Lecture Notes in Computer Science 6496

Commenced Publication in 1973
Founding and Former Series Editors:
Gerhard Goos, Juris Hartmanis, and Jan van Leeuwen

Peter F. Patel-Schneider   Yue Pan
Pascal Hitzler   Peter Mika
Lei Zhang   Jeff Z. Pan
Ian Horrocks   Birte Glimm (Eds.)

The Semantic Web – ISWC 2010

9th International Semantic Web Conference, ISWC 2010
Shanghai, China, November 7-11, 2010
Revised Selected Papers, Part I

 Springer

Volume Editors

Peter F. Patel-Schneider
Bell Labs Research, Murray Hill, NJ 07974, USA
E-mail: pfps@research.bell-labs.com

Yue Pan
IBM Research Labs, Beijing 100193, China
E-mail: panyue@cn.ibm.com

Pascal Hitzler
Wright State University, Dayton, OH 45435, USA
E-mail: pascal.hitzler@wright.edu

Peter Mika
Yahoo! Research, 08018 Barcelona, Spain
E-mail: pmika@yahoo-inc.com

Lei Zhang
IBM Research Labs, Shanghai 201203, China
E-mail: lzhangl@cn.ibm.com

Jeff Z. Pan
The University of Aberdeen, Aberdeen, AB24 3UE, UK
E-mail: jeff.z.pan@abdn.ac.uk

Ian Horrocks
University of Oxford, Oxford, OX1 3QD, UK
E-mail: ian.horrocks@comlab.ox.ac.uk

Birte Glimm
University of Oxford, Oxford, OX1 3QD, UK
E-mail: birte.glimm@comlab.ox.ac.uk

The cover photo was taken by Nicolas Rollier (flickr user nrollier).

Library of Congress Control Number: 2010940710

CR Subject Classification (1998): C.2, H.4, H.3, H.5, J.1, K.4

LNCS Sublibrary: SL 3 – Information Systems and Application, incl. Internet/Web
and HCI

ISSN   0302-9743
ISBN-10   3-642-17745-X Springer Berlin Heidelberg New York
ISBN-13   978-3-642-17745-3 Springer Berlin Heidelberg New York

springer.com

© Springer-Verlag Berlin Heidelberg 2010
Printed in Germany

Typesetting: Camera-ready by author, data conversion by Scientific Publishing Services, Chennai, India
Printed on acid-free paper   06/3180

Preface

The International Semantic Web Conferences (ISWC) constitute the major international venue where the latest research results and technical innovations on all aspects of the Semantic Web are presented. ISWC brings together researchers, practitioners, and users from the areas of artificial intelligence, databases, social networks, distributed computing, Web engineering, information systems, natural language processing, soft computing, and human–computer interaction to discuss the major challenges and proposed solutions, the success stories and failures, as well the visions that can advance research and drive innovation in the Semantic Web.

This volume contains the main proceedings of ISWC 2010, including papers accepted in the Research and Semantic-Web-in-Use Tracks of the conference, as well as long papers accepted in the Doctoral Consortium, and information on the invited talks.

This year the Research Track received 350 abstracts and 228 full papers from around the world. The Program Committee for the track was recruited from researchers in the field, and had world-wide membership. Each submitted paper received at least three reviews as well as a meta-review. The reviewers participated in many spirited discussions concerning their reviews. Authors had the opportunity to submit a rebuttal, leading to further discussions among the reviewers and sometimes to additional reviews. Final decisions were made during a meeting between the Track Chairs and senior Program Committee members. There were 51 papers accepted in the track, a 22% acceptance rate.

The Semantic-Web-in-Use Track, targeted at deployed applications with significant research content, received 66 submissions, and had the same reviewing process as the Research Track, except without the rebuttal phase. There were 18 papers accepted in this track, a 27% acceptance rate.

For the sixth consecutive year, ISWC also had a Doctoral Consortium Track for PhD students within the Semantic Web community, giving them the opportunity not only to present their work but also to discuss in detail their research topics and plans, and to receive extensive feedback from leading scientists in the field, from both academia and industry. Out of 24 submissions, 6 were accepted as long papers, and a further 7 were accepted for short presentations. Each student was assigned a mentor who led the discussions following the presentation of the work, and provided detailed feedback and comments, focusing on the PhD proposal itself and presentation style, as well as on the actual work presented.

The ISWC program also included four invited talks given by leading figures from both the academic and business world. This year talks were given by Li Xiaoming of Peking University, China; mc schraefel of the University of Southampton, UK; Austin Haugen of Facebook; and Evan Sandhaus of the New York Times.

The ISWC conference included the Semantic Web Challenge, as in the past. In the challenge, organized this year by Christian Bizer and Diana Maynard, practitioners and scientists are encouraged to showcase useful and leading-edge applications of Semantic Web technology, either on Semantic Web data in general or on a particular data set containing 3.2 billion triples. ISWC also included a large tutorial and workshop program, organized by Philippe Cudré-Mauroux and Bijan Parsia, with 13 workshops and 8 tutorials spread over two days. ISWC again included a Poster and Demo session, organized by Axel Polleres and Huajun Chen, for presentation of late-breaking work and work in progress, and a series of industry talks.

A conference as complex as ISWC requires the services of a multitude of people. First and foremost, we thank all the members of the Program Committees for the Research Track, the Semantic-Web-In-Use Track, and the Doctorial Consortium. They took considerable time, during summer vacation season for most of them, to read, review, respond to rebuttals, discuss, and re-discuss the submissions. We also thank the people involved in the other portions of the conference, particularly Birte Glimm, the Proceedings Chair; Lin Clark and Yuan Tian, the webmasters; Axel Polleres and Huajun Chen, the Posters and Demos Chairs, and their Program Committee; Yong Yu, the Local Arrangements Chair, Haofen Wang, who managed most aspects of the local arrangements, and Dingyi Han, Gui-Rong Xue and Lei Zhang, the Local Arrangements Committee; Sebastian Rudolph, the Publicity Chair; Jie Bao, the Metadata Chair; Anand Ranganathan and Kendall Clark, the Sponsor Chairs; and Jeff Heflin, the Fellowship Chair.

September 2010

Yue Pan and Peter F. Patel-Schneider
Program Chairs, Research Track Chairs

Pascal Hitzler, Peter Mika, and Lei Zhang
Semantic-Web-In-Use and Industry Track Chairs

Jeff Z. Pan
Doctoral Consortium Chair

Ian Horrocks
Conference Chair

Conference Organization

Organizing Committee

Conference Chair
Ian Horrocks   University of Oxford, UK

Program Chairs, Research Track Chairs
Yue Pan   IBM Research Labs, China
Peter F. Patel-Schneider   Bell Labs, USA

Semantic-Web-In-Use and Industry Chairs
Pascal Hitzler   Wright State University, USA
Peter Mika   Yahoo! Research, Spain
Lei Zhang   IBM Research Labs, China

Posters and Demos Chairs
Axel Polleres   National University of Ireland, Ireland
Huajun Chen   Shanghai Jiao Tong University, China

Doctoral Consortium Chair
Jeff Z. Pan   The University of Aberdeen, UK

Workshops and Tutorials Chairs
Philippe Cudré-Mauroux   Massachusetts Institute of Technology, USA
Bijan Parsia   University of Manchester, UK

Semantic Web Challenge Chairs
Chris Bizer   Freie Universität Berlin, Germany
Diana Maynard   University of Sheffield, UK

Metadata Chair
Jie Bao   Rensselaer Polytechnic Institute, USA

Local Organization Chair
Yong Yu   Shanghai Jiao Tong University, China

Local Organization Committee
Dingyi Han   Shanghai Jiao Tong University, China
Gui-Rong Xue   Shanghai Jiao Tong University, China
Haofen Wang   Shanghai Jiao Tong University, China
Lei Zhang   IBM Research Labs, China

Publicity Chair

Sebastian Rudolph   Karlsruher Institut für Technologie, Germany

Webmasters

Lin Clark   National University of Ireland, Ireland
Yuan Tian   Shanghai Jiao Tong University, China

Proceedings Chair

Birte Glimm   University of Oxford, UK

Sponsor Chairs

Anand Ranganathan   IBM T.J. Watson Research Center, USA
Kendall Clark   Clark & Parsia, LLC, USA

Fellowship Chair

Jeff Heflin   Lehigh University, USA

Senior Program Committee — Research

Hassan Ait-Kaci   Jeff Heflin
Abraham Bernstein   Aditya Kalyanpur
Paul Buitelaar   David Karger
Ciro Cattuto   Juanzi Li
Vinay Chaudhri   Li Ma
Bob DuCharme   Natasha Noy
Michel Dumontier   Jacco van Ossenbruggen
Tim Finin   Yuzhong Qu
Asunción Gómez-Pérez   Evren Sirin
Claudio Gutierrez

Program Committee — Research

Sudhir Agarwal   Mark Burstein
Harith Alani   Diego Calvanese
Paul André   Enhong Chen
Melliyal Annamalai   Key-Sun Choi
Kemafor Anyanwu   Philipp Cimiano
Knarig Arabshian   Lin Clark
Marcelo Arenas   Oscar Corcho
Jie Bao   Melanie Courtot
Michael Benedikt   Isabel Cruz
Chris Bizer   Claudia d'Amato
Eva Blomqvist   Mathieu d'Aquin
Kalina Bontcheva   David De Roure

Mike Dean
Stefan Decker
Ian Dickinson
Xiaoyong Du
Thomas Eiter
Robert H.P. Engels
Achille Fokoue
Enrico Franconi
Zhiqiang Gao
Nikesh Garera
Yolanda Gil
Stefan Gradmann
Michael Gruninger
Volker Haarslev
Harry Halpin
Siegfried Handschuh
Tom Heath
Nicola Henze
Martin Hepp
Nathalie Hernandez
Stijn Heymans
Kaoru Hiramatsu
Rinke Hoekstra
Andreas Hotho
Wei Hu
Zhisheng Huang
Jane Hunter
David Huynh
Eero Hyvönen
Zhi Jin
Lalana Kagal
Anastasios Kementsietsidis
Vladimir Kolovski
Markus Krötzsch
Ora Lassila
Georg Lausen
Faith Lawrence
Shengping Liu
Pankaj Mehra
Jing Mei
Riichiro Mizoguchi
Knud Moeller
Paola Monachesi
William Murray

Wolfgang Nejdl
Yuan Ni
Alexandre Passant
Chintan Patel
Alun Preece
Guilin Qi
Anand Ranganathan
Riccardo Rosati
Sebastian Rudolph
Uli Sattler
Ansgar Scherp
Daniel Schwabe
Yi-Dong Shen
Michael Sintek
Sergej Sizov
Kavitha Srinivas
Steffen Staab
Giorgos Stamou
Robert Stevens
Umberto Straccia
Heiner Stuckenschmidt
Mari Carmen Suárez-Figueroa
V.S. Subrahmanian
Xingzhi Sun
York Sure
Jie Tang
Christopher Thomas
Lieven Trappeniers
Tania Tudorache
Anni-Yasmin Turhan
Octavian Udrea
Michael Uschold
Haixun Wang
Haofen Wang
Fang Wei
Max Wilson
Katy Wolstencroft
Zhe Wu
Bin Xu
Peter Yeh
Yong Yu
Lei Zhang
Ming Zhang
Hai Zhuge

Program Committee — Semantic-Web-In-Use and Industry

Harith Alani
Sören Auer
Mathieu d'Aquin
Dave Beckett
Chris Bizer
Boyan Brodaric
Vinay Chaudri
Huajun Chen
Gong Cheng
Kendall Clark
John Davies
Leigh Dodds
Michel Dumontier
Aldo Gangemi
Paul Gearon
Mark Greaves
Stephan Grimm
Peter Haase
Michael Hausenblas
Manfred Hauswirth
Ivan Herman
Rinke Hoekstra
David Huynh
Eero Hyvönen

Renato Iannella
Krzysztof Janowicz
Atanas Kiryakov
Markus Krötzsch
Mark Musen
Knud Möller
Chimezie Ogbuji
Daniel Olmedilla
Eric Prud'hommeaux
Yuzhong Qu
Yves Raimond
Marta Sabou
Satya S. Sahoo
Andy Seaborne
Susie Stephens
Hideaki Takeda
Jie Tang
Jamie Taylor
Andraz Tori
Holger Wache
Haofen Wang
Jan Wielemaker
David Wood
Guo-Qiang Zhang

Program Committee — Doctoral Consortium

Abraham Bernstein
Meghyn Bienvenu
Huajun Chen
Ying Ding
Jianfeng Du
Jérôme Euzenat
Giorgos Flouris
Zhiqiang Gao
Marko Grobelnik
Siegfried Handschuh
Andreas Harth
Stijn Heymans
Wei Hu
Zhisheng Huang
Roman Kontchakov

Diana Maynard
Enrico Motta
Lyndon Nixon
Guilin Qi
Manuel Salvadores
Guus Schreiber
Pavel Shvaiko
Yi-Dong Shen
Amit Sheth
Elena Simperl
Giorgos Stamou
Giorgos Stoilos
Heiner Stuckenschmidt
Vojtech Svatek
Anni-Yasmin Turhan

Denny Vrandecic

Holger Wache

Haofen Wang

Shenghui Wang

Ming Zhang

Yuting Zhao

External Reviewers

Nor Azlinayati Abdul Manaf

Alessandro Adamou

Mark van Assem

Cosmin Basca

Sujoy Basu

Elena Botoeva

Jos de Bruijn

Carlos Buil-Aranda

Catherina Burghart

Jean Paul Calbimonte

Xiong Chenyan

DongHyun Choi

Alexandros Chortaras

Maria Copeland

Enrico Daga

Brian Davis

Renaud Delbru

Alexander DeLeon

Zhongli Ding

Laura Dragan

Fang Du

Liang Du

Alistair Duke

George Eadon

Jinan El-Hachem

Sean Falconer

Jun Fang

Nicola Fanizzi

Sébastien Ferré

Björn Forcher

Andrés García-Silva

Birte Glimm

Gunnar Aastrand Grimnes

Tudor Groza

Christian Hachenberg

Olaf Hartig

Norman Heino

Daniel Hienert

Aidan Hogan

Thomas Hornung

Matthew Horridge

Julia Hoxha

Gearoid Hynes

Robert Isele

Max Jakob

Martin Junghans

Aditya Kalyanpur

Kamal Kc

Malte Kiesel

Jörg-Uwe Kietz

Eun-Kyung Kim

Yoshinobu Kitamura

Pavel Klinov

Kouji Kozaki

Beate Krause

Thomas Krennwallner

Markus Krötzsch

Maurizio Lenzerini

Paea LePendu

Xuan Li

Yuan-Fang Li

Feiyu Lin

Maxim Lukichev

Sen Luo

Yue Ma

Frederick Maier

Theofilos Mailis

Michael Martin

Philipp Mayr

Anees ul Mehdi

Michael Meier

Pablo Mendes

Eleni Mikroyannidi

Fleur Mougin

Zhi Nie

Mathias Niepert

Nadejda Nikitina

Andriy Nikolov

Vit Novacek
Andrea Nuzzolese
Jasmin Opitz
Magdalena Ortiz
Raul Palma
Rafael Peñaloza
Jorge Pérez
Danh Le Phuoc
Axel Polleres
Freddy Priyatna
Jörg Pührer
Guilin Qi
Timothy Redmond
Yuan Ren
Achim Rettinger
Vinny Reynolds
Ismael Rivera
Mariano Rodriguez-Muro
Dmitry Ryashchentsev
Anne Schlicht
Florian Schmedding
Michael Schmidt
Thomas Schneider
mc schraefel
Floarea Serban
Wei Shen
Rob Shearer
Fuming Shih
Andrey Simanovsky
Mantas Simkus
Evren Sirin
Sebastian Speiser
Giorgos Stoilos
Cosmin Stroe
Mari Carmen Suárez-Figueroa

Kewu Sun
Xiaoping Sun
Martin Szomszor
Christer Thörn
VinhTuan Thai
Christopher Thomas
Despoina Trivela
Eleni Tsalapati
Dmitry Tsarkov
Alexander Ulanov
Natalia Vassilieva
Tasos Venetis
Kunal Verma
Boris Villazón-Terrazas
Denny Vrandecic
Bo Wang
Xiaoyuan Wang
Zhe Wang
Zhichun Wang
Jens Wissmann
Gang Wu
Kejia Wu
Linhao Xu
Yixin Yan
Fangkai Yang
Amapali Zaveri
Benjamin Zapilko
Maciej Zaremba
Lei Zhang
Xiao Zhang
Dmitriy Zheleznyakov
Hai-Tao Zheng
Qian Zhong
Ming Zuo

Sponsors

Platinum Sponsors	Gold Sponsors	Silver Sponsors
AI Journal	fluid Operations AG	IBM
Elsevier	LarKC	EMC^2
OntoText	SaltLux	W3C
	Yahoo!	Amiando

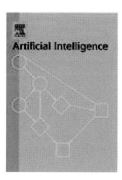

Table of Contents – Part I

Research Track

Table of Contents – Part II

Semantic-Web-In-Use Track

Doctoral Consortium

Invited Talks

Fusion – Visually Exploring and Eliciting Relationships in Linked Data

Samur Araujo[1], Geert-Jan Houben[1], Daniel Schwabe[2], and Jan Hidders[1]

[1] Delft University of Technology, PO Box 5031, 2600 GA Delft, The Netherlands
[2] PUC-Rio, Rua Marques de Sao Vicente, 225, Rio de Janeiro, Brazil
{s.f.cardosodearaujo,g.j.p.m.houben,a.j.h.hidders}@tudelft.nl,
dschwabe@inf.puc-rio.br

Abstract. Building applications over Linked Data often requires a mapping between the application model and the ontology underlying the source dataset in the Linked Data cloud. This mapping can be defined in many ways. For instance, by describing the application model as a view over the source dataset, by giving mappings in the form of dependencies between the two datasets, or by inference rules that infer the application model from the source dataset. Explicitly formulating these mappings demands a comprehensive understanding of the underlying schemas (RDF ontologies) of the source and target datasets. This task can be supported by integrating the process of schema exploration into the mapping process and help the application designer with finding the implicit relationships that she wants to map. This paper describes Fusion - a framework for closing the gap between the application model and the underlying ontologies in the Linked Data cloud. Fusion simplifies the definition of mappings by providing a visual user interface that integrates the exploratory process and the mapping process. Its architecture allows the creation of new applications through the extension of existing Linked Data with additional data.

Keywords: semantic web, data interaction, data management, RDF mapping, Linked Data.

1 Introduction

Nowadays, the Linked Data[1] cloud provides a new environment for building applications where many datasets are available for consumption. Although data in this cloud is ready to use, applications over the Linked Data cloud have currently an intrinsic characteristic: they consume RDF[2] data "as is", since designers do not have write permission over the data in the cloud which would enable them to change the data in any way. This fact raises an important issue concerning the development of applications over Linked Data: how to fill the gap between the ontology associated with the application model and the ontology used to represent the underlying data from the Linked Data cloud? The main benefit of mapping these two models is that then

---

[1] Linked Data - http://linkeddata.org/
[2] http://www.w3.org/TR/2004/REC-rdf-primer-20040210/

P.F. Patel-Schneider et al. (Eds.): ISWC 2010, Part I, LNCS 6496, pp. 1–15, 2010.
© Springer-Verlag Berlin Heidelberg 2010

Linked Data can be accessed through properties defined in the application model, which is more convenient for the designer, consequently simplifying considerably the development and maintenance of the application.

Although a number of techniques can be applied for mapping two RDF models, such as ontology matching, or inference rules, or views over RDF data, they often do not take into account that expressing the mapping rules themselves is a separate challenge, since in most cases Linked Data sources are represented using domain-specific ontologies that do not explicitly offer all common properties in the domain. Take for example DBLP[3] Linked Data, one of the best-known bibliography information sources available as Linked Data. Its ontology does not have an explicit property that connects directly co-authors, a common property in this domain. Although DBLP Linked Data contains paths that represent this relationship, it is not trivial to find them. Indeed, it requires understanding the schema behind the data and how this relationship is implicitly represented in this dataset. Similar examples can be found in any dataset in the Linked Data cloud, where the required information is implicitly encoded in the instance of data.

In this context, two specific and common scenarios often occur. The first is where the designer needs to express a mapping between a property in her application model (e.g. how a City is located in a Country) and a path in the RDF graph of the given dataset (e.g. a City belongs to a Province which belongs to a Country). Another example of this scenario can be given in the domain of government data. Suppose you are building an application over the GovTrack.Us[4] dataset and its application model requires a property *isSenatorOf* that directly connects instances of the class Politician to instances of the class State (e.g: Christopher Bond is a senator from Missouri). However, this relationship is not explicitly represented in the GovTrack.Us ontology. In order to obtain this relationship, the designer has to use the path *Senator -> has Role-> forOffice-> represents -> State* which, in this RDF graph, represents the relationship. Note that the designer needs a clear understanding of the GovTrack.Us schema in order to find the corresponding path to be mapped. The second scenario occurs when the mapping is in fact a computation over the existing data that produces a new explicit data value. For instance, a mapping between a property *screen resolution* from the application model and the concatenation of the properties *screen width* and *screen height* defined in the target dataset.

Note that in those scenarios for defining those mappings special attention should be paid to the exploratory process, especially when it demands from the designer to dive into the instances and the schema of the source dataset in order to find implicit relationships, which is not a straightforward task at all. Some authors have shown that visual exploration [1, 9] can help users to understand an unknown schema used to represent a known domain. Although those mechanisms help users to query an unknown schema, it will be always easier to explore a schema that is closer to the application models, often expressed in a specific application ontology. Although many tools are available for exploring Linked Data and for expressing mapping rules between RDF models, there is still a lack of tools that integrate these processes.

---

[3] http://dblp.l3s.de/d2r/
[4] http://www.govtrack.us/

This paper presents Fusion[5], a lightweight framework to support application designers in building applications over Linked Data. It supports designers in mapping the ontology of the used Linked Data sources to their application model by integrating the process of exploration of the target schema with the task of expressing a mapping rule itself. Fusion features a visual user interface that guides the designer in the process of specifying a mapping rule. It uses a standard RDF query language and allows Linked Data to be accessed using properties defined in the application model, consequently simplifying the use of Linked Data in a specific context.

The remainder of this paper is organized as follows. Section 2 presents relevant related work. Section 3 describes how Fusion supports the designer in deriving rules; while Section 4 describes Fusion's architecture. Section 5 presents some examples and shows how Fusion solves the problem of enriching access to Linked Data with application model properties. Finally, Section 6 presents the conclusion of this work.

2   Related Work

2.1   Ontology Mapping

The problem of mapping data models can also be conceived as an ontology-mapping problem, since it encompasses describing existing data in another vocabulary. In [8] a SPARQL extension is proposed to achieve that. Their solution merges SPARQL++ [3] and PSPARQL [9], two extensions of the SPARQL specification. The first extension adds some functions for enabling SPARQL to translate one vocabulary to another one by just using SPARQL *CONSTRUCT*. The second one adds path expressions to SPARQL, allowing a better navigation through the graph. Together they empower the SPARQL language to perform ontology mapping over two or more ontologies. Although the theory is given, the authors do not provide a concrete implementation especially because the proposed primitives have many implications for the performance of the query over the distributed environment of Linked Data.

2.2   SPARQL Construct Queries and Their Extensions

Another way to solve the problem of mapping RDF datasets is by specifying a *CONSTRUCT* query in SPARQL [2] that derives the triples in the target data set from the source data set. The resulting graph can then be stored in an arbitrary RDF repository. However, the *CONSTRUCT* query has limited expressive power, since some computation over the original RDF triples cannot be done, such as string manipulation and aggregation. For instance, using this approach it is not possible to generate the triple that would represent the mapping between the properties *screen width* and *screen height* (shown in Fig. 1) to a property *resolution* (shown in Fig. 2) that is their simple concatenation.

<http://sw.tv.com/id/2660> <http://sw.tv.com/screen_width> "128" .
<http://sw.tv.com/id/2660> <http://sw.tv.com/screen_height> "160" .

Fig. 1. A resource with predicates *screen width* and *screen height*

---

[5] http://www.wis.ewi.tudelft.nl/index.php/fusion

<http://sw.tv.com/id/2660> <http://sw.tv.com/resolution> "128x160" .

Fig. 2. A resource with predicate resolution

Polleres et al. [3] have proposed an extension of *CONSTRUCT* that overcomes such limitations, however this extension is limited to a specific RDF query engine that implements this SPARQL extension. Therefore, at this moment, such a solution is not feasible for the Semantic Web environment, which is very diverse in terms of query engines - the majority of data is stored in repositories that implement variations of the standard SPARQL specification that do not include the extensions discussed here.

2.3 Views over RDF Data

Another way to specify mappings between different representational models is by defining *views* [5, 6, 7]. This concept is well known in the field of database theory, and can be used to aggregate and personalize data. A *view* is a query accessible as a virtual table composed of the result of the query. Although *views* are frequently used in relational databases, building *views* over Linked Data presents many additional challenges. Issues such as view maintenance (including updates) and querying over virtual (non-materialized) views in the distributed environment of Linked Data are still open problems, besides several other performance issues that arise.

Volz et al. [4] have proposed a language based on RQL [5] for specifying views over RDF data. It defines views over RDF classes and views of RDF properties. Although this proposal presents a complex specification of views over RDF, it cannot solve the simple scenario described in Section 2.2, and its solution is based on RQL, which is not the standard RDF query language used nowadays. Magkanaraki et al. [7] have proposed a view specification language also based on RQL. Its processing model is based on materialized views. Chen et al. [6] present a scenario of accessing relational data using RDF views. In their approach a query over a view result in query rewriting that exploits the semantics of RDF primitives, such as, *subPropertyOf* or *subClassOf*. While their approach enriches the access to the relational data, it does not cover the transformations over the data that we are considering here, moreover it is focused on mapping relational schema to an RDF/S ontology.

2.4 SWRL Rules

Hassanpour et. al [12] proposes a tool for supporting the user on creating SWRL[6] rules. Their tool contains a visual interface that guides the user in visualizing, managing and eliciting SWRL specifications. Although this tool can be used to map two models using SWRL rules, it does not integrate the process of specifying the rules with the process of exploring an unknown schema, which is the main aim of Fusion.

2.5 RDF Exploration

RelFinder [1] is a visual tool for finding *n*-ary relationships between RDF resources. It contains a visual interface that allows the user to visualize the relationship in a directed graph layout. Basically, RelFinder issues a set of queries against a specific

---

[6] http://www.w3.org/Submission/SWRL/

SPARQL endpoint in order to find relationships between two or more RDF resources. RelFinder aims to be a better mechanism for finding relationships among data than any other exploratory mechanism. Explorator [9] is another tool that aims to facilitate the querying of instances of an unknown RDF schema, consequently allowing the user to discover relations between data instances even without previous knowledge of the domain. These tools re-enforce the idea that accessing RDF data is not a trivial task and demands a complex exploratory model behind it. In spite of the fact that they support users in finding relationships between data, they do not solve the problem of accessing the Linked Data through a schema associated with the application model.

2.6 Interlinking

From an operational point of view the mapping of two RDF models can be perceived as the addition of new triples to the original dataset for any new relationship expressed in the target ontology. Clearly this task requires some sort of automation. For instance, Silk [10] is a linking framework for discovering relationships between data items within different Linked Data sources. By specifying rules, the application designer can define how two distinct sets of resources, possibly belonging to distinct endpoints, can be interlinked, and as a result it produces a graph with all discovered connections. Although Silk automates the process of interlinking resources, Fusion goes one step further, since it supports also the process of specifying the rule. They solve two different problems: Silk interlinks two disconnected RDF graphs while Fusion extends the knowledge for a single endpoint. Although Silk's mapping language can be used for materializing the rules defined in Fusion, it does not support the full process supported by Fusion, which also includes, most notably, the discovery of a path in the schema to be mapped. While Silk allows the user to serialize a rule, it does not support her in finding it and expressing it.

3 Discovering and Deriving RDF Relationships

The main aim of Fusion is to help the designer in discovering relationships in RDF graphs that exist in the Linked Data cloud and specifying rules for the derivation of new properties for these relationships. We refer to this process as *relationship derivation*. The result of the *relationship derivation* process is a set of rules such that each produces RDF triples based on queries over an existing RDF graph. The evaluation of a rule results in a set of triples, each of which contains either a new *object property*[7] or a new *datatype property*. In the cases where it results in a new *object property*, the triples produced connect existing resources, while in the case where it derives a new *datatype property* the triples produced connect existing resources with values computed by a function over the RDF graph being queried. In the remainder of this section we describe how Fusion supports the designer in specifying these derivation rules.

3.1 Deriving Object Property Relationships

The main issue regarding the derivation of new *object property relationship* is to specify the correspondence between resources. For example, if a user wants to create

---

[7] http://www.w3.org/TR/owl-ref/#ObjectProperty-def

a new object property *locatedIn* that directly connects cities to their respective countries, she needs to specify the relationship between cities and countries in the existing data, i.e. which cities are located in which country. Such a correspondence can be obtained by following a certain path between two resources in the RDF graph of the source dataset. For example, Fig. 3 shows a sub-graph of Geonames Linked Data[8] that represents a path between the resource for the city of Delft and that of the country The Netherlands. By using the predicates in this example path and generalizing the intermediate nodes, it is possible to generalize such an example correspondence to apply to all cities and respective countries in this dataset, and thus bring the correspondence to the class level. By exploiting this resulting path, Fusion can map all cities to their corresponding country and thereby add a new object property to the original graph. Note that this process maps a newly added relationship that is defined in the application model, onto an implicit relationship that exists in the original graph. In the example above, the object property *locatedIn* defined in an application model could be mapped to the generalization of the path between Delft and the Netherlands *geo:parentFeature* — *Province of Zuid-Holland* — *geo:parentFeature*.

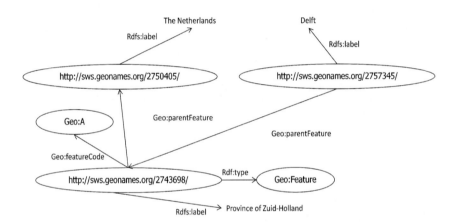

Fig. 3. A path between the resources Delft and The Netherlands in Geonames Linked Data

3.1.1   A Path Discovery Algorithm
The first step in the mapping process is to find the paths that could potentially be used in the mapping. Fusion automates this step by eliciting all possible paths in an RDF graph that connect two example RDF resources (e.g. Delft and The Netherlands) that have a specified maximum length. Thus, finding the relationship between two RDF resources becomes finding a path in the RDF graph that would allow navigating from the one resource to the other one. This process can be implemented as a modified version of the *breadth-first search algorithm (BFS)* with a maximum limit on the depth of the search and without the restriction that it should stop when the goal node is reached. Consequently, it can be used to retrieve all paths in the graph within a maximum length from the source to the target node. This algorithm is applied to the

---
[8] http://www.geonames.org/ontology/

RDF graph by interpreting each triple as an undirected edge between its subject and object. Since this algorithm is a small variation on the standard BFS and retrieves all possible paths from a to b of a maximum length d, its complexity is $O(c^d)$, where c is the maximum branching factor in the graph. This asymptotic complexity is in this case the theoretical optimum since it describes the size of the output.

3.1.2 Implementing the Path Discovery Algorithm over a SPARQL Endpoint

Considering that Fusion searches for paths in a Linked Data dataset, the path discovery algorithm needs to be implemented as a set of SPARQL queries, since the most direct way to search in an RDF graph in the Linked Data cloud is by issuing SPARQL queries over its remote SPARQL endpoint. In order to generate these queries we consider the RDF graph as an *undirected graph* as previously described. Thus, all paths with length n from node a to node b in this graph can be obtained with a set of SPARQL queries containing 2^n queries. Since we want to ignore the direction in the graph, we issue a distinct graph pattern for all possible choices of direction for each triple pattern in the path. Each query in this set contains n connected triple patterns, one for each edge in the path. For example, to obtain all paths between a and b with length 3, 8 ($= 2^3$) graph patterns, each containing 3 triple patterns, are generated. Fig. 4 shows all these 8 patterns.

```
1  (:a,:p1,:a2),(:a2,:p2,:a3),(:a3,:p3,:b)
2  (:a2,:p1,:a),(:a2,:p2,:a3),(:a3,:p3,:b)
3  (:a,:p1,:a2),(:a3,:p2,:a2),(:a3,:p3,:b)
4  (:a2,:p1,:a),(:a3,:p2,:a2),(:a3,:p3,:b)
5  (:a,:p1,:a2),(:a2,:p2,:a3),(:b,:p3,:a3)
6  (:a2,:p1,:a),(:a2,:p2,:a3),(:b,:p3,:a3)
7  (:a,:p1,:a2),(:a3,:p2,:a2),(:b,:p3,:a3)
8  (:a2,:p1,:a),(:a3,:p2,:a2),(:b,:p3,:a3)
```

Fig. 4. Graph patterns generated for path length 3

Each of these patterns will be transformed into a single SPARQL query, as shown in Fig. 5 for pattern 1 from Fig. 4, where a and b were specified as the resource Christopher Bond (http://www.rdfabout.com/rdf/usgov/congress/people/B000611) and the resource Missouri (http://www.rdfabout.com/rdf/usgov/geo/us/mo.), respectively. In this example, the path connects the US politician Christopher Bond with the state (Missouri) that he represents.

```
PREFIX Geo: <http://www.rdfabout.com/rdf/usgov/geo/us/>
PREFIX Gov:
<http://www.rdfabout.com/rdf/usgov/congress/people>
SELECT DISTINCT ?p1 ?a2 ?p2 ?a3 ?p3
WHERE {
    Gov:B000611 ?p1 ?a2 .
    ?a2 ?p2 ?a3 .
    ?a3 ?p3 Geo:mo .
}
```

Fig. 5. Query performed for graph pattern 1 from Fig. 4

Thus, when all graph patterns are executed, all paths of length n are retrieved. Fig. 6 shows a sample path found as result of the query from Fig. 5 being issued over the GovTrack.Us endpoint.

Fig. 6. A sample path found between the resources Christopher Bond (a senator) and Missouri (a US state) in the GovTrack.Us dataset

This algorithm is similar to RelFinder's, however RelFinder's algorithm considers the RDF graph as a *directed graph* and it searches *only* for 4 graph patterns, which does not cover all possible paths between *a* and *b*.

3.1.3 Derivation Process

The algorithm described previously is used in the first stage of the derivation process. The complete derivation process ends with the designer choosing one of the paths found in the first stage and generalizing it to find correspondences between two broader (more general) sets of resources. As a result of this procedure, a derivation rule is produced.

In order to apply the chosen path to a broader set of resource pairs the designer needs to generalize all nodes in the path. For instance, the path in Fig. 7 shows a generalization of the path in Fig. 6 that considers all sources and targets (their generalization is indicated by the ?) that are connected through the path *hasRole —R1 (blank node) —forOffice —Senators for MO —represents,* i.e., it will find the correspondence between all senators and the state of Missouri.

Fig. 7. A possible generalization between senator resources and the Missouri resource

Fig. 8 shows another (yet more general) generalization that can find the correspondence between all senators and the respective state that they represent, since now *all* intermediate nodes are variables.

Fig. 8. A possible generalization between senator resources and the respective US state that they represent

It should be noticed that the predicates in the path are not generalized and remain fixed. These generalizations define derivation rules, which select the resources that will be interconnected.

For the designer to control this generalization process, we provide a graphical user interface that will be shown later.

3.2 Deriving New Datatype Properties

Fusion also supports application designers to extend the original dataset with *datatype properties*. As Fusion's goal is to allow application designers to map a property in their application model to an existing Linked Data dataset, the values of the new *datatype properties* are computed over the existing values in the original dataset.

Formally, a derivation rule that produces *datatype properties* is defined by a tuple (q, p, f) with a query q, a predicate name p, and a function f. The query q defines a function that maps an RDF graph to a set of URIs in that graph, which defines the set of resources for which the new *datatype* property is defined. The predicate name p defines the predicate name of the new property. Finally the function f maps an RDF graph and a particular URI within that graph to an RDF value. The result of applying such a rule to an RDF graph G is the addition of all tuples (s, p, o) such that $s \in q(G)$ and $o = f(G, s)$.

4 Architecture Overview

Fusion's implementation architecture provides a complete environment to specify and execute a derivation rule. An overview of this architecture is shown in Fig. 9. The specification of the rules is supported in Fusion's user interface that will be explained further in the section 5. Fusion's server engine is responsible for executing the derivation rule itself. During the process of executing of a rule, it queries a source endpoint in the Linked Data, processes the result, and produces a set of new triples that will be added to the Fusion repository. Any RDF data store can be used as Fusion's repository. Currently, Fusion implements adapters for Sesame[9] and Virtuoso[10] data stores, although other adapters can be easily added to its architecture. All derived triples in Fusion contain as subject a resource that belongs to the queried dataset, so the derived data is intrinsically interlinked with the Linked Data cloud. For this reason, a query over a federation of endpoints, that includes the Fusion repository endpoint, will allow the designer to have a view over the Linked Data that also includes the properties defined in her application model.

There are other approaches to how to store the derived triples. For example, it is possible to use a user-defined namespace for the subjects of the derived triples, and add an owl:sameAs statement linking it to the original URIs, as opposed to using the original URIs directly as subject. The shortcoming of this alternative is that others who want to find out about the new derived properties would not look for them in Fusion's local repositories, but in the original URI, which doesn't know about these new derived properties. On the other hand, with the current approach, if the VoID

---

[9] http://www.openrdf.org/
[10] http://virtuoso.openlinksw.com/dataspace/dav/wiki/Main/

description of Fusion's local repository is updated to reflect the inclusion of information about the original URI, then others would still find its SPARQL endpoint when looking for endpoints containing information about that URI, thus having access to all derived triples in the Fusion's repository.

So it is really a modeling trade-off, with no clear advantage to either side, and our solution requires less involvement from third parties (e.g., for owl:sameAs processing).

Although Fusion does not serialize a rule before executing it, any serialization mechanism could be used in its architecture. For example, its rules could be serialized as inference rules, by using Spin Inference Notation[11], which contains a collection of RDF vocabularies enabling the use of SPARQL to define constraints and inference rules on Semantic Web models. Although this configuration is theoretically possible, executing inference rules, or even instantiating a *virtual view* over the Linked Data is still an open problem, since it raises many performance issues. Such issues do not occur in the current Fusion architecture because it materializes the result of the rules as new triples in the Fusion repository. The performance of querying data that is already materialized is always faster than querying data that needs to be processed at runtime. The main drawback with materializing the result of the rules is that once the original source is updated, the rules have to be executed again. Furthermore, detecting these changes in the Linked Data is not trivial. Such synchronizing or updating issues are actually the realm of research (and practice) in (database) view definitions and updates. The performance trade-offs in each case are well known, and are addressed by researchers working in that area, which while relevant, is not the research focus for Fusion at this state. Fusion is be able to benefit from whatever techniques are available on this topic.

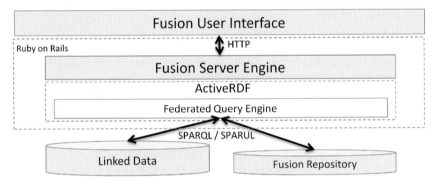

Fig. 9. Fusion's architecture overview

Fusion is implemented in Ruby on Rails[12] as a web application. It uses the ActiveRDF API [11] that allows an RDF graph to be accessed in the *object-oriented* paradigm. By using this API the properties of an RDF *resource* can be accessed as an attribute of its corresponding Ruby[13] object. For instance, the predicate

---

[11] http://www.spinrdf.org/
[12] http://rubyonrails.org/
[13] http://www.ruby-lang.org/en/

http://www.geonames.org/ontology#population   can   be   accessed   as   *<re-source>.population.* This architecture allows the designer to write complex functions for computing a new *datatype* property value using the full power of the Ruby language, which cannot be achieved simply by using the SPARQL language.

5   Examples of Use

This section describes two concrete scenarios that illustrate the use of Fusion to create an application by extending Linked Data sources with additional properties.

5.1   Scenario 1 – Adding the *isSenatorOf* Object Property to GovTrack.Us

In this example, we suppose that the designer wants to establish the relationship between US senators and the US state that they represent. Therefore she needs to construct a derivation rule that will find and define such a correspondence between politicians and states in GovTrack.Us's Linked Data repository. In the first step in the process, the designer provides an example of two resources in GovTrack.Us that she knows in advance that are actually related, for instance, politician Christopher Bond and the state of Missouri. Also, she needs to declare the GovTrack.Us endpoint to be queried and the maximum depth of the path. This step is shown in Fig. 10.

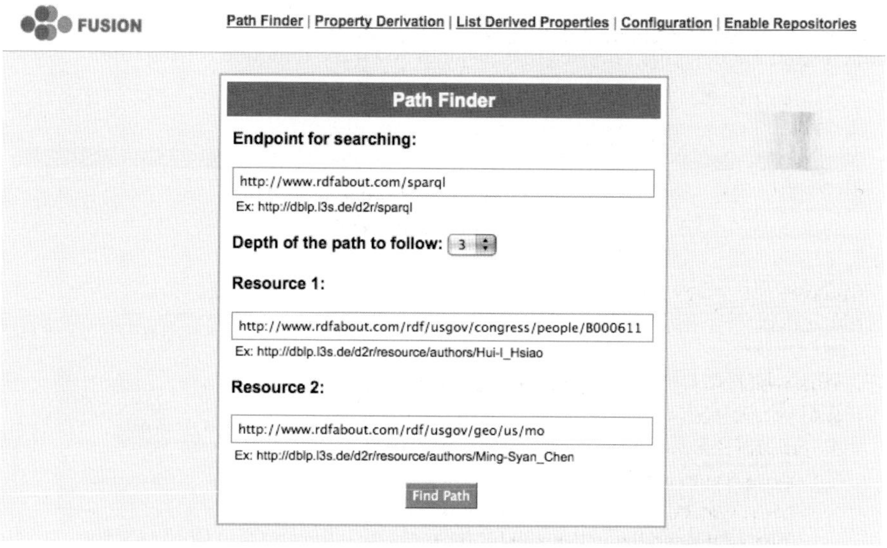

Fig. 10. Fusion's interface for finding a path between two known resources

As the result of this first step, Fusion shows all the paths that connect these two example resources satisfying the maximum path length. This result is shown in Fig. 11. In this example, the paths found have a maximum length of 3.

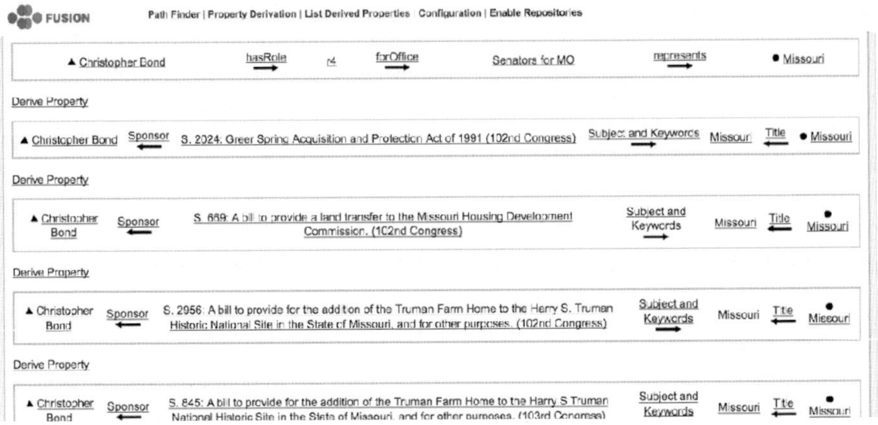

Fig. 11. Fusion's interface showing the discovered paths

In this view, the designer can now look for the path that has the intended semantics. Note that with this view the tool assists the designer in this discovery process,
since she does not need to query the schema manually in order to find these paths.
The first path shown in Fig. 11 indicates that the politician Christopher Bond has a
role as senator representing the state Missouri, and in our example case the designer
can now infer that this is an instance of the path that she is looking for. After this
conclusion, the designer chooses that instance to be the template for the rule.

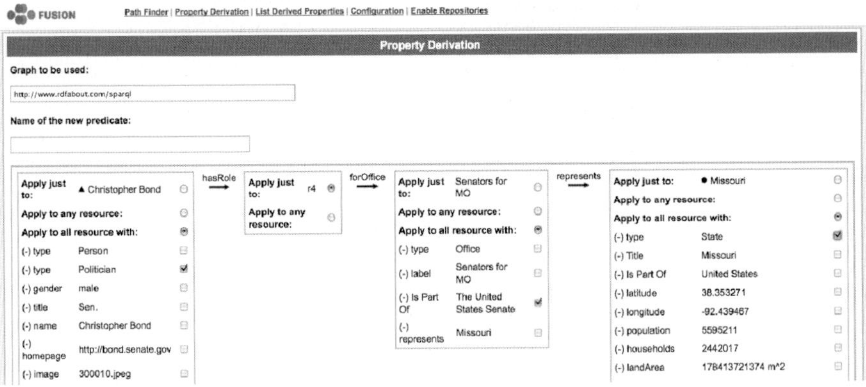

Fig. 12. Generalizing the path for the property isSenatorOf in GovTrack.Us

In the next step, shown in Fig. 12, the designer will define the derivation rule itself,
which means that she visually formulates a query, which generalizes the selected path
from the first step into a query that selects the elements to be connected through the
property *isSenatorOf*. To complete this operation she also needs to define the graph
where the derived triples will be stored and a specific URI to be used as the predicate
of the new triples, which in this example will be *http://example.org/isSenatorOf*. Note

that in this example 3 nodes were generalized such that only paths between resources of the RDF type *Politician* and RDF type *State* that contain an intermediate node that *is part of* the United States Senate will be considered during the derivation process. Consequently, Fusion will derive the new property *isSenatorOf* for all instances of the class Politician that are connected to an instance of the class State through the specified path. The whole process concludes with Fusion adding new triples to Fusion repository.

5.2 Scenario 2 – Adding a Datatype Property *citySize* to Geonames

In this example, we suppose that the designer wants to derive a new property *citySize* for cities in Geonames to distinguish small and large cities. Therefore she needs to create a derivation rule that will compute the appropriate values for this new property. She will want this property to have the value 'small' for cities with less than 1.000.000 inhabitants, and 'large' otherwise. As the first step in the process, she needs to provide the Geonames Linked Data endpoint to be queried and a resource (a city) in Geonames as an example. In this case she supplies the URI http://sws.geonames.org/2757345/, which represents the city of Delft. This step is shown in Fig. 13.

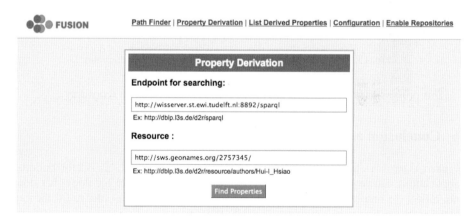

Fig. 13. Fusion's interface for deriving a data type property

In the next step, Fusion uses the city URI to construct the visual interface where the designer will express the derivation rule $R=(q,p,f)$. In this interface, she visually defines the query q, a URI for the new property p, and a Ruby expression that will be used as the function f. In Fig. 14 we show Fusion's datatype property derivation view.

In this view the designer specifies that the new property is to be defined for all resources for which the predicate *parentFeature* equals *Province Zuid-Holland*, i.e., it will compute it for all cities in the province of Zuid-Holland. Also she defines that the

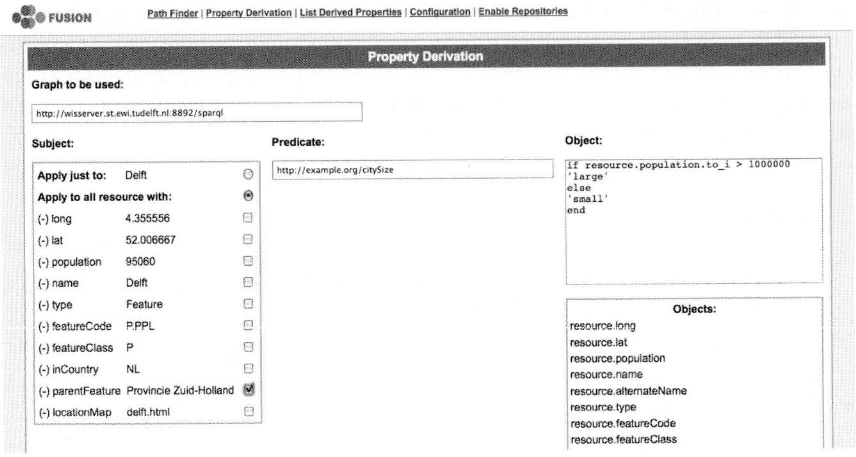

Fig. 14. Fusion's datatype property derivation interface

URI of the new property will be http://example.org/citySize. As Fusion was developed in Ruby, using the ActiveRDF API, the function *f* can be defined as a Ruby expression that for this example is shown in Fig. 15.

```
resource.population.to.i > 1.000.000 ? 'large':'small'
```

Fig. 15. Sample Ruby expression for computing the *citySize* value

This process ends with Fusion adding new triples to the Fusion repository.

6   Conclusion and Future Work

Linked Data is a cloud of distributed datasets that can be used "as is" for building applications. However, its data is often expressed in a low-level ontology that does not reflect the ontology associated with the application model. In order to fill the gap between these two representational models it is necessary to somehow map them. Although there exist approaches for solving this problem, such as ontology matching techniques, views over RDF and inference rules, they do not consider this task as a process that also involves Linked Data schema exploration. In others words, whatever strategy is used, it will demand from the designer to identify in both models exactly what to map and how to map it, which is not trivial, since it also demands a clear understanding of the underlying schema in the used Linked Data. Fusion integrates the exploratory task into the process of mapping, thereby helping the designer with identifying the relationships between her application model and the Linked Data schema, and also in providing a full architecture for expressing the mapping and extending the used Linked Data such that it implements the application model.

Fusion works by querying Linked Data and extending it by adding new data into the cloud without directly altering the original dataset. Fusion also provides a visual interface that allows the user to explore Linked Data, express the rules and derive new data, which in the end covers the whole process of mapping and extending. As Fusion materializes the result of the mapping as new triples in an extra endpoint in the cloud, it consequently allows the separation of the processes of building the application and managing the mapping between models.

References

1. Heim, P., Lohmann, S., Stegemann, T.: Interactive Relationship Discovery via the Semantic Web. In: Aroyo, L., Antoniou, G., Hyvönen, E., ten Teije, A., Stuckenschmidt, H., Cabral, L., Tudorache, T. (eds.) ESWC 2010. LNCS, vol. 6088, pp. 303–317. Springer, Heidelberg (2010)
2. SPARQL Specification, http://www.w3.org/TR/rdf-SPARQL-query/
3. Polleres, A., Scharffe, F., Schindlauer, R.: SPARQL++ for mapping between RDF vocabularies. In: Proceedings of the 6th International Conference on Ontologies, DataBases, and Applications of Semantics (ODBASE 2007), Vilamoura, Algarve, Portugal, November 27-29 (2007)
4. Volz, R., Oberle, D., Studer, R.: Views for light-weight Web ontologies. In: Proceedings of the 2003 ACM Symposium on Applied Computing (SAC 2003), Melbourne, Florida, March 9-12 (2003)
5. Karvounarakis, G., Alexaki, S., Christophides, V., Plexousakis, D., Scholl, M.: RQL: a declarative query language for RDF. In: Proceedings of the 11th International Conference on World Wide Web, Honolulu, Hawaii, USA, May 7-11 (2002)
6. Chen, C., Wu, Z., Wang, H., Mao, Y.: RDF/RDFS-based Relational Database Integration. In: Proceedings of the 22nd International Conference on Data Engineering (ICDE 2006), April 3-7 (2006)
7. Magkanaraki, A., Tannen, V., Christophides, V., Plexousakis, D.: Viewing the semantic web through RVL lenses. Web Semant. 1(4), 359–375 (2004)
8. Euzenat, J., Polleres, A., Scharffe, F.: Processing ontology alignments with SPARQL. In: Proceedings of the 2008 International Conference on Complex, Intelligent and Software Intensive Systems (CISIS 2008), Washington, DC, USA, March 4-7 (2008)
9. Araujo, S., Schwabe, D.: Explorator: a tool for exploring RDF data through direct manipulation. In: Proceedings of Linked Data on the Web (LODW 2009) Workshop at WWW 2009, Madrid, Spain, April 20 (2009)
10. Volz, J., Bizer, C., Gaedke, M., Kobilarov, G.: Silk – A Link Discovery Framework for the Web of Data. In: Proceedings of Linked Data on the Web (LODW 2009) Workshop at WWW 2009, Madrid, Spain, April 20 (2009)
11. Oren, E., Delbru, R., Gerke, S., Haller, A., Decker, S.: ActiveRDF: ObjectOriented Semantic Web Programming. Digital Enterprise Research Institute National University of Ireland, Galway Galway, Ireland (2007)
12. Hassanpour, S., O'Connor, M.J., Das, A.K.: A Software Tool for Visualizing, Managing and Eliciting SWRL Rules. In: Aroyo, L., Antoniou, G., Hyvönen, E., ten Teije, A., Stuckenschmidt, H., Cabral, L., Tudorache, T. (eds.) ESWC 2010, Part I. LNCS, vol. 6088, pp. 381–385. Springer, Heidelberg (2010)

Converting and Annotating Quantitative Data Tables

Mark van Assem[1], Hajo Rijgersberg[3], Mari Wigham[2,3], and Jan Top[1,2,3]

[1] VU University Amsterdam, The Netherlands
mark@cs.vu.nl
[2] Top Institute Food and Nutrition, The Netherlands
[3] Wageningen University and Research Centre, The Netherlands
first.last@wur.nl

Abstract. Companies, governmental agencies and scientists produce a large amount of quantitative (research) data, consisting of measurements ranging from e.g. the surface temperatures of an ocean to the viscosity of a sample of mayonnaise. Such measurements are stored in tables in e.g. spreadsheet files and research reports. To integrate and reuse such data, it is necessary to have a semantic description of the data. However, the notation used is often ambiguous, making automatic interpretation and conversion to RDF or other suitable format difficult. For example, the table header cell "f (Hz)" refers to frequency measured in Hertz, but the symbol "f" can also refer to the unit farad or the quantities force or luminous flux. Current annotation tools for this task either work on less ambiguous data or perform a more limited task. We introduce new disambiguation strategies based on an ontology, which allows to improve performance on "sloppy" datasets not yet targeted by existing systems.

1 Introduction

In this paper we study how to convert and annotate unstructured, "raw" quantitative data stored in tables into a semantic representation in RDF(S). Quantitative data are found in diverse sources, such as scientific papers, spreadsheets in company databases and governmental agencies' reports. The data consist of observations such as the heart rate of a patient measured in beats per minute, the viscosity of a sample of mayonnaise in pascal second, or the income of households in dollars in the US. Usually the tables consist of a header row that indicates which quantities and units are being measured and which objects; e.g. *Sample Nr. / Fat % / Visc. (Pa.s)*. Each content row then contains the values of one actual measurement.

Current reuse and integration of such data is not optimal, because a semantic description is not available. Researchers tend to write their data down in a "sloppy" way, because it is not anticipated that the data will ever be reused. This causes data to be "lost" and experiments to be needlessly repeated. To enable integration of data from different tables with each other, a complete description of all quantities and units in the table is necessary; annotation with

P.F. Patel-Schneider et al. (Eds.): ISWC 2010, Part I, LNCS 6496, pp. 16–31, 2010.

a few key concepts does not suffice. There are two main reasons why it is difficult to automatically convert the original data to a semantic description. Firstly, humans use different syntax for expressing quantities and units (e.g. separating the quantity from the unit with either brackets or a space). Secondly, the symbols and abbreviations used are highly ambiguous. For example, the symbol "g" can refer to at least ten different quantities and units.

This problem is not tackled by existing systems for conversion of tabular data to RDF, such as XLWrap [8]. These rely on a mapping specification constructed by a human analyst that is specific to the header of one table. Creating such a mapping is labour-intensive, especially if there are many differently structured tables involved. This is the case in government repositories such as Data.gov [4], and repositories of research departments of companies such as Unilever and DSM (from experience we know these contain thousands of different tables).

A solution is to include an automated annotation system into the conversion tool, as proposed by [9]. However, such an annotation system needs to tackle the ambiguity problem if it is to be succesfully used in the domain of quantities and units. We know of two existing annotation systems that target the domain of quantities and units [7,1], and our research can be seen as a continuation of these efforts. The results of these systems are good (over 90% F-measure), but they target "clean" datasets such as patent specifications, or focus on part of the total problem, such as detecting units only. In our work we focus on datasets with a high degree of ambiguity and attempt to detect quantities and units (including compound units).

The main contribution of our work is to show how ontology-based disambiguation can be used succesfully in several ways. Firstly, ambiguous quantity and unit symbols can be disambiguated by checking which of the candidate units/quantities are explicitly related to each other in the ontology. Secondly, ambiguous unit symbols may refer to units in specific application areas (e.g. nautical mile) or generic ones (meter). Some concepts act as indicators for a particular area (e.g. the unit nautical mile for "shipping"). After the area is identified by the presence or absence of indicators, we can disambiguate unit symbols. Thirdly, ambiguous compound unit expressions such as g/l can refer to gram per liter or gauss per liter. Only the former makes sense, as the ontology allows to derive that it refers to the quantity density, while the latter matches no known quantity. We show the benefits of ontology-based disambiguation by measuring precision and recall on two datasets and comparing with the performance achieved without these techniques. The datasets concerned are: (1) tables from the Top Institute Food and Nutrition; and (2) diverse scientific/academic tables downloaded from the Web.

The structure of this paper is as follows. We first present a detailed description of the problem, followed by related work (Sections 2 and 3). In Section 4 the datasets and ontology used in our experiment are described. Our approach is given in Section 5, which we evaluate in Section 6. We conclude with a discussion in Section 7.

2   Problem Description

Correct annotation of documents is faced with similar problems across many domains, including homonymy (a cause of low precision) and synonymy (a cause of low recall if the synonym is not known to the system). Below we discuss in what way these problems play a role in this domain.

Homonymy is caused in several ways. Firstly, it is not known beforehand whether cells contain a quantity (e.g. frequency), a unit (e.g. hertz), or both (e.g. f (Hz)). Secondly, homonymous symbols such as f are used, which can refer to quantities (frequency, force), units (farad) and prefixes (femto). The cell ms-1 might stand for either reciprocal millisecond or for meter per second (in the latter case m and s-1 should have been separated by a multiplication sign or space). This problem is aggravated because people often do not use official casing (e.g. f for force instead of the official F).

There are several types of synonymy involved in this domain: partial names (current for electric current), abbreviations (e.g. freq, Deg. C), plural forms (meters) and contractions (ms-1 instead of the correct form m s-1 for meter per second). Another type of synonym occurs when a quantity is prefixed with a term that describes the situation in more detail ("finalDiameter", "start_time", "mouthTemperature"). People also use colloquial names for quantities which overlap with other quantity names (i.e. confuse them). Two examples are weight (kg) and speed (1/s). The former should be mass (weight is measured in newton), the latter should be frequency.

A problem that is specific to this domain is the correct detection of compound units. The system has to detect the right compound unit instead of returning the units of which the unit is composed. For example, it should detect that km/h means kilometer per hour, instead of returning the units kilometer and hour separately (these should be counted as wrong results). This problem is aggravated by the fact that the number of compound units is virtually unlimited. For example, the quantity speed can be expressed in km/h, mm/picosecond, mile/year, etcetera. It is impractical to list them explicitly in an ontology. The interpretation of compound expressions is also difficult because of homonymy: g/l might stand for gram per liter or gauss per liter. The annotation process must somehow detect that gram per liter is the right compound unit (gauss per liter is not used), without gram per liter being present in the ontology. Returning gram, gauss and liter means returning three wrong results.

For correct detection of compound expressions, syntactic variations have to be taken into account (multiplication signs, brackets, etcetera). Compound expressions are also sometimes combined with substances, e.g. Conc. (g sugar/l water). Taken together this means a flexible matching process is needed instead of a strict grammar parser.

Particular to this domain is also that people tend to write down a quantity that is too generic or specific for the situation. For example, velocity (m/s) is too specific if the table contains scalar values only. The quantity velocity is only appropriate when a vector or a direction is indicated (e.g. "180 km/h north"). The other way around, the cell viscosity (stokes) should not be annotated with viscosity. The specific quantity kinematic viscosity (measured in stokes), is more precise. These "underspecifications" need to be corrected before successful data integration can take place.

3   Related Work

Annotation systems for quantitative data. As far as we know there are two existing systems that focus on automated annotation of tables with quantities and units. The system of [7] annotates table headers with both quantities and units, focusing on the biological domain (it contains generic physical quantities such as temperature and domain-specific ones such as colony count). The names and symbols are matched against their own ontology of 18 quantities with their associated unit symbols. Table headers and labels in the ontology are first lemmatized, turned into a vector space model, and compared using cosine similarity. Weights for terms are fixed beforehand: tokens that appear in the ontology get a weight of 1, stopwords and single letter tokens get weight zero. The advantage of this technique is that the order of tokens within terms is not important, so that "celsius temperature" matches "temperature celsius". This technique does not take abbreviations and spelling errors into account (e.g. "temp cels" will not match).

[1] present a system based on GATE/ANNIE for annotating measurements found in patent specifications (natural language documents). Symbols found in the documents are first tagged as possible unit matches using a flat list[1]. Domain-specific pattern matching rules then disambiguate the results, using the actual text plus detected types as input. For example, if a number is followed by letter(s) that match a unit symbol (e.g. 100 g), then the letter(s) are classified as a unit. It uses a similar rule to detect that 40-50mph refers to a range of numbers. Thirty of such rules were defined using the JAPE pattern language, but these cannot be inspected because the work is not open source. As far as we can tell no use is made of features of an ontology.

Both systems make simplifications. [1] only aim to identify units, not quantities. No techniques are provided to deal with homonymy and synonymy of unit symbols. The matching step is based on a list of units that does not contain homonymous symbols (e.g. uses "Gs" for gauss instead of the official "G"; fahrenheit has symbol "degF"). Matching using this list will miss correct matches (e.g. when "g" is used to refer to gauss).

Simplifications made by [7] include that they assume that quantities are only written with their full name, and units only written with their symbol. Both system's high performance (over 90% F-measure) are not likely to be reached on ambiguous data as found in repositories of research results. We conclude that existing systems do not sufficiently target the homonymy and synonymy problems. In the remainder of this section we discuss techniques used in other domains that may help solve these.

---

[1] Obtained from http://www.gnu.org/software/units/

Scoring functions. A usual technique for filtering out false positives and disambiguating between alternative candidates is to provide a *scoring function* and a threshold. The candidate with the highest score is accepted (if it scores above the threshold). We give two examples of scoring functions found in literature.

Firstly, the similarity of the whole document being annotated can be compared to already correctly annotated documents. Their vector representations are compared using cosine similarity. [5] uses this technique to disambiguate matches for the same text fragment, and to find matches missed earlier in the process (in the BioCreative effort where genes are detected in medical texts; a task similar to ours). Unfortunately, the "documents" in our domain usually contain little content (in natural language) to compare. Often there is no more information available beyond the text in the header row, which is already ambiguous itself. Secondly, an example of a scoring function specific to our domain is proposed by [7]. They observe that sometimes the data cells in a column contain units and can be used as evidence to disambiguate the column's quantity. Their function is composed of (1) cosine similarity of quantity to column header; and (2) average cosine similarity of units in that column to the quantity's units. Cosine similarity is computed on a vector representation of the terms; terms are first lemmatized. This function only works if the data cells in the column contains units, which is relatively rare in our datasets.

Ontology-based filtering and disambiguation. A useful *ontology-based scoring technique* is to use concepts related to the candidate concept. If these related concepts are detected in the text near to the candidate concept, this increases the likelihood that a candidate is correct. [5] implemented this technique so that the candidate genes for string "P54" are disambiguated by comparing the gene's species, chromosomal location and biological process against occurrences of species, location and process in the text surrounding "P54". We implement this technique for our domain through the relationship between units and their quantity listed in our ontology.

[7] use the value range of units stored in the ontology to filter out false positives. They look up the data values (numbers) in the column. If the values lie outside the unit's value range, the candidate is removed. This works on their data set and quantities, but this is not likely to work for large quantitative ontologies and varied datasets. For example, a temperature value of "-20" can only rule out the unit kelvin (its scale starts from 0), but leaves celsius and fahrenheit as possible interpretations. In case we are dealing with a relative temperature, then "-20" can even not strike kelvin from the list of candidates. Celsius and fahrenheit can only be disambiguated by a few actual values, which are unlikely to appear in actual measurements.

None of the techniques mentioned above addresses the problem of ambigous compound concepts (e.g. m/s might refer to meter per second or mile per siemens). We developed a solution that uses an ontology to determine whether the units together express a quantity that is defined in the ontology.

4   Materials

The data, annotator instructions, gold standard and ontology used in our evaluation are available online[2]. We start by giving a more detailed description of the problem.

4.1   Datasets

We use two datasets to develop and validate our approach. The first set is obtained from a data repository of researchers at the Dutch Top Institute Food and Nutrition.[3]

The second dataset was collected from the Web, especially from .edu, and .org sites and sites of scientific/academic organizations. The files were found through Google by querying for combinations of quantity names and unit symbols and filtering on Excel files, such as in "speed (m/s)" filetype:xls". Topics include: chemical properties of elements, throughput of rivers, break times and energy usage of motor cycles, length and weight of test persons.

Our datasets can be considered a "worst-case scenario". The dataset of [7] is simpler in that (1) quantities are always written in their full name and units with symbols only; (2) no abbreviations or misspellings occur; (3) no compound units appear; and (4) both data and ontology contain no ambiguous unit symbols. The dataset used by [1] may be simpler because the documents (patent specifications) are intended to be precise.

We make the assumption, like [7] and [1], that the header rows have already been identified and separated from the content rows. We have effectuated this assumption by deleting cells that do not belong to the table header from the Excel files used in our experiment.

4.2   Ontology

We use an ontology we developed, the *Ontology of Units of Measure and related concepts* (OUM) in the annotation process [11]. OUM's main classes are Quantity, Unit of Measure, Dimension and Application Area. (see Figure 1 for an overview). OUM currently consists of approximately 450 quantities and about 1,000 units. Concepts have English labels, an extension in Dutch is under development.

For each quantity the units in which it can be expressed are listed. For example, speed can be expressed in (amongst others) km/s and mm/s. Each unit belongs to one or more quantities. OUM groups similar quantities into classes. For example, Kinetic energy and Heat are subclasses of Energy.

Units can be split into singular units, multiples and submultiples, and compound units. *Singular units* (units with a special name) such as meter can be prefixed to create so-called multiples and submultiples (e.g. *kilo*meter, *milli*meter).

---

[2] See http://www.cs.vu.nl/~mark/iswc2010/. The food dataset was not included as it is commercially sensitive data.

[3] http://www.tifn.nl

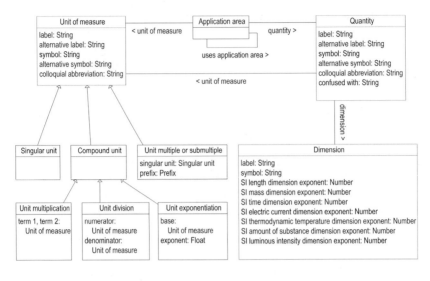

Fig. 1. UML diagram of main OUM classes and properties

Compound units are constructed by multiplying, dividing, or exponentiating units (e.g. m/s^2). Unit multiplications are linked to their constituent units through the properties term1 and term2, unit divisions are linked to their constituents through numerator and denominator.

Because units can be prefixed and composed, the number of possible units is almost endless. For example, units for the quantity velocity may be a combination of any unit for length (e.g. kilometer, centimeter, nordic mile) and any unit for time (hour, picosecond, sidereal year, etcetera). For practical reasons OUM only lists the more common combinations, but the analysis of what is "common" has not been finalised yet. As a consequence, for specific application areas some compound units may be missing. Each quantity or unit has one full name and one or more symbols. Each full name is unique, but words in the name can overlap (e.g. "magnetic field intensity", "luminous intensity").

Humans regulary confuse some quantities with each other (e.g. weight and mass). Our ontology records the concepts and their definitions as they are prescribed in standards, but for automated annotation it is useful to know which terms people use to denote these concepts. This dichotomy is well-known in the vocabulary world, and reflected in the SKOS standard through the skos:hidden-Label property[4]. It is used to record labels not meant for display but useful in searching. We introduce a property confused_with (subproperty of skos:hiddenLabel). By attaching the label "weight" to mass our annotation system will be able to generate mass as a candidate. In the same vein we introduce colloquial_abbreviation to denote often used abbreviations as "temp" and "freq" for temperature and frequency. Less than ten of such abbreviations and confusions are currently included.

---

[4] http://www.w3.org/TR/2009/NOTE-skos-primer-20090818/#sechidden

Quantities and units are sometimes used primarily in a particular *application area*. OUM specifies generic application areas, such as *space and time* (contains units such as mile and second). OUM also contains specific areas like *shipping* (contains nautical mile) and *astronomy* (contains sidereal second).

Quantities and units have dimensions, which are abstractions of quantities ignoring magnitude, sign and direction aspects. Analysis of dimensions is common practice in science and engineering [2]. It allows for example to detect errors in equations and to construct mathematical models of e.g. aircraft. OUM lists all dimensions which occur in practice, which can be used in disambiguation of compound units (see Section 5.4). The dimension of a quantity or unit can be viewed as a vector in a space spanned by an independent set of base vectors (i.e. base dimensions). For example, the quantity speed has a dimension that can be decomposed into base dimension *length* and base dimension *time* (with certain magnitudes as we show below). In principle we could also have expressed time in terms of base dimensions distance and speed. Each *system of units* used defines such a set of base dimensions to span the dimensional space. Each other dimension can be expressed as a combination of these base dimensions, each with a certain magnitude.

For example, the SI system of units has selected as its base dimensions *length (L), mass (M), time (T), electric current (I), thermodynamic temperature (Θ), amount of substance (N)* and *luminous intensity (J)*. Since all other dimensions can be computed by multiplication and division of one or more of these base dimensions, an arbitrary dimension can be expressed as multiplication $L^\alpha M^\beta T^\gamma I^\delta \Theta^\epsilon N^\zeta J^\eta$. If an exponent is 0, the respective basic quantity does not play a role. For example, the quantity velocity and unit cm/hr have SI-dimension $L^1 M^0 T^{-1} I^0 \Theta^0 N^0 J^0$, which is equivalent to $L^1 T^{-1}$ or length per time. A quantity or unit with a dimension for which all powers are 0 is said to be dimensionless. It is typically obtained as a ratio between quantities of equal dimension, such as strain or Reynolds number, and expressed as for example fractions or percentages.

5   Approach

We have divided the annotation process into the following steps: (0) table extraction; (1) tokenization; (2) basic matching; (3) matching compounds listed in OUM; (4) matching unknown compounds using dimensional analysis; (5) disambiguation. We do not treat the extraction step here; its output is a list of cells and their contents. Our main assumption is that the identification of the header row(s) has already been done.

5.1   Tokenization

The string value of a cell is separated into tokens by first splitting on spaces, underscores ("start_time") and punctuation marks (brackets, dots, stars, etc.). Number-letter combinations such as "100g" are separated, as are camel-cased tokens ("StartTime"). Basic classification of tokens into numbers, punctuation, and words is performed. Punctuation tokens that may represent multiplication (period, stars, dots), and division (slash) are also typed. Two other token types are detected: stopwords and a list of "modifiers" that are particular to this domain (e.g. mean, total, expected, estimated).

5.2   Basic Matching: Full Names and Symbols

Before matching takes place we generate several synonyms: plural forms of units (e.g. "meters"), contractions of compound unit symbols (e.g. "Pas" for pascal second), some alternative spellings (e.g. "C" for °C, s-1 vs. s^-1 vs. 1/s for reciprocal units, s^2 vs. s2 for exponentiated units). Because these can be generated systematically this is easier than adding them to the ontology.

Matching starts by comparing the input to full names of quantities and units, including confused_with and colloquial_abbreviations. The match with the highest score above a threshold is selected. We have used a string distance metric to overcome spelling mistakes, called Jaro-Winkler-TFIDF [3].

After full name matching is completed, a second matcher finds matches between input tokens and quantities/units based on their *symbols*, e.g. "f", "km", "s"). This is a simple exact match that ignores case. The outcome of this step will contain many ambiguous matches, especially for short unit and quantity symbols.

5.3   Matching: Compounds in OUM

The matches obtained in the basic matching in some cases represent compound units that are listed in OUM. For example, the previous step will return for the cell C.m the matches calorie, coulomb, meter, nautical mile. We detect that this is the compound coulomb meter by detecting that some of the unit matches are constituents of a compound listed in OUM. Comparison to a unit multiplication uses the properties term1 and term2, for comparing to unit division the properties numerator and denominator. In the latter case the additional constraint is that units have to appear in the input in the order prescribed (first numerator, then denominator). The punctuation used in the input determines whether we are dealing with a multiplication or a division. Notice that this step already helps to disambiguate matches; in this case calorie and nautical mile could be excluded.

A special case are compounds consisting of (sub)multiple units, e.g. μNm which stands for micronewton meter. Because OUM only lists newton meter, we have to first detect the prefix (in this case μ, other prefixes include m, M, k, T), remove it and then perform the compound check described above.

5.4   Matching: Compounds Not in OUM

The previous step will miss compound units not listed in OUM. If the unit symbols in the compound are not ambiguous, we can assume that this interpretation is correct. However, in many cases the symbols are ambiguous. For example, g/l can either denote gauss per liter or gram per liter. A way to disambiguate is to find out if the compound expresses a quantity that is listed in OUM. The quantity implied by the compound can be computed using the dimensional properties of the units (also listed in OUM).

The first step is to compute the overall dimension of the compound based on the individual units, the second step is to check whether a quantity with that dimension exists in OUM. Computing dimensions is a matter of subtracting the dimension exponents of the underlying elementary dimensions. Each unit is associated with an instance of Dimension, which in turn lists the dimension exponents through the properties SI length exponent, SI time exponent, etcetera. If, for example, we interpret g/l as gram per liter, we retrieve the units' dimensions (mass-dimension and volume-dimension, respectively). Then we divide the dimensional exponents of mass $L^0 M^1 T^0 l^0 \Theta^0 N^0 J^0$ by the dimensional exponents of volume $L^3 M^0 T^{-1} l^0 \Theta^0 N^0 J^0$ which gives $L^{-3} M^1 T^{-1} l^0 \Theta^0 N^0 J^0$ These dimensional exponents match exactly with the dimensions of the quantity density. On the other hand, viewing g as gauss would yield $L^{-3} M^1 T^{-2} l^{-1} \Theta^0 N^0 J^0$ for the dimension of the compound unit, which does not correspond to the dimension of any quantity in OUM.

This step is implemented by normalizing the input string, constructing a tree representation of the compound through a grammar parser, assigning the units to it, and sending it to a service that calculates the implied dimension components.

An interesting option in the future is to automatically enrich OUM with new compounds that pass the above test, and add them to OUM. This would be a valid way to continuously extend the set of compound units in OUM, not in an arbitrary manner, but learning from actual occurrences in practice. If we combine this with monitoring which compound units are never used in practice (but were added for theoretical reasons or just arbitrarily), a reliable mechanism for maintaining a relevant set of compound units in OUM would be created.

5.5   Disambiguation

The previous step will still contain ambiguous matches, e.g. for the cells f (Hz) and wght in g. We have developed a set of heuristics or "rules" to remove the remaining ambiguities.[5] First we list domain-specific pattern matching rules in the style of [1], then three disambiguation rules that make use of relations in the ontology (rules 7, 8 and 9).

Rule 1: SYMBOLS IN BRACKETS USUALLY REFER TO UNITS. For example, "s" in delay (s) refers to second and not area or entropy.

Rule 2: PREFER SINGULAR UNITS OVER (SUB)MULTIPLES. Symbols for singular units (e.g. pascal (Pa)) overlap with symbols for (sub)multiples (e.g. picoampere (pA)). In these cases, select the singular unit because it is more likely.

---

[5] Formulated as "rules" for reading convenience, but both the rules and previous "steps" can be implemented differently.

Rule 3: A SYMBOL THAT FOLLOWS A NUMBER USUALLY REFERS TO A UNIT. For example, 100 g refers to gram. This disambiguation deletes six potential quantity matches for "g", and retains units gram and gauss. (Rule also used by [1].)

Rule 4: TAKE LETTER CASE INTO ACCOUNT FOR LONGER SYMBOLS. People are sloppy in the correct letter case of symbols. One-letter symbols such as "t" may stand for temperature (T) or tonne (t). Two-letter symbols as "Km" may stand for kilometer (km) or maximum spectral luminous efficacy (Km). Casing used in the text cannot be trusted to disambiguate; the context usually does make clear which is meant. However, casing used in writing down units of three or more letters is usually reliable. For example, (sub)multiples such as mPa and MPa (milli/megapascal) are usually written correctly. Humans pay more attention to submultiples because errors are hard to disambiguate for humans too. We thus perform disambiguation based on case if the symbol is three letters or longer.

Rule 5: MODIFIER WORDS USUALLY APPEAR BEFORE QUANTITIES, NOT UNITS. For example, mean t or avg t is an indication that "t" stands for the quantity Time instead of the unit tonne. The idea of using specific types of tokens to improve correct concept detection is due to [6] in the gene annotation domain.

Rule 6: TOO MANY SYMBOL MATCHES IMPLIES IT IS NOT A QUANTITY OR UNIT. If previous steps were not able to disambiguate a symbol that has many candidate matches (e.g. "g" can match ten quantities and units), then the symbol probably does not refer to a quantity or unit at all (it might be a variable or e.g. part of the code of product). For such an ambiguous symbol, humans usually provide disambiguating information, such as the quantity. We therefore delete such matches. This rule can hurt recall, but has a greater potential to improve precision which will pay off in the F-measure. This rule should be executed after all other rules.

Rule 7: SYMBOLS THAT REFER TO RELATED QUANTITIES AND UNITS ARE MORE LIKELY THAN UNRELATED QUANTITIES AND UNITS. For example, T (C) is more likely to refer to temperature and celsius than to time and coulomb. The former pair is connected in OUM through property unit_of_measure (domain/range Quantity/Unit), while the latter pair is not. We filter out the second pair of matches. We first apply this rule on quantities and units in the same cell. This rule also allows to select the quantity mass for cell weight (g) instead of the erroneous weight. Mass was found in basic matching through its confused_with label. We repeat application of the rule on the whole table after application on single cells. A quantity mentioned in one cell (e.g. mass) can thus be used to disambiguate cells where the quantity was omitted (e.g. containing only "g"). During application of this rule we prefer matches on preferred symbols over matches on non-preferred ("alternative") symbols. For example, cell Length (m) matches length-meter (meter has symbol "m") which we prefer over length-nautical mile (mile has alternative_symbol "m").

Rule 8: CHOOSE THE MOST SPECIFIC QUANTITY THAT MATCHES THE EVIDENCE. Generic quantities such as Viscosity and Temperature have specific instances such as kinematic viscosity and celsius temperature. The user may have meant the specific quantity. If a unit is given, this can be disambiguated. For example, viscosity expressed in stokes means that kinematic viscosity was meant. When poise is used, dynamic viscosity was meant. In other cases, the units of the specific quantities overlap, so that the proper quantity cannot be determined (e.g. diameter and radius are forms of Length measured in units such as meter).

Rule 9: CHOOSE THE INTERPRETATION BASED ON THE MOST LIKELY APPLICATION AREA. Symbols such as "m" can refer to units from a generic application area or a specific application area (e.g. nautical mile in *shipping* or meter in *space and time*). If there is evidence that the table contains measurements in a specific area then all ambiguous units can be interpreted as a unit used in that area, instead of those in more generic areas. If there is no such evidence, the unit from the generic area is more likely. As evidence that the observations concern a specific area we currently accept that the table contains at least one unambiguous unit that is particular to that area (i.e. written in its full name). Other types of evidence can be taken into account in the future (e.g. column name "distance to star").

5.6 Implementation

We developed a prototype implementation of our annotation approach in Java. It provides a simple framework to implement matchers and disambiguation rules. Our matchers and disambiguation rules can probably also be implemented as JAPE rules on top of GATE; this is future work.

The Excel extractor uses the Apache POI library[6]. The prototype can emit the parsed and annotated tables as RDF files or as CSV files. For representing and manipulating the OUM ontology and the output as objects in Java we used the Elmo framework[7] with Sesame as RDF backend. For string metrics we use the SecondString[8] library developed by Cohen et al. The parser for compound units was built using YACC.

6 Evaluation and Analysis

6.1 Evaluation Type and Data Selection

We evaluate our approach by measuring recall and precision against a gold standard for two datasets. We could not measure the performance of our system on the data of [1] because it is not publicly available. Comparison against the data of [7] is not useful as they identify only a few (unambiguous) quantities and units.

---

[6] http://poi.apache.org/
[7] http://www.openrdf.org/doc/elmo/1.5/
[8] http://secondstring.sourceforge.net/

The tables were selected as follows. We randomly selected files from the food dataset and removed those that were unsuitable for our experiment because they were (1) written in Dutch; or (2) contained no physical quantities/units; or (3) had the same header as an already selected file (this occurs because measuring machines are used that produce the same table header each time). We kept selecting until we obtained 39 files. Selection of 48 Web tables was also random; no tables had to be removed.

The number of correctly and wrongly assigned URIs is counted on a per-document basis, by comparing the set of URIs returned by the system with the set of URIs of the human, ignoring the cell in which they were found. Based on the total number of correct/wrong/retrieved URIs, the macro-averaged precision and recall is calculated (each correct/wrong URI contributes evenly to the total score).[9]

6.2   Gold Standard Creation

The files were divided over three annotators (the authors). They used an Excel add-in [11] developed in earlier work that allows selection of concepts from OUM. Each cell could be annotated with zero or one quantity, and zero or one unit. The annotators were incouraged to use all knowledge they could deduce from the table in creating annotations. If the exact quantity was not available in OUM, a more generic quantity was selected. For example, the cell half-life (denoting the quantity for substance decay) was annotated with Time. After that, each file was checked on consistency by one of the authors.

Compound units that do not appear in OUM can not be annotated by assigning a URI to them (simply because they have no URI in OUM). They were put in a separate result file and were compared by hand.

6.3   Results

We have tested different configurations (Table 1). Firstly, a *baseline* system that only detects exact matches, including our strategies to enhance recall such as con-traction of symbols and generation of plural forms (comparable to [7]'s system). Secondly, with flexible string matching turned on. Thirdly, with pattern disam-biguation rules turned on (rules 1-6); this may be comparable to the GATE-based system [1]. We cannot be certain because their system is not open source. This indicated what can be achieved with pattern matching only. Fourthly, with also compound detection and ontology-based rules turned on (rules 7-9) .

The following points are of interest. Firstly, the baseline scores show that the extent of the ambiguity problem is different for quantities and units. Performance

---

[9] A comparison per cell would introduce a bias towards frequently occurring quanti-ties/units, which either rewards or punishes the system for getting those frequent cases right. Micro-averaging calculates precision/recall for each document and takes the mean over all documents. A single annotation may contribute more or less to the total precision or recall, depending on whether it appears in a document with little or a lot of annotations.

for quantities is not high (F-measure ranging from 0.09 to 0.20), while F-measure for units is already reasonable (around 0.40). It turns out that the datasets in our experiment relatively often use non-ambigous unit symbols, including "N" for newton and "sec" for second. Secondly, flexible string matching does not help to increase recall (threshold 0.90 was used but no clear increase was seen at 0.85 either). The results of the remaining two configurations are obtained with flexible matching turned off. Thirdly, pattern matching rules help considerable, improving F-measure with 0.15-0.60. Fourthly, ontology-based disambiguation increases the F-measure further for units: 0.16-0.25. The results for quantities are mixed: 0.07 increase in the Food dataset, no difference in the Web dataset. Fifthly, in the Web dataset unit scores are higher than quantity scores, and the other way around in the Food dataset.

Table 1. Results of evaluation. Separate precision (P), recall (R) and F-measure (F) are given for both datasets, based on macro-averaging. Best F-measures are in bold.

	Food						Web					
	Quantities			Units			Quantities			Units		
	P	R	F	P	R	F	P	R	F	P	R	F
baseline	0.11	0.84	0.20	0.30	0.61	0.40	0.05	0.70	0.09	0.29	0.61	0.40
flex. match	0.11	0.84	0.20	0.29	0.61	0.39	0.05	0.72	0.09	0.28	0.61	0.39
pat. rules	0.78	0.82	0.80	0.50	0.57	0.53	0.63	0.64	**0.63**	0.50	0.57	0.53
full	0.83	0.93	**0.87**	0.72	0.83	**0.78**	0.59	0.67	**0.63**	0.63	0.76	**0.69**

6.4 Qualitative Analysis

We analyzed the causes for false positives and false negatives in the results. The following should be highlighted. Firstly, the performance of the pattern rules is not improved upon as much as we had expected in the case of quantities. One explanation is that many of the symbols in the input did not represent a quantity, and the pattern rules successfully filter these false positives out through rule 6. In the future we will try our method on more varied datasets to determine if this effect is consistent or not.

Secondly, some quantities are simply missing in OUM, such as half life and resonance energy. The annotators used the more generic quantity (time and molar energy) to annotate the cells where they appear. The generic quantities are not not found because there is no lexical overlap. This can be solved by adding them or importing them from another ontology. Thirdly, a number of quantities is not found because they are not mentioned explicitly, but implied. For example, letters X and Y are used to indicate a coordinate system, and thus imply length. Failing to detect the quantity also causes loss of precision in unit detection: the quantity would help to disambiguate the units through rule 7. These issues points to the importance of a high-coverage ontology.

Fourthly, another cause for missed quantities is that the object being measured is stated, which together with the unit implies the quantity. For example, the cell Stock (g), refers to quantity mass as the word "stock" implies a food product (stock is a basis for making soup). This can be solved by using more ontologies in the matching step, and link concepts from those ontologies to OUM. For example, a class Food product could be linked to quantities that are usually measured on food products such as mass. Because field strength is not one of those quantities, the erroneous match gauss could be removed.

Fifthly, some of the problems are difficult to solve as very case-specific background knowledge would be required. For example, cells Lung (L) and Lung (R) produce false positive matches such as röntgen and liter and can only be solved with knowledge on human fysiology.

Lastly, analysis of the detection of compounds not in OUM shows that this step performed well at recognizing unit divisions (kilojoule per mole, newton per square millimeter). However, its performance is degraded considerably by false positives such as dP for decapoise and V_c for volt coulomb.

7   Discussion

In this paper we have studied annotation of quantitative research data stored in tables. This is relevant to today's world because scientists, companies and governments are accumulating large amounts of data, but these datasets are not semantically annotated. We presented several ways in which an ontology can help solve the ambiguity problems: (1) detection of compound units present in the ontology; (2) dimensional analysis to correctly interpret compound units not explicitly listed in the ontology; (3) identification of application areas to disambiguate units; and (4) identification of quantity-unit pairs to disambiguate them both. Especially the performance for unit detection is good. This is positive, as correct unit detection is more important than correct quantity detection: the quantity can be derived from the unit using the ontology. For example, time can be derived from millisecond. Even when the right specific quantity is not known (e.g. half-life), the more generic quantity that could be derived is a suitable starting point for data integration. For example, to integrate two datasets about the half-life of elements it is sufficient to know that columns are being merged that deal with time (if the units are not the same they can be automatically converted into each other).

However, performance is still far from perfect. We have suggested several ways in which performance may be improved, of which linking ontologies about the objects being measured is an attractive one. One promising line of future work is the application of machine learning (ML) techniques to the disambiguation problem. However, this is not straightforward since our domain lacks the typical features that ML approaches rely on, e.g. those based on the surrounding natural language text. We do see possibilities to use the properties of the candidate concepts as features and thus combine our rule-based approach with a machine learning approach – as e.g. proposed by [10]. This would require a larger annotated dataset to serve as training and test set.

An implication of this work for the Web of Data is that conversion tools need to be tuned to the domain at hand. Current tools target sources that are already structured to a large extent, but if the Web of Data is to grow, more unstructured sources should be targeted. The work of [9] already suggests to include an annotation system into a conversion tool, but the annotation system is generic. As shown a generic system will fail to capture the semantics of this domain. A system that can be configured for the domain is required.

Acknowledgements

This work was carried out within the Food Informatics subprogram of the Virtual Laboratory for e-Science, a BSIK project of the Dutch government. We thank Jeen Broekstra for implementation advice, Remko van Brakel for the Excel export tool, and Laura Hollink and Tuukka Ruotsalo for their comments.

References

1. Agatonovic, M., Aswani, N., Bontcheva, K., Cunningham, H., Heitz, T., Li, Y., Roberts, I., Tablan, V.: Large-scale, parallel automatic patent annotation. In: Conference on Information and Knowledge Management (2008)
2. Bridgman, P.: Dimensional Analysis. Yale University Press, New Haven (1922)
3. Cohen, W., Ravikumar, P., Fienberg, S.E.: A comparison of string distance metrics for name-matching tasks. In: Proc. of IJCAI 2003 Workshop on Inf. Integration, pp. 73–78 (2003)
4. Ding, L., DiFranzo, D., Magidson, S., McGuinness, D.L., Hendler, J.: The Datagov Wiki: A Semantic Web Portal for Linked Government Data. In: Bernstein, A., Karger, D.R., Heath, T., Feigenbaum, L., Maynard, D., Motta, E., Thirunarayan, K. (eds.) ISWC 2009. LNCS, vol. 5823. Springer, Heidelberg (2009)
5. Hakenberg, J., Royer, L., Plake, C., Strobelt, H., Schroeder, M.: Me and my friends: gene mention normalization with background knowledge. In: Proc. 2nd BioCreative Challenge Evaluation Workshop, pp. 1–4 (2007)
6. Hanisch, D., Fundel, K., Mevissen, H., Zimmer, R., Fluck, J.: ProMiner: rule-based protein and gene entity recognition. BMC bioinformatics 6(Suppl. 1), S14 (2005)
7. Hignette, G., Buche, P., Dibie-Barthélemy, J., Haemmerlé, O.: Fuzzy Annotation of Web Data Tables Driven by a Domain Ontology. In: Aroyo, L., Traverso, P., Ciravegna, F., Cimiano, P., Heath, T., Hyvönen, E., Mizoguchi, R., Oren, E., Sabou, M., Simperl, E. (eds.) ESWC 2009. LNCS, vol. 5554, p. 653. Springer, Heidelberg (2009)
8. Langegger, A., Woss, W.: Xlwrap - querying and integrating arbitrary spreadsheets with sparql. In: Bernstein, A., Karger, D.R., Heath, T., Feigenbaum, L., Maynard, D., Motta, E., Thirunarayan, K. (eds.) ISWC 2009. LNCS, vol. 5823, pp. 359–374. Springer, Heidelberg (2009)
9. Lynn, S., Embley, D.W.: Semantically Conceptualizing and Annotating Tables. In: Domingue, J., Anutariya, C. (eds.) ASWC 2008. LNCS, vol. 5367, pp. 345–359. Springer, Heidelberg (2008)
10. Medelyan, O., Witten, I.: Thesaurus-based index term extraction for agricultural documents. In: Proc. of the 6th Agricultural Ontology Service (AOS) Workshop at EFITA/WCCA (2005)
11. Rijgersberg, H., Wigham, M., Top, J.L.: How semantics can improve engineering processes - a case of units of measure and quantities (2010); accepted for publication in Advanced Engineering Informatics

JustBench: A Framework for OWL Benchmarking

Samantha Bail, Bijan Parsia, and Ulrike Sattler

The University of Manchester
Oxford Road, Manchester, M13 9PL
{bails,bparsia,sattler}@cs.man.ac.uk

Abstract. Analysing the performance of OWL reasoners on expressive OWL ontologies is an ongoing challenge. In this paper, we present a new approach to performance analysis based on *justifications* for entailments of OWL ontologies. Justifications are minimal subsets of an ontology that are sufficient for an entailment to hold, and are commonly used to debug OWL ontologies. In *JustBench*, justifications form the key unit of test, which means that *individual justifications* are tested for correctness and reasoner performance instead of entire ontologies or random subsets. Justifications are generally small and relatively easy to analyse, which makes them very suitable for transparent analytic micro-benchmarks. Furthermore, the JustBench approach also allows us to isolate reasoner errors and inconsistent behaviour. We present the results of initial experiments using JustBench with FaCT++, HermiT, and Pellet. Finally, we show how JustBench can be used by reasoner developers and ontology engineers seeking to understand and improve the performance characteristics of reasoners and ontologies.

1 Introduction

The Web Ontology Language (OWL) notoriously has very bad worse case complexity for key inference problems, at least, OWL Lite (EXPTIME-complete for satisfiability), OWL DL 1 & 2 (NEXPTIME-complete), and OWL Full (undecidable) (see [5] for an overview). While there are several highly optimised reasoners (FaCT++, HermiT, KAON2, Pellet, and Racer) for the NEXPTIME logics, it remains the case that it is frustratingly easy for ontology developers to get unacceptable or unpredictable performance from them on their ontologies. Reasoner developers continually tune their reasoners to user needs in order to remain competitive with other reasoners. However, communication between reasoner developers and users is tricky and, especially on the user side, often mystifying and unsatisfying.

Practical OWL DL reasoners are significantly complex pieces of software, even just considering the core satisfiability testing engine. The basic calculi underlying them are daunting given that they involve over a dozen inference rules with complex conditions to ensure termination. Add in the extensive set of optimisations

P.F. Patel-Schneider et al. (Eds.): ISWC 2010, Part I, LNCS 6496, pp. 32–47, 2010.

and it is quite difficult for non-active reasoner developers to have a reasonable mental model of the behaviour of reasoners. Middleware issues introduce additional layers of complexity ranging from further optimisations (for example, classification vs. isolated subsumption tests) to the surprising effects of different parsers on system performance.

In this paper, we present a new approach to analysing the behaviour of reasoners by focusing on justifications of entailments. Justifications—minimal entailing subsets of an ontology—already play a key role in debugging unwanted entailments, and thus are reasonably familiar to users. They are small and clearly defined subsets of the ontology that can be analysed manually if necessary, which reduces user effort when attempting to understand the source of an error in the ontology or unwanted reasoner behaviour. We present results from analysing six ontologies and three reasoners and argue that justifications provide a reasonable starting point for developing empirically-driven analytical micro-benchmarks.

2   Reasoner Behaviour Analysis

2.1   Approaches to Understanding Reasoner Behaviour

Consider five approaches to understanding the behaviour of reasoners on a given ontology, especially by ontology modellers:

1. **Training.** In addition to the challenges of promulgating detailed understanding of the performance implications of the suite of calculi and associated optimisations (remembering that new calculi or variants thereof are cropping up all the time), it is unrealistic to expect even sophisticated users to master the engineering issues in particular implementations. Furthermore, it is not clear that the requisite knowledge is available to be disseminated: New ontologies easily raise new performance issues which require substantial fundamental research to resolve.

2. **Tractable logics.** In recent years, there has been a renaissance in the field of tractable description logics which is reflected in the recent set of tractable OWL 2 profiles.[1] These logics tend to not only have good worst case behaviour but to be "robust" in their performance profile especially with regard to scalability. While a reasonable choice for many applications, they gain their performance benefits by sacrificing expressivity which might be required.

3. **Approximation.** Another approach is to give up on soundness or completeness when one or the other is not strictly required by an application, or, in general, when some result is better than nothing. Approximation [17,3,16] can either be external (e.g., a tool which takes as input an OWL DL ontology and produces an approximate OWL EL ontology) or internal (e.g., anytime computation or more sophisticated profile approximation). A notable difficulty of approximation approaches is that they require *more* sophistication

---

[1] http://www.w3.org/TR/2009/REC-owl2-profiles-20091027

on the part of users and sophistication of a new kind. In particular, they need to understand the semantic implications of the approximation. For example, it would be quite simple to make existing reasoners return partial results for classification—classification is inherently anytime. But then users must recognise that the absence of an entailment no longer reliably indicates non-entailment. In certain UIs (such as the ubiquitous tree representations), it is difficult to represent this additional state.

4. **Fixed rules of thumb.** These may occur as a variant or result of training or be embodied in so-called "lint" tools [12]. The latter is to be much preferred as such tools can evolve as reasoners do, whereas "folk knowledge" often changes slowly or promulgates misunderstanding. For example, the rules of thumb "inverses are hard" and "open world negation is less efficient than negation as failure"[2] do not help a user determine which (if either) is causing problems in their particular ontology/reasoner combination. This leads users to start ripping out axioms with the "dangerous" constructs in them which, e.g., for negation in the form of disjointness axioms, may in fact make things worse. Lint tools fare better in this case but do not support *exploration* of the behaviour of a reasoner/ontology combination, especially when one or the other does not fall under the lint tools coverage. Finally, rules of thumb lead to *manual* approximation which can distort modelling.

5. **Analytical tools.** The major families of analytical tools are profilers and benchmarks. Obviously, one can use standard software profilers to analyse reasoner/ontology behaviour, and since many current reasoners are open source, one can do quite well here. This, however, requires a level of sophistication with programming and specific code bases that is unreasonable to demand of most users. While there has been some work on OWL specific profilers [19], there are none, to our knowledge, under active development. Benchmarks, additionally, provide a common target for reasoner developers to work for, hopefully leading to convergence in behaviour. On the flip side, benchmarks cannot cover all cases and excessive "benchmark tuning" can inflate reasoner performance with respect to the benchmarks without improving general behaviour in real cases.

2.2   Benchmarks

Application and Analytical Benchmarks. For our current purposes, a *benchmark* is simply a reasoning problem, typically consisting of an ontology and an associated entailment. A *benchmark suite*, although often called a benchmark or benchmarks, is a set of benchmarks.

We can distinguish benchmark suites by three characteristics: their *focus*, their *realism*, and their *method of generation*. With regard to focus, the classic distinction is between *analytical* benchmarks and *application* benchmarks.

Analytical benchmarks attempt to determine the presence or absence of certain performance related features, e.g., the presence of a query optimiser in a

---

[2] This latter rule of thumb is actually *false* in general. Non-monotonic features generally *increase* worst case complexity, often quite significantly.

relational database can be detected[3] by testing a query written in sub-optimal form. More generally, they attempt to isolate particular behaviours of the system being analysed.

Application benchmarks attempt to predict the behaviour of a system on certain classes of application by testing an example (or select examples) of that class. The simplest form of an application benchmarking is retrospective recording of the behaviour of the application on the system in question in real deployment (i.e., performance measurement). Analytical benchmarks aim to provide a more precise understanding of the tested system, but that precision may not help predict how the system will perform in production. After all, an analytical benchmark does not say which part of the system will be stressed by any given application. Application benchmarks aim for better predictions of actual behaviour in production, but often this is at the expense of understanding. Accidental or irrelevant features might dominate the benchmark, or the example application may not be sufficiently representative.

In both cases, benchmark suites might target particular classes of problem, for example, conjunctive query answering at scale in the presence of $\mathcal{SHIQ}$ TBoxes.

Choice of Reasoning Problems. In order to be reasonably analytic, benchmarks need to be understandable enough so that the investigator can correlate the benchmark and features thereof with the behaviour observed either on theoretical grounds, e.g., the selectivity of a query, or by experimentation, e.g. by making small modifications to the test and observing the result. If we have a good theoretical understanding, then individual benchmarks need not be small. However, we do not have a good theoretical understanding of the behaviour of reasoners on real ontologies and, worse, real ontologies tend to be extremely heterogenous in structure, which makes sensible uniform global modifications rather difficult. While we we can measure and compare the performance of reasoners on real ontologies, we often cannot understand or analyse *why* some (parts of) ontologies are particularly hard for a certain reasoner—or even isolate these parts. Thus, we turn to subsets of existing ontologies. However, arbitrary subsets of an ontology are unlikely to be informative and there are too many for a systematic exploration of them all. Thus, we need a selection principle for subsets of the ontology. In JustBench, our initial selection principle is to select *justifications* of atomic subsumptions, which will be discussed in section 3.

Artificial subsets. Realism forms an axis with completely artificial problems at one pole, and naturally occurring examples at the other. The classic example of an artificial problem is the kSAT problem for propositional, modal, and description logics [7,18,10,11]. kSAT benchmark suites are presented in terms of how to generate random formulae (to test for satisfiability) according to certain parameters. Some of the parameters can be fixed for any test situation (e.g.

---

[3] The retrospective on the Wisconsin Benchmark [6] for relational databases has a good discussion of this.

clause length which is typically 3) and others are allowed to vary within bounds. Such benchmarks are comparatively easy to analyse theoretically[4] as well as empirically.

However, these problems may not even *look* like real problems (kSAT formulae have no recognisable subsumption or equivalence axioms), so extrapolating from one to the other is quite difficult. One can always use naturally occurring ontologies when available, but they introduce many confounding factors. This includes the fact that users tend to modify their ontologies to perform well on their reasoner of choice. Furthermore, it is not clear that existing ontologies will resemble future ontologies in useful ways. This is especially difficult in the case of OWL DL due to the fragility of reasoner behaviour: seemingly innocuous changes can have severe performance effects. Also, for some purposes, existing ontologies are not hugely useful—for example, for determining scalability, as existing ontologies can only test for scalability up to their actual size.

The realism of a benchmark suite can constrain its method of generation. While artificial problems (in general) can be hand built or generated by a program, naturally occurring examples have to be found (with the exception of naturally occurring examples which are generated e.g., from text or by reduction of some other problem to OWL). Similarly, application benchmarks must be at least "realistic" in order to be remotely useful for predicting system behaviour on real applications.

Modules. A module is a subset of an ontology which captures "everything" an ontology has to say about a particular subsignature of the ontology [4], that is, a subset which entails everything that the whole ontology entails which can be expressed in the signature of the module itself. Modules are attractive for a number of reasons including the fact that they capture all the relevant entailments and support a principled removal of "slow" parts of an ontology. However, most existing accounts of modularity are very fine grained with respect to signature choice, which preclude blind examination of all modules of an ontology.

If we restrict attention to modules for the signature of an atomic subsumption (which corresponds more closely to justifications for atomic subsumptions) we find that modules can be too big. First, at least by current methods, modules contain all justifications for *all* entailments expressible in their signature. As we can see in the Not-Galen ontology, this can lead to very large sets even just considering one subsumption. Second, current and prospective techniques involve various sorts of approximation which brings in additional axioms. While this excess is reasonable for many purposes, and might be more realistic as a model for a stand alone ontology, it interferes with the analysability of the derived benchmark. That being said, modules clearly have several potential roles for benchmarking, and incorporating them into *JustBench* is part of our future work.

---

[4] "Easy" in the sense of possible and feasible enough that analyses eventually emerge.

2.3   Existing OWL Benchmarks

The most popular reasoner benchmark, at least in terms of citation count, is the Lehigh University Benchmark (LUBM) [9]. LUBM is designed primarily to test the scalability of conjunctive query and consists of a small, simple, hand-built "realistic" ontology, a program for generating data conforming to that ontology, and a set of 14 hand-built "realistic" queries. LUBM is an application focused, realistic benchmark suite with artificial generation. LUBM's ontology and data were notoriously weak, for example, the ontology lacked coverage of many OWL features, a fact that the University Ontology Benchmark (UOBM) [13] was invented to rectify. For an extensive discussion and critique of existing synthetic OWL benchmarks see [20].

Several benchmarks suites, notable those described in [14,8], make use of naturally occurring ontologies, but do not attempt fine grained analysis of how the reasoners and ontologies interact. Generally, it can be argued that the area of transparent micro-benchmarks based on real (subsets of) OWL ontologies, as opposed to comprehensive (scalabiliy-, system-, or application) benchmarks is currently neglected.

3   Justification-Based Reasoner Benchmarking

Our goal is to develop a framework for benchmarking ontology TBoxes which is analytic, uses real ontologies, and supports the generation of problems. In order to be reasonably analytic, particular benchmarks need to be understandable enough so that the investigator can correlate the benchmark and features thereof with the behaviour observed either on theoretical grounds, e.g., the selectivity of a query, or by experimentation, e.g., by making small modifications to the test and observing the result. If we have a good theoretical understanding, then individual benchmarks need not be small. However, we do not have a good theoretical understanding of the behaviour of reasoners on (arbitrary) real ontologies and, worse, real ontologies tend to be extremely heterogenous in structure, which makes sensible uniform global modifications rather difficult. Thus, we turn to subsets of existing ontologies. However, arbitrary subsets of an ontology are unlikely to be informative and there are too many for a systematic exploration of them all. Thus, we need a selection principle for subsets of the ontology. In JustBench, our initial selection principle is to select *justifications* of entailments, e.g., of atomic subsumptions.

Definition 1 (Justification). A set of axioms $J \subseteq \mathcal{O}$ is a justification for $\mathcal{O} \models \eta$ if $J \models \eta$ and, for all $J' \subset J$, it holds that $J' \nvDash \eta$.

As an example, the following ontology[5] entails the atomic subsumption *C SubClassOf: owl:Nothing*, but only the first three axioms are necessary for the

---

[5] We use the Manchester OWL Syntax for all examples, omitting auxiliary declarations of entities for space and legibility reasons.

entailment to hold. Therefore, the set { *C SubClassOf: A and D, A SubClassOf: E and B, B SubClassOf not D and r some D*} is a justification for this entailment.

O = { *C SubClassOf: A and D,*
        A SubClassOf: E and B,
        B SubClassOf: not D and r some D,
        F SubClassOf: r only A,
        D SubClassOf: s some owl:Thing }

The size of a justification can range, in principle, from a single axiom to the number of all axioms in the ontology, with, in one study, an average of approximately 2 axioms per justification [1]. The number of justifications for an entailment can be exponential in the size of the ontology, and multiple (potentially overlapping) justifications for a single entailment occur frequently in ontologies used in practice.

An explanation framework that provides methods to exhaustively compute all justifications for a given entailment has been developed for the OWL API v3,[6] which we use in our benchmarking framework.

3.1   Limitations of this Selection Method

Justifications, while having several attractive features as benchmarks, also have drawbacks including: First, we can only generate test sets if computing at least some of the entailments and at least some of their justifications for them is feasible with at least one reasoner. Choice of entailment is critical as well, although, on the one hand, we have a standard set of entailments (atomic subsumptions, instantiations, etc.) and on the other hand we can analyse arbitrary sets of entailments (e.g., conjunctive queries derived from an application). As the test cases are generated by a reasoner, their correctness is determined by the correctness of the reasoner, which itself is often at issue. This problem is mitigated by checking individual justifications on all reasoners (for soundness) and using different reasoners to generate all entailments and their justifications (for completeness). The latter is very time consuming and infeasible for some ontologies.

Second, justification-based tests do not test scalability, nor do they test interactions between unrelated axioms, nor do they easily test non-entailment finding, nor do they test other global effects. With regard to scalability, we have two points: 1) Not every performance analysis needs to tackle scalability. For example, even if a reasoner can handle an ontology (thus, it scales to that ontology), its performance might be less than ideal. 2) Analysis of scalability problems needs to distinguish between reasoner weaknesses that are merely due to scale and those that are not. For example, if a reasoner cannot handle a particular two line ontology, it will not be able to handle that ontology with an additional 400 axioms. Thus, micro-benchmarks are still useful even if scalability is not relevant.

---

[6] http://owlapi.sourceforge.net

Finally, in the first instance, justification test successful entailment finding, but much of what an OWL reasoner does is find non-entailments. Non-entailment testing is a difficult matter to support analytically, however, even their justifications offer some promise. For example, we could work with repaired justifications.

3.2   JustBench: System Description

The *JustBench* framework is built in Java using the OWL API v3 and consists of two main modules that generate the justifications for an ontology and perform the benchmarks respectively. The generator loads an ontology from the input directory, finds entailments using the *InferredOntologyGenerator* class of the OWL API and generates justifications for these entailments with the explanation interface. The entailments in question are by adding specific *InferredAxiomGenerators* to the ontology generator. For example, one can add *InferredSubClassAxiomGenerator* to get all subsumptions between named classes and *InferredClassAssertionAxiomGenerator* to get all atomic instantiations. By default, we just look for atomic subsumptions and unsatisfiable classes. The justifications and entailments are saved in individual OWL files which makes them ready for further processing by the benchmarking module.

For each performance measurement, a new instance of the OWLReasoner class is created which loads the justification and checks whether it entails the subsumption saved as *SubClassOf* axiom in the entailment ontology. We measure the times to create the reasoner and load the ontology, the entailment check using the *isEntailed()* call to the respective reasoner, and the removal of the reasoner instance with *dispose()*. Regarding the small run-times of the entailment checks, there exists a trade-off between fine-grained, transparent micro-benchmarks and large test cases, where the results may be more robust to interference, but also harder to interpret for users. Limiting the impact that actions in the Java runtime and the operating system have on the measurements is an important issue when benchmarking software performance [2], which we take into account in our framework. In order to minimise measurement variation, the sequence of load, check, dispose is repeated a large number of times (1000 in our current setting) and the median of the values measured after a warm-up phase is taken as the final result. In preliminary tests it was detected that the mean value of the measurements was distorted due to a small number of outliers that differed from the majority of the measured values by several orders of magnitude, which was presumably caused by the JVM garbage collection. Basing the measurement on the median instead proved to yield stable and more reliable results.

We also experimented with a slightly different test involving a one-off call to *prepareReasoner()* is included before the measured entailment check. *prepareReasoner()* triggers a complete classification of the justification. Thus, we can isolate the time required to do a simple "lookup" for the atomic subsumption in the entailment. The times for loading, entailment checking and disposing are then saved along with the results of the entailment checks. Since the tested ontologies are justifications for the individual entailments, this should naturally return *true* for all entailment checks if the reasoner works correctly. As we will show in the next section, a *false* result here can indicate a reasoner error.

4   Experiments and Evaluation

4.1   Experimental Setup

The test sets were generated using JustBench and FaCT++ 1.4.0 on a Mac Pro desktop system (2.66 GHz Dual-Core Intel Xeon processor, 16 GB physical RAM) with 2GB of memory allocated to the Java virtual machine. The tested ontologies were Building, Chemical, Not-Galen (a modified version of the Galen ontology), DOLCE Lite, Wine and MiniTambis.[7] This small test set can already be regarded as sufficient to demonstrate our approach and show how its transparency and restriction to small subsets of the ontologies helps to isolate and understand reasoner behavior, as well as quickly trace the sources of errors.

Most test sets could be generated by our system within a few minutes, however, for the Not-Galen ontology the process was aborted after it had generated several hundred explanations for a single entailment. In order to limit the processing time, a reasoner time out was introduced, as well as a restriction on the number of justifications to be generated. Thus, the justifications for Not-Galen are not complete, and we assume that generating all explanations for all entailments of this particular ontology is not feasible in practical time. The number of justifications for each entailment ranged from 1 to over 300, as in the case of Not-Galen, with the largest containing 36 axioms.

The benchmarking was performed on the same system as the test set generation using three reasoners that are currently compatible with the OWL API version 3, namely FaCT++ 1.4.0, HermiT 1.2.3, and Pellet 2.0.1.

4.2   Results and Discussion

Reasoner Performance. The measurements for the justifications generated from our five test ontologies show a clear trend regarding the reasoner performance. Generally, it has to be noted that the performance of all three reasoners can be regarded as suitably on these ontolgoies, and there are no obvious hard test cases in this test set. On average, FaCT++ consistently performs best in almost all checks, with HermiT being slowest in most cases. Pellet exhibits surprising behaviour, as it starts out with a performance close to that of FaCT++ for smaller justifications and then approximates or even "overtakes" HermiT for justifications with a larger number of axioms. This behaviour, e.g., as shown in figure 1, is seen in all the ontologies tested.

Generally, the time required for an entailment check grows with the size of the justification for all three reasoners, as shown in figure 2—justifications with a size larger than 13 are all obtained from the Not-Galen ontology. Again, Pellet

---

[7] All ontologies that were used in the experiments may be found online:
http://owl.cs.man.ac.uk/explanation/justbenchmarks

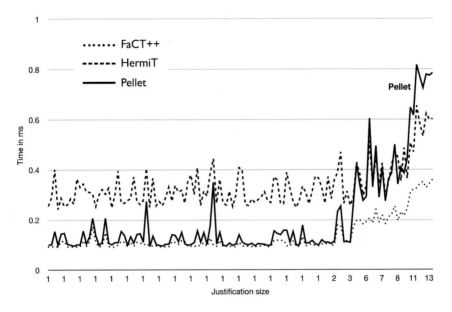

Fig. 1. Reasoner performance on justifications of the MiniTambis ontology

exhibits the behaviour mentioned above and eventually "overtakes" HermiT. The dip at size 16 is caused by the existence of only one justification of this size and can be neglected here.

HermiT in particular starts out with a higher baseline than the other reasoners, but only increases slowly with growing justification size. We are investigating further to pinpoint the exact factors in play.

For the atomic subsumptions in our examples, the expressivity—which seems quite wide ranging—does not significantly affect performance. The average time for each size group indicates that the hardest expressivities for all reasoners are $\mathcal{ALCN}$ and $\mathcal{ALCON}$. We expect that a analysis of the laconic versions of these justifications i.e., that only contain the axiom parts that are relevant to the entailment) will reveal to which extent the performance is affected by the use of expressive constructors.

Reasoner Errors. While the system returns *true* for nearly all entailment checks, a small subset of Not-Galen is wrongly identified as not entailed by Pellet after adding a call to *prepareReasoner()* to force a full classification of the justification. All the misclassified justifications have the DL expressivity $\mathcal{ALEH}$, indicating that they contain subproperty hierarchies. On closer inspection it can be found that Pellet produces an error for the justifications that have axioms of the form

Tears SubClassOf:
  NAMEDBodySubstance, isActedOnSpecificallyBy some
  (Secretion and (isFunctionOf some LachrymalGland))

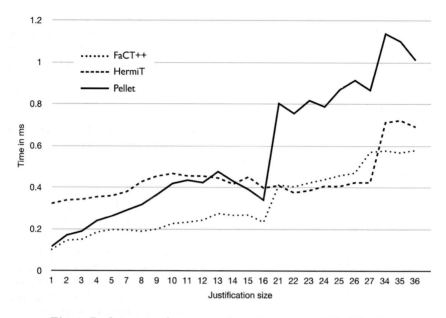

Fig. 2. Performance of reasoners depending on size of justifications

where *isActedOnSpecificallyBy* is a subproperty of some other property that is necessary for the entailment to hold. Communication with Pellet developers revealed that the problem is due to an error in the optimizations of the classification process and not in the core satisfiability check. This demonstrates the need to test all code paths.

By using justifications for the testing process, we detected and isolated and error which affects the correctness of the reasoner but was not otherwise visible. Performing an entailment check on the whole ontology would not exhibit this behaviour, as several other justifications for the entailment masked the entailment failure.

Errors Caused by Signature Handling. We also identified a problem in how FaCT++ handles missing class declarations when performing entailment checks. For some justifications FaCT++ aborts the program execution with an "invalid memory access error", which is not shown by Pellet and HermiT. We isolated the erroneous justifications and perform entailment checks outside the benchmarking framework to verify that the problem was not caused by any of the additional calls to the OWL API made by JustBench. We found that the subsumptions were all entailed due to the superclass being equivalent to *owl:Thing* in the ontology. Consider the following entailment:

NerveAgentSpecificPublishedWork
    SubClassOf: PublishedWork

and the justification for it consists of the following three axioms:

refersToPrecursor
  Domain: PublishedWork

NerveAgentRelatedPublishedWork
  SubClassOf: PublishedWork

VR_RelatedPublishedWork
  EquivalentTo: refersToPrecursor only VR_Precursor
  SubClassOf: NerveAgentRelatedPublishedWork

The subclass *NerveAgentSpecificPublishedWork* does not occur in the justification, as the entailment follows from *Class: PublishedWork EquivalentTo: owl:Thing* and therefore the subclass would not be declared in the justification. How should an OWL reasoner handle this case? Pellet and HermiT accept the ontologies and verify the entailment, whereas FaCT++ requires the signature of the entailment to be a subset of the signature of the justification. This causes FaCT++ to not even perform an entailment check and abort with the error "Unable to register 'NerveAgentSpecificPublishedWork' as a concept". While this is not a correctness bug per se, it is a subtle interoperability issue.

4.3   Additional Tests and Discussion

Performance for Full Classification. In order to compare our justification-based approach to typical benchmarking methods, we measure a full classification of each of our test ontologies. Therefore, an instance of the OWL API's *InferredOntologyGenerator* class is generated and the time required for a call to its *fillOntology()* method is measured to retrieve all inferred atomic subsumptions from the respective ontology. Surprisingly, the rankings based on the individual justifications are inverted here: FaCT++ performed worst for all tested ontologies (except for Wine, where the reasoner cannot handle a "PositiveInteger" datatype and crashes), with an average of 1.84 s to retrieve all atomic subsumptions. HermiT and Pellet do this form of classification in much shorter time (0.98 s and 1.16 s respectively), but the loading times for HermiT (a call to *createReasoner()*) are an order of magnitude larger than those of FaCT++.

Additional Entailments. Choice of entailments makes a big difference to the analysis of the ontology. For example, we examined an additional ontology which has a substantial number of individuals. For full classification of the ontology, both HermiT and Pellet performed significantly worse than FaCT++. Using JustBench with justifications for all entailed atomic subsumptions of the ontology did not lead to any explanation for this behaviour: all three reasoners performed well on the justifications and the sum of their justification reasoning times was much less than their classification time. However, after adding the justifications for inferred class assertions to the test set, the time HermiT takes for entailment checks for these justifications is an order of magnitude larger than for the other reasoners. The isolation of the classes of entailment, as well as shared characteristics of entailments in each class, falls out quite naturally by looking at their justification. In this case, it is evident that there is a specific issue with HermiT's instantiation reasoning and we have small test cases to examine and compare with each other.

An Artificially Generated Ontology. In an additional test with an artificially generated ontology we attempt to verify our claim about loading and classification times of the three reasoners. The ontology contains over 200 subsumptions of the form

A1EquivalentTo: A2
* and (p some (not (A2)))*

with entailments being atomic subsumptions of the type *A1 SubClassOf: A2, A2 SubClassOf: A3 ... A1 SubClassOf: A210.* Justifications for 62 of these entailments were generated before the system ran out of memory, and the entailments were checked against their respective justifications and the full ontology. The right chart of figure 3 shows clearly how the performance of both Pellet and FaCT++ for an entailment check worsens with growing ontology size, whereas HermiT has an almost consistently flat curve. FaCT++ in particular shows almost exponential growth. In contrast, the loading times for larger ontologies only grow minimally for Pellet and FaCT++, while HermiT's loading time increases rapidly, as can be seen in the left chart of figure 3.

All three reasoners perform much worse on the artificial ontology than on the "real-life" ones (except for Not-Galen). This is a bit surprising, considering that the expressivity of this ontology is only $\mathcal{ALC}$, as opposed to the more expressive $\mathcal{ALCF}(\mathcal{D})$, $\mathcal{SHIF}$, $\mathcal{SHOIN}(\mathcal{D})$, and $\mathcal{ALCN}$ respectively of the other ontologies. The justifications for its entailments however are disproportionally large (up to 209 axioms for *Class: A1 SubClassOf: A210*) whereas those occurring in the other "real" ontologies have a maximum size of only 13 axioms. This indicates that a complex justificatory structure with a large number of axioms in the justifications poses a more difficult challenge for the reasoners.

The measurements based on the artificial ontology indicate that HermiT performs more preparations in its *createReasoner()* method and has only minimal lookup times, which confirms our results from the entailment checks following a call to *prepareReasoner()*. We can conclude that, once the ontology is loaded and fully classified, HermiT performs well for larger ontologies, whereas FaCT++ suffers from quickly growing classification times. With respect to the lookup performance of FaCT++, is very likely that the JNI used in order to access the reasoner's native C++ code over the OWL API acts as a bottleneck that affects it negatively. The use of C++ code clearly affects the times for the calls to *dispose()*, as the FaCT++ framework has to perform actual memory management tasks in contrast to the two Java reasoners Pellet and HermiT which defer them indefinitely.

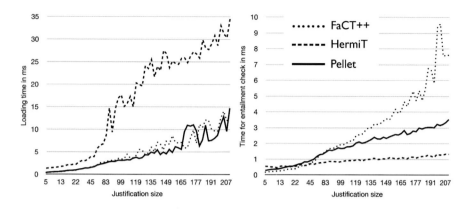

Fig. 3. Reasoner performance on an artificially generated ontology

4.4   Application of JustBench

We propose our framework as a tool that helps reasoner developers as well as ontology engineers check the correctness and performance of reasoners on specific ontologies. With respect to ontology development, the framework allows ontology engineers to carry out fine-grained performance analysis of their ontologies and to *isolate* problems. While measuring the time for a full classification can give the developer information about the overall performance, it does not assist in finding out *why* the ontology is easy or hard for a particular reasoner. JustBench isolates minimal subsets of the ontology, which can then be analysed manually to find out which particular properties are hard for which reasoners. One strategy for mitigating performance problems is to introduce redundancy. Adding the entailments of particularly hard justifications to the ontology causes them to "mask" other, potentially harder, justifications for the entailment. This leads to the reasoner finding the easier justifications first, which may improve its performance when attempting to find out whether an entailment holds in an ontology.

Reasoner developers also benefit from the aforementioned level of detail of justification-based reasoner analysis. By restricting the analysis to small subsets, developers can detect reasoner weaknesses and trace the sources by inspecting the respective justifications, which will help understanding and improving reasoner behaviour. Additionally, as shown in the previous section, the method also detects unsound reasoning which may not be exhibited otherwise.

5   Conclusion and Future Work

To our knowledge, JustBench is the first framework for analytic, realistic benchmarking of reasoning over OWL ontologies. Even though we have currently only examined a few ontologies, we find the general procedure of using meaningful subsets of real ontologies to be insightful and highly systematic. At the very least, it is a different way of interacting with an ontology and the reasoners.

Our current selection principle (i.e., selecting justifications) has proven fruitful: While justifications alone are not analytically complete (e.g., they fail to test non-entailment features), they score high on understandability and manipulability and can be related to overall ontology performance. Thus, arguably, justifications are a good "front line" kind of test for ontology developers.

Future work includes:

- **Improving the software:** While we believe we have achieved good independence from irrelevant system noise, we believe this can be refined further, which is critical given the typically small times we are working with. Furthermore, some OWL API functions (such as *prepareReasoner()*) do not have a tightly specified functionality. We will work with reasoner developers to ensure the telemetry functions we use are precisely described and comparable across reasoners.
- **Testing more ontologies:** We intend to examine a wide range of ontologies. Even our limited set revealed interesting phenomena. Working with substantively more ontologies will help refine our methodology and, we expect, support broader generalisations about ontology difficulty and reasoner performance.
- **More analytics:** Currently, we have been doing fairly crude correlations between "reasoner performance" and gross features of justifications (e.g., size). This can be considerably improved.
- **New selection principles:** As we have mentioned, modules are an obvious candidate, though there are significant challenges, not the least that the actual number of modules in real ontologies tends to be exponential in the size of the ontology [15]. Thus, we need a principle for determining and computing "interesting" modules. Other possible selection principles include "repaired" justifications and unions of justifications.

Furthermore, we intend to experiment with exposing users to our analysis methodology to see if this improves their experience of dealing with performance problems.

References

1. Bail, S., Parsia, B., Sattler, U.: The justificatory structure of OWL ontologies. In: OWLED (2010)
2. Boyer, B.: Robust Java benchmarking (2008),
   http://www.ibm.com/developerworks/java/library/j-benchmark1.html
3. Brandt, S., Küsters, R., Turhan, A.-Y.: Approximation and difference in description logics. In: Proc. of KR 2002. Morgan Kaufmann Publishers, San Francisco (2002)
4. Cuenca Grau, B., Horrocks, I., Kazakov, Y., Sattler, U.: Modular reuse of ontologies: Theory and practice. J. of Artificial Intelligence Research 31, 273–318 (2008)
5. Cuenca Grau, B., Horrocks, I., Motik, B., Parsia, B., Patel-Schneider, P., Sattler, U.: OWL 2: The next step for OWL. J. of Web Semantics 6(4), 309–322 (2008)

6. Dewitt, D.J.: The Wisconsin benchmark: Past, present, and future. In: Gray, J. (ed.) The Benchmark Handbook for Database and Transaction Processing Systems. Morgan Kaufmann Publishers Inc., San Francisco (1992)
7. Franco, J.V.: On the probabilistic performance of algorithms for the satisfiability problem. Inf. Process. Lett. 23(2), 103–106 (1986)
8. Gardiner, T., Tsarkov, D., Horrocks, I.: Framework for an automated comparison of description logic reasoners. In: Cruz, I., Decker, S., Allemang, D., Preist, C., Schwabe, D., Mika, P., Uschold, M., Aroyo, L.M. (eds.) ISWC 2006. LNCS, vol. 4273, pp. 654–667. Springer, Heidelberg (2006)
9. Guo, Y., Pan, Z., Heflin, J.: LUBM: A benchmark for OWL knowledge base systems. Web Semantics: Science, Services and Agents on the World Wide Web 3(2-3), 158–182 (2005)
10. Horrocks, I., Patel-Schneider, P.F.: Evaluating optimized decision procedures for propositional modal K(m) satisfiability. J. Autom. Reasoning 28(2), 173–204 (2002)
11. Hustadt, U., Schmidt, R.A.: Scientific benchmarking with temporal logic decision procedures. In: Fensel, D., Giunchiglia, F., McGuinness, D.L., Williams, M.-A. (eds.) KR, pp. 533–546. Morgan Kaufmann, San Francisco (2002)
12. Lin, H., Sirin, E.: Pellint - a performance lint tool for pellet. In: Dolbear, C., Ruttenberg, A., Sattler, U. (eds.) OWLED. CEUR Workshop Proceedings, vol. 432. CEUR-WS.org. (2008)
13. Ma, L., Yang, Y., Qiu, Z., Xie, G.T., Pan, Y., Liu, S.: Towards a complete OWL ontology benchmark. In: Sure, Y., Domingue, J. (eds.) ESWC 2006. LNCS, vol. 4011, pp. 125–139. Springer, Heidelberg (2006)
14. Pan, Z.: Benchmarking DL reasoners using realistic ontologies. In: OWLED (2005)
15. Parsia, B., Schneider, T.: The modular structure of an ontology: An empirical study. In: Lin, F., Sattler, U., Truszczynski, M. (eds.) KR. AAAI Press, Menlo Park (2010)
16. Rudolph, S., Tserendorj, T., Hitzler, P.: What is approximate reasoning? In: Calvanese, D., Lausen, G. (eds.) RR 2008. LNCS, vol. 5341, pp. 150–164. Springer, Heidelberg (2008)
17. Schaerf, M., Cadoli, M.: Tractable reasoning via approximation. Artificial Intelligence 74, 249–310 (1995)
18. Selman, B., Mitchell, D.G., Levesque, H.J.: Generating hard satisfiability problems. Artif. Intell. 81(1-2), 17–29 (1996)
19. Wang, T., Parsia, B.: Ontology performance profiling and model examination: first steps. In: Aberer, K., Choi, K.-S., Noy, N., Allemang, D., Lee, K.-I., Nixon, L.J.B., Golbeck, J., Mika, P., Maynard, D., Mizoguchi, R., Schreiber, G., Cudré-Mauroux, P. (eds.) ASWC 2007 and ISWC 2007. LNCS, vol. 4825, p. 595. Springer, Heidelberg (2007)
20. Weithöner, T., Liebig, T., Luther, M., Böhm, S.: What's wrong with OWL benchmarks? In: SSWS (2006)

Talking about Data: Sharing Richly Structured Information through Blogs and Wikis

Edward Benson, Adam Marcus, Fabian Howahl, and David Karger

MIT CSAIL
{eob,marcua,fabian,karger}@csail.mit.edu

Abstract. Several projects have brought rich data semantics to collaborative wikis, but blogging platforms remain primarily limited to text. As blogs comprise a significant portion of the web's content, engagement of the blogging community is crucial to the development of the semantic web. We provide a study of blog content to show a latent need for better data publishing and visualization support in blogging software. We then present DataPress, an extension to the WordPress blogging platform that enables users to publish, share, aggregate, and visualize structured information using the same workflow that they already apply to text-based content. In particular, we aim to preserve those attributes that make blogs such a successful publication medium: one-click access to the information, one-click publishing of it, natural authoring interfaces, and easy copy and paste of information (and visualizations) from other sources. We reflect on how our designs make progress toward these goals with a study of how users who installed DataPress made use of various features.

1 Introduction

Recent efforts to generate and curate high-value structured datasets have made great headway on several fronts, as exemplified by open government initiatives, Facebook's Open Graph project, and Freebase's structured wiki. While these centralized, top-down approaches are significant, we have yet to see wide adoption of structured data publication at the grass-roots level. Taking note that the development of hosted blogging platforms encouraged millions of web readers to become content authors as well, we aim to entice these users to publish data by building data-oriented features into their existing blogging software.

Large projects can rely on the promise of societal and technical benefits to justify the costs required to curate and publish structured data. We believe that for independent bloggers to take part in data publishing efforts of their own, the promise of later portability and reuse is not enough. Instead, end-user-focused data publishing tools should offer immediate gratification in the form of useful visualizations and interesting data aggregation before they focus on formal ontologies and namespaces. Only after the user has seen the benefit of data publishing as part of their content authoring workflow can we take steps to link, integrate, and further reuse the underlying data.

P.F. Patel-Schneider et al. (Eds.): ISWC 2010, Part I, LNCS 6496, pp. 48–63, 2010.

We also take inspiration from efforts such as the Semantic MediaWiki project, which has brought structured data publishing to wikis by exposing it in the Wiki-Text format already familiar to wiki users. We aim to similarly provide bloggers with data publishing tools that blend in with existing blogging environments. The popular blog publishing platforms that we target differ from wikis in that they depend more heavily on WYSIWYG editing, click-to-embed rich media, and an easy-access copy-and-paste culture. To facilitate the adoption of grass-roots data publishing, we must build tools that minimize the difference between traditional text-based blogging and the future of publishing, in which all content producers are data publishers.

To understand how to accommodate the data-blogger of the future, this paper first examines the properties of blogging platforms that led to their popularity among content authors on the web. We then demonstrate a latent need of, and great potential for, data-centric blogging tools with a content study of 210 blog entries on the web. This study quantifies the kinds of data-supported arguments that blog authors make and shows that bloggers are already using structured data in their content, but the tools they have to communicate it are limiting.

We then present DataPress, a plugin for the WordPress blogging platform which facilitates data visualization from minimally structured files, allows bloggers to point at other data presentations as a starting point for their own, and allows bloggers to publish and aggregate their own data sets. We build into DataPress the ability to easily link to external data from spreadsheets, RDF sources, and Semantic MediaWiki sites. We further demonstrate the ability to rely on Semantic MediaWiki as a community ontology server to encourage schema convergence across data feeds produced by DataPress bloggers.

Finally, we examine the log data and interview the authors of real DataPress deployments, one of which having seen over 55,000 page views. These users provide insight into how web authors are using DataPress to publish data today (not in their blogs, to our surprise) and where opportunities exist to improve data publishing tools for tomorrow.

2   Requirements for the 21st Century Blogger

Our goal is to bring structured data publishing to the blogging community, and to do so we must build tools appropriate to the environment that bloggers expect. This section examines the properties of blogging platforms that made them such a successful grass-roots medium of contribution to the web. We expect that by preserving these traits in a data-oriented blogging tool, we are more likely to gain traction from the blogging community. Of the many features of blogging systems that make them popular, we highlight:

One-click Publishing. Though publishing through a blogging platform is equivalent, from a technical standpoint, to uploading HTML documents over FTP, the increased usability and convenience that a web publishing interface provides encourages participation by a far wider audience.

Visual Authoring Environment. Blog platforms offer users familiar, word-processor-like WYSIWYG text editors, with HTML forms to guide them through more complicated tasks. Notably, the author does not have to co-ordinate her work through several distinct applications—all her authoring needs are met within the editing environment of the blogging tool.

Copy and Paste. Web blogs have developed a publishing culture that makes significant use of copy and paste, both to quote information found in other sources and also to replicate layout or visualization functionality that the author could not construct herself. Sometimes, such quoting is an end in itself; at other times, the goal is to use the original content as a starting point for publishing by modification. It is much easier to copy someone else's nicely formatted page and replace its content with your own than to understand how to create such a layout in the first place.

Pre-Packaged Widgets. Blogging systems make it easy to include rich media widgets—such as slide shows and video clips—in article text without having to manually write code or configuration. By simply uploading several pictures (in the case of a slide show) or adding a link (in the case of a YouTube video clip), the blogger benefits from the platform's ability to package up this simple data into a rich format that entertains visitors.

These traits help blogging platforms turn the technical task of publishing web content into an easy process accessible to the grassroots authors that provide so much of the web's "long tail" of content. If we wish to encourage these grassroots authors to provide data as well, we must give them tools that work from within these familiar environments and share their properties.

We use these traits as a guide to construct DataPress, and we argue that as a result DataPress can support, for rich structured data, the same behaviors that made the web's text authoring tools so effective:

- It has the same "click and you see it" immediacy that made hosted blogging such a big change over the FTP publishing workflow: it enables users to insert data into a blog post the same way they insert an image, offers readers embedded data visualizations inside article bodies, and it does so without leaving the metaphor provided by the blogging platform
- It does not require the author to understand complex data models, but instead can be based on concepts already familiar to end users: simple forms, embedded media, and links to data-laden websites
- It offers the same copy-and-paste workflow as text, making it easy for authors to "quote" both the data and the data visualizations authored by other users, either to be used unchanging or as a starting point for authors (who may not yet know how to author their own data or visualization) to make their own points by authoring changes in the acquired data or visualizations
- It inspects a user's data in order to better guide her through the creation of rich, interactive visualizations. Users can add faceted navigation, interactive maps and timelines, and search functionality all by selecting a few options in the blogging editor.

3   The Latent Potential for Grassroots Data

Blogging platforms facilitated the enormous growth of the web over the past two decades, but the capabilities of these tools are primarily limited to text. In contrast, professional publications often deal with rich, structured data: shopping sites offer faceted browsing across their product databases; product review sites let readers dynamically pit products against each other in feature-by-feature tables; and news sites such as the New York Times (which runs its own Visualization Lab[1]) publish interactive presentations of complex information. Arguably, these professionally managed web sites are significantly more expressive than grassroots authors' pages. One might think that this is because only large professional publishers care for such expressivity, but we observe that the desire to publish and present data extends far beyond large publishers.

In this section we present the results of a blog content study that indicates that bloggers are in fact *already* frequently talking about data; they are just doing it using text and static images, the best way that they can given their current publishing platforms. We believe this is a hopeful result for the semantic web community, for it suggests that grassroots bloggers would be eager to make use of structured data if their tools made this process easy and beneficial to their needs.

For the purposes of this study, we use the term *blog* to refer to any article-style publication on the web, including both personal journals and professional periodicals. A *semantic entity* refers to an object with one or more properties described in structured or unstructured (natural language) form. A *collection of semantic entities* refers to a sequence of semantic entities of the same type described in a document. For example, a semantic entity might be a paragraph of text or a table row that describes the technical specifications of a new camera. A collection of semantic entities would be a text document or full table comparing several cameras to each other.

We coded 210 blog articles across 21 blogs to measure the occurrence and nature of semantic entities and semantic entity collections within their text[2]. We generated this blog sample by selecting the 10 most recent entries (at time of study) from a semi-random list of blogs taken from the Technorati[5] blog indexing service. This list of blogs included:

- The top ten blogs according to Technorati's "authority rank"
- Eleven blogs selected at random from Technorati's list of "rising" articles

We used this selection method to attempt to capture both high quality, professional content (top ten blogs) and also blogs that varied in style and represented the "long tail" of the web (top rising posts). For each blog in our sample, we downloaded its RSS feed and coded each of its ten latest entries.

---

[1] http://vizlab.nytimes.com/

[2] The data for this survey can be found at
http://projects.csail.mit.edu/datapress/content_survey

Table 1. Number of occurrences overall and number of articles with various properties

	Lone Semantic Entities	Collections of Semantic Entities	Visual Collections	Referenced Datasets	Referenced Resources
Articles with one or more occurrences (of 210)	45 (21%)	64 (30%)	22 (10%)	67 (32%)	191 (91%)
Total Count	58	105	49	428	1061

Aggregating across the articles for each blog, we found that:

– 17 of 21 blogs contained at least one article in their latest 10 that enumerated the properties of a single semantic entity.
– 18 of 21 blogs contained at least one article in their latest 10 that enumerated the properties of a collection of semantic entities.
  • Half of these blogs used natural language text to describe the collection.
  • The other half used a table or a static image containing an info-graphic.

Aggregating across the articles for *all* blogs, we found that:
  Table 1 shows us that:

– 21% of articles surveyed contained at least one semantic entity.
– 30% of articles surveyed contained a collection of semantic entities (anecdotally, these were things such as polling results in different states, economic conditions in different countries, and professional sports records).
  • Two-thirds of these collections were presented in natural language text instead of a structured or visual format.

Finally, our data revealed that blog entries frequently refer to external sources of data rather than present original content. Authors made reference to some externally attributed datum or statistic in 91% of articles surveyed. In, 32% of articles, this reference was to an explicit data set, often given by name (e.g., "A 2008 Zogby Poll reported that..."), while in the other 59% it simply referred to a person or organization who had claimed the truth of the numerical fact. In all, we counted 428 total references across the 67 articles which mentioned data sets. These numbers are surprisingly high, perhaps influenced by the fact that our study was done in the midst of an electoral season, but they serve to reinforce the intuition that bloggers are in many respects serving as topic- or geo-localized journalists. They are writing about issues, and these issues involve data. We aim to make that data navigable, linked, and reusable.

Anecdotally, much of the presentation of semantic entities was inlined in text, rather than in a structured tabular format. Interactive data visualizations were rare—structured presentation tended to be either static tables or images. In fact, most of these collections were included in an HTML table or rudimentary list rather than a full-blown visualization. Data "links," if at all present, tended to be narrative references to a data set rather than resolvable URLs.

These results suggest significant latent potential for tools that allow bloggers to publish data with the same ease with which they already publish text. These authors are already interested in data, and at times they are publishing tables or images, indicating that they are willing to adopt non-prose presentation styles if they are available.

Currently the only non-prose data presentation tools that blogging platforms support are HTML tables and static images. We aim to fill this gap with DataPress, which provides both interactive visualization capabilities as well as raw data publishing and linking.

4   DataPress

DataPress[3] is our attempt to create a blogger's tool to publish, share, and copy data and data visualizations. We start from the premise that, from the standpoint of content authors, visualizations are an end in themselves—if a picture is worth a thousand words, surely a good interactive visualization is worth at least tens of pictures. They allow anyone encountering that data to understand it better by exploring it.

But DataPress is also a means to an end: making data more easily available for reuse. DataPress' rich data visualizations encourage authors to use it, but the tool also *exposes* the data it is showing off, making it easy to link to or snapshot, thus enabling the same reuse ecology already pervasive in textual blogs. With this in mind, we will describe DataPress in its four distinct roles:

1. Authoring data
2. Consuming data originating elsewhere
3. Authoring visualizations
4. Exposing data for consumption by other tools

For the authoring roles, our key goal is to fit data and visualization authoring naturally into the already existent workflow associated with WordPress. For the data sharing roles, we arrange for our tool to offer, with no extra user labor, JSON and RDF data "feeds" that can be consumed by others. We also facilitate easy linking of diverse content on the web for aggregation and visualization within a blog post.

DataPress is implemented as a plugin for the WordPress blogging platform. We chose WordPress because it has a large install base (over 3,816,965 downloads in 2007 alone [3]) and because it is a blogging tool used widely for both personal blogs and professional publications, including media outlets as large as the Washington Post online edition. Like other blogging platforms, WordPress places a high value on guided workflow and simple form-based configuration, so the DataPress plugin exposes all of its features as enhancements to the existing WordPress authoring interface.

This section describes the most recent version of DataPress. Our user study, presented in Section 6, was conducted across users of a previous release of DataPress. This previous release contained of all the features described below except for individual item publishing (in Section 4.1), data feeds (in Section 4.4), and Semantic MediaWiki hooks (Section 5).

---

[3] Downloadable source code, examples, and demo blog for testing available at:
http://projects.csail.mit.edu/datapress/

4.1   Authoring and Uploading Data

DataPress allows authors to create individual data items to publish with a post or upload entire datasets at a time. Both these option are accessible from buttons added to the WordPress post editor, seen in Figure 1.

Authoring a Data Item. By pressing the Data Item button, bloggers can enter key-valued information to associate with a typed semantic entity and publish this information as metadata with a blog post. This usage scenario fits the type of blogger who publishes

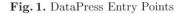

Fig. 1. DataPress Entry Points

similarly themed articles over time and would like to benefit from being able to aggregate their structured content for presentation purposes or export to the community.

Consider the practice of blogging one's academic reading list—some students and professors enjoy blogging summaries of papers they have read so that they can share their thoughts with others in the community. DataPress allows this temporal stream of activity to be published as structured data as well. While writing the blog post, the author clicks the "Data Item" button seen in Figure 1 and DataPress will bring up a selection of "data templates," shown in the first screenshot in Figure 2. A data template is simply a blank form derived from the schema of some item type.

Fig. 2. Choosing and filling in a Template

This list of data templates can draw from a variety of places. DataPress comes with a collection of built-in data templates, such as academic papers, books, and workouts, but it can also be configured to talk to Semantic Media Wiki installations or other data template repositories on the web, allowing the blog author to take advantage of communities that maintain such information and encouraging schema convergence across web sites (this idea will be expanded in Section 5). If no template fits the item, DataPress allows users to create their own by entering a custom class name and the properties that should be associated with its instances.

Once the user has selected a data template, DataPress loads the template schema (possibly from a remote repository) and transforms it into a web form for the user to fill out. One such form is shown in the right-hand side of Figure 2. From here, the data is stored in DataPress' back-end database, while a visual marker for the data is embedded into the text of the post as a small icon that allows editing or removal of the item from the post, shown in Figure 3.

The Data Item interface allows a blogger to follow their natural habit of writing a new article about each data item, while also producing an aggregate data set over time and across blog posts for rich visualization. Our reading-list blogger can place, sticky on their front page, a single rich "My Reading List" visualization showing all articles they have read, with links to the individual blog postings about the articles. This visualization becomes a new, non-chronological index into their blog content.

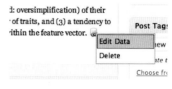

Fig. 3. Data Item in a Blog Post

Uploading Data Sets. DataPress also lets users associate entire data sets with a blog post. Using WordPress' built-in file upload tool, they can upload a file, and then using the Data Set button provided by DataPress, they can associate that file with a blog post. DataPress utilizes the data import mechanisms of the Exhibit framework [12] and the MIT Babel [8] data translation web service to accept a wide variety of formats, including RDF, JSON, CSV, XML, Microsoft Excel, and Bibtex.

Once a data file is associated with a blog post, DataPress stores this information in its database and provides the option of attaching *Data Footnote* links at the end of the blog post, allowing the reader to visually see links to the data that accompanies the writeup. These associated data sets are also used as inputs for data visualizations, shown later.

4.2 Data Linking

DataPress also lets authors link to remote data sets via URL. In addition to supporting a wide number of data formats that can be linked to directly, DataPress contains a special importer that handles what we call *approximate links*—URLs that point to web pages that talk *about* data, rather than links to the raw data itself. We currently support four such kinds of approximate links:

- URLs of DataPress-powered pages are automatically converted into data links to that page's data sources
- URLs of web pages containing an Exhibit-powered visualization are automatically converted into data links to that page's data sources
- URLs of Google Spreadsheet files are automatically converted into API calls into Google's JSON data service
- URLs of third-party JSON data files are converted into JSONP calls routed through a DataPress JSON-to-JSONP service

We expect to grow support for approximate linking as we believe that it supports, for data, the same copy-and-paste-ability that made blogging tools successful. If a user sees a page with data they want to use, they should only have to copy and paste that page's URL to be able to remix and republish its data. As we will show in the following section, we are currently also working on support for easy import of Semantic MediaWiki data via remote `ASK` queries.

4.3   Visualization Authoring

The "Visualization" button, shown above the post editor in Figure 1, provides access to a wizard which walks the user through the creation of a data visualization. DataPress uses the Exhibit framework for displaying interactive visualizations. This allows the plugin to benefit from the developer community that builds data importers and visualization plugins for Exhibit. DataPress' configuration wizard, shown in Figure 4, contains many of the various options Exhibit provides, as well as some blog-specific enhancements.

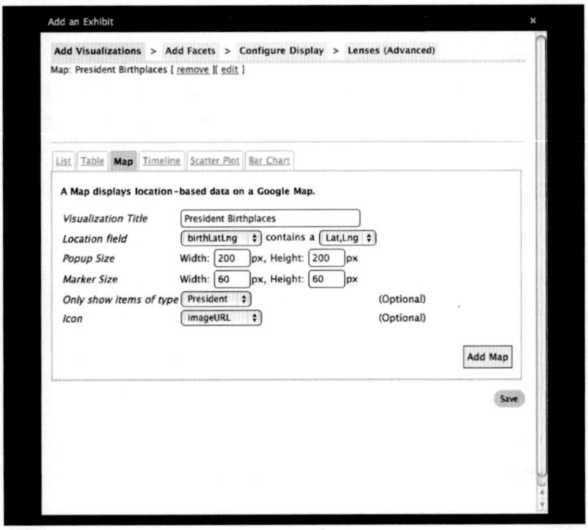

Fig. 4. Adding a Data Visualization

The wizard consists of four main steps:

Add Visualizations. Supported visualization types include lists, tables, maps, timelines, scatter plots, and bar charts.

Add Facets. Add faceted navigation to the visualization. Supported facet types include free-text search, list facets, range sliders, and tag clouds.

Configure Display. Many blogs follow a narrow-width article format while some rich visualizations are wide, so DataPress includes an "lightbox" option which presents visualizations as YouTube-style previews that expand to hover over the full web page when clicked. This step in the wizard also allows the blogger to link custom CSS files to the visualization.

Lenses. Lenses may be thought of as data style sheets—they are templates that define how items of a particular type should be displayed. DataPress provides a WYSIWYG lens editor that includes support for images whose URLs appear in the data.

Because DataPress is aware of the data that has already been associated with the blog post, it is able to suggest values for many of the configuration options required to create a visualization. When each new data item is added to the blog post, DataPress uses the Exhibit framework to parse the data in the background and update a running list of the item types and properties relevant to the visualization. This is particularly useful if a user is linking to data from another site on the web. Without even looking at the raw data or schema, the user is able to immediately begin crafting a visualization, with data-aware autosuggest fields providing the possible answers to necessary questions.

Once a visualization is configured, it can be inserted into the blog post by clicking a button in the wizard. The visualization appears in the blog text editor as the placeholder token {{Exhibit}} to mark its desired placement. Users can always re-edit their visualization by clicking on the toolbar button once again (we currently only support one visualization per blog post).

4.4 Data Sharing

After data has been associated with blog posts in DataPress, it can be shared with others in two different ways. The first is by nature of the fact that blog posts created with DataPress have links both visible (as optional data footnotes) and invisible (as links embedded in the markup of the page) that allow others to re-use the data associated with the post. Other bloggers with DataPress, for example, need only reference the URL of a data-laden blog post to automatically import all of its data and begin crafting visualizations to rebut, reinforce, or simply echo the message.

The second, and more intriguing, form of data sharing is made possible via data feeds. Just as WordPress allows RSS readers to fetch custom feeds specific to a particular tag or category of post, DataPress responds to requests to assemble data item feeds along similar lines. This feed generator creates aggregate collections of data items for a particular tag or category tracked by the blog. It does so by grouping together all data items published with posts that are marked with the specified tag or category. The following URL is an example of such a request:

```
/.../datapress/feed.php?tag=Research+Paper
```

Using data feeds web users may fetch a feed (in either JSON or RDF) of the structured data added to blog posts and incorporate that data into their own visualizations. A research group, for example, could aggregate the individual users' reading blogs into a group-wide record of readings.

If we accept that many bloggers blog out of the hope that others will consume what they blog, we can conclude that bloggers will be attracted to the idea of offering rich consumable data feeds with no additional effort on their part. We believe that such an access methodology will encourage increased *casual data curation*, as users who blog about similar items over time (trips, meals, workouts, papers, etc.) will value the data feed more than the sum of each individual annotated item once others can present the data in a visual, interactive manner.

5 DataPress in a Data Ecosystem

While the first step toward data publishing for bloggers is to give them value for using structured data, we keep in view the eventual goal of integrating linked data sources across the web. One natural link is the one between blogs run by individuals and wikis curated by communities. Projects like Semantic MediaWiki (SMW) and Freebase already offer several tools to support community-curated datasets. This section describes DataPress' features for integrating into such an ecosystem of data publishing.

To demonstrate the possibilities of such an ecosystem, we extend SMW with a plugin we have developed called Wibit[4]. Wibit enables interactive visualizations, data sharing, and schema sharing using the data contained in the SMW knowledge base. From a visualization perspective, Wibit provides a WikiText syntax that enables SMW users to create Exhibit visualizations that aggregate the results of an ASK query (to be contrasted with approaches like Project Halo [10] which make use of graphical interfaces). From a data perspective, our development version of Wibit provides a data API that permits external services to query the wiki knowledge base.

Working together with DataPress, the Wibit extension provides a number of integration points between SMW and data-aware blogs. Using the Wibit API, DataPress users can issue a remote ASK query and visualize its results from within a blog post. As DataPress allows multiple data sets to be combined, this means that a blogger can combine a wiki's data set with their own data feeds. This data flow also works in reverse: Wibit can aggregate data feeds across several blogs to display a visualization of blogged items.

As the data web evolves, we believe this blog-wiki connection is also a mechanism to encourage schema convergence within communities of interest. Users of a community can collaborate on the common definition of an item type on their community wiki, and then bloggers can use this schema to publish instances with their blog articles. Wibit's API exports SMW schemas in a JSON format for the DataPress template loader to read. When DataPress users are adding data items to their posts, they may pick from one of these community-defined item schemas instead of creating their own.

By facilitating the transfer of visualizations, data and schemas across blogs and wikis, data-oriented tasks can live closest to where they are natural: wikis for crowd curation and blogs for individual publication and reflection.

---

[4] A wiki running the introduced extensions and examples is available at:
http://projects.csail.mit.edu/wibit

6   Lessons Learned

DataPress is available as an open sourced plugin for WordPress. We now describe initial observations about how early adopters have used DataPress and provide lessons learned from phone interviews with three of these DataPress users. The users in this study are all running DataPress 1.2 or earlier, which lacks individual item publishing, data feeds, and Semantic MediaWiki hooks.

6.1   DataPress in the Wild

Since releasing DataPress, the tool has been downloaded 90 times. Of these downloads, 21 users chose to participate in a statistic collection study the software offers as an option, including one website whose DataPress-built exhibit has seen over 55,000 page views. We present some observations about this log data, though we stress that the number of users does not give our results statistical significance. In total, 56 visualizations were created and reported back to our servers, receiving 64,324 total page views. Of these, approximately half were created as permanent pages on their site, while the other half were embedded in blog entries.

Facets, or Lack Thereof. Facets are an important component for navigating structured data, and many Exhibits found online are heavily faceted to support deeper navigation of the data. One might expect highly-faceted Exhibit configurations through DataPress, but many DataPress-based Exhibits we found consisted of simply an unfaceted map or timeline for inline display of data. This suggests that even simple tools without the interactive features Exhibit provides—such as search, faceted navigation, and data lenses—are of great help to bloggers with data to display.

Lightboxing. Because of the narrow-width layout of many blogs, we assumed that the lightboxing feature of DataPress would be heavily used. However while lightboxed visualizations received more pageviews than "inline" visualizations in our data set, the lightbox setting was not frequently employed. User interviews indicated that some users might not have understood what the feature provided. Additionally, the reduced use of facets might have resulted in less space-constrained visualizations than we expected.

Data Footnotes. Finally, in our goal of exposing data for future reuse, we tried to make data footnotes simple to embed in a blog entry. While data footnotes are included by default as a textual token in any post that contains a DataPress-configured visualization, most visualization authors removed the footnotes from their entries; user interviews indicated the reasons are varied.

6.2   User Interviews

To better understand user motivations for seeking out DataPress and to learn how they used the tool, we conducted e-mail and phone interviews with three

DataPress users: a scientist managing a publications list, a large website owner wishing to add dynamic features to HTML tables, and a hobbyist/entrepreneur who maintains a website for his citys local music scene. None of the three users were technical: the most experienced felt confident enough to edit CSS, but not write JavaScript.

None of the three users indicated "data blogging" as their goal, and none of them published a visualization inside a blog post. Rather, each used DataPress to place visualizations on permanent, dedicated pages of their own, more like a content management system than a blog. Time will tell whether this contradicts our claims that blogs are a natural place for structured data publishing. Much as Flash animations or audio files were once a destination of their own, while today they are casually embedded in blog posts, one might expect a similar transition for rich data objects.

Latent Data Needs. We claimed in Section 3 that the prevalence of data in natural language blog posts indicates a latent need for better data publishing tools. For the three users we spoke with, the need was not so latent: each actively sought a way to present dynamic data visualizations on their site. One had heard of Exhibit, and installed WordPress and DataPress in order to create Exhibits without having to edit HTML. The other two had actively searched for a data visualization tool over the course of several months and eventually found Exhibit. After trying unsuccessfully to integrate Exhibit into their WordPress installations, they returned to the web looking for help and found DataPress.

The fact that some users are searching for months to find data visualization tools suggests that the many APIs offering data visualization services have an untapped audience of bloggers who want these services but dont know what to do with an API. For each of these authors who found our tool, there are surely more that have yet to find a tool to help them, and still more who haven't even thought to look for such tools because they believe data visualization to be outside the reach of bloggers.

Crossing the Structured Data Chasm. We learned that tools which provide a compelling reason for users to publish structured data can lead users to structure previously unstructured or poorly structured collections. One user said that she previously maintained a list of her publications in a MS Word document, but the ability to publish a faceted list of publications online motivated her to structure this list in a Google Spreadsheet. Another user initially maintained an HTML table to present a hand-curated collection of data, but moved this data into a separate data file so that he could provide users the ability to better explore the data.

Features without examples go unused. When we asked authors why they did not use some of DataPress' features, a common response was that our interface did not show examples of what the result of those features would be. This is food for thought from a design perspective: even though the high-level function of these feature was often clear ("Add a map", for example, or "Add a search box") the users still wanted to see usage examples first. After seeing (or hearing) these, they decided that the feature would be useful to them in many cases. Our high-traffic user who was only publishing a dynamic table was so enthused about the other configuration options after speaking with us that he added a map, timeline, and several dynamic facets to his visualization the day after our interview with him.

We must accommodate a spectrum of data ownership philosophies. We also learned that users are well-aware of the potential perils of publishing reusable datasets and easily-replicated visualizations. While not deeply technical, all three users understood that someone could copy their dataset by linking to it, and replicate their visualization by copying their Exhibit HTML. Their reactions varied. One author felt ownership over his data and would want any reuse negotiated beforehand, though he recognized that symbiotic relationships could be built around collaborative data editing. This author took the time to modify the CSS of his site to hide Exhibits bundled data copying interface. Despite that, he was happy see his entire visualization embedded in another site as long as the site drove traffic back to his. Another author was fine with reuse of either her data or her visualization, but felt that reuse of both together would be inexcusable copying. Finally, the third author was fine with his visualizations and data being reused by others, as long as proper attribution and links were provided. As tool builders, we must remember to try to accommodate both the information sharer and the businessperson who seeks benefit in exclusivity.

Users want more data tools. The authors we spoke to also understood— and requested—features related to the wider data ecosystem, apart from the visualization capabilities. One mentioned encouraging other site owners to collaboratively maintain complimentary data sets so that they could display the data in different ways on their sites. Two of the three authors specifically requested the ability to communally maintain data on a wiki and then display visualizations of it from within their blog environment. This is significant because these authors were using a version of DataPress without this feature and were unaware that it existed.

7   Related Work

The past few years have seen a great number of projects devoted to visualizing and cataloging structured data on the web. Many Eyes [13] allows users to upload data files and create interesting data visualizations via a web interface. These visualizations are both viewable on their site and embeddable into other sites. While Many Eyes facilitates data visualization, it requires the user to step outside their authoring tool of choice (such as a blog or wiki) and use a third-party service to create and host their visualization. DataPress instead enables authoring from within the blog environment and without third-party services. Further, while Many Eyes focuses on numerical data and content modeling, we target faceted navigation [9] across semi-structured data sets. Semi-structured data opens doors to visualizations involving multiple datasets, allowing authors to build on discussions with novel contributions from new data.

Sense.us [11] is a study in visualizations which facilitate asynchronous collaboration in a *centralized* fashion. We want to modify this model by *decentralizing* the visualizations and data references, allowing collaboration to occur in the native content publishing platform(s) of the user(s).

Exhibit [12] is a client-side web framework for creating rich visualizations of data. Exhibit combines textual data files (such as RDF or JSON) with an HTML-embedded configuration file to produce interactive faceted data displays. While they needn't be programmers, Exhibit authors must be comfortable editing raw HTML and often must be familiar with data formats such as JSON. DataPress relies on Exhibit to power its visualizations, but it relieves the need to understand Exhibit's configuration syntax by providing a wizard that integrates with the blogging platform. In doing so, we aim to bring Exhibit's effective visualization capabilities to the broader class of users.

The Google Visualization API [4] enables programming-savvy webmasters to create a variety of data visualizations. As we aim to bring such visualizations into the realm of blog and wiki content, we see tools like this as potential components to incorporate into our own framework.

While the New York Times Visualization Lab [6] does not appear to use a generally-available framework for authoring displays, it deserves mention as an organization which puts a lot of effort to embed rich information displays in online content. The fact that the interactive data-driven diagrams appearing in its online edition appear to be hand-coded only underscore the need for better general-purpose visualization tools accessible to web authors.

Several projects have also risen to prominence to provide entry and cataloguing of structured data on the web. DBpedia [7] curates the structured information already present in Wikipedia taxonomies (categories) and Info Boxes. DBpedia crawls Wikipedia weekly and coerces that information into an RDF database. Semantic MediaWiki [14] is a MediaWiki extension that enables users to embed key-value annotations about a wiki topic directly in its article text. An alternative approach to DBpedia, Semantic MediaWiki integrates awareness of the inherent structure and types of data into the wiki, and thus the authorship process, itself rather than attempting to recover structure from the natural-language oriented MediaWiki database. Other tools, such as Freebase [2] and Factual [1] provide many-to-many *data* authorship environments rather than attempting to interweave structured data curation along with natural language information repositories. These projects are an interesting new class of democratized data management tools by themselves, and we see them as being another important public data source in the connected data ecosystem that is evolving.

8   Conclusion

The design of DataPress reflects a belief that a data-aware web needs tools that make grassroots authors *want* to work with data. We show the need for such tools with a study of data-oriented blog content. DataPress makes progress on this goal by fitting portions of the semantic web vision into a tool crafted specifically for the blogging workflow. DataPress provides bloggers with an easy way to create, link, and publish data while preserving many of the properties that make blogging an attractive publication platform: one-click publishing, flexible format support, easy copy and paste, and immediate results. DataPress further demonstrates a possible ecosystem of grassroots semantic web publishing in which community wikis serve to centralize ontology management while bloggers use these definitions to create feeds of data over time. We reflect on conversations with our users to better understand how this need is manifested and how to build better tools to facilitate casual use of structured data on the web.

References

1. Factual, http://www.factual.com/ (accessed October 13, 2009)
2. Freebase, http://www.freebase.com/ (accessed October 13, 2009)
3. About Wordpress, http://wordpress.org/about/ (accessed October 29, 2009)
4. Google visualization API, http://code.google.com/apis/visualization/ (accessed October 29, 2009)
5. Technorati, http://technorati.com/ (accessed October 29, 2009)
6. The New York Times Visualization Lab, http://vizlab.nytimes.com/ (accessed October 29, 2009)
7. Auer, S., Bizer, C., Lehmann, J., Kobilarov, G., Cyganiak, R., Ives, Z.: Dbpedia: A nucleus for a web of open data. In: Aberer, K., Choi, K.-S., Noy, N., Allemang, D., Lee, K.-I., Nixon, L.J.B., Golbeck, J., Mika, P., Maynard, D., Mizoguchi, R., Schreiber, G., Cudré-Mauroux, P. (eds.) ASWC 2007 and ISWC 2007. LNCS, vol. 4825, pp. 722–735. Springer, Heidelberg (2007)
8. Butler, M., Gilbert, J., Seaborne, A., Smathers, K.: Data conversion, extraction and record linkage using XML and RDF tools in Project SIMILE. Technical report, HP Laboratories Bristol (2004)
9. Elliott, A.: Flamenco image browser: using metadata to improve image search during architectural design. In: CHI 2001 Extended Abstracts on Human Factors in Computing Systems, CHI 2001, pp. 69–70. ACM, New York (2001)
10. Friedland, N.S., Allen, P.G.: The Halo Pilot: Towards A digital Aristotle. Technical report, Vulcan (2009)
11. Heer, J., Viégas, F.B., Wattenberg, M.: Voyagers and voyeurs: supporting asynchronous collaborative information visualization. In: Proceedings of the SIGCHI Conference on Human Factors in Computing Systems, CHI 2007, pp. 1029–1038. ACM, New York (2007)
12. Huynh, D., Miller, R., Karger, D.: Exhibit: Lightweight structured data publishing. In: Proceedings of the 16th International Conference on World Wide Web, WWW 2007. ACM, New York (2007)
13. Viégas, F., Wattenberg, M.: Shakespeare, god, and lonely hearts: transforming data access with many eyes. In: Proceedings of the 8th ACM/IEEE-CS Joint Conference on Digital Libraries, JCDL 2008, pp. 145–146. ACM, New York (2008)
14. Völkel, M., Krötzsch, M., Vrandecic, D., Haller, H., Studer, R.: Semantic wikipedia. In: Proceedings of the 15th International Conference on World Wide Web, WWW 2006, pp. 585–594. ACM, New York (2006)

$\mathcal{EL}$ with Default Attributes and Overriding

Piero A. Bonatti, Marco Faella, and Luigi Sauro

Università di Napoli Federico II, Napoli, Italy
{bonatti,mfaella,sauro}@na.infn.it
http://people.na.infn.it/~bonatti

Abstract. Biomedical ontologies and semantic web policy languages based on description logics (DLs) provide fresh motivations for extending DLs with nonmonotonic inferences—a topic that has attracted a significant amount of attention along the years. Despite this, nonmonotonic inferences are not yet supported by the existing DL engines. One reason is the high computational complexity of the existing decidable fragments of nonmonotonic DLs. In this paper we identify a fragment of circumscribed $\mathcal{EL}^{\perp}$ that supports attribute inheritance with specificity-based overriding (much like an object-oriented language), and such that reasoning about default attributes is in P.

Keywords: Nonmonotonic description logics, Defeasible inheritance.

1   Introduction

The ontologies at the core of the semantic web — as well as ontology languages like RDF and OWL — are based on fragments of first-order logic and inherit strengths and weaknesses of this well-established formalism. Limitations include monotonicity, and the consequent inability to design knowledge bases (KBs) by describing prototypes whose general properties can be later refined with suitable exceptions. This natural approach is commonly used by biologists and calls for an extension of DLs with defeasible inheritance with overriding (a mechanism normally supported by object-oriented languages) [18, 19]. Another motivation for nonmonotonic DLs stems from the recent development of policy languages based on DLs [21,13,22,17]. DLs nicely capture role-based policies and facilitate the integration of semantic web policy enforcement with reasoning about semantic metadata (which is typically necessary in order to check policy conditions). However, in order to formulate standard default policies such as *open* and *closed* policies,[1] and authorization inheritance with exceptions, it is necessary to adopt a nonmonotonic semantics (see the survey [9] for more details).

Given the massive size of semantic web ontologies and RDF bases, it is mandatory that reasoning in nonmonotonic DLs be possible in polynomial time. Unfortunately, in general nonmonotonic DL, reasoning can be highly complex [11,12,8];

---

[1] If no explicit authorization has been specified for a given access request, then an open policy permits the access while a closed policy denies it.

P.F. Patel-Schneider et al. (Eds.): ISWC 2010, Part I, LNCS 6496, pp. 64–79, 2010.

the best approaches so far belong to the second level of the polynomial hierarchy [10, 7].

In this paper we identify a fragment of circumscribed DLs that extends $\mathcal{EL}$ with default attributes and inheritance with overriding. Informally, the extension allows us to express defeasbile inclusions such as "the instances of C are *normally* in D", for two concepts C and D. Such axioms can be overridden by more specific inclusions, according to a priority mechanism. Our strategy is preserving the classical semantics of $\mathcal{EL}$ as much as possible, in order to facilitate the adaptation of the existing monotonic ontologies. Our framework restricts nonmonotonic inferences to setting the default attributes of "normal" concept instances, without changing the extension of atomic concepts. We define two slightly nonstandard reasoning tasks to query the properties of normal instances. In general, these reasoning tasks are NP-hard. The main cause of intractability is the presence of conflicting defeasible inclusions, i.e., inclusions that give rise to an inconsistency when applied to the same individual. However, if for all pairs of conflicting inclusions δ_1 and δ_2, with non-comparable priority, there exists a disambiguating, higher priority inclusion that blocks at least one of δ_1 and δ_2, then the time complexity of the reasoning tasks becomes polynomial. We show that the identification of such δ_1 and δ_2 can be carried out in polynomial time; then the disambiguation can be left to the ontology engineer or performed automatically by generating a default that blocks both δ_1 and δ_2.

The paper is organized as follows. In Sec. 2, we recall the basics of circumscribed DLs with defeasible inclusions, using the notation adopted in [7]. In Sec. 2.1 we motivate and define a new reasoning task, tailored to inferring the default properties of concepts. Section 3 is devoted to the complexity analysis of this inference problem for the general case and for the restricted class of KBs outlined above. In Sec. 4 the new task and complexity results are extended to instance checking. A section on related work (Sec. 5) and one summarizing our results and discussing interesting future work (Sec. 6) conclude the paper.

2   Preliminaries

In DLs, *concept expressions* are inductively defined using a set of *constructors* (e.g. $\exists$, $\neg$, $\sqcap$), starting with a set $\mathsf{N_C}$ of *concept names*, a set $\mathsf{N_R}$ of *role names*, a set $\mathsf{N_I}$ of *individual names*, and the constants top $\top$ and bottom $\bot$. In what follows, we will deal with expressions

$$C, D ::= A \mid \top \mid \bot \mid C \sqcap D \mid \neg C \mid \exists R.C\,,$$

where A is a concept name and R a role name. In particular, the logic $\mathcal{EL}^\bot$ supports all of the above expressions except negation ($\neg C$). Knowledge bases consist in a (finite) set of concept inclusion assertions of the form $C \sqsubseteq D$ (TBox) and a (finite) set of instance assertions of the form $C(a)$, $R(a, b)$ with $a, b \in \mathsf{N_I}$ (ABox).

The semantics of the above concepts is defined in terms of *interpretations* $\mathcal{I} = (\Delta^{\mathcal{I}}, \cdot^{\mathcal{I}})$. The *domain* $\Delta^{\mathcal{I}}$ is a non-empty set of individuals and the *interpretation*

Name	Syntax	Semantics
negation	$\neg C$	$\Delta^{\mathcal{I}} \setminus C^{\mathcal{I}}$
conjunction	$C \sqcap D$	$C^{\mathcal{I}} \cap D^{\mathcal{I}}$
existential restriction	$\exists R.C$	$\{d \in \Delta^{\mathcal{I}} \mid \exists (d,e) \in R^{\mathcal{I}} : e \in C^{\mathcal{I}}\}$
top	$\top$	$\top^{\mathcal{I}} = \Delta^{\mathcal{I}}$
bottom	$\bot$	$\bot^{\mathcal{I}} = \emptyset$

Fig. 1. Syntax and semantics of some DL constructs

function $\cdot^{\mathcal{I}}$ maps each concept name $A \in \mathsf{N_C}$ to a set $A^{\mathcal{I}} \subseteq \Delta^{\mathcal{I}}$, each role name $R \in \mathsf{N_R}$ to a binary relation $R^{\mathcal{I}}$ on $\Delta^{\mathcal{I}}$, and each individual name $a \in \mathsf{N_I}$ to an individual $a^{\mathcal{I}} \in \Delta^{\mathcal{I}}$. The interpretation of arbitrary concepts is inductively defined as shown in Figure 1. An interpretation $\mathcal{I}$ is called a *model* of a concept C if $C^{\mathcal{I}} \neq \emptyset$. If $\mathcal{I}$ is a model of C, we also say that C is *satisfied* by $\mathcal{I}$.

An interpretation $\mathcal{I}$ *satisfies* (i) an inclusion $C \sqsubseteq D$ if $C^{\mathcal{I}} \subseteq D^{\mathcal{I}}$, (ii) an assertion $C(a)$ if $a^{\mathcal{I}} \in C^{\mathcal{I}}$, and (iii) an assertion $R(a,b)$ if $(a^{\mathcal{I}}, b^{\mathcal{I}}) \in R^{\mathcal{I}}$. Then, $\mathcal{I}$ is a *model* of a knowledge base $\mathcal{S}$ iff $\mathcal{I}$ satisfies all the elements of $\mathcal{S}$.

Here we consider *defeasible* $\mathcal{EL}^{\bot}$ knowledge bases $\mathcal{KB} = (\mathcal{S}, \mathcal{D})$ that consist of a (finite) set of classical axioms (inclusions and assertions) $\mathcal{S}$ and a (finite) set $\mathcal{D}$ of defeasible inclusions (DIs for short). Hereafter, with $C \sqsubseteq_{\mathcal{KB}} D$ we mean that D classically subsumes C, that is $\mathcal{S} \models C \sqsubseteq D$. A classical axiom can be either a normal form axiom [1] or an inclusion/disjointness of existential restrictions:

$$A \sqsubseteq B \qquad A_1 \sqsubseteq \exists R.A_2 \qquad A_1 \sqcap A_2 \sqsubseteq B$$

$$\exists P.A \sqsubseteq B \quad \exists R.A_1 \sqsubseteq \exists S.A_2 \quad \exists R.A_1 \sqcap \exists R.A_2 \sqsubseteq \bot$$

where letters of type A can be either a concept name or $\top$, whereas letters B either a concept name or $\bot$. Defeasible axioms take the form $A_1 \sqsubseteq_n \exists R.A_2$ and can be informally be read as *the instances of A_1 are normally in $\exists R.A_2$*.

Example 1. A well-known example of prototypical property in a biomedical domain is reported by Rector [18, 19]: *"In humans, the heart is usually located on the left-hand side of the body; in humans with situs inversus, the heart is located on the right-hand side of the body"*. A possible formalization in the above language is:

Human $\sqsubseteq_n$ ∃has_heart.LHeart
SitusInversus $\sqsubseteq$ Human $\sqcap$ ∃has_heart.RHeart
LHeart $\sqsubseteq$ Heart $\sqcap$ ∃position.Left
RHeart $\sqsubseteq$ Heart $\sqcap$ ∃position.Right .

In the absence of functional roles, we prevent humans to have both a LHeart and a RHeart with the disjointness axiom:

∃has_heart.LHeart $\sqcap$ ∃has_heart.RHeart $\sqsubseteq \bot$.   □

The nonmonotonic semantics summarized below follows the circumscriptive approach of [7].

Intuitively, a model of a knowledge base $\mathcal{KB}$ is a classical model of $\mathcal{S}$ that maximizes the set of individuals satisfying the defeasible inclusions in $\mathcal{D}$. Formally, for all defeasible inclusions $\delta = (A \sqsubseteq_n C)$ and all interpretations $\mathcal{I}$, the set of individuals *satisfying* δ is:

$$\mathsf{sat}_{\mathcal{I}}(\delta) = \{x \in \Delta^{\mathcal{I}} \mid x \notin A^{\mathcal{I}} \text{ or } x \in C^{\mathcal{I}}\}.$$

How such sets *can be maximized* depends on what is allowed to vary in an interpretation. Here we assume that only the extension of roles can vary, whereas the domain and the extension of concept names are assumed to be fixed. This semantics is called $\mathsf{Circ_{fix}}$.

The reason of this choice is rooted in the goal of having a minimal impact on the classical semantics of DLs. If a concept name A is allowed to vary and has exceptional properties, then A may become empty as illustrated in [8]; in most cases, however, it is undesirable to empty a concept only because it has non-standard properties. It should be possible to extend an existing ontology with default attributes without incurring in such side effects. With $\mathsf{Circ_{fix}}$, a subsumption $A \sqsubseteq B$ where A and B are atomic concepts is nonmonotonically valid iff it is classically valid. At the same time, it is possible to infer new inclusions like $A \sqsubseteq \exists R.B$ that specify default properties of A. In other words, $\mathsf{Circ_{fix}}$ supports default attributes without changing the extension of atomic concepts, as desired.

Maximizing defeasible inclusions may lead to conflicts between defeasible inclusions whose right-hand sides are mutually inconsistent. For this reason, it is useful to provide a means to say that a defeasible inclusion δ_1 has higher priority than another defeasible inclusion δ_2. This can be in general provided explicitly by any partial order over $\mathcal{D}$, but here we focus on an implicit way of defining priorities, namely *specificity*, which is based on classically valid inclusions.[2] For all DIs $\delta_1 = (A_1 \sqsubseteq_n C_1)$ and $\delta_2 = (A_2 \sqsubseteq_n C_2)$, we write

$$\delta_1 \prec \delta_2 \text{ iff } A_1 \sqsubseteq_{\mathcal{KB}} A_2 \text{ and } A_2 \not\sqsubseteq_{\mathcal{KB}} A_1.$$

Example 2. Consider the access control policy: "*Normally users cannot read project files; staff can read project files; blacklisted staff is not granted any access*". In circumscribed $\mathcal{EL}^{\perp}$:

```
Staff ⊑ Users
Blacklisted ⊑ Staff
UserRequest ≡ ∃subject.Users ⊓ ∃target.Projects ⊓ ∃action.Read
StaffRequest ≡ ∃subject.Staff ⊓ ∃target.Projects ⊓ ∃action.Read
UserRequest ⊑_n ∃decision.Deny
StaffRequest ⊑_n ∃decision.Grant
∃subject.Blacklisted ⊑ ∃decision.Deny
∃decision.Grant ⊓ ∃decision.Deny ⊑ ⊥.
```

---

[2] Since concept names are all fixed and retain their classical semantics, specificity can be equivalently defined using nonmonotonically valid inclusions instead. The result is the same, for all priority relations over defeasible inclusions.

As usual, $C \equiv D$ abbreviates $C \sqsubseteq D$ and $D \sqsubseteq C$. The two equivalences can be reformulated using normal form axioms (see Example 5). Clearly the two defeasible inclusions cannot be simultaneously satisfied for any staff member (due to the last inclusion above). According to specificity, the second defeasible inclusion *overrides* the first one and yields the intuitive inference that non-blacklisted staff members are indeed allowed to access project files.   □

We are finally ready to formalize the semantics of KBs with defeasible inclusions. The maximization of the sets $\mathsf{sat}_\mathcal{I}(\delta)$ is modelled by means of the following preference relation $<_\mathcal{D}$ over interpretations. Roughly speaking, $\mathcal{I} <_\mathcal{D} \mathcal{J}$ holds iff $\mathcal{I}$ improves $\mathcal{J}$ by extending the set of individuals that satisfy some defeasible inclusions. More precisely, if $\delta_1 \prec \delta_2$ (i.e., δ_1 has higher priority than δ_2), then the set of individuals satisfying δ_1 may be extended at the cost of restricting those that satisfy δ_2.

Definition 1. *For all interpretations $\mathcal{I}$ and $\mathcal{J}$, let $\mathcal{I} <_\mathcal{D} \mathcal{J}$ iff:*

1. *$\Delta^\mathcal{I} = \Delta^\mathcal{J}$;*
2. *$a^\mathcal{I} = a^\mathcal{J}$, for all $a \in \mathsf{N_I}$;*
3. *$A^\mathcal{I} = A^\mathcal{J}$, for all $A \in \mathsf{N_C}$;   (concept name extensions are fixed)*
4. *for all $\delta \in \mathcal{D}$, if $\mathsf{sat}_\mathcal{I}(\delta) \not\supseteq \mathsf{sat}_\mathcal{J}(\delta)$ then there exists $\delta' \in \mathcal{D}$ such that $\delta' \prec \delta$ and $\mathsf{sat}_\mathcal{I}(\delta') \supset \mathsf{sat}_\mathcal{J}(\delta')$;*
5. *there exists a $\delta \in \mathcal{D}$ such that $\mathsf{sat}_\mathcal{I}(\delta) \supset \mathsf{sat}_\mathcal{J}(\delta)$.*

The subscript $\mathcal{D}$ will be omitted when clear from the context. Now $\mathcal{I}$ is a model of $\mathsf{Circ_{fix}}(\mathcal{KB})$ iff $\mathcal{I}$ is a model of $\mathcal{S}$ that cannot be further improved (defeasible inclusions are satisfied "as much as possible").

Definition 2 (Model). *Let $\mathcal{KB} = (\mathcal{S}, \mathcal{D})$, an interpretation $\mathcal{I}$ is a model of $\mathsf{Circ_{fix}}(\mathcal{KB})$ iff $\mathcal{I}$ is a (classical) model of $\mathcal{S}$ and for all models $\mathcal{J}$ of $\mathcal{S}$, $\mathcal{J} \not< \mathcal{I}$.*

Example 3. Let $\mathcal{KB}$ be the knowledge base of Example 2. According to condition 2 in Def. 1, model improvements cannot change the extension of atomic concepts;[3] therefore, if Grant and Deny are empty in a model, then the two defeasible inclusions of $\mathcal{KB}$ cannot possibly force any request to satisfy $\exists\mathsf{decision.Grant}$ nor $\exists\mathsf{decision.Deny}$. In order to "enable" the two DIs, it suffices to assert that Grant and Deny are non-empty, by means of an auxiliary role aux and two simple inclusions:

$$\top \sqsubseteq \exists\mathsf{aux.Grant} \qquad \top \sqsubseteq \exists\mathsf{aux.Deny}\,.[4]$$

Now the two DIs can "fire" and, as a consequence, the models of $\mathsf{Circ_{fix}}(\mathcal{KB})$ are all the models of the classical inclusions of $\mathcal{KB}$ such that for all individuals x satisfying $\exists\mathsf{target.Projects} \sqcap \exists\mathsf{action.Read}$,

---

[3] Recall that this is one of our requirements, aimed at controlling the side effects of adding defeasible inclusions to existing classical ontologies.

[4] These axioms are usually harmless and can be inserted with the help of automated tools, that identify which concepts occurring in the right hand side of a DI can possibly be empty.

- if x satisfies $\exists$subject.Blacklisted, then x satisfies $\exists$decision.Deny;
- otherwise, if x satisfies $\exists$subject.Staff, then x satisfies $\exists$decision.Grant;
- otherwise, if x satisfies $\exists$subject.User, then x satisfies $\exists$decision.Deny. □

The above example shows the need for declaring the non-emptyness of default attribute ranges, such as B in $A \sqsubseteq_n \exists R.B$. In theory, such declarations may be inconsistent with the knowledge base; however, in practice, concept names are usually meant to be non-empty and, accordingly, concept consistency checking is a typical step in ontology validation. In other words, we only need to make explicit some assumptions that are sometimes left implicit; this can be done automatically for all default attribute ranges B. These additional axioms can be easily checked for consistency: In $\mathcal{EL}^{\perp}$, if all non-emptyness statements are individually consistent with the KB, then also the set of all non-emptyness statements is collectively consistent; consequently, no combinatorial problems arise and consistency checking remains polynomial. It is not difficult to extend this framework with nominals and concrete datatypes; when default attributes range over nominals or concrete domains, non-emptyness is implicit in the logic and no explicit declarations are needed.

2.1 A New Reasoning Task

Now that we have provided constructs for associating concepts to default properties, we need a suitable reasoning task to retrieve them. For example, from the formalization of human heart we would like to infer that typical humans satisfy $\exists$has_heart.LHeart. Subsumption queries, according to [8], are defined as follows: $\mathsf{Circ}_{\mathsf{fix}}(\mathcal{KB}) \models C \sqsubseteq D$ iff for all models $\mathcal{I}$ of $\mathsf{Circ}_{\mathsf{fix}}(\mathcal{KB})$, $C^{\mathcal{I}} \subseteq D^{\mathcal{I}}$. This reasoning method is not completely appropriate for our purposes, because a standard subsumption query $A \sqsubseteq \exists R.C$ considers not only the typical members of A, but also the typical members of A's subconcepts, where A's default properties may be overridden. In this way, some of A's default properties might not be included in the answer. For instance, in the context of the *situs inversus* example, it is generally not possible to entail Humans $\sqsubseteq \exists$has_heart.LHeart, because the members of Humans comprise all the members of SitusInversus, too, that are forced to satisfy $\exists$has_heart.RHeart, instead. For this reason, in this work we consider a slightly modified subsumption problem, according to which a query $A \sqsubseteq \exists R.C$ is interpreted as: "Do the individuals belonging to A *and no subconcepts of* A satisfy $\exists R.C$?". This is a sort of closed world assumption. It is equivalent to interpreting $A \sqsubseteq \exists R.C$ as $CWA_{\mathcal{KB}}(A) \sqsubseteq \exists R.C$, where $CWA_{\mathcal{KB}}(A) = A \sqcap \prod\{\neg B \mid B \in \mathsf{N_C} \text{ and } A \not\sqsubseteq_{\mathcal{KB}} B\}$. In $\mathcal{EL}^{\perp}$, this closure cannot introduce any inconsistency:

Theorem 1. *For all $\mathcal{EL}^{\perp}$ knowledge bases $\mathcal{KB}$, $CWA_{\mathcal{KB}}(A)$ is satisfiable w.r.t. $\mathcal{KB}$ iff A is satisfiable w.r.t. $\mathcal{KB}$.*

$CWA_{\mathcal{KB}}(A)$ can be equivalently defined in purely model theoretic terms not involving $\neg$ as the set $\lfloor A \rfloor^{\mathcal{I}}$ that denotes the set of all individuals $d \in A^{\mathcal{I}}$ such that, for all concept names B, $d \in B^{\mathcal{I}}$ holds only if $A \sqsubseteq_{\mathcal{KB}} B$. Then, we define the modified entailment problem as follows:

Definition 3. *Let* $\mathsf{Circ}_{\mathsf{fix}}(\mathcal{KB}) \models_{\mathsf{cw}} A \sqsubseteq D$ *hold if and only if for all models* $\mathcal{I}$ *of* $\mathsf{Circ}_{\mathsf{fix}}(\mathcal{KB})$, $\lfloor A \rfloor^{\mathcal{I}} \subseteq D^{\mathcal{I}}$.

Example 4. Extend the knowledge base of Example 1 with $\top \sqsubseteq \exists \mathsf{aux.LHeart}$, to ensure that there exists at least one normal heart.[5] Note that

$$CWA_{\mathcal{KB}}(\mathsf{Human}) = \mathsf{Human} \sqcap \neg \mathsf{SitusInversus} \sqcap \neg \mathsf{LHeart} \sqcap \neg \mathsf{RHeart} \sqcap$$
$$\neg \mathsf{Heart} \sqcap \neg \mathsf{Left} \sqcap \neg \mathsf{Right}.$$

It is not hard to see that $\mathsf{Circ}_{\mathsf{fix}}(\mathcal{KB}) \models_{\mathsf{cw}} \mathsf{Human} \sqsubseteq \exists \mathsf{has\_heart.LHeart}$ and $\mathsf{Circ}_{\mathsf{fix}}(\mathcal{KB}) \models_{\mathsf{cw}} \mathsf{SitusInversus} \sqsubseteq \exists \mathsf{has\_heart.RHeart}$, as desired. □

The reader may wonder whether in general the CWA can be too restrictive and miss valid default properties. This might happen if a concept A's extension could be completely covered by n subconcepts $A_1, \ldots, A_n$ sharing a same default property $\exists R.B$. In this case, it would be natural to require A's prototypical members to satisfy $\exists R.B$, as they must necessarily fall into some A_i. However, in $\mathcal{EL}^{\perp}$ such coverings cannot be defined, i.e. there is always a model $\mathcal{I}$ in which there exists $d \in A^{\mathcal{I}} \setminus \bigcup_{i=1}^{n} A_i^{\mathcal{I}}$. Such d need not satisfy $\exists R.B$, and hence it would be inappropriate to list $\exists R.B$ among the default properties of A.

3   Complexity

3.1   NP-Hardness of the General Case

In general, deciding whether $\mathsf{Circ}_{\mathsf{fix}}(\mathcal{KB}) \models_{\mathsf{cw}} A \sqsubseteq D$ holds is NP-hard. This can be proved by reducing SAT to our reasoning task. For each clause c_i in the SAT instance introduce two roles C_i and $\bar{C}_i$. Intuitively, the meaning of $\exists C_i$ and $\exists \bar{C}_i$ is: c_i is/is not satisfied, respectively. For each propositional symbol p_j introduce two roles P_j and $\bar{P}_j$. Intuitively, $\exists P_j$ and $\exists \bar{P}_j$ represent the truth of the complementary literals p_j and $\neg p_j$, respectively. Then, we need two concept names B_0 and B_1, and a role $\bar{F}$. Intuitively, $\exists \bar{F}$ represents the falsity of the set of clauses. The semantics of clauses is axiomatized by adding the inclusions

$$\exists P_j \sqsubseteq \exists C_i, \quad \exists \bar{P}_k \sqsubseteq \exists C_i,$$

for all disjuncts p_j and $\neg p_k$ in c_i. The space of possible truth assignments is generated by the following inclusions:

$$B_0 \sqsubseteq_n \exists P_j, \quad B_0 \sqsubseteq_n \exists \bar{P}_j, \quad \exists P_j \sqcap \exists \bar{P}_j \sqsubseteq \perp.$$

All of the above defaults have the same priority. The defeasible inclusions with the same index j "block" each other; we make at least one of them active by assuming B_0; this "forces" a complete truth assignment. Then we introduce a defeasible inclusion with lower priority:

$$B_0 \sqsubseteq B_1, \quad B_1 \sqsubseteq_n \exists \bar{C}_i.$$

---

[5] See Example 3 for an explanation of this kind of axioms.

This defeasible inclusion "assumes" that c_i is not satisfied. The first three groups of axioms may defeat this assumption (if the selected truth assignment entails $\exists C_i$) thanks to the following disjointness axiom:

$$\exists C_i \sqcap \exists \bar{C}_i \sqsubseteq \perp.$$

Finally, we need the inclusions $\exists \bar{C}_i \sqsubseteq \bar{F}$ to say that the set of clauses is not satisfied when at least one of the clauses is false. Now let $\mathcal{KB}$ denote the above set of inclusions. It can be proved that the given set of clauses is unsatisfiable iff:

$$\mathsf{Circ}_{\mathsf{fix}}(\mathcal{KB}) \models_{\mathsf{cw}} B_0 \sqsubseteq \exists \bar{F}.$$

Consequently:

Theorem 2. *Let $\mathcal{KB}$ range over $\mathcal{EL}^{\perp}$ knowledge bases. The problem of checking whether $\mathsf{Circ}_{\mathsf{fix}}(\mathcal{KB}) \models_{\mathsf{cw}} C \sqsubseteq D$ is NP-hard, even if C is a concept name and D an unqualified existential restriction.*

3.2   A Polynomial Case

The above reduction of SAT is based on concepts with equally specific, conflicting default properties. In our reference scenarios, we expect such situations to be symptoms of representation errors. For instance, in modelling prototypical entities, equally specific and conflicting default properties constitute a contradictory prototype definition. In the access control domain, a class of requests associated to conflicting decisions with the same priority constitutes an ambiguous policy, with potentially dangerous consequences. In this section, we focus on a class of KBs called *conflict safe*, where this kind of conflicts cannot occur. This restriction turns out to reduce the computational complexity of reasoning.

Intuitively, the idea is that it is possible to check efficiently whether two defaults δ_1 and δ_2 block each other and none of them is more specific than the other (as in the reduction from SAT). Such conflicts, that make the search space grow, can be solved (either manually or automatically) by adding more specific defaults that determine how to resolve the conflict (either in favor of one of the δ_is or blocking them both). In the following, let $\mathcal{KB} = (\mathcal{S}, \mathcal{D})$ be an arbitrary knowledge base. The next definitions are all relative to $\mathcal{KB}$.

We say that two defeasible inclusions are in conflict when they can be simultaneously activated (their premises are mutually consistent) and their conclusions are mutually inconsistent. The formal definition follows.

Definition 4. *Two defeasible inclusions $\delta_1 = A_1 \sqsubseteq_n \exists R.A_1'$ and $\delta_2 = A_2 \sqsubseteq_n \exists S.A_2'$ are in conflict, denoted by $\delta_1 \nleftrightarrow \delta_2$, iff $A_1 \sqcap A_2 \not\sqsubseteq_{\mathcal{KB}} \perp$ and $\exists R.A_1' \sqcap \exists S.A_2' \sqsubseteq_{\mathcal{KB}} \perp$.*

Since classical subsumption in $\mathcal{EL}^{\perp}$ knowledge bases can be computed in polynomial time [2], we have:

Proposition 1. *Given an $\mathcal{EL}^{\perp}$ knowledge base $\mathcal{KB} = (\mathcal{S}, \mathcal{D})$ and two defaults δ_1 and δ_2 in $\mathcal{D}$, the problem of checking whether $\delta_1 \nleftrightarrow \delta_2$ is in PTIME.*

A naive approach to listing all the conflicting pairs, consists in performing a quadratic number of $\mathcal{EL}^{\perp}$ subsumptions. Better strategies can be obtained by adapting the ideas behind efficient classification algorithms [6, Chap. 9] to reduce the number of comparisons (the details lie beyond the scope of this paper). In this section we assume that $\mathcal{KB}$ is *conflict safe* in the following sense:

Definition 5. $\mathcal{KB}$ *is* conflict safe *iff whenever two defeasible inclusions* $\delta_1 = A_1 \sqsubseteq_n \exists R.A_1'$ *and* $\delta_2 = A_2 \sqsubseteq_n \exists S.A_2'$ *are incomparable and in conflict (i.e.* $\delta_1 \not\prec \delta_2$, $\delta_2 \not\prec \delta_1$ *and* $\delta_1 \leftrightarrow \delta_2$), *then (i)* $A_1 \not\equiv_{\mathcal{KB}} A_2$, *(ii) there exists a concept name* A_3 *such that* $A_3 \equiv_{\mathcal{KB}} A_1 \sqcap A_2$, *and (iii) one of the following sets of inclusions belongs to* $\mathcal{KB}$:

- $A_3 \sqsubseteq_n \exists R.A_1'$;
- $A_3 \sqsubseteq_n \exists S.A_2'$;
- $A_3 \sqsubseteq_n \exists T$ *and* $\exists T \sqcap \exists R.A_1' \sqsubseteq \perp$ *and* $\exists T \sqcap \exists S.A_2' \sqsubseteq \perp$.   □

Note that the above three DIs (whose priority is higher than δ_1 and δ_2) correspond to three possible ways of resolving the conflict between δ_1 and δ_2, namely, supporting the conclusion of δ_1, supporting the conclusion of δ_2, or blocking both δ_1 and δ_2. The third option constitutes a possible default conflict resolution strategy that can be performed automatically by introducing fresh roles T and the corresponding disjointness axioms. Note also that our two running examples are conflict safe because all conflicting defaults are comparable and specificity resolves the conflict.

We proceed towards a PTIME algorithm for reasoning with conflict safe KBs. We first need some preliminary definitions. Given a concept C, $\mathsf{SupCls}(C)$ denotes the set of *superclasses* of C:

$$\mathsf{SupCls}(C) = \{B \mid C \sqsubseteq_{\mathcal{KB}} B\} \cup \{\exists R.A \mid C \sqsubseteq_{\mathcal{KB}} \exists R.A\}. \tag{1}$$

We write $C \rightsquigarrow A$ if $C \sqsubseteq_{\mathcal{KB}} \exists R.A$ for some R, and we denote by $\overset{*}{\rightsquigarrow}$ the transitive closure of $\rightsquigarrow$. Given a concept C, the operator $\mathsf{NE}(C)$ represents the set of concepts that are forced to be *non-empty* whenever C is. Notice that this set includes some concepts that are forced to be non-empty by the ABox in $\mathcal{KB}$, independently of C.

$$\mathsf{NG}(C) = \{C\} \cup \bigcup_{a \in \mathsf{N_I}} \{A \mid \mathcal{KB} \models A(a)\} \cup \bigcup_{a \in \mathsf{N_I}, R \in \mathsf{N_R}} \{A \mid \mathcal{KB} \models (\exists R.A)(a)\} \tag{2}$$

$$\mathsf{NE}(C) = \bigcup_{A \in \mathsf{NG}(C)} \{A' \mid A \overset{*}{\rightsquigarrow} A'\}. \tag{3}$$

When trying to satisfy a certain defeasible inclusion $A_1 \sqsubseteq_n \exists R.A_2$, we have to check two forms of consistency. First, the addition of an R edge to A_2 should be possible without modifying the interpretation of the concepts names, that are fixed. This check is realized by the following function $\mathsf{Comp}_{\mathrm{fix}}$. Second, the addition of $\exists R.A_2$ should not lead to classical inconsistencies, also considering other defeasible inclusions that were previously satisfied. This check is realized by the function Cons.

Algorithm 1:

Data: C, $\mathcal{KB} = \langle \mathcal{S}, \mathcal{D} \rangle$.

1    $\mathsf{X} \leftarrow \mathsf{SupCls}(C)$;
2   **while** $\mathcal{D} \neq \emptyset$ **do**
3   remove from $\mathcal{D}$ an inclusion $\delta = (A_1 \sqsubseteq_n \exists R.A_2)$ with maximal priority;
4   **if** $A_1 \in \mathsf{SupCls}(C)$ *and* $\delta \in \mathsf{Comp}_{\mathrm{fix}}(C) \cap \mathsf{Cons}(\mathsf{X})$ **then**
5           $\mathsf{X} \leftarrow \mathsf{X} \cup \mathsf{SupCls}(\exists R.A_2)$;

6 **return** X;

For a concept C, $\mathsf{Comp}_{\mathrm{fix}}(C)$ (for *fixed-atoms compatible*) is the set of defeasible inclusions whose r.h.s. agree with C on the inferred and non-empty concept names. That is, a defeasible inclusion $A_1 \sqsubseteq_n \exists R.A_2$ is in $\mathsf{Comp}_{\mathrm{fix}}(C)$ if and only if: *(i)* $\mathsf{NE}(\exists R.A_2) \subseteq \mathsf{NE}(C)$ and *(ii)* for all concept names $A \in \mathsf{SupCls}(\exists R.A_2)$, it holds $A \in \mathsf{SupCls}(C)$.

For a set of concepts X, $\mathsf{Cons}(\mathsf{X})$ is the set of defeasible inclusions whose r.h.s. is logically *consistent* with X. That is, a defeasible inclusion $A_1 \sqsubseteq_n \exists R.A_2$ is in $\mathsf{Cons}(\mathsf{X})$ if and only if $\bigsqcap_{D \in \mathsf{X}} D \sqcap (\exists R.A_2) \not\sqsubseteq_{\mathcal{KB}} \bot$.

We claim that Algorithm 1, when invoked over the concept C, returns the set of all concepts C' that are implied by C under the closed world assumption.

Theorem 3. *Let X be the result of Algorithm 1 on the concept C. If $\mathcal{KB}$ is conflict safe and assertion-free[6] then $\mathsf{X} = \{C' \mid \mathsf{Circ}_{\mathrm{fix}}(\mathcal{KB}) \models_{\mathrm{cw}} C \sqsubseteq C'\}$.*

Proof. ($\subseteq$) Let $C' \in \mathsf{X}$. If C' was inserted in line 1 of the algorithm, then C' is classically implied by C (i.e., $C \sqsubseteq_{\mathcal{KB}} C'$), and hence $\mathsf{Circ}_{\mathrm{fix}}(\mathcal{KB}) \models_{\mathrm{cw}} C \sqsubseteq C'$. Otherwise, C' was inserted in line 5. Hence, there is a defeasible inclusion $\delta = (A_1 \sqsubseteq_n \exists R.A_2)$ such that $A_1 \in \mathsf{SupCls}(C)$, $\delta \in \mathsf{Comp}_{\mathrm{fix}}(C) \cap \mathsf{Cons}(\mathsf{X}')$ and $C' \in \mathsf{SupCls}(\exists R.A_2)$, where X' is the value of the variable X when C' was inserted. By applying the definition of $\mathsf{Comp}_{\mathrm{fix}}$, we obtain that *(i)* $\mathsf{NE}(\exists R.A_2) \subseteq \mathsf{NE}(C)$, and *(ii)* for all concept names $A' \in \mathsf{SupCls}(\exists R.A_2)$, it holds $A' \in \mathsf{SupCls}(C)$.

Let $\mathcal{I}$ be a model of $\mathsf{Circ}_{\mathrm{fix}}(\mathcal{KB})$ with an individual $d \in \lfloor C \rfloor^{\mathcal{I}}$, we show that $d \in C'^{\mathcal{I}}$. Assume by contradiction that $d \notin C'^{\mathcal{I}}$. Since A_1 is a classical consequence of C, we have $d \in A_1^{\mathcal{I}}$. Since $C' \in \mathsf{SupCls}(\exists R.A_2)$, we have $d \notin (\exists R.A_2)^{\mathcal{I}}$. We show that there exists a classical model $\mathcal{J}$ of $\mathcal{KB}$ that improves $\mathcal{I}$, i.e., $\mathcal{J} <_{\mathcal{D}} \mathcal{I}$. To define $\mathcal{J}$, for all $\exists S.A_3 \in \mathsf{SupCls}(\exists R.A_2)$ (including $\exists R.A_2$ itself), we add to $\mathcal{I}$ an S-arc from d to an individual $x \in A_3^{\mathcal{J}}$. The existence of such an individual is guaranteed by the fact that $\mathsf{NE}(\exists R.A_2) \subseteq \mathsf{NE}(C)$. As a result, we have in particular that $d \in (\exists R.A_2)^{\mathcal{J}}$. By *(ii)*, all atomic concepts that are classical consequences of $\exists R.A_2$ are also consequences of C. This, together with the fact that $\delta \in \mathsf{Cons}(\mathsf{X}')$, ensures that $\mathcal{J}$ is a classical model of $\mathcal{KB}$. It remains to prove that $\mathcal{J} <_{\mathcal{D}} \mathcal{I}$. Since $\mathcal{I}$ and $\mathcal{J}$ only differ on the arcs outgoing from d, we have

---

[6] In DL jargon: the ABox is empty. The reason for considering ABox assertions in the definition of $\mathsf{NG}(C)$ will be clear in the next section, when we deal with instance checking.

$\mathsf{sat}_{\mathcal{I}}(\delta) \subset \mathsf{sat}_{\mathcal{J}}(\delta)$ and for all $\delta' \neq \delta$ in $\mathcal{D}$, we have $\mathsf{sat}_{\mathcal{I}}(\delta') = \mathsf{sat}_{\mathcal{J}}(\delta')$. Therefore, we obtain the thesis.

($\supseteq$) Let C' be a concept such that $\mathsf{Circ}_{\mathsf{fix}}(\mathcal{KB}) \models_{\mathsf{cw}} C \sqsubseteq C'$. Assume by contradiction that C' does not belong to the output X of the algorithm. Clearly, $C \not\sqsubseteq_{\mathcal{KB}} C'$, otherwise C' would have been added to X in step 1 of the algorithm. Since $\mathsf{Circ}_{\mathsf{fix}}(\mathcal{KB}) \models_{\mathsf{cw}} C \sqsubseteq C'$, there is a defeasible inclusion $A_1 \sqsubseteq_n \exists R.A_2 \in \mathcal{D}$ such that $A_1 \in \mathsf{SupCls}(C)$ and $\exists R.A_2 \sqsubseteq_{\mathcal{KB}} C'$ Let $\hat{\delta} \in \mathcal{D}$ be a defeasible inclusion with the above property and maximal priority. At some point, $\hat{\delta}$ is extracted from $\mathcal{D}$ at step 3 of the algorithm. Since C' is never added to X, we have that either $\hat{\delta} \notin \mathsf{Comp}_{\mathsf{fix}}(C)$ or $\hat{\delta} \notin \mathsf{Cons}(\mathsf{X}')$, where X' is the current value of the variable X. In both cases, it is possible to define a model $\mathcal{I}$ of $\mathsf{Circ}_{\mathsf{fix}}(\mathcal{KB})$ with an individual $d \in \Delta^{\mathcal{I}}$ such that $d \in \lfloor C \rfloor^{\mathcal{I}} \setminus C'^{\mathcal{I}}$, which is a contradiction.

We define $\mathcal{I}$ as follows.

- $\Delta^{\mathcal{I}} = \{d_C\} \cup \{d_A \mid A \in \mathsf{NE}(C)\} \cup \{d_a \mid a \in \mathsf{N_I}\}$;
- for each concept name A, $A^{\mathcal{I}} = \{d_X \mid X \sqsubseteq_{\mathcal{KB}} A\} \cup \{d_a \mid \mathcal{KB} \models A(a)\}$;
- for each role name R, we start by putting all edges that are classically required, i.e., $R^{\mathcal{I}} = \{(d_X, d_Y) \mid X \sqsubseteq_{\mathcal{KB}} \exists R.Y\} \cup \{(d_a, d_b) \mid R(a, b) \in \mathcal{KB}\} \cup \{(d_a, d_X) \mid \mathcal{KB} \models (\exists R.X)(a)\}$. Moreover, for each $\exists R.Y \in \mathsf{X}$, we add the edge (d_C, d_Y) to R^I. Extra edges starting from individuals other than d_C are not relevant.

By construction, $\mathcal{I}$ is a classical model of $\mathcal{KB}$ and, as $C' \notin \mathsf{X}$, $d_C \in \lfloor C \rfloor^{\mathcal{I}} \setminus C'^{\mathcal{I}}$. It remains to prove that there is no model $\mathcal{J}$ that improves $\mathcal{I}$ by making d_C satisfy $\hat{\delta}$.

If $\hat{\delta} \notin \mathsf{Comp}_{\mathsf{fix}}(C)$, then either $\mathsf{NE}(\exists R.A_2) \not\subseteq \mathsf{NE}(C)$ or there exists a concept name A' such that $A' \in \mathsf{SupCls}(\exists R.A_2)$ and $A' \notin \mathsf{SupCls}(C)$ (hence, $d_C \notin A'^{\mathcal{I}}$). Since any model $\mathcal{J}$ that is comparable with $\mathcal{I}$ has the same interpretation for the concept names, such model cannot have $d_C \in \exists (R.A_2)^{\mathcal{J}}$.

If instead $\hat{\delta} \notin \mathsf{Cons}(X')$, we have $\bigsqcap_{D' \in X'} D' \sqcap \exists R.A_2 \sqsubseteq_{\mathcal{KB}} \bot$. If this inconsistency derives from classical consequences of C (i.e., $\exists R.A_2 \sqcap \mathsf{SupCls}(C) \sqsubseteq_{\mathcal{KB}} \bot$), the thesis is obvious. Otherwise, the inconsistency is due to one or more defeasible inclusions δ that were chosen in the previous iterations of the loop, on line 3. For each such δ, either its priority is higher than the one of $\hat{\delta}$, or it is incomparable with it. In the first case, clearly it is not worth modifying δ in order to improve $\hat{\delta}$. In the latter case, we employ the assumption that $\mathcal{KB}$ is conflict safe. In particular, we have that δ and $\hat{\delta}$ are incomparable and in conflict. Let $\delta = (A_3 \sqsubseteq_n \exists R.A_4)$. There is a concept name A_5 such that $A_5 \equiv_{\mathcal{KB}} A_1 \sqcap A_3$ and the defeasible inclusion $\delta' = (A_5 \sqsubseteq_n \exists R.A_4)$ belongs to $\mathcal{KB}$. Then, the priority of δ' is higher than both δ and $\hat{\delta}$. Hence, it is not worth modifying δ' to improve $\hat{\delta}$. □

Theorem 4. *Algorithm 1 runs in polynomial time.*

Proof. The main cycle of the algorithm performs as many iterations as the number of defeasible inclusions in $\mathcal{KB}$. The polynomial complexity of the auxiliary operators NE, SupCls, $\mathsf{Comp}_{\mathsf{fix}}$ and Cons derive from the polynomial complexity of reasoning in $\mathcal{EL}$. □

The following example shows how to apply Algorithm 1 to the KB of Example 2.

Example 5. Assume that $\mathcal{KB}$ is the knowledge base of Example 3 and we want to check whether staff members can read project files. First, we have to reduce the $\mathcal{KB}$ in normal form as follows. We introduce six new concept names — SubUsers, SubStaff, TargProjects, AuxUsers, AuxStaff and ActRead — together with the following equivalences.

$$\exists\texttt{subject.Users} \equiv \texttt{SubUsers}$$
$$\exists\texttt{subject.Staff} \equiv \texttt{SubStaff}$$
$$\exists\texttt{target.Projects} \equiv \texttt{TargProjects}$$
$$\exists\texttt{action.Read} \equiv \texttt{ActRead}$$
$$\texttt{SubUsers} \sqcap \texttt{TargProjects} \equiv \texttt{AuxUsers}$$
$$\texttt{SubStaff} \sqcap \texttt{TargProjects} \equiv \texttt{AuxStaff}$$
$$\texttt{AuxUsers} \sqcap \texttt{ActRead} \equiv \texttt{UserRequest}$$
$$\texttt{StaffUsers} \sqcap \texttt{ActRead} \equiv \texttt{StaffRequest}$$

The above equivalences replace the original definitions of UserRequest and StaffRequest. The other inclusions remain unchanged. Recall that the $\mathcal{KB}$ contains

$$\top \sqsubseteq \exists\texttt{aux.Grant}$$
$$\top \sqsubseteq \exists\texttt{aux.Deny}$$

Algorithm 1 receives as input

$$C = \exists\texttt{subject.Staff} \sqcap \exists\texttt{target.Projects} \sqcap \exists\texttt{action.Read}.$$

On line 1, the superclasses of C are computed. At that point, X contains, among the others, StaffPolicy and $\mathsf{NE}(C)$ contains Grant. According to specificity, the first defeasible inclusion removed from $\mathcal{D}$ is $\texttt{StaffPolicy} \sqsubseteq_n \exists\texttt{decision.Grant}$. Since $\exists\texttt{decision.Grant}$ has no proper superclasses and $\mathsf{NE}(\exists\texttt{decision.Grant})$ contains only Grant, the condition on line 4 is satisfied and X becomes $X \cup \{\exists\texttt{decision.Grant}\}$. Thus, we have that

$$\mathsf{Circ}_{\mathsf{fix}}(\mathcal{KB}) \models_{\mathsf{cw}}$$
$$\exists\texttt{subject.Staff} \sqcap \exists\texttt{target.Projects} \sqcap \exists\texttt{action.Read} \sqsubseteq \exists\texttt{decision.Grant}.$$

Note that the second defeasible inclusion $\texttt{UsersPolicy} \sqsubseteq_n \exists\texttt{decision.Deny}$ does not belong to $\mathsf{Cons}(X)$ since $\exists\texttt{decision.Grant}$ and $\exists\texttt{decision.Deny}$ are inconsistent. □

4   Reasoning about Individuals

The ideas illustrated so far can be naturally extended to reasoning about individuals, that is, instance checking. This task suffers from the same problem as subsumption: given an assertion $A(a)$, the individual a might well be a member of any subclass of A, which may prevent the default properties of A from being inherited by a if the standard definition of instance checking [7] is used. Therefore, some form of closure similar to $CWA_{\mathcal{KB}}$ is needed. The closure, in this case, applies to the atomic concepts that contain the individuals in the ABox, as collected by the meta-function $AtCls_{\mathcal{KB}}(a) = \bigsqcap\{A \mid \mathcal{KB} \models A(a)\}$.

Definition 6. *Let $\mathcal{KB}$ be any defeasible $\mathcal{EL}^{\perp}$ KB. $CWA(\mathcal{KB})$ denotes the knowledge base obtained from $\mathcal{KB}$ by adding the assertions $CWA_{\mathcal{KB}}(AtCls_{\mathcal{KB}}(a))(a)$, for all individuals a occurring in $\mathcal{KB}$.*

Instance checking $\mathsf{Circ_{fix}}(\mathcal{KB}) \models_{\mathsf{cw}} C(a)$ is then defined as $\mathsf{Circ_{fix}}(CWA(\mathcal{KB})) \models C(a)$ or, in a model-theoretic view:

Definition 7. $\mathsf{Circ_{fix}}(\mathcal{KB}) \models_{\mathsf{cw}} C(a)$ *if and only if for all models $\mathcal{I}$ of $\mathsf{Circ_{fix}}(\mathcal{KB})$ if $\{A \in \mathsf{N_C} \mid a^{\mathcal{I}} \in A^{\mathcal{I}}\} = \{A \in \mathsf{N_C} \mid \mathcal{KB} \models A(a)\}$, then $a^{\mathcal{I}} \in C^{\mathcal{I}}$.*

Since $\mathsf{Circ_{fix}}$ preserves the classical semantics of atomic concepts and $\mathcal{EL}^{\perp}$ KBs behave like Horn theories in many respects, it can be proved that:

Proposition 2. *For all defeasible $\mathcal{EL}^{\perp}$ knowledge bases $\mathcal{KB}$, and all conjunctions of atomic concepts C, $\mathsf{Circ_{fix}}(\mathcal{KB}) \models_{\mathsf{cw}} C(a)$ iff $CWA(\mathcal{KB}) \models C(a)$ iff $\mathcal{KB} \models C(a)$.*

In other words, membership to atomic concepts and conjunctions thereof is fully classical. Therefore, in this paper, we focus on the more interesting problem of inferring the default properties of individuals. The reasoning task of our interest is the following: *Given an individual "a" and a concept $\exists R.A$, decide whether*

$$\mathsf{Circ_{fix}}(\mathcal{KB}) \models_{\mathsf{cw}} (\exists R.A)(a).$$

The NP-hardness proof for subsumption can be easily adapted to instance checking (using the same reduction plus assertion $B_0(a)$ and the query $\mathsf{Circ_{fix}}(\mathcal{KB}) \models_{\mathsf{cw}} (\exists \bar{F})(a)$). So we get:

Theorem 5. *Let $\mathcal{KB}$ range over $\mathcal{EL}^{\perp}$ knowledge bases. The problem of checking whether $\mathsf{Circ_{fix}}(\mathcal{KB}) \models_{\mathsf{cw}} C(a)$ is NP-hard, even if the existential restriction is unqualified (i.e., $A = \top$).*

For conflict safe knowledge bases, the instance checking problem can be decided using the same algorithm as for subsumption. What we need is to provide as input a concept which is the conjunction of all the atomic concepts and existential restrictions which a is classically an instance of. Let $GenCls_{\mathcal{KB}}(a)$ be such a conjunction:

$$GenCls_{\mathcal{KB}}(a) = \bigsqcap\{A \mid \mathcal{KB} \models A(a)\} \sqcap \bigsqcap\{\exists R.A \mid \mathcal{KB} \models (\exists R.A)(a)\}.$$

The proof of the following theorem is analogous to Theorem 3 and is left to the reader.

Theorem 6. *Let* X *be the result of Algorithm 1 on the concept GenCls$_{\mathcal{KB}}(a)$. If* $\mathcal{KB}$ *is conflict safe then* Circ$_{\text{fix}}(\mathcal{KB}) \models_{\text{cw}} (\exists R.A)(a)$ *iff* $(\exists R.A) \in$ X.

Example 6. Let $\mathcal{KB}$ be a knowledge base obtained by adding to Example 4 the assertions:

$$\text{Human(Mary)}$$
$$\text{SitusInversus(John)}$$

Recall that $\mathcal{KB}$ contains $\top \sqsubseteq \exists \text{aux.LHeart}$, where aux is a new role name.

We want to check that Circ$_{\text{fix}}(\mathcal{KB}) \models_{\text{cw}} (\exists \text{has\_heart.LHeart})(\text{Mary})$ and Circ$_{\text{fix}}(\mathcal{KB}) \models_{\text{cw}} (\exists \text{has\_heart.RHeart})(\text{John})$. Let consider the first query, the input of Algorithm 1 is the concept $C = \text{Human}$. As Human has no proper superclasses, at line 1 $X = \{\text{Human}\}$. The only defeasible inclusion to be checked in lines 2-5 is Human $\sqsubseteq_n \exists \text{has\_heart.LHeart}$. The set NE(Human) consists of all the concept names occurring in the knowledge base, $\exists \text{has\_heart.LHeart}$ is consistent with Human and it does not force other concept names to be locally true. Therefore, the condition in line 4 is satisfied and $\exists \text{has\_heart.LHeart}$ is added to X as expected.

For the second query, as seen before $\exists \text{has\_heart.RHeart}$ classically derives from SitusInversus and hence it is added to X directly in line 1. Note that, even if Human $\sqsubseteq_n \exists \text{has\_heart.LHeart}$ is *activated* by the fact that SitusInversus $\sqsubseteq_{\mathcal{KB}}$ Human, the defeasible inclusion Human $\sqsubseteq_n \exists \text{has\_heart.LHeart}$ is not in Cons(X) because $\exists \text{has\_heart.LHeart}$ and $\exists \text{has\_heart.RHeart}$ are inconsistent.   □

5   Related Work

DLs have been extended with nonmonotonic constructs such as default rules [20, 3,4], autoepistemic operators [11,12], and circumscription [10,8,7]. An advantage of circumscription is that nonmonotonic properties apply to all individuals, while the other approaches restrict nonmonotonic inferences to the individuals that are explicitly denoted in the ABox, as observed in [8]. While [8] focusses on expressive circumscribed description logics whose complexity may reach NEXPTIME^{NP}, [10] and [7] deal with lower-complexity DLs like $\mathcal{ALE}$, DL-lite, and $\mathcal{EL}$; however, upper complexity bounds are all at the second level of the polynomial hierarchy or harder, while here we have identified a tractable case. The two works [8, 7] consider more general forms of circumscription (with variable concept names) and reasoning tasks (satisfiability and KB consistency) that we do not consider here. However, they do not deal with the modified entailment $\models_{\text{cw}}$ on which this paper is focussed. Another recent attempt at low-complexity, nonmonotonic DL reasoning is based on a modal *typicality operator* [15, 14], whose extension is maximized to achieve nonmonotonic inferences. Unfortunately, reasoning is intractable in this approach.

6   Conclusions and Perspectives

The need for supporting prototypical reasoning and exceptions in DLs can be addressed by restricting the expressiveness of the underlying DL and by selecting an appropriate form of inference ($\models_{cw}$). We have shown how to encode a recurring example originated by the work on biomedical ontologies, and a representative example related to semantic web policies. The adoption of $Circ_{fix}$ makes it possible to add default attributes to the concepts of a given (classical) ontology in a controlled way, without affecting the extension of atomic concepts. For conflict safe KBs, the problem of reasoning about default attributes belongs to P; we provided an algorithm based on $\mathcal{EL}$ classification problems that enjoy efficient implementations [5]. This is a promising starting point for addressing the performance challenges posed by the semantic web.

In the full version of this paper we will provide more details on the strategies for making KBs conflict safe. We are also going to support more general queries and more constructs from $\mathcal{EL}^{++}$, identifying the tractability threshold.

An interesting direction for further research consists in studying the impact of variable concept names on the complexity of $\models_{cw}$.

Acknowledgements. This work is partially supported by the national project LoDeN (`http://loden.fisica.unina.it/`). The authors are grateful to the anonymous referees for their constructive comments that helped improving the paper.

References

1. Baader, F.: The instance problem and the most specific concept in the description logic EL w.r.t. terminological cycles with descriptive semantics. In: Günter, A., Kruse, R., Neumann, B. (eds.) KI 2003. LNCS (LNAI), vol. 2821, pp. 64–78. Springer, Heidelberg (2003)
2. Baader, F., Brandt, S., Lutz, C.: Pushing the EL envelope. In: Proc. of the Nineteenth International Joint Conference on Artificial Intelligence, IJCAI 2005, pp. 364–369. Professional Book Center (2005)
3. Baader, F., Hollunder, B.: Embedding defaults into terminological knowledge representation formalisms. J. Autom. Reasoning 14(1), 149–180 (1995)
4. Baader, F., Hollunder, B.: Priorities on defaults with prerequisites, and their application in treating specificity in terminological default logic. J. Autom. Reasoning 15(1), 41–68 (1995)
5. Baader, F., Lutz, C., Suntisrivaraporn, B.: CEL - a polynomial-time reasoner for life science ontologies. In: Furbach, U., Shankar, N. (eds.) IJCAR 2006. LNCS (LNAI), vol. 4130, pp. 287–291. Springer, Heidelberg (2006)
6. Baader, F., McGuiness, D.L., Nardi, D., Patel-Schneider, P.: The Description Logic Handbook: Theory, implementation and applications. Cambridge University Press, Cambridge (2003)
7. Bonatti, P.A., Faella, M., Sauro, L.: Defeasible inclusions in low-complexity DLs: Preliminary notes. In: Boutilier, C. (ed.) IJCAI, pp. 696–701 (2009)
8. Bonatti, P.A., Lutz, C., Wolter, F.: The complexity of circumscription in dls. J. Artif. Intell. Res. (JAIR) 35, 717–773 (2009)
9. Bonatti, P.A., Samarati, P.: Logics for authorization and security. In: Chomicki, J., van der Meyden, R., Saake, G. (eds.) Logics for Emerging Applications of Databases, pp. 277–323. Springer, Heidelberg (2003)

10. Cadoli, M., Donini, F., Schaerf, M.: Closed world reasoning in hybrid systems. In: Proc. of ISMIS 1990, pp. 474–481. Elsevier, Amsterdam (1990)
11. Donini, F.M., Nardi, D., Rosati, R.: Autoepistemic description logics. In: IJCAI (1), pp. 136–141 (1997)
12. Donini, F.M., Nardi, D., Rosati, R.: Description logics of minimal knowledge and negation as failure. ACM Trans. Comput. Log. 3(2), 177–225 (2002)
13. Finin, T.W., Joshi, A., Kagal, L., Niu, J., Sandhu, R.S., Winsborough, W.H., Thuraisingham, B.M.: ROWLBAC: representing role based access control in OWL. In: Ray, I., Li, N. (eds.) SACMAT, pp. 73–82. ACM, New York (2008)
14. Giordano, L., Gliozzi, V., Olivetti, N., Pozzato, G.L.: Prototypical reasoning with low complexity description logics: Preliminary results. In: Erdem, E., Lin, F., Schaub, T. (eds.) LPNMR 2009. LNCS, vol. 5753, pp. 430–436. Springer, Heidelberg (2009)
15. Giordano, L., Gliozzi, V., Olivetti, N., Pozzato, G.L.: Reasoning about typicality in ALC and EL. In: Grau, et al. (eds.) [16]
16. Grau, B.C., Horrocks, I., Motik, B., Sattler, U. (eds.): Proceedings of the DL Home 22nd International Workshop on Description Logics (DL 2009), Oxford, UK, July 27-30. CEUR Workshop Proceedings, vol. 477. CEUR-WS.org. (2009)
17. Kolovski, V., Hendler, J.A., Parsia, B.: Analyzing web access control policies. In: Williamson, C.L., Zurko, M.E., Patel-Schneider, P.F., Shenoy, P.J. (eds.) WWW, pp. 677–686. ACM, New York (2007)
18. Rector, A.L.: Defaults, context, and knowledge: Alternatives for OWL-indexed knowledge bases. In: Altman, R.B., Dunker, A.K., Hunter, L., Jung, T.A., Klein, T.E. (eds.) Pacific Symposium on Biocomputing, pp. 226–237. World Scientific, Singapore (2004)
19. Stevens, R., Aranguren, M.E., Wolstencroft, K., Sattler, U., Drummond, N., Horridge, M., Rector, A.L.: Using OWL to model biological knowledge. International Journal of Man-Machine Studies 65(7), 583–594 (2007)
20. Straccia, U.: Default inheritance reasoning in hybrid KL-ONE-style logics. In: IJCAI, pp. 676–681 (1993)
21. Uszok, A., Bradshaw, J.M., Jeffers, R., Suri, N., Hayes, P.J., Breedy, M.R., Bunch, L., Johnson, M., Kulkarni, S., Lott, J.: KAoS policy and domain services: Towards a description-logic approach to policy representation, deconfliction, and enforcement. In: 4th IEEE International Workshop on Policies for Distributed Systems and Networks (POLICY), Lake Como, Italy, pp. 93–96. IEEE Computer Society, Los Alamitos (June 2003)
22. Zhang, R., Artale, A., Giunchiglia, F., Crispo, B.: Using description logics in relation based access control. In: Grau, et al. (eds.) [16]

Supporting Natural Language Processing with Background Knowledge: Coreference Resolution Case

Volha Bryl, Claudio Giuliano, Luciano Serafini, and Kateryna Tymoshenko

Fondazione Bruno Kessler, via Sommarive 18, 38123 Trento, Italy
{bryl,giuliano,serafini,tymoshenko}@fbk.eu

Abstract. Systems based on statistical and machine learning methods have been shown to be extremely effective and scalable for the analysis of large amount of textual data. However, in the recent years, it becomes evident that one of the most important directions of improvement in natural language processing (NLP) tasks, like word sense disambiguation, coreference resolution, relation extraction, and other tasks related to knowledge extraction, is by exploiting semantics. While in the past, the unavailability of rich and complete semantic descriptions constituted a serious limitation of their applicability, nowadays, the Semantic Web made available a large amount of logically encoded information (e.g. ontologies, RDF(S)-data, linked data, etc.), which constitutes a valuable source of semantics. However, web semantics cannot be easily plugged into machine learning systems. Therefore the objective of this paper is to define a reference methodology for combining semantic information available in the web under the form of logical theories, with statistical methods for NLP. The major problems that we have to solve to implement our methodology concern (i) the selection of the correct and minimal knowledge among the large amount available in the web, (ii) the representation of uncertain knowledge, and (iii) the resolution and the encoding of the rules that combine knowledge retrieved from Semantic Web sources with semantics in the text. In order to evaluate the appropriateness of our approach, we present an application of the methodology to the problem of intra-document coreference resolution, and we show by means of some experiments on the standard dataset, how the injection of knowledge leads to the improvement of this task performance.

1 Introduction

The two key aspects of natural language applications based on machine learning techniques are the learning algorithm, and the feature extraction and representation of the documents, entities, or words that have to be manipulated. Reviewing the relevant literature of the last years, one realizes that, typically, the difference between the results obtained by different learning algorithms (e.g., support vector machines vs. decision trees) is significant when they are fed with the same information. On the other hand, the feature extraction and representation methods play a crucial role for the accuracy of the system. Simple representations, e.g., the bag-of-words, and more complex ones, e.g., tree kernels, have been exploited in different tasks and their difference has been proved to be significant as well. For example, in relation extraction approaches that

P.F. Patel-Schneider et al. (Eds.): ISWC 2010, Part I, LNCS 6496, pp. 80–95, 2010.

exploit deep syntactic parsing outperform the ones that represent only shallow syntactic analysis. Until now, the majority of the approaches focus on representing syntactic information while background knowledge extracted from knowledge bases has been restricted to WordNet and ad-hoc gazetteers [12,7]. The main reasons are due to the low coverage of the available knowledge resources and the difficulty to match text and ontology elements.

Nowadays, the Semantic Web made available a large amount of logically encoded information (e.g., ontologies, RDF(S)-data, linked data, etc.), which constitute a valuable source of semantic knowledge. However, extending the state-of-the-art natural language applications to use these resources is not a trivial task due to the following reasons: (i) The *heterogeneity* and the *ambiguity* of the schemes adopted by the different resources of the Semantic Web. This means, e.g., that the same relation can be encoded by different URIs, and that URIs are used by different resources for denoting different relations. (ii) The *irregular coverage* of the knowledge available in the Web. This means that for some "famous" entities the Semantic Web contains a large amount of knowledge, and only a little is relevant for solving a specific task (e.g., coreference resolution or relation extraction), while for other entities there is no knowledge at all. (ii) The *logical-statistical knowledge integration problem*, i.e., the fact that algorithms for coreference resolution are based on statistical feature models, while background knowledge in the Semantic Web is encoded in some logical form.

In this paper, we define a general methodology for supporting natural language processing by exploiting background knowledge available in the Web, by proposing practical solutions for the before mentioned problems. *First*, we map terms in text to URIs through Wikipedia mediation. Since most of the resources available in the Semantic Web are linked to Wikipedia, we can use it as a *semantic mediator*. So we propose to link text with Wikipedia entries and then to exploit the linking between Wikipedia and the other resources to access the knowledge encoded in them. Wikipedia represents a practical choice, as it is playing a central role in the development of the Semantic Web, given the large and growing number of resources linked to it, which makes Wikipedia one of the central interlinking hubs of the emerging Web of Data. *Second*, we query the Semantic Web using the URIs to obtain the background knowledge expressed in the RDF/OWL formalism and apply feature selection techniques to retrieve the relevant knowledge for the specific task. In this way we do not assume to have any a priori knowledge of the specific task but we delegate to the *feature selection* phase the responsibility of finding the relevant information from an arbitrary Semantic Web resource to model it. Differently, in our previous work [5] we experimented with the small predefined subset of properties from one specific knowledge source (YAGO ontology) to support the coreference resolution task. *Finally*, as we presented in more details in [5], we use the Alchemy tool [1] for the integration of uncertain knowledge, and facts expressed in first-order language. Alchemy provides both reasoning and learning functionalities, though we only use the reasoning part. The extension of this work, however, could require learning capabilities.

To evaluate the methodology, we run a number of experiments in coreference resolution, which are reported in Section 5. The experiments consist in selecting a set of features relevant for the given task from three large-scale Semantic Web resources and then testing the coreference resolution model extended with the selected features. The results show that our method performs in the order of the state-of-the-art coreference algorithms, and, importantly, that the use of background knowledge provides a tangible advantage for coreference resolution.

2   Coreference Resolution: Task Definition and Related Work

The task of coreference resolution consists in identifying mentions that refer to the same real-world entity. E.g., it is required to identify that the mentions *Barack Obama* and *president* are coreferent in the text *"Barack Obama will make an appearance on the TV show. The president is scheduled to come on Friday evening."* This constitutes an important subtask in many natural language processing (NLP) applications such as information extraction, textual entailment, and question answering. Machine learning (ML) is widely used to approach the coreference task. State-of-the-art coreference resolvers are mostly extensions of the Soon et al. approach in which a mention-pair classifier is trained using solely surface-level features to determine whether two mentions are coreferring or not [25]. In the last decade, two independent research lines have extended the Soon et al. approach yielding significant improvements in accuracy.

The first aims at defining a more sophisticated ML framework to overcome the limits of the mention-pair model. Entity-mention and mention-ranking models and their combination cluster-ranking are some of the relevant approaches proposed (e.g. [9,16]). An entity-mention model considers candidate pairs, which consist of a cluster of mentions, referring to the same entity, and a new mention. [18] motivate the entity-mention model using an example of a mention set such as "Mr. Obama", "Obama" and "she". A mention-pair model might first predict that "Mr. Obama" and "Obama" are coreferent, then it might predict that "Obama" and "she" are coreferent as well, and finally cluster all these mentions as referring to the same entity. An entity-mention model first classifies "Mr. Obama" and "Obama" as coreferent, and then immediately clusters them into an entity cluster. Then the model considers the *entity cluster* ("Mr. Obama", "Obama") and the mention "she" as the coreference candidates. In this case "she" is unlikely to be added to the given cluster, as there is a gender disagreement between "Mr." and "she". The mention-ranking models attempt to choose the most probable candidate antecedent for a mention, among *all* the preceding mentions within a given scope. E.g., if a text contains the mentions "she", "Barack Obama" and "Michele Obama", the set of candidate antecedents for the mention "Michele Obama" includes "she" and "Barack Obama". The models ranks them and chooses the most probable one.

The second research line investigates the usage of semantic knowledge sources to augment the feature space [25,20,17,27]. Here the majority of the approaches exploit WordNet [11] and, more recently, Wikipedia[1] or corpora annotated with semantic classes. E.g., in [25] a candidate pair of mentions was represented as a vector of twelve features, two of which, namely the semantic class agreement and alias, were of semantic nature. The alias feature contributed greatly to the performance of the system. It was obtained using a set of heuristics, e.g. it was considered true if one mention was an acronym of another. Therefore, its value could be evaluated only in a limited number of cases. The semantic

---

[1] http://wikipedia.org/

class agreement feature did not impact the performance of the system, which may be due to the fact that the most frequent sense of a mention in the WordNet lexical database was employed as its semantic class. Therefore, the possible ambiguity of a mention was not taken into account. In [19] a set of features from [25] was expanded, with the semantic relatedness features based on WordNet taxonomy. However, they did not perform the disambiguation as well, and the new semantic features did not impact the final performance of the system either. Recently, Wikipedia has also started to be exploited as a source of semantic knowledge for coreference resolution [20,27]. E.g., its category structure and article texts are used in [20] in order to obtain a set of six features based on the semantic relatedness of mentions. In order to find the Wikipedia articles which correspond to a mention, Wikipedia is queried for pages titled as the head lemma of the mention. If the disambiguation page is hit, an heuristic algorithm is employed. However, such approach is likely to return the Wikipedia page that corresponds to the most frequent sense of a mention. The problem of possible noun mention ambiguity was taken into account in [17]. In this work a special classifier was trained on the BBN entity corpus to assign one of five semantic classes to the mentions. Even though the set of semantic classes is not large, the features based on usage of these classes gave an improvement of the precision of the common noun resolution by 2-6% over [25]. These results show that taking into account the ambiguity of the mentions is crucial for obtaining the semantic knowledge relevant for coreference resolution. Knowledge representation format and the structure of the knowledge sources used by the above described approaches are different, therefore, in each specific case information from a resource has to be extracted and processed differently. In the following section we present an approach that allows us to overcome this issue and work with knowledge from heterogeneous sources with only minimal assumptions on their representation and structure.

3   Background Knowledge Acquisition

3.1   Sources of Background Knowledge

Our approach is concerned with using background knowledge from multiple resources in a unified way. We propose to acquire it from collections of RDF data, made available by the members of the Linked Data Community, e.g., DBpedia [2], Freebase [4], YAGO [26], and, perspectively, many others. In order to obtain semantic knowledge about a mention in plain text, we need to map it to a Linked Data resource entry. We benefit from the fact that some of the Linked Data resources are aligned with Wikipedia. Therefore, we link a mention to Wikipedia, using an approach described in Section 3.2, and then exploit this link to obtain data from the specific Linked Data resource. Moreover, Linked Data datasets are interconnected by means of RDF links and in future these inter-dataset links can be exploited as well. In the current work, we limit the scope of our research to the following resources, that can be directly accessed by using a Wikipedia link:

DBpedia is a structured twin of Wikipedia. Currently it describes more than 3.4 million entities. DBpedia resources bear the names of the Wikipedia pages, from which they have been extracted.

YAGO is an automatically created ontology, with taxonomy structure derived from WordNet, and knowledge about individuals extracted from Wikipedia. Therefore, the identifiers of resources describing individuals in YAGO are named as the corresponding Wikipedia pages. YAGO contains knowledge about more than 2 million entities and 20 million facts about them.

Freebase is a collaboratively constructed database. It contains knowledge automatically extracted from a number of resources including Wikipedia, MusicBrainz,[2] and NNDB,[3] as well as the knowledge contributed by the human volunteers. Freebase describes more than 12 million interconnected entities. Each Freebase entity is assigned a set of human-readable unique keys, which are assembled of a value and a namespace. One of the namespaces is the Wikipedia namespace, in which a value is the name of the Wikipedia page describing an entity.

3.2   Linking to Wikipedia

The linking problem is cast as a word sense disambiguation (WSD) exercise, in which each mention in the text (excluding pronouns) has to be disambiguated using Wikipedia to provide the sense inventory and the training data. The idea of using Wikipedia to train a supervised WSD system was first proposed in [6]. The proposed approach, called *The Wiki Machine*,[4] is summarized as follows.

Training Set. To create the training set, for each mention m, we collect from the English Wikipedia dump[5] all contexts where m is an anchor of an internal link, where a context corresponds to a line of text in the Wikipedia dump and it is represented as a paragraph in a Wikipedia article. The set of target articles represents the senses of m in Wikipedia and the contexts are used as labeled training examples. E.g., the proper noun *Bush* is a link anchor in $17,067$ different contexts that point to 20 different Wikipedia pages, `George_W._Bush`, `Bush_(band)`, and `Dave_Bush` are some example of possible senses. The set of contexts with their corresponding senses is then used to train the WSD system described below. E.g., the context "*Alternative Rock bands from the mid-90's , including Bush , Silverchair , and Sponge.*" is a training instance for the sense defined by the Wikipedia entry `Bush_(band)`.

Learning Algorithm. To disambiguate mentions in text, we implemented a kernel-based approach originally proposed in [13]. Different kernel functions are employed to integrate syntactic, semantic, and pragmatic knowledge sources typically used in the WSD literature. Kernel methods are theoretically well founded in statistical learning theory and shown good empirical results in many applications [24]. The strategy adopted consists in splitting the learning problem into two parts. They first embed the input data in a suitable feature space, and then use a linear algorithm (e.g., support vector machines) to discover nonlinear patterns in the input space. The kernel function is

---

[2] http://musicbrainz.org/
[3] http://www.nndb.com/
[4] http://thewikimachine.fbk.eu/
[5] http://download.wikimedia.org/enwiki/20100312

the only task-specific component of the learning algorithm. For each knowledge source a specific kernel has been defined. By exploiting the property of kernels, basic kernels are then combined to define the WSD kernel. Specifically, we used a linear combination of gap-weighted subsequences, bag-of-words, and latent semantic kernels .

Gap-weighted subsequences kernel. This kernel learns syntactic and associative relations between words in a local context. We extended the gap-weighted subsequences kernel to subsequences of word forms, stems, part-of-speech tags, and orthographic features (capitalization, punctuation, numerals, etc.). We defined gap-weighted subsequences kernels to work on subsequences of length up to 5. E.g., suppose we have to disambiguate the verb to score in the context "Maradona scored Argentina's third goal", given the labeled example "Ronaldo scored two goals in the second half" as training, a traditional approach, that only consider contiguous ngrams, has no clues to return the correct answer because the two contexts have no features in common. The use of gap-weighted subsequences allows us to overcame this problem and extract the feature "score goal", shared by the two examples.

Bag-of-words kernel. This kernel learns domain, semantic, topical information. Bag-of-words kernel takes as input a a wide context window around the target mention. Words are represented using stems. The main drawback of this approach is the need of a large amount of training data to reliably estimate model parameters. E.g., despite the fact that the examples "People affected by AIDS" and "HIV is a virus" express concepts related, their similarity is zero using the bag-of-words model since they have no words in common (they are represented by orthogonal vectors). On the other hand, due to the ambiguity of the word virus, the similarity between the contexts "the laptop has been infected by a virus" and "HIV is a virus is greater than zero", even though they convey very different messages.

Latent semantic kernel. To overcome the drawback of the bag-of-words, we incorporate semantic information acquired from English Wikipedia in an unsupervised way by means of latent semantic kernel. This kernel extracts semantic information through co-occurrence analysis in the corpus. The technique used to extract the co-occurrence statistics relies on a singular value decomposition of the term-by-document matrix. E.g., the similarity in the latent semantic space of the two examples "People affected by AIDS" and "HIV is a virus" is higher than in the bag-of-words representation, because the terms AIDS, HIV and virus very often co-occur in the medicine domain.

Implementation Details. The latent semantic model is derived from the 200,000 most visited Wikipedia articles. After removing terms that occur less than 5 times, the resulting dictionary contain about 300,000 and 150,000 terms respectively. We used the SVDLIBC package to compute the SVD, truncated to 400 dimensions.[6] To classify each mention in Wikipedia entries, we used a LIBSVM package.[7] No parameter optimization was performed.

---

[6] http://tedlab.mit.edu/~dr/svdlibc/
[7] http://www.csie.ntu.edu.tw/~cjlin/libsvm/

Evaluation. We evaluate The Wiki Machine on the ACE05-WIKI Extension [3]. This dataset extends the the English Automatic Content Extraction (ACE) 2005 dataset with ground-truth links to Wikipedia.[8] ACE 2005 is composed of 599 articles assembled from a variety of sources selected from broadcast news programs, newspapers, newswire reports, internet sources and from transcribed audio. It contains the annotation of a series of entities (person, location, organization) and their mentions. In the extension each nominal or named entity mention (in total 29,300 entity mentions) is manually assigned a Wikipedia link(s). The results of the evaluation are reported in the first line of Table 1. The training sets were collected from the English Wikipedia dump of March, 2010.

We have compared our approach with the state-of-the-art system described in [15]. In this approach, a plain text is *wikified*, i.e. terms in the text are linked to Wikipedia and then keywords are selected among them. We are interested only in the linking step. In this step a set of candidate Wikipedia pages (senses) for all terms in the text is collected as described in Section 3.2, when possible. The pages to which terms in the text can be linked unambiguously form the *context*. Different senses of an ambiguous term are evaluated using a classifier, based on three features, namely *commoness* of a sense, its *relatedness* to the context and the *context quality*.

The approach is implemented in the *Wikipedia Miner* tool.[9] We used it with the default parameters. The tool requires a Wikipedia dump preprocessed in a special way. We used the preprocessed Wikipedia dump of July, 2008, made available by the authors of the tool. The results are reported in the second line of Table 1. The Wikipedia Miner achieves six points better precision, however, its recall is considerably lower, thus making the F_1 13 points less than that of The Wiki Machine. The performance

Table 1. Comparisons of the two linking methods on the ACE05-WIKI Extension

Approach	Precision	Recall	F_1
The Wiki Machine	0.716	0.714	0.715
Wikipedia Miner	0.779	0.471	0.587

difference between the two systems could not be only due to the use of different version of Wikipedia, as the ACE corpus contains references to entities dated before 2005 and Wikipedia covered most of them in 2008. On the other hand, varying the Wiki Miner free parameters did not produce significant improvement.

4   Selecting Relevant Background Knowledge

The amount of information obtained from a Semantic Web resource even for a single named entity can be very big. For instance, DBpedia alone contains around 600 RDF triples describing *Barack Obama*. Most of this information is irrelevant to the NLP task at hand (e.g. Obama's website, residence, the name of his spouse, etc.), and only

[8] http://www.itl.nist.gov/iad/mig//tests/ace/ace05/index.html
[9] http://wikipedia-miner.sourceforge.net/

some of the triples can be useful to resolve coreferences (e.g. *type* properties stating that Obama is a politician and a president).

Indeed, many learning algorithms are originally not designed to deal with large amounts of irrelevant information, consequently, combining them with the *feature selection* techniques has become necessary in many applications. This is particularly true when the information needed is retrieved from heterogeneous knowledge sources as the ones made available on the Semantic Web. Recall that we do not assume any prior knowledge on the nature of the background knowledge that can be obtained, barring the availability in RDF.

We use the chi-square test to assess the relevance of background knowledge for the coreference resolution task by looking only at the intrinsic properties of the data. The chi-square test is a test for dependence between a feature and a class. Specifically, chi-square metric is calculated for each feature, and low-scoring features are removed. Afterwards, this subset of features is presented as input to the learning algorithm. Benefits of the chi-square test are that it easily scales to very high-dimensional data sets, it is computationally simple and fast, and the search in the feature space is separated from the search in the hypothesis space. The next sections describe the feature extraction and selection methods.

4.1   Feature Extraction

We obtain feature sets for coreference candidates, in which mentions are either a proper noun and a common noun (NAM-NOM), or both are common nouns (NOM-NOM). We denote a coreference candidate pair by (m_1, m_2). In the case of a NAM-NOM pair m_1 refers to the proper noun mention and m_2 to the common noun mention. As regards NOM-NOM, we consider two (m_1, m_2), pairs which differ by the order of the mentions, e.g. for the coreference candidate ("state", "country") we consider $(m_1$="state", m_2 = "country") and $(m_1$ = "country", m_2 = "state").

An (m_1, m_2) pair is processed as follows. We extract all RDF triples referring to m_1 from a knowledge source, using the methodology described in Section 3. In average we obtain 200 triples per mention. An RDF triple consists of subject, predicate and object. If m_1 is the object of the triple, we check if there is a string match between m_2 and the subject. In the other case, we check whether there is a string match between m_2 and the object. If the string match is observed, then the coreference candidate pair has a feature named as the predicate of the RDF triple, and the feature is included into the feature set. If for RDF-triples with a given predicate the string match never occurs in the entire training set, then the corresponding feature is not included into the feature set.

Examples of features for some of the mention pairs are presented in Table 2. Each mention is composed of the number of a document, the position in the document and the mention string itself. We select distinct sets of features for NAM-NOM and NOM-NOM mentions of person (PER) and geopolitical entities (GPE). Consequently from each of three background knowledge sources we extract four sets of features, namely NAM-NOM-GPE, NOM-NOM-GPE, NAM-NOM-PER, and NOM-NOM-PER. They typically contain 10-50 features. We apply the feature selection technique to each set.

Table 2. Feature examples

Mention pair	Feature
1-225-Clinton, 1-87-president	http://www.w3.org/2004/02/skos/core#subject
529-324-Yasser Arafat, 529-402-leader	http://www.w3.org/2004/02/skos/core#subject
410-23-state, 410-109-country	http://www.w3.org/2004/02/skos/core#subject
2-637-Kuwait, 2-956-city	http://rdf.Freebase.com/ns/location.country.capital
3-10-U.S.,3-892-States	http://www.w3.org/2002/07/owl#sameAs

4.2 Feature Selection

In machine learning coreference candidates are called instances. We say than an instance belongs to class 1 if the mentions in the candidate pair are coreferent; 0 otherwise. Let us introduce some notation.

n_{1f} number of instances in class 1 with feature f
$n_{1\bar{f}}$ number of instances in class 1 without feature f
n_{0f} number of instances in class 0 with feature f
$n_{0\bar{f}}$ number of instances in class 0 without feature f
n_1 total number of instances in class 1
n_0 total number of instances in class 0
n_f total number of instances with feature f
$n_{\bar{f}}$ total number of instances without feature f
n total number of instances

The chi-square feature selection metric, $\chi^2(f, c)$, measures the dependence between feature f and class $c \in \{0, 1\}$. If f and c are independent, then $\chi^2(f, c)$ is equal to zero. To select a relevant set of features, we utilized the following metric

$$\chi^2(f, c) = \frac{n(n_{1f}n_{0\bar{f}} - n_{0f}n_{1\bar{f}})^2}{n_1 n_f n_0 n_{\bar{f}}},$$

by averaging over the classes we obtain the metric for selecting a subset of features

$$\chi^2(f) = \sum_{i=0}^{1} Pr(c_i)\chi^2(f, c).$$

E.g., we extract from Freebase a set of 22 features for the NAM-NOM pairs of mentions which refer to a GPE entity. After feature selection, the scores of 9 features are near to zero, consequently only 13 features should be considered. The two top-scoring features in this case are http://www.w3.org/2002/07/owl#sameAs and http://www.w3.org/1999/02/22-rdf-syntax-ns#type. These features and their equivalents in other knowledge sources turned out to be highly relevant for other kinds of coreference as well.

5 Evaluation: Coreference Resolution with Background Knowledge

In this section we report on our experiments with the coreference resolution task. Namely, we give some hints on the implementation of the model we used as a baseline (more details can be found in [5]), explain how the background knowledge is plugged into the model, and present the results of the experiments.

5.1   Baseline Model Definition

Tool Selection. A recently introduced family of approaches to the task of coreference resolution try to represent the coreference task into some logical theory that supports the representation of uncertain knowledge. Among these approaches we can find a number of works [22,14,8] based on the formalism called Markov logic [10], which is a first-order probabilistic language which combines first-order logic with probabilistic graphical models.

In essence, Markov logic model is a set of first-order rules with weights associated to each rule. Weights can be learned from the available evidence (training data) or otherwise defined, and then inference is performed on a new (test) data. Such a representation of the model is intuitive and allows for the background knowledge be integrated naturally into it. It has been shown that the Markov logic framework is competitive in solving NLP tasks (see, for instance, [21,23], and [1] for more references). Another advantage of the weighted first-order representation is that the model can be easily extended with extra knowledge by simply adding logical axioms, thus minimizing the engineering effort and making the knowledge enrichment step more straightforward and intuitive.

Given the above, the inference tool we have selected to be used in the coreference resolution tasks is the inference module of the Alchemy system [1], with Markov logic as a representation language.

The Alchemy inference module takes as inputs (i) a Markov logic model, that is, a list of weighted first-order rules, and (ii) an evidence database, that is, the list of known properties (true of false values of predicates) of domain objects. In the case of coreference resolution, domain objects are the entity mentions, and the properties they might have are gender, number, distance, semantic class, etc. In the following we discuss how these two parts of input are constructed.

Markov Logic Model. In defining a model for coreference resolution, we were inspired by Soon et al baseline [25], which uses the following features: pairwise distance (in terms of number of sentences), string match, alias, number, gender and semantic class agreement, pronoun, definite/demonstrative noun phrase and both proper names feature. This approach achieves an F-measure of 62.2% in the MUC-6 coreference task and of 60.4% on the MUC-7 coreference task.

A Markov logic model consists of a list of predicates and a set of weighted first-order formulae. Some predicates in our model correspond to Soon et al features: binary predicates such as *distance* between two entity mentions (in terms of sentences) and string match, and unary predicates such as *proper name*, *semantic class*, number (*singular* or *plural*) and gender (*male*, *female* or *unknown*). Also, we use *string overlap* in addition to *string match* and define yet another predicate to describe *distance*, which refers to the number of named entities of the same type between two given ones (e.g. if there are no other named entities classified as "person" between "Obama" and "President", the distance is 0). The predicate *corefer(mention,mention)* describes the relation of interest,

and is called *query* predicate in Alchemy terminology, that is, we are interested in evaluating the probability of each grounding of this predicate given the known properties of
all the mentions.

The second part of the model definition concerns constructing the first-order rules
appropriate for a given task. We have defined the rules that connect the above properties
of the mentions with the coreference property. Some of the examples are given below[10].

String match is very likely to indicate coreference for proper names, while for common nouns it is still likely but makes more sense in combination with a distance property:

$$20\ match(x,y) \wedge proper(x) \wedge proper(y) \rightarrow corefer(x,y)$$

$$3\ match(x,y) \wedge noun(x) \wedge noun(y) \wedge dist0(x,y) \rightarrow corefer(x,y)$$

The number before a formula corresponds to the *weight* assigned to it.

Gender and number agreement between two neighboring mentions of the same type
provides a relatively strong evidence for coreference:

$$4\ male(x) \wedge male(y) \wedge singular(x) \wedge singular(y) \wedge follow(x,y) \rightarrow corefer(x,y)$$

We also define hard constraints, that is, crisp first-order formulae that should hold in
any given world. Fullstop after the formula refers to an infinite weight, which, in turn,
means that the formula holds with the probability equal to 1.

$$\neg corefer(x,x).$$

$$corefer(x,y) \wedge \rightarrow corefer(y,x).$$

In this paper we do not consider weight learning, so weights are assigned manually. We
do not consider pronoun mentions as the background knowledge is relevant for proper
name/common noun pairs in the first place.

Evidence Database. The second input to the Alchemy inference module is an evidence
database, i.e. the known values of non-query predicates listed in the previous section.
Normally, the coreference resolution task is performed on a document corpus, in which
each document is firstly preprocessed. Preprocessing consists in identifying the named
entities (persons, locations, organization, etc.), as well as their syntactic properties, such
as part of speech, number, gender, pairwise distance, etc.

The data corpus we use for the experiments is ACE 2005 data set, with around 600
documents from the news domain. We work on a corpus in which each word is annotated with around 40 features (token and document ID, Part of Speech tags by TextPro[11],
etc.). This allowed us to extract the syntactic properties of the mentions presented before. Note that for the gender property, we used male/female name lists to annotate
proper names in the corpus. For common nouns, we defined two lists of gender tokens
(which included "man","girl", "wife", "Mr.", etc.). Some examples of the properties we
obtained are given below.

---

[10] Full model is available at
    https://copilosk.fbk.eu/images/1/1f/Coreference2.txt
[11] TextPro – http://textpro.fbk.eu/

semclass ("2-83-Bob Dornan", "person")
neihgbourNouns ("2-82-Congressman","2-83-Bob Dornan")
propername ("2-83-Bob Dornan")
male ("2-83-Bob Dornan")
singular ("2-83-Bob Dornan")
pmatch ("2-740-Bob", "2-83-Bob Dornan")
match ("2-83-Bob Dornan", "2-942-Bob Dornan")
DBPedia_NAM-NOM_PER_2_type ("2-83-Bob Dornan", "2-62-Congressman")
YAGO_NAM-NOM_PER_1_type ("2-83-Bob Dornan", "2-86-Republican")

We worked on the gold standard annotation for named entities, and considered five named entity types: PERson, LOCation, GeoPoliticalEntity, FACility and ORGanization (although only the first two types were used in the experiments presented later in this section). Alchemy inference was performed separately for each named entity type. Note that the size of the document corpus does not impact the quality of the results as documents are processed independently, one by one.

The Alchemy inference module, which takes as input the weighted Markov logic model and the database containing the properties of mentions, produces as a result the probabilities of coreference for each of $N \times N$ possible pairs of mentions, where N is the number of mentions:

$$corefer(m_i, m_j) \quad p_{ij}, \quad 0 \leq p_{ij} \leq 1, \ i,j = \overline{1, N}$$

After having obtained this, we setup a probability threshold (e.g. $p = 0.9$) and consider only those pairs for which $p_{ij} \geq p$. On these pairs, we perform a transitive closure. Then the pairwise scores and, after a simple clustering step, MUC scores [28] are calculated. The resulting output consists of the list of coreference chains for each of the processed documents, and the measures of the efficiency, namely, recall, precision and their harmonic mean (F1).

5.2   Injecting Background Knowledge into Coreference Model

In the Markov logic model, in addition to the syntactic predicates and rules described above, a set of predicates and rules that deal with background knowledge were introduced. The predicates, or pairwise semantic properties of mentions, are the most relevant features selected according to the methodology described in Section 4 from the DBpedia, YAGO and Freebase knowledge sources. The list of the selected features is given in Table 3.

The Markov logic model is extended with the rules relating these semantic predicates with the coreference property. The arguments of a semantic predicate should be of the same named entity type (person or geopolicical entity), and the distance relation relation must hold between them.

For the experiments, the ACE data set was first ordered by the number of named entities linked to Wikipedia and split into two subsets of equal size (*ACE-SUBSET-1* and *ACE-SUBSET-2*): odd documents from the ordered list formed the first subset, even formed the second one. *ACE-SUBSET-1* was used for feature selection, while on *ACE-SUBSET-2* the Markov logic model extended with background knowledge was

Table 3. Selected features

KB name	NE type	Pair type	Property name
Freebase	GPE	NAM-NOM	http://www.w3.org/1999/02/22-rdf-syntax-ns#type
Freebase	GPE	NAM-NOM	http://www.w3.org/2002/07/owl#sameAs
Freebase	PER	NAM-NOM	http://www.w3.org/2002/07/owl#sameAs
Freebase	PER	NAM-NOM	http://rdf.freebase.com/ns/people.person.profession
Freebase	PER	NOM-NOM	http://www.w3.org/2002/07/owl#sameAs
YAGO	GPE	NAM-NOM	type
YAGO	GPE	NAM-NAM	means
YAGO	PER	NAM-NOM	type
DBPedia	GPE	NAM-NOM	http://dbpedia.org/property/reference
DBPedia	GPE	NAM-NOM	http://www.w3.org/2004/02/skos/core#subject
DBPedia	GPE	NAM-NOM	http://www.w3.org/1999/02/22-rdf-syntax-ns#type
DBPedia	PER	NAM-NOM	http://www.w3.org/2004/02/skos/core#subject
DBPedia	PER	NAM-NOM	http://www.w3.org/1999/02/22-rdf-syntax-ns#type
DBPedia	PER	NAM-NOM	http://dbpedia.org/property/title

tested. For the latter experiments, we have created yet another document set, *ACE-SUBSET-3*, which contains 50 documents from *ACE-SUBSET-2* with the highest background knowledge coverage (i.e. with the highest number of entity mentions linked to Wikipedia).

Tables 4 and 5 present MUC scores of the experiments for *ACE-SUBSET-2* and *ACE-SUBSET-3*, accordingly. Each table reports the values of MUC recall, precision and F1 for the models without and with the use of background knowledge extracted from DBpedia, YAGO and Freebase. Experiments were conducted for geopolitical entities (GPE) and persons (PER). Compared to the other three NE types (locations, organizations and facilities), persons and geopolitical entities constitute the major part of the corpus, so we do not report these results here. Also, we do not report the experiments for geopolitical entities with knowledge obtained from Freebase and DBpedia as the corresponding improvement for these cases was insignificant.

Table 4. MUC scores for GPE and PER NE types, *ACE-SUBSET-2* document set

NE type	KB	R	P	F1
GPE	none	0.7446	0.9371	0.8298
GPE	YAGO	0.8314	0.9308	**0.8783**
PER	none	0.7003	0.7302	0.7149
PER	DBpedia	0.7125	0.7196	0.7160
PER	Freebase	0.7178	0.7343	**0.7259**
PER	YAGO	0.7208	0.7348	**0.7277**

The improvement in $F1$ is 5% for GPE due to the use of YAGO on both datasets. The improvement in $F1$ for PER with the use of YAGO and Freebase is a bit higher for *ACE-SUBSET-3* (1.5% versus 2%) due to the increase of coverage in the latter. The

Table 5. MUC scores for GPE and PER NE types, *ACE-SUBSET-3* document set

NE type	KB	R	P	F1
GPE	none	0.7763	0.9380	0.8495
GPE	YAGO	0.8536	0.9335	**0.8918**
PER	none	0.7447	0.6946	0.7188
PER	DBpedia	0.7669	0.6852	**0.7238**
PER	Freebase	0.7749	0.7024	**0.7369**
PER	YAGO	0.7785	0.7039	**0.7393**

results for YAGO and Freebase are comparable to the ones presented in [5], while lower improvement for DBpedia is most probably due to the fact that this knowledge source is much less structured and polished with respect to YAGO and Freebase.

6 Conclusion and Future Work

In this paper we have defined a methodology for supporting a natural language processing task with semantic information available in the Web under the form of logical theories. In order to empower an NLP task with the knowledge from publicly available large scale knowledge sources, we map the terms in the text to concepts in Wikipedia and then, to other knowledge resources linked to Wikipedia (DBpedia, Freebase and YAGO). An important aspect of the mapping that was addressed in the paper is word sense disambiguation. We have applied the proposed approach to the task of intra-document coreference resolution. We have proposed a method for selecting a subset of knowledge relevant for a given text for solving the coreference task, which is based on feature selection algorithms. We have implemented the coreference resolution process with the help of the inference module of the Alchemy tool. The latter is based on Markov logic formalism and allows combining logical and statistical representation and inference. The results were evaluated on the ACE 2005 data set.

To the best of our knowledge, there are no approaches nor to coreference resolution, neither to other NLP tasks, which make use of structured semantic knowledge available in the Web. One of the key points in addressing this problem is combining the logic based representation of the model with statistical reasoning. Such model representation and the available Semantic Web knowledge resources "speak the same language", which is the language of logic. Another important point of our approach is that no prior assumptions on the structure of the Semantic Web knowledge sources are needed for them to be used to support an NLP task.

Future work directions include further exploiting the Linked Data resources (including the one not used in this paper, e.g. Cyc[12]) to extract more properties and rules to support coreference resolution, as well as using the links between different Linked Data resources to obtain more knowledge. Also, we are interested in experimenting with the full task, which includes named entity recognition module and learning the weights of the formulae of the model from the training data. Testing the proposed reference

---

[12] http://www.cyc.com

methodology on the other NLP task, like semantic relation extraction, is another challenging future work direction.

Acknowledgments

The research leading to these results has received funding from the ITCH project (http://itch.fbk.eu), sponsored by the Italian Ministry of University and Research and by the Autonomous Province of Trento, and the Copilosk project (http://copilosk.fbk.eu), a Joint Research Project under Future Internet – Internet of Content program of the Information Technology Center, Fondazione Bruno Kessler.

References

1. Alchemy, http://alchemy.cs.washington.edu/
2. Auer, S., Bizer, C., Kobilarov, G., Lehmann, J., Cyganiak, R., Ives, Z.G.: Dbpedia: A nucleus for a web of open data. In: Aberer, K., Choi, K.-S., Noy, N., Allemang, D., Lee, K.-I., Nixon, L.J.B., Golbeck, J., Mika, P., Maynard, D., Mizoguchi, R., Schreiber, G., Cudré-Mauroux, P. (eds.) ASWC 2007 and ISWC 2007. LNCS, vol. 4825, pp. 722–735. Springer, Heidelberg (2007)
3. Bentivogli, L., Forner, P., Giuliano, C., Marchetti, A., Pianta, E., Tymoshenko, K.: Extending English ACE 2005 Corpus Annotation with Ground-truth Links to Wikipedia. In: 23rd International Conference on Computational Linguistics, pp. 19–26 (2010)
4. Bollacker, K., Evans, C., Paritosh, P., Sturge, T., Taylor, J.: Freebase: a collaboratively created graph database for structuring human knowledge. In: Proceedings of the 2008 ACM SIGMOD International Conference on Management of Data, pp. 1247–1250. ACM, New York (2008)
5. Bryl, V., Giuliano, C., Serafini, L., Tymoshenko, K.: Using background knowledge to support coreference resolution. In: 19th European Conference on Artificial Intelligence (ECAI 2010), pp. 759–764 (2010)
6. Cucerzan, S.: Large-scale named entity disambiguation based on Wikipedia data. In: Proceedings of the 2007 Joint Conference on Empirical Methods in Natural Language Processing and Computational Natural Language Learning (EMNLP-CoNLL), pp. 708–716 (2007)
7. Culotta, A., Sorensen, J.: Dependency tree kernels for relation extraction. In: Proceedings of the 42nd Annual Meeting on Association for Computational Linguistics, pp. 423–431. Association for Computational Linguistics (2004)
8. Culotta, A., Wick, M.L., McCallum, A.: First-order probabilistic models for coreference resolution. In: Human Language Technology Conference of the North American Chapter of the Association of Computational Linguistics, pp. 81–88 (2007)
9. Denis, P., Baldridge, J.: Joint determination of anaphoricity and coreference resolution using integer programming. In: Human Language Technologies 2007: The Conference of the North American Chapter of the Association for Computational Linguistics, pp. 236–243 (2007), http://www.aclweb.org/anthology/N/N07/N07-1030
10. Domingos, P., Kok, S., Lowd, D., Poon, H., Richardson, M., Singla, P.: Markov logic. In: De Raedt, L., Frasconi, P., Kersting, K., Muggleton, S.H. (eds.) Probabilistic Inductive Logic Programming. LNCS (LNAI), vol. 4911, pp. 92–117. Springer, Heidelberg (2008)
11. Fellbaum, C., et al.: WordNet: An electronic lexical database. MIT Press, Cambridge (1998)
12. Giuliano, C., Lavelli, A., Pighin, D., Romano, L.: FBK-IRST: Kernel methods for semantic relation extraction. In: Proceedings of the 4th International Workshop on Semantic Evaluations, pp. 141–144. Association for Computational Linguistics (2007)

13. Giuliano, C., Gliozzo, A.M., Strapparava, C.: Kernel methods for minimally supervised wsd. Computational Linguistics 35(4), 513–528 (2009)
14. Huang, S., Zhang, Y., Zhou, J., Chen, J.: Coreference resolution using Markov Logic Networks. In: Proceedings of CICLing, pp. 157–168 (2009)
15. Milne, D., Witten, I.H.: Learning to link with Wikipedia. In: Proceedings of the 17th ACM Conference on Information and Knowledge Management, CIKM 2008, pp. 509–518. ACM, NY (2008)
16. Ng, V.: Learning noun phrase anaphoricity to improve coreference resolution: issues in representation and optimization. In: Proceedings of the 42nd Annual Meeting on Association for Computational Linguistics, ACL 2004, pp. 151–158 (2004)
17. Ng, V.: Semantic class induction and coreference resolution. In: Proceedings of the ACL, vol. 45, pp. 536–543 (2007)
18. Ng, V.: Supervised noun phrase coreference research: The first fifteen years. In: Proceedings of the 48th Annual Meeting of the Association for Computational Linguistics, Uppsala, Sweden, pp. 1396–1411 (July 2010),
    http://www.aclweb.org/anthology/P10-1142
19. Ng, V., Cardie, C.: Improving machine learning approaches to coreference resolution. In: Proceedings of the 40th Annual Meeting on Association for Computational Linguistics, pp. 104–111 (2002)
20. Ponzetto, S.P., Strube, M.: Exploiting semantic role labeling, wordnet and wikipedia for coreference resolution. In: Human Language Technology Conference of the North American Chapter of the Association of Computational Linguistics, pp. 192–199 (2006)
21. Poon, H., Domingos, P.: Joint inference in information extraction. In: Proceedings of the 22nd National Conference on Artificial Intelligence, AAAI 2007, pp. 913–918 (2007)
22. Poon, H., Domingos, P.: Joint unsupervised coreference resolution with Markov Logic. In: Proceedings of the 2008 Conference on Empirical Methods in Natural Language Processing, pp. 650–659 (2008)
23. Riedel, S., Meza-Ruiz, I.: Collective semantic role labelling with markov logic. In: Proceedings of the Twelfth Conference on Computational Natural Language Learning, pp. 193–197 (2008)
24. Shawe-Taylor, J., Cristianini, N.: Kernel Methods for Pattern Analysis. Cambridge University Press, Cambridge (2004)
25. Soon, W.M., Ng, H.T., Lim, D.C.Y.: A machine learning approach to coreference resolution of noun phrases. Computational Linguistic 27(4), 521–544 (2001)
26. Suchanek, F.M., Kasneci, G., Weikum, G.: Yago: a core of semantic knowledge. In: Proceedings of the 16th International Conference on World Wide Web, WWW 2007, pp. 697–706. ACM Press, New York (2007)
27. Versley, Y., Ponzetto, S.P., Poesio, M., Eidelman, V., Jern, A., Smith, J., Yang, X., Moschitti, A.: Bart: a modular toolkit for coreference resolution. In: Proceedings of the 46th Annual Meeting of the Association for Computational Linguistics on Human Language Technologies, pp. 9–12 (2008)
28. Vilain, M., Burger, J., Aberdeen, J., Connolly, D., Hirschman, L.: A model-theoretic coreference scoring scheme. In: Proceedings of the 6th Conference on Message Understanding, MUC6 1995, pp. 45–52 (1995)

Enabling Ontology-Based Access to Streaming Data Sources

Jean-Paul Calbimonte[1], Oscar Corcho[1], and Alasdair J.G. Gray[2]

[1] Ontology Engineering Group, Departamento de Inteligencia Artificial,
Facultad de Informática, Universidad Politécnica de Madrid,
Campus de Montegancedo s/n 28660, Boadilla del Monte, Spain
jp.calbimonte@upm.es, ocorcho@fi.upm.es
[2] School of Computer Science, The University of Manchester,
Oxford Road, Manchester M13 9PL, United Kingdom
a.gray@cs.man.ac.uk

Abstract. The availability of streaming data sources is progressively increasing thanks to the development of ubiquitous data capturing technologies such as sensor networks. The heterogeneity of these sources introduces the requirement of providing data access in a unified and coherent manner, whilst allowing the user to express their needs at an ontological level. In this paper we describe an ontology-based streaming data access service. Sources link their data content to ontologies through S_2O mappings. Users can query the ontology using SPARQL$_{Stream}$, an extension of SPARQL for streaming data. A preliminary implementation of the approach is also presented. With this proposal we expect to set the basis for future efforts in ontology-based streaming data integration.

1 Introduction

Recent advances in wireless communications and sensor technologies have opened the way for deploying networks of interconnected sensing devices capable of ubiquitous data capture, processing and delivery. Sensor network deployments are expected to increase significantly in the upcoming years because of their advantages and unique features. Tiny sensors can be installed virtually anywhere and still be reachable thanks to wireless communications. Moreover, these devices are inexpensive and can be used for a wide variety of applications including security surveillance, healthcare provision, and environmental monitoring.

As an example, consider a web application which aids an emergency planner to detect and co-ordinate the response to a forest fire in Spain. This involves retrieving relevant data from multiple sources, e.g. weather data from AEMET (Agencia Española de Meteorología)[1], sensor data from sensor networks deployed in the region, and any other relevant sources of data such as the ESA satellite

---

[1] http://www.aemet.es accessed 15 September 2010.

P.F. Patel-Schneider et al. (Eds.): ISWC 2010, Part I, LNCS 6496, pp. 96–111, 2010.

imagery providing fire risks[2]. Typically sources are managed autonomously and model their data according to the needs of the deployment. To integrate the data requires linking the sources to a common data model so that conditions that are likely to cause a fire can be detected, and presented to the user in terms of their domain, e.g. fire risk assessment. We propose that ontologies can be used as such a common model. For the scenario presented here, we use an ontology that extends ontologies from SWEET[3] and the W3C incubator group's semantic sensor network ontology[4].

The work presented in this paper considers advances done by the semantic web and database communities over the last decade. On the one hand, the semantic web research has produced mapping languages and software for enabling ontology-based access to stored data sources, e.g. R2O [1] and D2RQ [2]. These systems provide semantic access to traditional (stored) data sources by providing mappings between the elements in the relational and ontological models [3]. However, similar solutions for streaming data mapping and querying using ontology-based approaches have not been explored yet.

On the other hand, the database research community have investigated data stream processing where the data is viewed as an append-only sequence of tuples. Systems such as STREAM [4] and Borealis [5] have focused on query evaluation and optimisation over streams with high, variable, data rates. Other systems such as SNEE [6] and TinyDB [7], have focused on data generated by sensor networks, which tends to be at a lower rate, and query processing in the sensor network where resources are more constrained and energy efficiency is the primary concern. There have also been proposals for query processing over streaming RDF data [8,9]. However there is still no bridging solution that connects these technologies coherently in order to answer the requirements of i) establishing mappings between ontological models and streaming data source schemas, and ii) accessing streaming data sources through queries over ontology models.

In this paper we focus on providing ontology-based access to streaming data sources, including sensor networks, through declarative continuous queries. We build on the existing work of R2O for enabling ontology-based access to relational data sources, and SNEE for query evaluation over streaming and stored data sources. This constitutes a first step towards a framework for the integration of distributed heterogeneous streaming and stored data sources through ontological models. In Section 2 we provide more detailed descriptions of R2O and stream query processing in order to present the foundations of our approach in Section 3. In Section 4 we present the syntactic extensions for SPARQL to enable queries over RDF streams, and present S2O for stream-to-ontology mappings. The semantics of these extensions are detailed in Section 5 and a first implementation of the execution of the streaming data access approach is explained in Section 6. Related work is discussed in Section 7 and our conclusions in Section 8.

---

[2] http://dup.esrin.esa.int/ionia/wfa/index.asp accessed 15 September 2010.

[3] http://sweet.jpl.nasa.gov/ accessed 15 September 2010.

[4] http://www.w3.org/2005/Incubator/ssn/wiki/
Semantic_Sensor_Network_Ontology accessed 15 September 2010.

2 Background

This section describes the existing work upon which our approach for enabling ontology-based access to streaming data sources is based, *viz.* R2O which provides ontology-based access to stored relational data, and SNEE, a query processing engine over relational data streams. A full discussion of related work can be found in Section 7.

2.1 Ontology-Based Access to Stored Relational Data

The goal of ontology-based data access is to generate semantic web content from existing relational data sources available on the web [3]. The objective of these systems is to allow users to construct queries over an ontology (e.g. in SPARQL), which are then rewritten into a set of queries expressed in the query language of the data source (typically SQL), according to the specified mappings. The query results are then converted back from the relational format into RDF, which is returned to the user. ODEMAPSTER is one such system which uses R2O (Relational-to-Ontology) to express the mappings between the relational data source and the ontology [1].

The mapping definition language R2O defines relationships between a set of ontologies and relational schemas [1]. The mappings are expressed in terms of selections and transformations over database relations following a Global-as-View (GAV) approach [10], and can be created either manually or with the help of a mapping tool. The resulting mappings are saved as XML which enables them to be independent of any specific DBMS or ontology language.

Mapping relations to ontologies often requires performing operations on the relational sources. Several cases are handled by R2O and detailed below.

Direct Mapping. A single relation maps to an ontology class and the attributes of the relation are used to fill the property values of the ontology instances. Each row in the relation will generate a class instance in the ontology.

Join/Union. A single relation does not correspond alone to a class, but it has to be combined with other relations. The result of the join or union of the relations will generate the corresponding ontology instances.

Projection. Not all the attributes of a relation are always required for the mapping. The unnecessary attributes can simply be ignored. In order to do so, a projection on the needed attributes can be performed.

Selection. Not all rows of a relation correspond to instances of the mapped ontology class. A subset of the rows must be extracted. To do so, selection conditions can be applied to choose the desired subset for the mapping.

It is possible to combine joins, unions, projections and selections for more complex mapping definitions. R2O also enables the application of functions, e.g. concatenation, sub-string, or arithmetic functions, to transform the relational data into the appropriate form for the ontology.

2.2   Querying Relational Data Streams

A relational data stream is an append only, potentially infinite, sequence of timestamped tuples [11], examples of which include stock market tickers, heart rate monitors, and sensor networks deployed to monitor the environment. Data streams can be classified into two categories:

Event-streams. A tuple is generated each time an event occurs, e.g. the sale of shares, and can have variable, potentially very high, data rates.

Acquisitional-streams. A tuple is measured at a predefined regular interval, e.g. the readings made by a sensor network.

Users are typically interested in being informed continuously about the most recent stream values, with older tuples being less relevant. Classical database query processing is not adequate since data must first be stored and then queried with one-off evaluation. Hence, query languages [12,13] and data stream management systems (DSMS) [4,5,6,7] have been developed to process continuous long-lived queries over data streams as tuples arrive.

One existing approach is SNEEql, which has a well defined, unified semantics for declarative expressions of data needs over event-streams, acquisitional-streams, and stored data [12]. SNEEql can be viewed as extending SQL for processing data streams. The additional constructs are explained below.

Window. A window over a data stream transforms the infinite sequence of tuples into a bounded bag of tuples over which traditional relational operators can be applied. A window is specified as 'FROM *start* TO *end* [SLIDE *int unit*]', where *start* and *end* are of the form 'NOW − *literal*' and define the range of the window with respect to the evaluation time. The optional SLIDE parameter specifies how often windows are evaluated.

Window-to-Stream. Window-to-stream operators are used to convert a stream of windows into a stream of tuples. SNEEql supports three such operators: RSTREAM for all tuples appearing in the window, ISTREAM for tuples that have been added since the last window evaluation, and DSTREAM for tuples that have been deleted since the last window evaluation.

Queries expressed in the SNEEql language are optimized for evaluation within a sensor network over acquisitional-streams by the SNEE compiler [6]. SNEE has recently been extended to enable query evaluation over event-streams either within the sensor network (in-network query processing) or on computational hardware outside of the sensor network.

3   Ontology-Based Streaming Data Access

Our approach to enable ontology-based access to streaming data is depicted in Fig 1. The service receives queries specified in terms of the classes and properties[5] of the ontology using SPARQL$_{Stream}$, an extension of SPARQL that supports operators over RDF streams (see Section 4.1). In order to transform the

---

[5] We use the OWL nomenclature of classes, and object and datatype properties for naming ontology elements.

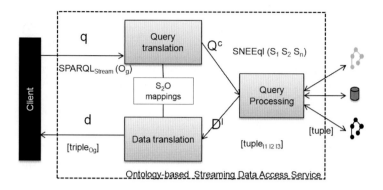

Fig. 1. Ontology-based streaming data access service

SPARQL$_{Stream}$ query, expressed in terms of the ontology, into queries in terms of the data sources, a set of mappings must be specified. These mappings are expressed in S$_2$O, an extension of the R$_2$O mapping language, which supports streaming queries and data, most notably window and stream operators (see Section 4.2). This transformation process is called *query translation*, and the target is the continuous query language SNEEql, which is expressive enough to deal with both streaming and stored sources.

After the continuous query has been generated, the query processing phase starts, and the evaluator uses distributed query processing techniques [14] to extract the relevant data from the sources and perform the required query processing, e.g. selection, projection, and joins. Note that query execution in sources such as sensor networks may include in-network query processing, pull or push based delivery of data between sources, and other data source specific settings. The result of the query processing is a set of tuples that the *data translation* process transforms into ontology instances.

This approach requires several contributions and extensions to the existing technologies for continuous data querying, ontology-based data access, and SPARQL query processing. This paper focuses on a first stage that includes the process of transforming the SPARQL$_{Stream}$ queries into queries over the streaming data sources using SNEEql as the target language. The following sections provide the syntax and semantics for the querying of streaming RDF data and the mappings between streaming sources and an ontology. We will then provide details of an actual implementation of this approach.

4   Query and Mapping Syntax

In this section we introduce the SPARQL$_{Stream}$ query language, an extension to SPARQL for streaming RDF data, which has been inspired by previous proposals such as C-SPARQL [9] and SNEEql [12]. However, significant improvements have been made that correct the types supported and the semantics of windowing operations, which can be summarised as: (i) we only support windows defined in time, (ii) the result of a window operation is a window of triples, not a stream, over which traditional operators can be applied, as such we have added window-to-stream operators, and (iii) we have adopted the SPARQL 1.1 definition for aggregates. We also present S$_2$O for the definition of stream-to-ontology mappings.

4.1 SPARQL$_{\text{Stream}}$

Just as in C-SPARQL we define an RDF *stream* as a sequence of pairs (T_i, τ_i) where T_i is an RDF triple $\langle s_i, p_i, o_i \rangle$ and τ_i is a timestamp which comes from a monotonically non-decreasing sequence. An RDF stream is identified by an IRI, which provides the location of the data source[6].

Window definitions are of the form 'FROM *Start* TO *End* [STEP] [*Literal*]', where the *Start* and *End* are of the form NOW or NOW − *Literal*, and *Literal* represents some number of time unit (DAYS, HOURS, MINUTES, or SECONDS)[7]. The optional STEP indicates the gap between each successive window evaluation. Note, if the size of the step is smaller than the range of the window, then the windows will overlap, if it coincides with the size of the window then every triple will appear in one and only one window, and if the step is larger than the range then the windows *sample* the stream. Also note that the definition of a window can be completely in the past. This is useful for correlating current values on a stream with values that have previously occurred.

The result of applying a window over a stream is a timestamped bag of triples over which conjunctions between triple patterns, and other "classical" operators can be evaluated. Windows can be converted back into a stream of triples by applying one of the window-to-stream operators in the SELECT clause: ISTREAM for returning all newly inserted answers since the last window, DSTREAM for returning all deleted answers since the last window, and RSTREAM for returning all answers in the window.

Listing 1 shows a complete SPARQL$_{\text{Stream}}$ query which, every minute, returns the average of the last 10 minutes of wind speed measurements for each sensor, if it is higher than the average speed from 2 to 3 hours ago.

Note, SPARQL$_{\text{Stream}}$ only supports time-based windows. C-SPARQL also has the notion of a triple-based window. However, such windows are problematic since the number of triples required to generate an answer may be greater than the size of the triple window. For example, consider a window size of 1 triple and the graph pattern from the example query in Listing 1. Only one of the triples that form the graph pattern would be kept by the window, and hence it would not be possible to compute the query answer.

---

[6] Note in our work the IRI's identify virtual RDF streams since they are derived from the streaming data sources.

[7] Note that the parser will also accept the non-plural form of the time units and is not case sensitive.

```
PREFIX  fire :  <http://www.semsorgrid4env.eu#>
PREFIX  rdf :  <http://www.w3.org/1999/02/22−rdf−syntax−ns#>
SELECT RSTREAM  ?WindSpeedAvg
FROM STREAM <www.semsorgrid4env.eu/SensorReadings.srdf> [FROM NOW − 10
       MINUTES TO NOW STEP 1 MINUTE]
FROM STREAM <www.semsorgrid4env.eu/SensorArchiveReadings.srdf> [FROM NOW − 3
       HOURS TO NOW −2 HOURS STEP 1 MINUTE]
WHERE {
    {
       SELECT AVG(?speed) AS ?WindSpeedAvg
       WHERE
       {
          GRAPH <www.semsorgrid4env.eu/SensorReadings.srdf> {
          ?WindSpeed  a  fire:WindSpeedMeasurement ;
                 fire:hasSpeed ?speed; }
       } GROUP BY ?WindSpeed
    }
    {
       SELECT AVG(?archivedSpeed) AS ?WindSpeedHistoryAvg
       WHERE
       {
         GRAPH <www.semsorgrid4env.eu/SensorArchiveReadings.srdf>  {
         ?ArchWindSpeed  a  fire:WindSpeedMeasurement ;
         fire:hasSpeed ?archivedSpeed ;    }
       } GROUP BY ?ArchWindSpeed
    }
    FILTER (?WindSpeedAvg > ?WindSpeedHistoryAvg )
}
```

Listing 1. An example SPARQL$_{Stream}$ query which every minute computes the average wind speed measurement for each sensor over the last 10 minutes if it is higher than the average of the last 2 to 3 hours

4.2   S$_2$O: Expressing Stream-to-Ontology Mappings

The mapping document that describes how to transform the data source elements to ontology elements is written in the S$_2$O mapping language, an extended version of R$_2$O [1]. An R$_2$O mapping document includes a section that describes the database relations, dbscehma-desc. In order to support data streams, R$_2$O has been extended to also describe the data stream schema. A new component called streamschema-desc has been created, as shown in the top part of Listing 2.

The description of the stream is similar to a relation. An additional attribute streamType has been added, it denotes the kind of stream in terms of data acquisition, i.e. event or acquistional. In the same way as key and non-key attributes are defined, a new timestamp-desc element has been added to provide support for declaring the stream timestamp attribute. Since S$_2$O extends R$_2$O, relations can also be specified using the existing R$_2$O mechanism. For the class and property mappings, the existing R$_2$O definitions can be used for stream schemas just as it was for relational schemas. This is specified in the conceptmap-def element as shown in the bottom part of Listing 2.

In addition, although they are not explicitly mapped, the timestamp attribute of stream tuples could be used in some of the mapping definitions, for instance in the URI construction (uri-as element). Finally, a SPARQL$_{Stream}$ streaming query requires an RDF stream to have an IRI identifier. S$_2$O creates a *virtual* RDF stream

```
streamschema−desc
    name MeteoSensors
    has−stream SensorWind
        streamType pushed
        documentation "Wind measurements"
        keycol−desc measurementId
            columnType integer
        timestamp−desc   measureTime
            columnType datetime
        nonkeycol−desc   measureSpeed
            columnType float
        nonkeycol−desc measureDirection
            columnType float
...
conceptmap−def Wind
    virtualStream <http://semsorgrid4env.eu/SensorReadings.srdf>
    uri−as
        concat(SensorWind.measurementID)
    applies−if
        <cond−expr>
    described−by
        attributemap−def hasSpeed
            virtualStream http://semsorgrid4env.eu/SensorReadings.srdf>
            operation constant
                has−column SensorWind.measureSpeed
```

Listing 2. An example S₂O declaration of a data stream schema and mapping from a stream schema to an ontology concept

and its IRI is specified in the S₂O mapping using the virtualStream element. It can be specified at the conceptmap-def level or at the attributemap-def level.

5 Semantics of the Streaming Extensions

Now that the syntax of SPARQL$_{\text{Stream}}$ and S₂O have been presented, we define their semantics.

5.1 SPARQL$_{\text{Stream}}$ Semantics

The SPARQL extensions presented here are based on the formalisation of Pérez *et al.* [15]. An RDF stream S is defined as a sequence of pairs (T, τ) where T is a triple $\langle s, p, o \rangle$ and τ is a timestamp in the infinite set of timestamps $\mathbb{T}$. More formally,

$$S = \{(\langle s, p, o \rangle, \tau) \mid \langle s, p, o \rangle \in ((I \cup B) \times I \times (I \cup B \cup L)), \tau \in \mathbb{T}\},$$

where I, B and L are sets of IRIs, blank nodes and literals. Each of these pairs can be called a *tagged triple*.

We define a stream of windows as a sequence of pairs (ω, τ) where ω is a set of triples, each of the form $\langle s, p, o \rangle$, and τ is a timestamp in the infinite set of timestamps $\mathbb{T}$, and represents when the window was evaluated. More formally, we define the triples that are contained in a time-based window evaluated at time $\tau \in \mathbb{T}$, denoted ω^τ, as

$$\omega^\tau_{t_s, t_e, \delta}(S) = \{\langle s, p, o \rangle \mid (\langle s, p, o \rangle, \tau_i) \in S, t_s \leq \tau_i \leq t_e\}$$

where t_s, t_e define the start and end of the window time range respectively, and may be defined relative to the evaluation time τ. Note that the rate at which windows get evaluated is controlled by the STEP defined in the query, which is denoted by δ.

We define the three window-to-stream operators as

$$\mathsf{RStream}((\omega^\tau, \tau)) = \{(\langle s, p, o \rangle, \tau) \mid \langle s, p, o \rangle \in \omega^\tau\}$$

$$\mathsf{IStream}((\omega^\tau, \tau), (\omega^{\tau-\delta}, \tau - \delta)) = \{(\langle s, p, o \rangle, \tau) \mid \langle s, p, o \rangle \in \omega^\tau, \langle s, p, o \rangle \notin \omega^{\tau-\delta}\}$$

$$\mathsf{DStream}((\omega^\tau, \tau), (\omega^{\tau-\delta}, \tau - \delta)) = \{(\langle s, p, o \rangle, \tau) \mid \langle s, p, o \rangle \notin \omega^\tau, \langle s, p, o \rangle \in \omega^{\tau-\delta}\}$$

where δ is the time interval between window evaluations. Note that RStream does not depend on the previous window evaluation, whereas both IStream and DStream depend on the contents of the previous window.

We have provided a brief explanation of the semantics of SPARQL$_{\text{Stream}}$. This is particularly useful in the sense that users may know what to expect when they issue a query using these new operators. However, as the actual data source is not an RDF stream but a sensor network or an event-based stream, e.g. exposed as a SNEEql endpoint, we need to transform the SPARQL$_{\text{Stream}}$ queries into SNEEql queries. The next section describes the semantics of the transformation from SPARQL$_{\text{Stream}}$ to SNEEql using the S$_2$O mappings.

5.2   S$_2$O Semantics

In this section we will present how we can use the S$_2$O mapping definitions to transform a set of conjunctive queries over an ontological schema, into the streaming query language SNEEql that is used to access the sources. This work is based on extensions to the ODEMAPSTER processor [1] and the formalisation work of Calvanese *et al.* [16] and Poggi *et al.* [17].

A conjunctive query q over an ontology $\mathcal{O}$ can be expressed as:

$$q(\boldsymbol{x}) \leftarrow \varphi(\boldsymbol{x}, \boldsymbol{y})$$

$$\varphi(\boldsymbol{x}, \boldsymbol{y}) : \bigwedge_{i=1\ldots k} P_i, \text{ with } P_i \begin{cases} C_i(x), C \text{ is an atomic class.} \\ R_i(x, y), R \text{ is an atomic property.} \\ x = y \end{cases}$$

$$x, y \text{ are variables either in } \boldsymbol{x}, \boldsymbol{y} \text{ or constants.}$$

where $\boldsymbol{x}$ is a tuple of distinct distinguished variables, and $\boldsymbol{y}$ a tuple of non-distinguished existentially quantified variables. The answer to this query consists in the instantiation of the distinguished variables [16]. For instance consider:

$$q_1(x) \leftarrow WindSpeedMeasurement(x) \wedge measuredBy(x, y) \wedge WindSensor(y)$$

It requires all instances x that are wind speed measurements captured by wind sensors. In this example x is a distinguished variable and y a non-distinguished one.

The query has three atoms: $WindSpeedMeasurement(x)$, $measuredBy(x,y)$, and $WindSensor(y)$.

Concerning the formal definition of the query answering, let $\mathcal{I} = (\Delta^{\mathcal{I}}, \cdot^{\mathcal{I}})$ be an interpretation, where $\Delta^{\mathcal{I}}$ is the interpretation domain and $\cdot^{\mathcal{I}}$ the interpretation function that assigns an element of $\Delta^{\mathcal{I}}$ to each constant, a subset of $\Delta^{\mathcal{I}}$ to each class and a subset of $\Delta^{\mathcal{I}} \times \Delta^{\mathcal{I}}$ to each property of the ontology. Given a query $q(\boldsymbol{x}) \leftarrow \varphi(\boldsymbol{x}, \boldsymbol{y})$ the answer to q is the set $q_{\boldsymbol{x}}^{\mathcal{I}}$ of tuples $\boldsymbol{c} \in \Delta^{\mathcal{I}} \times \cdots \times \Delta^{\mathcal{I}}$ that substituted to $\boldsymbol{x}$, make the formula $\exists \boldsymbol{y}.\varphi(\boldsymbol{x}, \boldsymbol{y})$ true in $\mathcal{I}$ [16,17,18]. Now we can introduce the definition of the mappings. Let $\mathcal{M}$ be a set of mapping assertions of the form:

$$\Psi \rightsquigarrow \Phi$$

where Ψ is a conjunctive query over the global ontology $\mathcal{O}$, formed by terms of the form $C(x), R(x,y), A(x,z)$, with C, R, and A being classes, object properties and datatype properties respectively in $\mathcal{O}$; x, y being object instance variables, and z being a datatype variable. Φ is a set of expressions that can be translated to queries in the target continuous language (e.g. SNEEql) over the sources.

A mapping assertion $C(f_C^{Id}(\boldsymbol{x})) \rightsquigarrow \Phi_{S_1,\ldots,S_n}(\boldsymbol{x})$ describes how to construct the concept C from the source streams (or relations) $S_1, \ldots, S_n$. The function f_C^{Id} creates an instance of the class C, given the tuple $\boldsymbol{x}$ of variables returned by the Φ expression. More specifically this function will construct the instance identifier (URI) from a set of attributes from the streams and relations. In this case the expression Φ has a declarative representation of the form:

$$\Phi_{S_1,\ldots,S_n}(\boldsymbol{x}) = \exists \boldsymbol{y}. p_{S_1,\ldots,S_n}^{Proj}(\boldsymbol{x}) \wedge p_{S_1,\ldots,S_n}^{Join}(\boldsymbol{v}) \wedge p_{S_1,\ldots,S_n}^{Sel}(\boldsymbol{v})$$

where $\boldsymbol{v}$ is a tuple of variables in either $\boldsymbol{x}, \boldsymbol{y}$. The term p^{Join} denotes a set of join conditions over the streams and relations S_i. Similarly the term p^{Sel} represents a set of condition predicates over the variables $\boldsymbol{v}$ in the streams S_i (e.g. conditions using $<, \leq, \geq, >, =$ operators).

A mapping assertion $R(f_{C_1}^{Id}(\boldsymbol{x_1}), f_{C_2}^{Id}(\boldsymbol{x_2})) \rightsquigarrow \Phi_{S_1,\ldots,S_n}(\boldsymbol{x_1}, \boldsymbol{x_2})$ describes how to construct instances of the object property R from the source streams and relations S_i. The declarative form of Φ is:

$$\Phi_{S_1,\ldots,S_n}(\boldsymbol{x_1}, \boldsymbol{x_2}) = \exists \boldsymbol{y}. \Phi_{S_1,\ldots,S_k}(\boldsymbol{x_1}) \wedge \Phi_{S_{k+1},\ldots,S_n}(\boldsymbol{x_2}) \wedge p_{S_1,\ldots,S_n}^{Join}(\boldsymbol{v})$$

where $\Phi_{S_1,\ldots,S_k}, \Phi_{S_{k+1},\ldots,S_n}$ describe how to extract instances of C_1 and C_2 from the streams $S_1, \ldots, S_k$ and $S_{k+1}, \ldots, S_n$ respectively. The term p^{Join} is the set of predicates that denotes the join between the streams and relations $S_1, \ldots, S_n$.

Finally an expression $A(f_C^{Id}(\boldsymbol{x}), f_A^{Trf}(\boldsymbol{z})) \rightsquigarrow \Phi_{S_1,\ldots,S_n}(\boldsymbol{x}, \boldsymbol{z})$ describes how to construct instances of the datatype property A from the source streams and relations $S_1, \ldots, S_n$. The function f_A^{Trf} executes any transformation over the tuple of variables $\boldsymbol{z}$ to obtain the property value (e.g. arithmetic operations, or string operations). The declarative form of Φ in this case is:

$$\Phi_{S_1,\ldots,S_n}(\boldsymbol{x}, \boldsymbol{z}) = \exists \boldsymbol{y}. \Phi_{S_1,\ldots,S_k}(\boldsymbol{x}) \wedge \Phi_{S_{k+1},\ldots,S_n}(\boldsymbol{z}) \wedge p_{S_1,\ldots,S_n}^{Join}(\boldsymbol{v})$$

The definition follows the same idea as the previous one. The variables of z will contain the actual values that will be used to construct the datatype property value using the function f_A^{Trf}.

When a conjunctive query is issued against the global ontology, the processor first parses it and transforms it into an abstract syntax tree and then uses the expansion algorithm described in [1] (that is based on the PerfectRef algorithm of [16]) to produce an expanded conjunctive query based on the TBox of the ontology. Afterwards the rewritten query can be translated to an extended relational algebra.

A query $Q_\mathcal{O}(\boldsymbol{x})[t_s, t_e, \delta]$ is a conjunctive query with a window operator (where t_s, t_e are the start and end points of the window range and δ is the slide) in order to narrow the data set according to a given criteria. For a query:

$$Q_\mathcal{O}(\boldsymbol{x})[t_s, t_e, \delta] = (C_1(x) \wedge R(x, y) \wedge A(x, z))[t_s, t_e, \delta]$$

the translation is given by $\lambda(\Phi)$, following the mapping definition:

$$\lambda(\Phi_{S_1, \dots, S_n}(\boldsymbol{x})[t_s, t_e, \delta]) = \pi_{pProj} (\bowtie_{pJoin} (\sigma_{pSel}(\omega_{t_s, t_e, \delta}S_1), \dots \\ , \sigma_{pSel}(\omega_{t_s, t_e, \delta}S_n)))$$

The expression denotes first a window operation $\omega_{t_s, t_e, \delta}$ over the relations or streams $S_1, \dots, S_n$, with t_s, t_e, and δ being the time range and slide. A selection σ_{pSel} is applied over the result, according to the conditions defined in the mapping. A multi-way join $\bowtie_{pJoin}$ is then applied to the selection, also based on the corresponding mapping definition. Finally a projection π_{pProj} is applied over the results. For any conjunctive query with more atoms, the construction of the algebra expression will follow the same direct translation using the GAV approach.

6   Implementation and Walkthrough

The presented approach of providing ontology-based access to streaming data has been implemented as an extension to the ODEMAPSTER processor [1]. This implementation generates SNEEql queries that can be executed by the streaming query processor.

```
windsamples: (sensorId INT PK, ts DATETIME PK, speed FLOAT, direction FLOAT)
sensors: (sensorId INT PK, sensorName CHAR(45), lat FLOAT, long FLOAT)
```

Listing 3. Relational schema of the data source

Consider the motivating example where a sensor network generates a stream `windsamples` of wind sensor measurements. The associated stored information about the sensors, e.g. location and type, are stored in a relation `sensors`. The schemas are presented in Listing 3. Also consider the following ontological view:

```
conceptmap−def  WindSpeedMeasurement
    virtualStream  <http :// semsorgrid4env . eu / SensorReadings . srdf>
    uri−as
        concat ( ' http :// semsorgrid4env . eu / WindSpeedMeasurement_ ' , windsamples .
            sensorId , windsamples . ts )
    described−by
        attributemap−def  hasSpeed
            operation  constant
                has−column  windsamples . speed
        dbrelationmap−def  isProducedBy
            toConcept  Sensor
            joins−via
                condition  equals
                    has−column  sensors . sensorId
                    has−column  windsamples . sensorId

conceptmap−def  Sensor
    uri−as
        concat ( ' http :// semsorgrid4env . eu / Sensor_ ' , sensors . sensorId )
    described−by
        attributemap−def  hasSensorId
            operation  constant
                has−column  sensors . sensorId
```

Listing 4. S_2O mapping from the data stream **windsamples** to the ontology concepts
WindSpeedMeasurement

$$SpeedMeasurement \sqsubseteq Measurement$$
$$WindSpeedMeasurement \sqsubseteq SpeedMeasurement$$
$$WindDirectionMeasurement \sqsubseteq Measurement$$
$$SpeedMeasurement \sqsubseteq \exists hasSpeed$$
$$Measurement \sqsubseteq \exists isProducedBy.Sensor$$
$$Sensor \sqsubseteq \exists hasName$$

We can define an S_2O mapping that splits the **windsamples** stream tuples into instances of two different concepts *WindSpeedMeasurement* and *WindDirection-
Measurement*. Listing 4 is an extract of the S_2O mapping document concerning
the *WindSpeedMeasurement*. The mapping extract defines how to construct
the *WindSpeedMeasurement* and *Sensor* class instances from the **windsamples**
stream and the **sensors** table: $\Psi_{WindSpeedMeasurement} \rightsquigarrow \Phi_{\texttt{windsamples}}$ and
$\Psi_{Sensor} \rightsquigarrow \Phi_{\texttt{sensors}}$. In the case of the *WindSpeedMeasurement* the function
$f^{Id}_{WindSpeedMeasurement}$ produces the URI's of the instances by concatenating the
sensorId and **ts** attributes. Now we can pose a query over the ontology using
SPARQL$_{Stream}$, for example to obtain the wind speed measurements taken in the
last 10 minutes (See the query in Listing 5).

A class query atom $WindSpeedMeasurement(x)$ and a datatype property
atom $hasSpeed(x, z)$ can be extracted from the SPARQL$_{Stream}$ query. The win-
dow specification $[t_s = \mathsf{NOW} - 10, t_e = \mathsf{NOW}, \delta = 1]$ is also obtained[8]. The S_2O
mapping defines that *WindSpeedMeasurment* instances are generated based on

---

[8] For the simplicity of presentation, we assume that the system rewrites all time
specifications to minutes. The implemented system uses milliseconds as the common
time unit.

```
PREFIX fire: <http://www.ssg4env.eu#>
PREFIX rdf: <http://www.w3.org/1999/02/22-rdf-syntax-ns#>
SELECT RSTREAM ?speed
FROM STREAM <www.ssg4env.eu/SensorReadings.srdf> [FROM NOW - 10 MINUTES TO
    NOW STEP 1 MINUTE]
WHERE {
  ?WindSpeed a fire:WindSpeedMeasurement;
      fire:hasSpeed ?speed;
}
```

Listing 5. SPARQL_Stream query which every minute returns the wind speed for the last ten minutes

```
SELECT RSTREAM concat('http://ssg4env.eu#WindSpeedMeasurement',windsamples.
    sensorId,windamples.ts) AS id, windsamples.speed AS speed
FROM windsamples[FROM NOW - 10 MINUTES TO NOW SLIDE 1 MINUTE];
```

Listing 6. The SNEEql query that is generated for the input query in Listing 5

the `sensorId` and `ts` attributes of the `windsamples` stream, using a concatenation function to generate each instance URI. Similarly the S$_2$O mapping defines that *hasSpeed* properties are generated from the values of the speed attribute of the `windsamples` stream. The processor will evaluate this as:

$$\lambda(\Phi_{\mathtt{windsamples}}(x_{\mathtt{sensorId}}, x_{\mathtt{ts}}, z_{\mathtt{speed}})[now - 10, now, 1]) =$$
$$\pi_{\mathtt{sensorId,ts,speed}}(\omega_{now-10,now,1}(\mathtt{windsamples}))$$

In this case no joins and other selection conditions are needed, and only one stream has to be queried to produce the results. The query generated in the SNEEql language is shown in Listing 6[9]. The relational answer stream that results from evaluating the query in Listing 6 are transformed by the **Data Transformation** module depicted in Figure 1 according to the S$_2$O mappings. This results in a stream of tagged triples which are instances of the class *WindSpeedMeasurement*.

7   Related Work

Several systems exist to provide ontology-based access to stored data, mainly in the form of relational databases, as described in [3].

A simple approach is to first generate a syntactical translation of the database schema to an ontological representation. Although the resulting ontology has no real semantics, it may be argued that this is a first step through an ontology model and could later be mapped to a real domain ontology [18]. Virtuoso [19] and D2RQ [2], like R$_2$O, use mappings between the source relational schema to RDF ontologies enabling users to issues queries over a semantically rich domain ontology. The expressiveness of the queries supported by these systems is limited to conjunctive queries, and none of the approaches takes into account streaming data and continuous queries.

---

[9] Although the current available implementation of the SNEE processor lacks the `concat` operator, we include the sample query in its complete form here.

Several stream processing and querying engines have been built in the last decade and can be grouped in two main areas: event stream systems (e.g. Aurora/Borealis [5], STREAM [4], TelegraphCQ [20]), and acquisitional stream systems (e.g. TinyDB [7], SNEE [6], Cougar [21]). For the first, the stream system does not have control over the data arrival rate, which is often potentially high and usually unknown and the query optimization goal is to minimize latency. For acquisitional streams, it is possible to control when data is obtained from the source, typically a sensor network, and the query optimisation goal is to maximize network lifetime. All these systems have their own continuous query language, generally based on SQL, although most of them share the same features. CQL (Continuous Query Language) [13] is the best known of these languages, but there is still no common standard language for stream queries. The SNEEql [12] language for querying streaming data sources is inspired by CQL, but it provides greater expressiveness in queries, including both event and acquisitional streams, and stored extents. Our work does not aim to improve on relational stream query processing, but to enable these systems to be accessible via ontology-based querying.

Finally, there are two existing proposals for extending SPARQL with stream-based operators: streamingSPARQL [8] and C-SPARQL [9]. Both languages introduce extensions for the support of RDF streams, and both define time-based and triple-based window operators where the upper bound is fixed to the current evaluation time. The SPARQL$_{Stream}$ windowing operator enables windows to be defined in the past so as to support correlation with historic data. We have not included triple-based windows in SPARQL$_{Stream}$ due to the problems with their semantics, discussed in Section 4.1. Window-to-stream operators are also missing in both existing approaches, which provides ambiguous semantics for the language. In SPARQL$_{Stream}$ the result of a window operator is a bag to triples over which traditional operators can be applied. We have introduced three window-to-stream operators inspired by SNEEql and CQL. The aggregate semantics introduced in C-SPARQL follow an approach of extending the data, which differs from standard aggregation semantics of summarising the data. We have opted to support the aggregation semantics being defined for SPARQL 1.1 [22], which summarise the data.

Table 1. Summary of key contributions

Extension	Base Approach	Summary
SPARQL$_{Stream}$	SPARQL 1.1	Window definitions with variable upper boundary
		Window-to-stream operators
S$_2$O	R$_2$O	Stream definitions in mapping
		Streaming data types
		Virtual RDF stream IRIs
	ODEMAPSTER	Translation of SPARQL$_{Stream}$ queries into SNEEql

8   Conclusions and Future Work

We have presented an approach for providing ontology-based access to streaming data, which is based on SPARQL$_{Stream}$, a SPARQL extension for RDF streams, and S$_2$O, an extension to R$_2$O for expressing mappings from streaming sources to ontologies. We have shown the semantics of the proposed extensions and the mechanism to generate data source queries from the original ontological queries using the mappings. The case presented here generated SNEEql queries but the techniques are independent of the target stream query language, although issues of stream data model and language evaluation semantics would need to be considered for each case. Finally the prototype implementation, which extends ODEMAPSTER, has shown the feasibility of the approach. This work constitutes a first effort towards ontology-based streaming data integration, relevant for supporting the increasing number of sensor network applications being developed and deployed in the recent years. The extensions presented in this paper can be summarised in Table 1.

Although we have shown initial results querying the underlying SNEE engine with basic queries, we expect to consider in the near future more complex query expressions including aggregates, and joins involving both streaming and stored data sources. Another important strand of future work is the optimization of distributed query processing [14] and the streaming queries [5,6]. It is also our goal to provide a characterization of our algorithms. In the scope of a larger streaming and sensor networks integration framework, we intend to achieve the following goals: i) integrating streaming and stored data sources through an ontological unified view; ii) combining data from event-based and acquisition-based streams, and stored data sources; iii) considering quality-of-service requirements for query optimization and source selection during the integration.

Acknowledgments. This work has been supported by the European Commission project SemSorGrid4Env (FP7-223913). We also thank Alvaro A. A. Fernandes, Ixent Galpin, and Norman W. Paton, from the University of Manchester, for their valuable ideas and suggestions.

References

1. Barrasa, J., Corcho, O., Gómez-Pérez, A.: R2O, an extensible and semantically based database-to-ontology mapping language. In: SWDB 2004, pp. 1069–1070 (2004)
2. Bizer, C., Cyganiak, R.: D2RQ. Lessons Learned. In: W3C Workshop on RDF Access to Relational Databases (October 2007)
3. Sahoo, S.S., Halb, W., Hellmann, S., Idehen, K., Thibodeau Jr, T., Auer, S., Sequeda, J., Ezzat, A.: A survey of current approaches for mapping of relational databases to RDF. W3C (January 2009)
4. Arasu, A., Babcock, B., Babu, S., Cieslewicz, J., Datar, M., Ito, K., Motwani, R., Srivastava, U., Widom, J.: Stream: The stanford data stream management system. In: Garofalakis, M., Gehrke, J., Rastogi, R. (eds.) Data Stream Management (2006)

5. Abadi, D.J., Ahmad, Y., Balazinska, M., Cetintemel, U., Cherniack, M., Hwang, J.H., Lindner, W., Maskey, A.S., Rasin, A., Ryvkina, E., Tatbul, N., Xing, Y., Zdonik, S.: The Design of the Borealis Stream Processing Engine. In: CIDR 2005 (2005)

6. Galpin, I., Brenninkmeijer, C.Y., Jabeen, F., Fernandes, A.A., Paton, N.W.: Comprehensive optimization of declarative sensor network queries. In: SSDBM 2009, pp. 339–360 (2009)

7. Madden, S.R., Franklin, M.J., Hellerstein, J.M., Hong, W.: TinyDB: an acquisitional query processing system for sensor networks. ACM Trans. Database Syst. 30(1), 122–173 (2005)

8. Bolles, A., Grawunder, M., Jacobi, J.: Streaming SPARQL - extending SPARQL to process data streams. In: Bechhofer, S., Hauswirth, M., Hoffmann, J., Koubarakis, M. (eds.) ESWC 2008. LNCS, vol. 5021, pp. 448–462. Springer, Heidelberg (2008)

9. Barbieri, D.F., Braga, D., Ceri, S., Grossniklaus, M.: An execution environment for C-SPARQL queries. In: EDBT 2010, Lausanne, Switzerland, pp. 441–452 (March 2010)

10. Lenzerini, M.: Data integration: a theoretical perspective. In: PODS 2002, pp. 233–246 (2002)

11. Golab, L., Özsu, M.T.: Issues in data stream management. SIGMOD Record 32(2), 5–14 (2003)

12. Brenninkmeijer, C.Y., Galpin, I., Fernandes, A.A., Paton, N.W.: A semantics for a query language over sensors, streams and relations. In: Gray, A., Jeffery, K., Shao, J. (eds.) BNCOD 2008. LNCS, vol. 5071, pp. 87–99. Springer, Heidelberg (2008)

13. Arasu, A., Babu, S., Widom, J.: The CQL continuous query language: semantic foundations and query execution. The VLDB Journal 15(2), 121–142 (2006)

14. Kossmann, D.: The state of the art in distributed query processing. ACM Comput. Surv. 32(4), 422–469 (2000)

15. Pérez, J., Arenas, M., Gutierrez, C.: Semantics and complexity of SPARQL. ACM Trans. Database Syst. 34(3), 1–45 (2009)

16. Calvanese, D., De Giacomo, G., Lembo, D., Lenzerini, M., Rosati, R.: DL-Lite: Tractable description logics for ontologies. In: AAAI 2005, pp. 602–607 (2005)

17. Poggi, A., Lembo, D., Calvanese, D., Giacomo, G.D., Lenzerini, M., Rosati, R.: Linking data to ontologies. J. Data Semantics 10, 133–173 (2008)

18. Lubyte, L., Tessaris, S.: Supporting the development of data wrapping ontologies. In: 4th Asian Semantic Web Conference (December 2009)

19. Erling, O., Mikhailov, I.: RDF support in the Virtuoso DBMS. In: Conference on Social Semantic Web. LNI, vol. 113, pp. 59–68. GI (2007)

20. Chandrasekaran, S., Cooper, O., Deshpande, A., Franklin, M.J., Hellerstein, J.M., Hong, W., Krishnamurthy, S., Madden, S.R., Reiss, F., Shah, M.A.: TelegraphCQ: continuous dataflow processing. In: SIGMOD 2003, p. 668 (2003)

21. Yao, Y., Gehrke, J.: The Cougar approach to in-network query processing in sensor networks. SIGMOD Rec. 31(3), 9–18 (2002)

22. Harris, S., Seaborne, A. (eds.): SPARQL 1.1 query language. Working draft, W3C (2010)

Evolution of *DL-Lite* Knowledge Bases

Diego Calvanese, Evgeny Kharlamov[*], Werner Nutt, and Dmitriy Zheleznyakov

KRDB Research Centre, Free University of Bozen-Bolzano, Italy
last_name@inf.unibz.it

Abstract. We study the problem of evolution for Knowledge Bases (KBs) expressed in Description Logics (DLs) of the *DL-Lite* family. *DL-Lite* is at the basis of OWL 2 QL, one of the tractable fragments of OWL 2, the recently proposed revision of the Web Ontology Language. We propose some fundamental principles that KB evolution should respect. We review known model and formula-based approaches for evolution of propositional theories. We exhibit limitations of a number of model-based approaches: besides the fact that they are either not expressible in DL-Lite or hard to compute, they intrinsically ignore the structural properties of KBs, which leads to undesired properties of KBs resulting from such an evolution. We also examine proposals on update and revision of DL KBs that adopt the model-based approaches and discuss their drawbacks. We show that known formula-based approaches are also not appropriate for *DL-Lite* evolution, either due to high complexity of computation, or because the result of such an action of evolution is not expressible in *DL-Lite*. Building upon the insights gained, we propose two novel formula-based approaches that respect our principles and for which evolution is expressible in *DL-Lite*. For our approaches we also developed polynomial time algorithms to compute evolution of *DL-Lite* KBs.

1   Introduction

Description Logics (DLs) provide excellent mechanisms for representing structured knowledge, and as such they constitute the foundations for the various variants of OWL, the standard ontology language of the Semantic Web[1]. DLs have traditionally been used for modeling at the intensional level the static and structural aspects of application domains [1]. Recently, however, the scope of ontologies has broadened, and they are now used also for providing support in the maintenance and evolution phase of information systems. Moreover, ontologies are considered to be the premium mechanism through which services operating in a Web context can be accessed, both by human users and by other services. Supporting all these activities, makes it necessary to equip DL systems with additional kinds of inference tasks that go beyond the traditional ones of satisfiability, subsumption, and query answering provided by current DL inference engines. The most notable one, and the subject of this paper, is that of *knowledge base evolution* [2], where the task is to incorporate new knowledge into an existing knowledge base (KB) so as to take into account changes that occur in the underlying domain of interest. In general, the new knowledge to incorporate is represented by a set of formulas denoting

---

[*] The author is co-affiliated with INRIA Saclay.
[1] http://www.w3.org/TR/owl2-overview/

P.F. Patel-Schneider et al. (Eds.): ISWC 2010, Part I, LNCS 6496, pp. 112–128, 2010.

those properties that should be true after the KB has evolved. In the case where the new knowledge interacts in an undesirable way with the knowledge in the KB, e.g., by causing the KB or relevant parts of it to become unsatisfiable, the new knowledge cannot simply be added to the KB. Instead, suitable changes need to be made in the KB so as to avoid the undesirable interaction, e.g., by deleting parts of the KB that conflict with the new knowledge. Different choices are possible, corresponding to different semantics for KB evolution [3–8].

In the literature, two main types of KB evolution have been considered: namely revision and update [4]. Both have a precise formal grounding in terms of *postulates* [4, 5] and a number of update and revision operators were proposed in the literature [5, 6]. This work has been carried out for propositional logic, providing a thorough understanding of the various options, both wrt semantics and wrt computational properties.

Work relevant to KB evolution has been carried out initially in schema evolution in databases, cf., [9], and more recently for expressive DLs [7, 8]. However, for such richer representation formalisms, the picture is much less clear, and the various possibilities are far from being completely explored. *(i)* The fundamental distinction in DLs between TBox (for *terminological*, or intensional knowledge) and ABox (for *assertional*, or extensional knowledge), calls for distinguishing these two components (both in the existing and in the new knowledge) also in the study of evolution. *(ii)* Going from propositional letters to first-order predicates and interpretations, on the one hand calls for novel principles underlying the semantics of evolution, and on the other hand broadens the spectrum of possibilities for defining such semantics. *(iii)* The combination of constructs of the considered DL will obviously affect the complexity of computing the result of evolution, independently of the chosen semantics. *(iv)* While in propositional logic the result of an evolution step is always expressible in the same formalism, this does not hold in general for DLs [10, 11].

In this paper we address several of the points raised by the above observations, thus contributing substantially to a clarification of the problem.

In line with Item (i), we carry out our investigation and establish our results by considering separately the role of the ABox and of the TBox in evolution.

Regarding Item (ii), we propose some fundamental principles that KB evolution should respect. We review known model and formula-based approaches for evolution of propositional theories [5, 6], and we lift them to the first-order case in two natural ways (by considering symmetric difference on symbol interpretations vs. interpretation atoms). Previous proposals for KB evolution, such as ABox updates under Winslett's semantics [10, 11], and the approaches proposed in [8], fit nicely into our classification.

Regarding Items (iii) and (iv), we concentrate our technical development on the *DL-Lite* family [12], which is at the basis of OWL 2 QL, one of the tractable profiles of OWL 2. We exhibit limitations of a number of model-based approaches for the logics of the *DL-Lite* family: besides the fact that evolution under such approaches is either not expressible in *DL-Lite* or hard to compute, they intrinsically ignore the structural properties of KBs, which leads to undesired properties of KBs resulting from such an evolution. We also examine proposals on update and revision of DL KBs that adopt the model-based approaches and discuss their drawbacks. We show that known formula-based approaches are also not appropriate for *DL-Lite* evolution, either due to high complexity of computation, or because the result of such an action of evolution is not expressible in *DL-Lite*. Building upon the insights gained, we propose two novel formula-based approaches that respect our principles and for which evolution is expressible in *DL-Lite*. For our approaches we also developed polynomial time algorithms to compute evolution of *DL-Lite* KBs.

2   Preliminaries and Problem Definition

Description Logics. We introduce some basic notions of Description Logics (DLs), more details can be found in [13]. A DL *knowledge base* (KB) $\mathcal{K} = \mathcal{T} \cup \mathcal{A}$ is the union of two sets of assertions, those representing the *intensional-level* of the KB, that is, the general knowledge, and constituting the *TBox* $\mathcal{T}$, and those providing information on the *instance-level* of the KB, and constituting the *ABox* $\mathcal{A}$. In our work we consider a family of DLs, *DL-Lite* [12], which form a tractable fragment of OWL 2.

All the logics of the *DL-Lite* family have the following constructs for (complex) *concepts* and *roles*: (i) $B ::= A \mid \exists R$, (ii) $C ::= B \mid \neg B$, (iii) $R ::= P \mid P^-$, where A and P stand for an *atomic concept* and *role*, respectively, which are just names. A *DL-Lite$_{core}$* TBox consists of concept inclusion assertions $B \sqsubseteq C$. *DL-Lite$_{\mathcal{FR}}$* extends *DL-Lite$_{core}$* by allowing in a TBox role inclusion assertions $R_1 \sqsubseteq R_2$ and functionality assertions (funct R), in a way that if $R_1 \sqsubseteq R_2$ appears in a TBox, then neither (funct R_2) nor (funct R_2^-) appears in the TBox. This syntactic restriction keeps the logic tractable. ABoxes in *DL-Lite$_{core}$* and *DL-Lite$_{\mathcal{FR}}$* consist of membership assertions of the form $B(a)$ and $P(a,b)$. When we write in this paper *DL-Lite* without a subscript, specifying a concrete language, we mean *any* language of this family. The *DL-Lite* family has nice computational properties, for example, KB satisfiability has polynomial-time complexity in the size of the TBox and logarithmic-space in the size of the ABox [14, 15].

The semantics of DL-Lite KBs is given in the standard way, using first order interpretations, all over the same infinite countable domain Δ. An *interpretation* $\mathcal{I}$ is a function $\cdot^{\mathcal{I}}$ that assigns to each concept C a subset $C^{\mathcal{I}}$ of Δ, and to each role R a binary relation $R^{\mathcal{I}}$ over Δ in such a way that $(\neg B)^{\mathcal{I}} = \Delta \setminus B^{\mathcal{I}}$, $(\exists R)^{\mathcal{I}} = \{a \mid \exists a'.(a, a') \in R^{\mathcal{I}}\}$, and $(P^-)^{\mathcal{I}} = \{(a_2, a_1) \mid (a_1, a_2) \in P^{\mathcal{I}}\}$. We assume that Δ contains the constants and that $c^{\mathcal{I}} = c$, i.e., we adopt *standard names*. Alternatively, we view an interpretation as a set of atoms and say that $A(a) \in \mathcal{I}$ iff $a \in A^{\mathcal{I}}$ and $P(a, b) \in \mathcal{I}$ iff $(a, b) \in P^{\mathcal{I}}$. An interpretation $\mathcal{I}$ is a *model* of a membership assertion $B(a)$ if $a \in B^{\mathcal{I}}$, and of $P(a, b)$ if $(a, b) \in P^{\mathcal{I}}$, of an inclusion assertion $D_1 \sqsubseteq D_2$ if $D_1^{\mathcal{I}} \subseteq D_2^{\mathcal{I}}$, and of a functionality assertion (funct R) if the relation $R^{\mathcal{I}}$ is a function.

As usual, we use $\mathcal{I} \models F$ to denote that $\mathcal{I}$ is a model of an assertion F, and $\mathcal{I} \models \mathcal{K}$ to denote that $\mathcal{I} \models F$ for each assertion F in $\mathcal{K}$. We use $Mod(\mathcal{K})$ to denote the set of all models of $\mathcal{K}$. A KB is *satisfiable* if it has at least one model and it is coherent[2] if for every concept and role S occurring in $\mathcal{K}$ there is an $\mathcal{I} \in Mod(\mathcal{K})$ such that $S^{\mathcal{I}} \neq \emptyset$. We use entailment on KBs $\mathcal{K} \models \mathcal{K}'$ in the standard sense. We say that an ABox $\mathcal{A}$ $\mathcal{T}$-*entails* an ABox $\mathcal{A}'$, denoted $\mathcal{A} \models_{\mathcal{T}} \mathcal{A}'$, if $\mathcal{T} \cup \mathcal{A} \models \mathcal{A}'$, and $\mathcal{A}$ is $\mathcal{T}$-*equivalent* to $\mathcal{A}'$, denoted $\mathcal{A} \equiv_{\mathcal{T}} \mathcal{A}'$, if $\mathcal{A} \models_{\mathcal{T}} \mathcal{A}'$ and $\mathcal{A}' \models_{\mathcal{T}} \mathcal{A}$. The deductive *closure* of a TBox $\mathcal{T}$ (of an

---

[2] Coherence is often called *full satisfiability*.

ABox $\mathcal{A}$), denoted $cl(\mathcal{T})$ (resp., $cl_{\mathcal{T}}(\mathcal{A})$), is the set of all TBox (resp., ABox) assertions F such that $\mathcal{T} \models F$ (resp., $\mathcal{T} \cup \mathcal{A} \models F$). It is easy to see that in *DL-Lite* $cl(\mathcal{T})$ (and $cl_{\mathcal{T}}(\mathcal{A})$) is computable in quadratic time in the size of $\mathcal{T}$ (resp., $\mathcal{T}$ and $\mathcal{A}$). In our work we assume that all TBoxes and ABoxes are closed.

Ontology Evolution. Let $\mathcal{K} = \mathcal{T} \cup \mathcal{A}$ be a *DL-Lite* KB and $\mathcal{N}$ a set of "new" (TBox and/or ABox) assertions. We want to study how to incorporate the assertions $\mathcal{N}$ into $\mathcal{K}$, that is, how $\mathcal{K}$ *evolves* [2] under $\mathcal{N}$. More practically, we want to develop *evolution operators* that take $\mathcal{K}$ and $\mathcal{N}$ as input and return, possibly in *polynomial time*, a *DL-Lite* KB $\mathcal{K}'$ that captures the evolution, and which we call *the evolution of $\mathcal{K}$ under $\mathcal{N}$*.

In the Semantic Web context, update and revision [4, 5], the two classical understandings of ontology evolution, are too restrictive from the intuitive and formal perspective: in many applications we know neither the status of the real world, nor how accurate $\mathcal{N}$ is wrt to the world. For example, if in $\mathcal{K}$ we store knowledge from Web sources, say, online newspapers that we collected using RSS feeds or Web crawling, then there is no chance to say how this information is related to the state of the real world. When a new portion of knowledge $\mathcal{N}$ arrives to $\mathcal{K}$ and conflicts with $\mathcal{K}$, then it might be unclear whether the conflict is due to outdated or wrong information in $\mathcal{K}$. This situation does not fit in the formalisms of update and revision and, therefore, we propose now some new postulates to be adopted in the context of evolution in the Semantic Web.

First, we assume that the KBs we are dealing with make sense, that is, they are coherent (and hence also satisfiable), and we want evolution to preserve this property:

EP1: Evolution should preserve coherence of the KB, that is, $\mathcal{K}'$ is coherent.

The same postulate is stipulated in [8]. Notice that in *DL-Lite*[3] coherence can be reduced to satisfiability. Moreover, when $\mathcal{N}$ may contain ABox assertions, one can enforce coherence by adding to $\mathcal{N}$ for each atomic concept A an assertion $A(d_A)$, and for each atomic role P an assertion $P(d_P, d'_P)$, where d_A, d_P, d'_P are fresh individuals.

For example, if our online newspapers KB $\mathcal{K} = \mathcal{T} \cup \mathcal{A}$ records that John is married to Mary and that a person can be married to at most one person, and if the new knowledge $\mathcal{N}$ says that John is married to Patty, then $\mathcal{K} \cup \mathcal{N}$ is unsatisfiable (and hence incoherent) and does not comply with **EP1**. This can be resolved by either *(i)* discarding the old information about John's marriage, that is, by changing $\mathcal{A}$, or *(ii)* weakening the constraint in $\mathcal{K}$ on the number of spouses, that is, by changing $\mathcal{T}$, or *(iii)* discarding $\mathcal{N}$. What to do depends on the application. In data-centric applications, the most valuable information is the (extensional) data and we would have to discard the constraint on the number of spouses from $\mathcal{T}$. In Web data integration, the constraints of $\mathcal{T}$ define the global schema and the data coming from different Web sources may be contradictive by nature. Thus, it makes more sense to discard one of the two assertions about John's spouses using, for example, the trust we have in the sources of the data. To formalize this consideration we introduce the notion of *protected part* of a KB, which is simply a subset $\mathcal{K}_{pr} \subseteq \mathcal{K}$ that is preserved by evolution. This is sanctioned by our second postulate:

EP2: Evolution should entail the new knowledge and preserve the protected part, that is, $\mathcal{K}' \models \mathcal{K}_{pr} \cup \mathcal{N}$.

---

[3] Actually, in all logics enjoying the disjoint-union model property.

This postulate is different from the classical ones of update and revision where it is only required that the new KB $\mathcal{K}'$ should entail the new knowledge $\mathcal{N}$. We observe, however, that evolution of $\mathcal{K}$ with a protected part $\mathcal{K}_{pr}$ wrt $\mathcal{N}$ is conceptually the same as evolution of $\mathcal{K}$ with the empty protected part wrt $\mathcal{K}_{pr} \cup \mathcal{N}$.

Another principle that is widely accepted [4, 5] is the one of *minimality of change*:

EP3: The change to $\mathcal{K}$ should be minimal, that is, $\mathcal{K}'$ is minimally different from $\mathcal{K}$.

There are different approaches to define minimality, suitable for particular applications, and the current belief is that there is no general notion of minimality that will "do the right thing" under all circumstances [6].

Based on these principles, we will study evolution operators. We will consider the classical update and revision operators coming from AI [5] and also operators proposed for DLs [8, 15], and try to adapt them to our needs. In the following we distinguish three types of evolution: *TBox evolution*, when $\mathcal{N}$ consists of TBox assertions only, and we denote it $\mathcal{N}_T$, *ABox evolution*, when $\mathcal{N}$ consists of ABox assertions only, and we denote it $\mathcal{N}_A$, *KB evolution*, when $\mathcal{N}$ includes both TBox and ABox assertions.

Running Example. In our online-newspapers KB we have structural knowledge that wives (W) are exactly those individuals who have husbands (hh) and that some wives are employed (E). Singles (S) cannot be husbands. Priests (P) are clerics (C) and clerics are singles. Both clerics and wives are receivers of rent subsidies (R). We also know that Adam (a) and Bob (b) are priests, Mary (m) is a wife who is employed and her husband is John (j). Also, Carl (c) is a catholic minister (M). This knowledge can be expressed in *DL-Lite* by the KB $\mathcal{K}_{ex}$, consisting of the following assertions:

$\mathcal{T}$: $W \sqsubseteq \exists hh$, $\exists hh \sqsubseteq W$, $E \sqsubseteq W$, $S \sqsubseteq \neg\exists hh^-$, $P \sqsubseteq C$, $C \sqsubseteq S$, $C \sqsubseteq R$, $W \sqsubseteq R$;
$\mathcal{A}$: $P(a)$, $P(b)$, $E(m)$, $hh(m, j)$, $M(c)$.

By crawling some Web sources we found out that John is now single (that is, $S(j)$), in the Oxford Dictionary we discovered that catholic ministers are superiors of some religious orders and hence clerics ($M \sqsubseteq C$), and from economic news we found out that the current crisis affects people receiving rent subsidies in that subsidies were canceled for wives ($W \sqsubseteq \neg R$) and for clerics ($C \sqsubseteq \neg R$), since the former may receive support from their husbands and the latter from their church. In the rest of the paper we will discuss how to incorporate this new knowledge into our KB.

3   Approaches to Evolution

A number of candidate semantics for evolution operators have been proposed in the literature [3, 6, 8, 15, 16]. They can be divided into two groups, *model-based approaches* (MBAs) and *formula-based approaches* (FBAs).

3.1   Model-Based Approaches

In model-based approaches (MBAs) the result of evolution of a KB $\mathcal{K}$ wrt new knowledge $\mathcal{N}$ is a set $\mathcal{K} \diamond \mathcal{N}$ of models. The general idea of MBAs is to choose as the result of evolution some models of $\mathcal{N}$ depending on their distance to the models of $\mathcal{K}$. Katsuno and Mendelzon [4] considered two ways of choosing these models of $\mathcal{N}$.

The idea of the first one, which we call *local*, is to go over all models $\mathcal{I}$ of $\mathcal{K}$ and for each $\mathcal{I}$ to take those models $\mathcal{J}$ of $\mathcal{N}$ that are minimally distant from $\mathcal{I}$. Formally,

$$\mathcal{K} \diamond \mathcal{N} = \bigcup_{\mathcal{I} \in Mod(\mathcal{K})} \arg\min_{\mathcal{J} \in Mod(\mathcal{N})} dist(\mathcal{I}, \mathcal{J}),$$

where $dist(\cdot, \cdot)$ is a function whose range is a partially ordered domain and $\arg\min$ stands for the *argument of the minimum,* that is, in our case, the set of models $\mathcal{J}$ for which the value of $dist(\mathcal{I}, \mathcal{J})$ reaches its minimum value, given $\mathcal{I}$. The distance function *dist* varies from approach to approach and commonly takes as values either numbers or subsets of some fixed set.

The idea of the second way, called *global*, is to choose those models $\mathcal{J}$ of $\mathcal{N}$ that are minimally distant from the entire set of models of $\mathcal{K}$. Formally,

$$\mathcal{K} \diamond \mathcal{N} = \arg\min_{\mathcal{J} \in Mod(\mathcal{N})} dist(Mod(\mathcal{K}), \mathcal{J}), \tag{1}$$

where $dist(Mod(\mathcal{K}), \mathcal{J}) = \min_{\mathcal{I} \in Mod(\mathcal{K})} dist(\mathcal{I}, \mathcal{J})$.

The classical MBAs were developed for propositional theories. In this context, an interpretation was identified with the set of propositional atoms that it makes true and two distance functions were introduced, respectively based on symmetric difference and on the cardinality of symmetric difference,

$$dist_\subseteq(\mathcal{I}, \mathcal{J}) = \mathcal{I} \ominus \mathcal{J} \qquad \text{and} \qquad dist_\sharp(\mathcal{I}, \mathcal{J}) = |\mathcal{I} \ominus \mathcal{J}|. \tag{2}$$

where the symmetric difference of two sets is defined as $\mathcal{I} \ominus \mathcal{J} = (\mathcal{I} \setminus \mathcal{J}) \cup (\mathcal{J} \setminus \mathcal{I})$. Distances under $dist_\subseteq$ are sets and are compared by set inclusion, that is, $dist_\subseteq(\mathcal{I}_1, \mathcal{J}_1) \leq dist_\subseteq(\mathcal{I}_2, \mathcal{J}_2)$ iff $dist_\subseteq(\mathcal{I}_1, \mathcal{J}_1) \subseteq dist_\subseteq(\mathcal{I}_2, \mathcal{J}_2)$. Distances under $dist_\sharp$ are natural numbers and are compared in the standard way.

One can extend these distances to DL interpretations in two different ways. One way is to consider interpretations $\mathcal{I}$, $\mathcal{J}$ as sets of *atoms*. Then $\mathcal{I} \ominus \mathcal{J}$ is again a set of atoms and we can define distances as in Equation (2). We denote these distances as $dist_\subseteq^a(\mathcal{I}, \mathcal{J})$ and $dist_\sharp^a(\mathcal{I}, \mathcal{J})$. While in the propositional case distances are always finite, this may not be the case for DL interpretations that are infinite. Another way is to define distances at the level of the concept and role *symbols* in the underlying signature Σ:

$$dist_\subseteq^s(\mathcal{I}, \mathcal{J}) = \{S \in \Sigma \mid S^\mathcal{I} \neq S^\mathcal{J}\}, \quad \text{and} \quad dist_\sharp^s(\mathcal{I}, \mathcal{J}) = |\{S \in \Sigma \mid S^\mathcal{I} \neq S^\mathcal{J}\}|.$$

Summing up across the different possibilities, we have three dimensions, which give eight possibilities to define a semantics of evolution according to MBAs by choosing: (1) the *local* or the *global* approach, (2) *atoms* or *symbols* for defining distances, and (3) *set inclusion* or *cardinality* to compare symmetric differences.

We denote each of these eight possibilities by a combination of three symbols, indicating the choice in each dimension. By $\mathcal{L}$ we denote local and by $\mathcal{G}$ global semantics. We attach the superscripts a or s to indicate whether distances are defined in terms of atoms or symbols. We use the subscripts $\subseteq$ or $\sharp$ to indicate whether distances are compared in terms of set inclusion or cardinality. For example, $\mathcal{L}_\sharp^a$ denotes the local semantics where the distances are expressed in terms of cardinality of sets of atoms.

Considering that in the propositional case a distinction between atom and symbol-based semantics is meaningless, we can also use our notation, without superscripts, to identify MBAs in that setting. Interestingly, the two classical local MBAs proposed by Winslett [6] and Forbus [17] correspond, respectively, to $\mathcal{L}_\subseteq$, and $\mathcal{L}_\sharp$, while the one by Borgida [18] is a variant of $\mathcal{L}_\subseteq$. The two classical global MBAs proposed by Satoh [5] and Dalal [19] correspond, respectively, to $\mathcal{G}_\subseteq$, and $\mathcal{G}_\sharp$.

Under each of our eight semantics, evolution results in a set of interpretations. In the propositional case each set of interpretations over finitely many symbols can be captured by a formula whose models are exactly those interpretations. In the case of DLs this is no more necessarily the case, since on the one hand, interpretations can be infinite and on the other hand logics may miss some connectives like disjunction or negation.

Let $\mathcal{D}$ be a DL and M one of the eight MBAs introduced above. We say $\mathcal{D}$ is *closed under evolution wrt* M (or evolution wrt M is *expressible* in $\mathcal{D}$) if for any KBs $\mathcal{K}$ and $\mathcal{N}$ written in $\mathcal{D}$, there is a KB $\mathcal{K}'$ written in $\mathcal{D}$ such that $Mod(\mathcal{K}') = \mathcal{K} \diamond \mathcal{N}$. We study now whether the logics of the *DL-Lite* family are closed under the various semantics.

Global Model-Based Approaches. We start with an example showing that wrt all four semantics $\mathcal{G}_\subseteq^s$, $\mathcal{G}_\sharp^s$, $\mathcal{G}_\subseteq^a$ and $\mathcal{G}_\sharp^a$, TBox evolution is not expressible in *DL-Lite*.

The observation underlying these results is that on the one hand, the minimality of change principle introduces implicit disjunction in the evolved KB. On the other hand, *DL-Lite* can be embedded into a slight extension of Horn logic [20] and therefore does not allow one to express genuine disjunction. Technically, this can be expressed by saying that every *DL-Lite* KB that entails a disjunction of *DL-Lite* assertions entails one of the disjuncts. The lemma gives a contrapositive formulation of this statement. Although *DL-Lite* does not have a disjunction operator, by abuse of notation we write $\mathcal{J} \models \phi \vee \psi$ as a shorthand for "$\mathcal{J} \models \phi$ or $\mathcal{J} \models \psi$" for *DL-Lite* assertions ϕ, ψ.

Lemma 1. *Let M be a set of interpretations. Suppose there are DL-Lite assertions ϕ, ψ such that (1) $\mathcal{J} \models \phi \vee \psi$ for every $\mathcal{J} \in M$; (2) there are $\mathcal{J}_1, \mathcal{J}_2 \in M$ such that $\mathcal{J}_1 \not\models \phi$ and $\mathcal{J}_2 \not\models \psi$. Then there is no DL-Lite KB $\mathcal{K}$ such that $M = Mod(\mathcal{K})$.*

Example 2. Consider the KB $\mathcal{K}_{ex}$ of our running example and assume that the new information $\mathcal{N}_T = \{ W \sqsubseteq \neg R \}$ arrived. We explore evolution wrt the $\mathcal{G}_\sharp^s$ semantics of $\diamond$, which counts for how many symbols the interpretation changes.

Consider three assertions, (derived) from $\mathcal{K}$, that are essential for this example: $E \sqsubseteq W$, $E \sqsubseteq R$, and $E(m)$. One can show that the minimum of $dist_\sharp^s(\mathcal{I}, \mathcal{J})$ for $\mathcal{I} \in Mod(\mathcal{K})$ and $\mathcal{J} \in Mod(\mathcal{N}_T)$ equals 1. Let $\mathcal{J} \in \mathcal{K} \diamond \mathcal{N}_T$. Then there exists $\mathcal{I} \in Mod(\mathcal{K})$ such that $dist_\sharp^s(\mathcal{I}, \mathcal{J}) = 1$. Hence, there is only one symbol $S \in \{E, W, R\}$ whose interpretation has changed from $\mathcal{I}$ to $\mathcal{J}$, that is $S^\mathcal{I} \neq S^\mathcal{J}$. Observe that S cannot be E. Otherwise, W and R would be interpreted identically under $\mathcal{I}$ and $\mathcal{J}$, and W and R would not be disjoint under $\mathcal{J}$, since m is an instance of both, thus contradicting $\mathcal{N}_T$. Now, assume that W has not changed. Then $\mathcal{J} \models E \sqsubseteq W$, since this held already for $\mathcal{I}$. However, $\mathcal{J} \not\models E \sqsubseteq R$, since $m \in E^\mathcal{J}$, but $m \notin R^\mathcal{J}$, due to the disjointness of W and R with respect to $\mathcal{J}$. Similarly, if we assume that R has not changed, it follows that $\mathcal{J} \models E \sqsubseteq R$, but $\mathcal{J} \not\models E \sqsubseteq W$. By Lemma 1 we conclude that $\mathcal{K} \diamond \mathcal{N}_T$ is not expressible in *DL-Lite*.

Analogously one can also show inexpressibility for $\mathcal{G}^s_{\sqsubseteq}$, $\mathcal{G}^a_{\sqsubseteq}$, and $\mathcal{G}^a_{\sharp}$. ∎

From the example we conclude our first inexpressibility result.

Theorem 3. *DL-Lite is not closed under TBox evolution wrt* $\mathcal{G}^s_{\sqsubseteq}$, $\mathcal{G}^s_{\sharp}$, $\mathcal{G}^a_{\sqsubseteq}$, *and* $\mathcal{G}^a_{\sharp}$.

With a similar argument one can show that the operator $\diamond_{M'}$ of Qi and Du [8] (and its stratified extension $\diamond_S$), is not expressible in *DL-Lite*. This operator is a variant of $\mathcal{G}^s_{\sharp}$ where in Equation (1) one considers only models $\mathcal{J} \in Mod(\mathcal{N})$ that satisfy $A^{\mathcal{J}} \neq \emptyset$ for every A occurring in $\mathcal{K} \cup \mathcal{N}$. The modification does not affect the inexpressibility, which can again be shown using Example 2. We note that $\diamond_{M'}$ was developed for KB revision with empty ABoxes and the inexpressibility comes from the non empty ABox.

Local Model-Based Approaches. We start with an example showing that both ABox and TBox evolution wrt the $\mathcal{L}^a_{\sqsubseteq}$ and $\mathcal{L}^a_{\sharp}$ semantics are not expressible in *DL-Lite*.

Example 4. We turn again to our KB $\mathcal{K}_{ex}$ and consider the scenario where we are informed that John is now a single, formally $\mathcal{N}_A = \{S(j)\}$. Suppose we want to perform ABox evolution where the TBox of $\mathcal{K}_{ex}$ is protected. The TBox assertions essential for this example are $W \sqsubseteq \exists hh$, $\exists hh \sqsubseteq W$, and $P \sqsubseteq \neg \exists hh^-$, that is, an individual is a wife iff she has a husband, and a priests is not a husband. The essential ABox assertions are $W(m)$, $P(a)$, and $P(b)$. We show the inexpressibility of evolution wrt $\mathcal{L}^a_{\sqsubseteq}$ using Lemma 1.

Under $\mathcal{L}^a_{\sqsubseteq}$, in every $\mathcal{J} \in \mathcal{K} \diamond \mathcal{N}_A$ one of four situations holds: *(i)* Mary is not a wife, that is, $\mathcal{J} \not\models W(m)$, and both Adam and Bob are priests, that is, $\mathcal{J} \models P(a) \wedge P(b)$. Hence, $\mathcal{J} \models P(a) \vee P(b)$. *(ii)* Mary is a wife and her husband is different from Adam and Bob. Due to minimality of change, both Adam and Bob are still priests, as in Case (i), and again $\mathcal{J} \models P(a) \vee P(b)$. *(iii)* Mary is a wife and her husband is Adam. Then Bob, due to mininality of change, is still a priest. Hence, $\mathcal{J} \models P(a) \vee P(b)$. Moreover, the new husband cannot stay priest any longer and $\mathcal{J} \not\models P(a)$. *(iv)* Mary is a wife and her husband is Bob. Analogously to Case (iii), we have $\mathcal{J} \models P(a) \vee P(b)$ and $\mathcal{J} \not\models P(b)$. We are in the conditions of Lemma 1, that is, for every model $\mathcal{J} \in \mathcal{K} \diamond \mathcal{N}_A$ it holds that $\mathcal{J} \models P(a) \vee P(b)$, and there are $\mathcal{J}' \in \mathcal{K} \diamond \mathcal{N}'_A$ s.t. $\mathcal{J}' \not\models P(a)$ and $\mathcal{J}'' \in \mathcal{K} \diamond \mathcal{N}_A$ s.t. $\mathcal{J}'' \not\models P(b)$. Consequently, the set of models $\mathcal{K} \diamond \mathcal{N}_A$ is not expressible in *DL-Lite*.

To show that the example works also for $\mathcal{L}^a_{\sharp}$, we need extra arguments. Intuitively, if a model $\mathcal{I} \models \mathcal{K}_{ex}$ contains individuals that are single, but not clerics, then the models $\mathcal{J} \models \mathcal{N}_A$ closest to $\mathcal{I}$ in terms of $dist_{\sharp}$ are such that Mary, if she remains a wife, marries one of these individuals and Adam and Bob remain priests, since this involves the fewest changes of atoms. However, this is no more the case if we consider a model $\mathcal{I}_0 \models \mathcal{K}_{ex}$ where everyone, except John, is a priest, that is $P^{\mathcal{I}_0} = \Delta \setminus \{j\}$. Reasoning as before, one can see that among the models $\mathcal{J}$ of $\mathcal{N}_A$ closest to $\mathcal{I}_0$, there are some such that $\mathcal{J} \not\models P(a)$ and others such that $\mathcal{J} \not\models P(b)$, while all of them satisfy $\mathcal{J} \models P(a) \vee P(b)$. Then Lemma 1 implies that $\mathcal{K} \diamond \mathcal{N}_A$ under $\mathcal{L}^a_{\sharp}$ is not expressible in *DL-Lite*.

Now, we consider TBox evolution, which means that the ABox of $\mathcal{K}_{ex}$ is protected. Suppose we found out that ministers are clerics, formally $\mathcal{N}_T = \{M \sqsubseteq C\}$. The assertions of $\mathcal{K}_{ex}$ essential for this example are $C \sqsubseteq S$ and $M(c)$. Assume there is a representation $\mathcal{K}'$ of the $\mathcal{K}_{ex} \diamond \mathcal{N}_T$ under $\mathcal{L}^a_{\sqsubseteq}$. Since $\mathcal{K}_1 = \mathcal{K}_{ex} \cup \mathcal{N}_T$ is fully satisfiable, one might expect that $\mathcal{K}' = \mathcal{K}_1$. It turns out this is not the case. Indeed, since every model $\mathcal{J} \in \mathcal{K}_{ex} \diamond \mathcal{N}_T$ is such that $\mathcal{J} \models \mathcal{N}_T \cup \{M(c)\}$, it holds that $c \in M^{\mathcal{J}} \subseteq C^{\mathcal{J}}$. Moreover, if $\mathcal{I} \in Mod(\mathcal{K})$ is such that $c \notin S^{\mathcal{I}}$, then $c \notin S^{\mathcal{J}'}$ for any $\mathcal{J}' \in Mod(\mathcal{N}_T)$ minimally different from $\mathcal{I}$. At the same time $\mathcal{K}_1 \models S(c)$, hence, such a $\mathcal{J}'$ is not a model of $\mathcal{K}_1$ and $\mathcal{K}_1$ cannot be $\mathcal{K}'$. Since the inclusion $C \sqsubseteq S$ caused the problem above, it might be the case that $\mathcal{K}'$ is $\mathcal{K}_2 = \mathcal{K}_1 \setminus \{C \sqsubseteq S\}$. It turns out this is not the case either, since $\mathcal{K}_2$ has models that are not in $\mathcal{K} \diamond \mathcal{N}_T$. Can we resolve this by adding some assertion to $\mathcal{K}_2$? No, again. If one adds *any DL-Lite* TBox assertion to $\mathcal{K}_2$ that is not entailed by $\mathcal{K}_2$ or not $C \sqsubseteq S$, one gets a KB with models not in $\mathcal{K} \diamond \mathcal{N}_T$. Hence, no representation $\mathcal{K}'$ of $\mathcal{K} \diamond \mathcal{N}_T$ exists. Analogously, one can show that $\mathcal{K} \diamond \mathcal{N}_T$ under $\mathcal{L}_{\sharp}^a$ is also not expressible in *DL-Lite*.   ∎

This example proves our second inexpressibility result, which follows.

Theorem 5. *DL-Lite is not closed under evolution wrt $\mathcal{L}_{\sqsubseteq}^a$ and $\mathcal{L}_{\sharp}^a$. This holds already for the special cases of TBox evolution and ABox evolution with protected TBox.*

De Giacomo et al. [21] considered ABox evolution with protected TBox wrt $\mathcal{L}_{\sqsubseteq}^a$ semantics. They presented an algorithm to compute *DL-Lite$_{\mathcal{FR}}$* KBs that represent $\mathcal{K} \diamond \mathcal{N}_A$ for *DL-Lite$_{\mathcal{FR}}$* KBs $\mathcal{K}$ and $\mathcal{N}_A$. As a consequence of Theorem 5, their algorithm is not complete.

A strange effect of evolution under $\mathcal{L}_{\sharp}^a$ semantics is that new information may "erase" completely the previous KB.

Proposition 6. *Let $\mathcal{K}$ be a KB with at least one finite model and let $\mathcal{N}$ be a satisfiable KB such that all its models are infinite. Then under $\mathcal{L}_{\sharp}^a$ we have that $\mathcal{K} \diamond \mathcal{N} = \mathcal{N}$.*

Since the *DL-Lite* logics without role functionality have the finite model property, that is, every satisfiable KB in these logics has a finite model, the above situation cannot occur for them. At the same time, in every *DL-Lite* logic with role functionality there are KBs all of whose models are infinite and such an erasure can take place.

The properties of the $\mathcal{L}_{\sqsubseteq}^s$ and $\mathcal{L}_{\sharp}^s$ semantics are still an open problem for us.

We now discuss conceptual problems with all the local semantics. Recall Example 4 for local MBAs $\mathcal{L}_{\sqsubseteq}^a$ and $\mathcal{L}_{\sharp}^a$. We note two problems. First, the divorce of Mary from John had a strange effect on the priests Bob and Adam. The semantics questions their celibacy and we have to drop the information that they are priests. This is counterintuitive, since Mary and her divorce have nothing to do with any of these priests. Actually, the semantics also erases from the KB assertions about all other people belonging to concepts whose instances are not married, since potentially each of them is Mary's new husband. Second, a harmless clarification introduced to the TBox that ministers are in fact clerics strangely affects the whole class of clerics. The semantics of evolution "requires" one to allow marriages for clerics. This appears also strange, because intuitively the clarification on ministers does not contradict by any means the celibacy of clerics.

Also the four global MBAs have conceptual problems that were exhibited in Example 2. The restriction on rent subsidies that cuts the payments for wives introduces a counterintuitive choice for employed wives. Under the symbol-based global semantics, they must either *collectively* get rid of their husbands or *collectively* lose the subsidy. Under atom-based semantics the choice is an individual one.

Summing up on both the global and the local MBAs that we have considered, they focus on minimal change of *models* of KBs and, hence, introduce choices that cannot be captured in *DL-Lite*, which owes its good computational properties to the absence of disjunction. This mismatch with regard to the structural properties of KBs leads to counterintuitive and undesired results, like inexpressibility in *DL-Lite* and erasure of the entire KB. Therefore, we think that these semantics are not suitable for the evolution of *DL-Lite* KBs, whether or not they satisfy **EP1-EP3**, and now study evolution according to formula-based approaches.

3.2 Formula-Based Approaches

Under formula-based approaches, the objects of change are sets of formulas. Given a KB $\mathcal{K}$ and new knowledge $\mathcal{N}$, a natural way to define the result of evolution seems to choose a maximal subset $\mathcal{K}_m$ of $\mathcal{K}$ that is consistent with $\mathcal{N}$. The result of evolution in this case is a set of formulas $\mathcal{K} \diamond \mathcal{N} = \mathcal{K}_m \cup \mathcal{N}$. However, a problem with this is that in general such a $\mathcal{K}_m$ is not unique.

Let $\mathcal{M}(\mathcal{K}, \mathcal{N})$ be the set of all such maximal $\mathcal{K}_m$. In the past, researchers have proposed a number of approaches to combine all elements of $\mathcal{M}(\mathcal{K}, \mathcal{N})$ into one set of formulas, which is then added to $\mathcal{N}$ [5, 6]. The two main ones are known as *Cross-Product*, or *CP* for short, and *When In Doubt Throw It Out*, or *WIDTIO* for short. The corresponding sets $\mathcal{K}_{CP}$ and $\mathcal{K}_{WIDTIO}$ are defined as follows:

$$\mathcal{K}_{CP} := \left\{ \bigvee_{\mathcal{K}_m \in \mathcal{M}(\mathcal{K}, \mathcal{N})} \left(\bigwedge_{\phi \in \mathcal{K}_m} \phi \right) \right\}. \qquad \mathcal{K}_{WIDTIO} := \bigcap_{\mathcal{K}_m \in \mathcal{M}(\mathcal{K}, \mathcal{N})} \mathcal{K}_m,$$

In CP one adds to $\mathcal{N}$ the disjunction of all $\mathcal{K}_m$, viewing each $\mathcal{K}_m$ as the conjunction of its assertions, while in WIDTIO one adds to $\mathcal{N}$ those formulas present in all $\mathcal{K}_m$. In terms of models, every model of $\mathcal{K}_{WIDTIO}$ is also a model of $\mathcal{K}_{CP}$, whose models are exactly the interpretations satisfying *some* of the $\mathcal{K}_m$.

Example 7. We consider again our running example. Suppose, we obtain the new information that priests no longer obtain rental subsidies. This can be captured by the set of TBox assertions $\mathcal{N}_T = \{P \sqsubseteq \neg R\}$. We now incorporate this information into our KB, under both CP and WIDTIO semantics. Clearly, $\mathcal{K}_{ex} \cup \mathcal{N}_T$ is not coherent and to resolve the conflict one can drop either $P \sqsubseteq C$ or $C \sqsubseteq R$. Hence, $\mathcal{M}(\mathcal{K}_{ex}, \mathcal{N}_T) = \{\mathcal{K}_m^{(1)}, \mathcal{K}_m^{(2)}\}$, where $\mathcal{K}_m^{(1)} = \mathcal{K} \setminus \{P \sqsubseteq C\}$, and $\mathcal{K}_m^{(2)} = \mathcal{K} \setminus \{C \sqsubseteq R\}$. Consequently, the results of evolving $\mathcal{K}$ with respect to $\mathcal{N}_T$ under the two semantics are

$$\mathcal{N}_T \cup \mathcal{K}_{CP} = \mathcal{N}_T \cup ((\mathcal{K} \setminus \{P \sqsubseteq C\}) \vee (\mathcal{K} \setminus \{C \sqsubseteq R\})) \qquad (3)$$

$$\mathcal{N}_T \cup \mathcal{K}_{WIDTIO} = \mathcal{N}_T \cup \left(\mathcal{K}_m^{(1)} \cap \mathcal{K}_m^{(2)}\right) = \mathcal{N}_T \cup \mathcal{K}_{ex} \setminus \{P \sqsubseteq C, C \sqsubseteq R\},$$

where in (3) we have combined DL notation with first order logic notation.　■

Intuitively, CP does not lose information, but the price to pay is that the resulting KB can be exponentially larger than the original KB, since there can exist exponentially many $\mathcal{K}_m$. In addition, as the example shows, even if $\mathcal{K}$ is a *DL-Lite* KB, the resulting $\mathcal{K}_{CP}$ may not be representable in *DL-Lite* anymore since it requires disjunction. This effect is also present if the new knowledge involves only ABox assertions.

WIDTIO, on the other extreme, is expressible in *DL-Lite*. However, it can lose many assertions, which may be more than one is prepared to tolerate. Even, if one deems this loss acceptable, one has to cope with the fact that it is generally difficult to decide whether an assertion belongs to $\mathcal{K}_{WIDTIO}$. This problem is already difficult if our KBs are TBoxes that are specified in the simplest variant of *DL-Lite*. We note that the following theorem can be seen as a sharpening of a result about WIDTIO for propositional Horn theories in [5], obtained with a different reduction than ours.

Theorem 8. *Given DL-Lite TBoxes $\mathcal{T}$ and $\mathcal{N}_T$ and an inclusion assertion $A \sqsubseteq B$, deciding whether $A \sqsubseteq B \in \bigcap_{\mathcal{K}_m \in \mathcal{M}(\mathcal{T}, \mathcal{N}_T)} \mathcal{K}_m$, is* coNP-*complete. Hardness holds already for DL-Lite$_{core}$.*

Against this backdrop we conclude that neither CP nor WIDTIO are good for practical solutions. As a pragmatic alternative we will explore the approach to nondeterministically choose *some* $\mathcal{K}_m^{(0)}$ among the $\mathcal{K}_m$. We call this semantics *bold semantics*.

4   Bold Semantics

We define as *bold semantics* the approach to evolution where, given a KB $\mathcal{K} = \mathcal{T} \cup \mathcal{A}$ and new knowledge $\mathcal{N}$, we add to $\mathcal{N}$ a *maximal compatible* subset $\mathcal{K}_m^{(0)} \subseteq cl(\mathcal{K})$, that is, a set such that $\mathcal{N} \cup \mathcal{K}_m^{(0)}$ is coherent and such that that $\mathcal{K}_m^{(0)}$ is maximal wrt to this property. Note that now we choose a subset of the *deductive closure* of $\mathcal{K}$ and not of $\mathcal{K}$ alone. By abuse of notation, we will use a binary operator to denote any result of bold evolution and write

$$\mathcal{K} \diamond_b \mathcal{N} = \mathcal{N} \cup \mathcal{K}_m^{(0)},$$

although $\mathcal{K} \diamond_b \mathcal{N}$ is not uniquely defined.

Example 9. Consider the KB and the update request from Example 7. As shown there, $\mathcal{M}(\mathcal{K}_{ex}, \mathcal{N}_T) = \{\mathcal{K}_m^{(1)}, \mathcal{K}_m^{(2)}\}$. According to bold semantics the result of the update is a KB $\mathcal{K}' = \mathcal{N} \cup \mathcal{K}_m^{(0)}$ for some $\mathcal{K}_m^{(0)} \in \mathcal{M}(\mathcal{K}_{ex}, \mathcal{N}_T)$. Thus, the result of the update is either $\mathcal{N}_T \cup \mathcal{K}_{ex} \setminus \{P \sqsubseteq C\}$ or $\mathcal{N}_T \cup \mathcal{K}_{ex} \setminus \{C \sqsubseteq R\}$. Whether to select one or the other of these two options depends on preferences, which we do not consider here.   ■

Choosing an arbitrary $\mathcal{K}_m$ has the advantage that $\mathcal{K} \diamond_b \mathcal{N}$ can be computed in polynomial time. In Fig. 1 we present a nondeterministic algorithm that, given a KB $\mathcal{K}$ and new knowledge $\mathcal{N}$, returns a set $\mathcal{K}_m \subseteq cl(\mathcal{K})$ that is a maximal compatible set of assertions for $\mathcal{K}$ and $\mathcal{N}$. The algorithm loops as many times as there are assertions in $cl(\mathcal{K})$. The number of such assertions is at most quadratic in the number of constants, atomic concepts, and roles. The crucial step is the check for coherence, which is performed once per loop. If this test is polynomial in the size of the input then the entire runtime of the algorithm is polynomial. For *DL-Lite$_{\mathcal{FR}}$* TBoxes $\mathcal{T}$, coherence can be checked in time quadratic in the number of assertions in the TBox, that is, $O(|\mathcal{T}|^2)$. Satisfiability of an ABox $\mathcal{A}$ with respect to $\mathcal{T}$ can be checked in time $O(|\mathcal{T}|^2 \times |\mathcal{A}|)$, where $|\mathcal{A}|$ is the number of assertions of $\mathcal{A}$. The $O(|\mathcal{T}|^2)$ complexity can be shown by reduction to satisfiability of sets of propositional Horn clauses (see [22] for details).

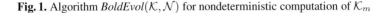

```
INPUT:      KBs K and N
OUTPUT:     a set Kₘ ⊆ cl(K) of TBox and ABox assertions
[1]   Kₘ := N; S := cl(K)
[2]   repeat
[3]      choose some φ ∈ S;  S := S \ {φ}
[4]      if {φ} ∪ Kₘ is coherent then Kₘ := Kₘ ∪ {φ}
[5]   until S = ∅
```

Fig. 1. Algorithm *BoldEvol*$(\mathcal{K}, \mathcal{N})$ for nondeterministic computation of $\mathcal{K}_m$

Theorem 10. *The algorithm BoldEvol runs in polynomial time and computes evolution wrt bold semantics, that is, $\mathcal{K} \diamond_b \mathcal{N} = BoldEvol(\mathcal{K}, \mathcal{N})$.*

This shows that bold semantics has the great advantage that evolution can be computed in polynomial time. However, its nondeterminism is a disadvantage. Clearly, we can *avoid* nondeterminism if we impose a linear order on the assertions in $cl(\mathcal{K})$, and let *BoldEvol* choose them in this order. The question how to define such an order depends on the characteristics of the application, and we cannot discuss it here.

One may wonder whether it is possible to efficiently compute a $\mathcal{K}_m$ with maximal *cardinality*. (Recall that our algorithm is only guaranteed to compute a $\mathcal{K}_m$ that is maximal wrt set inclusion.) Unfortunately, it turns out, using various reductions from the Independent Set problem, that under this requirement computation is hard, even for $\mathcal{K}$ and $\mathcal{N}$ that consist only of TBox or only of ABox assertions, except when both, $\mathcal{K}$ and $\mathcal{N}$, are ABoxes, in which case no conflicts can arise.

Theorem 11. *Given DL-Lite KBs $\mathcal{K}$ and $\mathcal{N}$ and an integer $k > 0$, to decide whether there exists a subset $\mathcal{K}_0 \subseteq \mathcal{K}$ such that $\mathcal{K}_0 \cup \mathcal{N}$ is coherent and $|\mathcal{K}_0| \geq k$ is NP-complete. NP-hardness already holds for DL-Lite$_{core}$ if (1) both $\mathcal{K}$ and $\mathcal{N}$ are TBoxes, or (2) $\mathcal{K}$ is an ABox and $\mathcal{N}$ is a TBox, or (3) $\mathcal{K}$ is a TBox and $\mathcal{N}$ is an ABox.*

In the next section we will see that nondeterminism is not present in ABox evolution with a protected TBox and that there is always a single maximal compatible ABox.

5   ABox Evolution

We study ABox evolution assuming that the new knowledge $\mathcal{N}_A$ is satisfiable with the old TBox $\mathcal{T}$, may only conflict with the old ABox $\mathcal{A}$, and that $\mathcal{T}$ is protected.

ABox Evolution under Bold Semantics. In *DL-Lite*, unsatisfiability of a KB is caused either by a single ABox assertion, which will be a membership assertion for an unsatisfiable concept or role, or by a pair of assertions contradicting either a disjointness or a functionality assertion of the TBox.

Lemma 12. *Let $\mathcal{T} \cup \mathcal{A}$ be a DL-Lite KB. If $\mathcal{T} \cup \mathcal{A}$ is unsatisfiable, then there is a subset $\mathcal{A}_0 \subseteq \mathcal{A}$ with at most two elements, such that $\mathcal{T} \cup \mathcal{A}_0$ is unsatisfiable.*

```
INPUT:      TBox T, and ABoxes A, D, each satisfiable with T
OUTPUT:     finite set of membership assertions A^w
[1]  A^w := A
[2]  for each B_1(c) ∈ D do
[3]      A^w := A^w \ {B_1(c)} and
[4]      for each B_2 ⊑ B_1 ∈ cl(T) do A^w := A^w \ {B_2(c)}
[5]          if B_2(c) = ∃R(c) then
[6]              for each R(c,d) ∈ A^w do D := D ∪ {R(c,d)}
[7]  for each R_1(a,b) ∈ D do
[8]      A^w := A^w \ {R_1(a,b)} and
[9]      for each R_2 ⊑ R_1 ∈ cl(T) do A^w := A^w \ {R_2(a,b)}
```

Fig. 2. Algorithm $Weeding(T, A, D)$ for $DL\text{-}Lite_{\mathcal{FR}}$

```
INPUT:      TBox T, and ABoxes A, N_A, each satisfiable with T
OUTPUT:     finite set of membership assertions A_m
[1]  A_0 := cl_T(A ∪ N_A), N_A := cl_T(N_A), CA := ∅
[2]  for each B ⊑ ¬B' ∈ cl(T) do
[3]      if {B(c), B'(c)} ⊆ A_0 then
[4]          if B(c) ∉ N_A then CA := CA ∪ {B(c)}
[5]          otherwise CA := CA ∪ {B'(c)}
[6]  for each (funct R) ∈ T do
[7]      if {R(a,b), R(a,c)} ⊆ A_0 then
[8]          if R(a,b) ∉ N_A then CA := CA ∪ {R(a,b)}
[9]          otherwise CA := CA ∪ {R(a,c)}
[10] A_m := Weeding(T, cl_T(A), CA)
```

Fig. 3. Algorithm $FastEvol(A, N_A, T)$ for $DL\text{-}Lite_{\mathcal{FR}}$

The lemma implies that if $T \cup N_A \cup A$ is unsatisfiable, then there are two assertions $\phi \in N_A$ and $\psi \in A$ such that $T \cup \{\phi, \psi\}$ is unsatisfiable. In other words, whether or not $\psi \in A$ needs to be eliminated from A as a result of evolution depends on ψ alone. As a consequence, ABox evolution wrt bold semantics is deterministic.

Theorem 13. *The result of ABox evolution* $(T \cup A) \diamond_b N_A$ *is uniquely defined.*

In principle, *BoldEvol* can be used to compute ABox evolution and regardless of the order in which it selects the assertions, it will always return the same result, due to Theorem 13. A drawback of *BoldEvol* is that it performs a coherence check during each loop, which is not needed in that form, since ABox evolution does not affect coherence of the TBox. We exhibit now a new algorithm *FastEvol* that replaces the coherence check with implicit satisfiability checks.

The algorithm *FastEvol* computes the set $A_m \subseteq cl_T(A)$ of all assertions that do not conflict with T and N and is based on Lemma 12. It exploits the algorithm *Weeding* (see Fig. 2), which takes as input T, A, and a set D of membership assertions to be "deleted" from A. For every assertion $\phi \in D$, *Weeding* deletes from A ϕ and also all the assertions that T-entail ϕ. The algorithm *FastEvol* (see Fig. 3) takes as input T, A, and N_A. It detects assertions in the closure of $\mathcal{A} \cup \mathcal{N}_A$ that *conflict* with the new data and stores them in CA. Finally, it resolves the conflicts by deleting CA from $\mathcal{A}$ using *Weeding*.

Theorem 14. *The algorithm FastEvol computes ABox evolution wrt bold semantics, that is, $(\mathcal{T} \cup \mathcal{A}) \diamond_b \mathcal{N}_A = \mathcal{T} \cup \mathcal{N}_A \cup FastEvol(\mathcal{K}, \mathcal{N}_A)$, and runs in polynomial time.*

Note that, although *FastEvol* may look similar to the algorithm *ComputeUpdate* in [21], it is actually different. Our algorithm always keeps at least as many assertions as *ComputeUpdate*. In some cases, however, *ComputeUpdate* drops an existential restriction of the form $\exists R(a)$, although it would not cause a contradiction.

Careful Semantics. We start with an example illustrating drawbacks of bold semantics. Apparently, the drawbacks come from the minimality of evolution principle **EP3**.

Example 15. Coming back to $\mathcal{K}_{ex}$, consider evolution wrt bold semantics for the news that John is getting single, formally, $\mathcal{N}_A = \{S(j)\}$. One can see that the only assertion to be dropped from $\mathcal{K}_{ex}$ is that John is the husband of Mary, that is, $\mathcal{K}_{ex} \diamond_b \mathcal{N}_A = \mathcal{K}_{ex} \cup \mathcal{N}_A \setminus \{hh(m, j)\}$. This implies that $\mathcal{K}_{ex} \diamond_b \mathcal{N}_A \models W(m)$ and, consequently, Mary *still has a husband* who is not John, despite the divorce with John, that is, $\mathcal{K}_{ex} \diamond_b \mathcal{N}_A \models \phi$, where $\phi = \exists x(hh(m, x) \wedge (x \neq j))$. The only option that bold semantics offers to Mary is to find another husband immediately after the divorce. It does not consider it an option for her to become single. We are interested in a semantics that allows for both possibilities. Note that the entailment $\mathcal{K}_{ex} \diamond_b \mathcal{N}_A \models \phi$ is *unexpected* in the sense that neither $\mathcal{K}_{ex}$ nor $\mathcal{N}_A$ entail ϕ, that is, $\mathcal{K}_{ex} \not\models \phi$ and $\mathcal{N}_A \not\models \phi$ hold. ∎

As the example shows, the situation when the result of evolution entails *unexpected information*, that is, information coming neither from the original KB, nor from the new knowledge, may be counterintuitive. In our example, the unexpected information is the formula $\exists x(hh(m, x) \wedge (x \neq j))$, which has a specific form: it restricts the possible values in the second component of the role hh. Our next semantics prohibits these role restrictions from being unexpectedly entailed from the result of evolution.

We say that a formula is *role-constraining*, or an *RCF* for short, if it is of the form $\exists x(R(a, x) \wedge (x \neq c_1) \wedge \cdots \wedge (x \neq c_n))$, where a and all c_i are constants. Let $\mathcal{T}$ be a TBox, and $\mathcal{A}$, $\mathcal{N}_A$ be ABoxes. A subset $\mathcal{A}_1 \subseteq \mathcal{A}$ is *careful* if for every RCF φ, whenever $\mathcal{A}_1 \cup \mathcal{N}_A \models_\mathcal{T} \varphi$ holds, either $\mathcal{A}_1 \models_\mathcal{T} \varphi$ or $\mathcal{N}_A \models_\mathcal{T} \varphi$ holds.

Theorem 16. *Let $\mathcal{T}$ be a DL-Lite TBox and $\mathcal{A}$, $\mathcal{N}_A$ DL-Lite ABoxes, and suppose that both $\mathcal{T} \cup \mathcal{A}$ and $\mathcal{T} \cup \mathcal{N}_A$ are satisfiable. Then, the set*

$$\{\mathcal{A}_0 \subseteq cl_\mathcal{T}(\mathcal{A}) \mid \mathcal{A}_0 \text{ is careful and } \mathcal{A}_0 \cup \mathcal{N}_A \text{ is } \mathcal{T}\text{-satisfiable}\}$$

has a unique maximal element wrt set inclusion.

We can exploit the maximal set $\mathcal{A}_m^c$ of assertions (where c stands for careful), whose uniqueness is guaranteed by Theorem 16, to define the *careful evolution*:

$$(\mathcal{T} \cup \mathcal{A}) \diamond_c \mathcal{N}_A := \mathcal{T} \cup \mathcal{N}_A \cup \mathcal{A}_m^c. \tag{4}$$

INPUT: TBox $\mathcal{T}$, and ABoxes $\mathcal{A}$, $\mathcal{N}_A$, each satisfiable with $\mathcal{T}$
OUTPUT: finite set of membership assertions $\mathcal{A}_m^c$
[1] $\mathcal{A}_m^c := FastEvol(\mathcal{T}, \mathcal{A}, \mathcal{N}_A)$, $UF := \emptyset$
[2]   **for each** $\exists R(a) \in Precl_{\mathcal{T}}(\mathcal{N}_A)$ **do**
[3]   **if** $R(a, b) \notin \mathcal{A}_m^c$ for every b **then**
[4]   **for each** $\exists R^- \sqsubseteq \neg C \in cl(\mathcal{T})$ **do**
[5]   **for each** $C(d) \in cl_{\mathcal{T}}(\mathcal{A}) \setminus cl_{\mathcal{T}}(\mathcal{N}_A)$ **do**
[6] $UF := UF \cup \{C(d)\}$
[7]   **for each** $\exists R(a) \in \mathcal{A}_m^c \setminus Precl_{\mathcal{T}}(\mathcal{N}_A)$ **do**
[8]   **if** $R(a, b) \notin \mathcal{A}_m^c$ for every b **then**
[9]   **if** there is a concept C in $\mathcal{T} \cup \mathcal{A} \cup \mathcal{N}_A$ s.t.
[10] $(\exists R^- \sqsubseteq \neg C) \in cl(\mathcal{T})$ **and** $C(d) \in cl_{\mathcal{T}}(\mathcal{N}_A) \setminus cl_{\mathcal{T}}(\mathcal{A})$ for some d **then**
[11] $UF := UF \cup \{\exists R(a)\}$
[12] $\mathcal{A}_m^c := Weeding(\mathcal{T}, \mathcal{A}_m^c, UF)$

Fig. 4. Algorithm $CarefulEvol(\mathcal{T}, \mathcal{A}, \mathcal{N}_A)$ for *DL-Lite*$_{\mathcal{FR}}$

One can see that, by its definition, careful semantics satisfies the principles **EP1**, **EP2**, and **EP3**, where for **EP3** the minimality should take into account carefulness. We exhibit now the algorithm *CarefulEvol*, which computes the uniquely determined set $\mathcal{A}_m^c$ of Equation (4). The *preclosure of $\mathcal{A}$ wrt $\mathcal{T}$*, denoted $Precl_{\mathcal{T}}(\mathcal{A})$, is a subset of $cl_{\mathcal{T}}(\mathcal{A})$ obtained as follows: one removes from $cl_{\mathcal{T}}(\mathcal{A})$ all the assertions of the form $\exists R(a)$, whenever there is an assertion of the form $R(a, c)$ in $cl_{\mathcal{T}}(\mathcal{A})$, for some constant c. The preclosure is needed to detect unexpected RCFs. The algorithm *CarefulEvol* (see Fig. 4) takes as input ABoxes $\mathcal{A}$, $\mathcal{N}_A$, and a TBox $\mathcal{T}$. It first computes the evolution wrt bold semantics. Then, it computes the set UF of assertions that cause *unexpectedness* in $FastEvol(\mathcal{T}, \mathcal{A}, \mathcal{N}_A)$ and belong to $cl_{\mathcal{T}}(\mathcal{A})$. Then it removes UF from $FastEvol(\mathcal{T}, \mathcal{A}, \mathcal{N}_A)$ by means of *Weeding*.

Theorem 17. *The algorithm CarefulEvol computes ABox evolution wrt careful semantics, that is, $(\mathcal{T} \cup \mathcal{A}) \diamond_c \mathcal{N}_A = \mathcal{T} \cup \mathcal{N}_A \cup CarefulEvol(\mathcal{T}, \mathcal{A}, \mathcal{N}_A)$ and runs in polynomial time.*

Again, *CarefulEvol* differs from *ComputeUpdate* in [21]. Sometimes the first may drop an existential restriction $\exists R(a)$ and the second keep it, while sometimes it may be the other way round.

We defer a detailed discussion on how bold and careful semantics are related to the classical update and revision postulates of [4] to an extended version of this paper.

6 Conclusion

We studied evolution of *DL-Lite* KBs. There are two main families of approaches to evolution: model-based and formula-based ones. We singled out and investigated a three-dimensional space of model-based approaches, and proved that most of them are not appropriate for *DL-Lite* due to their counterintuitive behavior and the inexpressibility of evolution results. Thus, we examined formula-based approaches, showed that the classical ones are again inappropriate for *DL-Lite*, and proposed a novel bold semantics. We showed that this semantics can be computed in polynomial time, but the result is, in general, non-deterministic. Then, we studied ABox evolution under bold semantics and showed that the result in this case is unique. We developed a polynomial time algorithm for DL-Lite KB evolution under this semantics, and an alternative optimized one for ABox evolution. We presented a conceptual drawback of ABox evolution under bold semantics and introduced careful semantics, which repairs the drawback. For this approach we proved that the evolution result is unique and developed a polynomial time algorithm to compute it.

Acknowledgements. The authors are partially supported by the EU projects ACSI (FP7-ICT-257593) and Ontorule (FP7-ICT-231875). The second author is also supported by the European Research Council grant Webdam (under FP7), agreement n. 226513.

References

1. Borgida, A., Brachman, R.J.: Conceptual modeling with description logics. In: [13], ch.10, pp. 349–372
2. Flouris, G., Manakanatas, D., Kondylakis, H., Plexousakis, D., Antoniou, G.: Ontology change: Classification and survey. Knowledge Engineering Review 23(2), 117–152 (2008)
3. Abiteboul, S., Grahne, G.: Update semantics for incomplete databases. In: Proc. of VLDB 1985 (1985)
4. Katsuno, H., Mendelzon, A.: On the difference between updating a knowledge base and revising it. In: Proc. of KR 1991, pp. 387–394 (1991)
5. Eiter, T., Gottlob, G.: On the complexity of propositional knowledge base revision, updates and counterfactuals. Artificial Intelligence 57, 227–270 (1992)
6. Winslett, M.: Updating Logical Databases. Cambridge University Press, Cambridge (1990)
7. Flouris, G.: On belief change in ontology evolution. AI Communications 19(4) (2006)
8. Qi, G., Du, J.: Model-based revision operators for terminologies in description logics. In: Proc. of IJCAI 2009, pp. 891–897 (2009)
9. Peters, R.J., Özsu, M.T.: An axiomatic model of dynamic schema evolution in objectbase systems. ACM Trans. on Database Systems 22(1), 75–114 (1997)
10. De Giacomo, G., Lenzerini, M., Poggi, A., Rosati, R.: On instance-level update and erasure in description logic ontologies. J. of Logic and Computation, Special Issue on Ontology Dynamics 19(5), 745–770 (2009)
11. Liu, H., Lutz, C., Milicic, M., Wolter, F.: Updating description logic ABoxes. In: Proc. of KR 2006, pp. 46–56 (2006)
12. Calvanese, D., De Giacomo, G., Lembo, D., Lenzerini, M., Rosati, R.: Tractable reasoning and efficient query answering in description logics: The *DL-Lite* family. J. of Automated Reasoning 39(3), 385–429 (2007)
13. Baader, F., Calvanese, D., McGuinness, D., Nardi, D., Patel-Schneider, P.F. (eds.): The Description Logic Handbook: Theory, Implementation and Applications. Cambridge University Press, Cambridge (2003)
14. Artale, A., Calvanese, D., Kontchakov, R., Zakharyaschev, M.: The *DL-Lite* family and relations. J. of Artificial Intelligence Research 36, 1–69 (2009)
15. Poggi, A., Lembo, D., Calvanese, D., De Giacomo, G., Lenzerini, M., Rosati, R.: Linking data to ontologies. J. on Data Semantics X, 133–173 (2008)

16. Ginsberg, M.L., Smith, D.E.: Reasoning about action I: A possible worlds approach. Technical Report KSL-86-65, Knowledge Systems, AI Laboratory (1987)
17. Forbus, K.D.: Introducing actions into qualitative simulation. In: Proc. of IJCAI 1989 (1989)
18. Borgida, A.: Language features for flexible handling of exceptions in information systems. ACM Trans. on Database Systems 10(4), 565–603 (1985)
19. Dalal, M.: Investigations into a theory of knowledge base revision. In: Proc. of AAAI 1988, pp. 475–479 (1988)
20. Calvanese, D., Kharlamov, E., Nutt, W.: A proof theory for *DL-Lite*. In: Proc. of DL 2007. CEUR, vol. 250, pp. 235–242 (2007), `ceur-ws.org`
21. De Giacomo, G., Lenzerini, M., Poggi, A., Rosati, R.: On the update of description logic ontologies at the instance level. In: Proc. of AAAI 2006, pp. 1271–1276 (2006)
22. Zheleznyakov, D., Calvanese, D., Kharlamov, E., Nutt, W.: Updating TBoxes in *DL-Lite*. In: Proc. of DL 2010. CEUR, vol. 573, pp. 102–113 (2010), `ceur-ws.org`

Ontology Similarity in the Alignment Space

Jérôme David[1], Jérôme Euzenat[1], and Ondřej Šváb-Zamazal[2]

[1] INRIA & LIG
Grenoble, France
{Jerome.David,Jerome.Euzenat}@inrialpes.fr
[2] University of Economics
Prague, Czech Republic
ondrej.zamazal@vse.cz

Abstract. Measuring similarity between ontologies can be very useful for different purposes, e.g., finding an ontology to replace another, or finding an ontology in which queries can be translated. Classical measures compute similarities or distances in an ontology space by directly comparing the content of ontologies. We introduce a new family of ontology measures computed in an alignment space: they evaluate the similarity between two ontologies with regard to the available alignments between them. We define two sets of such measures relying on the existence of a path between ontologies or on the ontology entities that are preserved by the alignments. The former accounts for known relations between ontologies, while the latter reflects the possibility to perform actions such as instance import or query translation. All these measures have been implemented in the OntoSim library, that has been used in experiments which showed that entity preserving measures are comparable to the best ontology space measures. Moreover, they showed a robust behaviour with respect to the alteration of the alignment space.

1 Introduction

There are many uses for measuring the proximity between ontologies, such as finding a representation in which some assertions can be translated or queried. In [1], we compared distances between ontologies based on ontology content. In this paper, we extend this work by distinguishing between measures in an ontology space, obtained by comparing the content of ontologies, and measures in an alignment space, obtained with regard to how the ontologies are related by alignments.

We call alignment space a structure populated by ontologies related by alignments. An alignment expresses relations between entities in the ontologies [2]. More specifically, a distance or similarity measure is alignment-based if it is computed without relying on the content of ontologies, but only on that of the alignments. So, such measures can only be applied when alignments are available, but we assume that the semantic web will have the characteristic of such a space with many ontologies already available and some alignments, sometimes competing, between them.

Alignment space measures may seem more remote from the true distance between ontologies because they do not directly consider ontology content. However, there are cases in which they can be very useful. This is obviously the case when ontologies are not available, e.g., because they are on a closed server, but alignments between

P.F. Patel-Schneider et al. (Eds.): ISWC 2010, Part I, LNCS 6496, pp. 129–144, 2010.

these ontologies and others exist. Such unavailable ontologies may be used as a target ontology or as an intermediate ontology (and then alignments may be composed).

This is also the case when the similarity between ontologies has to reflect the ability to transform a statement or a query from one ontology to another, e.g., in semantic peer-to-peer systems or dynamic composition of semantic web services. Since alignment spaces are structured by actual alignments, an alignment space measure is indeed reflecting to some extent the capacity to translate ontology expressions. Such measures are as useful as they can be computed quickly with respect to a particular query or formula. On the other hand, distances in an ontology space only provide a measure of closeness, and an alignment or a mediator remains to be produced.

In addition, even if ontologies are available, such measures may be useful as approximations of the "real distance" which are easier to compute than comparing the ontologies: alignment-based measures can quickly provide a hint on what are the most promising options. Indeed, because they already provide the structure to compute the measure, alignments are faster to compare than elaborate comparison of two ontologies as a whole.

In this paper we investigate the design of proximity measures in alignment spaces. We introduce two families of measures and evaluate them with regard to other measures in ontology spaces. We show that some of these are worth considering.

In the remainder, we first briefly consider the work designed for measuring a distance or a similarity between ontologies (§2) showing that it is exclusively based on the content of ontologies. We then provide general definitions about ontologies, alignments and similarities (§3). This introduces alignment spaces. We then define two families of alignment space measures: measures based on paths (§4) and measures based on coverage (§5). Finally, we provide an experimental evaluation of these measures (§6), showing in particular that coverage-based measures behave comparably to the best ontology-based measures and that they are reasonably robust to data alteration. Complements to this work can be found in [3].

2   Related Works

Most of the work dealing with ontology measures [4,5,6] is in reality concerned with concept distances. Such measures are widely used in ontology matching algorithms [2].

[4] introduced a concept similarity based on terminological and structural aspects of ontologies. This very precise proposal combines an edit distance on strings and a structural distance on hierarchies (the cotopic distance). The ontology similarity strongly relies on the terminological similarity. OLA [7] uses a concept similarity for ontology matching. This measure takes advantage of most of the ontological aspects (labels, structure, extension) and selects the maximum similarity. It is thus a good candidate for ontology similarity. The framework presented in [8] provides a similarity combining string similarity, concept similarity – considered as sets – and similarity across usage traces. [5] presents an elaborate framework for comparing concepts in a vector space in which dimensions are primitive concepts. It is said to be extensible to ontologies as well.

Finally, [6] more directly considered metrics evaluating ontology quality. This is nevertheless one step towards semantic measures since they introduce normal forms for ontologies which could be used for developing syntactically neutral measures.

These works generally rely on elaborate distance or similarity measures between concepts and they extend these measures to distances between ontologies. This extension is often considered as straightforward, although, there are many ways to do so. In [1], we have explicitly proposed and evaluated a collection of ontology distances and similarities based on the comparison of the content of ontologies.

[9] investigated ontology agreement which is used as a measure for choosing compatible ontologies. It can be seen as another kind of distance or similarity between ontologies. However, the way agreement/disagreement is computed is still based on ontology content; alignments are only used for identifying connected entities which are not immediately comparable, hence they are neutral. Link frequency – inverse dataset frequency [10] is a "popularity" measure which relies on references between datasets. Although it does not consider explicitly alignments and is not meant to be a similarity, it uses techniques related to our coverage-based measures.

The present paper provides and evaluates measures which, contrary to all the cited ones, are based on alignments between ontologies, hence the term "alignment space".

3   Ontologies, Alignment Spaces and Similarities

In this section, we introduce the ingredients which will be used for defining alignment space measures: ontologies and alignments, alignment spaces and finally the notion of similarity.

3.1   Ontologies and Alignments

We will use very simple definitions of ontologies and alignments. In particular, we will consider an ontology o represented as a set of named entities $Q_L(o)$. These entities could be classes (C), properties (P) or individuals (I): $Q_L(o) = C \cup P \cup I$.

Alignments express correspondences between entities belonging to different ontologies. Here we will only use a simplified version of alignments; a more complete definition and discussion can be found in [2]. Simple alignments contain correspondences in which entities are the ontology vocabulary and the relations between entities are equivalence $(=)$ or subsumption $(\sqsubseteq, \sqsupseteq)$.

Definition 1 (Simple alignment). *Given two ontologies o and o', a simple alignment is a set of triples $\langle e, e', r \rangle$, such as:*

- *$e \in Q_L(o)$ and $e' \in Q_{L'}(o')$ are named entities issued from the ontologies;*
- *$r \in \{=, \sqsubseteq, \sqsupseteq\}$.*

The correspondence $\langle e, e', r \rangle$ asserts that the relation r holds between the ontology entities e and e'.

Example 1. In Figure 1, the alignments are as follows:

$$A_{1,2} \text{ is } \{\langle a_1, a_2, = \rangle, \langle b_1, b_2, = \rangle, \langle c_1, c_2, = \rangle\}$$
$$A_{1,3} \text{ is } \{\langle d_1, d_3, = \rangle, \langle e_1, e_3, = \rangle\}$$
$$A_{2,3} \text{ is } \{\langle c_2, c_3, = \rangle, \langle d_2, d_3, \sqsupseteq \rangle, \langle e_2, e_3, \sqsubseteq \rangle\}$$
$$A_{2,4} \text{ is } \{\langle a_2, a_4, = \rangle, \langle b_2, b_4, = \rangle, \langle c_2, c_4, = \rangle\}$$
$$A_{3,4} \text{ is } \{\langle c_3, c_4, = \rangle, \langle d_3, d_4, = \rangle, \langle e_3, e_4, = \rangle\}$$

We use the notation $A(s)$ for the action of replacing any ontology entity of a set of entities s by the term it is in correspondence through A if any (otherwise, the entity is simply skipped). More precisely, the replacement is performed if there is a unique correspondence for each entity in s with a relation belonging to a set of relations θ. Depending on the task for which the measure is performed θ may be different. For instance, if the task is to transform a query, then taking $\theta = \{=\}$ provides exact transformations. However, if completeness is not a concern but correctness is, using $\theta = \{=, \sqsupseteq\}$ provides more options for transforming entities which remain correct (because it selects a subclass of the initial one). This is the value of θ used in the examples.

Definition 2 (Application of an alignment). *Given A a functional alignment, i.e., an alignment in which each entity appears at most once, θ a set of relations and s a set of ontology entities, the application of A to s denoted by $A(s)$ is[1]:*

$$A(s) = \{e' | \exists! \langle e, e', r \rangle \in A \text{ such that } e \in s \wedge r \in \theta\}$$

Example 2. Given the alignments of Example 1, $A_{1,2}(\{a_1, c_1, e_1\}) = \{a_2, c_2\}$ Alignments can be used in both ways through the inverse operation ($^{-1}$), such that $\sqsubseteq^{-1}$ is $\sqsupseteq$, and $=^{-1}$ is $=$. For instance, $A_{2,3}^{-1} = \{\langle c_3, c_2, = \rangle, \langle d_3, d_2, \sqsubseteq \rangle, \langle e_3, e_2, \sqsupseteq \rangle\}$ can be used for converting queries from o_3 to o_2.

3.2   Alignment Space

We call "alignment space" a set of ontologies and a set of alignments between these ontologies. Measuring proximity in a frozen alignment space allows for grounding the measure on actual alignments instead of non existing potential alignments.

Definition 3 (Alignment space). *An alignment space $\langle \Omega, \Lambda \rangle$ is made of a set Ω of ontologies and a set Λ of simple alignments between ontologies in Ω. We denote as $\Lambda(o, o')$ the set of alignments in Λ between o and o'.*

An alignment space can be represented as a multigraph[2] $G_{\Omega,\Lambda}$ in which ontologies are vertices and alignments are edges. Figure 2 (left) represents the graph corresponding to the alignments and ontologies of Figure 1.

---

[1] The notation $\exists!$ stands for "there exists a unique".

[2] A multigraph is needed, because there may be several alignments available between two ontologies.

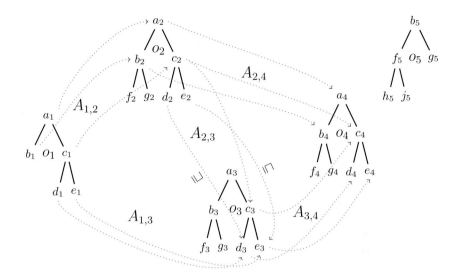

Fig. 1. Five ontologies (o_1, o_2, o_3, o_4 and o_5) and five alignments ($A_{1,2}$, $A_{1,3}$, $A_{2,3}$, $A_{2,4}$ and $A_{3,4}$)

It is possible to define the operation of inverse of an alignment ($^{-1}$), composition of two consecutive alignments ($\cdot$) and union of two alignments between the same ontologies ($\cup$) [11]. The inverse, composition or union closure of an alignment space is obtained by applying these operations to all possible (pairs of) alignments within the space until they do not generate any new alignments. The semantics of a closed space is the same as the initial space.

A path is simply defined as a path in $G_{\Omega,\Lambda}$.

Definition 4 (Path). *Given a set of alignments Λ, a path π in Λ is a finite sequence of alignments $A_1 \cdot \ldots A_n$ such that for each $i \in [1, n-1]$, $A_i \in \Lambda(o_i, o_i')$ and $A_{i+1} \in \Lambda(o_{i+1}, o_{i+1}')$, $o_i' = o_{i+1}$. The set of paths in an alignment space is named Π and the set of paths starting at an ontology o and ending at an ontology o' is identified by $\Pi(o, o')$.*

Example 3. For instance, $\Pi(o_1, o_4)$ contains the four following acyclic paths: $A_{1,2} \cdot A_{2,4}$, $A_{1,3} \cdot A_{3,4}$, $A_{1,2} \cdot A_{2,3} \cdot A_{3,4}$ and $A_{1,3} \cdot A_{2,3}^{-1} \cdot A_{2,4}$.

We extend the notation $A(s)$ to paths. If $\pi = A_1 \cdot \ldots A_n$, then $|\pi| = n$ and $\pi(s) = A_n(\ldots A_1(s) \ldots)$.

Definition 5 (Application of a path). *Given $\pi = A_1 \cdot \ldots A_n$ a functional path, θ a set of relations and s a set of ontology entities, the* application of π to s *denoted by $\pi(s)$ is:*

$$\pi(s) = \{e_n | \forall i \exists! \langle e_{i-1}, e_i, r \rangle \in A_i \text{ such that } e_0 \in s \wedge r \in \theta\}$$

By convention, we introduce the empty path ϵ from one ontology to itself, such that $\epsilon(s) = s$ and $|\epsilon| = 0$. We note $o \dot{\in} \pi$ if o is one of the ontologies involved in an alignment of the path π. There may be an infinite number of paths due to circuits in the graph.

3.3 Algebraic Similarity Properties

We consider ontology measures which are functions from two ontologies to a scalar domain. We use the term "measure" for both similarities and dissimilarity. A similarity is a a real positive function σ of two ontologies which is as large as ontologies are similar. It is defined as follows.

Definition 6 (Similarity). *A similarity $\sigma : \Omega \times \Omega \to \mathbb{R}$ is a function from a pair of entities to a real number expressing the similarity between two objects such that:*

$$\forall o, o' \in \Omega, \sigma(o, o') \geq 0 \qquad (positiveness)$$
$$\forall o \in \Omega, \forall o', o'' \in \Omega, \sigma(o, o) \geq \sigma(o', o'') \qquad (maximality)$$
$$\forall o, o' \in \Omega, \sigma(o, o') = \sigma(o', o) \qquad (symmetry)$$

Some authors consider a 'non symmetric (dis)similarity' [12]; we then use the term non symmetric measure or pre-similarity. All the measures presented in this paper are pre-similarities and labelled as such. However, if applied to a symmetrically closed space, they become similarities.

Very often, the measures are normalised. This is especially useful when the dissimilarity of different kinds of entities must be compared. Reducing each value to the same scale in proportion to the size of the considered space is the common way to normalise.

Definition 7 (Normalised measure). *A measure is said to be normalised if it ranges over the unit interval of real numbers [0 1].*

We consider only normalised measures and assume that a measure between two ontologies returns a real number between 0 and 1.

In the remainder, we define measures based on the structure of alignment spaces instead of relying directly on the ontology content. A first approach, considers alignment spaces as graphs and the proximity between ontologies is based on their topology (§4). Another family of measures is based on the capacity of alignments to cover a large proportion of the ontology entities as well as to keep them distinct (§5).

4 Path-Based Measures

The first kind of similarity between two ontologies may be based on paths between these ontologies in the graph $G_{\Omega, \Lambda}$. In fact, the existence of a path guarantees that it is possible to transform queries from one ontology to another. This can be refined by considering different values if the path is made of zero, one or several alignments:

Definition 8 (Alignment path pre-similarity)

$$\sigma_{ap}(o, o') = \begin{cases} 1 & \text{if } o = o' \\ 2/3 & \text{if } o \neq o' \text{ and } \Lambda(o, o') \neq \emptyset \\ 1/3 & \text{if } o \neq o' \text{ and } \Lambda(o, o') = \emptyset \text{ and } \Pi(o, o') \neq \emptyset \\ 0 & \text{otherwise} \end{cases}$$

Example 4. From the alignment space of Figure 1, we can see that $\sigma_{ap}(o_1, o_2) = 2/3$ because there is an alignment between o_1 and o_2, $\sigma_{ap}(o_1, o_4) = 1/3$ because there are paths between o_1 and o_4, and $\sigma_{ap}(o_4, o_1) = 1/3$ because there are also paths using inverse operations. All the values are given in Figure 2.

Such a measure is minimal between two non connected ontologies and it is normalised. It is symmetric as long as alignments are considered symmetric, i.e., as soon as an alignment A is available, it is assumed that A^{-1} is available as well. It is relatively easy to compute and it reflects the possibility to propagate information between two ontologies. However, it is not very precise in the number of transformations that may have to be performed to propagate this information.

So, a natural measure depends on the shortest path in the graph $G_{\Omega,\Lambda}$. Indeed, the fewer alignments are applied to a query, the more it is expected that it is an accurate translation (in first approximation).

Definition 9 (Shortest alignment path pre-similarity). *Given an alignment space* $\langle \Omega, \Lambda \rangle$, *the shortest alignment path pre-similarity* σ_{sap} *between two ontologies* $o, o' \in \Omega$ *is the complement to 1 of the length of the shortest path between o and o' in $G_{\Omega,\Lambda}$:*

$$\sigma_{sap}(o, o') = \begin{cases} 1 - \frac{\min_{\pi \in \Pi(o,o')} |\pi|}{\varnothing_{\Omega,\Lambda}} & \textit{if } \Pi(o, o') \neq \emptyset \\ 0 & \textit{otherwise} \end{cases}$$

In order to normalise the similarity, $\varnothing_{\Omega,\Lambda}$ can either be the size of $|\Omega|$, or the diameter of $G_{\Omega,\Lambda}$, i.e., the length of the longest shortest path, plus 1.

Example 5. From the alignment space of Figure 1, if we take the size of the network as ($\varnothing_{\Omega,\Lambda} = |\Omega| = 5$), $\sigma_{sap}(o_1, o_2) = 4/5$ because there is an alignment between o_1 and o_2 which is a path of length 1, $\sigma_{sap}(o_1, o_4) = 3/5$ because the shortest path between o_1 and o_4, e.g., through o_2, is of length 2, and $\sigma_{sap}(o_4, o_1) = 3/5$ because one can take the converse of the previous path. All the values are given in Figure 2.

The computation of this measure is not significantly more expensive than the computation of the alignment path pre-similarity. The shortest alignment path pre-similarity is more precise because it depends on the minimum necessary transformations between the two ontologies.

However, an alignment between two ontologies can be just empty: this does not mean that the ontologies are very close but rather that they are very different. Even if alignments are not empty, this measure does not tell how much of an ontology is preserved through the translation. Indeed, considering the alignment space of Figure 2, it shows that for both measures, o_4 is farther from o_1 than o_3, however, if one looks at the alignments in Figure 1, the composition of $A_{1,2}$ and $A_{2,4}$ preserves more information than the alignment $A_{1,3}$. This is the reason why we consider more precise measures.

5 Coverage-Based Measures

If we want to go further in measuring the precise proximity for querying applications, it may be useful to consider the ratio of elements of the ontology which are covered by

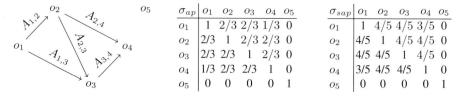

σ_{ap}	o_1	o_2	o_3	o_4	o_5
o_1	1	2/3	2/3	1/3	0
o_2	2/3	1	2/3	2/3	0
o_3	2/3	2/3	1	2/3	0
o_4	1/3	2/3	2/3	1	0
o_5	0	0	0	0	1

σ_{sap}	o_1	o_2	o_3	o_4	o_5
o_1	1	4/5	4/5	3/5	0
o_2	4/5	1	4/5	4/5	0
o_3	4/5	4/5	1	4/5	0
o_4	3/5	4/5	4/5	1	0
o_5	0	0	0	0	1

Fig. 2. Alignment space (left) corresponding to Figure 1 and the corresponding path-based measures (right). σ_{sap} is computed with $\varnothing_{\Omega,\Lambda} = |\Omega| = 5$ (using the length of the longest shortest path (2) plus 1 would have given the same results as σ_{ap} in this case).

an alignment. In fact this can be applied to any set of elements, not just an ontology. Hence the coverage can be given with regard to an ontology entity (the ratio is 1 or 0), to a query or to an ontology. It corresponds to the percentage of entities which have an image through the alignment.

Definition 10 (Alignment coverage). *Given a set of ontology entities s over an ontology o, a set of relations θ, and an alignment $A \in \Lambda(o, o')$, the coverage of s by A is given by:*

$$cov(s, A) = \frac{|\{e \in s | \exists \langle e, e', r \rangle \in A \wedge r \in \theta\}|}{|s|}$$

Example 6. In Figure 3, the coverage of alignment A_{0-4} is $2/3$ because out of a, b and c, only b and c are covered by the alignment.

There is a second important notion which is the ability for the alignment to preserve the difference between entities which are deemed different in the source ontology. The alignment distinguishability measure is the proportion of matched entities which are kept distinct. This could be considered as preservation of information.

Definition 11 (Alignment distinguishability). *Given a set of ontology entities s over an ontology o, a set of relations θ, and an alignment $A \in \Lambda(o, o')$, the distinguishability (or separability) of s by A is given by:*

$$sep(s, A) = \frac{|\{e' | \exists \langle e, e', r \rangle \in A \wedge e \in s \wedge r \in \theta\}|}{|\{e \in s | \exists \langle e, e', r \rangle \in A \wedge r \in \theta\}|}$$

Example 7. In Figure 3, the distinguishability of alignment A_{0-4} is $1/2$ because out of b and c covered by the alignment, there remain only one image in $A_{0-4}(\{b, c\})$.

For functional alignments, separability remains smaller than 1. These two notions are obviously tied to the concepts of existence and injectivity of a function. cov depends on $Q_L(o)$ alone, while sep also depends on $Q_{L'}(o')$, hence these measures cannot be reduced to one another.

In the following, we use a measure which accounts for both coverage and distinguishability at once: instead of making the count of ontology entities which have an image by the alignment, we only count those distinct images. Hence the lack of distinguishability automatically lowers the returned value.

Definition 12 (Alignment coverage distinguishability). *Given a set of ontology enti-*
ties s over an ontology o and an alignment $A \in \Lambda(o, o')$, the coverage distinguishability
of s by A is given by:

$$covdis(s, A) = cov(s, A) \times sep(s, A) = \frac{|A(s)|}{|s|}$$

Example 8. In Figure 3, the coverage distinguishability of alignment A_{0-4} is $1/3$ because out of a, b and c, there remain only one image in $A_{0-4}(\{a, b, c\})$. Other examples,
are provided in Figure 3.

This measure can easily be extended to paths. If we still retain functional paths, the
relation between cov, sep and $covdis$ of Definition 12 still holds for paths. Figure 3
shows the differences between the three measures.

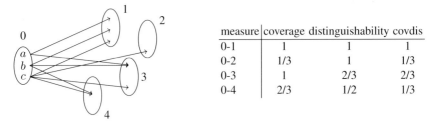

measure	coverage	distinguishability	covdis
0-1	1	1	1
0-2	1/3	1	1/3
0-3	1	2/3	2/3
0-4	2/3	1/2	1/3

Fig. 3. Simple alignments (left) and the corresponding coverage and distinguishability measures
(right)

5.1 Largest Coverage

The natural measure is that of largest coverage.

Definition 13 (Largest covering pre-similarity). *Given an alignment space $\langle \Omega, \Lambda \rangle$,*
the largest covering pre-similarity σ_{lc} *between two ontologies $o, o' \in \Omega$ is*

$$\sigma_{lc}(o, o') = \max_{A \in \Lambda(o, o')} covdis(o, A)$$

Such a measure is clearly not symmetric, even if the alignment is only made of equal-
ities: the ratio depends on the size of the source ontology, independently of the target
ontology. It is not definite either: if all information is preserved and distinguishable, the
similarity will be 1 though the ontologies are not the same.

We have applied this measure to direct alignments and not to paths. However, it may
be that a path better covers and preserves the ontology entities than a direct alignment.

For instance, if there were a direct alignment $A_{1,4} = \{\langle a_1, a_4, = \rangle\}$ from o_1 to o_4.
Then the coverage would be $1/5$, while the coverage provided by the path $A_{1,2} \cdot A_{2,4}$
is $3/5$. In that respect, o_4 is closer to o_1 than o_3 is.

Hence, it is necessary to apply the measure to the paths which lead to an ontology.
Composing the measures obtained by the alignments in order to get the measure for the
path is not sufficient. Indeed, if two alignments have a similarity of 80%, the similarity

of their compound alignment can be anything between 0% and 80%. We have computed the product of the similarity as the $\sigma_{\times lc}$ in Table 1.

It is thus necessary to evaluate path coverage distinguishability. In order to address this problem, we introduce measures which are based on path instead of simple alignments. The first one is the largest covering preservation pre-similarity:

Definition 14 (Largest covering preservation pre-similarity). *Given an alignment space* $\langle \Omega, \Lambda \rangle$, *the* largest covering preservation pre-similarity σ_{lcp} *between two ontologies* $o, o' \in \Omega$ *is:*

$$\sigma_{lcp}(o, o') = \max_{\pi \in \Pi(o,o')} covdis(o, \pi)$$

Example 9. From the alignment space of Figure 1, $\sigma_{lcp}(o_1, o_2) = 3/5$ because over 5 entities in o_1 the alignment $A_{1,2}$ preserves 3, $\sigma_{lcp}(o_1, o_4) = 3/5$ because the path $A_{1,2} \cdot A_{2,4}$ between o_1 and o_4 also preserves the same 3 entities (other paths of Example 3 preserve less entities). This time $\sigma_{lcp}(o_4, o_1) = 3/8$ because o_4 contains 8 entities and the $A_{2,4}^{-1} \cdot A_{1,2}^{-1}$ path preserves 3 entities. All the values of measures from o_1 are given in Table 1.

5.2   Union Path Coverage

So far, we only considered that a query would take one path at a time and that the query would be entirely evaluated through this path. In this case, the above measure is perfectly accurate. However, very often a query is split into parts which are sent to different peers and the results are composed through join or union depending on the query.

In this case, the measure above does not reflect the semantics of alignment spaces and does not provide a measure of the proximity of the two ontologies for evaluating queries. The meaning of alignment spaces can basically be rendered by the transitive and union closure of this alignment space[3]. In consequence, the coverage distinguishability should be computed not on the path that brings the maximal coverage but on the coverage provided by the combination of all the possible paths.

Definition 15 (Union path coverage pre-similarity).   *Given an alignment space* $\langle \Omega, \Lambda \rangle$, *the* union path coverage σ_{upc} *between two ontologies* $o, o' \in \Omega$ *is:*

$$\sigma_{upc}(o, o') = \frac{|(\bigcup_{\pi \in \Pi(o,o')} \pi)(s)|}{|s|}$$

The set of paths, eventually containing cycles, may be infinite; but what they preserve of s is necessarily finite, hence a finite subset of these paths is sufficient for computing σ_{upc}.

This measure takes full advantage of all the alignments provided within the alignment space. In particular, it is able to account for the fact that, in the example of Figure 1, any query expressed with regard to entities of ontology o_1 can be evaluated in ontology o_4, yet through different paths depending on the considered entities.

---

[3] We assume here that this alignment space is consistent.

Example 10. From the alignment space of Figure 1, $\sigma_{upc}(o_1, o_2) = 4/5$ because over 5 entities in o_1 the alignment $A_{1,2}$ preserves 3 but in addition the path $A_{1,3} \cdot A_{2,3}^{-1}$ preserves d_1. $\sigma_{upc}(o_1, o_4) = 1$ because the path $A_{1,2} \cdot A_{2,4}$ between o_1 and o_4 also preserves the same 3 entities and the path $A_{1,3} \cdot A_{3,4}$ preserves the two remaining ones. This time $\sigma_{upc}(o_4, o_1) = 5/8$ because out of the 8 entities in o_4, the $A_{2,4}^{-1} \cdot A_{1,2}^{-1}$ path preserves 3 entities and $A_{3,4}^{-1} \cdot A_{1,3}^{-1}$ preserves two other ones. All the values of measures from o_1 are given in Table 1.

Table 1. Coverage and distinguishability based similarities with regard to o_1 for the ontologies of Figure 1 (with $\theta = \{=, \sqsupseteq\}$)

measure	o_1	o_2	o_3	o_4	o_5
σ_{lc}	1	3/5	2/5	0	0
$\sigma_{\times lc}$	1	3/5	2/5	9/35	0
σ_{lcp}	1	3/5	2/5	3/5	0
σ_{upc}	1	4/5	3/5	1	0

5.3   OntoSim

OntoSim is a Java library for computing distance or similarity measures between ontologies[4]. It can be used by other tools, such as matchers, through its API.

OntoSim implements the measures described in [1] and here. The alignment space measures presented here usually rely on the sets of paths between two nodes in a graph which is a highly complex problem (the number of acyclic paths being $n!$ in a complete graph). However, because we have a quantity to optimise (the degree of coverage), this provides a ground for implementing branch-and-bound strategies (even for the union path coverage). In addition, we have developed a focussed search heuristics aiming at maximising the potentially preserved coverage (preservation can only decrease monotonously). Both approaches put together are really efficient in practice.

6   Comparison of Presented Measures

In order to better understand how these measures behave, we have performed experiments. These experiments follow those comparing measures in ontology spaces on the ontology alignment evaluation initiative (OAEI) benchmark ontologies [1]. They especially offered a separate evaluation of entity similarity measures and set similarity measures. The following experiment compares ontology space measures and alignment space measures on the OntoFarm data set (OAEI conference data set). Two experiments have been carried out for evaluating respectively the agreement between different measures and the robustness of alignment space measures.

6.1   Dataset Description

There are very few datasets available which have the structure of an alignment space: many ontologies and alignments. The OntoFarm dataset[5] [13] is made of a collection of

---

[4] http://ontosim.gforge.inria.fr
[5] http://nb.vse.cz/~svatek/ontofarm.html

15 ontologies dealing with the conference organisation domain. Ontologies are based upon three types of underlying resources:

- actual conference (series) and its web pages,
- actual software tool for conference organisation support,
- experience of people with personal participation in organisation of actual conference.

This dataset has been used several times in the OAEI evaluation campaigns. We have used those of 2009. For the experiment purpose, we have used a set of 105 alignments obtained as a majority vote between 7 matchers (Aroma, ASMOV, DSSim, Falcon, Lily, OLA, TaxoMap). We have suppressed empty alignments, resulting in 91 alignments containing 827 correspondences. Alignments are non-oriented: they can be traversed in both ways.

6.2 Agreement

The first experiment aims at comparing rank correlation between measures. Its goal is to compare if the proximity orders induced by alignment space measures can be correlated with the proximity orders induced by ontology space measures. We compare the alignment space measures with the measures that have been found the best in our previous study [1]. JaccardVM and CosineVM are measures between vectors determined by the terms used to describe entities in both ontologies, EntityLexicalMeasure computes a similarity between entities from their annotations, e.g., labels and comments, and extract a similarity between ontologies, while TripleBasedEntitySim compares entities on the basis of the RDF triples that involve them and extract a similarity between ontologies.

We use the standard Kendall τ_b rank correlation for computing the correlation between compared measures. In these experiments, the significance test at level of 5% gives a confidence interval of $[-0.09; 0.09]$.

Agreement results. The resulting agreement is shown in Table 2 using the Kendall τ_b correlation coefficient [14]. It ranges between -1 and 1, hence all these measures are correlated to some extent.

More interesting information is found when using these data for clustering the measures with respect to their agreement. Hierarchical clustering from agreement provides the dendrogram of Figure 4 (we have used single linkage, but the other linkage measures give the same results).

The two path measures, i.e., σ_{ap} and σ_{sap}, do not agree with other measures. This can be easily explained because the graph of alignments is very connected (91 alignments out of 105 possible ones) so these measures are not very informative: the ontologies come in few groups depending on how they are connected to the others, most of them being reachable through one alignment. This is not discriminating enough and it is penalised by the τ_b variant. As expected, this shows that these measures are very dependent on the topology of the alignment space.

The most interesting aspect of this test is that coverage-based measures, i.e., σ_{lcp} and σ_{upc}, are far more correlated with the content based measures than to the path-based measures. They are even more correlated to the vector-space measures than the

Table 2. Agreement results between measures

	σ_{sap}	σ_{lcp}	σ_{upc}	JaccardVM	CosineVM	EntityLexicalMeasure	TripleBasedEntitySim
Alignment path (σ_{ap})	0.881	0.147	0.147	0.418	0.315	0.117	0.115
Shortest path (σ_{sap})	-	0.138	0.138	0.414	0.32	0.099	0.092
Largest covering (σ_{lcp})	-	-	0.237	0.169	0.127	0.086	0.081
Union path coverage (σ_{upc})	-	-	-	0.169	0.127	0.086	0.081
JaccardVM	-	-	-	-	0.681	0.288	0.272
CosineVM	-	-	-	-	-	0.196	0.158
EntityLexicalMeasure	-	-	-	-	-	-	0.902

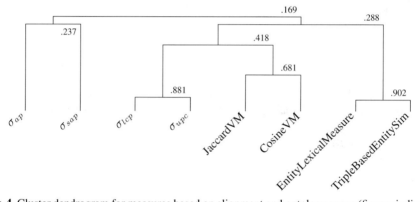

Fig. 4. Cluster dendrogram for measures based on alignment and ontology space (figures indicate agreement)

vector space measures agree with the entity-based measures. This is a very good sign especially that in our previous experiments we saw that JaccardVM and TripleBasedEntitySim were the best ontology space measures. This shows that these measures, which do not have access to the content of ontologies, are meaningful with regard to this content.

6.3   Robustness

The second experiment focuses on robustness of alignment space measures. For that purpose, alignment spaces are altered in a systematic manner. We have retained two variants for this degradation:

variant 1: Randomly remove $n\%$ of alignments in an alignment space
variant 2: Randomly remove $n\%$ of correspondences in all alignments

The experiment consisted of evaluating, for each measure, the agreement between the alignment space measure without degradation and the same measure computed on the altered alignment space. This experiment has been done with several levels of degradation, from 10% to 100% with a step of 10%. Since this procedure is based on random degradation, we repeated it 10 times for each level and averaged the results.

Agreement is still measured by the Kendall τ_b rank correlation between the measure obtained on the initial alignment space and that obtained on the degraded alignment space.

For the second variant, we only compare the two coverage measures because this type of degradation has no impact on path measures since it preserves the topology of alignment spaces.

We expect that the degradation obtained with the first variant will have a more negative impact on the robustness of measures.

Results of Variant 1. Results of this first variant are given in Figure 5 (left). Path-based measures do not have good results for the same reason as before: the graph being very connected, most ontologies are at the same distance to one another, then the τ_b coefficient penalises this behaviour. Still the correlation remains positive (0 means random).

Coverage-based measures have a linearly decreasing curves. This result shows the strong dependency of all these measures on available alignments. Both measures are very close, and indeed, we have observed this in other experiments as well.

Results of Variant 2. Results of the second variant are given in Figure 5 (right). Both measures show a sub-linear degradation: this shows that they are quite robust to correspondence degradation. We replicated these experiments with different datasets, different modus operandi and different agreement measures. The results are the same with a different amplitude of the robustness to the correspondence degradation (which is

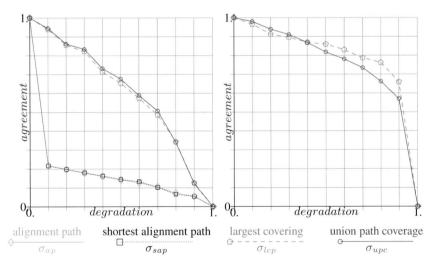

Fig. 5. Robustness of measures in function of the degree of degradation (Variant 1: alignment degradation and 2: correspondence degradation)

sometimes better and sometimes worse than the one observed here, but always more resistant than linear).

Results of σ_{lcp} (degree of agreement with non-degraded variant) seems higher, therefore we can conclude that it is less dependent on particular correspondences (this does not mean that they are better, just more robust).

The robustness tests show that alignment space measures are indeed correlated with the quality of the alignment space (so they are not random measures). In both cases, the measures are rather robust since their agreement with their initial behaviour decreases less than the degradation. The coverage-based measures shows some independence from correspondences degradation.

7   Conclusion

We have introduced a new way to measure similarity between ontologies adapted to a context in which alignments are available, such as the semantic web or semantic peer-to-peer systems [15]. Such measures rely on the available alignments instead of the content of the ontologies. They are useful when some ontologies are not available or when the proximity must denote the ability to transfer information from one ontology to another.

We have defined precisely some possible such measures. Path-based measures take into account the topology of alignment spaces. Coverage-based measures are based on the coverage and distinguishability of alignments and can account for combined alignment paths for transforming queries. This allows global reasoning on alignments alone which is something less easy in local environments.

The proposed measures have been implemented in the OntoSim library and compared to measures taking advantage of ontology content in order to detect similarity. Although not strongly correlated with the best measures, the coverage-based measures provide results comparable to these. Moreover, in addition to not depend on the ontology content, they have proved to be reasonably robust to errors in the alignments, especially if individual correspondences are missing. This is very encouraging.

The proposed measures have been designed with simplifying hypotheses that requires further investigation in order to relax them. This mostly concerns taking into account different alignment relations and alignment confidence, in the style of [11], as well as considering more closely non functional alignments. It would also be interesting to look further into the joint use of ontology space and alignment space measures.

Acknowledgements. This work has been partly supported by the European Commission IST project NeOn (IST-2006-027595). Ondřej Šváb-Zamazal has been partly supported by grant no. P202/10/1825 of the Grant Agency of the Czech Republic.

References

1. David, J., Euzenat, J.: Comparison between ontology distances (preliminary results). In: Sheth, A.P., Staab, S., Dean, M., Paolucci, M., Maynard, D., Finin, T., Thirunarayan, K. (eds.) ISWC 2008. LNCS, vol. 5318, pp. 245–260. Springer, Heidelberg (2008)

2. Euzenat, J., Shvaiko, P.: Ontology matching. Springer, Heidelberg (2007)
3. Euzenat, J., Allocca, C., David, J., d'Aquin, M., Le Duc, C., Svab-Zamazal, O.: Ontology distances for contextualisation. deliverable 3.3.4, NeOn (2009)
4. Mädche, A., Staab, S.: Measuring similarity between ontologies. In: Gómez-Pérez, A., Benjamins, V.R. (eds.) EKAW 2002. LNCS (LNAI), vol. 2473, pp. 251–263. Springer, Heidelberg (2002)
5. Hu, B., Kalfoglou, Y., Alani, H., Dupplaw, D., Lewis, P., Shadbolt, N.: Semantic metrics. In: Staab, S., Svátek, V. (eds.) EKAW 2006. LNCS (LNAI), vol. 4248, pp. 166–181. Springer, Heidelberg (2006)
6. Vrandečić, D., Sure, Y.: How to design better ontology metrics. In: Franconi, E., Kifer, M., May, W. (eds.) ESWC 2007. LNCS, vol. 4519, pp. 311–325. Springer, Heidelberg (2007)
7. Euzenat, J., Valtchev, P.: Similarity-based ontology alignment in OWL-lite. In: Proc. 16th European Conference on Artificial Intelligence (ECAI), Valencia (ES), pp. 333–337 (2004)
8. Ehrig, M., Haase, P., Hefke, M., Stojanovic, N.: Similarity for ontologies – a comprehensive framework. In: Proc. 13th European Conference on Information Systems, Information Systems in a Rapidly Changing Economy (ECIS), Regensburg, DE (2005)
9. d'Aquin, M.: Formally measuring agreement and disagreement in ontologies. In: Proc. 5th International Conference on Knowledge Capture (K-CAP), Redondo Beach (CA US), pp. 145–152 (2009)
10. Delbru, R., Toupikov, N., Catasta, M., Tummarello, G., Decker, S.: Hierarchical link analysis for ranking web data. In: Aroyo, L., Antoniou, G., Hyvönen, E., ten Teije, A., Stuckenschmidt, H., Cabral, L., Tudorache, T. (eds.) ESWC 2010. LNCS, vol. 6089, pp. 225–239. Springer, Heidelberg (2010)
11. Euzenat, J.: Algebras of ontology alignment relations. In: Sheth, A.P., Staab, S., Dean, M., Paolucci, M., Maynard, D., Finin, T., Thirunarayan, K. (eds.) ISWC 2008. LNCS, vol. 5318, pp. 387–402. Springer, Heidelberg (2008)
12. Tverski, A.: Features of similarity. Psychological Review 84(2), 327–352 (1977)
13. Šváb, O., Svátek, V., Berka, P., Rak, D., Tomášek, P.: Ontofarm: Towards an experimental collection of parallel ontologies. In: Gil, Y., Motta, E., Benjamins, V.R., Musen, M.A. (eds.) ISWC 2005. LNCS, vol. 3729. Springer, Heidelberg (2005)
14. Kendall, M.: Rank correlation methods, Griffin, London, UK (1970)
15. Bouquet, P., Giunchiglia, F., van Harmelen, F., Serafini, L., Stuckenschmidt, H.: Contextualizing ontologies. Journal of Web Semantics 1(1), 325–343 (2004)

SameAs Networks and Beyond: Analyzing Deployment Status and Implications of owl:sameAs in Linked Data

Li Ding, Joshua Shinavier, Zhenning Shangguan, and Deborah L. McGuinness

Tetherless World Constellation, Rensselaer Polytechnic Institute
{dingl,shinaj,shangz,dlm}@cs.rpi.edu

Abstract. Millions of owl:sameAs statements have been published on the Web of Data. Due to its unique role and heavy usage in Linked Data integration, owl:sameAs has become a topic of increasing interest and debate. This paper provides a quantitative analysis of owl:sameAs deployment status and uses these statistics to focus discussion around its usage in Linked Data.

Keywords: owl:sameAs, Linked Data, Network.

1 Introduction

The Web of Data is growing rapidly, with an ever-expanding set of inter-connected datasets depicted in the Linking Open Data (LOD) cloud diagram [1]. In the Web of Data, an increasing number of owl:sameAs statements have been published to support merging distributed descriptions of equivalent RDF resources. Although these statements are just binary relations, when all of these owl:sameAs statements are taken together, they form a very large directed graph connecting RDF resources to each other. We will refer to this large graph of RDF resources connected by sameAs statements as a **SameAs network**. SameAs networks are interesting both for their structural properties, e.g. size, diameter and in/out-degree and their semantic properties, e.g. reflexivity, symmetry and transitivity.

According to OWL semantics [2], all RDF resources in a single sameAs network are indistinguishable, such that they can be merged into one RDF resource and change the structure of RDF graph. However, recent literature [3-7], mainly from the Linked Data community, reports many issues related to owl:sameAs usage in the Web of Data: owl:sameAs is often used in ways that do not strictly agree with the official semantics of owl:sameAs in OWL. Some researchers [4, 6] further called for new ontological semantic relations to complement owl:sameAs in capturing similarity relations between RDF resources. To the best of our knowledge, most reported results on owl:sameAs are derived from very small sample datasets, and no statistically significant analysis has been conducted on the deployment status and implications of owl:sameAs in the Web of Data.

We conducted a large scale analysis on SameAs networks extracted from the Web of Data to answer two types of key questions: (i) *How is owl:sameAs actually deployed? How many SameAs networks have been published? Do these SameAs*

P.F. Patel-Schneider et al. (Eds.): ISWC 2010, Part I, LNCS 6496, pp. 145–160, 2010.

networks have interesting topological properties? (ii) *What are the implications of owl:sameAs inference in Linked Data integration? How can owl:sameAs be used to connect the ontologies of the datasets in the LOD cloud?* In order to reduce the bias caused by a small sample dataset, we use the Billion Triple Challenge (BTC) 2010 dataset which covers a significant portion of the Web of Data.

This work provides contributions related to the definition and analysis of SameAs networks. We highlight the practical value of our work in network settings focusing on (1) how Linked Open Datasets are connected and (2) how sameAs networks may be used for automated ontology mapping and error detection. The rest of this paper is structured as follows. Section 2 defines SameAs networks and identifies research problems. Section 3 describes the sample dataset extracted from the BTC 2010 dataset and experiment settings. Sections 4, 5 and 6 report the analytical discoveries on SameAs networks, along with two special classes of networks (Pay-Level Domains and Class-Level Similarity). Section 7 reviews related work. Section 8 concludes our work with future directions.

2   SameAs Networks

The importance of owl:sameAs in Linked Data integration is widely recognized, however there have not been many studies characterizing its usage in very large data-sets. The goal of our work was to review existing usage of owl:sameAs in a dataset that contains a significant number of sameAs statements and also to analyze usage in a practical Linked Data integration setting. We therefore will define the notion of a SameAs network and then show a selection of research problems derived from the motivating questions from Section 1.

2.1   Definitions and Notations

Definition 1. **SameAs statement.** A *SameAs statement* is an RDF triple which connects two RDF resources by means of an owl:sameAs predicate.

Definition 2. **Predicate-based Sub-graph Filter.** A *Predicate-based Sub-graph Filter* is a function $H = psf(G, P)$, where H and G are RDF graphs and P is a set of RDF properties. This function returns H which is a sub-graph of G, and the predicate of any triple in H is a member of P.

Definition 3. **SameAs network.** Given an RDF graph G, a *SameAs network SN* in G is a weakly connected component[1] of $psf(G, \{owl:sameAs\})$.

Figure 1 illustrates an example SameAs Network, where an RDF resource "dbpedia:Paul_Allen"[2] is denoted as a node, and a SameAs statement

---

[1] A weakly connected component is a maximum sub-graph where all pairs of nodes are by an undirected path. See
http://mathworld.wolfram.com/WeaklyConnectedComponent.html
[2] Throughout this paper, we use QName to encode URI reference and Turtle to encode RDF triples and RDF graphs. See http://www.w3.org/TeamSubmission/turtle/

"dbpedia:Paul_Allen owl:sameAs umbel:Paul_Allen" is denoted as a directed arc. This figure also exhibits additional interesting structural patterns: (i) two RDF resources could be linked by one-way or and reciprocal owl:sameAs statements; and (ii) there exist authority nodes (with high in-degree, e.g. dbpedia:Paul_Allen) and hub nodes (with high out-degree, e.g. freebase:guid.9202a8c04000641f800000000002e633).

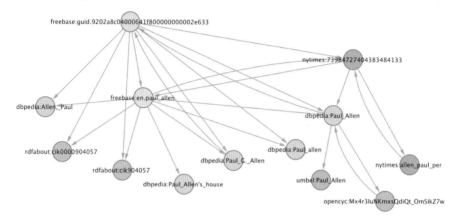

Fig. 1. An example SameAs network about "Paul Allen"

The official semantics of owl:sameAs is specified in OWL [2]: *"an owl:sameAs statement indicates that two URI references actually refer to the same thing."* Recent studies reported diverse usage that is NOT consistent with the official semantics:

- Is owl:sameAs symmetric? Vatant [7] suggested that owl:sameAs, when used in mashup, is not necessarily a symmetric property, i.e., "X owl:sameAs Y" does not imply "Y owl:sameAs X". Therefore, two RDF resources X and Y are considered to be strongly equivalent only when their owners make reciprocal SameAs statements.
- Is owl:sameAs transitive? Jaffri et al [5] reported that the equivalence relationship represented by owl:sameAs is often context-dependent, and is accurate only within the context of particular applications. While transitivity is automatically granted by OWL semantics, SameAs statements asserted in the Web of Data seldom guarantee transitivity.

2.2 SameAs Networks Analysis

In order to analyze the deployment status and implications of SameAs Networks, we identify the following three research problems:

How have SameAs Networks been deployed on the Web of Data? Since we are not the owners of the SameAs statements in the Web of Data, it would be quite subjective to speculate the intended semantics of owl:sameAs. In order to produce objective and convincible reports, we focus on the structural properties of SameAs networks. In order to avoid the bias caused by small sample datasets, we collected a large sample dataset from the real world Web of Data. Section 3 provides a quantitative analysis of the dataset.

What are the common interests among Linked Data publishers? Since there are many URIs using "dbpedia" for a namespace in the example SameAs network in Figure 1, it is possible to summarize SameAs statements to higher level connections to provide an overview of SameAs networks. We are particularly interested in "pay-level domain" (PLD)[3] as it is frequently used to identify Linked Data publishers and can often be connected to LOD datasets via one-to-one mappings. Now, with such summarization, users can analyze how and why Linked Data publishers (or LOD datasets) are inter-connected via SameAs statements.

How will Web ontologies be affected by owl:sameAs inference? Mapping Web ontologies is a well-known difficult problem due to the high cost of manually assert-ing mapping relations among ontological terms. Instance-based approaches have been used in mapping RDFS/OWL classes, i.e. two classes are considered "associated" if they share common instances. Now, with owl:sameAs inference, users may merge different RDF resources and thus find more associated classes.

3   Building ESameNet Dataset and Experiment Settings

In order to study the three problems identified in section 2.2, we will extend SameAs networks with additional information:

- *PLD statements*, each RDF resource can be connected to a literal name iden-tifying a PLD. PLD statements can be pre-computed before the creation of SameAs networks and stored in triples using ex:hasPLD as predicate.
- *Type statements*, each RDF is connected to zero-to-many RDFS/OWL classes via rdf:type. Type statements are already asserted in the RDF graph from which SameAs networks were obtained.

Definition 4: **Extended SameAs network.** Given an RDF graph *G*, an *extended SameAs network ESN* is constructed by extending a SameAs network *SN* of *G* with additional nodes and arcs. Besides the RES world, i.e. the world of all RDF resources in *SN*, two more worlds of nodes will be added: (i) the CLS world, i.e. a world of RDFS/OWL classes; (iii) the PLD world, i.e. a world of PLD names. A new node *n* will be added when there exists a PLD (or Type) statement *s* that links from a node in *SN* to *n*. Meanwhile, the corresponding statement *s* will be added as a new arc.

---

[3] A PLD is an internet domain that requires payment at a generic top-level domain (gTLD) or country code top-level domain (ccTLD) registrar. PLDs are usually one level below the cor-responding gTLD (e.g., dbpedia.org vs. org), with certain exceptions for cc-TLDs (e.g., ebay.co.uk, det.wa.edu.au) [8].

Figure 2 illustrates an example fragment of an extended SameAs network, including: RDF resources, e.g. *dbpedia:Virginia* and *fbase:en.virginia*; PLD statements, e.g. "*dbpedia:Virginia* ex:hasPLD "*dbpedia.org.*" and CLS statements, e.g. "*dbpedia:Virginia* rdf:type yago:StatesOfTheUnitedStates, dbpedia-owl:Place."

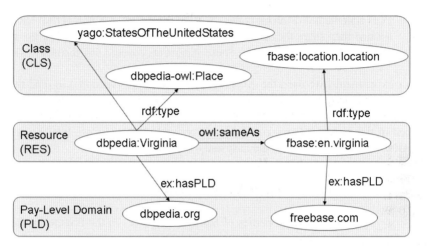

Fig. 2. An example fragment of an extended SameAs network

Our study is based on the **ESameNet** dataset (a collection of extended SameAs networks) extracted from the BTC 2010 dataset. We chose this dataset for two reasons: (i) With approximately 9 million SameAs statements, it constitutes a large-scale sample dataset which is suitable for providing statistical results with high confidence; (ii) Since the BTC 2010 dataset was gathered by crawling the Web based on seeding datasets provided by major Semantic Web search engines, it can be considered as a representative sample of the Web of Data, with relatively low sample distribution bias.

The ESameNet dataset is publicly available[4] in N-Quads[5] format and it consists of the following three subsets:

- **SameAs statements.** We copied all SameAs statements in the BTC 2010 dataset and removed invalid and duplicate statements. A few SameAs statements do not comply with Definition 1 (SameAs statement), e.g. some simply connect an RDF resource to a literal string[6]. From 9,358,227 valid SameAs statements, we obtained 8,711,398 triples after removing duplications. These statements covered 6,932,678 unique RDF resources with URI (aka. URI resource) and 645,400 blank nodes.
- **Type statements.** We copied all rdf:type statements for RDF resources mentioned in BTC 2010 dataset and found 552,622,105 such statements. These

---

[4] See http://tw.rpi.edu/2010/ESameNet
[5] http://sw.deri.org/2008/07/n-quads/
[6] E.g. http://sw.nokia.com/language-1/zh-CH owl:sameAs "zh_CH"^^xsd:lang.

statements covered 488,138,983 distinct RDF resources, and 168,503 distinct RDFS/OWL classes.

- **PLD statements:** We extracted PLD (pay-level domain) statements by parsing the URI of RDF resources in SameAs networks using regular expression. For RDF resource with HTTP URI, we can directly extract its PLD and create the PLD statement. For blank nodes (or RDF resources with non-HTTP URI), we assume they share the same PLDs as the named graphs which host the corresponding SameAs statements. These statements covered 967 distinct PLDs.

In our experiments, we used the AllegroGraph triple store (version 4.0)[7] and the Allegro Common Lisp (version 8.2)[8] programming environment to load the entire BTC 2010 dataset and extract the ESameNet dataset. All of the computational tasks described in this paper were executed on a server with 2x Quad-Core Intel Xeon CPU 2.33GHz CPU, 64GB physical memory and 4 TB hard disk space.

4   Basic Properties of SameAs Networks

We first analyze the basic properties of SameAs Networks in the ESameNet dataset. Each SameAs network is essentially a graph of URIs connected by non-redundant owl:sameAs statements. Due to the difficulties and limitations of automatic entity resolution, the creation of owl:sameAs statements is usually costly and requires manual efforts. Therefore, there are fewer owl:sameAs statements in the Web of Data than one might expect.

Weakly connected components. Overall, the ESameNet dataset contains 6,932,678 URI resources connected by 8,711,398 unique owl:sameAs statements. The graph consists of 2,890,027 weakly connected components, each of which covers on average 2.4 URI resources. The average path length of the graph is only 1.07, which is consistent with this very small average component size (see Figure 3); most components are simply pairs of nodes joined by (usually reciprocal) owl:sameAs links. There are a small number of larger components, including 41 components with hundreds of resources, and two components with thousands of resources. This observation implies that the typical size of SameAs networks is either a small constant or growing slowly; therefore, performing transitive inference on individual SameAs networks is not expensive and could be parallelized. A manual inspection of individual components revealed that the vast majority were star-like in structure, consisting of a single central resource connected to a number of peripheral resources. SameAs networks are not large and complex networks like those of foaf:knows, or even shallow tree-like structures like those of rdfs:subClassOf. Furthermore, SameAs networks tends to have small size components: 24,559 persons were found in the largest component of the foaf:knows network in 2005 [9] vs. 5000 resources were found in the

---

[7] http://www.franz.com/agraph/allegrograph/
[8] http://www.franz.com/products/allegrocl/

largest component in SameAs networks in 2010. One potential explanation could be that Linked Data principles are in favor of reusing URIs rather than duplicating resource decriptions in many distinct LOD datasets. An alternative explanation is that people simply haven't done enough large-scale linking yet[9] due to technology limitations.

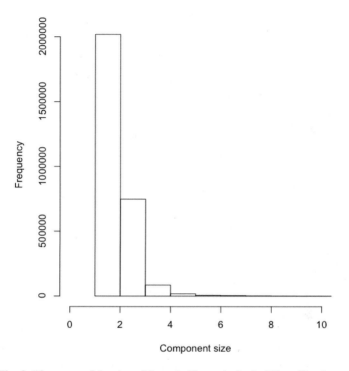

Fig. 3. Histogram of the size of SameAs Networks in the ESameNet dataset

Degree distribution. We investigated the overall in-degree distribution of ESameNet as it measures the popularity (or authority) of resources in sameAs networks[10]. Having plotted the in-degree distribution on a log-log scale, we can see that it exhibits the power law pattern characteristic of scale-free networks. We also noticed that there are slightly more resources with an owl:sameAs in-degree of 1 (that is, 2,974,914 resources) than one would expect of a power law distribution (see Figure 4), and there are also slightly more resources in the 10 to 20 range of in-degree than one would expect. The resources at the high end of the distribution contain on the order of 4,000 inbound owl:sameAs links. Note that we omitted resources with no inbound links.

---

[9] This alternative explanation is kindly suggested by reviewers of this paper.

[10] We skipped out-degree analysis to save space. The out-degree is typically controlled by the publishers for sameAs statements, but the in-degree shows how many publishers are willing to link to a resource using owl:sameAs.

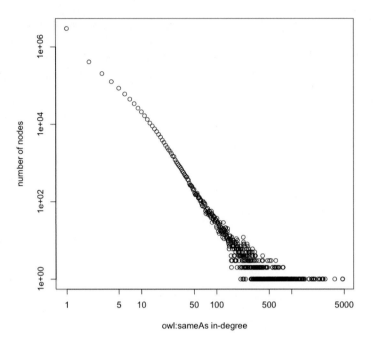

owl:sameAs in-degree

Fig. 4. The in-degree distribution of RDF resources in ESameNet

5 Pay-Level-Domain (PLD) Network Analysis

In order to gain deeper understanding of the common interests between different Linked Data publishers, users may demand a high-level meaningful network to abstract the SameAs networks. The PLD statements provide an ideal opportunity to meet this demand because a PLD can often be used to identify Linked Data publishers and millions of RDF resources in ESameNet can be reduced to hundreds of PLDs.

5.1 Definitions and Notations

Definition 5. **PLD network.** A *PLD network* is a weighted directed graph where (i) each node denotes a unique PLD (labeled by PLD name); (ii) each arc links two PLDs. The weight of an arc *<pld1, pld2>* is calculated by counting the unique SameAs statements between any possible pair of *u1* and *u2*, where (*u1* ex:hasPLD *pld1*) and (*u2* ex:hasPLD *pld2*), normalized by the out degree of *pld1*.

Intuitively, the PLD network is an abstraction of SameAs Networks where each PLD groups some RDF resources. Arcs in PLD network are created using the following SPARQL query:

```
SELECT ?pld1  ?pld2
WHERE { ?u1  ex:hasPLD ?pld1 .  ?u2  ex:hasPLD ?pld2 . ?u1 owl:sameAs? u2 . }
```

Figure 5 shows the largest (also the most interesting) cluster in the PLD network[11] generated from the ESameNet dataset, plotted using Cytoscape [10]. In this diagram, the size of a node is determined by the sum of the weights of both its incoming and outgoing arcs. The thickness of an arc is determined only by its weight. For the purpose of visual clarity, we omit arcs whose weight is less than a threshold (0.00001 in this study with 0.069 being the maximum weight), and self-loops (arcs linking from a node to the node itself). The color of a node is randomly assigned, with the guarantee that no two nodes have the same color. We adopt the "Organic" graph layout provided by Cytoscape to render this diagram to visually highlight clusters.

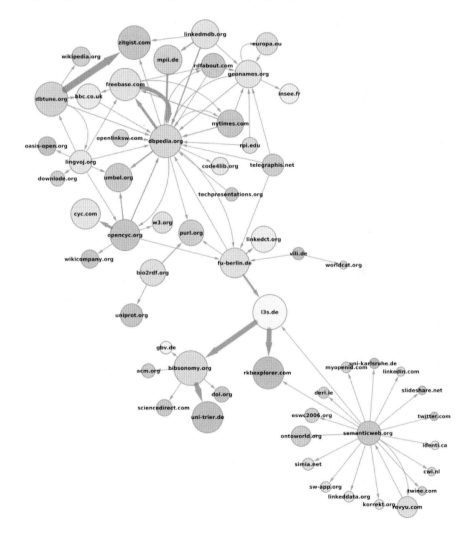

Fig. 5. The largest cluster in the PLD Network generated from ESameNet

---

[11] Due to space limitations, only the most significant cluster is shown. Other clusters can also be generated using the same method and tools as discussed.

5.2 Implications of the PLD Network

The PLD network is an abstraction of SameAs networks in that it establishes connections at PLD level based on instance-level SameAs statements, while retaining the basic structure of one-per-dataset nodes connected by links in a star-like fashion. It can help us gain better insights to the following research problems:

How are data publishers connected? The PLD network provides intuitive and straightforward insights into how publishers are connected via owl:sameAs assertions and what communities are potentially emerging. Figure 5 shows a clear depiction of the associations between different data publishers, in which thicker arcs reflect intensive occurrences of owl:sameAs assertions between corresponding domains. By using appropriate clustering algorithms which apply to any generic graph (e.g., social network), communities of data publishers can be easily identified by eyes. Nodes inside such a cluster can be considered as covering similar topics from possibly different perspectives. In Figure 5, some clusters can be visually identified. The cluster with the set of PLD nodes {*ls3.de, rkbexplorer.com, uni-tier.de, sciencedirect.com, acm.org, gbv.de, doi.org*} represents a community whose members publish data about scientific publications. Other clusters centering on bioinformatics and Semantic Web communities can also be easily identified. In general, we believe that applying novel clustering algorithms to this large-scale PLD network will facilitate detection of communities that share common knowledge and interests. We perceive this as an interesting future research direction.

Why are data publishers connected? After determining which Linked Data publishers are connected via owl:sameAs assertions, it is natural to further investigate why they are connected. Although it is possible to achieve this goal by manually analyzing Figure 5, it usually requires strong expertise in Linked Data, and thus is labor-intensive and error-prone. With the help of rdf:type information, semi-automatic or even automatic ways of explaining such connectivity is possible.

For all owl:sameAs statements between the source PLD *d1* and target PLD *d2*, we can retrieve the rdf:type information for *u* and *v* using the following SPARQL query:

```
SELECT ?subj_type ?obj_type
WHERE {
        ?s ex:hasPLD "d1".
        ?o ex:hasPLD "d2".
        ?s a ?sub_type.
        ?o a ?obj_type.
}
```

Then comparing the k-most frequently used types in *d1* with the k-most frequently used types in *d2* can help us to understand how the instance resources served by *d1* and *d2* are connected. Table 1 lists the top five (if exists) type labels for the source and target PLD of arcs.

Table 1. Top five most frequently used types for each arc in Table 1

Arc	Top-5 Types in Source PLD	Top-5 Types in Target PLD
\<dbtune.org, zitgist.com>	rdfs:Resource: 2864 mo:Track: 2382 mo:Record: 280 mo:MusicArtist: 202	mo:MusicArtist: 99515 mo:MusicGroup: 61368 foaf:Group: 61368 mo:Record: 58245 mo:SoloMusicArtist: 26058
\<l3s.de, bibsonomy.org>	foaf:Document: 366416 swrc:InProceedings: 254905 swrc:Article: 104295 swrc:Proceedings: 4164 swrc:Book: 550	N/A
\<l3s.de, rkbexplorer.com>	foaf:Document: 366073 swrc:InProceedings: 254567 swrc:Article: 104294 swrc:Proceedings: 4161 swrc:Book: 549	N/A
\<bibsonomy.org, uni-trier.de>	swrc:InProceedings: 308486 swrc:Article: 13339 swrc:Proceedings: 3216 swrc:InCollection: 1284 owl:Ontology: 89	N/A
\<freebase.com, dbpedia.org>	freebase:base.intellectualproperty. valuable_item: 240685 freebase:medicine.hospital: 51587 freebase:user.morrowjtm.default_ domain.sexuality: 46726 freebase:base.onlineadvertising.ad _pricing_model: 24968 freebase:user.ericqianli.default_do main.css: 24123	yago:NeighbourhoodsOfLewisham: 4312 RailwayStationsInLewisham: 638 dbpedia-owl:ProtectedArea: 564 yago:HighSchoolsInCentralPennsylvania: 524 yago:IndigenousPeoplesOfEurope: 519

The first row in Table 1 indicates that both PLD $d1$ = dbtune.org and PLD $d2$ = zitgist.com are publishing data about music, because the top five types related to all owl:sameAs links between them are generally well aligned and are using concepts from the Music Ontology[12]. Row 2, 3, and 4 are all missing the type information in the target PLD, which indicates that cross-PLD owl:sameAs links do not have type information for resources in the target PLD. Finally, the top five types in the source and target PLD do not align very well in the last row. This might be due to the vast amount of general human knowledge encoded by dbpedia.org and freebase.com, as well as the unique role of "knowledge hubs" that they have been playing on the Web. Actually, the top-k types discussed here can also be used to form a more complete view of either the source or the target PLD, in which case the owl:sameAs statements function as a clue for discovering more information for either side.

6   CLS Network Analysis

In order to show how Web ontologies are affected by owl:sameAs inference, we select an ontology mapping use-case: detecting the relations between two RDFS/OWL

---

[12] Music Ontology: http://musicontology.com/

classes. Two classes are considered overlapping when they share common instances. Classes inter-connected by such "class-overlap" relation form a Class-Level Similarity (CLS) network. With the CLS network, users can automatically detect clusters of classes and ontology mappings using machine learning techniques.

6.1 Definitions and Notation

Definition 6: **CLS network.** A *CLS network* is a weighted directed graph of classes where (i) each node denotes a unique RDFS/OWL class; (ii) each arc links two classes using one of the following relations: equivalence, subclass-of, disjointness and class-overlap. While the first three types of relations can be mapped to OWL properties, the last one cannot. The weight of an arc is calculated based on the number of common instances shared by the two classes linked by the arc.

As shown in Table 2, A CLS network can be constructed using SPARQL queries, namely Query A and Query B. Note that Query B leverages owl:sameAs inference to derive additional class-overlap relations, and it simply assumes that owl:sameAs is neither symmetric nor transitive. Other possible assumptions are left for future study.

Table 2. Two SPARQL queries for generating class-overlap relations

Query A	CONSTRUCT ?C1 ex:overlaps ?C2
	WHERE { ?s a ?C1, ?C2. filter (?C1!=?C2) }
Query B	CONSTRUCT ?C1 ex:overlaps ?C2
	WHERE { ?u1 a ?C1 . ?u2 a ?C2 . ?u1 owl:sameAs ?u2. filter (?C1 != ?C2) }

6.2 CLS Network and Enhancement

We executed Query A on all Type statements in ESameNet to build a CLS network CLS-ALL, which contains 168,503 unique nodes (RDFS/OWL classes) and hundreds of millions of arcs. Overall, the in-degree of classes (i.e. how many instances the classes have) follows a power-law distribution: about 45% (77 K) classes only have one instance, while a few have over 100 million instances each.

Focusing on the RDF resources connected by SameAs statements, we created a smaller CLS network CLS-SAME, which contains 6,555 unique nodes (RDFS/OWL classes) and 21,628 arcs (weighted differently) using Query B. Although CLS-SAME is much smaller than CLS-ALL, it helps users to quickly gather additional pairs of classes for determining class-level relations. Table 3 lists 20 class pairs in the CLS-SAME dataset. We found a couple of interesting observations:

- The rows with type [EQ] show that some class pairs could be mapped via equivalence relation because their URIs have the same local-name. This kind of class pairing can be used to guess equivalence relations.
- The rows with type [ERR] show that some class pairs may also be inappropriate mappings after checking their ontological definitions. Although this kind of class pairing does not help ontology mapping, it does help users to detect potential errors in Linked Data.

- The rows without a type label show that it is hard to determine the mapping relations between the class pairs by checking their URIs or ontological definitions. This kind of case usually involves a general-purpose class, such as <http://semantic-mediawiki.org/swivt/1.0#Subject>. This kind of class paring may be used to guess sub-class relations.

Table 3. List of 20 class pairs in CLS-SAME dataset

type	FROM	TO
	<http://semantic-mediawiki.org/swivt/1.0#Subject>	<http://xmlns.com/foaf/0.1/Person>
EQ	<http://www.w3.org/2002/07/owl#Class>	<http://www.w3.org/2000/01/rdf-schema#Class>
ERR	<http://www.w3.org/2002/07/owl#Class>	<http://www.w3.org/2002/07/owl#Thing>
	<http://www.geonames.org/ontology#Code>	<http://www.w3.org/2004/02/skos/core#Concept>
	<http://www.w3.org/2004/02/skos/core#Concept>	<http://www.geonames.org/ontology#Code>
EQ	<http://www.daml.org/2001/09/countries/iso-3166-ont#Country>	<http://rdf.geospecies.org/ont/geospecies#Country>
EQ	<http://www.geonames.org/ontology#Country>	<http://rdf.geospecies.org/ont/geospecies#Country>
	<http://semantic-mediawiki.org/swivt/1.0#Subject>	<http://referata.com/wiki/Special:URIResolver/Category-3APeople>
	<http://referata.com/wiki/Special:URIResolver/Category-3APeople>	<http://semantic-mediawiki.org/swivt/1.0#Subject>
	<http://www.w3.org/1999/02/22-rdf-syntax-ns#Property>	<http://www.w3.org/2002/07/owl#ObjectProperty>
	<http://semantic-mediawiki.org/swivt/1.0#Subject>	<http://xmlns.com/foaf/0.1/Agent>
	<http://semantic-mediawiki.org/swivt/1.0#Subject>	<http://discoursedb.org/wiki/Special:URIResolver/Category-3APositions>
	<http://discoursedb.org/wiki/Special:URIResolver/Category-3APositions>	<http://semantic-mediawiki.org/swivt/1.0#Subject>
EQ	<http://www.rdfabout.com/rdf/schema/usgovt/State>	<http://rdf.geospecies.org/ont/geospecies#State>
EQ	<http://data.linkedmdb.org/resource/movie/country>	<http://rdf.geospecies.org/ont/geospecies#Country>
	<http://xmlns.com/foaf/0.1/Person>	<http://semantic-mediawiki.org/swivt/1.0#Subject>
	<http://sw.opencyc.org/2008/06/10/concept/Mx4rqEYnNVMqEdaSKAACs0x8nw>	<http://www.w3.org/2002/07/owl#Class>
	<http://semantic-mediawiki.org/swivt/1.0#Subject>	<http://discoursedb.org/wiki/Special:URIResolver/Category-3ASources>
	<http://discoursedb.org/wiki/Special:URIResolver/Category-3ASources>	<http://semantic-mediawiki.org/swivt/1.0#Subject>
ERR	<http://xmlns.com/foaf/0.1/PersonalProfileDocument>	<http://xmlns.com/foaf/0.1/Person>

The above observations about the class pairs in the CLS network reflect that the BTC 2010 dataset is quite heterogenous and the current Semantic Web vocabularies are largely orthogonal. They also enlighten the potential use of the CLS network: with effective classification techniques, we may appropriately label class pairs in the CLS network and then support automated class alignment and error detection. In our future work will also try other combinations of assumptions including the assumption that owl:sameAs is transitive.

7 Related Work

Various recent literature [4-6] investigating pragmatic issues of owl:sameAs in the context of the Web of Data can be considered as directly related to our study. They provide valuable insights into the essential research problem of whether the ubiquitous use of owl:sameAs to inter-linked datasets is correct. Some of them identify incorrect usages of owl:sameAs in the Web of Data [5], leading to the explicit need for a co-reference management infrastructure for the Semantic Web. Others carry out in-depth discussions of the issues with the current semantics of owl:sameAs. For example, McCusker and McGuinness [6] discuss how and why using owl:sameAs could possibly result in confusions of provenance and ground truths in the bioinformatics context. Halpin and Hayes [4] view owl:sameAs statements as a special type of "identity link", and analyze the more general problem of identity links on the Semantic Web from a philosophical and knowledge representation perspective. They also outline four alternative interpretations of owl:sameAs, which all differ from the canonical OWL semantics as defined by W3C documents. Our work differs from all of the above in that, to the best of our knowledge, we are the first to conduct this type of large-scale empirical study on the deployment status of owl:sameAs using datasets from the Web of Data.

Another related research effort is the analysis of the graph structure of the Semantic Web. Some recent work [13-17] presents important graph metrics that reflect the basic shape, structure, and even dynamics of the whole Semantic Web viewed as a giant graph. It is reported in [14] that ontologies on the Semantic Web, like many natural and social networks, are scale-free. Some earlier [16, 17] and later [15] studies show more structural features of the Semantic Web, such as size, diameter and power-law degree distribution of the graph. In one of the more recent efforts that falls into this category, Ge *et al* [13] propose the notion of an Object Link Graph (OLG) for the Semantic Web, and show that it is also scale-free and has a small diameter. Our work is similar to these research efforts in the sense that we also present critical graph structure metrics. However, the subject of research focus, i.e., the owl:sameAs statements, and the scale are two major factors that differentiate our work with theirs.

Some of the existing endeavors, which make use of instance-level links to derive potential alignments and associations at the schema level, are also related to our work. Qu et al [18] propose the notion of a Class Association Graph (CAG), which is obtained from the Object Link Graph (OLG) defined in [13]. Similarly, Nikolov et al [19] illustrate how to establish schema-level mappings based on existing instance-level mappings in the Web of Data. Our study shares essentially the same idea of deriving schema-level relations using vast amounts of instance-level data.

8   Conclusion and Future Work

In order to better understand and use owl:sameAs in Linked Data, it is useful to study how owl:sameAs is actually deployed, which has implications for how data should be consumed. To the best of our knowledge, this work is the first study on SameAs networks extracted from the real world Web of Data, and it has reported statistically significant results based on the BTC 2010 dataset. The experiment results are the core of this work, and they support the goal of this paper – to highlight the uniqueness, interestingness and utility of SameAs networks to Linked Data researchers as well as practitioners.

- Section 4 shows that SameAs networks have unique graph properties in comparison with other networks in the Semantic Web. The graph properties also lead to nice computational properties of the SameAs network.

- Section 5 explains the interestingness of SameAs networks by showing the similarity between the PLD network and the LOD graph. We also showed that the PLD network could be used to explain how LOD datasets are actually linked by common topics.
- Section 6 shows one practical use of SameAs networks, where classes can be linked by means of common instances (derived by owl:sameAs inference). The CLS network has a great potential in detecting schema-level inconsistencies in interlinked datasets and supporting ontology alignment.

The results reported in this study can be easily extended with additional data, semantics and applications. For example, we can enrich the ESameNet dataset with SameAs statements generated using OWL inference on the entire BTC dataset (e.g. inferring owl:sameAs using owl:InverseFunctionalProperty) [11] and then evaluate the impact on the diameter of SameAs networks. Although this study does not assume the transitivity of owl:sameAs for the purpose of deriving the CLS network, future work may explore the alternative - evaluating the impact of transitive inference on SameAs networks. Another potential research direction is to follow up on our previous discussions on the operational semantics of owl:sameAs [12]. Last but not least, it is worth noting that owl:sameAs has implications not only for the two networks mentioned in this study, but rather, we can use BTC datasets from consecutive years to evaluate the evolution of SameAs Networks over time, and use owl:sameAs statements to compute property-level mappings.

References

[1] Bizer, C., Heath, T., Berners-Lee, T.: Linked Data - The Story So Far. International Journal on Semantic Web and Information Systems (IJSWIS) 5(3), 1–22 (2009)
[2] Bechhofer, S., van Harmelen, F., Hendler, J., Horrocks, I., McGuinness, D.L., Patel-Schneider, P., Stein, L.A.: OWL Web Ontology Language Reference. W3C Recommendation (February 2004)
[3] Cyganiak, R.: Linked data at the New York Times: Exciting, but buggy,
     `http://dowhatimean.net/2009/10/`
     `linked-data-at-the-new-york-times-exciting-but-buggy`
     (last retrieved September 2010)
[4] Halpin, H., Hayes, P.J.: When owl:sameAs isn't the same: An analysis of identity links on the semantic web. In: Proceedings of the International Workshop on Linked Data on the Web (2010)
[5] Jaffri, A., Glaser, H., Millard, I.: URI disambiguation in the context of linked data. In: Proceedings of the 1st International Workshop on Linked Data on the Web (2008)
[6] McCusker, J., McGuinness, D.L.: owl:sameAs considered harmful to provenance. In: Proceedings of the ISCB Conference on Semantics in Healthcare and Life Sciences (2010)
[7] Vatant, B.: Using owl:sameAs in linked data,
     `http://blog.hubjects.com/2007/07/`
     `using-owlsameas-in-linked-data.html` (last retrieved September 2010)
[8] Lee, H., Leonard, D., Wang, X., Loguinov, D.: IRLbot: scaling to 6 billion pages and beyond. In: Proceeding of the 17th International Conference on World Wide Web (2008)

[9] Ding, L., Zhou, L., Finin, T., Joshi, A.: How the Semantic Web is Being Used: An Analysis of FOAF Documents. In: HICSS38 (2005)

[10] Shannon, P., Markiel, A., Ozier, O., Baliga, N.S., Wang, J.T., Ramage, D., Amin, N., Schwikowski, B., Ideker, T.: Cytoscape: a software environment for integrated models of biomolecular interaction networks. Genome Res. 13(11), 2498–2504 (2003)

[11] Williams, G.T., Weaver, J., Atre, M., Hendler, J.A.: Scalable Reduction of Large Datasets to Interesting Subsets. Journal of Web Semantics: Science, Services and Agents on the World Wide Web 8 (2010)

[12] Ding, L., Shinavier, J., Finin, T., McGuinness, D.L.: owl:sameAs and Linked Data: An Empirical Study. In: Proceedings of the WebSci10: Extending the Frontiers of Society On-Line (2010)

[13] Ge, W., Chen, J., Hu, W., Qu, Y.: Object Link Structure in the Semantic Web. In: Aroyo, L., Antoniou, G., Hyvönen, E., ten Teije, A., Stuckenschmidt, H., Cabral, L., Tudorache, T. (eds.) ESWC 2010. LNCS, vol. 6089, pp. 257–271. Springer, Heidelberg (2010)

[14] Zhang, H.: The Scale-Free Nature of Semantic Web Ontology. In: Proceeding of the 17th International Conference on World Wide Web, WWW (2008)

[15] Theoharis, Y., Tzitzikas, Y., Kotzinos, D., Christophides, V.: On Graph Features of Semantic Web Schemas. IEEE Transactions on Knowledge and Data Engineering 20(5) (May 2008)

[16] Ding, L., Finin, T.: Characterizing the Semantic Web on the Web. In: Cruz, I., Decker, S., Allemang, D., Preist, C., Schwabe, D., Mika, P., Uschold, M., Aroyo, L.M. (eds.) ISWC 2006. LNCS, vol. 4273, pp. 242–257. Springer, Heidelberg (2006)

[17] Ding, L.: Enhancing Semantic Web Data Access. Ph.D Thesis. Department of Computer Science and Electrical Engineering, University of Maryland, Baltimore County (2006)

[18] Qu, Y., Ge, W., Cheng, G., Gao, Z.: Class Association Structure Derived From Linked Objects. In: Proceedings of the WebSci 2009: Society On-Line (2009)

[19] Nikolov, A., Uren, V., Motta, E.: Data Linking: Capturing and Utilising Implicit Schema-level Relations. In: Proceedings of the Linked Data on the Web Workshop, 19th International World Wide Web Conference, WWW (2010)

Deciding Agent Orientation on Ontology Mappings

Paul Doran[1], Terry R. Payne[1], Valentina Tamma[1], and Ignazio Palmisano[2]

[1] Department of Computer Science, University of Liverpool,
Liverpool L69 3BX, United Kingdom
{P.Doran,T.R.Payne,V.Tamma}@liverpool.ac.uk
[2] School of Computer Science, University of Manchester M13 9PL, UK
ignazio.palmisano@cs.manchester.ac.uk

Abstract. Effective communication in open environments relies on the ability of agents to reach a mutual understanding of the exchanged message by reconciling the vocabulary (ontology) used. Various approaches have considered how mutually acceptable mappings between corresponding concepts in the agents' own ontologies may be determined dynamically through argumentation-based negotiation (such as Meaning-based Argumentation, *MbA*). In this paper we present a novel approach to the dynamic determination of mutually acceptable mappings, that allows agents to express a private acceptability threshold over the types of mappings they prefer. We empirically compare this approach with the Meaning-based Argumentation and demonstrate that the proposed approach produces larger agreed alignments thus better enabling agent communication. Furthermore, we compare and evaluate the fitness for purpose of the generated alignments, and we empirically demonstrate that the proposed approach has comparable performance to the *MbA* approach.

1 Introduction

The problem of dynamic reconciliation of *ontologies* (vocabularies) used by agents during interactions has received significant attention [8,10,12], due to the growing adoption of mobile and service computing. In these scenarios, agents situated in open environments encounter unknown agents offering new services as a user's context or location changes. As the heterogeneity that permeates these environments increases, fewer assumptions on the vocabulary and content of these ontologies can be made, thus hindering seamless interaction between the agents.

The reconciliation of heterogeneous vocabularies has been investigated at length by research efforts in *ontology alignment* [7], which tries to determine suitable mappings between two ontologies. However, there are few traditional alignment approaches suitable for use in purely dynamic interaction scenarios as most require human intervention, or they align the ontologies at design time [11]. Although recent systems [6] have emerged that can generate alignments at

P.F. Patel-Schneider et al. (Eds.): ISWC 2010, Part I, LNCS 6496, pp. 161–176, 2010.

run time, these are often machine-learning based, requiring pre-labelled training data to guide the learning process.

Whilst this has been demonstrated to be effective when such data is available, it is not always suitable for all dynamic problems. Two agents may encounter each other for the first time with the aim of interacting to achieve some goal (where each agent may have its own preferences or policies over the terms and axioms used within a specific interaction). Whilst alignments may exist between the agents' ontologies, these may have been determined under different contexts or assumptions, and thus may not necessarily satisfy the current agents' preferences or policies. In order to address this limitation, and to consider the context within which the alignment is to be used, Laera *et. al.* [8] proposed in their Meaning-based Argumentation (*MbA*) approach the use of argumentation to select a set of mappings (i.e. an *alignment*) that is mutually acceptable to the negotiating agents, from the union of disparate, precomputed alignments where different alignments may have previously been generated (e.g. for previous agent-agent interactions) and then published or retained for future use.

Therefore, the problem can be cast as a search for a mutually acceptable set of mappings between two ontologies O_1 and O_2 (in the union of mappings previously computed), given the agents' individual, private preferences over the mapping type (i.e. terminological, extensional, etc.). Approaches such as those proposed by Laera *et al.* [8] and dos Santos *et al.* [10] assume that mappings have an associated confidence value, and based on this, utilise both an acceptance threshold, ϵ, and their preferences to determine whether or not a candidate mapping is suitable for a task.

The search is conducted collaboratively, through the use of argumentation. By specifying arguments that *support* (or *refute*) different mappings, the negotiating agents identify a subset of mappings that are considered mutually acceptable, which can subsequently be used to support further communication between the agents. The arguments are determined from the individual agent's preferences over the mapping types (which can vary, depending on the agents task or the expressive power of ontology it commits to) and its acceptance threshold. The argumentation converges on a set of *agreed* mappings, i.e. mappings that are mutually acceptable to the negotiating agents.

As the generation of arguments is directed by a single preference and acceptance threshold specified by each agent, this approach is susceptible to rejecting those mappings which, whilst not optimal, may still be considered acceptable to all the agents involved. This results in smaller alignments, which may fail to sufficiently support the agent's subsequent communication. This approach may also fail to reflect the true preference of an agent, as the different grounds supporting the choice or type of mapping may actually generate similar mappings in some cases.

In this paper, we demonstrate the effect of this limitation on the resulting alignment empirically, and propose a novel approach for generating arguments for each of the candidate mappings, utilising a weaker notion of *suitability* than that originally proposed. The *flexible approach for determining agents' orientation on ontology mappings* (*FDO*) proposed here provides a flexible mechanism for agents to decide whether they support or refute an argument about a mapping, and hence it allows agents to *compromise* over the suitable mappings; i.e by arguing in favour of an assertion that may not be amongst the preferred ones, but that facilitates the negotiation process in converging on a mutually acceptable solution. In this way, the agents create a larger consensus base, by increasing the number of arguments over which the agents negotiate, and that better reflect the agents' preferences over the type of mappings deemed to facilitate the exchange of messages. Whilst this approach results in agents relaxing some of their preferences over suitable mappings, we demonstrate that it produces a larger consensus over possible mappings due to the generation of a greater number of arguments in favour of the candidate mappings (compared to Laera *et al.*'s *MbA* approach), and better reflects the agents preferences than when only a single threshold and preference value is used. We also demonstrate that allowing the negotiation to take place over a larger set of arguments does not degrade the quality of the alignment produced, measured in terms of precision and recall over query answering tasks. Therefore, the contribution of this paper is twofold: we provide a novel approach to the determination of whether an agent supports or refutes an argument, and we provide an evaluation of this novel approach against the *MbA* approach.

The paper is organised as follows: the *MbA* approach is briefly summarised, followed by the description of our novel *FDO* approach for determining an agent's orientation on a mapping. This approach is then illustrated by means of an example, before being evaluated empirically. The results of the evaluation are then discussed, before concluding.

2   Arguing over Ontology Mappings

Meaning-based Argumentation (*MbA*), as proposed by Laera *et al.* [8], assumes that a number of precomputed alignments (i.e. sets of mappings) exist within some publicly available repository. A similar assumption is also made by dos Santos *et al.* [10], whereby such alignments are known (possibly computed *on-the-fly*) by different agents. Before presenting our *flexible approach for determining agent orientation*, we first give the formal definition of these alignments, and summarise the *MbA* approach[1].

A mapping between two agent ontologies O_1 and O_2 is described as a tuple: $m = \langle e, e', n, r \rangle$, where $e \in O_1$ and $e' \in O_2$ are the entities (concepts, properties or individuals) between which a relation, r, is asserted, such as equivalence, or subsumption, and n is a degree of confidence in this correspondence [7]. These mappings can either be computed offline and stored by a dedicated server, an *Ontology Alignment Service*, that provides the set of available *candidate mappings* the agents need to argue over [8], or they can be determined on the fly [10]. Whatever the approach used to generate the mappings, the argumentation process considers as input a set of pre-computed mappings, and a set

---

[1] We focus primarily on the *MbA* approach since the negotiation phase in dos Santos *et al.* is the same as the one used in *MbA*.

of justifications that motivate the existence of a mapping, that are provided by the mapping generation approach.

The Meaning-based Argumentation (*MbA*) process is based on the *Value-Based Argumentation Framework* (VAF) [3], which introduced the notions of *audience* and *preference values*. An audience represents a group of agents who share the same preferences over a set of values, with a single value being assigned to each argument. This framework extends the seminal work by Dung on the use of argumentation theory [5]. In Dung's framework, attacks always succeed; in essence they are all given equal value. For deductive arguments this suffices, but within the ontology alignment negotiation scenario [8] the persuasiveness of an argument could change depending on the audience, where an audience represents a certain set of preferences. Thus, the Value-Based Argumentation Framework (VAF) facilitates the assignment of different strengths to arguments on the basis of the values they promote and the ranking given to these values by the audience for the argument. Hence, it is possible to systematically relate strengths of arguments to their motivations and to accommodate different audience interests.

Definition 1. *A Value-Based Argumentation Framework (VAF) is defined as* $\langle AR, A, \mathcal{V}, \eta \rangle$, *where:*

- $\langle AR, A \rangle$ *is an argumentation framework;*
- $\mathcal{V}$ *is a set of k values which represent the types of arguments;*
- $\eta : AR \rightarrow \mathcal{V}$ *is a mapping that associates a value* $\eta(x) \in \mathcal{V}$ *with each argument* $x \in AR$.

The types of arguments represented by $\mathcal{V}$ typically varies, depending upon the application. Within the *MbA* process, the values of $\mathcal{V}$ correspond to five different types of ontological mismatches that can occur between ontologies, as represented in Table 1.

In order to model the notion of different agents having different perspectives on the same candidate mappings, we define an *audience*, i.e. the representation of a preference ordering of $\mathcal{V}$. The notion of *audience* is central to the VAF. Audiences are individuated by their preferences over the values. Thus, potentially, there are as many audiences as there are orderings[2] of $\mathcal{V}$. The set of arguments is assessed by each audience in accordance to its preferences. An audience is defined as follows:

Definition 2. *An* audience *for a VAF is a binary relation* $\mathcal{R} \subseteq \mathcal{V} \times \mathcal{V}$ *whose irreflexive transitive closure,* $\mathcal{R}^*$, *is asymmetric, i.e. at most one of* (v, v'), (v', v) *are members of* $\mathcal{R}^*$ *for any distinct* $v, v' \in \mathcal{V}$. *We say that* v_i *is preferred to* v_j *in the audience* $\mathcal{R}$, *denoted* $v_i \succ_{\mathcal{R}} v_j$, *if* $(v_i, v_j) \in \mathcal{R}^*$.

As this notion allows different agents (represented by an audience) to have different perspectives on the same candidate mapping, we need to model what it means for an argument to be acceptable relative to some audience. This is defined within the VAF as follows:

---

[2] Number of audiences corresponds to the different combinations of the elements in $\mathcal{V}$; i.e. *Number of audiences* $= |\mathcal{V}|!$

Table 1. The classification of different types of ontological alignment approaches

Semantic	M	These methods utilise model-theoretic semantics to determine whether or not there is a correspondence between two entities, and hence are typically deductive. Such methods may include propositional satisfiability and modal satisfiability techniques, or logic based techniques.
Internal Structural	IS	Methods for determining the similarity of two entities based on the internal structure, which may use criteria such as the range of their properties (attributes and relations), their cardinality, and the transitivity and/or symmetry of their properties to calculate the similarity between them.
External Structural	ES	Methods for determining external structure similarity may evaluate the position of the two entities within the ontological hierarchy, as well as comparing parent, sibling or child concepts.
Terminological	T	These methods lexically compare the strings (tokens or n-grams) used in naming entities, or in the labels and comments concerning entities. Such methods may employ normalisation techniques (often found in Information Retrieval systems) such as stemming or eliminating stop-words, etc.
Extensional	E	Extension-based methods which compare the extension of classes, i.e., their set of instances. Such methods may include determining whether or not the two entities share common instances, or may use alternate similarity based extension comparison metrics.

Definition 3. *Let $\langle AR, A, \mathcal{V}, \eta \rangle$ be a VAF, with R and S as subsets of AR, and an audience $\mathcal{R}$:*

(a) For $x, y \in AR$, x is a successful attack *on y with respect to $\mathcal{R}$ if $(x, y) \in A$ and $\eta(y) \not\succ_{\mathcal{R}} \eta(x)$.*

(b) $x \in AR$ is acceptable *with respect to S with respect to $\mathcal{R}$ if for every $y \in AR$ that successfully attacks x with respect to $\mathcal{R}$, there is some $z \in S$ that successfully attacks y with respect to $\mathcal{R}$.*

(c) S is conflict-free *with respect to $\mathcal{R}$ if for every $(x, y) \in S \times S$, either $(x, y) \notin A$ or $\eta(y) \succ_{\mathcal{R}} \eta(x)$*

(d) A conflict-free set S is admissible *with respect to $\mathcal{R}$ if every $x \in S$ is acceptable to S with respect to $\mathcal{R}$*

(e) S is a preferred extension *for the audience $\mathcal{R}$ if it is a maximal admissible set with respect to $\mathcal{R}$*

(f) $x \in AR$ is subjectively acceptable *if and only if x appears in the preferred extension for some specific audience.*

(g) $x \in AR$ is objectively acceptable *if and only if x appears in the preferred extension for every specific audience.*

(h) $x \in AR$ is indefensible *if it is neither subjectively nor objectively acceptable.*

Laera *et. al.* [8] subsequently adopted the VAF for the Meaning-based Argumentation (*MbA*) process, allowing agents to express preferences for different mapping types, and restricting the arguments to those concerning ontology mappings allowing agents to explicate their mapping choices. The definition of an agent and an argument are as follows:

Definition 4. *An agent, Ag_i, is characterised by the tuple $\langle O_i, VAF_i, Pref_i, \epsilon_i \rangle$ such that O_i is an ontology, VAF_i is a instance of a VAF, $Pref_i$ is an ordering over the possible values in $\mathcal{V}$ and ϵ_i is a private threshold between 0 and 1.*

Definition 5. *An argument* $x \in AR$ *is a triple* $x = \langle G, m, \sigma \rangle$ *where* m *is a mapping,* G *is the grounds justifying the* prima facie *belief that the mapping does or does not hold and* σ *is one of* $\{+, -\}$ *depending on whether the argument is that* m *does or does not hold.*

Thus, when arguing over ontology mappings using the VAF, an argument $x \in AR$ either supports or refutes a mapping m, depending on the value of σ. An agent determines this σ (i.e. decides whether to argue for or against a mapping) based on its preferences and threshold. Given the set of mappings $\mathcal{M} = \{m\}_{j=1,\ldots,p}$, such that p is the number of mappings, and the function[3] $\tau : M \to \mathcal{V} \mid \tau(m) = v \in \mathcal{V}$ then an agent can set the value of σ for an argument, x, about a mapping, m, as follows:

$$\sigma = \begin{cases} +, & \text{if } max(Pref_i) = \tau(m) \wedge n_m \geq \epsilon_i \\ -, & \text{otherwise} \end{cases} \quad (1)$$

The notion of an attack and counter-attack is also formally defined, whereby x is attacked by the assertion of its negation, $\neg x$.

Definition 6. *An argument* $x \in AR$ *attacks an argument* $y \in AR$ *if* x *and* y *are arguments for the same mapping,* m, *but with different* σ. *For example, if* $x = \langle G_1, m, + \rangle$ *and* $y = \langle G_1, m, - \rangle$, x *counter-argues* y *and vice-versa.*

The agents can now express, and exchange, their arguments about ontology mappings and decide from their perspective, audience, what arguments are in their preferred extension; but the agents still need to reach a mutually acceptable position with regards to what ontology alignment they actually agree upon. Laera *et. al.* define the notion of *agreed* and *agreeable* alignment as follows:

Definition 7. *An* agreed alignment *is the set of mappings supported by those arguments which are in every preferred extension of every agent.*

Definition 8. *An* agreeable alignment *extends the agreed alignments with those mappings supported by arguments in some preferred extensions of every agent.*

Thus, a mapping is *rejected* if it is in neither the agreed nor agreeable alignment. Given the context of agent communication it is rational for the agents to accept as many candidate mappings as possible [8], thus both sets of alignments are considered. The agents should only completely disagree when they want the opposite, indeed, the agents gain little by arguing and not reaching some kind of agreement.

The definition of audience is central to the notion of *acceptability* of an argument, since given a set of arguments, and their respective counter-arguments, the agents in an audience need to consider which of them they should accept. The acceptability of some arguments with respect to some audience, depends on the agents ability to determine a *preferred extension* that represents a consistent

---

[3] In some cases $\tau(m) = \eta(x_m)$, however in general this assumption does not hold.

position within an argumentation framework that can be defended against all attacks, and cannot be further extended without causing it to be inconsistent or open to attacks. The mappings supported in the preferred extensions form the mutually agreed set of mappings [8].

3   A Flexible Approach for Determining Agents' Orientation on Mappings

The meaning based negotiation approach by Laera *et al.* is the first attempt to tackle the problem of dynamic reconciliation of heterogeneous agent ontologies. Whilst the approach has the merit of having highlighted an important problem, the proposed solution presents a serious limitation, primarily due to the way σ is obtained.

In Laera's approach an agent argues only in favour of those arguments whose grounds have the *highest* ranking in the ordering of agent preferences, whilst all the other mappings are argued against. Hence, effectively the agents can only express one preference towards one type of mapping, and will argue against any other type of mapping, therefore greatly reducing the possibility to find a suitable agreement on a set of mappings. In other words, this approach fails to distinguish mappings that are *less preferred* from those mappings for which an agent is against.

In addition, this type of strict decision process could potentially increase the chance that inconsistent mappings are determined by the VAF. The walkthrough example presented in the next section illustrates an occurrence of this unlikely but possible event.

In this paper, we present an alternative approach that aims at recognising how agents can have different preferences over the types of mappings to use in interactions with other agents, and that these preferences can influence the decision making process behind the negotiation. An agent would ideally try to maximise the use of those types of mappings with the highest preferences, however, since it needs to interact with other agents (with their own preferences) then it might decide to *compromise*, i.e. to agree to use a less preferred mapping type if this facilitates communication.

This is the main motivation behind the novel approach to mapping selection that we present here. It builds on some of the notions presented in the previous section for the *MbA* approach, but gives agents more flexibility in deciding their orientation, i.e. whether to support or refute a mapping.

Given two agents ontologies O_1 and O_2, a mapping between $e \in O_1$ and $e' \in O_2$ is a tuple $m = \langle e, e', n, r \rangle$, as defined in the previous section. Analogously to *MbA* we define a VAF as a tuple $\langle AR, A, V, \eta \rangle$ that is similar to the definition given in the previous section (likewise for the definition of mapping m). In the *flexible approach for determining agents' orientation on a mapping* (*FDO*) proposed here, we define an agent as a tuple $Ag_i = \langle O_i, VAF_i, Pref_i, \phi_i \rangle$, where O_i is an ontology, VAF_i is a instance of a VAF, $Pref_i$ is an ordering of the values in V and $\phi_i : V \rightarrow [0, \cdots, 1]$ maps each v in V to a value $0 \leq z \leq 1$.

$\phi_i(v)$ represents the minimum confidence threshold for Ag_i to argue in favour of a mapping of type v.

Let us consider the function $\tau : \mathcal{M} \to \mathcal{V}$ that assigns a $v \in \mathcal{V}$ to every $m \in \mathcal{M}$, then the agent decides whether to be in favour or against the mapping as follows:

$$\sigma = \begin{cases} +, & \text{if } n_m \geq \phi_i(\tau(m)) \\ -, & \text{otherwise} \end{cases} \qquad (2)$$

In our approach, an agent determines its orientation on a mapping solely on the basis of the minimum confidence threshold for arguing in favour of a mapping type, and no longer on the ordering of preferences. In this way, the agents express how much they prefer each of the possible mapping types, and how willing they are to argue in their favour. The ordering of preferences is now only used by the VAF when dealing with arguments and their attacks.

4   Illustrative Example

The following example illustrates how the proposed *FDO* approach differs from the original *MbA* approach, assuming the two ontologies illustrated in Figure 1, with the mappings given with their relevant mapping types. Mapping m_1 is a *Terminological* equivalence mapping between concepts A and C, with a confidence of 0.75, whereas mappings m_2 and m_3 are *External Structural* equivalence mappings: m_2 between concepts B and D (confidence 0.85); and m_3 between concepts B and E (confidence 0.65). Note that concepts D and E are disjoint, and thus an alignment containing both mappings m_2 and m_3 would be inconsistent.

Given two agents that wish to communicate: Ag_1 has the preference ordering ES≻T; whereas Ag_2 has the preference ordering T≻ES. Table 2 shows the sets of mappings that will be argued in favour of (+) or against (-). With the *MbA* approach, we assume that the acceptance threshold $\epsilon_1 = \epsilon_2 = 0.5$. Ag_1 will argue in favour of m_2 and m_3, and against m_1; whereas Ag_2 will argue against m_2 and m_3, but in favour of m_1. This is due to the fact that, in the case of Ag_1, only

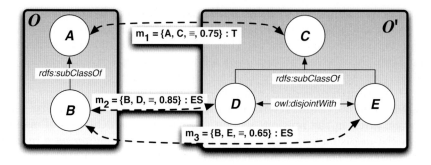

Fig. 1. An alignment between O and O'

Table 2. The arguments that support (+) or refute (-) different mappings, given thresholds and preferences

Approach	Mapping Type Preference	Acceptance Threshold	Arguments *in favor of +*	*against -*
MbA	ES ≻ T	0.5	$\{m_2, m_3\}$	$\{m_1\}$
	T ≻ ES	0.5	$\{m_1\}$	$\{m_2, m_3\}$
FDO	ES ≻ T	ES=0.5, T=0.7	$\{m_1, m_2, m_3\}$	$\{\}$
	T ≻ ES	T=0.5, ES=0.7	$\{m_1, m_2\}$	$\{m_3\}$

mappings of the first preference ordering were considered (subject to exceeding the acceptance threshold), and all other mappings were automatically refuted. The resulting attack graph is illustrated in Figure 2 (left), where each argument is assigned a label corresponding to its mapping, and the mapping type. These types are the values in the VAF, with each agent having a private preference ordering over them.

The *FDO* approach, however, assigns a separate acceptance threshold for each mapping type. Ag_1 assumes a 0.5 threshold for ES, but a 0.7 threshold for T, whereas Ag_2 assumes a 0.7 threshold for ES, and a 0.5 threshold for T. In this case, arguments are generated by Ag_1 in favour of all three mappings, whereas Ag_2 generates mappings in favour of m_1 and m_2, but against m_3. Although Ag_1 expresses a preference ordering for ES ≻ T, the confidence value of all three mappings exceeds the acceptance threshold for the different mapping types. The resulting attack graph is illustrated in Figure 2 (right).

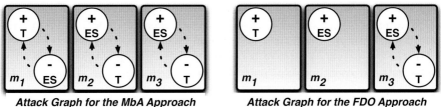

Attack Graph for the MbA Approach   **Attack Graph for the FDO Approach**

Fig. 2. Attack graphs for the *MbA* and *FDO* procedures

From the attack graphs shown in Figure 2 the preferred extensions for each audience can be computed for the *MbA* approach (see below). This does not produce an agreed alignment, but does produce an agreeable alignment, corresponding to $\{m_1, m_2, m_3\}$. However, as mentioned earlier, if this agreeable alignment were to be accepted by both agents, their ontologies would become inconsistent, thus making the ontologies and the resulting alignment unusable.

- T ≻ ES = $\{m_1+, m_2-, m_3-\}$
- ES ≻ T = $\{m_1-, m_2+, m_3+\}$

In contrast, the *FDO* approach produces an agreed alignment $\{m_1, m_2\}$, whereas mapping $\{m_3\}$ would only appear in an agreeable alignment. Thus, if the agreed alignment is accepted by both agents, they would be able to communicate with respect to concepts A, B, C, and D, but not with concept E.

5   Empirical Evaluation

The aim of the evaluation is to contrast the proposed *FDO* approach with the original *MbA* approach presented in [8]. Two hypotheses are explored: that the *FDO* approach generates a larger number of supporting arguments, resulting in more selected mappings that *MbA*; and that the increased number of mappings will better support communication tasks such as query answering (i.e. the resulting alignments are *fit for purpose*).

5.1   Evaluating the Generated Arguments

To explore the first hypothesis, the ratio of arguments in favour of mappings to those against was computed for both approaches, and the resulting mappings examined. This requires multiple candidate mappings based on different onto-logical grounds (and hence different mapping types) between ontologies of the same domain. Eleven ontologies were therefore taken from the *OAEI 2007 and 2008 Conference Track* repositories (with three exceptions[4]), as they represent different domain theories for the same, real-world domain (thus reflecting *real-world heterogeneity*) and can be used generate better pairwise alignments than ontologies from other tracks[5]. These ontologies (originally developed as part of the OntoFarm Project[6]) are listed in Table 3, complete with a brief characteri-sation in terms of the number of classes (named and anonymous) and properties (object and datatype), and their Description Logic expressivity[7].

For the evaluation, a total of 55 ontology pairs were identified[8]. The align-ments between each ontology pair were generated using the *Alignment API* [7], which only produces mappings of type internal structural (IS), external structural (ES) and terminological (T); thus for our evaluation, we assume $\mathcal{V} = \{ES, IS, T\}$.

In order to investigate the differences depending on the threshold, 4 thresh-olds have been identified for each mapping type. The first, $\epsilon_1 = 0$ corresponds to the case where the agent will argue in favour of all arguments. The remaining

---

[4] These ontologies have memory requirements of $>$1.5GB.

[5] http://oaei.ontologymatching.org/2007/conference/

[6] http://nb.vse.cz/~svatek/ontofarm.html

[7] The expressivity of an ontology (and hence complexity of a reasoner) for a De-scription Logic is indicated by the concatenation of letters representing different DL operators [1].

[8] Note that the ordering of the ontologies in each pair is irrelevant; thus rendering an evaluation on the symmetric pairs unnecessary. Therefore, a total of $N(N-1)/2$ ontology pairs were used, where N correspond to the 11 ontologies listed in Table 3.

Table 3. Characteristics for the ontology test set

Ontology	Named Classes	Object Prop.	Datatype Prop.	Anon. Classes	Expressivity
cmt	29	49	10	11	$\mathcal{ALCHIF(D)}$
Conf	59	46	18	33	$\mathcal{ALCHIF(D)}$
confOf	38	13	23	42	$\mathcal{SHIF(D)}$
crs_dr	14	15	2	0	$\mathcal{ALCHIF(D)}$
edas	103	30	20	30	$\mathcal{ALCHIF(D)}$
ekaw	73	33	0	27	$\mathcal{SHIN}$
MICRO	31	17	9	33	$\mathcal{ALCHOIF(D)}$
OpenConf	62	24	21	63	$\mathcal{ALCHOI(D)}$
paperdyne	45	58	20	109	$\mathcal{ALCHOIF(D)}$
PCS	23	24	14	26	$\mathcal{ALCHIF(D)}$
sigkdd	49	17	11	15	$\mathcal{ALCHI(D)}$

thresholds are generated by determining the mean $\bar{x}$ and standard deviations of the confidence values for all the mappings for each of the types in $\mathcal{V}$, generated for the evaluation. Thus, $\epsilon_2 = \bar{x} - stdev(x)$, $\epsilon_3 = \bar{x}$, and $\epsilon_4 = \bar{x} + stdev(x)$. Whilst the upper limit ($\epsilon = 1$) was considered, this would have resulted in the agents arguing against all the mappings, resulting in empty alignments. The four levels have been varied independently, producing four actual preferences for each ordering; this produces 144 preferences for each pair of ontologies (again, discarding duplicates). The total number of argumentation situations is, therefore, 7920.

Each experimental argumentation scenario (AS) is defined by the following tuple:

$$AS = (O_1, O_2, P_1, P_2, A^{\sigma+}, A^{\sigma-}, M_{acc})$$

where the set of mappings over which to argue is determined univocally by the ontologies O_1 and O_2, together with the alignment technique used, with P_1 and P_2 representing the actual sets used depending on the approach. For MbA, $P_1 = (Pref_1, \epsilon_1)$, $P_2 = (Pref_2, \epsilon_2)$, i.e. for each agent we use the pair composed of the preference ordering and the threshold. For FDO, $P_x = (Pref_x, \phi_x)$, but in this case the $Pref_x$ is used only by the VAF (*not* in determining the agent orientation). $A^{\sigma+}$ and $A^{\sigma-}$ represent the set of arguments in favour and against any of the mappings in the argumentation respectively, while M_{acc} represents the set of accepted mappings, i.e., the mappings belonging to at least one preferred extension of one agent. These latter three parameters are recorded for each evaluation.

To compare the results between different ontologies, an index relating $A^{\sigma-}$ to the total number of arguments used has been defined; $NegArgs(AS) : AS \rightarrow [0, 1]$, where:

$$NegArg = \frac{|A^{\sigma-}|}{(|A^{\sigma-}| + |A^{\sigma+}|)}$$

The results have been grouped into nine scenarios based on the first mapping type of each agent preference $Pref_x$. Thus, each row entry in Table 4 is labeled by an Argument Scenario (AS) pair, such that the two values correspond to the first preferred mapping type of Ag_1 and Ag_2 respectively. The results present the averages[9] over each of the subsequent preference values; i.e. the pair (ES,IS) averages values for Ag_1 preferences ES $\succ$ (IS $\succ$ T | T $\succ$ IS), whereas for Ag_2, IS $\succ$ (ES $\succ$ T | T $\succ$ ES), etc. To compare scenarios based on these pairs, a comparison was made between *FDO* and *MbA* by pairing same ordering and same thresholds, since the structure of the preferences is the same for both approaches.

Table 4. Average number of arguments for each scenario

Argument Scenario	FDO Approach				MbA Approach			
	$A^{\sigma+}$	$A^{\sigma-}$	M_{acc}	NegArgs	$A^{\sigma+}$	$A^{\sigma-}$	M_{acc}	NegArgs
(ES, ES)	5230	2591	1364	0.34	1498	6533	739	0.8
(ES, IS)	5685	2896	1325	0.35	2560	6310	33	0.72
(ES, T)	5720	2698	1358	0.33	1209	7680	92	0.84
(IS, ES)	4626	2216	1136	0.33	1802	4640	20	0.73
(IS, IS)	5230	2591	1364	0.34	3032	4752	439	0.64
(IS, T)	6413	3132	1490	0.33	2177	6388	175	0.76
(T, ES)	4416	2479	1050	0.36	987	5828	73	0.85
(T, IS)	4237	2170	1036	0.35	1488	5135	111	0.77
(T, T)	5230	2591	1364	0.34	700	6880	418	0.89

When using *MbA*, the proportion of arguments against mappings averaged 78%, significantly greater than the 34% average of arguments that were generated against mappings with *FDO*. This can be clearly seen when examining the number of mappings that were generated when using *FDO* (for example, 1325 mappings on average for (ES, IS), compared to only 32.67 with *MbA*). This higher number of negative arguments generated by *MbA* suggests that it may result in a higher probability of generating empty alignments, thus resulting in unnecessary communication failure. Whilst these results support our hypothesis, it raises questions as to the suitability and hence fitness of the accepted mappings for a given task, which is addressed below.

5.2   Fitness Evaluation

The above evaluation demonstrated that the *FDO* approach produced a greater number of arguments in favour of mappings being generated than when using *MbA*, resulting in a larger number of mutually acceptable mappings. However, it is unclear whether the increase in mappings will result in a better alignment between two ontologies. To address this, new alignments were generated and evaluated (in terms of precision and recall) for a typical *query-answering* task. An alignment was selected to answer simple queries against one of the ontologies

---

[9] Note that these results include the arguments generated by both agents over all the mappings considered.

involved in the alignment, and the results compared to that achieved when a set of hand-crafted *reference* mappings (from the OAEI Alignment Challenge) were used. To investigate how the availability of different alignments affects the task, four alignment systems (*Asmov, Falcon, Lily* and *OntoDNA* [13] were used to generate the alignments, and the evaluations were conducted over different alignment combinations.

Table 5. Precision(P), Recall (R) and F-Measure (FM) values for a selection of combinations of alignments (where each alignment system is referenced by their initials)

	O_1, O_2	Base			FDO			MbA		
		R	P	FM	R	P	FM	R	P	FM
A/L/O	(cmt, ekaw)	0.60	1	0.75	0.60	1	0.75	0.58	1	0.74
	(cmt, sigkdd)	0.19	1	0.32	0.19	1	0.32	0.10	0.81	0.18
	(confOf, ekaw)	0.55	1	0.71	0.55	1	0.71	0.43	1	0.60
A/F/L	(cmt, confOf)	0.83	0.94	0.88	0.83	0.99	0.91	0.77	1	0.87
	(confOf, ekaw)	0.90	0.93	0.91	0.9	0.99	0.94	0.75	1	0.85
	(confOf, sigkdd)	1	0.96	0.98	1	0.99	1	0.59	0.61	0.60
A/O	(cmt, ekaw)	0.60	1	0.75	0.60	1	0.75	0.58	1	0.74
	(cmt, sigkdd)	0.19	1	0.32	0.19	1	0.32	0.10	0.81	0.18
	(confOf, ekaw)	0.55	1	0.71	0.55	1	0.71	0.43	1	0.60
A/F	(cmt, confOf)	0.83	0.94	0.88	0.83	0.99	0.91	0.77	1	0.87
	(confOf, ekaw)	0.90	0.93	0.91	0.90	0.99	0.94	0.75	1	0.85
	(confOf, sigkdd)	1	0.96	0.98	1	0.99	1	0.59	0.61	0.60
L/O	(cmt, ekaw)	0.60	1	0.75	0.60	1	0.75	0.58	1	0.74
	(cmt, sigkdd)	0.19	1	0.32	0.19	1	0.32	0.10	0.81	0.18
	(confOf, ekaw)	0.55	1	0.71	0.55	1	0.71	0.43	1	0.60
F	(cmt, confOf)	0.83	0.94	0.88	0.83	0.99	0.91	0.77	1	0.87
	(confOf, ekaw)	0.90	0.93	0.91	0.90	0.99	0.94	0.75	1	0.85
	(confOf, sigkdd)	1	0.96	0.98	1	0.99	1	0.59	0.61	0.60
O	(cmt, ekaw)	0.60	1	0.75	0.60	1	0.75	0.58	1	0.74
	(cmt, sigkdd)	0.19	1	0.32	0.19	1	0.32	0.10	0.81	0.18
	(confOf, ekaw)	0.55	1	0.71	0.55	1	0.71	0.43	1	0.60
F/L	(cmt, confOf)	0.83	0.94	0.88	0.83	0.99	0.91	0.77	1	0.87
	(confOf, ekaw)	0.90	0.93	0.91	0.90	0.99	0.94	0.75	1	0.85
	(confOf, sigkdd)	1	0.96	0.98	1	0.99	1	0.59	0.61	0.60

The query-answering tasks were evaluated by querying instances from various knowledge-bases (KBs) defined using the different ontologies. In each case, queries were constructed by considering each named concept in one ontology O_1, and querying the KB for O_2. To overcome the ontological heterogeneity, the query was resolved using $O_2 \cup M$, where M was the alignment used. As the resulting instance set depends on the generated alignment, a reference "*gold standard*" instance set was constructed by using the hand-crafted *reference* alignment. To evaluate scenarios where alternate alignments were available from the different alignment systems used, alignments were generated by all of the systems, resulting in 12 different alignments, where each one was partitioned between three or five ontology pairs. Query answering tasks were performed for three cases: when all the mappings in the alignments were aggregated and used without any use of the argumentation process (i.e. *Base*); when *MbA* was used; and when *FDO* was used. In each case, the answers generated for each query were analysed and compared with that obtained when using the Gold Standard set, and the Precision (P), Recall (R) and F-measure (FM) results (using these classical Information Retrieval measures) are reported in Table 5[10].

The results suggest that in most cases, there is a slight improvement in the success of a task when *FDO* is used (compared to *Base*) for the scenarios listed in Table 5, with an average F-measure of 0.83 (compared to 0.82 for *Base*). This contrasts sharply with *MbA*, which achieves only an average F-measure of 0.72. In general, the precision of *FDO* is higher or comparable with that exhibited by *MbA*. Interestingly, when *FDO* is compared with the base case in general, a marked increase in precision is observed. *Base* already represents a best-case scenario, in which the different alignment systems are tuned in order to provide the best accuracy when computing the mappings, and therefore typically generate only those mappings for which the system has the highest level of confidence. These results suggest that the further filtering of results due to the use of *FDO* pays off in terms of the increase in precision.

6   Related Work

A number of solutions have been proposed that attempt to resolve ontological mismatches within open environments [14,4,8,9]. An ontology mapping negotiation [14] was proposed to establish a consensus between different agents using the MAFRA alignment framework. It was based on the utility and meta-utility functions used by the agents to establish if a mapping is accepted, rejected or negotiated, making it highly dependent on the MAFRA framework and unsuitable for other environments.

Bailin and Truszkowski [2] present an ontology negotiation protocol that enabled agents to exchange parts of their ontology, by a process of successive interpretations, clarifications, and explanations. The result was that each agent would converge on a single, shared ontology. However, within an open environment, agents may not always want to modify their own ontologies, as this may affect subsequent communication with other agents.

The work by van Diggelen *et al.* [4] dynamically generates a *minimal* shared ontology, where minimality is evaluated against the ability of the different components to communicate with no information loss. The agents can explain concepts to each other via the communication mechanism; either by defining the concept in terms already understood or by invoking an extensional learning mechanism.

---

[10] In eight cases, the recall and precision of the *Base* and *FDO* evaluations were of value 1 (i.e. they returned only those instances in the "*gold standard*" instance set), and thus have not been included in the Table. In these cases, the precision of *MbA* was also 1, but the recall varied between 0.9 and 0.99.

However, the ontological model used here is limited and non-standard, as its expressivity supports only simple taxonomic structures, with no properties and few restrictions other than disjointness and partial overlap, and does not correspond to any of the OWL flavours[11]. As a consequence, its applicability to the augmentation of existing real-world, published, OWL ontologies on the web is limited.

dos Santos *et al.* [9,10] address the problem of generating a canonical alignment using an extended version of the VAF, which considers both the strength and value of an argument. They do not consider the problem of dynamically aligning two agent ontologies to facilitate communication and fail to consider the preferences of the agents.

7   Conclusions

This paper presents a novel mechanism for determining whether agents are in favour or against ontology mappings during a process of dynamic selection of mutually acceptable alignements. The *flexible approach for determining agents' orientation on ontology mappings* (*FDO*) allows agents to express a minimum acceptability thresholds for each of the mapping types to include in the alignment used during communication. In this respect *FDO* provides a more flexible framework the Meaning-based argumentation (*MbA*) approach in order to decide whether agents support or refute a mapping.

A systematic evaluation has been presented, aiming at assessing the performance of this novel mechanism over the 11 ontologies used in the OAEI 2007 initiative. In particular, the evaluation investigated whether the *FDO* approach generates larger set of mutually acceptable mappings than the original *MbA* approach, thus improving the possibility of finding an alignment agents can use to interact. In addition, we investigated whether these mappings are *fit for purpose* for a query answering task.

The results obtained suggest that the *FDO* approach produces a considerably larger set of mutually acceptable mappings by reducing the number of mappings an agent argues against when compared with *MbA*. The fitness for purpose evaluation shows that the *FDO* approach has a comparable if not higher F-measure than the case when no argumentation is used, and definitely outperforms *MbA*.

References

1. Baader, F., Calvanese, D., McGuinness, D.L., Nardi, D., Patel-Schneider, P.F. (eds.): The Description Logic Handbook: Theory, Implementation, and Applications. Cambridge University Press, Cambridge (2003)
2. Bailin, S.C., Truszkowski, W.: Ontology negotiation: How agents can really get to know each other. In: Truszkowski, W., Hinchey, M., Rouff, C.A. (eds.) WRAC 2002. LNCS, vol. 2564, pp. 320–334. Springer, Heidelberg (2003)

---

[11] The authors mention a reformulation of their model using Description Logics, but provide no formal proof of its soundness [4].

3. Bench-Capon, T.: Value based argumentation frameworks. In: Proceedings of Non Monotonic Reasoning, pp. 444–453 (2002)
4. van Diggelen, J., Beun, R.J., Dignum, F., van Eijk, R., Meyer, J.J.: Ontology negotiation in heterogeneous multi-agent systems: The anemone system. Applied Ontology 2(3-4), 267–303 (2007)
5. Dung, P.: On the Acceptability of Arguments and its Fundamental Role in Non-monotonic Reasoning, Logic Programming and n-person Games. In: Artificial Intelligence, vol. 77, pp. 321–358 (1995)
6. Eckert, K., Meilicke, C., Stuckenschmidt, H.: Improving ontology matching using meta-level learning. In: Aroyo, L., Traverso, P., Ciravegna, F., Cimiano, P., Heath, T., Hyvönen, E., Mizoguchi, R., Oren, E., Sabou, M., Simperl, E. (eds.) ESWC 2009. LNCS, vol. 5554, pp. 158–172. Springer, Heidelberg (2009)
7. Euzenat, J., Shvaiko, P.: Ontology Matching. Springer, Heidelberg (2007)
8. Laera, L., et al.: Argumentation over ontology correspondences in mas. In: 6th International Joint Conference on Autonomous Agents and Multiagent Systems (AAMAS 2007), Honolulu, Hawaii, USA, May 14-18, p. 228 (2007)
9. dos Santos, C.T., Quaresma, P., Vieira, R.: Conjunctive queries for ontology based agent communication in mas. In: 7th International Joint Conference on Autonomous Agents and Multiagent Systems (AAMAS 2008), Estoril, Portugal, May 12-16, vol. 2, pp. 829–836 (2008)
10. dos Santos, C.T., et al.: A cooperative approach for composite ontology mapping. Journal of Data Semantics 10, 237–263 (2008)
11. dos Santos, C.T., Euzenat, J., Tamma, V., Payne, T.R.: Argumentation for reconciling agent ontologies. In: SASFA 2010. Springer, Heidelberg (2010) (in press)
12. Sensoy, M., Yolum, P.: A cooperation-based approach for evolution of service ontologies. In: 7th International Joint Conference on Autonomous Agents and Multiagent Systems (AAMAS 2008), Estoril, Portugal, May 12-16, vol. 2, pp. 837–844 (2008)
13. Shvaiko, P., Euzenat, J., Giunchiglia, F., He, B. (eds.): Proceedings of the 2nd International Workshop on Ontology Matching (OM-2007) Collocated with the 6th International Semantic Web Conference (ISWC 2007) and the 2nd Asian Semantic Web Conference (ASWC 2007), Busan, Korea, CEUR Workshop Proceedings, November 11, vol. 304. CEUR-WS.org (2008)
14. Silva, N., Maio, P., Rocha, J.: An approach to ontology mapping negotiation. In: Proceedings of the Workshop on Integrating Ontologies (2005)

One Size Does Not Fit All: Customizing Ontology Alignment Using User Feedback

Songyun Duan, Achille Fokoue, and Kavitha Srinivas

IBM T.J. Watson Research Center, NY, USA
{sduan,achille,ksrinivs}@us.ibm.com

Abstract. A key problem in ontology alignment is that different ontological features (*e.g.,* lexical, structural or semantic) vary widely in their importance for different ontology comparisons. In this paper, we present a set of principled techniques that exploit user feedback to customize the alignment process for a given pair of ontologies. Specifically, we propose an *iterative* supervised-learning approach to (i) determine the weights assigned to each alignment strategy and use these weights to combine them for matching ontology entities; and (ii) determine the degree to which the information from such matches should be propagated to their neighbors along different relationships for *collective* matching. We demonstrate the utility of these techniques with standard benchmark datasets and large, real-world ontologies, showing improvements in F-scores of up to 70% from the weighting mechanism and up to 40% from collective matching, compared to an unweighted linear combination of matching strategies without information propagation.

1 Introduction

Ontology alignment and the related problem of schema matching is a richly studied area [9, 10, 14], with significant advances of alignment techniques in recent years. There are a number of systems that perform pretty well on the ontology alignment evaluation initiative (OAEI) benchmarks (for most recent examples, see Lily [17], ASMOV [8], Anchor-Flood [11], and RiMOM [12]).

A common aspect of most alignment systems is that they combine semantic and lexical features of ontology entities with structural propagation (*e.g.,* as in similarity flooding [13] or in iterative structural propagation of QOM [6]). When such structural propagation is applied, two key assumptions dominate the literature: (i) Structural propagation is beneficial to ontology alignment; and (ii) The alignment results at the last iteration are the best to be produced as the final results. Due to the lack of a principled way to determine the optimal number of iterations, most systems perform structural propagation either to a fixed number of iterations, or until further propagation does not produce additional matchings [6, 7].

Our key observation, based on work with real-world ontologies, is that the importance of any of these features (lexical, semantic or structural) varies widely across different ontology alignments. Furthermore, the degree of structural propagation required for optimal performance also varies widely. More structural propagation does not necessarily lead to better alignment results; in some cases, *any* structural propagation actually *impairs* alignment quality.

P.F. Patel-Schneider et al. (Eds.): ISWC 2010, Part I, LNCS 6496, pp. 177–192, 2010.

More recently, collective matching approaches (*e.g.*, [1]) have been proposed to take structural information into account, in a principled manner, for matching ontology entities. These approaches typically use sophisticated statistical models such as Markov Networks [15] to explicitly represent interdependencies between various matching choices. In a sense, they do not optimize the quality of individual matching decisions (*i.e.*, matching between individual pairs of ontology entities); instead, they optimize the quality of the whole collection of matching decisions. However, a serious drawback with these approaches to ontology alignment using complex models is their high computational cost; thus making such systems hard to use with large, real-world ontologies.

In this paper, we propose a principled and scalable technique to incorporate lexical, semantic and structural features, using *iterative supervised structural propagation*. Our approach relies on customizing two key components of ontology alignment. First, at a lexical level, alignment depends on a number of different alignment strategies (*e.g.*, alignment based on the names of ontology entities as encoded in a URI, or the associated documentation in terms of *rdfs:label*, *rdfs:comment*). For a given pair of ontologies, empirical evaluation may find out that an alignment strategy based on name may be more appropriate than that based on documentation. Our approach uses user feedback to learn the relative importance of these different alignment strategies for a given ontology pair, which is similar to the approach taken by APFEL [7]. Specifically, we use logistic regression [2] to determine the weights assigned to different strategies based on user feedback.

Second, we address the issue of how to systematically propagate lexical-level and user-specified matches along structural relations in the ontology. Here, we diverge from previous iterative structural propagation approaches such as [13] and [6] in that we adopt *iterative supervised learning* to estimate the optimal number of iterations needed for a given ontology pair. Specifically, we use the training phase to observe exactly which iteration yields maximal benefits in alignment, and use this information to determine the stopping condition for structural propagation at test. Our experimental evaluation shows clear advantages of our approach over previous approaches (*e.g.*, APFEL [7]) that do not take user feedback as guidance across iterations during the structural propagation phase.

Our contributions in this paper are as follows:

- We use supervised learning to customize the weights for different alignment strategies for a given ontology pair, and to customize the degree to which those matches at an entity level get propagated to its neighbors for collective matching.
- We demonstrate the effectiveness of this approach on two benchmark datasets, and 6 other large, real-world ontology alignments. The experimental results show good scalability of our approach, and confirm the hypotheses about great variability in features across ontology alignments. Our results also show dramatic improvements in alignment from the weighting (up to 70% increase in F-scores), and collective matching (up to 40% increase in F-scores).
- We demonstrate that incorporating supervision into the process of structural propagation is key to the selection of the relevant features. Weighting features using supervision *after* the process of unsupervised structural propagation yields poor results in some cases.

The rest of the paper is organized as follows. Section 2 gives an overview of the framework for ontology alignment. Section 3 describes the ontological features and similarity metrics. Section 4 presents a supervised-learning technique for similarity aggregation and interpretation. Section 5 presents the technique of iterative supervised structural propagation. Section 6 presents experimental results. Section 7 discusses related work, and Section 8 concludes.

2 Overview of Ontology Alignment

In this section, we briefly introduce important notations, and present the overall structure of our approach to ontology alignment. We use the terms alignment/matching and element/entity interchangeably when there is no confusion.

An ontology $\mathcal{O}$ is represented as a labeled graph $G = (V, E, vlabel, elabel)$. The set of vertices V contains ontology entities such as concepts and properties. Edges in E ($E \subseteq V \times V$) represent structural relationships between entities. The edge labeling function $elabel$, which maps an edge $(v, v') \in E$ to a subset of the set SL of structural labels, which in turn specify the nature of the structural relationships between entities (*e.g.*, subclassOf). Let LL denote the set of lexical labels associated with entities (*e.g.*, name, documentation). Finally, the vertex labeling function, $vlabel : V \times LL \rightarrow String$, maps a pair $(e, l) \in V \times LL$ to a string corresponding to the value of the lexical label l (*e.g.*, name) associated with the entity e.

Given two ontologies $\mathcal{O}$ and $\mathcal{O}'$, the ontology alignment problem consists of finding a set of matchings (e, e'), where e and e' are entities in $\mathcal{O}$ and $\mathcal{O}'$, respectively. Additionally, a similarity measure, denoted sim_{agg}, which maps the pair of entities $(e, e') \in \mathcal{O} \times \mathcal{O}'$ to a real number in $[0, 1]$, provides the confidence in a matching. We assume that for any entity in $\mathcal{O}$, there is at most one matching entity in $\mathcal{O}'$.

The alignment approach presented in this paper is similar in its overall structure to the process adopted by many existing matching engines such as [6]:

1. **Generation of Candidate Matchings:** This step includes feature engineering (*i.e.,*, the extraction of the relevant characteristics of ontology entities in both the source and the target ontology) and the selection of candidate matchings (to avoid considering the Cartesian product of entities in the two ontologies).
2. **Similarity Aggregation and Interpretation:** This step computes various similarity metrics on candidate matchings identified in the previous step. Each individual similarity metric is a function of only the features extracted from the two ontology entities being compared. The similarity scores are then aggregated into a single similarity score for each candidate matching. Interpretation is then based on the aggregated similarity scores, and involves a decision about which candidate matchings should be selected as valid matchings — typically using a threshold.
3. **Structural Propagation.** This step propagates matching information along ontology structure, by repeating the previous steps, typically, either to a fixed number of iterations or until no additional matchings are produced.

Our approach significantly differs from previous work in two ways. First, our similarity aggregation step is not based on an unsupervised (thus ad-hoc) weighted combination of similarity scores. We use a fully supervised-learning approach (described in more details in Section 4) to learn, at each iteration, from user feedback an optimal combination of similarity scores. Second, our stopping condition for the structural propagation is more principled. Note that previous work stop propagation based on an arbitrary number of iterations or the absence of additional matchings, which assumes that matching quality monotonically improves over successive iterations (this assumption does not hold in many cases, as shown in the experiment section). We stop iterations when there is no significant improvement in information gain at training, and select only the matchings produced at the iteration where the matching result has the best consistency with user feedback (see Section 5 for more details).

3   Generation of Candidate Matchings

In this section, we describe the features that can be extracted from ontologies, and the lexical similarity metrics we consider in this paper (structural similarities are discussed in Section 5).

3.1   Feature Engineering

In our approach, the feature engineering step is essentially responsible for transforming models in various representations (*e.g.,* XML Schemas, UML models, OWL ontologies, etc) into an ontology $\mathcal{O}$ represented as the labeled graph $G = (V, E, vlabel, elabel)$. Structural features are represented as edge labels.

In this section, we present features extracted from models encoded as OWL ontologies or OBO ontologies.

Lexical features (*i.e.,* elements of the set LL) extracted from ontology entities (concepts or properties) are as follows:

- name, which corresponds to the last segment of the ontology entity's URI (*e.g.,* 'Person' for 'http://www.ibm.com/hr/Person').
- documentation, which consists of the concatenation of the values of rdfs:label, rdfs:comment, obo:def, obo:comment, and obo:synonym.

Structural features (*i.e.,* elements of the set SL) are shown in the first column of Table 1. The second column of Table 1 indicates the condition under which an edge (e_0, e_1) is assigned a given label. Note that, although these structural features do not capture all the structural and semantic constructs of OWL ontologies (*e.g.,* union, disjointWith, complementOf, and nested structures are not currently taken into account), they are sufficient to produce robust structural improvements on the ontologies we tested with (see Section 6 for more details).

3.2   Lexical Similarities and Initial Selection of Candidate Matchings

Similarity Metrics. Various similarity metrics can be employed to compare entities from different perspectives. In an abstract form, a similarity metric is a function that maps a pair of entities to a value between 0 and 1.

$$\mathtt{sim}(e, e') \rightarrow [0, 1] \qquad\qquad (1)$$

Table 1. Structural Labels

Label	Label $\in elabel(e_0, e_1)$ **iff.**
subclassOf	e_0 is a direct subclass of e_1.
superclassOf	e_0 is a direct superclass of e_1.
isRangeOf	The concept e_0 is the range of the property e_1.
isDomainOf	The concept e_0 is the domain of the property e_1.
subPropertyOf	e_0 is a direct subproperty of e_1
superPropertyOf	e_0 is a direct superproperty of e_1
hasRange	The range of the property e_0 is the concept e_1.
hasDomain	The domain of the property e_0 is the concept e_1.
hasExistRestrictionOnProperty	The property e_1 is used to define the concept e_0 in terms of an existential or minimal cardinality restriction (*e.g.*, e_0 is defined as $e_0 \sqsubseteq \exists e_1.C$)
hasForAllRestrictionOnProperty	The property e_1 is used to define the concept e_0 in terms of a universal restriction (*e.g.*, e_0 is defined as $e_0 \sqsubseteq \forall e_1.C$))
hasExistRestrictionOnClass	The concept e_1 is used to define e_0 in terms of an existential or minimal cardinality restriction (*e.g.*, assuming normalization to NNF, $e0$ is defined as $e_0 \sqsubseteq \exists R.e_1$).
hasForAllRestrictionOnClass	The concept e_1 is used to define e_0 in terms of a universal restriction (e.g, assuming normalization to NNF, $e0$ is defined as $e_0 \sqsubseteq \forall R.e_1$).
existRestrictionUsedFor	The concept e_1 is defined as an existential or minimum cardinality restriction using the property e_0 (*e.g.*, if e_1 is defined as $e_1 \sqsubseteq \exists e_0.C$)
forAllRestrictionUsedFor	The concept e_1 is defined as a universal restriction using the property e_0 (*e.g.*, if e_1 is defined as $e_1 \sqsubseteq \forall e_0.C$)

For a given pair of entities (e, e'), multiple similarity metrics can be applied. The similarity metrics are denoted as $\text{sim}_i(e, e')$ ($i = 1, 2, \ldots$). Note that the similarity metric can be as general as a matching technique.

For lexical similarity, standard similarity metrics exist for strings such as Levenshtein similarity or Jaccard similarity on n-grams. This works fine for lexical features, such as name, whose values are expected to consist of only a few words. However, for lexical features such as documentation, the values may consist of many paragraphs. Therefore, as explained in [3], we cast the problem into a classical information retrieval problem. We transform entities (*e.g.*, concepts and properties) into virtual documents. A virtual document consists of fields corresponding to the two lexical features described in the previous section, namely, name and documentation. These virtual documents are stored and indexed by a high-performance text search engine such as Lucene[1]. A Vector Space Model (VSM) [16] is adopted for comparison: each field F (name or documentation) of a virtual document is represented as a vector in a N_F-dimensional space, with N_F denoting the number of distinct words in field F of all documents.

---

[1] http://lucene.apache.org/java/docs/index.html

Traditional TF-IDF (Term Frequency-Inverse Document Frequency) values are used as the weights of coordinates associated with terms. The lexical similarity on a field $F \in \{\text{name}, \text{documentation}\}$ between two entities e and e' is referred to as $\text{sim}_F(e, e')$, and is computed as the *cosine* of the angle formed by their F vectors. We adjust for slight syntactic variations by using a term similarity metric (such as Levenshtein or Jaccard over n-grams) between terms as explained in [3].

Candidate Selection. In the first iteration (*i.e.,* before any structural propagation is performed), we use the text search engine, for each entity e in the source ontology $\mathcal{O}$, to select top-k candidate matchings of e in the target ontology $\mathcal{O}'$, by retrieving the virtual documents representing entities in $\mathcal{O}'$ that match well with e in terms of lexical similarity (*e.g.,* based on Lucene score).

4   Similarity Aggregation and Interpretation

4.1   User Feedback

In this paper, we assume that for any entity in $\mathcal{O}$, there is at most one matching entity in $\mathcal{O}'$. Also, we assume a simple format for user feedback (users specify which pairs of entities should be matched) that is fed to our system through a file of *gold standard* matchings. For a matching (e, e') specified by the user, we will label the matching (e, e') as `true`. For any candidate matching (e, e'') generated in Section 3, where e'' is not equal to e', we label it as `false`. Thus, we generate a set of training tuples in the following form:

$$\langle\, \text{sim}_1(e, e'), \ldots, \text{sim}_n(e, e'), \texttt{true}\rangle \tag{2}$$
$$\langle\, \text{sim}_1(e, e''), \ldots, \text{sim}_n(e, e''), \texttt{false}\rangle (\forall e'' \neq e')$$

4.2   Weighted Aggregation

To interpret the matching result, a common practice is to aggregate the similarity metrics with a linear combination and set a threshold to decide which matchings are estimated to be `true`. However, it is well accepted that linearly (unweighted) combining the similarity metrics (or matching strategies) may adversely affect the overall matching quality. With user feedback, we can infer which similarity metrics are more reliable than others, and assign higher weights to the more reliable ones. A natural extension is to get a weighted sum (with the weight vector $\overrightarrow{w}$) of the similarity measures and apply a threshold ω_0 to predict whether a matching is `true` or `false`. The prediction is done with a decision boundary $\text{f}\,(\overrightarrow{w}, \text{sim}) = 0$, where the function f is defined as follows:

$$\text{f}(\overrightarrow{w}, \text{sim}) = \omega_0 + \omega_1 \times \text{sim}_1 + \ldots + \omega_n \times \text{sim}_n \tag{3}$$

4.3   Probabilistic Matching

For a candidate matching, the above decision boundary produces a binary value indicating the matching is `true` or `false`. However, it is more important to also produce a probability (between 0 and 1) along with the binary prediction, such that the matching result can be easily incorporated in other matching strategies (we will see such an

---

Algorithm 1. Learning of Weights for Ontology Matching

Input: ontologies $\mathcal{O}$ and $\mathcal{O}'$, gold standard matchings M from user feedback, similarity metrics $\texttt{sim}_i$

Output: a list of matchings, $\langle (e, e'), P((e, e') = \texttt{true}) \rangle$

1. **for** *each matching* $m = (e, e')$ *in gold standard* M **do**
   > (i) Label candidate matchings for e: (e, e') as $\texttt{true}$ and (e, e'') $(\forall e'' \neq e')$ as $\texttt{false}$;
   > (ii) Compute the similarities of each candidate matching with given similarity metrics $\texttt{sim}_i$;
   > (iii) Generate the training tuples in the way described in Section 4.1;
2. Learn the weights for combining the similarities and the threshold to decide whether a matching should be produced or not;
3. Use the learned weights and the threshold to generate the matching result.

---

example in Section 5). In statistics, the output of the real-valued function $\texttt{f}$ can be mapped to a probability value, using the sigmoid function $P(t) = \frac{1}{1+e^{-t}}$. Specifically, given a candidate matching (e, e') with similarity measures $\texttt{sim}$, the probability of this matching is $\texttt{true}$ is:

$$P((e, e') = \texttt{true}) = \frac{1}{1 + e^{-f(\vec{w}, \texttt{sim})}} \qquad (4)$$

The probability that the matching is false is $P((e, e') = \texttt{false}) = 1 - P((e, e') = \texttt{true})$. The key issue is how to determine the weight vector $\vec{w}$ based on user feedback. Recall that the user feedback can be represented in the form of tuples $\langle \texttt{sim}, \texttt{true}/\texttt{false} \rangle$. The weight vector $\vec{w}$ that maximizes the likelihood of observing these tuples is the one that is most consistent with user feedback. In statistics, $\vec{w}$ can be determined using the MLE (maximum likelihood estimation) technique for logistic regression [2]. Algorithm 1 describes the key steps of the supervised-learning approach to ontology alignment.

5 Iterative Supervised Structural Propagation of User Feedback

In this section, we make the internal linkages of entities within ontologies explicit for learning. Specifically, for a candidate matching (e, e'), we take into account the matching results of e's neighbors in the ontology $\mathcal{O}$ when making the matching decision for (e, e'). The intuition is, for example, the matching of e's subclass with e''s subclass may add evidence that e and e' should be matched.

5.1 Structure-Based Similarity

For a candidate matching (e, e'), we extend the list of similarity metrics introduced in Section 3 with structure-based metrics as follows. Consider a structural label l (e.g., subclassOf) in the ontologies. Suppose there is a set of entities $\texttt{SE}(e, l)$ that are connected to e with the structural label l in $\mathcal{O}$ (i.e., $\texttt{SE}(e, l) = \{x | l \in elabel(e, x)\}$); correspondingly, $\texttt{SE}(e', l)$ for e' in $\mathcal{O}'$. It is important to aggregate the similarity values between the two sets, i.e., $\texttt{SE}(e, l)$ and $\texttt{SE}(e', l)$, and extend the list of similarity metrics for (e, e') with the aggregation metrics. Below we briefly describe two types of aggregation metrics. (We considered other types of aggregation metrics such as min and sum, but empirically observed that max and avg are more effective.)

- max(S_1, S_2, sim) is the maximum similarity between any pair of entities, from two sets of entities S_1 and S_2 respectively, in the Cartesian product of $S_1 \times S_2$. For instance, S_1 can be SE(e, l), S_2 can be SE(e', l), and sim can be a lexical similarity metric, as described in Section 3.

$$\text{max}(S_1, S_2, \text{sim}) = \max_{(e_1, e_2) \in S_1 \times S_2} \text{sim}(e_1, e_2)$$

- avg(S_1, S_2, sim) is the average similarity of pairs of entities in the Cartesian product $S_1 \times S_2$:

$$\text{avg}(S_1, S_2, \text{sim}) = \frac{\sum_{(e_1, e_2) \in S_1 \times S_2} \text{sim}(e_1, e_2)}{(|S_1| + |S_2|)/2}$$

For a candidate matching (e, e'), we can generate various structure-based similarity metrics based on their sets of neighbors SE(e, l) and SE(e', l). Concretely, the structure-based similarity metrics can be:

- max$(\text{SE}(e, l), \text{SE}(e', l), \text{sim}_{name})$
- avg$(\text{SE}(e, l), \text{SE}(e', l), \text{sim}_{name})$
- max$(\text{SE}(e, l), \text{SE}(e', l), \text{sim}_{doc})$
- avg$(\text{SE}(e, l), \text{SE}(e', l), \text{sim}_{doc})$
- max$(\text{SE}(e, l), \text{SE}(e', l), \text{sim}_{agg})$
- avg$(\text{SE}(e, l), \text{SE}(e', l), \text{sim}_{agg})$

In the above metrics, sim_{name} is lexical similarity on the name field of two entities, sim_{doc} is the lexical similarity on the documentation/comment field of two entities, and sim_{agg} can be the aggregated score of similarity metrics in Algorithm 1 (*i.e.*, $\text{sim}_{agg}(e, e') = P((e, e') = \text{true})$).

5.2   Determining the Degree of Structural Propagation

At the bootstrapping step, we generate the aggregated similarity for a candidate matching (e, e') in the following way: If (e, e') is part of the ground truth (*i.e.*, provided by user feedback), its value is 1; otherwise, its value is 0. At the following iterations, we can utilize the matching result from the previous iteration. Note that for the pairs of entities that appear as training tuples (see Formula 3), we replace their matching scores with the ground truth (1 for true, and 0 for false).

The above structural similarity metrics allow the propagation of information conveyed by user feedback along the structure of the two ontologies. We thus extend the initial set of similarity metrics (Section 3) with the six structure-based similarity metrics per relation type. As a result, the number of similarity metrics that can be used for ontology matching is large. Since the amount of user feedback is limited, we adopt dimensionality reduction techniques to avoid the overfitting problem in Section 5.4.

At each iteration, the selection of matching candidates is extended to include pairs of entities having at least one non-zero structural similarity measure. The impact of the neighbor matching scores on the candidate matching in consideration is learned based

---

Algorithm 2. Iterative Supervised Structural Propagation for Ontology Matching

---

Input: ontologies $\mathcal{O}$ and $\mathcal{O}'$
Output: a list of matchings, $\langle (e, e'), P((e, e') = \texttt{true}) \rangle$
1. Bootstrapping: Generate training tuples with basic similarity metrics and structure-based similarity metrics;
2. Learn a weight vector to integrate similarity metrics that maximize the likelihood of user feedback being correct;
3. Generate a new list of matchings by combining the similarity metrics using the newly learned weight vector;
4. Update the training tuples with the aggregated similarities from Step 3, and add candidate matchings whose structure-based similarity measures become nonzero;
5. If it does not meet stopping condition, go to Step 2.

---

on user feedback, as described in Section 4. This process iterates until some stopping condition is satisfied; the following describes a metric to define the stopping condition.

5.3 Determining the Right Number of Iterations

We observe that too many iterations may be detrimental to matching quality (see the experiment section). Therefore, we propose a metric G, which is the *training error*, to decide the optimal number of iterations. G is computed as the absolute difference of the matching result (in the form of $\langle (e, e'), P((e, e') = \texttt{true}) \rangle$) at each iteration with regard to the ground truth (*i.e.,* user feedback).

$$\texttt{G} = \sum_{(e,e') \in \texttt{Ground Truth}} (1 - P((e, e') = \texttt{true})) + \sum_{(e,e') \notin \texttt{Ground Truth}} P((e, e') = \texttt{true})$$

The hypothesis is that the smaller the value of G, the better the matching result. This hypothesis will be verified with experiments in the next section.

5.4 Techniques for Scalability

Dimensionality Reduction. For large ontologies, possibly with many edge-labels, the generated attribute list (of similarity metrics) can be huge. Due to the limited amount of user feedback, it is necessary to reduce the dimensionality of the attribute space, to avoid the well-known overfitting problem. We use a standard unsupervised dimensionality reduction technique, principle component analysis (PCA) [2], to extract the most important dimensions for learning from the originally high dimensional space.

Blocking Unreliable Information Propagation. In Algorithm 2, the number of candidate matchings will monotonically increase after each iteration, since new candidate matchings are generated if their neighbors have *confident* matchings. To avoid propagating noisy information from neighbors, we set a threshold on the matching scores to keep the low-confidence matchings from propagating to neighbors. A side benefit of such blocking is efficiency; the number of tuples in the training data generated based on user feedback will increase slowly, thus saving the time to learn the weight vector (in Section 4.2) for each iteration. Note that if there is no blocking of propagation, the number of tuples in the training data may increase exponentially during iterations.

6   Experimental Evaluation

The focus of our experimental evaluation is to determine whether the great variability in ontology alignments can be reduced by using (i) a supervised-learning technique to customize the weights assigned to lexical features, and (ii) an iterative supervised-learning approach to determine the appropriate degree of structural propagation for each ontology alignment.

6.1   Experimental Setting

We focused on parts of the OAEI benchmark suite that are most suited for evaluating the effects of structural propagation. Test 202 was selected because it modifies the original ontology by obfuscating all names and documentations, and is a test of alignment based on structural similarity. We also selected the anatomy segment of the benchmark because the pair of ontologies in that benchmark encode structural information within an extensive part-of hierarchy. We also added 6 other ontology alignments from Bio-Portal into the evaluation to ensure that our results generalize well to different types of ontology alignments. Table 2 shows the characteristics of these 6 additional ontology alignments, and the number of matchings manually discovered by domain experts.

To evaluate the effects of training on similarity combination and structural propagation[2], we performed random sampling to split the reference matchings in the following way: we assigned 50% of the matchings to the 'test' group, and from the rest we further sampled 50% of the matchings to create the 'training' group (*i.e.,*, training ratio was 25% of the total number of matchings for the ontology alignment). Note that the actual number of matchings used for learning is small with respect to ontology size. For both training and testing, we varied the number of iterations used for structural propagation to a maximum of about 10 iterations for each ontology alignment.

The experiments were performed on a server with 8 way machine with 4 dual-core Intel Xeon chips at 3.20 GHz, with 20 GB of memory. For all the experiments, we used a maximum Java heap size of 10 GB.

6.2   Evaluation Metric

In our experimental evaluation, we had a complete gold standard for Test 202; for all other ontology alignments, we only had partial reference alignments[3]. We therefore measured `F-scores` in the standard manner only on Test 202. For all other ontology matching tasks, we computed an `F-score` only on the partial alignments available to us, and only considered ontology entities that were in the reference alignments (all other matchings we produced for entities in the source ontology not present in the partial alignment were not taken into account for `precision` or `recall` estimates). We assumed that there is at most one matching entity in the target ontology for each entity in the source ontology.

---

[2] The threshold we used for blocking unreliable information (Section 5.4) is 0.5.

[3] The lack of complete reference alignments is a frequent problem in real world ontology alignments. The matchings in Bioportal, for example, are almost always partial because the ontologies are large, and cannot be perfectly aligned manually.

Table 2. BioPortal Ontology alignments

Ontology 1	#Classes	Ontology 2	#Classes	#matchings
Mosquito gross anatomy (TGMA)	2,404	Drosophila gross anatomy (FBbt)	8,742	324
Human devt. anatomy (EHDA)	11,575	Amphibian gross anatomy (AAO)	833	684
BRENDA tisse source (BTO)	4,950	Experimental Factor Ontology (EFO)	2,891	366
Experimental Factor Ontology (EFO)	2,891	Mouse Adult Gross Anatomy (MA)	3,504	212
ABA Adult Mouse Brain (ABA)	915	Mouse Adult Gross Anatomy (MA)	3,504	90
BIRNLex (birnlex)	3,582	UBER anatomy ontology (UBERON)	3,619	744

$$\text{precision} = \frac{|\text{M} \cap M_{GS}|}{|\text{M}|}, \text{recall} = \frac{|\text{M} \cap M_{GS}|}{|M_{GS}|}$$

$$\text{F-score} = \frac{2 \times \text{precision} \times \text{recall}}{\text{precision} + \text{recall}}$$

where M is the matchings discovered by our technique and the first ontology entity of each matching appears in the reference alignment M_{GS}.

In the following experimental results, we report F-score at the specific thresholds of (0.7, 0.8, 0.9), which are used to filter out low-confidence matchings, as users typically do not trust matchings with low matching confidence in practice.

6.3   Effect of Learning for Weighted Combination

Given the enormous variability in the importance of lexical and structural features to different ontology alignments, our hypothesis is that there is a principled way to weight these features appropriately using limited user feedback. We begin by examining the effect of learning to combine lexical features. Table 3 reports the F-score from our learning technique for weighted combination, compared with unweighted linear combination of similarity metrics for matching. For some ontology alignments (*e.g.*, BTO - EFO), there is a significant improvement in F-score (from 5% to around 70%); which clearly shows the effect of learning.

6.4   Effect of Iterative Supervised Structural Propagation

Figures 1 - 8 plot the changes in F-score as the structural propagation is iterated (in a supervised fashion), along with the corresponding training errors at iterations. These figures show:

- There is in fact a great deal of variability across ontologies, with lexical matches contributing to accuracy in the range of 10% to well above 90%.

Table 3. Effect of learning on F-scores at different thresholds

ontology alignment	Unweighted combination			Weighted combination		
	0.7 (%)	0.8 (%)	0.9 (%)	0.7 (%)	0.8 (%)	0.9 (%)
OAEI Anatomy	93	93	94	93	94	94
TGMA - FBbt	9	7	5	26	21	15
EHDA - AAO	99	99	99	99	99	99
BTO - EFO	5	1	1	74	72	68
EFO - MA	88	90	86	91	90	88
ABA - MA	93	90	85	94	92	90
BIRNLex - UBERON	77	66	46	83	81	73

- Structural propagation shows similar variability in its importance, with it improving accuracy by up to 40% in some cases (*e.g.,*, Figures 1, 3, 5, 8), but as shown in Figure 4, propagation of *any* structure in some ontologies causes a precipitous drop in accuracy by almost 25%, at high confidence thresholds (0.9).
- The number of iterations required to maximize the effects of structural propagation varies widely as well. In some cases (*e.g.,* Figure 1), a greater number of iterations of structural propagation is required, with peak matching quality being reached at about 5 iterations. In other cases (*e.g.,*, Figure 8), just one iteration is sufficient to maximize the benefits of structural propagation.

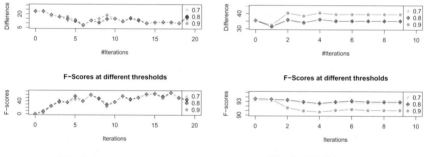

Fig. 1. OAEI 202   **Fig. 2.** OAEI Anatomy

Picking the Right Number of Iterations. For structural features, we hypothesized that the training error (*i.e.,*, the absolute difference between the matching results and reference matchings at training) can be used to estimate (i) whether structural propagation is useful, and (ii) to what degree structure needs to be propagated to maximize the overall matching quality. Because training error conceptually reflects *goodness of fit* [2], F-score at test is expected to be the best when training error is minimal. The general trend in Figures 1- 8 validated this hypothesis, therefore, we can pick the right number of iterations, in a principled way, to maximize the quality of matchings for a given pair of ontologies.

Comparison with Previous Work. We compare our approach with the technique proposed in [7] by simulating their process of matching in the following steps: (i) perform

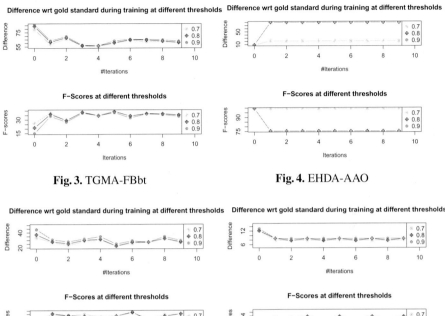

Fig. 3. TGMA-FBbt   **Fig. 4.** EHDA-AAO

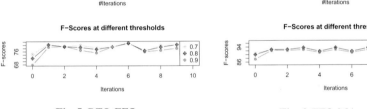

Fig. 5. BTO-EFO   **Fig. 6.** EFO-MA

iterative unsupervised structural propagation from iterations 1 to 8, and (ii) apply supervised learning to determine weighted combination of both lexical and structural similarity measures returned from the last iteration. The result of this matching approach is shown in Figures 9 and 10. Several points to note here include: (i) The unsupervised structural propagation actually affects the F-score adversely, thus highlighting the importance of supervised propagation; and (ii) At the last iteration (with supervised learning), we get mixed results; in the case of BTO-EFO the F-score at the last iteration improves over the matching results based on purely lexical similarity measures (from 2% to 71%), while in another case structural propagation hurts F-score compared to lexical similarity measures (from 85% to 78%). Note that this is in contrast to our result. For the same two cases, we observed (in Figures 5 and 6) a significant improvement in F-score. Specifically, with our approach, the F-score for BTO-EFO increases from 67% to 81%; and the F-score for EFO-MA increases from 87% to 92%. In any of the two cases, our approach outperforms that of the previous work, due to iterative supervised structural propagation. One lesson we learned here is iterative structural propagation without the guidance of user feedback is not reliable and can be harmful.

For OAEI 202, our best F-score (84%) across all thresholds makes our approach competitive to the top 5 matching engines with best F-score between 80% and 90%.

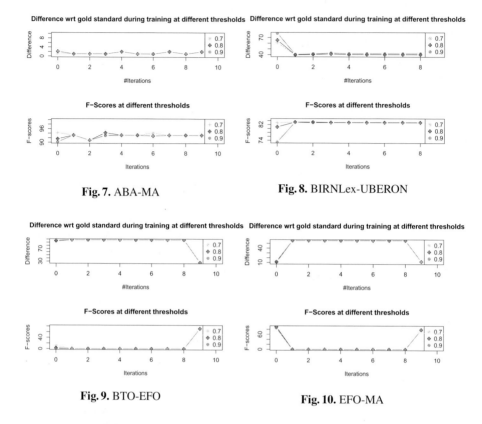

Fig. 7. ABA-MA

Fig. 8. BIRNLex-UBERON

Fig. 9. BTO-EFO

Fig. 10. EFO-MA

6.5 Discussion and Future Work

How much training data do we need to observe the beneficial results reported in this paper? We ran experiments with a smaller training ratio (10%) of the reference matchings, and observed a big variation in F-scores for the matching of BTO-EFO. The reason is that the absolute number of matchings (in this case, 36) used for training is too small considering the ontology size (in this case, 4,950). Our hypothesis is that when the sample sizes are too small (relative to the size of the ontology), careful selection of candidate matchings for user feedback is needed to ensure that enough structure is maintained for learning. Better sampling techniques (instead of random sampling) to reduce user feedback is an issue we leave for future work.

Another issue we observed is that for each ontology alignment in Table 2, our approach generates thousands of extra matchings with scores above 0.9, and these are not in the reference matchings. Based on the effectiveness of our technique on the reference alignments, we expect these extra matchings to be valuable to domain users, if only to recommend matchings for user validation.

One final point is about the scalability of our technique of iterative supervised structural propagation. The running time of each iteration was less than 3 minutes. Compared to existing collective matching based on sophisticated statistical models (*e.g.*, [1]), which have issues of scalability, our approach has a clear advantage in performance.

7 Related Work

Our approach, which applies an iterative supervised-learning technique to combine both lexical and structural similarities, can be contrasted with previous work that adopt either (unsupervised) iterative structural propagation technique (*e.g.,* similarity flooding [13] and its variants) or collective matching approaches (*e.g.,* [1]).

Similar to those systems (*e.g.,* [6] [12]) that apply variants of similarity flooding technique [13], our approach also iteratively propagates similarity metrics along ontology structures. However, our approach differs from them in two significant aspects. First, at each iteration, those systems aggregate various similarity metrics in an unsupervised (thus ad-hoc) fashion. In contrast, our approach applies supervised learning to learn from user feedback an optimal combination of both lexical and structural similarity metrics at each iteration; thus the information propagated across iterations is more reliable. Second, unlike those systems that assume matching result at the last iteration is the best (which is not necessarily true), we propose a novel and sound metric to estimate the matching quality at each iteration, based on the consistency of matching result with user feedback. Reference [7] views a matching engine (such as [6]) as a black box that returns its matching results and the similarity measures; it applies a supervised-learning technique to decide the optimal combination of the similarity measures returned by such a matching engine. Unlike our approach, the aggregation step occurring within the black-box matching engine remains unsupervised. As a result, the final structural similarities returned by the black-box engine may be less accurate; their iterative structural propagation misses the guidance from user feedback. Hence, the value of the supervised-learning approach to decide the weights of similarity measures is limited, resulting in sub-optimal matching results (we verified this point in the experiment section).

Recently, collective matching approaches (*e.g.,* [1]) have been proposed to take structural information into matching decisions using sophisticated statistical models. In a nutshell, those approaches use complex statistical models such as Markov Network [15] to explicitly represent interdependencies between matchings of interconnected ontology entities. Our approach is similar to this category of work in the sense that supervised learning techniques are applied to combine lexical and structural similarities in a principled way. However, due to the high computational complexity, in both learning and inference, of the complex statistical models used for encoding structural dependencies, those approaches based on sophisticated statistical models typically scale poorly to large ontologies[4].

Meta-learning (*i.e.,* integration of multiple alignment strategies) has also been implemented by GLUE [4] and other systems (*e.g.,* [5]). GLUE uses a supervised learning approach to build concept classifiers based on the associated instances (our approach does not assume instance information), but the way it combines inputs from various classifiers and performs structural propagation through relaxation labeling is unsupervised. Reference [5] also applies supervised learning to optimize the combination of multiple matching strategies, but it makes matching decisions for each entity independently, thus lacking the favor of collective matching.

---

[4] Simpler statistical models (*e.g.,* Markov Chain, Linear-chain Conditional Random Field, etc.) with scalable learning and inference algorithms are not sufficiently expressive to faithfully capture the structural dependencies.

8   Conclusion

To address the great variability in the importance of various features across ontology alignments, we have presented a principled and scalable technique to customize ontology alignment for a given pair of ontologies based on user feedback. We have shown how iterative supervised structural propagation, where each step is guided by user input, can optimally incorporate and propagate lexical-level and user-specified matches through the structure of the ontologies. Our experimental evaluation demonstrates the effectiveness of the new approach on both benchmark datasets and large, real-world bio-ontologies.

As future work, we plan to tackle the important, but orthogonal, problem of reducing user feedback by picking the most informative matches through active learning techniques.

References

1. Albagli, S., Ben-Eliyahu-Zohary, R., Shimony, S.E.: Markov network based ontology matching. In: IJCAI 2009 (2009)
2. Bishop, C.M.: Pattern Recognition and Machine Learning. Springer, Heidelberg (2007)
3. Byrne, B., Fokoue, A., Kalyanpur, A., Srinivas, K., Wang, M.: Scalable matching of industry models - a case study. In: OM (2009)
4. Doan, A., Madhavan, J., Domingos, P., Halevy, A.: Ontology matching: A machine learning approach. In: Handbook on Ontologies in Information Systems. Springer, Heidelberg (2003)
5. Eckert, K., Meilicke, C., Stuckenschmidt, H.: Improving ontology matching using meta-level learning. In: Aroyo, L., Traverso, P., Ciravegna, F., Cimiano, P., Heath, T., Hyvönen, E., Mizoguchi, R., Oren, E., Sabou, M., Simperl, E. (eds.) ESWC 2009. LNCS, vol. 5554, pp. 158–172. Springer, Heidelberg (2009)
6. Ehrig, M., Staab, S.: QOM – quick ontology mapping. In: McIlraith, S.A., Plexousakis, D., van Harmelen, F. (eds.) ISWC 2004. LNCS, vol. 3298, pp. 683–697. Springer, Heidelberg (2004)
7. Ehrig, M., Staab, S., Sure, Y.: Bootstrapping ontology alignment methods with APFEL. In: Gil, Y., Motta, E., Benjamins, V.R., Musen, M.A. (eds.) ISWC 2005. LNCS, vol. 3729, pp. 186–200. Springer, Heidelberg (2005)
8. Jean-Mary, Y.R., et al.: ASMOV: Results for OAEI 2009. In: OM (2009)
9. Euzenat, J., Shvaiko, P.: Ontology matching. Springer, Heidelberg (2007)
10. Giunchiglia, F., Shvaiko, P., Yatskevich, M.: S-match: an algorithm and an implementation of semantic matching. In: ESWC (2004)
11. Hanif, M.S., Aono, M.: Anchor-Flood: Results for OAEI 2009. In: OM (2009)
12. Li, J., Tang, J., Li, Y., Luo, Q.: RiMOM: A dynamic multistrategy ontology alignment framework. IEEE Trans. Knowl. Data Eng. (2009)
13. Melnik, S., Garcia-molina, H., Rahm, E.: Similarity flooding: A versatile graph matching algorithm. In: ICDE (2002)
14. Noy, N.F.: Semantic integration: a survey of ontology-based approaches. SIGMOD Rec. (2004)
15. Pearl, J.: Probabilistic reasoning in intelligent systems: networks of plausible inference (1988)
16. Raghavan, V.V., Wong, S.K.M.: A critical analysis of vector space model for information retrieval. Journal of the American Society for Information Science (1999)
17. Wang, P., Xu, B.: Lily: Ontology alignment results for OAEI 2009. In: OM (2009)

Compact Representation of Large RDF Data Sets for Publishing and Exchange*

Javier D. Fernández[1], Miguel A. Martínez-Prieto[1,2], and Claudio Gutierrez[2]

[1] Department of Computer Science, Universidad de Valladolid, Spain
{jfergar,migumar2}@infor.uva.es
[2] Department of Computer Science, Universidad de Chile, Chile
cgutierr@dcc.uchile.cl

Abstract. Increasingly huge RDF data sets are being published on the Web. Currently, they use different syntaxes of RDF, contain high levels of redundancy and have a plain indivisible structure. All this leads to fuzzy publications, inefficient management, complex processing and lack of scalability. This paper presents a novel RDF representation (HDT) which takes advantage of the structural properties of RDF graphs for splitting and representing, efficiently, three components of RDF data: *Header*, *Dictionary* and *Triples* structure. On-demand management operations can be implemented on top of HDT representation. Experiments show that data sets can be compacted in HDT by more than fifteen times the current naive representation, improving parsing and processing while keeping a consistent publication scheme. For exchanging, specific compression techniques over HDT improve current compression solutions.

1 Introduction and Related Work

The intended goal of the original RDF/XML representation design was to add small descriptions (metadata) to documents, to protocols, to mark web pages or to describe services. Representations like N3, Turtle and RDF/JSON, although having improved, in several respects, the original format, are still dominated by a *document-centric* view. Today, when one of the major trends in the development of the Web is RDF publishing at large scale, *i.e.* make RDF data publicly available for unknown purposes and users, the need to consider RDF under a *data-centric* view is becoming indispensable.

An analysis of published RDF data sets (the 2000 US Census, DBpedia, GeoNames, Uniprot, DBLP, etc.) reveals several undesirable features. First, the provenance and metadata about contents are barely present, and their information is neither complete nor systematic. Second, the files have neither internal structure nor a summary of their content. Basic data operations have to deal with the sequentiality of the information in the file, thus parsing the whole data and in most cases including human operation because the metadata is outside the file. For mashups of different sources, the situation is worse. Currently, the effort to prepare the data to be published is so costly, that

---

* Partially funded by MICINN (TIN2009-14009-C02-02), Millennium Institute for Cell Dynamics and Biotechnology (ICDB) (Grant ICM P05-001-F), and Fondecyt 1090565. The first author is granted by a fellowship from Erasmus Mundus, the Regional Government of Castilla y Leon (Spain) and the European Social Fund. Special thanks to Margaret Gagie.

P.F. Patel-Schneider et al. (Eds.): ISWC 2010, Part I, LNCS 6496, pp. 193–208, 2010.

files commonly have no design, no plan and no user in mind. They resemble unwanted creatures whose owners are keen to be rid of them.

This state of affairs does not scale to a Web where large data sets will soon, increasingly, be produced dynamically and automatically. Furthermore, most data would have to be *machine-understandable* in line with the aim of the original Semantic Web project. Thus, scaling the process of publishing and exchanging large RDF data sets should comply with some basic features. At the *logical level*, a large-scale data set should have standard metadata, like provenance (source, providers, publication date, etc.), editorial metadata (publisher, date, version, etc.), data set statistics (size, quality, type of data, basic parameters of the data) and intellectual property (types of copy[left|right]s). At the *physical level*, RDF representation at large scale should permit efficient processing, managing and exchanging (between systems and memory-disk movements). At the format level, desirable features include simple checks for triple existence, redundancy minimization and modular construction. Imagine a user who wants to publish or exchange a large data set from her preferred RDF data store. She would first need to dump the data into one RDF format, and then, due to the large size of the data, possibly compress it with a generic compressor. The resultant file has no structure, no metadata and it is hardly usable natively, *i.e.* without an appropriate external tool (another RDF data store, a visualization software, etc.).

This paper addresses these challenges, and shows that there are feasible and simple solutions. In particular, we introduce a new representation format (*Header-Dictionary-Triples*: HDT) that modularizes the data and uses the skewed structure of big RDF graphs [10,19,21] to achieve large spatial savings. It is based on three main components:

- A *header*, including logical and physical metadata describing the RDF data set. It serves as an entrance point to the information on the data set.
- A *dictionary*, organizing all the identifiers in the RDF graph. It provides a catalog of the information entities in the RDF graph with high levels of compression.
- A set of *triples*, which comprises the pure structure of the underlying RDF graph while avoiding the noise produced by long labels and repetitions.

We make use of succinct data structures and simple compression notions to approach a practical implementation for HDT. Our design, besides gaining modularity and compactness, also addresses other important features: 1) it allows on-demand indexed access to the RDF graph, and 2) it is used to develop a specific technique for RDF compression (referred to as HDT-Compress) able to outperform universal compressors.

Figure 1 shows a step-by-step description of the process to obtain an HDT representation of an RDF graph. The first three steps extract basic RDF features necessary to build the dictionary and the underlying graph, as well as information that will be included in the header. The fourth step covers some practical decisions in order to have the HDT concrete implementation for publication and exchange of RDF.

If we go back to the example of the user who wants to publish or exchange large RDF data, the advantages of HDT can be summarized as follows: 1) More compact and compressible: uses much less space, thus saving storing space and communication bandwidth and time; 2) Is clean and modular: it separates dictionary from triples (the graph structure), includes a header with metadata about the data; 3) Permits basic data operations by allowing access to parts of the graph without needing to process all of it.

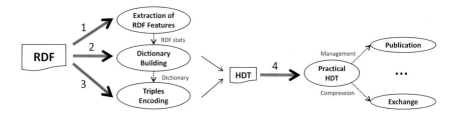

Fig. 1. A Step-by-step construction of the HDT format from a set of triples

The paper is organized as follows. Section 2 starts defining the set of metrics to characterize the structural RDF features used in HDT. Next, the HDT format is presented by an individual description of each component: Header, Dictionary, and Triples. Section 3, firstly details the practical implementation approach for HDT. Then, we detail the HDT management and compression. This section ends with an empirical study which analyzes the current HDT features on real-world data sets. Section 4 gives a brief discussion and addresses some future work. Finally, the appendix provides a study of the structural features of the data sets used in our experimentation, analyzing their impact on HDT. Additional resources and examples are available at `http://hdt.dcc.uchile.cl`.

Related Work. Today there are several representations for RDF data, *e.g.* RDF/XML [3], N3[1],Turtle[2], and RDF/JSON [1]. None of these proposals, though, seems to have considered data volume as a primary goal. RDF/XML, due to its verbosity is good for exchanging data, but only on a small scale. Turtle (a sub-language of N3) is a more compact and readable alternative. Although these formats present features to "abbreviate" constructions like URIs, groups of triples, common datatypes or RDF collections, the compactness of the representation definitively was not the main concern of their design. RDF/JSON resembles Turtle, with the advantage of being coded in a language easier to parse and more widely accepted in the programming world.

Regarding the structure of RDF real-world data, several studies point to the presence of power-law distribution, in term frequencies [10], resources [19] and schemas [21].

RDF compression capabilities have been studied [11] but have not been applied in a concrete format or implementation. The situation is not better for splitting RDF into components. Neither RDF/XML nor N3 (and their subsets Turtle and N-Triples) have the basic constructors to design modular files. To the best of our knowledge, none of these results have been applied in the design of RDF data sets. There is little work on the design of large RDF data sets. There have been projects discussing design issues of RDF[3], and a working group on design issues of translation from relational databases to RDF[4]. However, none of these works have touched the problem of RDF publication

---

[1] `http://www.w3.org/DesignIssues/Notation3`
[2] `http://www.w3.org/TeamSubmission/turtle/`
[3] Best Practices Publishing Vocab. W3C WG:
   `http://www.w3.org/2001/sw/BestPractices/`, and the Wordnet case `http://www.w3.org/2001/sw/BestPractices/WNET/wn-conversion.html`
[4] `http://www.w3.org/2001/sw/rdb2rdf/`

and exchange at large. The project that is currently systematically addressing the issue of publication of RDF at large, Linked Data, is starting to face some of these issues.

2   Compacting RDF with HDT: The Concepts

2.1   Taking Advantage of the Skewed Structure of Real-World RDF Data

Although power-law[5] distribution validation in RDF data remains an open field, in practice it is assumed as a common characteristic of RDF real-world data. Ding and Finn [10] reveal that Semantic Web graphs fit power-law distribution within some metrics such as the size of documents and term frequency use; most terms are described through few triples. Regarding the use of an RDF schema (RDFS[5]), the space of instances is sparsely populated, since most classes and properties have never been instantiated. By crawling the Web, Oren[19] comes to similar conclusions, showing that resources (URIs) in different documents fit to a power-law distribution. Theoharis [21] studies these properties for Semantic Web schemas, RDFS and OWL[16]. Similar distribution is found in the descendants of a class, as well as other schema features, such as the existence of few classes interconnecting schemas, or non-balanced hierarchies. The presence of star and chaining nodes has been also described in data and queries (star- and chain-shaped join queries) [17,18]. This schema analysis has contributed to synthetic schema generation for benchmarking [21]. These results motivate the adaptation to RDF of the well-known Web distribution, where power-law is present in successors list of a given domain, playing an important role in Web graphs compression [4,6].

For our purposes, a few indicators of the graph structure will be sufficient. RDF graph notation will follow [13,20], with no distinction between URIs, Blank nodes and Literals. A triple then, (s, p, o), is represented as a labeled graph $s \xrightarrow{p} o$. Let G be an RDF graph, and S_G, P_G, O_G be the sets of subjects, predicates and objects in G.

Definition 1 (out-degrees). *Let G be an RDF graph, and let $s \in S_G$ and $p \in P_G$.*

1. *The* out-degree *of s, denoted $deg^-(s)$, is defined as the number of triples of G in which s occurs as subject. Formally, $deg^-(s) = |\{(s, y, z)/(s, y, z) \in G\}|$. The maximum out-degree, $deg^-(G)$, and the mean out-degree, $\overline{deg^-}(G)$, are defined as the maximum and mean out-degrees of all subjects in S_G.*
2. *The* partial out-degree *of s respect to p, denoted $deg^{--}(s, p)$, is defined as the number of triples of G in which s occurs as subject and p as predicate. Formally, $deg^{--}(s, p) = |\{(s, p, z)/(s, p, z) \in G\}|$. The maximum partial out-degree of G, $deg^{--}(G)$, and mean partial out-degree, $\overline{deg^{--}}(G)$, are defined as the maximum (resp. the mean) partial out-degrees of all pairs of subject-predicates of G.*
3. *The* labeled out-degree *of s, denoted $degL^-(s)$, is defined as the number of different predicates (labels) of G with which s is related as a subject in a triple of G. Formally, $degL^-(s) = |\{p/p \in P_G, (s, p, z) \in G\}|$. The maximum labeled out-degree of G, $degL^-(G)$, and mean labeled out-degree, $\overline{degL^-}(G)$, are defined as the maximum (resp. the mean) labeled out-degrees of all subjects of G.*

---

[5] A power law is a function with scale invariance, which can be drawn as a line in the log-log scale with a slope equal to a scaling exponent, *e.g.* $f(x) = ax^{-\beta}$, thus $f(cx) \propto f(x)$, with a, c, β constants.

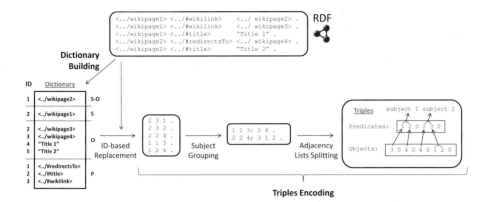

Fig. 2. Incremental representation of an RDF data set with HDT

Symmetrically, we define for objects the *in-degree*, denoted $deg^+(o)$, partial in-degree, $deg^{++}(o, p)$ and labeled in-degree, $degL^+(o)$. Their corresponding maximums and means are denoted as $deg^+(G)$, $deg^{++}(G)$, $degL^+(G)$, $\overline{deg^+}(G)$, $\overline{deg^{++}}(G)$ and $\overline{degL^+}(G)$. An additional property will be needed in what follows:

Definition 2 (subject-object ratio α_{s-o}). *It is defined as the proportion of common subjects and objects in the graph G. Formally,* $\alpha_{s-o} = \frac{|S_G \cap O_G|}{|S_G \cup O_G|}.$

Out-degree indicates the relevance of a subject node. A node with high out-degree, also called star, will have hundreds, or even thousands, of arcs (labeled edges in RDF). In conjunction with maximum and mean values, this constitutes good evidence of the existence of these types of nodes in a given graph. Similar reasoning can be made for in-degree, where the node is not a source, but is a common destination object node. Partial and labeled out- and in- degrees are meant to give information on the different types of edges coming out from (or going into) a node. Partial out-degree provides a metric of the multi evaluation of pairs (subject-predicate or predicate-object), while labeled degree refines the star-nodes categorization. Finally, subject-object ratio is a good measure of the percentage of nodes along which there are incoming and outgoing edges. These are the key edges to index, because of the different roles they play, either as subjects which are described elsewhere, or as objects describing other resources. In the final Appendix we illustrate these parameters for three real-world data sets.

In what follows, we will use these characteristics to provide a compact structure that represents, succinctly, the information of an RDF data set. Figure 2 outlines the incremental processing of our proposal. The result splits the RDF data set into three components that are represented and managed efficiently.

2.2 Header

The Header component is responsible for providing metadata information about the RDF collection. Although the most used RDF syntaxes consider the possibility of including metadata information, they present several drawbacks. Metadata is provided in

the same RDF syntax as the data set, inheriting some of its problems and making difficult the automatic distinction between data and metadata. The metadata of the collection
remains unclear and its management is inefficient.

We consider the Header as a flexible component in which the provider includes a
desired set of features. We distinguish four types of metadata:

- *Source and provider information.* This includes all kind of authority information
  about the source (or sources) of data and the provider of the data set, which can differ from the creator of data (*e.g.* in mashup applications). Note that this information
  can be shared between several data sets of a provider.
- *Publication data.* This collects the metadata about the publication act, such as the
  site of publication, dates of creation and modification, version of the data set (which
  could be useful for updating notification), language, encoding, namespaces, etc.
- *Data set statistics.* When managing huge collections, one could consider including
  some precomputed statistic about what follows in the data sets. For instance, it
  could be useful to include an estimation of the parameters presented in Section 2.1,
  or a subset of them used in the concrete design.
- *Other information.* A provider can take into account other metadata for the understanding and management of the data.

2.3   Dictionary

In general terms, a data dictionary is a centralized repository of information about data
such as meaning, relationships to other data, origin, usage, and format [15]. Current
RDF formats use elementary versions of dictionaries for namespaces and prefixes. This
allows for the abbreviation of long and repeated strings (URIs, Literals, etc.). A good example is "`http://www.w3.org/1999/02/22-rdf-syntax-ns#type`" re-
peated hundreds to thousands of times in the Billion Triple data set. Note that XML has
this functionality in the form of namespaces in conjunction with *XML Base*, and several
RDF formats allow abbreviations of this kind (`@base`, `@prefix` in N3 and Turtle).

Large RDF data sets are supposed to be managed by automatic processes, so that a
more effective replacement can be done. The Dictionary component assigns a unique
ID to each element in the data set. This way, the dictionary contributes to the goal of
compactness, by replacing the long repeated strings in triples by their short IDs. In fact,
the assignment of IDs, named as mapping [7], is usually the first step in RDF indexing.

The sets of subjects, predicates and objects in RDF are not disjoint. In order to assign
shorter IDs, we distinguish between four sets (in an RDF graph G):

- *Common subject-objects*, denoted as the set SO_G, are mapped to $[1, |SO_G|]$.
- The *non common subjects*, $S_G - SO_G$, are mapped to $[|SO_G| + 1, |S_G|]$.
- The *non common objects*, $O_G - SO_G$, are mapped to $[|SO_G| + 1, |O_G|]$.
- *Predicates* are mapped to $[1, |P_G|]$.

Figure 2 shows an example of these four sets within a dictionary building process. Note
that a given ID can belong to different sets, but the disambiguation of the correct set is
trivial when we know if the ID to search is a subject, a predicate or an object. A similar
partitioning is taken in some RDF indexing approaches [2].

The subject-object ratio defined in Section 2.1, α_{s-o}, characterizes the ratio of the subject-object set in the dictionary, composed of nodes with out-degree and in-degree greater than 0, $deg^-(a), deg^+(a) > 0$. We have noted that, in those data sets with a noticeable value of α_{s-o}, common subject-object identification has and advantage over a disjoint assignment, thus reducing the dictionary size. The set of predicates are treated independently because of their low number and the infrequent overlapping with other sets. Due to the sequential mapping of each set, the dictionary only has to include the strings, supposing an implicit order of IDs and some form of distinction between sets.

The Dictionary component allows multiple configurations. The order of the elements within each set could be random or sorted by some property, *e.g* the frequency of use or the alphabetical order. Prefixes and shared strings (specially for URIs) could be identified and written once and then reference the unshared portions incrementally. These design decisions should be declared in the Header component.

2.4 Triples

By means of the Dictionary component, an original RDF triple can now be expressed by three IDs, replacing each element in triples with the reference in the dictionary (ID-based replacement in Figure 2). As we transform a stream of strings into a stream of IDs, we can take advantage of some interesting properties.

Adjacency List is a compact data structure that facilitates managing and searching. For example, the set of triples:

$$\{(s, p_1, o_{11}), \cdots, (s, p_1, o_{1n_1}), (s, p_2, o_{21}), \cdots (s, p_2, o_{2n_2}), \cdots (s, p_k, o_{kn_k})\}$$

can be written as the adjacency list:

$$s \rightarrow [(p_1, (o_{11}, \cdots, o_{1n_1}), (p_2, (o_{21}, \cdots, o_{2n_2})), \cdots (p_k, (o_{kn_k}))].$$

Turtle (and hence N3) allows such generalized adjacency lists for triples. For example the set of triples $\{(s, p, o_j)\}_{j=1}^n$ can be abbreviated as $(s\ p\ o_1, \cdots, o_n)$.

The Triples component performs a subject ordered grouping, that is, triples are reorganized in adjacency lists, in sequential order of subject IDs. Due to this order, an immediate saving can be achieved by omitting the subject representation, as we know the first list corresponds to the first subject, the second list to the following, and so on.

In the notation above, all the data is represented by one stream, in which the list of objects associated with a subject (s) and a predicate (p) is represented just after the p. Instead, we decide to split this representation into two coordinated streams of *Predicates* and *Objects*. The first stream of *Predicates* corresponds to the lists of predicates associated with subjects, maintaining the implicit grouping order. The end of a list of predicates implies a change of subject, and must be marked with a separator, *e.g.* the non-assigned zero ID. The second stream (*Objects*) groups the lists of objects for each pair (s, p). These pairs are formed by the subjects (implicit and sequential), and coordinated predicates following the order of the first stream. In this case, the end of a list of objects (also marked in the stream) implies a change of (s, p) pair, moving forward in the first stream processing.

Figure 2 exemplifies the subject grouping and the final adjacency lists splitting into two coordinated streams. For instance, consider the list $[1, 2]$ in *Objects* stream. This is the fourth list in the stream, so it refers to the fourth predicate in *Predicates*: the ID 3. This predicate belongs to the second list in the stream; that is, it is related with the second subject. Thus, the considered list develops the triples $(2, 3, 1)$ and $(2, 3, 2)$.

The parameters defined in Section 2.1 characterize the streams. Labeled out-degree, $degL^-(s)$, indicates the size of the list of predicates for a given subject s. Therefore, the maximum size of any list in *Predicates* is limited by the maximum $degL^-(G)$, and the mean is given by $\overline{degL^-}(G)$. Symmetrically, partial out-degree, $deg^{--}(s, p)$ delimits the corresponding list in *Objects* for a given subject and predicate. Maximum and mean values, $deg^{--}(G)$ and $\overline{deg^{--}}(G)$ characterize the *Objects* stream.

This proposal leads to a compact dictionary-based triple representation in which the classical three-dimensional view of RDF has been reduced into two by the coordinated streams, considering implicit the third dimension of subjects. In the next section we introduce appropriate structures to effectively implement the HDT proposal.

3   Compacting RDF with HDT: Practical Aspects

HDT allows RDF data sets to be represented in a compact form, with no restriction on how it should be implemented. This feature allows HDT to be optimized in specific applications. In this section, we approach a practical HDT implementation focused on RDF publication and exchange. The optimization is based on high HDT compressibility and efficient management processes.

3.1   Implementation

The final HDT comprises the concrete implementation of the three complementary representations of the *header*, the *dictionary*, and the set of *triples*.

Header. Header information can include multiple types of metadata, and the selected configuration can vary between different data sets and different providers. In order to reach a mutual understanding between providers and consumers, in final implementation we restrict the Header to be one RDF-valid format and we provide a specific *hdt* vocabulary (http://hdt.dcc.uchile.cl/hdt#) to describe the Header through four top-level statements (containers): (1) *hdt:publicationInformation* describes publication, source and provider information, (2) *hdt:statisticalInformation* includes statistics about the data, *e.g.* the parameters defined in Section 2.1, (3) *hdt:formatInformation* groups the specification of the location and concrete Dictionary and Triples representation, and (4) *hdt:additionalInformation* contains further information given by the provider.

Dictionary. The final Dictionary configuration is encoded on a single stream in which all strings (ended with a reserved character, *e.g* '\2') are concatenated. Thus, the sequence represents the order of the strings in their correspondent vocabulary of subject-objects (S-O), subjects (S), objects (O), and predicates (P). An empty word (also ended with the reserved character) is appended to the end of each vocabulary in order to delimit its size.

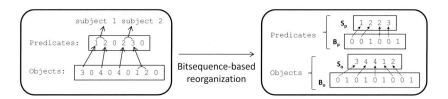

Fig. 3. Practical HDT Implementation

Triples. As we have previously explained, two coordinated ID-based streams, *Predicates* and *Objects*, *draw* the RDF graph, representing the triples with an implicit subject-grouping strategy. Both streams can be seen as sequences of non-negative integers in which 0-values mark the endings of predicate and object adjacency lists respectively. This means that positive integers represent predicates and objects, whereas 0's are auxiliary values embed in each stream to represent, implicitly, the graph structure. Our final implementation splits both parts in order to improve the HDT usability and to enhance its compactness. The graph structure is extracted from the original *Predicates* and *Objects* streams, so the 0-values can be deleted from them. The resultant sequences (respectively called S_p and S_o) hold the original ordering for the positive integers. In turn, the graph structure is indexed with two bitsequences (B_p and B_o, for predicates and objects) in which 0-bits mark IDs in the corresponding S_p or S_o sequence, whereas 1-bits are used to mark the end of an adjacency list.

Figure 3 shows a simple example of how the current approach reorganizes the original ID-based streams through the bitsequences. On the one hand, *Predicates* $= \{1, 2, 0, 2, 3, 0\}$ evolves to the sequence $S_p = \{1, 2, 2, 3\}$ and the bitsequence $B_p = \{001001\}$ whereas, on the other hand, *Objects*$= \{3, 0, 4, 0, 4, 0, 1, 2, 0\}$ is reorganized in $S_o = \{3, 4, 4, 1, 2\}$ and $B_o = \{010101001\}$.

The triples structure can be interpreted as follows. The i-*th* 1-bit in B_p marks the end of the predicate adjacency list for the i-*th* subject (it is referred to as P_i), whereas the length of the 0-bit sequences between two consecutive 1-bit represents the number of predicates in the corresponding list. For instance, the second 1-bit in B_p marks the end of the predicate adjacency list for the second subject (P_2). As we can see, a sequence of two 0-bit exists in between the previous and the current 1-bit. This means P_2 contains two predicates, which are represented by the third and fourth IDs in S_p by considering that the third and fourth 0-bit in B_p correspond to P_2. Thus, $P_2 = \{2, 3\}$.

Data in S_o and B_o are related in the same way. Hence, the j-*th* 1-bit in B_o marks the end of the object adjacency list for the j-*th* predicate. This predicate is represented by the j-*th* 0-bit in B_p and it is retrieved from the j-*th* position of S_p. For example, the third 1-bit in B_o refers the end of the object adjacency list for the third predicate in S_p which is related to the second subject as we have previously explained. Thus, this adjacency list stores all objects o in triples $(2, 3, o) \in G$.

Each element in S_p and S_o is encoded, respectively, with a fixed-length code of $\log(|P_G|)$ and $\log(|O_G|)$ bits, by considering that the data set comprises $|P_G|$ and $|O_G|$ different predicates and objects. The bitsequences used to represent B_p and B_o make use of *succinct structures*. They are able to support rank/select operations over a sequence S of length n drawn from an alphabet $\Sigma = \{0, 1\}$:

Algorithm 1. Check&Find operation for a triple (s, p, o)

---

1. $begin \leftarrow \textbf{select}_1(B_p, s - 1) + 2$;
2. $end \leftarrow \textbf{select}_1(B_p, s) - 1$;
3. $size_{P_s} \leftarrow end - begin$;
4. $P_s \leftarrow \textbf{retrieve}(S_p, 1 + \textbf{rank}_0(B_p, begin), size_{P_s})$;
5.
6. $plist \leftarrow \textbf{binary\_search}(P_s, p)$;
7. $pseq \leftarrow \textbf{rank}_0(B_p, begin) + plist$;
8.
9. $begin \leftarrow \textbf{select}_1(B_o, pseq - 1) + 2$;
10. $end \leftarrow \textbf{select}_1(B_o, pseq) - 1$;
11. $size_{O_{sp}} \leftarrow end - begin$;
12. $O_{sp} \leftarrow \textbf{retrieve}(S_o, 1 + \textbf{rank}_0(B_o, begin), size)$;
13.
14. $plist \leftarrow \textbf{binary\_search}(O_{sp}, o)$;

- $\texttt{rank}_a(\texttt{S}, \texttt{i})$ counts the occurrences of a symbol $a \in \{0, 1\}$ in $\mathcal{S}[1, i]$.
- $\texttt{select}_a(\texttt{S}, \texttt{i})$ finds the i-th occurrence of symbol $a \in \{0, 1\}$ in $\mathcal{S}$.

This problem has been solved using $n + o(n)$ bits of space while answering the queries in constant time [8]. We choose the *González, et al.* [12] approach to implement our bitsequences. This adds 5% of extra space to the original length of $\mathcal{B}_p$ and $\mathcal{B}_o$, and achieves constant time for the required $\texttt{select/rank}$ operations, which constitutes the basis for accessing to the structure of the graph.

3.2 HDT Management

A really huge RDF data set contains a volume of triples that becomes unmanageable when it is finally published. Let us suppose a very large data set has been published on any of the existing RDF syntaxes. Basic operations, *e.g.* check a triple existence or *access* to a subset of triples, are optimized in RDF storage systems, but this implies, firstly, configuring the system and then loading the full data set for the triple indexing. On the one hand, huge amounts of memory are required to render an efficient-time service when operating on the full data set. On the other hand, simple on-demand access to subsets of triples does not profit from the internal structure of RDF and suffers the cost of loading and searching the full data set.

Our current approach proposes an on-demand loading strategy able to take advantage of the structure indexed in $\mathcal{B}_p$ and $\mathcal{B}_o$ and accessible by fast $\texttt{rank/ select}$ operations. A functional prototype is implemented in order to test basic operations. An initial stage loads both the dictionary (in a hash table) and the bitsequences; we consider that these structures always fit into memory. The sequences $\mathcal{S}_p$ and $\mathcal{S}_o$ remain stored in disk, queried by using the Check&Find operation described in Algorithm 1.

Lines $1-4$ describe the steps performed to retrieve the predicate adjacency list for the subject s (P_s). First, we obtain its size by locating its begin/end positions in $\mathcal{B}_p$. Next, we $\texttt{retrieve}$ its sequence of predicate IDs from $\mathcal{S}_p$. This operation seeks the position where P_s begins in $\mathcal{S}_p$ (by using the $\texttt{rank}_0$ operation in line 4), and, next, retrieves the sequence of $size_{P_s}$ predicates that composed it. Once P_s is available in memory, we need to identify the position ($pseq$) where s and p are related in $\mathcal{S}_p$. Lines 6-7 describe it. First, p is located in P_s with a `binary_search`, and, next, this local position ($plist$) is used to obtain its global rank in $\mathcal{S}_p$. In this step, we are able to retrieve the object adjacency list for s, p (O_{sp}), by considering that it is indexed through the pseq$-th$ predicate. O_{sp} is retrieved (lines 9-12) similarly to P_s, considering $\mathcal{B}_o$ and $\mathcal{S}_o$. Finally, o is located with a `binary_search` on O_{sp}.

The cost of the `Check&Find` operation for a triple (s, p, o) is $O(size_{P_s} + size_{O_{sp}})$, assuming at most $size_{P_s} = degL^-(s)$ and $size_{O_{sp}} = deg^{--}(s,p)$. The distribution of lists assures an amortized cost in $(\overline{degL^-(G)} + \overline{deg^{--}(G)})$. Note that this operation does not just find the required triple (s, p, o), but also the triples $(s, p, z) \in G$. Besides, P_s contains all predicates from s, so the next operations on triples from s begin the `Check&Find` operation by identifying the position of p in $\mathcal{S}_p$ (from line 6).

Efficient access is obtained through `Check&Find`. If a triple $(s, p, o) \notin G$, it can be detected in step 6 (the predicate p is not in the predicate adjacency list for s: S_p) or in step 14 (the object o is not in the object adjacency list for s and p: O_{sp}). On the contrary, if $(s, p, o) \in G$, once the triple is found, the strings associated with s, p, and o are retrieved from the dictionary in time $O(1)$.

In addition, the `Check&Find` operation sets the basis for building efficient insertion and deletion. In both cases, the adjacency lists to be updated are already available in memory after the `Check&Find`. Thus, the changes can be performed in an efficient logarithmic time, and the final performance of the operations will depend on the strategy for writing the updated information on disk. Besides, as we explain, `Check&Find` checks the triple existence, $i.e$ the insertion is only performed if the triple does not exist and the deletion is carried out over the found triple.

We have assumed, in the initial step, that the dictionary fits into memory, a common assumption in the world of indexing regarding the size of the vocabulary. Our current development achieves reducing, in one order of magnitude, the original size by simply applying a prefix extraction. Other optimizations can be applied, such as a hierarchical treatment of URIs.

3.3 `HDT` Compression

RDF exchange is a common process in the global Web of data with the aim of sharing knowledge. The *data-centric* evolution of the Web will tend to demand even more exhaustive exchange processes in which efficiency is highly desirable. The performance of this task is directly related with the size of the data set, so large RDF data sets can overhead communication channels causing lengthy transmission delays. The use of universal compressors can alleviate this problem, although they are not able to detect and delete all the underlying redundancies of RDF.

We show, in Section 3.4, that our `HDT` representation (referred to as `Plain HDT` henceforth) is able to compact the RDF data set up to 15 times with respect to its original size. This compacting ability proves capable of achieving very large savings in communication bandwidth and transmission delays. However, `Plain HDT` is even more compressible with very little effort. `HDT-Compress` makes specific decisions for each component:

Table 1. Compression results

Data set	Triples (millions)	Size (MB)	HDT		Universal Compressors		
			Plain	Compress	gzip	bzip2	ppmdi
Billion Triples	106.9	15081.74	*31.87%*	**3.91%**	9.54%	6.83%	5.32%
Uniprot RDF	79.2	7083.22	*14.33%*	**3.23%**	8.71%	5.04%	3.99%
Wikipedia	47	6882.20	*6.62%*	**2.22%**	6.97%	5.11%	4.10%

Header: we keep this component in plain form as it should always be available to any receiving agent for processing.

Dictionary: it is compressed by considering that it stores all strings used in the RDF data set. Thus, we take advantage of repeated prefixes in URIs, specific n-gram distributions in literals, etc. This class of redundancy is able to be identified with a predictive high-order compressor. We chose PPM [9] to encode the dictionary.

Triples: the set of triples compression is independently attempted on each structure. On one hand, S_p comprises an integer sequence drawn from $[1, |P_G|]$. A Huffman [14] code is used to compress it. On the other hand, the compression of S_o (drawn from $[1, |O_G|]$) takes advantage of the power-law distribution of objects (see the right dispersion graph in Figure 5) through a second Huffman code. Finally, we hold a plain representation for bitsequences because of the small improvement obtained with specific techniques for bitsequence compression.

We chose shuff[6] and ppmdi[7] to implement, respectively, the Huffman and PPM-based encoding.

3.4 Experimental Results

This section shows the experimental results of the practical applications previously described for HDT. These tests were performed on a Debian 4.1.1 operating system, running on a computer with an AMD Opteron(tm) Processor 246 at 2 GHz and 4 GB of RAM. We used a g++ 4.1.2 compiler with -O9 optimization. This experimentation was run on the data sets described in the final Appendix.

First, we study the HDT performance with incremental size of the Uniprot data set, from 1 to 40 million triples. This is shown in Figure 4. The left table studies the HDT effectiveness evolution. As can be seen, it is distributed between $14 - 15\%$ for Plain HDT, and by around 3.5% for HDT-Compress (the percentage is always given with regard to the original file size). This proves the scalability of the HDT effectiveness by considering that the results do not directly depend on the size of the data set.

The right graph of Figure 4 shows relevant times for HDT. On the one hand, the *creation* time stands for the time required to transform an RDF data set (from plain N3) into HDT. This process is only performed once at publishing and shows a sublinear growth. On the other hand, after the *loading* time, the minimum information required for HDT management is in memory and available to be accessed with the Check&Find

---

[6] http://www.cs.mu.oz.au/~alistair/mr_coder/
[7] http://pizzachili.dcc.uchile.cl/experiments.html

Triples	Size	HDT	
(millions)	(MB)	Plain	Compress
1	89.07	15.11%	**3.73%**
5	444.71	14.54%	**3.48%**
10	893.39	14.04%	**3.27%**
20	1790.41	14.43%	**3.31%**
30	2680.51	14.39%	**3.27%**
40	3574.59	14.34%	**3.26%**

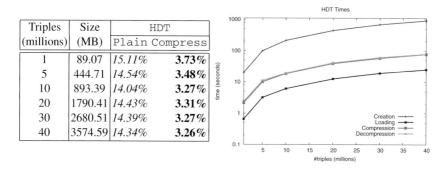

Fig. 4. Performance of HDT (Plain and Compress) with incremental size data sets from Uniprot. The left table shows effectiveness, whereas the right figure draws significative times.

mechanism (Algorithm 1). As can be seen, this time is only a very small fraction ($\approx 3\%$) of the creation one. Additionally, symmetrical *compression* and *decompression* times are achieved with HDT-Compress. This guarantees real time exchange processes by considering that the receiver is able to start the decompression as soon as the beginning of the compressed data set starts to arrive. In absolute terms, both compression and decompression times are slightly worse than the loading ones.

Table 1 compares HDT with respect to four well-known universal compressors. We choose gzip as a dictionary-based technique on LZ77, bzip2 based on the Burrows-Wheeler Transform, and ppmdi as a predictive high-order compressor. We consider a heterogeneous corpora of RDF data sets shown in the final Appendix: Billion Triples, Uniprot RDF and Wikipedia with 106.9, 79.2, and 47 million of triples respectively.

The most effective universal compressors for all data sets are ppmdi and bzip2 which achieve ratios of around 4% and 5% respectively. A very interesting result shows that Plain HDT is able to outperform gzip for the Wikipedia data set. This demonstrates the previously cited ability of HDT to obtain compact representations of RDF. HDT-Compress achieves the most effective results with ratios between $2-4\%$ for the considered data sets. This supposes reductions between $3-4$ times with respect to Plain HDT, and consequently proportional improvements on exchanging processes. In turn, HDT-Compress also outperforms universal compressors by improving the best results, achieved on ppmdi, of between $20-45\%$.

4 Conclusions and Future Work

RDF publication and exchange at large scale are seriously compromised by the scalability drawbacks of current RDF formats and the lack of modular structure, internal metadata information and native operations over the data. HDT addresses these problems by approaching a more compact representation format, decomposing an RDF data source into three main parts: Header, Dictionary, and Triples. Besides, this representation can be implemented by succinct structures and simple compression notions. This results in a very compact RDF representation able to support an on-demand Check&Find mechanism currently used to implement indexed access to the RDF graph. Our experimental results show the scalability of HDT for incremental data set sizes, being able to compact a data set up to 15 times current naive representations and providing efficient access to the data. In turn, a specific compression technique for RDF, HDT-Compress, outperforms universal compressors, which can serve as an essential choice in exchange processes involving huge data sets.

Current results open some interesting opportunities for future work. HDT compactness and the on-demand operations set the basis for developing an RDF storage system over HDT. The Check&Find mechanism and its ability to perform indexed access to the RDF graph will guide the design of efficient insertion and deletion (thus, also updating) which establish a full set of management operations. Additionally, we are currently analyzing S_p and S_o to be reorganized following a wavelet-tree-like strategy. This keeps the current spatial requirements of HDT but also provides indexed access inside both sequences, suggesting a full compressed index able to solve basic SPARQL queries natively. The resolution of a basic SPARQL join query can be performed through a series of wavelet-tree and bitsequences operations and dictionary accesses. Subject-object JOINs resolution can also profit from the common naming in the dictionary, as the elements are correctly and quickly localized in the top IDs.

References

1. Alexander, K.: RDF in JSON: A Specification for serialising RDF in JSON. In: SFSW 2008 (2008), http://www.semanticscripting.org/SFSW2008 (retrieved September 2010)
2. Atre, M., Chaoji, V., Zaki, M.J., Hendler, J.A.: Matrix "Bit" loaded: a scalable lightweight join query processor for RDF data. In: WWW 2010, pp. 41–50 (2010)
3. Beckett, D.: RDF/XML syntax specification (Revised). Technical report, W3C (February 2004)
4. Boldi, P., Vigna, S.: The webgraph framework I: compression techniques. In: WWW 2004, pp. 595–602 (2004)
5. Brickley, D.: RDF Vocabulary Description Language 1.0: RDF Schema. W3C Recomm. (2004), http://www.w3.org/TR/rdf-schema/ (retrieved September 2010)
6. Chierichetti, F., Kumar, R., Raghavan, P.: Compressed web indexes. In: WWW 2009, pp. 451–460 (2009)
7. Chong, E.I., Das, S., Eadon, G., Srinivasan, J.: An efficient sql-based rdf querying scheme. In: VLDB 2005, pp. 1216–1227 (2005)
8. Clark, D.: Compact PAT trees. PhD thesis, University of Waterloo (1996)
9. Cleary, J.G., Witten, I.H.: Data Compression Using Adaptive Coding and Partial String Matching. IEEE Transactions on Communications 32(4), 396–402 (1984)
10. Ding, L., Finin, T.: Characterizing the Semantic Web on the Web. In: Cruz, I., Decker, S., Allemang, D., Preist, C., Schwabe, D., Mika, P., Uschold, M., Aroyo, L.M. (eds.) ISWC 2006. LNCS, vol. 4273, pp. 242–257. Springer, Heidelberg (2006)
11. Fernández, J.D., Gutierrez, C., Martínez-Prieto, M.A.: RDF compression: basic approaches. In: WWW 2010, pp. 1091–1092 (2010)
12. González, R., Grabowski, S., Makinen, V., Navarro, G.: Practical implementation of rank and select queries. In: WEA 2005, pp. 27–38 (2005)
13. Gutierrez, C., Hurtado, C., Mendelzon, A.O.: Foundations of semantic web databases. In: PODS 2004, pp. 95–106 (2004)

14. Huffman, D.A.: A Method for the Construction of Minimum-Redundancy Codes. Proceedings of the IRE 40(9), 1098–1101 (1952)
15. IBM. IBM Dictionary of Computing. McGraw-Hill, New York (1993)
16. McGuinness, D., van Harmelen, F.: OWL Web Ontology Language Overview. W3C Recommendation (2004), http://www.w3.org/TR/owl-features/ (retrieved September 2010)
17. Neumann, T., Weikum, G.: RDF-3X: a RISC-style engine for RDF. Proceedings of the VLDB Endowment 1(1), 647–659 (2008)
18. Neumann, T., Weikum, G.: Scalable join processing on very large rdf graphs. In: COMAD 2009, pp. 627–640 (2009)
19. Oren, E., et al.: Sindice.com: a document-oriented lookup index for open linked data. International Journal of Metadata, Semantics and Ontologies 3(1), 37–52 (2008)
20. Pérez, J., Arenas, M., Gutierrez, C.: Semantics and Complexity of SPARQL. ACM Transactions on Database Systems 34(3), 1–45 (2009)
21. Theoharis, Y., Tzitzikas, Y., Kotzinos, D., Christophides, V.: On Graph Features of Semantic Web Schemas. IEEE Trans. on Know. and Data Engineering 20(5), 692–702 (2008)

Appendix: The Data Sets Used in the Study

This appendix comprises an experimental study on real-world data sets in order to characterize RDF structure and redundancy by applying the parameters presented in Section 2.1. We chose three data sets based on the huge amount of triples, different application domains and previous uses in benchmarking:

– *Billion Triples Challenge:* One of the largest RDF data sets (∼3.2 billion statements) available at the moment of writing, given as part of the Semantic Web Challenge[8]. Data is collected from Sindice, Swoogle, DBpedia and others.

– *Uniprot RDF*[9]: huge, freely-accessible RDF data set of protein sequence data, as part of the Uniprot project (∼0.7 billion statements).

– *Wikipedia triple-set*[10]: English Wikipedia in RDF (∼47 million statements).

A preprocessing step is first applied. Billion Triples data was parsed from N-Quads format[11] to NTriples by eliminating context information. We generated an N3 format from the original RDF/XML of Uniprot by using the tool SemWeb[12]. For Billion Triples and Uniprot, we used a random sample of the data, respecting the order of appearance and eliminating repeated triples.

Table 2 summarizes the statistical data, focusing on the most relevant parameters for our approach. Several comments are in order. First of all, note the high variability of values among the data sets. Billion Triple data set is a mashup of diverse sources, whereas Wikipedia triple-set and Uniprot are designed with one main purpose. The special condition of Billion Triple increments the number of different predicates, although they remain proportionally small to the number of triples, as well as decreasing the

---

[8] http://challenge.semanticweb.org/
[9] http://dev.isb-sib.ch/projects/uniprot-rdf/
[10] http://labs.systemone.at/wikipedia3
[11] http://sw.deri.org/2008/07/n-quads/
[12] http://razor.occams.info/code/semweb/semweb-current/doc/index.html

Table 2. Data sets statistic summary

Data Set	# triples	# pred.	deg^-	deg^-	deg^{--}	deg^{--}	$degL^-$	$degL^-$	α_{s-o}
Billion Triple	106.9M	50516	27387	2.74	27386	1.11	3293	2.46	6.54%
Uniprot RDF	79M	99	2030	4.80	2010	1.27	22	3.78	58.49%
Wikipedia	47M	9	7408	21.76	7400	3.95	7	5.51	17.61%

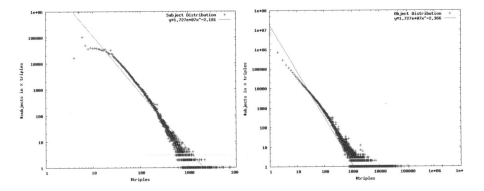

Fig. 5. Wikipedia triple-set distribution of subjects (left) and objects (right), *e.g.* a point (X,Y) in the rightmost graphic says that there are Y different objects each occurring in X triples. Both axis are logarithmic. The power laws have exponent -2.181 and -2.366 respectively.

subject-object ratio. In this case, out-degrees reveal that subjects are related with few predicates (a mean of 2.46) and each of these pairs match with a single object (a mean of 1.11). In turn, the design of Uniprot has more cohesion, with a high subject-object ratio and a smaller number of very frequent predicates (each subject is related with a mean of 3.78 predicates over a total of 99). This reveals a star chained design in which a subject is strongly characterize and interlinked with others. A similar interpretation could be done for the Wikipedia triple-set, although the number of predicates is extremely low. In this case, the high partial degree suggests that a pair *(subject,predicate)* is repeated within several objects (a mean of 3.95). This affirmation is consistent with the interlinked design of pages in Wikipedia.

Figure 5 shows the distribution of subjects and objects of the Wikipedia triple-set. As we expected, a power-law distribution is remarkably present in both cases, suggesting an implicit significant redundancy. The other data sets reveal the same distribution for subject and object as well as a lack of statistical distribution of predicates.

These results immediately point to possible compact design models of RDF. Our approach, HDT, exploits the significant correlation and the inherent redundancy in data and structure. In particular, the Dictionary component takes advantage of subject-object ratio characterization and groups the references to the same node. The Triples component represents the graph compacting the distribution with implicit and coordinated adjacency lists, parametrized by out-degree means.

Assessing Trust in Uncertain Information

Achille Fokoue[1], Mudhakar Srivatsa[1], and Rob Young[2]

[1] IBM Research, USA
{achille,msrivats}@us.ibm.com
[2] Defense Science and Technology Lab, UK
riyoung@dstl.gov.uk

Abstract. On the Semantic Web, decision makers (humans or software agents alike) are faced with the challenge of examining large volumes of information originating from heterogeneous sources with the goal of ascertaining trust in various pieces of information. While previous work has focused on simple models for review and rating systems, we introduce a new trust model for rich, complex and uncertain information.We present the challenges raised by the new model, and the results of an evaluation of the first prototype implementation under a variety of scenarios.

1 Introduction

Decision makers (humans or software agents alike) relying on information available on the web are increasingly faced with the challenge of examining large volumes of information originating from heterogeneous sources with the goal of ascertaining trust in various pieces of information. Several authors have explored various trust computation models (e.g., eBay recommendation system [14], NetFlix movie ratings [13], EigenTrust [10], PeerTrust [15], etc.) to assess trust in various entities. A common data model subsumed by several trust computation models (as succinctly captured in Kuter and Golbeck [11]) is the ability of an entity to assign a *numeric* trust score to another entity (e.g., eBay recommendation, Netflix movie ratings, etc.). Such pair-wise numeric ratings contribute to a (dis)similarity score (e.g., based on $\mathcal{L}_1$ norm, $\mathcal{L}_2$ norm, cosine distance, etc.) which is used to compute personalized trust scores (as in PeerTrust) or recursively propagated throughout the network to compute global trust scores (as in EigenTrust).

A pair-wise numeric score based data model may impose severe limitations in several real-world applications. For example, let us suppose that information sources $\{S_1, S_2, S_3\}$ assert axioms $\phi_1 = $ *all men are mortal*, $\phi_2 = $ *Socrates is a man* and $\phi_3 = $ *Socrates is not mortal* respectively. While there is an obvious conflict when all the three axioms are put together, we note that: (i) there is no pair-wise conflict, and (ii) there is no obvious numeric measure that captures (dis)similarity between two information sources.

This problem becomes even more challenging because of uncertainty associated with real-world data and applications. Uncertainty manifests itself in several diverse forms: from measurement errors (e.g., sensor readings) and stochasticity in physical processes (e.g., weather conditions) to reliability/trustworthiness of data sources; regardless of its nature, it is common to adopt a probabilistic measure for uncertainty. Reusing the

P.F. Patel-Schneider et al. (Eds.): ISWC 2010, Part I, LNCS 6496, pp. 209–224, 2010.
© Springer-Verlag Berlin Heidelberg 2010

Socrates example above, each information source S_i may assert the axiom ϕ_i with a certain probability $p_i = 0.6$. Further, probabilities associated with various axioms need not be (statistically) independent. In such situations, the key challenge is develop trust computation models for rich (beyond pair-wise numeric ratings) and uncertain (probabilistic) information.

The contributions of this paper are three fold. First, our approach offers a rich data model for trust. We allow information items to be encoded in inconsistency-tolerant extension of Bayesian Description Logics [3] (BDL)[1] with axioms of the form $\phi : X$[2] where ϕ is a classical axiom (in Description Logics (DL [1])) that is annotated with a boolean random variable from a Bayesian network [7]. Intuitively, $\phi : X$ can be read as follows: the axiom ϕ holds when the Boolean random variable X is true. Dependencies between axioms (e.g., $\phi_1 : X_1$ and $\phi_2 : X_2$) are captured using the Bayesian network that represents a set of random variables (corresponding to the annotations; e.g., X_1, X_2) and their conditional probability distribution functions (e.g., $Pr(X_2|X_1)$).

Second, our approach offers a trust computation model over uncertain information (encoded as BDL axioms). Intuitively, our approach allows us to compute a degree of inconsistency over a probabilistic knowledge base. We note that inconsistencies correspond to conflicts in information items reported by one or more information sources. Our approach assigns numeric weights to the degree of inconsistency using the *possible world* semantics (the formal semantics is given in section 3). Revisiting the *Socrates* example, three probabilistic axioms $\phi_i : p_i$[3] correspond to eight possible worlds (the power set of the set of axioms without annotations) corresponding to $\{\{\phi_1, \phi_2, \phi_3\}$, $\{\phi_1, \phi_2\}, \cdots, \emptyset\}$. Each possible world has probability measure that can be derived from p_i. For instance, the probability of a possible world $\{\phi_1, \phi_2\}$ is given by $p_1 * p_2 * (1-p_3)$. The degree of inconsistency of a knowledge base is then computed as the sum of the probabilities associated with possible worlds that are inconsistent.

In the presence of inconsistencies, our approach extracts justifications − minimal sets of axioms that together imply an inconsistency [9]. Our trust computation model essentially propagates the degree of inconsistency as blames (or penalties) to the axioms contributing to the inconsistency via justifications. This approach essentially allows us to compute trust in information at the granularity of an axiom. Indeed one may aggregate trust scores at different levels of granularity; e.g., axioms about a specific topic (e.g., birds), one information source (e.g., John), groups of information sources (e.g., all members affiliated with ACM), etc. Intuitively, our trust computation model works as follows. First, we compute a probability measure for each justification as the sum of the probabilities associated with possible worlds in which the justification holds (namely, all the axioms in the justification are present). Second, we partition the degree of inconsistency across all justifications; for instance, if a justification J_1 holds in 80% of the possible worlds then it is assigned four times the blame as a justification J_2 that holds in 20% of the possible worlds. Third, we partition the penalty associated with a

---

[1] BDL is a simple probabilistic extension of Description Logics, the foundation of OWL DL.

[2] This is a very simplified version of the BDL formulation given here for ease of the presentation. The complete and formal definition of BDL is presented in section 2.

[3] $\phi_i : p_i$ is a shorthand notation for $\phi_i : X_i$ and $Pr(X_i = true) = p_i$ for some independent random variable X_i.

justification across all axioms in the justification using a biased (on prior trust assessments) or an unbiased partitioning scheme. We note that there may be alternate approaches to derive trust scores from inconsistency measures and justifications; indeed, our approach is flexible and extensible to such trust computation models.

A naive implementation of our trust computation model requires *all* justifications. While computing a justification is an easy problem, exhaustively enumerating all possible justifications is known to be hard problem [9]. We formulate exhaustive enumeration of justifications as a tree traversal problem and develop an *importance sampling* approach to uniformly and randomly sample justifications without completely enumerating them. Unbiased sampling of justifications ensures that the malicious entities cannot game the trust computation model; say, selectively hide justifications that include axioms from malicious entities (and thus evade penalties) from the sampling process. For scalability reasons, our trust computation model operates on a random sample of justifications. A malicious entity may escape penalties due to incompleteness of justifications; however, across multiple inconsistency checks a malicious entity is likely to incur higher penalties (and thus lower trust score) than the honest entities.

Third, we have developed a prototype of our trust assessment system by implementing a probabilistic extension, PSHER, to our publicly available highly scalable DL reasoner SHER [6]. To avoid the exponential blow up due to the fact that the number of possible worlds in the worst case is exponential in the number of axioms, we use an error-bounded approximation algorithm to compute the degree of inconsistency of a probabilistic knowledge base and the weight of its justifications. Finally, we empirically evaluate the efficacy of our scheme (on a publicly available UOBM dataset) when malicious sources use an oscillating behavior to milk the trust computation model and when honest sources are faced with measurement errors (high uncertainty) or commit honest mistakes.

The remainder of the paper is organized as follows. After a brief introduction of Bayesian Description Logics (BDL) in Section 2, Section 3 describes an inconsistency-tolerant extension of BDL and presents solutions to effectively compute justifications (a proxy for (dis)similarity scores in our trust computation model). Section 4 describes our trust computation model. Section 5 presents an experimental evaluation of our system. We finally conclude in Section 6.

2   Background

In this section, we briefly describe our data model for uncertain information.

2.1   Bayesian Network Notation

We briefly recall notations for a Bayesian Network, used in the remainder of the paper. V: set of all random variables in a Bayesian network (e.g., $V = \{V_1, V_2\}$). $D(V_i)$ (for some variable $V_i \in V$): finite set of values that V_i can take (e.g., $D(V_1) = \{0, 1\}$ and $D(V_2) = \{0, 1\}$). v: assignment of all random variables to a possible value (e.g., $v = \{V_1 = 0, V_2 = 1\}$). $v|X$ (for some $X \subseteq V$): projection of v that only includes the random variables in X (e.g., $v|\{V_2\} = \{V_2 = 1\}$). $D(X)$ (for some $X \subseteq V$): Cartesian product of the domains $D(X_i)$ for all $X_i \in X$.

2.2   Bayesian Description Logics

Bayesian Description Logics [3] is a class of probabilistic description logic wherein each logical axiom is annotated with an event which is associated with a probability value via a Bayesian Network. In this section, we describe Bayesian DL at a syntactic level followed by a detailed example. A probabilistic axiom over a Bayesian Network BN over a set of variables V is of the form $\phi : e$, where ϕ is a classical DL axiom, and the probabilistic annotation e is an expression of one of the following forms: $X = x$ or $X \neq x$ where $X \subseteq V$ and $x \in D(X)$. Intuitively, every probabilistic annotation represents a scenario (or an event) which is associated with the set of all value assignments $V = v$ with $v \in D(V)$ that are compatible with $X = x$ (that is, $v|X = x$) and their probability value $Pr_{BN}(V = v)$ in the Bayesian network BN over V. Simply put, the semantics of a probabilistic axiom $\phi : X = x$ is as follows: when event $X = x$ occurs then ϕ holds. $\phi : p$, where $p \in [0, 1]$, is often used to directly assign a probability value to an classical axiom ϕ. This is an abbreviation for $\phi : X_0 = true$, where X_0 is a boolean random variable which is independent from all other variables and such that $Pr_{BN}(X_0 = true) = p$. We abbreviate the probabilistic axiom of the form $\top : e$ (resp. $\phi : \top$) as e (resp. ϕ).

A probabilistic knowledge base (KB) $K = (\mathcal{A}, \mathcal{T}, BN)$ consists of: 1) a Bayesian Network BN over a set of random variables V, 2) a set of probabilistic Abox axioms $\mathcal{A}$ of the form $\phi : e$, where ϕ is a classical Abox axiom, and 3) a set of probabilistic Tbox axioms $\mathcal{T}$ of the form $\phi : e$, where ϕ is a classical Tbox axiom. The following example illustrates how this formalism can be used to describe road conditions influenced by probabilistic events such as weather conditions:

$$\mathcal{T} = \{SlipperyRoad \sqcap OpenedRoad \sqsubseteq HazardousCondition,$$

$$Road \sqsubseteq SlipperyRoad : Rain = true\}$$

$$\mathcal{A} = \{Road(route9A), OpenedRoad(route9A) : TrustSource = true\}$$

In this example, the Bayesian network BN consists of three variables: $Rain$, a boolean variable which is true when it rains; $TrustSource$, a boolean variable which is true when the source of the axiom $OpenedRoad(route9A)$ can be trusted; and $Source$, a variable which indicates the provenance of the axiom $OpenedRoad(route9A)$. The probabilities specified by BN are as follows:

$$Pr_{BN}(TrustSource = true|Source = `Mary') = 0.8, \ Pr_{BN}(Rain = true) = 0.7$$

$$Pr_{BN}(TrustSource = true|Source = `John') = 0.5, \ Pr_{BN}(Source = `John') = 1$$

The first Tbox axiom asserts that a opened road that is slippery is a hazardous condition. The second Tbox axiom indicates that when it rains, roads are slippery. The Abox axioms assert that $route9A$ is a road and, assuming that the source of the statement $OpenedRoad(route9A)$ is trusted, $route9A$ is opened.

Informally, probability values computed through the Bayesian network 'propagate' to the 'DL side' as follows. Each assignment v of all random variables in BN (e.g.,$v = \{Rain = true, TrustSource = false, Source= `John'\}$) corresponds to a primitive event ev (or a scenario). A primitive event ev is associated, through BN, to a probability value p_{ev} and a classical DL KB K_{ev}[4] which consists of all classical axioms annotated with a compatible probabilistic annotation (e.g., $SlipperyRoad \sqcap OpenedRoad \sqsubseteq HazardousCondition, Road \sqsubseteq SlipperyRoad, Road(route9A)$). The probability value associated with the statement ϕ (e.g., $\phi = HazardousCondition(route9A)$) is obtained by summing p_{ev} for all ev such that the classical KB K_{ev} entails ϕ (e.g., $Pr(HazardousCondition(route9A)) = 0.35$).

3   Inconsistency and Justification

The ability to detect contradicting statements and measure the relative importance of the resulting conflict is a key prerequisite to estimate the (dis)similarity between information sources providing rich, complex and probabilistic assertions expressed as BDL axioms. Unfortunately, in the traditional BDL semantics [3], consistency is still categorically defined, i.e., a probabilistic KB is either completely satisfied or completely unsatisfied. In this section, we address this significant shortcoming by using a refined semantics which introduces the notion of degree of inconsistency. We start by presenting the traditional BDL semantics, which does not tolerate inconsistency.

For $v \in V$, we say that v is compatible with the probabilistic annotation $X = x$ (resp. $X \neq x$), denoted $v \models X = x$ (resp. $v \models X \neq x$), iff $v|X = x$ (resp. $v|X \neq x$).

Recall that BDL axioms $(\phi : e)$ are extensions of classical axioms (ϕ) with a probabilistic annotation (e). BDL semantics defines an annotated interpretation as an extension of a first-order interpretation by assigning a value $v \in D(V)$ to V. An annotated interpretation $\mathcal{I} = (\Delta^{\mathcal{I}}, .^{\mathcal{I}})$ is defined in a similar way as a first-order interpretation except that the interpretation function $.^{\mathcal{I}}$ also maps the set of variables V in the Bayesian Network to a value $v \in D(V)$. An annotated interpretation $\mathcal{I}$ satisfies a probabilistic axiom $\phi : e$, denoted $\mathcal{I} \models \phi : e$, iff $V^{\mathcal{I}} \models e \Rightarrow \mathcal{I} \models \phi$[5]. Now, a probabilistic interpretation is defined as a probabilistic distribution over annotated interpretations.

Definition 1. *(From [3]) A probabilistic interpretation Pr is a probability function over the set of all annotated interpretations that associates only a finite number of annotated interpretations with a positive probability. The probability of a probabilistic axiom $\phi : e$ in Pr, denoted $Pr(\phi : e)$, is the sum of all $Pr(\mathcal{I})$ such that $\mathcal{I}$ is an annotated interpretation that satisfies $\phi : e$. A probabilistic interpretation Pr satisfies (or is a model of) a probabilistic axiom $\phi : e$ iff $Pr(\phi : e) = 1$. We say Pr satisfies (or is a model of) a set of probabilistic axioms F iff Pr satisfies all $f \in F$.*

Finally, we define the notion of consistency of a probabilistic knowledge base.

Definition 2. *(From [3]) The probabilistic interpretation Pr satisfies (or is a model of) a probabilistic knowledge base $K = (\mathcal{T}, \mathcal{A}, BN)$ iff (i) Pr is a model of $\mathcal{T} \cup \mathcal{A}$ and (ii) $Pr_{BN}(V = v) = \sum_{\mathcal{I} \ s.t. \ V^{\mathcal{I}} = v} Pr(\mathcal{I})$ for all $v \in D(V)$. We say KB is consistent iff it has a model Pr.*

We note that condition (ii) in the previous definition ensures that the sum of probability values for annotated interpretations mapping V to $v \in D(V)$ is the same probability value assigned to $V = v$ by the Bayesian Network.

---

[4] K_{ev} was informally referred to as a 'possible world' in the introduction.

[5] This more expressive implication semantics differs from the equivalence semantics of [3].

3.1 Degree of Inconsistency

In the previously presented traditional BDL semantics, consistency is still categorically defined. We now address this significant shortcoming for our trust application using a refined semantics which introduces the notion of degree of inconsistency.

First, we illustrate using a simple example the intuition behind the notion of degree of inconsistency for a KB. Let K be the probabilistic KB defined as follows: $K = (\mathcal{T}, \mathcal{A} \cup \{\top \sqsubseteq \bot : X = true\}, BN)$ where $\mathcal{T}$ is a classical Tbox and $\mathcal{A}$ is a classical Abox such that the classical KB $cK = (\mathcal{T}, \mathcal{A})$ is consistent; BN is a Bayesian Network over a single boolean random variable X, and the probability $Pr_{BN}(X = true) = 10^{-6}$ that X is true is extremely low. Under past probabilistic extensions to DL, the K is completely inconsistent, and nothing meaningful can be inferred from it. This stems from the fact that when X is true, the set of classical axioms that must hold (i.e., $\mathcal{T} \cup \mathcal{A} \cup \{\top \sqsubseteq \bot\}$) is inconsistent. However, the event $X = true$ is extremely unlikely, and, therefore, it is unreasonable to consider the whole probabilistic KB inconsistent. Intuitively, the likelihood of events, whose set of associated classical axioms is inconsistent, represents the degree of inconsistency of a probabilistic KB.

We now formally define a degree of inconsistency and present an inconsistency-tolerant refinement of the semantics of a Bayesian DL.

Definition 3. *An annotated interpretation $\mathcal{I}$ is an annotated model of a probabilistic KB $K = (\mathcal{T}, \mathcal{A}, BN)$ where BN is a Bayesian Network over a set of variables V iff for each probabilistic axiom $\phi : e$, $\mathcal{I}$ satisfies $\phi : e$.*

In order, to measure the degree of inconsistency, we first need to find all primitive events v (i.e., elements of the domain $D(V)$ of the set of variables V) for which there are no annotated models $\mathcal{I}$ such that $V^{\mathcal{I}} = v$.

Definition 4. *For a probabilistic KB $K = (\mathcal{T}, \mathcal{A}, BN)$ where BN is a Bayesian Network over a set of variables V, the set of inconsistent primitive events, denoted $U(K)$, is the subset of $D(V)$, the domain of V, such that $v \in U(K)$ iff there is no annotated model $\mathcal{I}$ of K such that $V^{\mathcal{I}} = v$.*

Finally, the degree of inconsistency of a probabilistic knowledge base is defined as the probability of occurrence of an inconsistent primitive event.

Definition 5. *Let $K = (\mathcal{T}, \mathcal{A}, BN)$ be a probabilistic KB such that BN is a Bayesian Network over a set of variables V. The degree of inconsistency of K, denoted $DU(K)$, is a real number between 0 and 1 defined as follows:*

$$DU(K) = \sum_{v \in U(K)} Pr_{BN}(V = v)$$

A probabilistic interpretation Pr (as per Definition 1) satisfies (or is a model of) a probabilistic KB $K = (\mathcal{T}, \mathcal{A}, BN)$ to a degree d, $0 < d \leq 1$ iff.:

- *(i) Pr is a model as $\mathcal{T} \cup \mathcal{A}$ (same as in Definition 2)*
- *(ii) for $v \in V$,*

$$\sum_{\mathcal{I} \ s.t. \ V^{\mathcal{I}} = v} Pr(\mathcal{I}) = \begin{cases} 0 & \text{if } v \in U(K) \\ \frac{Pr_{BN}(V=v)}{d} & \text{if } v \notin U(K) \end{cases}$$

- *(iii) $d = 1 - DU(K)$*

A probabilistic knowledge base $K = (\mathcal{T}, \mathcal{A}, BN)$ *is consistent to the degree* d, *with* $0 < d \leq 1$, *iff there is a probabilistic interpretation that satisfies* K *to the degree* d. *It is completely inconsistent (or satisfiable to the degree 0), iff* $DU(K) = 1$.

Informally, by assigning a zero probability value to all annotated interpretations corresponding to inconsistent primitive events, (ii) in Definition 5 removes them from consideration, and it requires that the sum of the probability value assigned to interpretations mapping V to v for $v \notin U(K)$ is the same as the joint probability distribution Pr_{BN} defined by BN with a normalization factor d.

   In practice, computing the degree of inconsistency of a Bayesian DL KB can be reduced to classical description logics consistency check as illustrated by Theorem 1. First we introduce an important notation used in the remainder of the paper:

Notation 1. *Let* $K = (\mathcal{T}, \mathcal{A}, BN)$ *be a probabilistic KB. For every* $v \in D(V)$, *let* $\mathcal{T}_v$ *(resp.,* $\mathcal{A}_v$*) be the set of all axioms* ϕ *for which there exists a probabilistic axiom* $\phi : e$ *in* $\mathcal{T}$ *(resp.,* $\mathcal{A}$*), such that* $v \models e$. K_v *denotes the classical KB* $(\mathcal{T}_v, \mathcal{A}_v)$. *Informally,* K_v *represents the classical KB that must hold when the primitive event* v *occurs.* $K_{\mathcal{T}}$ *denotes the classical KB obtained from* K *after removing all probabilistic annotations:* $K_{\mathcal{T}} = (\cup_{v \in D(V)} \mathcal{T}_v . \cup_{v \in D(V)} \mathcal{A}_v)$.

Theorem 1. *A probabilistic KB* $K = (\mathcal{T}, \mathcal{A}, BN)$ *is consistent to the degree* d *iff.*

$$d = 1 - \sum_{v \ s.t. \ K_v \ inconsistent} Pr_{BN}(V = v)$$

The proof of Theorem 1 is a consequence of Lemma 1.

Lemma 1. *Let* K *be a probabilistic KB.* $v \in U(K)$ *iff* K_v *is inconsistent.*

3.2   Inconsistency Justification

A conflict or contradiction is formally captured by the notion of an inconsistency justification − minimal inconsistency preserving subset of the KB.

Definition 6. *Let* $K = (\mathcal{T}, \mathcal{A}, BN)$ *be a probabilistic KB consistent to the degree* d *such that* BN *is a Bayesian Network over a set of variables* V. $\mathcal{J}$ *is an inconsistency justification iff. 1)* $\mathcal{J} \subseteq (\mathcal{T}, \mathcal{A})$, *2)* $(\mathcal{J}, BN)$ *is probabilistic KB consistent to the degree* d' *such that* $d' < 1$, *and 3) for all* $\mathcal{J}' \subset \mathcal{J}$, $(\mathcal{J}', BN)$ *is probabilistic KB consistent to the degree 1 (i.e.* $(\mathcal{J}', BN)$ *is completely consistent). The degree* $DU(\mathcal{J})$ *of an inconsistency justification* $\mathcal{J}$ *is defined as the degree of inconsistency of the probabilistic KB made of its axioms:* $DU(\mathcal{J}) = DU((\mathcal{J}, BN))$.

Justification computation in a probabilistic KB reduces to justification computation in classical KBs as shown by the following theorem, which is a direct consequence of Theorem 1 and Definition 6:

Theorem 2. *Let* $K = (\mathcal{T}, \mathcal{A}, BN)$ *be a probabilistic KB, where* BN *is a Bayesian network over a set* V *of random variables.* $\mathcal{J}$ *is an inconsistency justification of* K *iff. there exists* $v \in D(V)$ *such that* $Pr_{BN}(V = v) > 0$ *and* $\mathcal{J}_{\mathcal{T}}$, *the classical KB obtained* *from $\mathcal{J}$ by removing all probabilistic annotations, is an inconsistency justification of K_v. Furthermore, the degree, $DU(\mathcal{J})$, of an inconsistency justification $\mathcal{J}$ is as follows:*

$$DU(\mathcal{J}) = \sum_{v \ s.t. \ \mathcal{J}_\top \subseteq K_v} Pr_{BN}(V = v)$$

Thus, once we have found a classical justification in a classical KB K_v for $v \in D(V)$ using, for example, the scalable approach described in our previous work [4], the degree of the corresponding probabilistic justification can be obtained through simple set inclusion tests.

Theorems 1 and 2 provide a concrete mechanism to compute degree of inconsistency of a probabilistic KB, and a degree of inconsistency of a justification. However, they are highly intractable since they require an exponential number, in the number of variables in BN, of corresponding classical tasks. We will address this issue in the next section.

3.3   Error-Bounded Approximate Reasoning

A Bayesian network based approach lends itself to fast Monte Carlo sampling algorithms for scalable partial consistency checks and query answering over a large probabilistic KB. In particular, we use a *forward sampling* approach described in [2] to estimate $pr = \sum_{v \in \Pi} Pr_{BN}(V = v)$ (recall theorem 1 and 2). The forward sampling approach generates a set of samples $v_1, \cdots, v_n$ from BN (each sample is generated in time that is linear in the size of BN) such that the probability pr can be estimated as $\widehat{pr_n}$ $= \frac{1}{n} * \sum_{i=1}^{n} I(v_i \in \Pi)$, where $I(z) = 1$ if z is true; 0 otherwise. One can show that $\widehat{pr_n}$ is an unbiased estimator of pr such that $\lim_{n \to \infty} \sqrt{n} * (\widehat{pr_n} - pr) \to \mathcal{N}(0, \sigma_z^2)$, where $\mathcal{N}(\mu, \sigma^2)$ denotes a normal distribution with mean μ and variance σ^2 and σ_z^2 denotes the variance of $I(z)$ for a boolean variable z. Hence, the sample size n which guarantees an absolute error of ϵ or less with a confidence level η is given by the following formula: $n = \frac{2*(erf^{-1}(\eta))^2 * \sigma_{z_{max}}^2}{\epsilon^2}$, where erf^{-1} denotes the inverse Gauss error function ($\sigma_{z_{max}}^2 = 0.25$ for a boolean random variable). For example, to compute the degree of consistency of a probabilistic KB within $\pm 5\%$ error margin with a 95% confidence, the sample size $n = 396$ is necessary.

3.4   Sampling Justifications in a Classical KB

Ideally, it is desirable to find all classical justifications. Computing a single justification can be done fairly efficiently by 1) using tracing technique to obtain a significantly small set S of axioms that is responsible for an inconsistency discovered by a single consistency test, and 2) performing additional $|S|$ consistency check on KBs of size at most $|S| - 1$ to remove extraneous elements from S. In our previous work [4], we presented a scalable approach to efficiently compute a large number of — but not all — justifications in large and expressive KBs through the technique of summarization and refinement [5]. The idea consists in looking for patterns of justifications in a dramatically reduced summary of the KB, and retrieve concrete instances of these patterns in the real KB.

Unfortunately, computing all justifications is well known to be intractable even for small and medium size expressive KBs [9]. [9] establishes a connection between the

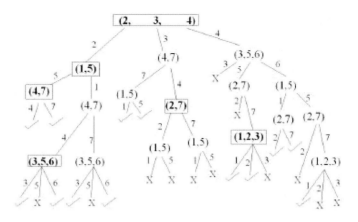

Fig. 1. Computing all justifications using Reiter's Hitting Set Tree Algorithm from [9]

problem of finding all justifications and the hitting set problem (i.e., given n sets S_i, find sets that intersect each S_i). The intuition behind this result is the fact that in order to make an inconsistent KB consistent at least one axiom from each justification must be removed. Therefore, starting from a single justification a Reiter's Hitting Tree can be constructed in order to get all justifications as illustrated in Figure 1 from [9]: Starting from the first justification $J = \{2, 3, 4\}$ computed in the KB K (J is set to be the root v_0 of the tree), the algorithm arbitrary selects an axiom in J, say 2, and creates a new node w with an empty label in the tree and a new edge $< v_0, w >$ with axiom 2 in its label. The algorithm then tests the consistency of the $K - \{2\}$. If it is inconsistent, as in this case, a justification J' is obtained for $K - \{2\}$, say $\{1, 5\}$, and it is inserted in the label of the new node w. This process is repeated until the consistency test is positive in which case the new node is marked with a check mark. As an important optimization, we stop exploring super set of path discovered earlier and marked the node with 'X'.

In order to avoid the high cost associated with exploring the whole Hitting Set Tree to find all conflicts. One can find the first K conflicts by exploring the Reiter's Hitting Set Tree (HST) until K distinct justifications are found. The problem with this approach is that nodes in the HST are not equally likely to be selected with such a scheme: the probability $\pi(v_d)$ of a node v_d in a path $< v_0 v_1 ... v_d >$ to be selected is $\pi(v_d) = \prod_{0 \leq i < d}(1/|v_i|)$, where $|v_i|$ denotes the number of axioms in the justification v_i. As a result, a malicious source can use the bias in the sampling to 'hide' its conflicts.

However, since the bias can be precisely quantified, one can obtain an unbiased sample as follows. We select K nodes in the HST by exploring the HST in the normal way, but each time a node v_i is encountered, it is selected iff. a random number r generated uniformly from [0,1] is such that $r \leq \min(\beta/\pi(v_i), 1)$, where β is a strictly positive real number. The following Proposition shows that, in this approach, for a sample of K HST nodes, if β is properly chosen, then the expected number of time a node is selected is identical for all nodes.

Proposition 1. *Let N_v denotes the random variable representing the number of time the node v appears in a HST sample of size K. The expected value $E(N_v)$ of N_v is:*

$$E(N_v) = \begin{cases} K * \pi(v) & \text{if } \beta \geq \pi(v) \\ K * \beta & \text{if } 0 < \beta < \pi(v) \end{cases}$$

Thus, if β is chosen such that $0 < \beta < \min_{v \in HST}(\pi(v))$, then we obtain an unbiased sample from the HST. Unfortunately, the minimum value of $\pi(v)$ depends on the tree structure (branching factor and maximum depth), and cannot be computed precisely without exploring the whole HST. In practice, we use the following sampling approach to select K nodes (the trade-off between computation cost and bias in the sample is controlled by a parameter of the algorithm, α):

1. Let $visited$ denote the set of visited nodes. Set $visited$ to $\emptyset$,
2. Traverse the HST in any order, and add the first $\max(K - |visited|, 1)$ nodes visited to $visited$
3. Let π_{min} be the minimum value of $\pi(v)$ for $v \in visited$.
4. Set $\beta = \pi_{min}/\alpha$, where $\alpha > 1$ is a parameter of the sampling algorithm which controls the trade-off between computation cost and biased in the sampling. Higher values of α, while reducing the bias in our sampling, increase the computation cost by reducing the probability of a node selection − hence, increasing the length of tree traversal.
5. For each $v \in visited$, add it to the result set RS with a probability of $\beta/\pi(v)$
6. If $|RS| < K$ and the HST has not been completely explored, then set $RS = \emptyset$ and continue the exploration from step 2; otherwise return RS

4   Trust Computation Model

We now briefly formalize the problem of assessing trust in a set IS consisting of n information sources. The trust value assumed or known prior to any statement made by an information source i is specified by a probability distribution $PrTV(i)$ over the domain $[0, 1]$. For example, a uniform distribution is often assumed for new information source for which we have no prior knowledge. Statements made by each source i is specified in the form of a probabilistic KB $\mathcal{K}^i = (\mathcal{T}^i, \mathcal{A}^i, BN^i)$. The knowledge function C maps an information source i to the probabilistic KB $\mathcal{K}^i$ capturing all its statements. The trust update problem is a triple $(IS, PrTV, C)$ whose solution yields a posterior trust value function $PoTV$. $PoTV$ maps an information source i to a probability distribution over the domain $[0, 1]$, which represents our new belief in the trustworthiness of i after processing statements in $\bigcup_{j \in IS} C(j)$.

In this paper, we only focus on trust computation based on direct observations, that is, on statements directly conveyed to us by the information sources. Inferring trust from indirect observations (e.g., statements conveyed to us from IS_1 via IS_2) is an orthogonal problem; one could leverage solutions proposed in [10], [15], [11] to infer trust from indirect observations.

4.1   Trust Computation

We model prior and posterior trust of a source i ($PrTV(i)$ and $PoTV(i)$) using a beta distribution $\mathcal{B}(\alpha, \beta)$ as proposed in several other trust computation models including [8]. Intuitively, the reward parameter α and the penalty parameter β correspond to good (non-conflicting) and bad (conflicting) axioms contributed to an information source respectively. The trust assessment problem now reduces to that of (periodically) updating the parameters α and β based on the axioms contributed by the information sources. One may bootstrap the model by setting $PrTV(i)$ to $\mathcal{B}(1,1)$ – a uniform and random distribution over $[0, 1]$, when we have no prior knowledge. In the rest of this section we focus on computing the reward (α) and penalty (β) parameters.

We use a simple reward structure wherein an information source receives unit reward for every axiom it contributes if the axiom is not in a justification for inconsistency[6]. We use a scaling parameter $\triangle$ to control the relative contribution of reward and penalty to the overall trust assessment; we typically set $\triangle > 1$, that is, penalty has higher impact on trust assessment than the reward. The rest of this section focuses on computing penalties from justifications for inconsistency.

Section 3.4 describes solutions to construct (a random sample of) justifications that explain inconsistencies in the KB; further, a justification J is associated with a weight $DU(J)$ that corresponds to the possible worlds in which the justification J holds (see section 3.2 for formal definition of $DU(J)$ and an algorithm to compute it). For each justification J_i we associate a penalty $\triangle(J_i) = \triangle * DU(J_i)$. The trust computation model traces a justification J_i, to conflicting information sources $\mathcal{S} = \{S_{i_1}, \cdots, S_{i_n}\}$ (for some $n \geq 2$) that contributed to the axioms in J_i. In this paper we examine three solutions to partition $\triangle(J_i)$ units of penalty amongst the contributing information sources as shown below. We use t_{i_j} to denote the expectation of $PrTV(i_j)$ for an information source i_j, that is, $t_{i_j} = \frac{\alpha_{i_j}}{\alpha_{i_j} + \beta_{i_j}}$.

$$\triangle(S_{i_j}) = \begin{cases} \frac{\triangle(J_i)}{n} & \text{unbiased} \\ \frac{\triangle(J_i)}{n-1} * (1 - \frac{t_{i_j}}{\sum_{k=1}^{n} t_{i_k}}) & \text{biased by trust in other sources} \\ \triangle(J_i) * \frac{\frac{1}{t_{i_j}}}{\sum_{k=1}^{n} \frac{1}{t_{i_k}}} & \text{biased by inverse self trust} \end{cases}$$

The unbiased version distributes penalty for a justification equally across all conflicting information sources; the biased versions tend to penalize less trustworthy sources more. One possible approach is to weigh the penalty for a source S_{i_j} by the sum of the expected prior trust values for all the other conflicting sources, namely, $\mathcal{S} - \{S_{i_j}\}$. For instance, if we have three information sources S_{i_1}, S_{i_2} and S_{i_3} with expected prior trust $t_{i_1} = 0.1$ and $t_{i_2} = t_{i_3} = 0.9$ then the penalty for source i_1 must be weighted by $\frac{1}{2} * \frac{0.9+0.9}{0.1+0.9+0.9} = 0.47$, while that of sources i_2 and i_3 must be weighted by 0.265. Clearly, this approach penalizes the less trustworthy source more than the trusted sources; however, we note that even when the prior trust in i_1 is arbitrarily close to zero, the penalty for the honest source i_2 and i_2 is weighted by 0.25. A close observation reveals that a malicious source (with very low prior trust) may penalize honest nodes (with high prior trust) by simply injecting conflicts that involve the honest nodes; for instance, if sources i_2 and i_3 assert axioms ϕ_2 and ϕ_3 respectively, then the malicious source i_1 can assert an axiom $\phi_1 = \neg\phi_2 \vee \neg\phi_3$ and introduce an inconsistency whose justification spans all the three sources. To overcome this problem, this paper uses a third scheme that weights penalties for justifications by the inverse value of prior trust in the information source.

---

[6] A preprocessing step weeds out trivial axioms (e.g., *sun rises in the east*).

5 Experimental Evaluation

To evaluate our approach, we have developed a prototype implementation, PSHER, that extends SHER reasoner [6] to support Bayesian $\mathcal{SHIN}$ (the core of OWL 1.0 DL) reasoning. SHER was chosen for its unique ability to scale reasoning to very large and expressive KBs [5], and to efficiently detect large number of inconsistency justifications in a scalable way [4]. PSHER uses the results of sections 3.1, 3.2 and 3.3 to reduce the problem of computing justifications on a probabilistic KB to detecting those justifications on classical KBs using SHER.

Axioms asserted by various information sources in our experiments were taken from the UOBM benchmark [12] which was modified to $\mathcal{SHIN}$ expressivity, and its Abox was modified by randomly annotating half of the axioms with probability values. Furthermore, we inserted additional Abox assertions in order to create inconsistencies involving axioms in the original UOBM KB. Note that not all axioms in the original UOBM KB end up being part of an inconsistency, which introduces an asymmetry in information source's knowledge (e.g., a malicious source is not assumed to have complete knowledge of all axioms asserted by other sources).

Due to space limitations, we only present an evaluation of our trust model under different scenarios. Scalability was already demonstrated in our previous work on SHER [4], where we presented a scalable approach to efficiently compute a large number of − but not all − justifications in large and expressive KBs through the technique

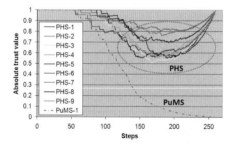

Fig. 2. Trust under single PuMS attack (No duplication)

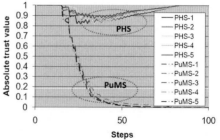

Fig. 3. Trust under 50% PuMS attack (No duplication)

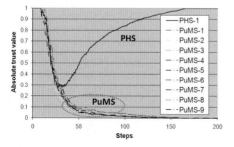

Fig. 4. Trust under 90% PuMS attack (No duplication)

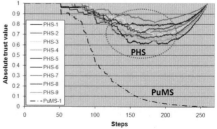

Fig. 5. Trust under single PuMS attack (25% duplication)

of summarization and refinement [5]. Scalability of PSHER is achieved through parallelism since each probabilistic reasoning task performed by PSHER is reduced to n corresponding classical tasks evaluated using SHER, where n depends on the desired precision as explained in Section 3.3. In the rest of this section, we report experiments conducted on UOBM1 (one department $\sim$ 74,000 axioms, including added inconsistent axioms and excluding duplication across sources).

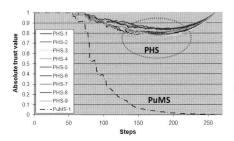

Fig. 6. Trust under single PuMS attack (50% duplication)

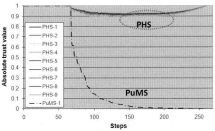

Fig. 7. Trust under single PuMS attack (100% duplication)

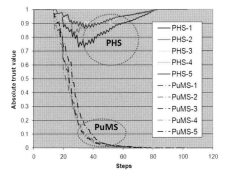

Fig. 8. Trust under 50% PuMS attack (25% duplication)

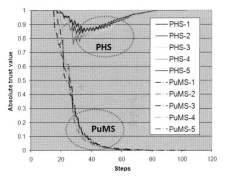

Fig. 9. Trust under 50% PuMS attack (50% duplication)

In our experiments, we considered 4 types of information sources:

- Perfect honest sources (PHS) whose axioms are taken from the UOBM KB before the introduction of inconsistencies.
- Purely malicious sources (PuMS) whose axioms are selected from the ones added to UOBM KB in order to create inconsistencies.
- Imperfect honest sources (IHS) have the majority of their axioms (more than 90%) from the UOBM KB before the introduction of inconsistencies. They allow us to simulate the behavior of our approach when honest sources are faced with measurement errors or commit honest mistakes.
- Partially malicious sources (PaMS) are such that between 10% to 90% of their axioms are selected from the axioms added to UOBM KB to create inconsistency. They are primarily used to simulate the behavior of our approach when malicious sources use an oscillating behavior to milk our trust computation scheme.

Axioms were randomly assigned to various sources without violating the proportion of conflicting vs. non-conflicting axioms for each type of source.

Our first experiment (Figure 2) measures the impact of a single purely malicious source (PuMS) on the trust values of 9 perfect honest sources. The PuMS asserts more and more incorrect axioms contradicting PHS's axioms (at each steps, each source asserts about 100 additional statements until all their axioms have been asserted) while the PHSs continue to assert more of what we consider as correct axioms. Axioms asserted by the PuMS do not necessarily yield an inconsistency in the same step in which they are asserted, but, by the end of the simulation, they are guaranteed to generate an inconsistency. For this experiment, there is no duplication of axioms across sources, and we do not assume any prior knowledge about the trustworthiness of the sources. Since each justification creates by the malicious source also involves at least one PuMS, initially, it manages to drop significantly the absolute trust value of some PHSs (up to 50% for PHS-3). However, a PuMS hurts its trust value significantly more than he hurts those of other sources. As a result of the fact that our scheme is such that less trustworthy sources get assigned a large portion of the penalty for a justification, the single PuMS eventually ends up receiving almost all the penalty for its inconsistencies, which allows the trust values of honest sources to recover. Due to information asymmetry (malicious sources do no have complete knowledge of informations in other sources and thus cannot contradict all the statements of an PHS), our scheme remains robust, in the sense that honest sources would recover, even when the proportion of PuMS increases (see Fig. 3 where 50% of the sources are PuMS and Fig. 4 where 90% of sources are PuMS).

In the previous experiments, although honest sources manage to recover from the attack, they can still be severely hurt before the credibility of the malicious sources decreased enough to enable a recovery for honest sources. This problem can be addressed in two ways: 1) by increasing the degree of redundancy between sources as illustrated in Figures 5, 6, 7, 8 and 9; and 2) by taking into account a priori knowledge of each source as illustrated in Figure 10.

In case of moderate to high redundancy between sources (Figures 5, 6, 7, 8 and 9), a justification generated by a malicious source to compromise a honest source is likely to hurt the malicious much more than the honest source because the axioms in the justification coming from the honest source are likely to be confirmed by (i.e. duplicated in) other honest sources. Therefore, the malicious source will be involved in as many justifications as there are corroborating honest sources, while each corroborating source will be involved in a single justification.

In Figure 10, we assume that we have a high a priori confidence in the trustworthiness of the honest sources: the prior distribution of the trust value of PHS in that experiment is a beta distribution with parameter $\alpha = 2000$ and $\beta = 1$. As expected, in Figure 10, the damage inflicted by the malicious source is significantly reduced compared to Figure 2 where no prior knowledge about the source trustworthiness was taken into account.

The next experiment evaluates the behavior of our scheme when partially malicious sources use an oscillating behavior. They alternate periods where they assert incorrect axioms, contradicting axioms asserted in the same period by other sources, with periods in which they assert only correct axioms. As opposed to previous experiments where malicious axioms asserted in a step were not guaranteed to yield an inconsistency in the

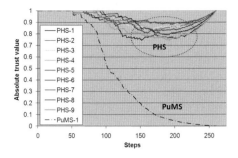

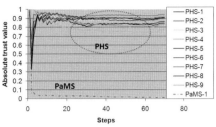

Fig. 10. Trust under single PuMS attack: No duplication - Prior = B(2000,1)

Fig. 11. Oscillating experiment - 90% PHS & 10% PaMS (No duplication)

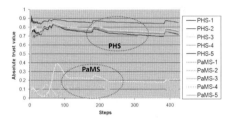

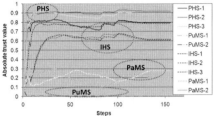

Fig. 12. Oscillating experiment - 50% PHS & 50% PaMS (No duplication)

Fig. 13. Oscillating experiment - 30% PHS, 20% PuMS, 30% IHS & 20% PaMS

same step, in the oscillation experiments, the inconsistency is observed at the same step. As shown in Figure 11 and 12, in absence of prior knowledge, the trust values of partially malicious sources (PaMS) and honest sources drop significantly at the first period in which incorrect axioms are stated. However, malicious sources, which due to information asymmetry, can only contradict limited set of statements from honest sources, never recover significantly, while honest sources quickly improve their trust values by asserting more axioms not involved in conflicts. As in the previous non-oscillating experiments, the negative impact on honest sources can be reduced considerably through axiom duplication and prior strong confidence in their trustworthiness.

The last experiment simulates an oscillating scenario where all four types of sources are present: 30% PHS, 20% PuMS, 30% IHS and 20%PaMS. Figure 13 shows how our scheme correctly separates the 4 types of sources as expected.

6   Conclusion

In this paper, we have introduced a new trust framework for rich, complex and uncertain information by leveraging the expressiveness of Bayesian Description Logics. We have demonstrated the robustness of the proposed framework under a variety of scenarios, and shown how duplication of assertions across different sources as well as prior knowledge of the trustworthiness of sources can further enhance it.

Acknowledgements. Research was sponsored by the U.S. Army Research Laboratory and the U.K. Ministry of Defence and was accomplished under Agreement Number W911NF-06-3-0001. The views and conclusions contained in this document are those of the author(s) and should not be interpreted as representing the official policies, either expressed or implied, of the U.S. Army Research Laboratory, the U.S. Government, the U.K. Ministry of Defence or the U.K. Government. The U.S. and U.K. Governments are authorised to reproduce and distribute reprints for Government purposes notwithstanding any copyright notation hereon.

References

[1] Baader, F., Calvanese, D., McGuinness, D., Nardi, D., Patel-Schneider, P.: The Description Logic Handbook. Cambridge University Press, Cambridge (2003)

[2] Cheng, J., Druzdzel, M.J.: AIS-BN: An Adaptive Importance Sampling Algorithm for Evidential Reasoning in Large Bayesian Networks. Journal of AI Research (2000)

[3] D'Amato, C., Fanizzi, N., Lukasiewicz, T.: Tractable reasoning with bayesian description logics. In: Greco, S., Lukasiewicz, T. (eds.) SUM 2008. LNCS (LNAI), vol. 5291, pp. 146–159. Springer, Heidelberg (2008)

[4] Dolby, J., Fan, J., Fokoue, A., Kalyanpur, A., Kershenbaum, A., Ma, L., Murdock, J.W., Srinivas, K., Welty, C.A.: Scalable cleanup of information extraction data using ontologies. In: Aberer, K., Choi, K.-S., Noy, N., Allemang, D., Lee, K.-I., Nixon, L.J.B., Golbeck, J., Mika, P., Maynard, D., Mizoguchi, R., Schreiber, G., Cudré-Mauroux, P. (eds.) ASWC 2007 and ISWC 2007. LNCS, vol. 4825, pp. 100–113. Springer, Heidelberg (2007)

[5] Dolby, J., Fokoue, A., Kalyanpur, A., Kershenbaum, A., Schonberg, E., Srinivas, K., Ma, L.: Scalable semantic retrieval through summarization and refinement. In: AAAI, pp. 299–304 (2007)

[6] Dolby, J., Fokoue, A., Kalyanpur, A., Schonberg, E., Srinivas, K.: Scalable highly expressive reasoner (sher). J. Web Sem. 7(4), 357–361 (2009)

[7] Hastie, T., Tibshirani, R., Friedman, J.: The Elements of Statistical Learning: Data Mining, Inference and Prediction. In: Springer Series in Statistics (2009)

[8] Josang, A., Ismail, R.: The beta reputation system. In: 15th Conference on Electronic Commerce (2002)

[9] Kalyanpur, A.: Debugging and Repair of OWL-DL Ontologies. PhD thesis, University of Maryland (2006), https://drum.umd.edu/dspace/bitstream/1903/3820/1/umi-umd-3665.pdf

[10] Kamvar, S., Schlosser, M., Garcia-Molina, H.: EigenTrust: Reputation management in P2P networks. In: WWW Conference (2003)

[11] Kuter, U., Golbeck, J.: SUNNY: A New Algorithm for Trust Inference in Social Networks, using Probabilistic Confidence Models. In: AAAI 2007 (2007)

[12] Ma, L., Yang, Y., Qiu, Z., Xie, G., Pan, Y.: Towards a complete owl ontology benchmark. In: Sure, Y., Domingue, J. (eds.) ESWC 2006. LNCS, vol. 4011, pp. 124–139. Springer, Heidelberg (2006)

[13] Netflix. Netflix Prize, http://www.netflixprize.com/

[14] Schafer, J.B., Konstan, J., Riedl, J.: Recommender Systems in E-Commerce. In: ACM Conference on Electronic Commerce (1999)

[15] Xiong, L., Liu, L.: Supporting reputation based trust in peer-to-peer communities. IEEE Transactions on Knowledge and Data Engineering (TKDE) 71, 16(7) (July 2004)

Optimising Ontology Classification

Birte Glimm, Ian Horrocks, Boris Motik, and Giorgos Stoilos

Oxford University Computing Laboratory, UK

Abstract. Ontology classification—the computation of subsumption hierarchies for classes and properties—is one of the most important tasks for OWL reasoners. Based on the algorithm by Shearer and Horrocks [9], we present a new classification procedure that addresses several open issues of the original algorithm, and that uses several novel optimisations in order to achieve superior performance. We also consider the classification of (object and data) properties. We show that algorithms commonly used to implement that task are incomplete even for relatively weak ontology languages. Furthermore, we show how to reduce the property classification problem into a standard (class) classification problem, which allows reasoners to classify properties using our optimised procedure. We have implemented our algorithms in the OWL HermiT reasoner, and we present the results of a performance evaluation.

1 Introduction

Ontology classification—the computation of subsumption hierarchies for classes and properties—is a core reasoning service provided by all OWL reasoners known to us. The resulting class and property hierarchies are used in ontology engineering, where they help users to navigate through the ontology and identify errors, as well as in tasks such as explanation and query answering.

Significant attention has been devoted to the optimisation of individual subsumption tests; however, most OWL reasoners solve the classification problem using an enhanced traversal (ET) classification algorithm similar to the one used in early description logic reasoners [1]. This can be inefficient when classifying large ontologies: even if each subsumption test is very efficient, the extremely large number of tests performed by ET can make classification an expensive operation. Moreover, with the exception of HermiT, all OWL reasoners we are aware of construct property hierarchies simply by computing the reflexive-transitive closure of the subproperty axioms occurring in the ontology—a procedure that is incomplete for each ontology language that supports existential restrictions (someValuesFrom), functional properties, and property hierarchies.

In order to address some of the problems of ET on large ontologies, an alternative classification algorithm, called KP, was proposed recently [9]. Unlike ET, KP does not construct the hierarchy directly; instead, it maintains the sets of known (K) and possible (P) subsumer pairs, and it performs subsumption tests to augment K and reduce P until the two sets coincide. To further reduce the number of tests, KP exploits the transitivity of the subclass relation to propagate (non-)subsumptions and thus speed up the convergence of K and P.

P.F. Patel-Schneider et al. (Eds.): ISWC 2010, Part I, LNCS 6496, pp. 225–240, 2010.

In this paper we address several issues that were left open in the work on KP, we present an optimised version of the resulting algorithm, and we evaluate its implementation in the HermiT reasoner. The new algorithm exhibits a consistent performance improvement over ET, and in some cases it reduces overall classification times by a factor of more than ten.

We then turn our attention to the classification of object and data properties. We show that merely computing the reflexive-transitive closure of the asserted hierarchies produces an incomplete hierarchy, and we discuss why the ET and KP algorithms do not perform well when applied to property classification. We then present a novel encoding of the property classification problem into a class classification problem, which allows us to exploit our new classification algorithm to correctly and efficiently compute property hierarchies. We have implemented our property classification algorithm in HermiT, thus making HermiT the only OWL reasoner we are aware of that correctly classifies object and data properties.

2   Preliminaries

An OWL 2 ontology consists of a set of axioms that describe the domain being modelled. For a full definition of OWL 2, please refer to the OWL 2 Structural Specification and Direct Semantics [7,6]; here we present only several examples of typical OWL axioms in the OWL 2 Functional Syntax:

$$\text{SubClassOf(Human Animal)} \qquad (1)$$

$$\text{DataPropertyAssertion}(\textit{age} \text{ Alex "27"}\hat{~}\hat{~}\text{xsd:integer}) \qquad (2)$$

$$\text{ObjectPropertyRange}(\textit{colour} \text{ ObjectOneOf(red green blue)}) \qquad (3)$$

Axiom (1) states that the class Human is a subclass of the class Animal (i.e., that all Humans are Animals); axiom (2) states that the individual Alex is related to the integer 27 by the data property *age* (i.e., that the age of Alex is 27); finally, axiom (3) states that the range of the object property *colour* consists of red, green, and blue (i.e., that the colour of an object can only be red, green, or blue). Concrete values such as the *literal* "27"$\hat{~}\hat{~}$xsd:integer in the above example are taken from the OWL 2 *datatype map*, which contains most of the XML Schema datatypes plus certain OWL-specific datatypes.

The interpretation of axioms in an OWL ontology $\mathcal{O}$ is given by means of two-sorted interpretations over the *object domain* and the *data domain*, where the latter contains concrete values such as integers, strings, and so on. An *interpretation* maps classes to subsets of the object domain, object properties to pairs of elements from the object domain, data properties to pairs of elements where the first element is from the object domain and the second one is from the data domain, individuals to elements in the object domain, a datatype to a subset of the data domain, and a literal (a data value) to an element in the data domain. For an interpretation to be a *model* of the ontology, several conditions have to be satisfied [6]. For example, if $\mathcal{O}$ contains SubClassOf(C D), then the interpretation of C must be a subset of the interpretation of D. If the axioms of $\mathcal{O}$ cannot be satisfied in any interpretation (i.e., if $\mathcal{O}$ has no model), then $\mathcal{O}$ is *inconsistent*; otherwise, $\mathcal{O}$ is *consistent*. If the interpretation of a class C is necessarily a subset of the interpretation of a class D in all models of $\mathcal{O}$, then we say that $\mathcal{O}$ entails $C \sqsubseteq D$ and write $\mathcal{O} \models C \sqsubseteq D$. If the interpretations of C and D necessarily coincide, we write $\mathcal{O} \models C \equiv D$. A class C is *satisfiable* if a model of $\mathcal{O}$ exists in which the interpretation of C is non-empty; otherwise, C is *unsatisfiable*. We use analogous notations for object and data properties. For full details of the OWL 2 Direct Semantics, please refer to the OWL 2 Direct Semantics specification [6]. We use $\mathbf{C}_\mathcal{O}$ to denote the set of classes that occur in $\mathcal{O}$ extended with owl:Thing and owl:Nothing; similarly, we use $\mathbf{OP}_\mathcal{O}$ (resp. $\mathbf{DP}_\mathcal{O}$) to denote the sets of object (resp. data) properties occurring in $\mathcal{O}$ extended with owl:TopObjectProperty and owl:BottomObjectProperty (resp. owl:TopDataProperty and owl:BottomDataProperty).

We next illustrate these definitions by means of an example. Let $\mathcal{O}$ be an ontology containing axioms (4) and (5); then, $\mathcal{O}$ entails $C \sqsubseteq E$ even though this is not stated explicitly. This is because axiom (4) ensures that in every model of $\mathcal{O}$, an instance i of C must be related to an instance of the class D with the property op. Since i has an op-successor, the property domain axiom (5) ensures that i is also an instance of the class E, and hence that C is contained in E.

$$\text{SubClassOf}(C\ \text{ObjectSomeValuesFrom}(op\ D)) \tag{4}$$

$$\text{ObjectPropertyDomain}(op\ E) \tag{5}$$

2.1 The KP Classification Algorithm

Classification of an ontology $\mathcal{O}$ computes all pairs of classes $\langle C, D \rangle$ such that $\{C, D\} \subseteq \mathbf{C}_\mathcal{O}$ and $\mathcal{O} \models C \sqsubseteq D$; similarly, object (resp. data) property classification of $\mathcal{O}$ computes all pairs of object (resp. data) properties $\langle R, S \rangle$ such that $\{R, S\} \subseteq \mathbf{OP}_\mathcal{O}$ (resp. $\{R, S\} \subseteq \mathbf{DP}_\mathcal{O}$) and $\mathcal{O} \models R \sqsubseteq S$. For example, given an ontology containing (4) and (5), a classification algorithm should compute

$$\{\langle \text{owl:Nothing}, C \rangle, \langle \text{owl:Nothing}, D \rangle, \langle C, E \rangle, \langle E, \text{owl:Thing} \rangle, \langle D, \text{owl:Thing} \rangle\}.$$

The recently proposed KP algorithm [9] extends the standard ET algorithm [1]. The KP algorithm maintains two binary relations K and P over $\mathbf{C}_\mathcal{O}$ such that, at any point during algorithm's execution, $\langle C, D \rangle \in K$ implies that $\mathcal{O} \models C \sqsubseteq D$ is known for certain, and $\langle C, D \rangle \in P$ implies that $\mathcal{O} \models C \sqsubseteq D$ is possible (i.e., no evidence to the contrary has been uncovered thus far). In particular, $\langle C, D \rangle \notin P$ means that $\mathcal{O} \not\models C \sqsubseteq D$ is known, so $P \setminus K$ contains all pairs $\langle C, D \rangle$ such that $C \sqsubseteq D$ is possible but not yet known. The algorithm expands K and reduces P until $K = P$, at which point $\mathcal{O} \models C \sqsubseteq D$ iff $\langle C, D \rangle \in K$. Roughly speaking, the algorithm chooses an unclassified class C (i.e., one where a class D exists such that $\langle C, D \rangle \in P \setminus K$), generates a partial hierarchy $\mathcal{H}_C$ of all unknown possible subsumers of C, and applies the standard ET procedure to insert C into $\mathcal{H}_C$. The newly computed subsumption and non-subsumption relations are then used to extend K and reduce P.

Algorithm 1. Prune Additional Possible Subsumptions

Algorithm: pruneNonPossible(P, K, V, N)
Input: P: a set of possible subsumptions to be pruned, K: a set of known subsumptions, V: a set of new positive subsumptions, N: a set of new non-subsumptions
1 **for each** $\langle C, D \rangle \in N$ **do**
2   **for each** E, F such that $\langle C, E \rangle \in K$ and $\langle F, D \rangle \in K$ remove $\langle E, F \rangle$ from P
3 **for each** $\langle C, D \rangle \in V$ **do**
4   **for each** $\langle D, E \rangle \in P$ **do**
5   **if** $\langle E, F \rangle \in K$ and $\langle C, F \rangle \notin P$ **then** remove $\langle D, E \rangle$ from P
6   **for each** $\langle E, C \rangle \in P$ **do**
7   **if** $\langle F, E \rangle \in K$ and $\langle F, D \rangle \notin P$ **then** remove $\langle E, C \rangle$ from P

The algorithm exploits the transitivity of $\sqsubseteq$ to reduce the number of subsumption tests needed to make K and P converge: whenever K is extended with fresh tuples it is also transitively closed, and a pruning strategy is used to remove tuples from P that correspond to obvious non-subsumptions. For example, if $\{\langle C, D \rangle, \langle E, F \rangle\} \subseteq K$, then $\langle D, E \rangle \in P$ implies $\langle C, F \rangle \in P$ since, by the transitivity of $\sqsubseteq$, adding $\langle D, E \rangle$ to K requires $\langle C, F \rangle$ to be added as well; but then, $\langle C, F \rangle \notin P$ implies $\langle D, E \rangle \notin P$. Analogously, if $\langle C, D \rangle \in P$, $\langle E, F \rangle \in K$ and $\langle C, F \rangle \notin P$, then $\langle C, D \rangle \in K$ implies $\langle D, E \rangle \notin P$. The complete pruning strategy of KP is shown in Algorithm 1. Note that this algorithm consists of several nested loops that iterate over potentially very large relations, which can make the algorithm inefficient in practice.

An important question when using KP is how to initialise K and P. The authors suggested to exploit the information generated by (hyper)tableau reasoners. In particular, when testing the satisfiability of a class A, (hyper)tableau algorithms usually initialise a node s_0 with the label $\mathcal{L}(s_0) = \{A\}$ and then apply expansion rules in order to try to construct a *pre-model*—an abstraction of a model for A; if a pre-model is constructed, then the (possibly expanded) label $\mathcal{L}(s_0)$ may provide information about subsumers and non-subsumers of A (if a pre-model cannot be constructed, then A is unsatisfiable and is equivalent to owl:Nothing). More precisely, if $\mathcal{L}(s_0)$ does not contain a class B, then we can infer the non-subsumption $A \not\sqsubseteq B$. Similarly, if B was deterministically added to $\mathcal{L}(s_0)$ (i.e., if no non-deterministic expansion was involved), then we can infer $A \sqsubseteq B$. Consequently, one can initially perform a satisfiability test for all the classes in $\mathbf{C}_\mathcal{O}$ and use the resulting pre-models to initialise K and P. It is not clear, however, whether it is generally efficient to perform all these tests.

3   Optimised Classification

We now present a new classification algorithm that we have implemented in the HermiT reasoner. Our algorithm is based on KP, but it addresses several open problems and incorporates numerous refinements and optimisations. The latter include, for example, a more efficient strategy for initialising K and P, a practical approach to pruning P, and several heuristics. We next describe our new algorithm and then contrast it with the relevant parts of KP.

Our approach is shown in Algorithm 2. Like KP, our algorithm maintains a set K of known and a set P of possible subsumption pairs. The algorithm uses an OWL reasoner to check satisfiability of classes (line 6) or subsumption between classes (line 25) using the well-known reduction of class subsumption to class satisfiability. In lines 2, 16, 24, 35 and 37, the algorithm manipulates K and P using operations that are defined next.

Definition 1. *Let U be a set of elements and let $R \subseteq U \times U$ be a binary relation over U. The set reachable(C, R) of elements reachable from $C \in U$ in R contains all $D \in U$ for which a path $\{\langle C, C_1 \rangle, \langle C_1, C_2 \rangle, \ldots, \langle C_n, D \rangle\} \subseteq R$ exists.*

Let $\sim$ be a relation over U defined as follows: $C \sim D$ if and only if $D = C$, or $D \in$ reachable(C, R) and $C \in$ reachable(D, R). Let $[C] := \{D \in U \mid D \sim C\}$ be the set of elements equivalent to C under $\sim$, and let $U_\sim := \{[C] \mid C \in U\}$. The relation $R_\sim$ induced by $\sim$ on R is defined as $R_\sim := \{\langle [C], [D] \rangle \mid \langle C, D \rangle \in R\}$.

The hierarchy in R is the triple hierarchy$(R) = (V, \mathcal{H}, \rho)$ where $V \subseteq U$ contains exactly one arbitrarily chosen element $C \in [D]$ for each $[D] \in U_\sim$, ρ maps each $C \in V$ into $\rho(C) = [C]$, and $\mathcal{H}$ is a transitively-reduced strict partial order over V such that $\langle C, D \rangle \in \mathcal{H}$ if and only if $\langle \rho(C), \rho(D) \rangle \in R_\sim$.

The projection project(R, S) of R to a set $S \subseteq U$, and the range $R[C]$ of an element $C \in U$ in R are defined as follows:

$$\text{project}(R, S) = \{\langle C, D \rangle \mid C, D \in S \text{ and } D \in \text{reachable}(C, R)\}$$
$$R[C] = \{D \mid \langle C, D \rangle \in R\}$$

Intuitively, hierarchy(K) extracts from K sets of classes for which $\mathcal{O} \models C \equiv D$ is known and then chooses one representative from each set to construct a transitively-reduced strict partial older.

Our algorithm can be roughly divided into two parts. Lines 1–15 are responsible for the initialisation of K and P using a novel heuristic, and lines 16–37 are responsible for extending K and reducing P using a mixture of the ET algorithm—as in KP—and a new technique for pruning P.

The Initialisation Phase. In KP, relations K and P are initialised by performing a satisfiability test for each atomic class in $\mathcal{O}$. Although modern reasoners can usually perform individual tests quite efficiently, the initialisation time can become large if there are many classes, so it is beneficial to avoid unnecessary tests whenever possible. For example, if $C \sqsubseteq D$ and C is satisfiable, then the pre-model constructed by a (hyper)tableau satisfiability test for C will also be a pre-model for D and for every other class occurring in the pre-model. We can thus avoid performing satisfiability tests for the classes outlined above, and from the pre-model for C we can read off information about the possible subsumers of all classes occurring in the pre-model. In order to maximise the effect of this optimisation, we first check the satisfiability of classes that are likely to be classified near the bottom of the hierarchy: such classes are likely to produce larger

Algorithm 2. New Classification Algorithm

Algorithm: Classify($\mathcal{O}$)

Input: $\mathcal{O}$: an ontology to be classified

1 $K := \mathsf{performStructuralSubsumption}(\mathcal{O})$
2 $(V, \mathcal{H}, \rho) := \mathsf{hierarchy}(K)$
3 Initialise a list $\mathsf{ToTest} := \{C \mid \langle \mathsf{owl:Nothing}, C \rangle \in \mathcal{H}\}$, $\mathsf{Unsat} := \emptyset$, and $P := \emptyset$
4 **while** $\mathsf{ToTest} \neq \emptyset$ **do**
5   Iteratively remove the head C from ToTest until C is found such that $P[C] = \emptyset$
6      $\mathcal{A} := \mathsf{buildModelFor}(C(s_0))$
7   **if** $\mathcal{A} = \emptyset$ **then**   // C is unsatisfiable
8   **for each** $\langle C, D \rangle \in \mathcal{H}$ **do** add D to the front of ToTest
9   **for each** descendant E of C in $\mathcal{H}$ that is not already in Unsat **do**
10   Add $\langle E, \mathsf{owl:Nothing} \rangle$ to K, add E to Unsat, and remove E from ToTest
11   **else**
12   **for each** $D \in \mathcal{L}(s_0)$ that was derived deterministically **do** add $\langle C, D \rangle$ to K
13   **for each** s in $\mathcal{A}$ and for each $D \in \mathcal{L}(s)$ **do**
14   **if** $P[D] = \emptyset$ **then** $P[D] := \mathcal{L}(s) \cap \mathbf{C}_{\mathcal{O}}$
15   **else** $P[D] := P[D] \cap \mathcal{L}(s)$
16 **for each** $D \in \mathbf{C}_{\mathcal{O}}$ and for each $E \in \mathsf{reachable}(D, K)$ **do** set $P[D] := P[D] \setminus \{E\}$
17 $\mathsf{UnClass} := \{C \in \mathbf{C}_{\mathcal{O}} \mid P[C] \neq \emptyset\}$
18 **while** $\mathsf{UnClass} \neq \emptyset$ **do**
19   Choose some $C \in \mathsf{UnClass}$ and set $B := P[C]$
20     $\mathcal{A} := \mathsf{buildModelFor}((C \sqcap \neg F)(s_0))$ with F the conjunction of all concepts in B
21   **if** $\mathcal{A} \neq \emptyset$ **then**   // all possible subsumers of C are non-subsumers
22   **for each** s in $\mathcal{A}$ and each $D \in \mathcal{L}(s)$ **do** set $P[D] := P[D] \cap \mathcal{L}(s)$
23   **else**
24        $(V, \mathcal{H}, \rho) := \mathsf{hierarchy}(\mathsf{project}(K, B \cup \{\mathsf{owl:Nothing}, \mathsf{owl:Thing}\}))$
25   Initialise a queue Q with $Q := \{\mathsf{owl:Thing}\}$
26   **while** $Q \neq \emptyset$ **do**
27   Remove the head H from Q
28   **for each** D such that $\langle D, H \rangle \in \mathcal{H}$ and $D \in P[C]$ **do**
29            $\mathcal{A} := \mathsf{buildModelFor}((C \sqcap \neg D)(s_0))$
30   **if** $\mathcal{A} \neq \emptyset$ **then**   // $C \sqcap \neg D$ was satisfiable—that is, $C \not\sqsubseteq D$
31   **for each** s in $\mathcal{A}$ and each $D \in \mathcal{L}(s)$ **do** set $P[D] := P[D] \cap \mathcal{L}(s)$
32   **else**
33   Add $\langle C, D \rangle$ to K, and add D to the end of Q
34        $P[C] := \emptyset$
35   **for each** $D \in \mathsf{UnClass}$ and $E \in \mathsf{reachable}(D, K)$ **do** set $P[D] := P[D] \setminus \{E\}$
36   Remove from $\mathsf{UnClass}$ each D such that $P[D] := \emptyset$
37 **return** $\mathsf{hierarchy}(K)$

---

pre-models that are richer in (non-)subsumption information and that can be used as pre-models for many other classes.

Our algorithm implements this idea as follows. First, it applies a simple structural subsumption algorithm to identify the obvious subsumptions in $\mathcal{O}$ and thus instantiate K. Then, it extracts a class hierarchy $\mathcal{H}$ from K and collects all classes C such that $\langle \text{owl:Nothing}, C \rangle \in \mathcal{H}$ (i.e., all 'leaves' of $\mathcal{H}$). Then, for each such C, the algorithm performs a satisfiability test; if C is satisfiable, then the constructed pre-model can be used to determine new known and possible subsumers as illustrated in lines 11–15. Note, however, that C is tested for satisfiability only if $P[C] = \emptyset$ (line 5), which avoids the test if a pre-model for C has been generated previously. The pre-model for C is used to update $K[C]$: if D was added to $\mathcal{L}(s_0)$ deterministically (which can easily be checked in reasoners that use dependency-directed backtracking), then D is guaranteed to be a subsumer of C [8], so $\langle C, D \rangle$ is added to K. The pre-model for C is also used to update $P[D]$ for *each* class D occurring in (any part of) the pre-model: if $D(s) \in \mathcal{A}$ and no possible subsumer for D is known yet, then $P[D]$ is initialised to $\mathcal{L}(s)$; otherwise, $P[D]$ is restricted to the elements in $\mathcal{L}(s)$. Note that $P[D]$ cannot become empty as it necessarily contains D.

Consider, for example, an ontology $\mathcal{O}$ containing axioms (4)–(7). Initially, structural subsumption initialises K by setting $K[X] = \{X, \text{owl:Thing}\}$ for each $X \in \mathbf{C}_\mathcal{O}$, and $K[\text{owl:Nothing}] = \mathbf{C}_\mathcal{O}$. At this point, ToTest contains C, D, E, F and G. Let us assume that C is chosen first, and a pre-model for $C(s_0)$ is generated. Due to axiom (4), s_0 must be related to an instance of D, say s_1, by property op. Since $D \in \mathcal{L}(s_1)$, the pre-model is also a pre-model for D. Due to axiom (6) and the ObjectUnionOf constructor, the reasoner can non-deterministically add E or F to $\mathcal{L}(s_1)$. Let us assume that the reasoner chooses E and then terminates returning $\mathcal{A}$; this pre-model can be used to infer that $P[C] = \{C\}$ and $P[E] = P[D] = \{D, E\}$. In the next iteration, D is chosen from the list, but $P[D] \neq \emptyset$ (information for D is already known), so no test is performed for D. At some point G is chosen and a model for $G(s_0)$ is constructed. Due to axiom (7), the reasoner relates s_0 with some fresh s_1 by property $op2$ such that $D \in \mathcal{L}(s_1)$. Let us assume, however, that to satisfy axiom (6), the reasoner now adds F to $\mathcal{L}(s_1)$. Since $P[D] \neq \emptyset$, $\mathcal{L}(s_1)$ can be used to prune $P[D]$; more precisely, since $E \notin \mathcal{L}(s_1)$, E is removed from $P[D]$.

$$\text{SubClassOf}(D \ \text{ObjectUnionOf}(E \ F)) \tag{6}$$

$$\text{SubClassOf}(G \ \text{ObjectSomeValuesFrom}(op2 \ D)) \tag{7}$$

Note that neither K nor P are updated if C is unsatisfiable, so little information is obtained from a satisfiability test for C. Hence, if $\mathcal{O}$ contains many unsatisfiable classes, initialisation might not provide enough initial information for K and P. Consequently, whenever our algorithm finds an unsatisfiable class C, it traverses $\mathcal{H}$ "upwards" until it finds a satisfiable class; furthermore, the unsatisfiability is propagated to all descendants of C in $\mathcal{H}$ (lines 7-10). Apart from making initialisation more robust, such an approach potentially identifies unsatisfiable classes without performing actual satisfiability tests (e.g., if D is discovered to be unsatisfiable and $\mathcal{O}$ contains $C \sqsubseteq D$). An example of such an ontology is FMA [2], which can be classified using our algorithm much more efficiently than with ET (see Section 6).

The Classification Phase. It is possible that all subsumers of a class D are identified after the initialisation phase, and this can happen even if the satisfiability of D had not been tested explicitly (in line 6). In our running example, all possible subsumers of D are already known (since $P[D] \subseteq K[D]$). For memory as well as for performance reasons, our algorithm next identifies only those classes for which there are unknown possible subsumers (lines 16-17), and operates only on them.

For these classes our algorithm proceeds as follows. It iteratively chooses a class C with $P[C] \neq \emptyset$ and checks $C \sqsubseteq D$ for each $D \in P[C]$. In order to perform these checks as efficiently as possible, the algorithm does not test each subsumption separately. Instead, inspired by the *clustering* optimisation [3], our algorithm tries to build a model for $C \sqcap \neg F$, where F is the conjunction of all possible subsumers of C (line 20). If a model exists, then $C \not\sqsubseteq F$ and so all concepts in $P[C]$ are non-subsumers of C.

If a model for $C \sqcap \neg F$ does not exist, then at least one concept in $P[C]$ is a subsumer of C, so a more detailed check is needed. The algorithm then proceeds as follows. It computes a transitively-reduced strict partial order $\mathcal{H}$ of the subsumers 'induced' by C. The standard ET algorithm is then applied to C over $\mathcal{H}$ in order to identify the (non-)subsumers of C. In contrast to KP, our algorithm introduces the following optimisation: if $C \sqcap \neg D$ is satisfiable for D a possible unknown subsumer of C (i.e., if $\mathcal{O} \not\models C \sqsubseteq D$), then the constructed pre-model can again be used to prune non-subsumers as was done in the initialisation phase. This process is performed in place of Algorithm 1, as it provides a more efficient pruning strategy. Another interesting and useful consequence of interleaving pruning with subsumption checking is that it can lead to the pruning of other possible subsumers of C that might otherwise be tested in a subsequent iteration. Therefore, the algorithm checks whether D is still a possible subsumer of C (line 28) before trying to construct a pre-model for $C \sqcap \neg D$ (line 29).

After the classification phase, all unknown possible subsumers will have been tested, and K contains all subsumption relations, so it is used to construct the final class hierarchy.

3.1   Further Comparisons with the KP Algorithm

We have already illustrated the major differences between Algorithm 2 and KP, such as the initialisation of K and P, and our new technique for pruning relations from P. In the following, we point out some additional differences, and we discuss further the pruning technique.

- **Memory Efficiency:** Our algorithm uses memory much more efficiently than KP. Recall that KP transitively closes K, which is not a good strategy on large ontologies such as FMA or SNOMED that contain thousands of classes. Furthermore, KP assumes that $P \supseteq K$—that is, all known subsumptions (including those derived by the transitive closure) are contained in P. In contrast, our algorithm uses a graph reachability algorithm to identify whether $\langle C, D \rangle$ belongs to the transitive closure of K, and removes the information about the classified classes from P, both of which can significantly reduce the algorithm's memory footprint.

- **Pruning:** Although our classification algorithm does not directly use Algorithm 1, it indirectly implements parts of Algorithm 1. For example, if $B \in P[A]$, but tests show that $\mathcal{O} \not\models A \sqsubseteq B$, then B can also be inferred to be a non-subsumer of all the subsumers of A as in the first loop of Algorithm 1. The second loop of Algorithm 1 prunes possible subsumptions when new positive subsumptions are inferred. However, our experience has shown that this strategy rarely identifies new non-subsumptions in practice. Consequently, the cost of applying such an expensive algorithm rarely outweighs the cost of performing a couple of additional subsumption tests.

- **Bottom-up Phase:** As in the ET algorithm, KP includes a bottom-up phase where the subsumees of an unclassified class C are identified in order to correctly place C into the class hierarchy. Our algorithm, however, does not include a bottom-up phase, which considerably simplifies the implementation as one does not need doubly-linked data structures for efficient retrieval of both successors and predecessors of C in K and P. Note that our algorithm is still complete since, if C is a possible but not yet known child of D, then $C \in P[D]$ and the relevant subsumption is tested when D is selected.

4   Object Property Classification

Classification of properties has, to the best of our knowledge, not been discussed in the literature. Apart from HermiT, all ontology reasoners that we are aware of construct the property hierarchy simply by computing the reflexive-transitive closure of the asserted property hierarchy. Such an algorithm is cheap to implement and requires no complex reasoning; however, it is incorrect for OWL as well as for considerably weaker ontology languages. Consider, for example, an ontology containing the following axioms:

$$\text{SubClassOf(ObjectSomeValuesFrom}(op_1 \; owl{:}Thing) \qquad (8)$$
$$\text{ObjectSomeValuesFrom}(op_2 \; owl{:}Thing) \;)$$

$$\text{SubObjectPropertyOf}(op_1 \; op_3) \qquad (9)$$

$$\text{SubObjectPropertyOf}(op_2 \; op_3) \qquad (10)$$

$$\text{FunctionalObjectProperty}(op_3) \qquad (11)$$

These axioms entail $op_1 \sqsubseteq op_2$: given $op_1(i_1, i_2)$, axiom (8) requires the existence of an op_2-successor for i_1; since both op_1 and op_2 are subproperties of op_3 and op_3 is functional, then i_2 must also be the op_2-successor for i_1, so we have $op_2(i_1, i_2)$.

Property chains and nominals can also imply implicit property subsumptions. The problems with property chains are demonstrated by the following example.

$$\text{SubClassOf}(owl{:}Thing \ \text{ObjectSomeValuesFrom}(op \ owl{:}Thing)) \qquad (12)$$

$$\text{SubObjectPropertyOf}(\text{ObjectPropertyChain}(\\ op_1 \ op \ \text{ObjectInverseOf}(op)) \ op_2) \qquad (13)$$

Whenever i_1 has an op_1-successor i_2, axiom (12) ensures that i_2 has an op-successor i_3; hence, we have $op_1(i_1, i_2)$, $op(i_2, i_3)$ and ObjectInverseOf(op)(i_3, i_2), and from axiom (13) we can infer $op_2(i_1, i_2)$, so the ontology implies $op_1 \sqsubseteq op_2$. Property classification in HermiT was initially based on the ET algorithm. Similarly to class subsumption testing, we concluded that $\mathcal{O} \models op_1 \sqsubseteq op_2$, for op_1 and op_2 object properties, iff $\mathcal{O} \cup \{op_1(a, b), \neg op_2(a, b)\}$ is not satisfiable, where a, b were individuals not occurring in $\mathcal{O}$. However, this is correct only for simple properties [7], where simple properties are roughly those that do not occur in property chains and transitivity axioms.

The problem with complex properties (i.e., non-simple ones) is that complex property assertions are not necessarily made explicit in the constructed pre-models. To ensure decidability, property chains and transitivity axioms are typically encoded into subclass axioms that propagate classes along paths in the pre-model in a way such that adding all missing property relationships does not violate any ontology axiom. Roughly speaking, given the property axiom SubObjectPropertyOf(ObjectPropertyChain($op \ op$) op) (which states that op is transitive), each axiom containing a universal quantifier over op is rewritten in a particular way; for example, axiom (14) is replaced with axioms (15)–(17)

$$\text{SubClassOf}(C \ \text{ObjectAllValuesFrom}(op \ D) \qquad (14)$$

$$\text{SubClassOf}(C \ \text{ObjectAllValuesFrom}(op \ D_{op})) \qquad (15)$$

$$\text{SubClassOf}(D_{op} \ D) \qquad (16)$$

$$\text{SubClassOf}(D_{op} \ \text{ObjectAllValuesFrom}(op \ D_{op})) \qquad (17)$$

where D_{op} is a fresh class. In order to compute all axioms required to eliminate all property inclusions, a non-deterministic finite automaton is constructed for each complex property, and subclass axioms are then extracted from automaton's transitions [4]. In order for the elimination to work as desired, negative property assertions with complex properties must be rewritten. For example, assertion (18) must be rewritten as (19)

$$\text{NegativeDataPropertyAssertion}(op \ a \ b) \qquad (18)$$

$$\text{ClassAssertion}(\text{ObjectAllValuesFrom}(op \\ \text{ObjectComplementOf}(\text{ObjectOneOf}(b))) \ a) \qquad (19)$$

where (19) states that a belongs to the class of individuals for which all op-successors are not b. The universal quantifier then triggers the generation of further axioms in the property chain elimination as described above.

Since complex property assertions are not necessarily made explicit in the pre-models, we cannot read off non-subsumptions from pre-models; that is, when $op_1(a,b)$ occurs but $op_2(a,b)$ does not occur in a pre-model, we cannot conclude $op_1 \not\sqsubseteq op_2$ if op_2 is a complex property. This significantly reduces the opportunities for pruning the search space, which makes property classification harder than standard (class) classification. We point out that, in the case described above, the publicly available 1.2.2 version of HermiT incorrectly concludes $op_1 \not\sqsubseteq op_2$. We corrected this error in the version of HermiT used for evaluation (see Section 6), which significantly decreased the performance of property classification.

In order to address these issues, we developed a new property classification technique that reduces property classification to standard (class) classification. Any classification algorithm, such as the one described in Section 3, can then be used to classify the property hierarchy, and it can use all relevant optimisations for pruning the search space. The reduction is defined as follows.

Definition 2. *Let $\mathcal{O}$ be an OWL 2 ontology and let* **OPE** *be the object properties and inverse object properties occurring in $\mathcal{O}$. An* object property to class mapping *w.r.t. $\mathcal{O}$ is a total and injective function τ from* **OPE** *to classes not occurring in $\mathcal{O}$. Let C_f be a class occurring neither in $\mathcal{O}$ nor in the range of τ. The* object property hierarchy induced by τ w.r.t. $\mathcal{O}$, *written $\mathcal{H}_{\mathcal{O}}^{\tau}$, is the transitive reduction of the relation $\{\langle op_1, op_2 \rangle \mid op_1, op_2 \in$* **OPE** *and $\mathcal{O}_{\tau} \models \tau(op_1) \sqsubseteq \tau(op_2)\}$, where $\mathcal{O}_{\tau}$ is an extension of $\mathcal{O}$ with axioms of the following form for each object property $op \in$* **OPE**.

$$Equivalent Classes(\tau(op) \ ObjectSome ValuesFrom(op \ C_f))$$

We write $(\mathcal{H}_{\mathcal{O}}^{\tau})^*$ to denote the reflexive-transitive closure of $\mathcal{H}_{\mathcal{O}}^{\tau}$.

Intuitively, to test $op_1 \sqsubseteq^? op_2$, we test $C_1 \sqsubseteq^? C_2$, where C_1 and C_2 are the representative classes introduced by τ for op_1 and op_2, respectively. As in standard classification, the reasoner checks this subsumption by trying to construct a pre-model containing $C_1(i)$ and $\neg C_2(i)$ for some individual i. The axioms in $\mathcal{O}_{\tau}$ then cause the addition of an op_1-successor of i, say i', with $C_f(i')$. If, due to other axioms in $\mathcal{O}$, i' is necessarily an op_2-successor of i as well, then the corresponding axiom in $\mathcal{O}_{\tau}$ for op_2 causes the addition of $C_2(i)$, which leads to a clash, which confirms the subsumption. Complex properties are handled using the transformation described earlier, so reading off non-subsumptions and pruning the set of possible subsumers works exactly as for classes.

The following theorem shows that this reduction of the object property classification problem to a standard classification problem is indeed correct.[1]

Theorem 1. *Let $\mathcal{O}$ be an OWL 2 ontology with $op_1, op_2 \in$* **OPE**, *let τ be an object property to class mapping w.r.t. $\mathcal{O}$, and let $\mathcal{H}_{\mathcal{O}}^{\tau}$ be the object property hierarchy induced by τ w.r.t. $\mathcal{O}$. Then $\mathcal{O} \models op_1 \sqsubseteq op_2$ iff $\langle op_1, op_2 \rangle \in (\mathcal{H}_{\mathcal{O}}^{\tau})^*$.*

---

[1] A complete proof is available in the accompanying technical report at
http://www.hermit-reasoner.com/2010/classification/Classification.pdf

5   Data Property Classification

Problematic constructors such as property chains do not apply to data properties, so one might think that data properties can be classified by just computing the reflexive-transitive closure of the asserted data property subsumptions. This, however, is not the case since we can easily adjust axioms (8)–(11) to work with data properties and *rdfs:Literal* instead of *owl:Thing*.

Another problem is that data property subsumption tests are difficult to implement. Since data properties are always simple, to test $\mathcal{O} \models dp_1 \sqsubseteq dp_2$ with dp_1 and dp_2 data properties, we might try to check whether $\mathcal{O} \cup \{dp_1(i,n), \neg dp_2(i,n)\}$ is unsatisfiable for i a fresh individual and n a fresh data value. We cannot, however, simply choose n to be any data value that does not occur in the input ontology. Assume, for example, that we selected an integer that does not occur in the input ontology $\mathcal{O}$; there are infinitely many integers, so there is always one not occurring in $\mathcal{O}$. This, however, might lead to conclusions that depend on the chosen integer: unlike for a fresh individual that can be interpreted as an arbitrary element of the object domain, the interpretation of a data value is fixed a priori. This problem can be solved by inventing a dummy datatype D that is considered to be non-disjoint with all datatypes in the OWL 2 datatype map (i.e., its value space can be intersected with any other data range without causing a contradiction); the only constraint for D is that a data value cannot belong to D and its complement. In order to check if $\mathcal{O} \models dp_1 \sqsubseteq dp_2$, the reasoner now checks the satisfiability of $\mathcal{O}$ extended with the following axioms, where i is a fresh individual:

$$\text{ClassAssertion}(\text{DataSomeValuesFrom}(dp_1 \ \mathsf{D}) \ i) \quad (20)$$

$$\text{ClassAssertion}(\text{DataAllValuesFrom}(dp_2 \ \text{DataComplementOf}(\mathsf{D})) \ i) \quad (21)$$

There is, however, still a problem with this approach. Datatype reasoning is typically implemented using a procedure such as the one presented by Motik and Horrocks [5]. If an individual i has a data property successor n, then one must check whether there are only finitely many values that n can take; if that is the case, one must find data values for n and the 'relevant' siblings of n that are related to the same individual i as n. A sibling n' is relevant if it can also have only finitely many possible data values and the assignment must be different from the one for n due to an inequality between n and n' (e.g., the inequality can be introduced by an at-least restriction). Thus, to handle D properly, an inequality must be generated between siblings n and n' if one of them must belong to D while the other must belong to the complement of D, which guarantees that the two nodes are not assigned the same data value in the procedure by Motik and Horrocks. Furthermore, note that even if n and n' must be assigned the same values, n and n' are not merged; for example, if an individual is required to have the integer 1 both as a dp_1- and a dp_2-successor, the two successors will be represented as separate objects in a pre-model. This again prevents the reading off of non-subsumptions between data properties. We should point out that this problem was also overlooked in HermiT 1.2.2, and correcting the error again significantly increased data property classification times.

We can, however, reduce data property classification to standard classification similarly as for object properties. This reduction allows us to read off subsumptions and non-subsumptions between data properties, because such (non-)subsumptions are reflected in the classes introduced by the encoding.

Definition 3. *Let $\mathcal{O}$ be an OWL 2 ontology and let* D *be a dummy datatype as discussed above. A* data property to class mapping w.r.t. $\mathcal{O}$ *is a total and injective function σ from* **DP** *to classes not occurring in $\mathcal{O}$. The* data property hierarchy induced by σ w.r.t. $\mathcal{O}$, *written $\mathcal{H}_{\mathcal{O}}^{\sigma}$, is the transitive reduction of the relation $\{\langle dp_1, dp_2 \rangle \mid dp_1, dp_2 \in$ **DP** *and* $\mathcal{O}_{\sigma} \models \sigma(dp_1) \sqsubseteq \sigma(dp_2)\}$, where $\mathcal{O}_{\sigma}$ is an extension of $\mathcal{O}$ with axioms of the following form for each data property $dp \in$ **DP**.*

$$EquivalentClasses(\sigma(dp) \; DataSomeValuesFrom(dp \; \mathsf{D}))$$

We write $(\mathcal{H}_{\mathcal{O}}^{\sigma})^{*}$ to denote the reflexive-transitive closure of $\mathcal{H}_{\mathcal{O}}^{\sigma}$.

The following theorem shows that the reduction is indeed correct. The proof is a straightforward adaptation of the proof of Theorem 1.

Theorem 2. *Let $\mathcal{O}$ be an OWL 2 ontology with $dp_1, dp_2 \in$ **DP**$_{\mathcal{O}}$, let σ be a data property to class mapping w.r.t. $\mathcal{O}$, and let $\mathcal{H}_{\mathcal{O}}^{\sigma}$ be the data property hierarchy induced by σ w.r.t. $\mathcal{O}$. Then $\mathcal{O} \models dp_1 \sqsubseteq dp_2$ iff $\langle dp_1, dp_2 \rangle \in (\mathcal{H}_{\mathcal{O}}^{\sigma})^{*}$.*

6   Evaluation

We have implemented Algorithm 2 and the property classification encodings in the HermiT 1.3 (hyper)tableau reasoner. To evaluate the effectiveness of our technique, we compared the performance of HermiT 1.3 against HermiT 1.2.2a (which implements the ET strategy, but with bugs related to property classification corrected as described in Sections 4 and 5). In our tests, we used two versions of the GALEN ontology, several ontologies from the Open Biological Ontologies (OBO) Foundry, the Food and Wine ontology from the OWL Guide, the Foundational Model of Anatomy (FMA), and ontologies from the Gardiner ontology suite. All ontologies and both HermiT versions are available online.[2] Table 1 summarises the numbers of classes and properties in each of the test ontologies.

The tests consisted of classifying the classes and properties of our test ontologies. We measured the classification time (in seconds) as well as the number of actual reasoning tests performed (including both satisfiability and subsumption tests). All experiments were performed on a UNIX machine of an Intel x86 64bit Cluster on one node with two quad core 2.8GHz processors and Java 1.5 allowing 2GB of heap memory. The results are summarised in Table 2. The upper part of the table contains all the deterministic ontologies (that is, the ontologies that do not use disjunctive constructors), while the lower part contains all the non-deterministic ontologies. For ontologies without data properties, we write '-' in Table 2 and OoM stands for Out of Memory.

---

[2] http://www.hermit-reasoner.com/2010/classification/Evaluation.zip

Table 1. Number of classes and properties in the evaluated ontologies

	classes	object prop.	data prop.		classes	object prop.	data prop.
GALEN-d	2 748	413	0	AEO	760	47	16
GALEN-und	2 748	413	0	substance	1 721	112	33
GO	19 528	1	0	ProPreO	482	30	0
GO_XP	27 883	5	0	OBI	2 638	77	6
chebi	20 979	10	0	Food-Wine	139	17	1
NCI	27 652	70	0	FMA 2.0	41 648	148	20

Table 2. Evaluation results for class and property classification (time in seconds)

Ontology	Classes				Object Properties				Data Properties			
	1.2.2a (ET)		1.3 (KP)		1.2.2a (ET)		1.3 (KP)		1.2.2a (ET)		1.3 (KP)	
	Tests	Time	Tests	Time	Tests	Time	Tests	Time	Tests	Time	Tests	Time
GALEN-d	2 744	3.6	3 380	2.9	6 073	439.2	197	< 1	-	-	-	-
GALEN-und	2 744	28.3	4 009	7.2	6 001	459.5	198	< 1	-	-	-	-
GO	19 260	43.0	14 288	3.7	4	< 1	3	< 1	-	-	-	-
GO_XP	27 880	119	20 029	14.4	9	10.4	6	4.8	-	-	-	-
chebi	20 693	69.8	13 484	7.6	26	59.9	12	18.1	-	-	-	-
NCI	27 652	71.1	21 367	10.5	71	< 1	72	< 1	-	-	-	-
AEO	2285	2.1	364	1.7	214	6.0	34	< 1	223	4.6	28	< 1
substance	4 569	15.9	2 730	12.8	962	23.6	107	< 1	957	22.5	40	< 1
ProPreO	1 441	7.3	1 157	6.8	518	3	33	< 1	-	-	-	-
OBI	12 444	254.7	3 047	170.1	2 278	310.5	52	3.4	39	6.0	7	< 1
Food-Wine	382	18.8	243	11.7	65	11.6	13	2.0	4	< 1	3	< 1
FMA 2.0	49 716	7 973.8	10 980	731.8	8 281	16 668.3	128	8.4	283	469.9	29	< 1

As Table 2 shows, the new classification strategy of HermiT 1.3 is in all cases significantly faster than the ET strategy of HermiT 1.2.2a, sometimes by one or even two orders of a magnitude. This is particularly the case for property classification where, as we have explained in the previous section, none of HermiT's standard optimisations can be applied, and one relies completely on the insertion strategy of ET to reduce the number of subsumption tests. In contrast, our property classification encoding can reuse the standard (class) classification optimisations, thus achieving a very good and robust performance. These results show that it is practically feasible to perform correct property classification through reasoning, instead of the cheap but incomplete transitive closure algorithms. The results for standard classification are similar: the new algorithm has significantly reduced the classification time in most cases. The significant performance gain in the classification of FMA is due in part to the heuristic implemented in lines 7–10 of Algorithm 2, which prevents HermiT from repeatedly performing class satisfiability tests for unsatisfiable classes.

The good performance results are also confirmed by the significant reduction in the number of required reasoning tests. The only case where HermiT 1.3 performs more tests is on GALEN, which is due to the fact that, on deterministic ontologies, HermiT 1.2.2a uses satisfiability tests and the pre-model reading technique [8] which identifies *all* subsumers of the tested class. In contrast, our method does not test the satisfiability of each class, so after the first phase there

Table 3. Number of tests performed by HermiT 1.3 compared to KP

	GO$^\sqcup$	GALEN$^\sqcup$	NCI$^\sqcup$
KP	32 614	4 657	48 389
HermiT 1.3	27 250	4 983	41 094

are unknown possible subsumers that need to be checked in the second phase. Especially in GALEN, most of them are subsumers, so the pruning step in lines 30–31 is rarely applicable. Nevertheless, such reasoning tests are usually very fast, so the overall system still performs better than HermiT 1.2.2a. On GALEN-und, where satisfiability tests are expensive, the benefits of not performing a satisfiability test for every class are particularly noticeable.

As a final experiment, we compared the performance of our system with the one that implements the KP algorithm [9]. We tested our system on three specially constructed ontologies that were used in [9] to evaluate the KP algorithm, and we compared the number of tests performed by our method with the number of tests published in [9]; Table 3 summarises the results. We can again see that for all ontologies but GALEN, our system performs fewer tests; furthermore, the same observations as above explain this difference. Unfortunately, the original implementation of KP was not available, so we were unable to compare the performance of HermiT with that of KP on the ontologies from Table 2.

7   Conclusions

In this paper, we considered the problem of efficiently classifying OWL ontologies. Unlike in previous approaches, we consider all classification tasks, including class, object and data property classification. To the best of our knowledge, property classification has not previously been discussed in the literature.

We presented a new classification algorithm that is based on KP [9], but that solves several open problems and that incorporates numerous refinements and optimisations. The latter include, for example, a novel heuristic strategy for initialising relations K and P, an efficient pruning strategy, and a novel heuristic for pruning unsatisfiable classes. Additionally, our new algorithm is more memory efficient than KP.

We presented examples that show why traditionally used algorithms based on the reflexive-transitive closure of the asserted property hierarchy are incomplete for property classification in OWL. We then discussed the difficulties in reusing well-known optimisations in the context of property classification, and we presented a novel reduction of the property classification problem to a standard classification problem. This reduction allows us to reuse all the optimisations applicable to the classification of classes.

Finally, we have implemented all our algorithms and reductions in version 1.3 of the HermiT reasoner, and have compared its performance with earlier versions using the standard classification method. Our results are very encouraging, showing significant improvements in classification times. Moreover, in the case of properties, our experiments show for the first time that complete property classification can be effectively implemented in practice.

We are currently working on extending our algorithm to handle realisation— the task of computing, for each individual i in an ontology, the most specific classes C such that i is an instance of C—and for realising property instances. Our preliminary results suggest that the performance of realisation can be significantly improved by applying the ideas outlined in this paper.

Acknowledgements. The presented work is funded by the EPSRC project HermiT: Reasoning with Large Ontologies. The evaluation has been performed on computers of the Oxford Supercomputing Centre.

References

1. Baader, F., Hollunder, B., Nebel, B., Profitlich, H.J., Franconi, E.: An empirical analysis of optimization techniques for terminological representation systems, or making kris get a move on. In: KR, pp. 270–281 (1992)
2. Golbreich, C., Zhang, S., Bodenreider, O.: The foundational model of anatomy in OWL: Experience and perspectives. Web Semantics 4(3), 181–195 (2006)
3. Haarslev, V., Möller, R.: High performance reasoning with very large knowledge bases: A practical case study. In: IJCAI, pp. 161–168 (2001)
4. Horrocks, I., Kutz, O., Sattler, U.: The even more irresistible $\mathcal{SROIQ}$. In: Proc. KR 2006, pp. 57–67 (2006)
5. Motik, B., Horrocks, I.: OWL datatypes: Design and implementation. In: Sheth, A.P., Staab, S., Dean, M., Paolucci, M., Maynard, D., Finin, T., Thirunarayan, K. (eds.) ISWC 2008. LNCS, vol. 5318, pp. 307–322. Springer, Heidelberg (2008)
6. Motik, B., Patel-Schneider, P.F., Cuenca Grau, B.: OWL 2 web ontology language direct semantics. W3C Recommendation (2009),
   http://www.w3.org/TR/owl2-direct-semantics/
7. Motik, B., Patel-Schneider, P.F., Parsia, B.: OWL 2 web ontology language structural specification and functional-style syntax. W3C Recommendation (2009),
   http://www.w3.org/TR/owl2-syntax/
8. Motik, B., Shearer, R., Horrocks, I.: Hypertableau Reasoning for Description Logics. Journal of Artificial Intelligence Research 36, 165–228 (2009)
9. Shearer, R., Horrocks, I.: Exploiting partial information in taxonomy construction. In: Bernstein, A., Karger, D.R., Heath, T., Feigenbaum, L., Maynard, D., Motta, E., Thirunarayan, K. (eds.) ISWC 2009. LNCS, vol. 5823, pp. 569–584. Springer, Heidelberg (2009)

SPARQL beyond Subgraph Matching

Birte Glimm and Markus Krötzsch

Oxford University Computing Laboratory, UK

Abstract. We extend the Semantic Web query language SPARQL by defining the semantics of SPARQL queries under the entailment regimes of RDF, RDFS, and OWL. The proposed extensions are part of the SPARQL 1.1 Entailment Regimes working draft which is currently being developed as part of the W3C standardization process of SPARQL 1.1. We review the conditions that SPARQL imposes on such extensions, discuss the practical difficulties of this task, and explicate the design choices underlying our proposals. In addition, we include an overview of current implementations and their underlying techniques.

1 Introduction

SPARQL provides a query language for querying RDF data that has gained significant popularity since its standardization by the World Wide Consortium (W3C) in January 2008 [12]. Almost all RDF stores support SPARQL either directly or via dedicated SPARQL wrappers. The main mechanism for computing query results in SPARQL is subgraph matching: RDF triples in both the queried RDF data and the query pattern are interpreted as nodes and edges of directed graphs, and the resulting query graph is matched to the data graph using variables as wild cards.

Various W3C standards, including RDF [3] and OWL [9], provide semantic interpretations for RDF graphs that allow additional RDF statements to be inferred from explicitly given assertions. It is desirable to utilize SPARQL as a query language in these cases as well, but this requires basic graph pattern matching to be defined using semantic entailment relations instead of explicitly given graph structures. Such extensions of the SPARQL semantics are known as *entailment regimes*.

The subject of this paper is to introduce SPARQL entailment regimes for RDF and RDFS entailment [3], OWL Direct Semantics [7], and OWL RDF-Based Semantics [14]. The proposed extensions are part of the SPARQL 1.1 Entailment Regimes specification, which is currently being developed by the W3C SPARQL working group.[1] The goal of this paper is to provide a detailed outline of these proposals that is valuable to practitioners and researchers alike. We provide extended discussions of the considerations that have led to our design, and we survey principal implementation techniques.

Although SPARQL has been designed to allow for the definition of entailment regimes, their precise definition is not straightforward. Naive approaches easily lead to infinite query results that are of no practical interest. Possible reasons include trivial renamings of blank nodes, RDFS's infinitely many axiomatic triples, and the entailment of arbitrary consequences from inconsistent inputs, each of which suggests different

---

[1] http://www.w3.org/2009/sparql/wiki/

P.F. Patel-Schneider et al. (Eds.): ISWC 2010, Part I, LNCS 6496, pp. 241–256, 2010.
© Springer-Verlag Berlin Heidelberg 2010

handling as discussed below. A second problem is that OWL is not primarily based on RDF triples but defines entailments in terms of ontological objects. Thus, triples can be genuine input data or merely part of the encoding of a complex object.

The paper is structured as follows. Section 2 gives a short introduction to RDF(S) and OWL, and Section 3 reviews the basics of SPARQL subgraph matching. In Section 4, we offer our interpretation of the conditions that SPARQL 1.0 defines for entailment regimes. The entailment regimes for RDF and RDFS are defined in Section 5, and the extensions of SPARQL with OWL's RDF-Based Semantics and the OWL Direct Semantics are presented in Section 6. Finally, Sections 7 and 8 explain basic implementation techniques for SPARQL entailment regimes and discuss further related work.

2 RDF Graphs and Their Semantics

SPARQL queries are evaluated over RDF graphs which remain the basic data structure even when adopting a more elaborate semantic interpretation. RDF is based on the set I of all *International Resource Identifiers* (IRIs), the set RDF-L of all *RDF literals*, and the set RDF-B of all *blank nodes*. The set RDF-T of *RDF terms* is I ∪ RDF-L ∪ RDF-B. We generally abbreviate IRIs using prefixes rdf, rdfs, owl, and xsd to refer to the RDF, RDFS, OWL, and XML Schema Datatypes namespaces, respectively. The prefix ex is used for an imaginary example namespace.

An *RDF graph* is a set of *RDF triples* of the form (subject, predicate, object) ∈ (I ∪ RDF-B) × I × RDF-T. We normally omit "RDF" in our terminology if no confusion is likely, and we use Turtle syntax [1] for all examples. The *vocabulary* Voc(G) of a graph G is the set of all terms that occur in G.

Semantically, RDF graphs can be interpreted in a number of ways based on various W3C recommendations. The *simple semantics* [3] considers only the graph structure of RDF, whereas more elaborate semantics such as RDFS entailment [3] or OWL Direct Semantics [7] provide a special meaning to certain RDF terms.

The common basis for all such semantics is that they were specified by defining a model theory: one defines a suitable kind of *interpretation*, and specifies necessary and sufficient conditions for one such interpretation to *satisfy* a given RDF graph. When defining a semantics E (such as RDF, RDFS, etc.) one often speaks of E-interpretations and E-satisfaction. The set of all E-interpretations that E-satisfy a graph G are called the E-*models* of G. Semantic entailment follows from this notion: a graph G E-*entails* a graph G', written G $\models_E$ G', if and only if every E-model of G is also an E-model of G'.

In this work, we encounter the *simple semantics*, *RDF semantics*, and *RDFS semantics* [3], as well as the *OWL Direct Semantics* [7] and *OWL RDF-Based Semantics* [14]. This order roughly mirrors the amount of entailments obtained under each of these semantics, e.g., all RDF-entailments are also RDFS-entailments. This ideal compatibility is not always given, especially since the OWL Direct Semantics is defined in the tradition of first-order logic, whereas the other semantics are based on a specific notion of interpretation introduced for RDF. The latter was found difficult to extend to expressive languages like OWL, and indeed entailment under the OWL RDF-Based Semantics is undecidable and is mostly used by tools that restrict to a sub-language of OWL.

On the other hand, the OWL Direct Semantics is only defined for graphs that respect certain additional conditions. This is so since this semantics is defined based on OWL objects of which RDF graphs are but an indirect representation. The OWL 2 functional-style syntax (FSS) directly corresponds to the OWL objects [8]. For example, the triple

ex:a owl:sameAs ex:b   corresponds to   SameIndividual(ex:a ex:b).

Since the mapping from RDF triples to OWL objects is not defined for arbitrary RDF graphs, the OWL 2 Direct Semantics makes restrictions on the well-formedness of RDF graphs that can be used with the semantics. *OWL 2 DL* describes the largest subset of RDF graphs for which the OWL 2 Direct Semantics is defined.

3   The SPARQL Query Language

We do not recall the complete surface syntax of SPARQL here but simply introduce the underlying algebraic operations using our notation. A detailed introduction to the relationship of SPARQL queries and their algebra is given in [4].

Queries are built using a countably infinite set V of *query variables* disjoint from RDF-T. SPARQL supports a variety of *filter expressions*, or just *filters*, built from RDF terms, variables, and a number of built-in functions and operators; see [12] for details.

Definition 1. *A* triple pattern *is member of the set* $(\text{RDF-T} \cup \text{V}) \times (\text{I} \cup \text{V}) \times (\text{RDF-T} \cup \text{V})$, *and a* basic graph pattern *(BGP) is a set of triple patterns. More complex graph patterns are inductively defined to be of the form* BGP, $\text{Join}(\text{GP}_1, \text{GP}_2)$, $\text{Union}(\text{GP}_1, \text{GP}_2)$, $\text{LeftJoin}(\text{GP}_1, \text{GP}_2, F)$, *and* $\text{Filter}(F, \text{GP})$, *where* BGP *is a BGP,* F *is a filter, and* $\text{GP}_{(i)}$ *are graph patterns that share no blank nodes.*[2] *The sets of* variables *and* blank nodes *in a graph pattern* GP *are denoted by* $\text{V}(\text{GP})$ *and* $\text{B}(\text{GP})$, *respectively.*

SPARQL allows literals to be used as triple subjects although RDF graphs cannot currently contain such triples. This is meant to support (future) extensions of RDF.

We exclude a number of SPARQL features from our discussion. First, we disregard any of the new SPARQL 1.1 query constructs since their syntax and semantics are still under discussion in the SPARQL working group. Second, we do not consider output formats (e.g., SELECT or CONSTRUCT) and solution modifiers (e.g., LIMIT or OFFSET) which are not affected by entailment regimes. Third, we exclude SPARQL datasets that allow SPARQL endpoints to cluster data into several named graphs and a default graph. For simpler presentation, we omit dataset clauses and assume that queries are evaluated over the default graph, called the *active graph* for the query.

Evaluating a SPARQL graph pattern results in a *solution sequence* that lists possible bindings of query variables to RDF terms in the active graph. Such bindings are represented by partial functions μ from V to RDF-T, called *solution mappings*. For a solution mapping μ – and more generally for any (partial) function – the set of elements on which μ is defined is the *domain* $\text{dom}(\mu)$ of μ, and the set $\text{ran}(\mu) := \{\mu(x) \mid x \in \text{dom}(\mu)\}$ is the *range* of μ. For a graph pattern GP, we use $\mu(\text{GP})$ to denote the pattern obtained by

---

[2] As in [12], disallowing GP_1 and GP_2 to share blank nodes is important to avoid unintended co-references. This was not needed in [10] where blank nodes were not considered.

Table 1. Evaluation of algebraic operators in SPARQL

$$[\![\mathsf{Union}(\mathsf{GP}_1, \mathsf{GP}_2)]\!]_G := \big\{(\mu, n) \mid n = M_1(\mu) + M_2(\mu) > 0\big\}$$

$$[\![\mathsf{Join}(\mathsf{GP}_1, \mathsf{GP}_2)]\!]_G := \big\{(\mu, n) \mid n = \textstyle\sum_{(\mu_1, \mu_2) \in J(\mu)} (M_1(\mu_1) * M_2(\mu_2)) > 0\big\} \text{ where}$$
$$J(\mu) := \{(\mu_1, \mu_2) \mid \mu_1, \mu_2 \text{ compatible and } \mu = \mu_1 \cup \mu_2\}$$

$$[\![\mathsf{Filter}(\mathsf{F}, \mathsf{GP})]\!]_G := \big\{(\mu, n) \mid M(\mu) = n > 0 \text{ and } [\![\mu(\mathsf{F})]\!] = \mathsf{true}\big\}$$

$$[\![\mathsf{LeftJoin}(\mathsf{GP}_1, \mathsf{GP}_2, \mathsf{F})]\!]_G := [\![\mathsf{Filter}(\mathsf{F}, \mathsf{Join}(\mathsf{GP}_1, \mathsf{GP}_2))]\!]_G \;\cup$$
$$\big\{(\mu_1, M_1(\mu_1)) \mid \text{ for all } \mu_2 \text{ with } M_2(\mu_2) > 0 : \mu_1 \text{ and } \mu_2 \text{ are}$$
$$\text{incompatible or } [\![(\mu_1 \cup \mu_2)(\mathsf{F})]\!] = \mathsf{false}\big\}$$

applying μ to all elements of GP in $\mathsf{dom}(\mu)$. This convention is extended in the obvious way to filter expressions, and to all functions that are defined on variables or terms.

The order of solution sequences is relevant for later processing steps in SPARQL, but not for obtaining the solutions for a graph pattern. To disregard the order formally, we use *solution multisets*. A *multiset* over an *underlying set* S is a total function $M : S \to \mathbf{N}^+ \cup \{\omega\}$ where $\mathbf{N}^+$ are the positive natural numbers, and $\omega > n$ for all $n \in \mathbf{N}^+$. The value $M(s)$ is the *multiplicity* of $s \in S$, and ω denotes a countably infinite number of occurrences. Infinitely many occurrences of individual solution mappings are indeed possible when considering SPARQL entailment regimes, and a major concern of this work is to avoid this for the entailment regimes we define.

We often represent a multiset M with underlying set S by the set $\{(s, M(s)) \mid s \in S\}$. Accordingly, we may write $(s, n) \in M$ if $M(s) = n$. Also, we assume that $M(s)$ denotes 0 whenever $s \notin S$. In some cases, it is also convenient to use a set-like notation where repeated elements are allowed, e.g. writing $\{a, b, b\}$ for the multiset M with underlying set $\{a, b\}$, $M(a) = 1$, and $M(b) = 2$.

To define the solution multiset for a BGP under the simple semantics, we still need to consider the effect of blank nodes. Intuitively, these act like variables that are projected out of a query result, and thus they may lead to duplicate solution mappings. This is accounted for using RDF instance mappings as follows:

Definition 2. *An* RDF instance mapping *is a partial function* $\sigma : \mathsf{RDF\text{-}B} \to \mathsf{RDF\text{-}T}$ *from blank nodes to RDF terms. We extend* σ *to pattern graphs and filters as done for solution mappings above. The* solution multiset $[\![\mathsf{BGP}]\!]_G$ *for a basic graph pattern* BGP *over the active graph* G *is the following multiset of solution mappings:*

$\{(\mu, n) \mid \mathsf{dom}(\mu) = \mathsf{V}(\mathsf{BGP}),$ *and n is the maximal number such that*
$\quad \sigma_1, \ldots, \sigma_n$ *are distinct RDF instance mappings such that, for all* $1 \le i \le n$,
$\quad \mathsf{dom}(\sigma_i) = \mathsf{B}(\mathsf{BGP})$ *and* $\mu(\sigma_i(\mathsf{BGP}))$ *is a subgraph of* G$\}$.

Note that the number n in the definition of $[\![\mathsf{BGP}]\!]_G$ is always finite.

The algebraic operators that are required for evaluating non-basic graph patterns correspond to operations on multisets of solution mappings which are the same for all entailment regimes. To take infinite multiplicities into account, we assume $\omega + n = n + \omega = \omega$ for all $n \ge 0$, $\omega * n = n * \omega = \omega$ for all $n > 0$ and $\omega * 0 = 0 * \omega = 0$. To

Table 2. Conditions for extending BGP matching to E-entailment (quoted from [12])

1. The scoping graph SG, corresponding to any consistent active graph AG, is uniquely specified and is E-equivalent to AG.
2. For any basic graph pattern BGP and pattern solution mapping P, P(BGP) is well-formed for E.
3. For any scoping graph SG and answer set $\{P_1, \ldots, P_n\}$ for a basic graph pattern BGP, and where $BGP_1, \ldots, BGP_n$ is a set of basic graph patterns all equivalent to BGP, none of which share any blank nodes with any other or with SG

$$SG \models_E (SG \cup P_1(BGP_1) \cup \ldots \cup P_n(BGP_n)).$$

4. Each SPARQL extension must provide conditions on answer sets which guarantee that every BGP and AG has a finite set of answers which is unique up to RDF graph equivalence.

incorporate the effect of filters, it suffices to know that SPARQL assigns to any filter F an effective truth value that we will denote by $[\![F]\!]$.

Definition 3. *Two solution mappings μ_1 and μ_2 are* compatible *if $\mu_1(v) = \mu_2(v)$ for all $v \in \text{dom}(\mu_1) \cap \text{dom}(\mu_2)$. If this is the case, a solution mapping $\mu_1 \cup \mu_2$ is defined by setting $(\mu_1 \cup \mu_2)(v) := \mu_1(v)$ if $v \in \text{dom}(\mu_1)$, and $(\mu_1 \cup \mu_2)(v) := \mu_2(v)$ otherwise.*

The evaluation *of a graph pattern over* G, *denoted $[\![\cdot]\!]_G$, is defined as in Table 1, where we abbreviate multisets $[\![GP]\!]_G / [\![GP_1]\!]_G / [\![GP_2]\!]_G$ by $M / M_1 / M_2$ for readability.*

Note that two mappings with disjoint domains are always compatible.Intuitively, $\text{Join}(GP_1, GP_2)$ represents all possible combinations of mappings from $[\![GP_1]\!]_G$ with compatible mappings from $[\![GP_2]\!]_G$, as accounted for by taking the product of multiplicities. One mapping in a join may result from various combinations of compatible mappings, so that we need to compute a sum of their multiplicities. The expression $\text{LeftJoin}(GP_1, GP_2, F)$ combines the filtered join of the inputs with all mappings of $[\![GP_1]\!]_G$ which are not represented in this filtered join.

4   Extending Basic Graph Pattern Matching

To extend SPARQL for entailment regimes like RDFS or OWL Direct Semantics, it suffices to modify the evaluation of BGPs accordingly, while the remaining algebra operations can still be evaluated as in Definition 3. When considering E-entailment, we thus define solution multisets $[\![BGP]\!]_G^E$. The SPARQL Query 1.0 specification [12] already envisages the extension of the BGP matching mechanism, and provides a set of conditions for such extensions that we recall in Table 2. We found these conditions hard to interpret since their terminology is not aligned well with the remaining specification. In the following, we discuss our reading of these conditions, leading to a revised clarified version presented in Table 3.[3]

Condition (1) forces an entailment regime to specify a so-called *scoping graph* based on which query answers are computed instead of using the active graph directly. Since

---

[3] The current SPARQL working group is not chartered to revise the existing specification, so the ongoing work on entailment regimes is based on the assumption that the conditions were meant to be in the revised form.

Table 3. Clarified conditions for extending BGP matching to E-entailment

An entailment regime E provides conditions on BGP evaluation such that for any evaluation $[\![\cdot]\!]_G^E$ that satisfies these conditions, any basic graph pattern BGP, and any graph G, the multiset of graphs $\{(\mu(\text{BGP}), n) \mid (\mu, n) \in [\![\text{BGP}]\!]_G^E\}$ is uniquely determined up to RDF graph equivalence.

1. For any consistent active graph AG, the entailment regime E uniquely specifies a *scoping graph* SG that is E-equivalent to AG.
2. A set of *well-formed* graphs for E is specified such that, for any basic graph pattern BGP, scoping graph SG, and solution mapping μ in the underlying set of $[\![\text{BGP}]\!]_{SG}^E$, the graph $\mu(\text{BGP})$ is well-formed for E.
3. For any basic graph pattern BGP, and scoping graph SG, if S denotes the underlying set of $[\![\text{BGP}]\!]_{SG}^E$, then there is a family of RDF instance mappings $(\sigma_\mu)_{\mu \in S}$ such that

$$\text{SG} \models_E \text{SG} \cup \bigcup_{\mu \in S} \mu(\sigma_\mu(\text{BGP})).$$

4. Entailment regimes *should* provide conditions to prevent trivial infinite solution multisets.

an entailment regime's definition of BGP matching is free to refer to such derived graph structures anyway, the additional use of a scoping graph does not increase the freedom of potential extensions. We assume, therefore, that the scoping graph is the active graph in the remainder. If the active graph is E-inconsistent, entailment regimes specify the intended behavior directly, e.g., by requiring that an error is reported.

Condition (2) refers to a "pattern solution mapping" though what is probably meant is a *pattern instance mapping* P, defined in [12] as the combination of an RDF instance mapping σ and a solution mapping μ where $P(x) = \mu(\sigma(x))$. We assume, however, that (2) is actually meant to refer to all solution mappings in $[\![\text{BGP}]\!]_G^E$. Indeed, even for simple entailment where well-formedness only requires P(BGP) to be an RDF graph, condition (2) would be violated when using *all* pattern instance mappings. To see this, consider a basic graph pattern BGP = {_:a ex:b ex:c}. Clearly, there is a pattern instance mapping P with P(_:a) = "1"^^xsd:int, but P(BGP) = {"1"^^xsd:int ex:b ex:c} is not an RDF graph. Similar problems occur when using all solution mappings. Hence we assume (2) to refer to elements of the computed solution multiset $[\![\text{BGP}]\!]_G^E$. The notion of *well-formedness* in turn needs to be specified explicitly for entailment regimes.

Condition (3) uses the term "answer set" to refer to the results computed for a BGP. To match the rest of [12], this has to be interpreted as the solution multiset $[\![\text{BGP}]\!]_G^E$. This also means mappings P_i are solution mappings (not pattern instance mappings as their name suggests). The purpose of (3), as noted in [12], is to ensure that if blank node names are returned as bindings for a variable, then the same blank node name occurs in different solutions only if it corresponds to the same blank node in the graph. To illustrate the problem, consider the following graphs:

G : ex:a ex:b _:c. G_1 : ex:a ex:b _:b_1. G_2 : ex:a ex:b _:b_2. G_3 : ex:a ex:b _:b_1.
    _:d ex:e ex:f.   _:b_2 ex:e ex:f.   _:b_1 ex:e ex:f.   _:b_1 ex:e ex:f.

Clearly, G simply entails G_1 and G_2, but not G_3 where the two blank nodes are identified. Now consider a basic graph pattern BGP = {ex:a ex:b ?x. ?y ex:e ex:f}. A solution multiset for BGP could comprise two mappings μ_1 : ?x $\mapsto$ _:b_1, ?y $\mapsto$ _:b_2 and

$\mu_2\colon \ ?x \mapsto \_{:}b_2, ?y \mapsto \_{:}b_1$. So we have $\mu_1(\text{BGP}) = G_1$ and $\mu_2(\text{BGP}) = G_2$, and both solutions are entailed. However, condition (3) requires that $G \cup \mu_1(\text{BGP}) \cup \mu_2(\text{BGP})$ is also entailed by G, and this is not the case in our example since this union contains G_3. The reason is that our solutions have unintended co-references of blank nodes that (3) does not allow. SPARQL's basic subgraph matching semantics respects this condition by requiring solution mappings to refer to blank nodes that actually occur in the active graph, so blank nodes are treated like (Skolem) constants.[4] The revised condition in Table 3 has further been modified to not implicitly require finite solution multisets which may not be appropriate for all entailment regimes. In addition, we use RDF instance mappings for renaming blank nodes instead of requiring renamed variants of the BGP.

Finally, condition (4) requires that solution multisets are finite and uniquely determined up to RDF graph equivalence, again using the "answer set" terminology. Our revised condition clarifies what it means for a solution multiset to be "unique up to RDF graph equivalence." We move the uniqueness requirement above all other conditions, since (2) and (3) do not make sense if the solution multiset was not defined in this sense. The rest of the condition was relaxed since entailment regimes may inherently require infinite solution multisets, e.g., in the case of the Rule Interchange Format RIF [6]. It is desirable that this only happens if there are infinite solutions that are "interesting," so the condition has been weakened to merely recommend the elimination of infinitely many "trivial" solution mappings in solution multisets. The requirement thus is expressed in an informal way, leaving the details to the entailment regime. Within this paper, we will make sure that the solution multisets are in fact finite (both regarding the size of the underlying set, and regarding the multiplicity of individual elements).

5 The RDF and RDFS Entailment Regimes

We focus on specifying the RDFS entailment regime, since the case of RDF is an obvious simplification of this entailment regime. The major problem for RDFS entailment is to avoid trivially infinite solution multisets as suggested by Table 3 (4), where three principal sources of infinite query results have to be addressed:

1. An RDF graph can be inconsistent under the RDFS semantics in which case it RDFS-entails all (infinitely many) conceivable triples.
2. The RDFS semantics requires all models to satisfy an infinite number of *axiomatic triples* even when considering an empty graph.
3. Every non-empty graph entails infinitely many triples obtained by using arbitrary blank nodes in triples.

We now discuss each of these problems, and derive a concrete definition for BGP matching in the proposed entailment regime at the end of this section.

---

[4] Yet, SPARQL allows blank nodes to be renamed when loading documents, so there is no guarantee that blank node IDs used in input documents are preserved.

5.1   Treatment of Inconsistencies

SPARQL does not require entailment regimes to yield a particular query result in cases where the active graph is inconsistent. As stated in [12], "[the] effect of a query on an inconsistent graph [...] must be specified by the particular SPARQL extension." One could simply require that implementations of the RDFS entailment report an error when given an inconsistent active graph. However, a closer look reveals that inconsistencies are extremely rare in RDFS, so that the requirement of checking consistency before answering queries would impose an unnecessary burden on implementations.

Indeed, graphs can only be RDFS-inconsistent due to improper use of the datatype rdf:XMLLiteral. A typical example for this is the following graph:

ex:a ex:b "<"^^rdf:XMLLiteral.   ex:b rdfs:range rdfs:Literal.

The literal in the first triple is *ill-typed* as it does not denote a value of rdf:XMLLiteral. This does not cause an inconsistency yet but forces "<"^^rdf:XMLLiteral to be interpreted as a resource that is not in the extension of rdfs:Literal, which in turn cannot be the case in any model that satisfies the second triple. Ill-typed literals are the only possible cause of inconsistency in RDFS and as such not a frequent problem.[5] Moreover, inconsistencies of this type are inherently "local" as they are based on individual ill-typed literals that could easily be ignored if not related to a given query.

It has thus been decided in the SPARQL working group that systems only have to report an error if they actually detect an inconsistency. Until this happens, queries can be answered as if all literals were well-typed. Our exact formalization corresponds to a behavior where tools simply assume that all strings are well-typed for rdf:XMLLiteral, and hence does not put additional burden on implementers.

5.2   Treatment of Axiomatic Triples

Every RDFS model is required to satisfy an infinite number of *axiomatic triples*. The reason is that the RDF vocabulary for encoding lists includes property names rdf:_i for all $i \geq 1$, with several (RDFS) axiomatic triples for each rdf:_i. For instance, we find a triple rdf:_i rdf:type rdf:Property for all $i \in \mathbf{N}$. Thus, the query ?x rdf:type rdf:Property could have infinitely many results. We consider such results trivial in the sense of Table 3 (4), and thus we want avoid them in the RDFS entailment regime.

We therefore propose that axiomatic triples with a subject of the form rdf:_i are only taken into account if the subject's IRI explicitly occurs in the active graph. This ensures that only finitely many axiomatic triples are considered, since there is only a finite number of axiomatic triples whose subjects do not have the form rdf:_i. To conveniently formalize this, Definition 5 below still refers to the standard RDFS entailment with all axiomatic triples, and restricts the range of solution mappings to an *answer domain* instead. Ignoring axiomatic triples for IRIs rdf:_i that occur only in a query but not in the active graph ensures that the total number of entailments that are relevant for query answering is finite. This would not be the case if new entailments would be required

---

[5] Implementations may support additional datatypes that can lead to similar problems. Such extensions go beyond the RDFS semantics we consider here, yet inconsistencies remain rare even in these cases.

whenever a given query contains a hitherto unused IRI. This distinguishes our approach from [5] where a *partial closure* algorithm is used to decide RDFS entailment for a set of axiomatic triples based on both the given graph and the query graph.

5.3   Treatment of Blank Nodes

Even if condition (3) in Table 3 holds, solution multisets could include infinitely many results that only differ in the identifiers for blank nodes. Simple entailment avoids this problem by restricting results to blank nodes that occur in the active graph. For entailment regimes, however, one must take entailed triples into account. This already leads to triples with different blank nodes, as illustrated in the graphs G_1 and G_2 in Section 4.

Restricting the range of solution mappings to blank nodes in the active graph would ensure finiteness but is not a satisfactory solution. To see why, consider the graph

$$G : \text{ex:a ex:b ex:c.} \quad \text{ex:d ex:e \_:f.}$$

The query $\mathsf{BGP} = \{\text{ex:a ex:b ?x}\}$ yields only one solution mapping $\mu : \text{?x} \mapsto \text{ex:c}$ under simple entailment. Yet, the mapping $\mu' : \text{?x} \mapsto \text{\_:f}$ uses only blank nodes from G, and satisfies $G \models \mu'(\mathsf{BGP})$ even under simple semantics. This shows that the latter two conditions are not sufficiently specific for handling blank nodes in entailment regimes. A more adequate approach is the use of *Skolemization*:

Definition 4. *Let the prefix* skol *refer to a namespace IRI that does not occur as the prefix of any IRI in the active graph or query. The* Skolemization $\mathsf{sk}(\_:b)$ *of a blank node* _:b *is defined as* $\mathsf{sk}(\_:b) := \text{skol:b}$. *We extend* $\mathsf{sk}(\cdot)$ *to graphs and filters just like other (partial) functions on RDF terms.*

Intuitively, Skolemization changes blank nodes into resource identifiers that are not affected by entailment. Clearly, we do not want Skolemized blank nodes to occur in query results, but it is useful to restrict to solution mappings μ for which $\mathsf{sk}(G) \models \mathsf{sk}(\mu(\mathsf{BGP}))$. In the above example, this condition is indeed satisfied by μ but not by μ'.

5.4   Defining the RDF(S) Entailment Regimes

The set of *well-formed* graphs for the RDFS entailment regime is simply the set of all RDF graphs. BGP matching for RDFS is defined as follows.

Definition 5. *Let* $\mathsf{Voc(RDFS)}$ *be the RDFS vocabulary,* G *an RDF graph, and* BGP *a basic graph pattern. The* answer domain w.r.t. G *under RDFS entailment, written* $\mathsf{AD_{RDFS}(G)}$, *is the set* $\mathsf{Voc(G)} \cup (\mathsf{Voc(RDFS)} \setminus \{\text{rdf:\_i} \mid i \in \mathbf{N}\})$. *The evaluation of* BGP *over* G *under RDFS entailment* $[\![\mathsf{BGP}]\!]_G^{\mathsf{RDFS}}$ *is the solution multiset*

$\{(\mu, n) \mid \mathsf{dom}(\mu) = \mathsf{V(BGP)}$, *and* n *is the maximal number such that*
  $\sigma_1, \ldots, \sigma_n$ *are distinct RDF instance mappings such that, for each* $1 \leq i \leq n$,
  $\mathsf{sk}(G) \models_{\mathsf{RDFS}} \mathsf{sk}(\mu(\sigma_i(\mathsf{BGP})))$ *and* $(\mathsf{ran}(\mu) \cup \mathsf{ran}(\sigma_i)) \subseteq \mathsf{AD_{RDFS}(G)}\}$.

Other types of graph patterns are evaluated as in Definition 3. If the active graph is RDFS-inconsistent, implementations may compute solution multisets based on the assumption that all literals of type rdf:XMLLiteral are well-typed, so that no inconsistency occurs. When the inconsistency is detected, implementations should report an error.

Since computing a partial RDFS closure for an RDF graph can be done in polynomial time [5] and BGP evaluation then amounts to subgraph matching over the partial closure, it follows that the complexity of the evaluation problem under the RDFS regime is the same as for standard SPARQL. For set semantics instead of multiset semantics this is known to be PSPACE-complete [10].

The entailment regime for RDF is defined similarly, but using RDF entailment and the RDF vocabulary instead. Note that the above definition can also be restricted to simple entailment, yielding the same solution multisets as Definition 2.

6   The OWL Entailment Regimes

In contrast to the RDFS semantics, a graph does no longer admit a unique canonical model that can be used to compute answers under the RDF-Based Semantics (RBS) and Direct Semantics (DS) of OWL, i.e., we can no longer imagine queries to act on a unique "completed" version of the active graph. This affects reasoning algorithms (see Section 7), but has only little effect on our definitions. The main new challenges for OWL are its expressive datatype constructs that may lead to infinite answers, and the fact that the OWL DS is defined in terms of OWL objects to which a given RDF graph and query must first be translated. The problems discussed for RDF(S) also require slightly different solutions for OWL:

1. Inconsistent input ontologies are required to be rejected with an error.
2. The axiomatic triples of RDFS are used only by the RBS and can again be handled by suitably restricting solutions to an answer domain.
3. The problem of blank nodes occurs for both semantics and can again be addressed by Skolemization, but for DS the blank nodes that are used to encode OWL objects must not be Skolemized.

The main difference to RDFS is the stricter first item which no longer permits deferred inconsistency detection. Inconsistencies in RDFS were easy to ignore since they always related to single literals. Neither OWL semantics suggests such simple reasoning under inconsistencies. Although proposals exists for addressing this, they disagree on the inferred entailments and tend to require complex computations. On the other hand, typical OWL reasoning algorithms are model building procedures which detect inconsistencies as part of their normal operation. Hence, reporting errors in this case can usually be done without additional effort.

6.1   Infinite Entailments in Datatype Reasoning

In order to see how datatype reasoning in OWL can cause infinite entailments, consider the graph and query in Table 4. Recall that a abbreviates rdf:type, [...] denotes an implicit blank node, and (...) denotes an RDF list. G states that all data values to which Peter is related via ex:dp are in the singleton set of the integer 5. The query asks for all data values to which ex:Peter cannot be related with ex:dp. Without suitable restrictions, all (infinitely many) integers other than 5 could be used in solution mappings for ?x.

Table 4. A query with infinitely many entailed solutions

G : ex:Peter a [a owl:Restriction;
    owl:onProperty ex:dp;
    owl:allValuesFrom [a rdfs:Datatype;
        owl:oneOf ("5"^^xsd:integer)]]

BGP : ex:Peter a [a owl:Restriction;
    owl:onProperty ex:dp;
    owl:allValuesFrom [a rdfs:Datatype;
        owl:datatypeComplementOf [
            a rdfs:Datatype; owl:oneOf (?x)]]]

Moreover, it is currently unknown how to compute all mappings for literal variables even for cases where there number is finite – testing all literals is clearly not an option.[6]

We therefore restrict the answer domain for the OWL entailment regimes to include only literals that are explicitly mentioned in the input graph. Like for the IRIs rdf:_i, this may lead to unexpected behavior, since mentioning a literal in the input may lead to new query results even for queries not directly related to this literal. Yet, we think this problem is so rare in practice that a more detailed analysis of the problematic datatype expressions is not worthwhile, even if it could further limit unintuitive behavior.

6.2 The OWL 2 RDF-Based Semantics Entailment Regime

The OWL 2 RDF-Based Semantics treats classes as individuals that refer to elements of the domain. Each such element is then associated with a subset of the domain, called the class extension. This means that semantic conditions on class extensions are only applicable to those classes that are actually represented by an element of the domain which can lead to less consequences than expected. An example is given by the following graph and BGP:

G : ex:a rdf:type ex:C   BGP : ?x rdf:type [rdf:type owl:Class ;
                                    owl:unionOf (ex:C ex:D)]

G states that ex:a has type ex:C, while BGP asks for instances of the complex class denoting the union of ex:C and ex:D. One might expect μ: ?x ↦ ex:a to be a solution, but this is not the case under the OWL 2 RDF-Based Semantics (see also [14, Sec. 7.1]). It is guaranteed that the union of the class extensions for ex:C and ex:D exists as a subset of the domain; no statement in G implies, however, that this union is the class extension of any domain element. Thus, μ(BGP) is not entailed by G.

The entailment holds, however, when the statement ex:E owl:unionOf (ex:C ex:D) is added to G. In the OWL Direct Semantics, in contrast, classes denote sets and not domain elements, so G entails μ(BGP) under DS where, formally, G must first be extended with an ontology header to become well-formed for DS. Note that a similar situation occurs for the example in Section 6.1, but the problem still occurs if the necessary expressions are introduced.

Summing up, the RBS handles blank nodes just like RDFS, even in cases where they are needed for encoding OWL class expressions. This allows us to use Skolemization just like in the case of RDFS in the next definition.

---

[6] Hence one cannot call such solutions "trivial" in the sense of Table 3. Indeed, our restrictions are motivated by pragmatic considerations, not by formal requirements of SPARQL.

Table 5. Grammar extension for extended OWL objects

Class := IRI | Var	ObjectProperty := IRI | Var	DataProperty := IRI | Var
Individual := NamedIndividual | AnonymousIndividual | Var		
Literal := typedLiteral | stringLiteralNoLanguage | stringLiteralWithLanguage | Var		

Definition 6. *Let* Voc(OWL2) *be the OWL 2 vocabulary,* G *a graph, and* BGP *a basic graph pattern. We write* $\models_{RBS}$ *to denote the OWL 2 RDF-Based Semantics entailment relation. The* answer domain *w.r.t.* G *under RDF-Based Semantics entailment, written* $AD_{RBS}(G)$*, is the set* $Voc(G) \cup (Voc(OWL2) \setminus \{rdf:\_i \mid i \in \mathbf{N}\})$*. The evaluation of* BGP *over* G *under RDF-Based Semantics entailment* $[\![BGP]\!]_G^{RBS}$ *is the solution multiset*

$\{(\mu, n) \mid dom(\mu) = V(BGP)$, *and* n *is the maximal number such that*
$\qquad \sigma_1, \ldots, \sigma_n$ *are distinct RDF instance mappings such that, for each* $1 \leq i \leq n$,
$\qquad sk(G) \models_{RBS} sk(\mu(\sigma_i(BGP)))$ *and* $(ran(\mu) \cup ran(\sigma_i)) \subseteq AD_{RBS}(G)\}$.

6.3 The OWL 2 Direct Semantics Entailment Regime

The OWL 2 Direct Semantics is not defined in terms of triples, but in terms of OWL objects that constitute an *ontology*. The OWL 2 recommendation specifies how to construct an ontology O_G from a graph G that satisfies some further conditions [9]. Thus G is *well-formed for the OWL DS entailment regime* if O_G is defined. In the following, we conveniently identify ontologies with their unique canonical representation in Functional-Style Syntax [8]. Some RDF triples are mapped to so-called *non-logical axioms* such as annotations, declarations, or import directives. Such axioms can only have indirect effect on DS entailment, e.g., since imported axioms are taken into account, but they do not directly lead to entailments. In particular, annotations do not contribute query results under DS.

Like the active graph, also the BGP of the query is mapped into an OWL 2 DL ontology, extended to allow variables in place of class names, object property names, datatype property names, individual names, or literals. Table 5 shows how productions of the OWL 2 functional-style syntax grammar [8] are extended to allow variables as defined by the Var production from the SPARQL grammar [12]. Solution mappings in a query result are applied to such extended ontologies to obtain a set of OWL DL axioms that is compatible with the queried ontology and also entailed by it under DS.

The construction of ontologies from graphs requires type declarations for properties, classes, and (custom) datatypes to avoid ambiguities, and we need similar typing information for terms *and* variables in BGPs. For example, the BGP {?s ?p ?o} could refer to DataPropertyAssertion(?p ?s ?o) or ObjectPropertyAssertion(?p ?s ?o) if the type of ?p is not given. We take type declarations from the queried ontology into account, so that only variables may require further typing.

Formally, an extended ontology O_{BGP}^G is constructed for a basic graph pattern BGP and graph G using the parsing process for RDF graphs as defined in [9] with three modifications: variable identifiers are allowed in place of IRIs and literals in all parsing steps, an ontology header may be added to BGP if not given, and the type declarations given in BGP are augmented with the declarations in G (denoted AllDecl(G) in [9]). The complete parsing process is detailed in the latest entailment regimes working draft.[7] BGP is *well-formed for the OWL DS entailment regime and a graph* G if O_{BGP}^{G} can be obtained in this way and is an extended OWL DL ontology.

We can now define the evaluation of graph patterns. Skolemization is now applied to O_G, which ensures that only blank nodes that represent anonymous OWL individuals are Skolemized, not blank nodes used for encoding complex OWL syntax in RDF.

Definition 7. *Consider a graph* G *that is well-formed for the OWL 2 DS entailment regime, and a basic graph pattern* BGP *that is well-formed for DS and* G*. With* $sk(O_G)$ *we denote the result of replacing each blank node b in* O_G *with* $sk(b)$. *The answer domain w.r.t.* G *under OWL 2 Direct Semantics entailment, written* $AD_{DS}(G)$, *is* $Voc(O_G)$. *If* O_G *is inconsistent, queries must be rejected with an error. Otherwise, we write* $\models_{DS}$ *for the OWL 2 Direct Semantics entailment relation and define the* evaluation of BGP *over* G *under OWL 2 Direct Semantics entailment* $[\![BGP]\!]_{G}^{DS}$ *as the solution multiset*

$\{(\mu, n) \mid dom(\mu) = V(BGP)$, *and n is the maximal number such that*
$\quad \sigma_1, \ldots, \sigma_n$ *are distinct RDF instance mappings such that, for each* $1 \leq i \leq n$,
$\quad O_G \cup \mu(\sigma_i(O_{BGP}^{G}))$ *is an OWL 2 DL ontology, and*
$\quad sk(O_G) \models_{DS} sk(\mu(\sigma_i(O_{BGP}^{G})))$ *and* $(ran(\mu) \cup ran(\sigma_i)) \subseteq AD_{DS}(G)\}$.

Since $AD_{DS}(G)$ is finite, clearly the solution multiset and each multiplicity is finite too. Although the restriction to $AD_{DS}(G)$ avoids infinite results as discussed in Section 6.1, reasoners may have to consider a large number of literals as potential variable bindings and we expect that not all systems will provide a complete implementation for queries with literal variables.

The complexity of standard reasoning problems in OWL are well-understood and BGP evaluation can be implemented using the standard reasoning techniques. The complexity of OWL reasoning usually outweighs that of the SPARQL algebra operations, i.e., checking whether a solution mapping is a solution is complete for nondeterministic exponential time in OWL DL and undecidable for the RDF-Based semantics.

7   Implementations of SPARQL Entailment Regimes

We now discuss how the interplay between SPARQL query processing and semantic inference can be implemented in practice. Three principal approaches for this task are reviewed below. An overview of optimized implementation techniques for SPARQL algebra operators or specific reasoning algorithms is beyond the scope of this work.

Materialization and Partial Closure. One can often extend the input graph with all relevant semantic consequences, pre-computed at load time, and evaluate SPARQL queries on this extended graph under the simple semantics. The approach is not applicable to entailment regimes for which one cannot pre-compute all relevant consequences, e.g., for OWL DS entailment where arbitrarily complex class expressions may be required. In the case of RDF(S) and OWL RDF-Based Semantics, however, our definitions ensure that the relevant consequences are finite and depend on the input graph only.[8]

---

[7] http://www.w3.org/TR/2010/WD-sparql11-entailment-20100601/
[8] Computing all such consequences for OWL RBS is of course still undecidable.

Materialization is the most common implementation technique, supported in systems such as AllegroGraph, Jena, BigOWLIM and SwiftOWLIM, Mulgara, OntoBroker, or Virtuoso.[9] The partial closure algorithm proposed in [5] for checking RDF(S) entailment can be adapted to implement the RDF(S) regime: Blank nodes in the initial graph have to marked since only they can be used in solution and instance mappings, whereas new blank nodes introduced by the partial closure algorithm cannot be used for variable bindings. Blank nodes in the query are treated as variables that are projected out immediately after BGP evaluation; the multiplicity of a solution is then given by the number of original solutions from which it can be obtained through this projection.

Query Rewriting. These techniques change the query rather than the queried graph. One or more, possibly more complex queries are then evaluated over the original graph. More expressive query features may be needed, e.g., by using regular expressions to capture the transitivity of rdfs:subClassOf. To the best of our knowledge a pure query rewriting techniques has so far only be proposed for a subset of RDFS [11]. A combination with materialization, however, is also possible and successfully used, e.g., to realize RDFS entailment in Sesame [17].

Modified Query Evaluation. The most direct approach for implementing our definitions is to modify existing SPARQL processors to evaluate BGPs differently. This can be accomplished, e.g., with the free ARQ library (http://jena.sourceforge.net/ARQ/). While this offers much flexibility, computing BGP matches on demand may preclude many optimizations for evaluating algebra operators. Yet, this method is a good approach for adding SPARQL support to systems that perform complex inferencing. The Hermit OWL reasoner (http://hermit-reasoner.com/) is currently being extended accordingly to support the proposed DS entailment regime. This work also includes the modification of the OWL API for parsing BGPs into extended OWL ontologies.

8   Related Work

Section 7 listed various efforts that are closely related to the implementation of our proposals. Here we focus on alternative proposals for querying expressive semantic data sources, especially for OWL.

OWL DS queries that ask for individuals and literals only are closely related to *conjunctive queries* (CQs) on description logic (DL) knowledge bases; see [4] for a basic introduction. An important difference is that CQs admit full existential variables that can represent any domain element which can be (indirectly) inferred to exist. In contrast, variables and blank nodes under OWL DS entailment may only bind to individuals that are represented by a given blank node or IRI in the input, corresponding to so-called *distinguished variables* in CQs. As of today, decidability of CQ entailment has only been established for a sublanguage of OWL 2 [13]. Restricted CQ answering still is the most common query service provided by OWL reasoners today. For example, KAON2 (http://kaon2.semanticweb.org/) and the TrOWL system

---

[9] See http://en.wikipedia.org/wiki/Triplestore for more information on the mentioned systems.

(`http://trowl.eu/`) support the CQ subset of the OWL DS regime, whereas Rac-erPro (`http://racer-systems.com/`) has its proprietary query language for CQs, called nRQL [2]. Similarly, OWLgres [16] and Quonto[10] support the CQ fragment, but they implement the OWL QL profile, which restricts the expressivity of the input ontology to allow for a more efficient implementation based on standard database techniques.

We are not aware of a complete implementation of the DS entailment regime. As of today, the Pellet OWL 2 DL reasoner (`http://clarkparsia.com/pellet`) is the most advanced system. The subset of SPARQL that Pellet supports – called SPARQL-DL [15] – consists of queries that can be translated into a pre-defined set of query atoms in an abstract syntax; with the semantics defined per abstract query atom.

Explicitly listing admissible queries has the advantage that one can focus on queries that are well supported by OWL reasoners. Our definition of OWL DS entailment, in contrast, uses a more general approach based on a direct mapping of BGPs to extended OWL ontologies. This allows for queries that are not typically supported by reasoners, e.g., when using variables to represent class names in complex class expressions.

Furthermore, SPARQL-DL treats blank nodes in queries like non-distinguished CQ variables with full existential meaning, whereas the DS regime treats such blank nodes like SPARQL variables that are projected out after BGP evaluation. Blank nodes under DS entailment thus are largely like distinguished CQ variables, though we allow blank nodes in the input to occur in results via Skolemization. Our design choice makes the treatment of blank nodes more uniform across all SPARQL entailment regimes, and it avoids the computational problems with non-distinguished variables in OWL.

9   Conclusions

We have presented extensions for SPARQL to incorporate RDF, RDFS, OWL RDF-Based semantics, and OWL Direct Semantics entailment. When comparing the individual entailment regimes, we find that a surprisingly high level of compatibility can be achieved between the different formalisms.

The presented regimes are closely related to the SPARQL Entailment Regimes document currently developed in the W3C SPARQL working group and we believe that our extended discussions and the resulting definitions provide a useful resource for implementers and users of SPARQL.

Our work also provides a basis for further extensions of SPARQL. Entailment regimes such as D-entailment can easily be added. A RIF entailment regime is also currently under development in the SPARQL Working Group, although some preliminaries still have to be clarified, e.g., how an RDF graph can import or encode a RIF rule set. An integration of new SPARQL operators, which are defined algebraically such as the *minus* operator currently under discussion, is straightforward. SPARQL modifications that introduce extension points besides BGP matching, in contrast, would require more considerations. Depending on the outcome of current discussions, this might be the case for *path expressions* in SPARQL 1.1. Yet, our overall impression is that SPARQL is ready – both theoretically and practically – for taking the step beyond sub-graph matching.

---

[10] `http://www.dis.uniroma1.it/quonto/`

Acknowledgements. This work was supported by EPSRC in the project *HermiT: Reasoning with Large Ontologies* and *RInO: Reasoning Infrastructure for Ontologies and Instances*, and by DFG in the project *ExpresST*. We thank the members of the SPARQL working group for valuable comments and suggestions.

References

1. Beckett, D., Berners-Lee, T.: Turtle – Terse RDF Triple Language. W3C Team Submission (January 14, 2008), http://www.w3.org/TeamSubmission/turtle/
2. Haarslev, V., Möller, R., Wessel, M.: Querying the semantic web with Racer + nRQL. In: Proc. KI 2004 International Workshop on Applications of Description Logics (2004)
3. Hayes, P. (ed.): RDF Semantics. W3C Recommendation (February 10, 2004), http://www.w3.org/TR/rdf-mt/
4. Hitzler, P., Krötzsch, M., Rudolph, S.: Foundations of Semantic Web Technologies. Chapman & Hall/CRC (2009)
5. ter Horst, H.J.: Completeness, decidability and complexity of entailment for RDF Schema and a semantic extension involving the OWL vocabulary. J. of Web Semantics 3(2-3), 79–115 (2005)
6. Kifer, M., Boley, H. (eds.): RIF Overview. W3C Working Group Note (June 22, 2010), http://www.w3.org/TR/rif-overview/
7. Motik, B., Patel-Schneider, P.F., Cuenca Grau, B. (eds.): OWL 2 Web Ontology Language: Direct Semantics. W3C Recommendation (October 27, 2009), http://www.w3.org/TR/owl2-direct-semantics/
8. Motik, B., Patel-Schneider, P.F., Parsia, B. (eds.): OWL 2 Web Ontology Language: Structural Specification and Functional-Style Syntax. W3C Recommendation (October 27, 2009), http://www.w3.org/TR/owl2-syntax/
9. Patel-Schneider, P.F., Motik, B. (eds.): OWL 2 Web Ontology Language: Mapping to RDF Graphs. W3C Recommendation (October 27, 2009), http://www.w3.org/TR/owl2-mapping-to-rdf/
10. Pérez, J., Arenas, M., Gutierrez, C.: Semantics and complexity of SPARQL. ACM Transactions on Database Systems 34(3), 1–45 (2009)
11. Pérez, J., Arenas, M., Gutierrez, C.: nSPARQL: A navigational language for RDF. J. of Web Semantics (to appear, 2010), http://web.ing.puc.cl/~jperez/papers/jws2010.pdf
12. Prud'hommeaux, E., Seaborne, A. (eds.): SPARQL Query Language for RDF. W3C Recommendation (January 15, 2008), http://www.w3.org/TR/rdf-sparql-query/
13. Rudolph, S., Glimm, B.: Nominals, inverses, counting, and conjunctive queries. J. of Artificial Intelligence Research 39, 429–481 (2010), http://www.comlab.ox.ac.uk/files/2175/paper.pdf
14. Schneider, M. (ed.): OWL 2 Web Ontology Language: RDF-Based Semantics. W3C Recommendation (October 27, 2009), http://www.w3.org/TR/owl2-rdf-based-semantics/
15. Sirin, E., Parsia, B.: SPARQL-DL: SPARQL query for OWL-DL. In: Golbreich, C., Kalyanpur, A., Parsia, B. (eds.) Proc. OWLED 2007 Workshop on OWL: Experiences and Directions. CEUR Workshop Proceedings, vol. 258. CEUR-WS.org (2007)
16. Stocker, M., Smith, M.: Owlgres: A scalable OWL reasoner. In: Dolbear, C., Ruttenberg, A., Sattler, U. (eds.) Proc. OWLED 2008 Workshop on OWL: Experiences and Directions. CEUR Workshop Proceedings, vol. 432. CEUR-WS.org (2008)
17. Stuckenschmidt, H., Broekstra, J., Amerfoort, A.: Time – space trade-offs in scaling up RDF Schema reasoning. In: Dean, M., Guo, Y., Jun, W., Kaschek, R., Krishnaswamy, S., Pan, Z., Sheng, Q.Z. (eds.) WISE 2005 Workshops. LNCS, vol. 3807, pp. 172–181. Springer, Heidelberg (2005)

Integrated Metamodeling and Diagnosis in OWL 2

Birte Glimm[1], Sebastian Rudolph[2], and Johanna Völker[3]

[1] Oxford University Computation Laboratory, UK
birte.glimm@comlab.ox.ac.uk
[2] Institute AIFB, Karlsruhe Institute of Technology, DE
rudolph@kit.edu
[3] KR & KM Research Group, University of Mannheim, DE
voelker@informatik.uni-mannheim.de

Abstract. Ontological metamodeling has a variety of applications yet only very restricted forms are supported by OWL 2 directly. We propose a novel encoding scheme enabling class-based metamodeling inside the domain ontology with full reasoning support through standard OWL 2 reasoning systems. We demonstrate the usefulness of our method by applying it to the OntoClean methodology. En passant, we address performance problems arising from the inconsistency diagnosis strategy originally proposed for OntoClean by introducing an alternative technique where sources of conflicts are indicated by means of marker predicates.

1 Introduction

Applications of metamodeling in Ontology Engineering are manifold, including the representation of provenance or versioning information as well as the documentation of modeling decisions. Roughly speaking, metamodeling allows for referring to *predicates* (classes and properties in OWL) as if they were domain individuals. This way it is possible to assert the membership of classes in metaclasses and interconnect them via metaroles.

Consider, for example, the following extract of a knowledge base about animals and the respective species they belong to.

$$(\mathsf{GoldenEagle} \sqcup \mathsf{HaastsEagle})(\mathit{harry}) \quad \mathsf{HouseMouse}(\mathit{jerry})$$

Intuitively, we specify that the individual *harry* is a golden or a Haast's eagle and *jerry* is a common house mouse. Now, assume the knowledge base also expresses taxonomic relationships assigning species to orders of animals.[1]

$$\mathsf{GoldenEagle} \sqsubseteq \mathsf{Falconiformes} \quad \mathsf{HouseMouse} \sqsubseteq \mathsf{Rodentia}$$
$$\mathsf{HaastsEagle} \sqsubseteq \mathsf{Falconiformes}$$

If, additionally, we were to specify which of the zoological terms actually denote species and which denote orders, we could introduce the classes `Species` and `Order`.

---

[1] Species is the most specific level within the biological classification and order is a more general one, e.g., Golden Eagle (A. chrysaetos) is a *species*, whereas Falconiformes is the *order* of Golden Eagle. In Europe the Falconiformes order is commonly split into Falconiformes and Accipitriformes, but we neglect that here.

P.F. Patel-Schneider et al. (Eds.): ISWC 2010, Part I, LNCS 6496, pp. 257–272, 2010.
© Springer-Verlag Berlin Heidelberg 2010

Treating those classes on a level with Rodentia etc. by subclass statements like Rodentia ⊑ Order leads to consequences that are doubtful (like Order(*jerry*)) or outright unwanted (like HouseMouse ⊑ Order). Therefore, species and order should be treated as *metaclasses* the members of which are themselves classes, i.e., we would like to make statements such as

Species(GoldenEagle)   Order(Falconiformes)
Species(HaastsEagle)   Order(Rodentia)
Species(HouseMouse)

Likewise we may think of *metaroles* that interrelate classes instead of individuals. In particular, the subclass relationship between classes can be seen as such a metarole (one with a built-in meaning instead of one that can be freely defined). In fact, many evaluation or design criteria for ontologies [12,4] directly refer to the hierarchy of classes in an ontology. Considering our example, one obvious design criterion would be that for A a species and B an order, $B ⊑ A$ must not hold as this would contradict the conventional organization of zoological taxonomies.

Current ontology languages differ with respect to their support for metamodeling. While it is supported by OWL Full, this high expressivity leads to undecidability as shown by Motik [9], who also discusses milder variants of metamodeling. One variant, which is also supported by OWL 2 DL, is called *punning*. Punning allows for using the same identifier, e.g., for an individual and a class. The class and its corresponding individual are, however, treated as entirely independent, which disallows many of the intended usage scenarios of metamodeling. As another lightweight metamodeling feature, OWL 2 allows for annotation properties, which may associate information to classes, roles, and even axioms. In OWL 2 DL and all its subprofiles, these properties do not carry any semantics and are not used for reasoning.

One way to facilitate more expressive metamodeling while still supporting the use of off-the-shelf reasoning tools for OWL is to maintain two (or more) ontologies, keeping the basic domain knowledge separate from the meta knowledge. In that case, the two ontologies must be kept in sync by additional external mechanisms. Thereby, information obtained from reasoning in the basic ontology (like its subclass hierarchy) is fed into the metaontology as explicit statements. Based on this, reasoning in the enriched metaontology can be carried out. Clearly, this approach comes with increased maintenance efforts. Examples for this strategy are [10] and [14].

We extend this state of the art in two ways, which are independent from each other, but can be combined:

1. We introduce a technique that enables class-based metamodeling within one ontology. Thereby, subclass relationships between classes are axiomatically synchronized with role memberships of class-representing individuals. Meta-level constraints on classes and their subsumption relationships can then be expressed as OWL axioms in the same ontology as the actual content.
2. We propose a way of expressing meta-level constraints in a way that does not lead to inconsistency, but rather indicates constraint violations by auxiliary classes or roles. Thus, the origins of these violations can be localized by comparably cheap instance retrieval operations instead of costly debugging strategies.

We proceed as follows: The next section introduces the necessary preliminaries of the description logic $\mathcal{SROIQ}$ underlying the OWL 2 standard. We use the DL notation for its brevity. Section 3 introduces our technique enabling ontology-inherent metamodeling. Section 4 sketches the OntoClean methodology as one possible metamodeling use case. Section 5 describes the original OWL-based OntoClean constraint checking approach as well as our metamodeling-based modification of it. Section 6 introduces another modification of the methodology by suggesting to use marker predicates instead of explanations. Finally, Section 7 provides an evaluation of the proposed techniques before we conclude in Section 8. A more detailed treatise can be found in the extended version of the paper [3].

2   Preliminaries

We just recall the basic definitions for the description logic $\mathcal{SROIQ}$ [6]. For further details on DLs we refer interested readers to the Description Logic Handbook [1]. As our definitions are based on DLs, we use the terms *ontology* and *knowledge base* interchangeably.[2]

Definition 1. *Let* N_R, N_C, *and* N_I *be three disjoint sets of* role names *containing the* universal role $U \in N_R$, class names, *and* individual names, *respectively. A* $\mathcal{SROIQ}$ *RBox for* N_R *is based on a set* **R** *of* roles *defined as* $\mathbf{R} := N_R \cup \{R^- \mid R \in N_R\}$, *where we set* $\text{Inv}(R) := R^-$ *and* $\text{Inv}(R^-) := R$ *to simplify notation. In the sequel, we will use the symbols* R, S, *possibly with subscripts, to denote roles.*

A generalised role inclusion axiom *(RIA) is a statement of the form* $S_1 \circ \ldots \circ S_n \sqsubseteq R$, *and a set of such RIAs is a generalised* role hierarchy. *A role will be called* non-simple *for some role hierarchy if it can be implied by some role chain, otherwise it is* simple.

A role disjointness assertion *is a statement of the form* $\text{Dis}(S, S')$, *where* S *and* S' *are simple. A* $\mathcal{SROIQ}$ *RBox is the union of a set of role disjointness assertions together with a role hierarchy. A* $\mathcal{SROIQ}$ *RBox is regular if its role hierarchy is regular.*

For brevity, we omit a precise definition of *simple* roles and role hierarchy *regularity*, and refer interested readers to [6]. Note that number restrictions (defined below) can only be formed with simple roles to guarantee the decidability of the standard reasoning tasks such as checking knowledge base consistency.

Definition 2. *Given a* $\mathcal{SROIQ}$ *RBox* $\mathcal{R}$, *the set of* class expressions **C** *is defined as follows:*

- $N_C \subseteq \mathbf{C}$, $\top \in \mathbf{C}$, $\bot \in \mathbf{C}$,
- *if* $C, D \in \mathbf{C}$, $R \in \mathbf{R}$, $S \in \mathbf{R}$ *a simple role,* $a \in N_I$, *and* n *a non-negative integer, then* $\neg C$, $C \sqcap D$, $C \sqcup D$, $\{a\}$, $\forall R.C$, $\exists R.C$, $\exists S.\text{Self}$, $\leq n\, S.C$, *and* $\geq n\, S.C$ *are also class expressions.*

---

[2] Moreover, we use the term *classes* instead of concepts for unary predicates, whereas we refer to binary predicates as *roles* instead of properties in order to avoid confusion with the term *metaproperties* introduced by OntoClean.

Table 1. Semantics of class expressions in $\mathcal{SROIQ}$ for an interpretation $\mathcal{I} = (\Delta^{\mathcal{I}}, \cdot^{\mathcal{I}})$

Name	Syntax	Semantics
inverse role	R^-	$\{\langle x, y\rangle \in \Delta^{\mathcal{I}} \times \Delta^{\mathcal{I}} \mid \langle y, x\rangle \in R^{\mathcal{I}}\}$
universal role	U	$\Delta^{\mathcal{I}} \times \Delta^{\mathcal{I}}$
top	$\top$	$\Delta^{\mathcal{I}}$
bottom	$\bot$	$\emptyset$
negation	$\neg C$	$\Delta^{\mathcal{I}} \setminus C^{\mathcal{I}}$
conjunction	$C \sqcap D$	$C^{\mathcal{I}} \cap D^{\mathcal{I}}$
disjunction	$C \sqcup D$	$C^{\mathcal{I}} \cup D^{\mathcal{I}}$
nominals	$\{a\}$	$\{a^{\mathcal{I}}\}$
univ. restriction	$\forall R.C$	$\{x \in \Delta^{\mathcal{I}} \mid \langle x, y\rangle \in R^{\mathcal{I}} \text{ implies } y \in C^{\mathcal{I}}\}$
exist. restriction	$\exists R.C$	$\{x \in \Delta^{\mathcal{I}} \mid \text{ for some } y \in \Delta^{\mathcal{I}}, \langle x, y\rangle \in R^{\mathcal{I}} \text{ and } y \in C^{\mathcal{I}}\}$
Self construct	$\exists S.\mathsf{Self}$	$\{x \in \Delta^{\mathcal{I}} \mid \langle x, x\rangle \in S^{\mathcal{I}}\}$
qualified number restriction	$\leq n\, S.C$	$\{x \in \Delta^{\mathcal{I}} \mid \#\{y \in \Delta^{\mathcal{I}} \mid \langle x, y\rangle \in S^{\mathcal{I}} \text{ and } y \in C^{\mathcal{I}}\} \leq n\}$
	$\geq n\, S.C$	$\{x \in \Delta^{\mathcal{I}} \mid \#\{y \in \Delta^{\mathcal{I}} \mid \langle x, y\rangle \in S^{\mathcal{I}} \text{ and } y \in C^{\mathcal{I}}\} \geq n\}$

In the remainder, we use C *and* D *to denote class expressions. A* $\mathcal{SROIQ}$ *TBox is a set of* general class inclusion axioms *(GCIs) of the form* C $\sqsubseteq$ D. *We use* C $\equiv$ D *to abbreviate* C $\sqsubseteq$ D *and* D $\sqsubseteq$ C. *An* individual assertion *can have the form* C(a) *or* R(a, b) *with* $a, b \in \mathsf{N}_I$ *individual names. A* $\mathcal{SROIQ}$ *ABox is a set of individual assertions.*

A $\mathcal{SROIQ}$ *ontology* O *is the union of a regular RBox* $\mathcal{R}$, *an ABox* $\mathcal{A}$ *and TBox* $\mathcal{T}$ *for* $\mathcal{R}$. *The* vocabulary *of an ontology, denoted* voc(O), *is a triple* (O_C, O_R, O_I) *with* O_C *the set of class names occurring in* O, O_R *the set of role names occurring in* O, *and* O_I *the set of individual names occurring in* O.

The semantics of $\mathcal{SROIQ}$ ontologies is given by means of interpretations.

Definition 3. *An* interpretation $\mathcal{I}$ *consists of a set* $\Delta^{\mathcal{I}}$ *called* domain *(the elements of it being called* individuals*) together with a function* $\cdot^{\mathcal{I}}$ *mapping individual names to elements of* $\Delta^{\mathcal{I}}$, *class names to subsets of* $\Delta^{\mathcal{I}}$, *and role names to subsets of* $\Delta^{\mathcal{I}} \times \Delta^{\mathcal{I}}$.

The function $\cdot^{\mathcal{I}}$ *is inductively extended to role and class expressions as shown in Table 1. An interpretation* $\mathcal{I}$ *satisfies an axiom* φ *if we find that* $\mathcal{I} \models \varphi$:

- $\mathcal{I} \models S \sqsubseteq R$ *if* $S^{\mathcal{I}} \subseteq R^{\mathcal{I}}$,
- $\mathcal{I} \models S_1 \circ \ldots \circ S_n \sqsubseteq R$ *if* $S_1^{\mathcal{I}} \circ \ldots \circ S_n^{\mathcal{I}} \subseteq R^{\mathcal{I}}$ (∘ *being overloaded to denote the standard composition of binary relations here),*
- $\mathcal{I} \models \mathsf{Dis}(R, S)$ *if* $R^{\mathcal{I}}$ *and* $S^{\mathcal{I}}$ *are disjoint,*
- $\mathcal{I} \models C \sqsubseteq D$ *if* $C^{\mathcal{I}} \subseteq D^{\mathcal{I}}$.

An interpretation $\mathcal{I}$ *satisfies* C(a) *if* $a^{\mathcal{I}} \in C^{\mathcal{I}}$ *and* R(a, b) *if* $(a^{\mathcal{I}}, b^{\mathcal{I}}) \in R^{\mathcal{I}}$. *An interpretation* $\mathcal{I}$ *satisfies an ontology* O *(we then also say that* $\mathcal{I}$ *is a* model *of* O *and write* $\mathcal{I} \models O$*) if it satisfies all axioms of* O. *An ontology* O *is* satisfiable *if it has a model. An ontology* O entails *an axiom* φ, *if every model of* O *is a model of* φ.

Further details on $\mathcal{SROIQ}$ can be found in [6]. We have omitted here several syntactic constructs that can be expressed indirectly, especially RBox assertions for transitivity, reflexivity of simple roles, and symmetry.

In the remainder, we use the following notational convention: individual names are written in italic, e.g., *jerry*. Class names are written in sans serif font, e.g., HouseMouse and role names are written in normal serif font, e.g., eats, unless they are used to denote metaclasses or metaroles for which we use typewriter font, e.g., Species or subClassOf.

3   Ontology-Inherent Metamodeling for Classes

We will now show how to define a metamodeling-enabled version O^{meta} for a given ontology O. The converted ontology O^{meta} will be such that each model of the converted ontology has two different kinds of individuals: the *class individuals* are individuals that represent classes and each such individual is an instance of the newly introduced metaclass Class. On the other hand, the model also contains *proper individuals* and all these are instances of the newly introduced class Inst. Subclass relationships between a class C and a class D in the given ontology O are materialized as role instances: the individual that represents the class C, say o_C, and the individual that represents D, say o_D, are interconnected by the newly introduced metarole subClassOf. Similarly, a class membership of an individual a in a class C in the given ontology becomes manifest in a type relationship between a and o_C, for type also a freshly introduced role in O^{meta}. We further introduce an auxiliary role R_{Inst}, which is used to localize the universal role. These correspondences can then be used to check for modeling errors and to examine quality properties of the ontology.

Definition 4. *Let O be a domain ontology with vocabulary* $\text{voc}(O_C, O_R, O_I)$. *The vocabulary of the metamodeling-enabled version O^{meta} of O is:*

$$O^{\text{meta}}_C := O_C \cup \{\text{Inst}, \text{Class}\}$$
$$O^{\text{meta}}_R := O_R \cup \{\text{type}, \text{subClassOf}, R_{\text{Inst}}\}$$
$$O^{\text{meta}}_I := O_I \cup \{o_C \mid C \in O_C\}$$

where all the newly introduced names are fresh, i.e., they are not part of $\text{voc}(O)$.
    We define the functions bound($\cdot$), **SepDom**($\cdot$), **Typing**($\cdot$), *and* **MatSubClass**($\cdot$), *which take an ontology, i.e., a set of axioms, and return a set of axioms. The function* bound($\cdot$) *returns its input after rewriting it as follows: first, every occurrence of X having one of the forms* $\top, \neg C, \forall R.C, \leq n\,R.C, \exists U.\text{Self}$ *is substituted by* $\text{Inst} \sqcap X$, *where we explicitly allow for complex classes* C. *Next, the universal role is localized by substituting every* $\forall U.C$ *by* $\forall U.(\neg\text{Inst} \sqcup C)$ *and every U occurring on the left hand side of a role chain axiom by* $R_{\text{Inst}} \circ U \circ R_{\text{Inst}}$ *where R_{Inst} is axiomatized via* $\exists R_{\text{Inst}}.\text{Self} \equiv \text{Inst}$.[3] *We extend* bound($\cdot$) *in the obvious way to also rewrite an axiom or a class expression. The functions* **SepDom**, **Typing**, *and* **MatSubClass** *return a set of axioms as specified in Table 2. The metamodeling-enabled version O^{meta} of O is*

$$\text{bound}(O) \cup \textbf{SepDom}(O) \cup \textbf{Typing}(O) \cup \textbf{MatSubClass}(O)$$

---

[3] It is not hard to check that none of these transformations harms the global syntactic constraints.

Table 2. Returned axioms for an ontology O by **SepDom**, **Typing**, and **MatSubClass**

SepDom(O):	`Inst ≡ ¬Class`		(1)
	`Class(o_C)`	for all $C \in O_C$	(2)
	`Inst(i)`	for all $i \in O_I$	(3)
	`∃R.⊤ ⊑ Inst`	for all $R \in O_R$	(4)
	`⊤ ⊑ ∀R.Inst`	for all $R \in O_R$	(5)
	`∃type.⊤ ⊑ Inst`		(6)
	`⊤ ⊑ ∀type.Class`		(7)
	`∃subClassOf.⊤ ⊑ Class`		(8)
	`⊤ ⊑ ∀subClassOf.Class`		(9)
Typing(O):	`C ≡ ∃type.{o_C}`	for all $C \in O_C$	(10)
MatSubClass(O):	`Class ⊓ ∀type⁻.∃type.{o_C} ≡ Class ⊓ ∃subClassOf.{o_C}`		
		for all $C \in O_C$	(11)

Roughly speaking, given an ontology O, the function bound(O) ensures that the complete domain of O is "squeezed" into the class `Inst` and also class construction is forced to only involve individuals from `Inst`. The axioms constructed by **SepDom(O)** have the following purpose: Axiom (1) makes sure that the newly established metalayer does not interfere with the instance layer. Axiom (2) ensures that all class-representative individuals lie in the metalayer. Axiom (3) forces every named individual of the original ontology to be in the instance layer. Axioms (4) and (5) state that every role of the original ontology is forced to start and end only in the instance layer. Axioms (6) and (7) stipulate that the `type`-role starts in the instance layer and ends in the metalayer. Finally, Axiom (8) and (9) specify that the subClassOf(O) role is allowed to interconnect only individuals from the metalayer. The axioms from **Typing(O)** ensure that class members of `C` are exactly those domain individuals which are connected to `C`'s representative o_C via the `type` role, while axioms from **MatSubClass(O)** finally synchronize actual subclass relationships in the instance layer with the subClassOf links between the corresponding representatives in the metalayer.

Note that the size of O^{meta} is linearly bounded by the size of O. Intuitively, the conversion from O to O^{meta} realizes a model conversion: given a model of O, the transformation endows the model with a metalayer containing reified atomic classes o_C. As in RDF, class membership of the original individuals is now indicated by the newly introduced `type` role and class subsumption by the subClassOf role which is axiomatically synchronized with the actually valid subclass relation in the considered model. Thereby, we materialize the hierarchy among classes of a particular model in the metalayer. Figure 1 depicts the established correspondences in a schematic way.

Note that no original model is ruled out by this process which also ensures that the conversion does not cause new (unwanted) consequences. In the sequel, we characterize the above mentioned properties of O^{meta} more formally:

Theorem 1. *Let O be an OWL ontology and O^{meta} its metamodeling-enabled version as specified in Definition 4. Then the following properties hold:*

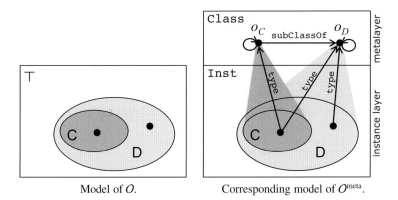

Model of O.   Corresponding model of O^{meta}.

Fig. 1. Sketch of the established interdependencies in the models of O^{meta}

1. *For any OWL axiom a containing only names from* O_C, O_R *and* O_I, *we have that* $O \models a$ *iff* $O^{\mathrm{meta}} \models \mathrm{bound}(a)$.
2. *For any class name* $\mathsf{C} \in O_C$ *and instance name* $i \in O_I$, *we have that* $O \models \mathsf{C}(i)$ *iff* $O^{\mathrm{meta}} \models \mathrm{type}(i, o_\mathsf{C})$.
3. *For any two named classes* $\mathsf{C}, \mathsf{D} \in O_C$, *we have that* $O \models \mathsf{C} \sqsubseteq \mathsf{D}$ *iff* $O^{\mathrm{meta}} \models$ $\mathrm{subClassOf}(o_\mathsf{C}, o_\mathsf{D})$.

Proof. For the first claim, given a model $\mathcal{I}$ of O, we construct a model $meta(\mathcal{I}) = \mathcal{J}$ of O^{meta} as follows:

$$\Delta^{\mathcal{J}} = \Delta^{\mathcal{I}} \cup \{\delta_\mathsf{C} \mid \mathsf{C} \in O_C\}$$
$$\varsigma^{\mathcal{J}} = \varsigma^{\mathcal{I}} \quad \text{for all } \varsigma \in O_C \cup O_R \cup O_I$$
$$\mathrm{type}^{\mathcal{J}} = \{\langle \delta, \delta_\mathsf{C} \rangle \mid \delta \in \mathsf{C}^{\mathcal{I}}\} \qquad \mathrm{Inst}^{\mathcal{J}} = \Delta^{\mathcal{I}}$$
$$\mathrm{Class}^{\mathcal{J}} = \{\delta_\mathsf{C} \mid \mathsf{C} \in O_C\} \qquad \mathrm{subClassOf}^{\mathcal{J}} = \{\langle \delta_\mathsf{C}, \delta_\mathsf{D} \rangle \mid \mathsf{C}^{\mathcal{I}} \subseteq \mathsf{D}^{\mathcal{I}}\}$$

By construction $\mathcal{J}$ satisfies all axioms from **SepDom**$(O) \cup$ **Typing**$(O) \cup$ **MatSubClass**(O). By induction, we obtain, for every class C containing only names from $\mathrm{voc}(O)$, that $\mathrm{bound}(\mathsf{C})^{\mathcal{J}} = \mathsf{C}^{\mathcal{I}}$ (claim †). This in turn guarantees that, for every axiom Ax using only terms from $\mathrm{voc}(O)$, we obtain $\mathcal{I} \models Ax$ iff $\mathcal{J} \models \mathrm{bound}(Ax)$. In particular, $\mathcal{J}$ also satisfies $\mathrm{bound}(O)$, whence it is a model of O^{meta} as claimed.

Using this transformation, we can show that $O^{\mathrm{meta}} \models \mathrm{bound}(Ax)$ implies $O \models Ax$. We demonstrate the case for GCIs. Suppose $O^{\mathrm{meta}} \models \mathrm{bound}(\mathsf{C}) \sqsubseteq \mathrm{bound}(\mathsf{D})$ but $O \not\models \mathsf{C} \sqsubseteq \mathsf{D}$. Then there is a model $\mathcal{I}$ of O with $\mathsf{C}^{\mathcal{I}} \not\subseteq \mathsf{D}^{\mathcal{I}}$. But then there is a model $\mathcal{J} = meta(\mathcal{I})$ with $\mathrm{bound}(\mathsf{C})^{\mathcal{J}} \not\subseteq \mathrm{bound}(\mathsf{D})^{\mathcal{J}}$ (according to †) contradicting our assumption. For the other axiom types, the correspondence can be shown along the same lines.

The other direction ($O \models Ax$ implying $O^{\mathrm{meta}} \models \mathrm{bound}(Ax)$) is shown analogously using the transformation converting models $\mathcal{J}$ of O^{meta} to models $\mathcal{I}$ of O as follows:

$$\Delta^{\mathcal{I}} = \mathrm{Inst}^{\mathcal{J}} \qquad \varsigma^{\mathcal{I}} = \varsigma^{\mathcal{J}} \text{ for all } \varsigma \in O_C \cup O_R \cup O_I$$

Note that the additional axioms of O^{meta} ensure that only individuals from $\Delta^{\mathcal{I}}$ occur in every $\varsigma^{\mathcal{J}}$, whence $\mathcal{I}$ is well-defined. Again we can establish $\mathrm{bound}(\mathsf{C})^{\mathcal{J}} = \mathsf{C}^{\mathcal{I}}$ by induction and use this to show $O^{\mathrm{meta}} \models \mathrm{bound}(Ax)$ implying $O \models Ax$.

Table 3. An example ontology O and its metamodeling-enabled version O^{meta}

Ontology O:	
HouseMouse $\sqsubseteq \exists$eats$^-$.GoldenEagle	Prey $\equiv \exists$eats$^-$.$\top$
HouseMouse(*jerry*)	(GoldenEagle $\sqcup$ HaastsEagle)(*harry*)

	Metaontology O^{meta}:
bound(O):	HouseMouse $\sqsubseteq \exists$eats$^-$.(Inst $\sqcap \neg$GoldenEagle) Prey $\equiv \exists$eats$^-$.Inst
	HouseMouse(*jerry*) (GoldenEagle $\sqcup$ HaastsEagle)(*harry*)
SepDom(O):	Inst $\equiv \neg$Class
	Class($o_{\mathsf{HouseMouse}}$) Class($o_{\mathsf{GoldenEagle}}$) Class(o_{Prey}) Class($o_{\mathsf{HaastsEagle}}$)
	Inst(*jerry*) Inst(*harry*)
	$\exists$eats.$\top \sqsubseteq$ Inst $\top \sqsubseteq \forall$eats.Inst $\exists$type.$\top \sqsubseteq$ Inst $\top \sqsubseteq \forall$type.Class
	$\exists$subClassOf.$\top \sqsubseteq$ Class $\top \sqsubseteq \forall$subClassOf.Class
Typing(O):	HouseMouse $\equiv \exists$type.$\{o_{\mathsf{HouseMouse}}\}$ Prey $\equiv \exists$type.$\{o_{\mathsf{Prey}}\}$
	GoldenEagle $\equiv \exists$type.$\{o_{\mathsf{GoldenEagle}}\}$ HaastsEagle $\equiv \exists$type.$\{o_{\mathsf{HaastsEagle}}\}$
MatSubClass(O):	Class $\sqcap \forall$type$^-$.$\exists$type.$\{o_{\mathsf{HouseMouse}}\} \equiv$ Class $\sqcap \exists$subClassOf.$\{o_{\mathsf{HouseMouse}}\}$
	Class $\sqcap \forall$type$^-$.$\exists$type.$\{o_{\mathsf{Prey}}\} \equiv$ Class $\sqcap \exists$subClassOf.$\{o_{\mathsf{Prey}}\}$
	Class $\sqcap \forall$type$^-$.$\exists$type.$\{o_{\mathsf{GoldenEagle}}\} \equiv$ Class $\sqcap \exists$subClassOf.$\{o_{\mathsf{GoldenEagle}}\}$
	Class $\sqcap \forall$type$^-$.$\exists$type.$\{o_{\mathsf{HaastsEagle}}\} \equiv$ Class $\sqcap \exists$subClassOf.$\{o_{\mathsf{HaastsEagle}}\}$

For the second claim, given $O \models \mathsf{C}(i)$ we can conclude $O^{\mathrm{meta}} \models \mathsf{bound}(\mathsf{C}(i))$ and hence $O^{\mathrm{meta}} \models \mathsf{C}(i)$ from which by **Typing**(O) follows $\mathsf{type}(i, o_\mathsf{C})$. The argument holds in both directions.

For the third claim, we have that from $O \models \mathsf{C} \sqsubseteq \mathsf{D}$ follows $O^{\mathrm{meta}} \models \mathsf{bound}(\mathsf{C}) \sqsubseteq \mathsf{bound}(\mathsf{D})$ and, therefore, $O^{\mathrm{meta}} \models \mathsf{C} \sqsubseteq \mathsf{D}$. Considering a model $\mathcal{J}$ of O^{meta}, **MatSubClass**(O) ensures that $\mathcal{J} \models \mathsf{subClassOf}(o_\mathsf{C}, o_\mathsf{D})$ iff $o_\mathsf{C} \in ($Class $\sqcap \forall$type$^-$.$\exists$type.$\{o_\mathsf{D}\})^{\mathcal{J}}$. This can be simplified to $\{\delta \mid \langle \delta, o_\mathsf{C}^{\mathcal{J}} \rangle \in \mathsf{type}^{\mathcal{J}}\} \subseteq \{\delta \mid \langle \delta, o_\mathsf{D}^{\mathcal{J}} \rangle \in \mathsf{type}^{\mathcal{J}}\}$ which, by **Typing**(O), coincides with $\mathsf{C}^{\mathcal{J}} \subseteq \mathsf{D}^{\mathcal{J}}$ and is true by assumption. Again, the argument holds in both ways. $\square$

As an example, consider the ontology O and its metamodeling-enabled version O^{meta} from Table 3. We find that HouseMouse $\sqsubseteq$ Prey is a consequence of O, whence O^{meta} entails $\mathsf{subClassOf}(o_{\mathsf{HouseMouse}}, o_{\mathsf{Prey}})$. In O^{meta} we can further make statements such as $\mathtt{ExtinctSpecies}(o_{\mathsf{HaastsEagle}})$ where $\mathtt{ExtinctSpecies}$ is a metaclass used to state that Haast's eagle is an extinct species. If we then add the axiom $\mathtt{ExtinctSpecies} \sqsubseteq \forall$type$^-$.$\bot$ to say that extinct species cannot have instances, an OWL 2 DL reasoner can deduce GoldenEagle(*harry*).

In the following, we will illustrate the benefits of our approach on the basis of a more concrete application scenario: the evaluation of ontologies with respect to the OntoClean methodology.

4   OntoClean

This section gives a brief introduction to OntoClean (for a more thorough description refer, e.g., to Guarino and Welty [4]), a methodology developed in order to ensure the correctness of taxonomies with respect to the philosophical principles of Formal Ontology. Central to OntoClean are the notions of **rigidity**, **unity**, **dependence** and **identity**, commonly known as *metaproperties*. Note that in the OntoClean terminology, *properties* are what is called *classes* in OWL. Metaproperties are, therefore, "properties of properties." Consequently, OntoClean can be considered a very natural application of metamodeling in ontology engineering and evaluation.

In the following, we will explain the process of applying the OntoClean methodology by making reference to the OntoClean example ontology introduced by Guarino and Welty ([4], Figure 1). This ontology, which consists of 22 classes such as Apple, Food, Person or Agent, illustrates some of the most frequent modeling errors in terms of OntoClean.

The process of applying the OntoClean methodology to a given ontology consists of two essential phases:

Phase 1: First, every single class of the ontology to be evaluated or redesigned is tagged with respect to the aforementioned metaproperties. This way, every class gets assigned a particular tagging such as +R-D+I+U, denoting the fact that this class is rigid (+R), non-dependent (-D), a sortal (+I, i.e. it carries an identity criterion) and that it has unity (+U).

Phase 2: In the second phase, after the metaproperty tagging has been completed, all the subsumption relationships of the ontology are checked according to a predefined set of OntoClean *constraints*. Any violation of such a constraint potentially indicates a fundamental misconceptualization in the subsumption hierarchy.

Hence, after performing the two steps, the result is a tagged ontology and a (potentially empty) list of misconceptualizations by whose means an ontology engineer can "clean" the ontology. In a nutshell, the key idea underlying OntoClean is to constrain the possible taxonomic relationships by disallowing subsumption relations between specific combinations of tagged classes. Welty et al. [13] show that analyzing and modifying an ontology according to the quality criteria defined by OntoClean can have a positive impact on the performance of an ontology-based application.

Metaproperties. As mentioned above, the original version of OntoClean is based on four metaproperties: rigidity, unity, identity and dependence – abbreviated as R, U, I and D, respectively. For brevity, we focus on rigidity omitting detailed explanations of the other metaproperties and referring the interested reader to, e.g., [4].

Rigidity is based on the notion of *essence*. A class is essential for an individual *iff* the individual is necessarily a member of this class, in all worlds and at all times. *Iff* a class is essential to all of its individuals, the class is called *rigid* and is tagged with +R. *Non-rigid* classes, i.e., classes which are not essential to some of their individuals, are tagged with -R. An anti-rigid class is one that is not essential to all of its individuals and thus tagged with ~R. Hence, every anti-rigid class is also a non-rigid class. Apple is a typical example for a rigid class, because the property of being an apple is *essential* to all of its individuals, or to put it differently: an apple is necessarily an apple and cannot stop being one. In this respect, Apple differs from classes such as Food which is mostly considered anti-rigid. Note, however, that the tagging of Food crucially hinges on the intended semantics of this class. Welty [14] nicely illustrates this by an example: If Food is the class of all things edible by humans then it should be tagged as rigid (+R). If, in contrast, Food is a role that can be played by any individual while it is being eaten, we must consider it anti-rigid (~R). The latter sense was assumed by Guarino and Welty when they designed the aforementioned example ontology.

Constraints. The following formulation of the OntoClean constraints is literally taken from Welty [14]. We adhere to this version rather than to the more stringent formulation provided by Guarino and Welty [4] as it directly maps to the axiomatization in the original OntoClean metaontology (cf. Section 5).

1. A rigid class (+R) cannot be a subclass of an anti-rigid class (~R).
2. A class with unity (+U) cannot be a subclass of a class with anti-unity (~U).
3. All subclasses of a sortal are sortals (+I).
4. All subclasses of a dependent class are dependent (+D).

What seems a matter of merely philosophical consideration can in fact have practical implications. Imagine, for instance, that a rigid class Apple is subsumed by Food which is tagged as anti-rigid. Thus, Apple(a) would imply Food(a). It might also appear reasonable to model the class Poisoned as disjoint to Food. But now, as the ontology evolves and further class instantiations are added, we could state, for example, that a has been poisoned (formally, Poisoned(a)) – and the ontology turns logically inconsistent.

5   OWL-Based Constraint Checking

Despite the fact that OntoClean is the single most well-known and theoretically founded methodology for evaluating the formal correctness of subsumption hierarchies, there has always been a lot of criticism regarding the high costs for tagging and constraint checking. To address this criticism and to make the OntoClean methodology more easily applicable in practical ontology engineering settings, Welty suggested an OWL-based formalization of metaproperty assignments and constraints [14], which leverages logical inconsistencies as indicators of constraint violations. We discuss this framework next and then introduce our novel metamodeling in this setting.

5.1   The Original OntoClean Metaontology

Welty's formalization, occasionally referred to as *OntOWLClean*, axiomatizes the aforementioned OntoClean constraints as domain-range restrictions on a transitive object property subClassOf, which serves as a replacement for the normal subclass relation ($\sqsubseteq$) and enables a reification of every subsumption relationship in a given domain ontology. For example, instead of writing Apple $\sqsubseteq$ Food to state that Apple is a subclass of Food, we write subClassOf($o_{\text{Apple}}, o_{\text{Food}}$) introducing fresh individuals for the classes (e.g., o_{Apple} for Apple). Metaproperty assignments then become class membership assertions, e.g., we write RigidClass(o_{Apple}). Similarly, we can state that o_{Apple} is a class by adding the assertion Class(o_{Apple}). Note that in order to have true metamodeling one would have to state that subClassOf is the same as $\sqsubseteq$ and that Class really implies that its instances are classes. This is possible in OWL Full, but not in OWL DL. Since the reasoning support for OWL Full is limited, the axioms required to equate subClassOf and Class with their OWL modeling constructs are available

Table 4. Fragment of the OntoClean metaontology [14]

$$
\begin{aligned}
\text{RigidClass} &\sqsubseteq \text{Class} \\
\text{NonRigidClass} &\sqsubseteq \text{Class} \\
\text{AntiRigidClass} &\sqsubseteq \text{NonRigidClass} \\
\text{Class} &\equiv (\text{NonRigidClass} \sqcup \text{RigidClass}) \sqcap \\
&\quad (\text{NonDependentClass} \sqcup \text{DependentClass}) \sqcap \\
&\quad (\text{SortalClass} \sqcup \text{NonSortalClass}) \sqcap \\
&\quad (\text{UnityClass} \sqcup \text{NonUnityClass}) \\
\text{NonRigidClass} \sqcap \text{RigidClass} &\sqsubseteq \bot \\
\text{RigidClass} &\sqsubseteq \forall \text{subClassOf}.\neg\text{AntiRigidClass}
\end{aligned}
$$

...

as a complementary OWL Full ontology, which is kept separate from the core of the OntOWLClean ontology.[4] Table 4 shows an excerpt from the OWL DL part of the ontology focussing on the axioms for rigidity.

Note that if we now state that **Apple** is rigid and **Food** is anti-rigid the metaontology becomes inconsistent as witnessed by the following set of axioms:

$$\text{RigidClass} \sqsubseteq \forall \text{subClassOf}.\neg\text{AntiRigidClass} \qquad \text{RigidClass}(o_{\text{Apple}})$$
$$\text{subClassOf}(o_{\text{Apple}}, o_{\text{Food}}) \qquad \text{AntiRigidClass}(o_{\text{Food}})$$

OntOWLClean has been a great step forward when it comes to the practical applicability of the OntoClean methodology, because it enables the use of standard DL reasoning for detecting constraint violations (see also [11]). However, the axiomatization of metaproperty assignments and constraints suggested by Welty has at least two drawbacks which can be overcome now that OWL 2 is available:

First, while syntactically metaproperty assignments and constraints could be part of the same ontology as classes and subsumption axioms, there is no semantic link between a class and its corresponding individual (e.g., the class **Apple** and the individual o_{Apple} being a member of **RigidClass** in the metaontology). Hence without the OWL Full part of the axiomatization, OntOWLClean does not allow for integrated reasoning over classes and their metaproperties. Furthermore, any changes to the subsumption hierarchy (in case of constraint violations, for example) involve modifications of two logically unrelated taxonomies, possibly maintained in two different files.

Second, the computational costs of determining the reasons for logical inconsistencies and thus constraint violations can be very high. The typical way of debugging an inconsistent ontology is to compute minimal subsets of the ontological axioms, which preserve the inconsistency. These subsets are called *explanations, justifications*, or *minAs*. In order to compute the explanations, axioms are removed from the original ontology in a step by step manner, while after each removal a reasoner is used to check whether the remaining set of axioms is still inconsistent. This process is repeated until no further axiom can be removed without turning the ontology consistent. Users presented with

---

[4] Note that the last axiom in Table 4 is specified as an equivalence in http://www.ontoclean. org/ontoclean-dl-v1.owl. We assumed this to be a mistake as equivalence is not used for modeling any of the constraints in Welty's paper [14], and corrected the ontology accordingly.

Table 5. An explanation for the conflict caused by Apple being rigid and Food being anti-rigid

$$\text{Apple} \sqsubseteq \text{Food}$$
$$\text{RigidClass} \sqsubseteq \forall \text{subClassOf}.(\neg \text{AntiRigidClass})$$
$$\exists \text{type}.\top \sqsubseteq \text{Inst}$$
$$\text{Apple} \equiv \exists \text{type}.\{o_{\text{Apple}}\}$$
$$\text{Food} \equiv \exists \text{type}.\{o_{\text{Food}}\}$$
$$\text{Class} \sqcap \forall \text{type}^-.\exists \text{type}.\{o_{\text{Food}}\} \equiv \text{Class} \sqcap \exists \text{subClassOf}.\{o_{\text{Food}}\}$$
$$\text{RigidClass}(o_{\text{Apple}})$$
$$\text{AntiRigidClass}(o_{\text{Food}})$$
$$\text{Class}(o_{\text{Apple}})$$

the explanations can then decide how to fix the ontology. In particular computing *all* such explanations is a computationally hard task. Due to the high costs, only limited tool support for inconsistency diagnosis in OWL is available.

5.2   Towards OntOWL2Clean

In order to address the first issue, we extend O^{meta} from Table 3 by a set of classes and axioms which enable us to express all of the OntoClean metaproperty assignments and constraints. In particular, we have the following axioms for the constraints:

$$\text{RigidClass} \sqsubseteq \forall \text{subClassOf}.\neg \text{AntiRigidClass} \qquad \text{(C1)}$$
$$\text{UnityClass} \sqsubseteq \forall \text{subClassOf}.\neg \text{AntiUnityClass} \qquad \text{(C2)}$$
$$\text{SortalClass} \sqsubseteq \forall \text{subClassOf}^-.\text{SortalClass} \qquad \text{(C3)}$$
$$\text{DependentClass} \sqsubseteq \forall \text{subClassOf}^-.\text{DependentClass} \qquad \text{(C4)}$$

We can now add the OntoClean taggings to the classes making use of the class individuals in O^{meta}. For example, since we assume Food to be anti-rigid while Apple is rigid, we add the following facts to O^{meta}:

$$\text{AntiRigidClass}(o_{\text{Food}}) \qquad \text{RigidClass}(o_{\text{Apple}})$$

Note that we do not have to add $\text{subClassOf}(o_{\text{Apple}}, o_{\text{Food}})$ explicitly since the subClassOf role between o_{Apple} and o_{Food} is implied in O^{meta}. Since constraint (C1) prevents an anti-rigid class from being a subclass of a rigid one, the ontology becomes inconsistent, witnessed by the explanation shown in Table 5.

One should be aware that adding constraints might have a "backward" impact on the semantics of the "original part" of the ontology in that they could rule out certain models thereby leading to additional consequences. To see this, assume our ontology has been corrected by removing Apple $\sqsubseteq$ Food. Axiom (C1) still enforces that Apple must not be a subclass of Food and we have as a consequence that the extension of Apple must be nonempty in every model. This is because the empty set is trivially a subset of every set and, in particular, a subset of the extension of Food. Depending on the concrete scenario, these ramifications might be unwanted or intended. They can be avoided by using the approach based on marker predicates described next.

6   Marker Predicates for Pinpointing Constraint Violations

The above method implements Welty's approach of specifying the constraints in a way that their violation results in an inconsistent knowledge base. In order to actually find and identify the reasons for these inconsistencies, diagnosis techniques [7,5] have to be employed. Typically these diagnosis techniques are rather costly as they require numerous calls to a reasoning system.

We argue that in certain cases, an alternative approach can be employed, wherein violations of OntoClean constraints do not cause the ontology to become inconsistent but lead to the creation of *marker classes* or *roles* that indicate which ontology elements are involved in a constraint violation. This alternative method can be combined with the original two-ontology approach as well as with our metamodeling technique.

Consider a constraint that prohibits $C \sqsubseteq D$ whenever C is endowed with the metaproperty T_1 and D is endowed with T_2. By specifying the axiom $T_1 \sqsubseteq \forall \mathsf{subClassOf}.\neg T_2$, we would turn an ontology inconsistent whenever it entails $T_1(o_C)$, $T_2(o_D)$, and $\mathsf{subClassOf}(o_C, o_D)$. Consequently, diagnosis would be required to locate the violated constraint. Instead, we propose to establish an auxiliary marker role $\mathsf{conflictsWith}$ between o_C and o_D in this case. Thereby, all conflicts can be readily spotted by simply retrieving all entailed $\mathsf{conflictsWith}$ role memberships. This wanted correspondence can be logically enforced in OWL 2 using an encoding introduced independently in [8] and [2] which makes use of additional auxiliary roles t_1, t_2 as well as some of the advanced features of $\mathcal{SROIQ}$:

$$T_1 \sqsubseteq \exists t_1.\mathsf{Self} \qquad T_2 \sqsubseteq \exists t_2.\mathsf{Self} \qquad t_1 \circ \mathsf{subClassOf} \circ t_2 \sqsubseteq \mathsf{conflictsWith}$$

In order to axiomatize the OntoClean constraints, we introduce a fresh role mp for each OntoClean metaproperty $\mathsf{mpClass}$ (e.g., rigid for $\mathsf{RigidClass}$) and an axiom

$$\mathsf{mpClass} \sqsubseteq \exists \mathsf{mp}.\mathsf{Self} \tag{M1}$$

where mp is the fresh role associated with $\mathsf{mpClass}$. We can then axiomatize the four OntoClean constraints (C1) to (C4) with the following role chain axioms, where we use one marker per conflict type:

$$\mathsf{rigid} \circ \mathsf{subClassOf} \circ \mathsf{antiRigid} \sqsubseteq \mathsf{rigidityConflict} \tag{M2}$$
$$\mathsf{unity} \circ \mathsf{subClassOf} \circ \mathsf{antiUnity} \sqsubseteq \mathsf{unityConflict} \tag{M3}$$
$$\mathsf{nonDependent} \circ \mathsf{subClassOf} \circ \mathsf{dependent} \sqsubseteq \mathsf{dependencyConflict} \tag{M4}$$
$$\mathsf{nonSortal} \circ \mathsf{subClassOf} \circ \mathsf{sortal} \sqsubseteq \mathsf{sortalConflict} \tag{M5}$$

For each role r on the right-hand side of Axioms (M2) to (M5)

$$r \sqsubseteq \mathsf{conflictsWith} \tag{M6}$$

Note that for (C3) and (C4) we use an equivalent formulation which is better suited for using the marker properties. In order to allow for retrieving all conflicts at once, we further introduce $\mathsf{conflictsWith}$ as a superrole of all the roles on the right-hand side of the above axioms.

7   Evaluation

We used the OntoClean example ontology [4] to test the different approaches. This leaves us with four settings: we first test Welty's metamodeling (see Section 5) and our metamodeling (see Section 3) with Explanations for discovering the modeling mistakes in settings Ex1 and Ex2, then we test the two approaches with the new Marker predicates (see Section 6) in settings Ma1 and Ma2.

Ex1   Our baseline is the metamodeling part of the example OntoClean ontology, i.e., we use the metamodeling as proposed by Welty. It is worth noting that we use the weakened OWL DL version of the original OWL Full meta ontology to be able to use OWL DL reasoners. For each axiom, e.g., Apple $\sqsubseteq$ Food in the original ontology, the meta version contains an assertion $\texttt{subClassOf}(o_{\mathsf{Apple}}, o_{\mathsf{Food}})$ with o_{Apple} and o_{Food} individuals and $\texttt{subClassOf}$ a role. Furthermore, the meta ontology contains the taggings such as $\texttt{RigidClass}(o_{\mathsf{Apple}})$ and the OntoClean constraints from Axioms (C1) to (C4).

Ex2   The second ontology is the metamodeling version of the OntoClean ontology according to our novel metamodeling technique. Since we have no separation between the metamodeling part and the original axioms, the ontology contains assertions of either kind: adjusted axioms from the source ontology as well as the taggings and the OntoClean constraints from Axioms (C1) to (C4).

Ma1   In this setting, we use the marker predicates to manifest modeling errors instead of inconsistencies. The ontology uses Welty's metamodeling as in setting Ex1 and contains the taggings, but instead of causing inconsistencies by adding Axioms (C1) to (C4), we use marker predicates as described in Section 6 and add Axioms (M1) to (M6).

Ma2   The last setting uses our novel metamodeling approach in combination with the marker predicates, i.e., we use an ontology as in setting Ex2, but instead of causing inconsistencies by adding Axioms (C1) to (C4), we again use marker predicates and add Axioms (M1) to (M6).

In order to find all potential modeling errors in the settings Ex1 and Ex2, we use the explanation framework by Horridge et al. [5] and we generate all minimal subsets O' of the ontology such that O' is inconsistent. For the settings Ma1 and Ma2, we retrieve instances of the roles that indicate a conflict. All tests have been performed using the OWL 2 DL reasoner HermiT. The ontologies, HermiT 1.2.4, the obtained results, and the program used to produce the results are available online.[5] The tests have been performed on a MacBook Air with Java 1.6, assigning 1GB memory to Java.

Both explanation approaches ran out of memory after 2.5 days, generating 51 explanations for setting Ex1 and 46 explanations for setting Ex2, making the approach not really feasible in practice. Although the first explanations are generated quickly, the later ones can take significant time and memory. We repeated the tests of setting Ex1 and Ex2 on a node of the Oxford Supercomputing Centre, assigning 24GB of main memory to Java. We terminated the programs after one week, getting 53 explanations for setting Ex1 and 66 for setting Ex2. By analyzing the ontology manually, we find that

---

[5] http://www.hermit-reasoner.com/2010/metamodeling/metamodeling.zip

Table 6. Time in seconds for retrieving instances of the marker properties

setting	rigidity	unity	conflict dependency	sortal	all
Ma1	< 1	< 1	< 1	< 1	< 1
Ma2	21	355	20	56	452

there should be 53 explanations for setting Ex1 and we assume that the code attempted to find more explanations without success.

There are more explanations for the new metamodeling approach since an explanation might contain the meta axioms for the involved classes only partially. In such a case, the explanation contains additional axioms that are not directly related to the conflict, but which contain enough meta axioms to cause the clash for the only partially axiomatized real inconsistency cause.

The novel marker approaches (Ma1 and Ma2) both find 40 conflicts: 10 rigidity conflicts, 16 unity conflicts, 12 dependency conflicts, and 2 identity conflicts. The timings are given in Table 6 and were averaged over 3 runs of the reasoner. It can be observed that the times for our new metamodeling approach are significantly slower than the ones for the original approach. This is a consequence of the more complex axiomatization that is required in order to achieve real metamodeling in OWL 2 DL, whereas in setting Ma1, the reasoner only works on a part of the ontology that suffices to detect these conflicts. Full metamodeling in the settings Ex1 and Ma1 requires an OWL Full reasoner and, as Welty states [14], a satisfactory implementation that can handle the full meta ontology is not (yet) available.

The number of marker conflicts is lower than the number of explanations because for several indirect subclass relationships, there are different ways of deriving the subsumption. E.g., the ontology contains:

$$\text{Organisation} \sqsubseteq \text{SocialEntity} \qquad \text{SocialEntity} \sqsubseteq \text{Agent}$$
$$\text{Organisation} \sqsubseteq \text{LegalAgent} \qquad \text{LegalAgent} \sqsubseteq \text{Agent}$$

In both cases, we have that Organisation is a subclass of Agent. Since we further have that Organisation is a rigid class, while Agent is anti-rigid, we have one explanation using the first set of subclass axioms and another explanation using the second set of subclass axioms, whereas the marker approach does not distinguish the two cases.

8   Conclusion

We have presented a novel approach to ontology-inherent metamodeling for classes in OWL based on an axiomatization of class reification. This approach allows for associating information to classes and asserting constraints on the subclass hierarchy in a way that allows for the usage of standard OWL reasoning tools. We demonstrated our approach by applying it to the OntoClean methodology. We found that the benefits of our approach in terms of maintenance and tight logical integration may come at the cost of runtime performance. On the other hand, we showed that performance can be increased by several orders of magnitude if the explanation-based diagnosis originally proposed for OntoClean is substituted by a novel consistency-preserving approach working with marker predicates that indicate potential modeling flaws in the ontology. The runtime improvements thus obtained outweigh the metamodeling-induced slowdown by far.

Acknowledgements. Birte Glimm is funded by the EPSRC project HermiT: Reasoning with Large Ontologies. Sebastian Rudolph is supported by the German Research Foundation (DFG) under the ExpresST project. Johanna Völker is financed by a Margarete-von-Wrangell scholarship of the European Social Fund (ESF) and the Ministry of Science, Research and the Arts Baden-Württemberg. The evaluation has been performed on computers of the Oxford Supercomputing Centre.

References

1. Baader, F., Calvanese, D., McGuinness, D., Nardi, D., Patel-Schneider, P. (eds.): The Description Logic Handbook: Theory, Implementation and Applications. Cambridge University Press, Cambridge (2007)
2. Gasse, F., Sattler, U., Haarslev, V.: Rewriting rules into $\mathcal{SROIQ}$ axioms. In: Poster at 21st International Workshop on Description Logics, DL (2008)
3. Glimm, B., Rudolph, S., Völker, J.: Integrated metamodeling and diagnosis in owl 2. Tech. Rep. 3006, Institut AIFB, KIT, Karlsruhe (September 2010), http://www.aifb.kit.edu/web/Techreport3006
4. Guarino, N., Welty, C.A.: An Overview of OntoClean. In: International Handbook on Information Systems, 2nd edn., pp. 201–220. Springer, Heidelberg (2009)
5. Horridge, M., Parsia, B., Sattler, U.: Explaining inconsistencies in OWL ontologies. In: Godo, L., Pugliese, A. (eds.) SUM 2009. LNCS, vol. 5785, pp. 124–137. Springer, Heidelberg (2009)
6. Horrocks, I., Kutz, O., Sattler, U.: The even more irresistible $\mathcal{SROIQ}$. In: Proc. 10th Int. Conf. on Principles of Knowledge Representation and Reasoning (KR 2006), pp. 57–67. AAAI Press, Menlo Park (2006)
7. Ji, Q., Haase, P., Qi, G., Hitzler, P., Stadtmüller, S.: RaDON – Repair and diagnosis in ontology networks. In: Proceedings of the European Semantic Web Conference (ESWC), pp. 863–867. Springer, Heidelberg (2009)
8. Krötzsch, M., Rudolph, S., Hitzler, P.: Description logic rules. In: Ghallab, M., Spyropoulos, C.D., Fakotakis, N., Avouris, N.M. (eds.) ECAI. Frontiers in Artificial Intelligence and Applications, vol. 178, pp. 80–84. IOS Press, Amsterdam (2008)
9. Motik, B.: On the properties of metamodeling in OWL. Journal of Logic and Computation 17(4), 617–637 (2007)
10. Tran, T., Haase, P., Motik, B., Grau, B.C., Horrocks, I.: Metalevel information in ontology-based applications. In: Fox, D., Gomes, C.P. (eds.) AAAI, pp. 1237–1242. AAAI Press, Menlo Park (2008)
11. Völker, J., Vrandečić, D., Sure, Y., Hotho, A.: AEON – An approach to the automatic evaluation of ontologies. Journal of Applied Ontology 3(1-2), 41–62 (2008)
12. Vrandečić, D.: Ontology Evaluation. In: International Handbook on Information Systems, 2nd edn., pp. 293–313. Springer, Heidelberg (2009)
13. Welty, C., Mahindru, R., Chu-Carroll, J.: Evaluating ontology cleaning. In: McGuinness, D.L., Ferguson, G. (eds.) Proc. 19th National Conf. on AI (AAAI) and 16th Conf. Innovative Applications of AI (IAAI). MIT Press, Cambridge (July 2004)
14. Welty, C.: OntOWLClean: Cleaning OWL ontologies with OWL. In: Bennet, B., Fellbaum, C. (eds.) Proceedings of Formal Ontologies in Information Systems (FOIS), pp. 347–359. IOS Press, Amsterdam (2006)

Semantic Recognition of Ontology Refactoring

Gerd Gröner, Fernando Silva Parreiras, and Steffen Staab

WeST — Institute for Web Science and Technologies
University of Koblenz-Landau
{groener,parreiras,staab}@uni-koblenz.de

Abstract. Ontologies are used for sharing information and are often collaboratively developed. They are adapted for different applications and domains resulting in multiple versions of an ontology that are caused by changes and refactorings. Quite often, ontology versions (or parts of them) are syntactical very different but semantically equivalent. While there is existing work on detecting syntactical and structural changes in ontologies, there is still a need in analyzing and recognizing ontology changes and refactorings by a semantically comparison of ontology versions. In our approach, we start with a classification of model refactorings found in software engineering for identifying such refactorings in OWL ontologies using DL reasoning to recognize these refactorings.

1 Introduction

Ontologies share common knowledge and are often developed in distributed environments. They are combined, extended and reused by other users and knowledge engineers in different applications. In order to support reuse of existing ontologies, remodeling and changes are unavoidable and lead to different ontology versions. Quite often, ontology engineers have to compare different versions and analyze or recognize changes. In order to improve and ease the understandability of changes, it is more beneficial for an engineer to view a more abstract and high-level change description instead of a large number of changed axioms (elementary changes) or ontology version logs like in [1]. Combinations of elementary syntactic changes into more intuitive change patterns are described as refactorings [2] or as composite changes [3].

However, the recognition of refactorings (or changes in general) is difficult due to the variety of possible changes that may be applied to an ontology. Especially if the comparison of different ontology versions is not only realized by a pure syntactical comparison, e.g. a comparison of triples of an ontology, but rather by a semantic comparison of entities in an ontology and their structure.

The need to detect high-level changes is already stated in [1,4,5]. High-level understanding of changes provides a foundation for further engineering support like visualization of changes and extended pinpointing focusing on entailments of refactorings rather than individual axiom changes. In order to tackle the described problem, the following issues need to be thoroughly investigated: (i) A high-level categorization of ontology changes like the well established refactoring

P.F. Patel-Schneider et al. (Eds.): ISWC 2010, Part I, LNCS 6496, pp. 273–288, 2010.
© Springer-Verlag Berlin Heidelberg 2010

patterns in software engineering. (ii) An automatic recognition of refactorings for OWL ontologies that goes beyond mere syntactic comparisons.

The recognition of refactorings is a challenging task due to the variety of possible changes and insufficient means for a semantic comparison of ontology versions. In particular, we identify the issue that we require a semantic comparison of different versions of classes rather than their syntactical comparison. Semantic comparison allows for taking available background knowledge into account while abstracting from elementary changes.

There are different approaches that detect ontology changes by a syntactical comparison like in [4,6] or the combination of adding and deleting RDF-triples to high-level changes in [5]. A structural comparison using matching algorithms is considered in [7]. More related to our research is the work on version reasoning for ontologies in [8,9]. However, their focus is on integrity checking, entailment propagation between versions and consistency checking of ontology mappings.

In this paper, we tackle the problem of refactoring recognition using description logics (DL) reasoning in order to semantically compare different versions of an OWL DL ontology. We apply the semantic comparison in heuristic algorithms to recognize refactoring patterns. Extrapolating from [2,3] we have defined different refactoring patterns of how OWL ontologies may evolve.

We organize this paper as follows. Sect. 2 motivates the problem of schema changes and describes shortcomings of existing approaches. In Sect. 3, we give an overview of the considered refactoring patterns and describe in detail two of them. The comparison of ontology versions and the recognition of the refactoring patterns using DL reasoning is demonstrated in Sect. 4 and 5. The evaluation is given in Sect. 6. We analyze related work in Sect. 7, followed by the conclusion.

2   An Ontology Refactoring Scenario

In order to clarify the problem we tackle, we start with a motivating example that highlights the problem followed by some argumentation in favor of a semantic version comparison for recognizing refactorings.

2.1   Motivating Example

In this section, we consider an ontology change from version V to $V\prime$ including multiple elementary changes. An example is displayed in Fig. 1 and 2. Snippets of the corresponding ontology versions are depicted below. In order to highlight the changed axioms in the example, we mark axioms that are deleted from version V with (d) and axioms that are added are marked with (a) at the end of the line. *Person, Employee, Project, ContactData* and *Department* are OWL classes, *Employee* is a subclass of *Person*. The properties *name, SSN, telephone* and *address* are datatype properties with range *string* and *project, department* and *contact* are object properties.

We recognize three refactorings from version V to $V\prime$. First: the pattern Move Of Property moves a datatype property restriction SSN from class *Employee* to

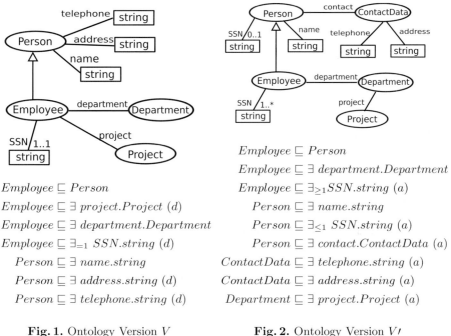

$$Employee \sqsubseteq Person$$
$$Employee \sqsubseteq \exists\, project.Project \; (d)$$
$$Employee \sqsubseteq \exists\, department.Department$$
$$Employee \sqsubseteq \exists_{=1} SSN.string \; (d)$$
$$Person \sqsubseteq \exists\, name.string$$
$$Person \sqsubseteq \exists\, address.string \; (d)$$
$$Person \sqsubseteq \exists\, telephone.string \; (d)$$

$$Employee \sqsubseteq Person$$
$$Employee \sqsubseteq \exists\, department.Department$$
$$Employee \sqsubseteq \exists_{\geq 1} SSN.string \; (a)$$
$$Person \sqsubseteq \exists\, name.string$$
$$Person \sqsubseteq \exists_{\leq 1} SSN.string \; (a)$$
$$Person \sqsubseteq \exists\, contact.ContactData \; (a)$$
$$ContactData \sqsubseteq \exists\, telephone.string \; (a)$$
$$ContactData \sqsubseteq \exists\, address.string \; (a)$$
$$Department \sqsubseteq \exists\, project.Project \; (a)$$

Fig. 1. Ontology Version V **Fig. 2.** Ontology Version $V\prime$

its superclass *Person*. In version V there are implicitly two cardinality restrictions in the property restriction $\exists_{=1} SSN.string$. This is semantically equivalent with the restrictions $\exists_{\leq 1} SSN.string$ and $\exists_{\geq 1} SSN.string$. The datatype property restriction with the maximal cardinality restriction is moved to the superclass *Person*. Second: Extract Class moves the datatype properties *address* and *telephone* to a newly created class *ContactData* that does not contain further properties. In version $V\prime$ the class *Person* has a further object property restriction on *contact* with range *ContactData*. Third: Move Of Property moves an object property *project* from the class *Employee* to the class *Department*.

As demonstrated in the ontology excerpt below the diagrams, the refactorings are syntactically represented by a number of added and deleted axioms from version V to $V\prime$. E.g., the movement of the property SSN from *Employee* to its superclass is represented in the ontology by the deleted axiom $Employee \sqsubseteq \exists_{=1} SSN.string$ and the added axioms $Person \sqsubseteq \exists_{\leq 1} SSN.string$ and $Employee \sqsubseteq \exists_{\geq 1} SSN.string$. In order to improve the understanding and recognition of changes between ontology versions, we argue that it is more intuitive for the ontology engineer to characterize changes at a higher abstraction level like by the recognition of refactorings instead of indicating a large collection of added and deleted axioms. For instance, consider the second mentioned refactoring which extracts the datatype properties *address* and *telephone* to the newly created class *ContactData*. Obviously, such a high-level change characterization is more intuitive for an ontology engineer than a listing of changed axioms. In this refactoring at least two axioms are deleted and three axioms are added to the ontology.

2.2   Discussion of Shortcomings

We already argued for the need of a semantic comparison of the versions rather than a syntactic or a purely structural comparison. This is mainly due to the various possibilities of defining classes in OWL compared to RDF(S) like class definitions using intersection, union or property restrictions. We give two examples of shortcomings for syntactical and structural comparisons.

Consider again the third refactoring (Move Of Property) from Fig. 1 and 2. Breaking down this refactoring to axiom changes, we would delete the axiom $Employee \sqsubseteq \exists\ project.Project$ and add the axiom $Department \sqsubseteq \exists\ project.Project$. Now, we slightly extend this refactoring. Suppose there are two subclasses of $Department$, $InternalDepartment$ and $ExternalDepartment$ and the property restriction $\exists\ project.Project$ is moved to both subclasses $InternalDepartment$ and $ExternalDepartment$ rather than to the superclass $Department$. In this case, the ontology contains two new axioms and one is still removed. If there is a further axiom in the ontology describing that each department is either an internal or an external department ($Department \equiv InternalDepartment \sqcup ExternalDepartment$) and there is no other department, we can conclude that after the refactoring $project$ is still a property of $Department$. Hence, we identify a refactoring that moves a property ($project$) from a class to another class ($Department$) but without changing an axiom that contains the class itself.

As a second example, we demonstrate shortcomings of structural (and frame-based) comparisons which compare classes and their connections, i.e. domain and range of properties. Consider again the move of the datatype property SSN with maximal cardinality restriction from the class $Employee$ to $Person$. Here, we compare the class $Employee$ in both versions. The cardinality restriction that restricts the class $Employee$ to exactly one SSN is explicitly stated in version V. Semantically, in version $V\prime$ the restriction for class $Employee$ is exactly the same due to inheritance and the conjunction of the minimal and maximal restrictions which also results in exactly one SSN property. This equivalence of the class $Employee$ in both versions is not detected by a purely structural comparison.

3   Modeling and Categorizing Refactoring Patterns

A first step towards the recognition of refactoring patterns is the categorization of well-known patterns, adopted from [2] and also presented as *composite changes* for ontology evolution in RDF(S) [3]. Hereafter, we demonstrate two such refactoring patterns in detail.

3.1   Modeling Foundations and Assumptions

For a more compact notation, we describe a class in version V with C and we use $C\prime$ to refer to this class in version $V\prime$. The class of the range of an object property restriction is called the referenced class. A refactoring pattern is an abstract description of an ontology change or evolution that is applied to realize a certain ontology remodeling. The kind of remodeling is mainly a collection of *best practise* ontology remodeling steps. A refactoring is an instantiation of a refactoring pattern, i.e. a concrete change of an ontology.

Our recognition approach works correctly for a slightly restricted subset of OWL DL where we add two restrictions (Def. 1). The second restriction is also known from OWL Lite (cf. [10]). Both restrictions are necessary in order to avoid exponential computation complexity or even infinite computations in the proposed algorithms that are used in Sect. 4.2 like the ExtractReferenceClasses-Algorithm, e.g., if there are further object property restrictions that appear in the range of another object property restriction.

Definition 1 (Language Restrictions). *We restrict OWL DL ($\mathcal{SHOIN}(D)$) by the following additional conditions:*

1. *In each property restriction $\exists p.E$ and $\forall p.E$, E is a named class. The same condition is also required for cardinality restrictions.*
2. *Individuals are not allowed in class definitions, i.e. no oneOf constructor.*

3.2 Overview of Refactoring Pattern

We start with an overview of the analyzed refactoring patterns and describe how they change an ontology (cf. Table 1). They are adopted from [2,3].

The first group of refactorings (No. 1-6) extract or merge classes and move properties to or from the extracted or deleted class. Extract Subclass and Extract Superclass are specializations of Extract Class. The refactorings in the second group (No. 7-9) move properties between existing classes. In No. 8 and 9, the properties are moved within a class hierarchy. Finally, the third group collects refactorings that add, delete or modify object property restrictions. Either the inverse object property is added or removed to a class description (No. 10, 11) or in No. 12 cardinality restrictions are modified.

3.3 Detailed Refactoring Descriptions

In this subsection, we give detailed descriptions of two refactoring patterns (Extract Class and Move of Property) and example representations in OWL in order to substantiate our approach. The recognition algorithms and the results for these two examples are given later on in Sect. 4. A comprehensive description of the other considered patterns from Table 1 and the recognition of them is presented in [11]. A refactoring pattern consists of the following elements:

1. Each pattern has a *Name* (cf. pattern overview in Table 1).
2. The *Problem Description* characterizes a modeling structure of an ontology and indicates when this pattern is applicable.
3. The *Solution* describes how the problem is (or could be) solved. This contains the required remodeling steps in order to realize the refactoring.
4. The *Example* demonstrates the technical details of this refactoring.

Table 1. Analyzed Refactoring Patterns

No.	Pattern Name	Description
1.	Extract Class	Properties of a class are extracted to a newly created class.
2.	Extract Subclass	Properties of a class are extracted to a newly created subclass.
3.	Extract Superclass	Properties of a class are extracted to a newly created superclass.
4.	Collapse Hierarchy	A subclass and its superclass are merged to one class.
5.	Extract Hierarchy	A class is divided into a class hierarchy. Properties are extracted to the newly created sub- and superclasses.
6.	Inline Class	A class that is referenced by another class is deleted and all its properties are moved to the class that had referenced this class.
7.	Move Of Property	At least one property is moved from a class to a referenced class.
8.	Pull-Up Property	At least one property is moved from a class to its superclass.
9.	Push-Down Property	At least one property is moved from a class to its subclass.
10.	Unidirectional to bidirectional Reference	An object property restriction is added to the target class of an existing object property restriction, where the object property is the inverse property.
11.	Bidirectional to unidirectional Ref.	The inverse property restriction of an object property restriction is removed.
12.	Cardinality Change	The cardinality restriction of a property restriction is changed.

Extract Class Refactoring Pattern. An example of the Extract Class refactoring and the corresponding DL representation is already given in the running example from Fig. 1 and 2 in Sect. 2.

Problem Description. In version V, there is a named class C with property restrictions containing the properties $p_1, \ldots, p_n$. An ontology engineer identifies some of the properties $p_{i_1}, \ldots, p_{i_n}$ ($\{p_{i_1}, \ldots, p_{i_n}\} \subseteq \{p_1, \ldots, p_n\}$) that are related to this class but should be grouped together and extracted into a new class D. Finally, a property restriction from class C to the new class D is needed.

Solution. A new class D is created and all the selected property restrictions on $p_{i_1}, \ldots, p_{i_n}$ are moved from C to D. An axiom for the object property restriction on p to the new class D is added, e.g. the axiom $C \sqsubseteq \exists p.D$.

Example. In the example of Fig. 1 and 2, the engineer identifies the property restrictions containing the properties *address* and *telephone* of the class *Person* in V that should be extracted to a new class. The new class *ContactData* is created in version $V\prime$ and the identified property restrictions are added by adding axioms to the new class like *ContactData* $\sqsubseteq \exists$ *address.string*. The corresponding axioms of the moved properties are removed from the class definition of the class *Person*. Finally, the object property restriction to the new class is added to *Person*, e.g., by the new axiom *Person* $\sqsubseteq \exists$ *contact.ContactData*.

Move of Property Refactoring Pattern. An example of the Move Of Property refactoring and the corresponding description of the ontologies in OWL are already described in the running example of Sect. 2 (Fig. 1 and 2).

Problem Description. A named class C has a property restriction on the property p and on the object property r, with the named class D in the range of the definition. The ontology engineer would like to move this property restriction from the class C to the referenced class D.

Solution. The identified property restriction on the property p is moved to the class D. The range of this moved property p is unchanged.

Example. In the example of Fig. 1 and 2, the property *project* should be moved from the class *Employee* to *Department*. The class *Department* is already referenced by *Employee* with the object property *department*. In version $V\prime$, the corresponding axiom *Employee* $\sqsubseteq \exists project.Project$ is deleted and the axiom *Department* $\sqsubseteq \exists project.Project$ is added to the ontology.

4   DL-Reasoning for Ontology Comparison

In this section, we describe the usage of DL reasoning in order to semantically compare ontology versions. We distinguish between three types of comparisons: (i) A *syntactic* comparison checks whether for a class or property in the ontology V there is an entity with the same name in $V\prime$. (ii) The *structural* comparison compares classes and their structure, i.e. sub- and superclass relations and object property restrictions of this class. Hence, a class with all "connected" classes is compared in both versions. (iii) In a *semantic* comparison, classes of both versions are compared using subsumption checking, testing the equivalence, sub- and superclass relations between a class by comparing the interpretations.

4.1   Combining Knowledge Bases

The first step towards a semantic version comparison (Sect. 4.2) is to allow reasoning on two versions of an ontology, e.g., by checking class subsumption of classes from two versions. This requires a renaming of classes that appear in both versions with the same name, otherwise we can not compare them by reasoning. Hence, we start with comparing the names of classes and properties of both versions and rename them. We build a combined, additional knowledge base that captures both, the original version V and the new version $V\prime$. This combined ontology is only a technical mean that is used in order to enable a semantic comparison of classes that appear in both versions. The ontology versions V and $V\prime$ remain unchanged and the semantics given by both versions is also not affected.

 For each named class C that occurs in both versions V and $V\prime$, we build the combined knowledge base as follows: (i) The class C is renamed, e.g., C_1 for the class in version V and a class C_2 for the class in version $V\prime$. (ii) Both classes C_1 and C_2 are subclasses of the superclass C. With this step, we guarantee that C_1 and C_2 are still related to each other. C_1 and C_2 are not disjoint. (iii) In every class expression (anonymous class) if C_i occurs as a class in the range of a property restriction, the class C_i is replaced by its superclass C.

4.2 Semantic Version Comparison

We distinguish between the name or label of a class (C) and the intensional description of the class, i.e. the object and datatype properties that describe the class. The extension of a class, i.e. the set of inferred instances of this class, is denoted using semantic brackets $[\![C]\!]$. A statement like $[\![C]\!] \sqsubseteq A$ means the subsumption $C \sqsubseteq A$ can be inferred.

We use $\hat{C}$ as a representation of the class C in a conjunctive normal form, i.e. $\hat{C} \equiv C_1 \sqcap \ldots \sqcap C_n$ where $\forall i = 1, \ldots, n$ there is an axiom in the ontology $C \sqsubseteq C_i$ and C_i is a class expression. Hence, C is subsumed by each C_i. In order to ease the comparison of classes in two versions, we apply a normalization and reduction of $\hat{C}$ resulting in a reduced conjunctive normal form $\tilde{C}$.

Definition 2 (Reduced Conjunctive Normal Form). *A class definition in conjunctive normal form $\hat{C}$ is reduced to $\tilde{C}$ by the following steps:*

1. *Nested conjunctions are flattened, i.e. $A \sqcap (B \sqcap C)$ becomes $A \sqcap B \sqcap C$.*
2. *Negations are normalized such that in all negations $\neg C$, C is a named class.*
3. *If $B \sqsubseteq A$ can be inferred and $A \sqcup B$ is a class expression in $\hat{C}$, the expression is replaced by A in $\tilde{C}$.*

The main advantage of the normalization is a unique representation that can be assumed for the class definition C which is exploited in the comparison later on. This unique representation is ensured by Lemma 1. The reduced conjunctive normal form $\tilde{C}$ is used in the comparison algorithms later on. We will see, that we are only interested in class expressions C_i that are either property restrictions or named superclasses.

Lemma 1 (Uniqueness of the Reduced Conjunctive Normal Form). $\hat{C} \equiv C_1 \sqcap \ldots \sqcap C_n$ *is a class in conjunctive normal form and $\tilde{C}$ is the reduced conjunctive normal form of the class C. For each class expression C_i $(i = 1, \ldots, n)$ one of the following conditions hold: (i) C_i is a named class, (ii) C_i is a datatype or object property restriction or (iii) C_i is a complex class definition that can neither be a named superclass of C nor a property restriction.*

Proof. It is easy to see whether C_i satisfies the first or second condition, i.e. either C_i is a named class or a property restriction (including qualified property restrictions). In the following, we prove the third condition, assumed that C_i is neither a named class nor a property restriction. We consider the remaining possible class constructors that are allowed according to the language restriction from Def. 1. We show that either the third condition is satisfied or the expression is not allowed after the reduction:

- if $C_i \equiv \neg D$ then C_i cannot be a named superclass of C and (iii) is satisfied.
- $C_i \equiv \neg \forall R.D$ or $C_i \equiv \neg \exists R.D$ is not allowed after the reduction according No. 2 in Def. 2
- $C_i \equiv D \sqcap E$ is not allowed as restricted in No. 1 in Def. 2 (flattening).
- $C_i \equiv D \sqcup E$ then C_i cannot be a named superclass of C. Trivial equivalent representations like $C_i \equiv D \sqcup E$ and $E \sqsubseteq D$ are not allowed (cf. No. 3). $\square$

Algorithm: Diff(Class C, Ontology versions V, $V\prime$)
Input: Class C and two ontology versions $(V, V\prime)$
Output: Set of class expressions that subsume $C\prime$ in $V\prime$ but not C in V

1: /* Compute the new additional class expressions in $C\prime$ of $V\prime$ */
2: $\mathcal{D} := \emptyset$
3: **for** each asserted class expression A of $\tilde{C}\prime$ $(\tilde{C}\prime \sqsubseteq A$ is asserted in $V\prime)$ **do**
4:     **if** $[\![C]\!] \not\sqsubseteq A$ in V **then**
5:        $\mathcal{D} := \mathcal{D} \cup \{A\}$
6:     **end if**
7: **end for**
8: **Return** $\mathcal{D}$.

Fig. 3. The Diff-Algorithm

We use two algorithms to compare versions V and $V\prime$. The Diff-Algorithm (Fig. 3) computes all class expressions that subsume the class $C\prime$ in version $V\prime$, but not C in V. To compute the difference[1], the Diff-Algorithm is used twice. $Diff(C, V, V\prime)$ returns all class expressions that subsume $C\prime$ in $V\prime$. Class expressions that subsume C of V are the result of $Diff(C, V\prime, V)$. $\tilde{C}\prime$ is a class in reduced conjunctive normal form, the expression A is a conjunct that appears in $\tilde{C}\prime$. We can extract the conjuncts due to the normal form representation. In line 4, it is checked whether the subsumption is inferred in version V.

The Common-Algorithm in Fig. 4 extracts the common class expressions of a class C in both versions. Therefore, the subsumption of the class expressions from one version compared with the other is checked in both directions, i.e. $\mathcal{D}_1$ are class expressions from version V that are subsumed by $V\prime$ and $\mathcal{D}_1\prime$ vice versa. $\mathcal{D}$ is the intersection of $\mathcal{D}_1$ and $\mathcal{D}_1\prime$ and consists of all class expressions A from C in both versions. As in the Diff-Algorithm, A is a conjunct of the reduced conjunctive normal forms $(\tilde{C}, \tilde{C}\prime)$.

We use the ExtractReferenceClasses-Algorithm from Fig. 5 to obtain the classes that are referenced by a class, i.e. we are looking for the class in the range of a property restriction in a class definition. The algorithm uses set operations and returns a set of classes. However, in the considered refactoring patterns, only one class is extracted. If multiple classes are extracted from one class, this is considered as multiple refactorings in succession. The input class expression C is an object property restriction like $\exists contact.ContactData$ (line 2). The result is the class that is referenced (R in line 4), e.g., *ContactData*.

The method *getProperty* returns the object property (object property name) of the given object property restriction (class expression) C. Such methods are provided by OWL-APIs like [13]. The referenced class can not directly be extracted from the expressions using API operations, since in general the expression could be more complex than just a single OWL class as in our applications with language restrictions. Therefore, we have to implement this algorithm. Methods

---

[1] This definition is different from the stronger definition of DL difference of [12], where the difference requires that the minuend is subsumed by the subtrahend.

Algorithm: Common(Class C, Ontology versions V, $V\prime$)
Input: Class C and two ontology versions (V, $V\prime$)
Output: Set of class expressions that subsume C in V and $C\prime$ in $V\prime$

1: /* Common class expressions $\mathcal{D}$ of C and $C\prime$ in both ontology versions V, $V\prime$ */
2: /* $\mathcal{D}_1$ are class expressions of C in V subsumed in $V\prime$, and $\mathcal{D}_1\prime$ are class expressions of $C\prime$ in $V\prime$ subsumed in V. */
3: $\mathcal{D}_1 := \emptyset$ and $\mathcal{D}_1\prime := \emptyset$
4: **for** each asserted class expression A of $\tilde{C}$ ($\tilde{C} \sqsubseteq A$ is asserted in V) **do**
5:     **if** $[\![C\prime]\!] \sqsubseteq A$ in $V\prime$ **then**
6:         $\mathcal{D}_1 := \mathcal{D}_1 \cup \{A\}$
7:     **end if**
8: **end for**
9: **for** each asserted class expression A of $\tilde{C}\prime$ ($\tilde{C}\prime \sqsubseteq A$ is asserted in $V\prime$) **do**
10:     **if** $[\![C]\!] \sqsubseteq A$ in V **then**
11:         $\mathcal{D}_1\prime := \mathcal{D}_1\prime \cup \{A\}$
12:     **end if**
13: **end for**
14: **Return** $\mathcal{D} := \mathcal{D}_1 \cap \mathcal{D}_1\prime$.

Fig. 4. The Common-Algorithm

Algorithm: ExtractReferenceClasses(Class expression C, Ontology version V)
Input: Class expression C that is an object property restriction,
            e.g., $\exists contact.ContactData$ and an ontology version (V)
Output: Set of classes which are referenced by the object property restriction C
            (e.g., the class $ContactData$)

1: $\mathcal{D} := \emptyset$ /* for the referenced classes */
2: **if** IsObjectPropertyRestriction(C) **then**
3:     **for** each class R of version V **do**
4:         **if** $[\![C]\!] \sqsubseteq \exists getProperty(C). R$ **then**
5:             $\mathcal{D} := \mathcal{D} \cup \{R\}$
6:         **end if**
7:     **end for**
8: **end if**
9: **Return** $\mathcal{D}$.

Fig. 5. The ExtractReferenceClasses-Algorithm

like *IsObjectPropertyRestriction* or *IsPropertyRestriction* are provided by APIs as well. For property restrictions with universal quantifiers, the referenced class can be extracted likewise, but this is not required in our approach.

The Diff- and Common-Algorithm compute for a class C, the class expressions C_i that subsume C. These class expressions are expressions C_i of the reduced conjunctive normal form $\tilde{C}$. Hence, all class expressions of the result of the Diff- and Common-Algorithm are in reduced conjunctive normal form too.

The focus of our approach is to recognize the introduced refactorings rather than identifying arbitrary ontology changes. Hence, we can neglect some of the class expressions that are in the result of the Diff- and Common-Algorithm. All the considered refactoring patterns only change sub- and superclass relations and property restrictions in class definitions. Therefore, the only relevant class expressions in the result set of the Diff- and Common-Algorithm are those class expressions that are named classes (representing superclasses) and property restrictions. According to Lemma 1, we can easily determine whether a class expression C_i of the result of the algorithms is a superclass, a property restriction or another complex class expression that can be neglected in the comparison.

5   Refactoring Pattern Recognition

In this section, we demonstrate the recognition of the already introduced refactoring patterns Extract Class and Move of Property. The recognition description of the other patterns can be found in [11].

Extract Class. This refactoring is illustrated in Fig. 1 and 2. One recognizes the refactoring according to the algorithm in Fig. 6.

The algorithm in Fig. 6 returns the extracted class if the refactoring is successfully recognized, otherwise the result is the empty class ($\perp$). The algorithm works as follows. All named classes C and $C\prime$ that exist in both versions and are different are compared (line 2). In line 3 the difference is computed. For instance, the set $\mathcal{D}_1$ consists of all class expressions which are only in $[\![C\prime]\!]$ of $V\prime$ but not in V. $C\prime$ of $V\prime$ contains exactly one additional object property restriction to another class, i.e. a change only extracts one class. Therefore, we require that $\mathcal{D}_1$ is a singleton (line 4) and that D_1 is an object property restriction (line 6).

Algorithm: Recognize-ExtractClass(Ontology versions V, $V\prime$)
Input: Ontology versions V and $V\prime$
Output: Extracted Class E

```
 1: E := ⊥
 2: for  all classes C and C′ that are different in version V and V′ do
 3:     D₁ := Diff(C, V, V′)  AND  D₂ := Diff(C, V′, V)
 4:     if |D₁| = 1 then
 5:         D₁ ∈ D₁:
 6:         if IsObjectPropertyRestriction(D₁) then
 7:             RC := ExtractReferenceClasses(D₁, V′)
 8:             if |RC| = 1 AND ∀ D₂ ∈ D₂ : ∃ RC ∈ RC : [[RC]] ⊑ D₂ AND
                   ∀ D₂ ∈ D₂ : IsPropertyRestriction(D₂) then
 9:                 E := RC
10:             end if
11:         end if
12:     end if
13: end for
14: Return E
```

Fig. 6. Algorithm for Recognizing *Extract Class*

Algorithm: Recognize-MoveOfProperty(Ontology versions V, $V\prime$)
Input: Ontology versions V and $V\prime$
Output: Set of moved property restrictions $\mathcal{P}$

1: $\mathcal{P} := \emptyset$
2: **for** all classes A and $A\prime$ in version V and $V\prime$ that are different **do**
3:    **for** all referenced classes B and $B\prime$ **do**
4:      **if** B and $B\prime$ are also different in version V and $V\prime$ **then**
5:        $\mathcal{C}_1 := Common(A, V, V\prime)$ AND $\mathcal{C}_2 := Common(B, V, V\prime)$ AND
6:        $\mathcal{A}_1 := Diff(A, V, V\prime)$ AND $\mathcal{A}_2 := Diff(A, V\prime, V)$ AND
7:        $\mathcal{B}_1 := Diff(B, V, V\prime)$ AND $\mathcal{B}_2 := Diff(B, V\prime, V)$
8:        **if** $\mathcal{A}_1 = \emptyset$ AND $\mathcal{B}_2 = \emptyset$ AND $\mathcal{A}_2 = \mathcal{B}_1$ AND $\forall E \in \mathcal{A}_2 : IsPropertyRestriction(E)$ **then**
9:           $\mathcal{P} := \mathcal{A}_2$
10:        **end if**
11:      **end if**
12:    **end for**
13: **end for**
14: **Return** $\mathcal{P}$

Fig. 7. Algorithm for Recognizing *Move of Property*

In line 7, the new class that is referenced by C is extracted. In line 8, we ensure that property restrictions are only moved to one class, i.e. $\mathcal{RC}$ is a singleton. Finally, it is required that all property restrictions are moved correctly to the new class RC (subsumption in line 8). The second and third conditions in line 8 ensure that only property restrictions and no other class expressions are moved and that they are moved to the correct class RC. The result is the referenced class RC ($\mathcal{RC}$ is a singleton).

The recognition algorithms for other extract and merge class refactorings work in the same way. E.g., to recognize an Extract Subclass refactoring, we just replace the referenced class (RC) by the corresponding subclass. The recognition result for the example in Fig. 1 and 2 is as follows:

$\mathcal{D}_1 = \{\exists contact.ContactData\}$ (object property restriction in $V\prime$)
$\mathcal{D}_2 = \{\exists address.string, \exists telephone.string\}$ (property restrictions in V)
$\mathcal{RC} = \{ContactData\}$ (only one restriction in $\mathcal{D}_1$ ($\exists contact.ContactData$))
$RC = ContactData$ and $[\![RC]\!] \sqsubseteq D_2$ is inferred for all $D_2 \in \mathcal{D}_2$

Move of Property. The algorithm in Fig. 7 recognizes the Move of Property refactoring by the following steps. In lines 2-4, it is checked for all classes whether the classes A and $A\prime$ are different in both versions V and $V\prime$ and the referenced classes (range of property) B and $B\prime$ are also different in V and $V\prime$. The common and different class expressions of class A and B in both versions are computed (lines 5-7). If all property restrictions are moved correctly from class A to B the four conditions of line 8 have to be satisfied. Finally, the moved property restrictions are the result of the algorithm (line 9 and 14). Algorithms to detect the other move refactorings like the movement of property restrictions within a class hierarchy work similarly. The recognition of the Move of Property example from Fig. 1 and 2 is as follows:

Common property restrictions of the classes *Employee* and *Department*:
$C_1 = \{\exists_{=1} SSN.string, \exists department.Department\}$, $C_2 = \{\}$
Department is referenced by *Employee*: $\exists department.Department \in C_1$
Moved property restrictions (from *Employee* to *Department*):
$A_1 = \{\}$, $A_2 = \{\exists project.Project\}$, $B_1 = \{\exists project.Project\}$ and $B_2 = \{\}$

6   Evaluation and Discussion

Analysis: We evaluated refactorings for the described refactoring patterns on two ontologies. The DOLCE Lite Plus ontology[2] is the smaller ontology with an average version size of 240 classes and 360 subclass axioms. For each pattern, 8 concrete refactorings were applied. The second ontology is a bio-medical ontology OBI[3] with an average size of 1200 classes, 1700 subclass axioms, and 14 concrete refactorings for each pattern. For both ontologies, we changed the original ontology V by adding and deleting classes, properties and axioms according to the pattern description and applied our recognition algorithms. All recognized refactorings were correctly recognized. The performance result is depicted in Table 2.

For the evaluation, we used the Pellet 2.0.0 reasoner in Java 1.6 on a computer with 2.5 GHz CPU and 2 GB RAM. In Table 2 only the time for the recognition is displayed. The time for matching and combining the ontologies (first step of the comparison) is on average 570 msec for the models with about 240 classes and 2900 msec for models with an average size of 1200 classes.

Limitations: We identified the following limitations that are further challenges for future work. (i) The refactoring patterns are adopted from existing work on ontology evolution (cf. [3]), but also on object-oriented modeling (cf. [2]). Therefore, we only recognize those elementary ontology changes that are specified in the refactoring recognition. However, there might be a couple of further ontology changes that are not considered in our approach. For instance, we do not consider changes of the property range yet which would lead to difficulties in the current approach in the combination step (cf. Sect. 4). (ii) We need a language restriction as described in Definition 1 and reduction according to Definition 2. Otherwise, we can not ensure the recognition.

Lessons Learned: Although the proposed semantic comparison between classes of different versions is the main benefit of our work, the comparison is rather a structural-semantic comparison than a purely semantical comparison. The Diff- and Common-Algorithms iterate and compare class expressions that are either superclasses or property restrictions which is a structural class comparison. The

---

[2] http://www.loa-cnr.it/DOLCE.html
[3] http://obi-ontology.org/page/Main_Page

Table 2. Analyzed Refactoring Patterns

No.	Refactoring	Recognition (Avg. 240)		Recognition (Avg. 1200)	
		Avg.[msec]	Max.[msec]	Avg.[msec]	Max.[msec]
1.	Extract Class	493	605	2050	2520
2.	Extract Subclass	412	480	1910	2430
3.	Extract Superclass	473	580	1860	2540
4.	Collapse Hierarchy	1062	1154	2260	2480
5.	Extract Hierarchy	886	1042	2170	2410
6.	Inline Class	1042	1075	2330	2590
7.	Move Attribute	1085	1240	2680	3230
8.	Pull-Up Attribute	864	1065	2150	2840
9.	Push-Down Attribute	840	957	2820	3360
10.	Unidirectional to bidirectional Ref.	1170	1254	1820	2140
11.	Bidirectional to unidirectional Ref.	1135	1174	1950	2280
12.	Cardinality Change	1180	1265	1740	1870

algorithms work properly even for more expressive OWL languages that do not satisfy the restrictions and reductions. However, we need these restrictions in order to guarantee a correct recognition.

7   Related Work

We group the related work into three categories. Firstly, the syntactical comparisons are analyzed. Secondly, related work on structural comparisons is presented. Finally, we consider OWL reasoning for ontology comparison.

The detection of changes of RDF knowledge bases is considered in [14]. High-level changes of RDF-graphs and version differences (RDF triples) are represented and detected in [5]. They categorize elementary changes like add and delete operations to high-level changes which are similar to refactoring patterns. Basically, they analyze the difference of RDF-triples of two RDF-graphs instead of OWL ontologies and the detection is based on a (syntactical) triple comparison, i.e. the high-level change is detected if all its required low-level changes (RDF-triples) are recognized.

Related work on ontology mappings and the computation of structural differences between OWL ontologies is given in [7,15,16]. In [7] a fix-point algorithm is used for comparing and mapping related classes and properties based on their names and structure (references to other entities). A heuristic matching is applied to detect structural differences. Benefits of the heuristics are mainly the identification of related classes and properties if their names have changed.

A framework for tracking ontology changes is introduced in [17]. It is implemented as a plug-in for Protégé [18] that creates a change and annotation ontology to record the changes and meta information on changes. This change ontology is used to display the applied changes to the user. Similarly, change logs are used to manage different ontology versions in [1]. The change logs are realized by a version ontology that represents instances for each class, property and individual of the analyzed ontology. The usage of version ontologies (meta ontology) for change representation is also proposed in [19].

More closely related to our work are the approaches on DL reasoning applying semantic comparison for versioning and ontology changes in OWL. OWL ontology evolution is analyzed in [20]. However, the focus of this work is not on detecting changes. They tackle inconsistency detection caused by already identified changes and in case of an inconsistency, additional changes are generated to result again in a consistent ontology. In [9] and [21], OWL reasoning on modular ontologies is considered in order to tackle the problem of consistency on mappings between ontologies. While the focus in [21] is on reasoning for consistency of ontology mappings and different from our work, in [9] the problem of consistency management for ontology modules is considered. The ontology modules are connected by conjunctive queries instead of merging based on syntactic matching as in our work. Although, subsumption checking is used to compare classes of versions, a classification and especially a recognition of refactoring pattern or complex changes is missing. The main difference to the related work on semantic comparison is the ability of our approach on recognizing ontology refactoring patterns based on change operations in OWL ontologies.

8   Conclusion and Future Work

In this paper, we have demonstrated a structural-semantic comparison approach to recognize specified refactoring patterns using standard DL reasoning. We provide technical information on the version comparison and recognition algorithms. One can apply the results of this work for schema versioning, semantic difference and conflict detection. Additionally, it paves the way for application of reasoning technologies in change prediction of ontologies as well as for guidance in versioning and evolution of ontologies. In future, we plan to cover additional refactoring patterns and plan to extend our approach by a heuristic mapping between classes and properties to handle name changes.

Acknowledgements. This work has been supported by the EU Project MOST (ICT-FP7-2008 216691).

References

1. Plessers, P., Troyer, O.D.: Ontology Change Detection Using a Version Log. In: Gil, Y., Motta, E., Benjamins, V.R., Musen, M.A. (eds.) ISWC 2005. LNCS, vol. 3729, pp. 578–592. Springer, Heidelberg (2005)
2. Fowler, M., Beck, K., Brant, J., Opdyke, W.: Refactoring: Improving the Design of Existing Code. Addison-Wesley, Reading (1999)

3. Stojanovic, L., Maedche, A., Motik, B., Stojanovic, N.: User-Driven Ontology Evolution Management. In: Gómez-Pérez, A., Benjamins, V.R. (eds.) EKAW 2002. LNCS (LNAI), vol. 2473, pp. 285–300. Springer, Heidelberg (2002)
4. Klein, M., Fensel, D., Kiryakov, A., Ognyanov, D.: Ontology Versioning and Change Detection on the Web. In: Gómez-Pérez, A., Benjamins, V.R. (eds.) EKAW 2002. LNCS (LNAI), vol. 2473, pp. 197–212. Springer, Heidelberg (2002)
5. Papavassiliou, V., Flouris, G., Fundulaki, I., Kotzinos, D., Christophides, V.: On Detecting High-Level Changes in RDF/S KBs. In: Bernstein, A., Karger, D.R., Heath, T., Feigenbaum, L., Maynard, D., Motta, E., Thirunarayan, K. (eds.) ISWC 2009. LNCS, vol. 5823, pp. 473–488. Springer, Heidelberg (2009)
6. Noy, N.F., Kunnatur, S., Klein, M.C.A., Musen, M.A.: Tracking Changes During Ontology Evolution. In: McIlraith, S.A., Plexousakis, D., van Harmelen, F. (eds.) ISWC 2004. LNCS, vol. 3298, pp. 259–273. Springer, Heidelberg (2004)
7. Noy, N.F., Musen, M.A.: PROMPTDIFF: A Fixed-Point Algorithm for Comparing Ontology Versions. In: AAAI/IAAI, pp. 744–750 (2002)
8. Meilicke, C., Stuckenschmidt, H., Tamilin, A.: Repairing ontology mappings. In: AAAI, pp. 1408–1413 (2007)
9. Stuckenschmidt, H., Klein, M.: Reasoning and Change Management in Modular Ontologies. Data & Knowledge Engineering 63(2), 200–223 (2007)
10. Horrocks, I., Patel-Schneider, P.F., Harmelen, F.V.: From SHIQ and RDF to OWL: The Making of a Web Ontology Language. J. of Web Semantics 1, 7–26 (2003)
11. Gröner, G., Staab, S.: Categorization and Recognition of Ontology Refactoring Pattern. Technical Report 9/2010, University of Koblenz-Landau (2010), http://www.uni-koblenz.de/~groener/documents/TR092010.pdf
12. Teege, G.: Making the Difference: A subtraction Operation for Description Logics. In: 4th Int. Conference on Knowledge Representation (KR), pp. 540–550 (1994)
13. The OWL API (2010), http://owlapi.sourceforge.net
14. Zeginis, D., Tzitzikas, Y., Christophides, V.: On the Foundations of Computing Deltas between RDF Models. In: Aberer, K., Choi, K.-S., Noy, N., Allemang, D., Lee, K.-I., Nixon, L.J.B., Golbeck, J., Mika, P., Maynard, D., Mizoguchi, R., Schreiber, G., Cudré-Mauroux, P. (eds.) ASWC 2007 and ISWC 2007. LNCS, vol. 4825, pp. 637–651. Springer, Heidelberg (2007)
15. Klein, M., Noy, N.: A Component-Based Framework for Ontology Evolution. In: Proc. of the IJCAI 2003 Workshop on Ontologies and Distributed Systems. CEUR-WS, vol. 71. Citeseer (2003)
16. Ritze, D., Meilicke, C., Sváb-Zamazal, O., Stuckenschmidt, H.: A Pattern-based Ontology Matching Approach for Detecting Complex Correspondences. In: Proc. of Int. Workshop on Ontology Matching, OM (2009)
17. Noy, N., Chugh, A., Liu, W., Musen, M.: A Framework for Ontology Evolution in Collaborative Environments. In: Cruz, I., Decker, S., Allemang, D., Preist, C., Schwabe, D., Mika, P., Uschold, M., Aroyo, L.M. (eds.) ISWC 2006. LNCS, vol. 4273, pp. 544–558. Springer, Heidelberg (2006)
18. Protégé - Ontology Editor (2010), http://protege.stanford.edu
19. Palma, R., Haase, P., Wang, Y., dAquin, M.: D1.3.1 Propagation Models and Strategies. Technical report, NeOn Project Deliverable 1.3.1 (2007)
20. Haase, P., Stojanovic, L.: Consistent Evolution of OWL Ontologies. In: Gómez-Pérez, A., Euzenat, J. (eds.) ESWC 2005. LNCS, vol. 3532, pp. 182–197. Springer, Heidelberg (2005)
21. Meilicke, C., Stuckenschmidt, H., Tamilin, A.: Reasoning Support for Mapping Revision. J. Log. Comput. 19(5), 807–829 (2009)

Finding the Achilles Heel of the Web of Data: Using Network Analysis for Link-Recommendation

Christophe Guéret, Paul Groth, Frank van Harmelen, and Stefan Schlobach

VU University Amsterdam
De Boelelaan 1081a, 1081 HV, Amsterdam, The Netherlands
{cgueret,pgroth,Frank.van.Harmelen,schlobac}@few.vu.nl

Abstract. The Web of Data is increasingly becoming an important infrastructure for such diverse sectors as entertainment, government, e-commerce and science. As a result, the *robustness* of this Web of Data is now crucial. Prior studies show that the Web of Data is strongly dependent on a small number of central hubs, making it highly vulnerable to single points of failure. In this paper, we present concepts and algorithms to analyse and repair the brittleness of the Web of Data. We apply these on a substantial subset of it, the 2010 Billion Triple Challenge dataset. We first distinguish the *physical structure* of the Web of Data from its *semantic structure*. For both of these structures, we then calculate their robustness, taking *betweenness centrality* as a robustness-measure. To the best of our knowledge, this is the first time that such robustness-indicators have been calculated for the Web of Data. Finally, we determine which links should be added to the Web of Data in order to *improve* its robustness most effectively. We are able to determine such links by interpreting the question as a very large optimisation problem and deploying an evolutionary algorithm to solve this problem. We believe that with this work, we offer an effective method to analyse and improve the most important structure that the Semantic Web community has constructed to date.

1 Introduction

The rapidly growing Web of Data increasingly resembles the Web in its network properties. It resembles a small world network that relies on central hubs to provide connectivity between resources on the Web of Data [10]. Such central hubs are potential points of failure. This is particularly dangerous for the Web of Data, which, unlike the Web, is designed to be used by automated agents that have less capability to recover from lack of access to resources than human users might have on the regular Web.

Current approaches to ensure robustness of the Web of Data are based on anecdotal observations. In this work, we propose a systematic approach for analysing the Web of Data and recommending where links can be added to

P.F. Patel-Schneider et al. (Eds.): ISWC 2010, Part I, LNCS 6496, pp. 289–304, 2010.

help ensure robustness against both infrastructure failure and semantic deviation. An example of the first is: how can we ensure that automated agents can still traverse the network if DBpedia is down? An example of the second is: if the SIOC ontology is updated, where do links need to be introduced to re-establish connectivity?

Our systematic approach uses well known network properties to characterise the robustness of both the infrastructure and semantic networks within the Web of Data. Based on these properties, we present an optimisation algorithm that produces recommendations about where links should be added to the Web of Data. The algorithm takes into account whether additional links would be semantically meaningful.

The contributions of this paper are (i) a characterisation of the strength of the current Web of Data in terms of its infrastructure and semantic network.; (ii) a recommendation algorithm for adding links to the Web of Data to increase its robustness; and (iii) applying this algorithm in order to determine how many (and which) links are required to obtain different levels of robustness.

Our main findings are that (a) the current Web of Data is indeed highly sensitive to failure of individual nodes, both at the infrastructure level and as a semantic network, and (b) this situation can be remedied by adding a surprisingly small number of links, provided that these links are chosen well, as calculated by our recommendation algorithm.

The paper is organised as follows. In Section 2, we discuss related work and argue why it is useful to distinguish infrastructural connectivity and semantic connectivity. This leads to Section 3 where the robustness of the current Web of Data is measured, followed by Section 4, which presents an algorithm to recommend how best to increase that robustness. Section 5 concludes.

2   Background

2.1   Related Work

The use of network properties to study complex systems has grown in a wide range of fields (*e.g.* biology, social science and web science) because it provides a mechanism to extract global properties of systems [12]. In terms of robustness, the classic result is from Barabasi, which shows that scale-free networks are robust against random failure, but not against targeted attacks [1]. The robustness of scale-free networks is important because they are widely seen in nature including power grids, the World Wide Web and social networks [2].

The application of such network analysis to the Web of Data has until now been limited, and has been performed on a wide variety of graph-structures: [10] analysed the 2009 BTC dataset[1] and showed that, interpreted as a sample of the Web of Data, it is scale-free and that semanticdesktop.org and purl.org are central in it. The same paper also analysed the well-known "bubble-graph" of the Web of Data, consisting of the datasets published and interlinked by the

---

[1] http://vmlion25.deri.ie/index.html

Linking Open Data project[2]. It showed the existence of topic-oriented hubs, with DBPedia connected to 50% of all the other datasets, and over 50% of all the shortest paths in the graph being routed through either DBPedia or DBLP.

In recent work, [8] analysed the "object link graph": the Web of Data restricted to its object-to-object links, i.e. after removing all links from objects to classes, and all class- and property-hierarchies. They found that this object-link graph also has a scale-free nature, with a diameter value of 12, which is small compared to the size of the graph, although the link density is rather low. Such a small diameter of a large but low density graph again points to the presence of central hubs that provide the main connectivity between many resources.

Other work, such as [9], also use network analysis tools, but apply them only to networks of ontologies, and do not consider the much more substantial collection of instances that form the real content of the Web of Data. At an even smaller scale, [13] applies concepts from network analysis to individual ontologies.

Summarising, only a handful of analyses have been performed on the network properties of the Web of Data. Furthermore, all these works have only *analysed* the Web of Data, but nobody has used the results of their analysis to effectively compute *improvements* to the Web of Data.

2.2   Infrastructure Failure and Semantic Failure

Connectivity on the Web of Data can be disrupted in two different ways: infrastructural failure or semantic failure. For the infrastructure, the problem is server unavailability, *e.g.* the `dbpedia.org` server is down. In the semantic network, the problem is robustness against change, for example still using `sioc:User` instead of the current `sioc:UserAccount`.

The robustness of an infrastructure is commonly improved by the use of mirrors and caches. Our approach is complimentary to using these techniques. In order to detect hosts that function as infrastructure hubs, and whose unavailability would hence break many paths, we aggregate the Web of Data into a *hostname graph*:

Definition 1 (hostnames graph). *The hostname graph $\mathcal{H}$ is a $\langle V, E \rangle$ where $h \in V$ is a node of $\mathcal{H}$ iff h is used as a hostname in any URI on the Web of Data, and $e \in E, e = \langle h_1, h_2 \rangle$ is an edge of $\mathcal{H}$ from node h_1 to node h_2 iff there is a triple $\langle s, p, o \rangle$ anywhere on the Web of data with h_1 the hostname referred to in the URI of s and h_2 the hostname referred to in the URI of o.*

Thus, the hostname graph has as many nodes as there are hostnames mentioned in all the triples on the Web of Data.

Similarly, the namespace graph is an aggregation of the semantic structure of the Web of Data:

Definition 2 (namespaces graph). *The namespace graph $\mathcal{S}$ is a tuple $\langle V, E \rangle$ where $n \in V$ is a node of $\mathcal{S}$ iff n is used as a namespace anywhere on the Web of*

---

[2] http://esw.w3.org/SweoIG/TaskForces/CommunityProjects/LinkingOpenData

Data, and $e \in E, e = \langle n_1, n_2 \rangle$ is an edge of $\mathcal{S}$ from node n_1 to node n_2 iff there is a $\langle s, p, o \rangle$ anywhere on the Web of Data with n_1 the namespace of s and n_2 the namespace of o.

Thus, the namespace graph has as many nodes as there are namespaces mentioned in all the triples on the Web of Data.

Definition 3 (content of nodes). *The content $cont(n)$ of a node n is defined as the set of URI such that there is a $\langle r, p, o \rangle$ anywhere on the Web of Data and n is the namespace of r for a namespaces graph or n is the hostname of r for an hostnames graph.*

3   Analysing the Web of Data

The networks and the programs described in this section are all publicly available at http://linkeddata.few.vu.nl/wod_analysis/.

3.1   Measures of Robustness

By robustness of a graph, we mean the degree to which connectivity in a graph is maintained after a node is removed from the graph. There are a number of network measures that can be used for measuring the robustness of a graph. For example, the diameter of a graph[3] provides information about connectivity. A smaller diameter implies that there are a large number of connections within the network while a larger diameter means that the network is less connected. While the diameter provides a reasonable global summary statistic, centrality statistics allow one to investigate the graph on a per node basis. In particular, betweenness centrality measures how often a node occurs on a shortest path any pair of nodes:

Definition 4 (Betweenness centrality). *For a graph $G = (N, E)$ with a set of nodes N and a set of edges E, the* betweenness centrality $B(n)$ *of a node $n \in N$ is defined as*

$$B(n) = \sum_{s \neq n \neq t \in N} \frac{S(s, n, t)}{S(s, t)}$$

where $S(s, t)$ is the number of shortest paths from s to t, and $S(s, n, t)$ is the number of shortest paths from s to t that pass through node n. Instead of $B(n)$ we will often report on its non-normalised version, $B'(n)$.

$$B'(n) = \sum_{s \neq n \neq t \in N} S(s, n, t),$$

Instead of "betweenness centrality" we will often simply speak of "betweenness".

---

[3] The diameter of a graph is the longest shortest path in the graph.

Betweenness is a measure of the importance of a node for the connectivity between other nodes. The intuition is that if a node lies on many shortest paths it is an important node, since removal of such a node will directly influence the cost of the connectivity between other nodes, as other (i.e., longer) shortest paths will have to be followed.[4]

A completely connected network has the maximal robustness, and correspondingly the lowest betweenness centrality: $B(n) = 0$ for every $n \in N$, and removing one node does not impact the overall connectivity of the network greatly.

If we want to improve the robustness of the Web of Data, we will want to lower the number of nodes that have high betweenness centrality, since these are important potential points of failure. For this, we will first need to analyse which nodes actually have a high betweenness centrality. This is obviously computationally intensive, since it involves calculating the shortest paths between all pairs of nodes on the Web of Data. This *robustness analysis* will be topic of the remainder of this section. Deciding how to *improve the robustness* will be tackled in Section 4.

3.2   Dataset

The 2010 Billion Triple Challenge (BTC) Dataset[5] was used as a representative sample of the Web of Data. It contains roughly 3.2 billion statements. From this dataset, the hostname graph and namespace graph were constructed. Given that namespaces cannot be systematically identified given a URL alone, we used a predefined list of widely used namespaces as defined by the `prefix.cc` service. Out of the 330 namespaces registered on the services, 198 were found to be used in the snapshot used to create the networks.

We removed from the BTC all triples where the object was a literal, all triples containing blank nodes, and all triples that refer only to URI's from the same dataset, since none of these triples would contribute to the objects of our study, namely the hostname graph and the namespace graph. Surprisingly, this reduced the BTC dataset to 530 million triples, showing that the vast majority of the Web of Data (or at least the BTC snapshot of it) does not contribute to it being a "web". Of those remaining 530 million triples, the vast majority (389 million) were covered by the namespace list built from `prefix.cc`. This gives us some confidence that the namespace list is sufficiently representative set of namespaces for building our namespace graph.

As a further characterisation of our dataset, Figure 1 shows the degree distribution of both the hostname graph (infrastructure) and the namespace graph (semantic links). Both distributions exhibit a pattern that is not linear. The

---

[4] Of course, if we are interested in connectivity, it is only an approximation to assume that connections only happen along shortest paths; variations of betweenness centrality such as "flow betweenness" and "random walk betweenness" have been proposed to allow for this. In many practical cases however, the simple (shortest path) betweenness centrality gives quite informative answers [12].

[5] http://km.aifb.kit.edu/projects/btc-2010/

Table 1. Size of the two studied networks

Network name	Number of nodes	Number of edges
Hostnames	558841	656012
Namespaces	198	936

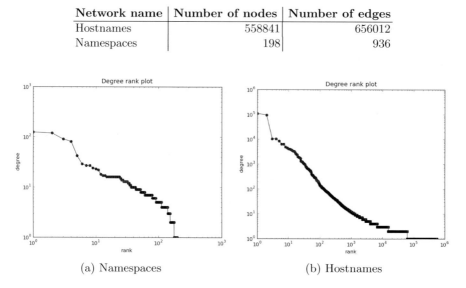

(a) Namespaces   (b) Hostnames

Fig. 1. Degree distribution of the namespaces and hostnames networks

degree shown in distribution does not follow a power law. From this we can conclude that that these two networks are not scale-free. However, they still have a few strongly connected hubs.

3.3   Robustness Results

Based on the extracted graphs, we calculated the betweenness centrality for all nodes in both graphs using the Small-world Network Analysis and Partitioning software (SNAP) [4]. Given the size of the hostname graph, we used an approximation algorithm implemented by SNAP and set the sampling percentage to 10% of all nodes. This is double the 5% percentage suggested for use in [4]. For more details on the algorithm used, see [3].

Infrastructure Analysis. Table 2(a) shows the non-normalised betweenness distribution ($B'(n)$) among the hostnames on the Web of Data, in ten bins starting from the maximal centrality and working down to zero. We note that the distribution does not follow a power-law curve but is in fact more extreme: essentially, almost all infrastructural connectivity on the Web of Data is mediated by only 3 servers. Table 2(b) reveals which hosts these are: `xmlns.org`, `dbpedia.org` and `purl.org`. All this points to an extreme brittleness of the infrastructure underlying the Web of Data: only taking out a handful of servers would completely cripple the entire network.

Figure 2 provides a good example of the potential impact that the dominance of hubs could have on the Web of Data. Recently, Radar Networks which owned www.twine.com was sold to another company Evri[6]. While the transition of twine.com to Evri was smooth, it is entirely possible that www.twine.com could have ceased to exist or no longer supported Web of Data content as a result of this takeover. Our analysis shows that this would have had a substantial impact on the infrastructural connectivity of the Web of Data.

Table 2. Histogram of betweenness for hostnames and the top ten hostnames with the highest betweenness

$B'(n)^7$	#Nodes
$5 - 6 \times 10^9$	2
$4 - 5 \times 10^9$	0
$3 - 4 \times 10^9$	0
$2 - 3 \times 10^9$	1
$1 - 2 \times 10^9$	0
$0.5 - 1 \times 10^9$	4053
$0 - 0.5 \times 10^9$	554785

(a) Distribution of the betweenness results

Hostname	$B'(n)$
xmlns.com	5 693 379 049
dbpedia.org	5 432 125 038
purl.org	2 163 504 423
www.kanzaki.com	532 149 372
www.w3.org	470 113 796
dbtune.org	323 796 691
identi.ca	318 896 524
www.twine.com	299 237 555
semanticweb.org	277 374 029
dblp.l3s.de	225 602 575

(b) Top 10 hostnames and their betweenness result

The 554 785 hostnames with a betweenness of 0 are dead ends in the network. Some of these hosts may be used to serve only non semantic content, such as images. Thus, they do not provide resources that can be interlinked and used to walk through the network. The 4056 other hosts are more representative of the interlinkage status of the graph. This number is much higher than the 198 nodes in the namespaces network (these namespaces account for 60 different hostnames).

Semantic Network analysis. Similar to the infrastructure network analysis, Table 3a shows the betweenness distribution of the namespaces, again arranged in 10 bins. The majority of nodes are not in-between at all and the overall distribution mirrors that of the hostnames graph. The semantic network of the Web of Data, like its infrastructure network, also relies heavily on hubs. Table 3b shows these hubs. These are indeed the hubs one would expect, perhaps with the exception of example.org, which, by definition, can provide no connectivity to other namespaces because it is reserved for examples[8].

---

[6] http://www.novaspivack.com/uncategorized/evri-ties-the-knot-with-twine
[7] Non-normalised betweenness.
[8] See RFC2606, http://www.rfc-editor.org/rfc/rfc2606.txt

Table 3. Histogram of betweenness for namespaces and the top ten namespaces with the highest betweenness

B'(n)	#Nodes
8001-9000	1
7001-8000	1
6001-7000	0
5001-6000	2
4001-5000	0
3001-4000	1
2001-3000	0
1001-2000	6
1-1000	70
0	117

(a) Distribution of the betweenness results

Namespace	$B'(n)$
www.w3.org/1999/02/22-rdf-syntax-ns#	8783
example.org/	7191
dbpedia.org/resource/	5428
xmlns.com/foaf/0.1/	5030
www.w3.org/2002/07/owl#	3926
sw.opencyc.org/concept/	1764
www.w3.org/2007/uwa/ context/deliverycontext.owl#	1737
www.w3.org/2003/01/geo/wgs84_pos#	1609
www.semanticdesktop.org/ ontologies/2007/11/01/pimo#	1300
ontologies.ezweb. morfeo-project.org/eztag/ns#	1225

(b) Top 10 namespaces and their betweenness result

4   Improving the Web of Data

The previous section has shown that the Web of Data is extremely brittle, and relies on a very small number of hubs that are crucial to its connectivity. Both the infrastructure network and the semantic network could be be strengthened by judiciously adding links to the network The expected impact of such new links is to reduce the *variation of the centrality* among the nodes of a graph, thereby diminishing the importance of hubs. The variation of betweenness centrality within a graph is termed the *centralisation betweenness index* [7]:

Definition 5 (centralisation betweenness index). *Given a graph, $G = (N, E)$ with a set of nodes N and a set of edges E, the centralisation betweenness index $C(G)$ of G is defined as*

$$C(G) = \sum_{i=1}^{G} \frac{[max_{n \in N}(B(n)) - B(i)]}{(|N| - 1)}$$

where $B(n)$ is the betweenness of node n in the graph.

4.1   The Cost of Fixing the WoD

The simplest way of reducing $C(G)$ would be to make G a fully connected graph, resulting in an optimal value of $C(G) = 0$. Of course, for the Web of Data this is neither feasible nor desirable, because only semantically meaningful links should be added. Besides, the creation of new edges has a cost. As is well known from the ontology mapping domain, establishing new relations between two ontologies is no easy task. Similarly, finding equivalent instances that can be related by a sameAs triple is challenging.

We have therefore chosen to characterise the problem of recommending where to introduce edges in the Web of Data as an optimisation problem that minimises the centralisation index $C(G)$ while at the same time minimising the cost of introducing an edge.

In the following, we estimate the cost of adding an edge as the inverse of the overlap between the used vocabularies. This estimates the chances of finding pairs of concepts or resources based on the shared usage of predicates by the respective nodes. Intuitively, this cost measure favours "meaningful" edges, i.e. edges between nodes with overlapping vocabularies. Of course, this is a very rough estimation, that could be changed for a more accurate one without impairing the applicability of our algorithms.

Definition 6 (vocabulary of a node). *The vocabulary of a node n from either a hostnames graph $\mathcal{H}$ or a namespaces graph $\mathcal{S}$ is the set of predicates used to describes the resources contained in the node.*

$$vocab(n) = \{p \mid \exists \langle r, p, o \rangle, r \in cont(n)\}$$

Our semantic cost for a link between two nodes will be based on the similarity of the vocabularies used in the nodes. We used the standard Jaccard measure to quantify the similarity between vocabularies. This is a measure commonly used in the ontology mapping domain.

Definition 7 (Vocabulary Similarity). *The similarity $S(n_1, n_2)$ between two nodes n_1 and n_2 from either the hostname graph or the namespace graph is defined as:*

$$S(n_1, n_2) = \frac{|vocab(n_1) \cap vocab(n_2)|}{|vocab(n_1) \cup vocab(n_2)|}$$

The corresponding cost of the edge, $\langle n_1, n_2 \rangle$, is defined as the complementary of the similarity between the nodes:

$$cost(\langle n_1, n_2 \rangle) = 1 - S(n_1, n_2)$$

Of course, we could use any other measure for semantic overlap from work in ontology alignment [6], and again these could be easily plugged into the algorithms we will describe next.

Using these calculations as our basis we now define the optimisation problem as follows:

$$\text{minimize } B(< N, E' >) \text{ subject to } min \sum_{e \in E'} cost(e), \text{ where } E' = E \cup (N \times N)$$

Note, that E' is the union of the existing edges with some set of newly introduced edges from the space of all possible edges in the graph.

4.2   Strategies for Adding New Edges

In order to put this strengthening of the WoD into a reasonable setting, the recommended fixes proposed hereby give an answer to the following question:

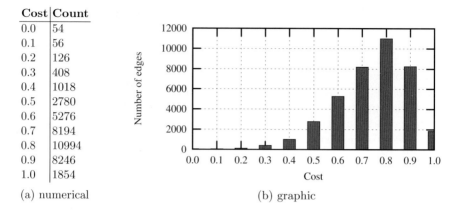

Cost	Count
0.0	54
0.1	56
0.2	126
0.3	408
0.4	1018
0.5	2780
0.6	5276
0.7	8194
0.8	10994
0.9	8246
1.0	1854

(a) numerical   (b) graphic

Fig. 2. Distribution of the costs for the edges than can be added to the namespaces graph

If I have computing resource available to create X new edges, what is the best way to spend them? In the following sections, we highlight some of the strategies that can be considered to give an answer. All the strategies have in common that they work on a finite set of edges that can be added. A graph made of n nodes can only have, at most, $n * (n - 1)$ directed edges if loops are avoided. When computing the list of edges to add, those already existing are not considered.

Note that, in this paper, we focus on the identification of links rather than their publication. The publication of links could be done by the data publishers or by a third party service (as done by Jaffri, Glaser and Millard in [11]).

Greedy Strategies

Start with the cheapest. This first strategy consists in sorting all the edges by their increasing cost and adding them one by one, stopping after X edges have been added. The rationale is that focusing on the cheapest connections will get the best reward/cost ratio for spending the resources available. This nodes are estimated to share the same vocabulary to describe their resources, linking them should increase the density of the clusters they are part of.

The implementation of this strategy requires the enumeration of all of the possible edges and sorting of them according to their cost. We implemented this as a greedy algorithm that computes the cost of all new edges, sorting them and inserting them one by one, measuring the centrality gain after each insertion.

Start with the most expensive. This second strategy is the exact inverse of the previous one. Instead of adding the new edges by increasing cost, the most expensive are added first. Linking nodes which are dissimilar should create bridges among different clusters, thereby diminishing the importance of the existing hubs already connecting these islands. The algorithm implementing this strategy is similar to the previous one and has the same scalability constraint.

Selective Strategies

Choose randomly. Rather than focusing on the cheapest or the most expensive nodes, it could be interesting to select a sample of X of them with different costs. The expected result is to mix bridging some clusters and increasing the density of others. The easiest most straightforward approach is then to randomly select the set of edges to create.

The algorithm implementing this strategy simply creates a set of new edges by sampling two random values between 1 and n. If the drawn edge is already present in the graph or in the set of edges to add, the process is repeated.

Choose wisely. This last strategy accounts for a property ignored by all other strategies: the fact that some edges could be nice to add *in combination with* others. Indeed, the centrality gain is likely not to depend only on how many new edges are created but also on *which ones*. The idea then is not to only select the edges to add one by one but to focus on a group of edges of size X, all at once.

Instead of creating only one set of edges like in the random selection, several sets are investigated in parallel and iteratively improved. This search strategy is done by an evolutionary algorithm, a population based class of algorithm known to perform well on combinatorial optimisation problems [5]. The outline of the evolutionary algorithm, a standard one, is detailed in Algorithm 1. It is a generational evolution with an elitism of 1: every new generation replaces the previous set of candidates with the exception of the best one which is kept.

4.3 Repair of the Namespaces Network

The namespaces network contains 198 nodes for 936 edges, leaving room for $198 * 197 - 936 = 38070$ new edges. The Figures 3 and 4 reports the result of the previously introduced strategies on that network.

The two greedy strategies are compared in figure 3. It can be observed that none of these baselines perform very well in two aspects: (1) many links must be added before obtaining a reasonable improvement of the centrality. 2500 links have to be added to halve the centrality. (2) both strategies first create more damage than improvements. The centrality first increases before going down again. Also, this behaviour is monotonic only after a minimum of edges have been added meaning that these strategies are only applicable if a minimum amount of resources are available. There is however a clear winner on this picture: adding edges by increasing cost is the best approach, damaging less of the network and decreasing its centrality starting at 125 edges. It can thus be concluded that focusing on the easiest pairs is best idea when one can not do better and X is large enough.

Choosing which edges to add is one way to do better than the greedy strategies. The results from the two selective strategies are reported in Figure 4. Our first observation is that both strategies outperform the greedy approaches: they are less damaging and reduce centrality faster. The random choice technique has some uncertain behaviour when less than 250 edges are added but is guaranteed

Algorithm 1. Main loop of evolutionary search strategy. The $\oplus$ is a "one-point crossover" operation than mixes two candidate solutions.

Initialise population P;
while *not terminated* **do**

    /* Evaluation of current sets   */

    foreach *Candidate set of edges s in P* **do**

        compute $\frac{C_B(<N, E \cup s>)}{C_B(<N, E>)}$

    /* Creation of new sets   */

    $P' \leftarrow$ best individual from P;

    while *Size of P' different than size of P* **do**

        switch *with a probability of 0.1* **do**

            $s \leftarrow$ tournament selection from P;

            $s' \leftarrow$ tournament selection from P;

            $P' \leftarrow P' \cup s \oplus s''$

        switch *with a probability of 0.8* **do**

            $s \leftarrow$ tournament selection from P;

            foreach *edge s_i of s* **do**

                switch *with a probability of 0.1* **do**

                    $s_i \leftarrow$ randomly created new edge

            $P' \leftarrow P' \cup s$;

    /* Generation replacement   */

    $P \leftarrow P'$;

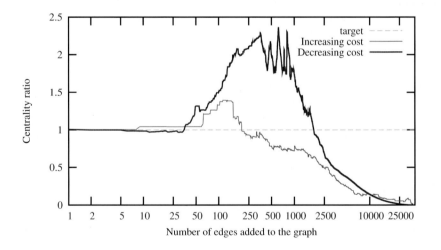

Fig. 3. Comparison of the two greedy strategies that consist in sorting all the edges according to their cost and insert them one by one, by (in/de)creasing cost

to decrease the centrality by almost 60% if at least 1000 edges are created (*e.g.* 2% of the amount of possible new edges). Both algorithms monotonically improve the centrality as soon as more than 250 edges are added. That is around 30% of the existing 936 edges. Above 10000 new edges, there is no difference in the results. For less than 250 new edges, the evolutionary algorithm finds the best sets. It achieves the best performance, decreasing the centrality by almost 60%, with a set of only 64 edges.

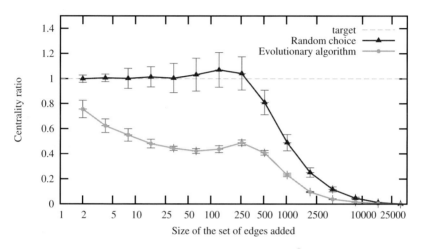

Fig. 4. Comparison of the two selective strategies applied to the namespaces network. They consist in creating a set of edges to add, either by random choice or iterative construction (evolutionary algorithm). The goal is to bring the ratio, at least, below 1.0 and, at best, close to 0.

Table 4 shows the four links recommended to create in order to decrease the centrality of the network by 30%. We now discuss whether the addition of the suggested links is feasible. Row ❶ suggests creating a link from the Lifecycle Schema to Freebase. The Lifecycle Schema describes the specification of a generic lifecycle for a resource. It defines notions such state, transition and task. Links could easily be created from this schema to descriptions of the corresponding concepts in Freebase. For example, one could link to the definition of Finite-state machine in Freebase (i.e. `http://rdf.freebase.com/ns/finite_state_machine`). Row ❷ recommends creating a link between annotations about papers from ISWC 2004 to the Ubiquitous Applications Location Ontology. This seems reasonable since one could describe the papers as having been presented at a particular geolocation, which this location ontology supports. An important note is that the given link for ISWC 2004 annotations is no longer operative. It should probably be updated to the Semantic Web Dogfood site. This is another example where old links cause the Web of Data to break. The third recommendation, Row ❸, suggests adding a link between a site describing labels for about 1 million commodities to SKOS-XL (an ontology for describing labels). A connection

302   C. Guéret et al.

between these sites again seems reasonable as one could possibly describe these commodity labels as subclasses of skosxl:Label. Finally, the recommendation, Row ❹, to link the Dublin Core types to the Cyc Ontology also could be done given that the Dublin Core types describe generic types such as Event, Image, Sound, which also appear in Cyc.

Table 4. When added all together to the namespaces graph, these 4 edge brings the centrality to 70% of its original value

From namespace	To namespace	Cost
❶ http://purl.org/vocab/ lifecycle/schema#	http://rdf.freebase.com/ns/	0.999803
❷ http://annotation.semanticweb. org/2004/iswc#	http://www.w3.org/2007/uwa/ context/location.owl#	0.892857
❸ http://openean.kaufkauf.net/id/	http://www.w3.org/2008/05/ skos-xl#	1.0
❹ http://purl.org/dc/dcmitype/	http://sw.opencyc.org/concept/	1.0

4.4 Repair of the Hostnames Network

The hostnames network contains 558784 nodes for 656012 edges, leaving room for $558784 * 558784 - 656012 = 312238902644$, 312 Billions, new edges. Unfortunately, such a huge number of edges makes search by enumeration impossible and the greedy approaches inapplicable. Instead, we only apply the selective strategies.

For the random strategy, as long as the number of edges added reaches 100M (that is, 0.03% of the 312B possibilities), it does not matter which ones are added. In every case, the centrality is diminished by at least 90%, going to 10% of the original value. This applies similarly for the evolutionary strategy, however, that strategy performs slightly better than the random strategy. Unfortunately, both strategies have a significant adverse impact on the hostname network before any improvement is seen for less than 100M edges added and no impact for less than 10k edges.

5 Conclusion

We can divide the conclusions of this paper into two categories: (i) *generic methods* for analysing the Web of Data, and (ii) *specific observations* on the state of the current Web of Data.

Generic methods for analysing the Web of Data

– We have defined two useful abstractions over the Web of Data, the hostname graph and the namespace graph, allowing us to analyse both the infrastructural and its semantic connectivity of the Web of Data.

- Following insights from network analysis, we have proposed betweenness centrality as the key metric for measuring network robustness (= the ability to maintain connectivity after removal of nodes).
- We have phrased the problem of improving the robustness as an optimisation problem, aiming to minimise the graph's centrality index under minimal cost of adding links. We proposed as a cost-function the Jaccard distance measure based on vocabulary overlap, but our approach is neutral as to the choice of the cost-function.
- We investigated the feasibility of a number of algorithms to solve this optimisation problem, and showed that, in particular, the use of an evolutionary algorithm was successful in identifying a small number of links that substantially increase the robustness of the graph.

Observations on the state of the current Web of Data. Assuming that the BTC dataset is indeed a representative snapshot, the following facts have been revealed by our analysis:

- The vast majority of triples on the Web of Data do not contribute to it being a web, but instead point to literals or blank nodes, or refer only to URI's internal to the same dataset. This concerns as much as 80% of all triples.
- The Web of Data is currently not a scale-free network. It shows a more extreme distribution, although it has some of the typical properties of a scale free network, in particular the presence of hub-nodes.
- Almost all infrastructural connectivity on the WoD is mediated by 3 servers, `xmlns.com`, `dbpedia.org` and `purl.org`, making the system very brittle.
- Similarly, almost all semantic connectivity is provided through a small number of namespaces, again a very brittle structure.
- On the positive side, the robustness of the Web of Data can be improved drastically: the centrality of the namespace graph can be improved by a factor of 2 by adding just 4 edges to the namespace graph.
- For the hostnames graph, we were not able to find any such easy fixes. In fact, it seems that the hostnames graph will need substantial (and hence automated) extension for it to become more robust.

Future Work. A first task would of course be to extend this work to larger snapshots of the Web of Data, to see if our methods scale and if our findings generalise. Currently, the hostname graph is already at the limits of what is computationally feasible to solve the link-optimisation problem. In particular, repeatedly testing the centrality index of candidate graphs that are generated by our evolutionary algorithm is very expensive. An incremental algorithm calculating the centrality index of a slightly modified graph would be helpful here.

A more fundamental extension to our work would be to change our analysis into a real-time monitoring engine that would constantly monitor the state of the Web of Data, *e.g.* by taking as input a stream of modifications, and produce as output a set of suggestions for useful links to add in order to maintain or improve robustness. Unlike the regular Web, where failure is tolerated, the Web of Data is meant for machine consumption, implying that it is more in need of constant and machine-assisted upkeep. In this paper, we have provided the necessary abstractions for such quality control, and we have shown that the Web of Data in its current form has severe vulnerabilities. We have also proposed effective algorithms for determining repairs. With these results our paper opens the way towards continuous and machine-assisted repairs to the Web of Data.

In some cases adding a link may be less expensive than deploying a mirror. While studying the cost of adding links versus that of deploying mirrors goes beyond the scope of this work, we plan to work on the automated identification and connection to cached data.

References

1. Albert, R., Jeong, H., Barabási, A.L.: Error and attack tolerance of complex networks. Nature 406(6794), 378–382 (2000)
2. Amaral, L.a., Scala, A., Barthelemy, M., Stanley, H.E.: Classes of small-world networks. Proceedings of the National Academy of Sciences of the USA 97(21), 11149–11152 (2000)
3. Bader, D., Kintali, S., Madduri, K., Mihail, M.: Approximating betweenness centrality. In: Bonato, A., Chung, F.R.K. (eds.) WAW 2007. LNCS, vol. 4863, pp. 124–137. Springer, Heidelberg (2007)
4. Bader, D., Madduri, K.: SNAP, Small-world Network Analysis and Partitioning: an open-source parallel graph framework for the exploration of large-scale networks. In: IEEE International Symposium on Parallel and, pp. 1–12. IEEE, Los Alamitos (April 2008)
5. Eiben, A., Smith, J.: Introduction to evolutionary computing. Springer, Heidelberg (2003)
6. Euzenat, J., Shvaiko, P.: Ontology matching. Springer, Heidelberg (2007)
7. Freeman, L.C.: A Set of Measures of Centrality Based on Betweenness. Sociometry 40(1), 35 (1977)
8. Ge, W., Chen, J., Qu, Y.: Object Link Structure in the Semantic Web. In: Aroyo, L., Antoniou, G., Hyvönen, E., ten Teije, A., Stuckenschmidt, H., Cabral, L., Tudorache, T. (eds.) ESWC 2010, Part I. LNCS, vol. 6088, pp. 257–271. Springer, Heidelberg (2010)
9. Gil, R., Garcia, R.: Measuring the semantic web. In: Advances in Metadata Research, Proceedings of MTSR 2005. Rinton Press (2006) ISBN 1-58949-053-3
10. Guéret, C., Wang, S., Schlobach, S.: The web of data is a complex system - first insight into its multi-scale network properties. In: Proceedings of the European Conference on Complex Systems, ECCS (2010) (to appear)
11. Jaffri, A., Glaser, H., Millard, I.: Uri identity management for semantic web data integration and linkage. In: 3rd International Workshop On Scalable Semantic Web Knowledge Base Systems. Springer, Heidelberg (2007)
12. Newman, M.E.J.: The Structure and Function of Complex Networks. SIAM Review 45(2), 167–256 (2003)
13. Zhang, X., Cheng, G., Qu, Y.: Ontology summarization based on rdf sentence graph. In: Proceedings of the 16th International Conference on World Wide Web, WWW 2007, pp. 707–716. ACM, New York (2007)

When owl:sameAs Isn't the Same: An Analysis of Identity in Linked Data

Harry Halpin[1], Patrick J. Hayes[2], James P. McCusker[3],
Deborah L. McGuinness[3], and Henry S. Thompson[1]

[1] School of Informatics
University of Edinburgh
10 Crichton St. EH8 9LW Edinburgh, UK
{hhalpin,ht}@inf.ed.ac.uk
[2] Institute for Human and Machine Cognition
40 South Alcaniz St.
Pensacola, FL 32502 USA
phayes@ihmc.us
[3] Tetherless World Constellation
Department of Computer Science
Rensselaer Polytechnic Institute
110 8th Street, Troy, NY 12180 USA
{mccusj,dlm}@cs.rpi.edu

Abstract. In Linked Data, the use of *owl:sameAs* is ubiquitous in interlinking data-sets. There is however, ongoing discussion about its use, and potential misuse, particularly with regards to interactions with inference. In fact, *owl:sameAs* can be viewed as encoding only one point on a scale of similarity, one that is often too strong for many of its current uses. We describe how referentially opaque contexts that do not allow inference exist, and then outline some varieties of referentially-opaque alternatives to *owl:sameAs*. Finally, we report on an empirical experiment over randomly selected *owl:sameAs* statements from the Web of data. This theoretical apparatus and experiment shed light upon how *owl:sameAs* is being used (and misused) on the Web of data.

Keywords: linked data, identity, coreference.

1 Introduction

As large numbers of independently developed data-sets have been introduced to the Web as Linked Data, the vexing problem of identity has returned with a vengeance to the Semantic Web. As the ubiquitous *owl:sameAs* property is used as the RDF property to connect these data-sets, it has been dubbed the '*owl:sameAs* problem' by publishers and users of Linked Data. However, the problem of identity lies not within Linked Data *per se*, but is a long-standing and well-known issue in philosophy, the problem of identity and reference. What precisely *is* new in the recent appearance of this problem on the Web of Linked Data is that this is the first time the problem is being encountered by different

P.F. Patel-Schneider et al. (Eds.): ISWC 2010, Part I, LNCS 6496, pp. 305–320, 2010.

individuals attempting to *independently* knit their knowledge representations together using the same standardized language. Much of the supposed "crisis" over the proliferation of *sameAs* in Linked Data can be traced to the fact that many mutually incompatible intuitions motivate the use of *owl:sameAs* in Linked Data. These intuitions almost always violate the rather strict logical semantics of identity demanded by *owl:sameAs* as officially defined.

To review, the *owl:sameAs* (abbreviated from hereon simply *sameAs*) construct is defined as stating "that two URI references actually refer to the same thing" [3]. For example, the city of Paris is referenced in a number of different Linked data-sets: ranging from OpenCyc to the New York Times. For example, we find that *dbpedia:Paris* is asserted to be *sameAs* both *cyc:CityOfParisFrance* and *cyc:Paris_DepartmentFrance* (and five other URIs). Yet OpenCyc explicitly states (in English!) that these two are distinct. Is there a contradiction here? Is DBPedia misusing *sameAs*? In this paper we will explore the origins of this (very common) situation, and suggest some ways forward.

As the Semantic Web is a project in development, it is always possible to specify anew various constructs. The project of inspecting alternative readings of *sameAs* has been begun by us in the past by looking at context [9] and proposed ontologies [12]. In this work we bring our research together and validate it empirically. We begin by reviewing the philosophical origin of the problem of identity from Leibnitz's Law in Section 2 and its implementation as *sameAs* in Section 3. In Section 4 we demonstrate a number of theoretically-motivated distinctions that are 'kind of close' to *sameAs* and then systematize these into an ontology in Section 5. Finally, test see if humans can reliably use these distinctions in Section 6, and conclude with recommendations for the future development of RDF in Section 7.

2   What Is Identity?

The father of knowledge representation, Leibnitz, was also the first to phrase a coherent and formalizable definition of identity, often called 'Leibnitz's Law' or the 'The Identity of Indiscernables,' namely that that if x is not identical to y, then there must be some property that they do not share [11]. Or put another way, if x and y share all properties (i.e. if they are indiscernable) then they are identical. This law can then stated logically as $\forall x \forall y \exists P. x \neq y \rightarrow P(x) \wedge \neg P(y)$. The inverse of this is the more trivial law of substitutivity, which can then be stated as $\forall x \forall y. P(x) \wedge P(y) \rightarrow x = y$. Leibnitz's law and the law of substitutivity, which are obvious from a logical perspective, have a number of very practical engineering reprecussions in a distributed knowledge representation system such as the Semantic Web.

A number of classical problems already crop up in this analysis of identity. For example consider changes over time. Should things with different temporal-spatial co-ordinates be counted as different, even if they share the rest of their properties? While that sounds like a common-sense distinction, is Tim Berners-Lee as an adult is the same as Tim Berners-Lee five minutes ago? Or as a child?

Or if he lost his arm? This leads straight in to arguments about perdurance and endurance in philosophy. In any engineering discipline such as knowledge representation (as opposed to say, metaphysical thought experiments), we can *never* enumerate all possible properties.

Instead, we consider only a subset of possible properties. As a result identity based on propery matching is under-determined. One solution is to have *some* properties count as those necessary for identity, namely an explicit *theory of identity criteria*. Are there two different kinds of properties, properties that are somehow *intrinsic* to identity and others that are *extrinsic*, i.e. purely relative to other things?[1] However, this does not mean that all such criteria-based theories are compatible. One can imagine theories of identity based on different criteria, where some theories of identity subsume weaker or stronger ones, but others are simply incommensurable. Problems also arise with respect to (comparing) property *values*, for example when values are vague (is "purple" the same as "rgb(255,0,255)") or imprecise (is "2 inches" the same as "5 cm").

Regardless of these well-known issues, the point of a logical analysis of identity is clear in terms of inference: When someone says two things are the same, *the two things share all the same properties* and so every property of one thing can be *inferred* to be a property of the other. The quesion is: Does such a definition of identity work in a decentralized environment such as the Web of Linked Data?

3   The Identity Crisis of Linked Data

Just because a construct in a knowledge representation language is explicitly and formally defined does not necessarily mean that people will follow that definition when actually using that construct 'in the wild.' This can be for a wide variety of reasons. In particular, the language may not provide the facilities needed by people as they actually try to encode knowledge, so they may use a construct that appears to be *close enough* to what they need. A combination of not reading specifications—especially formal semantics, which even most software developers and engineers lack training in—and the labeling of constructs with "English-like" mnemonics, will naturally lead the use of a knowledge representation language by actual users to vary from what its designers intended. In decentralized systems such as the Semantic Web, this problem is amplified. Far from being a sign of abuse, it is a sign of success, as it means that the Semantic Web is actually being deployed outside academia and research labs.

At first glance, *sameAs* seems to be harmless. Its informal definition is that "the built-in OWL property *owl:sameAs* links an individual to an individual" and "Such an *owl:sameAs* statement indicates that two URI references actually refer to the same thing: the individuals have the same identity" [1]. OWL states that "It is unrealistic to assume everybody will use the same name to refer to

---

[1] For example, using a *single* pre-defined criterion to define identity has been a success in terms of primary keys in databases. OWL also allows us to deploy such a property using the *owl:inverseFunctionalProperty* construct, although this is a rather simple approximation of a full-fledged theory of identity criterion.

individuals. That would require some grand design, which is contrary to the spirit of the web" [1]. The problems with *sameAs* start when we apply the principle of substitution to it, by inferring from a *sameAs* assertion that its subject and object share all the same properties.

Despite efforts such as OKKAM which attempt to get the Semantic Web to re-use URIs [4], with the distributed growth of Linked Data projects new URIs are often being minted for new data-sets independently and then *sameAs* links are added manually or automatically. Furthermore, the entire transitive closure of *all* individuals that are connected by *sameAs* share *all* the same properties, if the official (substitutive) definition is respected.

There is the possibility that *sameAs* could turn the Semantic Web from a web of interconnected data to the semantic equivalent of mushy peas. Of course identity is transitive and substitutive. If all the uses of *sameAs* are semantically correct, all these entailments would be exactly correct. The problem is not that *sameAs* itself is mushing up *Linked Data*, but that it's being used to mean other things than what the specification says it means.

While there have been heroic efforts to deal with these 'co-reference' bundles by the KnoFuss architecture [15] and the Consistent Reference Service [7], these have both been deployed only in certain domains. While there has been much related work in the database community on assessing information quality from uncertain sources of information [16], and some work in the Semantic Web community such as the work of WIQA [2] and Inference Web [13], this work has yet to be widely deployed for Linked Data. As imaginable, this has led to considerable discussion in the Linked Data community that such use of *sameAs* is dangerous and potentially 'wrong' as regards the formal semantics of OWL 1.0. However, since inference is rarely used with Linked Data, these problems are not always noticed. Does the possibility of incorrect inferences even matter if one's application does not use inference? With frameworks such as SiLK increasing the number of *sameAs* [17] statements, is the use of *sameAs* a potential time-bomb for Linked Data, or just a harmless convention?

4   Varieties of Identity and Similarity

What kinds of uses of *sameAs* inconsistent with its strict logical definition may be found in the wild? The kind of uses we find suggest that in some cases the context (which can be given on the Semantic Web as a named graph) of the use of name of is *referentially opaque* despite both names denoting a single thing. In other cases the two things are just similar. In neither case is it implied that either name can be freely substituted for the other (the Principle of Substitution does not hold), nor can all the properties of either name be inferred to hold of the other.

4.1   Identical but Referentially Opaque

The first case is when things are **identical**, *that is the two names do identify to the same thing, but all the properties ascribed to one name are not necessarily* *appropriate for the other*, so their names can not be substituted. In this case, the context of use, like a named graph on the Semantic Web, is referentially opaque. While this may appear to violate the very definition of identity, there are two general cases where this may hold.

The first case is when indeed the two names do identify the same thing, but not all properties asserted using one of the names may be asserted using the other name. This is the case when the particular name used to refer to an object matters in some important way. A typical example of referential opacity arises when we have an explicit representation of an agent's knowledge or belief, and the agent doesn't know that the names co-refer. If the agent believes that the 'Morning Star' refers to Venus, but does not know that the 'Evening Star' also refers to Venus, then an equality substitution (such as using *sameAs*) between the 'Evening Star' and 'Morning Star' will give a false representation of their beliefs, even though this equation is factually true.

Another case is when two names may refer to the same thing and all properties do hold of both names, but it is socially inappropriate to re-use the name in a different context (a context can be given as a named graph in RDF). The central intuition here is there are 'forms of reference' appropriate to a context, especially in social contexts. To use an informal example, when at an event of the Royal Society, Tim Berners-Lee is *Professor Sir* Tim Berners-Lee of MIT and Southhampton, not *timbl* on IRC. This does not mean that in an IRC chat Tim Berners-Lee is somehow *not* a professor, but that within that context those properties do not matter. This property is exceedingly important for Linked Data, as contrary to popular doctrine, URIs are uused often as kinds of names and it is possible that the Web is full of referentially opaque contexts.

4.2   Identity as Claims

One could attempt to avoid the entire problem by simply treating all statements of identity as **claims**, *where the statement of identity is not necessarily true, but only stated by a particular agent.* As different agents may have different sets of claims they accept, different agents may accept different identity statements and so have different inferences. These issues also apply to the Semantic Web insofar as it uses any kind of inference as once an agent accepts an identity claim, the agent is bound to all its valid inferences. Informally, it is one thing for me to link to your URI, but its another thing for me to believe what you say about it as though you were talking about my URI. Put another way, one should be wary of accepting conclusions *over here* that could have been drawn *over there*, so to speak.

In particular, this issue comes into play when different agents describe the world at different levels of granularity. For example, different sources of Linked Data may make subtley different claims about some common-sense term like 'sodium.' This occurs in the case of the concept of sodium in DBPedia, which has a *sameAs* link to the concept of sodium in OpenCyc. The OpenCyc ontology says that an element is the set of all pieces of the pure element, so that sodium in Cyc has a member which is a lump of pure metallic sodium with exactly twenty-three neutrons. On the other hand, sodium as defined by DBPedia includes all isotopes, which have different number of neutrons than 'standard' sodium, and in this particular case are unstable. So, one should not state the number of neutrons in DBPedia's use of sodium, but one *can* with OpenCyc. At least in web settings with little inference or reliance on detailed structures, it is unlikely that most deployers of Linked Data actually check whether or not *all* the properties and their associated inferences are shared amongst linked data-sets.

4.3   Matching

As inspired by *skos:exactMatch*, which states "indicates a high degree of confidence that two concepts can be used interchangeably across a wide range of information retrieval applications." [14], one can imagine a kind of strong similiarity relationship called **matching** where *different things share enough properties enough to substitute for each other*, at least for some purposes. Unlike *skos:exactMatch* this property would apply to things themselves, not just concepts of things. Two descriptions of things can share all the same properties due to only a finite and incomplete number of these properties being described. For example, while a wine-glass is identical to itself, it would match another wine-glass from the same set in a Semantic Web description...at least for the purposes of laying a table. We should also be careful not to mix up names and things. The "Department of Paris" and "District of Paris" may share the same geographical extent, but by what act of civil engineering on a grand scale or legal act in court could such things actually be substituted for each other? Obviously they are not identical and only strongly similar, even if the knowledge representation of them lists all the same properties by virtue of being incomplete.

4.4   Similar

Another relationship is a kind of weaker notion of being **similar**, which is when *two different things share some but not all properties* in their given incomplete description. A wine glass and a coffee-cup are similar as regards holding liquids, but they hold entirely different kinds of liquid usually and are different shapes, so Leibnitz's Law would not hold obviously as they are different things. A real-world example from Linked Data would be the relationship between two biospecimens coming from the same cell line in an experiment [12]. We have observed scientists inclination in practice to connect them with *sameAs*, as the two biospecimens are part of the same cell line. However, this creates inferential problems including causing the specimen to be derived from itself, and important experimental properties to be duplicated! Therefore, it makes more sense to have an identifier that only causes some (but perhaps not all) properties to be shared.

4.5   Related

The final relationship is **related**, when *two different things share no properties in common in a given description but are nonethless closely aligned in some fashion.*

For example, the relationship between a wine-glass and wine. Such complex, structured, yet hard-to-specify relationships between things that are 'kind of close to identity' often arise, such as the relationship between a quantity and a measurement of a quantity and between sodium and a isotope. One example of this from Linked Data is the use of a drug in a clinical trial and the drug itself, which is currently connected via *sameAs* in a Linked Data drug study [10]. Although on some trivial level 'everything is related', there are degrees of relatedness. A drug may be related to many things (such as certain plants it derived from), that fact may have little relevance to, much less identity with, the clinical trial that tested its properties, as these properties could also be synthetically brought about. One is also tempted to engage with some sort of "fuzzy" or numerical weighted uncertainty measure between one and zero of identity, but the real hard questions of precisely where these real values come from and their relationship to actual probability theory muddy these conceptual waters very quickly. It seems that beneath these predicates there is likely to be a whole family of heterogeneous and semi-structured relationships that should be studied more carefully and empirically observed before any hasty judgments are made.

5 The Similarity Ontology

Although in Section 4 we demonstrate a need for a notion of identity that does not have any entailments and the possibility that various forms of similarity are being confused with the notion of identity, we did not explicitly explore the details. One possibility as originally proposed and discussed in [12] would be to propose a number of new relationships of identity based on permutations around each of the properties of transitivity, symmetry, and reflexivity. A new ontology called the *Similarity Ontology* (SO) has been defined that separates each of these out as a new kind of relationship.[2] While one could use these properties to make inferences about the relationship in certain domain-specific cases, one would not thereby necessarily be claiming that any two objects having this new kind of relationship would share properties.

The properties of the Similarity Ontology are shown in Table 1. Unlike identity, similarity properties are not necessarily transitive and symmetric. Note that non-symmetric is not equivalent to asymmetric, but simply not necessarily symmetric. The same applies to non-reflexivity and irreflexivity, and non-transitivity and in-transitivity. Domain-specific properties can be created as sub-properties of one of the eight SO properties in order to maximize interoperability while maintaining distinctions among future concepts of similarity. We have also defined a mapping ontology that shows examples of mappings with existing similarity properties from RDFS, OWL, and SKOS[3] and show the sub-property relationship among the new and existing similarity properties in Fig. 1. These properties cover the wide range of relationships from "a is the same thing as b" to "b has more information about a" and allow the expression of precise concepts of similarity.

---

[2] http://purl.org/twc/ontologies/similarity.owl
[3] http://purl.org/twc/ontologies/similarity-mapping.owl

Table 1. The proposed Identity Ontology. Eight new identity properties derived from the original meta-properties of *sameAs*: Reflexivity, Symmetry, and Transitivity. The prefix "sim" is used for the ontology.

		Transitive	Non-transitive
Reflexive	Symmetric	*so:identical*	*so:similar*
	Non-Symmetric	*so:claimsIdentical*	*so:claimsSimilar*
Non-Reflexive	Symmetric	*so:matches*	*so:related*
	Non-Symmetric	*so:claimsMatches*	*so:claimsRelated*

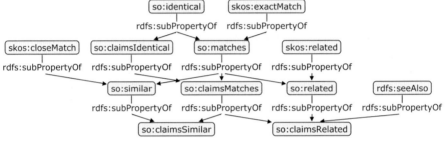

Fig. 1. Sub-property relationships between the properties of the Similarity Ontology and existing properties from OWL, RDFS, and SKOS

so:identical: Two URIs refer to the same thing and so share all the properties, but the reference is opaque. This is the most restrictive property of similarity in SO. It follows the same definition as *sameAs*, which "indicates that two URI references actually refer to the same thing: the individuals have the same identity", but it is referentially opaque and so does not follow Leibnitz's Law [1] As this is the most restrictive property, no other SO properties are sub-properties of it. *sameAs* is defined to be a sub-property so that existing valid assertions of identity are preserved.

so:claimsIdentical: Since this property is transitive and reflexive, but not necessarily symmetric, it serves as a way for one agent to claim two URIs are identical, without the inverse needing to be true. As a super-property of *so:identical*, everything that is actually identical makes the claim of identity, with both sides of the claim being made due to the symmetry of *so:identical*. This property is transitive because if an entity a claims to be entity b and b claims to be entity c, then a cannot deny that it is claiming to be c as well.

so:matches: Two URIs refer to possibly distinct things that share all the properties needed to substitute for each other in some graphs. This property is symmetric but not necessarily reflexive. *so:matches* is a super-property of *so:identical*.

so:claimsMatches: This is the same as *so:matches*, but is not necessarily symmetric, so that things can be claimed to match without reciprocation.

so:similar: Two URIS refer to possibly different things that share some properties but not enough to substitute for each other. *so:similar* is a super-property of *so:matches*. This is a super-property of *so:identical* since everything that is identical is also similar. It is also a super-property of *skos:closeMatch*[14].

so:claimsSimilar: This is the same as *so:similar* but is not necessarily symmetric. Agents can therefore use this property to claim similarity without reciprocation. As a statement of similarity is in actuality two claims of similarity, so *so:claimsSimilar* is a super-property of *so:similar*. In symmetry with *so:similar*, claims of identity and matching imply a claim of similarity.

so:related: Two URIS refer to possibly distinct things, and share no properties necessarily but are associated somehow. As it is only symmetric, there are no claims to any sort of similarity, matching, or identity. Because of this, *so:related* is a super-property of only *so:matches*, as *so:similar* and *so:identical* are reflexive, which would make *so:related* reflexive by proxy. This property is closely related to *skos:related* [14].

so:claimsRelated: This is the loosest sense of identity in SO. It is a similar property to *rdfs:seeAlso*, which is "used to indicate a resource that might provide additional information about the subject resource." [5] We define *rdfs:seeAlso* to be a sub-property of *so:claimsRelated*. *so:related* and *so:claimsMatches* are both super-properties of *so:claimsRelated*.

5.1   Inference

There is a real opportunity here for doing inference. How is this done? It can be said that a particular property or set of properties are isomorphic across a particular kind of similarity. This kind of entailment can be performed through introduction of a property chain, introduced in OWL 2. What people obviously want to express is 'same cell line as,' or more generally, 'same relevant property as' (One could imagine a number of relevant properties and sub-properties). This is much more structured than a vague notion of matching and similarity, and probably more useful. We could do this in OWL now by having a class of identity-restrictions, along these lines:

sameAsClass a IDRestriction.
samePropertyAs relevantProperty P.
A samePropertyAs B.
A P X. B P Y. →
X sameAs Y.

6   Experiment

We have carried out an empirical study of *sameAs* "in the wild". Examples of *sameAs* were taken from the Linked Data Web in order to determine how robust the distinctions offered above are in practice. That is, do people actually use *sameAs* in the different ways that are outlined in the Similarity Ontology? Can people recognize these kinds of distinctions reliably? If at least some of the distinctions between similarity relationships that are currently conflated by *sameAs* can be made in a robust manner, then these distinctions may be candidates for standardization.

6.1   Data

For our experiment we retrieved all *sameAs* triples from the copy of the Linked Open Data Cloud hosted by OpenLink, which totalled 58,691,520 *sameAs* triples from 1,202 unique domain names. The top eight providers of triples show a heavy slant towards biology, being in order: *bio2rdf* (26 million), *uniprot* (6 million), DBPedia (4.3 million), Freebase (3.2 million), Max Planck Institute (.85 million), OpenCyc (.2 million), Geonames (.1 million), Zemanta (.05 million). As shown in Figure 2, when the domain of each URI in the subject and object is plotted by rank-frequency in log-log space, these triples display what appears to be power-law behavior. This is in line with earlier results [8] that show that Linked Data does not necessarily follow a power-law, but something relatively close that does exhibit a somewhat fore-shortened long-tail and nearly exponential behavior in the head. When we used the standard method of Clauset et al. to detect a power-law, the exponent was estimated to be 2.42, but the Monte-Carlo generation of synthetic distributions showed that the distribution failed significantly ($p = .08, p \leq .1$, no power-law found) to be a power-law. Nonetheless, it is seemingly exponential and almost certainly non-parametric.

In order to select a subset for an initial experiment, we first eliminated some classes of triples, and then took a weighted random sample. As the data was to be rated by non-specialists, all biomedical data with *bio2rdf* and *uniprot* links was excluded from the random sampling. Furthermore, the two linked data-sets

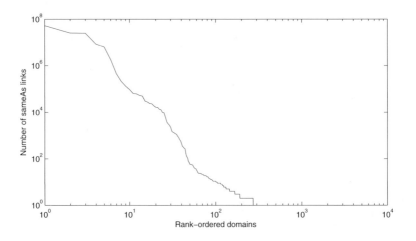

Fig. 2. Frequency of domains in *sameAs* statements in rank-order, logarithmic (base 10) scale

that just copied data (DBPedia) blindly, namely *zemanta* and *Freebase*, were also excluded.

We then drew approximately 500 sample *sameAs* statements at random from the remaining 2.3 million triples. In order to prevent the major data-providers from unfairly dominating the sample, the samples were chosen so that the frequency of URIs in the resulting triples from major providers (those in the exponential head of the distribution) was scaled down by the logarithm of their raw frequency. This down-weighting is intended to result in a balanced and diverse sample of *sameAs* statements. Finally, we attempted to retrieve RDF triples whose subjects were the subject and object URIs of those statements. The 250 cases where this retrieval was successful provided the material for our initial evaluation experiment.

6.2   Experimental Design

We used the *Amazon Mechanical Turk*[4] as a platform for a pilot experiment. Tasks that require some amount of human judgement (such as the judgement about identity) are broken into what are termed Human Intelligence Tasks (HITs) for presentation via the Web to three of the authors. Each HIT covered 10 *sameAs* pairs, as shown in Figure 3, with a standard sample of properties and values from each retrieved RDF triple displayed in two side-by-side tables. We hope to later repeat this experiment on a larger scale using crowd-sourcing via this platform.

The following instructions were given for the forced choice response: **The same:** clearly intended to identify the same thing, without necessarily using the same properties e.g. two different descriptions of a live performance by Queen of *Bohemian Rhapsody*. **Matches:** identifies two copies or versions of the same thing, with the same fundamental properties and differing only with regards to incidental properties, e.g. descriptions of two live performances by Queen of *Bohemian Rhapsody*, but at different locations. **Similar:** Identifies two fundamentally distinct things, but with some properties in common e.g. descriptions of two live performances of *Bohemian Rhapsody*, by two different bands. **Related:** not intended to identify the same thing, but related. E.g. descriptions of the band Queen and of a live performance by Queen of *Bohemian Rhapsody*. **Unrelated:** None of the above. Also, a 'Can't tell' response was available.

As a step towards creating a gold standard, three of the authors assessed all 250 samples. We plotted the results for each judge per category in Figure 4, revealing what appears to be substantial disagreement with respect to some categories. Merging the results of each judge, a table is given in Table 2 that gives raw agreement and disagreement frequencies.

First of all, the vast majority of *sameAs* statements were indeed judged to be correctly identical, and only a relatively small amount were judged to be incorrect. Interestingly enough, a relatively large amount were unknown. Only a small amount were judged as similar, while the amount judged to be matches and

---

[4] https://www.mturk.com/

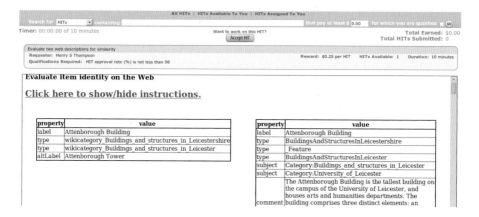

Fig. 3. Mechanical Turk Interface for identity rating

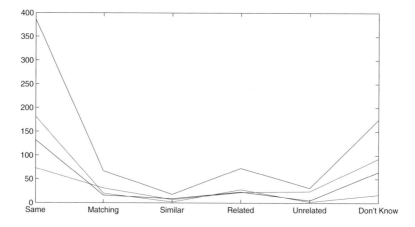

Fig. 4. Number of category assignments per judge. Total across all judges blue, each individual judge is red (1), black (2), and green (3). Y-axis is their frequency in the data-set.

related were modest. To return to Figure 4, it is very clear that the judges have different styles of judgement, with one judge preferring *sameAs* where another judge would be much more strict by usually answering that they can't tell. The remaining judge is in between these two extremes. The amount of disagreement shows that the categories are fairly unstable. However, there is clearly *something* in between not knowing if two URIs are identical and knowing that they are.

Since each question could be considered a binary response over nominal data, we employed the κ statistic to determine agreement between the judges. The κ statistic takes into account agreement between annotators that is greater than chance, and is only valid over nominal data (although our data could be considered ordinal, it is strictly speaking nominal, as each choice is a different relationship rather than a single identity gradient).[5] The κ for the six-way forced choice is 0.158, which is non-accidental but considered 'poor' agreement. Notice that while there was substantial disagreement, there were elements (particularly of identical) where nearly half the data-set was labelled in agreement, likewise for the 'related' category and a substantial portion of 'don't know'. However, the rest of the categories appear to be terminally prone to error. By optimizing and recombining categories, we were able to reach a κ of .319, which indicates 'fair' agreement. This was accomplished by merging the 'similar', 'matching', and 'related' categories, and then merging the 'can't tell' with 'not related' categories, and leaving the 'same as' category to itself. The results, as given per judge in Figure 5, are much more clear. However, there is still substantial disagreement. The main disagreement seems to consist of, rather surprisingly, an inability to agree on 'same as' versus 'can't tell'.

Table 2. Raw numbers of Similarity Categories before optimization

Categories-Rater	Rater 1	Rater 2	Rater 3
Identical	73	132	181
Matching	31	16	20
Similar	7	9	2
Related	22	23	28
Not Related	24	5	2
Can't Tell	93	65	17

The differing habits of the raters in this regard are actually more unstable than their ability to link something using a 'sort of similar or related' category, as shown by inspection of Table 3. It is not in the categories themselves that the problem surfaces, but in the lack of appropriate knowledge for use in determining whether two things are in some context-free manner actually identical. This brings into some doubt the concept of whether or not two things can be declared identical in a context-free manner, and also highlights the importance of background knowledge in determining accurate *sameAs* statements. In this regard, it should not be surprising that there was such high disagreement on manual judging of identity and similarity in Linked Data. However, there are a number of positive results that we can make a guess at by taking the mean of the collapsed categories per rater (and their standard deviation):

- The most postive result is that approximately 51% ($\pm$ 21%) percent of the usage of *sameAs* seems correct.
- While the distinctions made in the Similarity Ontology likely require special training beyond that of even RDF experts, a relatively coarse-grained referentially opaque 'kind-of-similar-and-related-to' relationship can be reliably

---

[5] The derivation of the κ statistic is described in mathematical detail elsewhere [6].

used instead of *sameAs* for intermediate cases (around 21% (± 3%) of our data);
– Approximately 27% (± 19%) of the *sameAs* cannot be reliably judged based only on the RDF retrieved.

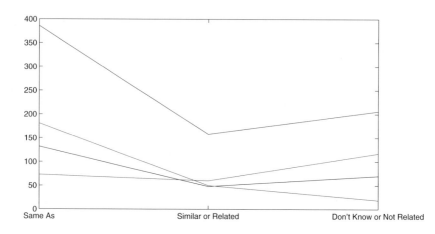

Fig. 5. Frequency of categories in trained expert judges after optimization. Total across all judges blue, each individual judge is red (1), black (2), and green (3). *X*-axis is categories, *Y*-axis is their frequency in the data-set.

Table 3. Raw numbers of Similarity Categories after optimization

Categories-Rater	Rater 1	Rater 2	Rater 3
Identical	73	132	181
Similar, Matching, Related	60	48	50
Can't Tell or Not Related	117	70	19

7   Conclusion

The issue of how to express relationships of identity and similarity on Linked Data is more complex than just applying *sameAs*. We believe the extent of disagreement and inaccurate usage as observed in practice at least calls for additional documentation providing clearer guidance on when to use *sameAs*. Further studies on much larger scales using crowd-sourcing need to be employed to see if the 'default' behaviors of the judges in our experiment generalizes. A further extension of our experiment will test whether the closures of *sameAs* produce surprising and incorrect inferences. This can be done by merging inferred triples with the *sameAs* statements used in the current experiment.

The proposed Similarity Ontology solution has a number of distinctions that may be difficult to deploy consistently in open-ended domains. In fact, like many ontologies, the initial distinctions proposed capture an important intuition, namely that there is a nuanced heterogeneous structure of similarity instead of a strict notion of identity in the use of *sameAs* on the Web, one that will likely result in an asymmetric flow of inference. However, the Similarity Ontology explores too large of a design space to be reliably deployed. A simple similarity property would be quite useful to add to RDF, such as sub-property of *rdfs:seeAlso*. Further study of approaches beyond *sameAs* would be useful if not provocative for the Linked Data community. Solving the issue of identity in Linked Data may require a certain refactoring of some core constructs of RDF, including relating identity to a fully-worked out semantics for named graphs. Furthermore, individuals could be thought of as being composed of differing aspects at different levels of granularity rather than the notion of individuals traditionally used in semantics. In future work, we will also continue investigations into the notion of aspects and named graphs and continue to be inspired by the use cases presenting themselves from the current abundance of misuse of *sameAs* in Linked Data space. The (ab)use of *sameAs* in Linked Data is not a threat, it's an opportunity.

Acknowledgements

We would like to thank reviewers of earlier versions of this work in OWLED 2010, LDOW 2010, and RDF Next Steps for their helpful feedback. Also, special thanks to Kingsley Idehen for helping provide the data-set.

References

1. Bechhofer, S., van Harmelen, F., Hendler, J., Horrocks, I., McGuinness, D.L., Patel-Schneider, P.F., Stein, L.A.: OWL Web Ontology Language Reference (2004)
2. Bizer, C., Cyganiak, R.: Quality-driven information filtering using the wiqa policy framework. Web Semantics: Science, Services and Agents on the World Wide Web 7(1), 1–10 (2009)
3. Bizer, C., Cygniak, R., Heath, T.: How to publish Linked Data on the Web (2007), http://www4.wiwiss.fu-berlin.de/bizer/pub/LinkedDataTutorial/ (last accessed on May 28, 2008)
4. Bouquet, P., Stoermer, H., Giacomuzzi, D.: OKKAM: Enabling a Web of Entities. In: I3: Identity, Identifiers, Identification. Proceedings of the WWW 2007 Workshop on Entity-Centric Approaches to Information and Knowledge Management on the Web, Banff, Canada, May 8. CEUR Workshop Proceedings (2007) ISSN 1613-0073, http://CEUR-WS.org/Vol-249/submission_150.pdf
5. Brickley, D., Guha, R.V.: RDF Vocabulary Description Language 1.0: RDF Schema (2004)
6. Carletta, J.: Assessing agreement on classification tasks: The kappa statistic. Computational Linguistics 22, 249–254 (1996)

7. Glaser, H., Millard, I., Jaffri, A.: RKBExplorer.com: A knowledge driven infrastructure for Linked Data providers. In: Bechhofer, S., Hauswirth, M., Hoffmann, J., Koubarakis, M. (eds.) ESWC 2008. LNCS, vol. 5021, pp. 797–801. Springer, Heidelberg (2008)
8. Halpin, H.: A query-driven characterization of linked data. In: Proceedings of the Linked Data Workshop at the World Wide Web Conference, Madrid, Spain (2009)
9. Halpin, H., Hayes, P.: When owl: sameas isn't the same. In: Proceedings of the WWW 2010 Workshop on Linked Data on the Web, Raleigh, USA (April 25, 2010), http://events.linkeddata.org/ldow2010/papers/ldow2010_paper09.pdf
10. Jentzsch, A., Hassanzadeh, O., Bizer, C., Andersson, B., Stephens, S.: Enabling tailored therapeutics with linked data. In: Proceedings of the WWW 2009 Workshop on Linked Data on the Web, April 20th, 2010, Madrid, Spain (April 2009), http://events.linkeddata.org/ldow2009/papers/ldow2009_paper9.pdf
11. Leibniz, G., Loemker, L.: Philosophical papers and letters. Springer, Heidelberg (1976)
12. McCusker, J., McGuinness, D.: Towards identity in linked data. In: Proceedings of OWL: Experience and Directions, San Francisco, USA (June 21-22, 2010), http://www.webont.org/owled/2010/papers/owled2010_submission_12.pdf
13. Mcguinness, D.L., Silva, P.P.: Explaining answers from the semantic web: The inference web approach. Journal of Web Semantics 1, 397–413 (2004)
14. Miles, A., Bechhofer, S.: SKOS Simple Knowledge Organization System Reference (2009)
15. Nikolov, A., Uren, V., Motta, E.: Knofuss: a comprehensive architecture for knowledge fusion. In: Proceedings of the 4th International Conference on Knowledge Capture, K-CAP 2007, pp. 185–186. ACM, New York (2007)
16. Pipino, L.L., Lee, Y.W., Wang, R.Y.: Data quality assessment. Communications of the ACM 45(4), 211–218 (2002)
17. Volz, J., Bizer, C., Gaedke, M., Kobilarov, G.: Discovering and maintaining links on the web of data. In: Bernstein, A., Karger, D.R., Heath, T., Feigenbaum, L., Maynard, D., Motta, E., Thirunarayan, K. (eds.) ISWC 2009. LNCS, vol. 5823, pp. 650–665. Springer, Heidelberg (2009)

Semantic Need: Guiding Metadata Annotations by Questions People #ask

Hans-Jörg Happel

FZI Research Center for Information Technology, Karlsruhe, Germany
happel@fzi.de

Abstract. In its core, the Semantic Web is about the creation, collection and interlinking of metadata on which agents can perform tasks for human users. While many tools and approaches support either the creation *or* usage of semantic metadata, there is neither a proper notion of metadata need, nor a related theory of guidance *which* metadata should be created. In this paper, we propose to analyze structured queries to help identifying missing metadata. We conduct a study on Semantic MediaWiki (SMW), one of the most popular Semantic Web applications to date, analyzing structured "ask"-queries in public SMW instances. Based on that, we describe *Semantic Need*, an extension for SMW which guides contributors to provide semantic annotations, and summarize feedback from an online survey among 30 experienced SMW users.

1 Introduction

Berners-Lee et al. [3] envisioned a *Semantic Web* populated by machine-understandable metadata based on which agents can reason and act to fulfill tasks for human users. Accordingly, one can distinguish two different roles: the *users* and the *providers* of semantic metadata.

Semantic Web research has addressed both roles and their corresponding work processes to a considerable extent. The usage of semantic metadata is supported by various tools ranging from semantic web service frameworks to ontology-based information retrieval systems. The creation and provision of semantic metadata has been studied in terms of manual and (semi-)automatic annotation systems (e.g. [8]) and with respect to exposing existing structured content on the Semantic Web (e.g. [4]). Surprisingly, only few research has studied topics such as incentives or methods for guiding the creation of semantic metadata so far. Since the provision of semantic metadata remains a costly process, several authors thus call for better means to "support users in the creation of metadata" [6][p. 148] and "to create incentives for annotations" [8][p. 198].

In this paper we propose to guide metadata provision by actual *metadata needs*. In previous research [11], we coined the term *Need-driven Knowledge Sharing* (NKS) to outline a framework connecting the usage and provision of information. We describe how NKS can be applied on the Semantic Web, taking *Semantic MediaWiki* (SMW) as a concrete example.

P.F. Patel-Schneider et al. (Eds.): ISWC 2010, Part I, LNCS 6496, pp. 321–336, 2010.
© Springer-Verlag Berlin Heidelberg 2010

After introducing NKS, we present two heuristics for identifying missing annotations in SMW and describe their application in an exploratory empirical study with SMW installations running on the public internet. Then we present the implementation of *Semantic Need*, an extension for SMW which uses structured queries to guide users in contributing semantic metadata. In a second study, we asked 30 experienced SMW administrators to provide feedback on our general concept and on Semantic Need in particular. Based on the analysis of this data, we discuss further improvements and application scenarios of our approach.

2   Need-Driven Knowledge Sharing

Ultimately, the Semantic Web can be seen as a specialized system for sharing codified information. As when sharing texts and documents, users and providers of information are separated due to the asynchrony of the technology, resulting in reduced motivation and contribution [12]. To address this, we developed the concept of *Need-driven Knowledge Sharing* (NKS) [11].

It is based on the assumption that information needs re-occur over time and across different information seekers, and can thus be used as a means to guide the creation and improvement of information. NKS rejects viewing information sharing as a linear process where all information has to be created prior to any request. In turn, it embraces that an information repository is never 100% *complete*, but grows and evolves over time. This perspective acknowledges the real world experience that individual requests might even fail to deliver any appropriate result, if some information is not yet known to the repository [13].

In a similar fashion, the logical formalisms underlying the Semantic Web share that "information [...] is in general viewed as being incomplete" [2, p. 68] and thus make a so-called *Open World Assumption* (OWA). In opposite to "closed world"-systems such as relational databases, facts that cannot be derived are not considered false but (yet) unknown under the OWA. Thus, a semantic knowledge base (*KB*) usually describes only a limited subset of what is considered true in a domain (see Fig. 1) and might grow over time.

A *KB* can be generally considered as a set of logical statements or axioms[1]. Such axioms might be used to state so-called terminological knowledge which describes classes and properties of the domain (i.e. "Professor is a subclass of Teacher") or about named individuals (i.e. "Rudi Studer is a Professor")[2]. If a *KB* cannot answer a request that is considered to have true results, this can either be due to missing assertions [2, p. 68] but also due to an incomplete specification of the terminology.

Although this evolutionary nature of captured knowledge is a fundamental principle underlying the Semantic Web, there do not exist appropriate methods providing guidance on how a knowledge base should evolve – i.e. which axioms should be added to satisfy information needs. We thus propose to use

---

[1] In RDF these axioms are called triples [16].

[2] The terminological and assertional part of a *KB* are usually referred to as TBox, respectively ABox [2, p. 46].

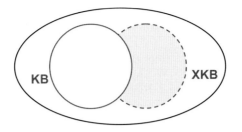

Fig. 1. *KB* denotes the set of all axioms in the knowledge base. *XKB* denotes the set of all axioms which have to be added to the KB to satisfy all structured queries.

structured queries for this purpose. While there is no universal definition of structured queries, we consider so-called *conjunctive queries*[3] [14, p. 294] which are composed of conjunctive query atoms. These atoms may contain variables (i.e. "*Professor(x) ∧ worksAt(x, y)*") which will be assigned concrete instance values from the *KB* if suitable results can be derived from the axioms in *KB*. Formally, a query q can be *satisfied* by a knowledge base KB, if $\exists \mu : KB \models \mu(q)$. The function μ maps every variable of the query to the name of an individual, ensuring that only known individuals are returned by a query [14, p. 295].

We choose *QBox* as the set of all structured queries that have been formulated against a knowledge base *KB*. Due to the inherent incompleteness of the *KB*, we expect that there is a set of *unsatisfied* queries UQ ($UQ \subseteq QBox$) for which holds: $\neg \exists \mu : \forall q \in UQ : KB \models \mu(q)$. UQ' is the subset of UQ, for which true results can be assumed[4]. We thus choose a set of logical axioms XKB such that $\exists \mu : \forall q \in UQ' : XKB \cup KB \models \mu(q)$. We assume that *KB* and $KB \cup XKB$ are *consistent* knowledge bases. Note that *XKB* thus loosely corresponds to the set of axioms filling the "semantic gap between supply and demand on the Semantic Web" as described by [15].

Finally, we choose a set of *partially unsatisfied* queries PUQ' ($PUQ' \subseteq UQ'$) by requiring that $\exists \mu : \forall q \in PUQ' : \exists atom \in q : KB \models \mu(atom)$. We consider PUQ' a particularly relevant subset of the *QBox*, since in opposite to queries in $UQ \setminus PUQ'$, queries in PUQ' ("PUQs") have at least one query atom that can be satisfied from *KB*. Using PUQs, axioms contributing yet missing knowledge can thus be related to existing *KB* statements.

3   Semantic Need Heuristics

We now want to investigate if "semantic gaps" as described in the previous section really occur on the Semantic Web. Since query data is not widely available

---

[3] In particular, we only consider the case of "DL-safe" conjunctive queries in this paper – i.e. we do not allow for non-distinguished variables in query atoms.

[4] For instance, a query for "All volcanoes in Karlsruhe" would not be contained in UQ' since there cannot be any true result (at least if we consider the real world as our domain).

for analysis, we decided to analyze Semantic MediaWiki (SMW) installations, since they contain persisted structured queries on Wiki pages (so called "inline queries").

In the following section, we thus give a brief introduction to SMW. We then apply the NKS concept to SMW, describing *incomplete* and *sparse result sets* as two heuristics for identifying PUQs. Finally, we apply these heuristics to structured queries extracted from public SMW installations.

3.1   Semantic MediaWiki

Semantic MediaWiki (SMW)[5] is an extension to the widespread MediaWiki engine[6]. It allows users to semantically annotate content on Wiki pages such that data can be exported and queried in a structured way.

- A list of conditions (basically categories and property values but also named instances) which should be matched against the knowledge base to constrain the result set
- A list of printout statements from which values should be contained in the result set

PUQs, as introduced in section 2, may either lead to incomplete or sparse result sets for queries in Semantic MediaWiki. We will now elaborate both cases in more detail and introduce heuristics for identifying axioms that should be added to the *KB* to satisfy PUQs.

3.2   Incomplete Result Set

An incomplete result set denotes the situation that an *expected* result is not returned by a structured query. There are several reasons why this can be the case. First, a result instance might no yet be captured – e.g. a query for all instances of the class employee ([[Category:Employee]]) would not yield employees that are not yet known to the system. Second, instance annotations might be incomplete – e.g. a query for all employees with a salary >40.000 ([[Category:Employee]][[salary::>40000]]) would not yield Employee instances that lack any information about their salary.

Clearly, it is not obvious to decide if a given result set is incomplete. One option could be to leverage ontological background knowledge such as cardinality statements on properties. However, such statements are only possible in more feature-rich formalisms such as OWL, but not in RDF or SMW. Thus, another option is to heuristically infer missing results. In the following we present one particular heuristic for this purpose.

---

[5] http://www.semantic-mediawiki.org
[6] http://www.mediawiki.org

Country	Area	Population	Capital	Currency
Burundi	28,000 km²	8,700,000	Bujumbura	
Central African Republic		4,400,400		Central African CFA franc
Mauritius	2,040 km2		Port Louis	Mauritian rupee
Nigeria			Abuja	
Seychelles	455 km2	87,476		Seychellois rupee
Somalia	637,661 km2	9,133,000		Somali shilling
South Africa	1,221,037 km2		Pretoria Bloemfontein Cape Town	Rand
Zimbabwe		12,521,000	Harare	US dollar

Fig. 2. An example of a sparse result set

Near matches. As stated before, structured queries often contain multiple conditions to select particular subsets of an ontology class. The previously mentioned query for employees with a certain salary is an example for this. We define *near matches* as instances in the knowledge base which are potentially relevant results for a given query, but which do not appear in its result set due to missing semantic metadata.

To identify such cases, we only consider queries with at least two conditions. Technically, a candidate "near match" has to match at least one condition of a query and must not match at least one other condition, for which it lacks any annotation. This is to avoid considering instances which are properly described (e.g. an employee with a salary of 30.000, which does not match the query by purpose).

Near matches can thus help indicating missing annotations that prevent instances from appearing as a query result. The underlying assumption is, that these instances potentially could match the information need if metadata would have been properly annotated. Accordingly, we consider them *near matches* and assume that this might offer valuable insights on required metadata to people contributing to a knowledge base.

3.3 Sparse Result Set

In SMW-QL, a cell in the result set will remain empty by default if there is no appropriate binding for that variable (see Fig. 2)[7]. We define *sparse result set* as a case, when at least one cell in a result set remains empty.

Missing Result Values. For SMW queries, empty cells can be considered an unsatisfied information need, since the query requests a variable binding which

---

[7] Note that e.g. in SPARQL the default behaviour will not show the entire result set tuple, if at least one variable can not be bound. This default behaviour can be changed using the *OPTIONAL* modifier [16]. However, in this case, we would end up with an *incomplete result set* as discussed before.

can not be satisfied from the knowledge base. Thus, we define *missing result values* as a heuristic to infer missing annotations. They can easily be derived by simply counting empty cells in query result sets. For the maintainer of a Wiki page it might be interesting to know which printout statements are missing on a particular page in order to help delivering additional information for queries.

3.4 Public SMW Analysis

We now investigate if "missing result values" and "near matches" can be useful heuristics to identify PUQs in real-world settings. Since SMW is a popular MediaWiki extension, there exists a large number of publicly accessible installations which we could use for this purpose.

Design. To check our heuristics we follow the basic research interests how many *missing result values* respectively *near matches* exist for real world structured queries. In terms of information need indicators, we will rely on the analysis of inline queries, since these are the only information needs in SMW which currently have a persistent representation. To select public SMW instances for analysis, we derived an initial list by consulting overview pages and search engines. By dismissing Wikis with only few semantic data (less than 3 queries and 250 annotations), we cut down our list from around 200 to 100. We then ruled out Wikis which were not accessible via a public API or difficult to crawl due to connection problems during the test runs of our evaluation tooling. Out of these, we randomly selected 26 Wikis, which we crawled. Due to the massive amount of data we decided to carry out deeper investigations on eight Wikis described in Table 1.

Table 1. Overview of surveyed SMW installations

Sitename	Pages	ANN^8	$PG_{ANN}{}^9$	IQ^{10}	IQ_{EC}	IQ_{ECPO}	IQ_{ECCJ}
CS Wiki (CS)	195	1.591	67	5	5	5	4
Eroge Wiki (ER)	340	1.853	182	3	1	0	0
HAR2009 (HA)	2.892	3.468	940	38	0	0	0
Historiographus (HI)	998	2.724	390	19	14	10	8
Mount Wiki (MN)	2.662	1.422	833	199	0	0	0
Protege Wiki (PR)	1.545	253	367	11	10	6	4
Sharing Buttons (SH)	122	590	18	7	0	0	0
territoile (TR)	1.801	3.135	502	3	1	1	1
Σ	10.564	15.036	3.299	285	31	22	17

---

[8] Overall number of semantic annotations.

[9] Number of pages containing at least one semantic annotation.

[10] Number of inline queries. Further columns indicate subsets of IQ constrained by evaluation conditions as described in the text.

Process. In order to retrieve data for our analysis, we wrote a crawler[11] which accesses the MediaWiki API. It extracts all semantic annotations and structured queries from the pages and stores it into a database. After retrieving the data, we applied further processing in order to restrict the number of queries for analysis. First, we chose an evaluation condition ("EC") which selects queries that a) are "ask"-queries (ruling out "show" queries) and that b) have either "table" (=default) or "broadtable" as output format (ruling out, e.g., RSS exports of query results). The number of queries satisfying the evaluation condition is shown in Table 1 as IQ_{EC}. In order to further align the set of queries to our analysis, we applied a final selection step. For the analysis of missing result values, we selected those queries that actually contain printout statements (IQ_{ECPO}). Accordingly, we selected only conjunctive queries for the analysis of near matches (IQ_{ECCJ}).

Overall, this processing resulted in 22 queries satisfying IQ_{ECPO} and 17 queries satisfying IQ_{ECCJ}. Due to overlaps of both sets (see Table 2 and 3) this results in 25 distinct queries[12]. As a first step of analysis, we derived the number of results for all queries. Since many queries were located on Template pages, the corresponding fields in Table 2 and 3 denote "n.a.", since the number of results would depend on the page embedding the template. Instead, we computed the number of instances for the [[Category:]] part of the query ($Results_{CAT}$).

Results

Missing result values. Table 2 summarizes the analysis of the IQ_{ECPO} query set. We computed the number of missing result values (e.g. empty cells) in the result set. For queries on normal pages, this is the actual number. For queries on template pages, we summarized the number of missing result values across all instances ($Results_{CAT}$).

As it can be seen from the results, all queries on normal pages provide a complete result set. However, for queries on template pages, up to 63% of cells in the query result set were empty. To estimate if these empty cells were really due to missing information (instead of consciously ommitted), we manually investigated three empty cells for each of five different queries. It turned out, that only two of the 15 empty cells could not be considered missing information. This shows that queries lack result values to a considerable extent. In average, 16% of cells remained empty across all queries surveyed.

Near matches. For the conjunctive queries, we first observed that all 17 queries under consideration consisted of exactly two conjunctions. In most cases, this is a category stametement combined with a restriction on one property (e.g. PR2: [[Category:Plugin]] [[For Application::PAGENAME]]). In order to derive near matches, we computed the number of instances which completly lack the

---

[11] Available at http://www.teamweaver.org/wiki/index.php/MediaWikiTools
[12] See http://www.teamweaver.org/downloads/data/sneed/sneed-smw-queries.pdf

Table 2. Empty/missing result values for the surveyed queries

ID	Results	$Results_{CAT}$[13]	Empty cells	Printout requests	% Empty cells
CS1	n.a.	8	19	4	59%
CS2	n.a.	7	0	3	0%
CS3	n.a.	1	0	1	0%
CS4	n.a.	16	0	2	0%
CS5	7	7	0	4	0%
HI1	1	18	0	3	0%
HI2	28	65	0	2	0%
HI4	n.a.	18	27	3	50%
HI5	n.a.	65	22	2	17%
HI7	n.a.	24	60	4	63%
HI8	n.a.	4	1	3	8%
HI9	n.a.	35	6	4	4%
HI10	n.a.	15	3	4	5%
HI11	n.a.	14	2	4	4%
HI12	n.a.	15	9	4	15%
PR1	72	80	0	1	0%
PR2	n.a.	80	13	1	16%
PR3	n.a.	91	1	1	1%
PR5	n.a.	91	75	2	41%
PR6	n.a.	91	1	1	1%
PR4	n.a.	91	57	1	63%
TR1	70	102	0	1	0%
		Σ 938	Σ 296	Ø2,5	Ø16%

annotation of the restricted property. The rationale behind this is, that these instances might qualify to appear in the query result set, once a correct value for the property is annotated.

As described in the last column of Table 3, up to 94% of instances lacked the annotation on the selection property in extreme cases. Again, we performed a deeper investigated on three near matches for each of five different queries. Out of these 15, five turned out be "false positives" - i.e. were lacking annotations by purpose. While near matches might thus not be a strict indicator for "missing" annotations, they are nevertheless a strong hint. On average, across all queries, a value of 22% turns out. This is a rather high number, considered that this rules out the instances from appearing in the results of the surveyed queries.

Although our analysis is based on a rather small set of queries, this selection can already help to identify up to 296 missing printout statements and up to 147 missing selection properties within the surveyed Wikis. Given the fact that we only analyzed around 9% of the overall inline queries (due to our evaluation conditions), this stresses the potential for using "missing result values" and "near matches" as heuristics for guiding semantic annotations.

---

[13] Number of instances for the [[Category:]] part of the query.

Table 3. Near matches for the surveyed queries

ID	Results	$Results_{CAT}$[13]	Missing selection property	% Missing selection property
CS1	n.a.	8	6	75%
CS2	n.a.	7	0	0%
CS3	n.a.	1	0	0%
CS4	n.a.	16	4	25%
HI1	1	18	17	94%
HI2	28	65	10	15%
HI3	1	3	1	33%
HI4	n.a.	18	17	94%
HI5	n.a.	65	10	15%
HI6	n.a.	3	0	0%
HI7	n.a.	24	13	54%
HI8	n.a.	4	2	50%
PR1	72	80	8	10%
PR2	n.a.	80	9	11%
PR3	n.a.	91	0	0%
PR4	n.a.	91	18	20%
TR1	70	102	32	31%
		Σ 676	Σ 147	Ø22%

4   Semantic Need Implementation

In this section, we first discuss how semantic gaps in a knowledge base can be resolved. Then we present our prototypical implementation of Semantic Need as an extension for SMW and summarize results of a survey conducted among 30 SMW administrators.

4.1   Resolving Semantic Gaps

Semantic gaps in the knowledge base, as indicated in the previous paragraphs, can either be resolved by capturing or by sharing knowledge.

Capturing is necessary, if knowledge is not yet formalized at all. This can involve both, schema-level knowledge or data/annotations. Concerning annotations, "near matches" and "missing result values" can help identifying concrete properties which are not yet annotated for a knowledge base instance. Thus, users can be provided with an interface denoting all missing properties for a given instance as derived by these heuristics. Similarly, one can try to identify if "near matches" and "missing result values" are due to missing schema mappings. This denotes the case if query atoms do not correspond to existing categories or properties in the Wiki. This can either imply that parts of the domain knowledge are missing in the ontology of the Wiki, or it can be an indictor of synonyms – e.g. if a user asks for [[Category:Worker]] instead of [[Category:Employee]]. Thus, the system might assist users in finding candidate mappings to improve the ontology schema and thus help satisfying information needs.

Sharing knowledge can be done if information *is* already formally captured, but not available at query time, since it is hidden in a yet unknown or not accessible knowledge base. Information needs might thus be satisfied by either sharing (i.e. copying) semantic information into the queried knowledge base, or by introducing suitable mappings, which allow the query engine to retrieve semantic information from distributed spaces.

4.2 Semantic Need for MediaWiki

Our current implementation adresses the *capturing* of semantic *annotations*, while the *sharing* of semantic information and the provision *schema-level knowl-edge* are foreseen in the system design, but not yet realized. We also currently focus on so-called "inline queries" embedded in Wiki pages. We consider them the most relevant, since many end users might not be able or willing to formulate ad hoc structured queries on their own. Basic information about inline queries is stored in a "semantic query log" which includes the conditions and printout statements of the query. Due to space restrictions, we will skip details on the storage by now[14]. Based on the query log, a so-called *Need API* offers *metadata need* information such as "near matches" and "missing result values".

One consumer of such need information is the *Capturing UI*, a special user interface which allows *knowledge engineers, domain experts* or *end users* to contribute potentially missing facts to the knowledge base. We realized two different types of implementation so far. First, we provide "global" overview pages which list all queries – in particular those without results – and a Wiki-wide overview of pages and their missing annotations. Second, the same feature is applied to individual pages, resulting in an overview of missing annotations for a specific Wiki page. This can be considered a semantic counterpart to the MediaWiki page `Special:WhatLinksHere`, which helps users to find out how a Wiki page is *syntactically* embedded (i.e. linked).

Although usability issues are not a core focus of this paper, we also thought about how to address actual end users who might contribute to the Wiki more directly (see Fig. 3). Besides this, several other ways to inform users about contribution possibilities can be imagined – including integration in Java-Script based annotation UIs[15], game-based interfaces (e.g. [17]) or identifying and approaching potential contributors directly (e.g., by E-Mail).

4.3 Semantic Need Survey

While an initial implementation of the Semantic Need extension is already available, it is not yet robust enough for an evaluation in the field. We thus decided to evaluate the current version based on an expert survey among experienced SMW administrators, which we describe in the following.

---

[14] Initial (but more general) ideas have been presented in [9].
[15] Such as http://smwforum.ontoprise.com/smwforum/index.php/ Help:Introduction_to_Advanced_Annotation_Mode

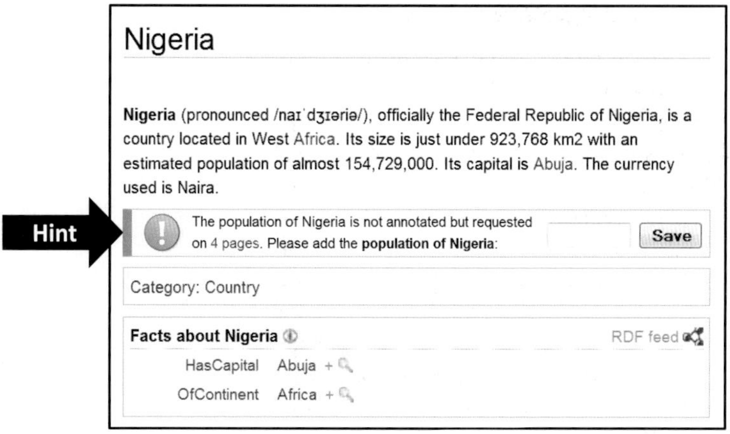

Fig. 3. In-page display and input form for missing annotations

Design and Process. The main goal of the survey was to gather feedback on our current concept and its realization. We thus decided to include a small example scenario with screenshots of SMW and our extension. Since this requires a) prior knowledge of SMW, b) a holistic view of an existing SMW installation and its usage and c) results in a rather large questionnaire, our main target group consists of experienced SMW administrators rather than end users.

The questionnaire consists of five major components. Two parts address the problems of a *sparse* and *incomplete result set*, asking respondents about the frequency and severity of these issues. Another part deals with semantic annotation practices. People are asked how they find out missing annotations in a standard SMW. Afterwards, screenshots of Semantic Need are shown (including Fig. 3) and people are asked if they agree that Semantic Need might be effective to a) generally help maintaining annotations, b) focus annotation effort and c) motivate users to provide contributions. Two other parts of the survey address the usage context of SMW. We asked about the knowledge domain captured in the Wiki and the structure and content of the knowledge base.

The final questionnaire has 34 questions[16]. It was pre-tested by 5 persons resulting in some minor modifications and clarifications. To gather participants for the survey, we followed two strategies. Since we were interested in frequent SMW users, we advertised our survey on the official SMW user and developer mailinglists. Furthermore, we directly contacted 15 persons which are known to drive own SMW projects.

Results. We received 30 complete answers. A majority of 15 answers came from Germany, 7 from the US while the remaining participants are scattered across eight different (mostly European) countries. Concerning their experience with

---

[16] See http://www.teamweaver.org/downloads/data/sneed/sneed-survey.pdf

SMW, 15 respondents describe themselves as "intermediate", 11 as "expert" and 4 as "novice". On average, they are using SMW for 2.3 years.

The knowledge domain captured in SMW is characterized as "fixed/standardized" in 8 cases, as a "generally open domain without many predetermined entities and properties" in 6 cases and as a mix of both options in 15 cases. Accordingly, the semantic data model is largely prescribed by Semantic Forms/Templates in 19 cases. Only 7 SMWs have an equal level of prescribed and ad hoc structure and another 4 rely mostly on free-form annotations. None of the Wikis surveyed do *not* use Semantic Forms/Templates at all. 12 people answered that no particular methodologies, practices or tools are used to maintain the semantic data, while 5 people claim to follow simple informal practices and 7 people implement changes based on more advanced measures such as scripts, documentation and team decisions. In 7 cases, the data stored in SMW is driven by the structure from external data and systems.

The problem of *sparse result set* was observed "often" or "sometimes" by 18 people, while 12 indicated "rarely" or "never". 15 people rate the issue as "not problematic" while 12 answered "somehow problematic". No one rated query result sparseness as "very problematic". In their free text justification, people made the point that the application context (4 answers) and the nature of the data itself (5) have an impact on if query result sparseness is an actual problem.

For *incomplete result sets*, 19 people answered to have observed the issue "often" or "sometimes" while 9 observed it "rarely" or "never". Furthermore, only 5 people consider the issue "not problematic", while 18 answered "somehow problematic" and 5 even "very problematic". This is stressed by the free text justifications in which 16 respondents repeated that query result incompleteness is a problematic issue. Key aspects are the "invisiblility" of the issue (which makes it worse than sparse query results) which is quantified if the dataset grows large: *"due to the nature of our wiki (IT company) it is hard to know when a query is incomplete. For example, there are hundreds of pages on servers so impossible to know when one or several are missing."*

We also asked how people would deal with finding out missing annotations for a particular Wiki page and clustered the free-form answers in four main categories. 6 answers suggest to make a comparison with annotations on similar Wiki pages. Related to that, 7 people would check the schema (i.e. properties) and forms related to that page. Another 4 people would do an analysis of the page text to identify additional content that could be formalized. Finally, 10 answers suggested to create specific ask-queries for this purpose. It turns out that *decisions* are a core part of this process – as one answer puts it: *"Write down a list of all the quantifiable data on the page. - Then decide if any of these are excessive in depth for most users. - In this case I would add part of africa, size, population, and currency."*

The global overview about Wiki pages and their missing annotations is generally appreciated in the survey. On a 5-point scale ranging from "strongly diagree" to "strongly agree", most respondents agree that this feature can be effective to maintain semantic annotations in SMW $(8/18/2/2/0^{17})$. The agreement is slightly less on if it can help to guide annotation efforts towards most crucial information needs $(5/18/6/1/0)$ and on if it can motivate users to provide missing annotations $(9/13/5/2/1)$. The page-specific features of Semantic Need are even more appreciated. 15 respondents strongly agree that it can be effective to maintain semantic annotations in SMW $(15/11/2/2/0)$. Concerning annotation guidance and user motivation, 26 respondents at least chose "agree" in both cases $(12/14/3/1/0)$. Finally, 20 participants (66%) are interested in using the Semantic Need extension on their own Wiki.

Summarizing we can observe that SMW usage differs largely – ranging from prescribed data structures to more open, Semantic Web-inspired scenarios. While the first group argues that data quality and completeness is crucial in their case and thus considers missing annotations a serious problem, others stress the evolving nature of Semantic Web applications: *"I don't see this is a 'problem' - it's the way things are, always in flux, always perfecting, always coming to stasis. Law of Thermodynamics."* Semantic Need however, was considered helpful by both groups – either to help raising data quality or to provide guidance in less predefined settings.

5   Design Implications

In this section we reflect on our overall approach, the Semantic Need prototype and the data we have gathered to validate it. We identified a number of design parameters which we consider useful for our own future work but also for other people developing Semantic Web applications.

The need for need specification: Surprisingly the Semantic Web, which is all about expressing knowledge in a formal way, has not yet done much in terms of *expressing information needs*. We thus consider an ontology which helps users to characterize their information needs more precisely (e.g., duration or urgency) helpful. This should be complemented by appropriate *semantic query log* standards and storage mechanisms.

Data quality modeling: Several people in our survey argued that some of the identified "problems" might just be intended states: *"I have seen instances in which sparseness was intentional, i.e., a query is created specifically to show the absence of data - it can be useful in the right circumstances.".* Thus, the precision of information need heuristics depends on assumptions and background knowledge of the application domain. While some people suggested to specify properties e.g. as *mandatory*, these features are either not part of the knowledge representation formalism[18] or hardly used (such as cardinality constraints).

---

[17] Amount of answers stating: strongly agree/agree/neutral/disagree/strongly disagree.
[18] In the case of SMW, some are artificially enforced by the Semantic Forms extension.

A scattered Semantic Web: While we focus on single SMW instances in this paper, we think that our concepts are also useful on a larger scale. The Semantic Web is decentralized and heterogeneous and so are the "semantic intranets" of some of our survey partcipants: *"because my data comes from ExternalData, I wouldn't ENTER those properties on the Wiki itself, but this extension would help us to go back to the source and add it.".* A Semantic Need-enabled SMW could thus pull data (and information needs) from external systems, capture mappings and share data in external places. While an interconnected set of SMW instances would be a straightforward idea, we also think that our approach could be implemented in other Semantic Web applications, given a set of standards for information need description and exchange.

Ontology evolution vs. maturing: Much research on ontology dynamics has a technical spin under the label of *ontology evolution.* However, our survey results show that data integrity is not the only concern in this field. While the Semantic Forms extension, which helps to "freeze" parts of the data structure, has been quickly adopted by many SMW administrators, the process for dealing with emerging entities is not yet well adressed. We thus argue that methodological considerations such as the *ontology maturing* concept [5] should be given more attention in the design of Semantic Web applications.

6   Related Work

Since our main goal is to guide the creation of semantic metadata, work in the area of *semantic annotation* is partly relevant for us. However, most systems, such as CREAM [8] are inspired by the linear perspective of the information foodchain [7] and thus drive the annotation process by the pre-defined ontology structure. While guidance and incentives for annotation are considered major open issues ([8][p. 198], [6][p. 148]), we are not aware of other approaches considering queries for guiding the annotation process.

The probably most directly related work to ours is a recent study by Mika et al. [15]. Similar to our NKS framework, they contrast and connect the perspectives of semantic metadata provision and usage. However, they use a slightly different approach by taking keyword queries from Yahoo query logs and mapping those to entity/property pairs, which they compare to actual semantic metadata from DBPedia. Thus, the work is primarily of descriptive nature and does not suggest actual technical solutions. The evaluation track of the SemSearch workshop[19] and the evaluation campaign of the SEALS project [18] are recently emerging initiatives to capture and analyze structured query data.

The work presented in this paper might also be considered related to approaches for maintaining or *gardening* semantic knowledge bases. A particular example is the Semantic Gardening Extension[20] for SMW. However, it is focused

---

[19] http://km.aifb.kit.edu/ws/semsearch10/

[20] http://smwforum.ontoprise.com/smwforum/index.php/
    Help:Semantic_Gardening_Extension

on knowledge base instances without properly defined ontology classes or ontology classes without instances. A *need dimension*, taking into account the actual usage of semantic data is currently not part of this work.

As for the core idea of driving knowledge sharing by user requests, the seminal Answer Garden system [1] deserves credit. While Answer Garden uses experts to filter and answer requests, so called "Collaborative Question Answer Systems" (such as Yahoo Answers[21]) and our Woogle system [10] embrace all users as potential contributors.

7 Summary

This paper has described three major contributions. First, we have argued for considering information needs – and in particular structured queries – as drivers for the process of creating semantic metadata. To this end we introduced the *Semantic Need* approach which guides contributors to create metadata which is of the most value for other users in the Semantic Web. Second, we introduced an extension for SMW as a proof-of-concept realization of this approach. While this stresses the general feasibility of our ideas, we think that a realization within a larger Semantic Web scope is possible as well (see also [9]).

Third, we conducted two empirical studies to validate our claims. Our analysis of public SMW installations shows, that the current application areas of Semantic Need – *missing result values* and *near matches* – occur in the surveyed dataset to a considerable extent and are thus of practical relevance. This is also stressed by the result of an expert survey among 30 experienced SMW administrators. Their feedback provides initial evidence that Semantic Need can be an effective tool to support the guided growth of semantic knowledge bases.

Beyond that, our framework and our empirical data will enable us to pursue further studies of that kind. Obvious directions would be to guide ontology schema evolution and mapping or query refinements based on information needs.

Acknowledgements

This work was partially supported by the THESEUS project, which is funded by the German Federal Ministry of Economics (BMWi) under grant 01MQ07019, and the GlobaliSE project, which is funded by the Baden-Württemberg Stiftung. Thanks go to Andreas Abecker, Markus Krötzsch, Sebastian Rudolph, Stephan Grimm, Athanasios Mazarakis and Heiko Haller for helpful feedback and to Paul Hübner and Hristo Valev for their implementation of the MediaWiki crawler and the Semantic Need extension.

References

1. Ackerman, M.S., Malone, T.W.: Answer garden: a tool for growing organizational memory. In: Proceedings of the ACM SIGOIS Conference on Office Information Systems, pp. 31–39. ACM, New York (1990)

---

[21] http://answers.yahoo.com/

2. Baader, F., Nutt, W.: Basic description logics. In: Baader, F., Calvanese, D., McGuinness, D.L., Nardi, D., Patel-Schneider, P.F. (eds.) Description Logic Handbook, pp. 43–95. Cambridge University Press, Cambridge (2003)
3. Berners-Lee, T., Hendler, J., Lassila, O.: The semantic Web. Scientific American 284(5), 34–43 (2001)
4. Bizer, C., Cyganiak, R.: D2r server-publishing relational databases on the semantic web (poster). In: International Semantic Web Conference (2006)
5. Braun, S., Schmidt, A., Walter, A., Nagypal, G., Zacharias, V.: Ontology maturing: a collaborative web 2.0 approach to ontology engineering. In: Proceedings of the Workshop on Social and Collaborative Construction of Structured Knowledge (CKC 2007), CEUR Workshop Proceedings, vol. 273 (2007)
6. Decker, S.: Semantic web methods for knowledge management. Ph.D. thesis, University of Karlsruhe (2002)
7. Decker, S., Jannink, J., Melnik, S., Mitra, P., Staab, S., Studer, R., Wiederhold, G.: An information food chain for advanced applications on the www. In: Borbinha, J.L., Baker, T. (eds.) ECDL 2000. LNCS, vol. 1923, pp. 490–493. Springer, Heidelberg (2000)
8. Handschuh, S.: Creating ontology-based metadata by annotation for the semantic web. Ph.D. thesis, University of Karlsruhe (2005)
9. Happel, H.J.: Growing the semantic web with inverse semantic search. In: 1st Workshop on Incentives for the Semantic Web (INSEMTIVE 2008), pp. 1–12 (2008)
10. Happel, H.J.: Social search and need-driven knowledge sharing in wikis with woogle. In: Proceedings of the 5th International Symposium on Wikis and Open Collaboration, WikiSym 2009, pp. 1–10. ACM, New York (2009)
11. Happel, H.J.: Towards need-driven knowledge sharing in distributed teams. In: Proceedings of the 9th International Conference on Knowledge Management, pp. 128–139. JUCS (2009)
12. Happel, H.J.: Semantic need: An approach for guiding users contributing metadata to the semantic web. Int. J. Knowledge Engineering and Data Mining (to appear, 2010)
13. Happel, H.J., Mazarakis, A.: Considering information providers in social search. In: Proceedings of the 2nd International Workshop on Collaborative Information Seeking (CIS 2010), pp. 1–5 (2010)
14. Hitzler, P., Krötzsch, M., Rudolph, S.: Foundations of Semantic Web Technologies. Chapman & Hall/CRC (2009)
15. Mika, P., Meij, E., Zaragoza, H.: Investigating the semantic gap through query log analysis. In: Bernstein, A., Karger, D.R., Heath, T., Feigenbaum, L., Maynard, D., Motta, E., Thirunarayan, K. (eds.) ISWC 2009. LNCS, vol. 5823, pp. 441–455. Springer, Heidelberg (2009)
16. Prud'Hommeaux, E., Seaborne, A.: SPARQL query language for RDF. World Wide Web Consortium, Recommendation REC-rdf-sparql-query-20080115 (January 2008)
17. Siorpaes, K., Hepp, M.: Ontogame: weaving the semantic web by online games. In: Bechhofer, S., Hauswirth, M., Hoffmann, J., Koubarakis, M. (eds.) ESWC 2008. LNCS, vol. 5021, pp. 751–766. Springer, Heidelberg (2008)
18. Wrigley, S.N., Reinhard, D., Elbedweihy, K., Bernstein, A., Ciravegna, F.: Methodology and Campaign Design for the Evaluation of Semantic Search Tools. In: Proceedings of the Semantic Search 2010 Workshop, SemSearch 2010 (2010)

SAOR: Template Rule Optimisations for Distributed Reasoning over 1 Billion Linked Data Triples[*]

Aidan Hogan[1], Jeff Z. Pan[2], Axel Polleres[1], and Stefan Decker[1]

[1] Digital Enterprise Research Institute, National University of Ireland, Galway
{firstname.lastname}@deri.org
[2] Dpt. of Computing Science, University of Aberdeen
jeff.z.pan@abdn.ac.uk

Abstract. In this paper, we discuss optimisations of rule-based materialisation approaches for reasoning over large static RDF datasets. We generalise and re-formalise what we call the "partial-indexing" approach to scalable rule-based materialisation: the approach is based on a separation of terminological data, which has been shown in previous and related works to enable highly scalable and distributable reasoning for specific rulesets; in so doing, we provide some completeness propositions with respect to semi-naïve evaluation. We then show how related work on template rules – T-Box-specific dynamic rulesets created by binding the terminological patterns in the static ruleset – can be incorporated and optimised for the partial-indexing approach. We evaluate our methods using LUBM(10) for RDFS, pD* (OWL Horst) and OWL 2 RL, and thereafter demonstrate pragmatic distributed reasoning over 1.12 billion Linked Data statements for a subset of OWL 2 RL/RDF rules we argue to be suitable for Web reasoning.

1   Introduction

More and more structured data is being published on the Web in conformance with the Resource Description Framework (RDF) for disseminating machine-readable information, forming what is often referred to as the "Web of Data". This data is no longer purely academic: in particular, the Linked Data community – by promoting pragmatic best-practices and applications – has overseen RDF exports from, for example, corporate bodies (e.g., BBC, New York Times, Freebase), community driven efforts (e.g., Wikipedia, GeoNames), the biomedical domain (e.g., DrugBank, Linked Clinical Trials) and governmental bodies (e.g., data.gov, data.gov.uk). At a conservative estimate, there now exists tens of billions of RDF statements on the Web.

Sitting atop RDF are the RDF Schema (RDFS) and Web Ontology Language (OWL) standards. Primarily, RDFS and OWL allow for defining the relationships between the classes and properties used to organise and describe entities, providing a declarative and extensible domain of discourse through use of rich formal semantics. One could thereafter view the Web of Data as a massive, heterogeneous, collaboratively edited

---

[*] The work presented in this paper has been funded in part by Science Foundation Ireland under Grant No. SFI/08/CE/I1380 (Lion-2), by the EU MOST project, the EPSRC LITRO project, and by an IRCSET Scholarship.

P.F. Patel-Schneider et al. (Eds.): ISWC 2010, Part I, LNCS 6496, pp. 337–353, 2010.

knowledge-base amenable for reasoning: however, the prospect of applying reasoning over (even subsets of) the Web of Data raises unique challenges, the most obvious of which are the need for scale, and tolerance to noisy, conflicting and impudent data [6].

Inspired by requirements for the Semantic Web Search Engine (SWSE) project [9] – which aims to offer search and browsing over Linked Data – in previous work we investigated pragmatic and scalable reasoning for Web data through work on the Scalable Authoritative OWL Reasoner (SAOR) [7,8]; we discussed the formulation and suitability of a set of rules inspired by pD* [16] for materialisation over Web data. We gave particular focus to scalability and Web tolerance showing that by abandoning completeness, materialisation over a diverse Web dataset – in the order of a billion statements – is entirely feasible wrt. a significant fragment of OWL semantics. From the scalability perspective, we introduced a partial-indexing approach based on a separation of terminological data from assertional data in our rule execution model: terminological data – the most frequently accessed segment of the knowledge base for reasoning which in our scenario represents only a small fraction of the overall data [8] – is stored and indexed in-memory for fast access, whereas the bulk of (assertional) data is processed by file-scans. Related approaches have since appeared in the literature which use a separation of terminological data for applying distributed RDFS and pD* reasoning over datasets containing hundreds of millions, billions and hundreds of billions of statements [19,18,17]. However, each of these approaches has discussed completeness and implementation/optimisation based on the specific ruleset at hand.

In this paper, we reformulate the partial-indexing approach – generalising to arbitrary rulesets – and discuss when it is (i) complete with respect to standard rule closure; and (ii) appropriate and scalable. We then introduce generic optimisations based on "template rules" – where terminological data is bound by the rules prior to accessing the A-Box – and provide some initial evaluation over a small LUBM dataset for RDFS, pD*, and OWL 2 RL/RDF. Thereafter, we look to apply our optimisations for scalable and distributed Linked Data reasoning, initially reintroducing our *authoritative reasoning* algorithm which incorporates provenance, detailing distribution of our approach, and then providing evaluation for reasoning over 1.12b Web triples.

2   Preliminaries

Before we continue, we briefly introduce some concepts prevalent throughout the paper. We use notation and nomenclature as is popular in the literature (cf. [4,8]). Herein, we denote infinite sets by $\mathbf{S}$ and corresponding finite subsets by $\mathcal{S}$.

2.1   RDF and Rules

RDF Constant. Given the set of URI references $\mathbf{U}$, the set of blank nodes $\mathbf{B}$, and the set of literals $\mathbf{L}$, the set of *RDF constants* is denoted by $\mathbf{C} := \mathbf{U} \cup \mathbf{B} \cup \mathbf{L}$.

RDF Triple. A triple $t := (s, p, o) \in (\mathbf{U} \cup \mathbf{B}) \times \mathbf{U} \times \mathbf{C}$ is called an *RDF triple*, where s is called subject, p predicate, and o object. A triple $t := (s, p, o) \in \mathbf{G}$, $\mathbf{G} := \mathbf{C} \times \mathbf{C} \times \mathbf{C}$ is called a *generalised triple*, which allows any RDF constant in any triple position: hence forth, we assume generalised triples [2]. We call a finite set of triples $\mathcal{G} \subset \mathbf{G}$ a *graph*. (For brevity, we sometimes use $\mathtt{r}$: for the RDFS namespace, $\mathtt{o}$: for OWL namespace, and $\mathtt{f}$: for the well-known FOAF namespace; we use 'a' as a shortcut for $\mathtt{rdf:type}$.)

Triple Pattern, Basic Graph Pattern. A *triple pattern* is a generalised triple where variables from the set $\mathbf{V}$ are allowed; i.e.: $t^v := (s^v, p^v, o^v) \in \mathbf{G}^\mathbf{V}$, $\mathbf{G}^\mathbf{V} := (\mathbf{C} \cup \mathbf{V}) \times (\mathbf{C} \cup \mathbf{V}) \times (\mathbf{C} \cup \mathbf{V})$. We call a set (to be read as conjunction) of triple patterns $\mathcal{G}^\mathbf{V} \subset \mathbf{G}^\mathbf{V}$ a *basic graph pattern*. We denote the set of variables in graph pattern $\mathcal{G}^\mathbf{V}$ by $\mathbf{V}(\mathcal{G}^\mathbf{V})$.

Variable Bindings. Let $\mathbf{M}$ be the set of endomorphic *variable binding mappings* $\mathbf{V} \cup \mathbf{C} \to \mathbf{V} \cup \mathbf{C}$ which map every constant $c \in \mathbf{C}$ to itself and every variable $v \in \mathbf{V}$ to an element of the set $\mathbf{C} \cup \mathbf{V}$. A triple t is a *binding* of a triple pattern $t^v := (s^v, p^v, o^v)$ iff there exists $\mu \in \mathbf{M}$, such that $t = \mu(t^v) = (\mu(s^v), \mu(p^v), \mu(o^v))$. A graph $\mathcal{G}$ is a binding of a graph pattern $\mathcal{G}^\mathbf{V}$ iff there exists a mapping $\mu \in \mathbf{M}$ such that $\bigcup_{t^v \in \mathcal{G}^\mathbf{V}} \mu(t^v) = \mathcal{G}$; we use the shorthand $\mu(\mathcal{G}^\mathbf{V}) = \mathcal{G}$. We use $\mathbf{M}(\mathcal{G}^\mathbf{V}, \mathcal{G}) := \{\mu \mid \mu(\mathcal{G}^\mathbf{V}) \subseteq \mathcal{G}, \mu(v) = v \text{ if } v \notin \mathbf{V}(\mathcal{G}^\mathbf{V})\}$ to denote the set of variable binding mappings for graph pattern $\mathcal{G}^\mathbf{V}$ in graph $\mathcal{G}$ which map variables outside $\mathcal{G}^\mathbf{V}$ to themselves.

Inference Rule. We define an *inference rule* r as the pair $(\mathcal{A}nte_r, \mathcal{C}on_r)$, where the *antecedent* (or *body*) $\mathcal{A}nte_r \subset \mathbf{G}^\mathbf{V}$ and the *consequent* (or *head*) $\mathcal{C}on_r \subset \mathbf{G}^\mathbf{V}$ are basic graph patterns such that $\mathbf{V}(\mathcal{C}on_r) \subseteq \mathbf{V}(\mathcal{A}nte_r)$ (range restricted) – rules with empty antecedents model axiomatic triples. We write inference rules as $\mathcal{A}nte_r \Rightarrow \mathcal{C}on_r$.

Rule Application and Standard Closure. A *rule application* is the immediate consequences $T_r(\mathcal{G}) := \bigcup_{\mu \in \mathbf{M}(\mathcal{A}nte_r, \mathcal{G})} (\mu(\mathcal{C}on_r) \setminus \mu(\mathcal{A}nte_r))$ of a rule r on a graph $\mathcal{G}$; accordingly, for a ruleset $\mathcal{R}$, $T_\mathcal{R}(\mathcal{G}) := \bigcup_{r \in \mathcal{R}} T_r(\mathcal{G})$. Now, let $\mathcal{G}_{i+1} := \mathcal{G}_i \cup T_\mathcal{R}(\mathcal{G}_i)$ and $\mathcal{G}_0 := \mathcal{G}$; the *exhaustive application* of the $T_\mathcal{R}$ operator on a graph $\mathcal{G}$ is then the least fixpoint (the smallest value for n) such that $\mathcal{G}_n = T_\mathcal{R}(\mathcal{G}_n)$. We call $\mathcal{G}_n$ the *closure* of $\mathcal{G}$ wrt. ruleset $\mathcal{R}$, denoted as $Cl_\mathcal{R}(\mathcal{G})$, or succinctly $\overline{\mathcal{G}}$ where the ruleset is obvious.

The above closure takes a graph and a ruleset and recursively applies the rules over the union of the original graph and the inferences until a fixpoint. Usually, this would consist of indexing all input and inferred triples; however, the cost of indexing and performing query-processing over large graphs can become prohibitively expensive. Thus, in [7] we originally proposed an alternate method based on a separation of terminological data, which we now generalise and discuss.

3   Partial Indexing Approach: Separating Terminological Data

In the field of Logic Programming, the notion of a 'linear program' refers loosely to a ruleset where only one pattern in each rule is recursive [12]. Our partial indexing approach is optimised for linear rules, where the non-recursive segment of the data is identified, separated and prepared, and thereafter each recursive pattern can then be bound via a triple-by-triple stream: we cater for non-linear rules, but as the number of recursive rules, the amount of recursion, and the amount of recursive data involved increases, our approach performs worse than the "full-indexing" approach.

Specifically regarding RDFS and OWL, the terminological segment of the data presents itself as relatively small and 'non-recursive' (or at least, mostly only recursive within itself), which can be leveraged for partial indexing. Herein, we define our notion of RDF(S)/OWL terminological data. (To generalise the following, the reader can consider terminological data as the RDFS/OWL archetype for any non-recursive and sufficiently small element of the data commonly required during rule application.)

Meta-class. We consider a *meta-class* as a class specifically of classes or properties; i.e., the members of a meta-class are themselves either classes or properties. Herein, we restrict our notion of meta-classes to the set defined in RDF(S) and OWL specifications, where examples include `rdf:Property`, `rdfs:Class`, `owl:Restriction`, `owl:-DatatypeProperty`, `owl:TransitiveProperty`, etc.; `rdfs:Resource`, `rdfs:-Literal`, e.g., are not meta-classes.

Meta-property. A *meta-property* is one which has a meta-class as its domain; again, we restrict our notion of meta-properties to the set defined in RDF(S) and OWL specifications, where examples include `rdfs:domain`, `rdfs:subClassOf`, `owl:hasKey`, `owl:inverseOf`, `owl:oneOf`, `owl:onProperty`, `owl:unionOf`, etc.; `rdf:type`, `owl:sameAs`, `rdfs:label`, e.g., do *not* have a meta-class as domain.

Terminological Triple. We define the set of *terminological triples* $\mathbf{T} \subset \mathbf{G}$ as the union of (i) triples with `rdf:type` as predicate and a meta-class as object; (ii) triples with a meta-property as predicate; (iii) triples forming a *valid* RDF list whose head is the object of a meta-property (e.g., a list used for `owl:unionOf`, etc.).

Terminological/Assertional Pattern. We refer to a *terminological -triple/-graph pattern* as one whose instance can only be a terminological triple or, resp., a set thereof. An *assertional pattern* is any pattern which is not terminological.

Given the above notions of terminological data/patterns, we now define a $\mathcal{T}$-split inference rule where part of the rule body is strictly matched by terminological data.

Definition 1. $\mathcal{T}$-**split inference rule:** *Given a rule* $r := (\mathcal{A}nte_r, \mathcal{C}on_r)$, *we define a* $\mathcal{T}$-split rule r^τ *as the triple* $(\mathcal{A}nte_{r^\tau}^{\mathcal{T}}, \mathcal{A}nte_{r^\tau}^{\mathcal{G}}, \mathcal{C}on)$ *where* $\mathcal{A}nte_{r^\tau}^{\mathcal{T}}$ *is the set of terminological patterns in* $\mathcal{A}nte_r$, *and* $\mathcal{A}nte_{r^\tau}^{\mathcal{G}} := \mathcal{A}nte_r \setminus \mathcal{A}nte_{r^\tau}^{\mathcal{T}}$. *We denote the set of all* $\mathcal{T}$-split rules by $\mathbf{R}^\tau$, *and the mapping of a rule to its* $\mathcal{T}$-split version as $\tau : \mathbf{R} \to \mathbf{R}^\tau; r \mapsto r^\tau$. *We additionally give the convenient sets* $\mathbf{R}^\emptyset := \{r^\tau \mid \mathcal{A}nte_{r^\tau}^{\mathcal{T}} = \emptyset, \mathcal{A}nte_{r^\tau}^{\mathcal{G}} = \emptyset\}$, $\mathbf{R}^{\mathbf{T}\emptyset} := \{r^\tau \mid \mathcal{A}nte_{r^\tau}^{\mathcal{T}} \neq \emptyset, \mathcal{A}nte_{r^\tau}^{\mathcal{G}} = \emptyset\}$, $\mathbf{R}^{\emptyset\mathbf{G}} := \{r^\tau \mid \mathcal{A}nte_{r^\tau}^{\mathcal{T}} = \emptyset, \mathcal{A}nte_{r^\tau}^{\mathcal{G}} \neq \emptyset\}$, $\mathbf{R}^{\mathbf{T}\mathbf{G}} := \{r^\tau \mid \mathcal{A}nte_{r^\tau}^{\mathcal{T}} \neq \emptyset, \mathcal{A}nte_{r^\tau}^{\mathcal{G}} \neq \emptyset\}$, $\mathbf{R}^{\mathbf{G}} := \mathbf{R}^{\mathbf{T}\mathbf{G}} \cup \mathbf{R}^{\emptyset\mathbf{G}}$ *and* $\mathbf{R}^{\mathbf{T}} := \mathbf{R}^{\mathbf{T}\emptyset} \cup \mathbf{R}^{\mathbf{T}\mathbf{G}}$ *as the set of all* $\mathcal{T}$-split rules with an empty antecedent, only terminological patterns, only assertional patterns, both types of patterns, some terminological patterns, and some assertional pattern respectively, where $\mathbf{R}^\tau = \mathbf{R}^\emptyset \cup \mathbf{R}^{\mathbf{T}\emptyset} \cup \mathbf{R}^{\emptyset\mathbf{G}} \cup \mathbf{R}^{\mathbf{T}\mathbf{G}} = \mathbf{R}^\emptyset \cup \mathbf{R}^{\mathbf{T}} \cup \mathbf{R}^{\mathbf{G}}$. *We also give the sets* $\mathbf{R}^{\mathbf{G}^1} := \{r^\tau \in \mathbf{R}^{\mathbf{G}} : |\mathcal{A}nte_{r^\tau}^{\mathcal{G}}| = 1\}$, $\mathbf{R}^{\mathbf{G}^n} := \{r^\tau \in \mathbf{R}^{\mathbf{G}} : |\mathcal{A}nte_{r^\tau}^{\mathcal{G}}| > 1\}$, *denoting the set of linear and non-linear rules respectively. Given a* $\mathcal{T}$-split ruleset $\mathcal{R}^\tau$, *herein we may use, e.g.,* $\mathcal{R}^{\mathbf{G}}$ *to denote* $\mathcal{R}^\tau \cap \mathbf{R}^{\mathbf{G}}$.

Example 1. For the rule $r := (\texttt{?c1,r:subClassOf,?c2}) \wedge (\texttt{?x,a,?c1}) \Rightarrow (\texttt{?x,a,?c2})$, $Ante^T := \{(\texttt{?c1,r:subClassOf,?c2})\}$ and $Ante^{\mathcal{G}} := \{(\texttt{?x,a,?c1})\}$. Underlining $Ante^T$, we write $\tau(r) := r^{\tau} := \underline{(\texttt{?c1,r:subClassOf,?c2})} \wedge (\texttt{?x,a,?c1}) \Rightarrow (\texttt{?x,a,?c2})$.

We then define our T-Box as the set of terminological triples in a given graph which are required by the terminological patterns of a given ruleset.

Definition 2. T-Box/A-Box: *Given a graph $\mathcal{G}$ and a $\mathcal{T}$-split ruleset $\mathcal{R}^{\tau}$, let $\mathcal{R}^{\mathbf{T}} := \mathcal{R}^{\tau} \cap \mathbf{R}^{\mathbf{T}}$ represent the subset of rules in $\mathcal{R}^{\tau}$ which require terminological data; the T-Box of $\mathcal{G}$ wrt. $\mathcal{R}^{\tau}$ is then $\mathbf{T}(\mathcal{G}, \mathcal{R}^{\tau}) := \bigcup_{r^{\tau} \in \mathcal{R}^{\mathbf{T}}} \bigcup_{t^v \in Ante_{r_{\tau}}^T} \bigcup_{\mu \in \mathbf{M}(\{t^v\}, \mathcal{G})} \mu(t^v)$, representing the subset of terminological triples in $\mathcal{G}$ which satisfy a terminological pattern of a rule antecedent ($Ante_{r_{\tau}}^T$) in $\mathcal{R}^{\tau}$; where ruleset and graph are obvious, we may abbreviate $\mathbf{T}(\mathcal{G}, \mathcal{R}^{\tau})$ to simply $\mathcal{T}$. Our A-Box is synonymous with $\mathcal{G}$: i.e., we also consider our T-Box as part of our A-Box in a form of unidirectional meta-modelling.*

Given the notion of a $\mathcal{T}$-split rule and our T-Box, we can now define how $\mathcal{T}$-split rules are applied, and how $\mathcal{T}$-split closure is achieved wrt. a static T-Box.

Definition 3. $\mathcal{T}$-split rule application and closure. *We define a $\mathcal{T}$-split rule application for a $\mathcal{T}$-split rule r^{τ} wrt. a graph $\mathcal{G}$ to be:*

$$T_{r^{\tau}}(\mathcal{T}, \mathcal{G}) := \bigcup_{\mu_0 \in \mathbf{M}(Ante_{r^{\tau}}^T, \mathcal{T})} \bigcup_{\mu_1 \in \mathbf{M}(\mu_0(Ante_{r^{\tau}}^{\mathcal{G}}), \mathcal{G})} (\mu_0 \circ \mu_1)(Con_{r^{\tau}}) \qquad (1)$$

here formalising the notion that the terminological patterns of the rule are strictly instantiated from a separate T-Box $\mathcal{T}$. Again, for a $\mathcal{T}$-split ruleset $\mathcal{R}^{\tau}$, $T_{\mathcal{R}^{\tau}}(\mathcal{T}, \mathcal{G}) := \bigcup_{r^{\tau} \in \mathcal{R}^{\tau}} T_{r^{\tau}}(\mathcal{T}, \mathcal{G})$. Now, let Ax denote the set of axiomatic triples given by $\mathcal{R}^{\tau}$ (the same set as for $\mathcal{R}$), and $\mathcal{T}_0 := \mathbf{T}(\mathcal{G} \cup Ax, \mathcal{R}^{\tau})$ be our initial T-Box derived from $\mathcal{G}$ and axiomatic triples, and $\mathcal{T}_{i+1} := \mathcal{T}_i \cup \mathbf{T}(T_{\mathcal{R}^{\mathbf{T}\emptyset}}(\mathcal{T}_i, \emptyset), \mathcal{R}^{\tau})$; we define our closed T-Box as $\mathcal{T}_n$ for the least value of n such that $\mathcal{T}_n = \mathcal{T}_n \cup T_{\mathcal{R}^{\mathbf{T}\emptyset}}(\mathcal{T}_n, \emptyset)$, denoted $\overline{\mathcal{T}}^{\tau}$, representing the closure of our initial T-Box wrt. rules requiring only terminological knowledge. Finally, let $\mathcal{G}_0^{\tau} := \mathcal{G} \cup \overline{\mathcal{T}}^{\tau} \cup Ax$ and $\mathcal{G}_{i+1}^{\tau} := \mathcal{G}_i^{\tau} \cup T_{\mathcal{R}^{\mathbf{G}}}(\overline{\mathcal{T}}^{\tau}, \mathcal{G}_i^{\tau})$; we now define the exhaustive application of the $T_{\mathcal{R}^{\tau}}$ operator on a graph $\mathcal{G}$ wrt. a static T-Box $\mathcal{T}$ as being upto the least fixpoint such that $\mathcal{G}_n^{\tau} = T_{\mathcal{R}^{\mathbf{G}}}(\overline{\mathcal{T}}^{\tau}, \mathcal{G}_n^{\tau})$. We call $\mathcal{G}_n^{\tau}$ the $\mathcal{T}$-split closure of $\mathcal{G}$ with respect to the $\mathcal{T}$-split ruleset $\mathcal{R}^{\tau}$, denoted as $Cl_{\mathcal{R}^{\tau}}(\mathcal{T}, \mathcal{G})$ or simply $\overline{\mathcal{G}}^{\tau}$.

The $\mathcal{T}$-split closure algorithm consists of two main steps: (i) deriving the closed T-Box from axiomatic triples, the input graph, and recursively applied $\mathbf{R}^{\mathbf{T}\emptyset}$ rules; (ii) applying 'A-Box' reasoning for all triples wrt. the $\mathbf{R}^{\mathbf{G}}$ rules and the static T-Box. We now give some propositions relating the $\mathcal{T}$-split closure with the standard rule application closure described in the preliminaries; firstly, we must give an auxiliary proposition which demonstrates how mappings for sub-graphs-patterns can be combined to give the mappings for the entire graph pattern, which relates to the $\mathcal{T}$-split rule application.

Proposition 1. *For any graph $\mathcal{G}$ and graph pattern $\mathcal{G}^{\mathbf{V}} := \mathcal{G}_a^{\mathbf{V}} \cup \mathcal{G}_b^{\mathbf{V}}$, it holds that $\mathbf{M}(\mathcal{G}^{\mathbf{V}}, \mathcal{G}) = \{\mu_b \circ \mu_a \mid \mu_a \in \mathbf{M}(\mathcal{G}_a^{\mathbf{V}}, \mathcal{G}), \mu_b \in \mathbf{M}(\mu_a(\mathcal{G}_b^{\mathbf{V}}), \mathcal{G})\}$.*

Proof. Firstly, $\mu_b \circ \mu_a \in \mathbf{M}$ since μ_a and μ_b are endomorphic. By definition, $(\mu_b \circ \mu_a)(c) = c$ for $c \in \mathbf{C}$. Next, we need to show that $(\mu_a \circ \mu_b)(v) = v$ if $v \notin \mathbf{V}(\mathcal{G}^{\mathbf{V}})$:

since by definition $\mu_a(v) = v$ if $v \notin \mathcal{G}_a^\mathbf{V}$ and $\mu_b(v) = v$ if $v \notin \mu_a(\mathcal{G}_b^\mathbf{V})$, and since $\mathbf{V}(\mu_a(\mathcal{G}_b^\mathbf{V})) \subseteq \mathbf{V}(\mathcal{G}_b^\mathbf{V})$ and $\mathbf{V}(\mathcal{G}^\mathbf{V}) = \mathbf{V}(\mathcal{G}_a^\mathbf{V}) \cup \mathbf{V}(\mathcal{G}_b^\mathbf{V})$, then $(\mu_b \circ \mu_a)(v) = v$ if $v \notin \mathbf{V}(\mathcal{G}^\mathbf{V})$. By definition, $\mu_a(\mathcal{G}_a^\mathbf{V}) \subseteq \mathcal{G}$ and thus we have $\mathbf{V}(\mu_a(\mathcal{G}_a^\mathbf{V})) = \emptyset$, and $\mu_a(\mathcal{G}_a^\mathbf{V}) = (\mu_b \circ \mu_a)(\mathcal{G}_a^\mathbf{V})$; again by definition we have $(\mu_b \circ \mu_a)(\mathcal{G}_b^\mathbf{V}) \subseteq \mathcal{G}$, and so $(\mu_b \circ \mu_a)(\mathcal{G}_a^\mathbf{V}) \cup (\mu_b \circ \mu_a)(\mathcal{G}_b^\mathbf{V}) = (\mu_b \circ \mu_a)(\mathcal{G}_a^\mathbf{V} \cup \mathcal{G}_b^\mathbf{V}) = (\mu_b \circ \mu_a)(\mathcal{G}^\mathbf{V}) \subseteq \mathcal{G}$. We now have $\mu_b \circ \mu_a \in \mathbf{M}(\mathcal{G}^\mathbf{V}, \mathcal{G})$ for every $\mu_a \in \mathbf{M}(\mathcal{G}_a^\mathbf{V}, \mathcal{G}), \mu_b \in \mathbf{M}(\mu_a(\mathcal{G}_b^\mathbf{V}), \mathcal{G})$, and need to show that for every $\mu \in \mathbf{M}(\mathcal{G}^\mathbf{V}, \mathcal{G})$, there exists a $(\mu_b \circ \mu_a)$ such that $(\mu_b \circ \mu_a)(\mathcal{G}^\mathbf{V}) = \mu(\mathcal{G}^\mathbf{V})$; by definition, we know that there exists a μ_a such that $\mu_a(\mathcal{G}_a^\mathbf{V}) = \mu(\mathcal{G}_a^\mathbf{V})$ for any μ as defined, and that for every such μ_a there exists a μ_b such that $(\mu_b \circ \mu_a)(\mathcal{G}_b^\mathbf{V}) = (\mu \circ \mu_a)(\mathcal{G}_b^\mathbf{V}) = \mu(\mathcal{G}_b^\mathbf{V})$, and hence the proposition holds. $\square$

Theorem 1. Soundness: *For any given ruleset* $\mathcal{R} \subset \mathbf{R}$*, its* $\mathcal{T}$*-split version* $\mathcal{R}^\tau := \tau(\mathcal{R})$*, and any graph* $\mathcal{G}$*, it holds that* $\overline{\mathcal{G}}^\tau \subseteq \overline{\mathcal{G}}$*.*

Proof. Clearly, $\mathcal{A}x$ gives the same set of triples for $\mathcal{R}^\tau$ and $\mathcal{R}$, and thus $\mathcal{T}_0 \subseteq \overline{\mathcal{G}}$ since $\mathbf{T}(\mathcal{G} \cup \mathcal{A}x, \mathcal{R}^\tau) \subseteq \mathcal{G} \cup \mathcal{A}x \subseteq \overline{\mathcal{G}}$. From Proposition 1, it follows that $\mathbf{M}(Ante_r, \mathcal{G}) = \mathbf{M}(Ante_r^\mathcal{T} \cup Ante_r^\mathcal{G}, \mathcal{G}) = \{\mu_0 \circ \mu_1 \mid \mu_0 \in \mathbf{M}(Ante_r^\mathcal{T}, \mathcal{G}), \mu_1 \in \mathbf{M}(\mu_0(Ante_r^\mathcal{G}), \mathcal{G})\}$; we can then show that $T_r(\mathcal{G}) = T_{r^\tau}(\mathcal{G}, \mathcal{G})$ by replacing $\mathcal{T}$ with $\mathcal{G}$ in Equation 1, from which follows $T_\mathcal{R}(\mathcal{G}) = T_{\mathcal{R}^\tau}(\mathcal{G}, \mathcal{G})$. Given that $T_{\mathcal{R}_a^\tau}(\mathcal{G}, \mathcal{G}) \subseteq T_{\mathcal{R}^\tau}(\mathcal{G}, \mathcal{G})$ if $\mathcal{R}_a^\tau \subseteq \mathcal{R}^\tau$, and $T_{\mathcal{R}^\tau}(\mathcal{G}_a, \mathcal{G}_b) \subseteq T_{\mathcal{R}^\tau}(\mathcal{G}, \mathcal{G})$ if $\mathcal{G}_a \subseteq \mathcal{G}$ and $\mathcal{G}_b \subseteq \mathcal{G}$ – i.e., that our rule applications are monotonic – we can show by induction that $\overline{\mathcal{T}}^\tau \subseteq \overline{\mathcal{G}}$: given $\mathcal{T}_0 \subseteq \overline{\mathcal{G}}$ from above, we can say that $\mathcal{T}_{i+1} \subseteq \overline{\mathcal{G}}$ iff $\mathcal{T}_i \subseteq \overline{\mathcal{G}}$ since $\mathbf{T}(T_{\mathcal{R}^{T\emptyset}}(\mathcal{T}_i, \emptyset)) \subseteq T_{\mathcal{R}^{T\emptyset}}(\mathcal{T}_i, \mathcal{T}_i) \subseteq T_\mathcal{R}(\mathcal{T}_i) \subseteq \overline{\mathcal{G}}$. Now, clearly $\mathcal{G}_0^\tau \subseteq \overline{\mathcal{G}}$, and since $T_{\mathcal{R}^\mathbf{G}}(\overline{\mathcal{T}}^\tau, \mathcal{G}_i^\tau) \subseteq T_{\mathcal{R}^\tau}(\mathcal{G}_i^\tau, \mathcal{G}_i^\tau) = T_\mathcal{R}(\mathcal{G}_i^\tau) \subseteq \overline{\mathcal{G}}$, we can say that if $\mathcal{G}_i^\tau \subseteq \overline{\mathcal{G}}$, then $\mathcal{G}_{i+1}^\tau \subseteq \overline{\mathcal{G}}$; by induction, $\overline{\mathcal{G}}^\tau \subseteq \overline{\mathcal{G}}$. $\square$

Theorem 2. Conditional Completeness: *If* $\overline{\mathcal{T}}^\tau = \mathbf{T}(\overline{\mathcal{G}}^\tau, \mathcal{R}^\tau)$*, then* $\overline{\mathcal{G}}^\tau = \overline{\mathcal{G}}$*.*

Proof. First, $T_{\mathcal{R}^\tau}(\mathbf{T}(\mathcal{G}, \mathcal{R}^\tau), \mathcal{G}) = T_{\mathcal{R}^\tau}(\mathcal{G}, \mathcal{G})$ since by definition $\mathbf{T}(\mathcal{G}, \mathcal{R}^\tau)$ only removes triples from $\mathcal{G}$ that cannot be bound by terminological patterns in $\mathcal{R}^\tau$. Given the criteria $\overline{\mathcal{G}}^\tau = \overline{\mathcal{G}}^\tau \cup T_{\mathcal{R}^\mathbf{G}}(\overline{\mathcal{T}}^\tau, \overline{\mathcal{G}}^\tau)$ – or, rephrasing, $T_{\mathcal{R}^\mathbf{G}}(\overline{\mathcal{T}}^\tau, \overline{\mathcal{G}}^\tau) \subseteq \overline{\mathcal{G}}^\tau$ – we first know that $\mathcal{A}x \cup \overline{\mathcal{T}}^\tau \cup \mathcal{G} = \mathcal{G}_0^\tau \subseteq \overline{\mathcal{G}}^\tau$. Thus, $T_{\mathcal{R}^\tau}(\overline{\mathcal{T}}^\tau, \overline{\mathcal{G}}^\tau) = T_{\mathcal{R}^\mathbf{G}}(\overline{\mathcal{T}}^\tau, \overline{\mathcal{G}}^\tau) \subseteq \overline{\mathcal{G}}^\tau$. If $\overline{\mathcal{T}}^\tau = \mathbf{T}(\overline{\mathcal{G}}^\tau, \mathcal{R}^\tau)$, then $T_{\mathcal{R}^\tau}(\overline{\mathcal{T}}^\tau, \overline{\mathcal{G}}^\tau) = T_{\mathcal{R}^\tau}(\mathbf{T}(\overline{\mathcal{G}}^\tau, \mathcal{R}^\tau), \overline{\mathcal{G}}^\tau) = T_{\mathcal{R}^\tau}(\overline{\mathcal{G}}^\tau, \overline{\mathcal{G}}^\tau) = T_\mathcal{R}(\overline{\mathcal{G}}^\tau)$, which gives $\mathcal{G}_0 \subseteq \overline{\mathcal{G}}^\tau \subseteq \overline{\mathcal{G}}$: i.e., $\overline{\mathcal{G}}^\tau$ is known to be the partial closure of $\mathcal{G}$. Given the fixpoint condition $\overline{\mathcal{G}} = \overline{\mathcal{G}} \cup T_\mathcal{R}(\overline{\mathcal{G}})$, then $\overline{\mathcal{G}}^\tau$ must be the fixpoint: $\overline{\mathcal{G}}^\tau = \overline{\mathcal{G}}$. $\square$

Proposition 2. *A triple* $t \in \mathbf{T}(\overline{\mathcal{G}}^\tau, \mathcal{R}^\tau) \setminus \overline{\mathcal{T}}^\tau$ *can only be produced for* $\overline{\mathcal{G}}^\tau$ *through an inference for a rule in* $\mathbf{R}^\mathbf{G}$*.*

Proof. Any T-Box triples in the original graph, or T-Box triples produced by the 'closure' of $\mathbf{R}^\emptyset$ rules are added to the initial T-Box $\mathcal{T}_0$. Any T-Box triples produced by the closure of $\mathbf{R}^{\mathbf{T}\emptyset}$ rules over $\mathcal{T}_0$ are added to the closed T-Box $\overline{\mathcal{T}}^\tau$. Since $\mathbf{R}^\tau := \mathbf{R}^\emptyset \cup \mathbf{R}^{\mathbf{T}\emptyset} \cup \mathbf{R}^\mathbf{G}$, the only new triples – terminological or not – that can arise in the computation of $\overline{\mathcal{G}}^\tau$ after deriving $\overline{\mathcal{T}}^\tau$ are from rules in $\mathbf{R}^\mathbf{G}$. $\square$

We have shown that for an arbitrary ruleset and graph, the $\mathcal{T}$-split closure is sound wrt. the standard closure, and that if no T-Box triples are produced by rules requiring assertional knowledge, then $\mathcal{T}$-split closure is complete wrt. the standard closure. So, when are T-Box triples produced by $\mathbf{R}^\mathbf{G}$ rules? Analysis must be applied per ruleset,

---

Algorithm 1. Partial indexing approach for $\mathcal{T}$-split closure

---

Required: $\mathcal{R}, \mathcal{G}$

1 $\mathcal{R}^\tau := \tau(\mathcal{R}); \mathcal{T}_0 := \mathbf{T}(\mathcal{A}x, \mathcal{R}^\tau); n := 0;$ /* get t-split rules & ax. T-Box triples */

2 **for** $t \in \mathcal{G}$ **do** $\mathcal{T}_0 := \mathcal{T}_0 \cup \mathbf{T}(\{t\}, \mathcal{R}^\tau);$ /* SCAN 1: extract T-Box from main data */

3 **while** $\mathcal{T}_{n+1} \neq \mathcal{T}_n$ **do** $\mathcal{T}_{n+1} := \mathcal{T}_n \cup \mathbf{T}(T_{\mathcal{R}\mathbf{T}\emptyset}(\mathcal{T}_i, \emptyset), \mathcal{R}^\tau); n++;$ /* do T-Box reasoning */

4 $\overline{\mathcal{T}}^\tau := \mathcal{T}_{n+1}; \overline{\mathcal{G}}^\tau := \mathcal{G}_0^\tau := \mathcal{G} \cup \overline{\mathcal{T}}^\tau \cup \mathcal{A}x; \mathcal{A} := \emptyset;$ /* initialise A-Box structures */

5 **for** $t^I \in \mathcal{G}_0^\tau$ **do**   /* SCAN 2: A-Box reasoning over all data */

6 $\quad$ $\mathcal{G}_0^I := \emptyset; \mathcal{G}_1^I := \{t^I\}; n := 1;$ /* initialise set to hold inferences from t^I */

7 $\quad$ **while** $\mathcal{G}_n^I \neq \mathcal{G}_{n-1}^I$ **do**   /* while we find new triples to reason over */

8 $\quad\quad$ **for** $t \in \mathcal{G}_n^I \setminus \mathcal{G}_{n-1}^I$ **do**   /* scan new triples */

9 $\quad\quad\quad$ $\mathcal{G}_{n+1}^I := \mathcal{G}_n^I \cup T_{\mathcal{R}\mathbf{G}\mathbf{1}}(\overline{\mathcal{T}}^\tau, \{t\});$ /* do all 'no A-Box join' rules for t */

10 $\quad\quad$ **for** $r \in \mathcal{R}^{\mathbf{G}^\mathbf{n}}$ **do**   /* for each 'A-Box join' rule */

11 $\quad\quad\quad$ **for** $t^v \in \mathcal{A}nte_r^{\mathcal{G}}$ **do**   /* for each assertional pattern */

12 $\quad\quad\quad\quad$ **if** $\exists \mu \in \mathbf{M} : \mu(t^v) = t$ **then** $\mathcal{A} := \mathcal{A} \cup \{t\};$ /* index t if needed */

13 $\quad\quad$ $\mathcal{G}_{n+1}^I := \mathcal{G}_{n+1}^I \cup T_r(\overline{\mathcal{T}}^\tau, \mathcal{A});$ /* apply 'A-Box join' rule over index */

14 $\quad\quad$ $n++;$ /* recurse */

15 $\quad$ $\overline{\mathcal{G}}^\tau := \overline{\mathcal{G}}^\tau \cup \mathcal{G}_n^I;$ /* write set of recursive inferences for t^I to output */

Return : $\overline{\mathcal{G}}^\tau$

---

but for RDFS, pD* and OWL 2 RL/RDF, we informally posit that by inspection, one can show that such a condition can only arise through so called *non-standard usage* [8]: the assertion of terminological triples which use meta-classes and meta-properties in positions other than the object of `rdf:type` triples or predicate position respectively – e.g., `my:subPropertyOf rdfs:subPropertyOf rdfs:subPropertyOf`.

The $\mathcal{T}$-split approach can be implemented through partial indexing using two scans of the data: the first separates and builds the T-Box and the second reasons over the A-Box – note that the first scan can be over a separate T-Box graph. Algorithm 1 details this approach, which largely follows the formalisms in Definition 3: the major variance consists of the application of rules in $\mathcal{R}^\mathbf{G}$, which one can convince themselves is equivalent since *all* triples encountered are passed through every rule in $\mathcal{R}^\mathbf{G}$. For brevity, we omit some implementational details such as partial duplicate detection implemented using an LRU locality cache. The "non-trivial" aspects of the implementation include the indexing of the T-Box $\overline{\mathcal{T}}^\tau$, and the indexing of the A-Box $\mathcal{A}$. Again, as $\mathcal{A}$ is required to store more data, the two-scan approach becomes more inefficient than the "full-indexing" approach; in particular, a rule in $\mathcal{R}^{\mathbf{G}^\mathbf{n}}$ with an open pattern – e.g., OWL 2 RL/RDF rule **eq-rep-s**: $(?s, o\!:\!\mathtt{sameAs}, ?s') \wedge (?s, ?p, ?o) \Rightarrow (?s', ?p, ?o)$ – will require indexing of all data, negating the benefits of the approach. Again, partial-indexing performs well if $\mathcal{A}$ remains small and performs best if $\mathcal{R}^{\mathbf{G}^\mathbf{n}} = \emptyset$ – i.e., no rules require A-Box joins and thus A-Box indexing is not required.

4 Template Rules

We now discuss optimisations for deriving $\mathcal{T}$-split closure based on template rules, which are currently used by DLEJena [13] and also used in RIF for supporting OWL 2 RL/RDF [15]; however, instead of manually specifying a set of template rules, we leverage our general notion of terminological data to create a generic template function: after separating and closing the T-Box, we bind the T-Box patterns of rules before accessing the A-Box to create a set of new *templated rules* (or $\mathcal{T}$-ground rules) which themselves 'encode' the T-Box, thus avoiding repetitive T-Box pattern bindings during the A-Box reasoning process. We now formalise these notions.

Definition 4. Template Function: *For a $\mathcal{T}$-split rule r^{τ}, the template function is given as $\alpha : \mathbf{R}^{\tau} \times 2^{\mathbf{G}} \to 2^{\mathbf{R}}; (r^{\tau}, \mathcal{T}) \mapsto \{(\mu(Ante_{r^{\tau}}^{\mathcal{G}}), \mu(Con_{r^{\tau}})) \mid \mu \in \mathbf{M}(Ante_{r^{\tau}}^{\mathcal{T}}, \mathcal{T})\}.$*

Example 2. Given a simple T-Box $\mathcal{T} := \{(\texttt{f:Person,r:subClassOf,f:Agent})\}$ and a rule $r^{\tau} := (\texttt{?c1,r:subClassOf,?c2}) \wedge (\texttt{?x,a,?c1}) \Rightarrow (\texttt{?x,a,?c2})$, then the template function is given as $\alpha(r^{\tau}, \mathcal{T}) := \{(\texttt{?x,a,f:Person}) \Rightarrow (\texttt{?x,a,f:Agent})\}$.

Templated rule application is synonymous with standard rule application. We may use α as intuitive shorthand to map a *set* of $\mathcal{T}$-split rules to the union of the set of resulting templated rules. We now (i) propose that applying a $\mathcal{T}$-split rule gives the same result as applying the respective set of templated rules wrt. arbitrary graphs $\mathcal{T}$ & $\mathcal{G}$; (ii) describe the closure of a graph using templated rules; (iii) show that the templated-rule closure equals the $\mathcal{T}$-split closure previously outlined.

Proposition 3. *For any graphs $\mathcal{T}, \mathcal{G}$ and for any rule r with a $\mathcal{T}$-split rule $r^{\tau} = \tau(r)$, it holds that $T_{r^{\tau}}(\mathcal{T}, \mathcal{G}) = T_{\alpha(r^{\tau}, \mathcal{T})}(\mathcal{G})$.*

Proof. $T_{r^{\tau}}(\mathcal{T}, \mathcal{G}) = \bigcup_{\mu_0 \in \mathbf{M}(Ante_{r^{\tau}}^{\mathcal{T}}, \mathcal{T})} \bigcup_{\mu_1 \in \mathbf{M}(\mu_0(Ante_{r^{\tau}}^{\mathcal{G}}), \mathcal{G})} (\mu_0 \circ \mu_1)(Con_{r^{\tau}}) = \bigcup_{r \in \alpha(r^{\tau}, \mathcal{T})} \bigcup_{\mu \in \mathbf{M}(Ante_r, \mathcal{G})} \mu(Con_r) = T_{\alpha(r^{\tau}, \mathcal{T})}(\mathcal{G})$. □

Definition 5. Templated rule closure: *Given a ruleset $\mathcal{R}$, its $\mathcal{T}$-split version $\mathcal{R}^{\tau} := \tau(\mathcal{R})$, and a graph $\mathcal{G}$, let $\overline{\mathcal{T}}^{\tau}$ represent the closed T-Box as derived in the $\mathcal{T}$-split closure, and let $\mathcal{R}^{\alpha} := \alpha(\mathcal{R}^{\mathbf{G}}, \overline{\mathcal{T}}^{\tau})$. Again, let $\mathcal{G}_0^{\alpha} := \mathcal{G} \cup \overline{\mathcal{T}}^{\tau} \cup Ax$, but this time $\mathcal{G}_{i+1}^{\alpha} := \mathcal{G}_i^{\alpha} \cup T_{\mathcal{R}^{\alpha}}(\mathcal{G}_i^{\alpha})$; as before, the templated rule closure is $\mathcal{G}_n$ for the smallest value of n such that $\mathcal{G}_n^{\alpha} = T_{\mathcal{R}^{\alpha}}(\mathcal{G}_n^{\alpha})$, denoted as $Cl_{\mathcal{R}^{\alpha}}(\mathcal{T}, \mathcal{G}^{\alpha})$, or simply $\overline{\mathcal{G}}^{\alpha}$.*

Theorem 3. *For any graph $\mathcal{G}$, and any ruleset $\mathcal{R} \subset \mathbf{R}$, its $\mathcal{T}$-split version $\mathcal{R}^{\tau}$, and the respective templated ruleset $\mathcal{R}^{\alpha}$, we can say that $\overline{\mathcal{G}}^{\alpha} = \overline{\mathcal{G}}^{\tau}$.*

Proof. The only divergence between the $\mathcal{T}$-split closure and templated-rule closure is in the fixpoint calculation: $\mathcal{G}_{i+1}^{\alpha} := \mathcal{G}_i^{\alpha} \cup T_{\mathcal{R}^{\alpha}}(\mathcal{G}_i^{\alpha})$ versus $\mathcal{G}_{i+1}^{\tau} := \mathcal{G}_i^{\tau} \cup T_{\mathcal{R}^{\mathbf{G}}}(\overline{\mathcal{T}}^{\tau}, \mathcal{G}_i^{\tau})$. Using induction, by def. $\mathcal{G}_0^{\tau} = \mathcal{G}_0^{\alpha}$; if $\mathcal{G}_i^{\tau} = \mathcal{G}_i^{\alpha}$, then $\mathcal{G}_{i+1}^{\tau} = \mathcal{G}_i^{\alpha} \cup \bigcup_{r^{\tau} \in \mathcal{R}^{\mathbf{G}}} T_{r^{\tau}}(\overline{\mathcal{T}}^{\tau}, \mathcal{G}_i^{\alpha}) = \mathcal{G}_i^{\alpha} \cup \bigcup_{r \in \alpha(\mathcal{R}^{\mathbf{G}}, \overline{\mathcal{T}}^{\tau})} T_{r^{\tau}}(\mathcal{G}_i^{\alpha}) = \mathcal{G}_i^{\alpha} \cup T_{\mathcal{R}^{\alpha}}(\mathcal{G}_i^{\alpha}) = \mathcal{G}_{i+1}^{\alpha}$. □

The templated rules can be applied in lieu of the $\mathcal{R}^{\mathbf{G}}$ rules in Algorithm 1. Indeed, a large number of rules can be templated for a sufficiently complex T-Box, and naïve application of all such rules on all triples could worsen performance; however, the templated rules are more amenable to further optimisations, which we now discuss.

4.1 Merging Equivalent Template Rules

The templating procedure may result in rules with equivalent antecedents – which can be aligned by variable rewriting – being produced; these rules can subsequently be merged. We formalise such notions here.

Definition 6. Equivalent Graph Patterns: *Let* **N** *be the set of automorphic* variable rewrite mappings *containing all* ν *as follows:*

$$\nu : \mathbf{V} \cup \mathbf{C} \rightarrowtail \mathbf{V} \cup \mathbf{C}; \; x \mapsto \begin{cases} x & if \; x \in \mathbf{C} \\ v \in \mathbf{V} & otherwise \end{cases} \qquad (2)$$

(Note: **N** $\subset$ **M***). We denote by* $\sim_\nu$ an equivalence relation for graph patterns *such that* $\mathcal{G}_i^\mathbf{V} \sim_\nu \mathcal{G}_j^\mathbf{V}$ *iff there exists a mapping* $\nu \in \mathbf{N}$ *such that* $\nu(\mathcal{G}_i^\mathbf{V}) = \mathcal{G}_j^\mathbf{V}$.

Proposition 4. *The relation* $\sim_\nu$ *is reflexive, symmetric and transitive.*

Proof. Reflexivity is trivially given by the identity morphism $\nu(x) = x$, symmetry is given by the inverse morphism $\nu^{-1}(\mathcal{G}_j^\mathbf{V})$ where $\nu^{-1} \in \mathbf{N}$ if $\nu \in \mathbf{N}$ since ν is automorphic, and transitivity is given by the presence of the composite morphism $(\nu_a \circ \nu_b)(\mathcal{G}^\mathbf{V})$ where again $\nu_a \circ \nu_b \in \mathbf{N}$ since ν_a and ν_b are automorphic.   □

Definition 7. Rule Merge: *Let* $\sim_\mathbf{R}$ *be an equivalence relation – slightly abusing notation – which holds between two rules such that* $r_i \sim_\mathbf{R} r_j$ *iff* $Ante_{r_i} \sim_\nu Ante_{r_j}$. *Given an equivalence class* $[r]$ *– a set of rules between which* $\sim_\mathbf{R}$ *holds – select a canonical rule* $r \in [r]$*; we can now describe the merge of the equivalence class as* $\beta([r]) := (Ante_r, Con_{[r]})$ *where* $Con_{[r]} := \bigcup_{r_i \in [r]} \nu_i(Con_{r_i})$ *for some* $\nu_i \in \mathbf{N}$ *such that* $\nu_i(Ante_{r_i}) = Ante_r$. *Now letting* $\mathcal{R}/\sim_\mathbf{R} := \{[r] \mid r \in \mathcal{R}\}$ *denote the quotient set of* $\mathcal{R}$ *by* $\sim_\mathbf{R}$ *– the set of all equivalent classes* $[r]$ *wrt.* $\sim_\mathbf{R}$ *in* $\mathcal{R}$ *– we can generalise the rule merge function for a set of rules as* $\beta : 2^\mathbf{R} \to 2^\mathbf{R}, \; \mathcal{R} \mapsto \bigcup\{\beta([r]) \mid [r] \in \mathcal{R}/\sim_\mathbf{R}\}$.

Example 3. Taking the two templated rules: $(?x,f\!:\!img,?y) \Rightarrow (?x,a,f\!:\!Person)$ and $(?s,f\!:\!img,?o) \Rightarrow (?s,f\!:\!depicts,?o)$; they can be merged by ν where $\nu(?s) = ?x$, $\nu(?o) = ?y$, giving $(?x,f\!:\!img,?y) \Rightarrow (?x,a,f\!:\!Person) \wedge (\; ?x,f\!:\!depicts,?y\;)$.

The choice of canonical rule is unimportant since ν is automorphic; we now show that the application of any ruleset and the respective merged ruleset are extensionally equal.

Proposition 5. *For any graph* $\mathcal{G}$ *and* $\sim_\mathbf{R}$ *equivalence class* $[r]$*,* $T_{[r]}(\mathcal{G}) = T_{\beta([r])}(\mathcal{G})$*; for any ruleset* $\mathcal{R}$*,* $T_\mathcal{R}(\mathcal{G}) = T_{\beta(\mathcal{R})}(\mathcal{G})$*; wrt. closure,* $Cl_{\mathcal{R}^a}(\mathcal{T}, \mathcal{G}) = Cl_{\beta(\mathcal{R}^a)}(\mathcal{T}, \mathcal{G})$.

Proof. We denote $\beta([r])$ as $(Ante_\beta, Con_\beta)$. If $\mathcal{G}_i^\mathbf{V} \sim_\nu \mathcal{G}_j^\mathbf{V}$, then by def. $\nu(\mathcal{G}_i^\mathbf{V}) = \mathcal{G}_j^\mathbf{V}$, and for any graph $\mathcal{G}$ and any mapping $\mu \in \mathbf{M}$, $\mu(\nu(\mathcal{G}_i^\mathbf{V})) = \mu(\mathcal{G}_j^\mathbf{V})$; i.e., if $\mathcal{G}_i^\mathbf{V} \sim_\nu \mathcal{G}_j^\mathbf{V}$, $\mathbf{M}(\nu(\mathcal{G}_i^\mathbf{V}), \mathcal{G}) = \mathbf{M}(\mathcal{G}_j^\mathbf{V}, \mathcal{G})$. Thus we give $\mathcal{M}_\beta := \{\mu \mid \mu(Ante_\beta) \subseteq \mathcal{G}\} = \bigcup_{r_i \in [r]}\{\mu \mid \mu(\nu_i(Ante_{r_i})) \subseteq \mathcal{G}\}$. Let $\mathcal{M}_i := \{\mu \mid \mu(Ante_{r_i}) \subseteq \mathcal{G}\}$; now, it follows that $T_{\beta([r])}(\mathcal{G}) = \bigcup_{\mu \in \mathcal{M}_\beta} \mu(Con_\beta) = \bigcup_{r_i \in [r]} \bigcup_{\mu \in \mathcal{M}_\beta} \mu(\nu_i(Con_{r_i})) = \bigcup_{r_i \in [r]} \bigcup_{\mu \in \mathcal{M}_i} \mu(Con_{r_i}) = T_{[r]}(\mathcal{G})$. The rest of the proposition follows naturally.   □

4.2   Rule Index

We have reduced the amount of templated rules through merging; however, given a sufficiently complex T-Box, we may still have a prohibitive number of rules for efficient recursive application. We now look at the use of a rule index which maps a triple t to rules containing an antecedent pattern which t is a binding for, thus enabling the efficient identification and application of only relevant rules for a given triple.

Definition 8. Rule Lookup: *Given a triple* t *and ruleset* $\mathcal{R}$, *the* rule lookup function *is* $\omega : \mathbf{G} \times 2^{\mathbf{R}} \rightarrow 2^{\mathbf{R}}$, $(t, \mathcal{R}) \mapsto \{r \in \mathcal{R} \mid \exists \mu \in \mathbf{M} : \exists t^v \in \mathcal{A}nte_r : (\mu(t^v) = t)\}$.

Example 4. Given a triple $t := (\text{ex:me,a,f:Person})$, and a simple example ruleset $\mathcal{R} := \{(\text{?x,f:img,?y}) \Rightarrow (\text{?x,a,f:Person}); (\text{?x,a,f:Person}) \Rightarrow (\text{?x,a,f:Agent}); (\text{?x,a,?y}) \Rightarrow (\text{?y,a,r:Class})\}$, $\omega(t, \mathcal{R})$ returns a set containing the latter two rules.

A triple pattern has $2^3 = 8$ possible forms: $(?, ?, ?)$, $(s, ?, ?)$, $(?, p, ?)$, $(?, ?, o)$, $(s, p, ?)$, $(?, p, o)$, $(s, ?, o)$, (s, p, o). Thus, we require eight indices for antecedent triple patterns, and eight lookups to perform $\omega(t, \mathcal{R})$ – to find all relevant rules for a triple. We use seven in-memory hashtables storing the constants of the rule antecedent patterns as key, and a set of rules containing such a pattern as value; e.g., $\{(\text{?x,a,f:Person})\}$ is put into the $(?, p, o)$ index with $\{\text{a,f:Person}\}$ as key. Rules containing patterns without constants are stored in a set, as they are relevant to all triples.

4.3 Rule Dependency – Labelled Rule Graph

Within our rule index, there may exist rule dependencies: the application of one rule may/will lead to the application of another. Thus, instead of performing lookups for rules for each recursively inferred triple, we can model dependencies in our rule index using a rule graph. In Logic Programming, a rule graph is defined as a directed graph $\mathcal{H} := (\mathcal{R}, \Omega)$ where $(r_i, r_j) \in \Omega$ (i.e., $r_i \Omega r_j$, read "r_j follows r_i") iff there exists a mapping $\mu \in \mathbf{M}$ such that $\mu(t^v) \in \mathcal{C}on_{r_i}$ for $t^v \in \mathcal{A}nte_{r_j}$ (cf. [14]).

By building and encoding such a rule graph into our index, we can "wire" the recursive application of rules for a given triple. However, from the merge function (or otherwise) there may exist rules with large consequent sets. We therefore extend the notion of the rule graph to a directed labelled graph with inclusion of the labelling function $\lambda : \mathbf{R} \times \mathbf{R} \rightarrow 2^{\mathbf{G}^V}$; $(r_i, r_j) \mapsto \{t^v \in \mathcal{C}on_{r_i} \mid \exists \mu \in \mathbf{M} : \mu^{-1}(t^v) \in \mathcal{A}nte_{r_j}\}$; in simpler terms, $\lambda(r_i, r_j)$ gives the set of consequent triple patterns in r_i that would be matched by patterns in the antecedent of r_j, labelling the edges Ω of the rule graph with the consequent patterns that give the dependency.

Example 5. For a rule $r_i := (\text{?x,f:img,?y}) \Rightarrow (\text{?x,a,f:Person}) \wedge (\text{?y,a,f:Image})$, and a rule $r_j := (\text{?s,a,f:Person}) \Rightarrow (\text{?s,a,f:Agent})$, we say that $r_i \Omega r_j$, where $\lambda(r_i, r_j) = \{(\text{?x,a,f:Person})\}$.

In practice, our rule index stores sets of elements of a linked list, where each element contains a rule and links to rules which are relevant for that rule's consequent patterns. Thus, for each input triple, we can retrieve all relevant rules for all eight possible patterns, apply those rules, and if successful, follow the respective labelled links to recursively find relevant rules without re-accessing the index until the next input triple.

4.4 Rule Saturisation

We very briefly describe one final and intuitive optimisation technique we investigated – which later evaluation demonstrates to be mostly disadvantageous – involving the *saturisation of rules*; we say that a subset of dependencies in the rule graph are *strong*

Algorithm 2. Partial-indexing approach using templated rule optimisations

Required: $\mathcal{R}, \mathcal{G}$

1 derive $\overline{\mathcal{T}}^\tau$ and $\mathcal{R}^\tau$ as in Algorithm 1;   /* SCAN 1: See Algorithm 1 */
2 $\mathcal{R}^\alpha := \alpha(\mathcal{R}^G); \mathcal{R}^\beta := \beta(\mathcal{R}^\alpha);$ /* template and merge $\mathcal{T}$-split rules */
3 build ω index for $\mathcal{R}^\beta$ encoding graph $\mathcal{H}$ with edges $\lambda;$ /* build rule index w/ dependencies */
4 $\overline{\mathcal{G}}^\tau := \mathcal{G}_0^\tau := \mathcal{G} \cup \overline{\mathcal{T}}^\tau \cup \mathcal{A}x; \mathcal{A} := \emptyset;$ /* init A-Box structures */
5 **for** $t^I \in \mathcal{G}_0^\tau$ **do**   /* SCAN 2: A-Box reasoning over all data */
6    $\quad \mathcal{R}\mathcal{G}_0^I := \emptyset; \mathcal{R}\mathcal{G}_1^I := \{(r, t^I) \mid r \in \omega(t^I, \mathcal{R}^\beta)\}; n := 1;$ /* initialise relevant rules for t^I */
7    $\quad$ **while** $\mathcal{R}\mathcal{G}_n^I \neq \mathcal{R}\mathcal{G}_{n-1}^I$ **do**   /* while we find new rule/triple pairs to reason over */
8    $\qquad$ **for** $(r, t) \in \mathcal{R}\mathcal{G}_n^I \setminus \mathcal{R}\mathcal{G}_{n-1}^I$ **do**   /* scan new rule/triple pairs */
9    $\qquad\quad$ $\mathcal{G}_{rt} := \emptyset; \mathcal{R}\mathcal{G}_{n+1}^I := \mathcal{R}\mathcal{G}_n^I;$ /* initialise state for rule triple pair */
10   $\qquad\quad$ **if** $|Ante_r| > 1$ **then**   /* if rule requires A-Box join */
11   $\qquad\qquad$ **for** $t^v \in Ante_r^\mathcal{G}$ **do**   /* for each assertional pattern */
12   $\qquad\qquad\quad$ **if** $\exists \mu \in \mathbf{M} : \mu(t^v) = t$ **then** $\mathcal{A} := \mathcal{A} \cup \{t\};$ /* index t if needed */
13   $\qquad\qquad$ $\mathcal{G}_{rt} := T_r(\overline{\mathcal{T}}^\tau, \mathcal{A});$ /* apply 'A-Box join rule' over index */
14   $\qquad\quad$ **else**
15   $\qquad\qquad$ $\mathcal{G}_{rt} := T_r(\overline{\mathcal{T}}^\tau, \{t\});$ /* apply 'non A-Box join rule' for t */
16   $\qquad\quad$ **if** $\mathcal{G}_{rt} \neq \emptyset$ **then**   /* if rule creates inference */
17   $\qquad\qquad$ **for** $r^+ : (r, r^+) \in \Omega$ **do**   /* find successive rules in graph */
18   $\qquad\qquad\quad$ **for** $t_n^v \in \lambda(r, r^+)$ **do**   /* for the consequent patterns bound */
19   $\qquad\qquad\qquad$ $\mathcal{R}\mathcal{G}_{n+1}^I := \mathcal{R}\mathcal{G}_{n+1}^I \cup \{(r^+, t_n) \mid t_n \in \mathcal{G}_{rt}\};$ /* add rule/triple pair */

20   $\quad$ n++;   /* recurse for unique rule/triple pair */

21 $\overline{\mathcal{G}}^\tau := \overline{\mathcal{G}}^\tau \cup \{t \mid (r, t) \in \mathcal{R}\mathcal{G}_n^I\};$ /* write recursive inferences for t^I to output */

Return : $\overline{\mathcal{G}}^\tau$

dependencies, where the successful application of one rule *will* always lead to the successful application of another. For linear rules, we can saturate rules by pre-computing the recursive rule application of its dependencies; we give the gist with an example:

Example 6. Take rules $r_i := (?\mathtt{x},\mathtt{f:img},?\mathtt{y}) \Rightarrow (?\mathtt{x},\mathtt{a},\mathtt{f:Person}) \wedge (?\mathtt{y},\mathtt{a},\mathtt{f:Image})$, $r_j := (?\mathtt{s},\mathtt{a},\mathtt{f:Person}) \Rightarrow (?\mathtt{s},\mathtt{a},\mathtt{f:Agent})$, $r_k := (?\mathtt{x},\mathtt{a},?\mathtt{y}) \Rightarrow (?\mathtt{y},\mathtt{a},\mathtt{r:Class})\}$. We can see that $r_i \, \Omega \, r_j$, $r_i \, \Omega \, r_k$, $r_j \, \Omega \, r_k$. We can remove the links from r_i to r_j and r_k (and similarly from r_j to r_k) by saturating r_i to $(?\mathtt{x},\mathtt{f:img},?\mathtt{y}) \Rightarrow (?\mathtt{x},\mathtt{a},\mathtt{f:Person}) \wedge$ $(?\mathtt{y},\mathtt{a},\mathtt{f:Image}) \wedge (?\mathtt{x},\mathtt{a},\mathtt{f:Agent}) \wedge (\mathtt{f:Person},\mathtt{a},\mathtt{r:Class}) \wedge (\mathtt{f:Image},\mathtt{a},\mathtt{r:Class})$ $\wedge \, (\mathtt{f:Agent},\mathtt{a},\mathtt{r:Class})\}$.

As we will see in Sections 4.6 & 5.2, saturation produces more duplicates and thus puts more load on the duplicate-removal cache, negatively affecting performance.

4.5   Optimised Partial Indexing Approach Using Template Rules

We now integrate the above notions as optimisations for the partial indexing approach, with the new procedure detailed in Algorithm 2. We no longer need to bind T-Box patterns during A-Box access; we mitigate the cost of extra templated rules by first merging rules, and instead of brute-force applying all rules to all triples in the A-Box reasoning scan, we use our linked rule index to retrieve only relevant rules for a given triple and to find recursively relevant rules. We now initially evaluate our methods.

Table 1. Details of reasoning for LUBM(10) given different reasoning configurations

input	LUBM(10) - 1.27M data triples, 295 ontology triples																	
fragment	RDFS						pD*						OWL 2 RL					
inferred	748k						1,328k						1,597k					
tmpl. rules	149						175						378					
after merge	87						108						119					
config.	N	NI	T	TI	TIM	TIMS	N	NI	T	TI	TIM	TIMS	N	NI	T	TI	TIM	TIMS
time (s)	99	117	404	89	81	69	365	391	734	227	221	225	858	940	1,690	474	443	465
rule apps (m)	16.5	15.5	308	11.3	9.9	7.8	62.5	50	468	22.9	21.1	13.9	149	110	1,115	81.8	78.6	75.6
% success	43.4	46.5	2.4	64.2	62.6	52.3	18.8	23.4	2.6	51.5	48.7	61.3	4.2	5.6	0.8	10.5	6.8	15
cache hit (m)	10.8	10.8	8.2	8.2	8.2	8.1	19.1	19.1	15.1	15.1	14.9	38.7	16.5	16.5	13.1	13	12.7	34.4

4.6 Preliminary Performance Evaluation

In order to initially evaluate the above optimisations, we applied small-scale reasoning for RDFS (minus the infinite rdf:_n axiomatic triples [4]), pD* and OWL 2 RL/RDF over LUBM(10) [3], consisting of about 1.3m triples – note that we do exclude **lg/gl** rules for RDFS/pD* since we allow generalised triples [2]. All evaluation in this paper has been run on single-core 2.2GHz Opteron x86-64 machine(s) with 4GB of main memory. Table 1 gives the performance for the following partial-indexing configurations: (i) N: 'normal' $\mathcal{T}$-split closure; (ii) NI: normal $\mathcal{T}$-split closure with linked rule index; (iii) T: $\mathcal{T}$-split closure wrt. templated rules; (iv) TI: $\mathcal{T}$-split closure wrt. linked templated rule index; (v) TIM: $\mathcal{T}$-split closure wrt. linked & merged templated rule index; (vi) TIMS: $\mathcal{T}$-split closure wrt. linked, merged & saturated templated rule index.

In all approaches, exhaustively applying templated rules demonstrates the worst performance; after indexing the approach becomes the most efficient. RDFS gains little in the way of improvement, but in fact only contains 8 rules requiring A-Box data: the reduction in rule applications given by templating and indexing is modest. OWL 2 RL and pD* take just over half the time for TI* vs. N* approaches. A correlation between increased rule applications and increased inferencing time is evident, but sometimes fails: e.g., for pD*, TIMS gives less rule applications than TIM, but takes more time – in such cases, we see the cache encountering more duplicates – as mentioned, saturated rules can immediately produce a batch of duplicates that would otherwise halt a chain of inferences mid-way. OWL 2 RL creates more templated rules than pD* due to expanded T-Box level reasoning, but these are merged to a number just above pD*: OWL 2 RL supports intersection-of inferencing used by LUBM and not in pD*. LUBM does not contain OWL 2 constructs, but redundant rules are factored out during templating.

Although we improve the performance of pD* and OWL 2 RL/RDF inferencing, we perform A-Box joins in-memory, and in fact cannot scale much beyond the limited scale above for these fragments: again our optimisations focus on linear rules. We now reunite with our original use-case of Linked Data reasoning, focussing on the application of linear rules and shifting up three orders of magnitude.

5   Reasoning for Linked Data

Again, we aim at reasoning over Linked Data for the SWSE project. In previous works, we have investigated the unique challenges of reasoning over the open Web, and identified the need for scale, incompleteness, and consideration of the source of data.

In [8], we applied reasoning over 1 billion Linked Data triples *using T-Box optimisations specific to a subset of pD**; we (i) demonstrated that aside from equality reasoning, pD* rules which do not require A-Box joins covered 99% of inferences possible in our Web dataset, based on the observation that the most commonly instantiated vocabularies on the Web typically use lightweight RDFS and OWL terms supportable by linear rules; (ii) discussed the dangers of applying materialisation over open Web data, which can naïvely lead to an explosion of inferences: for example, one document[1] defines owl:Thing to be a member of 55 union classes, another defines nine *properties* as the domain of rdf:type[2], etc. Observation (i) ties in with our linear-rule optimisations; however, equality reasoning requires A-Box joins: we see owl:sameAs related inferencing as very important for data integration within the Linked Data use-case, but prefer a decoupling of such reasoning – which entails its own requirements and challenges – and have analysed the issue separately in previous works [10]. Observation (ii) motivates our next discussion: we now reintroduce our notion of *authoritative reasoning*.

5.1   Authoritative Reasoning

In order to curtail the possible side-effects of open Web data publishing, we include the source of data in inferencing. Our methods are based on the view that a publisher instantiating a vocabulary's term (class/property) thereby accepts the inferencing mandated by that vocabulary (and recursively referenced vocabularies) for that term. Thus, once a publisher instantiates a term from a vocabulary, only that vocabulary and its references should influence what inferences are possible through that instantiation.

Firstly, we must define the relationship between a term and a vocabulary. We view a term as an RDF constant, and a vocabulary as a Web document: we give the function $http : \mathbf{U} \rightarrow 2^{\mathbf{G}}$ as the mapping from a URI (a Web location) to an RDF graph it may provide by means of a given HTTP lookup. In Linked Data principles, dereferencable URIs are encouraged; dereferencing can be seen as a function $deref : \mathbf{U} \rightarrow \mathbf{U}$ which maps one URI to another by means of HTTP dereferencing mechanisms (this may include removal of a URI fragment identifier and recursive but finite redirects, and maps a URI to itself in case of failure; such functions are fixed to the time the data was crawled).

We then give the authoritative function:

$$auth : \mathbf{U} \rightarrow 2^{\mathbf{C}}; \ u \mapsto \{c \mid c \in \mathbf{B}, c \in t \in http(u) \text{ or } c \in \mathbf{U}, deref(c) = u\} \quad (3)$$

where a Web document is authoritative for URIs which dereference to it and the blank nodes it contains; e.g., the FOAF vocabulary is authoritative for terms in its namespace.

To negate the effects of non-authoritative axioms on reasoning over Web data, we apply restrictions to the $\mathcal{T}$-split rule application of rules in $\mathbf{R}^{\mathrm{TG}}$, whereby, for the mapping μ of the rule application as before, there must additionally exist a $\mu(v)$ such that $v \in \mathbf{V}(Ante^{\mathcal{T}}) \cap \mathbf{V}(Ante^{\mathcal{G}})$, $\mu(v) \in auth(u)$, $\mu(Ante^{\mathcal{T}}) \subseteq http(u)$.[3]

---

[1] http://lsdis.cs.uga.edu/~oldham/ontology/wsag/wsag.owl

[2] http://www.eiao.net/rdf/1.0

[3] Note here that we restrict the T-Box segment of a $\mathbf{R}^{\mathrm{TG}}$ rule to be instantiated by one document; this is not so restrictive where in OWL 2 RL/RDF, all such rules contain one 'T-Box axiom', possibly described using multiple triples; cf. [8]. Also, we do not consider the results of T-Box level reasoning as authoritative.

Example 7. Take rule $r^\tau := (?c1,r:subClassOf,?c2) \wedge (?x,a,?c1) \Rightarrow (?x,a,?c2)$. Here, $\mathbf{V}(\mathcal{A}nte_{r^\tau}^{\mathcal{T}}) \cap \mathbf{V}(\mathcal{A}nte_{r^\tau}^{\mathcal{G}}) = \{?c1\}$. Take an A-Box triple $(ex:me,a,f:Person)$; $\mu(?c1) = f:Person$. Let $deref(f:Person) = f:$ the FOAF spec; now, $\{u \mid \mu(?c1) \in auth(u)\} = \{f:\}$. Any triple of the form $(f:Person,r:subClassOf,?c2)$ must come from $f:$ for the rule to be authoritatively applied. Note that $?c2$ can be arbitrarily bound; i.e., FOAF can *extend* any classes they like.

We refer the reader to [8] for more detail on authoritative reasoning. Note that the previous two examples from documents in Footnotes 1 & 2 are ignored by the authoritative reasoning. Since authoritativeness is on a T-Box level, we can apply the above additional restriction to our templating function when binding the terminological patterns of the rules to derive a set of *authoritative templated rules*.

5.2 Linked Data Reasoning Evaluation

We now give evaluation over 1.12b quads (947m unique triples) of Linked Data crawled for SWSE in May 2010. Note that we use a GZip compressed file of quadruples as input to the reasoning process: the fourth element element encodes the provenance (Web source) of the contained triple; we also require information about redirects encountered in the crawl to reconstruct the *deref* function. We output a flat file of GZipped triples. We perform reasoning over a subset of OWL 2 RL/RDF containing 42 rules: firstly, we omit datatype reasoning which can lead to the inference of near-infinite triples (e.g., $1.000^{\wedge\wedge}xsd:float$ $owl:sameAs$ $1.00^{\wedge\wedge}xsd:float$); secondly, we currently omit inconsistency checking rules (we will examine use-cases for these rules in later work); thirdly, we omit rules which infer 'tautologies' – statements that hold for every term in the graph, such as reflexive $owl:sameAs$ statements (we also filter these from the output). Given our use-case SWSE, we wish to infer a circumspect amount of data with utility for query-answering – completeness is not a requirement (cf. [5] for related discussion). For reasons of efficiency as described, we omit rules which require A-Box joins. Thus, our subset consists of the OWL 2 RL/RDF axiomatic rules, 'schema rules'[2, Table 9], and rules with one assertional pattern which we give in Table 3.

Reasoning over the dataset described inferred 1.58b raw triples, which were filtered to 1.14b triples removing non-RDF generalised triples and 'tautological statements' – post-processing revealed that 962m ($\sim 61\%$) were unique and had not been asserted (roughly a 1:1 *reasoned:asserted* ratio). The first step – extracting 1.1m T-Box triples from the dataset – took 8.2 hrs. Subsequently, Figure 1 gives the results for reasoning on one machine for each approach as before. T-Box level processing – e.g., templating, rule indexing, etc. – took roughly the same time. For A-Box reasoning, saturation causes the same problems with extra duplicate triples as before, and so the fastest approach is TIM, which takes $\sim 15\%$ of the time for the naïve $\mathcal{T}$-split closure algorithm; we also show the linear performance of TIM in Figure 1 (we would expect all methods to be similarly linear). 301k templated rules with 2.23m links are merged to 216k with 1.15m links; after saturation, each rule has an average of 6 consequent patterns and all links are successfully removed. Note that with 301k templated rules without indexing, applying all rules to all statements would take approx. 19 years.

Since all of our rules are linear, we can also distribute our approach by flooding the templated rules to all machines. In Table 2, we give the performance of such an

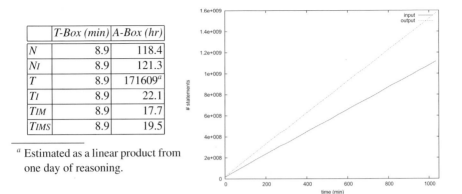

	T-Box (min)	A-Box (hr)
N	8.9	118.4
N_I	8.9	121.3
T	8.9	171609^a
T_I	8.9	22.1
T_{IM}	8.9	17.7
T_{IMS}	8.9	19.5

^{a} Estimated as a linear product from
one day of reasoning.

Fig. 1. Performance for reasoning over 1.1B statements on one machine for all approaches *(left)*, and detailed throughput performance for A-Box reasoning using the fastest approach T_{IM} *(right)*

Table 2. Distributed reasoning in minutes using T_{IM} for 1, 2, 4 & 8 machines

Machines	Extract T-Box	Build T-Box	Reason A-Box	Total
1	492	8.9	1062	1565
2	240	10.2	465	719
4	131	10.4	239	383
8	67	9.8	121	201

approach for 1, 2, 4, and 8 machines using a simple RMI architecture [9]. Note that the most expensive aspects of the reasoning process – extracting the T-Box from the dataset and reasoning over the A-Box – can be executed independently in parallel. The only communication required between machines is the aggregation of the T-Box, and creation of the shared templated-rule index: this takes ∼10 mins, and becomes the lower bound for time taken for distributed evaluation with arbitrary machine count. In summary, we perform reasoning over 1.12b Linked Data triples in 3.35 hours using 8 machines, deriving 1.58b inferred triples, of which 962m are novel and unique.

6 Related Work

We have discussed our previous work on SAOR throughout the paper. Following initial work on SAOR – which had not yet demonstrated distribution – a number of scalable distributed reasoners adopted a similar approach to partial indexing herein reformalised. Weaver et al. [19] discuss a similar approach for distributed reasoning over RDFS; however, their experiments were solely over LUBM and their discussion was specific to RDFS. Urbani et al. [18] use MapReduce for distributed reasoning for RDFS over 850m Linked Data triples; they do not consider authority and produce 30b triples which is too much for our SWSE use-case – interestingly, they also tried pD* on 35m Web triples and stopped after inferring 3.8b inferences in 12 hours, lending strength to our arguments for authoritative reasoning. In very recent work, the same authors [17] apply incomplete but comprehensive pD* to 100b LUBM triples, discussing rule-specific optimisations for performing join rules over pD*: however, we feel that materialisation wrt. rules over 1b triples of arbitrary Linked Data is still an open research goal.

A viable alternative approach to Web reasoning employed by Sindice [1] – the relation to which is discussed in depth in [8] – is to consider merging small "per-document" closures which quarantines reasoning to a given document and the related documents it either implicitly or explicitly imports. Works on LDSR select clean subsets of Linked Data $\sim$0.9b triples and apply reasoning using the proprietary BigOWLIM reasoner [11].

With respect to template rules, DLEJena [13] uses the Pellet DL reasoner for T-Box level reasoning, and uses the results to template rules for the Jena rule engine; they only demonstrate methods on synthetic datasets up to a scale of $\sim$1M triples. We take a somewhat different direction, discussing optimisations for partial-indexing.

Table 3. OWL 2 RL/RDF rules we apply *for Web reasoning* with exactly one assertional pattern. Authoritative variable positions are given in bold. Not shown are axiomatic and schema rules [2].

R^{G^1} : only one assertional pattern in antecedent			
OWL2RL	**Antecedent**		**Consequent**
	terminological	*assertional*	
eq-sym	-	?x owl:sameAs ?y .	?y owl:sameAs ?x .
prp-dom	**?p** rdfs:domain ?c .	?x ?p ?y .	?x a ?c .
prp-rng	**?p** rdfs:range ?c .	?x ?p ?y .	?y a ?c .
prp-symp	**?p** a owl:SymmetricProperty .	?x ?p ?y .	?y ?p ?x .
prp-spo1	**?p$_1$** rdfs:subPropertyOf ?p$_2$.	?x ?p$_1$?y .	?x ?p$_2$?y .
prp-eqp1	**?p$_1$** owl:equivalentProperty ?p$_2$.	?x ?p$_1$?y .	?x ?p$_2$?y .
prp-eqp2	?p$_1$ owl:equivalentProperty **?p$_2$** .	?x ?p$_2$?y .	?x ?p$_1$?y .
prp-inv1	**?p$_1$** owl:inverseOf ?p$_2$.	?x ?p$_1$?y .	?y ?p$_2$?x .
prp-inv2	?p$_1$ owl:inverseOf **?p$_2$** .	?x ?p$_2$?y .	?y ?p$_1$?x .
cls-int2	**?c** owl:intersectionOf (?c$_1$ ... ?c$_n$) .	?x a ?c .	?x a ?c$_1$...?c$_n$.
cls-uni	?c owl:unionOf (?c$_1$ ... **?c$_i$** ... ?c$_n$) .	?x a ?c$_i$.	?x a ?c .
cls-svf2	?x owl:someValuesFrom owl:Thing ; owl:onProperty **?p** .	?u ?p ?v .	?u a ?x .
cls-hv1	**?x** owl:hasValue ?y ; owl:onProperty **?p** .	?u a ?x .	?u ?p ?y .
cls-hv2	?x owl:hasValue **?y** ; owl:onProperty **?p** .	?u ?p ?y .	?u a ?x .
cax-sco	**?c$_1$** rdfs:subClassOf ?c$_2$.	?x a ?c$_1$.	?x a ?c$_2$.
cax-eqc1	**?c$_1$** owl:equivalentClass ?c$_2$.	?x a ?c$_1$.	?x a ?c$_2$.
cax-eqc2	?c$_1$ owl:equivalentClass **?c$_2$** .	?x a ?c$_2$.	?x a ?c$_1$.

7   Conclusion

We have introduced the notion of terminological data for RDF(S)/OWL, and have generalised and formalised the notion of partial indexing techniques which are optimised for application of linear rules and which rely on a separation of terminological data – a non-recursive segment of the data; we then related the derived closure to semi-naïve evaluation. We subsequently discussed inclusion of a template function in such an algorithm, showing that naïvely, templated rules worsen performance, but with rule merging, indexing and linking techniques, templated rules outperform the base-line $\mathcal{T}$-split closure esp. for a complex T-Box. This work, along with DLEJena, supports uncited claims within the recently standardised RIF working group that rule templating offers a more efficient solution for supporting OWL 2 RL than a direct translation of OWL 2 RL/RDF rules [15, Section 1]. We then reintroduced some discussion relating to reasoning over Linked Data, including our notion of authoritativeness, and demonstrated scalable distributed reasoning over a subset of OWL 2 RL for 1.1b quads (without need for manual

T-Box massaging or pre-selection). The SAOR system is actively used to provide reasoned data to the SWSE system [9] for live search and browsing over Linked Data: `http://swse.deri.org/`.

References

1. Delbru, R., Polleres, A., Tummarello, G., Decker, S.: Context Dependent Reasoning for Semantic Documents in Sindice. In: Proc. of 4th SSWS Workshop (October 2008)
2. Grau, B.C., Motik, B., Wu, Z., Fokoue, A., Lutz, C.: OWL 2 Web Ontology Language: Profiles. W3C Recommendation (October 2009)
3. Guo, Y., Pan, Z., Heflin, J.: LUBM: A benchmark for OWL knowledge base systems. J. Web Sem. 3(2-3), 158–182 (2005)
4. Hayes, P.: RDF semantics. W3C Recommendation (February 2004)
5. Hitzler, P., van Harmelen, F.: A Reasonable Semantic Web. Semantic Web Journal 1(1) (to appear 2010), `http://www.semantic-web-journal.net/`
6. Hogan, A., Harth, A., Passant, A., Decker, S., Polleres, A.: Weaving the Pedantic Web. In: Proc. of 3rd Workshop (April 2010)
7. Hogan, A., Harth, A., Polleres, A.: SAOR: Authoritative Reasoning for the Web. In: Domingue, J., Anutariya, C. (eds.) ASWC 2008. LNCS, vol. 5367, pp. 76–90. Springer, Heidelberg (2008)
8. Hogan, A., Harth, A., Polleres, A.: Scalable Authoritative OWL Reasoning for the Web. Int. J. Semantic Web Inf. Syst. 5(2) (2009)
9. Hogan, A., Harth, A., Umbrich, J., Kinsella, S., Polleres, A., Decker, S.: Searching and Browsing Linked Data with SWSE: the Semantic Web Search Engine. Technical Report DERI-TR-2010-07-23 (2010)
10. Hogan, A., Polleres, A., Umbrich, J., Zimmermann, A.: Some entities are more equal than others: statistical methods to consolidate Linked Data. In: Proc. of NeFoRS Workshop (2010)
11. Kiryakov, A., Ognyanoff, D., Velkov, R., Tashev, Z., Peikov, I.: LDSR: a Reason-able View to the Web of Linked Data. In: Proc. of 7th Semantic Web Challenge (2009)
12. Lloyd, J.W.: Foundations of Logic Programming, 2nd edn. Springer, Heidelberg (1987)
13. Meditskos, G., Bassiliades, N.: DLEJena: A practical forward-chaining OWL 2 RL reasoner combining Jena and Pellet. J. Web Sem. 8(1), 89–94 (2010)
14. Ramakrishnan, R., Srivastava, D., Sudarshan, S.: Rule Ordering in Bottom-Up Fixpoint Evaluation of Logic Programs. In: Proc. of 16th VLDB, pp. 359–371 (1990)
15. Reynolds, D.: OWL 2 RL in RIF. W3C Working Group Note (June 2010)
16. ter Horst, H.J.: Completeness, decidability and complexity of entailment for RDF Schema and a semantic extension involving the OWL vocabulary. J. Web Sem. 3, 79–115 (2005)
17. Urbani, J., Kotoulas, S., Maassen, J., van Harmelen, F., Bal, H.E.: OWL reasoning with WebPIE: Calculating the closure of 100 billion triples. In: Aroyo, L., Antoniou, G., Hyvönen, E., ten Teije, A., Stuckenschmidt, H., Cabral, L., Tudorache, T. (eds.) The Semantic Web: Research and Applications. LNCS, vol. 6088, pp. 213–227. Springer, Heidelberg (2010)
18. Urbani, J., Kotoulas, S., Oren, E., van Harmelen, F.: Scalable Distributed Reasoning Using MapReduce. In: Bernstein, A., Karger, D.R., Heath, T., Feigenbaum, L., Maynard, D., Motta, E., Thirunarayan, K. (eds.) ISWC 2009. LNCS, vol. 5823, pp. 634–649. Springer, Heidelberg (2009)
19. Weaver, J., Hendler, J.A.: Parallel Materialization of the Finite RDFS Closure for Hundreds of Millions of Triples. In: Bernstein, A., Karger, D.R., Heath, T., Feigenbaum, L., Maynard, D., Motta, E., Thirunarayan, K. (eds.) ISWC 2009. LNCS, vol. 5823, pp. 682–697. Springer, Heidelberg (2009)

Justification Oriented Proofs in OWL

Matthew Horridge, Bijan Parsia, and Ulrike Sattler

School of Computer Science,
The University of Manchester

Abstract. Justifications — that is, minimal entailing subsets of an ontology — are currently the dominant form of explanation provided by
ontology engineering environments, especially those focused on the Web
Ontology Language (OWL). Despite this, there are naturally occurring
justifications that can be very difficult to understand. In essence, justifications are merely the premises of a proof and, as such, do not articulate
the (often non-obvious) reasoning which connect those premises with the
conclusion. This paper presents justification oriented proofs as a potential solution to this problem.

1 Introduction and Motivation

Modern ontology development environments such as Protégé-4, the NeOn Toolkit,
Swoop, and Top Braid Composer, allow users to request explanations for entailments (inferences) that they encounter when editing or browsing ontologies. Indeed, the provision of explanation generating functionality is generally seen as
being a vital component in such tools. Over the last few years, *justifications* have
become the dominant form of explanation in these tools. This paper examines
justifications as a kind of explanation and highlights some problems with them.
It then presents *justification lemmatisation* as a non-standard reasoning service,
which can be used to augment a justification with *intermediate inference steps*,
and gives rise to a structure known as a *justification oriented proof*. Ultimately, a
justification oriented proof could be used as an input into some presentation device to help a person step though a justification that is otherwise too difficult for
them to understand.

1.1 Justifications as Explanations

A justification is a minimal subset of an ontology (a set of axioms) that is
sufficient for a given entailment to hold. As an example, consider the small
ontology $\mathcal{O} = \{A \sqsubseteq B, A(i), C \sqsubseteq D\}$, which *entails* $B(i)$, written $\mathcal{O} \models B(i)$[1]. A
justification $\mathcal{J}$ for $\mathcal{O} \models B(i)$ is a *minimal* subset of $\mathcal{O}$ that entails $B(i)$, in this
case $\mathcal{J} = \{A \sqsubseteq B, A(i)\}$.

The major benefit of justifications is that they pinpoint and isolate the hand-
fuls of axioms, in what could be a very large ontology, that cause the entailment

---

[1] $B(i)$ means i is an instance of B.

P.F. Patel-Schneider et al. (Eds.): ISWC 2010, Part I, LNCS 6496, pp. 354–369, 2010.

InverseProperties(hasPet, isPetOf)
isPetOf(Rex, Mick)
Domain(hasPet, Person)
Male(Mick)
reads(Mick, DailyMirror)
drives(Mick, Q123ABC)
Van(Q123ABC)
Van ⊑ Vehicle
WhiteThing(Q123ABC)
Driver ≡ Person ⊓ ∃drives.Vehicle
Driver ⊑ Adult
Man ≡ Adult ⊓ Male ⊓ Person
WhiteVanMan ≡ Man ⊓ ∃drives.(Van ⊓ WhiteThing)
WhiteVanMan ⊑ ∀reads.Tabloid
Tabloid ⊑ Newspaper

Person ⊑ ¬Movie
RRated ⊑ CatMovie
CatMovie ⊑ Movie
RRated ≡ (∃hasScript.ThrillerScript)
⊔ (∀hasViolenceLevel.High)
Domain(hasViolenceLevel, Movie)

Fig. 1. A justification for Person ⊑ ⊥   **Fig. 2.**   A   justification   for
Newspaper(DailyMirror)

to hold. For example, the SNOMED medical ontology contains roughly 400,000 axioms, but a justification for an entailment in this ontology is on average less than ten axioms in size [2].

Unlike full blown proofs, justifications are conceptually simple structures with a natural relation to the ontology development process—they are directly related to what has been asserted or stated in an ontology. Justifications require very little additional knowledge beyond the semantics of the language. This conceptual simplicity, coupled with the fact that the computation of justifications for real ontologies tends to be practical [10], and the fact that off-the-shelf implementations of justification finding services exist, has most likely lead to the large uptake of justifications as a type of explanation.

1.2 Problems with Justifications

However, despite the fact that justifications are a popular form of explanation in the OWL world, observations show there are justifications that people find difficult or impossible to understand. Indeed, the justifications shown in Figure 1 and Figure 2, both from real ontologies, gave many users trouble when trying to understand how they lead to their respective entailments. Indeed, some people questioned whether the justification shown in Figure 1 was a justification at all.

In the case of the justification shown in Figure 1, which is a justification for Person ⊑ ⊥[2] spotting that the justification entails Movie ≡ ⊤[3] is key to understanding how the justification works. Since everything is entailed to be a Movie, and Person is disjoint with Movie, Person is disjoint with ⊤, hence Person is

---
[2] ⊥ is read as "bottom" and is the same as owl:Nothing.
[3] ⊤ is read as "top" and is the same as owl:Thing.

unsatisfiable. People who fail to realise that Movie $\equiv \top$ is also entailed generally fail to understand how the justification gives rise to the entailment.

Similarly, the justification shown in Figure 2, is also rather difficult for people to work through. There are fifteen axioms of many different types in the justification. It is far from obvious how these axioms interplay with each other to result in the entailment Tabloid(DailyMirror). When a user works through this justification, they have to spot intermediate entailments, for example, WhiteVanMan(Mick) and Person(Mick), in order to arrive at the conclusion Tabloid(DailyMirror)).

In a exploratory study [5], it was observed that many justifications *for entailments of interest in real ontologies* can be understood by people with a variety of backgrounds, and these kinds of justifications serve extremely well as explanations. However, it was also observed that there are justifications that are difficult or impossible for people to work through. Two obvious reasons for this are: (1) People do not spot key entailments within justifications, that are necessary for them to understand how the justification works (as is the case with the justification in Figure 1, and (2) People find large justifications, with many types of axioms, tedious and therefore difficult to work through (as is the case with the justification in Figure 2). In other words, when people fail, or find it difficult, to spot *intermediate entailments, conclusions or steps* they can fail to understand why a justification supports the entailment in question, and hence fail to understand why the entailment in question holds in their ontology.

1.3   From Justifications towards Proofs

The above notion of "intermediate steps" that could guide a person through understanding a justification, raises the question of whether full blown proofs, such as natural deduction style proofs with inference rules, should be used for explaining entailments in OWL ontologies.

One of the typical claims about natural deduction is that it mimics human reasoning—that is, it has a *strong cognitive adequacy* [18]. However, there is ongoing debate in the field of cognitive psychology about how human reasoning actually works. Some camps favour a "logic" or rule based account [16], while others favour a "model" based account [8]. Even for simple cases of natural language based deduction, it is unclear which account is correct. Moreover, other research [14] shows that relatively untrained people—clearly without having a complete set of deduction rules at their disposal—can successfully work through surprisingly complex reasoning puzzles. It is therefore impossible to say whether or not natural deduction and similar proof systems mimic human reasoning. What is clear, is that representations that have a strong cognitive adequacy are not necessarily useable. Hence, even if natural deduction has a strong cognitive adequacy, there is no guarantee that it is usable as a form of explanation for entailments in ontologies.

In summary, it is likely that natural deduction style proofs are not necessarily the best form of explanation. On the other hand, justifications are an appealing type of explanation. It is known that that a wide range of people can cope with justifications [5]. This includes domain experts who have very little training or background with the Description Logics that underpin OWL. Justifications appeal to these kinds of people because they are conceptually simple—very little training is needed in order to understand how justifications work. The same cannot be said about natural deduction style proofs. Additionally, people are used to seeing axioms, albeit in a frame-based style of presentation, and justifications reflect this familiarity. If natural deduction style proofs were presented to people such as domain experts, they would require special training in order to read the proofs.

1.4   Justification Oriented Proofs

What is needed, is something that lies between justifications and proofs. Given the popularity and conceptual simplicity of justifications, the work presented in this paper uses them as building blocks for structures that begin to look like proofs, but are independent of any calculus or deduction rules. In essence, intermediate steps are introduced into a justification, which are themselves explained with justifications. This results in a directed acyclic proof graph of the form shown in Figure 3. Ultimately a justification is extended with "lemmas" into a *justification oriented proof.*

The main idea behind a justification oriented proof is depicted in Figure 3. The numbered lozenges represent axioms, with the leftmost lozenge, labelled η, representing the entailment of interest. The white lozenges labelled with "1" – "6" represent exactly the axioms that appear in the original justification $\mathcal{J}$ for the entailment (and are therefore in the ontology as asserted axioms). Grey shaded lozenges represent *lemmas* that are entailed by the deductive closure of $\mathcal{J}$ but are not in $\mathcal{J}$ as asserted axioms. For a given node, its direct predecessors constitute a justification for that node. This produces a weakly connected directed acyclic graph, with one sink node that represents the entailment of interest and a source node for each axiom in the justification. Hence, in the example shown in Figure 3, $\mathcal{J} = \{1, 2, 3, 4, 5, 6\}$ is a justification for η with respect to the ontology that entails η. Axiom 7 is a lemma for axioms 1, 2 and 3 (conversely, axioms 1, 2 and 3 are a justification for axiom 7). Axiom 8 is a lemma for axioms 3, 4 and 5 (conversely axioms 3, 4 and 5 are a justification for axiom 8). Together axioms 6, 7 and 8 constitute a justification for η i.e. the entailment. Notice that axiom 3 participates in different justifications for different lemmas.

Ultimately, the justification oriented proof guides a person through the understanding of the original justification. In essence, lemmas are intermediate steps that may be non-obvious, but may be significant to understanding how the justification results in the entailment. They also provide a chunking mechanism, which can help guide a user through a large and tedious to understand justification.

1.5   Contributions

The *main contribution* of this paper is the novel *framework* that is presented for *constructing* justification oriented proofs. This framework is rather different to

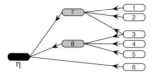

Fig. 3. A schematic of a Justification Oriented Proof — Predecessors of a node represent a justification for that node

other approaches: First, the framework *does not use any deduction rules* per se to derive the intermediate steps or conclusions. The choice of steps is ultimately governed by a *pluggable* justification complexity model which is used to choose one justification over another during the proof construction. Details of a *practical* model are supplied in this paper, but it is important to realise that this paper shows that the idea of using *a model* to select intermediate steps works well in practice. Second, the framework is entirely *black-box based*. Any reasoner, such as FaCT++, HermiT, Pellet or Racer, that implements a decision procedure for entailment checking for OWL 2 (or any other monotonic logic) may be used for generating the proofs. In other words, the internals of the reasoner need not be modified to extract some kind of intermediate proof. Third, and finally, the presented framework ought to be easily adaptable to deal with other fragments of First Order Logic that may or may not overlap the fragment that corresponds to OWL 2.

2   Related Work

The idea of using proofs as forms of explanation is obviously not new. Indeed, in some camps [3,11], proofs are essentially regarded as *the* main form of explanation. However, the work that is presented in this paper is based on the intuitions mentioned in the introduction. That is, it is arguably more practical and more helpful to *not* show full blown proofs because (1) users already know and understand justifications, and (2) it avoids having to teach users a new calculus or deduction rules.

In [7], Huang acknowledges that Natural Deduction proofs are too fine-grained to be used as explanations, and introduces Natural Deduction Style Proofs at the assertional level, where trivial steps are eliminated from proofs. Parallels may be drawn with the basic motivations presented here. The main difference here is that the proofs here are arguably targeted at an even higher level of abstraction, and that an entirely black-box complexity model based approach is used to generate the proofs rather than extracting them from a theorem prover.

Finally, in [17] Schlobach introduces optimal interpolants, and so called *illustrations* that are intended to bridge the gap between subsumee and subsumer class expressions. The notion of lemmas and justifications oriented proofs as presented here are in the spirit of Schlobach's illustrations. However, the main

Table 1. OWL 2 Class, Object Property and Individual Axioms

$C \sqsubseteq D$	$C \equiv D$	$\mathsf{DisjointClasses}(C_1, \ldots, C_n)$
$\mathsf{DisjointUnion}(C, D_1, \ldots, D_n)$		
$R \sqsubseteq S$	$R \equiv S$	$\mathsf{DisjointProperties}(R_1, \ldots, R_n)$
$\mathsf{InverseProperties}(R, S)$	$\mathsf{Domain}(R, C)$	$\mathsf{Range}(R, C)$
$\mathsf{Functional}(R)$	$\mathsf{InverseFunctional}(R)$	$\mathsf{Transitive}(R)$
$\mathsf{Symmetric}(R)$	$\mathsf{Asymmetric}(R)$	$\mathsf{Reflexive}(R)$
$\mathsf{Irreflexive}(R)$		
$C(a)$	$R(a, b)$	$\mathsf{DifferentIndividuals}(a, ..., a_n)$
$\mathsf{SameIndividual}(a_1, \ldots, a_n)$		

difference is that Schlobach's work primarily deals with subsumption between two class expressions in isolation, whereas the work presented here deals with arbitrary entailments that arise from a sets of axioms.

3   Preliminaries

OWL 2 and Description Logics The work presented in this paper focuses on OWL 2. OWL 2 [12] is the latest standard in ontology languages from the W3C. An OWL 2 ontology roughly corresponds to a $\mathcal{SROIQ}(D)$ [6] knowledge base. For the purposes of this paper, an *ontology* is regarded as a finite set of axioms $\{\alpha_0, \ldots, \alpha_n\}$ of the form shown in Table 1[4], where C and D are (possibly complex) class expressions, R and S are (possibly inverse or complex) properties, and a and b are individuals. (Note that subscripts are used to represent different occurrences, or class expressions, properties etc.).

Definition 1 (Justification). *$\mathcal{J}$ is a justification for $\mathcal{O} \models \eta$ if $\mathcal{J} \subseteq \mathcal{O}$, $\mathcal{J} \models \eta$ and for all $\mathcal{J}' \subsetneq \mathcal{J}$ $\mathcal{J}' \not\models \eta$.*

By a slight abuse of notation, the nomenclature used in this paper also refers to a minimally entailing set of axioms (that is not necessarily a subset of an ontology) as a justification.

The Structural Transformation — δ Much of the work presented in the remainder of the paper uses the "well known" structural transformation — referred to here as δ. This transformation takes a set of axioms and flattens out each axiom by introducing names for sub-concepts, transforming the axioms into an equi-satisfiable set of axioms. The structural transformation was first described in Plaisted and Greenbaum [15], with a version of the rewrite rules for description logics given in [13]. For the sake of brevity, the structural transformation is not defined here — the interested reader is referred to [13,4] for a full definition.

---

[4] For the sake of brevity, axioms involving data properties and data ranges are not presented here. However, the framework extends to these axioms in the obvious way.

4   Proof Generation Framework

In what follows the framework for generating justification oriented proofs is presented. The framework consists of two main ideas: (1) The notion of justification lemmatisation. Subsets of a justification may be replaced with simple summarising axioms, which are known as *lemmas*. One justification is lemmatised into another justification. (2) The notion of stitching a series of lemmatised justifications into a justification oriented proof. First a definition of justification lemmatisation is presented and then a definition for justification oriented proofs is given.

4.1   Justification Lemmatisation

Given a justification $\mathcal{J}$ for an entailment η, the aim is to lemmatise $\mathcal{J}$ into $\mathcal{J}'$, so that $\mathcal{J}'$ *is less complex by some measure* and for *some purpose* than $\mathcal{J}$. With this notion in hand, lemmas for justifications can now be defined. First, an informal description is given, then a more precise definition is given in Definition 3.

Informally, a set of lemmas $\Lambda_{\mathcal{S}}$ for a justification $\mathcal{J}$ for η is a set of axioms that is entailed by $\mathcal{J}$ which can be used to replace some set $\mathcal{S} \subseteq \mathcal{J}$ to give a new justification $\mathcal{J}' = (\mathcal{J} \setminus \mathcal{S}) \cup \Lambda_{\mathcal{S}}$ for η. If, additionally, $\mathcal{J}'$ is less complex, by some measure, than $\mathcal{J}$. $\mathcal{J}'$ is called a *lemmatisation of $\mathcal{J}$.*

Various restrictions are placed on the generation of the set of lemmas $\Lambda_{\mathcal{S}}$ that can lemmatise a justification $\mathcal{J}$. These restrictions prevent "trivial" lemmatisations, an example of which will be given below. Before these restrictions are discussed, it is useful to introduce the notion of a *tidy* set of axioms.

Intuitively, a set of axioms is *tidy* if it is consistent, contains no synonyms of $\perp$ (where a class name is a synonym of $\perp$ with respect to a set of axioms $\mathcal{S}$ if $\mathcal{S} \models A \sqsubseteq \perp$), and contains no synonyms of $\top$ (where a class name is a synonym of $\top$ with respect to a set of axioms $\mathcal{S}$ if $\mathcal{S} \models \top \sqsubseteq A$).

Definition 2 (Tidy sets of axioms). *A set of axioms $\mathcal{S}$ is tidy if $\mathcal{S} \not\models \top \sqsubseteq \perp$, $\mathcal{S} \not\models A \sqsubseteq \perp$ for all $A \in Signature(\mathcal{S})$, and $\mathcal{S} \not\models \top \sqsubseteq A$ for all $A \in Signature(\mathcal{S})$.*

The definition of lemmatisation that follows, mandates that a set of lemmas $\Lambda_{\mathcal{S}}$ must only be drawn from (i) the deductive closure of *tidy* subsets of the set $\mathcal{S} \subseteq \mathcal{J}$, (ii) from the *exact* set of synonyms of $\perp$ or $\top$ over $\mathcal{S}$.

Without the above restrictions on the axioms in $\Lambda_{\mathcal{S}}$, it would be possible to lemmatise a justification $\mathcal{J}$ to produce a justification $\mathcal{J}'$ that, in isolation, is simple to understand, but otherwise bears little or no resemblance to $\mathcal{J}$. For example, consider $\mathcal{J} = \{A \sqsubseteq \exists R.B, \ B \sqsubseteq E \sqcap \exists S.C, \ B \sqsubseteq D \sqcap \forall S.\neg C\}$ as a justification for $A \sqsubseteq \perp$. Suppose that *any* axioms entailed by $\mathcal{J}$, could be used as lemmas (i.e. there are no restrictions on the axioms that make up $\Lambda_{\mathcal{S}}$). In this example, A is unsatisfiable in $\mathcal{J}$, meaning that it would be possible for $\mathcal{J}' = \{A \sqsubseteq E, A \sqsubseteq \neg E\}$ to be a lemmatisation of $\mathcal{J}$. Here, $\mathcal{J}'$ is arguably easier to understand than $\mathcal{J}$, but bears little resemblance to $\mathcal{J}$. In other words, $A \sqsubseteq E$

and $A \sqsubseteq \neg E$ are not helpful lemmas for $\mathcal{J} \models A \sqsubseteq \bot$. Similarly unhelpful results arise if lemmas are drawn from *inconsistent* sets of axioms, or sets of axioms that contain synonyms for $\top$.

Given the above intuitions and the notion of tidy sets of axioms, the notion of justification lemmatisation is defined as follows:

Definition 3 (Justification Lemmatisation). *Let* $\mathcal{J}$ *be a justification for* η *and* $\mathcal{S}$ *a set of axioms such that* $\mathcal{S} \subseteq \mathcal{J}$. *Let* $\Theta_{\mathcal{S}}$ *be the set of tidy subsets of* $(\mathcal{S} \cup \delta(\mathcal{S}))$. *Recall that* $\mathcal{T}^{\star}$ *is the deductive closure of a set of axioms* $\mathcal{T}$. *Let*

$$\Lambda_{\mathcal{S}} \subseteq \bigcup_{\mathcal{T} \in \Theta_{\mathcal{S}}} \mathcal{T}^{\star} \cup \{\alpha \mid \alpha \text{ is of the form } A \sqsubseteq \bot \text{ or } \top \sqsubseteq A, \\ \text{and } \exists \mathcal{K} \subseteq (\mathcal{S} \cup \delta(\mathcal{S})) \text{ that is consistent and } \mathcal{K} \models \alpha\}$$

$\Lambda_{\mathcal{S}}$ *is a set of lemmas for a justification* $\mathcal{J}$ *for* η *if, for* $\mathcal{J}' = (\mathcal{J} \setminus \mathcal{S}) \cup \Lambda_{\mathcal{S}}$

1. $\mathcal{J}'$ *is a justification for* η *over* $\mathcal{J}^{\star}$, *and,*
2. *Complexity*$(\eta, \mathcal{J}') <$ *Complexity*$(\eta, \mathcal{J})$.

The ability to lemmatise one justification into another justification *is a key process* in constructing a justification oriented proof. Given a regular justification $\mathcal{J}$ for η, $\mathcal{J}$ can be lemmatised into $\mathcal{J}_1$ for η. The axioms in $\mathcal{J}_1$ may then be inspected to determine which of them are lemmas – lemmas are axioms that are not in $\mathcal{J}$. Given a lemma $\alpha \in \mathcal{J}_1$ ($\alpha \notin \mathcal{J}$) a new justification $\mathcal{J}_2 \subseteq \mathcal{J}$ for α can be identified. If necessary, $\mathcal{J}_2$ can then be lemmatised into a simpler justification for α. Axioms in $\mathcal{J}_2$ can then be inspected and the process can be repeated as necessary. Ultimately the process builds up a justification oriented proof. Justification oriented proofs are defined as follows:

Definition 4 (Justification Oriented Proof). *A justification oriented proof for a justification* $\mathcal{J}$ *for an entailment* η *in* $\mathcal{O}$ *is a weakly connected directed acyclic graph* $G = (V, E)$ *such that* $\mathcal{J} \subseteq V \subset \mathcal{J}^{\star}$ *and either,* $G = (\{\eta\}, \{\langle \eta, \eta \rangle\})$ *or,*

1. η *is the one and only sink node in* G,
2. $\mathcal{J}$ *is the exact set of source nodes in* G, *and*
3. *For a given node, the set of predecessor nodes are a justification for the node over* $\mathcal{J}^{\star}$.

In summary, as shown in Figure 3, a node in a justification oriented proof that has incoming edges, is either a lemma or the entailment (sink node) itself. Source nodes (nodes with no predecessors) are the axioms in the original justification. Finally, given one justification $\mathcal{J}$ for η, there may be *multiple* justification oriented proofs, even if the set of lemmas in the proof is fixed.

It should be noted that, in the same way that raw unordered justifications are not presented directly to end users, it is unlikely the graph which constitutes a justification oriented proof should be presented *directly* to end users. Instead, the graph can be used as an input into some interactive presentation device.

4.2 Complexity Models

As can be seen from Definition 3, justification lemmatisation depends upon the notion of justification complexity. More specifically, it depends upon whether one justification is more complex, by some measure and for some purpose, than another justification. In this framework, *complexity models* are used to assign complexity scores to justifications and determine whether one justification is more complex, than another. The framework makes *no commitment to a particular complexity model*. Indeed, models are intended to be pluggable. A model may depend upon the application in question and the intended audience. In the work presented here, the primary aim is to produce justification oriented proofs, which pick out difficult to spot lemmas, and chunk and summarise sets of heterogeneous axioms in justifications. With these goals in mind, a simple model is presented later in this paper. However, before this model is presented, models that deal with special use cases are first discussed. The main intention here, is to give a feel for how different models can be appropriate for different applications, and how different models may be plugged into the framework.

A Model for Deriving Proofs for Laconic Justifications. A laconic justification [4] is a justification whose axioms have no superfluous parts and whose parts are as weak as possible. Given $\mathcal{O} \models \eta$, a laconic justification oriented proof consists of a sink node η, and predecessors of η which are either (1) leaf nodes representing axioms contained in $\mathcal{O}$, or (2) are nodes representing axioms entailed by $\mathcal{O}$, for which each one has a predecessor representing an axiom contained in $\mathcal{O}$. Given a justification $\mathcal{J}$ for η, a simple complexity model for computing such proofs assigns a score of zero to $(\mathcal{J}', \eta)$ if $\mathcal{J}'$ is a laconic justification for η, a score of zero to $(\mathcal{J}', \alpha)$ if $\alpha \neq \eta$ and α is in the laconic justification in question, and $\mathcal{J}'$ is a singleton set containing an axiom from the original ontology, and otherwise, a score of one.

A Model for Deriving Proofs for Root/Derived Unsatisfiable Classes. Given an ontology $\mathcal{O}$ which contains unsatisfiable classes ($\mathcal{O} \models A \sqsubseteq \bot$ for some class name A in the signature of $\mathcal{O}$), a root unsatisfiable class [9] is a class in the signature of $\mathcal{O}$ whose unsatisfiability does not depend on the unsatisfiability of any other class in the signature of $\mathcal{O}$. A derived unsatisfiable class is a class whose unsatisfiability depends on the unsatisfiability of some other class in the signature of $\mathcal{O}$. More precisely, given $\mathcal{O} \models A \sqsubseteq \bot$, A is a derived unsatisfiable class if there exists some class B such that $\mathcal{O} \models B \sqsubseteq \bot$ and there is a justification $\mathcal{J}_A \models A \sqsubseteq \bot$ and another justification $\mathcal{J}_B \models B \sqsubseteq \bot$ such that $\mathcal{J}_B \subsetneq \mathcal{J}_A$, otherwise, A is a root unsatisfiable class.

A suitable model that will lemmatise and "collapse" a subset that corresponds to a justification for a root unsatisfiable class (corresponding to $\mathcal{J}_B$ above) is as follows: Given $\mathcal{O} \models A \sqsubseteq \bot$, the model assigns a score of 1 to a justification $\mathcal{J}_A$ for $\mathcal{O} \models \eta$ if there exists a justification $\mathcal{J}' \subset \mathcal{J}$ for $\mathcal{J} \models B \sqsubseteq \bot$, where $\mathcal{J}' \neq \{B \sqsubseteq \bot\}$ and $\mathcal{J}'' = \mathcal{J} \setminus \mathcal{J}' \cup \{B \sqsubseteq \bot\}$ is a justification for $A \sqsubseteq \bot$ over the deductive closure of $\mathcal{O}$, the model otherwise assigns a score of 0.

4.3   A General Model for Deriving Justification Oriented Proofs

For the purposes of introducing non-obvious and summarising intermediate steps into justifications, a simple justification complexity model is presented in Table 2. This model was derived partly from intuitions on what makes justifications difficult to understand, and partly from the observations made during a pilot/exploratory study [5] in which people attempted to understand justifications from real ontologies. The model uses various components to produce complexity scores which are summed to produce an overall complexity score for a justification. Broadly speaking, there are two types of components: (1) Structural components, such as C1, which require a syntactic analysis of a justification, and (2) Semantic components, such as C4, which require entailment checking to reveal non-obvious phenomena. Although the model presented in Table 2 is rather simple, it is surprisingly effective in that it produces pleasing justification oriented proofs.

5   An Algorithm for Generating Proofs

Given the above definitions, the main algorithms for generating proofs are presented below. There are three main algorithms: 1) GenerateProof, which takes a justification as an input and outputs a proof; 2) LemmatiseJustification, which takes a justification as an input and outputs either a lemmatised justification or the justification itself; 3) ComputeJPlus, which takes a justification and computes a set of axioms that are in the deductive closure of tidy subsets of the justification from which lemmas may be drawn. The GenerateProof algorithm uses the LemmatiseJustification as a sub-routine, and the LemmatiseJustification algorithm uses the ComputeJPlus algorithm as a sub-routine. Note that due to space limitations, the ComputeJPlus algorithm is not specified line by line in this paper—instead, a definition of $\mathcal{J}^+$ (Definition 5) is given below, and it is assumed that the algorithm simply computes $\mathcal{J}^+$ in accordance with this definition.

5.1   GenerateProof

The GenerateProof algorithm for computing justification oriented proofs is depicted in Figure 4. The basic idea is that, given an input of a justification $\mathcal{J}$ for η, a lemmatised justification $\mathcal{J}'$ for η is computed. $\mathcal{J}'$ is then used to initialise a justification oriented proof $\mathcal{P}$. For each node λ in the proof corresponding to an axiom in $\mathcal{J}'$, if λ is not in $\mathcal{J}$ then it is a lemma and a justification needs to be computed for it. In this case a new justification $\mathcal{J}''$ is computed for α' over $\mathcal{J}$. Next, $\mathcal{J}''$ is lemmatised to give $\mathcal{J}'''$ which is inserted into the proof $\mathcal{P}$. The process then repeats for lemmas in $\mathcal{P}$ that do not have any predecessors until none of the leaves in the proof are lemmas. Although not depicted in Figure 4, it is important to note that, in order to comply with Definition 4, there is a test in step 6 to determine whether inserting $\mathcal{J}'''$ as a result of the lemmatisation process into $\mathcal{P}$ would result in a cyclic graph instead of a DAG. If this is the

Table 2. A Simple Complexity Model for Generating Justification Oriented Proofs

Name	Description
C1 AxiomTypes	Counts the axiom types in $\mathcal{J}$ and η. The count is multiplied by a weighting (10.0) and added to the overall complexity score.
C2 ClassConstructors	Counts the class constructors in $\mathcal{J}$ and η. The count is multiplied by a weighting (10.0) and added to the overall complexity score.
C3 UniversalImplication	If $\alpha \in \mathcal{J}$ and α is of the form $\forall R.C \sqsubseteq D$ or $D \equiv \forall R.C$ a constant (50.0) is added to the overall complexity score.
C4 SynonymOfThing	If $\mathcal{J} \models \top \sqsubseteq A$ for some $A \in$ Signature($\mathcal{J}$) and $\top \sqsubseteq A \notin \mathcal{J}$ and $\top \sqsubseteq A \neq \eta$ then a constant (50.0) is added to the complexity score.
C4 SynonymOfNothing	If $\mathcal{J} \models A \sqsubseteq \bot$ for some $A \in$ Signature($\mathcal{J}$) and $A \sqsubseteq \bot \notin \mathcal{J}$ and $A \sqsubseteq \bot \neq \eta$ then a constant (50.0) is added to the complexity score.
C5 DomainAndNoExistential	If Domain$(R, C) \in \mathcal{J}$ and $\mathcal{J} \not\models E \sqsubseteq \exists R.D$ for some class expressions E and D then a constant (50.0) is added to the complexity score.
C6 ModalDepth	The maximum modal depth of all class expressions in $\mathcal{J}$ is multiplied by a weighting (50.0) and added to the overall complexity score
C7 SignatureDifference	For each $A \in$ Signature(η), where $A \notin$ Signature($\mathcal{J}$) a weighting (50.0) is added to the overall complexity score
C8 AxiomTypeDifference	If the axiom type of η is not the set of axiom types of $\mathcal{J}$ then a weighting (50.0) is added to the overall complexity score
C9 ClassConstructorDifference	For each class constructor in η that is not in the set of class constructors of $\mathcal{J}$, a weighting (50.0) is added to the overall complexity score

case, then an alternative lemmatisation of $\mathcal{J}''$ must be chosen (or if there are no alternatives then $\mathcal{J}''$ itself must be chosen) to insert into $\mathcal{P}$. This enforcement of non-cyclical proofs is also part of the mechanism that ensures the GenerateProof algorithm terminates. A discussion on termination is presented later.

5.2   LemmatiseJustification

The LemmatiseJustification algorithm is presented in Algorithm 1. The algorithm takes a justification J for η as its input and returns a justification L as its output. Either L is a lemmatisation of J or L is equal to J. In essence, the algorithm produces a lemmatised justification by computing a *filter* S on the deductive closure of tidy subsets of J, which obviously includes axioms that could lemmatise J. Justifications for η are then computed with respect to S.

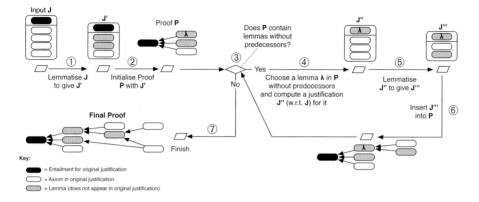

Fig. 4. GenerateProof – A Depiction of an Algorithm for Generating Justification Oriented Proofs. Justification Lemmatisation is used as a Sub-routine.

A complexity score is computed for each justification $L \subseteq S$, which is compared to the complexity of J. If the difference between the score for J and the score for L is positive then L is selected as a lemmatisation of J. Algorithm 1 always terminates due to the fact that S is finite in size and hence there are a finite number of justifications for η with respect to S.

5.3 ComputeJPlus

Definition 3 mandates that, for a justification $\mathcal{J}$, lemmas must be drawn from the deductive closure of tidy subsets of $\mathcal{J}$. However, the deductive closure of a set of axioms is *infinite*. For practical purposes it is necessary to work with a finite representative of the deductive closure that suffices for computing pleasing lemmatisations and pleasing justification oriented proofs. In addition to these practicalities, a finite representation of the deductive closure is needed because the ability to draw lemmas from an infinite set of axioms could lead to non-termination of the GenerateProof algorithm. In order to ensure termination, not only is it necessary to disallow cycles in the proof, but it is also necessary to introduce a filter on the deductive closure that produces a *finite* set of axioms, $\mathcal{J}^+$ from which lemmas may be drawn. In essence, $\mathcal{J}^+$ is some finite subset of the deductive closure of $\mathcal{J}$.

Algorithm 1. LemmatiseJustification(J, η)

Function-1: LemmatiseJustification(J, η)

1: $S \leftarrow$ ComputeJPlus$(J, \eta) \setminus \{\eta\}$
2: $justs \leftarrow$ ComputeJustifications(S, η)
3: $c_1 \leftarrow$ ComputeComplexity(J, η)
4: $L \leftarrow J$
5: **for** $J' \in justs$ **do**
6:     $c_2 \leftarrow$ ComputeComplexity(J', η)
7:     **if** $c_2 < c_1$ **then**
8:         $L \leftarrow J'$
9: **return** L

The question is, given a justification $\mathcal{J}$, what axioms should $\mathcal{J}^+$ contain? Although there is no definitive answer to this, it must be remembered that the ultimate goal is to include enough in $\mathcal{J}^+$ so that it is possible to produce a series of candidate lemmatised justifications, from which a "nice" one may be chosen using a complexity model. With this in mind, there are a number of possible options for $\mathcal{J}^+$ generation:

Generation with Sub-Concepts. One possibility is to specify $\mathcal{J}^+$ so that it contains axioms of the forms specified in Table 1, which are build up from sub-concepts of axioms in $\mathcal{J}$. However, while such a strategy can go a long way to producing a set of axioms containing lemmas that could result in pleasing proofs, there could be axioms, which might be lemmas of choice, that are not be contained in the set. For example, given $\mathcal{O} = \{A \sqsubseteq \exists R.B, \exists R.B \sqsubseteq C, \mathsf{Trans}(R)\} \models A \sqsubseteq C$, a lemma of choice might be $\exists R.A \sqsubseteq \exists R.\exists R.B$ (entailed by $A \sqsubseteq \exists R.B$). However, with the above schema, based on sub-concepts, the class expression on the right hand side of the axiom ($\exists R.\exists R.B$) does not exists as a sub-concept in $\mathcal{J}$ and so the axiom would never be generated. What is needed is a set of class expressions that is rich enough so as to be able to build a rich set of axioms that constitute candidate lemmas. This is achieved using nested sub-concepts:

Generation with Nested Sub-Concepts

Definition 5 ($\mathcal{J}^+$). *For a justification $\mathcal{J}$ for η, let $\mathcal{S}$ be the set of sub-concepts occurring in the axioms in $\mathcal{J} \cup \{\eta\}$ plus $\top$ and $\bot$. Let $\mathcal{S}'$ be the smallest set of class expressions such that $\mathcal{S}' \supseteq \mathcal{S}$ and $\mathcal{S}'$ contains class expressions of the form:*

- *$\neg C$ where $C \in \mathcal{S}'$ and C is not negated.*
- *$C_1 \sqcap \cdots \sqcap C_i$ or $C_1 \sqcup \cdots \sqcup C_i$ for $2 \leq i \leq |\mathcal{S}|$ and for any $C_j \in \{C_1, \ldots, C_i\}$ it is the case that $C_j \in \mathcal{S}$ or $C_j = \neg C$ for some $C \in \mathcal{S}$ where C is not negated.*

Now, let $d = |\mathcal{J}| \times c$ where c is the maximum modal depth [1] of the class expressions in $\mathcal{S}$. Let R be a property in the signature of $\mathcal{J}$ and m be the sum of all numbers occurring in cardinality restrictions. Let $\mathcal{S}''$ be the smallest set of class expressions such that $\mathcal{S}'' \supseteq \mathcal{S}'$ and $\mathcal{S}''$ contains class expressions of the form:

- *$\exists R.C, \forall R.C, \geq nR.C$ or $\leq nR.C$, where $C \in \mathcal{S}''$, the modal depth of C is no greater than d, and $n \leq m$.*
- *$\exists R.\{a\}$, where a and R are in the signature of $\mathcal{J}$ or η.*
- *$\neg C$ where $C \in \mathcal{S}''$ and C is not negated.*

Given $\mathcal{S}''$, $\mathcal{J}^+$ is now defined as the set of axioms of the form given in Table 1, where C and D are substituted for class expressions in $\mathcal{S}''$, R and S are substituted for property expressions in $\mathcal{J}$, a and b are substituted for individuals in the signature of $\mathcal{J}$, and for each axiom $\alpha \in \mathcal{J}^+$, there exists a tidy subset $\mathcal{J}' \subseteq \mathcal{J}$ such that $\mathcal{J}' \models \alpha$.

The ComputeJPlus algorithm in now defined to compute $\mathcal{J}^+$ in accordance with Definition 5. Since $\mathcal{S}$ is finite, $\mathcal{S}''$ is also finite and therefore $\mathcal{J}^+$ is also finite. Therefore, there are finite number of justifications for an entailment η with respect to $\mathcal{J}^+$, hence GenerateProof algorithm is guaranteed to terminate.

6   The Feasibility of Computing Justification Oriented Proofs

The GenerateProof algorithm and its sub-routines, and the complexity model shown in Table 2 were implemented in Java using the OWL API. The algorithm has two basic, but necessary, optimisations. First, $\mathcal{J}^+$ is computed incrementally and the number of entailment checks is minimised in the obvious way, for example, if $\mathcal{J} \not\models A \sqsubseteq B$ then an entailment test is not performed for $A \sqsubseteq B \sqcap C$. Second, justifications in the LemmatiseJustification algorithm are computed one by one rather than all at once. This means that if a justification $\mathcal{J}'$ is found as a lemmatisation of $\mathcal{J}$ this justification is selected rather than continuing to look for one of lower complexity. If necessary, $\mathcal{J}'$ could be lemmatised to produce a justification of possibly lower complexity.

The implemented algorithm, with the Pellet reasoner, was tested against the ontologies listed in Table 3. For each ontology, a maximum of 5 justifications for entailments of the form $A \sqsubseteq B$, $A \sqsubseteq \bot$ and $A(a)$ were computed. Proofs were then computed for these justifications. Times for computing the justifications, and times for computing proofs were measured and averaged.

The implementation, although naive, with plenty of room for further optimisation, shows that it can be practical to compute proofs for entailments in real ontologies. Generally speaking, if it is possible to compute a justification for an entailment, it is possible to compute a justification oriented proof for that justification and entailment. In all cases, the time required to compute the proof is *at least* an order of magnitude higher than the time required to compute a justification. The difference is particularly striking for the Tambis ontology, where there were several justifications for which it took a significant time to perform entailment checking while computing $\mathcal{J}^+$ and then compute justifications over $\mathcal{J}^+$.

7   Examples

A selection of videos showing examples of justification oriented proofs may be found online at http://www.cs.man.ac.uk/~horridgm/2010/iswc/proofs/examples/. The examples illustrate the kinds of lemmas that get introduced into proofs and illustrate what is possible using the complexity model presented in Table 2. Figure 5 shows a justification oriented proof for the justification shown in 1. It should be noted that the presentation style used for the examples is merely for illustrative purposes. In the tree presentation used, the children of an axiom represent a justification for that axiom.

Table 3. Mean Times for Computing Justifications and Proofs

Ontology Expressivity/Axioms	Just. Size (Mean/SD/Max)	Just. Time (mean ms)	Proof Time (mean ms)
Generations ($\mathcal{ALCOIF}$/38)	4 / 2.1 / 8	31	2034
Economy ($\mathcal{ALCH}$/1625)	2 / 0.6 / 6	32	144
People+Pets ($\mathcal{ALCHOIN}$/108)	4 / 2.5 / 16	31	801
Tambis ($\mathcal{SHIN}$/595)	8 / 4.1 / 21	1047	244987
Nautilus ($\mathcal{ALCF}$/38)	3 / 2.0 / 6	20	758
Transport ($\mathcal{ALCH}$/1157)	5 / 2.1 / 9	19	469
University ($\mathcal{SOIN}$/52)	5 / 2.1 / 9	21	1738
PeriodicTable ($\mathcal{ALU}$/100)	4 / 9.9 / 36	72	1026
Chemical ($\mathcal{ALCHF}$/114)	8 / 1.2 / 11	38	3690

Entailment : Person $\sqsubseteq \bot$

Person $\sqsubseteq$ **¬Movie**

$\top \sqsubseteq$ Movie

    $\forall$hasViolenceLevel. $\bot \sqsubseteq$ Movie

        $\forall$hasViolenceLevel. $\bot \sqsubseteq$ RRated

            RRated $\equiv$ ($\exists$**hasScript**. **ThrillerScript**) $\sqcup$ ($\forall$hasViolenceLevel. **High**)

      RRated $\sqsubseteq$ Movie

           RRated $\sqsubseteq$ **CatMovie**

          CatMovie $\sqsubseteq$ **Movie**

  $\exists$hasViolenceLevel. $\top \sqsubseteq$ Movie

      Domain(hasViolenceLevel, Movie)

Fig. 5. A schematic of a justification oriented proof for the justification shown in Figure 1

8   Conclusions and Future Work

This paper has presented justification oriented proofs as possible solution to the problem of people understanding justifications. Justification lemmatisation has been introduced as a new non-standard reasoning service, which is a key component of for producing justification oriented proofs. Justification lemmatisation is based on the notion of a justification having a certain complexity for a given task. In the approach taken here, a simple complexity model based on various structural and non-structural phenomena was used as a basis for producing justification oriented proofs for entailments in real ontologies. Although, there is plenty of room for optimisation, some initial experiments on a series of published ontologies indicate that it is practical to compute justification oriented proofs for entailments in real ontologies.

It must be emphasised that the main contribution of this paper has been to formalise the notions of justification lemmatisation, justification oriented proofs and using complexity models to generate pleasing proofs. Preliminary user feedback, garnered from poster presentations at various conferences, has been very positive. However, as future work, a series of detailed user studies will be carried out to ascertain the specific benefit of justification oriented proofs to end users. Smooth presentation and interaction mechanisms will be designed to support this evaluation.

References

1. Baader, F., Calvanese, D., McGuinness, D.L., Nardi, D., Patel-Schneider, P.F.: The Description Logic Handbook (2003)
2. Baader, F., Suntisrivaraporn, B.: Debugging SNOMED CT using axiom pinpointing in the description logic $\mathcal{EL}^+$. In: KR-MED 2008 (2008)
3. Borgida, A., Calvanese, D., Rodriguez, M.: Explanation in DL-Lite. In: DL 2008 (2008)
4. Horridge, M., Parsia, B., Sattler, U.: Laconic and precise justifications in OWL. In: Sheth, A.P., Staab, S., Dean, M., Paolucci, M., Maynard, D., Finin, T., Thirunarayan, K. (eds.) ISWC 2008. LNCS, vol. 5318, pp. 323–338. Springer, Heidelberg (2008)
5. Horridge, M., Parsia, B., Sattler, U.: Lemmas for justifications in OWL. In: DL 2009 (2009)
6. Horrocks, I., Kutz, O., Sattler, U.: The even more irresistible $\mathcal{SROIQ}$. In: KR 2006 (2006)
7. Huang, X.: Reconstructing proofs at the assertion level. In: Bundy, A. (ed.) CADE 1994. LNCS, vol. 814, pp. 738–752. Springer, Heidelberg (1994)
8. Johnson-Laird, P.N., Byrne, R.M.J.: Deduction. Psychology Press, San Diego (1991)
9. Kalyanpur, A.: Debugging and Repair of OWL Ontologies. PhD thesis, The Graduate School of the University of Maryland (2006)
10. Kalyanpur, A., Parsia, B., Horridge, M., Sirin, E.: Finding all justifications of OWL DL entailments. In: Aberer, K., Choi, K.-S., Noy, N., Allemang, D., Lee, K.-I., Nixon, L.J.B., Golbeck, J., Mika, P., Maynard, D., Mizoguchi, R., Schreiber, G., Cudré-Mauroux, P. (eds.) ASWC 2007 and ISWC 2007. LNCS, vol. 4825, pp. 267–280. Springer, Heidelberg (2007)
11. Kwong, F.K.H.: Practical approach to explaining $\mathcal{ALC}$ subsumption. Technical report, The University of Manchester (2005)
12. Motik, B., Patel-Schneider, P.F., Parsia, B.: OWL 2 Web Ontology Language structural specification and functional style syntax. W3C Recommendation, W3C – World Wide Web Consortium (October 2009)
13. Motik, B., Shearer, R., Horrocks, I.: Optimized reasoning in description logics using hypertableaux. In: Pfenning, F. (ed.) CADE 2007. LNCS (LNAI), vol. 4603, pp. 67–83. Springer, Heidelberg (2007)
14. Newstead, S.E., Brandon, P., Handley, S.J., Dennis, I., Evans, J.S.B.: Predicting the Difficult of Complex Logical Reasoning Problems, vol. 12. Psychology Press, San Diego (2006)
15. Plaisted, D.A., Greenbaum, S.: A structure-preserving clause form translation. Journal of Symbolic Computation (1986)
16. Rips, L.J.: The Psychology of Proof. MIT Press, Cambridge (1994)
17. Schlobach, S.: Explaining subsumption by optimal interpolation. In: Alferes, J.J., Leite, J. (eds.) JELIA 2004. LNCS (LNAI), vol. 3229, pp. 413–425. Springer, Heidelberg (2004)
18. Strube, G.: The role of cognitive science in knowledge engineering. In: Proc. of Contemporary Knowledge Engineering and Cognition (1992)

Toponym Resolution in Social Media

Neil Ireson and Fabio Ciravegna

University of Sheffield, UK

Abstract. Increasingly user-generated content is being utilised as a source of information, however each individual piece of content tends to contain low levels of information. In addition, such information tends to be informal and imperfect in nature; containing imprecise, subjective, ambiguous expressions. However the content does not have to be interpreted in isolation as it is linked, either explicitly or implicitly, to a network of interrelated content; it may be grouped or tagged with similar content, comments may be added by other users or it may be related to other content posted at the same time or by the same author or members of the author's social network. This paper generally examines how ambiguous concepts within user-generated content can be assigned a specific/formal meaning by considering the expanding context of the information, i.e. other information contained within directly or indirectly related content, and specifically considers the issue of toponym resolution of locations.

Keywords: Concept Disambiguation, Social networks, Information Extraction.

1 Introduction

The growth in the use of social media for sharing content (text, images or video) to other individuals who can be close personal associates or random strangers, is staggering. If the latest statistics from Facebook[1] are to be believed, around 7% of the world's population are active users and they spend on average 40 minutes per day on this one site. Whilst the value of this User-Generated Content (UGC) is being realised, utilising the information it contains poses a number of challenges. Contributions to social media sites (blogs, forums, Twitter, etc.) are conversational in nature and thus tend to be brief and informal, containing imprecise, subjective and ambiguous information. The provider of the content may make assumptions about the receivers' ability to interpret the meaning, despite the fact that the message (i.e. content and any associated metadata) may imperfectly represent their intended meaning. For example, incidental finding in a recent study on photo retrieval [1] indicated that people are unable to retrieve their own content due to their inconsistent descriptions.

One solution to this issue would be to facilitate the user in providing clear semantics defining any potential ambiguous concept they use. Recently a number

---

[1] http://www.facebook.com/press/info.php?statistics

P.F. Patel-Schneider et al. (Eds.): ISWC 2010, Part I, LNCS 6496, pp. 370–385, 2010.

of services (such as OpenCalais[2]) attempt to guide the user providing content to link concepts in their text to some URI, e.g. to Wikipedia or IMDB[3] articles. If such approaches are to be useful they must make suggestion which match the content provider's intentions. In order to determine the correct resolution of an ambiguous concept it is necessary to consider its context, whilst this context is most readily provided by the other information contained in the content (i.e. other text, image features, tags, etc.) the conversational nature of social media means that this information might well be limited and imperfect. However the interrelated nature of social media means that the disambiguation process may be able to use more distant but still related context, for example content posted around the same time or by the same user or by members of the user's social network.

In this paper we examine the use of this expanding context to resolve ambiguous concepts in social media and specifically consider the issue of toponym resolution, i.e. the allocation of specific geolocations to target location terms. Section 2 considers the related work; the disambiguation of both generic and location concepts in text and social media. Section 3 then outlines the methodology used to determine the expanding information context and discusses measurements for determining the degree of term ambiguity. Section 4 describes the experiment; the data used and generation of the disambiguation classification model. Section 5 presents the results and Section 6 discusses the short-comings of the work and how these might be addressed in the future. Section 7 then summarises the findings of the paper.

2   Related Work

The automatic disambiguation of concepts in social media has concentrated on the issue of ambiguous textual tags, this work can be broadly divided into two areas. The first approach disambiguates a target concept (tag) by creating clusters a frequently co-occurring tags, where each cluster is assumed to provide a separate meaning, defined by the tags it contains [2]. Such an approach has the advantage of being applicable to any tag, however as no specific meaning is assigned to each tag cluster it limits its usefulness and the ability to evaluate the approach. Although further processing has been used to assign a unique URI to the clusters based on the co-occurrence between cluster tags and terms found in an ontology [3], the work suffers from a limited evaluation of the techniques performance.

The second approach attempts to identify the correct meaning of tag given its use for a specific resource (i.e. image, web page). The co-occurring tags are used to provide context, these tags are compared with some tag-concept model to determine the most likely meaning. Angeletou [4] used WordNet [5] to identify ambiguous tags, however other work claims that WordNet tends to produce

---

[2] http://www.opencalais.com/
[3] Internet Movie Database: http://www.imdb.com

overly generic concepts [3]. Other approaches use purpose built tag-concept models based on Wikipedia/DBpedia [6,7,8].

The principal difficulty with these approaches is the issue of evaluating whether the disambiguation processes used actually assigns the correct meaning. All the studies perform post-experiment, human review of the results, and in general do not specify the nature of the evaluation (i.e. number of reviewers, inter-annotator agreement, etc.). The generation of "Gold-Standards" for the disambiguation of generic concepts in (multilingual) text has been undertaken by the SENSEVAL[4] evaluation exercises. However, this data is concerned with the identification of concepts in natural language, whilst in social media text, and particularly with tags, concepts have little or no grammatical context.

Toponym resolution has the advantage over general concept resolution that user-generated, gold standard data is available, and especially in social media data. This is due to the ability to, and interest in, geotagging UGC. In addition, in a limited context it is highly likely that a given location will have only one meaning [9], a hypothesis which is shown to be true for the data used in this experiment in respect to a given user context. A number of researchers have examined the disambiguation of locations in text, for example; in news articles [10], in Wikipedia articles [11] and general Web pages [12,13,14,15]. The disambiguation processes generally combine a number of techniques, including; statistical likelihood (selecting the most probable location), textual context (considering the surrounding text of a location) and co-referent locations within the document and, for web pages, hypertext-linked documents. The use of co-occurring locations to provide disambiguating context is stressed as a key technique which generally provides high precision, but low recall, when compared with other techniques. This is due to the requirement for related locations to occur, which may not be satisfied for a given document. One of the key issues in this technique is deriving a function to determine how co-occurrence of locations affects the disambiguation. Most frequently this involves using some heuristics to propagate toponym likelihood (based on location similarity, i.e. the spatial distance or the relative distance in some location taxonomy, between co-occurring locations and possible toponym resolutions), but more recent work forms a feature vector based on the co-occurrences and uses machine learning to calculate the most predictive function [11].

Crandall, et al. [16] combine image features and temporal context to geolocate Flickr images, and in their conclusion they indicate the potential advantage which could be derived from also considering social context. Davis, et al. [17] explore the combination of user and temporal context to determine the location of an uploaded photograph, but unfortunately do not provide enough description of the experimental results to determine the relative effects of these two contexts.

The recent work by Serdyukov, et al. [18], examines the issue of geolocating Flickr photos using the associated tags. Although their work does use the GeoNames[5] database to boost the importance of location names, the predictive model

---

[4] http://www.senseval.org/
[5] http://www.geonames.org

incorporates all the photo tags. Their aim is not specifically toponym resolution, instead they attempt to calculate the actual latitude/longitude of a photo. The resultant model provides an association between a tag (or set of tags) and a location. This is similar to previous work looking at Flickr data to determine the location (and event) related semantics of a tag [19], i.e. the degree to which a given tag could be associated with a given location, and Wang et al.'s work on finding the relationships between news/blog tags and countries [20].

Weinberger, et al. [21] explore the general issue of tag ambiguity. The work defines ambiguity in terms of the probability of observing a given tag in a given context (i.e. set of tags). They then determine the two tags, if added to the set of tags, that give rise to maximally different probability distributions. For example the tags "UK" AND "MA" significantly effect the probability distributions of the tag "Cambridge". It is interesting to note that the research, which considers tag from 100 million images, indicates that 16% of tag ambiguity is explained by other geographic metadata. This emphasises the importance of determining the correct location associated with tags not only to geolocate UGC but also to disambiguate other tags. Indeed there has been work exploring the use of the known location (and time) of UGC to build a recommendation classifier to suggest other tags to the user [22,23].

Perhaps a surprising feature of the previous work on toponym resolution in social media, is that the social context has generally not been exploited. User models have been utilised in the disambiguation of general tags [6], and recent work, on tag recommendations [24] and determining the quality of reviews [25], have demonstrate that using social contextual information, i.e. information relating to a user and their social network, can help improve prediction especially overcoming the issue of data sparsity. The work in this paper examines how such information can be applied to improve the toponym resolution process.

3   Methodology

This section describes the techniques applied in toponym resolution (and applicable to concept disambiguation in general) and how social context can be used to improve performance.

3.1   Information Context

Similar to previous work the surrounding context is used to disambiguate the target concept. This context is provided by the tags associated with the UGC or the users themselves. There are both advantages and disadvantages in using tags over actual textual content; in text you can exploit grammatical structures and term proximity, whilst tags are, in effect, a "bag-of-words". However tags are intended to provide an overall description of the content so, if efficacious tagging is performed, tags should provide a valuable source of descriptive information. However not all tags will be equally informative and the degree of relevance of a given tag will depend upon the application. For toponym resolution the target concepts (tags) are locations and therefore it is necessary to determine those tags which influence the location description of content. The approach adopted was to limit the tags by only considering location names, whilst there are a number of freely-available geographic resources, Yahoo! GeoPlanet[6] was selected due to it providing a semantically structured lexical database. Its specified aim is to provide "geo-referencing data on the Internet", which it does by providing a common naming convention (each location is allocated a URI in the form of a Where-On-Earth Identifier (WOEID)) and a framework or taxonomy describing the relative geography of these locations. The version[7] used contains over 5 million locations/toponyms, but more importantly the data provides an analogous structure to that found in general concept resources, such as WordNet. Each location is a node in a hierarchy from which it is possible to determine the parental locations (hypernyms) which contain the location, the child locations (hyponyms) that the location contains and also locations that share the same parent (coordinate terms). In addition it is possible to extract neighbouring locations, which are coordinate terms which are adjacent to the target location. A further attractive feature of Yahoo! GeoPlanet is the work on namespace concordance[8] which maps between the WOEID and a variety of other namespaces (e.g. location identifiers from; Geonames[9], OpenStreetMap[10], Wikipedia[11]). This means that it become possible to link content identified by a WOEID to information from multifarious providers including that from the Linked Data community.

The specific application scenario is a classic Information Retrieval problem whereby a user wishes to retrieve all the UGC which relates to a given instance of a concept, e.g. Sheffield, South Yorkshire, UK (WOEID:34503). The user can apply three strategies to retrieve the desired information:

1. Query for the ambiguous term and sift through the results. The effectiveness of this strategy is dependent upon the likelihood of content being tagged with the desired instance. If the user is looking for an obscure location, i.e. one which is relatively infrequently tagged, or the search term is highly ambiguous then many of the results will be irrelevant. In addition, if the location can be tagged with several synonyms (e.g. New York, NY, Big Apple) then relevant results may be missed.

2. Rely on the user to have tagged the content with the actual location URI. Note that with location it is also possible to use the geocoding coordinates of the content, if they are provided, although this does not necessarily uniquely identify the location, for example a point location (latitude/longitude) may refer to the immediate surroundings or it may simply be the central point of some wider area, e.g. city, county, country.

---

[6] http://developer.yahoo.com/geo/geoplanet/
[7] Version 7.5.1 released 2010-06-03
[8] http://developer.yahoo.com/geo/geoplanet/guide/
   api-reference.html#api-concordance
[9] http://www.geonames.org
[10] http://www.openstreetmap.org
[11] http://en.wikipedia.org

3. Form a query which is likely to return content relating to the desired instance. This strategy is reliant on the user's ability to form the complex query and the content being tagged with the disambiguating information. Weinberger, et al.'s [21] work on determining the most disambiguating tag would be relevant to directing the content provider tagging by suggesting the most effective tags.

The approach adopted, in effect, applies the third strategy by automatically constructing the complex query. In practice the process is applied offline allocating a location URI to every occurrence of a target location tag by considering all the co-occurring related location name tags: used to tag content, used by the tagging user and used by the users in their social network. Note that is is also possible to allocate a "non-location" URI, indicating that the target location term does not relate to any of the possible toponyms. Thus, information context is provided by a vector of related term frequencies:

$$IC = freq(T)_1, freq(T)_2, \ldots, freq(T)_l \tag{1}$$

and the meaning of any given term is provided by some function combining all the term's information contexts:

$$M(T) = f(IC_1, IC_2, \ldots, IC_n) \tag{2}$$

3.2 Ambiguity

The traditional types of ambiguity include lexical, syntactic, semantic, and pragmatic ambiguity (for a detailed discussion of ambiguity in natural language see [26]). This current work is only concerned with lexical ambiguity, that is where a term (i.e. a text string) has several different meanings. Lexical ambiguity can be subdivided into homonymy and polysemy. Homonymy occurs when a term can have a number of unrelated meanings, whilst polysemy occurs when a term has several related meanings. However this distinction is subtle and it is often unclear which type of ambiguity to apply, and is not considered in this work.

Mich [27] provides two measures for lexical ambiguity:

lexical ambiguity of a term T:

$$a(T) = \text{the number of meanings of T} \tag{3}$$

frequency-weighted lexical ambiguity of a term T:

$$a^*(T) = \sum_{i=1}^{a(T)} log_2 freq(M_i(T)), \tag{4}$$

where $M_i(T)$ is the ith meaning of T, and freq(m) is the observed frequency of that meaning.

The meanings are provided by some lexical resource (e.g. WordNet) and the weighted function is calculated from the frequency of occurrences of meanings found in some text corpus. However the use of frequency seems erroneous; as a frequently used term with a single meaning is still deemed ambiguous. A preferred measure would be to use Shannon's information entropy, which more accurately measures the degree to which the occurrence of a term determines its meaning.

$$H(T) = -\sum_{i=1}^{a(T)} P(M_i(T))log_2 P(M_i(T)), \qquad (5)$$

where $P(M_i(T))$ is the probability of observing the ith meaning of T.

Whilst a term may appear to be highly ambiguous due to a multitude of possible meanings (i.e. a high value according to Equation 3), in a given usage context only a limited set of those possibilities may be likely to occur. For example whilst there are 54 possible toponym resolutions of the location name Cambridge, any given user is only likely to refer to a single one of those possibilities. Although users may be unlikely to refer to multiple toponyms with the same name, they may use the term for meanings other than location names. For example, the term Barry can be associated with: 14 distinct toponyms, a common first name and can be used to describe a particular striped pattern in heraldry. Thus an individual user may use one of the non-geographic meanings in addition to using a single locational meaning for a given term. In the experiments report below the relative effect of term ambiguity on performance is considered.

4   Experiment

4.1   Data

The experiments were performed using Flickr data, a summary of the data is provided in Table 1. Three location areas were chosen; Cambridge, including Ely, Newmarket and Haverhill (as a classical example of an ambiguous location); Sheffield, including, Chesterfield, Barnsley, Hope Valley and Rotherham (for which an accurate local geographic database is available which can be used to assess the quality of the Geoplanet database); and Cardiff, including Barry, Ferndale, Sully, Penarth, Porth, Bridgend, Aberdare, Mountain Ash, Pentre, Cowbridge (which offers a number of highly non-ambiguous location names, and location names which are ambiguous due to also being common terms, namely Barry, Sully and Mountain Ash). These 20 target location names can be resolved into 268 toponyms. In total 1,143,529 photos were tagged with at least one of the these terms (after removing duplicates), of which 123,124 (10.8%) have an associated geolocation (latitude/longitude), these were uploaded by 12,326 users (approximately 10 photos per user). These geolocated photos are used to provide the gold standard.

The geolocated photos contain 165,389 target location name tags, note that each photo must contain at least one target location tag to be retrieved. The users' 580,296 contacts produce 1,140,668 target location tags and the contacts' 5,700,749 contacts produce 3,998,763 target location tags. Whilst all the collected data was limited to an upload date before the end of 2009, all the contact and tag values are up-to-date at the time of retrieve (March 2010).

Each geolocation was then assigned to its nearest toponym, or, if it is greater than 30km away from any toponym it was assigned a null (i.e. non-location) value, this resulted in 99,215 photos assigned to toponyms and 23,909 "other" non-location meanings. When compared to Overell's [28] work on geolocation of Wikipedia articles, where the data contained 1,395 locations and 7,660 non-locations, the Flickr data contains over 22 times the proportion of location to non-location references. This may well be due to the fact tags are less likely to contain proper names (e.g. Person and Organisation names) when compared to free-text, as they are intended as a generic label. It is worth noting that in the experimental data for a given user all the occurrences of a specific target location term (e.g. Sheffield) resolve to a single toponym (e.g. WOEID:34503). However 1,229 (10%) of the users use the same term for both location and non-location meanings.

Table 1 shows the number of photos and users for each target location, note that the row values are not mutually exclusive, as a single document can contain to multiple location tags, therefore the totals are less than the sum of the rows. The final three columns provide measures for the term ambiguity, the first column, *Num*, gives the total number of meanings (toponyms) provided by the lexical database. The next two columns give the information entropy measures, computed from Equation 5, for the term ambiguity, the *Location* column considers the ambiguity with respect to the toponyms, whilst the *Term* column also includes the occurrence of non-location meanings. In general the inclusion of the non-location meaning increases the term ambiguity, however for the term Barry it is reduced due to the fact 85.1% of the occurrences of Barry refer to non-location meanings.

4.2   Classification

In order to resolve the location names it is necessary to determine their contextual information. As stated above this is provided by the co-occurrence of related location names, which are gleaned from the Yahoo! Geoplanet API. For each toponym the related location names are determined by: their *ancestors* (hypernyms), *children* (hyponyms) and *neighbours* (adjacent coordinate terms). However, whilst all the 268 toponyms have ancestors, only 36 possess children and 203 possess neighbours. In general the larger and more populous locations have a highly number of children, this is in part due to the fact such locations actual contain more child locations and also possible due to them being more accurately represented in the Geoplanet data. As a relatively accurate resource was available for the Sheffield area this was used to provide a basic analysis of the Geoplanet data. For Sheffield Geoplanet provides 43 child (suburbs) locations, whilst the more accurate resource provides 99 possibilities. In addition two of the suburb names provided by Geoplanet have incorrect spellings. Whilst

Table 1. Summary of location term data (number of photos, users and term ambiguity)

Location Name	Photos		Users		Ambiguity		
	All	Geo	All	Geo	Num	Location	Term
Cambridge	159969	29467	11881	2200	54	1.408	1.574
Ely	5953	1515	1608	301	13	1.388	1.852
Newmarket	4940	1020	664	154	16	2.135	2.384
Haverhill	3637	210	286	43	7	1.378	1.670
Cardiff	255012	36546	14337	2080	19	0.141	0.389
Barry	225629	29503	39337	3588	14	1.559	0.839
Ferndale	29722	6795	1953	299	30	1.515	1.801
Sully	26905	4450	6347	718	10	1.275	1.394
Penarth	12980	2652	1011	212	2	0.000	0.068
Porth	10060	2284	1785	384	2	0.392	1.154
Bridgend	5626	1109	654	140	14	0.857	1.109
Aberdare	5528	527	394	64	2	0.105	0.909
Mountain Ash	4222	392	1923	257	2	0.000	0.236
Pentre	1657	287	454	93	6	1.413	1.526
Cowbridge	1195	224	184	43	2	0.060	0.354
Sheffield	290253	39368	13424	2015	26	0.209	0.717
Chesterfield	40799	5907	4137	591	30	1.812	2.056
Barnsley	29589	6824	1460	240	7	0.022	0.554
Hope Valley	15692	1216	981	198	10	1.537	1.584
Rotherham	13970	2068	971	170	2	0.007	0.460
Total	1143529	123124	96109	12326	268		

such missing and erroneous data will adversely effect the absolute performance of the disambiguation process, the aim of the current work is to examine the relative performance of using an expanding context, rather than maximise the performance on the given data.

The co-occurrence between each of the 20 target location names and their related (i.e. ancestor, child and neighbour) locations is calculated. The document context is provided by all the related tags assigned to the photo. The user context is provided by all the related tags added by the user who uploads the photo, these tags are weighted by their frequency. The (uploading) user contacts' context is provided by all the related location tags added by the contacts, with tags weighted by the number of contacts who have used that tag. Similarly each of the contacts' contacts' tag usage provides further context. Although tags can be assigned any user, the vast majority of tags are provided by the uploading user, of the 8,193,877 location tags observed in the data only 23,534 (0.28%) are provided by other users. Thus four experiments are performed:

D : using only the related tags in the immediate document (photo) context
U : as D, including all the (uploading) user related tags as context

C : as U, including all the (uploading) user contacts' related tags as context

CC : as C, including all the (uploading) user contacts' contacts' related tags as context

For each experiment the set of co-occurring related location name frequencies provides a feature vector, from which a classification model is constructed. A Support Vector Machine (SVM) classifier used is (LibSVM [29]), applying a Radial-Basis Function kernel. For each experiment the feature vector values are normalised (between [0,1]) and a ten-fold cross-validation was performed. The photos uploaded by a specific user are placed in a single fold to prevent the classifier learning a user specific rather than generic classification model. For each fold the cost parameter was optimised using a three-fold cross-validation experiment on the training data. Note that along with the possible toponyms associated to a location term the classifier also learns to predict non-location references.

5 Results

Table 2 provides an overview of the experimental results, showing the location names, ordered according to increasing term ambiguity and the f-measure for the four experiments. Note that the reported f-measure for the locations is the micro-average of all the classes (toponyms), calculated by summing the one-versus-all matrices, as a result precision equals recall equals f-measure. The final row provides the macro-average of these micro averages.

The current approaches to concept disambiguation tend to rely on related information found solely within the context of the document, shown by the results in the second column. From these results it can be seen that including information from the creator of the content can significantly improve the disambiguation (paired t-test confidence <0.004), in addition including information from their social network contacts does produce some advantage but not highly significant (paired t-test confidence <0.42), whilst including information from the contacts' social network produces a slight detrimental effect over just using contacts' information. This trend can be observed in the macro-average values in the final row.

Note that the three location names where solely using the document context produced the best results have by far the three highest proportions of non-location meanings (i.e. Mountain Ash (0.961), Barry (0.851) and Sully (0.643)). Therefore the performance of such disambiguation techniques, which rely on the co-occurrence of related location names, are likely to be adversely affected by the presence of a significant proportion of non-location meanings.

Figure 1 and 2 examine the relationship between disambiguation performance and term ambiguity for the four experiments. Figure 1 shows the actual performance which indicates the expected decrease in performance with increasing ambiguity, this effect can be more clearly in Figure 2 which shows the trend lines for the data. The graphs indicate that for the experiments including user

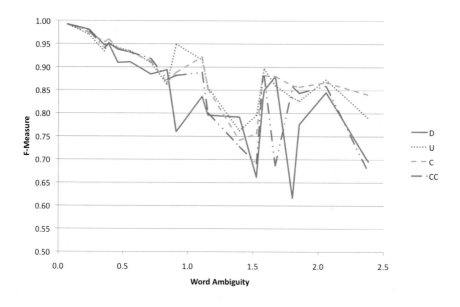

Fig. 1. Performance in relation to Word Ambiguity

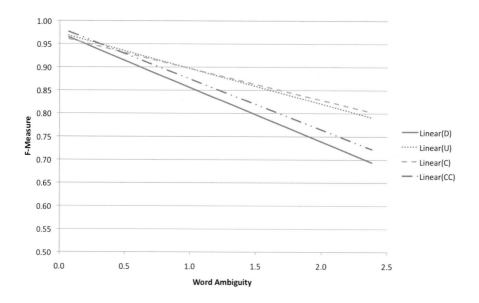

Fig. 2. Linear Trend in Performance in relation to Word Ambiguity

(U) and their social network contacts (C) as context the fall in performance is less sensitive to increase in term ambiguity.

Finally the overall results are depicted in Figure 3, which shows a generalised Precision-Recall curve for all classes (toponyms). The curve is formed using

Table 2. Performance measure for expanding social context

Location Name	Term Ambiguity	Performance (F-Measure)			
		Document	User	Contacts	Contacts' Contacts
Penarth	0.068	**0.992**	**0.992**	**0.992**	**0.992**
MountainAsh	0.236	**0.981**	0.970	0.974	0.979
Cowbridge	0.354	0.947	0.934	**0.954**	0.941
Cardiff	0.389	0.949	0.952	**0.960**	0.951
Rotherham	0.460	0.910	**0.941**	**0.941**	0.938
Barnsley	0.554	0.911	**0.935**	0.931	0.932
Sheffield	0.717	0.885	0.909	0.913	**0.918**
Barry	0.839	**0.894**	0.863	0.867	0.873
Aberdare	0.909	0.760	**0.949**	0.889	0.882
Bridgend	1.109	0.836	0.915	**0.922**	0.889
Porth	1.154	0.796	**0.856**	0.852	0.801
Sully	1.394	**0.792**	0.762	0.742	0.728
Pentre	1.526	0.662	**0.796**	0.756	0.692
Cambridge	1.574	0.828	**0.882**	0.879	0.880
HopeValley	1.584	0.850	**0.896**	0.882	0.880
Haverhill	1.670	0.880	0.860	**0.880**	0.687
Ferndale	1.801	0.617	0.833	**0.860**	0.858
Ely	1.852	0.776	0.827	**0.857**	0.844
Chesterfield	2.056	0.845	**0.873**	0.867	0.859
Newmarket	2.384	0.696	0.789	**0.840**	0.673
Macro-Average		0.832	0.891	**0.892**	0.860

a similar generalisation approach as outlined by Hand and Till [30] for ROC curves, whereby each of the individual Precision-Recall curves are combined with a weight according to the number of instances they represent, and missing points on each curve are calculate by linear interpolation.

The curve indicates that at low-recall values, where only highly probable instances are classified, considering more proximate contextual information (i.e. in the document rather than user tags, or user rather than contact tags) produces higher precision, which would be intuitively expected. However as recall increases utilising more distant context becomes more beneficial.

6 Discussion

The experiment and results described above indicate the importance of considering the user context when disambiguating location terms used to tag their content. In addition a potential benefit in considering the information provided by the user's social network contacts is shown. However the major limitation of the work is that it only uses a single data source, namely Flickr. Within this single domain an attempt was made to apply the techniques employed to a wide variety of concept types, varying the levels of term ambiguity and contextual

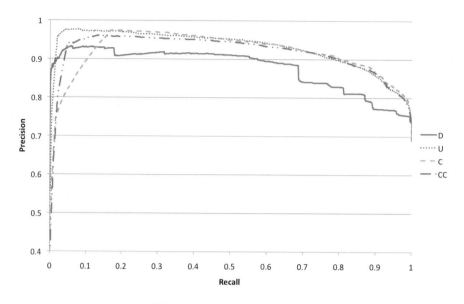

Fig. 3. Precision-Recall Curve

information available, and analyse the resultant performance. However drawing any conclusions outside the domain of toponym resolution in photo sharing location tags must be made with care, in particular for concepts where the user is more likely to use multiple meanings. Recently Twitter[12] have introduced the ability to geotag user content, which could be used to provide a comparison for toponym resolution in a different form of social media. For other, more generic concept types, the issue of creating an experimental gold standard to provide an objective evaluation needs to be addressed.

Although the experimental data described above provides a large number of labelled instances, a number of assumptions have been made. It is presumed that the user has accurately geocoded their content and any erroneous geocoding will provide a low level of noise, not significantly affecting the results. Although the experiment was mainly concerned with determining the relative probability of resolving a location term to competing toponyms, the limit of 30km set to determine that a geocoded photo was related to a given toponym is arbitrary and in practice this limit should be related to the toponym concept type, i.e. country, county, city, town, suburb, etc. In addition it should be noted that in their usage location terms do not have strict boundaries and the allocation of a given point to a toponym depends upon user context [31].

Whilst the main performance variability can be explained by the ambiguity of a given term (shown in the Figures 1 and 2), there is certain variation from the general trend. Other variables which potentially influence performance are the number of documents/users available for each target location, the number/type of related location tags (children, neighbours, ancestors) provided by

---

[12] http://twitter.com/

the GeoPlanet resource and the number of non-location references for each target location name. In addition, whilst Yahoo! Geoplanet was used as a semantically structured lexical database to provide a set of terms related to a concept, the resource is shown to contain missing and erroneous information. Further research should examine how such data and resource variables and imperfections influence the disambiguation process. The recent work on linking the various namespaces used to identify geolocations means it is possible to combine the information contained in these linked resources.

In this experiment related information is provided by the user and their social context, however a fairly naïve approach was adopted whereby all the user's information is assumed to relate equally to all their content. Similarly all the user's social network is assumed to have an equal impact. A more realistic approach would be to consider a temporal dimension, where information relatedness is dependent on temporal proximity, which previous work has shown to be effective [16]. If it is possible to determine the relative strength of social ties, e.g. with some measure of the degree of mutual interaction, this may also prove significant in determining the relative impact of information provided by the user's contacts. While the experiments showed that considering the information from the user's contacts' social network is detrimental to performance, considering the impact of social ties may allow information to be utilised from more distance social context.

7   Conclusion

This paper considers the issue of disambiguating concepts in social media. The approach adopted links the location concepts, found in user-generated content, to a URI defining its intended meaning; enabling the content to be retrieved according to a specific semantic query. Due to the nature of social media, the content containing the ambiguous concept may possess limited contextual information with which to disambiguate, however the interrelationship between content and users in social media means it is possible to exploit more distant, related contextual information. The paper shows the importance of considering user context when disambiguating the location terms they use to describe their content, and indicates that this is more important for terms with a higher degree of ambiguity. The information provided by a user's social network contacts can also provide some advantage although further work is required to determine if a more sensitive consideration of context, i.e. considering a temporal aspect or the strength of social ties, might improve the significance of using such social context.

Acknowledgments

This work has been supported by the European Commission as part of the WeKnowIt project (FP7-215453).

References

1. Whittaker, S., Bergman, O., Clough, P.: Easy on that trigger dad: a study of long term family photo retrieval. Personal Ubiquitous Comput 14(1), 31–43 (2010)
2. Yeung, C.m.A., Gibbins, N., Shadbolt, N.: Tag meaning disambiguation through analysis of tripartite structure of folksonomies. In: Web Intelligence/IAT Workshops, pp. 3–6. IEEE, Los Alamitos (2007)
3. Specia, L., Motta, E.: Integrating folksonomies with the semantic web. In: Franconi, E., Kifer, M., May, W. (eds.) ESWC 2007. LNCS, vol. 4519, pp. 624–639. Springer, Heidelberg (2007)
4. Angeletou, S.: Semantic enrichment of folksonomy tagspaces. In: Sheth, A.P., Staab, S., Dean, M., Paolucci, M., Maynard, D., Finin, T., Thirunarayan, K. (eds.) ISWC 2008. LNCS, vol. 5318, pp. 889–894. Springer, Heidelberg (2008)
5. Fellbaum, C.: WordNet: An Electronic Lexical Database. Bradford Books (1998)
6. Tesconi, M., Ronzano, F., Marchetti, A., Minutoli, S.: Semantify del.icio.us: Automatically turn your tags into senses. In: Social Data on the Web (2008)
7. Garcia, A., Szomszor, M., Alani, H., Corcho, O.: Preliminary results in tag disambiguation using dbpedia. In: Knowledge Capture (K-Cap 2009) - First International Workshop on Collective Knowledge Capturing and Representation - CKCaR 2009 (September 2009)
8. Overell, S., Sigurbjörnsson, B., van Zwol, R.: Classifying tags using open content resources. In: Proceedings of the Second ACM International Conference on Web Search and Data Mining, WSDM 2009, pp. 64–73. ACM, New York (2009)
9. Yarowsky, D.: One sense per collocation. In: Proceedings of the workshop on Human Language Technology, Morristown, NJ, USA, Association for Computational Linguistics, HLT 1993, pp. 266–271 (1993)
10. Garbin, E., Mani, I.: Disambiguating toponyms in news. In: Proceedings of the conference on Human Language Technology and Empirical Methods in Natural Language Processing, HLT 2005, Morristown, NJ, USA, Association for Computational Linguistics, pp. 363–370 (2005)
11. Overell, S., Rüger, S.: Using co-occurrence models for placename disambiguation. International Journal of Geographical Information Science 22, 265–287 (2008)
12. Ding, J., Gravano, L., Shivakumar, N.: Computing geographical scopes of web resources. In: Proceedings of the 26th International Conference on Very Large Data Bases, VLDB 2000, pp. 545–556. Morgan Kaufmann Publishers Inc., San Francisco (2000)
13. Amitay, E., Har'El, N., Sivan, R., Soffer, A.: Web-a-where: geotagging web content. In: Proceedings of the 27th Annual International ACM SIGIR Conference on Research and Development in Information Retrieval, SIGIR 2004, pp. 273–280. ACM, New York (2004)
14. Clough, P., Sanderson, M., Joho, H.: Extraction of semantic annotations from textual web pages. Deliverable D15 6201, EU Project: SPIRIT (2004)
15. Zong, W., Wu, D., Sun, A., Lim, E.P., Goh, D.H.L.: On assigning place names to geography related web pages. In: Proceedings of the 5th ACM/IEEE-CS Joint Conference on Digital libraries, JCDL 2005, pp. 354–362. ACM, New York (2005)
16. Crandall, D.J., Backstrom, L., Huttenlocher, D., Kleinberg, J.: Mapping the world's photos. In: Proceedings of the 18th International Conference on World Wide Web, WWW 2009, pp. 761–770. ACM, New York (2009)

17. Davis, M., King, S., Good, N., Sarvas, R.: From context to content: leveraging context to infer media metadata. In: Proceedings of the 12th Annual ACM International Conference on Multimedia, MULTIMEDIA 2004, pp. 188–195. ACM, New York (2004)

18. Serdyukov, P., Murdock, V., van Zwol, R.: Placing flickr photos on a map. In: Proceedings of the 32nd international ACM SIGIR Conference on Research and Development in Information Retrieval, SIGIR 2009, pp. 484–491. ACM, New York (2009)

19. Rattenbury, T., Naaman, M.: Methods for extracting place semantics from flickr tags. ACM Trans. Web 3(1), 1–30 (2009)

20. Wang, C., Wang, J., Xie, X., Ma, W.Y.: Mining geographic knowledge using location aware topic model. In: Proceedings of the 4th ACM Workshop on Geographical Information Retrieval, GIR 2007, pp. 65–70. ACM, New York (2007)

21. Weinberger, K.Q., Slaney, M., Van Zwol, R.: Resolving tag ambiguity. In: Proceeding of the 16th ACM International Conference on Multimedia, MM 2008, pp. 111–120. ACM, New York (2008)

22. Naaman, M., Paepcke, A., Garcia-Molina, H.: From where to what: Metadata sharing for digital photographs with geographic coordinates. In: Meersman, R., Tari, Z., Schmidt, D.C. (eds.) CoopIS 2003, DOA 2003, and ODBASE 2003. LNCS, vol. 2888, pp. 196–217. Springer, Heidelberg (2003)

23. Sarin, S., Nagahashi, T., Miyosawa, T., Kameyama, W.: On the design and exploitation of user's personal and public information for semantic personal digital photograph annotation. Adv. MultiMedia 2008(2), 1–16 (2008)

24. Rae, A., Sigurbjrnsson, B., van Zwol, R.: Improving tag recommendation using social networks. In: RIAO 2010, Paris, France (2010)

25. Lu, Y., Tsaparas, P., Ntoulas, A., Polanyi, L.: Exploiting social context for review quality prediction. In: 19th International World Wide Web Conference, WWW 2010 (April 2010)

26. Ceccato, M., Kiyavitskaya, N., Zeni, N., Mich, L., Berry, D.M.: Ambiguity identification and measurement in natural language texts. Technical Report Technical Report DIT-04-111, Univeristy of Trento (December 2004)

27. Mich, L.: On the use of ambiguity measures in requirements analysis. In: Proceedings of the 6th International Workshop on Applications of Natural Language to Information Systems, NLDB 2001, pp. 143–152. GI (2001)

28. Overell, S.: Geographic Information Retrieval: Classification, Disambiguation and Modelling. PhD thesis, Imperial College London (2009)

29. chung Chang, C., Lin, C.J.: Libsvm: a library for support vector machines (2001) Software available at, http://www.csie.ntu.edu.tw/~cjlin/libsvm

30. Hand, D.J., Till, R.J.: A simple generalisation of the area under the roc curve for multiple class classification problems. Mach. Learn. 45(2), 171–186 (2001)

31. Jones, C.B., Purves, R.S., Clough, P.D., Joho, H.: Modelling vague places with knowledge from the web. Int. J. Geogr. Inf. Sci. 22(10), 1045–1065 (2008)

An Expressive and Efficient Solution to the Service Selection Problem

Daniel Izquierdo, María-Esther Vidal, and Blai Bonet

Departamento de Computación
Universidad Simón Bolívar
Caracas 89000, Venezuela
{idaniel,mvidal,bonet}@ldc.usb.ve

Abstract. Given the large number of Semantic Web Services that can be created from online sources by using existing annotation tools, expressive formalisms and efficient and scalable approaches to solve the service selection problem are required to make these services widely available to the users. In this paper, we propose a framework that is grounded on logic and the Local-As-View approach for representing instances of the service selection problem. In our approach, Web services are semantically described using LAV mappings in terms of generic concepts from an ontology, user requests correspond to conjunctive queries on the generic concepts and, in addition, the user may specify a set of preferences that are used to rank the possible solutions to the given request. The LAV formulation allows us to cast the service selection problem as a query rewriting problem that must consider the relationships among the concepts in the ontology and the ranks induced by the preferences. Then, building on related work, we devise an encoding of the resulting query rewriting problem as a logical theory whose models are in correspondence with the solutions of the user request, and in presence of preferences, whose best models are in correspondence with the best-ranked solutions. Thus, by exploiting known properties of modern SAT solvers, we provide an efficient and scalable solution to the service selection problem. The approach provides the basis to represent a large number of real-world situations and interesting user requests.

1   Introduction

Existing Web infrastructures support the publication and access to a tremendous amount of Web data sources, some of which can be labeled and converted into Semantic Web Services by using existing annotation tools like the one proposed by Ambite et al. [3]. Once a large dataset of Semantic Web Services become available, users require techniques to effectively select the services that meet their requirements. In order to achieve this goal, the services in the dataset must be tagged with their functional and non-functional properties, and the user preferences and requirements must be formally described as well. In this paper, we extend an approach traditionally used in the area of data integration

P.F. Patel-Schneider et al. (Eds.): ISWC 2010, Part I, LNCS 6496, pp. 386–401, 2010.

to solve the problem of selecting the best services that meet a user request, a problem that we call in this paper the Service Selection Problem (SSP).[1]

As in other approaches, we use domain ontologies for describing the services in the dataset, yet we differ in how the services are described. In this paper, we use the recent approach of Ambite et al. [3] that describes services as views on the generic concepts of the ontology following the Local-As-View (LAV) approach that is widely used in integration systems [21], instead of the traditional Global-As-View (GAV) approach where the generic concepts are expressed in terms of the services. The adoption of the LAV approach instead of GAV is not accidental. LAV descriptions are tailored towards systems with constantly changing datasets and a relatively stable set of generic concepts, while GAV descriptions are tailored towards systems with a constantly changing set of generic concepts but a relatively stable dataset of services.

As it is shown below, LAV descriptions of services correspond to mappings that define services as conjunctive queries involving the generic concepts in the ontology. Thus, every time that a service changes or a new one becomes available, only a tiny fraction of the mappings must be updated, usually just one mapping. Likewise, user requests can be modeled as conjunctive queries over the generic concepts in a way that the SSP can be cast as the problem of rewriting a query in terms of a set of views, the so-called Query Rewriting Problem (QRP) that is well-known in the area of data integration [8,16], query optimization and data maintenance [1,21], and for which several scalable approaches have been proposed [4,12,13,21,23]. Furthermore, user preferences and constraints on the possible solutions for a given request may be specified with a simple yet expressive language for preferences. These preferences and constraints refine and rank the set of valid rewritings of the posed query so that the best solution to the SSP corresponds to the best-ranked rewritings of the QRP.

Our solution extends the recent approach of Arvelo et al. [4] for QRPs that is based on the efficient enumeration of models for a propositional logic theory. In our case, an input instance of the SSP is converted into an instance of a QRP with preferences and constraints that is further translated into a (weighted) logical theory for which its models are in correspondence with the solutions of the SSP, and the rank of the models induced by the propositional weights corresponds to the rank of the solutions induced by the user preferences and constraints. These translations, from SSP to QRP to logic, are performed efficiently, in (low) polynomial time, and the best models are found using off-the-shelf SAT tools. Thus, we are able to exploit the benefits of modern SAT techniques such as conflict-directed backtracking and caching and decomposition of common subproblems to perform the necessary search in the combinatorial space of solutions.

In summary, we make the following crisp contributions to the problem of service selection and composition: (1) advocate the LAV approach as it provides an scalable solution for describing the continuously changing set of available Web Services, (2) propose a simple yet powerful language for expressing preferences

---

[1] We assume that a discovery service previously crawled the Web and located the services, and that an annotation tool stored their descriptions in our catalog.

and constraints on the valid solutions of the SSP, (3) describe how to transform the SSP to the QRP extended with preferences and constraints, and (4) describe how to change an efficient and scalable solution to the QRP, based on propositional logic and SAT tools, to handle preferences and constraints. The rest of this paper is as follows. The next two sections describe the SSP and the language of preferences and constraints, and the proposed solution to the SSP. Then, we present preliminary experiments, related work and finish with a discussion.

2   Service Selection Problem

An SSP consists of a description IS of the integration framework and a user request R. Formally, the integration framework is a tuple $IS = \langle D, S, M \rangle$ where D is the ontology of generic concepts, S is the set of available services, and M is the collection of LAV mappings that semantically describe the services in terms of the ontology. On the other hand, a user request is a tuple $R = \langle Q, P \rangle$ that is made of a query Q expressed as a conjunctive query over the generic concepts and a set of preferences P. In the following, we describe all these elements in detail and illustrate the framework through a number of examples.

2.1   Domain Ontology

The domain ontology D is a tuple $\langle \sigma, A \rangle$ where σ is a *signature* for a logical language and A is a collection of axioms describing the ontology. A signature σ is a set of relational and constant symbols from which logical formulas can be constructed; it corresponds to a tuple $\langle R_1^{r_1}, \dots, R_n^{r_n}, c_1, \dots, c_m \rangle$ where each R_i is a relational symbol of arity $r_i,$[2] and each c_j is a constant symbol. The axioms describe the ontology by defining the relationships between the ontology concepts. For the present work, we only consider subsumption relationships between concepts that are expressed with rules of the form:

$$R(\bar{x}, \bar{y}, \bar{a}) \sqsubseteq P(\bar{x}, \bar{b}), \tag{1}$$

where R and P are predicates in σ (of appropriate arity), $\bar{x}$ and $\bar{y}$ are lists of variables (repetitions allowed), and $\bar{a}$ and $\bar{b}$ are lists of constant symbols (repetitions allowed). All these lists may be empty except $\bar{x} \cup \bar{b}$.

Although limited in appearance, subsumption rules are quite expressive as they allow us to specify diverse relationships between concepts; e.g.,

- **Hierarchy of classes and subclasses** (or types and subtypes): classes are specified with unary predicates. A subclass relationship can be specified with a simple rule; e.g., $penguin(x) \sqsubseteq bird(x)$ tells that penguins are birds.
- **Subrelations via specialization:** A subrelation of R^r can be specified by constraining another relation P^s $(r < s)$. For example, the rule:

$$descendant(Elizabeth\ II, x) \sqsubseteq noble(x) \tag{2}$$

tells that the descendants of Queen Elizabeth II are noble.

---

[2] We use the notation R^r to say that R is a relational symbol of arity r.

- **Indirect subsumption:** it is even possible to specify a subrelation via another seemingly unrelated predicate. For example, the rule:

$$citizen\text{-}of(x, Montreal) \sqsubseteq lives\text{-}in(x, Canada)$$

says that when the second argument of '*citizen-of*' is fixed to the constant '*Montreal*', the tuples in the relation '*citizen-of*' are contained in the set of tuples in the relation '*lives-in*' whose second component is '*Canada*'.

However, we require that the *dependency graph* $G(D)$ of the ontology to be a *forest of trees*. The dependency graph is a labeled directed graph that is constructed as follows: the nodes of the graph are the relational symbols in the signature, and there is an edge (R, P) in the graph iff there is a rule of the form (1). The edge is labeled with the *bindings* induced by the rule; e.g., if $descendant(x, y)$ and $noble(z)$ are two predicates in the signature and there is the rule (2), then there is an edge from $descendant$ to $noble$ labeled with the bindings $\{x = Elizabeth\ II, y = z\}$.

2.2 Services and Mappings

The available services are represented by means of another signature $S = \tau = \langle S_1^{s_1}, \ldots, S_k^{s_k} \rangle$ called the *services signature*, where each symbol S_i represents a concrete service in the Internet that "offers" some information.

The semantic description of services is expressed with the LAV paradigm in terms of mappings that describe the services in terms of concepts in the domain ontology [26]: for each service S_i, there is a mapping that describes S_i as a *conjunctive query* on the concepts in the ontology that also distinguish input and output attributes of the service. For example, a service $S(x, y)$ that returns information about flights originating at a given US city can be described as:

$$S(\$x, y) \ :- \ flight(x, y), uscity(x).$$

where $flight^2$ and $uscity^1$ are relational symbols in the ontology. The symbol '$\$$' denotes that x is an input attribute. The semantic interpretation of a mapping like this one enforces the following:

- the service represented by S provides information in the form of tuples (x, y),
- the service is called with x as input attribute and returns (x, y),
- each tuple (x, y) returned by the service satisfies the rhs of the view; i.e., $flight(x, y)$ and $uscity(x)$, and
- the views are not necessarily *complete*; i.e., there may be other tuples (x, y) that satisfy the rhs of the view but which are not available through S.

The LAV approach is commonly used in integration systems because it permits the scalability of the system as new services become available [26]. Under LAV, the appearance of a new service only causes the addition of a new mapping describing the service in terms of the concepts in the ontology. Under GAV, on the other hand, the ontology concepts are semantically described using views in terms of the sources of information. Thus, the extension or modification of the ontology is an easy task in GAV as it only involves the addition or local modification of few descriptions [26]. Therefore, the LAV approach is best suited for applications with a stable ontology but with changing data sources whereas the GAV approach is best suited for applications with stable data sources and a changing ontology. For the Semantic Web, we assume that the ontology of concepts is the stable component. We believe that this is a reasonable assumption since, once a common language is agreed upon to describe Web resources, the only changing characteristic is the number and nature of resources which constantly pop up or disappear from the Web.

Up to here, we have described all elements in the integration framework $IS = \langle D, S, M \rangle$ where $D = \langle \sigma, A \rangle$ is an ontology of concepts, $S = \tau$ represent the available services in the Web and M is a collection of LAV mappings describing the services in terms of the concepts in D. The integration framework can be thought as the "knowledge base" (KB) in a system designed for answering requests about the selection and composition of Web services. Ideally, the KB should support the efficient processing of user requests.

2.3   User Requests

A user request is a tuple $R = \langle Q, P \rangle$ where Q is a conjunctive query in terms of concepts in the ontology that describes how these concepts must be combined to resolve a given task, and a set P of preferences. For example, the query:

$$Q(x) \; :- \; \mathit{flight}(\mathrm{LA}, x), \mathit{flight}(x, \mathrm{Paris}).$$

finds all cities on which a two-leg flight from Los Angeles to Paris stop. This query can be answered using the view $S(\$x, y)$ as $I(x) :- S(\mathrm{LA}, x), S(x, \mathrm{Paris})$. Observe that this rewriting is correct yet not *necessarily* complete because there may be two-leg flights from Los Angeles to Paris that stop at non-US cities (which are not available through the service $S(\$x, y)$) or because there may be a two-leg flight from Los Angeles to Paris that stops at a US city that is unknown to $S(\$x, y)$.

The preferences are used to rank the collection of valid rewritings. Once this ranking is obtained, the solution for the request R is any best-ranked valid rewriting. In this work, we consider a simple yet expressive language for preferences in which preferences are "soft constraints" on valid rewritings.

A soft constraint is a tuple $\pi = \langle \varphi, c \rangle$ where φ is a propositional formula and c is the cost associated with φ. The idea is that each valid rewriting is associated with a cost equal to the sum of the costs of the preferences violated by the rewriting, and that these costs induce a ranking on valid rewritings. Thus, a best-ranked valid rewriting is one that has minimum cost.

It only remains to say what type of propositional formulas φ are allowed and when a preference is violated by a rewriting. The set of propositions for constructing preferences is $\mathcal{L}(IS) = \{R : R \in \sigma\} \cup \{S : S \in \tau\}$ that corresponds to the relational symbols either in the ontology signature or in the services signature. Elements of $\mathcal{L}(IS)$ are propositional symbols that should not be confused with their relational interpretation in IS; indeed, if the reader is more comfortable, he may think on a different symbol altogether such as P_R, $[R]$, or other. The validity of a preference is defined with respect to the propositional model $\mathcal{M}(I)$ (truth assignment for the symbols in $\mathcal{L}(IS)$) constructed from a valid rewriting $I(\bar{x})$: $\mathcal{M}(I) \vDash S$ for $S \in \tau$ iff the service S appears in $I(\bar{x})$, and $\mathcal{M}(I) \vDash R$ for $R \in \sigma$ iff the concept R appears in the unique path from a concept R' to the root in the dependency graph where R' is a concept in a service $S(\bar{y})$ used in $I(\bar{x})$. That is, the model makes true the service symbols used in $I(\bar{x})$, or the ontology symbols used in services in $I(\bar{x})$, or the ontology symbols that can be reached from the latter in the dependency graph $G(D)$. For example, the rewriting $I(x) :{-}S(\text{LA}, x), S(x, \text{Paris})$ defines the model $\mathcal{M}(I) = \{S = \textbf{true}, flight = \textbf{true}, uscity = \textbf{true}\}$. Finally, a preference φ holds in an answer $I(\bar{x})$ iff $\mathcal{M}(I) \vDash \varphi$.

This simple language permits us to express interesting preferences such as:

- **Hard constraints:** a soft constraint of the type $\pi = \langle \varphi, \infty \rangle$ can be thought as a hard constraint that must be satisfied by every rewriting because if the best rewriting has infinite cost, we know that there is no valid rewritig that satisfies φ.
- **QoS preferences:** this type of preferences can be used to assign absolute quantities of reward/cost to single services as the one used for integrated QoS parameters. For example, if each service S_i is associated with a QoS reward of r_i, then the collection of preferences $\pi_i = \langle \neg S_i, -r_i \rangle$ selects a valid rewriting with services that have the highest combined QoS,
- **Conditional preferences:** a user preference of the type 'if service S is used, then service R should be used as well' can be modeled with the 'hard' constraint $S \Rightarrow R$,
- **Preferences of the type at-least-one:** a user preference of the type that at least one of the services $S_1, \ldots, S_n$ should be used in the rewriting, can be modeled with the 'hard' constraint $S_1 \vee \cdots \vee S_n$, and
- **Preferences of the type at-most-one:** a user preference of the type that at most one of the services $S_1, \ldots, S_n$ should be used in the rewriting, can be modeled with the collection $\{\neg S_i \vee \neg S_j : i \neq j\}$ of 'hard' constraints.

2.4 Examples

Consider a travel-information system that contains information about flight and train trips between cities and information about which cities are in the US. The domain ontology is comprised of the predicates $trip^2$, $flight^3$, $train^3$ and $uscity^1$, and the constants AA, UA, AT, UP, LA, NY, and Paris. The first predicate relates cities (x, y) if there is a direct trip either by plane or train between them. The flight predicate relates (x, y, t) whenever there is a direct flight from x to y operated by airline t, and similarly for $train$, and $uscity$ indicates when a given city is a US city or not. The ontology axioms capture two subsumption relations:

$$flight(x, y, t) \sqsubseteq trip(x, y),$$

$$train(x, y, t) \sqsubseteq trip(x, y).$$

For the services, assume that the available data sources on the Internet contain the following information:

- *national-flight*(x, y) relates two US cities that are connected by a direct flight,
- *AA-flight*(x, y) relates cities that are connected by American flights,
- *UA-flight*(x, y) relates cities that are connected by United flights,
- *one-way-flight*(x, y) relates two cities that are connected by a one-way flight,
- *one-stop*(x, y) relates two cities that are connected by a one-stop flight,
- *to-pa*(x) tells if there is a direct flight from x to Paris,
- *from-la*(x) tells if there is a flight from Los Angeles to x,
- *national-train*(x, y) relates US cities that are connected by a direct train,
- *AT-train*(x, y) relates cities that are connected by Amtrak trains, and
- *UP-train*(x, y) relates cities that are connected by Union Pacific Railway trains.

These services are semantically described using the concepts in the ontology by the following LAV mappings:

$$
\begin{aligned}
\textit{national-flight}(\$x, y) \;&:-\; \textit{flight}(x, y, t),\ \textit{uscity}(x),\ \textit{uscity}(y)\,, \\
\textit{AA-flight}(\$x, y) \;&:-\; \textit{flight}(x, y, \text{AA})\,, \\
\textit{UA-flight}(\$x, y) \;&:-\; \textit{flight}(x, y, \text{UA})\,, \\
\textit{one-way-flight}(x, y) \;&:-\; \textit{flight}(x, y, t)\,, \\
\textit{one-stop}(x, z) \;&:-\; \textit{flight}(x, y, t),\ \textit{flight}(y, z, t)\,, \\
\textit{to-pa}(\$x) \;&:-\; \textit{flight}(x, \text{Paris}, \text{AA})\,, \\
\textit{from-la}(\$x) \;&:-\; \textit{flight}(\text{LA}, x, \text{UA})\,, \\
\textit{national-train}(\$x, y) \;&:-\; \textit{train}(x, y, t),\ \textit{uscity}(x),\ \textit{uscity}(y)\,, \\
\textit{AT-train}(\$x, y) \;&:-\; \textit{train}(x, y, \text{AT})\,, \\
\textit{UP-train}(\$x, y) \;&:-\; \textit{train}(x, y, \text{UP})\,.
\end{aligned}
$$

Observe that each tuple produced by each service satisfies the semantic description given in the body of the rule; e.g., the tuples that satisfy *national-flight*(x, y) meet the conjunctive formula:

$$\exists t(\textit{flight}(x, y, t) \,\wedge\, \textit{uscity}(x) \,\wedge\, \textit{uscity}(y))\,.$$

However, there may be tuples that satisfy this formula that are not produced by *national-flight*(x, y), i.e., this service is not necessarily complete.

Consider now a user who is interested in identifying the services able to retrieve one-stop round trips from a US city x to any city y in the world. Notice that the trip from x to y stops at a city u, that the back trip from y to v stops at a city v, and that u may not be equal to v. This request can be modeled with the conjunctive query:

$$Q(x, y) \;:-\; \textit{uscity}(x),\ \textit{trip}(x, u),\ \textit{trip}(u, y),\ \textit{trip}(y, v),\ \textit{trip}(v, x)\,.$$

Any rewriting of the ontology predicates in terms of the services that respect the input/output constraints on the parameters correspond to a *composition of services* that implements the request. For example, the following rewriting is a valid solution to the request:

$$I(x, y) \;:\!-\; \textit{national-flight}(x, u), \; \textit{to-pa}(u),$$
$$\textit{one-way-flight}(\text{Paris}, v), \; \textit{national-flight}(v, x) \,.$$

But, the following two rewritings are not valid solutions:

$$I'(x, y) \;:\!-\; \textit{national-flight}(x, u), \; \textit{to-pa}(u), \; \textit{from-la}(v), \; \textit{national-flight}(v, x) \,,$$
$$I''(x, y) \;:\!-\; \textit{one-stop}(x, y), \; \textit{one-way-flight}(y, v), \; \textit{national-flight}(v, x) \,.$$

The first is not valid because it maps the query variable y into two different constants Paris and LA that denote different cities, and the second rewriting is not valid because the service $\textit{one-stop}(x, y)$ does not receive as input, or produce as output, the middle city u where the flight from x to y must stop.

As shown, one can use a system for rewriting queries in terms of views for computing solutions to the SSP, since the valid solutions correspond to the valid rewritings of the query. However, in the presence of user preferences, the solutions must be ranked according to the preferences and the best solutions should be returned. To illustrate the use of preferences, consider the following request:

$$Q(x, y) \;:\!-\; \textit{trip}(\text{LA}, x), \; \textit{trip}(x, \text{NY}), \; \textit{trip}(\text{NY}, y), \; \textit{trip}(y, \text{LA}).$$

that looks for round-trips between Los Angeles and New York such that each direction is a one-stop trip. Observe that the query is posed in a way that there are no restrictions whatsoever on the use of planes or trains for any leg of the trip. However, users typically have preferences about using planes or trains. For this example, we study four different scenarios for user preferences and show how to model them in the proposed framework:

P1. The user prefers to fly rather than to travel by train. This can be modeled by assigning a high reward to the symbol *flight*. Likewise, a preference of trains over airplanes is obtained by assigning a high reward to *train*.

P2. The user is indifferent with respect to trains or airplanes, yet she does not want to mix both. This preference is an at-most-one preference over the set $\{\textit{flight}, \textit{train}\}$ that corresponds to the formula $\neg\textit{flight} \vee \neg\textit{train}$ and a cost for the violation of the preference.

P3. If the user travels by airplane, she prefers to always use the same airline. This preference can be modeled with the formula $\neg AA\textit{-flight} \vee \neg UA\textit{-flight}$ together with a cost. Additionally, the other means of air transportation should be 'disabled' since they may return flights operated by any airline; e.g., add the constraint $\neg\textit{national-flight}$ with a high cost.

P4. Finally, if the user travels by airplane, she prefers to use UA. This is a non-trivial preference that can be modeled with the formula:

$$(\textit{flight} \Rightarrow UA\textit{-flight}) \wedge (\neg UA\textit{-flight} \vee \neg AA\textit{-flight}) \,.$$

The first part says that if a leg of the trip is done by plane, then UA must be used, while the second part says that whenever UA is used, AA should not be used. Also, the services that do not guarantee airline operators should be disabled as in the previous case.

All these preferences correspond to formulas over the propositional language $\mathcal{L}(IS)$. The formulas for all but the first case involve preferences that can be treated as hard constraints if they are associated with infinite cost, or soft constraints meaning the user prefers, but is not limited to, solutions that do not violate the preferences.

3   Solution and System Architecture

We extend the McDSat system of Arvelo et al. [4] for QRP. An instance of QRP consists of a collection of views and a query on abstract concepts. The problem consists in rewriting the query in terms of the views such that each tuple produced by the rewriting is a tuple of the solution [26]. McDSat reduces QRP to the problem of finding the models of a propositional logic theory that satisfies the following properties: (1) there is a 1-1 correspondence between the valid rewritings of the query and the models of the logical theory, (2) given a model of the theory, one can recover the corresponding rewriting in linear time, and (3) the theory can be constructed in polynomial time from the QRP instance. Once the logical theory is constructed, one can be interested in finding all minimal rewritings of the query as done in data integration systems with incomplete sources, or just one rewriting as done when sources are complete [26]. For the former, off-the-shelf model enumeration tools such as c2d [9] and Relsat [5] can be used, while off-the-shelf SAT solvers such as Minisat [14] or Rsat [22] can be used in the latter case.

In this section, we have just enough space to explain how the logical theory constructed by McDSat can be extended to capture the features associated with SSPs that are not present in QRPs; namely, handling constant symbols, input and output attributes, the ontology of concepts with subsumption relationships, and user preferences. The result is an extended theory whose models are in correspondence with the valid solutions of the SSP and, in the presence of preferences, whose *best models* are in correspondence with the best-ranked valid solutions of the SSP.

Constant Symbols. McDSat does not provide support for constant symbols, yet incorporating this functionality is straightforward. Basically, one only has to track the unification of variables with constants using new propositional symbols, and to propagate such unifications transitively using implications in order to avoiding the unification of different constant symbols. This modification involves the addition of a small number of propositional symbols and clauses to the CNF generated by McDSat.

Input and Output Attributes. The general principle for properly handling input and output attributes is that every input attribute of a service must unify either with a constant symbol or with an output attribute of another service, while avoiding the cycles in the dependencies among the services that 'produce' (output) and 'consume' (input) attributes.

This principle can be enforced by adding propositional symbols to the theory of the form $In(z, R)$ and $Out(z, R)$, where z is a variable and R is a service, and symbols $Prec(R, S)$ for each pair of services R and S. The intended interpretation for these symbols is that $In(z, R)$ holds iff attribute z is an input attribute of R, that $Out(z, R)$ holds iff attribute z is an output attribute of R, and that $Prec(R, S)$ holds iff the service R produces an attribute that is consumed by S. Accordingly, the theory is extended with clauses that enforce this interpretation plus rules of the form $Prec(R, S) \land Prec(S, T) \Rightarrow Prec(R, T)$ that propagate the precedence relation via transitivity, and $\neg Prec(S, S)$ that prunes rewritings containing cycles.

Ontology. The subsumption relationships make the rewriting process more complex as now one needs to consider unification among predicate symbols of different name and arity. Indeed, consider the following four concepts, where a, b, c and d are constant symbols, and x, y and z are variables:

$$P(b, y, z) \sqsubseteq R(a, y),$$
$$R(a, y) \sqsubseteq T(c, y),$$
$$P(d, x, z) \sqsubseteq M(a, x),$$
$$M(a, x) \sqsubseteq N(d, x),$$

and the user request $Q(x, y)$ with services $S1$, $S2$ and $S3$:

$$Q(x, y) \; :\!- \; T(z, y), N(z, x),$$
$$S_1(y) \; :\!- \; R(a, y),$$
$$S_2(x, z) \; :\!- \; N(z, x),$$
$$S_3(x, z) \; :\!- \; P(d, x, z).$$

Then, the system must be able to infer that the query can be rewritten as $I(x, y) :\!-S_1(y), S_2(x, c)$ since $R(a, y)$ unifies with $T(z, y)$ producing the *binding* $\{z = c\}$, and $S_2(x, z)$ unifies with $N(z, x)$ and becomes $S_2(x, c)$ once the binding is propagated. On the other hand, the system must also infer that $Q(x, y)$ cannot be rewritten as $I(x, y) :\!-S_1(y), S_3(x, z)$ because $R(a, y)$ unifies with $T(z, y)$ with binding $\{z = c\}$, $P(d, x, z)$ unifies with $N(z, x)$ with binding $\{z = d\}$, and these two bindings are non-unifiable since constants denote unique objects.

We incorporate the subsumption relation into MCDSAT by means of the dependency graph $G(D)$. Once the graph is built using the subsumption rules, its transitive closure is computed along with the bindings associated with each edge: edges generated by the transitive closure have labels that correspond to the union of the bindings along the edges that generate this edge (if the set of bindings is inconsistent, then the label is assigned the binding {**false**}). These labels are unique and well defined as $G(D)$ is assumed to be a forest of trees. Once the transitive closure $G(D)^*$ is computed, all edges with inconsistent labels can be dropped. The transitive closure is then used to extend the rules in the logical theory that permit the cover of relational symbols in the query with symbols in the views: a predicate P is allowed to cover a predicate R whenever there is an edge from P to R in $G(D)^*$, and when this covering becomes active, the bindings associated with it become active as well.

Preferences. To incorporate preferences, we use the concepts of literal-ranking function and best-ranked models for propositional logic. A literal ranking function r is a function that assign ranks (weights) to literals. Given a literal-ranking function r, the rank $r(\omega)$ of a model ω is the aggregation of the ranks for each literal made true by the model; i.e., $r(\omega) = \sum_{\omega \models \ell} r(\ell)$ [11]. Thus, the models can be ordered by their rank and the best-ranked models are the models with minimum rank. Some model enumerators like c2d can be used to compute all the best-ranked models of a propositional theory. Likewise, Weighted-Max-SAT solvers such as MiniMaxSAT [15] can be used to find a best ranked model.

For SSPs, we accommodate the preferences by using a suitably defined literal-ranking function r^* and by computing best-ranked models. First, a new propositional variable is created for each relational symbol in the ontology and services signatures along with clauses that turn this proposition true whenever the corresponding symbol become active (true). Second, for each preference $\pi = \langle \varphi, c \rangle$, a new propositional symbol p_π is created along with the formula $p_\pi \Leftrightarrow \varphi$. Thus, p_π is true iff φ is satisfied in the model (rewriting). Finally, the literal-ranking function r^* is defined as $r^*(\neg p_\pi) = c$ for each such preference. Clearly, the rank of a model corresponds to the sum of the costs associated with the preferences violated by the model, and thus a best-ranked model corresponds to a rewriting of minimum regret.

3.1   System Architecture

We define an architecture for solving SSPs that is comprised of a Catalog of service descriptions, an Ontology Reasoner, the Encoder, the best model Finder, and the Decoder. Figure 1 depicts the overall architecture of the system. In this framework, an instance of SSP consists of an integration framework IS and a user request R. The Catalog of the system is populated with the components of the integration system, i.e., the domain ontology including the subsumption rules, the services and the LAV mappings between them.

The input instance is then translated into a CNF theory and a literal-ranking function r^* by the Encoder module. The Encoder makes use of the transitive closure $G(D)^*$ that is calculated by the Ontology Reasoner together with the bindings associated with the edges. Once the theory is obtained, it is fed to the Finder that returns a best model. The model is given to the Decoder that reconstructs the solution to the input instance.

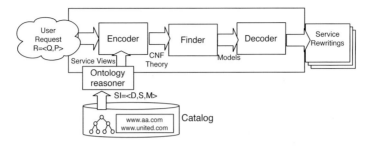

Fig. 1. System Architecture

4   Preliminary Experiments

We have developed a system prototype that implements the above ideas except for the support to distinguish input and output attributes of services (all attributes are assumed to be output), and with partial support for handling preferences; a complete implementation is ongoing work. With this prototype, we conducted experiments on two type of domains: airline domains of the type seen before and random domains. The Finder module is built using c2d (http://reasoning.cs.ucla.edu/c2d) that compiles (transforms) the CNF formula for the propositional theory into deterministic and decomposable negation normal form (d-DNNF) from which all models or just the best models can be efficiently enumerated in linear time [10]. The compilation process from CNF into d-DNNF is intractable in the worst case, yet this is not always the case as the experiments below show.

The objective of the experiments is to test several features of the approach and to see the scalability of the approach. The main benefit of the approach is that one can compile the logical theory for a problem instance and then calculate all the rewritings, or the best ones, any number of times, and the cost/rewards associated with the preferences can also be changed without the need to recompile the theory. Therefore, the time complexity of our approach is basically the time to compile the CNF theory into d-DNNF since calculating the CNF from the SSP and decoding the models is negligible. Thus, we only report the time to compile the CNF into d-DNNF.

4.1   Airline Experiments

The first benchmark consists of problems for air-travel queries. Service views are of the form $V_i(x, y) :- flight(x, y, \mathrm{AL}_i)$ where AL_i is a constant that denotes the name of an airline and the view is assumed to return flights between two cities served by the airline AL_i. For the query, we consider a request to find trips between Paris and New York with a number of stops. The query returns the stops and has the form:

$$Q(x_1, \ldots, x_n) \;\; :- \;\; flight(\mathrm{Paris}, x_1, t), \; flight(x_1, x_2, t), \; \ldots, \; flight(x_n, \mathrm{NY}, t) \,.$$

Observe that the existentially quantified variable, t, is the same for each flight meaning that it can only be unified with the same constant; i.e., each leg of the

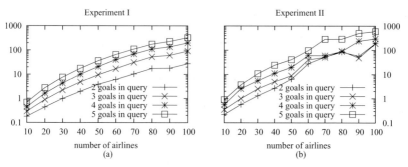

Fig. 2. Compilation times for experiments I and II for different number of goals and different number of views. Experiment II involves $n(n-1)/2$ preferences of a problem with n views. The plots are in logarithmic scale, and the time is in seconds.

flight is served by the same airline. We solved several instances for this type of query with a number of stops from 2 to 5 and a number of services from 10 to 100. Fig. 2(a) shows the results of the compilations: the vertical axis refers to the time in seconds in logarithmic scale and the horizontal axis to the number of views in the benchmark. The results show good performance since realistic instances of the problem (sets of 100 airlines with 5-stop flights) can be compiled in 328 seconds. The size in disk of the d-DNNF for 100 airlines and 5-stop flights is 3.4Mb from which the best model can be computed in 0.29 seconds, and the enumeration of all models can be done in 0.47 seconds.

In the second experiment, we test our system with user preferences. The query is the same except that the existentially quantified variables are all distinct for each flight meaning that any combination of airlines can fulfill the user request. As user preferences, we consider the set $\{\neg V_i \vee \neg V_j : 1 \leq i \neq j \leq n\}$ of $n(n-1)/2$ constraints each with cost $c_{i,j}$, for a problem with n services. Thus, a best model is one that violates the minimum number of preferences and this is equivalent to using the same airline for each leg of the flight. Fig.2(b) shows the result for the compilation also in logarithmic scale. As it can be seen, the compilation times are very similar for the Experiment I where there are no preferences. The largest instance is a complex problem involving 100 views, 5 subgoals in the query and 4,950 user preferences; the total number of rewritings is 100^5, yet it can be compiled in 600 seconds.

In the third experiment, we test the system with ontologies of different sizes. Ontologies corresponding to full binary trees of depth 2 to 7 were generated with the predicate $trip(x, y)$ at the root. Then, for each node in the tree, there is a view that is described by the predicate at that node. The user request is:

$$Q(x_1, \ldots, x_4) \; :- \; uscity(x_1), \; trip(x_1, x_2), \; trip(x_2, x_3), \; trip(x_3, x_4), \; trip(x_4, x_1).$$

In all cases, the compilation time was always less than 13 seconds.

4.2   Random Experiments

For the last experiments, we generated random unstructured instances of SSPs as follows: each user request contains 6 subgoals, 10 distinct variables and 10

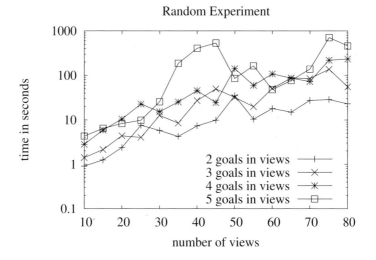

Fig. 3. Compilation times for random experiments with different number of views. The plots are in logarithmic scale and the time is in seconds.

distinct constant symbols, while each service view contains between 2 and 5 subgoals. The constants were randomly placed on the subgoals arguments with a 50% probability. Fig. 3 shows the compilation time for these instances. The size of the compiled theories and number of models does not increase monotonically with the number of views given the random nature of the instances. As it can be seen, these are complex instances and the approach is able to solve them in reasonable time.

5   Related Work

The problem of selecting the services that satisfy a user request is a combinatorial optimization problem and several heuristics have been proposed to find a good solution in a reasonably period of time [2,6,17,18,19,20,24,25,27].

In a series of papers, Berardi and others [6,7] describe services and user requests in terms of deterministic finite-state machines that are encoded using Description Logics theories whose models correspond to solutions of the problem, yet there are no efficient methods to compute these models as in the case of SAT.

Ko et al. [18] propose a constraint-based approach that encodes the non-functional permissible values as a set of constraints whose violation needs to be minimized. Alrifai and Risse [2] develop a two-fold solution that uses a hybrid integer programming algorithm to find the decomposition of global QoS into local constraints, and then, selects the services that best meet the local constraints.

Recently, two planning-based approaches have been proposed. Kuter and Golbeck [19] extend the SHOP2 planning algorithm to select the trustworthy composition of services that implement a given OWL-S process model, while Sohrabi and McIlraith [25] propose a HTN planning-based solution where user preference metrics and domain regulations are used to guide the planner into the space of relevant compositions. Finally, Lécué [20] develops a genetic-based algorithm to identify the composition of services that best meet the quality criteria for a set of QoS parameters.

These existing solutions scale up to a number of abstract concepts. In addition to scalability, our approach provides a more expressive framework where services are semantically described in terms of domain ontology concepts, user preferences restrict the space of solutions, and ontology relationships augment the space of possible solutions. Finally, our approach is sound and complete in the sense that every solution produced by the system is a valid solution, that every valid solution can be produced by the system, and that the best-ranked valid solution is the best solution in terms of the user preferences.

6   Discussion

We proposed a novel formalism for expressing Service Selection Problems involving an ontology of generic concepts, services described using views in terms of the concepts, following the LAV approach, and user preferences. This is a general, well-defined and scalable framework since it is based on logic, the LAV approach, and permits the modeling of real-life scenarios and preferences.

We also showed how the propositional theory used in MCDSAT for solving the QRP can be extended to handle SSPs. This formulation allows us to exploit the properties of modern SAT solvers to provide an efficient and scalable solution.

The preliminary experiments show that the approach can be applied to real-sized problems. We are currently working on a complete implementation of the formalism in order to offer its full expressiveness. In the future, we plan to use other off-the-shelf SAT tools such as MiniMaxSat that is able to find a best model without the need to compile the CNF into d-DNNF.

References

1. Afrati, F.N., Li, C., Ullman, J.D.: Using views to generate efficient evaluation plans for queries. J. Comput. Syst. Sci. 73(5), 703–724 (2007)
2. Alrifai, M., Risse, T.: Combining global optimization with local selection for efficient qos-aware service composition. In: WWW, pp. 881–890 (2009)
3. Ambite, J.L., Darbha, S., Goel, A., Knoblock, C.A., Lerman, K., Parundekar, R., Russ, T.A.: Automatically constructing semantic web services from online sources. In: Bernstein, A., Karger, D.R., Heath, T., Feigenbaum, L., Maynard, D., Motta, E., Thirunarayan, K. (eds.) ISWC 2009. LNCS, vol. 5823, pp. 17–32. Springer, Heidelberg (2009)
4. Arvelo, Y., Bonet, B., Vidal, M.-E.: Compilation of query-rewriting problems into tractable fragments of propositional logic. In: AAAI (2006)
5. Bayardo, R.: Relsat: A Propositional Satisfiability Solver and Model Counter, http://code.google.com/p/relsat/
6. Berardi, D., Cheikh, F., Giacomo, G.D., Patrizi, F.: Automatic Service Composition via Simulation. Int. J. Found. Comput. Sci. 19(2), 429–451 (2008)

7. Berardi, D., Giacomo, G.D., Mecella, M., Calvanese, D.: Composing Web Services with Nondeterministic Behavior. In: ICWS, pp. 909–912 (2006)
8. Chen, H., Wu, Z., Mao, Y.: Rewriting queries using views for rdf-based relational integration. In: ICTAI, pp. 260–264 (2005)
9. Darwiche, A.: The c2d compiler, http://reasoning.cs.ucla.edu/c2d/
10. Darwiche, A.: New advances in compiling cnf into decomposable negation normal form. In: ECAI, pp. 328–332 (2004)
11. Darwiche, A., Marquis, P.: Compiling propositional weighted bases. Artif. Intell. 157(1-2), 81–113 (2004)
12. Duschka, O.M., Genesereth, M.R.: Answering recursive queries using views. In: PODS, pp. 109–116 (1997)
13. Duschka, O.M., Genesereth, M.R.: Query planning in infomaster. In: SAC, pp. 109–111 (1997)
14. Een, N., Sorensson, N.: Minisat, http://minisat.se/
15. Heras, F., Larrosa, J., Oliveras, A.: MiniMaxSAT: An efficient Weighted Max-SAT Solver. Journal of Artificial Intelligence Research 31, 1–32 (2008)
16. Jaudoin, H., Petit, J.-M., Rey, C., Schneider, M., Toumani, F.: Query rewriting using views in presence of value constraints. In: Description Logics (2005)
17. Junghans, M., Agarwal, S., Studer, R.: Towards practical semantic web service discovery. In: Aroyo, L., Antoniou, G., Hyvönen, E., ten Teije, A., Stuckenschmidt, H., Cabral, L., Tudorache, T. (eds.) ESWC 2010, Part II. LNCS, vol. 6089, pp. 15–29. Springer, Heidelberg (2010)
18. Ko, J.M., Kim, C.O., Kwon, I.-H.: Quality-of-Service Oriented Web Service Composition Algorithm and Planning Architecture. Journal of Systems and Software 81(11), 2079–2090 (2008)
19. Kuter, U., Golbeck, J.: Semantic web service composition in social environments. In: Bernstein, A., Karger, D.R., Heath, T., Feigenbaum, L., Maynard, D., Motta, E., Thirunarayan, K. (eds.) ISWC 2009. LNCS, vol. 5823, pp. 344–358. Springer, Heidelberg (2009)
20. Lécué, F.: Optimizing qos-aware semantic web service composition. In: Bernstein, A., Karger, D.R., Heath, T., Feigenbaum, L., Maynard, D., Motta, E., Thirunarayan, K. (eds.) ISWC 2009. LNCS, vol. 5823, pp. 375–391. Springer, Heidelberg (2009)
21. Levy, A.Y., Rajaraman, A., Ordille, J.J.: Querying heterogeneous information sources using source descriptions. In: VLDB, pp. 251–262 (1996)
22. Pipatsrisawat, K., Darwiche, A.: A lightweight component caching scheme for satisfiability solvers. In: Marques-Silva, J., Sakallah, K.A. (eds.) SAT 2007. LNCS, vol. 4501, pp. 294–299. Springer, Heidelberg (2007)
23. Pottinger, R., Halevy, A.Y.: Minicon: A scalable algorithm for answering queries using views. VLDB J 10(2-3), 182–198 (2001)
24. Rahmani, H., GhasemSani, G., Abolhassani, H.: Automatic Web Service Composition Considering User Non-functional Preferences. Next Generation Web Services Practices 0, 33–38 (2008)
25. Sohrabi, S., McIlraith, S.A.: Optimizing web service composition while enforcing regulations. In: Bernstein, A., Karger, D.R., Heath, T., Feigenbaum, L., Maynard, D., Motta, E., Thirunarayan, K. (eds.) ISWC 2009. LNCS, vol. 5823, pp. 601–617. Springer, Heidelberg (2009)
26. Ullman, J.D.: Information integration using logical views. Theor. Comput. Sci. 239(2), 189–210 (2000)
27. Wada, H., Champrasert, P., Suzuki, J., Oba, K.: Multiobjective Optimization of SLA-aware Service Composition. In: IEEE Congress on Services, Workshop on Methodologies for Non-functional Properties in Services Computing (2008)

Ontology Alignment for Linked Open Data

Prateek Jain[1], Pascal Hitzler[1], Amit P. Sheth[1], Kunal Verma[2], and Peter Z. Yeh[2]

[1] Kno.e.sis Center, Wright State University, Dayton, OH
[2] Accenture Technology Labs, San Jose, CA

Abstract. The Web of Data currently coming into existence through the Linked Open Data (LOD) effort is a major milestone in realizing the Semantic Web vision. However, the development of applications based on LOD faces difficulties due to the fact that the different LOD datasets are rather loosely connected pieces of information. In particular, links between LOD datasets are almost exclusively on the level of instances, and schema-level information is being ignored. In this paper, we therefore present a system for finding schema-level links between LOD datasets in the sense of ontology alignment. Our system, called BLOOMS, is based on the idea of bootstrapping information already present on the LOD cloud. We also present a comprehensive evaluation which shows that BLOOMS outperforms state-of-the-art ontology alignment systems on LOD datasets. At the same time, BLOOMS is also competitive compared with these other systems on the Ontology Evaluation Alignment Initiative Benchmark datasets.

1   Introduction

The Linked Open Data (LOD) community effort is a cornerstone in the realization of the Semantic Web vision [1]. So far it has resulted in an openly available "Web of Data" comprising several billion RDF triples. LOD captures knowledge from diverse domains and is constantly growing. Some of the domains include: information from Wikipedia, governmental and geospatial data, entertainment, bio-informatics and publications. However, in terms of practical usability, LOD is still in its infancy. Several central issues remain to be investigated and solved, and discussions of these are ongoing among researchers (see, e.g., [2,3,4]). Our own preliminary investigations into LOD querying [2] in particular exposed a need for schema-level integration of LOD datasets, an issue which has also been pointed out in [1], and elsewhere, as a core challenge.

While LOD datasets are well interlinked on the instance level, they are very loosely connected on the schema level (see also Table 3). Since our work involves schema alignment for our work, we investigated the most competitive state-of-the-art ontology alignment systems available in order to use them for the integration task. However, it turned out that the performance of these systems on LOD schema datasets was rather poor, even though they performed fine on established benchmarks. We were thus left with finding our own solution to LOD schema alignment, on which we report here. Our resulting system, BLOOMS, in fact outperforms state-of-the-art ontology alignment systems in LOD schema alignment, while is roughly on par with these systems on established ontology alignment benchmarks (see Section 4).

P.F. Patel-Schneider et al. (Eds.): ISWC 2010, Part I, LNCS 6496, pp. 402–417, 2010.

Conceptually a key strength of BLOOMS is that we utilize a bootstrapping approach (see Section 3). The system computes alignments with the help of noisy community-generated data available on the Web. Currently, BLOOMS uses Wikipedia and the Wikipedia category hierarchy for this purpose. However there is no conceptual reason why one would not be able to use other inputs (or even existing upper-level ontologies or upper-level domain-specific ontologies) instead. This would simply result in a different bias for the alignment, which could potentially be exploited, e.g., for alignment tasks on narrower thematic domains (see also our discussion of future work, Section 6). Furthermore BLOOMS utilizes the Alignment API [5] as a base system by exploiting its capabilities which complement the native BLOOMS bootstrapping approach.

The structure of the paper is as follows. In Section 2, we clarify some notions and explain the background. In Section 3, we give details about our bootstrapping approach. In Section 4, we give a detailed quantitative evaluation of BLOOMS by comparing it with state-of-the-art ontology alignment systems, for LOD schema alignment and for the Ontology Alignment Evaluation Initiative benchmark. In Section 5, we discuss related work. In Section 6, we conclude with a summary and ideas for future work.

Further details on the evaluation, and the BLOOMS system for download, can be found at http://wiki.knoesis.org/index.php/BLOOMS.

2   Preliminaries

An overview of LOD appears in [1]. Although different LOD datasets are interlinked, it should be noted that interlinks are still relatively scarce. Interlinks are mainly on the instance level (using owl:sameAs), and are clustered within three major thematic domains which are hardly connected by links—see [6]. Schema-level information, by which we mean taxonomies built using rdfs:subClassOf (possibly enriched with further RDF Schema or OWL axioms not involving instance data), is also relatively scarce. In particular there is a lack of interlinks between the different schemas.

DBpedia [7] is an LOD dataset which is based on Wikipedia infoboxes. Our bootstrapping approach employs noisy community-generated data, and we have chosen to use Wikipedia and DBPedia. A central role is played by the Wikipedia category hierarchy, which is a user-generated class hierarchy for Wikipedia pages. It is important to notice that this category hierarchy is not a taxonomy in any reasonable sense. In particular, many of the "sub-category" relations are semantically not rdfs:subClassOf relations [8]. We will discuss our reasons for choosing Wikipedia/DBpedia later on in Section 3.

BLOOMS is a system for schema alignment. For the purpose of this paper, we mean by schema alignment the generation of links between class hierarchies (taxonomies), which are rdfs:subClassOf relations. For an example, if "Human" occurs in some dataset and "Woman" occurs in some other dataset, then we would expect BLOOMS (or any other ontology alignment system) to create a relation between these two classes in the form of an RDF triple "Woman rdfs:subClassOf Human". Note that two classes A and B will always be related by one out of four relationships: A rdfs:subClassOf B, B rdfs:subClassOf A, A owl:equivalentClass[1] B, or none of the previous three.

---

[1] This is semantically equivalent to stating *both* A rdfs:subClassOf B and B rdfs:subClassOf A, and we abstract from the (syntactic) difference.

3   The BLOOMS Approach

At the core of the BLOOMS bootstrapping approach is the utilization of the Wikipedia category hierarchy. In essence, BLOOMS constructs a forest (i.e., a set of trees) T_C (which we call the *BLOOMS forest* for C) for each matching candidate class name C, which roughly corresponds to a selection of supercategories of the class name. Comparison of the forests T_C and T_B for matching candidate classes C and B then yields a decision whether or not (and with which of the candidate relations) C and B should be aligned. We next spell this out in detail.

BLOOMS accepts as input two ontologies which are assumed to contain schema information. It then proceeds with the following steps.

1. **Pre-processing of the input ontologies** in order to (i) remove property restrictions, individuals, and properties, and to (ii) tokenize composite class names to obtain a list of all simple words contained within them, with stop words removed.
2. **Construction of the BLOOMS forest** T_C for each class name C, using information from Wikipedia.
3. **Comparison of constructed BLOOMS forests**, which yields decisions which class names are to be aligned.
4. **Post-processing** of the results with the help of the Alignment API and a reasoner.

We now give more details and examples on the key steps just described. As a running example, we use the class names Event and JazzFestival taken from the LOD datasets DBpedia and Music Ontology, respectively.

Pre-processing of the input ontologies. This involves a straightforward algorithm which normalizes each input class name C into a string C' obtained by replacing underscores and hyphens[2] by spaces, splitting at capital letters, and the like.[3] For stop word removal we used the 319 stop words defined by the Information Retrieval Research Group of Glasgow University.[4]

For our running example, JazzFestival is transformed to "Jazz Festival", whereas Event is not modified at all.

Construction of the BLOOMS forest T_C from C. The first step in constructing T_C is to invoke a call to the Wikipedia Web service using C' as input. This Web service returns a set of Wikipedia pages[5] W_C as results of a search on Wikipedia for the words in the string. If a returned result is a Wikipedia disambiguation page, it is then removed from W_C and replaced by all Wikipedia pages mentioned in the disambiguation page. We call the elements of the resulting set W_C *senses* for C.

Concerning our running example, for Event, the Web service returns Event, Eventing, Sport, NFL Draft, News, Festival, Event-driven programming, Rodeo, Athletics at the Summer Olympics, and Extinction event.

---

[2] We actually did the hyphens manually, because they occurred only in one of our test ontologies, namely the AKT Portal Ontology (see Section 4).

[3] There was no need to make use of a dictionary, mainly because the resulting strings are used as input to Wikipedia search, which works well without stemming etc.

[4] `http://ir.dcs.gla.ac.uk/resources/linguistic_utils/stop_words`

[5] More precisely, their URLs.

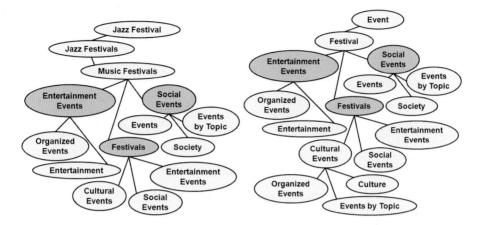

Fig. 1. BLOOMS trees for Jazz Festival with sense Jazz Festival and for Event with sense Event. To save space, some categories are not expanded to level 4.

In the next step, for each sense $s \in W_C$, a tree $T_s \in T_C$, called the *BLOOMS tree for C with sense s*, is constructed, as follows.

- The root of the tree is s.
- Children of s are exactly all the Wikipedia categories into which the Wikipedia page s is categorized.
- Subsequently, for each category c which is a node in the tree, its children are exactly all Wikipedia categories of which c is a subcategory.
- T_s is the resulting tree, which is cut at level 4 (i.e., branches of T_s have maximally 5 nodes, including the root).

We decided to cut the tree at level 4 because we found that deeper levels involve Wikipedia categories which are very general, like "Humanities", and thus would be ineffective for our purposes.

Figure 1 shows the BLOOMS tree for Event with sense Event and for Jazz Festival with sense Jazz Festival.

Comparison of constructed BLOOMS forests. Any concept name C in the one input ontology is now matched against any concept name D in the other input ontology. This is done by comparing each $T_s \in T_C$ with each $T_t \in T_D$. For this, we define a function o, which assigns a real number in the unit interval to each (ordered) pair of BLOOMS trees. The value $o(T_s, T_t)$, called the *overlap of T_t with T_s*, is defined as follows.

1. Remove from T_s all nodes for which there is a parent node which occurs in T_t. All leaves of the resulting tree T'_s are either of level 4 or occur in T_t. Note that due to the way BLOOMS trees are constructed, we removed only nodes from T_s which actually occur in T_t—we remove them because they do not give us any essential additional information for comparing T_s with T_t.
2. $o(T_s, T_t) = \frac{n}{k-1}$, where n is the number of nodes in T'_s which occur also in T_t, and k is the total number of nodes in T'_s (we do not count the root).

In our running example, the BLOOMS trees in Figure 1 are pruned beneath the dark gray nodes. We obtain $o(T_{\text{Event}}, T_{\text{Jazz Festival}}) = \frac{3}{4}$ and $o(T_{\text{Jazz Festival}}, T_{\text{Event}}) = \frac{3}{5}$.

The decision on an alignment is then made as follows.

- If, for any choice of $T_s \in T_C$ and $T_t \in T_D$, we have that $T_s = T_t$, then we set C owl:equivalentClass D.
- If $\min\{o(T_s, T_t), o(T_t, T_s)\} \geq x$ for any choice of $T_s \in T_C$ and $T_t \in T_D$, and for some pre-defined threshold x,[6] then set C rdfs:subClassOf D if $o(T_s, T_t) \leq o(T_t, T_s)$, and set D rdfs:subClassOf C if $o(T_s, T_t) \geq o(T_t, T_s)$.

For our running example, we have $o(T_{\text{Event}}, T_{\text{Jazz Festival}}) > o(T_{\text{Jazz Festival}}, T_{\text{Event}})$, and therefore obtain Jazz Festival rdfs:subClassOf Event.

Post-processing. For post-processing, we first invoke the Alignment API for finding alignments between the original input ontologies. Those alignments returned with a confidence value of at least 0.95 are kept, and added to the results previously obtained.[7] We then invoke a reasoner (in fact, Jena) which finds inferred alignments. E.g., if A is a subclass of B in one of the input ontologies, and an alignment B rdfs:subClassOf C has already been found, then the alignment A rdfs:subClassOf C is also added, and finding these alignments is done using a reasoner. We finally output the alignment results, in the Alignment API format.

The BLOOMS approach as just described makes heavy use of Wikipedia/DBPedia for bootstrapping. It is natural to ask, if Wikipedia could be replaced with something else. In general, we think so. In fact, any upper level ontology or thesaurus could be used, and perhaps there are even more options we did not think of. BLOOMS currently uses Wikipedia because it seemed an intuitive choice due to a number of reasons.

- Wikipedia provides wide thematic coverage.
- The Wikipedia category hierarchy is community-built and thus seemed a natural choice for an alignment system for community-built LOD datasets.
- Wikipedia provides a search feature which we could exploit. This search feature makes it possible to naturally include trees in BLOOMS forests which would be difficult to associate with the input concept name in a more controled setting, e.g., when using an upper level ontology.

We have not systematically investigated any alternatives yet. The evaluation in Section 4 shows that the current approach using Wikipedia is already rather strong. It is left for future work to investigate to what extent alternatives would bring an increase in performance. We hypothesize that alternatives should indeed be very helpful for alignment in more specialized thematic domains, e.g., for life science data in the LOD Cloud. Potential alternatives include the following: Ontologies such as Cyc or SUMO, as used, e.g., in [9]; Thesauri such as WordNet;[8] Taxonomies created from Wikipedia, such as the one reported in [8]; or efforts like the Open Directory Project[9] or YAGO [10].

---

[6] This threshold was typically 0.8 or 0.6 in our experiments in Section 4, where we will discuss how to set suitable thresholds.

[7] 0.95 seems to be the lowest threshold generally giving indisputable results.

[8] WordNet is used by the Alignment API [5], and thus is indirectly utilized by our approach.

[9] http://www.dmoz.org/

4   Evaluation

We have implemented our approach in the BLOOMS[10] system in Java on top of the Alignment API framework [11]. We utilize the Jena Framework[11] for parsing the ontologies, extracting the concepts and for the mentioned reasoning step. The input for BLOOMS is two different ontologies serialized using RDF/XML or OWL.

We performed a comprehensive evaluation of BLOOMS using third party datasets and other state-of-the-art systems in ontology matching. More specifically, we evaluated BLOOMS in two different ways. Firstly, we examined the ability of BLOOMS to serve as a general purpose ontology matching system, by comparing it with other systems on the Ontology Alignment Evaluation Initiative (OAEI) benchmarks.[12] Secondly, we evaluated BLOOMS for the purpose of LOD schema integration and compared it with other systems for ontology matching on LOD schema alignment.

Established in 2004 by leading researchers in the area of ontology matching, the OAEI aims at forging consensus on methods available for schema matching/ontology integration. As a part of this initiative various datasets and reference alignments between these datasets have been made available for evaluating the performance of the participating systems. The systems are evaluated on various parameters such as precision, recall, endurance to lack of structure in the ontologies and absence of properly named concepts.

The initiative consists of various tracks such as a benchmark track, instance matching and oriented matching. The datasets mainly belong to the very narrow domain of bibliographic information with a number of alternative ontologies of the same domain for which alignments are provided. We decided to evaluate BLOOMS on both the benchmark track and the oriented matching track. In the former the task is to identify (only) equivalence relationships. In the latter the task is to identify subclass relationships. The objective of the BLOOMS system is naturally aligned with these two tracks. Furthermore, the OAEI provides us with baselines, and results from the previous version of the oriented matching track are available on the web.[13]

In the 2009 initiative, there were five major systems in the oriented matching track: ASMOV [12], CSR [13], RiMOM [14], AROMA [15] and TaxoMAP [16]. We picked RiMOM and AROMA, for the following reasons: (1) RiMOM was the top system in the oriented track in terms of f-measure and available for download. It was one of the consistent performers in the past two years. (2) AROMA ranked second in the 2008 event. (3) Another important factor was the availability of systems for download in order to run experiments on LOD datasets using them.[14] (4) RiMOM and AROMA

---

[10] BLOOMS is available from http://wiki.knoesis.org/index.php/BLOOMS

[11] http://www.openjena.org/

[12] http://oaei.ontologymatching.org/

[13] http://oaei.ontologymatching.org/2009/results/oriented/

[14] In the OAEI 2009 initiative there were other systems which performed better than RiMOM, namely ASMOV, Lily and CSR. However, ASMOV is a commercial system and the free version runs only on OAEI 2009 datasets and therefore we cannot use it on LOD datasets. CSR is not available for download and our requests for an evaluation copy remained unanswered. TaxoMAP and Lily we could not get working due to platform incompatibility issues, and our support requests were not answered in time.

utilize different techniques and hence this gives good variety in the techniques utilized for the purpose of matching. RiMOM, in fact, automatically determines which ontology alignment methods to use for a particular matching task, and what kinds of information to use in the similarity calculation and how to combine multiple methods as necessary. AROMA is an ontology matcher which utilizes association rule mining.

In order to achieve more breadth in our evaluation, we also included recent systems which have not participated in the OAEI. OMViaUO [9] utilizes upper level ontologies such as SUMO and DOLCE as semantic bridges in the ontology matching process. S-Match [17] is another novel approach in which semantic correspondences are discovered by computing and returning, (as a result) the semantic information implicitly or explicitly codified in the labels of nodes and arcs.

Some of the systems had tunable parameters. As mentioned in Section 3, we used BLOOMS with a threshold value of 0.8 for the ontologies belonging to the same domain, and used a value of 0.6 where one of the ontologies was an abstract ontology such as DBpedia or SUMO. This was done for the following reasons: (1) We expect BLOOMS trees for concepts belonging to the same domain to have higher overlap. (2) Relations between an abstract and a domain specific ontology can be found using a lower overlap. This is because BLOOMS trees constructed for concepts in the domain specific ontology will usually require more nodes to become generic enough in order to match a concept of the more generic ontology.

For RiMOM, while evaluating on LOD datasets, based on our understanding we specified a number of thresholds in the "MatchThreshold" parameter, which range from 0.3 to 0.8. However, the execution with the different parameters always resulted in the same output. On inspection of the results, we found that there were entries with threshold values as low as 0.01 in the output file.

For AROMA, we utilized a threshold of 0.6 for "lexicalThreshold". While parameters below 0.5 were too low and resulted in very poor precision, higher thresholds such as 0.8 resulted in identification of very few results. If guidelines were available for deciding the thresholds, we might have been able to tune the system in a better way.

We could not tune S-Match since the S-Match GUI does not provide for this.

We consulted the OMViaUO literature to get information related to setting suitable thresholds. However, we found no discussion related to this. Further, with respect to the Alignment API and OMViaUO, altering the threshold values (even to 0) did not result in any significant improvement of results on LOD datasets. For the Alignment API and OMViaUO we kept the threshold at 0.5 to achieve an optimum balance between precision and recall.

4.1   Evaluation: Ontology Alignment Evaluation Initiative Oriented Track

In order to test the quality of mappings generated using BLOOMS, we ran our system on the oriented datasets using the reference alignment and compared its performance with the other systems mentioned above. Table 1 presents our results on the oriented matching track of the OAEI. The different tests 1XX, 2XX, and 3XX comprise of matching a single source ontology (101) to other ontologies beginning with the prefix digit of the test. Thus, test 1XX comprises of matching ontology 101 to ontologies 101, 103, and so forth. Similarly 2XX comprises of matching ontology 101 to ontologies 201, 202, and

Table 1. Results on the oriented matching track. Results for RiMOM and AROMA have been taken from the OAEI 2009 website. Legends: Prec=Precision, Rec=Recall, A-API=Alignment API, OMV=OMViaUO, NaN=division by zero, likely due to empty alignment.

						Ontology Alignment Initiative—Oriented Matching Track						
	A-API		OMV		S-Match		AROMA		RiMoM		BLOOMS	
Test	Prec	Rec	Prec	Rec	Prec	Rec	Prec	Rec	Prec	Rec	Prec	Rec
1XX	0	0	0.02	0.06	0.01	0.71	NaN	0	1	1	1	1
2XX	0	0	0.01	0.03	0.05	0.30	0.84	0.08	0.67	0.85	0.52	0.51
3XX	0.01	0.03	0.02	0.047	0.01	0.14	0.72	0.11	0.59	0.81	1	0.84
Avg.	0.00	0.01	0.02	0.04	0.03	0.38	0.63	0.07	0.75	0.88	0.84	0.78

so forth. Unlike the ontologies used in the tests 1XX and 2XX which are created by the organizers, the test 3XX comprises of ontologies which have been created by other organizations and are used in the real world. We computed the precision and recall figures using the baselines and results made available on the OAEI website.

In the oriented matching track, BLOOMS along with RiMOM provided superior results in the test 1XX. For the test 2XX, all systems including BLOOMS show a drop in the performance. We fathom that the reasons for this drop are the following. (1) Some ontologies in test 2XX contain concepts from French. Thus systems which rely on lexico-syntactic tools obviously have difficulties with these ontologies.[15] (2) Some of these ontologies consist of concepts with random names where the matching has to be done on the basis of structure alone.

For the test 3XX, BLOOMS outperforms the other systems in its recall without comprising on its precision. The reasons for the superior performance of BLOOMS could be the following: (1) Wikipedia has a large number of articles with a rich category hierarchy in which the articles and categories summarize the concepts mentioned in the real world ontologies. (2) The ontologies in these tests are of related domains (e.g. Scientific Publishing) and therefore, require a higher overlap between the BLOOMS trees for two concepts to be related. A higher overlap threshold enforces that the concepts and their corresponding BLOOMS trees have to be very similar. This reduces the number of false positives. (3) The mentioned invocation of a reasoner allows us to identify some of the concepts which otherwise have to be found using the structure of the ontology.

The other systems (besides RiMOM) suffer from poor precision and recall due to a variety of reasons. (1) A number of systems such as OMViaUO generate only equivalence mappings. In the oriented matching track, the provided reference alignments consist mainly of subsumption relationships. (2) While S-Match provides good results for the recall, its precision is affected by a plethora of results which are generated for the ontologies. S-Match produces two different output files. We utilized the "default results" file, since it gives a larger number of results. The other file "minimal results" produces a very small set of results, which one could expect to have a higher precision but lower recall, but this is not necessarily the case. For example, for matching

---

[15] In future investigations, one could attempt to exploit the fact that Wikipedia is available in many languages, and that the different-language versions are in fact interlinked.

Table 2. Comparison of various systems on the benchmark track. Results for RiMOM and AROMA have been reused from the OAEI 2009 website. Legends: Prec=Precision, Rec=Recall.

	Ontology Alignment Initiative—Benchmark Track											
	S-Match		OMViaUO		Alignment API		BLOOMS		AROMA		RiMoM	
Test	Prec	Rec	Prec	Rec	Prec	Rec	Prec	Rec	Prec	Recall	Prec	Rec
1XX	0.11	1	0.26	0.37	0.59	0.96	0.71	1	1	1	1	1
2XX	0.1	0.2	0.21	0.31	0.3	0.54	0.38	0.49	0.88	0.65	0.93	0.81
3XX	0.1	0.2	0.28	0.28	0.45	0.77	0.62	0.84	0.80	0.76	0.81	0.82
Avg.	0.1	0.46	0.25	0.33	0.45	0.76	0.57	0.78	0.88	0.81	0.91	0.88

ontologies 101 and 103, S-Match produced 267 results in the default file (precision: 0.46; recall: 0.50), and 57 in the minimal file (precision: 0; recall: 0). (3) OMViaUO could not produce satisfactory results due to poor matching performance. We believe the reason for this is the absence of required ontological concepts in WordNet and in the upper level ontologies utilized by OMViaUO. (4) The Alignment API also suffered from poor precision and recall due to reasons similar to those for OMViaUO. (5) We think AROMA suffers from poor results due to difficulties in identifying association rules related to the ontologies.

4.2 Evaluation: Ontology Alignment Evaluation Initiative Benchmark Track

To test the quality of mappings generated using BLOOMS, we ran it on the benchmark datasets using the reference alignment and compared its performance with the other systems mentioned above. Table 2 presents our results on the benchmark track of the ontology alignment initiative. As in the oriented matching track, the different tests 1XX, 2XX and 3XX comprise of matching a source ontology to other ontologies beginning with the prefix digit of the test. This test utilizes a larger number of ontologies than the oriented matching track. However, to a large extent the ontologies involved are identical.

In the benchmark track, BLOOMS is able to retrieve all results in 1XX, however, it results in loss of precision. In the 1XX track, the other systems gave varying performances. RiMOM and AROMA are impressive with their excellent precision and recall, whereas S-Match and OMViaUO suffer from retrieval of few and incorrect results.

BLOOMS does a better job in 3XX than 2XX due to the involvement of real world ontologies. It ranks right behind RiMOM and AROMA in its recall and does a decent job with respect to precision. The Alignment API does a significantly better job in retrieving the results and matching the ontologies, probably due to the fact, that this track involves finding equivalence relations between ontological concepts. The reasons for poor performance of the other systems are identical to those in the oriented track.

For the 3XX test, BLOOMS outperforms RiMOM and the other systems in finding the correct results. However, the increase in recall goes with a dip in precision. AROMA performs the best in terms of precision.

4.3   Evaluation: LOD Schema Alignment

For a comparative evaluation of BLOOMS on LOD schema alignment, it was necessary to provide a baseline for the alignment task. Since there are no established benchmarks or available baselines for measuring precision and recall for LOD schema alignment, we asked human experts familiar with the domains to create reference alignments.[16] The experts were asked to identify if the concepts belonging to the to-be-matched pairs of schemas are related to each other via a *subclass* or an *equivalence* relationship. In case the experts were not familiar with the terms they utilized descriptions of the concepts (if available in the ontology) or other appropriate references for identifying the relationships. The experts identified all possible subclass and equivalence mappings between the concepts of different ontologies. This process may result in some redundancy if equivalence has already been established at a top level concept C1 and C2. As a result of this mapping, subclass relationships between can be inferred between C1 and subclasses of C2 using this equivalence relationship automatically. This process is obviously subjective to some extent, but in the absence of a community agreed reference alignment, there is no other way to identify the accuracy of any of the systems. However, this phenomenon A similar methodology has been utilized previously in [9].

Table 3. LOD datasets=LOD datasets utilizing this schema, D=taxonomic depth, # C=number of classes, Linked datasets=LOD datasets they are linked to at the instance level

Schema	LOD datasets	D	# C	Linked datasets
DBpedia[17]	DBpedia	4	204	Geonames, US Census, Freebase
Geonames[18]	Geonames, Geospecies	2	11	DBpedia, Jamendo, FOAF Profiles
Music Ontology[19]	Jamendo, Music Brainz, DBTunes	4	136	GovTrack, DBpedia, Geonames
BBC Program[20]	BBC Programs, BBC Music	4	100	BBC Music, BBC Playcount Data
FOAF Profiles[21]	FOAF, Music Brainz	3	16	Crunch Base, QDOS, SIOC Sites
SIOC[22]	DBpedia, LinkedMDB	2	14	Virtuoso Sponger, FOAF Profiles, SemanticWeb.org
AKT Reference Ontology[23]	ACM, DBLP	5	17	Pisa, IEEE, eprints
Semantic Web Conference Ontology[24]	SW Conference Corpus	5	177	SemanticWeb.org, Revyu

[16] The reference alignments and the other material related to this work are available for download at http://wiki.knoesis.org/index.php/BLOOMS
[17] http://wiki.dbpedia.org/Downloads351/#dbpediaontology
[18] http://www.geonames.org/ontology/
[19] http://musicontology.com/
[20] http://purl.org/ontology/po/
[21] http://xmlns.com/foaf/spec/
[22] http://rdfs.org/sioc/ns#
[23] http://www.aktors.org/ontology/support
[24] http://data.semanticweb.org/ns/swc/ontology

Table 4. Results of various systems for LOD Schema Alignment. Legends: Prec=Precision, Rec=Recall, M=Music Ontology, B=BBC Program Ontology, F=FOAF Ontology, D=DBpedia Ontology, G=Geonames Ontology, S=SIOC Ontology, W=Semantic Web Conference Ontology, A=AKT Portal Ontology, err=System Error, NA=Not Available.

Linked Open Data Schema Ontology Alignment												
	Alignment API		OMViaUO		RiMoM		S-Match		AROMA		BLOOMS	
Test	Prec	Rec	Prec	Rec	Prec	Rec	Prec	Rec	Prec	Rec	Prec	Rec
M,B	0.4	0	1	0	err	err	0.04	0.28	0	0	0.63	0.78
M,D	0	0	0	0	err	err	0.08	0.30	0.45	0.01	0.39	0.62
F,D	0	0	0	0	err	err	0.11	0.40	0.33	0.04	0.67	0.73
G,D	0	0	0	0	err	err	0.23	1	0	0	0	0
S,F	0	0	0	0	0.3	0.2	0.52	0.11	0.30	0.20	0.55	0.64
W,A	0.12	0.05	0.16	0.03	err	err	0.06	0.4	0.38	0.03	0.42	0.59
W,D	0	0	0	0	err	err	0.15	0.50	0.27	0.01	0.70	0.40
Avg.	0.07	0.01	0.17	0	NA	NA	0.17	0.43	0.25	0.04	0.48	0.54

Since the LOD cloud consists of more than 100 datasets, we had to make a selection for the purpose of evaluation. Table 3 gives the brief core data about the various LOD datasets which we used. We decided to use the schemas mentioned above due to the following reasons: (1) As specified in the second column of table 3, instance data on LOD are often created using a few well known schemas. For example, Jamendo, Music Brainz and AudioScrobbler primarily utilize concepts from the Music Ontology. The chosen datasets give significant coverage of the LOD cloud. (2) These datasets cover different domains such as Music, Publication and the Web. Thus they allow us to identify connections between ontologies belonging to different thematic domains. (3) Some of the dataset providers such as LinkedMDB have not made their schema publicly available. To eliminate any unfair advantage we might obtain as a result of "self created schemas", we decided not to include the datasets where the schema was not explicitly provided. Please note that the choice of the datasets was made a priori. The selection is not tailored to favor any specific system.

Discussion of Results: LOD Schema Alignment. The precision and recall values of the various systems performed on combinations of the various LOD schemas are listed in Table 4. The reasons for picking these combinations are the following: (1) The combinations correspond to datasets which are of related domains. For example, the Music Ontology and BBC Program both belong to the entertainment domain. Similarly, the Semantic Web Conference Ontology and the AKT Reference Ontology are for the scientific publication domain. (2) DBpedia schema is generic enough to encapsulate various kinds of domains and it can be matched to a large number of schemas. In a sense DBpedia can be understood as having an "umbrella" function.[25]

On LOD schema matching, BLOOMS outperforms the other state of the art systems, as seen in Table 4. In the following paragraph, we examined each of the individual pairs and discuss possible reasons for the performance of the various systems next.

---

[25] This mirrors the central position which DBPedia currently has in the LOD Cloud.

(1) **Music Ontology and BBC Program:** These two datasets are very closely related to each other due to the reuse of concepts and similarity in domain. Unfortunately, RiMOM failed to work on these two ontologies, possibly because of their size.[26] AROMA did not find any relevant relations. OMViaUO finds only a few correct answers. OMViaUO was only able to match identical concepts being used across the two ontologies. The Alignment API also found same concepts, but there were also other concepts which were wrongly matched, which lead to a lower precision than that of OMViaUO. S-Match retrieves some of the results, but again due to the vast number of results computed by S-Match, its precision is low. BLOOMS performed significantly better, we think for the following reasons: (1) Concepts used in these ontologies are not commonly used English language terms (e.g., "DAT"). They belong to the domain of signal recording and cannot be found in common thesauri. However, since Wikipedia contains many domains, they are properly categorized as well. (2) It is hard to match domains just on the basis of linguistic and structural matching.

(2) **Music Ontology and DBpedia:** For the matching of the Music Ontology with DBpedia, other than S-Match and BLOOMS all systems fail to deliver any noteworthy results. While S-Match identifies plenty of relevant relations, again due to the large number of erroneous results its precision is low. The Alignment API and OMViaUO fail to find any matching concepts besides alignments to owl:Thing. For the purpose of this evaluation, this particular correspondence (of concepts to owl:Thing) was excluded due to its highly obvious nature and lack of usefulness. Again we could not get RiMOM to work. AROMA identifies a few correspondences but few of them are correct.

(3) **FOAF Ontology and DBpedia:** The results in this case are similar to that of Music Ontology and DBpedia. S-Match registers a slight increase in both precision and recall. Similarly for AROMA there is an increase in the recall, but its precision drops. BLOOMS registers an increase in both precision and recall due to (a) the fact that concepts in FOAF are very specific and limited, such as "PersonalProfileDocument", and (b) the knowledge about these concepts is very well categorized in Wikipedia.

(4) **Geonames and DBpedia:** This is an interesting category, since apart from S-Match, all systems fail to deliver any results. This is partly due to the modeling of the Geonames ontology which is (a) very limited in nature, and (b) consists of concepts which are hard to understand and identify using only their names—examples are "Code" and "Feature". It is hard to relate them to concepts in DBpedia due to their ambiguous meaning and absence of corresponding concepts in DBpedia.

(5) **FOAF and SIOC:** The FOAF and SIOC Ontologies consist of data related to on-line communities and social networks. RiMOM, which outperforms all the other systems on the ontology alignment initiative benchmarks, suffers from poor precision and recall on these ontologies. The performance of AROMA is identical to that of RiMOM. While they identify about a fifth of the relevant relations, they do so at the cost of low precision. S-Match, on the other hand, does a significantly good job at picking a few correct relations. The other two systems, OMViaUO and the Alignment API, fail to find any relevant relations.

---

[26] We contacted the authors for assistance in resolving the issue, but at the time of writing of this paper, they had not replied to our request with a solution for the problem.

(6) **The Semantic Web Conference Ontology and the AKT Reference Ontology:** These two datasets belong to the domain of scientific publications and hence consist of terms which are somewhat related to some of the terms in the OAEI. Correspondingly, BLOOMS performs well by retrieving more than half of the relations while S-Match retrieves about 40% of the relations. Again due to the vast number of correspondences retrieved by S-Match, its precision is low. AROMA retrieves largely correct but few relations. Due to its excellent performance in the OAEI track, we expected RiMOM to do well in this category. However, because of similar problems as above, RiMOM did not execute on this pair of ontologies. The Alignment API delivers better results than AROMA, whereas OMViaUO retrieves few results with a large number of wrong ones.

(7) **The Semantic Web Conference Ontology and DBpedia:** Using this pair we created alignment between the domain of scientific publication and general information. The overlapping concepts in the two ontologies consist of terms describing professions of people, events, and places. BLOOMS retrieves close to half of the correspondences with about 70% precision. S-Match retrieves half of the relations with a low precision. The other systems either do not retrieve any results or their results are insignificant.

Summary of the Results on LOD Schema Alignment. The results illustrate that on an average BLOOMS performs significantly better (40% better recall with at least twice the precision of the other systems) than the other state-of-the-art systems when it comes to ontology matching on the LOD cloud. Even individually, BLOOMS gives one of the highest recalls in 5 out of 7 pairs utilized for the purpose of evaluation. BLOOMS is a close second in one of the pairs. Of the other systems S-Match is impressive with its consistency in retrieving correct relations, however it comes at the cost of low precision. With regards to precision, BLOOMS leads in 6 out of the 7 pairs. Hence, by providing high recall, with high precision, BLOOMS makes it easy to curate the results for high quality mappings, thus making the results useful for practical purposes. [27]

Although S-Match gives decent recall, its low precision makes it difficult to work with the output. This is due to the vast number of mappings retrieved by S-Match, containing only relatively few correctly found ones. For example, S-Match found 3120 relations between concepts of BBC Program and the Music Ontology, of which only 4% were correct.

Of the other systems, AROMA gives decent precision but suffers from poor recall on LOD datasets. OMViaUO and the Alignment API suffer from both poor precision and recall on LOD datasets.

To summarize, these results indicate that state-of-the-art systems fail to provide the support required for a practically useful alignment of ontologies on the LOD cloud. On the other hand, BLOOMS provides significantly better precision and recall. The reasons for this significantly better performance lie in the fact that the BLOOMS approach is much better suited to handling the diverse domains of the LOD cloud datasets.

---

[27] A reviewer pointed out the advantage which BLOOMS might be obtaining due to the use of reasoner. Possibly other systems could also be improved by adding suitable post-processing by a reasoner. However, our experiments demonstrated that even without the reasoner BLOOMS is superior than other system on LOD schema alignment. For example, for Music Ontology and BBC Program schema alignment without using a reasoner results in precision and recall figures of 0.63 and 0.60 which are still significantly better than those obtained by other systems.

Thus, it significantly utilizes its advantage of using Wikipedia, a community-created data source, in dealing with community-created LOD datasets.

5   Related Work

To the best of our knowledge, this is the first work which exploits a generic and noisy categorization system such as Wikipedia in the context of ontology matching. In [18,19] the authors present a survey in the area of ontology matching.[28] Previously, Wikipedia categorization has been utilized for creating and restructuring taxonomies [8,20]. A taxonomy that covers popular approaches in database schema matching was presented in [21]. A generic algorithm for the same was presented in [22]. In [23] ontology schema matching was used to improve instance co-reference resolution. This helps in cleaning up the data and improving the quality of links at the instance level, but the issue of identifying appropriate relationships at the schema level has not been addressed. The voiD Framework [24] along with the SILK Framework [25] automate the process of link discovery between LOD datasets at the instance level. At the schema level, a notable effort for creating a unified reference point for LOD schemas is UMBEL [26], which is a coherent framework for ontology development which can serve as a reference framework.

6   Conclusion and Future Work

We have presented our approach—BLOOMS—for bootstrapping ontology alignment using the LOD cloud. Our results demonstrate that BLOOMS does not only significantly outperform state-of-the-art ontology alignment systems in LOD schema alignment; it also outperforms most other systems on the Ontology Alignment Initiative benchmark, and is roughly on par with the other best performing other system, Ri-MoM. We believe that BLOOMS draws its strength from (1) bootstrapping noisy data, and (2) the richness of Wikipedia which is used for the bootstrapping.

The fact that BLOOMS does so well on the benchmark is a rather pleasing result, in particular since it was developed solely with LOD schema alignment in mind. Indeed the initial motivation for BLOOMS came from a bottleneck in the LOD querying approach we are currently following—as outlined in [2]—which requires significant tool support for LOD schema matching in order to be scalable, in order to keep up with the growth of the LOD cloud. With BLOOMS we have made a major step towards solving this bottleneck; progress on our LOD querying approach will be reported elsewhere.

We can only hypothesize that BLOOMS will be able to keep up in the future, with the expanding LOD cloud. Our optimism is based on two core observations:

– The high precision and recall values which BLOOMS achieves on the LOD cloud, have in fact been achieved without significant fine-tuning. This gives ample optimism that there is room for improvement and future development, e.g., through adjusting the system to large thematic subdomains of the LOD cloud (e.g., life sciences), or by substituting the current use of DBPedia in our system by some other dataset which may only come into existence in the future.

---

[28] The ontology matching portal at http://www.ontologymatching.org/ gives a good review of the state-of-the-art research in this area.

– Due to the central use of DBPedia, i.e., of information exported from Wikipedia, it can be expected that the performance of BLOOMS will increase with the expansion of DBPedia/Wikipedia, since we will then have more data to bootstrap. So in a sense, assuming that DBPedia/Wikipedia will keep growing as the LOD cloud keeps growing, one might be tempted to say that BLOOMS has a bit of what some people call the "Google property"—namely that a system is getting better the more data it gets.

As future work we intend to identify other kinds of relationships such as partonomical relationships or disjointness on the LOD cloud. We therefore focused on the performance evaluation. We also intend to publicly release an upper level ontology for LOD based on existing upper level ontologies such as SUMO and DOLCE, created with significant but curated input of BLOOMS. We would also like to evaluate BLOOMS using other platforms such as OWL-API and other reasoners besides Jena. This will be the next few steps in our quest to LOD querying as outlined in [2]. Further, we would like to evaluate BLOOMS w.r.t scalability on ontologies larger than used for evaluation presented in Table 3.

Acknowledgement. This work is funded primarily by NSF Award:IIS-0842129, titled "III-SGER: Spatio-Temporal-Thematic Queries of Semantic Web Data: a Study of Expressivity and Efficiency". Pascal Hitzler acknowledges support by the Wright State University Research Council. We would like to sincerely thank Jérôme Euzenat for his insightful comments about the work which were extremely helpful in refining our manuscript.

References

1. Bizer, C., Heath, T., Berners-Lee, T.: Linked data – the story so far. International Journal On Semantic Web and Information Systems 5(3), 1–22 (2009)
2. Jain, P., Hitzler, P., Yeh, P.Z., Verma, K., Sheth, A.P.: Linked Data is Merely More Data. In: Brickley, D., Chaudhri, V.K., Halpin, H., McGuinness, D. (eds.) Linked Data Meets Artificial Intelligence, pp. 82–86. AAAI Press, Menlo Park (2010)
3. Polleres, A., Hogan, A., Harth, A., Decker, S.: Can we ever catch up with the Web? Semantic Web—Interoperability, Usability, Applicability (to appear), http://www.semantic-web-journal.net/
4. Hitzler, P., van Harmelen, F.: A reasonable Semantic Web. Semantic Web—Interoperability, Usability, Applicability (to appear), http://www.semantic-web-journal.net/
5. Euzenat, J.: An API for ontology alignment. In: McIlraith, S.A., Plexousakis, D., van Harmelen, F. (eds.) ISWC 2004. LNCS, vol. 3298, pp. 698–712. Springer, Heidelberg (2004)
6. Guéret, C., Wang, S., Schlobach, S.: The Web of Data is a complex system—first insight into its multi-scale network properties. In: Proceedings of the ECCS 2010 European Conference on Complex Systems, Lisbon, Portugal (September 2010)
7. Bizer, C., et al.: DBpedia—A crystallization point for the Web of Data. Journal of Web Semantics 7(3), 154–165 (2009)
8. Ponzetto, S.P., Strube, M.: Deriving a large scale taxonomy from Wikipedia. In: Proceedings of the 22nd National Conference on Artificial Intelligence, pp. 1440–1445. AAAI Press, Menlo Park (2007)

9. Mascardi, V., Locoro, A., Rosso, P.: Automatic Ontology Matching via Upper Ontologies: A Systematic Evaluation. IEEE Trans. on Knowledge and Data Engr. 22(5), 609–623 (2010)

10. Suchanek, F.M., Kasneci, G., Weikum, G.: Yago: A Core of Semantic Knowledge. In: Williamson, C.L., et al. (eds.) Proceedings of the 16th International Conference on World Wide Web, WWW 2007, Banff, Alberta, Canada, May 8-12. ACM Press, New York (2007)

11. David, J., Euzenat, J., Scharffe, F., dos Santos, C.T.: The Alignment API 4.0 Semantic Web—Interoperability, Usability, Applicability (to appear),
    http://www.semantic-web-journal.net/

12. Jean-Mary, Y.R., Shironoshita, E.P., Kabuka, M.R.: Ontology matching with semantic verification. Journal of Web Semantics 7(3), 235–251 (2009)

13. Spiliopoulos, V., Valarakos, A.G., Vouros, G.A.: CSR: Discovering Subsumption Relations for the Alignment of Ontologies. In: Bechhofer, S., et al. (eds.) ESWC 2008. LNCS, vol. 5021, pp. 418–431. Springer, Heidelberg (2008)

14. Li, J., Tang, J., Li, Y., Luo, Q.: RiMOM: A dynamic multistrategy ontology alignment framework. IEEE Transactions on Knowledge and Data Engineering 21, 1218–1232 (2009)

15. David, J., Guillet, F., Briand, H.: Matching directories and OWL ontologies with AROMA. In: Proceedings of the 15th ACM International Conference on Information and Knowledge Management, CIKM 2006, pp. 830–831. ACM, New York (2006)

16. Hamdi, F., Safar, B., Niraula, N.B., Reynaud, C.: Taxomap in the OAEI 2009 Alignment Contest. In: Shvaiko, P., et al. (eds.) Proceedings of the 4th International Workshop on Ontology Matching (OM 2009) at the 8th International Semantic Web Conference (ISWC 2009) Chantilly, USA, October 25 (2009)

17. Giunchiglia, F., Shvaiko, P., Yatskevich, M.: S-Match: an algorithm and an implementation of semantic matching. In: Kalfoglou, Y., et al. (eds.) Semantic Interoperability and Integration. Number 04391 in Dagstuhl Seminar Proceedings, Dagstuhl, Germany (2005)

18. Euzenat, J., Shvaiko, P.: Ontology matching (DE). Springer, Heidelberg (2007)

19. Choi, N., Song, I.Y., Han, H.: A survey on ontology mapping. SIGMOD Rec 35(3), 34–41 (2006)

20. Ponzetto, S.P., Navigli, R.: Large-scale taxonomy mapping for restructuring and integrating wikipedia. In: Boutilier, C. (ed.) Proceedings of the 21st International Joint Conference on Artificial Intelligence, Pasadena, California, USA, July 11-17, pp. 2083–2088 (2009)

21. Rahm, E., Bernstein, P.A.: A survey of approaches to automatic schema matching. The VLDB Journal 10(4), 334–350 (2001)

22. Madhavan, J., Bernstein, P.A., Rahm, E.: Generic schema matching with Cupid. In: Proceedings of the 27th International Conference on Very Large Data Bases, VLDB 2001, pp. 49–58. Morgan Kaufmann Publishers Inc., San Francisco (2001)

23. Nikolov, A., Uren, V.S., Motta, E., Roeck, A.N.D.: Overcoming schema heterogeneity between linked semantic repositories to improve coreference resolution. In: Gómez-Pérez, A., Yu, Y., Ding, Y. (eds.) ASWC 2009. LNCS, vol. 5926, pp. 332–346. Springer, Heidelberg (2009)

24. Alexander, K., Cyganiak, R., Hausenblas, M., Zhao, J.: Describing Linked Datasets – On the Design and Usage of voiD, the 'Vocabulary of Interlinked Datasets'. In: WWW 2009 Workshop on Linked Data on the Web (LDOW 2009), Madrid, Spain (2009)

25. Volz, J., Bizer, C., Gaedke, M., Kobilarov, G.: Discovering and maintaining links on the web of data. In: Bernstein, A., Karger, D.R., Heath, T., Feigenbaum, L., Maynard, D., Motta, E., Thirunarayan, K. (eds.) ISWC 2009. LNCS, vol. 5823, pp. 650–665. Springer, Heidelberg (2009)

26. Bergman, M.K., Giasson, F.: UMBEL ontology, volume 1, technical documentation,
    http://umbel.org/doc/UMBELOntology_vA1.pdf

SPARQL Query Optimization on Top of DHTs*

Zoi Kaoudi, Kostis Kyzirakos, and Manolis Koubarakis

Dept. of Informatics and Telecommunications
National and Kapodistrian University of Athens, Greece

Abstract. We study the problem of SPARQL query optimization on top
of distributed hash tables. Existing works on SPARQL query processing
in such environments have never been implemented in a real system, or
do not utilize any optimization techniques and thus exhibit poor performance. Our goal in this paper is to propose efficient and scalable algorithms for optimizing SPARQL basic graph pattern queries. We augment
a known distributed query processing algorithm with query optimization
strategies that improve performance in terms of query response time and
bandwidth usage. We implement our techniques in the system Atlas and
study their performance experimentally in a local cluster.

1 Introduction

With interest in the Semantic Web rising rapidly, the problem of SPARQL query
processing and optimization has received a lot of attention. This paper concen-
trates on the optimization of SPARQL queries over RDF data stored on top
of distributed hash tables (DHTs). The first such implemented P2P system is
RDFPeers [1] where only a restricted query class is supported (conjunctive multi-
predicate queries). In [11], we have extended the work of RDFPeers and presented
two algorithms for the distributed evaluation of conjunctions of triple patterns.
The algorithms in [11] have been evaluated only by simulations and no query optimization techniques or an implemented system has been presented. Motivated
by [1, 11], our group has been developing Atlas (http://atlas.di.uoa.gr),
a full-blown open source P2P system for the distributed processing of RDF(S)
data stored on top of DHTs. The RDFS reasoning functionality, the architecture
and various applications of Atlas are presented in [7–9].

In this paper, we present for the first time the *query optimization techniques*
we have developed in Atlas, and evaluate them experimentally. Although query
optimization has been extensively studied and is widely used in the database
area, SPARQL query optimization has been addressed only recently even in
centralized environments [12, 13, 23]. The first works that dealt with distributed
query optimization of SPARQL queries are [17, 24]. However, the architecture
proposed in these papers is very different from the one offered by a DHT. In [10]
a DHT-based system is presented which supports a SPARQL-like query language
and utilizes optimization techniques complementary to the ones we propose.

---

* This work was partially supported by the European project SemsorGrid4Env.

P.F. Patel-Schneider et al. (Eds.): ISWC 2010, Part I, LNCS 6496, pp. 418–435, 2010.

In this work, we address SPARQL query optimization over RDF data stored on top of DHTs and target the minimization of the time required to answer a query and the network bandwidth consumed during query evaluation. Our work starts from the QC algorithm of [11] which we enhance with a query graph representation to avoid Cartesian products and with a distributed mapping dictionary (Section 3). Although mapping dictionaries are by now standard in centralized RDF stores [2, 12, 27], our paper is the first that discusses how to implement one in a DHT environment. In addition, we fully implement and evaluate a DHT-based optimizer which is used to find the best ordering of a query's triple patterns. We describe three greedy optimization algorithms for this purpose: two static and one dynamic. These algorithms utilize selectivity estimates to determine the order with which triple patterns should be evaluated in order to improve query response time and network bandwidth consumption (Section 4). We also propose methods for estimating selectivities of SPARQL basic graph pattern queries utilizing techniques from relational databases (Section 5). We discuss which statistics should be kept at each peer and use histograms for estimating data distributions. We demonstrate that it is sufficient for a peer to create and maintain local statistics, i.e., statistics about the data values for which it is responsible. These statistics can be obtained by other peers by sending low cost messages (Section 6). We implement all our techniques in the system Atlas and present an extensive experimental evaluation in a local cluster using the widely used LUBM benchmark [4] (Section 7).

2   System and Data Model

System Model. We assume a structured overlay network where peers are organized according to a DHT protocol. DHTs are structured P2P systems which try to solve the *lookup problem*; given a data item x, find the peer which holds x. Each peer and each data item are assigned a unique m-bit identifier by using a hash function. The identifier of a peer can be computed by hashing its IP address. For data items, we first have to determine a *key* k and then hash this key to obtain an identifier id_k. A LOOKUP(id_k) operation returns a pointer to the peer responsible for the identifier id_k. Atlas uses the Bamboo DHT [18] but our algorithms can be implemented on top of any DHT network. When a peer receives a LOOKUP request, it efficiently routes the request to a peer with an identifier that is numerically closest to id_k. This peer is responsible for storing the data item with key k and we will call it *responsible peer for key k*.

Data Model and Query Language. We assume that the reader is familiar with the notions of RDF *triple* and *triple pattern*. We deal with RDF triples with no blank nodes and SPARQL queries of basic graph patterns (BGP). A SPARQL query with filter expressions involving equality operators can be easily rewritten to a BGP query. In the following, we define an internal representation of a query extending the graph-based approach used in [12, 23].

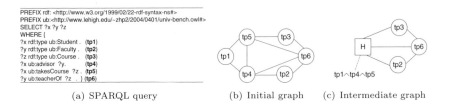

(a) SPARQL query   (b) Initial graph   (c) Intermediate graph

Fig. 1. Query example

Definition 1. *A query graph g is a tuple (N, H, E), where N is the set of nodes in g, H is the set of hypernodes in g and E is the set of undirected edges in g. Each node in N denotes a single triple pattern and each node in H denotes a conjunction of triple patterns. Two nodes from $N \cup H$ are connected with an edge in E if and only if the triple pattern or the conjunction of triple patterns represented by these two nodes share at least one variable.*

Initially, the query graph of a query consists only of simple nodes. During query processing, evaluated triple patterns are merged into hypernodes. In the rest of the paper, we focus only on connected graphs. The evaluation of unconnected graphs is straightforward since each connected subgraph can be evaluated independently, and then the union of the results can be created at the peer that posed the query. A *query plan* q_g for graph g is a total order of the nodes in N of g. In Fig. 1(a), we present an example SPARQL query which will be used throughout the paper (LUBM Q9 [4]). The initial query graph is shown in Fig. 1(b). Figure 1(c) shows an intermediate query graph where the hypernode H represents the conjunction of triple patterns $tp1 \wedge tp4 \wedge tp5$ (i.e., triple patterns $tp1$, $tp4$ and $tp5$ have already been evaluated).

3   Query Evaluation

We start by first explaining the triple indexing scheme we have adopted from [1], where each triple is indexed in the DHT three times. The hash values of the subject, predicate and object of each triple are used to compute the identifiers that will indicate the peers responsible for storing the triple. The peer that receives a store request for a set of RDF triples uses a MULTISEND message as in [11] to distribute the triples among the peers. Each peer keeps its triples in a local database consisting of a single relation with four columns (*triple relation*). The first three columns correspond to the three components of the triples stored, while the fourth column indicates which of the three components is the key that led the triple to this peer.

Query processing. The algorithm we use (QC*) is based on algorithm QC of [11]. Unlike [11] that uses lists of triple patterns, we employ the query graph representation, which ensures that no Cartesian product will be computed and transferred through the network.

When a peer receives a query request, it translates it into a query graph g. Based on the query plan generated by the optimizer, the peer also marks the node of the query graph which represents the triple pattern that should be evaluated first and sends a `QEval` message to the peer that will start the query evaluation. Figure 2 shows the pseudocode when such a message arrives at a peer p. Keyword **event** is used for handling messages also indicating the peer where the handler is executed. Keyword **sendto** prefixed by an identifier declares that the message should be sent to the peer which is responsible for this identifier. In this case, a LOOKUP operation is performed first to discover the peer responsible for this identifier and then the message is sent directly to this peer.

```
Algorithm 1: QC*
1 event p.QEval(id, g, interRes, vars, retIP)
2     IR:=MATCH (g.marked_node().triplepattern());
3     if interRes = {} then interRes':=IR;
4     else interRes':=IR join interRes;
5     if interRes' = {} then
6          sendto retIP.queryResp({ });
7          return;
8     end
9     g':=g.MERGE(g.hypernode, g.marked_node);
10    if g'.N={} then        //all triple patterns are evaluated
11         answer := project interRes' on vars;
12         sendto retIP.queryResp(answer);
13         return;
14    end
15    project out unnecessary vars from interRes';
16    g'.MARKNEXTNODE();
17    key':=FINDKEY(g'.marked_node().triplepattern());
18    id':=HASH(key');
19    sendto id'.QEval(id',g',interRes',vars, retIP);
20 end event
```

Fig. 2. QC* algorithm

First, peer p evaluates the triple pattern which correspond to the marked node of query graph g and forms a temporary relation lR by posing a selection query to its triple relation. If relation $interRes$, which holds the intermediate results so far, is empty, peer p is the first peer of the query evaluation and assigns lR to $interRes'$. Otherwise, p assigns to $interRes'$ the natural join of lR and $interRes$. If the result of the join is an empty relation, p returns an empty set to the peer that posed the query (peer with IP address $retIP$) and query evaluation terminates. Otherwise, query evaluation continues and peer p merges the marked node with the hypernode in g creating a new query graph g'. In case peer p is the first peer participating in the query evaluation, p just transforms the marked node n into a hypernode. If the new graph g' consists only of a hypernode, all triple patterns have been evaluated and p computes the projection of $interRes'$ on the answer variables $vars$ and sends the answer to the peer with IP address $retIP$. Otherwise, query evaluation continues and p projects out from $interRes'$ variables that neither appear in $vars$ nor in the rest of the triple patterns. Then, a new node in g' is marked as the next triple pattern that should be evaluated and p sends a `QEval` message to the next peer[1]. Local procedure MARKNEXTNODE ensures that the chosen node is connected with the hypernode of g', so that no Cartesian product will be computed.

Mapping dictionary. QC* utilizes a distributed mapping dictionary which replaces long strings (URIs and literals) by unique integer values. Triple storage and query evaluation is, then, performed more efficiently using these integers.

---

[1] We assume that each triple pattern has at least one bound component. The case where all three components of a triple pattern are variables requires a slightly different implementation which we do not discuss here.

The uniqueness of the integer values used in the mapping dictionary could be ensured in various ways. We propose the following scheme which is fully distributed (thus scalable and fault tolerant) and does not require any kind of coordination between the peers. Each peer keeps a local integer counter consisting of l bits which is initially set to 0. l is incremented by 1 everytime a new integer value needs to be generated. Each peer that joins the network is assigned a unique m-bit identifier by hashing its IP address. We create an n-bit identifier for a triple component by concatenating the m bits of the peer's unique identifier with the l bits of the current local counter. Depending on the network setting and the application requirements, we can determine an appropriate value for l so that each n-bit identifier is of reasonable space.

When a peer receives a store request, it transforms the given triples into new triples containing integers. The peer, then, sends the new set of triples to be stored in the network using MULTISEND. Together with the new triples, it also sends the mapping from strings to integers that created these triples. Note that we use the *string values* of the triple's components as keys to create the identifiers. Each peer that receives a MULTISEND message, stores in its triple relation the triples it is responsible for. Each such peer also maintains a two-column relation which serves as a local dictionary which holds the mappings for *all* the components of its local triples (*dictionary relation*).

During query evaluation, each string appearing in the triple patterns of a query is transformed into the corresponding unique integer. This transformation is performed during the lookup operation as follows. Whenever peer y wants to send a QEval message for triple pattern tp, it first sends a LOOKUP request to determine the peer responsible for this triple pattern (peer p). Peer p, which receives the LOOKUP request, retrieves the integers corresponding to the strings of tp from its local dictionary relation and sends them to y together with its IP address. Then, peer y replaces the strings of tp with the integers sent from peer p and continues query processing. In case any of the strings has no assigned integer, the answer to the query is empty and query processing terminates. Finally, the peer which computes the answer to the query is responsible for replacing integers in the triples of the answer set with their string values. To achieve this, it contacts the least possible number of peers that have already participated in the query evaluation and have the appropriate values in their dictionary relation. During query evaluation, the IP address of these peers is appended within the QEval message (using an extra parameter).

4   Query Optimization Algorithms

The goal of query optimization is to find a query plan that optimizes the performance of a system with respect to a metric of interest. In our work, we are interested in improving the time required to answer a query (*query response time*) and the *network bandwidth* consumed. The query response time of our algorithm can be improved if the time spent for query evaluation locally by each peer and the time required for network messages to reach relevant peers is improved. One way to accomplish this is by minimizing the size of intermediate relations produced during query evaluation (*interRes'* in QC*). In this case, we benefit in two ways: first, we achieve lower bandwidth consumption and second, we accomplish the computation of joins with smaller intermediate relations locally at peers. Lessons learned from earlier versions of Atlas persuaded us to concentrate on optimizing these metrics to improve the scalability of our system.

In the following, we present three greedy optimization algorithms which try to minimize the size of intermediate relations produced by the query processing algorithm utilizing selectivity-based heuristics. We describe both static and dynamic optimization algorithms. The two static query optimization algorithms are completely executed by the peer that receives the initial query request and output a fully specified query plan (an ordered list of triple patterns). In the dynamic query optimization algorithm, optimization decisions take place at each step of the query processing algorithm. Using standard terminology from relational systems, the *selectivity of a triple pattern tp*, $sel(tp)$, is the fraction of the total number of triples in the network that match tp. Similarly, if H is a conjunction of triple patterns, the *selectivity of the conjunction of triple patterns*, $sel(H)$, is the fraction of total number of triples in the network that match H. We later discuss how we can estimate these selectivities (Section 5).

Naive static algorithm. The naive algorithm (NA) orders triple patterns based on their selectivity (from the most selective to the least selective) and in a fashion where a Cartesian product computation will not be required. The optimization algorithm works as follows. Using the initial query graph representation, each node is assigned with the selectivity of the corresponding triple pattern. The algorithm firstly selects the query graph node n_0 with the minimum selectivity and adds it to the query plan. Then, it marks the nodes that are connected with n_0 and removes n_0 from the graph. The algorithm iteratively chooses the node n_{min} with the minimum selectivity, selecting only from the marked ones, adds it to the query plan, marks the nodes connected with n_{min} and removes n_{min} from the graph. The algorithm terminates when no nodes are left in the graph. NA is based on the assumption that after joining two very selective triple patterns, the joining result will also be selective. Certainly, this assumption is not always true, but the algorithm often performs in a satisfactory way, as we will see in the experimental section.

Semi-naive static algorithm. The semi-naive algorithm (SNA) is a variation of the minimum selectivity algorithm [22] and has also been used in [23]. SNA goes beyond NA by taking into account the selectivity of *pairs* of triple patterns. Besides assigning each node of the graph with the selectivity of its triple pattern, each *edge* of the graph is also assigned with the selectivity of the conjunction of the connected triple patterns. The algorithm begins by selecting the edge with the minimum selectivity, orders its nodes based on their selectivity and adds them to the query plan. Then, SNA iteratively chooses the edge that has the minimum selectivity, but also has one of its nodes in the query plan, and adds the other node to it. SNA terminates when all nodes have been added to the

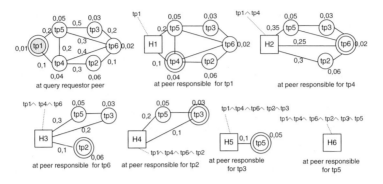

Fig. 3. Dynamic query optimization example

query plan. In case of a tie between the selectivities of two edges, the algorithm chooses the one that has the node with the smaller selectivity.

Dynamic algorithm. Finally, we propose a dynamic optimization algorithm (DA) which seeks to construct query plans that minimize the number of intermediate results during query evaluation. Initially, the peer that received the query request, assigns all edges and nodes of the query graph with the corresponding selectivities and chooses the first triple pattern to be evaluated as in SNA. Then, the optimization step is carried out at each peer p which receives a QEval message. After the new query graph g' with the new hypernode H' has been created at peer p, p selects the triple pattern that should be evaluated next. The candidate triple patterns are the triple patterns of the query graph nodes which are directly connected with the hypernode H'. In this way, the computation of a Cartesian product is avoided. Peer p estimates the selectivity of the join between the intermediate results so far (which correspond to H') and each candidate triple pattern and assigns the corresponding edges. Then, peer p selects the node which is connected to H' with the minimum edge selectivity. In case a tie between the selectivities of two edges emerges, p chooses the node with the smaller selectivity. Figure 3 shows an example execution of DA. The query requestor peer assigns the edges and nodes of the query graph and chooses $tp1$ as the first node. At each query processing step, each peer finds the edge with the minimum selectivity from the set of edges connected to the hypernode and marks the corresponding node (shown with a double circle). In the last step, the query graph consists only of the hypernode.

5   Selectivity Estimation

In this section, we propose methods for estimating the selectivity of single triple patterns as well as the selectivity of a conjunction of triple patterns. To achieve this, we need to compute statistics from the data stored in the network. Section 6 describes how these statistics are generated.

5.1   Single Triple Patterns

We present two ways to estimate the selectivity of a single triple pattern; one based on a simple heuristic also presented in [23] and one based on an analytical estimation technique using the attribute value independence assumption [20].

Bound-is-easier heuristic. We consider a simple variation of the standard bound-is-easier heuristic of relational and datalog query processing [25], also used in [23], and assume that the more bound components a triple pattern has, the more selective it is. We further enrich this heuristic by considering the position of the bound components of a triple pattern, if two triple patterns have the same number of bound components. In this case, we assume that subjects are more selective than objects, which in turn are more selective than predicates.

Analytical estimation. Given a triple pattern $tp = (s, p, o)$, where s, p, o are variables or constants, the selectivity of tp using the attribute value independence assumption [20] is computed by the formula $sel(tp) = sel(s) \times sel(p) \times sel(o)$, where $sel(s), sel(p), sel(o)$ are the selectivities of the triple pattern's components. We assume a selectivity of 1.0 for the triple pattern components which are variables as well as for the predicate value `rdf:type`. The selectivity of the other components depends on the frequency with which their value appears in the set of triples stored in the network. We define the frequency of a triple component c with value v (denoted by $freq_c(v)$ where $c \in \{S, P, O\}$) as the total number of occurrences of value v as a triple component c in the set of triples stored in the network. For example, $freq_S(\text{ub:zoi})$ is the number of occurrences of value `ub:zoi` as a subject, while $freq_O(\text{ub:zoi})$ denotes the number of occurrences of value `ub:zoi` as an object in the set of triples stored in the network.

The selectivity of a triple pattern component $c \in \{S, P, O\}$ with value v can now be computed by the formula $sel_c(v) = \frac{freq_c(v)}{T}$, where $freq_c(v)$ is the frequency of value v as a component c and T is the total number of triples. Although in [23], the attribute value independence assumption is also used, their method assumes a uniform distribution for subjects and requires a bound predicate for the objects. In Section 6, we describe how $freq_c(v)$ is computed. For the computation of the total number of triples indexed in the network (T), we use a broadcast protocol. More elegant solutions for distributed counting in P2P are proposed in [14], but adopting such a method is out of the scope of the paper.

5.2   Conjunction of Triple Patterns

The selectivity of the conjunction of *two* triple patterns tp_1 and tp_2 is $\frac{joinCard(tp_1, tp_2)}{T^2}$, where $joinCard$ is the number of tuples (cardinality) of the relation that results from joining tp_1 and tp_2 and T is the number of triples stored in the network.

To compute the expression $joinCard$, we adopt a method proposed for relational systems in [25]. Assume that we have two triple patterns tp_1 and tp_2 and the corresponding relations R_1 and R_2 which contain all the tuples formed with values existing in the triples stored in the network that satisfy tp_1 and tp_2. Relations R_1 and R_2 have as attributes the variables of triple patterns tp_1 and tp_2, respectively. Since we deal with triple patterns that have at least one constant component, two triple patterns can share at most two variables. The cardinality of joining R_1 with R_2 is computed by the formula:

$$joinCard(R_1, R_2) = \frac{T_{R_1} \times T_{R_2}}{max(I_{R_1(?x_1)}, I_{R_2(?x_1)}) \times max(I_{R_1(?x_2)}, I_{R_2(?x_2)})}$$

where T_{R_1} and T_{R_2} are the number of tuples of R_1 and R_2 respectively, $?x_1$ and $?x_2$ are the variables shared by tp_1 and tp_2, and $I_{R_i(?x_j)}$ is the size of the domain of attribute $?x_j$ of relation R_i. In other words, T_{R_i} is the number of triples that match tp_i, and $I_{R_i(?x_j)}$ is the number of distinct values that variable $?x_j$ has in the bindings of tp_i. This formula can easily be adapted to the case that tp_1 and tp_2 share less than two variables. In [23], the authors propose to precompute the join cardinality by executing the actual SPARQL queries which can become a very expensive operation, especially in a distributed environment.

We can easily determine the number of triples T_R that match a triple pattern tp. If tp has one bound component c with value v, then the number of triples that match tp is equal to the number of occurrences of value v as a component c, i.e., $T_R = freq_c(v)$. If a triple pattern tp has two bound components, then we compute the number of triples that match tp using the selectivity of the triple pattern as explained earlier, i.e., $T_R = sel(tp) \times T$, where T is the total number of triples stored in the network. For the computation of the size of the domain of a variable $?x$ in a triple pattern tp, namely $I_{R(?x)}$, we distinguish two cases. If tp has one variable, then $I_{R(?x)}$ is equal to the number of bindings of variable $?x$. Since no duplicate triples exist in the network and $?x$ is the only variable in the triple pattern, each binding will be unique. In this case, we have $I_{R(?x)} = T_R$. In the case where tp has two variables, the corresponding domain size for the shared variable can be determined using the techniques of Section 6.

We now discuss the use of the above selectivity estimation techniques by the optimization algorithms of Section 4. While NA requires only the selectivity of single triple patterns and thus both the bound-is-easier heuristic and the analytical estimation can be applied, SNA and DA require also the selectivity of conjunctions of triple patterns and hence only the analytical estimations will be used. Especially in DA, the estimation of the selectivity of the join between the intermediate results (R_1) and one triple pattern (R_2) is required. The formulas are the same as described above with the exception that relation R_1 is already formed. Therefore, the number of tuples of R_1 and the domain size of any variable in the attributes of R_1 can be computed on the fly by examining relation R_1.

6   Statistics for RDF

In this section, we present an efficient DHT-based scheme for collecting and using the statistics that enable the estimation of the selectivities described in Section 5. These statistics are the frequency of a triple component and the size of the domain of a variable in a triple pattern. Peers keep statistics only from their *local data* and specifically for the data values for which they are responsible (i.e., values that are the *keys* that led a triple to a specific peer). These turn out to be *global statistics* required by the optimization algorithms and can be obtained by sending low cost messages. This is a very good property of the indexing scheme and has not been pointed out in the literature before.

subject	predicate	object	object-class
$freq_s$	$freq_p$	$freq_o$	$freq_c$
dp_s	ds_p	ds_o	–
do_s	dp_p	dp_o	–

Fig. 4. Statistics kept at each peer

Creating statistics. Let us first introduce some new notation which is useful for keeping statistics for the sizes of the domain of the variables appearing in a query. We will denote by $ds_c(v)$ ($dp_c(v)$ and $do_c(v)$, respectively) the total number of distinct subject (predicate and object, respectively) values that exist in the triples stored in the network which contain value v as component c. For example, let tp be the triple pattern (?x,ub:advisor,?y). Then, ds_p(ub:advisor) denotes the number of the distinct subject values in the triples with predicate ub:advisor, i.e., the size of the domain of variable ?x. Similarly, do_p(ub:advisor) denotes the number of the distinct object values in the triples with predicate ub:advisor, i.e., the size of the domain of variable ?y.

The following observation allows us to collect local statistics at each peer of a DHT. *Let v be the value of a bound subject of a triple pattern and p_v the peer responsible for key v. Then, peer p_v is capable of computing the exact frequency of v as a subject, ($freq_s(v)$), the exact number of the distinct predicate values with subject v ($dp_s(v)$), and the exact number of the distinct object values with subject v ($do_s(v)$) in the set of triples stored by looking* only *in its local database.* Given that peer p_v is responsible for key v, our indexing scheme forces all triples that contain v either as a subject, predicate or object to be stored at peer p_v. Therefore, peer p_v can retrieve from its local triple relation all triples that contain v as a subject and hence it can compute the occurrences of v as a subject in the set of all triples stored in the network, i.e., $freq_s(v)$. In addition, peer p_v can compute the number of the distinct predicate values and object values for subject v by projecting the triples that contain v as subject on the predicate and object attribute respectively. The same holds for a triple pattern's bound predicate or object. Following that, it is sufficient for each peer to create and maintain statistics of its *locally* stored RDF data only, and more precisely only of the triple components' values which are the *keys* that led to the peer.

For each triple component, a peer keeps the statistics shown in Fig. 4. A data structure which would keep the exact distribution of each triple component would require excessive memory space for very large amount of data. A commonly used method dating back from relational systems is estimating the frequency distribution of an attribute by creating *histograms* [16]. Given a space budget B for each statistical structure, each peer decides if the exact distribution can be kept in memory or an estimation of the distribution is required by creating a histogram. We use v-optimal-end-biased histograms [16] which we experimentally found to be more suitable. As shown in Fig. 4, we differentiate between objects of triples of the form (s,rdf:type,o), which are classes, and objects of triples (s,p,o) with p $\neq$ rdf:type since we discovered that a more accurate estimation for the objects statistics can be achieved in this way.

Retrieving statistics. Whenever the query optimizer of peer x needs statistics for the selectivity estimation of one triple pattern or a conjunction of triple patterns, it sends a STATSREQ(v_i, c_i) message for each bound component value v_i appearing in the triple patterns to the peer responsible for key v_i specifying also the type c_i of the component value (i.e., subject, predicate, object or class). The peer that receives such a message retrieves the required statistics for value v_i from the corresponding statistical structure (depending on the value c_i) and sends them back to peer x. The time required to retrieve the statistics is negligible compared to the time required for evaluating a query, as we will see in the experimental section. The cost of sending these messages is very low since: (a) messages are small in size, (b) messages are sent in parallel, (c) each message requires only $O(logn)$ hops to reach the destination peer, and (d) the statistics at the destination peer are kept in main memory.

7   Experimental Evaluation

In this section, we present an experimental evaluation of our optimization techniques. All algorithms have been implemented as an extension to our prototype system Atlas. In the latest version of Atlas, we have adopted SQLite as the local database of each peer since the Berkeley DB included in the Bamboo implementation was inefficient. For our experiments, we used as a testbed both the PlanetLab network as well as a local shared cluster (http://www.grid.tuc.gr/). Although we have extensively tested our techniques on both testbeds, here we present results only from the cluster where we achieve much better performance. The cluster consists of 41 computing nodes, each one being a server blade machine with two processors at 2.6GHz and 4GB memory. We used 30 of these machines where we run up to 4 peers per machine, i.e., 120 peers in total.

For our evaluation, we use the Lehigh University benchmark (LUBM) [4]. In each experiment we first infer all triples and then store them in the network. We present results only for queries with more than 4 triple patterns so that the benefits of the proposed optimizations can be clearly demonstrated. We omit query Q12 since it always produces an empty result set and does not exhibit interesting results. All measurements are averaged over 10 runs using the geometric mean which is more resilient to outliers. In the following, QG denotes that the query graph is used to avoid Cartesian products but no other optimization is utilized. The naive algorithm using the bound-is-easier heuristic is denoted by NA^-, while the naive and semi-naive algorithm using the analytical estimation is denoted by NA and SNA, respectively. Finally, DA denotes the dynamic optimization algorithm.

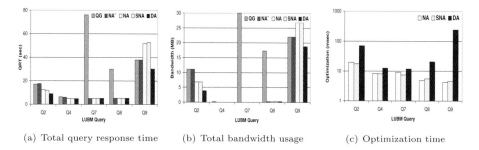

(a) Total query response time   (b) Total bandwidth usage   (c) Optimization time

Fig. 5. Query optimization performance for LUBM-50

7.1 Comparing the Optimization Algorithms

In this section, we compare and evaluate the optimization algorithms described in Section 4. For this set of experiments, we store all the inferred triples of the LUBM-50 dataset ($9,437,221$ triples) in a network of 120 peers. Then, using each optimization algorithm, we run the queries. In all graphs of Fig. 5, the x-axis shows the LUBM queries while the y-axis depicts the metric of interest. Figure 5(a) shows the query response time (QRT) for the different LUBM queries. QRT is the total time required to answer a query and it also includes the time required by the query optimizer for determining a query plan (optimization time). Figure 5(b) shows the total bandwidth consumed during query evaluation.

Queries Q2 and Q9 consist of 6 triple patterns having only their predicates bound. In both queries, there exists a join among the last three triple patterns (in the order given by the benchmark) and the combination of all three triple patterns is the one that yields a small result set. *DA* finds a query plan that combines these three triple patterns earlier than the other algorithms. This results in producing smaller intermediate result sets, as it is also shown by the bandwidth consumption in Fig. 5(b), and thus results in better QRT. Although *NA* and *SNA* perform close to DA for query Q2, they fail to choose a good query plan for Q9 affecting both the QRT and the bandwidth consumption. At this point, we should note that *QG* and NA^- depend on the initial order of a query's triple patterns. For this reason, both algorithms choose a relatively good query plan for query Q9 since the order in which its triple patterns are given by the benchmark is a good one. Q4 is a star-shape query with all its triple patterns sharing the same subject variable, while only the first two triple patterns have two bound components. Therefore, since these two triple patterns are the more selective ones, all optimization algorithms choose the same query plan and perform identically in terms of both QRT and bandwidth. The same holds for query Q7 where QRT is significantly reduced when using either optimization technique compared to *QG*. Q8 is a query similar to Q7.

In Fig. 5(c), we show the total optimization time in msec on a logarithmic scale. For *QG* and NA^- the optimization overhead is negligible and is not shown in the graph. The optimization time contains the time for retrieving the required statistics from the network, the time for the selectivity estimation and the time

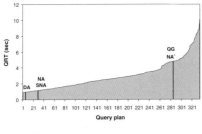

LUBM Query	Min QRT	Max QRT	DA QRT
Q2	0.91 s	10.06 s	0.96 s
Q4	2.54 s	17.39 s	2.89 s
Q7	1.53 s	16.62 s	1.55 s
Q8	2.88 s	10.20 s	3.06 s
Q9	3.30 s	64.06 s	4.41 s

(a) LUBM Q2 query plan space   (b) QRT of LUBM query plans

Fig. 6. Exploring the query plan space for LUBM-10

spent by the optimization algorithm. As expected, DA spends more time than the other optimization algorithms since it runs at each query processing step. However, the optimization time is still one order of magnitude smaller than the time required by the query evaluation process. Therefore, although DA requires more time than the other optimization algorithms, the system manages to perform efficiently for all queries when DA is used. We observe similar results for the bandwidth consumed by the query optimizer. NA and SNA consume $\sim 2KB$ while DA consumes $\sim 7KB$, still one order of magnitude less than the bandwidth spent during query evaluation. We omit this graph due to space limitations.

7.2   Effectiveness of Query Optimization

In this section, we explore the query plan space of the LUBM queries to show how effective the optimization algorithms are. The size of the query plan space of a query consisting of N triple patterns is $N!$. Since query plans that involve Cartesian products are very inefficient to evaluate in a distributed environment, we consider only triple pattern permutations which do not produce any Cartesian product. In this experiment, we store the LUBM-10 dataset in a network of 120 peers and run all possible query plans for several LUBM queries. In Fig. 6(a), we depict the QRT of all possible query plans for query Q2 in ascending order. The query plan space of Q2 consists of 335 query plans which do not involve any Cartesian product. In this figure, we highlight the position of the query plans chosen by the different optimization algorithms. We observe that DA chooses one of the best query plans, while NA and SNA perform worse choosing the 27th best query plan. NA^- performs poorly choosing one the worst query plans. Similar results are observed for the other queries as well. In Fig. 6(b), we list the QRT for all queries of the best and the worst query plan together with the QRT when using DA. We observe that the QRT when using DA is very close to the QRT of the optimal query plan for all queries. Note that without the query plans that involve Cartesian products, the difference between the min and the max QRT of all queries is not very large.

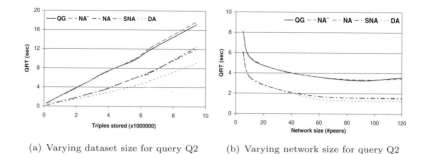

(a) Varying dataset size for query Q2   (b) Varying network size for query Q2

Fig. 7. Varying dataset and network size

7.3   Varying the Dataset and Network Size

In these sets of experiments, we study the performance of our system when varying the number of triples stored in the network and the number of peers. We show results only for Q2 which involves a join among three triple patterns.

Figure 7(a) shows the behavior of our system using each optimization algorithm as the dataset stored in the network grows. In a network of 120 peers, we stored datasets from LUBM-1 to LUBM-50. Every time we measured the QRT of query Q2 using each optimization algorithm. As expected, QRT increases as the number of triples stored in the network grows. This is caused by two factors. Firstly, the local database of each peer grows and as a result local query processing becomes more time-consuming. Secondly, the result set of query Q2 varies as the dataset changes. For example, for LUBM-1 the result set is empty, while for LUBM-50 the result set contains 130 answers. This results in transferring larger intermediate result sets through the network which also affects the QRT of the query. Besides, this experiment brings forth an interesting conclusion regarding the optimization techniques. While query plans chosen by NA, SNA and DA perform similarly up to approximately $1.8M$ triples stored (i.e., LUBM-10), we observe that for bigger datasets the query plan chosen by DA outperforms the others. This shows that the system becomes more scalable with respect to the number of triples stored in the network when using DA. Similar results are observed for Q9, while for the rest queries all optimization algorithms choose the same query plan independently of the dataset size.

In the next set of experiments, we start networks of 5, 10, 30, 60, 90 and 120 peers and store the LUBM-10 dataset. We then run the queries using all optimization techniques. In Fig. 7(b), we show the QRT for Q2 as the network size increases. We observe that QRT improves significantly as the network size grows up to 60, while it remains almost the same for bigger network sizes. The decrease in the QRT for small networks is caused by the fact that the more peers join the network the less triples are stored in each peer's database and thus local processing load is reduced. The same result was observed in other queries where QRT either improved or remained unaffected as the number of peers increased.

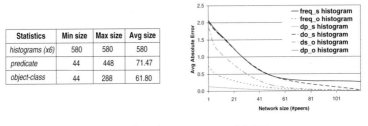

Statistics	Min size	Max size	Avg size
histograms (x6)	580	580	580
predicate	44	448	71.47
object-class	44	288	61.80

(a) Size of statistics per peer (bytes)   (b) Histograms error

Fig. 8. Statistics

7.4 Statistics

We present measurements concerning the size of the statistics kept by each peer. We set a space budget of 500 bytes per statistical structure per peer leading to using histograms of 10 buckets only for values appearing in the subjects and objects of triples. Each statistical structure for the subjects and objects is kept in a separate histogram resulting in a total of six histograms per peer. For the predicates and object classes, we keep the exact distributions of their values. This is typical of a large DHT network where a peer is responsible for very few predicate or class values. Figure 8(a) shows the size of the generated statistics for each peer for the LUBM-50 dataset in a network of 120 peers. Histograms always occupy the same amount of space, while the exact statistics for predicate and object-class vary depending on the amount of values the peer is responsible for. The total statistics kept at each peer result in a total amount of memory of $4K$ in average which is negligible compared to today's powerful machines.

In order to show that it is sufficient to maintain local statistics at each peer and only for the values for which the peer is responsible, we have computed the average absolute error for each histogram for different network sizes ranging from 1 to 120 peers for LUBM-5. A network consisting of a single peer resembles a centralized system where a global histogram is created from all data stored. For every value v_i appearing in the dataset as a component c, we have measured its real frequency $freq_c(v_i)$ and the estimated frequency $\widehat{freq}_c(v_i)$ taken from the corresponding histogram of the peer responsible for value v_i. The absolute error for v_i equals to $e^{abs}(v_i) = |freq_c(v_i) - \widehat{freq}_c(v_i)|$. If N is the total number of distinct values of component c in the dataset, the average absolute error is computed as $\frac{1}{N}\sum_{i=1}^{N} e^{abs}(v_i)$. The same holds for the estimated number of distinct subjects, predicates and objects. Results for each statistical structure that is estimated by a histogram are shown in Fig. 8(b). We observe that as the number of peers increases the error drops significantly. This shows that the values of each triple component are independent and hence the more peers join the network (i.e., more histograms created), the better the estimation becomes.

7.5 Discussion

We have also experimented with different datasets using the SP^2B benchmark [19] as well as a real world dataset of the US Congress vote results presented in [26]. The results were similar to the ones observed using LUBM. For all datasets, DA consistently chooses a query plan close to the optimal regardless of the query type or dataset stored and without posing a significant overhead neither to the total time for answering the query nor to the bandwidth consumed. On the contrary, the static optimization methods are dependent on the type of the query and the dataset, which make them unsuitable in various cases (as shown earlier for query Q9). In addition, we have also tested indexing all possible combinations of the triples' components, as proposed in [11]. In this case, we have used histograms at each peer for combinations of triples' components as well. However, we did not observe any difference in the choice of the query plan and thus, showed results only with the triple indexing algorithm. This results from the nature of the LUBM queries which mostly involve bound predicates and object-classes for which we kept an exact distribution in both cases.

8 Related Work

Earlier works that consider SPARQL query processing on top of DHTs such as [1, 6, 7, 11] lack optimization techniques resulting in handling very small datasets (only thousands of triples). Another DHT-based system is UniStore where a triple-based model and a SPARQL-like query language is supported [10]. In UniStore, a cost-based optimizer is implemented which estimates the cost of physical operators in terms of the number of hops and messages required for each operator. The evaluation presented in [10] is conducted in PlanetLab and hence only small datasets are used. The work of [10] is complementary to ours and the two approaches could actually be combined by an appropriate cost model. Early works that studied query optimization in a distributed environment, although not a DHT, are [17, 24]. In [17], the authors present an engine for federated SPARQL databases and make use of query rewriting and cost-based optimization techniques. For the cost-based optimization, they use iterative dynamic programming but fail to estimate the selectivity of conjunctions of triple patterns and set it to a fixed value instead. Other works in the area of distributed SPARQL query processing are studied in [3, 5, 15]. However, these papers focus on distributed computing platforms based on powerful clusters and do not discuss any optimization techniques.

Finally, a lot of attention has been given to SPARQL query optimization in centralized environments [12, 13, 23]. In [23], the authors present a selectivity-based framework for optimizing SPARQL BGP queries. In RDF-3X [12], the authors propose two kinds of statistics for the selectivity estimation of the joins: specialized histograms which can handle both triple patterns and joins by leveraging the aggregated indexes built, and the computation of frequent join paths in the RDF graphs. In [13], the authors of RDF-3X go one step further to propose more accurate selectivity estimations by precomputing exact join cardinalities for all possible choices of one or two constants in a triple pattern and material-izing the results in additional indexes. This can be a very expensive operation in a distributed setting such as a DHT. Finally, a method for the cardinality esti-mation of SPARQL queries using a probability distribution is presented in [21].

9   Conclusions and Future Work

We studied the problem of distributed SPARQL query optimization on top of DHTs. We discussed the query optimization techniques we have developed in our system Atlas, and presented an experimental evaluation. Our current re-search is focused on the implementation and evaluation of algorithm SBV [11], which achieves a better load balancing, in the presence of the query optimization framework that we have developed.

References

1. Cai, M., Frank, M.R., Yan, B., MacGregor, R.M.: A Subscribable Peer-to-Peer RDF Repository for Distributed Metadata Management. Journal of Web Semantics (2004)
2. Chong, E.I., Das, S., Eadon, G., Srinivasan, J.: An Efficient SQL-based RDF Querying Scheme. In: VLDB 2005
3. Erling, O., Mikhailov, I.: Towards Web Scale RDF. In: SSWS 2008
4. Guo, Y., Pan, Z., Heflin, J.: LUBM: A Benchmark for OWL Knowledge Base Systems. Journal of Web Semantics (2005)
5. Harth, A., Umbrich, J., Hogan, A., Decker, S.: YARS2: A Federated Repository for Querying Graph Structured Data from the Web. In: Aberer, K., Choi, K.-S., Noy, N., Allemang, D., Lee, K.-I., Nixon, L.J.B., Golbeck, J., Mika, P., Maynard, D., Mizoguchi, R., Schreiber, G., Cudré-Mauroux, P. (eds.) ASWC 2007 and ISWC 2007. LNCS, vol. 4825, pp. 211–224. Springer, Heidelberg (2007)
6. Heine, F.: Scalable P2P based RDF Querying. In: InfoScale 2006
7. Kaoudi, Z., Koubarakis, M., Kyzirakos, K., Magiridou, M., Miliaraki, I., Papadakis-Pesaresi, A.: Publishing, Discovering and Updating Semantic Grid Resources using DHTs. In: CoreGRID Workshop on Grid Programming Model, Grid and P2P Sys-tems Architecture 2006
8. Kaoudi, Z., Koubarakis, M., Kyzirakos, K., Miliaraki, I., Magiridou, M., Papadakis-Pesaresi, A.: Atlas: Storing, Updating and Querying RDF(S) Data on Top of DHTs. Journal of Web Semantics (System paper) (2010)
9. Kaoudi, Z., Miliaraki, I., Koubarakis, M.: RDFS Reasoning and Query Answering on Top of DHTs. In: Sheth, A.P., Staab, S., Dean, M., Paolucci, M., Maynard, D., Finin, T., Thirunarayan, K. (eds.) ISWC 2008. LNCS, vol. 5318, pp. 499–516. Springer, Heidelberg (2008)
10. Karnstedt, M.: Query Processing in a DHT-Based Universal Storage - The World as a Peer-to-Peer Database. PhD thesis (2009)
11. Liarou, E., Idreos, S., Koubarakis, M.: Evaluating Conjunctive Triple Pattern Queries over Large Structured Overlay Networks. In: Cruz, I., Decker, S., Alle-mang, D., Preist, C., Schwabe, D., Mika, P., Uschold, M., Aroyo, L.M. (eds.) ISWC 2006. LNCS, vol. 4273, pp. 399–413. Springer, Heidelberg (2006)

12. Neumann, T., Weikum, G.: RDF-3X: a RISC-style engine for RDF. In: VLDB 2008
13. Neumann, T., Weikum, G.: Scalable Join Processing on Very Large RDF Graphs. In: SIGMOD 2009
14. Ntarmos, N., Triantafillou, P., Weikum, G.: Distributed Hash Sketches: Scalable, Efficient, and Accurate Cardinality Estimation for Distributed Multisets. ACM TOCS (2009)
15. Owens, A., Seaborne, A., Gibbins, N., schraefel, m.: Clustered TDB: A Clustered Triple Store for Jena. Technical Report (2008) (Unpublished)
16. Poosala, V., Ioannidis, Y., Haas, P., Shekita, E.: Improved Histograms for Selectivity Estimation of Range Predicates. In: ACM SIGMOD 1996
17. Quilitz, B., Leser, U.: Querying Distributed RDF Data Sources with SPARQL. In: Bechhofer, S., Hauswirth, M., Hoffmann, J., Koubarakis, M. (eds.) ESWC 2008. LNCS, vol. 5021, pp. 524–538. Springer, Heidelberg (2008)
18. Rhea, S., Geels, D., Roscoe, T., Kubiatowicz, J.: Handling Churn in a DHT. In: USENIX Annual Technical Conference 2004
19. Schmidt, M., Hornung, T., Lausen, G., Pinkel, C.: SP^{2}Bench: A SPARQL Performance Benchmark. In: ICDE 2009
20. Selinger, P.G., Astrahan, M.M., Chamberlin, D.D., Lorie, R.A., Price, T.G.: Access Path Selection in a Relational Database Management System. In: SIGMOD (1979)
21. Shironoshita, E.P., Ryan, M.T., Kabuka, M.R.: Cardinality Estimation for the Optimization of Queries on Ontologies. SIGMOD Record (2007)
22. Steinbrunn, M., Moerkotte, G., Kemper, A.: Heuristic and Randomized Optimization for the Join Ordering Problem. VLDB Journal (1997)
23. Stocker, M., Seaborne, A., Bernstein, A., Kiefer, C., Reynolds, D.: SPARQL Basic Graph Pattern Optimization using Selectivity Estimation. In: WWW 2008
24. Stuckenschmidt, H., Vdovjak, R., Broekstra, J., jan Houben, G., Eindhoven, T., Amersfoort, A.: Towards Distributed Processing of RDF Path Queries. Int. J. Web Eng. and Tech. (2005)
25. Ullman, J.D.: Principles of Database and Knowledge-Base Systems, vol. I and II. Computer Science Press, Rockville (1988)
26. Vidal, M.-E., Ruckhaus, E., Lampo, T., Martínez, A., Sierra, J., Polleres, A.: Efficiently Joining Group Patterns in SPARQL Queries. In: Aroyo, L., Antoniou, G., Hyvönen, E., ten Teije, A., Stuckenschmidt, H., Cabral, L., Tudorache, T. (eds.) ESWC 2010. LNCS, vol. 6088, pp. 228–242. Springer, Heidelberg (2010)
27. Weiss, C., Karras, P., Bernstein, A.: Hexastore: Sextuple Indexing for Semantic Web Data Management. In: VLDB 2008

Optimizing Enterprise-Scale OWL 2 RL Reasoning in a Relational Database System

Vladimir Kolovski[1], Zhe Wu[2], and George Eadon[1]

Oracle
[1] 1 Oracle Drive, Nashua, NH 03062 USA
[2] 400 Oracle Parkway, Redwood City, CA 94065 USA
{vladimir.kolovski,alan.wu,george.eadon}@oracle.com

Abstract. OWL 2 RL was standardized as a less expressive but scalable subset of OWL 2 that allows a forward-chaining implementation. However, building an enterprise-scale forward-chaining based inference engine that can 1) take advantage of modern multi-core computer architectures, and 2) efficiently update inference for additions remains a challenge. In this paper, we present an OWL 2 RL inference engine implemented inside the Oracle database system, using novel techniques for parallel processing that can readily scale on multi-core machines and clusters. Additionally, we have added support for efficient incremental maintenance of the inferred graph after triple additions. Finally, to handle the increasing number of owl:sameAs relationships present in Semantic Web datasets, we have provided a hybrid in-memory/disk based approach to efficiently compute compact equivalence closures. We have done extensive testing to evaluate these new techniques; the test results demonstrate that our inference engine is capable of performing efficient inference over ontologies with billions of triples using a modest hardware configuration.

1 Introduction

As part of the OWL 2 [9] standardization effort, three new, less expressive OWL subsets were proposed that have polynomial (or less) complexity and are suitable for efficient and scalable reasoning over large datasets [12]. These profiles are OWL 2 EL, based on the EL++ description logic [7], OWL 2 QL based on DL-Lite [5] and OWL 2 RL, which was designed with rule-based implementations in mind.

Since it is described as a collection of positive Datalog rules, OWL 2 RL can be theoretically implemented on top of semantic stores that already provide rule-based reasoning. One of these semantic inference engines is Oracle's Semantic Technologies offering [10], which has supported inference over scalable rule-based subsets of OWL since Oracle Database 11g Release 1. Oracle's inference engine pre-computes and materializes all inferences using forward chaining, and later uses the materialized graph for query answering[1].

---

[1] Note that our focus is not on query time inference; therefore we have not incorporated techniques such as magic sets rewriting.

P.F. Patel-Schneider et al. (Eds.): ISWC 2010, Part I, LNCS 6496, pp. 436–452, 2010.

There are several challenges in handling enterprise-scale OWL 2 RL reasoning:

- OWL 2 RL supports equivalence relations such as owl:sameAs or owl:equivalentClass. With the emergence of inter-connected Linked Data and its heavy use of owl:sameAs, it becomes increasingly difficult to fully materialize owl:sameAs closures. A naïve representation of the closure could be $O(N^2)$ in the size of the original triple set; we have observed these owl:sameAs blowups using UniProt[2] and OpenCyc [24] data.
- New RDF data is being published at an increasing rate; efficiently reasoning through such updates becomes a bottleneck if the inference closure needs to be maintained. There exists previous work on optimizing Datalog reasoning through updates using semi-naïve evaluation (see [21] for a survey); however it has neither been applied nor evaluated in an OWL setting using large-scale datasets and complex rule sets.
- Since OWL 2 RL has more than 70 rules, performing RL inference on billion-triple sized datasets could take hours to finish. With the proliferation of multi-core and multi-CPU machines, an approach is needed that could efficiently parallelize OWL 2 RL inference so that it could readily scale by adding more processors to the inference engine.

In this paper, we present a new[2] version of the inference engine built inside Oracle Database that supports OWL 2 RL and addresses the above challenges. The main contributions are the following:

Compact Materialization of Equivalence Closures – We address the challenge of efficiently computing owl:sameAs equivalence closures on massive scales by providing a hybrid (memory and disk-based) algorithm for generating compressed closures and integrating it with the general forward chaining inference engine.

Incremental Maintenance of Inferred Closure – We have developed a technique to efficiently update the inferred graph after triple additions to the underlying data model. Our technique is based on semi-naïve evaluation, with additional optimizations such as lazy duplicate elimination and dynamic semi-naïve evaluation.

Parallel Inference – We have parallelized the rules engine by leveraging Oracle's support for parallel SQL execution [17], which scales well with modern multi-core and multi-CPU architectures. To this end, we developed novel rule optimization techniques specifically aiming at parallel execution of queries. We also developed a source table design to align the structure of the table that stores semantic models with the table that stores intermediate temporary data generated during inference. Finally, we developed optimizations to reduce the data storage footprint of inference to reduce memory and I/O consumption.

Note that no knowledge of Oracle internals is needed to apply the techniques presented in this paper. Thus, they should be applicable to any RDBMS-based OWL 2 RL implementation (except for parallel inference, which assumes that the underlying database has support for parallel query evaluation).

---

[2] The algorithms described in this paper along with full support for the OWL 2 RL/RDF entailment and validation rules are available in an Oracle Database 11g Release 2 patch and will be part of the next release.

We evaluated the new features using datasets with billions of triples, including versions of the LUBM ontology benchmark, UniProt ontology and various other real world datasets. With the optimized handling of equivalence closures, inference over owl:sameAs-heavy datasets that was extremely time and space consuming in previous versions of Oracle can now be done in minutes. We also show that our incremental OWL 2 RL inference over graphs of 1 billion triples takes less than 30 seconds to update the inferred graph. Finally, in the empirical evaluation section, we demonstrate the advantages of using parallel inference; this allows us to perform inference faster on less powerful hardware than well-known triple store vendors [3].

2 Preliminaries

2.1 OWL 2 RL

OWL 2 RL is a profile of OWL 2 aiming at applications that require scalable reasoning, efficient query answering, and more expressiveness than RDF(S), without needing the full expressive power of OWL 2. The specification [12] provides a partial axiomatization of the OWL 2 RDF-Based Semantics in the form of first order implications, called the OWL 2 RL/RDF rules.

The OWL 2 RL/RDF rule set is a superset of the non-trivial RDF(S) rules [22]; total number of rules in the partial axiomatization of OWL 2 RL is 78, compared to the 14 rules defined for RDF(S). In addition to supporting all of the RDF(S) constructs (except for axiomatic triples which are omitted for performance reasons), OWL 2 RL also supports inverse and functional properties, keys, existential and value restrictions, and owl:intersectionOf, owl:unionOf to some extent. For a lack of space, we will not enumerate all of the OWL 2 RL rules; we refer the reader to the standard specification [12] for more information. Inference and query answering has polynomial data complexity for OWL 2 RL.

2.2 Oracle Semantic Technologies

Oracle Semantic Technologies [23] provides a semantic data management framework in Oracle Database that supports storing, querying, and inferencing of RDF/OWL data via either SQL or Java APIs. It allows users to create one or more semantic models to store an RDF dataset or OWL ontology. The built-in native inference engine allows inference on semantic models using OWL, SKOS, RDF(S), and user-defined rules. The semantic model (and/or *entailed semantic model*, that is, model data plus inferred data) is materialized and can be queried using either SPARQL query patterns embedded in SQL or standard SPARQL query interface in Java. Oracle also supports ontology-assisted querying over enterprise relational data and semantic indexing of documents.

Inference Engine: The semantic inference engine [10] in Oracle 11g Database is based on forward chaining. It compiles entailment rules directly to SQL and uses Oracle's native cost-based SQL optimizer to choose an efficient execution plan for each rule. Various optimizations were added to improve performance and scalability:

- Dependency Graph – We developed a dependency graph such that we only apply a rule in round *n* if in round *n-1* there have been new inferences for at least one of the predicates contained in the rule's body.
- Using a Partitioned, Un-indexed Table – A temporary table is used to materialize all inferences while applying the inference rules. This table is partitioned by predicate to allow efficient queries, but is not indexed, since inserting inferred triples in an indexed table significantly slows down total inference time.
- Optimized Transitive Closure Evaluation – this optimization is critical for predicates such as rdfs:subClassOf. Instead of using hierarchical queries natively provided by Oracle Database, we discovered that implementing semi-naïve evaluation [21] to compute transitive closure results in better performance.

The following notation is used throughout this paper to refer to various data structures maintained in the semantic store: *M* refers to a single semantic model, i.e., an RDF graph containing asserted instance and schema triples. I(M), or *I* for short, refers to the entailed OWL 2 RL graph for M which contains only the materialized inferred triples. *PTT* is the partitioned, un-indexed temporary table that stores inferred triples during inference. *D* and *DI* are related to incremental inference: D stores the triples added to M since the last inference call, and DI contains the triples inferred in the current inference round.

3   Optimized Equivalence Reasoning

For equivalence relations such as owl:sameAs, owl:equivalentProperty or owl:equivalentClass, fully materializing the equivalence closures can be problematic for large datasets. In general, given a connected RDF graph with N resources using only owl:sameAs relationship, there will be $O(N^2)$ inferred owl:sameAs triples. Note that the alternative of searching the RDF graph at query time to determine if two URIs are equivalent is not feasible because of the interactions among owl:sameAs inferences with other rules in OWL 2 RL. This will require a query rewriting approach, which given the large number of rules in OWL 2 RL will slow down queries.

Each group of owl:sameAs-connected resources represents a *clique*; when doing full materialization the cliques' sizes (number of owl:sameAs triples) can grow quite large. For instance, the Oracle 11g inference engine [10] exhausts disk space (500GB) before completing the owl:sameAs closure for the benchmark ontology UniProt 80M [2]. Note that this version of UniProt80 contains a clique of size 22,000+ individuals so that a full materialization generates more than 480 million triples.

Our approach to handling equivalence closures is based on partial materialization. Instead of materializing the cliques, we choose one resource (individual) from each clique as a representative and all of the inferences for that clique are consolidated using that representative. The idea behind this partial materialization has been explored in previous work [3, 6, 8]; our novel contribution is in developing a hybrid (memory and disk-based), scalable approach for building the owl:sameAs[3] cliques.

---

[3] For brevity, we will only be discussing owl:sameAs closures in the rest of this paper, but our approach is applicable to other equivalence relations such as (owl:equivalentClass).

Following, we discuss how we solve the technical challenge of large scale clique building, that is: given an arbitrarily large input of owl:sameAs pairs, efficiently build a map $\rho : ID \to ID$ which will take an ID^4 of a subject, a property or an object as input and return the corresponding clique representative ID. Note while building ρ, we maintain an invariant that $x \geq \rho(x)$.

3.1   Large Scale Clique Building

The main challenges in building owl:sameAs cliques are that 1) a pure memory based approach does not scale due to memory size limitations, and 2) a pure SQL based approach is not efficient because of the performance implications of many joins on input required to build ρ.

Our proposed solution uses a hybrid approach – we load batches of owl:sameAs assertions (where the batch size is a tunable parameter) from the input table, merge each batch in memory and then append the generated cliques to ρ, which is stored as a table. After all batches are processed, there may be owl:sameAs relationships across different cliques. To capture these cases, we again employ batch processing on the ρ table itself, merging where needed, until we reach a fixpoint.

The flow of the algorithm is as follows:

```
function build_cliques (I)
I  :     input table containing owl:sameAs pairs
ρ :      empty map (resource_id -> clique_id)

1. Read batch B from I
        a. ρ B = Merge(B);  b. Append ρ B to ρ
2. Repeat 1 until no batches left in I
3. Loop
        a. Select batch of merge candidates B from ρ
        b. ρ B = Merge(B)
        c. Update ρ with ρ B
4. Repeat 3 until no more merges possible in ρ
5. return ρ
```

Merge is done in memory using the Union-Find algorithm [8, 27]. Given an input of equivalence relations (i.e., owl:sameAs assertions), Merge builds a map of resources to clique representatives such that given a resource, retrieving its representative is done using only one lookup. The algorithm has time complexity of O(N log N) and polynomial space complexity, however using path compression [8] we achieved almost linear performance in our testing.

After steps 1 and 2, ρ is not fully merged since there may be inter-clique merges remaining. For instance, if one clique contains A owl:sameAs B and another contains A owl:sameAs C, then B and C should belong to the same clique and they will be

---

[4] Note that in our internal storage structures, URIs and literals are mapped to number-based IDs for performance reasons.

selected as merge candidates. Additionally, if one clique contains A owl:sameAs B and another contains B owl:sameAs C, then A and C should be in the same clique and they will also be selected as merge candidates. In step 3c, ρ is updated with the merged in-memory map ρ_B. This is done using an OUTER JOIN where, for each key x in ρ_B, ρ (x) is replaced by ρ_B(x).

After ρ has been built, we update the asserted and inferred graph with the new information, replacing resources x with their clique representatives ρ (x).

Performance: On a UniProt 80 million triple sample, the optimized owl:sameAs approach took 26 minutes to finish inference, producing 61 million consolidated triples. More than 100,000 cliques were generated with an average membership size of 5.6; the largest owl:sameAs clique had 22,064 resources. The storage savings compared to a full materialization of the owl:sameAs closure are more than 95%. More evaluation results are shown in Section 6.

4 Parallel OWL Inference

An extensive performance evaluation of the previous version of Oracle's inference engine (11g Release 1) on a server class machine with solid-state disk based storage revealed that the inference process is CPU-bound in such a setup. Thus, the native OWL inference engine needs to be parallelized to fully leverage hardware configurations that have multiple CPUs (cores) and high I/O throughput.

We explored several schemes to parallelize the native OWL inference process. Simply applying Oracle SQL engine's parallel execution capability to each inference rule (which is translated to a SQL query) without any modification to the inference algorithm did not produce any performance benefits. In the following subsections, we propose several new inference optimization techniques that successfully leverage Oracle's parallel execution engine. We believe they are general enough to be applied to any database supporting parallel query executions.

4.1 Query Simplification for Efficient Parallel Inference

After comparing the performance difference of all the rules running in serial and parallel mode, we observed that rules with smaller number of patterns in the body tend to have bigger performance gains when run in parallel. This observation leads to a new optimization to *simplify complex, multi-pattern rules*. Next, we provide an example of how this rule simplification by break up technique is used to optimize the parallel execution of the OWL 2 RL rule CLS-SVF1 (listed below):

```
T(?x,owl:someValuesFrom,?y)
T(?x,owl:onProperty,?p)
T(?u,?p,?v)
T(?v, rdf:type, ?y)    →        T(?u, rdf:type, ?x).
```

The first two patterns T(?x, owl:someValuesFrom, ?y) and T(?x, owl:onProperty, ?p) are much more selective compared to the rest. Intuitively, execution of this rule can be divided into two parts, where one part focuses on the selective patterns, and the other part focuses on the rest. After the selective sub-query is executed, the variable bindings are then used to further constraint the rest of the patterns. Putting this idea into context, a query can be executed to find all bindings for ?x, ?y, and ?p that satisfy the first two patterns T(?x, owl:someValuesFrom, ?y) and T(?x, owl:onProperty, ?p). For each binding tuple (x, y, p) coming from the query result set, we execute the following rule in parallel:

```
T(?u, p, ?v). T(?v, rdf:type, y)  →  T(?u, rdf:type, x)
```

Executing CLS_SVF1 in parallel mode using the hybrid approach described above is five times faster than running this rule as a single SQL statement.

This idea of breaking up a rule in two parts can easily be generalized to complex rules containing selective and unselective patterns in the rule body[5]. The pseudo code of the algorithm is as follows:

```
function find_sel_patterns (I, R) returns C
I   :   Input RDF graph containing asserted data
R   :   Set of triple patterns belonging to an OWL 2
        RL rule body
C   :   Candidate selective subset that is returned,
        initially empty
1. Estimate average out- and in- degree for subjects
   and property nodes respectively for each property
   in I
2. Estimate selectivity for each property in I by sam-
   pling
3. For each subset S of R
     If est(S, I) < threshold[6] then
        If cardinality(S) > cardinality(C) then C := S
        Else if cardinality(S) == cardinality(C) and
                est(S, I) < est(C, I) then C := S
4. Return C
```

Note that all of the rules in OWL 2 RL have less than 10 triple patterns in the body, so the search space for selective subsets is fairly small.

In the pseudo-code above, est(S, I) estimates the selectivity (size of return set) of a set of triple patterns S against a triple dataset I. We use a simple, conservative estimation technique where we start with the property count estimates and then we iteratively multiply by the average in- or out- degrees (depending on the position of the join variables). These property count and average in/out degree estimates are done once, when the first time find_sel_patterns is executed. Note that more sophisticated SPARQL selectivity estimation methods like [26] could be used here.

---

[5] Note that we are not simply reordering the patterns; instead we find a selective subset of rule body patterns to be used as the driving query.

[6] Currently, we set the threshold for the selective triple pattern estimate to 1000. We do not use a larger number because we need to re-evaluate the second part of the rule for each binding produced by the selective part.

The idea of query simplification also applies to those rules, including CLS-INT1 [12], with recursive/hierarchical structures that use rdf:list. Using CLS-INT1 as an example, instead of using a single complex SQL to find all ?y that satisfy `T(?y, rdf:type, ?C1) ... T(?y, rdf:type, ?Cn)`, a series of simpler SQLs are used to first find all matches for the `T(?y, rdf:type, ?C1)` and then join this result set with the next pattern `T(?y, rdf:type, ?C2)`. This kind of operation is repeated until the last pattern `T(?y, rdf:type, ?Cn)` is processed. Apart from the query simplification, another benefit is that for ontologies containing tens of thousands or more owl:intersectionOf axioms, it is feasible to process all the axioms together in an iterative fashion. Details are omitted here due to space limitations.

4.2 Compact Data Structures

An examination of the underlying table design shows a discrepancy between the table that holds the original semantic model(s) and the partitioned temporary table (PTT) that holds the intermediate inference results. Namely, PTT is partitioned using predicates while the semantic models are not.

To allow efficient parallel execution, we designed a single *source table* with the *same* structure as PTT; this source table contains all data of the original semantic model(s). Then, queries executed during inference only use this source table and PTT. This design change produced critical performance improvements for parallel inference. For example, rules that tend to generate many new inferred triples including RDFS2, RDFS3, RDFS9, RDFS11, RDFS7, PRP-INV1 [12], PRP-INV2 are running 30% ~ 60% faster when Oracle SQL engine's parallel query execution is turned on and the degree of parallelism (**DOP**[7]) is set to 4[8].

As an additional storage optimization, we use an 8-byte binary RAW type as a column type for the PTT and source table instead of a generic numeric type (NUMBER). RAW is an Oracle-specific native datatype which is returned as a hexadecimal string. This column type change saves more than 12% disk storage size using typical benchmark ontologies and this space saving directly translates into better inference performance.

As a final optimization, we also use *perfect reverse hashing*, based on the fact that the set of all generated resource IDs for even a large-scale ontology tends to be sparse (imagine 1 billion = 10^9 unique IDs scattered across a space consisting of 2^{64} which is roughly $1.8*10^{19}$ different values). Perfect reverse hashing provides additional storage savings by mapping the sparse ID values into a sequential set of values starting with 1. For example, assume the original data model has the following set of unique ID values: {10, 1009123, 834132227519661324, 76179824290317, 621011710366788}, where some of them require multiple bytes for storage. If we map them to this sequence {1, 2, 3, 4, 5}, then one byte for each ID is sufficient. In our algorithm, we get the set of unique integer IDs out from the semantic models, map them into a set of sequential integer values, which are then stored in a variable length data type. Then, the RDBMS determines the number of bytes needed for storage. Note that the more compact table structures provided by perfect reverse hashing will improve serial inference as well.

---

[7] Degree of parallelism (DOP) is an Oracle setting that specifies the number of parallel processes that should be used to execute a SQL statement.

[8] On a PC with dual-core CPU, three 1TB disks and 8GB RAM running 64-bit Linux.

5   Incremental Inference

Incremental inference tackles the following problem: Given a model M with a materialized inference graph I, how can we efficiently update I after a new set of triples D is added to M?

Our algorithm for incremental inference is based on semi-naïve evaluation [21]. The goal is to avoid re-deriving existing facts in I after an update. The following example illustrates the basic idea using the rule:

$$X\ rdf{:}type\ C1.\quad C1\ rdfs{:}subClassOf\ C2\ => \ X\ rdf{:}type\ C2$$
$$\quad p_1 \qquad\qquad\qquad p_2 \qquad\qquad\qquad\qquad h$$

In "naïve" inference, the patterns p1 and p2 are both selected from M UNION I. For shorthand, we use $p^{A,B}$ to indicate that pattern p selects from relation A UNION B. After adding D to M, we know that the join $p_1^{M,I} \times p_2^{M,I}$ was already evaluated. Joining $p_1^{M,I,D} \times p_2^{M,I,D}$ means mostly re-deriving the same inferences.

To avoid redundant derivation, at least one predicate should select from the new set of triples D. The semi-naïve rule evaluation is done in two steps:

$$1) \quad h \leftarrow p_1^{D} \times p_2^{M,I,D}$$
$$2) \quad h \leftarrow p_1^{M,I,D} \times p_2^{D}$$

Given the assumption that D is small relative to M and I, this divide and conquer approach has the potential for significant performance improvements. We implemented two custom optimizations on top of this well-known evaluation algorithm in order to further improve performance.

5.1   Lazy Duplicate Elimination

During inference, inferred triples are checked to see if they already exist in M, I or PTT before the triples are inserted in PTT. This check usually involves a hash join which essentially scans through the M, I, and PTT tables. Given a small size of D, we assume that the number of triples inferred will be relatively small compared to M and I, so we allow duplicates to accumulate by not removing them after firing each rule. Instead, we perform the join to remove duplicates only once, at the end of each inference round.

Lazy duplicate elimination will introduce duplicates in DI during an inference round. However, our results (see Table 1) indicate that the duplicate overhead is acceptable since we do not have to perform duplicate elimination after each rule.

Table 1. Duplicate Triples in Incremental Inference. The number of newly asserted triples (i.e., delta size) is 10,000.

Model name (#triples)	Yago (19.9 million)	WordNet (1.9 million)	LUBM8000 (1.06 billion)
#Duplicate/#Unique triples	83,583 / 17,180	123,123 / 23,410	20,944 / 2,453

5.2 Dynamic Semi-naïve Evaluation

The semi-naïve evaluation technique described in this section can also be used when performing inference from scratch, by treating the inferred triples in each round as delta D.

However, we observed that using semi-naïve evaluation for each inference round (we refer to it as *static* semi-naïve evaluation) is not always the optimal choice. This is because in the initial inference rounds the number of inferred triples |DI| could be quite large compared to the size of the asserted model(s) |M|; in such cases, when splitting and evaluating each rule the same execution plan (usually consisting of hash joins) might be used and it might be slower than evaluating the rule in one step. On the other hand, if |DI| is small enough, the SQL optimizer will select a different plan where a nested loop join with index is used instead of hash join, which could dramatically improve performance when the driving table |DI| << |M|.

Thus, we *selectively* use semi-naïve evaluation depending on the number of triples inferred in an inference round. At the end of each round r, we use the following heuristic formula to determine whether to use semi-naïve evaluation in round $r+1$:

$$\frac{|DI|}{|PTT| + \sum_i |M_i|} < t \tag{1}$$

where the threshold t is set to 0.1 by default. In other words, if the number of triples inferred is less than 10% of the overall triple count (including cumulative inferences and asserted models), then we use semi-naïve evaluation in the following round.

Below, we demonstrate the benefits of dynamic semi-naïve evaluation compared to naïve evaluation and to "static" semi-naïve evaluation (running times in seconds).

Table 2. Evaluation results for dynamic semi-naïve evaluation

Dataset	Model Size	#Inf. Rounds	Dynamic Semi-Naive	Naïve	Static Semi-Naive
LUBM1000	133M	3	**3,628**	4,497	4,996
Yago	19.9M	2	**981**	1,230	2,049
Wordnet	1.9M	6	**335**	427	605

6 Evaluation

This section presents the results of a performance study of the techniques presented in this paper using various real-world and synthetic semantic datasets.

6.1 Experimental Setup

We used 2 commodity PCs for our experiment; we refer to them as S1 and S2. Each runs Redhat Enterprise Linux v5 64 Bit (2.6.18-128). Each PC has Oracle Database 11.2 installed and three disks attached. We used Automatic Storage Management (ASM) to spread the I/O load across multiple disks.

	CPU	Memory	Disk
S1	Intel Core 2 Duo 2.13 GHz	6GB	750GB
S2	Intel Core 2 Quad 2.4 GHz	8GB	3TB

The databases were setup with a block size of 8k bytes. S1 had 2400M memory allocated to system global area (SGA), and 3200M to aggregated program global area (PGA) whereas S2 had allocated 3400M and 4400M to SGA and PGA.

In addition to the two commodity PC setups, we also use two server-class machines. S3 is a Sun 4150 server with dual quad core CPUs and Sun Storage 5100 Flash Array. S4 is a Sun Oracle Database Machine and Exadata Storage Server (Full Rack[9] with 8 nodes).

	CPU	Memory	Disk
S3	Intel Xeon CPU E5440 2.83GHz	32GB	1TB
S4[4]	Intel Xeon CPU E5540 2.53GHz	72GB each node	100 TB+

We used various real-world and synthetic datasets to evaluate our inference engine. Lehigh University Benchmark [4] is used frequently to evaluate performance of semantic stores; we evaluated against LUBM1000, LUBM8000, LUBM25K and LUBM50K where each has 133M, 1.1B, 3.3B and 6.6B triples respectively. We used Yago (20M), OpenCyc [24] (1.5M), Wordnet (1.9M) and UniProt [2] (two versions, one of 80M, another of 740M) as real-world datasets.

6.2 Parallel Inference Evaluation

We evaluated parallel inference on UniProt 740M and various sizes of LUBM. On server class machine S3, we measure the performance improvement as DOP changes from 1 to 8. Figure 1 shows that performance improves drastically as we move from a serial execution to parallel execution with DOP set to 8.

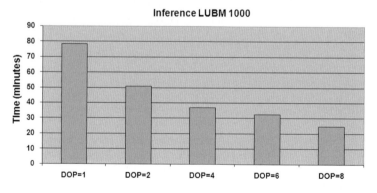

Fig. 1. Inference Performance on server S3

The evaluation results achieved using machine S2 results are shown in Figure 2. In the case of LUBM8000, inference time drops from 42 hours to 11 hrs when using parallel inference. We observe similar improvements with LUBM25000. In all cases, the parallel inference is run with DOP=4.

Using the parallel inference optimizations, we are able to achieve comparable performance to other triple stores while using much weaker hardware: e.g., BigOWLIM uses a server machine while reporting similar inference performance numbers to ours for LUBM25000 and LUBM8000 [15].

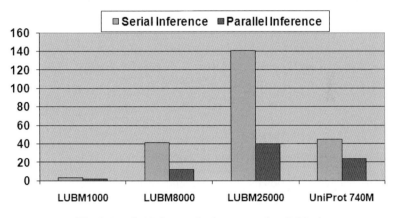

Fig. 2. Parallel Inference Performance when DOP=4

We also evaluated parallel inference performance on server-class machine S4. Due to time limitations, we only collected a few data points. The results nonetheless prove the effectiveness of the parallel inference engine inside Oracle database and the scalability of the particular server-class machine tested.

Benchmark	Parallel Inference Time with S4
LUBM 8000, DOP = 64	46 minutes and 23 seconds
LUBM 25000, DOP = 32	247 minutes and 9 seconds

6.3 Incremental Inference Evaluation

This section contains the incremental inference performance results. The evaluation was done on server S1. For each dataset, we removed a number of triples, performed inference on the remaining dataset and then added back the removed triples in batches of various sizes while measuring the time needed to update the inference graph. We performed this three times and measured the average incremental inference time for each batch. Results are presented in Figure 3.

Our evaluation shows that updating the inference graph is orders of magnitude faster using the incremental inference techniques. For instance, even in the case of a 1 billion triple dataset like LUBM8000, we are able to update the inference graph in less than 20 seconds if the delta is less than 100. Even when adding 10,000 triples the inference update time takes only a few minutes (compared to 11 hours needed to build LUBM8000 inference graph from scratch).

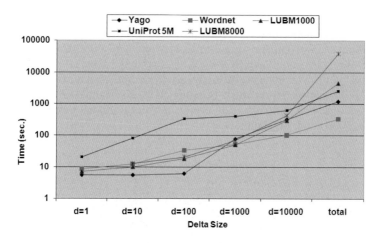

Fig. 3. Incremental Inference Evaluation. As a reference, we also show total (non-incremental) inference time when building the inference graph from scratch.

6.4 Optimized owl:sameAs Handling Evaluation

To evaluate our optimized owl:sameAs handling techniques, we used the following datasets: UniProt 80 million, UniProt 740M and OpenCyc. All three of these datasets have more than 100,000 asserted owl:sameAs triples. Performance results are shown in Table 3. Without the optimized owl:sameAs handling, UniProt 80M and OpenCyc did not finish inference: they exhausted disk space (500GB) after running for 40+ hrs, and UniProt 740M took 24 hours to finish inference.

Table 3. Performance Results for Optimized owl:sameAs Handling

Dataset (#triples)	UniProt (80 Million)	UniProt (740 Million)	OpenCyc (1.5 Million)
owl:sameAs Closure # of Triples	2,129,166,152	42,159,397	295,540,812
owl:sameAs Compressed Closure # of Triples	766,905	12,282,537	395,527
Inferred Triple Size	63,161,568	740,269,215	91,192,106
Cliques Building Time	29 sec	6 min	39 sec
Total Inference Time	25min 48sec	4hr 14min	13 hr 47min

Figure 4 shows the distribution of number of cliques across bins of various clique sizes. As expected, most of the cliques in all three datasets have less than 10 members. Interestingly, UniProt80m and OpenCyc have a surprising number of cliques larger than 1000. In the case of UniProt80m, the largest clique is of size 22,065 and can blow up to 486 million triples when fully materialized. In the more recent, updated version of UniProt (with 740million triples), the modeling issues leading to these large cliques seems to have been fixed; almost all cliques have less than 1000 resources.

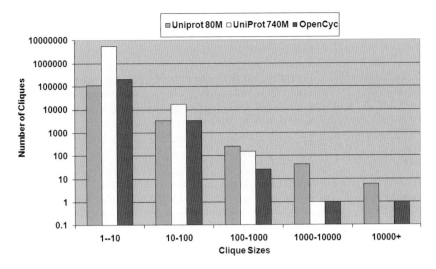

Fig. 4. Distributions of Clique Sizes

The latest version of OpenCyc [24], on the other hand, seems to contain some modeling issues. Apart from the largest clique containing 17,030 resources, many of the resources in that clique are plain literals with distinct values.

7   Related Work

Due to a lack of space, in this section we only provide a survey of the semantic stores and inference engines most closely related to our work.

Jena [11,14] is a Java framework for Semantic Web Applications. In addition to providing an API for RDF, RDFS, OWL and SPARQL, it includes a rule-based inference engine; the inference engine can use both forward and backward chaining, and it supports the most common OWL constructs. Additionally it allows users to define their own custom rules, however it does not natively support any constructs introduced in OWL 2. Like our inference engine Jena supports incremental maintenance (when the forward-chaining RETE-based engine is used); unlike our engine Jena does not optimize owl:sameAs handling.

Sesame [13] is a semantic data repository for RDF and RDFS. Inference wise, it does not support OWL and OWL 2 constructs as we do. It provides an inference engine for RDFS that uses forward chaining and materialization of the data. In [13], an algorithm is proposed for truth maintenance of RDFS data, which could be used to optimize reasoning after updates.

BigOWLIM [3] is a semantic repository that is fully compatible with the Sesame RDF framework. It supports RDFS, some OWL constructs, and extensions with user-defined rules. BigOWLIM's inference engine materializes inferred triples using forward chaining.

BigOWLIM has reported results for the LUBM benchmark [15]. For example, BigOWLIM 3.1 can load, infer, and store the LUBM 8000 dataset in 14.4 hours on a desktop machine. However, their approach seems to require a much larger memory footprint when operating against large ontologies [15] compared to ours.

AllegroGraph [1] is a persistent triple store that can handle large RDF knowledge bases. It has inferencing capabilities that extend beyond RDFS, including custom rules and some OWL constructs, but does not natively support any constructs introduced in OWL 2. Virtuoso Open Link Server [18] is a persistent triple store that scales well on multiple machines but it also provides limited inference support (rdfs:subClassOf, owl:sameAs and rdfs:subPropertyOf constructs).

The Web-Scale Parallel Inference Engine (WebPIE) - although not a true RDF repository because it lacks query capabilities - shows the power of massive parallelism for OWL reasoning. WebPIE [25] is able to infer 4.97 billion triples from a 10 billion-triple LUBM data set in 4.06 hours, using a 64-node cluster. As a comparison, in an inference run using the server-class machine S4, Oracle's parallel inference engine is able to infer, in one inference round, 5.5 billion triples from a 13 billion-triple LUBM data set in 1.97 hours, using DOP=32. The same university ontology and the same OWL Horst semantics were chosen to make the comparison meaningful. We plan to do more testing using high performance platforms like S4.

While our inference engine is to able to cover the whole OWL 2 RL profile, for applications that need additional expressiveness there has been recent work in coupling OWL 2 DL reasoner Pellet [19] with the OWLPrime inference engine in Oracle Database 11g. The scalable-yet-expressive engine PelletDB [20] uses Pellet to compute the class hierarchy and Oracle for Abox reasoning and instance query answering.

Recently a system has been proposed for parallel inference in shared-nothing clusters using existing systems for local computation on each node [16]. The parallelism within our system could work with this to take advantage of multi-core machines within a shared-nothing cluster.

8   Conclusions and Future Work

This paper described the next generation OWL 2 RL inference engine, implemented in the Oracle Database, capable of handling ontologies with billions of triples. We described a number of techniques that we developed to make this engine enterprise-scale, incremental and parallelized. Additionally, to accommodate the high degree of owl:sameAs interlinking between semantic datasets, we implemented a novel scalable, hybrid in-memory/disk-based approach that can compute compact equivalence closures. Using this owl:sameAs approach we were able to discover some modeling issues in real world datasets (e.g., OpenCyc). Our final contribution consists of a thorough evaluation of all our techniques on large-scale real world and synthetic RDF/OWL datasets.

As part of future work, we plan to develop an efficient technique to update inference graphs in presence of deletions. Additionally, we plan to investigate how we can generalize our approach and extend our inference engine to cover the remaining OWL 2 profiles (EL and QL). Finally, the optimization techniques described in this paper are only applied to the axiomatic rules of OWL 2 RL, OWLPrime and RDF(S). We plan to generalize the approach to cover user-specified rules and evaluate it using the OpenRuleBench suite [28].

Acknowledgement. We thank Jay Banerjee for his continuous support and suggestions. We thank Tim Cline for his help in providing server-class machines S3 and S4.

References

[1] AllegroGraph, http://www.franz.com/products/allegrograph/
[2] The UniProt Consortium. The Universal Protein Resource (UniProt). Nucleic Acids Res. 36, D190–D195 (2008), http://www.UniProt.org/
[3] Kiryakov, A., Ognyanov, D., Manov, D.: OWLIM – a Pragmatic Semantic Repository for OWL. In: Proc. International Workshop on Scalable Semantic Web Knowledge Systems (SSWS 2005), New York City, USA (2005)
[4] Guo, Y., Pan, Z., Heflin, J.: LUBM: A Benchmark for OWL Knowledge Base Systems. Journal of Web Semantics 3(2), 158–182 (2005)
[5] Calvanese, D., de Giacomo, G., Lembo, D., Lenzerini, M., Rosati, R.: Tractable Reasoning and Efficient Query Answering in Description Logics: The DL-Lite Family. J. of Automated Reasoning 39(3), 385–429 (2007)
[6] Stocker, M., Smith, M.: Owlgres: A Scalable OWL Reasoner. In: Proc. of OWL Experiences and Directions EU, OWLED-EU (2008)
[7] Baader, F., Brandt, S., Lutz, C.: Pushing the EL Envelope. In: Proc. of the 19th Joint Int. Conf. on Artificial Intelligence, IJCAI 2005 (2005)
[8] Tarjan, R.: A Class of Algorithms which Require Nonlinear Time to Maintain Disjoint Sets. Journal of Computer and System Sciences 18(2), 110–127 (1979)
[9] Motik, B., Patel-Schneider, P.F., Parsia, B. (eds.): OWL 2 Web Ontology Language: Structural Specification and Functional-Style Syntax. Latest version available at, http://www.w3.org/TR/owl2-syntax/
[10] Wu, Z., Eadon, G., Das, S., Chong, E.I., Kolovski, V., Annamalai, M., Srinivasan, J.: Implementing an Inference Engine for RDFS/OWL Constructs and User-Defined Rules in Oracle. In: IEEE 24th International Conference on Data Engineering, ICDE, pp. 1239–1248 (2008)
[11] Jena Framework, http://jena.sourceforge.net/
[12] Motik, B., Grau, B.C., Horrocks, I., Wu, Z., Fokoue, A., Lutz, C. (eds.): OWL 2 Web Ontology Language: Profiles. Latest version available at, http://www.w3.org/TR/owl2-profiles/
[13] Broekstra, J., van Harmelen, F., Kampman, A.: Sesame: A Generic Architecture for Storing and Querying RDF and RDF Schema. In: Horrocks, I., Hendler, J. (eds.) ISWC 2002. LNCS, vol. 2342, p. 54. Springer, Heidelberg (2002)
[14] Wilkinson, K., Sayers, C., Kuno, H., Reynolds, D.: Efficient RDF storage and retrieval in Jena. In: Proc. VLDB Workshop on Semantic Web and Databases (2003)
[15] OWLIM: LUBM Tests, http://ontotext.com/owlim/benchmarking/lubm.html
[16] Narayanan, S., Catalyurek, U., Kurc, T., Saltz, J.: Parallel materialization of large ABoxes. In: Proc. of the 2009 ACM Symposium on Applied Computing, Honolulu, Hawaii, pp. 1257–1261 (2009)
[17] Oracle SQL Parallel Execution, http://www.oracle.com/technology/products/bi/db/11g/pdf/twp_bidw_parallel_execution_11gr1.pdf
[18] Virtuoso Universal Server Platform, http://virtuoso.openlinksw.com/
[19] Pellet – Open Source OWL DL Reasoner, http://clarkparsia.com/pellet/

[20] PelletDB. More information, `http://clarkparsia.com/pelletdb`

[21] Ceri, S., Gottlob, G., Tanca, L.: What you always wanted to know about Datalog (and never dared to ask). IEEE Transactions on Knowledge and Data Engineering 1(1) (1989)

[22] Hayes, P. (ed.): RDF Semantics, W3C Recommendation. Latest version available at, `http://www.w3.org/TR/rdf-mt/`

[23] Oracle Semantic Technologies, `http://www.oracle.com/technology/tech/semantic_technologies/index.html`

[24] OpenCyc, `http://www.opencyc.org/downloads`

[25] Urbani, J., Kotoulas, S., Maassen, J., van Harmelen, F., Bal, H.: OWL reasoning with WebPIE: calculating the closure of 100 billion triples. In: Aroyo, L., Antoniou, G., Hyvönen, E., ten Teije, A., Stuckenschmidt, H., Cabral, L., Tudorache, T. (eds.) ESWC 2010. LNCS, vol. 6088, pp. 213–227. Springer, Heidelberg (2010)

[26] Stocker, M., Seaborne, A., Bernstein, A., Kiefer, C., Reynolds, D.: SPARQL Basice Graph Pattern Optimization Using Selectivity Estimation. In: Proc. of the World Wide Web Conference (WWW 2008), Beijing, China, April 21-15 (2008)

[27] Fiorio, C., Gustedt, J.: Memory Management for Union-Find Algorithms. In: Reischuk, R., Morvan, M. (eds.) STACS 1997. LNCS, vol. 1200, pp. 67–79. Springer, Heidelberg (1997)

[28] Liang, S., Fodor, P., Wan, H., Kifer, M.: OpenRuleBench: An Analysis of the Performance of Rule Engines. In: WWW 2009, pp. 601–610. ACM Press, New York (2009)

Linked Data Query Processing Strategies

Günter Ladwig and Thanh Tran

Institute AIFB, Karlsruhe Institute of Technology, 76128 Karlsruhe, Germany
{guenter.ladwig,duchthanh.tran}@kit.edu

Abstract. Recently, processing of queries on linked data has gained attention. We identify and systematically discuss three main strategies: a bottom-up strategy that discovers new sources during query processing by following links between sources, a top-down strategy that relies on complete knowledge about the sources to select and process relevant sources, and a mixed strategy that assumes some incomplete knowledge and discovers new sources at run-time. To exploit knowledge discovered at run-time, we propose an additional step, explicitly scheduled during query processing, called *correct source ranking*. Additionally, we propose the adoption of *stream-based query processing* to deal with the unpredictable nature of data access in the distributed Linked Data environment. In experiments, we show that our implementation of the mixed strategy leads to early reporting of results and thus, more responsive query processing, while not requiring complete knowledge.

1   Introduction

The amount of Linked Data on the Web is large and ever increasing. This development is exciting, paving new ways for next generation applications on the Web. We contribute to this development by investigating the problem of how to process queries against Linked Data. Linked Data query processing can be seen as a special case of federated query processing, i.e., to process queries against data that resides in different data sources. However, the highly distributed structure and evolving nature of Linked Data presents unique challenges.

- **Volume of the Source Collection:** According to the Linked Data principles [2], each URI can be dereferenced and the document returned represents a virtual "data source". This dramatically increases the *number of Linked Data sources* that need to be considered for query processing.
- **Dynamic of the Source Collection:** Linked Data sources are added and removed and sources' content changes rapidly over time. Due to this dynamic, it is no longer safe to assume that information about all sources can be obtained. In particular, sources might be a priori *unknown* and can only be discovered at run-time.
- **Heterogeneity of Sources, Source Descriptions and Access Options:** Sources *vary in size*. There might be large sources, corresponding to Web databases today. Sources could also just comprise several RDF statements obtained via URI lookup. Further, there is *no standard for describing sources*

P.F. Patel-Schneider et al. (Eds.): ISWC 2010, Part I, LNCS 6496, pp. 453–469, 2010.

yet. Not all sources are accompanied with a voiD[1] description and even if so, they are often incomplete. Also, the *range of access options is vast*. Sources can be obtained via HTTP lookup, retrieved from SPARQL endpoints or directly loaded from a local repository or cache. Even using the same access method, the time required to obtain the same amount of data might vary greatly due to network latency.

Recently, Harth et al. [5] proposed a probabilistic data structure that aims to improve the efficiency of Linked Data query processing. In order to deal with a large number of sources, rich statistics about them are acquired and stored locally. These statistics are used to determine relevant sources and to optimize query processing. Hartig et al. [6] proposed a method for dealing with the dynamic aspect of Linked Data query processing. As opposed to [5], the strategy employed here does not assume information about sources to be available. Sources are discovered via lookups of URIs found during query processing. We follow the direction of this line of work and make the following contributions:

- For Linked Data query processing, we identify the challenges, discuss concrete tasks, and derive three main strategies. There is a *top-down strategy* corresponding to the approach implemented by [5], a *bottom-up strategy* implemented by [6], and a *mixed strategy* that as opposed to [5], does not assume complete but only partial knowledge about the sources and unlike [6], have to discover only some but not all sources at run-time.
- We propose an implementation of the mixed strategy that is able to use runtime information for *corrective source selection and ranking*. The proposed ranking scheme can deal with different types of source descriptions containing knowledge at varying levels of granularity.
- As an alternative to the pull-based non-blocking iterator [6], we propose the use of *push- and stream-based query processing* where source data is treated as finite streams that can arrive at any time in any order. This approach is better suited to deal with network latency as it is driven by incoming data and does not require temporary rejection of answers.

We implement the proposed approach and perform an evaluation where we compare the mixed strategy with the bottom-up [6] and top-down [5] strategies. The results suggest that the implemented mixed strategy is able to report results much earlier than the bottom-up strategy, while not relying on the assumption that complete knowledge is available, as opposed to the top-down strategy. First results (25% of total results) were on average reported 42% faster than for the bottom-up strategy.

Outline: In Section 2 we discuss techniques for Linked Data query processing. In Sections 3 & 4 we present our approach to stream-based query processing and corrective source ranking. Finally, we present related work in Section 5 before the discussion on evaluation results in Section 6 and the conclusions in Section 7.

---

[1] `http://vocab.deri.ie/void/guide`

2   Linked Data Query Processing

We begin with a discussion on Linked Data. For Linked Data query processing, we discuss the tasks (a) source discovery, (b) source ranking and for (c) query evaluation, we discuss the (1) top-down, (2) bottom-up and (3) mixed strategy.

2.1   Linked Data

In this work, we simply conceive Linked Data sources as sets of RDF triples [10].

Definition 1. *A source s is a set of RDF triples $\langle s, p, o \rangle \in T^s$ where s is the subject, p the predicate and o the object. It is uniquely identified by an URI and can be retrieved by dereferencing that URI. A source s_i links to another source s_j if the URI of s_j appears as the subject or object in at least one triple $t^s \in T^{s_i}$.*

The standard language for querying RDF data is SPARQL [14]. An important part of SPARQL queries are basic graph patterns (BGP). In this work we are concerned with answering BGP queries.

Definition 2. *A* basic graph pattern *is a set of* triple patterns $\langle s, p, o \rangle \in T^q$ *where every s, p and o is either a* variable *or a constant. Variables may interact in an arbitrary way such that the triple patterns $t^q \in T^q$ may form a graph.*

An answer to a BGP query is given by μ which maps patterns $t^q \in T^q$ to triples $t^s \in T^s$. By applying such a mapping, each variable in T^q is replaced by the corresponding subject, predicate or object of triples in T^s (called a binding). When processing queries over a set of Linked Data sources, the query is not evaluated on a single source, but on the graph formed by the union of all retrieved sources. A BGP query is evaluated by performing a series of joins between RDF triples that match the triple patterns in the query. In particular, two triple patterns that share a variable form a *join pattern*.

There are several types of source descriptions the system might be able to obtain for a source: A *metadata description* is like a voiD description of the content. It captures basic information such as the size of the source, the RDF predicate it contains etc. *Statistics* capture detailed information that can be derived from the source data such as triple pattern cardinality, join pattern cardinality, histograms, etc. A representative *sample* of the source data might be available.

2.2   Source Discovery

There are multiple ways for sources to be discovered: Sources can be explicitly set in the *query* using special syntax or can be part of a triple pattern. The query engine can maintain a list of *known sources*. This list can either be entered manually or be compiled from previously executed queries. Sources can be *discovered* during query processing by following links mentioned in the content of retrieved sources.

In the first two cases, sources are known before the execution of the query. Compile-time optimization decisions concerning source ranking and query optimization (discussed in the following) are based exclusively on information derived from these sources. In the last case, sources are dynamically added at runtime. New information derived from these sources has an impact on the compile-time optimization plan. This information might render the plan no longer optimal. It is used in our work for corrective query optimization.

2.3   Source Ranking

A source is *relevant* if it contains data that can contribute to the final answers. The standard optimization goal is to (1) obtain all results as fast as possible. However, given the volume and dynamic of the Linked Data collection, it is often infeasible to retrieve and process all sources. It is important to rank sources by their relevancy to the query and more fine-grained optimization goals. In particular, it might be desirable to (2) report results as early as possible, (3) to optimize the time for obtaining the first k results, or (4) to maximize the number of total results, given a fixed amount of time.

Source ranking uses available source descriptions that may vary in quality and completeness, i.e., they may lack information important for ranking. This means that it is essential to incorporate not only a priori available knowledge, but also knowledge discovered obtained query execution.

2.4   Query Evaluation Strategies

Top-Down Query Evaluation. Linked Data comprises heterogeneous data that comes from different sources. Typically, a federated database system is used to integrate multiple sources and systems into one single federated database. The goal is to obtain a fully-integrated virtual database that provides transparent access to data of all its constituent sources.

Typically, sources and databases are geographically decentralized in a federated system. However, a system, which discovers, retrieves and stores Linked Data sources centrally, also falls into the category of a federated system. In fact, no matter the physical location (and other characteristics) of the sources, a source is considered if and only if the federated system knows about it. The federated system assumes that *all source descriptions are available* and based on that, compiles a *query evaluation plan* that specifies the relevant sources, and the order for retrieving and processing these sources. Thus, query planning and optimization is a one-off process performed in a top-down fashion based on complete information.

Harth et al. [5] implement this top-down evaluation. The main focus is on using a data structure capturing rich statistics that can be used to improve query planning and optimization. In approaches that fall into this category, source discovery is performed offline and source ranking is not part of the process. In order to deal with the large amount of sources, source ranking based on approximative triple and join pattern cardinality estimation is used to consider only a fixed number of top-ranked sources.

Bottom-Up Query Evaluation. As opposed to top-down query processing, this strategy does not assume source descriptions to be available beforehand and computes results in a bottom-up fashion. Without planning and optimization, it directly evaluates the query. During this process, it (1) retrieves the sources that are mentioned in the query, (2) discovers further sources based on source URIs and links found in the data of the retrieved sources, (3) incorporates the content of these discovered sources into query evaluation and (4) terminates when all sources found to be relevant have been processed.

Systems that implement this strategy do not rely on sources or source descriptions being managed centrally but discover and retrieve sources from external locations. *Source discovery* and *retrieval* are an integral part of the online process. These online tasks make this approach to query processing different from traditional database approaches. They might be needed due to the Linked Data specific challenges we have discussed. The large volume and the dynamic of the sources and source collection render the traditional top-down approach impracticable. In particular, it cannot be applied when there are sources that are not known beforehand and can only be discovered during online processing.

Another aspect distinct to this approach is *completeness*. As opposed to traditional query processing, it might not be possible to obtain complete knowledge about all sources. In particular, processing queries against Linked Data where sources have to be discovered online might not yield all results. Results to the query cannot be found when they are part of sources that are unknown and cannot be discovered during online processing. This is the case when a link between two sources is only stored in one of the sources, meaning that the link cannot be discovered from the other source.

This strategy is implemented in [6], using non-blocking iterators to avoid blocking due to network delay (see Section 3.2).

Mixed Strategy Query Evaluation. This strategy combines the two other strategies by assuming that knowledge about some sources is available (the sources' data themselves are not necessarily locally available), and more knowledge can be obtained during online query processing. Compared to the top-down strategy, it does not rely on complete knowledge. Similar to the bottom-up strategy, online source discovery is an integral part of query processing. As opposed to that strategy, it makes use of knowledge available beforehand to do query planning and optimization. However, the plan built at compile time might be corrected according to newly acquired knowledge about sources. In particular, the additional optimization tasks that have to be performed online are *corrective source ranking* and *join order optimization*. Source ranking is not a by-product of query optimization [5], but explicitly scheduled as an integral task.

For processing queries on Linked Data, this strategy begins with (1) "best-effort" query planning, and based on this plan, evaluates the query. During this process, (2) sources are retrieved, (3) new sources are discovered, (4) new sources' content are incorporated into evaluation and in a continuous fashion, (5) new sources' descriptions are used for corrective source ranking and optimization. The evaluation proceeds with the continuously refined plan and (6) terminates when all relevant sources have been processed.

This mixed strategy explicitly addresses two of the challenges discussed previously. It uses online discovery to deal with Linked Data *volume* and *dynamic*. Also targeting the aspect of volume, compile-time combined with evaluation-time corrective source ranking and optimization are employed to make processing the large amount of sources affordable.

In the following, we discuss our implementation of this strategy that addresses also the remaining challenges. It features a novel approach for corrective source ranking that is designed to deal with Linked Data heterogeneity by exploiting the *different types of source descriptions* discussed previously. A stream-based query processing is employed to deal with the unpredictable nature of Linked Data resulting from *different source access options*, and to report results early.

3   Stream-Based Linked Data Query Processing

We provide an overview of our approach to Linked Data query processing and then discuss stream-based evaluation based on push-based symmetric hash joins.

3.1   Overview of the Process

Query Planning. A query plan is constructed during query compilation. We only consider left-deep query plans in this implementation, while in principle, query plans with other shapes such as bushy plans are possible. Depending on available source descriptions, basic information or detailed statistics as discussed before can be used to plan the order of operators to be executed, and to perform other kinds of database optimizations that might consider indexes, materialized views, or the concrete join implementations [12]. Apart from joins, for Linked Data query processing, the operators we consider additionally include source discovery, source retrieval and source ranking. In this work we do not consider the general case of operator order optimization (and join order optimization) but focus on the specific aspect of corrective source ranking at run-time.

Query Evaluation. For evaluating the query according to the query plan, we run each operator in a separate thread. Communication between operators is based on bounded message queues to enable parallel query processing. After query planning, threads for all operators are started. As a first step at run-time, local indexes are probed using the query triple patterns to obtain an initial list of possibly relevant sources, which is then sent to the source ranker. Fig. 1 shows an overview of the operators involved in the query execution.

Source Ranking. The source ranker also runs in its own thread and receives source URIs, either obtained through discovery or from local indexes. It ranks the sources according to the methods described in Section 4. Ranking is performed only when necessary. The source ranker checks this continuously, using the parameters given in Section 4.4. If ranking is to be performed, the scores of all sources are calculated and normalized. The source ranker keeps track of the source retrieval threads and assigns them the top-ranked sources.

Source Retrieval. Because of network delay it is usually necessary to request data from several sources at once, which is accomplished by running more than one source retrieval thread [6]. They filter the incoming data using the triple patterns of the query and push matching triples to the join operators as soon as they are decoded from the incoming data. This push-based join processing and the join operator are discussed in Section 3.2.

Source Discovery. In addition to retrieving sources, the retrieval threads perform discovery of new sources based on the content of the source currently being processed. They notify the source ranker of all sources linked from the source just found.

Termination. Several termination conditions can be configured: (1) maximum discovery distance, (2) maximum number of sources to load and (3) number of results to produce. If any of these conditions are reached, the source ranker notifies the join operators so that query execution is terminated as soon as all remaining intermediate results have been processed.

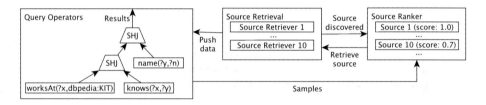

Fig. 1. Join, source ranking and retrieval operators

3.2   Push-Based Symmetric Hash Join

Query processing in highly distributed environments, where data is often stored at remote locations, presents unique challenges. These environments require flexible scheduling: operators should not block, so that the query plan can make progress when input is delayed for another part [8]. In query plans using iterator-based ("pull-based") operators, the `next` method blocks until it is able to produce a result. Non-blocking iterators [6] were proposed to address this problem in the context of Linked Data queries. A non-blocking iterator is able to temporarily reject input from iterators lower in the operator tree when it would otherwise block because of unavailable data on the other input. On the next call to its `next` method, the lower operator randomly either returns a new intermediate result or one of the previously rejected results, for which data might now be available in the upper iterator. This ensures that query processing can progress even if data for a particular triple pattern is not yet available. The advantage of this solution is that it can be used in existing query engines. However, while waiting for input to become available the query engine essentially performs busy-waiting in a loop by alternately asking for new results and then rejecting them. Even if no new data arrives the query engine is active, consuming CPU time.

To alleviate this problem, we propose the adoption of a stream-based approach where source data is treated as a (finite) stream that can arrive in any order. To process such streams, pipelined operators are required that produce results even before the whole input has been read. Query plans using these operators can be implemented using threads and message queues, taking advantage of multi-core and parallelization capabilities of modern CPUs.

One such operator is the symmetric hash join (SHJ), which, in constrast to traditional hash joins, can start reporting results as soon as input tuples arrive in the operator and does not have to wait until one of the input has been completely read [16]. This is achieved by maintaing one hash table on each input. Instead of a pull-based iterator, we employ the SHJ in a push-based mode where operators are driven by their inputs. Instead of a `next` method that is called by operators higher in the operator tree, the join operator has a `push` method for each of its inputs. Algorithm 1 shows the operation of this method in a SHJ operator. First, the arriving tuple t is inserted into the hash table H_m that corresponds to the input where the tuple belongs to (i.e., $H_m = H_1$ or H_2). Then, t is used to probe hash table H_n for valid join combinations. All such valid join combinations are then immediately reported to subsequent operator out by calling its `push` method. Pushing is done from the operators corresponding to the leaf nodes of the operator tree to the root operator. The root operator pushes early results to the caller of the query evaluator. Compared to blocking operators such as the hash join, the SHJ produces results as soon as input tuples are available and input tuples can arrive on all inputs in any order.

Algorithm 1. SHJ: $push(in, t)$

Input: Operator in from which input tuple t was pushed
Data: Hash tables H_1 and H_2; current operator $this$; subsequent operator out
1 **if** in is $left$ $input$ **then** $m = 1$, $n = 2$
2 **else** $m = 2$, $n=1$
3 Insert t into hash table H_m
4 Probe H_n with join keys of t
5 **forall** $valid$ $join$ $combinations$ j **do** out.push($this,j$)

---

4   Corrective Source Ranking

The relevance of a source depends on several factors and is measured based on the current query, any available intermediate results and an overall optimization goals as discussed in Section 2.3. In this section, we elaborate on the source features that are taken into account, concrete metrics derived from them, the indexes used to compute the metrics, newly discovered information used to refine and correct previously computed metrics, and how they are incorporated into source ranking.

4.1   Source Features and Metrics

Triple Pattern Results. A source is more relevant if it contains data that contributes to answers of the query. Thus, a source is relevant if it contains triples matching a query triple pattern. The estimation of triple pattern results is based on the metrics triple pattern cardinality and triple pattern specificity.

Definition 3 (Triple Pattern Cardinality and Specificity). *The* triple pattern cardinality $card(s,t)$ *gives the number of triples in source s that match the triple pattern t. The* triple pattern specificity $spec(t)$ *gives the number of constants that occur in the triple pattern t.*

Clearly, the higher the cardinality and the more specific the triple pattern, the more relevant is a source matching that pattern. However, these two metrics alone are yet no good indicator for the relevance of a source. Given the power-law distribution in the Web of Data [4], some triple patterns might have a high cardinality for all or many sources. These patterns do not discriminate sources, just like words that frequently occur in all documents of a collection. One example is $\langle ?x, rdf{:}type, ?y \rangle$, which can be found in most Linked Data sources. To alleviate this problem, we adopt the TF-IDF concept to obtain weights for triple patterns (capturing their *importance*). Similarly to words in IR, the importance of a triple pattern positively correlates with how often bindings to this pattern occur in a source as measured by its cardinality, and negatively correlates with how often its bindings occur in all sources of the collection. Higher weight is thus given to discriminative triple patterns.

Definition 4 (TF-ISF). *Given a source s and a triple pattern t, the* triple frequency - inverse source frequency *(TF-ISF) measure is defined as* $\text{TF-ISF}(s,t) = card(s,t) \cdot \log \frac{|S|}{|\{r \in S | card(r,t) > 0\}|}$ *where S is the set of all sources to be ranked.*

Join Pattern Results. A source containing data matching larger parts of the query is more relevant. Thus, a source that contains data matching a join pattern is considered highly relevant. However, not containing data for a join does not render a source irrelevant as its data might be joined with data from other sources. The join pattern cardinality estimates the results of a join pattern.

Definition 5 (Join Pattern Cardinality). *Given the join pattern* $t_i \bowtie_v t_j$ *on the shared variable v, the* join pattern cardinality *of a source s denoted* $card(s, t_i, t_j, v)$ *gives the number of results a join on the variable v between triples retrieved from s for t_i and t_j produces.*

Links to Results. A source containing many links coming from relevant sources is more useful. The relevance of such sources is even higher when these links match query predicates. Note that unlike triple pattern results that can be computed given a source, links can only be discovered by processing several sources. A source at first considered irrelevant based on triple pattern results might become relevant during the process. For measuring links to results, links to other sources are extracted from sources discovered during the process.

Definition 6 (Links to Results). *Let S be the set of sources already processed,* $links(s_i, s_j)$ *be a function that return all links between a source* $s_i \in S$ *and the source* s_j*, the* links to results *of* s_j *is defined as* $links(s_j) = \bigcup_{s_i \in S} links(s_i, s_j)$*.*

Retrieval cost. Sources are more useful the faster they can be retrieved.

Definition 7 (Retrieval Cost). *The* retrieval cost *of a source s is a monotonic aggregation of the* size *of s and the* bandwidth *of a host h, defined as* $cost(s) = Agg(size(s), bandwidth(h))$*.*

Source size is available in the source description. Bandwidth is approximately derived for a particular host based on past experiences or, when available, average performance recorded during the process for sources retrieved from this host.

4.2 Metric Computation

In the mixed strategy, some of the source metrics are available locally. We store these metadata in specialized indexes (1) to select relevant sources and (2) to compute cardinalities for these sources.

Indexes for Source Selection. Given a triple pattern, these indexes return a set of sources that contain triples matching the pattern. The only "interesting" patterns are those with one or two variables. Patterns with no variables match only themselves and pattern with no constants match all triples and thus, match all sources. Three indexes are sufficient to support all patterns with one variable. In particular, we create the indexes SP, PO and OS (where S, P, O stand for subject, predicate and object). Each maps the indexed pattern to a set of sources. For example, to find sources for $\langle ?x, rdf{:}type, foaf{:}Person \rangle$, we use the PO index to retrieve relevant sources. Using prefix lookup, the same indexes can be used to cover all patterns with two variables.

Index for Cardinality Computation. In [5], a probabilistic index structure is used to support triple and join pattern cardinality estimation of individual sources. A different technique based on aggregation indexes is presented in [13]. We adopt this method, but extend it to support lookup of triple pattern cardinalities and estimation of join cardinalities for individual sources. Instead of calculating the statistics and indexes for the whole dataset, we treat each source as its own dataset and create the aggregation indexes accordingly. While we lose the ability to perform selectivity and cardinality estimation over the indexed data as a whole, we can now calculate estimates for individual sources, which is what is necessary for source ranking.

4.3 Metric Correction and Refinement

During query processing as sources are retrieved and their data is processed, more information becomes available to compute new or to refine and correct previously computed metrics. This is especially important in the case of very general "non-discriminative" triple patterns, such as $\langle ?x, rdf:type, ?y \rangle$. When such a pattern is joined with another pattern, it is more or less by chance that matching join combinations are found.

When processing queries over data that is stored and indexed locally, this problem can be alleviated by performing index nested-loop joins. An index nested-loop join between two triple patterns t_1, t_2 uses triples that match t_1 to instantiate triple pattern t_2 by replacing variables with bindings of the join variables in triples matching t_1. This creates more specific triple patterns which are then used perform index lookups to retrieve further data that is guaranteed to create join combinations.

In the case where data is not locally available, we cannot perform such joins. However, we employ a similar technique to estimate join pattern cardinalities, taking into account current intermediate results and information in the cardinality indexes. In particular, a triple pattern of a join is instantiated with intermediate results and then used to perform lookups on the triple pattern cardinality indexes to calculate better join cardinality estimates:

Definition 8 (Join Pattern Cardinality Estimate). *Let t_i, t_j be two triple patterns, T_i^s a set of triples in s matching the pattern t_i, and $T_i^s(v)$ denotes the set of bindings to the variable v of the triple pattern t_i. Based on triple pattern cardinalities, a cardinality estimate of a join $t_i \bowtie_v t_j$ is calculated as $card(s, t_i, t_j, v) = \sum_{b \in T_i^s(v)} card(s, t_j.inst(v, b))$, where $t_j.inst(v, b)$ denotes the instantiation of the variable v of the triple pattern t_j with the binding b.*

The SHJ operators maintain hash tables on both of their inputs, storing data by the join attribute. The data of a source indexed in a hash table is used to instantiate the triple patterns of the join to obtain more specific triple patterns. Then, the cardinality of these more specific patterns is looked up using the index and aggregated to obtain an estimate for the size of the join. In order to reduce the cost of this process, we perform *sampling* to estimate the join cardinality by instantiating the triple pattern with only a random subset of the triples. Sampling has been used in database research to perform estimation of join cardinalities, see Section 5 for related work on this topic.

4.4   Source Ranking at Run-Time

In our implementation we prioritized early result reporting, i.e., producing results as early as possible is the optimization goal. First, for every indexed source, we calculate the TF-ISF measure for all query triple patterns. In order to produce early results the join cardinality is important. We employ both methods for join cardinality estimation: using join pattern indexes and sampling from join states obtained during query processing. Less information is available for sources that are not indexed and were only discovered during query processing. No join cardinality estimation is performed for these sources. For all sources, however, the count and type of incoming links are available. In particular, we follow *owl:sameAs* and *rdfs:seeAlso* links as well as links that have a predicate that occurs in a query triple pattern. Links with query predicates receive a higher weight than others as these are more likely to deliver results. Finally, all scores are normalized separately and then combined using a monotonic aggregation function, in this case a weighted summation.

Ranking of sources is not a one-off process but needs to be done continuously during query processing as new sources and more information about already known sources are discovered. However, ranking also represents an overhead, and therefore should be executed only when "necessary". We define several parameters that are used to influence the behavior and cost of the ranking process: (1) *Invalid Score Threshold*: the score of a source is invalid if it has not been calculated before, or if new information about the source is available. A ranking is performed when the number of invalid scores passes a threshold. (2) *Sample Size*: using larger samples for join size estimation will give better estimates, but are also more costly to use. (3) *Resampling Threshold*: results of previous join size estimates are cached for each indexed source. Only when the corresponding hash table maintained by the join operator grows over a given threshold, join size re-estimation is performed using a new sample.

5   Related Work

Seminal work on Linked Data query processing [6,5] and some concrete techniques related to our work have been discussed throughout the paper. Here, we summarize the relation between the proposed corrective ranking and steam-based processing techniques to database work on query optimization and processing in an distributed environment.

Query Optimization. One main problem of query optimization is finding the optimal join order. To do that, it is necessary to estimate their selectivity. Histograms [15] and more complex probabilistic data structures have been suggested to store and estimate selectivity information of RDF triples. In [13], aggregation indexes are used to improve the accuracy of selectivity estimation for joins between triple patterns. As discussed in Section 4.2, we extend these indexes to estimate the cardinality of joins for individual sources (instead of the entire source collection).

Compared to these approaches, [9] does not perform compile-time join ordering, but optimizes the query at run-time by using chain sampling to estimate the selectivity of joins that were not yet performed. In our work, we use sampling combined with triple pattern cardinality indexes to estimate the cardinality of joins given data in a particular source.

Sideways information passing has been employed to complement compile-time optimization with a run-time decision-making technique for reusing intermediate states from one query part to prune and reduce computation of other parts [13,8]. The feedback process between query execution and source ranking employed in our approach for metric refinement can be seen as a case of sideways information passing.

Query Processing in Distributed Environments. In distributed environments data is often stored in remote locations, causing delays in data access.

Much research has been focused on compensating for these delays. Widely used for this are pipelineable query operators that operate on streams. As discussed in Section 3.2, the symmetric hash join is one such operator. Another aspect of stream-based query processing is adaptivity. Query processing techniques have been proposed to adapt the query plan at run-time to deal with changing characteristics of the data. One technique is to switch among query plans at run-time [7]. Other techniques use special operators, such as Eddies [1] and STAIRs [3] that adaptively route incoming tuples through a series of query operators.

Comparison. Our work is the first to provide a systematic overview of Linked Data query processing. The specific techniques proposed extend related work in database research to deal with the specific aspects of Linked Data. In particular, whereas selectivity information has been used for query optimization [13,15,5], it is incorporated in this work into a framework for source ranking, a task that is novel and specific to Linked Data query processing. Likewise, the ideas behind stream-based and adaptive processing [7] and sideways information passing techniques [8] are adopted to address the specific challenges of Linked Data, to enable corrective source ranking on Linked Data streams.

6   Evaluation

In the experiments, we systematically compare the three strategies and examine the impact of various parameters on corrective source ranking. A more extensive description of the evaluation, including the queries, can be found in [11].

Queries and Data. We create a set of eight queries that can all be executed using a discovery-only approach (i.e. results can be discovered by exploring from sources mentioned in the query). These queries use popular datasets from the Linked Open Data Project, such as DBPedia, Geonames, DBLP, Semantic Web Dog Food, data.gov, Freebase and others. Overall, during answering these queries, 6200 sources were retrieved containing 500k triples in total.

Systems. We compare the approaches proposed in [6] for bottom-up evaluation (BU), [5] for top-down (TD), and our implementation of the mixed (MI) strategies. All approaches were implemented on top of the same stream-based query engine. We randomly chose 25% of the sources from the complete index of TD to construct a partial index for MI. Note that these indexes are used for obtaining source descriptions, but the actual data used for query processing comes from a local proxy server. Because local access has lower latency than remote, we applied a configurable delay to the proxy server. For this evaluation this was set to 2s, wherease under real conditions this can be much higher.

Comparison of Strategies. The strategies under investigation vary w.r.t completeness of results. The bottom-up strategy finds only sources and results that can be discovered by following links, the mixed strategy usually finds some more, and the top-down strategy finds all of them. To make the approaches comparable, we restrict the sources to those that can be considered by all strategies, i.e., those discovered by the BU strategy.

Table 1 shows the results for six queries, capturing the times needed to obtain (some percentage of the) results, and the specific times needed for source selection and ranking. The results show that for all queries, the MI and TD approaches report results earlier than BU. The benefit lies in the use of prior knowledge about sources, which helps to retrieve more relevant sources first. Less expected, MI outperformed TD in some cases (Q1,Q3,Q5,Q6,Q7,Q8) in terms of early reporting. The cause lies in the higher source selection times resulting from the use of a larger index. On average the time to retrieve 25% and 50% of the results was 8.7s and 12.8s for MI and 15.1s and 22.0s for BU, respectively. This is an improvement of about 42% in both cases, which may increase with higher, more realistic latencies where the impact of ranking will be higher.

In terms of total execution time, MI and BU are comparable, while TD is significantly faster in most cases. While TD incurs more overhead for the initial source selection because of the larger index, it enables the exclusion of sources. Due to the high network cost, not retrieving irrelevant sources results in a significant performance gain. Using only a partial index, MI is not able to restrict the number of sources that have to be retrieved. This means that in the end MI processes almost the same sources, same data and thus does the same work as BU. The additional overhead incurred by source selection, ranking and sampling lead to execution times worse than BU in some cases (Q1,Q2,Q6). However, MI was able to process more useful sources and results earlier.

To better illustrate the behavior of the different approaches, Fig. 2 shows the arrival of results over time for query Q4. The first result for TD was produced after less than 10s and all results were reported after 33s. The difference to overall execution time of about 90s given in Table 1 is due to the fact that even after the final result was reported other relevant sources had to be processed, but did not contribute to the final result. This indicates that early result reporting resulting in better responsiveness is very important in some cases, where processing all

Table 1. Execution times for six of the evaluation queries. Times in ms.

	BU	MI	TD	BU	MI	TD	BU	MI	TD
	Q1			Q2			Q3		
25% res.	24810.5	10300.0	11038.0	10464.5	10162.0	8096.5	9207.0	7900.0	11166.0
50% res.	43464.5	40782.0	15787.0	13080.5	17974.5	8327.0	10568.0	8048.5	11391.5
Total	84066.5	86895.5	44323.5	21623.5	23273.0	21428.0	22711.0	21944.0	21733.5
Src. sel.	0.0	853.0	1444.5	0.0	805.0	1280.0	0.0	1211.0	1717.0
Ranking	25.5	2404.0	411.0	32.5	358.0	196.5	32.0	575.5	523.0
#Sources	622.0	612.0	154.0	120.0	120.0	67.0	134.0	134.0	67.0
	Q4			Q5			Q6		
25% res.	56800.5	26025.5	10969.5	16837.5	6580.5	4177.0	8222.5	4743.5	5545.0
50% res.	56804.5	26047.0	13605.0	21578.5	11855.5	9186.0	10961.5	7650.5	5634.0
Total	98129.0	98931.0	91352.0	29562.0	30603.5	20074.0	24086.0	20711.0	16469.0
Src. sel.	0.0	270.0	351.0	0.0	203.0	292.0	0.0	1331.0	1863.5
Ranking	31.0	3173.5	1358.5	25.5	283.5	414.5	23.5	292.5	335.0
#Sources	392.0	390.0	342.0	119.0	117.0	70.0	236.0	92.0	49.0

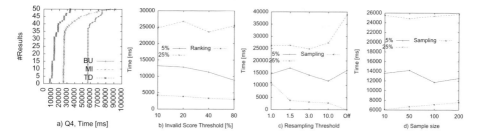

a) Q4, Time [ms]   b) Invalid Score Threshold [%]   c) Resampling Threshold   d) Sample size

Fig. 2. a: Result arrival times for query Q4; b-d: Effects of invalid score threshold, sample size and resampling threshold

sources might be very costly and not needed. Clearly, TD produced results earlier than MI, which was better than BU.

Corrective Source Ranking. In this part we examine the influence of various parameter configurations on sampling and ranking. To separate the effect of each parameter, we vary one while setting the other parameters to default values 40% for invalid score threshold, 3 for resampling and 50 for sample size.

Invalid Score Threshold. Fig. 2b shows average query times for computing 5% and 25% of the results and for sampling at different invalid score thresholds from 10%-80%. With increasing threshold, ranking is performed less often, and correspondingly, times for ranking decreased. The effect of performing ranking less often was positive for computing 5% results, but no clear trend could be observed for 25% results, where the best time was observerd for a threshold of 40%. Ranking is beneficial as query execution is more guided and sources that directly contribute to join results are preferred, especially by using join cardinality estimation with sampling.

Resampling Threshold. Fig 2c shows that times for sampling decrease with higher resampling thresholds, as sampling is performed less often. Times for 5% and 25% results are best for a threshold of 1.5 and 3, respectively. Clearly, sampling is better than no sampling, because the time to reach 25% of results is the highest when sampling is off.

Sample Size. Fig 2d shows that times for sampling increased as the sample grows larger. While sampling creates an overhead, it also provides benefits. Larger sample sizes can lead to more accurate cardinality estimates. Thus, total effect on result computation times varies. While the time for 25% results stayed largely the same, time for 5% results was clearly best for a sample size of 100.

7   Conclusion

We provided a systematic analysis of the challenges and tasks, and discussed concrete strategies for linked data query processing. We proposed an implementation of the mixed strategy that mimics a realistic linked data scenario where some partial knowledge of linked data sources are available. The implementation exploits different types of knowledge available beforehand, and also, incorporates information gained during query processing to perform corrective source ranking. The proposed ranking scheme specifies various types of metrics, which can be combined to reach different optimization goals. A stream-based processing technique is adopted to deal with the unpredictable nature of linked data access. Experiments showed that the proposed implementation leads to early reporting of results and thus, more responsive query processing. On average early results were reported 42% faster than for the bottom-up strategy. In the Linked Data scenario where response times are very high due to the large number of sources and network latency, the capability to produce early results is essential.

As future work, we aim to use information discovered at run-time not only for source ranking but for optimizing the entire evaluation process. In particular, we will target the problem of run-time corrective query optimization to refine the query determined at compile-time.

Acknowledgements. Research reported in this paper was supported by the German Federal Ministry of Education and Research (BMBF) under the CollabCloud project (grant 01IS0937A-E).

References

1. Avnur, R., Hellerstein, J.M.: Eddies: continuously adaptive query processing. SIGMOD Rec. 29(2), 261–272 (2000)
2. Bizer, C., Heath, T., Berners-Lee, T., Heath, T., Hepp, M., Bizer, C.: Linked data - the story so far. International Journal on Semantic Web and Information Systems (IJSWIS) (2009)
3. Deshpande, A., Hellerstein, J.M.: Lifting the burden of history from adaptive query processing. In: Proceedings of the Thirtieth International Conference on Very Large Data Bases, Toronto, Canada, vol. 30, pp. 948–959 (2004)
4. Ge, W., Chen, J., Hu, W., Qu, Y.: Object link structure in the semantic web. In: The Semantic Web: Research and Applications, pp. 257–271 (2010)
5. Harth, A., Hose, K., Karnstedt, M., Polleres, A., Sattler, K., Umbrich, J.: Data summaries for on-demand queries over linked data. In: Proceedings of the 19th International Conference on World Wide Web (2010)
6. Hartig, O., Bizer, C., Freytag, J.: Executing SPARQL queries over the web of linked data. In: Bernstein, A., Karger, D.R., Heath, T., Feigenbaum, L., Maynard, D., Motta, E., Thirunarayan, K. (eds.) ISWC 2009. LNCS, vol. 5823, pp. 293–309. Springer, Heidelberg (2009)
7. Ives, Z.G., Halevy, A.Y., Weld, D.S.: Adapting to source properties in processing data integration queries. In: Proceedings of the 2004 ACM SIGMOD International Conference on Management of Data, Paris, France. ACM, New York (2004)
8. Ives, Z.G., Taylor, N.E.: Sideways information passing for Push-Style query processing. In: Proceedings of the 2008 IEEE 24th International Conference on Data Engineering, pp. 774–783. IEEE Computer Society, Los Alamitos (2008)
9. Kader, R.A., Boncz, P., Manegold, S., van Keulen, M.: ROX: run-time optimization of XQueries. In: Proceedings of the 35th SIGMOD International Conference on Management of Data, Providence, Rhode Island, USA, pp. 615–626. ACM, New York (2009)

10. Klyne, G., Carroll, J.J., McBride, B.: Resource description framework (RDF): concepts and abstract syntax (2004)
11. Ladwig, G., Tran, T.: Linked data query processing strategies – technical report. Technical report (2010), `http://people.aifb.kit.edu/gla/tr/ldqp_report.pdf`
12. Neumann, T., Weikum, G.: RDF-3X: a RISC-style engine for RDF. Proc. VLDB Endow. 1(1), 647–659 (2008)
13. Neumann, T., Weikum, G.: Scalable join processing on very large RDF graphs. In: Proceedings of the 35th SIGMOD International Conference on Management of Data, Providence, Rhode Island, USA, pp. 627–640. ACM, New York (2009)
14. Prud'hommeaux, E., Seaborne, A.: SPARQL Query Language for RDF. W3C Recommendation (2008)
15. Stocker, M., Seaborne, A., Bernstein, A., Kiefer, C., Reynolds, D.: SPARQL basic graph pattern optimization using selectivity estimation. In: Proceeding of the 17th International Conference on World Wide Web, Beijing, China (2008)
16. Wilschut, A.N., Apers, P.M.G.: Dataflow query execution in a parallel main-memory environment. Distributed and Parallel Databases 1(1), 103–128 (1993)

Making Sense of Twitter

David Laniado[1] and Peter Mika[2]

[1] DEI, Politecnico di Milano
Via Ponzio 34/5, 20133 Milan, Italy
david.laniado@elet.polimi.it
[2] Yahoo! Research
Diagonal 177, 08018 Barcelona, Spain
pmika@yahoo.inc.com

Abstract. Twitter enjoys enormous popularity as a micro-blogging service largely due to its simplicity. On the downside, there is little organization to the Twitterverse and making sense of the stream of messages passing through the system has become a significant challenge for everyone involved. As a solution, Twitter users have adopted the convention of adding a hash at the beginning of a word to turn it into a *hashtag*. Hashtags have become the means in Twitter to create threads of conversation and to build communities around particular interests.

In this paper, we take a first look at whether hashtags behave as strong identifiers, and thus whether they could serve as identifiers for the Semantic Web. We introduce some metrics that can help identify hashtags that show the desirable characteristics of strong identifiers. We look at the various ways in which hashtags are used, and show through evaluation that our metrics can be applied to detect hashtags that represent real world entities.

1   Introduction

Twitter, a service for publishing short messages has been growing nearly exponentially in the past two years. Twitter handled over 600 messages every second by January, 2010[1], and has become a cultural phenomenon in many parts of the world. This success can be attributed in a large part to the simplicity of system, and the resulting cleanliness of its web site and its APIs. The ease of publishing also means that Twitter inspires timely contributions and has become an important source of information for late-breaking news, and it is already being exploited by major search engines. While appealing to publishers, the simplicity of Twitter has its downsides for anyone consuming and processing Twitter data, especially when it comes to aggregating messages. Aggregation is a necessary first step for many applications of Twitter mining, including news and trend detection, brand management and customer service, and it is also a crucial first step in separating personal communications from public discussions.

Within the current system, however, the aggregation functions are limited to filtering tweets by users or restricting by keywords. Even in the latter case,

---

[1] http://blog.twitter.com/2010/02/measuring-tweets.html

P.F. Patel-Schneider et al. (Eds.): ISWC 2010, Part I, LNCS 6496, pp. 470–485, 2010.

tweets are organized by time, and not by relevance as is common for search engines. Without formal organization, aggregating tweets that belong to the same conversation or discuss the same topic is daunting. Table 1 shows ten consecutive messages retrieved for the keyword *banana*. These messages are not only posted in different languages, but are part of different ongoing conversations and refer to very different topics (the plant, a chain store, a dance, a club, and others). Keyword search is not only imprecise in aggregation, but is also missing out on a number of messages that do not contain the particular keyword. As Twitter messages are unusually short, keyword search is likely to fail in recall. As an example, during a January, 2010 earthquake in the San Francisco Bay Area, search engines have been criticized in showing only tweets that explicitly mentioned the word *earthquake*. A second, related problem is separating personal communication and news publishing, the two main cases of Twitter usage [12]. This is a crucial function for aggregators that are interested only in the conversations that concern topics of broader interests such as news or current events.

As a community solution to these problems, Twitter users have adopted the convention of adding a hash at the beginning of a word to turn it into a *hashtag*. Hashtags are meant to be identifiers for discussions that revolve around the same topic. By including hashtags in a message, users indicate to which conversations their message is related to. When used appropriately, searching on these hashtags would return messages that belong to the same conversation (even if they don't contain the same keywords), and thereby solving the aggregation problem. Coincidentally, this is the same function that strong identifiers (URIs) play in the Semantic Web. The questions we ask then is which hashtags behave as strong identifiers (if any), and could they be mapped to concept identifiers in the Semantic Web?

In this paper, we propose a set of metrics to measure the extent to which hashtags exhibit the desirable properties of strong identifiers. Our first contribution is thus formalizing the characteristic properties of strong identifiers in terms of usage in social media systems. We give a general description of hashtag usage according to these metrics (Section 2). Using a manually collected data set, we evaluate how well our metrics can identify those hashtags that represent named entities and concepts found in Freebase, a large and broad-coverage knowledge base (Section 3). Our contribution is in measuring the quality of hashtags as identifiers and selecting the hashtags that are candidate concept identifiers, a necessary first step in mapping hashtags to Semantic Web knowledge bases and identifying hashtags that are candidates for extending knowledge bases. We discuss related work in Section 4 and point to future work in Section 5.

2   Metrics for Hashtag Evaluation

There is no special support for tagging in Twitter, and new tags are simply introduced by prefixing a word with the hash sign. Hashtags may be used for personal categorization, but in the vast majority of cases the intention of those who introduce a new hashtag is to evolve it into a symbol that is used by a

Table 1. A consecutive sequence of Twitter message for the query 'banana'

Boo368	@AvenLantz OMG I WANT A BANANA HAMMOCK XD
Endivisual	Got my dress..from banana republic..uhh im wearing dis dress once..? Thx..i dont need it to be so expensive -_-"
DevvonTerrell	World_of_Lala Fuh Sure!!RT @_RosettaStone_: Real talk DevvonTerrell grandmother needs to open up a bakery. Her Banana Pudding is on. HAHA!!
makalovesbieber	RT @bieberhechos: RT si te gusta la banana de Justin (? JAJAJA no mentira.
reidnwrite	@EDHMovement Unforgettable goes SUPER hard...he slipped like banana peels for not having you know you know on the album!
jojoserquina	Chicken Tinola with bitter melon, hot long horn and banana pepper, ginger and spices http://twitgoo.com/14sosn
Vol_Sus	RT @So_Delicious: Hot Fudge-Dipped Frozen Banana Bites wa recipe for Coconut Peanut Butter Hot Fudge Sauce! http://bit.ly/aknbRe YUM!
Markaw00	Eating a banana sandwich and watching Hero.
LauraRogers13	Mom asks me if I want a banana and I start doing the banana dance...I've been at cheer too much!
MissRicaRica	RT @philthyrichFOD: @MissRiCaRiCa *PHILTHY RICH* Coming Home Party And Video Shoot July 4th @ Banana Joes 950 10th St Modesto http://twitpic.com/1oh6ji PLZ RT.

community of users interested in and discussing a particular topic. The goal of such a hashtag is to help search and aggregation of messages related to the same topic, a function that is similar to the role of (shared) URIs in the Semantic Web.

There are a number of desirable criteria that a hashtag should fulfill in this role, similar to how 'cool URIs' are differentiated from poor URIs. In the following, we formalize some of these characteristics.

1. **Frequency.** The hashtag is used by a community of users with some frequency. We measure frequency both in number of users and number of messages sent, and explore the correlations between the two ways of measuring frequency.
2. **Specificity.** The extent to which the usage of a hashtag deviates from the usage of the word without a hash.
3. **Consistency in usage.** The hashtag is used consistently by different users and in different messages to indicate a single topic or concept.
4. **Stability over time.** The hashtag should become a part of the persistent vocabulary of Twitter users, i.e. it should have sustained levels of usage and should have a stable meaning over a period time.

In the following, we formalize these notions based on a Vector Space Model (VSM) for hashtags.

2.1   A Vector Space Model for Hashtags

The basic model of Twitter can be represented by a set of tuples $S \subset M \times U \times \mathcal{P}(H) \times T$ where M is the set of all sequences of not more than 140 characters, U is the set of registered Twitter users, H is the set of hashtags and T is a set of discrete timestamps with a total order. The set of hashtags is the set of possible words that start with a hash. Hashtags form part of the message in the raw data, and we extract them using a regular expression `"#[a-zA-Z0-9_]+"`. The size limitation imposed on messages puts an upper bound on the potential length of hashtags, the number of possible hashtags as well as the number of hashtags that may appear in a single message.

In line with previous works on the analysis of folksonomy systems [5], we capture the semantics of the hashtags by their usage in the social media system. In particular, we will represent the meaning of hashtags using a Vector Space Model (VSM) [20]. VSMs are commonly used in information retrieval as a representation of documents, where each dimension corresponds to a term in the collection and each value measures the weight of that term for the document. In our case, we form virtual documents for each hashtag by considering all messages where the hashtag appears. We don't filter messages by language, but it would be possible to build language specific representations this way.[2]

Formally, each hashtag h_j can be represented by a vector $\mathbf{h_j} = w_{1,j}, w_{2,j}..w_{N,j}$ where $w_{i,j} \in W, N = |W|$ and W is the set of unique terms in all of M. The simplest method for assigning weight is to consider term frequencies, i.e. $w_{i,j}$ is the number of messages in which term i co-occurs with hashtag h_j. In order to account for the different levels of specificity of terms with respect to hashtags, and to reduce the importance of the most common words, we obtain a more accurate model by applying *tf-idf* normalization: $w_{i,j} = tf_{i,j} \cdot idf_i$ where $tf_{i,j} = \frac{w_{i,j}}{\sum_{i=0}^{N} w_{i,j}}$ is the relative frequency of term i with respect to hashtag h_j; $idf_i = \log \frac{|H|}{|\{\mathbf{h_j} : w_{i_j} > 0\}|}$ is inversely proportional to the logarithm of the relative number of hashtags which term i appears with. For reasons of efficiency, we set elements $w_{i,j}$ lower than a threshold k to zero. In particular, this allows efficient indexing of the vectors using inverted indices.

We also introduce a bigram language model for hashtags; to do this, we define as *bigram* each pair of consecutive terms in a message, and as $\mathbf{b_j}$ the vector of all bigrams coocurring with tag h_j, $b_{i,j}$ being the number of messages in which bigram i and tag h_j co-occur. We apply tf-idf normalization in the same way as we compute it for single word co-occurrence.

Finally, we represent hashtags on a social dimension by means of their user occurrence vector $\mathbf{u_j}$, where $u_{i,j}$ is the number of messages tweeted by user u_i and containing hashtag h_j.

---

[2] Based on previous experience, languages can be detected with good accuracy despite the short length of messages. The Twitter Search API also allows restricting tweets by language.

474   D. Laniado and P. Mika

2.2   Frequency of Usage

The **frequency of a hashtag** $h_j \in H$ in terms of the number of users and messages can be defined as

$$F_u(h_j) = |\{u : \exists (m, u, H_i, t) \in S \land h_j \in H_i\}| \tag{1}$$
$$F_m(h_j) = |\{m : \exists (m, u, H_i, t) \in S \land h_j \in H_i\}| \tag{2}$$

where H_i is the set of tags used in message i.

2.3   Specificity

While in most tagging systems tags are added as external metadata to describe the content, in Twitter tags are just words making part of the message, high-lighted by means of a hash to assign them a special function. Often, the hash is added as a form of emphasis (e.g.: "I'm so #happy!"), and the user may not be aware that the word as a hashtag has a more specific or otherwise different meaning than the word itself. A hashtag can often just refer to the meaning of the corresponding word, but in some cases it can assume a very different usage. For example, the hashtag "#milan" seems to be prominently used to refer to the Italian town, while the word "Milan" is much more frequently used in the context of the football team.

It is thus interesting to observe if a hashtag has a meaning close to the one of the corresponding word without hash, that we will call a *non-tag*. As with URIs on the Semantic Web, we assume that hashtags that closely match the meaning of the corresponding non-tag will be used more frequently. On the other hand, we also expect that words that are used mostly as hashtags, or hashtags that are used with a different semantics than their non-tag, will be used more consistently, because they are re-used intentionally.

Similarly to our previous definitions, we define $\mathbf{n_j}$ as the term vector of the non-tag n_j derived from h_j by removing the hash. When building the term vector $\mathbf{n_j}$, we only consider non-tag n_j occurring in a message when the corresponding hashtag h_j is not used inside the same message. The intuition is that when a non-tag appears in a message where the corresponding hashtag has already been used, the semantics of the two are probably the same. We apply tf-idf normalization to non-tags analogously to the one described in Section 2.1 for hashtags.

We compute the **specificity of a hashtag** as the similarity between the vec-torial representation of the hashtag and the corresponding non-tag. For comput-ing similarity, we use the well-known cosine similarity of the two co-occurrence vectors [21].

$$wsim(h_j, n_j) = \frac{\mathbf{h_j} \cdot \mathbf{n_j}}{\|\mathbf{h_j}\| \, \|\mathbf{n_j}\|} \tag{3}$$

Analogously, we define $\mathbf{\bar{u}_j}$ as the model of the users of the non-tag u_j, where $\bar{u}_{i,j}$ is the number of messages in which user i used non-tag n_j. We measure *social specificity* by comparing the model of the users of hashtag h_j to the model of the users of non-tag n_j:

$$usim(h_j, n_j) = \frac{\mathbf{u_j} \cdot \mathbf{\bar{u}_j}}{\|\mathbf{u_j}\| \|\mathbf{\bar{u}_j}\|} \tag{4}$$

To be able to compare tags and non-tags also according to frequency, we define $\bar{F}_u(n_i)$ and $\bar{F}_m(n_i)$ the frequency of a non-tag in terms of users and messages, respectively.

2.4   Consistency of Usage

An important requirement for strong identifiers on the Semantic Web is that they need to be used consistently across documents and users. As a measure of the variety of usage contexts of a hashtag, we study the *entropy* of our vectorial representations of hashtags. Entropy measures the amount of uncertainty associated with the value of a random variable, in other words how uniformly the probabilities are distributed across possible values of the variable.

We define the entropy of a hashtag h_j as:

$$H(h_j) = -\sum_{i=1}^{n} p(w_{i,j}) \log p(w_{i,j}) \tag{5}$$

Higher values of entropy point to more even distributions of probabilities, corresponding to tags being used in a variety of contexts, while lower values of entropy signifies more restricted usage of a tag.

Similarly, we measure entropy of bigrams co-occurring with a tag as

$$Hb(h_j) = -\sum_{i=1}^{n} p(b_{i,j}) \log p(b_{i,j}) \tag{6}$$

Non-tag entropy is measured like tag entropy: $\bar{H}(j) = -\sum_{i=1}^{n} p(\bar{w}_{i,j}) \log p(\bar{w}_{i,j})$.

2.5   Stability over Time

To study the evolution of hashtags on a temporal dimension, we chose to analyze them day by day. First of all, to be able to identify new tags emerging, we define as *new* on day d a tag not appearing in the previous k days. We will define *longevity* of a new tag $l_{d,k}(h_j)$ as the number of days in which tag h_j appears at least once, over the k days after its first occurrence on day d.

We then define $\mathbf{h_j^d}$ the vector of words appearing with tag h_j in some message on day d, and we measure similarity of a hashtag h_j on day d with respect to the previous day as

$$wsim_d(h_j) = \frac{\mathbf{h_j^d} \cdot \mathbf{h_j^{d-1}}}{\left\|\mathbf{h_j^d}\right\| \left\|\mathbf{h_j^{d-1}}\right\|} \tag{7}$$

Analogously, $\mathbf{u_j^d}$ is the vector of users who used tag h_j on day d, and $usim_d(h_j)$ is the similarity among users on day d and $d-1$.

The intuition behind these measures is that a stable tag should endure over time and its meaning should not deviate much from one day to the other.

3   Evaluation

3.1   Dataset

For this study we relied on a dataset of 539,432,680 messages, collected over the whole month of November 2009 (about 18 million per day). Slightly less than 50% of tweets are in English; to filter out messages in non-latin encoding, that we are not able to parse and study, we discarded all messages containing non-ASCII characters, reducing the size of the dataset by about 28%.

Twitter user interfaces allow for forwarding of messages; the original message is so "retweeted" with a special string "rt" at the beginning. As our study is based on the co-occurrence of words inside the same message, and massive retweeting that characterizes several tags might have a strong impact biasing the results, we decided to filter out all retweets. Retweets constitute 5.4% of messages, so the actual dimension of our dataset, after filtering, is of about 369 million messages.

To compute words co-occurring with a hashtag, we filtered out from the messages all Web links and Twitter usernames (words starting with "@"). To reduce the size of co-occurrence vectors, discarding items having a very low tf-idf, we used a threshold $k = 0.01$.

3.2   Descriptive Statistics

Figure 1 shows the distribution of the number of hashtags per message; overall, only 31.5 million messages, corresponding to 8.5%, have at least one hashtag. The percentage of users using at least a hashtag is higher, around 20%. Figure 2 shows that the number of users per tag follows a power low distribution, with some outlier tags used by hundreds of thousands of users. Both the distribution of the number of messages and of distinct tags tweeted by each user also follow a heavy tailed distribution, with a few extremely active users, tweeting up to 10 thousand messages or one thousand distinct tags in a month. The total number of distinct tags encountered is over 2 million; however, only about 93 thousand, corresponding to 4.14%, appeared in more than 20 messages over the whole month: for our study, we considered only these tags, and discarded all the others.

3.3   Evaluating Hashtags

In this Section we will illustrate some results obtained by applying the metrics described in Section 2 to evaluate hashtags contained in our dataset.

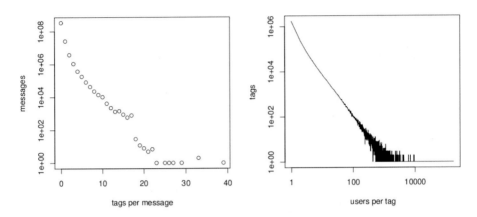

Fig. 1. Representation of the number of messages having a given number of hashtags, on a logarithmic scale

Fig. 2. Distribution of the number of users using a hashtag, on a log-log scale

Frequency of usage. A first interesting question about hashtags is whether the corresponding non-tags also appear; about 73.5% of hashtags have the corresponding non-tag appearing at least once in our dataset. Among these, 57.8% are more frequent as hashtags than as non-tags. A "map" representing the frequency F_m of each hashtag in function of the frequency $\bar{F}_m$ of the corresponding non-tag in shown in Figure 3. The graphic exhibits a *glove* shape, which seems to point out the distinction between two kinds of tags: those corresponding to common words, that appear only sometimes preceded by a hash, and those on the "thumb", Twitter specific tags which are more often used with hash, and do usually not correspond to any commonly used word. Examples of this second kind of tags are #tagtuesday, #iranelection, #sextips and #tcot (acronym for "top conservatives on Twitter"). We obtained a very similar shape for user frequencies F_u and $\bar{F}_u$.

Specificity. Figure 4 shows the similarity between tags and the corresponding non-tags, both in terms of co-occurrence vectors and of users. About a half of tags have null values of *usim*, meaning no user in common with the corresponding non-tag, while *wsim* is null for about one third of tags; while considering this second result, it must be taken into account the fact that we have cut all values of tf-idf below a threshold of 0.01.

Among tags having the highest values of *wsim* we find for example #daylight, almost always used in the context of "daylight savings", #lady, mostly referred to the singer Lady Gaga both as a tag and as a non-tag, and #comofaz, which is a Portuguese slang word for "How do I do?" Among those having null or very low similarity we find tags like #tweetphoto, mainly found in messages generated by an application, and #li, that corresponds to a common word in several languages, like Portuguese, Italian and Chinese, but as a hashtag is mainly used to refer to the social network platform LinkedIn.

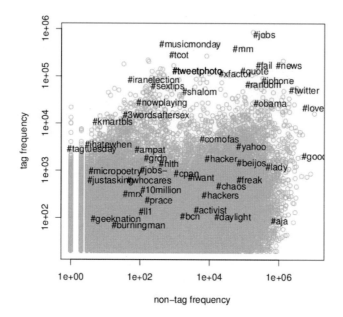

Fig. 3. Frequency of each hashtag in function of the frequency of the corresponding word with no hash

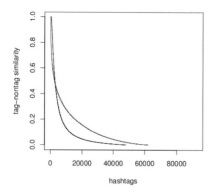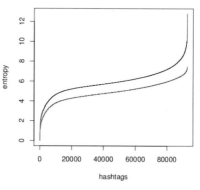

Fig. 4. Similarities *wsim* (red) and *usim* (blue), in descending order

Fig. 5. Entropies H (red) and Hb (blue) of tags, in ascending order

In Figures 6 and 7 similarity *wsim* is plotted in function of tag and non-tag frequency, respectively. Apart from a tendency of very frequent tags to have a lower similarity, no precise relationship can be detected between *wsim* and F_m. On the other hand, high values of similarity seem to be more likely for tags corresponding to words having a frequency in the order of a few thousands, with a peak between 1e+04 and 1e+05.

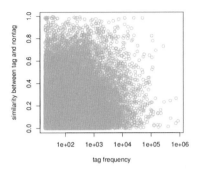
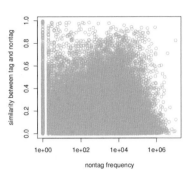

Fig. 6. Similarity between each tag and the corresponding non-tag, in function of tag frequency

Fig. 7. Similarity between each tag and the corresponding non-tag, in function of non-tag frequency

Consistency of usage. In Figure 5 we plotted the entropies of tags, in descending order. Most of the tags have values of H lying in the range between 4 and 6; entropy based on bigram co-occurrence tends to be higher, with values ranging mostly between 5 and 7.

Among tags having very high entropy we find especially tags expressing sentiments, like #whocares, #argh, #_#, beyond some words used in a variety of contexts, like #freak. Tags with a very low entropy are typically generated by applications, like #dongdongdong (a tweeting church), #tweetphoto or #iphonebabes.

3.4   Stability over Time

While until here we have studied tags as static entities for the whole period of observation, in this Section we will illustrate some results based on the observation of tags over different days.

As an example, we report some statistics observed for tags appearing on November 10th, 2009; to identify new tags we used a temporal window of $k = 9$ days. The total number of distinct hashtags observed on November 10th is over 160 thousand, about 50% of which were not appearing in any of the 9 previous days. We looked for these *new tags* in the messages from the 9 following days to evaluate their longevity l. Most of the tags have $l = 0$ and only 36 tags (about 0.045%) appear in all days until November 19th. This is an interesting indicator of how off-handedly users add hashes to words.

In this way, we have selected for each day very few new tags, that are potentially new trending topics; we can now illustrate the results obtained by applying the measures defined in Section 2.5 to two of these tags, to characterize them.

Tag #ampat stands for "American patriot", and seems to have been adopted by a well defined community. Frequency of messages and users (Figure 8) exhibit a slow decreasing trend, after starting with about 300 messages in the first day, tweeted by 50 users; entropy tends to decrease in time (Figure 10) pointing out a convergence towards some context; both the meaning and the community behind the tag seem to be quite stable, though users tend to differentiate a bit in the last observed days (Figure 9).

#kmartbls stands for Kmart's blue light special offers; the extremely high similarity between consecutive days in terms of co-occurrences (Figure 12), together with the very low entropy (Figure 13), is a signal of the scarce variety of information carried by the messages; these data, contrasted with the very high frequency (Figure 11), can easily bring to the conclusion that the tag has been massively promoted by some automatic application, retweeting almost identical messages from different accounts.

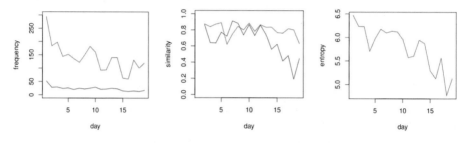

Fig. 8. Frequency F_m (red) and F_u (blue) of tag #ampat by day (November 12th-30th)

Fig. 9. Values of $wsim_d$ (red) and $usim_d$ (blue) of tag #ampat (November 12th-30th)

Fig. 10. Entropy of tag #ampat over days (November 12th-30th)

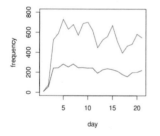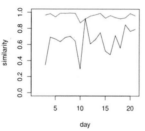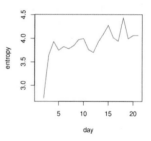

Fig. 11. Frequency F_m (red) and F_u (blue) of tag #kmartbls by day (November 10th-30th)

Fig. 12. Values of $wsim_d$ (red) and $usim_d$ (blue) for tag #kmartbls (November 10th-30th)

Fig. 13. Entropy of tag #kmartbls over days (November 10th-30th)

3.5 Manual Assessment

In order to assess how well our metrics are able to indicate which hashtags represent stable concepts with a unique identity, we have performed a manual evaluation on a random sample of 257 hashtags, relying on 7 evaluators, experts in the field of NLP. For each tag, we collected a random sample of 100 messages with that hashtag, and asked our evaluators to answer the following questions:

1. whether they could guess the meaning of the tag just by looking at it;
2. whether the hashtag represented:
   - an event, person, organization, product, or other named entity;
   - messages generated by an application (e.g. spam);
   - messages with a common sentiment;
   - other;
   - not clear;
3. whether the tag referred to the same meaning in all messages or not.

Furthermore, the evaluators were asked to choose the closest matching concept from Freebase[3], by means of the Freebase Suggest tool[4].

In roughly 39% of cases, the messages were found to refer to a named entity; for 20% of the tags the messages were characterized by a common sentiment (e.g. `#thankfulfor`, `#grrr` or `#youknowyouareuglyif`), while 12% of the times they were recognized as generated automatically by some application (e.g. `#soundcloud`, an audio distribution platform that relies on Twitter to spread notifications about users' activities, or `#shop`, massively used by spammers). In 26% of the cases, the hashtag did not represent a named entity, a sentiment or an application, but was created for some other reason, typically to discuss a general topic (e.g. `#tv`, `#politics`, `#immigration`). The meaning of the tag remained unclear in 6.7% of the cases. Among named entities, organizations were the most common (27%), followed by products, events, persons and other entities (16%, 12%, 6%, 29%).

Slightly more than half of the tags (137) could be associated to a Freebase entry; this is higher than the number of named entities because Freebase contains also some general terms, like domains or common words, which are not named entities. As expected, most application and sentiment tags could not be mapped to Freebase. Only 33% of application and 14% of sentiment tags could be resolved, and many of these mappings are rough approximations of the intended meaning (e.g. the protest tag `#freegary` mapped to `gary_mckinnon`). We have also explicitly measured agreement on this task by reevaluating 31 judgments. 18 out of the 31 tags in this sample could be mapped to Freebase. The inter-annotator agreement on the task of determining if a hashtag can be mapped to Freebase is very high (Cohen's κ of 0.79). The judges agreed on the exact target in 12 out of 18 cases, and 4 of the 6 instances of disagreements were simply due to the same topic appearing in multiple hierarchies within Freebase. One of the other two cases was a close match (`technician` vs `technology` for the tag `#tech`), the other a broader match (`bacon` vs `food` for `#bacon`).

Using the whole set of judgments, we have also performed a logistic regression on the binary variable indicating whether there was a mapping to Freebase for a given hashtag. We have normalized the input variables by a linear transformation to the [0,1] interval, so that we obtain coefficients that are comparable

---

[3] http://freebase.com

[4] http://code.google.com/p/freebase-suggest/

in magnitude. Table 2 shows the coefficients of the resulting model. This model
shows that tag frequency, non-tag frequency, the number of users are negatively
correlated with the success of mapping to Freebase, because these frequency
measures are indicators of Twitter-specific usage. Entropy is also negatively correlated, because the higher the entropy, the less consistently the tag is used. The
number of non-tag users is positively correlated, because it indicates common
words/sentiments. Similarities are also positively correlated, but to a smaller
extent. Altogether our model achieves a 66% accuracy, a relative improvement
of 25% over the baseline of choosing the majority class.

Table 2. Logistic regression coefficients of the input variables reported, for predicting
output variable FBID (i.e., whether a hashtag can be mapped onto a Freebase entry)

Variable	F_m	$\bar{F}_m$	F_u	$\bar{F}_u$	Hb	H	$\bar{H}$	$wsim$	$usim$	Intercept
Coefficient	-2.00	-3.45	-6.80	5.45	3.56	-3.68	0.11	0.78	0.34	-0.01

4   Related Work

After the appearance of the first social bookmarking applications, a considerable
effort has been spent in the study of tag semantics. Work in this field is strongly
related to ours, different in that tagging is explicit and often serves personal
categorization. Classifications of tags based on their usage are proposed in [8]
and [22]; an insight into the use of non subject related tags is offered in [11]. Motivations and incentives behind tagging have been investigated in [16] and [2]. In
[7] some metrics are introduced to evaluate tags, based on user behaviour. The
authors of [1] evaluate the potential of folksonomies to generate semantic metadata; an assessment of delicious tag vocabulary efficiency from an information
theory perspective is provided in [6]. Among the studies aiming at extracting
emergent semantics from folksonomies, the work described in [24] relies on a
metric of tag entropy to evaluate the ambiguity of tags.

   While in our work we could represent hashtags as virtual documents, based
on messages in which they appear, in traditional social tagging applications the
context in which a tag can be analyzed is usually just constitued by other tags
used concurrently; a tripartite model of tags, users and resources is the basis for
most works [17]. In [5] some measures to compute tag relatedness are presented,
and delicious tags are grounded to WordNet synsets in order to contrast semantic
relations with the results of the different metrics proposed; the best semantic
precision was achieved with metrics based on the cosine between each tag's
context, represented as a vector of co-occurring tags. Also the study described
in [4] resonates with our work for the use of information retrieval techniques to
compare tags with each other. In [13] a classification of users according to their
tagging behaviour is leveraged to improve the effectiveness of algorithms for
emergent semantics extraction from folksonomies. The idea of integrating tags
into the Semantic Web is pursued in FLOR [3], a framework for the enrichment

of folksonomies with semantic information from existing ontologies. Models have been proposed to anchor tags to Semantic Web URIs, such as MOAT [19] and CommonTag[5]; NiceTag ontology allows for the representation of different kinds of tagging actions, by means of named graphs [15].

Twitter's social network and information diffusion dynamics have been studied in [10] and [12]; the authors of [14] investigate the use of Twitter during conferences, identifying classes of hashtags and finding out a prevalence of technical terms, and a general tendency to address especially people belonging to the same community. In [9] tagging behaviour in Twitter is compared with the one in delicious, and it is described as *conversational*; the authors in particular study the phenomenon of memes emerging around hashtags that are often abandoned after a short time, and introduce statistical metrics to detect them. A tripartite model of users, hashtags and messages is introduced in [23] to turn Twitter into a folksonomy, and to extract emergent semantics. Special syntaxes have been proposed to allow users express structured information inside a tweet; among these we mention twitlogic[6] and HyperTwitter[7], which allows users specify relationships among hashtags (equivalent, subtag) and express arbitrary properties; an alternative distributed platform for microblogging, based on Semantic Web principles, is described in [18].

5   Conclusions and Future Work

Since their introduction, hashtags have shown to be a popular feature of microblogging platforms as a practical solution to the problem of aggregating content in the disorganized and fragmented stream of information that characterizes these systems. However, not all hashtags are used in the same way, not all of them aggregate messages around a community or a topic, not all of them endure in time, and not all of them have an actual meaning. In this work we have addressed the issue of evaluating Twitter hashtags as strong identifiers, as a first step in order to bridge the gap between Twitter and the Semantic Web.

The first contribution of this paper stands in the formalization of the problem, and in the elaboration of a number of desired properties for a good hashtag to serve as a URI. We have proposed a Vector Space Model for hashtags, representing them as virtual documents; in parallel we have introduced the notion of *non-tag*, to be able to compare each tag with the corresponding word without hash. We have defined several metrics, based both on the messages containing a hashtag and on the community adopting it, to characterize hashtag usage on a variety of dimensions: *frequency*, *specificity*, *consistency*, and *stability* over time. We have applied these metrics to a dataset of more than half a billion messages, collected over the whole month of November 2009. Beyond qualitatively illustrating the results, showing how the metrics proposed tend to correspond to actual properties of the data, we have performed manual classification of a

---

[5] http://commontag.org

[6] http://twitlogic.fortytwo.net/

[7] http://semantictwitter.appspot.com/

sample of tags. Based on these data, we have tested the results obtained with the algorithms described in the paper, showing how a combination of the proposed measures can help in the task of assessing which tags are more likely to represent valuable identifiers. These results are promising, with respect to the perspective of anchoring Twitter hashtags to Semantic Web URIs, and to detect concepts and entities valuable to be treated as new identifiers. Also spam detection tasks can benefit from the metrics we have illustrated.

This work is only a first step in the direction of the investigation of hashtag semantics, and of automatic hashtag classification. Different machine learning algorithms can be used to improve the performances; cleaner results might be obtained by taking into account the different languages of tweets. A more complete analysis may result by considering also links, usernames and emoticons, and by comprising retweet dynamics in the investigation. As a further step, we plan to study similarity between hashtags, based both on word and user co-occurrence vectors, in order to find clusters and study emergent semantics.

References

1. Al-Khalifa, H.S., Davis, H.C.: Exploring the value of folksonomies for creating semantic metadata. International Journal on Semantic Web and Information Systems (2007)
2. Ames, M., Naaman, M.: Why we tag: motivations for annotation in mobile and online media. In: Proc. of CHI (2007)
3. Angeletou, S.: Semantic enrichment of folksonomy tagspaces. In: Sheth, A.P., Staab, S., Dean, M., Paolucci, M., Maynard, D., Finin, T., Thirunarayan, K. (eds.) ISWC 2008. LNCS, vol. 5318, pp. 889–894. Springer, Heidelberg (2008)
4. Benz, D., Grobelnik, M., Hotho, A., Jaschke, R., Mladenic, D., Servedio, V.D.P., Sizov, S., Szomszor, M.: Analyzing tag semantics across collaborative tagging systems. In: Dagstuhl Seminar 08391 - Working Group Summary (2008)
5. Cattuto, C., Benz, D., Hotho, A., Stumme, G.: Semantic grounding of tag relatedness in social bookmarking systems. In: Sheth, A.P., Staab, S., Dean, M., Paolucci, M., Maynard, D., Finin, T., Thirunarayan, K. (eds.) ISWC 2008. LNCS, vol. 5318, pp. 615–631. Springer, Heidelberg (2008)
6. Chi, E.H., Mytkowicz, T.: Understanding the efficiency of social tagging systems using information theory. In: Proc. of HT (2008)
7. Farooq, U., Kannampallil, T.G., Song, Y., Ganoe, C.H., Carroll, J.M., Giles, L.: Evaluating tagging behavior in social bookmarking systems: metrics and design heuristics. In: Proc. of GROUP (2007)
8. Golder, S.A., Huberman, B.A.: Usage patterns of collaborative tagging systems. J. Inf. Sci. (2006)
9. Huang, J., Thornton, K.M., Efthimiadis, E.N.: Conversational tagging in twitter. In: Proc. of HT (2010)
10. Huberman, B.A., Romero, D.M., Wu, F.: Social networks that matter: Twitter under the microscope. First Monday (2009)
11. Kipp, M.E.: @toread and cool: Subjective, affective and associative factors in tagging. In: Proc. of CAIS/ACSI (2008)
12. Kwak, H., Lee, C., Park, H., Moon, S.: What is Twitter, a social network or a news media? In: Proc. of WWW (2010)

13. Krner, C., Benz, D., Strohmaier, M., Hotho, A., Stumme, G.: Stop thinking, start tagging - tag semantics emerge from collaborative verbosity. In: Proc. of WWW (2010)
14. Letierce, J., Passant, A., Breslin, J., Decker, S.: Understanding how twitter is used to widely spread scientific messages. In: Proc. of WebSci. (2010)
15. Limpens, F., Monnin, A., Gandon, F., Laniado, D.: Speech acts meet tagging: NiceTag ontology. In: Proc. of I-SEMANTICS (2010)
16. Marlow, C., Naaman, M., Boyd, D., Davis, M.: Ht06, tagging paper, taxonomy, flickr, academic article, to read. In: Proc. of HT (2006)
17. Mika, P.: Ontologies are us: A unified model of social networks and semantics. In: Gil, Y., Motta, E., Benjamins, V.R., Musen, M.A. (eds.) ISWC 2005. LNCS, vol. 3729, pp. 522–536. Springer, Heidelberg (2005)
18. Passant, A., Hastrup, T., Bojars, U., Breslin, J.: Microblogging: A semantic web and distributed approach. In: Proc. of SFSW (2008)
19. Passant, A., Laublet, P.: Meaning of a tag: A collaborative approach to bridge the gap between tagging and linked data. In: Proc. of LDOW (2008)
20. Raghavan, V.V., Wong, S.K.M.: A critical analysis of vector space model for information retrieval. Journal of the American Society for Information Science (1986)
21. Salton, G.: Automatic Text Processing – The Transformation, Analysis, and Retrieval of Information by Computer. Addison-Wesley, Reading (1989)
22. Sen, S., Lam, S.K., Rashid, A.M., Cosley, D., Frankowski, D., Osterhouse, J., Harper, F.M., Riedl, J.: Tagging, communities, vocabulary, evolution. In: Proc. of CSCW (2006)
23. Wagner, C., Strohmaier, M.: The wisdom in tweetonomies: Acquiring latent conceptual structures from social awareness streams. In: Proc. of SemSearch (2010)
24. Wu, X., Zhang, L., Yu, Y.: Exploring social annotations for the semantic web. In: Proc. of WWW (2006)

Optimize First, Buy Later: Analyzing Metrics to Ramp-Up Very Large Knowledge Bases

Paea LePendu, Natalya F. Noy, Clement Jonquet, Paul R. Alexander,
Nigam H. Shah, and Mark A. Musen

Stanford University, Stanford, California USA
{plependu,noy,jonquet,palexander,nigam,musen}@stanford.edu

Abstract. As knowledge bases move into the landscape of larger ontologies and have terabytes of related data, we must work on optimizing the performance of our tools. We are easily tempted to buy bigger machines or to fill rooms with armies of little ones to address the scalability problem. Yet, careful analysis and evaluation of the characteristics of our data—using metrics—often leads to dramatic improvements in performance. Firstly, are current scalable systems scalable enough? We found that for large or deep ontologies (some as large as 500,000 classes) it is hard to say because benchmarks obscure the load-time costs for materialization. Therefore, to expose those costs, we have synthesized a set of more representative ontologies. Secondly, in designing for scalability, how do we manage knowledge over time? By optimizing for data distribution and ontology evolution, we have reduced the population time, including materialization, for the NCBO Resource Index, a knowledge base of 16.4 billion annotations linking 2.4 million terms from 200 ontologies to 3.5 million data elements, from one week to less than one hour for one of the large datasets on the same machine.

1 Introduction

Researchers are using ontologies extensively to annotate their data, to drive decision-support systems, and to perform natural language processing and information extraction. As a result, we have an abundance of information across many domains making their way into knowledge-based systems. For example, annotation databases that link terms from biomedical ontologies to clinical data reach well into the tens of billions of records and help scientists discover new connections among genes and diseases, or drugs, treatments, and patient outcomes [11,20].

At the same time, ontologies are diverse; they are evolving; and they are getting larger. Many have over 25,000 classes. A few have over 200,000 classes. Some change on a daily basis. As we move into this abundant landscape, we are tempted to meet the computational challenges either by scaling-up and purchasing bigger machines, or by scaling-out and renting armies of little ones from the various compute clouds. Here, we study how a careful analysis and evaluation of the characteristics of our ontologies and data—using metrics—leads to dramatic improvements in performance, without spending on new infrastructure.

We focus on the domain of biomedicine, which has some of the largest, actively used, and actively evolving ontologies today. In our laboratory, as part of the National Center for Biomedical Ontology (NCBO), we have developed BioPortal [16]—the largest

P.F. Patel-Schneider et al. (Eds.): ISWC 2010, Part I, LNCS 6496, pp. 486–501, 2010.

repository of publicly available biomedical ontologies. It currently contains more than 200 ontologies, which comprises over 2.4 million classes.

BioPortal includes the NCBO Resource Index [19], which is a searchable database of semantic annotations for biomedical resources using all BioPortal ontologies. In this context, a biomedical *resource* is a repository of elements that may contain patient records, gene expression data, scholarly articles, and so on. A *data element* is unstructured text describing elements in the resource. An *annotation*—a central component—links an ontology term to a data element, indicating that the element refers to the term. To generate the Resource Index, we use a concept-recognition tool to find ontology terms and their synonyms in data elements, and to store these associations in the index. The Resource Index currently includes 22 different data resources, comprising over 3.5 million data elements resulting in 16.4 billion annotations stored in a 1.5 terabyte MySQL database. We are ramping-up the system to include nearly 100 different data resources, 50 million data elements, and well over 100 billion annotations.

Many large-scale knowledge bases will pay an amortized penalty up-front by materializing inferences (e.g., forward chaining, materializing views, computing transitive closures) so that queries will run much faster, but at what cost? Knowing these tradeoffs and performance limitations helps us make critical decisions on which systems will work best for our needs, or when—and how—to build something entirely new. As one example of a critical problem we encountered for the Resource Index: what happens when ontology evolution outpaces materialization?

Therefore, we approach the scalability problem for a knowledge base of annotations, like the Resource Index, first by examining existing, scalable systems. Our goal is to incorporate a large variety of ontologies as well as a large amount of data. However, we found that benchmarks fall short in illuminating the fundamental tradeoffs between query-time and load-time costs precisely because they do not account for variability among ontologies. The size and depth of an ontology hierarchy significantly affects the cost-curve for materialization. Furthermore, ontologies are not stagnant. Hence, in building our own tool to handle a variety of ontologies and datasets, we found that optimizing primarily for data distribution and ontology evolution significantly improves the performance of the system.

Besides providing metrics for the most comprehensive set of ontologies and annotations available in biomedicine (Section 3), we offer the following contributions:

- we used clustering algorithms on the size and depth of ontologies to identify some characteristic distributions (4.2).
- we synthesized a set of representative ontologies as a new benchmark and demonstrate that these ontologies illuminate previously opaque materialization costs (4.3).
- we analyzed the distribution of annotations together with ontology evolution metrics to determine partitioning schemes that streamline our workflow (5.2).
- we improved the performance time of the Resource Index by several orders of magnitude by applying our analysis toward optimization strategies (5.3).

2   Related Work

Some of the related areas for this study include work on annotation indexes, benchmarking tools, ontology metrics, and the motivating biomedical uses cases.

Annotation indexes. An annotation assigns a tag to some media as a whole, but it also records the context in which the tag applies. Text annotations differ from multimedia annotations mainly in terms of dimensionality. For text strings, the context can be as simple as a position-offset value. For images or video (such as radiology x-rays, EKG data, CT scans, MRI images, or even YouTube videos), additional dimensions, including temporal offsets are required to localize the annotation context [18]. Large-scale annotation systems often resort to scale-out architectures. Annotation databases and ontology-based indexes often exemplify this: GoPubMed [5] indexes all of the PubMed[1] articles using terms from the Gene Ontology; Sindice [6] is another example of a more general ontology-based index; Yahoo! is adopting the Healthline[2] ontology-based search engine. Some of the underlying technologies include databases such as MySQL, indexing tools like Lucene, and frameworks like Hadoop and Solr.[3]

Benchmarking tools. Researchers use the Lehigh University Benchmark (LUBM) [7] to evaluate the scalability of knowledge bases. LUBM provides developers, engineers and architects with methods for quantifying the load-time and query-time for computationally intensive tasks such as computing massive RDF closures [23,26]. Some researchers have indicated that the benchmark can be improved by reflecting real-world workloads based on various dimensions such as reasoning complexity [13], data distribution [24], and even ontology variation [21]. However, to our knowledge, still no benchmark takes the diversity of ontologies into account; therefore, materialization costs remain inadequately characterized.

Ontology metrics. Researchers aiming to characterize various dimensions of large numbers of ontologies in RDF(S), OWL or DAML format have performed several surveys of the Semantic Web landscape [4,14,21,25]. Ontologies vary considerably along many dimensions: by size and shape [4,25], by expressiveness and complexity [13,25], by feature selection [21,25], or by instance density [4].

Biomedical use cases. The NCBO annotations have been used to interpret high-throughput biomedical data (e.g., gene expression) using generalized (ontology neutral) enrichment analysis [22] to discover significant gene–disease relationships that are implicitly embedded in scientific literature. We have also used annotations to discover context-specific mappings among biomedical terminologies, which gives insights into the relationship between, for example, the liver (Minimal Anatomical terminology) and transduction (Gene Ontology) within the context of cancer (Human Disease Ontology).

One of the gaps in current research involves using multiple ontologies at once in large-scale knowledge base systems, such as the Resource Index, for scientific analysis. Not only must we keep pace with the growing abundance of biomedical data, we must also account for the number, variety and evolution of ontologies being used in practice.

---

[1] http://www.ncbi.nlm.nih.gov/pubmed
[2] http://www.healthline.com
[3] http://www.apache.org

3 Data

We have collected metrics for the most comprehensive set of publicly available ontologies and annotations in biomedicine. In this section, we outline what ontologies and annotations we used and how we gathered metrics on them. Our collected data represents a snapshot taken in May, 2010.

Ontologies. The BioPortal repository [16] stores biomedical ontologies developed in various formats—OWL, OBO, RRF, Lexgrid-XML and Protégé Frames—and ranging in subject matter from representation of anatomy and phenotype to diseases. Researchers in biomedicine actively contribute their ontologies to BioPortal. Users can submit new versions of their ontologies; visualize, browse and search them; make comments on and get notifications about ontology changes; or create mappings between terms from different ontologies.

BioPortal makes all data accessible programmatically via RESTful Web services. Almost all of the BioPortal ontologies, including earlier versions, can be downloaded.[4] We used these services to collect the ontologies used in our study. Of the 200 ontologies, we incorporated 145. We skipped the remainder due to limitations on what was available for parsing at the time.

Annotations. The Resource Index workflow, as illustrated by Figure 1, is composed of two main steps: First, *direct annotations* are created from the text metadata of a resource element using an off-the-shelf concept recognition tool, which in our case is MGREP [3]. Second, we use subclass relations to traverse ontology hierarchies to create new, *expanded annotations*.[5]

The ontology terms play a vital role because of the subsumption hierarchy. Users who search for a general term like "cancer" will find results for documents that have been annotated with, say, "melanocytic neoplasm" because it is defined as a kind of cancer in the NCI Thesaurus, one of the ontologies in BioPortal.

The Resource Index currently has 22 resources indexed. We use the following sample of 4 resources in this study because they are representative in terms of size, type of content, frequency of updates, and quantity of data per element:

Biositemaps (BSM) represent a mechanism that researchers in biomedicine use to publish and retrieve metadata about biomedical resources (1.5K elements).

ArrayExpress (AE) is a public repository of microarray data and gene-indexed expression profiles from a curated subset of experiments (10K elements).

ClinicalTrials.gov (CT) provides regularly updated information about federally and privately supported clinical trials (89K elements).

GRANTS combines three different funding databases: Research Crossroads, CRISP, and the Explorer of the NIH Reporter (1,400K elements).

---

[4] http://www.bioontology.org/wiki/index.php/
BioPortal_REST_services#Download_an_ontology_file

[5] This workflow is also available as a web service called the NCBO Annotator [10], which provides researchers with an easy mechanism to employ ontology-based annotation using BioPortal ontologies in their respective pipelines.

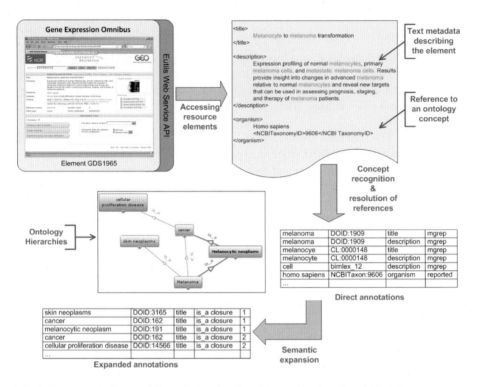

Fig. 1. Resource Index workflow: Using the Gene Expression Omnibus (GEO) data resource, we illustrate how direct annotations are associated with a data element, then we expand those annotations to account for the ontology hierarchy

Metrics. BioPortal maintains a limited set of ontology metrics, such as the number of classes or siblings (Table 1), which we gathered using REST services. For every ontology in either OWL or OBO format, we also used the ontology download service and used the OWL API Metrics tool [8] to complement those statistics. We followed imports during all calculations and consider only the asserted hierarchy.

For annotation metrics, we directly download the statistical data on the number and kinds of annotations per resource and per ontology kept by the Resource Index. Table 1 lists the specific metrics we used in this study. All the metrics are available online.[6]

Table 1. Ontology and annotation metrics that we used

Ontology Metrics	Annotation Metrics
Number of classes	Number of data elements
Number of versions	Number of direct annotations
Maximum depth of the class hierarchy	Number of expanded annotations
Maximum and average number of siblings	
Average number of adds, deletes and changes per version	

---

[6] http://www.bioontology.org/wiki/index.php/Metrics_Study

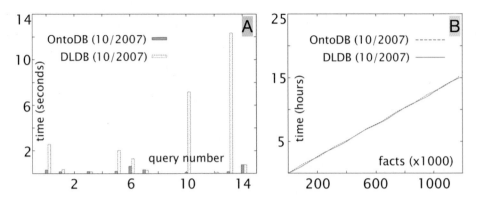

Fig. 2. Materialized versus non-materialized KBs: OntoDB outperforms DLDB on LUBM query-time performance (A). Yet, OntoDB *appears* to add no additional cost for load-time performance (B). (Note: we reported these figures previously [12]).

4   Are Existing Systems Scalable Enough?

The set of ontologies in BioPortal, which we use to generate the Resource Index, is extremely diverse, both in terms of size and depth of the class hierarchy. In order to analyze and improve the query performance of the Resource Index, we must first analyze the effects of the ontology characteristics on both load-time and query-time performance. We started by using popular ontology benchmarks to compare the performance of different approaches. However, as we show in this section, these benchmarks do not account for the diversity in size and depth of ontologies. We then discuss the complementary set of benchmarks that we synthesized based on the ontologies in our repository.

4.1   A Tale of Two KBs: Is Materialization Really Free?

Because materializing inferences is a large part of high-performance KBs, we would expect the obvious tradeoff: systems that perform materialization should obtain faster query time at the cost of slower load time. However, upon comparing two dichotomous KB systems, we were baffled to find that materialization seems to cost nothing at all.

DLDB [17] is a knowledge base system developed at Lehigh University that uses database views to assist with query answering on large sets of data. OntoDB [12] was developed at the University of Oregon using other database features. These two systems are very similar: they take an ontology and a set of instances as input; they create and load a relational database schema based on predicates from the ontology; they store the instances in database tables; and they use intrinsic database features to maintain the knowledge model and answer queries. OntoDB differs from DLDB by materializing inferences at load time—using triggers—rather than by unfolding views at query time.

Comparing OntoDB with DLDB using LUBM, we clearly see the expected gains in query-response time, as illustrated by Figure 2–A: materializing the inferences yields marked gains because of the pre-computation. However, suspiciously, the gains come at

Table 2. EM clustering: Cluster 3 characterizes 38 percent of ontologies

Cluster	Num. Classes	Max. Depth	Max. Siblings	Avg. Siblings
0 (8%)	19628	13.7	249.9	39.5
	(+/- 29189)	(+/- 5.6)	(+/- 118.6)	(+/- 13.4)
1 (26%)	1264	10.7	68.8	5.1
	(+/- 791)	(+/- 4.7)	(+/- 35.5)	(+/- 6.0)
2 (3%)	13338	37.0	2252.9	10.4
	(+/- 10483)	(+/- 5.3)	(+/- 2335.9)	(+/- 10.0)
3 (38%)	**181**	**7.5**	**13.6**	**3.4**
	(+/- 170)	**(+/- 3.4)**	**(+/- 8.4)**	**(+/- 2.6)**
4 (21%)	34401	18.8	385.1	1.1
	(+/- 66037)	(+/- 11.0)	(+/- 380.9)	(+/- 0.5)
5 (2%)	344095	29.7	9939	1
	(+/- 158541)	(+/- 10.3)	(+/- 100.4)	(+/- 18.0)
6 (3%)	45303	21.2	2226.3	76.8
	(+/- 20784)	(+/- 15.1)	(+/- 790.2)	(+/- 49.8)

Table 3. K-means clustering: The characteristics of small, medium and large ontologies

Cluster	Num. Classes	Max. Depth	Max. Siblings	Avg. Siblings
Small (80%)	4925	9.6	110.6	3.8
Medium (8%)	26062	14.7	654.7	55.8
Large (12%)	96502	33.2	2571.4	8.6

no apparent cost as shown in Figure 2–B: the slope of the overlapping lines indicates the same *constant cost* per assertion for both systems. Would larger or deeper ontologies demonstrate the expected load-time cost for materialization?

4.2   Is the LUBM Ontology Too Small and Shallow?

To determine whether larger and deeper ontologies would expose the expected load-time costs, we need ontologies that vary in size and depth. Motivated by results that are practical and relevant to biomedicine, we analyzed the 145 ontologies from BioPortal by running clustering algorithms on the ontology metrics that we gathered for them (Section 3, Table 1). We specifically considered the number of classes, maximum depth and maximum and average number of siblings. We used two clustering algorithms: the Expectation-Maximization clustering algorithm (EM) with an unspecified number of clusters, which uses an iterative mechanism to find an optimal clustering distribution (Table 2); and the simple K-means clustering algorithm with 3 clusters, identifying small, medium, and large ontologies (Table 3).

We found that the most representative ontology has 181 classes and depth 8 (Table 2). In addition, the three clusters with the largest number of ontologies (3, 1, 4) cover 85% of the ontologies. The characteristics in the K-means results are highly skewed, with 80% of ontologies falling into the "small" category. The previous EM results (cluster 3, 38%) suggest that there are a significant number of clustered ontologies having smaller characteristics—so, we might consider introducing a smaller division than K-means suggests. Furthermore, the medium and large categories can be collapsed into cluster 4.

Reasoning in this way, we combined the results of EM clusters 3, 1 and 4 with the K-means clusters to extrapolate a division between small, medium and large ontologies:

Table 4. Ontologies: nine synthetic ontologies representing biomedicine

	Ontology Parameters		Schema Load Time		Instance Load Time	
	Size	Depth	Mean		Mean	
Small	78	5	6.95 ms	7.57 ms	3.65 ms	4.78 ms
	81	10	7.12 ms		4.99 ms	
	72	20	8.64 ms		5.69 ms	
Medium	1623	5	8.21 ms	8.61 ms	6.51 ms	10.55 ms
	1555	10	8.93 ms		9.45 ms	
	1827	20	8.68 ms		15.69 ms	
Large	19992	5	9.57 ms	9.82 ms	11.84 ms	24.93 ms
	22588	10	9.59 ms		22.80 ms	
	19578	20	10.28 ms		40.14 ms	

100, 2,000 and 25,000 classes. Similarly, we defined the range of shallow, mid-range (mid) and deep ontologies as having depths of 5, 10, and 20. We use these results to generate a set of ontologies that account for the variety of depths and sizes.

4.3   Accounting for Size and Depth: New Benchmark Ontologies

The clustering results give us the characteristics of nine possible ontologies varying along the size and depth dimensions: small–medium–large and shallow–mid–deep. Our goal is to determine whether this benchmark produces enough variability to compare load times for use cases that rely on different ontologies.

We developed OntoGenerator, a tool that generates a synthetic ontology given the following parameters: a seed, the maximum number of classes, maximum number of siblings (i.e., span), density, and number of individuals. The density parameter introduces a degree of randomization in the fullness of the tree structure. It denotes the probability that the maximum number of siblings or the maximum depth will be reached along any path to a leaf node. We use the seed value to prime the randomizing function which allows us to reproduce the same ontology given the same parameters, or, conversely, to construct a new ontology (of the same kind) by using a different seed. Finally, the tool creates the given number of individuals as instances of randomly chosen classes in the ontology, distributing data uniformly at various depths in the hierarchy.

We used OntoGenerator to synthesize the nine different ontologies whose specific size and depth are outlined in Table 4. We also had OntoGenerator create one million individuals for each of the ontologies. The data is available online.[7]

4.4   Conclusion: Materialization Costs Depend on the Size and Depth

Our hypothesis that materialization costs will be exposed for larger and deeper ontologies was confirmed: the results demonstrate that cost is *not constant* per assertion, but it depends on ontology size and depth. We obtained the load-time results in Table 4 by populating OntoDB (the KB that explicitly materializes inferences) with the nine ontologies we generated, i.e., we loaded the ontology and the data. We measured load time in two phases: the time to transform the ontology into a schema and load it into

---

[7] http://www.bioontology.org/wiki/index.php/Metrics_Study

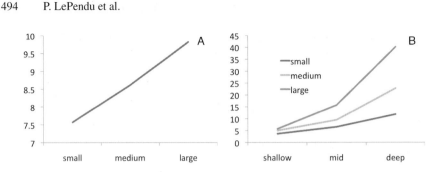

Fig. 3. New benchmark performance: For OntoDB, the average time in milliseconds to load a single assertion for small, medium, and large ontologies increases with larger-sized ontologies (A). Load time also increases with depth (B).

the database (averaged per class), and the average time for loading a single instance assertion (taken 1,000 at a time).

The positive slope of the lines displayed in Figure 3 show a clear cost dependency. Interestingly, size and depth have both super-linear and interactive effects on the cost. The crooked slope of each line indicates that size (and depth) independently yields a super-linear effect on load time. Furthermore, size and depth have a cumulative effect: the larger the ontology, the larger the role of depth (note increasing crookedness).

In conclusion, we should include ontologies of varying size and depth in KB benchmarking suites. We have proposed a set of ontologies that can be used to enhance the LUBM for characterizing materialization costs. This study clearly points out that, depending on their ontologies, system designers should worry about materialization.

5   Managing Large-Scale Annotation Databases

Based on the results above (cf. Table 4), neither OntoDB nor DLDB could handle data on the scale of the Resource Index: they would take several hours just to create the schema (let alone process any data!) for the NCI Thesaurus, which has 74,646 classes. Furthermore, the Resource Index uses not just one ontology, but over 200 of them (over 2.4 million classes) for annotation purposes. Finally, not only must we abate the costs of materializing inferences for large-scale KBs, but we must also consider how to manage that knowledge over time for various, evolving ontologies.

Whereas ontology size and depth affect materialization costs in the stagnant scenarios described above, we demonstrate below that data distribution and ontology evolution significantly also affect how we manage a very large KB over time. By optimizing for these metrics, we have streamlined the Resource Index population workflow by several orders of magnitude: from taking over one week for loading to less than one hour for one of our larger datasets on the same machine.

5.1   The NCBO Resource Index

Annotation databases such as the NCBO Resource Index take the structure of an ontology into account to provide enhanced search and retrieval functionality for documents.

Table 5. Annotations: A sample of four biomedical resources from the Resource Index shows the number of elements, the number of direct annotations, and the number of expanded annotations.

Resource	No. Elements	No. Direct Annotations	No. Expanded Annotations	Avg. I-Density
BSM	1.5 K	0.5 M	3 M	1.9
AE	11 K	13 M	115 M	68.6
CT	89 K	181 M	1,300 M	434.4
GRANTS	1,400 K	1,900 M	13,800 M	7621.6

As mentioned previously, a direct annotation "tags" a data element with a class from an ontology. If a document is annotated with a class from an ontology, then we infer that it is also annotated by the superclasses of that class (Figure 1).

In terms of load time versus query time, the tradeoff has to do with materializing those superclass annotations. We can perform the inference at query time by unwinding the hierarchical structure and issuing a union of sub-queries, one for each subclass (recursively), to retrieve annotations. However, unfolding queries is probably not a viable option because users expect split-second response times and the number of subclasses for a given class can reach into the thousands. The query-time results in Figure 2 testify to the slowness of query unfolding, even for a small ontology. But, as our benchmarking study further illustrates (Figure 3), we can expect that for very large and deep ontologies (e.g., the NCBI Organismal Classification Ontology, which has 513,248 classes) a user could potentially wait for several minutes to get answers on a very simple query as it unfolds into tens of thousands of sub-queries.

The other option is to materialize the expanded annotations by forward propagating (i.e., copying) them up the class hierarchy when they are created. As a result, no unfolding occurs during query time—we can directly look up the answers for each class quickly by using an index. However, materializing inferred annotations results in very large expanded annotation databases (Table 5). In general, the ratio of annotations to elements is 1000:1 and the ratio of expanded to direct annotations is 8:1.

Prior to optimization, the Resource Index population workflow took approximately one week to generate and materialize annotations for the CT resource (the largest processable resource), which has 89,000 data elements. We extrapolated that it would take several months to process a repository of a million documents (e.g., the GRANTS resource), which puts repositories serving nearly 20 million documents (e.g., PubMed) completely out of reach. Therefore, our goal is to reduce the load time by optimizing the workflow so that it will scale to handle these large data resources—and to keep up with them as each one grows.

5.2 Optimizing for Data Distribution and Ontology Evolution

Since our goal is to manage very large sets of annotations and expanded annotations, one obvious choice is to partition the data in a way that supports a streamlined and efficient execution. Ideally, a good partitioning mechanism will also distribute easily over multiple nodes (if necessary). In order to determine how to partition the data, we look at metrics on data distribution and ontology evolution:

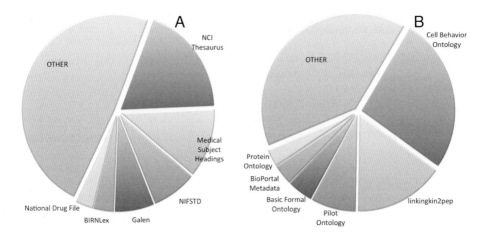

Fig. 4. Annotations per ontology: The number of annotations per ontology obviously preferences some of the larger ontologies like NCIT, MeSH and Galen (A). However, I-density (B) shows that small but generic ontologies have higher percentages of their terms used for annotation.

Data Distribution. Instance density tells us how the data is distributed. If we treat annotations as a kind of data instance, the instance density (I-density) of an ontology measures the number of instances for each class in an ontology [4]. As Table 5 shows, larger resources have larger I-densities because there is simply more data to be distributed. This indicates that resource size is an important consideration. By the same token, larger ontologies entail more annotations (Figure 4–A) because more terms are available for annotation matches, so ontology size is another factor. Interestingly, accounting for ontology size, the actual I-density of a resource per ontology preferences very small but very general ontologies such as the Cell Behavior Ontology (6 terms), Basic Formal Ontology (an upper ontology), and BioPortal Metadata Ontology (Figure 4–B) because there is a higher chance that a high percentage of terms are used in those ontologies.

Ontology Evolution. Ontologies evolve by growing in size and changing in structure. We used the structural differences (diffs) that BioPortal provides for 15 of its ontologies to understand their evolution. The diffs record the additions, deletions and changes to classes in the ontologies for consecutive versions. Figure 5 shows that users modify ontologies (change or delete terms) more often than they add new terms. This implies that materialized inferences must be updated regularly as well. Moreover, the number of versions per ontology in BioPortal indicates that the frequency of updates depends on the ontology in question: new versions of ontologies accumulate frequently for a few but rarely for most, following a power law distribution (Figure 6).

The metrics on data distribution and ontology evolution help to decide how to partition the database so that we can streamline the workflow. Our goal is not only to materialize the expanded annotations efficiently, but also to manage them as the ontologies

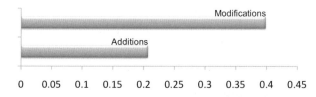

Fig. 5. Version differences: Based on a sample of 15 ontologies, normalized by ontology-size, the percentage of modifications (updates and deletes) are double the percentage of additions

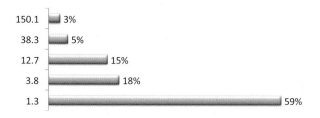

Fig. 6. Ontology versions: Ontologies loaded in BioPortal evolve according to a power law distribution: most ontologies (59%) are rarely modified (1.3 versions on average) but some (3%) are revised weekly (150 versions). The Gene Ontology changes the most (daily, 212 versions).

evolve. In developing an optimization strategy, we need to determine which partitioning criteria are most important, in what order.

Partitioning the Database. First of all, we can see from Table 5 and Figure 4 that data is naturally distributed along resource size and ontology size by examining instance density metrics—therefore, those attributes make good partitioning candidates. We confirmed this by using information theory on the annotation metrics to determine feature selection: we ran a basic information gain algorithm (ID3). However, we still need more information to decide which should be the primary partitioning criteria: resource size or ontology size.

Hence, we looked at ontology evolution metrics next. Evolution influences how we keep the database up-to-date. Whereas data is always added, metrics confirm that ontologies are mostly (but infrequently) updated. From these evolution metrics, we concluded that resources should be the primary partitioning criterion because data additions will far outweigh ontology updates in importance.

Therefore, we first partition the database by resource, which means that each set of annotations for a resource will be kept in its own database file. Next, we partition again by ontology. Partitioning first by resource allows us to process data elements in a pipelined fashion: it enables adding new data elements over time extremely quickly by merely appending to the end of the file (minimizing disk seeks). Furthermore, adding a new resource is as simple as adding a new partition: it has no detrimental effect on existing annotations. Finally, as we show in the next section, this partitioning helps with expanding annotations very rapidly.

Sub-partitioning by ontology allows us to drop annotations efficiently for only the ontology that has changed: it has no effect on annotations associated with other ontologies. Moreover, since sub-partitioning subdivides the files into smaller chunks, we also gain speed in reading, indexing and updating annotations by ontology.

Speeding-up Expanded Annotations. Expanding annotations along the subsumption hierarchy is—by far—the most computationally intensive, storage expensive and time consuming phase of the population workflow. Partitioning by resource makes it possible to compute the expansions extremely efficiently. Again, the goal is to take a direct annotation using a class and to use the ontology hierarchy to materialize (i.e., expand) the annotations to superclasses of that class (cf. Figure 1). There are 2.4 million classes in BioPortal ontologies, which, if we compute the transitive closure on subclass assertions, result in 20.4 million subclass relationships to consider during expansion. However, the real challenge comes when we have to cross-reference these 20.4M subclass relationships against 2 billion direct-annotation records (e.g., the GRANTS data set, Table 5).

Stored in a database, we essentially have two tables to join: [term, superclass] and [term, data-element]. The former contains the subclass relationships and the latter contains the direct annotations. The efficient way to compute this join—called a merge-join—is to sort both tables first by the join condition (term) and then scan them both sequentially, outputting the join result in a linear fashion in the size of each table. However, sorting 2 billion records on disk is unrealistic: it would take days if not weeks; moreover, data is constantly being added, so maintaining sorted order would slow down inserts. Joining them using just indexes causes thrashing and unnecessary seeking and re-reading of records (billions of times over!). The database is unable to automatically optimize the join because it lacks the appropriate metrics.

The solution relies on two important facts: (1) the data grows rapidly but is partitioned by resource, and (2) the ontology subclass relations will not grow quickly beyond the 20.4 million (or, say, a 100 million) mark any time soon. Therefore, we stream annotations by resource into the [term, data-element] table as rapidly as the disk can write them (which is many thousands of records a second). We also force the [term, superclass] table into main memory using the MySQL MEMORY engine. (Forcing the annotations into main memory is not an option because they are simply too large.) By having the entire hierarchy in main memory, we can achieve the optimal performance of the simple merge-join query without sorting any records. The result of applying these key decisions yields remarkable performance gains.

Table 6. Optimized workflow: We achieve highly scalable performance using partitioning and merge-join optimization based on density and evolution metrics

Resource	No. Elements	Old Population Time	Optimized Population Time
BSM	1.5 K	<< 1 day	0.4 min
AE	11 K	~ 1 day	3 min
CT	89 K	~ 7 days	49 min
GRANTS	1,400 K	N/A	492 min

5.3   Conclusion: Memory-Based Merge-Join with Partitioning Performs Well

Table 6 shows that we have achieved extraordinary performance gains—on a single machine. In the worst case, we could repopulate the entire database (all resources and all ontologies) in less than a day. Extrapolating these results, we anticipate that indexing PubMed (20M articles) with all BioPortal ontologies will take only 5 days, as opposed to nearly a year based on previous estimates.

In conclusion, by analyzing how data is distributed, how it grows, and how it evolves, we can determine how to partition the data, how to streamline population speed, what to keep in main memory, and how to update data without slowing down population.

6   Discussion

We would like to see previous results on large-scale knowledge base systems [17,23,26] re-evaluated using the nine new benchmark ontologies we created. For example, we believe that anomalies described when evaluating the BibTeX benchmark [24]—a phenomenon similar to that in our "tale of two KBs"—would be explained by using a variety of ontologies, rather than merely a variety of data. Furthermore, we believe that differences in materialization heuristics will be easier to comprehend for systems that compute massive RDF closures [23,26].

We would like to see metrics used more often for query optimization on Semantic Web queries [1]. As our study illustrates, database optimization techniques should be reused. Just as query optimizers use catalogs, which are metrics kept on the data distribution in each database table, so can metrics improve knowledge bases. For example, we can use ontology metrics for improving query federation over SPARQL endpoints: we can easily optimize a query intersecting two ontologies by taking the smaller one first. The same goes for annotations. What makes optimization in this area different from database optimization, is that ontologies are just like data—they are not treated as higher class citizens the way schemas are treated in databases.

As future work, we plan to develop better metrics on resource evolution: how frequently and how drastically do resources change. Likewise, we are working on computing the differences between all versions of all ontologies. What will be most interesting is to analyze those differences to come up with a viable change-propagation model that could increase the performance of our systems even further. Considering that only fractions of an ontology change during a revision, this approach should save a considerable amount time updating annotations.

Our future research directions include using metrics for SPARQL query optimization. Also, we can analyze differences between ontology revisions to determine an incremental, change-propagation model for updating materialized inferences as ontologies evolve. Finally, we will consider providing annotations as Linked Open Data in a scalable way.

7   Conclusion

We collected metrics for the most comprehensive ontologies available in biomedicine. By analyzing those ontology metrics, we have created a set of representative ontologies that improve upon existing benchmarks. These new ontologies help to illuminate previously opaque load time costs. By using these new evaluation tools, researchers can improve upon KB systems that use materialization strategies.

We followed up on our evaluation by studying additional metrics on instance density and ontology evolution to reduce load time for the NCBO Resource Index as much as possible. As a result of our analyses, we improved performance time by several orders of magnitude, without investing in new infrastructure. In the worst case, the current database of 16.4 billion annotations can be re-materialized overnight. More importantly, now we can annotate resources that were previously impossible, such as PubMed.

Acknowledgments. We thank Matthew Horridge for assisting with the OWL API Metrics tool. Benchmarking work was performed at the University of Oregon under the direction of Professor Dejing Dou, supported in part by grant R01 EB007684 from the National Institutes of Health. This work was also supported largely in part by the National Center for Biomedical Ontology, under roadmap-initiative grant U54 HG004028 from the National Institutes of Health.

References

1. Bernstein, A., Kiefer, C., Stocker, M.: OptARQ: A SPARQL Optimization Approach based on Triple Pattern Selectivity Estimation. Technical report, University of Zurich (2007)
2. Bodenreider, O., Smith, B., Kumar, A., Burgun, A.: Investigating Subsumption in SNOMED CT: An Exploration into Large Description Logic-Based Biomedical Ontologies. Artificial Intelligence in Medicine 39(3), 183–195 (2007)
3. Dai, M., Shah, N., Xuan, W.: An Efficient Solution for Mapping Free Text to Ontology Terms. In: AMIA Summit on Translational Bioinformatics (2008)
4. d'Aquin, M., Baldassarre, C., Gridinoc, L., Angeletou, S., Sabou, M., Motta, E.: Characterizing Knowledge on the Semantic Web with Watson. In: Evaluation of Ontology-based Tools Wkshp (ISWC), pp. 1–10 (2007)
5. Doms, A., Schroeder, M.: GoPubMed: exploring PubMed with the Gene Ontology. Nucleic Acids Research 33(Web Server issue), 783–786 (2005)
6. Tummarello, G., Oren, E., Delbru, R.: Sindice.com: Weaving the Open Linked Data. In: Int'l Sem. Web Conf. (ISWC), pp. 547–560 (2007)
7. Guo, Y., Pan, Z., Heflin, J.: LUBM: A Benchmark for OWL Knowledge Base Systems. Journal of Web Semantics 3(2), 158–182 (2005)
8. Horridge, M., Bechhofer, S.: The OWL API: A Java API for Working with OWL 2 Ontologies. In: OWL Experiences and Directions Wkshp, OWLED (2009)
9. Jonquet, C., Musen, M., Shah, N.: Building a Biomedical Ontology Recommender Web Service. Biomedical Semantics 1(S1) (2010)
10. Jonquet, C., Shah, N., Youn, C., Callendar, C., Storey, M., Musen, M.: NCBO Annotator: Semantic Annotation of Biomedical Data. In: Int'l Sem. Web Conf., ISWC (2009)
11. Krallinger, M., Leitner, F., Valencia, A.: Analysis of biological processes and diseases using text mining approaches. In: Methods in Molecular Biology, pp. 341–382 (2010)
12. LePendu, P., Dou, D.: Using Ontology Databases for Scalable Query Answering, Inconsistency Detection, and Data Integration. Intelligent Info. Sys., JIIS (2010)
13. Ma, L., Yang, Y., Qiu, Z., Xie, G., Pan, Y., Liu, S.: Towards a Complete OWL Ontology Benchmark. In: Sure, Y., Domingue, J. (eds.) ESWC 2006. LNCS, vol. 4011, pp. 125–139. Springer, Heidelberg (2006)

14. Magkanaraki, A., Sofia, A., Christophides, V., Plexousakis, D.: Benchmarking RDF Schemas for the Semantic Web. In: Horrocks, I., Hendler, J. (eds.) ISWC 2002. LNCS, vol. 2342, pp. 132–146. Springer, Heidelberg (2002)

15. Marwah, K., Katzin, D., Alterovitz, G.: Context-Specific Ontology Integration. In: (under review) Pacific Symp. on Biocomputing, PSB (2011)

16. Noy, N., Shah, N., Whetzel, P., Dai, B., Dorf, M., Griffith, N., Jonquet, C., Rubin, D., Storey, M.A., Chute, C., Musen, M.: BioPortal: Ontologies and Integrated Data Resources at the Click of a Mouse. Nucleic Acids Research 1(37) (2009)

17. Pan, Z., Heflin, J.: DLDB: Extending Relational Databases to Support Semantic Web Queries. In: Practical and Scalable Sem. Web Systems Wkshp (ISWC), pp. 109–113 (2003)

18. Rubin, D., Supekar, K., Mongkolwat, P., Kleper, V., Channin, D.: Annotation and Image Markup: Accessing and Interoperating with the Semantic Content in Medical Imaging. IEEE Intelligent Systems 24(1), 57–65 (2009)

19. Shah, N., Jonquet, C., Chiang, A., Butte, A., Chen, R., Musen, M.: Ontology-driven Indexing of Public Datasets for Translational Bioinformatics. BMC Bioinformatics 10 (2009)

20. Stenson, P., Ball, E., Howells, K., Phillips, A., Mort, M., Cooper, D.: The Human Gene Mutation Database: providing a comprehensive central mutation database for molecular diagnostics and personalized genomics. Hum. Genomics 4(2), 69–72 (2009)

21. Tempich, C., Volz, R.: Towards a benchmark for Semantic Web reasoners - an analysis of the DAML ontology library. In: Evaluation of Ontology-based Tools Wkshp, ISWC (2003)

22. Tirrell, R., Evani, U., Berman, A., Mooney, S., Musen, M., Shah, N.: An Ontology-Neutral Framework for Enrichment Analysis. In: AMIA Annual Symposium (2010)

23. Urbani, J., Kotoulas, S., Oren, E., Harmelen, F.: Scalable Distributed Reasoning using MapReduce. In: Bernstein, A., Karger, D.R., Heath, T., Feigenbaum, L., Maynard, D., Motta, E., Thirunarayan, K. (eds.) ISWC 2009. LNCS, vol. 5823, pp. 634–649. Springer, Heidelberg (2009)

24. Wang, S., Guo, Y., Qasem, A., Heflin, J.: Rapid Benchmarking for Semantic Web Knowledge Base Systems. In: Gil, Y., Motta, E., Benjamins, V.R., Musen, M.A. (eds.) ISWC 2005. LNCS, vol. 3729, pp. 758–772. Springer, Heidelberg (2005)

25. Wang, T., Parsia, B., Hendler, J.: A Survey of the Web Ontology Landscape. In: Int'l Sem. Web Conf. (ISWC), pp. 682–694 (2009)

26. Weaver, J., Hendler, J.: Parallel Materialization of the Finite RDFS Closure for Hundreds of Millions of Triples. In: Bernstein, A., Karger, D.R., Heath, T., Feigenbaum, L., Maynard, D., Motta, E., Thirunarayan, K. (eds.) ISWC 2009. LNCS, vol. 5823, pp. 682–697. Springer, Heidelberg (2009)

Using Reformulation Trees to Optimize Queries over Distributed Heterogeneous Sources

Yingjie Li and Jeff Heflin

Department of Computer Science and Engineering, Lehigh University
19 Memorial Dr. West, Bethlehem, PA 18015, U.S.A.
{yil308,heflin}@cse.lehigh.edu

Abstract. In order to effectively and quickly answer queries in environments with distributed RDF/OWL, we present a query optimization algorithm to identify the potentially relevant Semantic Web data sources using structural query features and a term index. This algorithm is based on the observation that the join selectivity of a pair of query triple patterns is often higher than the overall selectivity of these two patterns treated independently. Given a rule goal tree that expresses the reformulation of a conjunctive query, our algorithm uses a bottom-up approach to estimate the selectivity of each node. It then prioritizes loading of selective nodes and uses the information from these sources to further constrain other nodes. Finally, we use an OWL reasoner to answer queries over the selected sources and their corresponding ontologies. We have evaluated our system using both a synthetic data set and a subset of the real-world Billion Triple Challenge data.

Keywords: information integration, query optimization, query reformulation, source selectivity.

1 Introduction

In the Semantic Web, the definitions of resources and the relationship between resources are described by ontologies. The resources in the Web are independently generated and distributed in many locations. Under such an environment, there is often the need to integrate the ontologies and their data sources and access them without regard to the heterogeneity and the dispersion of the ontologies. Although recent research has led to the development of knowledge bases (KBs) and/or triple stores to support this need, such systems have many disadvantages. First, centralized knowledge bases will become stale unless they are frequently reloaded with fresh data; this can be especially expensive if the knowledge-bases rely on forward-chaining. Second, they can require significant disk space, especially for triple stores that use multiple triple indices to optimize queries. For example, Hexastore [16] replicates each triple six times. Finally, there may be legal or policy issues that prevent one from copying data or storing it in a centralized place. For this reason, we believe it is important to investigate algorithms that allow data to reside in its original location, and that use summary indexes to

P.F. Patel-Schneider et al. (Eds.): ISWC 2010, Part I, LNCS 6496, pp. 502–517, 2010.

determine which locations contain data relevant to a particular query. In particular, we have proposed an inverted index-based mechanism that indicates which documents mention certain URIs and/or literal strings [5]. This simple mechanism is clearly space efficient, but also surprisingly selective for many queries. However, because this index only indicates if URIs or literal strings are present in a document, specific answers to a subgoal of the given query cannot be calculated until the sources are physically accessed - an expensive operation given disk/network latency. To further complicate matters, ontology heterogeneity can lead to query answers being expressed in ontologies different from those used to express the query. To solve these issues, we further proposed a flat-structure query optimization algorithm that selects and processes sources given a set of conjunctive query rewritings [4]. For each rewriting, this algorithm employs a source selection strategy that prioritizes selective subgoals of the query and uses the sources that are relevant to these subgoals to provide constraints that could make other subgoals more selective. However, there are two key problems with this algorithm: 1) the number of rewrites can be exponential in the size of the query, especially when there are complex ontology axioms and 2) the selectivity of the algorithm is inhibited by its reliance on local information.

To solve the above issues, we present a novel algorithm for optimizing the selection of sources in ontology-based information integration systems. Like our prior work, this approach relies on an inverted index; however, the main contributions of this paper are:

- We present a tree-structure algorithm that performs optimizations that consider the structure of a reformulation tree. Using this tree and the term index, it estimates the most selective subgoals, incrementally loads the relevant sources, and uses the data from the sources to further constrain related subgoals.
- We demonstrate that this new algorithm outperforms the algorithms proposed in [4] and [5] on both a synthetic data set with 20 ontologies having significant heterogeneity and a real world data set with 73,889,151 triples distributed in 21,008,285 documents.

The remainder of the paper is organized as follows: Section 2 reviews related work. In Section 3, we describe the tree-structure source selection algorithm for ontology-based information integration. Section 4 describes our experiments and in Section 5, we conclude and discuss future work.

2   Related Work

Currently, there are mainly three areas of work related with our paper: database query optimization, RDF query optimization and query answering over distributed ontologies.

Query optimization has been extensively studied by traditional database researchers since the classic work by Selinger et al. [9]. Variations of these ideas are still common practice in relational optimizers: use statistics about the database instance to estimate the cost of a query plan; consider plans with binary joins in which the inner relation is a base relation (left-deep plans); and postpone Cartesian product after joins with predicate. Following this, a number of optimization techniques for databases systems were proposed. The representatives include join-ordering strategies, and techniques that combine a bottom-up evaluation with top-down propagation of query variable bindings in the spirit of the Magic-Sets algorithm [8]. Join-ordering strategies may be heuristic-based or cost-based; some cost-based approaches depend on the estimation of the join selectivity; others rely on the fan-out of a literal [12]. All of these database query optimization techniques are designed for situations where data of different database relations are stored in the same file. However, in the Semantic Web, it is very common that data from the same relation is spread among many files. If the available indices do not completely specify the triples contained in a document, then high latency makes determining the extensions of the relations in these files very expensive. In such situations, query plans need to be developed incrementally.

In RDF query optimization, RDF data can be serialized and stored in a database and a SPARQL query can be executed as an SQL join, hence recently a lot of database join query optimization techniques such as creating indexes have been applied to improve the performance of SPARQL queries. In recent years, many researchers have proposed ways of optimizing SPARQL join queries. MonetDB [11] exploits the fact that RDF data typically has many fewer predicates than triples, thereby vertically partitioning the data for each unique predicate and sorting each predicate table on subject, object order. RDF-3X [6] and Hexastore [16] attempt to achieve scalability by replicating each triple six times (SPO, SOP, PSO, POS, OPS, OSP): one for each sorting order of subject, predicate and object. It has been demonstrated that this strategy results in good response time for conjunctive queries. The major disadvantages of both of these approaches are that they rely on centralized knowledge bases and that the indexes (or replication) are quite expensive in terms of space. YARS2 [3] is another native RDF store and query answering system where index structures and query processing algorithms are designed from scratch and optimized for RDF processing. The novelty of the approach proposed by YARS2 lies in the use of multiple indexes to cover different access patterns. However, in this way, if more efficient query processing can be achieved, more disk space will be needed. GRIN [15] is a novel index developed specially for RDF graph-matching queries and focuses on path-like queries that cannot be expressed using existing SPARQL syntax. This index identifies selected central vertices and the distance of other nodes from these vertices. However, it is still not clear how GRIN could be adapted for a distributed context.

In query answering over distributed ontologies, T. Tran et al. [14] proposed Hermes, which translates a keyword query provided by the user into a federated query and then decomposes this into separate SPARQL queries that are issued to web data sources. A number of indexes are used, including a keyword index, mapping index, and structure index. The most significant drawback to the approach is that it does not account for rich schema heterogeneity (mappings are basically of the subclass/equivalent class variety). Stuckenschmidt et al. [13] proposed a global data summary for locating data matching query answers in different sources and optimizing the query. However, this method does not consider the heterogeneity of schemas of the distributed ontologies.

Most of the research on query answering over distributed schemas or ontologies are based on the P2P architecture. Piazza [2] proposes a language (based on XQuery/XPath) to describe the semantic mapping between two different ontologies. In this work, a peer reformulates a query by using the semantic mapping and forwards the reformulated query to another peer related by the semantic mapping. DRAGO [10] focuses on a distributed reasoning based on the P2P-like architecture. In DRAGO, every peer maintains a set of ontologies and the semantic mapping between its local ontologies and remote ontologies located in other peers. The semantic mapping supported in DRAGO is only the subsumption relationship between two atomic concepts and ABox reasoning is not supported. KAONP2P [1] also suggests the P2P-like architecture for query answering over distributed ontologies. KAONP2P supports more extended semantic mapping which describes the correspondence between views of two different ontologies, where each view is represented by a conjunctive query. To support federated query answering, it generates a virtual ontology including a target ontology to which the query is issued and the semantic mapping between the target and the other ontologies. Then, the query evaluation is performed against the virtual ontology. However, all of these P2P systems have a drawback in that each node must install system specific P2P software, presenting a barrier to adoption.

3   Query Optimization

In this section, we present some preliminary definitions regarding the distributed environment for our algorithm, the inverted term index, and the algorithms of our prior work. We then describe the novel tree-structure algorithm in detail.

3.1   Preliminaries

In the Semantic Web, there exist many ontologies, which can contain classes, properties and individuals. We assume that the assertions about the ontologies are spread across many data sources, and that mapping ontologies have been defined to align the classes and properties of the domain ontologies. For convenience of analysis, we separate ontologies (i.e. the class/property definitions and axioms that relate them) and data sources (assertions of class membership or property values). Formally, we treat an ontology as a set of axioms and a data source as a set of RDF triples. A collection of ontologies and data sources constitute what we call a *semantic web space*:

Definition 1. *(Semantic Web Space) A Semantic Web Space SWS is a tuple* $\langle D, o, s \rangle$, *where D refers to the set of document identifiers, o refers to an ontology function that maps D to a set of ontologies and s refers to a source function that maps D to a set of data sources.*

We have chosen to focus on *conjunctive queries*, which provide the logical foundation of many query languages (SQL, SPARQL, Datalog, etc.). A conjunctive query has the form $\langle \overline{X} \rangle \leftarrow B_1 \left(\overline{X}_1 \right) \wedge \ldots \wedge B_n \left(\overline{X}_n \right)$ where each variable appearing in $\langle \overline{X} \rangle$ is called a distinguished variable and each $B_i(\overline{X}_i)$ is a query triple pattern (QTP) $\langle s_i, p_i, o_i \rangle$, where s_i is a URI or variable, p_i is a predicate URI, and o_i is a literal, URI, or variable. Given a Semantic Web Space SWS, the answer set $ans(SWS, \alpha)$ for a conjunctive query α is the set of all substitutions θ for all distinguished variables in α such that: SWS $\models \alpha\theta^1$. In this definition, the entailment relation $\models$ is defined in the usual way, albeit with respect to the conjunction of every ontology and data source in the Semantic Web Space.

Our problem of interest is given a Semantic Web Space, how do we *efficiently* answer a conjunctive query? Recall, we are assuming that we do not have a local repository for the full content of data sources and due to network latency, we need to minimize the number of sources that we will load to ascertain their actual content. Therefore we need to prune sources that are clearly irrelevant and focus on those that might contain useful information for answering the query. Here, we consider a system architecture where an **Indexer** is periodically run to create an index for all of the data sources and to collect the axioms from domain and mapping ontologies. Given a conjunctive query, the **Reformulator** uses the domain and mapping ontologies to produce a set of query rewritings. A **Selector** takes these rewritings and uses the index to identify which sources are potentially relevant to the query (note, since the index is an abstraction, we cannot be certain that a source is relevant until we load it). Then the **Loader** reads the selected sources together with their corresponding ontologies and inputs them into a sound and complete OWL **Reasoner**, which is then queried to produce results. Since the selected sources are loaded in their entirety into a reasoner, any inferences due to a combination of these selected sources will also be computed by the reasoner.

In our prior work [5], we showed that a term index could be an efficient mechanism for locating the documents relevant to queries over distributed and heterogeneous semantic web resources. Basically, the term index is an inverted index, where each term is either a full URI (taken from the subject, predicate or object of a triple) or a string literal value. Formally, for a given document d, the terms contained in d can be expressed as following:

$terms(d) \equiv \{x | \langle s, p, o \rangle \in d \wedge [x \equiv s \vee x \equiv p \vee (o \in U \wedge x \equiv o) \vee (o \in L \wedge x \in \text{lit}-\text{terms}(o))]\}$,

where $\langle s, p, o \rangle$ stands for a triple contained in document d, U is the set of URIs, L is the set of Literals and *lit-terms*() is a function that extracts terms from literals, and may involve typical IR techniques such as stemming and stopwords. The term index can then be defined as follows:

Definition 2. *(Term Index) Given a Semantic Web Space $\langle D, o, s \rangle$, the term index is a function $I : T \rightarrow \mathcal{P}(D)$, where $T = \bigcup_{d \in D} terms(s(d))$.*

---

¹ $\alpha\theta$ is a shorthand for applying θ to the body of α, i.e., $B_1\theta \wedge B_2\theta \ldots \wedge B_n\theta$.

Using the term index we can define two functions that together determine how to select potentially relevant sources using the term index. Note that the sources for a QTP are basically those sources that contain each constant (URI or literal term) in the QTP.

Definition 3. *(Term Evaluation) Given the set of possible query triple patterns Q and a set of constant terms T (that appear as subjects, predicates or objects of any $q \in Q$), the term evaluation function qterms: $Q \to \mathcal{P}(T)$ maps QTPs to the (non-variable) terms that appear in them.*

Definition 4. *(Source Evaluation) Given the set of possible query triple patterns Q and a set of document identifiers D, the source evaluation function is qsources: $Q \to \mathcal{P}(D)$. Given a QTP q and a term index I, qsources$(q) = \bigcap_{c \in qterms(q)} I(c)$.*

Subsequently, we proposed a *flat-structure* query optimization algorithm [4] where the **Selector** took a set conjunctive query rewrites as input and then locally optimized each of them. Since loading sources is the primary bottleneck of this type of system, we focused on optimizing the *source selectivity* – the total number of sources loaded. We define the source selectivity of a selection procedure *sproc* for a query α as the number of sources not selected divided by the total number of sources available:

$$Sel_{sproc}(\alpha) = \frac{|D| - |sproc(\alpha)|}{|D|} \tag{1}$$

The flat-structure algorithm is based on the simple observation that the join selectivity of a pair of QTPs is often higher than the overall selectivity of these two QTPs treated independently. Consider two QTPs q_1 and q_2 from the same conjunctive query that share a variable x, in database parlance this situation is called a *join condition* and x is the *join variable*. We note that the number of sources required to answer the query are often less than $qsources(q_1) \cup qsources(q_2)$. If we load the sources for q_1 first, we can find a set rs of variable bindings for the QTP from the triples contained in the sources. We can then apply each substitution $\theta \in rs$ to q_2 to generate a set of queries and get a set of sources for q_2 by doing index lookups for each: $\bigcup_{\theta \in rs} qsources(q_2\theta)$. It should be clear that by adding an additional constant to each QTP, this join approach often has a higher source selectivity than naively applying *qsources* to each QTP in the query, although note that the *join selectivity* depends on which QTP is processed first. The flat-structure algorithm iteratively loads the most selective QTP and uses its substitutions to calculate the join selectivity for all remaining QTPs.[2] The two main problems with this algorithm are:

---

[2] Here, we assume that all data sources are relatively small; the presence of very large data sources may lead to an issue where a QTP that has high source selectivity actually has low answer selectivity. Such problems could be addressed by keeping additional size statistics in the index.

- In order to avoid complications with inference impacting the number of sources for each QTP, it repeats the source selection procedure for each possible query rewrite. However, when there is significant heterogeneity in the ontologies, synonymous ontology expressions can lead to an explosion in the number of query rewrites. Processing a large number of rewrites can slow the system down, even if we cache the results of index lookups and are careful not to load the same source multiple times.
- The inability to use the full structure of query rewrites reduces the possible source selectivity of the query process. Since source selection is independently executed for each query rewriting, selectivity is based only on local information, and does not account for the possibility that a subgoal that initially appears selective actually is not selective once all of its rewrites are taken into consideration.

3.2 Tree Structure Query Optimization Algorithm

To address the issues discussed in the previous section, we propose to replace the **Selector** component of our previous architecture with a *tree-structure* query optimization algorithm that takes a rule-goal tree expressing the query reformulation as its input and performs a greedy, bottom-up analysis of which sources to load. A rule-goal tree is basically an AND-OR graph, where goal nodes are labeled with QTPs (or their rewritings), and rule nodes are labeled with ontology axioms [2]. The purpose of the rule-goal tree is to encapsulate all possible ways the required information could be represented in the sources. See Figure 1 for an example of a rule goal tree for query with three QTPs. In this example, a property composition axiom $r0$ from a mapping ontology has been used to rewrite *swat:makerAffiliation* as the conjunction of *swrc:affiliation* and *foaf:maker*. Qasem et al. [7] have shown how to produce such rule-goal trees when all ontologies are expressed in OWLII, a subset of OWL DL that is slightly more expressive than Description Horn Logic.

We begin with an example to provide the intuition for our algorithm, and then discuss its details subsequently. Consider the rule-goal tree in Fig. 1 for the query Q, which asks for the publications affiliated with Lehigh University (*"lehigh-univ"*), complete with the ids and names of their authors. In the diagram, each goal node has three associated costs: the *initial-cost* is the number of sources relevant to that goal if we do not consider any axioms, the *local-optimal-cost* is the number of relevant sources after applying available constant constraints and the *total-cost* is the number of sources after applying available constant constraints and collecting sources from the descendants. Additionally, the order in which we process goal nodes is indicated by the parenthesized numbers. The first step is to use the term index to initialize the tree with source selectivity information, represented by initial-costs next to each goal node. We start with the QTP leaf node that selects the fewest sources: $< ?m, akt:has-affiliation, "lehigh-univ">$ (labeled with (1)). Since this is an OR node, we simply propagate its sources up to its parent goal. Thus, the total-cost for $< ?m, swrc:affiliation, "lehigh-univ">$ is updated to 60 (40 sources from its child plus 20 sources of itself; for simplicity of

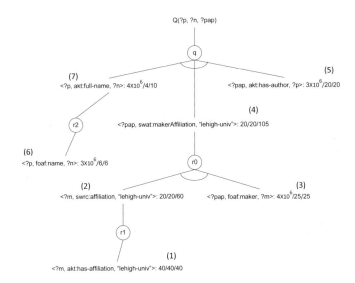

Fig. 1. Query resolution of one sample query with notations in form of initial-cost/local-optimal-cost/total-cost

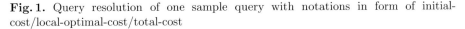

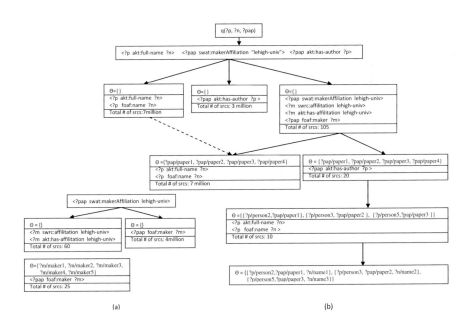

Fig. 2. AND-optimization. At each level of the tree a QTP is chosen greedily, its sources loaded and queried, and the answers applied to sibling QTPs.

exposition we are assuming that the sets of sources are disjoint, but this is not a requirement for the algorithm). Since all children of $<?m, swrc:affiliation, ``lehigh-univ">$ have been processed, it joins the leaf nodes as a candidate for processing, and since it's total cost is 60, which is less than the initial costs of all other candidates, it is the next node to be processed. Since it is a child of $r0$, an AND rule node (indicated by the arc), we can use it to constrain its sibling *foaf:maker* as shown in Fig. 2(a). First, we load all sources associated with the goal node and issue the goal as a query for these sources. This query results in the substitutions for $?m$: $\{?m/maker1, ?m/maker2, ...\}$. Each of these substitutions is then applied to $<?pap, foaf:maker, ?m>$, an index lookup is performed for each resulting QTP, and the total set of sources (in this case 25 of them) is used to update the total cost of this node in Figure 1, step (3). In step (4), the total cost of these nodes (60+25=85) is propagated to their parent *swat:makerAffiliation*, and is added to its initial cost (20), resulting in a total cost of 105. Since this node now has the best selectivity and is the child of an AND rule node (the original query), we need to perform another AND optimization as show in Fig. 2(b). As shown, once this node is selected, there are two siblings to choose from. However, before we can determine the cost of these nodes, we must repeat the tree process on the subtrees rooted at these nodes, thus the number of sources for $<?p, akt:full-name, ?n>$ is 7 million, the sum of its sources and the sources of its child $<?p, foaf:name, ?n>$. We apply the substitutions from *swat:makerAffiliation* to each sibling, resulting in the number of sources of *akt:has-author* being reduced to 20 (updating its local-optimal-cost in Fig. 1), but not changing the sources of *akt:full-name*. In step (5) of Fig. 1, we select *akt:has-author*, load its sources, issue a combined query with the previous goal, and get a new set of substitutions. These substitutions are then applied to the subtree of *akt:full-name*, changing the local-optimal-costs of *foaf:name* and *akt:full-name* to 6 and 4, respectively, and changing the total-cost of *akt:full-name* to 10. As a result, the total number of collected sources for the given conjunctive query is $105 + 20 + 10 = 135$, compared to over 11 million if no optimization was done. Once all sources are loaded, we can ask the original query of the reasoner in order to get a final set of substitutions.

The pseudo code for our algorithm is shown in Figure 3. Algorithm 1 processes a rule-goal tree, where the parameter rs, which provides a set of substitutions, is $\emptyset$ when first called, but instantiated in recursive calls. We use *frontier* to maintain a set of deepest, unprocessed goal nodes in the rule-goal tree; this is initialized to be the set of leaf nodes. In Lines 2-4, we use the term index to determine the initial selectivity of all goal nodes in the rule-goal tree. Then, the most selective node n is chosen from the *frontier* (Line 6). We check if n is a child of an AND rule, and if so Algorithm 2 is called to collect sources by using the greedy strategy (Lines 7-8). If the rule is an OR mapping, the sources from the rule children are directly broadcast upward to the rule parent goal node p (Lines 9-10). Since this completes the processing of n, we remove it from our *frontier* node set (Line 11) and if p currently has no descendants in *frontier*,

Algorithm 1 Source selection for structure-based query optimization
function getSourceList(*rtree*, *rs*) **returns** a list of sources
inputs: *rtree*, a given rule goal tree *rtree*
rs, a list of substitutions
1: Let *frontier* = leaf nodes of *rtree*,
srcs[] = array of sets of sources, indexed by goal nodes
2: **for each** goal node *n* in *rtree* **do**
3:     **for each** $\theta \in rs$ **do**
4:         *srcs[n]* = qsources($n\theta$)
5: **do**
6:     Let $n = min_{node \in frontier}$ ($|srcs[node]|$), p = getParent(*n*)
7:     **if** *n* is a child of an AND rule node *r* **then**
8:         *srcs[p]* = *srcs[p]* ∪ OptimizeANDNode(*n*,
siblings of *n*, *srcs*)
9:     **else**
10:         *srcs[p]* = *srcs[p]* ∪ *srcs[n]*
11:     remove *n* from *frontier*
12:     **if** *p* has no descendants on *frontier* **then**
13:         add *p* to *frontier*
14: **while** $(frontier \neq \{rtree.root\})$
15: **return** *srcs[rtree.root]*

Algorithm 2 Node optimization
function OptimizeANDNode(*on*, *sibs*, *srcs*) **returns** a list of sources
inputs: *on*, a given goal node in the rule-goal tree
sibs, a set of *on*'s sibling nodes
srcs, an array of sets of sources, indexed by goal nodes
1: Let *allsrcs* = ∅, *query* = true
2: *allsrcs* = *allsrcs* ∪ *srcs[on]*
3: load(*srcs[on]*, *KB*)
4: **do**
5:     Let *query* = *query* ∧ *on*
6:     Let *rs* = askReasoner (*KB*, *query*)
7:     **for each** *qtp* ∈ *sibs* **do**
8:         *srcs[qtp]* = getSourceList(subtree rooted at *qtp*, *rs*)
9:     Let $on = min_{t \in sibs\ that\ join\ with\ query}$ ($srcs[t]$)
10:     Remove *on* from *sibs*
11:     *allsrcs* = *allsrcs* ∪ *srcs[on]*
12:     load(*srcs[on]*, *KB*)
13: **while** $(sibs \neq \emptyset)$
14: **return** *allsrcs*

Fig. 3. Pseudo code of tree-structure source selection algorithm

we add p to the *frontier* (Lines 12-13). When the *frontier* contains only the root of the given rule-goal tree, the *while* loop terminates and our source collection ends (Line 14). Finally, all collected sources are returned (Line 15).

In Algorithm 2 we optimize an AND node, given a most selective goal node *on*, its siblings *sibs*, and an array of the sources for each node in the tree (the latter is used as an output parameter to update the log of sources found for each node). We start by loading *on*'s sources into the knowledge base KB. Then, we evaluate *on* by asking the reasoner to get the substitutions of the variables contained in *on* (Lines 5-6). These substitutions are then applied to *on*'s siblings to enhance their individual selectivity (Lines 7-8). Note the recursive call to *getSourceList*() in line 8; this ensures that any new constraints specified by rs are effectively applied to the subtree rooted at each sibling. Based on the new selectivity estimations, we choose the next most selective node that shares a join variable with the partial query to be the next *on* (Line 9). Then we remove *on* from *sibs*, add its sources to the sources retrieved so far, and load any newly selected sources (Lines 10-12). In the next iteration, *on* is conjuncted with the partial query *query*, the reasoner is queried, and the substitutions applied again to the siblings. This process is repeated until all sibling nodes of the initial given goal node are processed (Line 13). Finally, the sources collected by the current AND mapping rule are returned (Line 14). As an aside, the flat-structure algorithm essentially executes a variation of Algorithm 2 for every conjunctive query rewrite.

4   Evaluation

To evaluate our query optimization algorithm, we have conducted two experiments based on a synthetic data set and a real world data set respectively. The first experiment compares our tree-structure source selection algorithm to our previous non-structure [5] and flat-structure [4] source selection algorithms using a synthetic dataset with significant ontology heterogeneity. The second experiment tests the scalability and practicality of our algorithm using a subset of the real world Billion Triple Challenge (BTC) data set. For both experiments, we use a graph-based synthetic query generator to produce a set of queries that are guaranteed to have at least one answer each. These queries range from one to thirteen triples, have at most nine variables each, and each QTP of each query satisfies the join condition with at least one sibling QTP. All of our experiments are done on a workstation with a Xeon 2.93G CPU and 6G memory running UNIX. Our **Indexer** component is implemented using Lucene while our **Reasoner** is KAON2.

4.1   Heterogeneity Evaluation Using a Synthetic Data Set

Our first experiment compares the tree-structure algorithm to the non-structure and the flat-structure algorithms using a synthetic data generator that is designed to approximate realistic conditions. First, we ensure that each generated file is a connected graph, which is typical of most real-world RDF files. Based on a random sample of 200 semantic web documents, we set the average number of triples in a generated document to be 50. In order to achieve a very heterogeneous environment, we conducted experiments with 20 ontologies, 8000 data source sources, and a diameter of 6, meaning that the longest sequence of mapping ontologies between any two domain ontologies was six. In this configuration, the average number of sources committing to each ontology is 400. This configuration resulted in an index size of 75.3MB, which was built in 21.6 seconds. We issued 240 random queries, grouped by the number of unconstrained QTPs (from 0 to 10), where an unconstrained QTP is one with variables for both its subject and object or with an *rdf:type* predicate paired with a variable subject. For each group, we computed the average query response time, average number of selected sources and average number of index accesses. Due to the exponential increase in query response time, we only executed queries with up to 5 unconstrained QTPs for both the non-structure and flat-structure algorithms.

   Fig. 4(a) shows how each algorithm's average query response time is affected by increasing the number of unconstrained QTPs. From this result, we can see that the tree-structure algorithm and flat-structure algorithm are faster than the non-structure algorithm. The reason is that unconstrained QTPs are typically the least selective; thus, the more unconstrained QTPs there are, the more opportunities there are for the two optimization algorithms to use constraints to enhance the selectivity of goals. However, the benefits of the tree-structure algorithm become really noticeable for 5 or more unconstrained QTPs; in this situation the flat-structure algorithm begins to reveal exponential behavior while the tree-structure algorithm remains linear. This is because complex mapping ontologies can lead to a number of conjunctive query rewrites that is exponential in the size of the query.

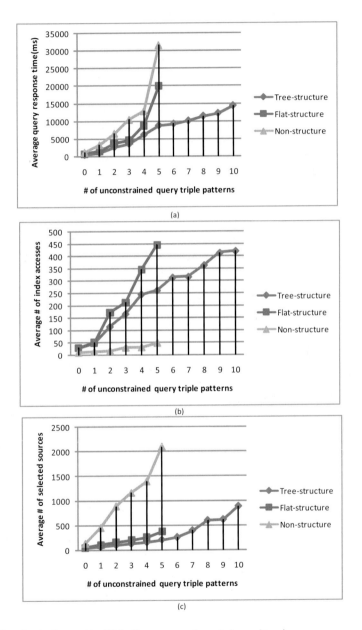

Fig. 4. Synthetic Semantic Web Space experimental results. Average query response time (a), index accesses (b) and number of selected sources (c) as the number of unconstrained QTPs varies.

Fig. 4(b) shows how each algorithm's average number of index accesses is affected by the number of unconstrained QTPs. Note the index is stored on disk and is optimized for fast lookups, but a large number of accesses can have a noticeable impact on performance. From this result, we can see that the tree-structure and flat-structure algorithms require more index accesses than the non-structure algorithm: for 5 unconstrained QTPs they require 5.3x and 9.1x more accesses, respectively. This is because both algorithms take into account the query structure information while solving the original query and might need several index lookups for the same query subgoal but using different substitutions. However, the tree-structure algorithm has 58% fewer index accesses than the flat-structure algorithm. The reason is that when using the flat-structure algorithm, one QTP can appear in multiple query rewritings and receive constraints from different sets of siblings representing different rewrites, while in the tree-structure algorithm the constraints of a sibling already consider all possible rewrites of the sibling.

Fig. 4(c) shows how the number of unconstrained QTPs impact the average number of selected sources for each algorithm. From this result, we can see the selectivity of the tree-structure and the flat-structure algorithms are roughly linear, while the non-structure algorithm is exponential in the number of unconstrained QTPs. Furthermore, the tree-structure algorithm has a gentler slope for its source selectivity than the flat-structure algorithm. Note, loading sources is the primary bottleneck of the system, since it requires that triples be read from the disk or network. The similar trends in Fig. 4(a) and Fig. 4(c) reflect the importance of source selectivity to overall query response time.

4.2   Scalability Evaluation Using the BTC Data Set

In this section, we evaluate our algorithm's scalability by using a subset of the BTC 2009 data set (much of which comes from the Linking Open Data Project Cloud). We have chosen four collections, as summarized in Table 1, with a total of 73,889,151 triples. Using the provenance information in the BTC, we re-created local N3 versions of the original files from the BTC resulting in 21,008,285 data sources. The size of these data sources varies from roughly 5 to 50 triples each. In order to integrate the four heterogeneous collections, we manually created some mapping ontologies, primarily using *rdf:subClassOf* and *rdf:subPropertyOf* axioms (these schemas do not have any meaningful alignments that are more complex). Since our algorithm does not yet select all relevant sources with *owl:sameAs* information, we assume an environment where any relevant *owl:sameAs* information is already supplied to the reasoner. We do this by initializing the KB with the necessary *owl:sameAs* statements. Our index construction time is around 58 hours and its size is around 18GB. Each document takes around 10ms on average to be indexed. The Lucence configurations are 1500MB for RAMBufferSize and 1000 for MergeFactor, which are the best tradeoff between index building and searching for our experiment.

Table 1. Data sources selected from the BTC 2009 dataset

Data Source	Namespace	# of Sources	# of Triples
http://data.semanticweb.org/	swrc	41,974	174,816
http://sws.geonames.org/	geonames	2,324,253	14,866,924
http://dbpedia.org	dbpedia	10,615,260	48,694,372
http://dblp.rkbexplorer.com	akt	8,026,878	10,153,039
Total		21,008,285	73,889,151

Because the non-structure algorithm does not refine goals with constraint information from related goals, it cannot scale to the BTC data set. In fact, most of our synthetic queries cannot be solved by this algorithm. For example, consider the query $Q:\{\langle<?x_0, swrc:affiliation, "lehigh-univ"\rangle.\langle?x_2, akt:has-title, "Hawkeye"\rangle.\langle?x_2, foaf:maker, ?x_0\rangle.\langle?x_0, akt:full-name, ?x_1\rangle\}$. For the non-structure algorithm, the number of sources that can potentially contribute to solving $\langle?x_2, foaf:maker, ?x_0\rangle$ is 3,485,607, which is far too many to load into a memory-based reasoner. However, the tree-structure and flat-structure algorithms can deal with it because the number of sources for the same QTP becomes 114 after variable constraints are applied. For this reason, we only compare the tree-structure and flat-structure algorithms here.

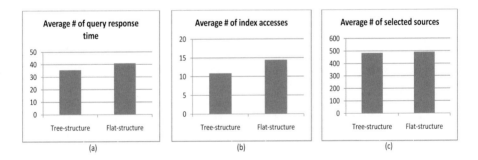

Fig. 5. BTC data set experimental results

We executed 150 synthetic queries with at most 10 QTPs and computed the same metrics as for the prior experiment. As shown in Fig. 5(a), the average query response time of the tree-structure algorithm is 35 seconds, which is a 13% improvement over the flat-structure algorithm. At the same time, it has 25% fewer index accesses as shown in Fig. 5(b). Fig. 5(c) shows that both algorithms select on average between 450 and 500 sources, and the tree-structure algorithm only shows a 1.6% improvement over the flat-structure algorithm here. We attribute this to the fact that the semantic mappings of the BTC experiment are not as complex as those for the synthetic data set, which leads to a small number of rewrites for each query. when there are potentially many rewrites for a query. We posit that in real-world settings where more ontologies are involved, that the superiority of the tree-structure algorithm will be more pronounced.

5   Conclusions, Limitations and Future Work

We have proposed a tree-structure optimization algorithm for integrating millions of data sources that commit to different ontologies. Given a reformulation tree, this algorithm uses a bottom-up process to select sources and uses the selectivity of each goal node as a heuristic to optimize and plan the query execution. Our experiments have demonstrated that this new algorithm is better than both of our prior algorithms [4] [5] in that not only does it demonstrate query response time performance that is linear with respect to the number of unconstrained QTPs, it also has better source selectivity and requires fewer index accesses than the flat-structure algorithm. Meanwhile, we have also shown that our algorithm scales well, allowing many complex randomly generated queries against 20 million heterogeneous data sources to complete in 35 seconds.

Despite showing initial promise there are a number of limitations to the work in its present form. First, the algorithm focuses on conjunctive queries, and does not consider richer features of SPARQL such as OPTIONALs. In addition, in order to avoid the computational challenges of higher-order logics, it does not allow variables in the predicate position. Second, the implementation only works with OWLII, a subset of OWL DL, although any rewriting algorithm that produces an AND-OR reformulation tree could be used. Since finite reformulation trees cannot express rewrites of a query whose reformulation involves cyclic rules, completeness is only guaranteed for acyclic OWLII axioms. We note that this algorithm is designed for a setting where there are large numbers of small RDF files, and that it is not intended to issue queries to large SPARQL end points. Fortunately, due to Linked Data guidelines, we note that most large SPARQL end points expose an interface where a URL can be dereferenced to retrieve a small set of RDF triples describing each instance. The algorithm assumes that a correct set of mapping ontologies has been provided, and we note that any errors in these mappings can result in a loss of "semantic fidelity." Finally, the current algorithm is not guaranteed to find all relevant sources if there are *owl:sameAs* statements in the Semantic Web Space.

Our future work includes attempting to improve the selectivity of our algorithm even further and addressing many of its limitations. We believe it is possible to make better estimates about the selectivity of a node by maintaining upper and lower bounds and we will also look at storing additional statistics in our index. With respect to the limitations mentioned in the previous paragraph, we think that the most critical need is to adapt the algorithm to locate relevant *owl:sameAs* statements, which must necessarily be an iterative process in order to find their transitive closure. We believe that this paper provides a major step towards a pragmatic solution for querying a large, distributed, and ever changing Semantic Web.

References

1. Haase, P., Wang, Y.: A decentralized infrastructure for query answering over distributed ontologies. In: Proceedings of the 2007 ACM Symposium on Applied Computing, SAC 2007, pp. 1351–1356. ACM, New York (2007)
2. Halevy, A.Y., Ives, Z.G., Madhavan, J., Mork, P., Suciu, D., Tatarinov, I.: The Piazza peer data management system. IEEE Trans. Knowl. Data Eng. 16(7), 787–798 (2004)
3. Harth, A., Umbrich, J., Hogan, A., Decker, S.: YARS2: A federated repository for querying graph structured data from the web. In: The Semantic Web, pp. 211–224 (2008)
4. Li, Y., Heflin, J.: Query optimization for ontology-based information integration. In: CIKM 2010. ACM, New York (2010)
5. Li, Y., Qasem, A., Heflin, J.: A scalable indexing mechanism for ontology-based information integration. In: IEEE/WIC/ACM International Conference on Web Intelligence and Intelligent Agent Technology (2010)
6. Neumann, T., Weikum, G.: Scalable join processing on very large RDF graphs. In: Proceedings of the 35th SIGMOD International Conference on Management of Data, SIGMOD 2009, pp. 627–640. ACM, New York (2009)
7. Qasem, A., Dimitrov, D.A., Heflin, J.: Efficient selection and integration of data sources for answering semantic web queries. In: International Conference on Semantic Computing, pp. 245–252 (2008)
8. Ramakrishnan, R., Ullman, J.D.: A survey of research on deductive database systems. Journal of Logic Programming 23, 125–149 (1993)
9. Selinger, P.G., Astrahan, M.M., Chamberlin, D.D., Lorie, R.A., Price, T.G.: Access path selection in a relational database management system. In: Proceedings of the 1979 ACM SIGMOD International Conference on Management of Data, SIGMOD 1979, pp. 23–34. ACM, New York (1979)
10. Serafini, L., Tamilin, A.: Drago: Distributed reasoning architecture for the semantic web. In: Gómez-Pérez, A., Euzenat, J. (eds.) ESWC 2005. LNCS, vol. 3532, pp. 361–376. Springer, Heidelberg (2005)
11. Sidirourgos, L., Goncalves, R., Kersten, M.L., Nes, N., Manegold, S.: Column-store support for rdf data management: not all swans are white. PVLDB 1(2), 1553–1563 (2008)
12. Staudt, M., Soiron, R., Quix, C., Jarke, M.: Query optimization for repository-based applications. In: Proceedings of the 1999 ACM Symposium on Applied Computing, SAC 1999, pp. 197–203. ACM, New York (1999)
13. Stuckenschmidt, H., Vdovjak, R., Broekstra, J., Houben, G.: Towards distributed processing of RDF path queries. Int. J. Web Eng. Technol. 2(2/3), 207–230 (2005)
14. Tran, T., Wang, H., Haase, P.: Hermes: Data web search on a pay-as-you-go integration infrastructure. Web Semantics 7(3), 189–203 (2009)
15. Udrea, O., Pugliese, A., Subrahmanian, V.S.: Grin: A graph based RDF index. In: AAAI, pp. 1465–1470 (2007)
16. Weiss, C., Karras, P., Bernstein, A.: Hexastore: sextuple indexing for semantic web data management. PVLDB 1(1), 1008–1019 (2008)

AnQL: SPARQLing Up Annotated RDFS

Nuno Lopes[1], Axel Polleres[1], Umberto Straccia[2], and Antoine Zimmermann[1]

[1] Digital Enterprise Research Institute,
National University of Ireland Galway, Ireland
{nuno.lopes,axel.polleres,antoine.zimmermann}@deri.org
[2] Istituto di Scienza e Tecnologie dell'Informazione (ISTI - CNR), Pisa, Italy
straccia@isti.cnr.it

Abstract. Starting from the general framework for Annotated RDFS which we presented in previous work (extending Udrea et al's Annotated RDF), we address the development of a query language – AnQL – that is inspired by SPARQL, including several features of SPARQL 1.1. As a side effect we propose formal definitions of the semantics of these features (subqueries, aggregates, assignment, solution modifiers) which could serve as a basis for the ongoing work in SPARQL 1.1. We demonstrate the value of such a framework by comparing our approach to previously proposed extensions of SPARQL and show that AnQL generalises and extends them.

Introduction

RDF (Resource Description Framework) [14] is the widely used representation language for the Semantic Web and the Web of Data. RDF exposes data as triples, consisting of *subject*, *predicate* and *object*, stating that *subject* is related to *object* by the *predicate* relation. Several extensions to RDF were proposed in order to deal with time [7,19,24], truth or imprecise information [15,22], trust [10,20] and provenance [4]. All these proposals share a common approach of extending the RDF language by attaching meta-information about the RDF graph or triple. RDF Schema (RDFS) [3] is the specification of a restricted vocabulary that allows to deduce further information from existing triples. In our previous work [23], we presented a general extension to RDFS, improving on Udrea et al's Annotated RDF [25], that is capable of encapsulating the mentioned RDF extensions as specific domains for RDF annotations. For this general extension, we present a generic RDFS reasoning procedure which can be formulated independently of the annotation domain by being parameterised with operations any domain needs to provide. An overview of Annotated RDFS is presented in Sect. 1.

SPARQL [21] is the W3C-standardised query language for RDF. In this paper we present an extension of SPARQL for querying annotated RDFS. SPARQL shares similarities with SQL although several features, such as aggregates, nested queries and variable assignments, are still missing from the current SPARQL specification. Our SPARQL extension presented here also deals with these missing features thus going beyond the features of SPARQL, heading towards the currently under development SPARQL 1.1 specification [9]. Our extension of SPARQL, called AnQL, is presented in Sect. 2. Furthermore, Sect. 3 presents a discussion of some of the most important issues with the design of our query language along with the comparison to some of the related works.

P.F. Patel-Schneider et al. (Eds.): ISWC 2010, Part I, LNCS 6496, pp. 518–533, 2010.

Related Work. The basis for Annotated RDF were first established by Udrea *et al.* [25, 26], in which their query language is restricted to conjunctive queries. SPARQL is compared to the presented conjunctive queries but excludes the possibility of querying annotations. Furthermore, OPTIONAL, UNION and FILTER SPARQL queries are not considered which results in a subset of SPARQL that can be directly translated into their previously presented conjunctive query system.

In [7], Gutiérrez *et al.* present conjunctive queries with built-in predicates for querying temporal RDF, neither considering full SPARQL. Pugliese *et al.* [19] also have a temporal framework where they only define conjunctive queries, thus ignoring some of the more advanced features of SPARQL. Tappolet and Bernstein [24] present temporal extensions for RDF and SPARQL. A storage format for temporal RDF is presented where each time interval is stored as a named graph. The τ-SPARQL query language allows to query the temporal RDF representation using an extended SPARQL syntax that can match the graph pattern against the snapshot of a temporal graph at any given time point and allows to query the start and endpoints of a temporal interval, whose values can then be used in other parts of the query. The RDF extensions towards uncertain or fuzzy information [15, 22] so far do not address SPARQL, presenting only extensions for RDFS reasoning but [22] formalises conjunctive queries.

SPARQL extensions towards querying trust have been presented by Hartig [10]. Hartig introduces a trust aware query language, tSPARQL, that includes a new constructor to access the trust value of a graph pattern. This value can then be used in other statements such as FILTER*s* or ORDER.

Another extension to query meta-knowledge in RDF, mostly considering provenance and uncertainty is presented by Dividino *et al.* [4]. In this work, the meta-information is stored using named graphs and the syntax and semantics of SPARQL are extended to consider an additional expression that enables querying the named graphs representing the meta-information.

Our present work can also be related to annotated relational databases, especially Green *et al.* [6] who provides a similar framework for the relational algebra. After presenting a generic structure for annotations, they focus more specifically on the provenance domain. The specificities of the relation algebra, especially Closed World Assumption, allows them to define a slightly more general structure for annotation domains, namely semiring (as opposed to residuated lattice in our approach).

1   Annotated RDFS

For the sake of making the paper self-contained, we recap essential parts from [23], where we only considered ground graphs, while here we do allow blank nodes as well.

1.1   Syntax

Consider pairwise disjoint alphabets **U**, **B**, and **L** denoting, respectively, *URI references*, *blank nodes* (i.e., *variables*, denoted x, y, z)[1] and *Literals*.[2] We call the elements

---

[1] We will often use the term blank node and variable synonymously in this paper.

[2] We assume **U**, **B**, and **L** fixed, and for ease we will denote unions of these sets simply concatenating their names.

in **UBL** (**B**) *terms*. An *RDF triple* is $\tau = (s, p, o) \in$ **UBL** $\times$ **U** $\times$ **UBL**.[3] We call s the *subject*, p the *predicate*, and o the *object*. An *annotated triple* is an expression $\tau\colon \lambda$, where τ is a triple and λ is an *annotation value* (defined below). An *annotated graph* G is a finite set of *annotated triples*. The *universe* of G, $universe(G)$, is the set of elements in **UBL** that occur in the triples of G. A *vocabulary* is a subset of **UL**.

As in our previous work, for presentation purposes, we rely on a fragment of RDFS, called ρdf [16], that covers essential features of RDFS.[4] ρdf is defined as the following subset of the RDFS vocabulary: ρdf $= \{$sp, sc, type, dom, range$\}$. Informally, (i) (p, sp, q) means that property p is a *subproperty* of property q; (ii) (c, sc, d) means that class c is a *subclass* of class d; (iii) (a, type, b) means that a is of *type* b; (iv) (p, dom, c) means that the *domain* of property p is c; and (v) (p, range, c) means that the *range* of property p is c. Annotations are added to triples to attach meta information such as temporal validity, trust or fuzzy value, provenance.

Example 1. For instance, the following annotated triple:

$$(\text{:Alain}, \text{:livesIn}, \text{:Paris})\colon \texttt{[1980, 1991]}$$

in a temporal setting [7] has intended meaning "Alain lives in Paris from 1980 to 1991", while in the fuzzy setting [22]:

$$(\text{audiTT}, \mathsf{type}, \text{SportsCar})\colon 0.8$$

has intended meaning "AudiTT is a sports car to degree not less than 0.8"; considering provenance as annotations:

$$(\text{Person}, \mathsf{sc}, \text{Agent})\colon \{\texttt{http://xmlns.com/foaf/0.1/}\}$$

would mean that the subclass relationship between persons and agents is defined by – or, "belongs to" – the document $\texttt{http://xmlns.com/foaf/0.1/}$.

1.2   RDFS Annotation Domains

Consider a lattice $\langle L, \preceq \rangle$. Elements in L are our annotation values. The order $\preceq$ is used to express redundant/entailed/subsumed information. For instance, for temporal intervals, an annotated triple $(s, p, o)\colon \texttt{[2000,2006]}$ entails $(s, p, o)\colon \texttt{[2003,2004]}$, as $\texttt{[2003,2004]} \subseteq \texttt{[2000,2006]}$ (here, $\subseteq$ plays the role of $\preceq$). Informally, an interpretation will map statements to elements of the annotation domain. Our semantics generalises the one of standard RDFS by using an algebraic structure that is well-known for Many-Valued FOL [8]. We say that an *annotation domain* for RDFS is a residuated bounded lattice $D = \langle L, \preceq, \wedge, \vee, \otimes, \Rightarrow, \bot, \top \rangle$.[5] That is,

1. $\langle L, \preceq, \wedge, \vee, \bot, \top \rangle$ is a bounded lattice, where $\bot$ and $\top$ are bottom and top elements, and $\wedge$ and $\vee$ are meet and join operators;
2. $\langle L, \otimes, \top \rangle$ is a commutative monoid;

---

[3] As in [16] we allow literals for s.

[4] Just as in [16] our annotation framework can be extended to full RDFS, adding additional semantic conditions and respective inference rules [13].

[5] We correct here an imprecision in the definition given in [23], in which we did not mention that the structure should be a residuated lattice.

3. $\Rightarrow$ is the so-called residuum of $\otimes$, *i.e.*, for all $\lambda_1, \lambda_2, \lambda_3$, $\lambda_1 \otimes \lambda_3 \preceq \lambda_2$ iff $\lambda_3 \preceq (\lambda_1 \Rightarrow \lambda_2)$.

Remark 1. Note that $\lambda_1 \Rightarrow \lambda_2$ can be determined uniquely as $\lambda_1 \Rightarrow \lambda_2 = \sup \{\lambda \mid \lambda_1 \otimes \lambda \preceq \lambda_2\}$ (see [11]). Furthermore, in the remaining of this paper, we do not use the $\wedge$ which is implicitly defined by the order $\preceq$. For these reasons, we represent a domain succinctly as 6-tuple $\langle L, \preceq, \vee, \otimes, \bot, \top \rangle$.

In what follows we define a *map* as a function $\mu : \mathbf{UBL} \rightarrow \mathbf{UBL}$ preserving URIs and literals, *i.e.*, $\mu(t) = t$, for all $t \in \mathbf{UL}$. Given a graph G, we define $\mu(G) = \{(\mu(s), \mu(p), \mu(o)) \mid (s, p, o) \in G\}$. We speak of a map μ from G_1 to G_2, and write $\mu : G_1 \rightarrow G_2$, if μ is such that $\mu(G_1) \subseteq G_2$.

1.3 Semantics

Fix an annotation domain $D = \langle L, \preceq, \vee, \otimes, \bot, \top \rangle$. Informally, an interpretation $\mathcal{I}$ will assign to a triple τ an element of the annotation domain $\lambda \in L$, dictating that under $\mathcal{I}$, the annotation of τ is greater or equal than (i.e., $\succeq$) λ. Formally, an *annotated interpretation* $\mathcal{I}$ over a vocabulary V is a tuple $\mathcal{I} = \langle \Delta_R, \Delta_P, \Delta_C, \Delta_L, P[\![\cdot]\!], C[\![\cdot]\!], \cdot^{\mathcal{I}} \rangle$ where $\Delta_R, \Delta_P, \Delta_C, \Delta_L$ where $\Delta_R, \Delta_P, \Delta_C, \Delta_L$ are the interpretation domains of $\mathcal{I}$, which are finite non-empty sets, and $P[\![\cdot]\!], C[\![\cdot]\!], \cdot^{\mathcal{I}}$ are the interpretation functions of $\mathcal{I}$. These have to satisfy:

1. Δ_R are the resources (the domain or universe of $\mathcal{I}$);
2. Δ_P are property names (not necessarily disjoint from Δ_R);
3. $\Delta_C \subseteq \Delta_R$ are the classes;
4. $\Delta_L \subseteq \Delta_R$ are the literal values and contains $\mathbf{L} \cap V$;
5. $P[\![\cdot]\!]$ maps each property name $p \in \Delta_P$ into a function $P[\![p]\!] : \Delta_R \times \Delta_R \rightarrow L$, *i.e.*, assigns an annotation value to each pair of resources;
6. $C[\![\cdot]\!]$ maps each class $c \in \Delta_C$ into a function $C[\![c]\!] : \Delta_R \rightarrow L$, *i.e.*, assigns an annotation value representing class membership in c to every resource.
7. $\cdot^{\mathcal{I}}$ maps each $t \in \mathbf{UL} \cap V$ into a value $t^{\mathcal{I}} \in \Delta_R \cup \Delta_P$, and such that $\cdot^{\mathcal{I}}$ is the identity for plain literals and assigns an element in Δ_R to each element in $\mathbf{L}$;

Intuitively, a triple $(s, p, o) : \lambda$ is satisfied by an annotated interpretation $\mathcal{I}$ if (s, o) belongs to the extension of p to a "wider" extent than λ. We formalise this intuition in terms of semantic conditions on the use of the RDFS vocabulary. That is, an interpretation $\mathcal{I}$ is a *model* of an annotated ground graph G, denoted $\mathcal{I} \models G$, iff $\mathcal{I}$ is an interpretation over the vocabulary $\rho df \cup universe(G)$ that satisfies the conditions in Table 1. Here, considering a set $\Delta \subseteq \Delta_R \cup \Delta_P$, we say that a function $p \colon \Delta \times \Delta \rightarrow L$ is sup-$\otimes$ *transitive* (or simply *transitive*) over Δ iff for all $x, z \in \Delta$, $\sup_{y \in \Delta}\{p(x, y) \otimes p(y, z)\} \preceq p(x, z)$.[6]

  Finally, entailment among annotated ground graphs G and H is as usual. Now, $G \models H$, where G and H may contain blank nodes, iff for any grounding G' of G there is a grounding H' of H such that $G' \models H'$.[7]

---

[6] As Δ is finite, sup-$\otimes$ transitivity is well defined.

[7] A grounding G' of graph G is obtained by replacing variables in G with terms in $\mathbf{UL}$.

Table 1. The conditions for annotated interpretations

Simple:
  1. (s, p, o): $\lambda \in G$ implies $p^{\mathcal{I}} \in \Delta_P$ and $P[\![p^{\mathcal{I}}]\!](s^{\mathcal{I}}, o^{\mathcal{I}}) \succeq \lambda$;
Subproperty:
  1. $P[\![\mathsf{sp}^{\mathcal{I}}]\!]$ is sup-$\otimes$ transitive over Δ_P;
  2. $P[\![\mathsf{sp}^{\mathcal{I}}]\!](p, q) = \inf_{(x,y) \in \Delta_R \times \Delta_R} P[\![p]\!](x, y) \Rightarrow P[\![q]\!](x, y)$;
Subclass:
  1. $P[\![\mathsf{sc}^{\mathcal{I}}]\!]$ is sup-$\otimes$ transitive over Δ_C;
  2. $P[\![\mathsf{sc}^{\mathcal{I}}]\!](c, d) = \inf_{x \in \Delta_R} C[\![c]\!](x) \Rightarrow C[\![d]\!](x)$;
Typing I:
  1. $C[\![c]\!](x) = P[\![\mathsf{type}^{\mathcal{I}}]\!](x, c)$;
  2. $P[\![\mathsf{dom}^{\mathcal{I}}]\!](p, c) = \inf_{(x,y) \in \Delta_R \times \Delta_R} P[\![p]\!](x, y) \Rightarrow C[\![c]\!](x)$;
  3. $P[\![\mathsf{range}^{\mathcal{I}}]\!](p, c) = \inf_{(x,y) \in \Delta_R \times \Delta_R} P[\![p]\!](x, y) \Rightarrow C[\![c]\!](y)$;
Typing II:
  1. For each $e \in \rho df$, $e^{\mathcal{I}} \in \Delta_P$
  2. $P[\![\mathsf{dom}^{\mathcal{I}}]\!](p, c)$ is defined only for $p \in \Delta_P$ and $c \in \Delta_C$
  3. $P[\![\mathsf{range}^{\mathcal{I}}]\!](p, c)$ is defined only for $p \in \Delta_P$ and $c \in \Delta_C$
  4. $P[\![\mathsf{type}^{\mathcal{I}}]\!](s, c)$ is defined only for $c \in \Delta_C$.

Remark 2. In [16], the authors define two variants of the semantics of ρdf: the default one includes reflexivity of the subclass and subproperty relations but in the present paper, we extend the alternative semantics presented in [16, Definition 4] which omits this requirement.

Remark 3. Note that we always have that $G \models \tau : \bot$. Clearly, triples of the form $\tau : \bot$ are uninteresting and, thus, in the following we not consider them as part of the language.

As for the crisp case, it can be shown [23] that *any annotated RDFS graph has a finite model* (modulo Remark 3) and, thus, we do not have to care about consistency.

1.4   Deductive System

The important feature of the annotation framework is that we are able to provide a deductive system in the style of the one for classical RDFS. Moreover, *the schemata of the rules are the same for any annotation domain* (only support for the domain dependent $\otimes$ and $\vee$ operations has to be provided) and, thus, are amenable to an easy implementation on top of existing systems. Specifically, the rule set contains the rules presented in Table 2. Please note that rule 6 from Table 2 is destructive, *i.e.*, this rule removes the premises as the conclusion is inferred, intuitively meaning that only "maximal" annotations are preserved.

Remark 4. We point out that rules $2 - 5$ from Table 2 can be represented concisely using the following inference rule:

$$(A) \; \frac{\tau_1 : \lambda_1, \, ..., \, \tau_n : \lambda_n, \{\tau_1, \ldots \tau_n\} \vdash_{\mathsf{RDFS}} \tau}{\tau : \bigotimes_i \lambda_i}$$

Table 2. Inference rules for annotated RDFS

1. Simple:

    $(a)\ \frac{G}{G'}$ for a map $\mu: G' \to G$ $(b)\ \frac{G}{G'}$ for $G' \subseteq G$

2. Subproperty:

    $(a)\ \dfrac{(A,\mathsf{sp},B):\lambda_1,(B,\mathsf{sp},C):\lambda_2}{(A,\mathsf{sp},C):\lambda_1 \otimes \lambda_2}$ $(b)\ \dfrac{(D,\mathsf{sp},E):\lambda_1,(X,D,Y):\lambda_2}{(X,E,Y):\lambda_1 \otimes \lambda_2}$

3. Subclass:

    $(a)\ \dfrac{(A,\mathsf{sc},B):\lambda_1,(B,\mathsf{sc},C):\lambda_2}{(A,\mathsf{sc},C):\lambda_1 \otimes \lambda_2}$ $(b)\ \dfrac{(A,\mathsf{sc},B):\lambda_1,(X,\mathsf{type},A):\lambda_2}{(X,\mathsf{type},B):\lambda_1 \otimes \lambda_2}$

4. Typing:

    $(a)\ \dfrac{(D,\mathsf{dom},B):\lambda_1,(X,D,Y):\lambda_2}{(X,\mathsf{type},B):\lambda_1 \otimes \lambda_2}$ $(b)\ \dfrac{(D,\mathsf{range},B):\lambda_1,(X,D,Y):\lambda_2}{(Y,\mathsf{type},B):\lambda_1 \otimes \lambda_2}$

5. Implicit Typing:

    $(a)\ \dfrac{(A,\mathsf{dom},B):\lambda_1,(D,\mathsf{sp},A):\lambda_2,(X,D,Y):\lambda_3}{(X,\mathsf{type},B):\lambda_1 \otimes \lambda_2 \otimes \lambda_3}$

    $(b)\ \dfrac{(A,\mathsf{range},B):\lambda_1,(D,\mathsf{sp},A):\lambda_2,(X,D,Y):\lambda_3}{(Y,\mathsf{type},B):\lambda_1 \otimes \lambda_2 \otimes \lambda_3}$

6. Generalisation:

    $\dfrac{(X,A,Y):\lambda_1,(X,A,Y):\lambda_2}{(X,A,Y):\lambda_1 \vee \lambda_2}$

Essentially, this rule says that if a classical RDFS triple τ can be inferred by applying a classical RDFS inference rule to triples $\tau_1, \ldots \tau_n$ (denoted $\{\tau_1, \ldots, \tau_n\} \vdash_{\mathsf{RDFS}} \tau$), then the annotation term of τ will be $\bigotimes_i \lambda_i$, where λ_i is the annotation of triple τ_i. It follows immediately that, using rule schema (A), the annotated framework extends to the whole RDFS rule set as well. We also assume that rule schema (A) or rule 6 of Table 2 are not applied if the consequence is of the form $\tau: \perp$ (see Remark 3).

Finally, like for the classical case, the *closure* is defined as $cl(G) = \{\tau: \lambda \mid G \vdash^* \tau: \lambda\}$, where $\vdash^*$ is as $\vdash$ for the annotated framework without rule $(1a)$. Notice that the size of the closure of G is polynomial in $|G|$ and can be computed in polynomial time, provided that the computational complexity of operations $\otimes$ and $\vee$ are polynomially bounded (from a computational complexity point of view, it is as for the classical case, plus the cost of the operations $\otimes$ and $\vee$ in L).

Proposition 1 (Soundness and completeness). *For an annotated graph, the proof system $\vdash$ is sound and complete for $\models$, that is, (1) if $G \vdash \tau: \lambda$ then $G \models \tau: \lambda$ and (2) if $G \models \tau: \lambda$ then there is $\lambda' \succeq \lambda$ with $G \vdash \tau: \lambda'$.*

1.5   Examples of Domains

Here, we instantiate the definition with several domains that have been discussed in the literature. The interested reader can find more details about the temporal and fuzzy domains in our previous work [23] and additional information in our accompanying technical report [13]. Furthermore, domains can be combined into a multi-dimensional annotation domain as explained in [23].

Crisp. Note that with the domain $D_{01} = \langle \{0,1\}, \leqslant, \max, \min, 0, 1 \rangle$, Annotated RDFS turns out to be the same as standard RDFS.

Fuzzy. The fuzzy domain has been presented in [15, 22] and to model fuzzy RDFS in our framework is easy: the annotation domain is $D_{[0,1]} = \langle [0,1], \leqslant, \max, \otimes, 0, 1 \rangle$ where $\otimes$ is any continuous t-norm on $[0,1]$ and $\vee$ is $\max$.

Temporal. For modelling the temporal domain we generalise the notions presented in [7,19,24] and consider that *time points* are elements of xsd:dateTimeStamp [18] value space $\cup \{-\infty, +\infty\}$.[8] A *temporal interval* is a non-empty interval $[\alpha_1, \alpha_2]$, where α_i are time points. An empty interval is denoted as $\emptyset$. We define a partial order on intervals as $I_1 \leqslant I_2$ iff $I_1 \subseteq I_2$ and L as (where $\bot = \{\emptyset\}$, $\top = \{[-\infty, +\infty]\}$). Therefore, a *temporal term* is a finite set of pairwise disjoint time intervals. Furthermore, on L we define the following partial order:

$$t_1 \preceq t_2 \text{ iff } \forall I_1 \in t_1 \exists I_2 \in t_2, \text{ such that } I_1 \leqslant I_2 .$$

The join and t-norm $\otimes$ operators are defined as:

$$t_1 \vee t_2 = \inf\{t \mid t \succeq t_i, i = 1, 2\}$$
$$t_1 \otimes t_2 = \sup\{t \mid t \preceq t_i, i = 1, 2\} .$$

Remark 5. Although we represent time points as dateTimeStamps, for presentation purposes in this paper we will only use years.

Provenance. We also generalise the representation of *provenance* as described, e.g., in [4, 5]. In this case, we start from a countably infinite set of *atomic provenances* **P**. We consider the propositional formulae made from symbols in **P** (atomic propositions), logical *or* ($\vee$) and logical *and* ($\wedge$), for which we have the standard entailment $\models$. A *provenance value* is an equivalent class for the logical equivalence relation, *i.e.*, the set of annotation values is the quotient set of **P** by the logical equivalence. The order relation is $\models$, t-norm and join are $\wedge$ and $\vee$ respectively. We set $\top$ to *true* and $\bot$ to *false*.

Trust. For the trust domain we rely on previous work by Schenk [20] that defines a *bilattice* structure to model some form of *trust*. We can directly use this algebraic structure as an annotation domain in our framework.

2   AnQL: Annotated SPARQL

We now present an extension of the SPARQL [21] query language, made for querying annotated graphs, which we call AnQL. For the rest of this section we fix a specific annotation domain, $D = \langle L, \preceq, \vee, \otimes, \bot, \top \rangle$.

---

[8] Note that we have a continuous set of time points as opposed to Gutiérrez *et al.* [7].

2.1 Syntax

A *simple AnQL query* is defined – analogously to a SPARQL query – as a triple $Q = (P, G, V)$ where P is an *annotated graph pattern*, the *dataset* G is an annotated graph and V is a set of variables, called the *result form*. We restrict ourselves to **SELECT** queries in this work so it is sufficient to consider the result form V as a list of variables to be projected.

Remark 6. Note that, for presentation purposes, we simplify the notion of datasets by excluding named graphs and thus GRAPH queries. Our definitions can be straight-forwardly extended to named graphs and we refer the reader to the SPARQL W3C specification [21] for details.

Triple patterns in annotated AnQL are defined the same way as in SPARQL. A *triple pattern* is a triple (s, p, o) where $s, o \in \mathbf{UBL}$ and $p \in \mathbf{UB}$. We denote variables from **B** in triple patterns by '?' prefixed names.[9] Let **V** be a distinct set of variables, called *annotation variables*. For a triple pattern τ and λ either an annotation term from D or an annotation variable, we call $\tau \colon \lambda$ an *annotated triple pattern*; sets of annotated triple patterns are called *basic annotated patterns* (BAP). An *annotated graph pattern* is defined in a recursive manner: any BAP is an annotated graph pattern; if P and P' are annotated graph patterns, R is a filter expression (see [21], and later on), then $(P\ \mathsf{AND}\ P')$, $(P\ \mathsf{OPTIONAL}\ P')$, $(P\ \mathsf{UNION}\ P')$, $(P\ \mathsf{FILTER}\ R)$ are annotated graph patterns.

Example 2. Suppose we are looking for people who live near Paris during some time period and optionally owned a car during that period. This query can be posed as follows:

```
SELECT ?p ?c ?l
WHERE {(?p :basedNear :paris):?l OPTIONAL{(?p :hasCar ?c):?l}}
```

Assuming the following input data:

$$(\text{:alain, :livesIn, :paris}) \colon [2007, 2010]$$
$$(\text{:alain, :hasCar, :peugeot}) \colon [2004, 2009]$$
$$(\text{:alain, :hasCar, :renault}) \colon [2010, 2010]$$
$$(\text{:livesIn, } \mathsf{sp}, \text{:basedNear}) \colon [-\infty, +\infty]$$

we will get the following answers:

$$\theta_1 = \{?p/\text{:alain}, ?l/[2007, 2010]\}$$
$$\theta_2 = \{?p/\text{:alain}, ?c/\text{:peugeot}, ?l/[2007, 2009]\}$$
$$\theta_3 = \{?p/\text{:alain}, ?c/\text{:renault}, ?l/[2010, 2010]\}.$$

The first answer corresponds to the answer in which the OPTIONAL pattern is not satisfied, so we get the annotation value $[2007, 2010]$ that corresponds to the time Alain lives in Paris. In the second and third answers, the OPTIONAL pattern is matched and, in this case, the annotation value is restricted to the time when Alain lives in Paris and has a car.   □

---

[9] Note that we do not consider blank nodes in triple patterns separately, since they can be treated just as other variables.

Note that – as we will see – this first query will return as a result for the annotation variable the periods where a car was owned.

Example 3. A slightly different query can be people who lived near Paris during some time period and optionally owned a car at some point during their stay. This query – which will rather return the time periods of employment – can be written as follows:

```
SELECT ?p ?c ?l WHERE {(?p :basedNear :paris):?l
                OPTIONAL {(?p :hasCar ?c):?l2 FILTER (?l2 ≼ ?l)} }
```

Using the input data from Example 2, we obtain the following answers:

$$\theta_1 = \{?p/:\text{alain}, ?l/[2007, 2010]\}$$
$$\theta_2 = \{?p/:\text{alain}, ?c/:\text{renault}, ?l/[2007, 2010]\}$$

In this example the FILTER behaves as in SPARQL by removing from the answer set the mappings that do not make the FILTER expression true. This query also exposes the issue of unsafe filters, noted in [2].   □

2.2   Semantics

We denote by $var(P)$ the set of variables and annotation variables present in a BAP P. A *substitution* θ for a BAP P is a mapping with domain $var(P)$ (annotation) variables into (annotation) terms occurring in G. We denote the *domain* of a substitution θ, i.e. the variables for which θ is defined, by $dom(\theta)$. For convenience, sometimes we will use the notation $\theta = \{x_1/t_1, \ldots, x_n/t_n\}$ to indicate that $\theta(x_i) = t_i$, *i.e.*, variable x_i is assigned to term t_i. Note that we do not allow any assignment of an annotation variables to $\bot$ (of the domain D). An annotation value of $\bot$, although it is a valid answer for any triple, does not provide any additional information and thus is of minor interest.

For a BAP P, and a substitution θ we denote by $\theta(P)$ the triples obtained by replacing the variables in P according to θ. By $G \models \theta(P)$ we denote the fact that $\theta(P)$ is entailed by G.

For the extension of the SPARQL relational algebra to the annotated case we introduce – inspired by the definitions in [17] – definitions of compatibility and union of substitutions: given two substitutions θ_1 and θ_2, θ_1 and θ_2 are $\otimes$-*compatible* if and only if (i) $\theta_1(x) = \theta_2(x)$ for any non-annotation variable $x \in dom(\theta_1) \cap dom(\theta_2)$; (ii) $\theta_1(\lambda) \otimes \theta_2(\lambda) \neq \bot$ for any annotation variable $\lambda \in dom(\theta_1) \cap dom(\theta_2)$. Further, for two $\otimes$-*compatible* substitutions θ_1 and θ_2 the $\otimes$-*union*, denoted $\theta_1 \otimes \theta_2$, is as $\theta_1 \cup \theta_2$, with the exception that for any annotation variable $\lambda \in dom(\theta_1) \cap dom(\theta_2)$, $\theta_1 \otimes \theta_2(\lambda) = \theta_1(\lambda) \otimes \theta_2(\lambda)$.

We are now ready to present the notion of evaluation for generic AnQL graph patterns. Let P be a BAP, P_1, P_2 annotated graph patterns, G an annotated graph and R a filter expression, then the evaluation $[\![\cdot]\!]_G$, *i.e.*, set of *answers*,[10] is recursively defined as:

---

[10] Strictly speaking, we consider *sequences* of answers – note that SPARQL allows duplicates and imposing and order on solutions, cf. Sect. 2.3 below for more discussion – but we stick with set *notation* representation here for illustration. Whenever we mean "real" sets where duplicates are removed we write $\{\ldots\}_{\text{DISTINCT}}$.

$$[\![P]\!]_G \qquad\qquad = \{\theta \mid dom(\theta) = var(P) \text{ and } G \models \theta(P)\}$$
$$[\![P_1 \text{ AND } P_2]\!]_G \qquad = \{\theta_1 \otimes \theta_2 \mid \theta_1 \in [\![P_1]\!]_G, \theta_2 \in [\![P_2]\!]_G, \theta_1 \text{ and } \theta_2 \otimes\text{-compatible}\}$$
$$[\![P_1 \text{ UNION } P_2]\!]_G \qquad = [\![P_1]\!]_G \cup [\![P_2]\!]_G$$
$$[\![P_1 \text{ FILTER } R]\!]_G \qquad = \{\theta \mid \theta \in [\![P_1]\!]_G \text{ and } R\theta \text{ is true}\}$$
$$[\![P_1 \text{ OPTIONAL } P_2[R]]\!]_G = \{\theta \mid \text{ and } \theta \text{ meets one of the following conditions:}$$

1.) $\theta = \theta_1 \otimes \theta_2$ if $\theta_1 \in [\![P_1]\!]_G, \theta_2 \in [\![P_2]\!]_G, \theta_1$ and θ_2 $\otimes$-compatible, and $R\theta$ is true

2.) $\theta = \theta_1 \in [\![P_1]\!]_G$ and $\forall \theta_2 \in [\![P_2]\!]_G$ such that θ_1 and θ_2 $\otimes$-compatible,
   $R(\theta_1 \otimes \theta_2)$ is true, and for all annotation variables λ in $dom(\theta_1) \cap dom(\theta_2)$,
   $\theta_2(\lambda) \prec \theta_1(\lambda)$

3.) $\theta = \theta_1 \in [\![P_1]\!]_G$ and $\forall \theta_2 \in [\![P_2]\!]_G$ such that θ_1 and θ_2 $\otimes$-compatible,
   $R(\theta_1 \otimes \theta_2)$ is false$\}$

Remark 7. For practical convenience, we retain in $[\![\cdot]\!]_G$ only "domain maximal answers". That is, let us define $\theta' \preceq \theta$ iff (i) $\theta' \neq \theta$; (ii) $dom(\theta) = dom(\theta')$; (iii) $\theta(x) = \theta'(x)$ for any non-annotation variable x; and (iv) $\theta'(\lambda) \preceq \theta(\lambda)$ for any annotation variable λ. Then, for any $\theta \in [\![P]\!]_G$ we remove any $\theta' \in [\![P]\!]_G$ such that $\theta' \preceq \theta$.

Remark 8. Please note that the cases for the evaluation of the OPTIONAL are compliant with the SPARQL specification [21], covering the notion of unsafe FILTERs as presented in [2]. However, there are some peculiarities inherent to the annotated case. More specifically case 2.) introduces the side effect that annotation variables that are compatible between the mappings may have different values in the answer depending if the OPTIONAL is matched of not. This is the behaviour demonstrated in Example 2.

The notion of *true filter* is defined as follows: for a FILTER expression R, the valuation of R on a substitution θ, denoted $R\theta$ is *true* iff:[11]

    (1) $R = \text{BOUND}(v)$ with $v \in dom(\theta)$;
    (2) $R = \text{isBLANK}(v)$ with $v \in dom(\theta)$ and $\theta(v) \in \mathbf{B}$;
    (3) $R = \text{isIRI}(v)$ with $v \in dom(\theta)$ and $\theta(v) \in \mathbf{U}$;
    (4) $R = \text{isLITERAL}(v)$ with $v \in dom(\theta)$ and $\theta(v) \in \mathbf{L}$;
    (5) $R = (u = v)$ with $u, v \in dom(\theta) \cup \mathbf{UBL}$ and $\theta(u) = \theta(v)$;
    (6) $R = (\neg R_1)$ with $\theta(R_1)$ is false;
    (7) $R = (R_1 \vee R_2)$ with $\theta(R_1)$ is true or $\theta(R_2)$ is true;
    (8) $R = (R_1 \wedge R_2)$ with $\theta(R_1)$ is true and $\theta(R_2)$ is true;
    (9) $R = (x \preceq y)$ with $x, y \in dom(\theta) \cup L$ and $\theta(x) \preceq \theta(y)$;
   (10) $R = p(\bar{z})$ with $p(\bar{z})\theta = $ true iff $p(\theta(\bar{z})) = $ true, where p is a built-in predicate.

In the FILTER expressions above, a built-in predicate p is any n-ary predicate p, where p's arguments may be variables (annotation and non-annotation ones), domain values of D, values from $\mathbf{UL}$, p has a fixed interpretation and we assume that the evaluation of the predicate can be decided in finite time. Annotation domains may define their own built-in predicates that range over annotation values as in the following query:

Example 4. Consider we want to know where Alain was living before 2009. This query can be expressed in the following way:

```
SELECT ?city
WHERE {(:alain :livesIn ?city):?l FILTER(before(?l, [2009]))}
```

---

[11] We consider a simple evaluation of filter expressions where the "error" result is ignored, see [21, Sect. 11.3] for details.

The following proposition shows that we have a conservative extension of SPARQL:

Proposition 2. *Let $Q = (P, G, V)$ be a SPARQL query over an RDF graph G. Let G' be obtained from G by annotating triples with $\top$. Then $[\![P]\!]_G$ under SPARQL semantics is in one-to-one correspondence to $[\![P]\!]_{G'}$ under AnQL semantics such that for any $\theta \in [\![P]\!]_G$ there is a $\theta' \in [\![P]\!]_{G'}$ with θ and θ' coinciding on $var(P)$.*

2.3   Further Extensions of AnQL

In this section we include various features from SPARQL 1.1[12] such as variable assignments, projection (i.e. sub-SELECTs), aggregates and solution modifiers to AnQL. We succinctly present both syntax and semantics of the constructs. The evaluation of a ASSIGN statement is defined as:

$$[\![P \text{ ASSIGN } f(\bar{\mathbf{z}}) \text{ } AS \text{ } z]\!]_G = \{\theta \mid \theta_1 \in [\![P]\!]_G, \theta = \theta_1[z/f(\theta_1(\bar{\mathbf{z}}))]\} \text{ ,}$$

where

$$\theta[z/t] = \begin{cases} \theta \cup \{z/t\} & \text{if } z \notin dom(\theta) \\ (\theta \setminus \{z/t'\}) \cup \{z/t\} & \text{otherwise .} \end{cases}$$

Essentially, we assign to the variable z the value $f(\theta_1(\bar{\mathbf{z}}))$, which is the evaluation of the function $f(\bar{\mathbf{z}})$ with respect to a substitution $\theta_1 \in [\![P]\!]_G$.

Example 5. Using a built-in function we can retrieve for each employee the length of employment for any company:

```
SELECT ?x ?y ?z
WHERE {(?x :worksFor :?y):?l ASSIGN length(?l) AS ?z }
```

Here, the *length* built-in predicate returns, given a set of temporal intervals, the overall total length of the intervals.   □

Remark 9. Note that this definition is more general than "SELECT *expr* AS *?var*" project expressions in current SPARQL 1.1 [9] due to not requiring that the assigned variable be unbound.

We introduce the ORDERBY clause where the evaluation of a $[\![P \text{ ORDERBY } ?x]\!]_G$ statement is defined as the ordering of the solutions – for any $\theta \in [\![P]\!]_G$ – according to the values of $\theta(?x)$. Ordering for non-annotation variables follows the rules in [21, Section 9.1]. In case the variable x is an annotation variable, the order is induced by $\preceq$. In case, $\preceq$ is a partial order then we may use some linearisation method for posets, such as [12]. Likewise, the SQL-like statement LIMIT(k) can be added straightforwardly.

We can further extend the evaluation of AnQL queries with aggregate functions

$$@ \in \{\text{SUM}, \text{AVG}, \text{MAX}, \text{MIN}, \text{COUNT}, \wedge, \vee, \otimes\}$$

as follows: the evaluation of a GROUPBY statement is defined as:[13]

---

[12] These features are currently being defined by W3C, see [9] for the latest draft.

[13] In the expression, $\bar{@}\bar{\mathbf{f}}(\bar{\mathbf{z}})$ AS $\bar{\alpha}$ is a concise representation of n aggregations of the form $@_i f_i(\bar{\mathbf{z}}_i)$ AS α_i.

$$[P \text{ GROUPBY}(\bar{\mathbf{w}}) \ \bar{@}\bar{\mathbf{f}}(\bar{\mathbf{z}}) \ AS \ \bar{\alpha}]_G = \{\theta \mid \theta_1 \text{ in } [P]_G, \theta = \theta_1|_{\bar{\mathbf{w}}}[\alpha_i/@_i f_i(\theta_i(\bar{\mathbf{z}}_i))]\}_{\text{DISTINCT}} ,$$

where the variables $\alpha_i \notin var(P)$, $\bar{\mathbf{z}}_i \in var(P)$ and none of the GROUPBY variables $\bar{\mathbf{w}}$ are included in the aggregation function variables $\bar{\mathbf{z}}_i$. Here, we denote by $\theta|_{\bar{\mathbf{w}}}$ the restriction of variables in θ to variables in $\bar{\mathbf{w}}$. Using this notation, we can also straightforwardly introduce projection, *i.e.*, sub-SELECTs as an algebraic operator in the language covering another new feature of SPARQL 1.1:

$$[\text{SELECT } \bar{\mathbf{V}} \ \{P\}]_G = \{\theta \mid \theta_1 \text{ in } [P]_G, \theta = \theta_1|_{\bar{\mathbf{v}}}\} .$$

Remark 10. Please note that the aggregator functions have a domain of definition and thus can only be applied to values of their respective domain. For example, SUM and AVG can only be used on numeric values, while MAX, MIN are applicable to any total order. Resolution of type mismatches for aggregates is currently being defined in SPARQL 1.1 [9] and we aim to follow those, as soon as the language is stable. The COUNT aggregator can be used for any finite set of values. The last three aggregation functions, namely $\wedge, \vee$ and $\otimes$, are defined by the annotation domain and thus can be used on any annotation values.

Remark 11. Please note that, unlike the current SPARQL 1.1 syntax, assignment, solution modifiers (ORDER BY, LIMIT) and aggregation are stand-alone operators in our language and do not need to be tied to a sub-SELECT but can occur nested withinin any pattern. This may be viewed as syntactic sugar allowing for more concise writing than the current SPARQL 1.1 [9] draft.

Example 6. Suppose we want to know, for each employee, the average length of their employments with different employers. Then such a query will be expressed as:

```
SELECT ?x ?avgL
WHERE{(?x :worksFor :?y):?l GROUPBY(?x) AVG(length(?l)) AS ?avgL}
```

Essentially, we group by the employee, compute for each employee the time he worked for a company by means of the built-in function $length$, and compute the average value for each group. That is, if $g = \{\langle t, t_1\rangle, \ldots, \langle t, t_n\rangle\}$ is a group of tuples with the same value t for employee x, and value t_i for y, where each length of employment for t_i is l_i (computed as $length(\cdot)$), then the value of $avgL$ for the group g is $(\sum_i l_i)/n$. □

Proposition 3. *Assuming the built-in predicates are computable in finite time, the answer set of any AnQL is finite and can also be computed in finite time.*

This proposition can be demonstrated by induction over all the constructs we allow in AnQL.

3   Twisting AnQL – Issues and Pitfalls

In this section we discuss some practical issues arising in formulating real-life questions in AnQL like the treatment of non-annotated queries, combination of domains in queries and some domain specific issues while highlighting problems in some related works.

3.1   Uniform Treatment of Annotated and Non-annotated Queries

We aim at providing a uniform treatment for queries, *i.e.*, it should be allowed to ask annotated queries against non-annotated graphs and vice-versa. There are two distinct situations where a default value must be determined, viz., in the RDF data or in SPARQL queries. The treatment of non-annotated triples in the data has been discussed in [23] and here we just use the meta-variable Ω_D to represent the default value for domain D. We consider a similar solution for evaluating a SPARQL query over an annotated RDFS dataset. We allow that any non-annotated triple pattern τ be considered a BAP by assigning it a default annotation. We consider that a graph pattern P, is in *Annotated Normal Form (ANF)* if it does not contain any non-annotated triple patterns. Any graph pattern P can be transformed into *ANF* by replacing each non-annotated triple pattern $\tau_i \in P$ by using one of the following approaches:

1. adding a single annotation variable for each triple: $\tau_i : \lambda$, where λ is a new annotation variable not occurring in P; or
2. adding a different annotation variable for each non-annotated triple: $\tau_i : \lambda_i$ s.t. each λ_i is a new annotation variable not occurring in P and different from any other generated variable; or
3. adding the $\top$ element from the domain: $\tau_i : \top$.

In later discussions, we will use the meta-variable Θ_D to represent the default value of domain D assigned to annotations in the query triples.

Example 7. For instance, if we again consider the query (excluding the annotation variables) and input data from Example 2, the query would look like:

```
SELECT ?p ?c
WHERE {(?p :basedNear :paris) OPTIONAL{(?p :hasCar ?c)}}
```

Now, given the 3 approaches for transforming this query into ANF we would get the following answers:

Approach 1	$?p/$:alain	-
	$?p/$:alain	$?c/$:peugeot
	$?p/$:alain	$?c/$:renault
Approach 2	$?p/$:alain	$?c/$:peugeot
	$?p/$:alain	$?c/$:renault
Approach 3	$\emptyset$	

3.2   Querying Multi-dimensional Domains

Similarly to the discussion in the previous subsection, we can encounter mismatches between the Annotated RDFS dataset and the AnQL query. In case the AnQL query contains only variables for the annotations, the query can be answered on any Annotated RDFS dataset. From a user perspective, the expected answers may differ from the actual annotation domain in the dataset, *e.g.*, the user may be expecting temporal intervals in the answers when the answers actually contain a fuzzy value. For this reason some built-in predicates to determine the type of annotation should be introduced, like isTEMPORAL, isFUZZY, etc.

If the AnQL query contains annotation values and the Annotated RDFS dataset contains annotations from a different domain, one option is to not provide any answers. Alternatively, we can consider combining the domain of the query with the domain of the annotation into a multi-dimensional domain, as illustrated in the next example.

Example 8. Assuming the following input data:
$$(:alain, :livesIn, :paris): \{ex.org\}$$
When performing the following query:

```
SELECT ?p ?c WHERE { (?p :livesIn ?c):[2009, 2010] }
```

we would interpret the data to the form:
$$(:alain, :livesIn, :paris): (\{ex.org, \Omega_{temporal}\})$$
while the query would be interpreted as:

```
SELECT ?p ?c WHERE  (?p :livesIn ?c):(Θprovenance, [2009, 2010])
```

where $\Omega_{temporal}$ and $\Theta_{provenance}$ are annotations corresponding to the default values their respective domains, as discussed in Section 3.1. The semantics of combining different domains into one multi-dimensional domain has been discussed in [23].

3.3 Temporal Issues

Let us highlight some specific issues inherent to the temporal domain. Considering queries using Allen's temporal relations [1] (before, after, overlaps, etc.) as allowed in [24], we can pose queries like "find persons who lived in Paris before Alain". this query raises some ambiguity when considering that persons may have lived in the same city at different disjoint intervals. We can model such situations – relying on sets of temporal intervals modelling the temporal domain. Consider the following input data:

$$(:betty, :livesIn, :paris): \{[1990, 1995]\}$$
$$(:alain, :livesIn, :paris): \{[1980, 2000], [2002, 2010]\}$$

Tappolet and Bernstein [24] consider the latter triple as two triples with disjoint intervals as annotations. For the following query in their language τSPARQL:

```
SELECT ?p WHERE {
[?s1,?e1] ?p :livesIn :paris . [?s2,?e2] :alain :livesIn :paris .
[?s1,?e1] time:intervalBefore [?s2,?e2] }
```

we would get :betty as an answer although Alain was already living in Paris when Betty moved there. This is one possible interpretation of "before" over a set of intervals. In AnQL we could add different domain specific built-in predicates, representing different interpretations of "before". For instance, we could define binary built-ins (i) beforeAny($?A1, ?A2$) which is true if there exists *any* interval in annotation $?A1$ before an interval in $?A2$, or, respectively, a different built-in beforeAll($?A1, ?A2$) which is only true if *all* intervals in annotation $?A1$ are before any interval in $?A2$. Using the latter, an AnQL query would look as follows:

```
SELECT ?p WHERE {(?p :livesIn :paris):?l1 .
   (:alain :livesIn :paris):?l2 . FILTER(beforeAll(?l1,?l2))}
```

This latter query gives no result, which might comply with people's understanding of "before" in some cases, while we also have the choice to adopt the behaviour of [24] by use of beforeAny instead. Our report [13] provides more details on this issue.

3.4 Constraints vs Filters

Considering the previous section, please note that FILTERs do not act as *constraints* over the query. It could be expected that, given the data from the previous section, and for the following query:

```
SELECT ?l1 ?l2 WHERE { (?p :livesIn :paris):?l1 .
  (:alain :livesIn :paris):?l2 }
```

with an additional *constraint* that requires $?l1$ to be "before" $?l2$. We could expect the answer $\{?l1/[1990, 1995], ?l2/[1996, 2000]\}$ that matches the query with regards to the data and satisfies the proposed *constraint*. However, we require maximality of the annotation values in the answers, which in general, do not exist in presence of *constraints*. For this reason, we do not allow general *constraints*.

Conclusions

Based on our previous work on Annotated RDFS [23], we presented a semantics for an extension of the SPARQL query language, AnQL, that enables querying RDF with annotations. Queries are specified with regards to a specific domain, from which we presented some of the more common ones. Queries exemplified in related literature for specific extensions of SPARQL can be expressed in AnQL.

Noticeably, our semantics goes beyond the expressivity of the current SPARQL specification and includes some features from SPARQL 1.1 such as aggregates, variable assignments and sub-queries.

A prototype implementation, including the annotated RDFS inferencing and annotated SPARQL query engine is available at `http://anql.deri.org`.

Acknowledgement. The work presented in this report has been funded in part by Science Foundation Ireland under Grant No. SFI/08/CE/I1380 (Líon-2) and supported by COST Action IC0801 on Agreement Technologies. We thank Jürgen Umbrich for his useful comments.

References

1. Allen, J.F.: Maintaining knowledge about temporal intervals. Communications of the ACM 26(11), 832–843 (1983)
2. Angles, R., Gutierrez, C.: The Expressive Power of SPARQL. In: Sheth, A.P., Staab, S., Dean, M., Paolucci, M., Maynard, D., Finin, T., Thirunarayan, K. (eds.) ISWC 2008. LNCS, vol. 5318, pp. 114–129. Springer, Heidelberg (2008)
3. Brickley, D., Guha, R.V.: RDF Vocabulary Description Language 1.0: RDF Schema. W3C Recommendation (2004), `http://www.w3.org/TR/rdf-schema/`
4. Dividino, R.Q., Sizov, S., Staab, S., Schueler, B.: Querying for Provenance, Trust, Uncertainty and other Meta Knowledge in RDF. Journal of Web Semantics 7(3), 204–219 (2009)
5. Flouris, G., Fundulaki, I., Pediaditis, P., Theoharis, Y., Christophides, V.: Coloring RDF Triples to Capture Provenance. In: Bernstein, A., Karger, D.R., Heath, T., Feigenbaum, L., Maynard, D., Motta, E., Thirunarayan, K. (eds.) ISWC 2009. LNCS, vol. 5823, pp. 196–212. Springer, Heidelberg (2009)
6. Green, T.J., Karvounarakis, G., Tannen, V.: Provenance semirings. In: Proc. of 26th ACM SIGACT-SIGMOD-SIGART Symposium on Principles of Database Systems (PODS 2010), pp. 31–40 (2007)

7. Gutiérrez, C., Hurtado, C.A., Vaisman, A.A.: Introducing Time into RDF. IEEE Transactions on Knowledge and Data Engineering 19(2), 207–218 (2007)
8. Hájek, P.: Metamathematics of Fuzzy Logic. In: Trends in Logic, Kluwer, Dordrecht (1998)
9. Harris, S., Seaborne, A.: SPARQL 1.1 Query Language. W3C Working Draft (2010),
   http://www.w3.org/TR/2010/WD-sparql11-query-20100601/
10. Hartig, O.: Querying Trust in RDF Data with tSPARQL. In: Aroyo, L., Traverso, P., Ciravegna, F., Cimiano, P., Heath, T., Hyvönen, E., Mizoguchi, R., Oren, E., Sabou, M., Simperl, E. (eds.) ESWC 2009. LNCS, vol. 5554, pp. 5–20. Springer, Heidelberg (2009)
11. Klement, E.P., Mesiar, R., Pap, E.: Triangular Norms. In: Trends in Logic, Kluwer, Dordrecht (2000)
12. Labrador, N.M., Straccia, U.: Monotonic Mappings Invariant Linearisation of Finite Posets. Technical report, Computing Research Repository (2010), Available as CoRR technical report at, http://arxiv.org/abs/1006.2679
13. Lopes, N., Lukácsy, G., Polleres, A., Straccia, U., Zimmermann, A.: A General Framework for Representing, Reasoning and Querying with Annotated Semantic Web Data. Technical report, DERI (2010),
   http://www.deri.ie/fileadmin/documents/DERI-TR-2010-03-29.pdf
14. Manola, F., Miller, E.: RDF Primer. W3C Recommendation (2004),
   http://www.w3.org/TR/rdf-primer/
15. Mazzieri, M., Dragoni, A.F.: A Fuzzy Semantics for the Resource Description Framework. In: da Costa, P.C.G., d'Amato, C., Fanizzi, N., Laskey, K.B., Laskey, K.J., Lukasiewicz, T., Nickles, M., Pool, M. (eds.) URSW 2005 - 2007. LNCS (LNAI), vol. 5327, pp. 244–261. Springer, Heidelberg (2008)
16. Muñoz, S., Pérez, J., Gutiérrez, C.: Minimal Deductive Systems for RDF. In: Franconi, E., Kifer, M., May, W. (eds.) ESWC 2007. LNCS, vol. 4519, pp. 53–67. Springer, Heidelberg (2007)
17. Pérez, J., Arenas, M., Gutiérrez, C.: Semantics and complexity of SPARQL. ACM Transactions on Database Systems 34(3) (2009)
18. Peterson, D., Gao, S., Malhotra, A., Sperberg-McQueen, C.M., Thompson, H.S.: W3C XML Schema Definition Language (XSD) 1.1 Part 2: Datatypes. W3C Working Draft (2009),
   http://www.w3.org/TR/xmlschema11-2/
19. Pugliese, A., Udrea, O., Subrahmanian, V.S.: Scaling RDF with time. In: Proc. of 17th International Conference on World Wide Web (WWW 2008), pp. 605–614 (2008)
20. Schenk, S.: On the Semantics of Trust and Caching in the Semantic Web. In: Sheth, A.P., Staab, S., Dean, M., Paolucci, M., Maynard, D., Finin, T., Thirunarayan, K. (eds.) ISWC 2008. LNCS, vol. 5318, pp. 533–549. Springer, Heidelberg (2008)
21. Seaborne, A., Prud'hommeaux, E.: SPARQL Query Language for RDF. W3C Recommendation (2008), http://www.w3.org/TR/rdf-sparql-query/
22. Straccia, U.: A Minimal Deductive System for General Fuzzy RDF. In: Polleres, A., Swift, T. (eds.) RR 2009. LNCS, vol. 5837, pp. 166–181. Springer, Heidelberg (2009)
23. Straccia, U., Lopes, N., Lukacsy, G., Polleres, A.: A General Framework for Representing and Reasoning with Annotated Semantic Web Data. In: Proc. of 24th AAAI Conference on Artificial Intelligence (AAAI 2010). AAAI Press, Menlo Park (2010)
24. Tappolet, J., Bernstein, A.: Applied Temporal RDF: Efficient Temporal Querying of RDF Data with SPARQL. In: Aroyo, L., Traverso, P., Ciravegna, F., Cimiano, P., Heath, T., Hyvönen, E., Mizoguchi, R., Oren, E., Sabou, M., Simperl, E. (eds.) ESWC 2009. LNCS, vol. 5554, pp. 308–322. Springer, Heidelberg (2009)
25. Udrea, O., Recupero, D.R., Subrahmanian, V.S.: Annotated RDF. In: Sure, Y., Domingue, J. (eds.) ESWC 2006. LNCS, vol. 4011, pp. 487–501. Springer, Heidelberg (2006)
26. Udrea, O., Recupero, D.R., Subrahmanian, V.S.: Annotated RDF. ACM Transactions on Computational Logic 11(2), 1–41 (2010)

Using Semantics for Automating the Authentication of Web APIs

Maria Maleshkova[1], Carlos Pedrinaci[1], John Domingue[1],
Guillermo Alvaro[2], and Ivan Martinez[2]

[1] Knowledge Media Institute (KMi)
The Open University, Milton Keynes, United Kingdom
{m.maleshkova,c.pedrinaci,j.b.domingue}@open.ac.uk
[2] Intelligent Software Components (iSOCO), Madrid, Spain
{galvaro,imartinez}@isoco.com

Abstract. Recent technology developments in the area of services on the Web are marked by the proliferation of Web applications and APIs. The implementation and evolution of applications based on Web APIs is, however, hampered by the lack of automation that can be achieved with current technologies. Research on semantic Web services is therefore trying to adapt the principles and technologies that were devised for traditional Web services, to deal with this new kind of services. In this paper we show that currently more than 80% of the Web APIs require some form of authentication. Therefore authentication plays a major role for Web API invocation and should not be neglected in the context of mashups and composite data applications. We present a thorough analysis carried out over a body of publicly available APIs that determines the most commonly used authentication approaches. In the light of these results, we propose an ontology for the semantic annotation of Web API authentication information and demonstrate how it can be used to create semantic Web API descriptions. We evaluate the applicability of our approach by providing a prototypical implementation, which uses authentication annotations as the basis for automated service invocation.

1 Introduction

Web services provide means for the development of open distributed systems, based on decoupled components, by overcoming heterogeneity and enabling the publishing and consuming of functionalities of existing pieces of software. Recently the world around services on the Web, thus far limited to "classical" Web services based on SOAP and WSDL, has been enriched by the proliferation of Web applications and APIs, also referred to as RESTful services [1], when conforming to the REST architectural principles. Web APIs are characterised by their relative simplicity and their natural suitability for the Web, relying directly on the use of URIs, for both resource identification and interaction, and HTTP for message transmission. Many popular Web 2.0 applications like Facebook,

P.F. Patel-Schneider et al. (Eds.): ISWC 2010, Part I, LNCS 6496, pp. 534–549, 2010.

Google, Flickr and Twitter offer easy-to-use, publicly available APIs, which not only provide simple access to different resources but also enable combining heterogeneous data coming from diverse services, in order to create data-oriented service compositions called mashups.

Despite their success, Web APIs are currently facing a number of limitations. While the development, publication and use of Web services is guided by standards and specifications, the Web API landscape is much more heterogeneous and diverse. This heterogeneity is especially present in the forms and structure of the documentation, since currently most Web API descriptions are given directly in text/HTML as part of a webpage. Providers publish the APIs in a way that they see fit, following no particular guidelines and conforming to no particular standards. As a consequence, in order to use Web APIs, developers are obliged to manually locate, retrieve and interpret heterogeneous documentation, and subsequently develop custom tailored software, which has a very low level of reusability. The diversity of the Web APIs is accompanied by a wide range of used authentication approaches, which hinder the automated Web API invocation. The lack of a common structured language for describing Web APIs is addressed by some initial proposals [2], [3], while lightweight annotations over Web API descriptions [4], [5] have been developed as means for overcoming the existing heterogeneity and providing basic support for service task automation. Still, up to date, the importance of authentication as part of the invocation process has been neglected. As our study points out (see Section 4.2) the majority of the Web APIs require some form of authentication but none of the existing formalisms and annotation approaches deal with this. Moreover, none of the available tools, which provide developer support for creating mashups, such as Yahoo Pipes and DERI Pipes[1], handle authentication in an integrated way and it still needs to be addressed separately. As a result, the invocation of individual Web APIs and their use within mashups, requires extensive manual development work, independently of whether they are semantically annotated or not.

In order to support the automated invocation of Web APIs, we propose the use of semantic annotations over the existing heterogeneous HTML descriptions. As shown by our study, the invocation of Web APIs requires authentication in more than 80% of the cases, but currently there is no description formalism or semantic annotation approach, which addresses this, and commonly the need for authentication support is simply neglected. Therefore, we provide an ontology for the annotation of authentication information on top of Web API descriptions, which covers all authentication mechanisms identified by our study and is extendable to cover further ones. We show how semantic descriptions of authentication details can be created and validate our approach by providing a prototype of an invocation system, which effectively uses the created annotations to support the automated invocation of Web APIs.

The remainder of this paper is structured as follows: Section 2 provides a motivating example that illustrates the challenges related to Web API authentication, while Section 3 gives an overview of previous work in the area of semantic Web

---

[1] http://pipes.yahoo.com/pipes/, http://pipes.deri.org/

API descriptions and Web service security. A detailed analysis of current API authentication approaches is given in Section 4. Based on the Web API survey, in Section 5 we propose an ontology and an implementation for supporting the automated authentication, by using lightweight semantic annotations. Section 6 presents an overview of related work and Section 7 concludes the paper.

2   Motivating Example

In this section we present a simple example, which demonstrates the necessity of authentication information during the invocation of Web APIs, and use it throughout the paper to illustrate the here proposed annotation approach. In particular, we describe one of the operations of the Last.fm Web API[2]. The Last.fm API enables developers to access and use the Last.fm data. This popular website for music claims more than 40 million active users and provides details about artists, albums, events and user-specific information, such as playlists.

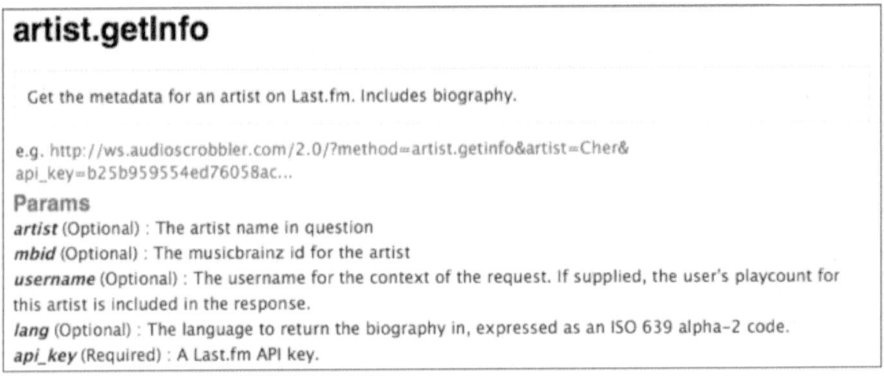

Fig. 1. Extract from the Last.fm API

Figure 1 shows the Web API operation for getting the details for a particular artist. The provided data can be used directly or as part of a mashup, where artists details are combined with latest charts news, for example. However, since the Web API is described solely in HTML, the discovery and interpreting of the documentation have to be done manually. Moreover, even if the Last.fm API were semantically annotated or had a machine-processable description, for example in WADL, the automated invocation would still not be possible because of the necessity of providing an authentication API key, which cannot be captured with existing description forms. The work presented in this paper is targeted at addressing precisely this problem.

---

[2] http://www.last.fm/api

3   Background

Since the advent of Web service technologies, research on semantic Web services (SWS) has been devoted to reduce the extensive manual effort required for manipulating Web services. The main idea behind this research is that tasks such as discovery, negotiation, composition and invocation can have a higher level of automation, when services are enhanced with semantic descriptions of their properties. Similarly to "classical" Web services, Web API-related tasks also require a lot of developer involvement and face even further difficulties, since there is no established common formalism for describing Web APIs. In order to address this, lightweight annotations over API descriptions have been proposed as means for achieving a higher-level of automation.

Currently, there are two main contributions aiming at using semantics to support the automation of common Web API service-related tasks. Both approaches rely on marking service properties within the HTML description and subsequently liking these to semantic entities. MicroWSMO [4] is a formalism for the semantic description of Web APIs, which is based on adapting the SAWSDL [6] approach for enhancing service properties with semantic information. MicroWSMO uses microformats for adding semantic information on top of HTML service documentation, by relying on hRESTS [7] for marking service properties. Listing 1.1 shows the hRESTS annotation of the Last.fm API, where the different service properties are identified via HTML tags.

Listing 1.1. Last.fm Web API

```
1   <div class="service" id="service1"><h1>Last.fm Web Services</h1>
2   <div class="operation" id="op1"><h2><span class="label">artist.getInfo</span></h2>
3   <div>Get the metadata for an artist on Last.fm. Includes biography.</div>
4   <span class="address">http://ws.audioscrobbler.com/2.0/?method=artist.getinfo&amp;
5   artist =Cher&amp;api_key=xxx</span>
6   <div class="input" id="input1">
7   <span>artist</span> (Optional) : The artist name in question<br>
8   <span>lang</span> (Optional) : The language to return the biography in.<br>
9   <span>api_key</span> (Required) : A Last.fm API key.<br></div>
10  <div class="output" id="output1">Artist</div></div></div>
```

Another formalism is SA-REST [5], which also applies the grounding principles of SAWSDL but instead of using hRESTS relies on RDFa [8] for marking service properties. Similarly to MicroWSMO, SA-REST enables the annotation of existing HTML service descriptions by identifying service elements and linking these to semantic entities. The main differences between the two approaches are not the underlying principles but rather the implementation techniques.

Both MicroWSMO and SA-REST, provide a solid foundation for the use of semantics as the basis for automating common service tasks. However, they are very lightweight and the automation support that they provide is limited. More importantly, all existing approaches neglect the need for addressing the automation of Web API authentication. Therefore, we use existing models for the semantic description of Web APIs as the basis for an incremental approach for reflecting authentication information.

3.1   WS-Security

The issues of authentication and security have already been tackled in the context of WSDL and SOAP-based Web services. The result is a unified Web service security standard. WS-Security [9] specifies a set of feature extensions to SOAP messaging, in order to provide message integrity and confidentiality. In addition, it also provides a mechanism for associating security tokens with message content and allows for a variety of signature formats and encryption algorithms. As a result, the defined enhancements provide support for ensuring that the sent message is not altered by a third party (message integrity), that its content cannot be read by anyone but the designated client or server (confidentiality) and that the user has the necessary credentials in order to access particular resources (authentication).

WS-Security addresses the main security issues in the context of Web services. However, in contrast to WSDL-based services, Web APIs are proliferating autonomously without the creation of standards and independently from Web services. The result is a very heterogeneous world of Web APIs, where security issues such as confidentiality and message integrity, guaranteed by the WS-Security standard, are not considered as crucial. In fact, as our study shows, security in the context of Web APIs is reduced only to authentication, which in turn serves mainly the purposes of access control, where providers rather want to restrict and track the number of requests, instead of providing data integrity.

4   Investigating Authentication for Web APIs

In order to become aware of currently used Web API authentication approaches, we conducted a study, analysing 222 Web APIs from the ProgrammableWeb[3] directory. ProgrammableWeb is a popular API directory, providing information about 2002 APIs and 4827 mashups (visited June 2010). For easier search and browsing, the APIs are sorted in categories and our analysis covered all 51 categories, including on average 4 APIs per category. The analyzed Web APIs from each category were randomly chosen, however, since some categories have only one or two entries, the analyzed number of Web APIs per category varies. As a result the survey covered 18% of the REST ProgrammableWeb APIs (1235 APIs at the time of the study, February 2010). Therefore, we consider the following results to be representative for the directory and in general, since ProgrammableWeb is currently the biggest directory[4].

In the following sections we first provide an overview of common authentication approaches, as identified by our Web API analysis, and then layout the results and conclusions of our Web API survey.

---

[3] http://www.programmableweb.com
[4] Webmashup.com (http://www.webmashup.com) contains around 1800 Web APIs and 3100 mashups, while APIFinder (http://www.apinder.com) provides around 1100 Web APIs.

4.1   Common Authentication Approaches

Currently, as our survey shows, most Web APIs use one of five authentication mechanisms. We differentiate between approaches based only on authentication credentials (API key or username and password), approaches using a transmission security protocol (HTTP Basic Authentication, HTTP Digest Authentication and OAuth), and approaches that use different parts of the HTTP request in order to transmit the authentication information. We start by describing authentication mechanisms relying only on the input credentials.

API Key. Currently, the most common way of Web API authentication is via API key (also called "developer key", "developer token", "token Id", "user Id", "user key"). Web APIs using this mechanism include Last.fm (`http://www.last.fm/api`) and Remember the Milk (`http://www.rememberthemilk.com/services/api/`). This authentication mechanism does not have any security measures for the message integrity and confidentiality but is rather only based on the necessary credentials. The user only needs to provide the API key, which is received by signing up for the particular Web API. The key is transmitted either as a parameter in the Web API URI or directly in the HTTP request. Each client that provides a valid API key is permitted access to the requested resources. This approach is very simple to use and to implement. However, since the API key is not protected in any way during the message transmission, but is rather sent directly as plain text, this method is suitable for Web API providers, who only want to somehow restrict the access to the available resources.

Username and Password. Similarly, to authentication via API key, authentication via username and password is also only based on the required credentials. It provides no message encryption or signature and the login details are transmitted as parameters of the request URI or are included in the HTTP request. Example Web APIs include Happenr (`http://www.happenr.com/webservices/`, for example `http://happenr.com/webservices/getEvents.php?username=xxx&password=xxx&town=London`) and FileSocial (`http://filesocial.com/api/docs`). The user only needs to create an account for the particular Web API and can use the username and password (in some cases email and password, telephone number and pin, username and token or API key and private key) to access resources. Similarly to the authentication via API key, this approach is only suitable for providers who want to restrict the traffic and the number of requests to their APIs.

The first two authentication mechanisms are only based on the required credentials, while the following approaches, including HTTP Basic Authentication and HTTP Digest Authentication, are transmission security protocols. These, provide a higher level of security for the login details and the client's message.

HTTP Basic Authentication. [10] provides a simple way for user authentication. It is based on a challenge-response model, where the HTTP server requests and validates the authentication of the Web client. Example Web APIs include Assembla (`https://www.assembla.com/wiki/show/breakoutdocs/Assembla_REST_API`) and Basecamp (`http://developer.37signals.com/`

basecamp/). In order to access a Web API operation, which requires authentication, the client needs to provide the corresponding username and password in the form of an authentication header (with value Base64encode [11] of the string `username+":"+password`). The Base64-encoded string is transmitted and decoded by the receiver, resulting in the colon-separated user name and password strings, which are checked against the expected values.

This type of authentication is very simple and is supported by all popular Web browsers. However, although it uses Base64 encoding, it does no encryption and the username and password can directly be decoded from the transmitted message. Therefore, this type of authentication is only suitable for Web APIs with lower data security demands.

HTTP Digest Authentication. [10] follows the same process as the HTTP Basic one – request, credentials challenge and response. However, it only transmits a digest of the username and password, which cannot be directly decoded. Example Web APIs include Talis (`http://n2.talis.com/wiki/Platform_API`) and AdSpeed   (`http://www.adspeed.com/Knowledges/830/AdSpeed_API/AdSpeed_API_Overview.html`, for example `http://api.adspeed.com/?method=METHOD NAME&param1=VALUE&param2=VALUE&md5=`**SIGNATURE**). The first time a client sends a request to the server, the server responds with a nonce (a random string) and the realm (typically a description of the computer or system being accessed). The client uses the username and password, to compute the digest response (result of `MD5(username:realm:password)`) and the digest of the nonce (usually by using MD5 [12]), which are put in the response. The server processes the response by retrieving the stored password for the user and testing the nonce. If the nonce is correct, the response digest is checked by using the nonce, username and password to compute a digest and compare it to the received one. If the two digests match, the client is allowed access to the resources.

OAuth. [13] is a protocol for making authenticated HTTP requests by using a token. It enables users to share private resources stored on one website with another one by using a token, which is an identifier denoting an access grant with specific scope, duration, and other attributes, instead of the username and password. Therefore, OAuth supports the interoperability and the combining of resources coming from different websites, in a way that is transparent for the user. Example Web APIs include Fire Eagle (`http://fireeagle.yahoo.net/developer/documentation`) and Delicious (`http://delicious.com/help/api`).

Whenever a Web API (or a website) needs to access resources from another Web API, the user is asked to provide his/her access information for the host Web API, while in the background, OAuth creates a token, which can be used by other APIs to gain access to the resources. As a result the username and password are kept private and unavailable for third-party websites and APIs, while the interoperability is still facilitated. This authentication approach is extremely important in the context of mashups, since it does not require that the user provides credentials for every Web API included in the composition, but rather relies on token-based user-transparent handling of authentication.

So far we have described authentication based on different credentials and on using different authentication mechanisms. In addition, there are also two main ways of transmitting the authentication information. One very common way is to directly provide the API key or username and password as parameters in the request URI. For example Last.fm (`http://ws.audioscrobbler.com/2.0/?method=artist.getinfo&artist=Cher&apikey=XXX`) and Fire Eagle (`https://fireeagle.yahooapis.com/api/0.1/`) use this approach. Since the authentication credentials are not protected in any way, this way of sending data is suitable for openly available resources, where providers want to restrict the access to the API but are not concerned with enforcing access rights. The other common way of sending authentication credentials is directly in the HTTP request. This method is somewhat more complex because the client needs to construct the request, instead of only calling a parameterized URI. However, it enables a higher level of security, since the information can be encrypted and signed. For example, this way of sending information is commonly used by the HTTP digest authentication approach.

In summary, current authentication approaches have three main characteristics: 1) the required credentials, 2) the used authentication protocol, and 3) the way of sending the authentication information.

4.2   Web API Survey Results

After introducing the most common authentication approaches, in this section we focus on describing the results and the main findings of our Web API study. Table 1 shows the results of our analysis in terms of Web API authentication approaches. As it can be seen, using an API key is by far the most common way of authentication (38%). It is followed by 19% of APIs, which do not require any authentication. HTTP Basic and HTTP Digest are not used as often (14%, 5%), while about 6% of the APIs use OAuth and 5% implement their own operations, which need to be called, before being able to invoke other operations. There are

Table 1. Common Web API Authentication Approaches

Authentication Mechanisms	Number	In %
API Key	89	38%
No Authentication	46	19%
HTTP Basic	32	14%
Username and Password	19	8%
OAuth	14	6%
Web API Operation	12	5%
HTTP Digest	11	5%
API Key in Combination with Other Credentials	5	2%
Session Based	5	2%
Other	2	1%
Authentication Only for Data Modification	4	2%
Offer Alternative Authentication Mechanisms	16	7%

also some APIs, which require authentication only for operations, which perform data modification but require no authentication for only reading resources. The sum of all APIs is greater than 222 because, APIs that offer more that one authentication mechanism were counted more than once.

It is important to point out that currently 81% of the Web APIs require some form of authentication. Therefore, providing support for the annotation and automation of authentication is crucial for Web API use. In addition, there is no established approach for Web API authentication but rather a landscape of different approaches. Also, about only a quarter of the APIs use a mechanism that protects the user credentials and does not transmit them directly in plain text. This shows that providers are not so much concerned with verifying the user identity and do not invest implementation work in securing the message transfer but rather prefer to employ simple measures for access control. This is verified by the fact that less than 10% of the Web APIs use signatures and encryption.

Table 2. Way of Transmitting Credentials

Transmission Medium	Number	In %
URI	117	70%
HTTP Header	45	27%
URI or HTTP Header (Depending on the Type of Authentication and HTTP Method)	6	3%

Table 2 shows the most commonly used ways of transmitting authentication credentials. As it can be seen, 70% of the Web APIs send authentication information directly in the URI, while less than one third require that the HTTP header is constructed. This means that even if Web APIs require authentication, most of them do not need a custom client but can rather be invoked directly from a Web browser.

The survey also delivered some important information about the Web API description forms. In particular, none of the analyzed APIs used WSDL [3] or WADL [2] and the majority of the APIs are documented directly in HTML Web pages. The main conclusions of the Web API survey can be summarised as follows:

1. More that 80% of the Web APIs require authentication. Therefore authentication is a vital part of the invocation process and any approach disregarding authentication information has very limited support.
2. The currently used authentication approaches are very heterogeneous and there is no commonly accepted way for addressing Web API authentication.
3. Only about 25% of the Web APIs use an authentication approach that protects the user credentials and/or the content of the message.

5   Supporting the Automation of Web API Authentication

As highlighted by the Web API survey, authentication is essential and any general purpose solution aiming at supporting the invocation of Web APIs or mashups would necessarily be restricted to only 20% of the APIs or requires customisation. In addition, the authentication approaches are very diverse and similarly to Web API descriptions in general, authentication details are not described in a machine processable way but are given as part of documentation websites. In order to overcome the heterogeneity and provide means for the automatic recognition and processing of authentication details, we propose that Web API descriptions are annotated with semantic information about authentication. We next present the main design principles followed when deriving the proposed authentication ontology and continue by describing it in detail.

5.1   Design Principles

Guided by the results of the Web API study, we analyzed the collected data and derived an authentication ontology, which enables the annotation of authentication information as part of a semantic Web API description. The process of defining this ontology was guided by a number of competency questions and design principles. First, we started by identifying the cases, in which authentication is required, and the information that is needed. Relevant information in this respect is: *"Does the service require authentication?"*, *"Which operations require authentication?"*, *"What kind of authentication is used?"*, and *"What is the required information to complete the authentication?"*. As we concluded, based on the analysis of common authentication approaches, authentication has three main characteristic including the required credentials, the used authentication protocol, and the way of sending the authentication information. Therefore, we can identify the information necessary for supporting a particular authentication mechanism by determining: *"What are the required credentials?"*, *"What is the used authentication protocol?"* and *"How is the authentication information transmitted?"*. In addition to the competency questions, used for identifying the information that needs to be captured by the authentication ontology, we implemented some complementary requirements, which are specified in the form of design principles. The principles are as follows:

1. The ontology should cover all common authentication approaches identified by the Web API study.
2. The ontology should be extendable to cover further mechanisms.
3. The ontology should capture the information required for the automation of authentication as part of the invocation process.
4. The ontology should be compatible with existing semantic annotation approaches, such as MicroWSMO and SA-REST.
5. The ontology should support simplicity of use for making annotations.
6. The ontology should aim to be minimal but capture the necessary information for supporting the authentication.

The so designed ontology is not bound to any particular annotation formalism, but can be used as an extension by simply attaching it to the service and operation elements. In the following section we introduce the authentication ontology and show how it supports the automation of the invocation of Web APIs by using it as part of an authentication engine.

5.2 Authentication Ontology

Figure 2 depicts the Web API authentication ontology with namespace *waa* (Web API Authentication), which consists of three main classes– *AuthenticationMechanism*, *Credentials* and *TransmissionMedium*[5].

The *AuthenticationMechanism* class has six subclasses, corresponding to common authentication mechanisms, where the *Direct* subclass is used to describe approaches, which rely on using only credential details and employ no authentication protocol. The *Credentials* class has a number of instances including *API_Key*, *Username*, *Password* and *OAuth Credentials*, which can be combined to produce composite credentials, such as authentication through username and password. The *composedOf* relationship as well as the class *AuthenticationMechanism*, which can have further subclasses, represent points of extensibility for the ontology. The *Service* class has a relationship to the *ServiceAuthentication* class, which has three instances including *All*, *Some* and *None* that are used to point out that the service requires authentication for all its operations, for only some of them or for none of them. The *TransmissionMedium* has two instances (*ViaHTTPHeader* and *ViaURI*), used to describe that the credentials are sent by using only the URI or through constructing an HTTP header.

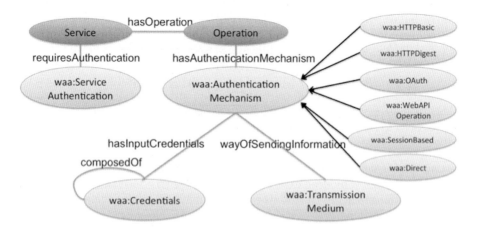

Fig. 2. Web API Authentication Ontology

The *Service* and *Operation* classes lack a namespace, because they serve as placeholders that can be replaced by the service and operation elements of any Web API model whether it is semantic, such as MicroWSMO or SA-REST, or not. In this way, the ontology can be used as an extension to existing formalisms or independently from them.

---

[5] The ontology is available at `http://purl.oclc.org/NET/WebApiAuthentication`

In order to show how the authentication ontology can be used to annotate Web APIs, we have taken the Last.fm motivating example from Section 2 and provide its HTML annotation (Listing 1.2) and semantic description (Listing 1.4). We apply the annotation approach presented in [14], where Web API descriptions are enhanced with semantic information by using MicroWSMO and SWEET as a supporting tool. In the example we use the *wsl* (`http://www.wsmo.org/ns/wsmo-lite`) namespace for WSMO-Lite, which is used in MicroWSMO annotations for the service model. However, as pointed out, the authentication annotations can be assigned to any operation and service model elements.

Listing 1.2. Example MicroWSMO Authentication Annotation

```
1   <div class="service" id="service1"><h1>Last.fm Web Services</h1>
2   <a rel="model" href="http://purl.oclc.org/NET/WebApiAuthentication#All">
3   <div class="operation" id="op1"><h2><span class="label">artist.getInfo</span></h2>
4   <a rel="model" href="http://purl.oclc.org/NET/WebApiAuthentication/LastFm">
5   <span class="address">http://ws.audioscrobbler.com/2.0/?method=artist.getinfo...</span>
6   <div class="input" id="input1">...</div>
7   <div class="output" id="output1">Artist</div></div></div>
```

Listing 1.2 shows the annotated HTML of the Last.fm API. The Web API requires authentication for all its operations (Line 2) and has authentication information for the *artist.getInfo* operation reflected in Line 4. The model reference contains a URI pointing to a particular instance of the *AuthenticationMechanism* class, which contains details about the operation requiring an API key, which is sent in the URI without the use of any authentication protocols.

Listing 1.3. Example Instance of the AuthenticationMechanism Class

```
1   @prefix rdf:      <http://www.w3.org/1999/02/22-rdf-syntax-ns#> .
2   @prefix waa:   <http://purl.oclc.org/NET/WebApiAuthentication#> .
3   <http://purl.oclc.org/NET/WebApiAuthentication/LastFm> rdf:type waa:Direct ;
4     waa:hasInputCredentials waa:API_Key ;
5     waa:wayOfSendingInformation waa:ViaURI .
```

Listing 1.3 shows how this instance of the *AuthenticationMechanism* class looks like. As it can be seen, the capturing of authentication information with the provided Web API authentication ontology is very simple and easy to apply.

Listing 1.4. Example RDF Authentication Annotation

```
1    : service1  rdf:type wsl: Service  ;
2         rdfs:isDefinedBy <http://www.last.fm/api/show?service=267> ;
3         waa:requiresAuthentication waa:All  .
4    : operation1 rdf:type wsl:Operation ;
5         rdfs: label " artist . getInfo" ;
6         hr: hasAddress "http://ws.audioscrobbler.com/2.0/?method=artist.getinfo&..." ;
7         waa:hasAuthenticationMechanism <http://purl.oclc.org/NET/WebApiAuthentication/LastFm> .
8    <http://purl.oclc.org/NET/WebApiAuthentication/LastFm> rdf:type waa:Direct ;
9         waa:hasInputCredentials waa:API_Key ;
10        waa:wayOfSendingInformation waa:ViaURI .
11   : service1 wsl:hasOperation : operation1 .
```

Based on the annotated HTML, the authentication information can easily be extracted in RDF (Listing 1.4) by using a simple XML transformation. All examples are available at `http://sweet.kmi.open.ac.uk/examples/`.

5.3   Authentication Engine

In this section we show how the authentication Web API annotations can be used to support the automated invocation of services. The contribution described here is implemented as part of SPICES[6] [15] (Semantic Platform for the Interaction and Consumption of Enriched Services), a platform for the easy consumption of services based on their semantic descriptions. In particular, SPICES supports both the end-user interaction with services and the invocation process itself, via the generation of appropriate user interfaces. Typically, the user is presented with a set of fields, which must be completed to allow the service to execute, and these fields cove input parameters as well as authentication credentials.

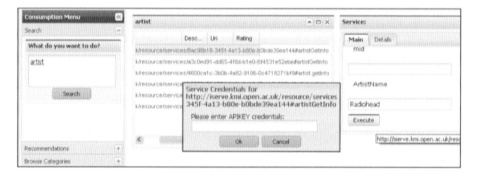

Fig. 3. Invoking the Last.fm API

Dealing with the different types of credentials and the way they have to be used, is the purpose of an Authentication Engine, part of SPICES, developed as a REST service, which is capable of handling the storage and retrieval of credentials for different Web APIs. This engine has the necessary logic to support the user in his/her interaction with services. In particular, if the engine has the credentials for a given service, thanks to the authentication annotations described previously, it is able to create a suitable request including the credentials. If the authentication credentials are not available yet, based on the authentication annotations, the authentication engine will be aware of the missing credentials and will prompt the user to provide them. Currently, the authentication engine plays the role of a trusted party, since it accesses and stores all user credentials. However, SPICES is only a prototypical implementation, with the main focus on supporting the invocation of Web APIs, and the authentication engine represents an initial practical application of the here presented approach. Therefore some

---

[6] `http://soa4all.isoco.net/spices`

issues such as appropriately storing and managing user credentials, still need to be addressed. However, since more than 70% of the authentication approaches do not protect the user credentials, this issue is not so crucial.

Figure 3 shows how the authentication engine prompts the user for the Last.fm API key, based on the API annotation, during the process of invoking the *artist.getInfo* operation. In this way the authentication engine can collect the required credentials and compose an API request, which together with the provided input information, supports the automated Web API invocation.

6   Related Work

In this section we describe further existing Web API authentication approaches, which address different challenges in the context of authentication but have not yet reached greater popularity.

Web-key [16] is an authentication mechanism, especially designed to tackle the difficulties arising in the context of mashup authentication. Web-key is an https URL convention for representing a transferable permission in a Web application. It binds each permission issued from the Web application to a randomly generated bit string (key), which is transmitted in the fragment segment of the URL (for example https://www.emaple.com/app/**#mhbqcmmva5ja3**). The keys are generated on behalf of the user for every Web API that is part of the composition. In this way, each Web API has its unique key, instead of the user having to share his username and password across all composite APIs. However, this approach is fairly new and its adoption is still limited.

Another authentication mechanism is **FOAF+SSL** [17]. FOAF+SSL is a simple protocol for RESTful authentication, which enables a one-click signing into websites by using a browser as the client application. It requires the user to enter neither a password nor an identifier but rather uses SSL and a custom trust protocol. The custom trust protocol is based on authentication certificates, which contain semantic descriptions of the authentication information in the subject alternative Name URI. FOAF+SSL presents a novel authentication approach, which includes Semantic Web principles in the authentication process. It remains to be seen how well it will be adopted for Web API implementations.

OpenID[7] targets to solve the problem of one user being forced to have many different Web application and API accounts, in order to be able to execute a mashup. It is a method based on using a single login at a trusted provider to automatically gain access to other websites. In this way the user can log into different services with the same digital identity, where these services trust the authentication body. Website providers, which use OpenID include AOL, IBM, Microsoft and others. OpenID is often seen as a complimentary approach to OAuth, where OpenID credentials can be used for generating OAuth tokens.

XAuth[8] provides an approach for extending authenticated user services throughout the Web by issuing user browser tokens for each of the participating

---

[7] http://openid.net
[8] http://xauth.org

services. In this way the provider can recognize, which users are logged into the services and not only give access to resources but also give additional relevant options. A different approach is followed by **Yadis**[9], who instead of suggesting a new authentication mechanism, propose means for automatically detecting, which authentication protocol a particular system is most likely to use. Therefore, Yadis addresses the question of how do we know, what authentication needs to be used, by providing a service discovery system that determines automatically, without end-user intervention, the most appropriate protocol to use.

The approach, which we propose in this paper, differs from Web-key, FOAF+ SSL, XAuth and Yadis and other authentication mechanisms because we are not suggesting to alter the current Web API authentication landscape by introducing a common standard. Instead, based on a study of current Web API authentication mechanisms, we provide a lightweight model and an approach for the annotation of APIs. The resulting semantic descriptions serve as the basis for automating the Web API invocation process.

In addition to the here listed authentication mechanisms, there is also one approach that uses semantic Web service descriptions in order to capture authorisation and privacy service properties [18]. The authors suggest that privacy and authentication policies should be incorporated into the OWL-S Web service descriptions. This additional information can then be integrated into the service matchmaking process. Similarly to our approach, this approach uses semantic descriptions for capturing authentication information. However, it is suitable only for WSDL-based services annotated in OWL-S.

7 Conclusion and Future Work

Nowadays, finding, interpreting and invoking Web APIs requires extensive human involvement due to the lack of API machine-processable descriptions. Efforts like SA-REST and MicroWSMO aim to overcome this difficulty and provide basic support for the automation of common service tasks such as discovery and composition. However, currently none of the existing approaches support the automated authentication as part of the Web API invocation process. As a result, developers are required to manually retrieve and interpret the HTML documentation, to signup with API providers, in order to receive access credentials, and to implement support for the different authentication protocols. In addition, none of the existing frameworks for supporting the creation of mashups, such as Yahoo Pipes and DERI Pipes, enable the handling of authentication in an integrated way and it has to be addressed with additional manual effort.

Our Web API study shows that more than 80% of the APIs require authentication, which makes authentication a vital part of the invocation process and any invocation approach disregarding authentication information has very limited support. Therefore, we propose the annotation of authentication information by using an authentication ontology, which overcomes Web API heterogeneity and

---

[9] http://yadis.org

provides the basis for automated authentication handling. We base the annotation approach on a thorough study of current Web API authentication mechanisms and show how it can be used as input to SPICES and the authentication engine, in order to support the automated invocation of Web APIs. Future work will focus on further developing the authentication engine.

Acknowledgments. The work presented in this paper is partially supported by EU funding under the project SOA4All (FP7 - 215219).

References

1. Richardson, L., Ruby, S.: RESTful Web Services. O'Reilly Media, Sebastopol (May 2007)
2. Hadley, M. J.: Web Application Description Language (WADL). Technical report, Sun Microsystems (November 2006), https://wadl.dev.java.net
3. Web Services Description Language (WSDL) Version 2.0. W3C Recommendation (June 2007), http://www.w3.org/TR/wsdl20/
4. Kopecký, J., Vitvar, T., Fensel, D., Gomadam, K.: hRESTS & MicroWSMO. Technical report (2009), http://cms-wg.sti2.org/TR/d12/
5. Sheth, A.P., Gomadam, K., Lathem, J.: SA-REST: Semantically Interoperable and Easier-to-Use Services and Mashups. IEEE Internet Computing 11(6), 91–94 (2007)
6. Kopecký, J., Vitvar, T., Bournez, C., Farrel, J.: SAWSDL: Semantic Annotations for WSDL and XML Schema. IEEE Internet Computing 11(6), 60–67 (2007)
7. Kopecký, J., Gomadam, K., Vitvar, T.: hRESTS: an HTML Microformat for Describing RESTful Web Services. In: Proceedings of International Conference on Web Intelligence, WI 2008 (2008)
8. RDFa in XHTML: Syntax and Processing. Proposed Recommendation, W3C (September 2008), http://www.w3.org/TR/rdfa-syntax/
9. Nadalin, A., Kaler, C., Monzillo, R., Hallam-Baker, P.: Web Services Security: SOAP Message Security 1.1, WS-Security 2004 (2006)
10. Franks, J., Hallam-Baker, P., Hostetler, J.: HTTP Authentication: Basic and Digest Access Authentication RFC 2617. The Internet Society (1999)
11. Freed, N., Borenstein, N.: Multipurpose Internet Mail Extensions (MIME) Part One: Format of Internet Message Bodies, http://tools.ietf.org/html/rfc2045
12. The MD5 Message-Digest Algorithm, http://tools.ietf.org/html/rfc1321 (visited June 2010)
13. Atwood, M., et al.: OAuth Core 1.0 Specification, http://oauth.net/core/1.0/
14. Maleshkova, M., Pedrinaci, C., Domingue, J.: Supporting the creation of semantic RESTful service descriptions. In: Service Matchmaking and Resource Retrieval in the Semantic Web (SMR2) at 8th International Semantic Web Conference (2009)
15. Álvaro, G., Martínez, I., Gómez, J.M., Lecue, F., Pedrinaci, C., Villa, M., Di Matteo, G.: Using SPICES for a Better Service Consumption. Poster at the 7th Extended Semantic Web Conference, ESWC (2010)
16. Close, T.: Web-key: Mashing with Permission. In: Proceedings of Web 2.0 Security and Privacy (2008)
17. Story, H., Harbulot, B., Jacobi, I., Jones, M.: FOAF+SSL: RESTful Authentication for the Social Web. In: SPOT 2009 European Semantic Web Conference (2009)
18. Kagal, L., Paolucci, M., Srinivasan, N., Denker, G., Finin, T., Sycara, K.: Authorization and Privacy for Semantic Web Services. IEEE Intelligent Systems 19(4) (July 2004)

Representing and Querying Validity Time in RDF and OWL: A Logic-Based Approach

Boris Motik

Oxford University Computing Laboratory, Oxford, UK

Abstract. RDF(S) and OWL 2 currently support only static ontologies. In practice, however, the truth of statements often changes with time, and Semantic Web applications often need to represent such changes and reason about them. In this paper we present a logic-based approach for representing validity time in RDF and OWL. Unlike the existing proposals, our approach is applicable to entailment relations that are not deterministic, such as the Direct Semantics or the RDF-Based Semantics of OWL 2. We also extend SPARQL to temporal RDF graphs and present a query evaluation algorithm. Finally, we present an optimization of our algorithm that is applicable to entailment relations characterized by a set of deterministic rules, such RDF(S) and OWL 2 RL/RDF entailment.

1 Introduction

RDF(S) and OWL 2 currently support only *static* ontologies. In practice, however, the truth of statements often changes with time, and Semantic Web applications often need to represent such changes and reason about them. We discuss these issues on an example derived from the author's collaboration with ExperienceOn (abbreviated EO)—an IT start-up company from Barcelona, Spain.

EO aims to improve search in the tourism domain by providing an advanced system that can answer complex queries such as "trips to the second week of Oktoberfest." Users will input their questions in natural language, and NLP technology will translate such questions into one or more queries over a knowledge base containing information about flights, lodging, events, geography, and so on. EO's system must be able to represent statements that are not universally true, but are associated with *validity times*. For example, "Oktoberfest is being held in Munich" is true only while the festival is being held; similarly, statements describing airline flight schedules are valid only in certain time intervals. Validity time must be tightly integrated with reasoning; for example, from the knowledge about Oktoberfest and German geography, EO's system should conclude that "Oktoberfest is being held in Bavaria" is true for the duration of the festival. Validity time should also be integrated with a query language, allowing one to retrieve "flights from London to Munich during Oktoberfest." Validity time thus affects virtually all aspects of knowledge representation and reasoning in scenarios such as EO's. Some applications also need to represent *transaction times*, which specify when facts were added to the database. In this paper we focus on validity time since it is more relevant to knowledge modeling.

P.F. Patel-Schneider et al. (Eds.): ISWC 2010, Part I, LNCS 6496, pp. 550–565, 2010.

Validity time has been extensively studied in databases and artificial intelligence [4,19]. Neither RDF nor OWL, however, supports validity time, and SPARQL does not provide temporal query primitives. These deficiencies have been recognized by the community, and several proposals have emerged. A comprehensive framework for representing validity time in RDF was presented in [7], and it encompasses notions of temporal graphs and entailment, a characterization of temporal entailment via closures [6], an encoding of temporal graphs into regular RDF graphs, and a sketch of a temporal query language. This approach was extended in [9] with more general temporal constraints. In [15], the authors extended the approach from [7] with unknown time points, defined a temporal query language based on graph matching, and presented a way for indexing temporal graphs. A general framework for annotating RDF data was presented in [17]; the very general notion of annotations can be used to represent validity time. A temporal extension of SPARQL was presented in [18]. Approaches to extending *description logics* (DLs) [3]—the family of formalisms underpinning OWL 2 DL—with temporal features were surveyed in [2]. A temporal extension of OWL based on concrete domains was presented in [10].

None of these proposals is applicable to all variants of RDF and OWL. For example, the notion of closures from [7] relies on the fact that the inference rules of RDF(S) are deterministic—an assumption that does not hold in expressive languages such as OWL. In this paper we present a novel approach for representing validity time that is applicable to all Semantic Web languages, including RDF(S) and all profiles of OWL 2. In particular, in Section 3 we develop a first-order interpretation of temporal graphs, which we use to define temporal graph entailment. Our approach coincides with the one from [7] on RDF(S), but it is applicable to all languages of the RDF and OWL family.

In Section 4, we argue that a temporal query language defined in the obvious way would allow for queries that have very large and often even infinite answers. We present a query language whose queries always have finite answers, we integrate our query primitives into the formalization of SPARQL from [14], and we present a general query evaluation algorithm. In Section 5 we optimize our general evaluation algorithm for the case of deterministic inference rules.

We implemented our approach in EO's system. Given the nature of EO's business, we cannot make the system publicly available; however, EO is successfully using our approach to answer temporal queries, which we take as indication that our approach is suitable for practice.

The proofs of all technical results can be found in the extended version of this paper available from the author's online publication list.

2   Preliminaries

We assume the reader to be familiar with the syntax and semantics of OWL 2 DL [12]; for simplicity, we write OWL 2 DL axioms using the description logic syntax [3]. We use the standard definitions of constants, variables, terms, predicates, atoms, multi-sorted first-order logic, and skolemization [5]. For α an

OWL 2 DL axiom or an ontology, let $\theta(\alpha)$ be the translation of α into a first-order formula. We assume that the equality predicate $\approx$ is treated in $\theta(\alpha)$ as a standard first-order predicate explicitly axiomatized as a congruence; this does not affect the consequences of $\theta(\alpha)$ [5]. Moreover, we assume that θ maps the blank nodes (also called anonymous individuals) in α into free first-order variables, so the semantics of α is $\exists y_1, \ldots, y_n : \theta(\alpha)$ where $y_1, \ldots y_n$ are the blank nodes of α.

Let $\mathcal{U}$, $\mathcal{B}$, and $\mathcal{L}$ be infinite sets of *URI references*, *blank nodes*, and *literals*, respectively, and let $\mathcal{UBL} = \mathcal{U} \cup \mathcal{B} \cup \mathcal{L}$. A *triple* is an assertion of the form $\langle s, p, o \rangle$ with $s, p, o \in \mathcal{UBL}$.[1] An *RDF graph* (or just *graph*) G is a finite set of triples. The semantics of RDF is determined by *entailment relations*.

Simple entailment, *RDF entailment*, *RDFS entailment*, and *D-entailment* are defined in [8], and *OWL 2 RL/RDF entailment* and *OWL 2 RDF-Based entailment* are defined in [16]. The logical consequences of each entailment relation X from this list can be characterized by a (possibly infinite) set of first-order implications Γ_X. For example, for RDF entailment, Γ_{RDF} contains the rules in [8, Section 7], and for OWL 2 RL/RDF entailment, Γ_{RL} contains the rules in [11, Section 4.3]. The semantics of a graph G w.r.t. X can be defined by transforming G into a first-order theory as follows. We assume that each blank node corresponds to a first-order variable (i.e., for simplicity, we do not distinguish blank nodes from variables). Let $\mathsf{b}_X(G)$ be the set of all blank nodes in G. For a triple $A = \langle s, p, o \rangle$, let $\pi_X(A) = T(s, p, o)$, where T is a ternary first-order predicate. For a graph G, let $\pi_X(G) = \bigwedge_{A \in G} \pi_X(A)$. The first-order theory corresponding to G is then $\nu_X(G) = \{\exists \mathsf{b}_X(G) : \pi_X(G)\} \cup \Gamma_X$. Let $\xi_X(G)$ be obtained from $\nu_X(G)$ by skolemizing the existential quantifiers $\exists \mathsf{b}_X(G)$—that is, by removing $\exists \mathsf{b}_X(G)$ and replacing each blank node in $\pi_X(G)$ with a fresh URI reference. Theory $\nu_X(G)$ is equisatisfiable with $\xi_X(G)$. A graph G_1 *X-entails* a graph G_2, written $G_1 \models_X G_2$, if and only if $\nu_X(G_1) \models \exists \mathsf{b}_X(G_2) : \pi_X(G_2)$; the latter is the case if and only if $\xi_X(G_1) \models \exists \mathsf{b}_X(G_2) : \pi_X(G_2)$.

We next define *OWL 2 Direct entailment* (written DL due to its relationship with description logic). A graph G *encodes an OWL 2 DL ontology* if G can be transformed into an OWL 2 DL ontology $\mathcal{O}(G)$ as specified in [13]. For such G, let $\mathsf{b}_{DL}(G)$ be the set of blank nodes occurring in $\mathcal{O}(G)$; let $\pi_{DL}(G) = \theta(\mathcal{O}(G))$; let $\nu_{DL}(G) = \exists \mathsf{b}_{DL}(G) : \theta(\mathcal{O}(G))$; and let $\xi_{DL}(G)$ be obtained from $\nu_{DL}(G)$ by skolemizing the existential quantifiers $\exists \mathsf{b}_{DL}(G)$. Formula $\nu_{DL}(G)$ is equisatisfiable with $\xi_{DL}(G)$. For G_1 and G_2 graphs that encode OWL 2 DL ontologies, G_1 *DL-entails* G_2, written $G_1 \models_{DL} G_2$, iff $\nu_{DL}(G_1) \models \exists \mathsf{b}_{DL}(G_2) : \pi_{DL}(G_2)$; the latter is the case if and only if $\xi_{DL}(G_1) \models \exists \mathsf{b}_{DL}(G_2) : \pi_{DL}(G_2)$.

SPARQL is the standard W3C language for querying RDF graphs, and the 1.1 version (currently under development) will support different entailment relations. In this paper we focus on *group patterns*—the core of SPARQL that deals with pattern matching and is largely independent from constructs such as aggregates and sorting. We formalize group patterns as in [14], and we treat answers as sets rather than multisets as this simplifies the presentation without changing

---

[1] RDF actually requires $s \in \mathcal{U} \cup \mathcal{B}$, $p \in \mathcal{U}$, and $o \in \mathcal{UBL}$, but this is not important in our framework so we assume $s, p, o \in \mathcal{UBL}$ for the sake of simplicity.

the nature of our results. Let $\mathcal{V}$ be an infinite set of *variables* disjoint from $\mathcal{UBL}$. A *mapping* is a partial function $\mu : \mathcal{V} \to \mathcal{UBL}$. The domain (resp. range) of μ is written $\mathsf{dm}(\mu)$ (resp. $\mathsf{rg}(\mu)$). We define $\mu(t) = t$ for $t \in \mathcal{UBL} \cup \mathcal{V} \setminus \mathsf{dm}(\mu)$. Mappings μ_1 and μ_2 are *compatible* if $\mu_1(x) = \mu_2(x)$ for each $x \in \mathsf{dm}(\mu_1) \cap \mathsf{dm}(\mu_2)$; in such a case, $\mu_1 \cup \mu_2$ is also a mapping. The following algebraic operations on sets of mappings Ω_1 and Ω_2 are used to define the semantics of group patterns.

$$\Omega_1 \bowtie \Omega_2 = \{\mu_1 \cup \mu_2 \mid \mu_1 \in \Omega_1, \mu_2 \in \Omega_2, \text{ and } \mu_1 \text{ and } \mu_2 \text{ are compatible}\}$$
$$\Omega_1 \setminus \Omega_2 = \{\mu_1 \in \Omega_1 \mid \text{each } \mu_2 \in \Omega_2 \text{ is not compatible with } \mu_1\}$$

A *built-in expression* is constructed using the elements of $\mathcal{V} \cup \mathcal{U} \cup \mathcal{L}$ as specified in [14]; furthermore, for each built-in expression R and each mapping μ, we can determine whether R evaluates to true under μ, written $\mu \models R$, as specified in [14]. A *basic graph pattern* (BGP) is a set of triples of the form $\langle s, p, o \rangle$ where $s, p, o \in \mathcal{UBL} \cup \mathcal{V}$. A *group pattern* (GP) is an expression of the form B, P_1 and P_2, P_1 union P_2, P_1 opt P_2, or P_1 filter R, where B is a BGP, P_1 and P_2 are group patterns, and R is a built-in expression. For A a built-in expression or a group pattern and μ a mapping, $\mathsf{var}(A)$ is the set of variables occurring in A, and $\mu(A)$ is the result of replacing each variable x in A with $\mu(x)$.

The *answer* to a group pattern P on a graph G depends on an entailment relation X. For each X, we assume that a function exists that maps each graph G to the set $\mathsf{ad}_X(G) \subseteq \mathcal{UBL}$ called the *answer domain* of G; this set determines the elements of $\mathcal{UBL}$ that can occur in answers to group patterns on G under X-entailment. To see why this is needed, let $B = \{\langle x, rdf{:}type, rdf{:}Property \rangle\}$; due to the axiomatic triples [8], $\emptyset \models_{RDF} \mu(B)$ whenever $\mu(x) \in \{rdf{:}\_1, rdf{:}\_2, \ldots\}$. Without any restrictions, the answer to B under RDF entailment would thus be infinite even in the empty graph. To prevent this, $\mathsf{ad}_{RDF}(G)$ excludes $rdf{:}\_1$, $rdf{:}\_2, \ldots$ that do not occur in G, which makes $\mathsf{ad}_{RDF}(G)$ finite and thus ensures finiteness of answers. Similar definitions are used for X other than RDF.

SPARQL treats blank nodes as objects with distinct identity. To understand this, let $G = \{\langle a, b, c \rangle, \langle d, e, \_{:}1 \rangle\}$ where $\_{:}1$ is a blank node, let $P = \langle a, b, x \rangle$, and let $\mu = \{x \mapsto \_{:}1\}$. Even though $G \models_{RDF} \mu(P)$, the answer to P on G under RDF entailment does not contain μ. Roughly speaking, $\_{:}1$ is distinct from c even though $\_{:}1$ is semantically a "placeholder" for an arbitrary URI reference. We capture this idea using skolemization: we replace the blank nodes in G with fresh URI references, thus giving each blank node a unique identity. Our answers are isomorphic to the answers of the official SPARQL specification, so skolemization allows us to simplify the technical presentation without losing generality. We formalize this idea by evaluating group patterns in $\xi_X(G)$ instead of $\nu_X(G)$. Table 1 defines the answer $[\![P]\!]_G^X$ to a group pattern P in a graph G w.r.t. X.

3   Representing Validity Time in RDF and OWL

To incorporate validity time into RDF, one could simply equip each triple with a validity time instant; however, it would be impractical or even impossible to explicitly list all such time instants. To this end, Chomicki distinguishes an

Table 1. Semantics of Group Patterns

$$[\![B]\!]_G^X = \{\mu \mid \mathsf{dm}(\mu) = \mathsf{var}(B), \mathsf{rg}(\mu) \subseteq \mathsf{ad}_X(G), \xi_X(G) \models \exists \mathsf{b}_X(\mu(B)) : \pi_X(\mu(B))\}$$
$$[\![P_1 \text{ and } P_2]\!]_G^X = [\![P_1]\!]_G^X \bowtie [\![P_2]\!]_G^X$$
$$[\![P_1 \text{ union } P_2]\!]_G^X = [\![P_1]\!]_G^X \cup [\![P_2]\!]_G^X$$
$$[\![P_1 \text{ opt } P_2]\!]_G^X = [\![P_1]\!]_G^X \bowtie [\![P_2]\!]_G^X \cup [\![P_1]\!]_G^X \setminus [\![P_2]\!]_G^X$$
$$[\![P_1 \text{ filter } R]\!]_G^X = \{\mu \in [\![P_1]\!]_G^X \mid \mu \models R\}$$

abstract from a *concrete* temporal database [4]. The former is a sequence of "static" databases each of which contains the facts true at some time instant. Since the time line is unbounded, an abstract temporal database is infinite, so a concrete temporal database is used as a finite specification of one or more abstract temporal databases. We next apply thus approach to RDF and OWL.

We use a discrete notion of time, since the ability to talk about predecessors/successors of time instants is needed in Section 4. Thus, the set $\mathcal{TI}$ of *time instants* is the set of all integers, $\leq$ is the usual total order on $\mathcal{TI}$, and $+1$ and -1 are the usual successor and predecessor functions on $\mathcal{TI}$. The set of *time constants* is $\mathcal{TC} = \mathcal{TI} \cup \{-\infty, +\infty\}$; we assume that $\mathcal{UBL} \cap \mathcal{TC} = \emptyset$. Time constants $-\infty$ and $+\infty$ are special in that they can occur in first-order formulae only in atoms of the form $-\infty \leq t$, $-\infty \leq +\infty$, and $t \leq +\infty$ for t a time instant or a variable; all such atoms are syntactic shortcuts for true. This allows us to simplify the notation for bounded and unbounded time intervals; for example, to say that the interval described by formula $t_1 \leq x^t \leq t_2$ has no lower bound, we write $t_1 = -\infty$, which makes the formula equivalent to $x^t \leq t_2$.

Definition 1. *A* temporal triple *has the form* $\langle s, p, o\rangle[t]$ *or* $\langle s, p, o\rangle[t_1, t_2]$, *such that* $s, p, o \in \mathcal{UBL}$, $t \in \mathcal{TI}$, $t_1 \in \mathcal{TI} \cup \{-\infty\}$, *and* $t_2 \in \mathcal{TI} \cup \{+\infty\}$. *A temporal graph* G *is a finite set of temporal triples.*

In this work, we focus mainly on the conceptual aspects of temporal graphs and we do not discuss practical issues such as serialization syntax. We interpret temporal graphs in multi-sorted first-order logic. Let t be a distinct *temporal sort* interpreted over $\mathcal{TI}$; we write x^t to stress that a variable x ranges over $\mathcal{TI}$. For each n-ary predicate P, let $\hat{P}$ be the $n+1$-ary predicate where positions 1–n have the same sort as in P, and position $n+1$ is of sort t. For t a term of sort t and $P(u_1, \ldots, u_n)$ an atom, let $P(u_1, \ldots, u_n)\langle t\rangle = \hat{P}(u_1, \ldots, u_n, t)$, and let $\varphi\langle t\rangle$ be obtained by replacing each atom A with $A\langle t\rangle$ in a first-order formula φ.

Intuitively, atom $\hat{P}(u_1, \ldots, u_n, t)$ encodes the truth of atom $P(u_1, \ldots, u_n)$ at time instant t: the former is true iff the latter is true at time t, so our approach is similar to the temporal arguments approach [19]. Similarly, $\varphi\langle t\rangle$ determines the truth of φ at time instant t. As explained in Section 2, $\approx$ is an ordinary predicate with an explicit axiomatization, so $\hat{\approx}$ is well defined and it gives us a notion of equality that changes with time. Finally, to understand why a multi-sorted interpretation is needed, consider a graph G that encodes the OWL 2 DL axiom $\top \sqsubseteq \{c\}$. Such G is satisfiable only in first-order interpretations consisting of a single object, which contradicts the requirement that a domain should contain $\mathcal{TI}$. Multi-sorted logic cleanly separates temporal instants from other objects in the domain, so axioms such as $\top \sqsubseteq \{c\}$ do not quantify over time instants, which solves the problem. We next define the semantics of temporal graphs.

Definition 2. *Let X be an entailment relation from Section 2 other than DL, and let Γ_X be the first-order theory that characterizes X. For G a temporal graph, $\mathsf{u}_X(G)$, $\mathsf{b}_X(G)$, and $\mathsf{tc}_X(G)$ are the subsets of $\mathcal{U} \cup \mathcal{L}$, $\mathcal{B}$, and $\mathcal{TC}$, respectively, that occur in G. Mappings π_X and ν_X are extended to temporal graphs as shown below, where O is a fresh unary predicate. Furthermore, $\xi_X(G)$ is obtained from $\nu_X(G)$ by skolemizing the existential quantifiers in $\exists\mathsf{b}_X(G)$, and $\mathsf{ub}_X(G)$ is $\mathsf{u}_X(G)$ extended with the URI references introduced via skolemization.*

$$\pi_X(\langle s, p, o\rangle[t]) = \hat{T}(s, p, o, t)$$

$$\pi_X(\langle s, p, o\rangle[t_1, t_2]) = \forall x^{\mathsf{t}} : (t_1 \leq x^{\mathsf{t}} \leq t_2) \rightarrow \hat{T}(s, p, o, x^{\mathsf{t}})$$

$$\pi_X(G) = \bigwedge_{u \in \mathsf{b}_X(G)} O(u) \wedge \bigwedge_{A \in G} \pi_X(A)$$

$$\nu_X(G) = \{\exists\mathsf{b}_X(G) : \bigwedge_{u \in \mathsf{u}_X(G)} O(u) \wedge \pi_X(G)\} \cup \{\forall x^{\mathsf{t}} : \varphi\langle x^{\mathsf{t}}\rangle \mid \varphi \in \Gamma_X\}$$

A temporal graph G_1 entails a temporal graph G_2 under entailment relation X, written $G_1 \models_X G_2$, if and only if $\nu_X(G_1) \models \exists\mathsf{b}_X(G_2) : \pi_X(G_2)$.

Intuitively, predicate O in $\nu_X(G)$ "contains" all elements of $\mathsf{u}_X(G) \cup \mathsf{b}_X(G)$ that occur in G, which ensures that, whenever $G_1 \models_X G_2$, all blank nodes in G_2 can be mapped to $\mathsf{u}_X(G_1) \cup \mathsf{b}_X(G_1)$. We discuss the rationale behind such a definition at end of this section; for the moment, we just note that, when applied to RDF(S), our definition of entailment coincides with the one from [7].

We next present a small example. Let G_1 be the temporal graph containing temporal triples (1)–(3). Triples in (1) state that there is a flight from LHR to MUC; this information may have been gathered from two distinct sources, so validity times of the two triples overlap. Triple (2) states that Munich hosts Oktoberfest. Finally, triple (3) states that, if x hosts y, then x has y as an attraction; that this statement is not universally true might be due to the fact that attractions are relevant only during holiday seasons. One can easily verity that $G_1 \models_{RDFS} \langle :Munich, :hasAttraction, :Oktoberfest\rangle[130, 180]$.

$$\langle :LHR, :flightTo, :MUC\rangle[50, 120] \qquad \langle :LHR, :flightTo, :MUC\rangle[100, 150] \qquad (1)$$

$$\langle :Munich, :hosts, :Oktoberfest\rangle[80, 180] \qquad (2)$$

$$\langle :hosts, rdfs:subPropertyOf, :hasAttraction\rangle[130, 300] \qquad (3)$$

OWL 2 Direct entailment is not characterized by a fixed set of first-order implications, so we define temporal OWL 2 Direct entailment separately.

Definition 3. *A temporal OWL 2 DL axiom has the form $\alpha[t]$ or $\alpha[t_1, t_2]$ for α an OWL 2 DL axiom, $t \in \mathcal{TI}$, $t_1 \in \mathcal{TI} \cup \{-\infty\}$, and $t_2 \in \mathcal{TI} \cup \{+\infty\}$. A*

temporal OWL 2 DL ontology $\mathcal{O}$ *is a finite set of temporal OWL 2 DL axioms.*
Temporal axioms and ontologies are mapped into formulae as $\theta(\alpha[t]) = \theta(\alpha)\langle t\rangle$,
$\theta(\alpha[t_1,t_2]) = \forall x^t : (t_1 \leq x^t \leq t_2) \rightarrow \theta(\alpha)\langle x^t\rangle$, *and* $\theta(\mathcal{O}) = \bigwedge_{A\in\mathcal{O}} \theta(A)$.

A temporal graph G *encodes a temporal OWL 2 DL ontology* $\mathcal{O}(G)$ *if* $\mathcal{O}(G)$
can be extracted from G *using the mapping from [13] modified as follows:*

- *Each* $\langle s,p,o\rangle$ *in Tables 3–8 and 10–15 is replaced with* $\langle s,p,o\rangle[-\infty,+\infty]$.
- *Each triple pattern from Tables 16 and 17 without a main triple[2] producing*
  an axiom α *is changed as follows: each* $\langle s,p,o\rangle$ *in the pattern is replaced with*
  $\langle s,p,o\rangle[-\infty,+\infty]$, *and the triple pattern produces* $\alpha[-\infty,+\infty]$.
- *Each triple pattern from Tables 16 and 17 with a main triple* $\langle s_m,p_m,o_m\rangle$
  producing an axiom α *is replaced with the following two triple patterns.*
  - *The first one is obtained by replacing each triple* $\langle s,p,o\rangle$ *in the pattern*
    other than the main one with $\langle s,p,o\rangle[-\infty,+\infty]$, *replacing the main triple*
    with $\langle s_m,p_m,o_m\rangle[t]$, *and making the triple pattern produce* $\alpha[t]$.
  - *The second one is obtained by replacing each triple* $\langle s,p,o\rangle$ *in the pattern*
    other than the main one with $\langle s,p,o\rangle[-\infty,+\infty]$, *replacing the main triple*
    with $\langle s_m,p_m,o_m\rangle[t_1,t_2]$, *and making the triple pattern produce* $\alpha[t_1,t_2]$.

For G *encoding a temporal OWL 2 DL ontology* $\mathcal{O}(G)$, $\mathsf{u}_{DL}(G)$, $\mathsf{b}_{DL}(G)$, *and*
$\mathsf{tc}_{DL}(G)$ *are the sets of named individuals, blank nodes, and temporal constants,*
respectively, in $\mathcal{O}(G)$. *Mappings,* π_{DL} *and* ν_{DL} *are extended to* G *as shown be-*
low, where O *is a fresh unary predicate. Furthermore,* $\xi_{DL}(G)$ *is obtained from*
$\nu_{DL}(G)$ *by skolemizing the existential quantifiers in* $\exists\mathsf{b}_{DL}(G)$, *and* $\mathsf{ub}_{DL}(G)$ *is*
$\mathsf{u}_{DL}(G)$ *extended with the named individuals introduced via skolemization.*

$$\pi_{DL}(G) = \bigwedge_{u\in\mathsf{b}_{DL}(G)} O(u) \wedge \theta(\mathcal{O}(G))$$

$$\nu_{DL}(G) = \exists\mathsf{b}_{DL}(G) : \bigwedge_{u\in\mathsf{u}_{DL}(G)} O(u) \wedge \pi_{DL}(G)$$

For G_1 *and* G_2 *temporal graphs that encode temporal OWL 2 DL ontologies, we*
have $G_1 \models_{DL} G_2$ *if and only if* $\nu_{DL}(G_1) \models_{DL} \exists\mathsf{b}_{DL}(G_2) : \pi_{DL}(G_2)$.

Definition 3 allows us to attach validity time to axioms (but not to parts of
axioms such as class expressions), which provides us with a flexible language
that can represent, for example, class hierarchies that change over time.

We next explain the intuition behind the predicate O in Definitions 2 and
3. Note that $\exists\mathsf{b}_X(G)$ occurs in ν_X before the universal quantifiers over $\mathcal{TI}$,
so blank nodes in G are interpreted *rigidly*—that is, they represent the same
objects throughout all time. For example, let $G_2 = \{\langle s,p,\_:1\rangle[-\infty,+\infty]\}$, so
$\pi_{DL}(G_2) = \exists\_:1 : O(\_:1) \wedge \forall x^t : \hat{p}(s,\_:1,x^t)$; since $\exists\_:1$ comes before $\forall x^t$, blank
node $\_:1$ refers to the same object at all time instants. In contrast, the existen-
tial quantifiers in $\varphi\langle x^t\rangle$ and $\theta(\mathcal{O}(G))$ are *not rigid*—that is, they can be satisfied by different objects at different time instants. For example, let G_3 be such
that $\mathcal{O}(G_3) = \{\exists p.\top(s)[-\infty,+\infty]\}$, so $\pi_{DL}(G_3) = \forall x^t : \exists y : \hat{p}(s,y,x^t)$; since $\exists y$

---

[2] Please refer to [13] for the definition of a main triple.

comes after $\forall x^t$, the value for y can be different at different time instants. Consequently, G_2 is *not* DL-equivalent to G_3; in fact, G_2 DL-entails G_3, but not vice versa. Blank nodes can thus be understood as unnamed constants, which we believe to be in the spirit of RDF and OWL. In line with this intuition, conjuncts $\bigwedge_{u \in \mathsf{b}_X(G)} O(u)$ and $\bigwedge_{u \in \mathsf{u}_X(G)} O(u)$ in Definitions 2 and 3 ensure that, if $G_2 \models_X G_3$, then the blank nodes in $\mathsf{b}_X(G_3)$ can be mapped to the rigid objects in $\mathsf{u}_X(G_2)$, but not to the nonrigid objects whose existence is implied by existential quantifiers. Without this restriction, G_3 would DL-entail $G_4 = \{\langle s, p, \_:1 \rangle [1,1]\}$ (since the triple in G_4 refers only to a single time instant, the nonrigidity of $\exists p.\top$ is irrelevant), which seems at odds with the fact that G_3 does not DL-entail G_2. Under our semantics, G_3 does not DL-entail G_4 due to the O predicate, which seems more intuitive and it is also easier to implement.

4   Querying Temporal Graphs

The first step in designing a query language is to identify the types of questions that the language should support. The language of first-order logic readily reveals the following natural types of questions:

Q1. Is BGP B true in G at some time instant t?
Q2. Is BGP B true in G at all time instants between t_1 and t_2?
Q3. Is BGP B true in G at some time instant between t_1 and t_2?

Such questions can be easily encoded in first-order formulae, and an answer to a formula Q over a graph G under entailment relation X can be defined as the set of mappings μ of the free variables of Q such that $G \models_X \mu(Q)$. Such an approach, however, has an important drawback. Let $G_5 = \{\langle a, b, c \rangle [5, 12], \langle a, b, c \rangle [9, +\infty]\}$ and let $Q(x_1, x_2) = \forall x : x_1 \leq x \leq x_2 \to \langle a, b, c \rangle [x]$ be a question of type Q2. Evaluating $Q(x_1, x_2)$ on G_5 is not a problem if x_1 and x_2 are concrete time instants. Note, however, that $Q(x_1, x_2)$ does not ask for *maximal* x_1 and x_2 for which the formula holds. Thus, the answer to $Q(x_1, x_2)$ on G_5 is infinite since it contains each mapping μ such that $5 \leq \mu(x_1) \leq \mu(x_2) \leq +\infty$.

One can restrict answers to mappings that refer to time instants explicitly occurring in G, but this is also problematic. First, answers can contain redundant mappings. For example, $\mu_1 = \{x_1 \mapsto 5, x_2 \mapsto +\infty\}$ is the "most general mapping" in the answer to $Q(x_1, x_2)$ on G_5, but the answer also contains a "less general" mapping $\mu_2 = \{x_1 \mapsto 9, x_2 \mapsto 12\}$. Second, answers can differ on syntactically different but semantically equivalent temporal graphs. For example, $G_6 = \{\langle a, b, c \rangle [5, 10], \langle a, b, c \rangle [7, +\infty]\}$ is equivalent to G_5 under simple entailment; however, μ_2 is not contained in the answer to $Q(x_1, x_2)$ on G_6, and $\mu_3 = \{x_1 \mapsto 7, x_2 \mapsto 10\}$ is not contained in the answer to $Q(x_1, x_2)$ on G_5. Third, computing redundant answers can be costly: an answer to a formula such as $Q(x_1, x_2)$ in a graph with n overlapping intervals consists of mappings that refer to any two pairs of interval endpoints, so the number of mappings in the answer is exponential in n. One might try to identify the "most general" mappings, but this would be an ad hoc solution without a clear semantic justification.

We deal with these problems in two stages. First, we introduce primitives that support questions of types Q1–Q3, as well as of types Q4–Q5, thus explicitly introducing a notion of maximality into the language.

Q4. Is $[t_1, t_2]$ the maximal interval such that BGP B holds in G for each time instant in the interval?

Q5. Is t the smallest/largest instant at which BGP B holds in G?

We define our notion of answers w.r.t. $\mathcal{TC}$, which makes the answers independent from the syntactic form of temporal graphs. To ensure finiteness, we then define a syntactic notion of *safety*, which guarantees that only questions of type Q4 and Q5 can "produce" a value.

Practical applications will often need to express constraints on time points and intervals retrieved via Q1–Q5. For example, to retrieve "hotels with vacancy during Oktoberfest," we must require the duration of Oktoberfest to be contained in the hotels' vacancy period. Such conditions can be expressed, for example, using Allen's interval algebra [1], and they can be integrated into our query language via built-in expressions; for example, we can provide a built-in expression that takes two pairs of interval end-points and that is true iff the first interval is contained in the second. Such extensions of our query language are straightforward, so we do not discuss them further in the rest of this paper.

Definition 4. *A temporal group pattern (TGP) is an expression defined inductively as shown below, where B is a BGP, P_1 and P_2 are TGPs, R is a built-in expression, $t_1 \in \mathcal{TI} \cup \{-\infty\} \cup \mathcal{V}$, $t_2 \in \mathcal{TI} \cup \{+\infty\} \cup \mathcal{V}$, and $t_3 \in \mathcal{TI} \cup \mathcal{V}$. TGPs from the first two lines are called* basic.

B at t_3	B during $[t_1, t_2]$	B occurs $[t_1, t_2]$	
B maxint $[t_1, t_2]$	B mintime t_3	B maxtime t_3	
P_1 and P_2	P_1 union P_2	P_1 opt P_2	P_1 filter R

We redefine a mapping *as a partial function $\mu : \mathcal{V} \to \mathcal{UBL} \cup \mathcal{TC}$. Let X be an entailment relation and G a temporal graph. Let $\mathrm{ad}_X(G) = \mathrm{ad}_X(G')$, where G' is the nontemporal graph obtained by replacing all triples in G of the form $\langle s, p, o \rangle [u]$ and $\langle s, p, o \rangle [u_1, u_2]$ with $\langle s, p, o \rangle$. The answer to a basic TGP P in G under X is the set of mappings defined as specified below, where $\delta_X(\mu(P))$ is a condition from Table 2. Answers to all other TGP types are defined in Table 1.*

$$\llbracket P \rrbracket_G^X = \{\mu \mid \mathrm{dm}(\mu) = \mathrm{var}(P), \mathrm{rg}(\mu) \subseteq \mathrm{ad}_X(G) \cup \mathcal{TC}, \text{ and } \delta_X(\mu(P)) \text{ holds}\}$$

We next present several TGPs that could be used in our running example. TGP (4) returns the maximal intervals $[y, z]$ during Oktoberfest in which a flight from airport x to the Munich airport exists; the answer to (4) on G_1 is $\{\{x \mapsto LHR, y \mapsto 80, y \mapsto 150\}\}$. TGP (5) retrieves all events z in London that have at least one time instant in common with Oktoberfest; if occurs were changed to during, the TGP would retrieve all events z in London whose duration is contained in the duration of Oktoberfest. TGP (6) retrieves the first time instant at which Munich hosted Oktoberfest; the answer to (6) on G_1 is

Table 2. Semantics of Temporal Graph Patterns

P	$\delta_X(P)$
B at t_3	$\xi_X(G) \models \exists b_X(B) : \pi_X(B)\langle t_3 \rangle$
B during $[t_1, t_2]$	$\xi_X(G) \models \exists b_X(B) \, \forall x^t : [t_1 \leq x^t \leq t_2] \rightarrow \pi_X(B)\langle x^t \rangle$
B occurs $[t_1, t_2]$	$\xi_X(G) \models \exists b_X(B) \, \exists x^t : [t_1 \leq x^t \leq t_2 \wedge \pi_X(B)\langle x^t \rangle]$
B maxint $[t_1, t_2]$	a function $\sigma : b_X(B) \rightarrow ub_X(G)$ exists such that $\xi_X(G) \models \forall x^t : [t_1 \leq x^t \leq t_2] \rightarrow \pi_X(\sigma(B))\langle x^t \rangle$, and $t_1 = -\infty$ or $\xi_X(G) \not\models \pi_X(\sigma(B))\langle t_1 - 1 \rangle$, and $t_2 = +\infty$ or $\xi_X(G) \not\models \pi_X(\sigma(B))\langle t_2 + 1 \rangle$
B mintime t_3	a function $\sigma : b_X(B) \rightarrow ub_X(G)$ exists such that $\xi_X(G) \models \pi_X(\sigma(B))\langle t_3 \rangle$ and $\xi_X(G) \not\models \pi_X(\sigma(B))\langle x^t \rangle$ for each $x^t \in \mathcal{TI}$ with $x^t \leq t_3 - 1$
B maxtime t_3	a function $\sigma : b_X(B) \rightarrow ub_X(G)$ exists such that $\xi_X(G) \models \pi_X(\sigma(B))\langle t_3 \rangle$ and $\xi_X(G) \not\models \pi_X(\sigma(B))\langle x^t \rangle$ for each $x^t \in \mathcal{TI}$ with $t_3 + 1 \leq x^t$

Note: $\delta_X(P)$ does not hold if P is malformed (e.g., if it is of the form B at t_3 and $t_3 \notin \mathcal{TI}$); and $\sigma(B)$ is the result of replacing each blank node v in B with $\sigma(v)$.

$\{\{x \mapsto 80\}\}$. Finally, TGP (7) returns all rooms x that have price y during an event z in Munich within the time interval $[50, 100]$.

$$\{\langle x, \mathit{:flightTo}, \mathit{:MUC}\rangle, \langle \mathit{:Munich}, \mathit{:hosts}, \mathit{:Oktoberfest}\rangle\} \text{ maxint } [y, z] \qquad (4)$$

$$\begin{aligned} \{\langle \mathit{:Munich}, \mathit{:hosts}, \mathit{:Oktoberfest}\rangle\} \text{ maxint } [x, y] \text{ and} \\ \{\langle \mathit{:London}, \mathit{:hosts}, z\rangle\} \text{ occurs } [x, y] \end{aligned} \qquad (5)$$

$$\{\langle \mathit{:Munich}, \mathit{:hosts}, \mathit{:Oktoberfest}\rangle\} \text{ mintime } x \qquad (6)$$

$$\{\langle x, \mathit{:hasPrice}, y\rangle, \langle \mathit{:Munich}, \mathit{:hosts}, z\rangle\} \text{ occurs } [50, 100] \qquad (7)$$

We next turn our attention to the formal properties of TGPs. By Definition 4, $ad_X(G)$ does not contain time constants occurring in G and answers are defined w.r.t. $\mathcal{TC}$, which ensures that answers do not depend on the syntactic form of temporal graphs. For example, temporal graphs G_5 and G_6 mentioned earlier are equivalent and $ad_X(G_5) = ad_X(G_6)$, so $[\![P]\!]^X_{G_5} = [\![P]\!]^X_{G_6}$ for each TGP P.

Proposition 1. *Let X be an entailment relation, and let G_1 and G_2 be temporal graphs such that $G_1 \models_X G_2$, $G_2 \models_X G_1$, and $ad_X(G_1) = ad_X(G_2)$. Then, for each temporal group pattern P, we have $[\![P]\!]^X_{G_1} = [\![P]\!]^X_{G_2}$.*

Since the answers are defined w.r.t. the entire set $\mathcal{TC}$, temporal basic graph patterns can have infinite answers. We next define a notion of safe TGPs and later show that such group patterns always have finite answers.

Definition 5. *For P a temporal group pattern, $\mathsf{uns}(P)$ is the set of variables as shown in Table 3. Pattern P is safe if and only if $\mathsf{uns}(P) = \emptyset$.*

Table 3. The Definition of Safety

P	$\mathsf{uns}(P)$	P	$\mathsf{uns}(P)$
B at t_3	$\{t_3\} \cap \mathcal{V}$	B maxint $[t_1, t_2]$	$\emptyset$
B during $[t_1, t_2]$	$\{t_1, t_2\} \cap \mathcal{V}$	B mintime t_3	$\emptyset$
B occurs $[t_1, t_2]$	$\{t_1, t_2\} \cap \mathcal{V}$	B maxtime t_3	$\emptyset$
P_1 and P_2	$\mathsf{uns}(P_1) \cup [\mathsf{uns}(P_2) \setminus \mathsf{var}(P_1)]$	P_1 union P_2	$\mathsf{uns}(P_1) \cup \mathsf{uns}(P_2)$
P_1 opt P_2	$\mathsf{uns}(P_1) \cup [\mathsf{uns}(P_2) \setminus \mathsf{var}(P_1)]$	P_1 filter R	$\mathsf{uns}(P_1)$

Intuitively, $x \in \mathsf{uns}(P)$ means that there is no guarantee that $\mu(x) \subset \mathcal{TC}$ implies $\mu(x) \in \mathsf{tc}_X(G)$ for each $\mu \in \llbracket P \rrbracket_G^X$. Thus, B at t_3, B during $[t_1, t_2]$, and B occurs $[t_1, t_2]$ are safe iff t_1, t_2, and t_3 are not variables: B can hold at potentially infinitely many time intervals, which could give rise to infinite answers if t_1, t_2, or t_3 were a variable. In contrast, B maxint $[t_1, t_2]$, B mintime t_3, and B maxtime t_3 are always safe as there are finitely many maximal intervals in which B holds. The nontrivial remaining cases are P_1 and P_2 and P_1 opt P_2, in which we assume that P_1 is evaluated "before" P_2—that is, that the values for variables obtained by evaluating P_1 are used to bind unsafe variables in P_2; this will be made precise shortly in our algorithm for TGP evaluation. Thus, $(B_1$ occurs $[x, y])$ and $(B_2$ maxint $[x, y])$ is not safe while $(B_2$ maxint $[x, y])$ and $(B_1$ occurs $[x, y])$ is, which may seem odd given that conjunction is commutative. Without a predefined evaluation order, however, we would need to examine every possible order of conjuncts in a conjunction to find an "executable" one, which could be impractical.

We next present an algorithm for evaluating TGPs. We start by showing how to decide three types of temporal entailment that are used as basic building blocks of our evaluation algorithm. We first present some auxiliary definitions. Let G be a temporal graph and X an entailment relation. A pair of time constants (t_1, t_2) is *consecutive* in G if $t_1, t_2 \in \mathsf{tc}_X(G)$, $t_1 < t_2$, and no $t \in \mathsf{tc}_X(G)$ exists with $t_1 < t < t_2$. The *representative* of such (t_1, t_2) is defined as $t_1 + 1$ if $t_1 \neq -\infty$, $t_2 - 1$ if $t_1 = -\infty$ and $t_2 \neq +\infty$, and 0 otherwise. Furthermore, $\mathsf{ti}_X(G) \subseteq \mathcal{TI}$ is the smallest set that contains $\mathsf{tc}_X(G) \cap \mathcal{TI}$ and the representative of each consecutive pair of time constants in G. Finally, note that by Definitions 2 and 3, $\xi_X(G)$ contains $\bigwedge_{u \in \mathsf{ub}_X(G)} O(u) \wedge \Lambda$ and zero or more formulae of the form $\forall x^{\mathsf{t}} : \varphi_i \langle x^{\mathsf{t}} \rangle$, and that Λ is a conjunction of formulae of the form $\psi_i \langle t_i \rangle$ and $\forall x^{\mathsf{t}} : (t_i^1 \leq x^{\mathsf{t}} \leq t_i^2) \to \kappa_i \langle x^{\mathsf{t}} \rangle$; then, for $t \in \mathcal{TI}$, $\Xi_X(G, t)$ is the set of all $O(u)$, all ψ_i such that $t_i = t$, all κ_i such that $t_i^1 \leq t \leq t_i^2$, and all φ_i.

Proposition 2. *Let G be a temporal graph, let X be an entailment relation, let B be a BGP such that $\mathsf{var}(B) = \emptyset$, and let $t_1 \in \mathcal{TI} \cup \{-\infty\}$, $t_2 \in \mathcal{TI} \cup \{+\infty\}$, and $t_3 \in \mathcal{TI}$. Then, the following claims hold:*

1. *$\xi_X(G)$ is satisfiable iff $\Xi_X(G, t)$ is satisfiable for each $t \in \mathsf{ti}_X(G)$.*
2. *$\xi_X(G) \models \exists \mathsf{b}_X(B) : \pi_X(B) \langle t_3 \rangle$ iff $\xi_X(G)$ is unsatisfiable or some function $\sigma : \mathsf{b}_X(B) \to \mathsf{ub}_X(G)$ exists such that $\Xi_X(G, t_3) \models \pi_X(\sigma(B))$.*

Table 4. Evaluation of Temporal Group Patterns

$\text{eval}_X(P, G)$ is the set of mappings defined as follows depending on the type of P:

$P = B$ at t_3 or $P = B$ during $[t_1, t_2]$ or $P = B$ occurs $[t_1, t_2]$:
  $\{\mu \mid \text{dm}(\mu) = \text{var}(B), \text{rg}(\mu) \subseteq \text{ad}_X(G), \text{ and } \delta_X(\mu(P)) \text{ holds}\}$

$P = B$ maxint $[t_1, t_2]$:
  $\{\mu \mid \text{dm}(\mu) = \text{var}(P), \text{rg}(\mu) \subseteq \text{ad}_X(G) \cup \text{ti}_X(G) \cup \{-\infty, +\infty\}, \text{ and } \delta_X(\mu(P)) \text{ holds}\}$

$P = B$ mintime t_3 :
  $\{\mu \mid \text{dm}(\mu) = \text{var}(P), \text{rg}(\mu) \subseteq \text{ad}_X(G) \cup \text{ti}_X(G), \delta_X(\mu(B \text{ at } t_3)) \text{ holds, and}$
    $\delta_X(\mu(B \text{ at } t')) \text{ does not hold for all } t' \in \text{ti}_X(G) \text{ such that } t' \leq \mu(t_3) - 1\}$

$P = B$ maxtime t_3 :
  $\{\mu \mid \text{dm}(\mu) = \text{var}(P), \text{rg}(\mu) \subseteq \text{ad}_X(G) \cup \text{ti}_X(G), \delta_X(\mu(B \text{ at } t_3)) \text{ holds, and}$
    $\delta_X(\mu(B \text{ at } t')) \text{ does not hold for all } t' \in \text{ti}_X(G) \text{ such that } \mu(t_3) + 1 \leq t'\}$

$P = P_1$ and P_2 :
  $\{\mu_1 \cup \mu_2 \mid \mu_1 \in \text{eval}_X(P_1, G) \text{ and } \mu_2 \in \text{eval}_X(\mu_1(P_2), G)\}$

$P = P_1$ union P_2 :
  $\text{eval}_X(P_1, G) \cup \text{eval}_X(P_2, G)$

$P = P_1$ opt P_2 :
  $\text{eval}_X(P_1 \text{ and } P_2, G) \cup \{\mu \in \text{eval}_X(P_1, G) \mid \text{eval}_X(\mu(P_2), G) = \emptyset\}$

$P = P_1$ filter R :
  $\{\mu \in \text{eval}_X(P_1, G) \mid \mu \models R\}$

---

3. $\xi_X(G) \models \exists b_X(B) \, \forall x^t : [t_1 \leq x^t \leq t_2] \rightarrow \pi_X(B)\langle x^t \rangle$ *iff* $\xi_X(G)$ *is unsatisfiable or some* $\sigma : b_X(G) \rightarrow \text{ub}_X(G)$ *exists such that* $\Xi_X(G, t) \models \pi_X(\sigma(B))$ *for each* $t \in \text{ti}_X(G)$ *with* $t_1 \leq t \leq t_2$.

4. $\xi_X(G) \models \exists b_X(B) \, \exists x^t : [t_1 \leq x^t \leq t_2 \wedge \pi_X(B)\langle x^t \rangle]$ *iff* $\xi_X(G)$ *is unsatisfiable or some* $\sigma : b_X(B) \rightarrow \text{ub}_X(G)$ *exists such that* $\Xi_X(G, t) \models \pi_X(\sigma(B))$ *for some* $t \in \text{ti}_X(G)$ *with* $t_1 \leq t \leq t_2$.

Proposition 2 reduces temporal entailment to standard entailment problems that can be solved using any decision procedure available. This provides us with a way to check conditions $\delta_X(\mu(P))$ needed to evaluate safe TGPs. Furthermore, note that Claim 3 can be straightforwardly extended to general temporal graph entailment. We use these results as building blocks for the function shown in Table 4 that evaluates safe temporal group patterns. For P a basic TGP, $\text{eval}_X(P, G)$ can be computed by enumerating all mappings potentially relevant to P and then eliminating those mappings that do not satisfy the respective conditions; optimizations can be used to quickly eliminate irrelevant mappings.

Proposition 3. *Let G be a temporal graph, let X be an entailment relation such that $\text{ad}_X(G)$ is finite, and let P be a safe temporal group pattern. Then $\text{eval}_X(P, G) = [\![P]\!]_G^X$ and $[\![P]\!]_G^X$ is finite.*

5   Optimized Query Answering

The algorithm from Table 4 checks temporal entailment using a black box decision procedure, which can be inefficient. In this section we first present an optimization of this algorithm that is applicable to simple entailment, and then we extend this approach to any entailment relation that can be characterized by deterministic rules, such as RDF(S) and OWL 2 RL.

5.1   Simple Entailment

Simple entailment is the basic entailment relation in which BGPs can be evaluated in nontemporal graphs by simple graph lookup. Such an approach provides the basis of virtually all practical RDF storage systems and has proved itself in practice, so it would be beneficial if similar approaches were applicable to TGPs and temporal graphs. As the following example demonstrates, however, this is not the case. Let $P = \{\langle :LHR, :flightTo, :MUC\rangle\}$ maxint $[x, y]$; then $\llbracket P \rrbracket_{G_1}^{simple} = \{\{x \mapsto 50, y \mapsto 150\}\}$. Note, however, that G_1 does not contain temporal triple $\alpha = \langle :LHR, :flightTo, :MUC\rangle[50, 150]$, so $\llbracket P \rrbracket_{G_1}^{simple}$ cannot be computed via lookup. Temporal graph G_1 is, however, equivalent to the normalized temporal graph $\mathsf{nrm}(G_1)$ obtained from G_1 by replacing (1) with α; then, P can be evaluated in $\mathsf{nrm}(G_1)$ via lookup, which simplifies query processing. TGPs of other types can additionally require adequate interval comparisons.

We next formalize this idea. We say that temporal triples $\langle s, p, o\rangle[t_1, t_2]$ and $\langle s', p', o'\rangle[t_1', t_2']$ *overlap* if $s = s'$, $p = p'$, $o = o'$, and $\max(t_1, t_1') \leq \min(t_2, t_2')$; this definition is extended to triples of the form $\langle s, p, o\rangle[t_1]$ by treating them as abbreviations for $\langle s, p, o\rangle[t_1, t_1]$. Let G be a temporal graph and let $A \in G$ be a temporal triple. The *maximal subset* of G w.r.t. A is the smallest set $G_A \subseteq G$ such that $A \in G_A$ and, if $\beta \in G_A$, $\gamma \in G$, and β and γ overlap, then $\gamma \in G_A$ as well. The *normalization* of G is the temporal graph $\mathsf{nrm}(G)$ that, for each $A \in G$ of the form $\langle s, p, o\rangle[t_1, t_2]$ or $\langle s, p, o\rangle[t_1]$, contains the temporal triple $\langle s, p, o\rangle[t_1', t_2']$ where t_1' and t_2' are the smallest and the largest temporal constant, respectively, occurring in the maximal subset G_A of G w.r.t. A.

Let G' be the list of the temporal triples in G of the form $\langle s, p, o\rangle[t_1, t_2]$ and $\langle s, p, o\rangle[t_1]$ sorted by s, p, o, t_1, and t_2. For each $A \in G$, the triples that constitute the maximal subset G_A of G occur consecutively in G', so $\mathsf{nrm}(G)$ can be computed by a simple sequential scan through G'.

We next show how to use $\mathsf{nrm}(G)$ to evaluate temporal group patterns. Let $B = \{\langle s_1, p_1, o_1\rangle, \ldots, \langle s_k, p_k, o_k\rangle\}$ be a BGP, let $x_1, \ldots, x_k$ and $y_1, \ldots, y_k$ be variables not occurring in B, and let G be a temporal graph. Then $\langle B\rangle^G$ is the set of all mappings μ such that $\mathsf{dm}(\mu) = \mathsf{var}(B) \cup \{x_1, y_1, \ldots, x_k, y_k\}$, $\mu(\langle s_i, p_i, o_i\rangle[x_i, y_i]) \in \mathsf{nrm}(G)$ for each $1 \leq i \leq k$, and $\mu^\downarrow \leq \mu^\uparrow$, where the latter are defined as $\mu^\downarrow = \max\{\mu(x_1), \ldots, \mu(x_k)\}$ and $\mu^\uparrow = \min\{\mu(y_1), \ldots, \mu(y_k)\}$. Furthermore, $\nu|_B$ if the restriction of a mapping ν to $\mathsf{var}(B)$. Table 5 then shows how to evaluate basic TGPs under simple entailment in a normalization.

Table 5. Evaluation of Temporal Group Patterns under Simple Entailment

---

$\mathsf{eval}_{simple}(P, G)$ is the set of mappings defined as follows:

$P = B$ at t_3 :
$$\{\nu \mid \mathsf{dm}(\nu) = \mathsf{var}(P) \text{ and } \exists \mu \in \langle B \rangle^G : \nu|_B = \mu|_B \wedge \mu^\downarrow \leq t_3 \leq \mu^\uparrow\}$$

$P = B$ during $[t_1, t_2]$:
$$\{\nu \mid \mathsf{dm}(\nu) = \mathsf{var}(P) \text{ and } \exists \mu \in \langle B \rangle^G : \nu|_B = \mu|_B \wedge \mu^\downarrow \leq t_1 \leq t_2 \leq \mu^\uparrow\}$$

$P = B$ occurs $[t_1, t_2]$:
$$\{\nu \mid \mathsf{dm}(\nu) = \mathsf{var}(P) \text{ and } \exists \mu \in \langle B \rangle^G : \nu|_B = \mu|_B \wedge \max(\mu^\downarrow, t_1) \leq \min(\mu^\uparrow, t_2)\}$$

$P = B$ maxint $[t_1, t_2]$:
$$\{\nu \mid \mathsf{dm}(\nu) = \mathsf{var}(P) \text{ and } \exists \mu \in \langle B \rangle^G : \nu|_B = \mu|_B \wedge \nu(t_1) = \mu^\downarrow \wedge \nu(t_2) = \mu^\uparrow\}$$

$P = B$ mintime t_3 :
$$\{\nu \mid \mathsf{dm}(\nu) = \mathsf{var}(P) \text{ and } \exists \mu \in \langle B \rangle^G :$$
$$\mu^\downarrow \in \mathcal{TI} \wedge \nu|_B = \mu|_B \wedge \nu(t_3) = \mu^\downarrow \wedge \forall \lambda \in \langle B \rangle^G : \mu|_B = \lambda|_B \to \mu^\downarrow \leq \lambda^\downarrow\}$$

$P = B$ maxtime t_3 :
$$\{\nu \mid \mathsf{dm}(\nu) = \mathsf{var}(P) \text{ and } \exists \mu \in \langle B \rangle^G :$$
$$\mu^\uparrow \in \mathcal{TI} \wedge \nu|_B = \mu|_B \wedge \nu(t_3) = \mu^\uparrow \wedge \forall \lambda \in \langle B \rangle^G : \mu|_B = \lambda|_B \to \lambda^\uparrow \leq \mu^\uparrow\}$$

Proposition 4. *For each temporal graph G and each safe temporal group pattern P, we have $\mathsf{eval}_{simple}(P, G) = [\![P]\!]_G^{simple}$.*

We explain the intuition behind this algorithm for $P = B$ maxint $[t_1, t_2]$. First, we compute $\langle B \rangle^G$ by evaluating the conjunctive query $\bigwedge \langle s_i, p_i, o_i \rangle [x_i, y_i]$ in $\mathsf{nrm}(G)$ via simple lookup. Consider now an arbitrary $\mu \in \langle B \rangle^G$. By the definition of normalization, each $\mu(x_i)$ and $\mu(y_i)$ determine the maximal validity interval of $\langle s_i, p_i, o_i \rangle$ so, to answer P, we must intersect all intervals $[\mu(x_i), \mu(y_i)]$. Note that $\mu^\downarrow$ and $\mu^\uparrow$ give the lower and the upper limit of the intersection, provided that $\mu^\downarrow \leq \mu^\uparrow$. Thus, what remains to be done is to convert μ into ν by setting $\nu(x) = \mu(x)$ for each $x \in \mathsf{var}(B)$ and ensuring that $\nu(t_1) = \mu^\downarrow$ and $\nu(t_2) = \mu^\uparrow$. Based on these ideas, EO's system translates TGPs into SQL, which allows us to use a standard database query planner to optimize query execution.

5.2 Entailments Characterized by Deterministic Rules

Let X be an entailment relation that can be characterized by a set Γ_X of deterministic rules of the form (8).

$$A_1 \wedge \ldots \wedge A_n \to B \tag{8}$$

To evaluate a SPARQL group pattern in a graph under X-entailment, most existing (nontemporal) RDF systems first compute the closure of the graph w.r.t. Γ_X. We next show how to compute the temporal closure $\mathsf{cls}_X(G)$ of a temporal graph G using the rules from Γ_X. After computing the closure, we can normalize it and then apply the algorithm from Section 5.1.

Definition 6. *For X and Γ_X as stated above, let Σ_X be the set containing the rule (9) for each rule (8) in Γ_X.*

$$A_1[x_1, y_1] \wedge \ldots \wedge A_n[x_n, y_n] \wedge \max(x_1, \ldots, x_n) \leq \min(y_1, \ldots, y_n) \rightarrow \\ B[\max(x_1, \ldots, x_n), \min(y_1, \ldots, y_n)] \qquad (9)$$

Let G be a temporal graph consisting of triples of the form $\langle s, p, o \rangle [t_1, t_2]$.[3] The skolemization of G is a temporal graph obtained from G be replacing each blank node with a fresh URI reference. Furthermore, the temporal closure of G is the (possibly infinite) temporal graph $\mathsf{cls}_X(G)$ obtained by exhaustively applying the rules in Σ_X to the skolemization of G.

By applying this approach to G_1 under RDFS entailment, one can see that (2) and (3) produce $\langle :Munich, :hasAttraction, :Oktoberfest \rangle [130, 180]$.

The following proposition shows that, instead of evaluating TGPs in G under X-entailment, one can evaluate them in $\mathsf{cls}_X(G)$ under simple entailment.

Proposition 5. *Let X and G be as stated in Definition 6. For each temporal group pattern P, we have $[\![P]\!]_G^X = [\![P]\!]_{G'}^{simple}$, where $G' = \mathsf{cls}_X(G)$.*

6   Implementation and Outlook

In this paper we presented an approach for representing validity time in RDF and OWL, an extension of SPARQL that allows for querying temporal graphs, and two query answering algorithms. We implemented our approach in EO's knowledge representation system. The system is based on a proprietary extension of RDF that supports n-ary relations; it uses an ontology language based on OWL 2 RL; and it implements a proprietary query language based on the primitives and the notion of safety outlined in Section 4. The PostgreSQL database is used for data persistence and query processing. Ontology reasoning is implemented by translating the ontology into a datalog program, which is then compiled into a plSQL script that implements the seminaïve datalog evaluation strategy. Datalog rules are modified as described in Section 5.2 in order to deal with validity time; furthermore, the resulting set of facts obtained by applying the rules is normalized to allow for efficient query answering. Finally, temporal queries are translated into SQL and then evaluated using PostgreSQL's query engine; the translation essentially encodes the query evaluation algorithm from Section 5.1. The source of the system is not open, and EO has no plans for licensing the system to third-party developers. Therefore, we do not present a performance evaluation since such results could not be validated by the community. We merely note that EO is successfully using our approach with datasets consisting of tens of millions of triples, which we take as confirmation that our approach is amenable to practical implementation.

An important open theoretical question is to determine the worst-case complexity bounds of the query answering problem for our query language. Furthermore, one should see whether the general algorithm from Section 4 can be

---

[3] For simplicity we assume that G does not contain triples of the form $\langle s, p, o \rangle [t_1]$.

successfully used with expressive languages such as OWL 2 DL. We believe this to be possible provided that the algorithm is adequately optimized.

References

1. Allen, J.F.: Maintaining Knowledge about Temporal Intervals. Communications of the ACM 26(11), 832–843 (1983)
2. Artale, A., Franconi, E.: A survey of temporal extensions of description logics. Annals of Mathematics and Artificial Intelligence 30(1-4), 171–210 (2000)
3. Baader, F., Calvanese, D., McGuinness, D., Nardi, D., Patel-Schneider, P.F. (eds.): The Description Logic Handbook: Theory, Implementation and Applications, 2nd edn. Cambridge University Press, Cambridge (August 2007)
4. Chomicki, J.: Temporal Query Languages: A Survey. In: Proc. ICTL, pp. 506–534 (1994)
5. Fitting, M.: First-Order Logic and Automated Theorem Proving, 2nd edn. Texts in Computer Science. Springer, Heidelberg (1996)
6. Gutiérrez, C., Hurtado, C.A., Mendelzon, A.O.: Foundations of Semantic Web Databases. In: Proc. PODS, pp. 95–106 (2004)
7. Gutierrez, C., Hurtado, C.A., Vaisman, A.A.: Introducing Time into RDF. IEEE Transactions on Knowledge and Data Engineering 19(2), 207–218 (2007)
8. Hayes, P.: RDF Semantics, W3C Recommendation (February 10, 2004)
9. Hurtado, C.A., Vaisman, A.A.: Reasoning with Temporal Constraints in RDF. In: Alferes, J.J., Bailey, J., May, W., Schwertel, U. (eds.) PPSWR 2006. LNCS, vol. 4187, pp. 164–178. Springer, Heidelberg (2006)
10. Milea, V., Frasincar, F., Kaymak, U.: Knowledge Engineering in a Temporal Semantic Web Context. In: Proc. ICWE, pp. 65–74 (2008)
11. Motik, B., Cuenca Grau, B., Horrocks, I., Wu, Z., Fokoue, A., Lutz, C.: OWL 2 Web Ontology Language: Profiles. W3C Recommendation (October 27, 2009)
12. Motik, B., Patel-Schneider, P.F., Parsia, B.: OWL 2 Web Ontology Language: Structural Specification and Functional-Style Syntax, W3C Recommendation (October 27, 2009)
13. Patel-Schneider, P.F., Motik, B.: OWL 2 Web Ontology Language: Mapping to RDF Graphs, W3C Recommendation (October 27, 2009)
14. Pérez, J., Arenas, M., Gutierrez, C.: Semantics and complexity of SPARQL. ACM Transactions on Database Systems 34(3) (2009)
15. Pugliese, A., Udrea, O., Subrahmanian, V.S.: Scaling RDF with Time. In: Proc. WWW, pp. 605–614 (2008)
16. Schneider, M.: OWL 2 Web Ontology Language: RDF-Based Semantics, W3C Recommendation (October 27, 2009)
17. Straccia, U., Lopes, N., Lukácsy, G., Polleres, A.: A General Framework for Representing and Reasoning with Annotated Semantic Web Data. In: Proc. AAAI (2010)
18. Tappolet, J., Bernstein, A.: Applied Temporal RDF: Efficient Temporal Querying of RDF Data with SPARQL. In: Aroyo, L., Traverso, P., Ciravegna, F., Cimiano, P., Heath, T., Hyvönen, E., Mizoguchi, R., Oren, E., Sabou, M., Simperl, E. (eds.) ESWC 2009. LNCS, vol. 5554, pp. 308–322. Springer, Heidelberg (2009)
19. Vila, L.: A Survey on Temporal Reasoning in Artificial Intelligence. AI Communications 7(1), 4–28 (1994)

Enhancing the Open-Domain Classification of Named Entity Using Linked Open Data

Yuan Ni, Lei Zhang, Zhaoming Qiu, and Chen Wang

IBM Research, China
{niyuan,lzhangl,qiuzhaom,chenwang}@cn.ibm.com

Abstract. Many applications make use of named entity classification. Machine learning is the preferred technique adopted for many named entity classification methods where the choice of features is critical to final performance. Existing approaches explore only the features derived from the characteristic of the named entity itself or its linguistic context. With the development of the Semantic Web, a large number of data sources are published and connected across the Web as Linked Open Data (LOD). LOD provides rich a priori knowledge about entity type information, knowledge that can be a valuable asset when used in connection with named entity classification. In this paper, we explore the use of LOD to enhance named entity classification. Our method extracts information from LOD and builds a type knowledge base which is used to score a (named entity string, type) pair. This score is then injected as one or more features into the existing classifier in order to improve its performance. We conducted a thorough experimental study and report the results, which confirm the effectiveness of our proposed method.

Keywords: Named Entity Classification, Linked Open Data.

1 Introduction

Automatically classifying named entities from text is a crucial step in several applications. For example, in the Question Answering system [17], one important step is to determine whether a candidate answer is of the correct type given the question, and in the Information Extraction system [9], the first step is to identify the named entities and their types from the text. As devices linked to the Internet proliferate, the amount of information available on the Web rapidly grows. At the same time, there is a trend to scaling applications to the Web. For example, the MULDER Question Answering system [19] is designed to answer open-domain questions from the Web, and the TextRunner system is designed to extract open information from the Web [6]. To satisfy the requirements of such Web-scale applications, named entity classification is evolving from a simple three-category classification system to one in which a large number of classes are specified by an ontology or by a taxonomy [8], clearly a more challenging task.

P.F. Patel-Schneider et al. (Eds.): ISWC 2010, Part I, LNCS 6496, pp. 566–581, 2010.

Most proposed approaches for named entity classification adopt machine-learning techniques. The choice of features is critical to obtaining a good classification of named entities. Traditional classification approaches focus on the word-level characters or the context information of the named entities [20] without exploiting in any way the valuable classification information available for the large number of entities provided by Linked Open Data (LOD) on the Web. As of this writing, the Linked Data project [1] includes more than 100 datasets which together contain about 4.2 billion triples, and the datasets cover such various domains as Wikipedia, IMDb, and Geonames. Given a named entity, it is highly likely that some type assertions can be found in LOD. Thus, knowledge from LOD can provide good features for named entity classification. Features used in existing methods can be considered as context-dependent linguistic features, and the machine-learning technique makes use of statistic information based on these features to obtain a posteriori knowledge, while the features from LOD are considered as context-independent features that explore a priori knowledge. Our proposed method is to integrate the a priori knowledge with the a posteriori knowledge to improve named entity classification.

In this work we examine how to make use of the type information from LOD to improve named entity classification. It is not a trivial task due to the following challenges. First, because LOD may contain noisy information and it may be incomplete with respect to the information required to determine named entities and their potential types, we need to improve the precision and completeness of the type information. Second, for the extracted and cleaned type information for named entities, we need a mechanism to store it and to guarantee the efficiency of the method used to process the large, unwieldy LOD for use in real applications. Third, LOD from various sources have various taxonomies, and named entity classification also has its own taxonomies; therefore, we need a mechanism to solve the type similarity problem among these multiple taxonomies. The final challenge is to provide a scoring strategy given the multiple possible type information for a named entity.

Organization. The remainder of this paper is organized as follows. Section 2 gives an overview of our proposed method; Section 3 presents the method we use to prepare the linked data and store the potential type information; the scoring method is discussed in Section 4; Section 5 presents our experimental evaluation of the feature; related work is discussed in Section 6, and the conclusion and future work are discussed in Section 7.

2   Overview

The key idea of our approach is to leverage the knowledge base from the LOD to generate a score to measure the probability that the given named entity could be classified as the target type. The scores can then be used by existing machine-learning methods as additional features to improve named entity classification. Let us use $< f_1, \ldots, f_n >$ to denote the feature vector for a named entity in the existing approach. Our method appends a set of features derived from LOD to the end of that vector to obtain a new feature vector $< f_1, \ldots, f_n, F_1, \ldots, F_m >$. The new generated feature vectors are used by the classifiers for named entity classification. The rationale behind this method is to combine context-independent, a priori knowledge from LOD with linguistic contextual information in a single feature vector and let the machine-learning algorithm determine how to best combine them statistically.

Fig 1 shows the architecture of the component to compute the LOD feature. In order to efficiently compute the additional feature, our system is divided into online and offline components. For the offline components, the *LOD Preprocessor* is used to extract the type information from the various LOD sources and precompute the type knowledge base in a format that best suits our algorithms. For the online components, the *Type Retrieval* is used to retrieve all type information of a given named entity string from the type knowledge base. The LOD scorer takes charge of computing the probability scores for the retrieved types of the named entity and the target type. An intermediate taxonomy is used to calculate the similarity between each type that is provided by LOD and the target type. Finally, the obtained score, together with other feature scores, are given to the classifier.

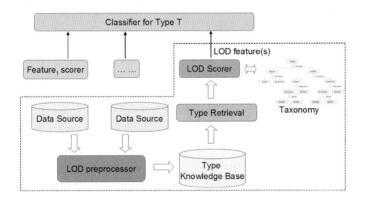

Fig. 1. The Architecture of the Component for Computing LOD feature

3   Type Knowledge Base Generation

The type knowledge base is needed for two reasons. First, the LOD may contain noise and may be incomplete; thus, we need to remove the noisy information and make the type information more complete. Second, to ensure the efficiency of the online scoring, the type information should be pre-computed and indexed in a format that supports fast retrieval. This section describes the method used to generate the type knowledge base. The goal is to enumerate all possible (name string, type) pairs from the LOD.

LOD uses a Uniform Resource Identifier (URI) to identify each instance, and type assertions are provided for each URI. For example, for an instance having the URI dbpedia:John_F._Kennedy[1], one triple from DBpedia indicates that the instance dbpedia:John_F._Kennedy belongs to type *President*:

(dbpedia:John_F._Kennedy, rdf:type, President).

For the instance dbpedia:John_F._Kennedy, we have another triple from DBpedia to indicate one of its names:

(dbpedia:John_F._Kennedy, rdfs:label, "John F. Kennedy").

Based on these two triples, we obtain the (name string, type) pair (John F. Kennedy, President) to specify that the name string "John F. Kennedy" belongs to category "President". From DBpedia, we have another triple which indicates another possible name for the instance dbpedia:John_F._Kennedy:

(dbpedia:John_F._Kennedy, dbpedia:birthName, "John Fitzgerald Kennedy").

From this we obtain another (name string, type) pair, (John Fitzgerald Kennedy, President).

As the type knowledge base requires the type information for each named entity, one problem we needed to solve is how to generate all possible name variants for each instance. Traditional approaches leverage the string similarity to determine whether some name variants correspond to one instance. However, the name variants of one instance may not always be similar. For example, ping-pong and table-tennis are the name variants for the same instance. In our method, we propose to use various name properties and certain relationships in LOD to enumerate all possible names variants for an instance.

3.1   Leverage the Name Properties

Thanks to the broad coverage of LOD, most name variants for an instance that may be mentioned in some text are likely to be specified by some name properties. We analyze the properties used in LOD sources and identify the ones that may describe the name information. For example, in terms of the definition from RDF schema, the property rdfs:label provides a human-readable description of a resource, making it a good candidate for name properties. We observe that there are many name properties in LOD, e.g., DBpedia has 106 properties about such names as dbpedia:name and dbpedia:fullname. To get maximal coverage on all possible names of an entity, we tried to use most of these properties. However, experimental results showed that they lead to many errors due to noisy LOD. For example, from DBpedia we have the following triples:

(dbpedia:Chrysler_New_Yorker, dbpedia:name, 1982)
(dbpedia:Chrysler_New_Yorker, rdf:type, Automobile)

We can then obtain the pair (1982, Automobile) which is not correct. Based on these experiments, we make use of only the properties that exactly describe the names, such as rdfs:label and foaf:name.

---

[1] The dbpedia: stands for the prefix for the URI from DBpedia.

3.2   Leverage the Relationships

In LOD, some relationships may connect two data instances where one data instance could be considered as providing another name variant of the other instance. We have identified three relationships and make use of them to enrich the name variants of the instances.

Redirects Relationship. The *redirects* relationship is used in DBpedia and links one URI, which has no description, to another URI that has a description. The purpose of creating and maintaining the *redirects* relationship is because the former URI has the relationships, including alternative name, less- and more-specific forms of names, abbreviations, etc. [3] with the later URI. If URI_1 redirects to URI_2, then the name of URI_1 can be considered as a name variant of the instance of URI_2. Therefore, for each (name string, type) pair, i.e., (name$_2$, type), that is derived from the type assertions for URI_2, we generate another pair (name$_1$, type).

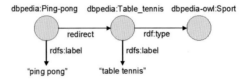

Fig. 2. The Example RDF Data for Redirect Relationship

Example 3.1. Let us use an example to illustrate this. Suppose we have a set of RDF triples, and their graph representation is shown in Fig 2. There is a redirect relationship from the URI dbpedia:Ping-pong to dbpedia:Table_tennis, meaning that ping-pong is a name variant of dbpedia:Table_tennis. From the description of dbpedia:Table_tennis, we obtain a pair (table tennis, Sport), and we then generate another pair (ping-pong, Sport) due to the redirects relationship.   □

owl:sameAs Relationship. According to the OWL specification, the owl:sameAs indicates that two URI references actually refer to the same thing. Therefore, if URI_1 owl:sameAs URI_2, we combine them as one instance. The instance has the name variants from both URI_1 and URI_2 and has the types from both URI_1 and URI_2.

Disambiguates Relationship. The *disambiguates* relationship is used in DBpedia. A disambiguation URI has many *disambiguates* relationships with other different URIs which could, in principle, have the same name as that of the disambiguation URI. For example, we have (dbpedia:Joker disambiguates dbpedia:Joker_butterfly) and (dbpedia:Joker disambiguates dbpedia:The_ Joker's_ Wild). It means that dbpedia:Joker_butterfly and dbpedia:The_Joker's_Wild can have the same name Joker. Joker is then a name variant of dbpedia:Joker_butterfly and is also a name variant of dbpedia:The_Joker's_Wild. For all type assertions about dbpedia:Joker_butterfly and dbpedia:The_Joker's_Wild, we generate corresponding (name string, type) pairs for the name Joker. For example, we know that dbpedia:Joker_butterfly belongs to the type *Insect*, then we generate a pair (Joker, Insect).

The enrichment using these relationships may not be 100% reliable because there may exist incorrect relationships due to the current quality level of LOD. However, we have conducted experiments that verify that the above enrichment helps improve the scoring accuracy.

3.3   Structure of the Type Knowledge Base

Given the (name string, type) pairs extracted from LOD, we need a mechanism to store and index them in order to guarantee the efficient retrieval for online scoring. The inverted list is used to store such information where the name string is the key. Different data instances may have the same name, but the scoring mechanism needs to distinguish the types for different instances (which will be introduced in detail in Section 4). Thus, the type information for a single name string is separated in terms of the data instance, i.e., the URI to which it corresponds. Fig. 3 shows the structure of the inverted list for storing the type information for name strings where the element $(t^1_{j1}, \ldots, t^1_{jm(j)})$ for entry N_1 stores all possible types of the jth instance S^1_j that has the name string N_1.

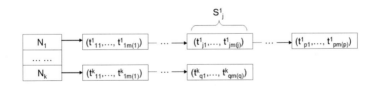

Fig. 3. The Inverted List for Type Knowledge Base

We could have used services like Sindice and Sigma[2] instead of building the inverted index on our own, but considering the complicated preprocessing we need to perform and the networking latency to use these services, we decided to build our own indexes.

In our work, we have generated a type knowledge base that includes LOD from DBpedia, IMDb, and GeoNames. The statistical information is shown in Table 1.

4   Scoring Method

Given a named entity string and a target type, the scoring is performed by computing a metric that measures the probability that the named entity can be classified as the target type using the type information from the type knowledge

---

[2] Sindice : http://sindice.com, Sigma:http://sig.ma

Table 1. Statistic Information for the Type Knowledge Base

Dataset	#instances (millions)	#name properties	#name variants (millions)	#type assertions (millions)
DBpeida	3.2	4	4.7	7.9
IMDb	24.5	8	12.0	24.4
GeoNames	6.7	3	8.4	6.5

base. There are three main challenges to doing this: (1) given a named entity string, how to find the matched names in the type knowledge base and get all possible type information for them; (2) because the types from LOD and the target type may be from different taxonomies, we need a strategy to precisely compute the similarity between these types; (3) because one named entity may correspond to multiple instances with multiple types, we need a mechanism to determine a final score. In the following sections, we introduce the details of the techniques used to meet these challenges.

4.1 Retrieving Types for the Named Entity String from Type Knowledge Base

To retrieve the possible type information, one simple method is to use the given named entity string as the key to find the corresponding types from the inverted lists. However, for the same entity, the given name may not be exactly the same as the names indexed in the type knowledge base. There are two reasons that can cause a name mismatch.

The entity itself has various names. For example, for President John F. Kennedy, one may use the full name John Fitzgerald Kennedy. As mentioned in Section 3, during the generation of the type knowledge base, we make use of the properties of names of instances to generate all possible names of an instance. Additionally, we make use of three types of relationships in LOD to enrich the possible names. With the help of all name properties of an instance and the relationships, our type knowledge base is likely to have a broad coverage of all possible names of an instance. Then, given a named entity string, a simple index lookup is enough to find its type information.

The names are presented in different format. For example, the indexed name is *tomato* while the given name is *tomatoes*. To solve this problem, we conduct the normalization on both the indexed names and the given names using the following rules: (1) perform word stemming on the names; (2) convert the names to lowercase; (3) remove any articles from names.

4.2 Matching the Target Type with the Retrieved Type

The types provided by LOD are considered as from open-domain. According to data publishers' requirements, new types can be added to describe new instances. Meanwhile, the types generated by data publishers are more flexible; for instance, a type can be represented by a phrase, such as the category *jewish american film directors* from DBpedia. Even more, each data source in LOD may have its own type system. On the other side, the target type would also be considered as from open-domain because of the requirement of scaling information extraction and question answering to the Web. It is very difficult to match various types from open-domain taxonomies. We propose an intermediate ontology (denoted as **O**) to compute the similarity. First, the target type and the retrieved type are linguistically matched to some nodes in **O**, and we then compute the semantic distance between the two matched nodes in **O** and use this distance measurement as the similarity score.

Intermediate Ontology. One simple method for the intermediate ontology is to leverage an existing, well-defined general taxonomy. WordNet [11] is a well-defined general taxonomy widely used by the natural-language processing community. However, WordNet lacks a formal constraint on classes; for example, WordNet does not provide information about disjoint classes which could help us determine that a named entity does not belong to a type. Additionally, WordNet contains word senses that are too diverse, and a very rarely used word sense may incur a negative effect on the similarity computation. The AI community has also built general-purpose ontologies, such as Cyc [2], with formal logic theories. However, the ontology is very complex and lacks linguistic features. Considering the drawbacks of existing taxonomies, in our work we built a simple intermediate ontology. The ontology is designed in terms of the following principles: (1) the ontology covers the most frequently used concepts; (2) the ontology captures the disjoint axioms between classes such that we can obtain a score to measure how the named entity does not conform to the target type. For example, *people* and *organization* are disjointed classes. Our ontology is relatively small so if some type cannot be matched, we revert back to using WordNet.

Calculating the Similarity Score. The created ontology **O** is used as an intermediate ontology to link the target type and the retrieved type. The first step is to find the corresponding types of the target type and the retrieved type in **O**. If T denotes the target type/retrieved type, and the corresponding type in **O** is denoted as T', then T' should stand for the same concept as T. If a node in **O** exactly matches the type T, then the node is considered the best match for T. However, because of the flexibility of types from the open-domain, especially types from linked data, some types can be a phrase with adjective qualifiers that are not covered in **O**. To match these types for which no exactly matched nodes exist in **O**, we perform a normalization on the type phrase to get the headword. By analyzing type phrases from the linked data, we observed that the qualifiers are mainly presented in three ways: (1) an adjective is used to qualify a noun, for example, "Australian anthropologists"; (2) a qualifier phrase beginning with *of xxx* is used, such as "people of the French and Indian war"; (3) a qualifier phrase with *from xxx* is used, such as "people from St. Louis, Missouri". Given the type T, we remove the qualifiers for the above three cases to get the headword of

T, denoted as T_{root}. Finally, the node in **O** that matches T_{root} exactly is the corresponding type of T in **O**.

Suppose, T'_{target} and $T'_{retrieve}$ are the corresponding types of the target type and the retrieved type, respectively. The similarity score (denoted as s) is computed as follows:

- if T'_{target} and $T'_{retrieve}$ are the same node, then $s = 1$.
- if T'_{target} and $T'_{retrieve}$ are disjointed in terms of the ontology, then $s = -1$.
- if T'_{target} and $T'_{retrieve}$ are on the same path and there exists n steps between them, then $s = 1/n$.
- if T'_{target} and $T'_{retrieve}$ are in different paths and n_1 and n_2 are the number of steps to their lowest common ancestors, then $s = 1/(n_1 + n_2)$.

It is possible that T_{head} cannot be exactly matched by some node in the ontology **O** because our created ontology cannot cover all possible kinds of types. For these cases, we make use of the online resource WordNet to calculate the similarity score. Given a word, WordNet provides an API to get matched nodes. We then use the method proposed in [5] to calculate the semantic similarity between nodes in WordNet.

4.3   Determining the Final Score

Given a named entity $N(i)$, it may correspond to multiple data instances S_1^i, ..., S_j^i, ...,$S_n^i(i)$, and for each data instance S_j^i, it may belong to multiple types T_{j1}^i, ..., T_{jk}^i, ..., $T_{jm(j)}^i$. For each type T_{jk}^i, we calculate a score with respect to the target type T using the mechanism discussed in the previous section. For the named entity string $N(i)$, we then obtain multiple scores, which are divided into subsets according to the instances they describe. In this paper, we propose a two-step strategy to determine the final score. The first step is to compute the score for each data instance S_i given the scores for $\{T_{j1}^i, \ldots, T_{jk}^i, \ldots, T_{jm(j)}^i\}$; after that, we compute the score for the named entity $N(i)$ given the scores for $\{S_1^i, \ldots, S_j^i, \ldots, S_n^i(i)\}$. The advantage of the two-step strategy is as follows. By considering the characteristic of a single instance, we can avoid some noisy scores, making the score for a single instance more precise. The precise score for each instance is indispensable to obtaining a precise final score.

Determining the score for an instance S_i. Given a certain instance S_i, it can be stated that there should not be any conflicts within all of the type information for that instance. Therefore, for all scores of an instance, it is unlikely that some are positive (i.e., conform to the target type to some degree) and some are negative (i.e., conflict with the target type to some degree). However, due to the fuzzy match in the type-matching step and possible noise in the linked data, conflicts may occur. We propose the use of a vote strategy to solve this problem. For an instance S_i, if most of its types get positive scores, then the largest score is picked as the score for S_i; otherwise, if most of its types get negative scores, then the smallest score is picked as the score for S_i.

Determining the score for $N(i)$**.** Given multiple data instances for a named entity, if we know its URI, then the score for the data instance with the matched URI is used as the final score. This situation may occur in some Web-based question-answering applications in which Wikipedia is used as the corpora [4]. When the title of the Wikipedia page is selected as a candidate answer, the URI of the page is considered as the URI of the named entity and corresponds directly to a DBpedia URI.

In cases where there is no URI for the named entity and we cannot know which instance is indicated for this named entity, then we can use the following strategies to determine the final score. (1) *The aggressive strategy*: Considering that the named entity could be any one of the indexed data instances, one aggressive heuristic is that if the maximum score is larger than 0, then we pick the largest score; else if the maximum score is smaller than 0, then we pick the smallest score; otherwise the final score is 0. The aggressive strategy tends to give an exact score, either exactly matched or exactly unmatched. The score will be distinguishable. (2) *The average strategy*: We assume that the named entity has the same probability to match each indexed data instance, then the average of the total scores is used as the final score. An experimental study is provided in Section 5 to compare the above two strategies.

4.4   Applying the Score in Machine Learning

This section introduces how to use the score (which measures the probability of whether the named entity belongs to the target type) as a feature in the machine-learning step.

We can simply add one feature, which is called TyCorLOD (Type Coercion using LOD) in machine learning, and the generated score is used as the feature score. The score range for the TyCorLOD feature would then be [-1,1]. The higher score indicates that the named entity is more likely to belong to the target type. However, the problem with this method is that we cannot give different weights on the positive effect and negative effect. Therefore, we developed another method. In the machine-learning step, we split the score into two features (we call them TyCorLOD and AnTyCorLOD). The TyCorLOD feature indicates the likelihood that the named entity **conforms** to the target type, while the AnTyCorLOD feature indicates the likelihood that the named entity **conflicts** with the target type. These two feature scores are generated in terms of the final score S using the following strategy: If $S >= 0$, then we give the score S for TyCorLOD and 0 for AnTyCorLOD; otherwise (i.e., $S < 0$), we give the score 0 for TyCorLOD and $|S|$ for AnTyCorLOD. The comparison of the above two options is discussed in Section 5.

5   Experimental Study

To verify our proposed feature using LOD for named entity classification, we conducted extensive experiments to demonstrate the effectiveness of our proposed method.

5.1   Experimental Setup

Datasets. We have tested our proposed method on two datasets. The first dataset (denoted as $Data_Q$) is extracted from an IBM question-answering system for open-domain factoid questions. We randomly selected 400 questions. For each question we manually labeled the type of entity that was expected as an answer to the question (i.e., the target type) and we also extracted the top 10 candidate answers that were generated by the system. We asked one test person to determine whether the candidate answer belongs to the target type. For all candidate answers that belonged to the target types, we generated a list of (candidate answer, type) pairs, called *ground-truth* (denoted as $Data_Q^{true}$); for all candidate answers that did not belong to the target types, we generated a list of (candidate answer, type) pairs, called *ground-wrong* (denoted as $Data_Q^{wrong}$). We obtained 1,967 pairs for ground-truth and 3,053 pairs for ground-wrong. There are 114 distinguishing target types, which reflects the fact that the data was from open-domain. The second dataset (denoted as $Data_P$) is the *People Ontology*, which is a benchmark in [16], [15]. It was extracted from WordNet and contained 1,657 distinct person instances arranged in a multilevel taxonomy having 10 fine-grained categories, such as chemist or actor. Each instance and its category could also be considered as a pair in ground-truth. From this dataset, we obtained 1,618 pairs for ground-truth. The obtained dataset is denoted as $Data_P^{true}$.

Evaluation Metric. We conducted two types of evaluations. First, we measured how our feature and scoring method performed on ground-truth/ground-wrong. The three metrics used here, i.e., accuracy, false-rate, and unknown-rate, are described in Table 2, where N stands for total number of pairs.

Table 2. Descriptions for Used Metrics

Metric Name	Description	Measurement	Remarks
accuracy	#(correctly scored pairs)/N	correctness percentage of the scoring method	for ground-truth (resp. ground-wrong) dataset, the positive scores (resp. negative scores) are considered as correct
unknown-rate	#(pairs with score 0)/N	the coverage of the linked data	if the named entity is not indexed in the type knowledge base, we give the score 0
false-rate	#(incorrectly scored pairs)/N	incorrectness percentage of the scoring method	for ground-truth (resp. ground-wrong) dataset, the negative scores (resp. positive scores) are considered as incorrect

Second, we wanted to illustrate how our proposed feature helps to improve the performance of named entity classification. To do this, we compared the precision/recall of the classification with and without our feature.

5.2   Scoring Accuracy

This section demonstrates the scoring accuracy of LOD feature on the ground-true/wrong dataset. As introduced in Section 4.3, we have two strategies to determine the final score, i.e., the aggressive strategy and the average strategy. We compared the performance of these two strategies on the ground-true/wrong of $Data_Q$ and $Data_P$ using the metrics introduced in Table 2. The results for the *aggressive* strategy and for the *average* strategy are shown in Table 3.

Table 3. The Scoring Performance on the Ground-Truth/Wrong Dataset

Data Set	Aggressive			Average		
	accuracy	unknown-rate	false-rate	accuracy	unknown-rate	false-rate
$Data_Q^{true}$	83.4%	10.2%	6.35%	71.9%	12.6%	15.5%
$Data_Q^{wrong}$	50.6%	25.9%	23.5%	53.9%	26.2%	19.9%
$Data_P^{true}$	91.5%	7.85%	0.65%	88.3%	9.88%	1.83%

We first observed that the aggressive strategy resulted in high accuracy, i.e., 83.4% for $Data_Q^{true}$ and 91.5% for $Data_P^{true}$ on ground-true data. This indicates that the linked data provides high-quality type information with good coverage and that our scoring method measures the type similarity quite well. Second, it shows that the accuracy for the ground-wrong data from $Data_Q$ is a little lower. For the higher unknown-rate, although in a question-answering project the correct response should be a fact, the extracted candidate answers may not all be fact entities, for example, "Germany history" or "the best movie." Therefore, these candidates are not covered by the linked data. Actually, named entity classification does not target for this kind of entity. For the higher false-rate, there are two main reasons: (1) our strategy for final score gives higher preference on the positive score. One name may correspond to multiple instances in our type knowledge base, and we do not know which instance the named entity corresponds to. Giving a relaxed score is safer for our linked data feature. With the additional context information of the named entity, information that can be used as additional features in named entity classification, the false-rate could be reduced; (2) the type knowledge base contains only more general types than the target type. For example, the named entity has a type *person*, and the target type is *actor*. The named entity may not be an *actor*, but as the type *person* is not disjoint with the type *actor*, we could not give a negative score for this named entity with respect to the type *actor*. This may incur the false-positive cases.

To compare the aggressive strategy with the average strategy, we observed that the aggressive strategy outperforms the average strategy for the ground-true data while the average strategy outperforms the aggressive strategy for the ground-wrong data. Because the aggressive strategy gives a higher priority on the positive score, the ground-true pairs benefit more from this strategy. For ground-wrong data, as mentioned in the previous paragraph, the false-positive instances may occur due to the more general types in the type knowledge base. Using the aggressive strategy, this false-positive instance will affect the final score; while using the average strategy, the false-positive instance may be compensated by other correctly scored instances. That is why the average strategy is better than

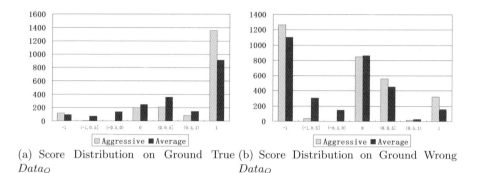

(a) Score Distribution on Ground True
$Data_Q$

(b) Score Distribution on Ground Wrong
$Data_Q$

Fig. 4. Score Distribution on the Dataset

the aggressive strategy for ground-wrong data. When combining the ground-true and ground-wrong data together for $Data_Q$, the aggressive strategy is better in general.

Fig. 4(a) and (b) illustrate the distribution of the scores between the range [-1,1] for the ground-true and ground-wrong datasets from $Data_Q$. We can observe that for correct scores (i.e., positive scores for ground-true and negative scores for ground-wrong), most are exactly correct (i.e., score 1 for ground-true and score -1 for ground-wrong). This indicates the accuracy of our scoring. To compare the aggressive strategy and the average strategy, it shows that the aggressive strategy gives more exact scores and the average aggressive tends to be more balanced. The reason is that the average strategy computes the average of all instance scores, so the exact score may be reduced by other scores. Considering both the accuracy and the distribution, we suggest using the aggressive strategy.

5.3   Impact on the Machine Learning for Classification

This section reports the results of adding our proposed feature into the feature space of an existing named entity classification method using latent semantic kernels (LSI) [15]. We use LSI to denote the existing approach and LSI+LOD to denote the approach with our feature. Given each instance from the people ontology, the multi-context is derived by collecting 100 English snippets by querying $Google^{TM}$. The proximity matrices are derived from 200,000 Wikipedia articles. and the features are then generated using jLSI code [14]. With respect to our LOD feature, for each target class i, one feature is added to measure the probability of whether the instance belongs to the class i. Therefore, for 10 classes, we need to add 10 features. We use the KNN (k=1) method to do the classification. Fig. 5(a) compares the precision/recall/f-measure between the approaches LSI and LSI+LOD. It is shown that LSI+LOD outperforms the LSI on both precision and recall, and then f-measure. Specifically, the precision is improved from 81% to 84.3%, the recall is improved from 80.3% to 84.3%, and the f-measure is improved from 80.5% to 84.3%.

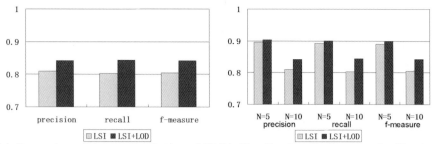

(a) Comparison of Approaches between LSI (b) Classification Performance by Varying
and LSI+LOD $Data_P$ the Number of Classes $Data_P$

Fig. 5. Effectiveness of Linked Data Feature

Fig. 5(b) illustrates the performance on different numbers of target classes. The bars with $N = 5$ are the results for the 5-class dataset where we select 5 classes from the 10 classes and extract the corresponding instances to generate the dataset. The bars with $N = 10$ are the results for the 10-class dataset. It shows that the improvement of LSI+LOD over LSI is larger on the dataset with $N = 10$ than the dataset with $N = 5$. Specifically, for $N = 5$, the f-measure is improved by 1.1% using LSI+LOD, and for $N = 10$, the f-measure is improved by 4.7% using LSI+LOD. This indicates that as the number of target classes grows larger, the improvement using our LOD feature becomes greater. The reason is that as the number of classes becomes larger, the classification accuracy using traditional features, such as word characteristic and word context, becomes lower. However, as the linked data knowledge base provides more fine-grained type information, the scoring is still accurate for fine-grained classes. Therefore, the improvement of the method using the LOD features becomes greater.

We also conducted the experiments on a dataset from an IBM question-and-answering system to compare the performance of using feature TyCorLOD only or using both TyCorLOD and AnTyCorLOD, as discussed in Section 4.4. The results verify that the strategy to use both TyCorLOD and AnTyCorLOD outperforms the strategy of using only TyCorLOD. Due to space limitations, we omit the detailed results here.

6   Related Work

Named entity classification has been studied by many researchers [13], [10], [15], [12], [16]. The early named entity classification task considered a limited number of classes. For example, the MUC named entity task [18] distinguishes three classes, i.e., PERSON, LOCATION and ORGANIZATION, and the CoNLL-2003 adds one more class, i.e., MISC. The dominant technique for addressing this task is supervised learning where the labeled training data for the set of classes is provided [7]. Recently, a more fine-grained categorization of named entities has been studied. Fleischman and Hovy [12] examined different features and learning algorithms to automatically subcategorize person names into eight fine-grained classes. Cimiano and Völker [8] have leveraged the context of named entities and used unsupervised learning to categorize named entities with respect to an ontology. In short, machine-learning approaches are widely adopted by the proposed approaches and the features used for the learning method can be classified into three categories: (1) the word-level feature [7], such as the word case or digit pattern; (2) handcrafted resources, such as gazetteers or lists [13], [12]; (3) the context of the named entities [8], [15].

The feature proposed in this paper is different from these existing approaches. It exploits a resourceful knowledge base, i.e., linked open data. This knowledge base is different from precomplied lists for classifying certain categories, as used in [13]. The linked data is published by various data providers and has a broad coverage. Because information in the linked data is still growing, more type information will be available in the future. The linked data feature is orthogonal with existing features and can be combined with them in order to improve the performance of named entity classification, including both the three-class task and fine-grained classification.

7 Conclusion and Future Work

In this paper, we proposed to explore the extensive type information provided by LOD to generate additional features, and these new features, together with existing features, can be used in machine-learning techniques for named entity classification. Specifically, in the first step, we proposed a mechanism to generate a type knowledge base that precisely and completely captures the type information for all possible named entity strings. We then proposed scoring strategies to generate feature scores based on the precomputed type information. Our experimental results verified the effectiveness of the proposed method and indicated that the improvement margin becomes larger as the number of target classes grows. In the future, we plan to investigate more data sources for LOD in order to provide better coverage.

Acknowledgements

We thank Christopher Welty and Aditya Kalyanpur for their valuable suggestions regarding this paper.

References

1. Linked data, http://linkeddata.org/
2. Opencyc, http://www.cyc.com/opencyc
3. Redirects, http://en.wikipedia.org/wiki/Redirects_on_wikipedia
4. Ahn, D., Jijkoun, V., Mishne, G., Muller, K., de Rijke, M., Schlobach, S.: Using wikipedia at the trec qa track. In: Proceedings of the 13rd Text REtrieval Conference, TREC 13 (2004)

5. Banerjee, S., Pedersen, T.: Extended gloss overlaps as a measure of semantic relatedness. In: Proceedings of the 18th International Joint Conference on Artificial Intelligence (2003)
6. Banko, M., Cafarella, M.J., Soderland, S., Boardhead, M., Etzioni, O.: Open information extraction from the web. Communications of the ACM (2008)
7. Bikel, D.M., Miller, S., Schwartz, R., Weischedel, R.: Nymble: High-performance learning name-finder. In: Proceedings of the 5th Conference on Applied Natural Language Processing (1997)
8. Cimiano, P., Volker, J.: Towards large-scale, open-domain and ontology-based named entity classification. In: Proceedings of the International Conference on Recent Advances in Natural Language Processing, RANLP (2005)
9. Etzioni, O., Cafarella, M., Downey, D., Kok, S., Popescu, A.-M., Shaked, T., Soderland, S., Yates, D.S.W.S.A.: Web-scale information extraction in knowitall. In: Proceedings of the 13th International Conference on World Wide Web, WWW (2004)
10. Evans, R.: A framework for named entity recognition in the open domain. In: Proceedings of the Recent Advances in Natural Language Processing, RANLP (2003)
11. Fellbaum, C. (ed.): Wordnet: An electronic lexical database. MIT Press, Cambridge (1998)
12. Fleischman, M., Hovy, E.: Fine-grained classification of named entities. In: Proceedings of the 19th International Conference on Computational Linguistics, Coling (2002)
13. Ganti, V., Konig, A.C., Vernica, R.: Entity categorization over large document collections. In: Proceedings of the 14th ACM SIGKDD International Conference On Knowledge Discovery & Data Mining (2008)
14. Giuliano, C.: jLSI a for latent semantic indexing (2007) Software available at, http://tcc.itc.it/research/textec/tools-resources/jLSI.html
15. Giuliano, C.: Fine-grained classification of named entities exploiting latent semantic kernels. In: Proceedings of the 13rd Conference onCcomputational Natural Language Learning, CoNLL (2009)
16. Giuliano, C., Gliozzo, A.: Instance-based ontology population exploiting named-entity substitution. In: Proceedings of the 22nd International Conference on Computational Linguistics, Coling (2008)
17. Harabagiu, S., Moldovan, D., Pasca, M., Mihalcea, R., Surdeanu, M., Bunescu, R., Girju, R., Rus, V., Morarescu, P.: Falcon: Boosting knowledge for answer engines. In: Proceedings of 9th Text REtrieval Conference, TREC 9 (2000)
18. Hirschman, L., Chinchor, N.: Muc-7 named entity task definition. In: Proceedings of the 7th Message Understanding Conference, MUC-7 (1997)
19. Kwok, C.C.T., Etzioni, O., Weld, D.S.: Scaling question answering to the web. In: Proceedings of the 10th World Wide Web Conference, WWW (2001)
20. Nadeau, D., Sekine, S.: A survey of named entity recognition and classification. In: Linguisticae Investigationes (2007)

Forgetting Fragments from Evolving Ontologies

Heather S. Packer, Nicholas Gibbins, and Nicholas R. Jennings

Intelligence, Agents, Multimedia Group,
School of Electronics and Computer Science,
University of Southampton,
Southampton SO17 1BJ, UK
{hp07r,nmg,nrj}@ecs.soton.ac.uk

Abstract. Ontologies underpin the semantic web; they define the concepts and their relationships contained in a data source. An increasing number of ontologies are available on-line, but an ontology that combines information from many different sources can grow extremely large. As an ontology grows larger, more resources are required to use it, and its response time becomes slower. Thus, we present and evaluate an on-line approach that forgets fragments from an OWL ontology that are infrequently or no longer used, or are cheap to relearn, in terms of time and resources. In order to evaluate our approach, we situate it in a controlled simulation environment, RoboCup OWLRescue, which is an extension of the widely used RoboCup Rescue platform, which enables agents to build ontologies automatically based on the tasks they are required to perform. We benchmark our approach against other comparable techniques and show that agents using our approach spend less time forgetting concepts from their ontology, allowing them to spend more time deliberating their actions, to achieve a higher average score in the simulation environment.

1 Introduction

Evolving ontologies enable the completion of tasks and queries that were unforeseen during the design phase. Ontologies may evolve in use, by incorporating information from other ontologies. Due to the abundance of available ontologies it is possible that the uncontrolled evolution of an ontology may lead to an ontology that is large in size. Large ontologies require increasing amounts of resources to host, manage, and use. In time-critical environments where a fast response time is required, large ontologies can critically degrade response times. By forgetting concepts from an ontology in order to reduce its size the resources required to host, manage, and use the ontology can be reduced, therefore improving response times. However, forgetting concepts is a trade-off, because if too many concepts are forgotten, the quality of the answers will degrade, while forgetting too few concepts degrades the response time. For example, consider a fire brigade that uses an ontology to describe fire vehicle capability information on a portable device. A fire requires immediate use of vehicles which can remove rubble. A nearby building site has suitable vehicles, although information about their capabilities

P.F. Patel-Schneider et al. (Eds.): ISWC 2010, Part I, LNCS 6496, pp. 582–597, 2010.

and operational requirements is not present in the fire brigade's ontology. As such, this information can be reused from the construction vehicles manufacturers' ontologies by the fire brigade portable device, using a low-bandwidth mobile internet connection. Response time is critical so that damage to surrounding areas can be minimised, while inferring which construction vehicles are appropriate for the situation and will best protect the vehicles' operators.

In this paper, we focus on reducing the size of evolving ontologies to improve their response time by removing fragments, sets of axioms that represent concepts [6]. Our approach removes fragments according to the frequency and recency with which they are used, and the cost of their acquisition, in terms of time and resources. We reduce the size of an ontology when it becomes too large to complete a task within a given period of time. By removing a fragment we hypothesise that the associated costs of using the ontology will be reduced.

We illustrate our work within a controlled environment so that, for now, we can regulate the information available to software agents, and use a standard success metric to compare state of the art approaches. We use an extension of RoboCup Rescue (RCR), a widely used multi-agent platform for agent research that simulates an emergency response scenario. RCR agents learn about concepts that enable them to rescue targets that have been victim to an earthquake in a virtual city. A team of agents has five second time limit in which to determine their actions, or else automatically forfeit their turn, and thus their ontology must enable them to use the information it contains and perform actions given the timeframe. The agents' ontologies evolve as they encounter tasks that require additional information to complete. By reusing the additional information in their ontologies, they do not need to incur the cost of re-learning it in future. However, with each additional fragment they learn, the performance of using their ontology degrades. Thus, to ensure that agents can use their ontology efficiently, our approach forgets concepts from their ontologies. While we situate our approach using a specific multi-agent system exemplar, our algorithm is a general approach to selecting concepts to forget from an ontology, and could therefore be applied outside of our framework. For example, agents could learn and forget concepts from ontologies on the Semantic Web, although doing so might bring about challenges in managing inconsistencies and issues regarding trust. Likewise, our approach is not specific to the search and rescue domain; it can be applied to any domain. It should be noted that our forgetting approach does not guarantee that reasoning from an evolving ontology is sound or complete. However, this is not always a requirement, and a 'good enough' answer is appropriate in many cases where a response is required quickly, as discussed in the example above. Our approach is agnostic to a specific ontology language, however for simplicity we describe our approach in the context of OWL-Lite ontologies.

Against this background, we advance the state of the art in automatic ontology evolution, with our main contribution, a technique that selects a fragment which represents the least useful concept in the ontology to remove. We contribute a technique that rates all concepts in an ontology and weights according to their use and acquisition cost. Then, in our empirical evaluation, we compare our forgetting approach to other state of the art approaches and evaluate the outcome using RCR's scoring system. Agents using our approach save 99.3% more civilians and 12.4% more of the city, compared with the next best approach.

In Section 2 we introduce related work. In Sections 3 and 4, we introduce RCR and highlight the benefits of forgetting. Section 5 describes our approach and Section 6 discusses the empirical evaluation by describing our benchmark techniques. Section 7 presents our results and Section 8 concludes.

2   Related Work

The agent community focuses primarily on augmenting an agent's ontology, instead of pruning it. In particular, Bailin and Truszkowski [7], Afsharchi et al. [8], Wiesman and Roos [9], and Soh [10] enable their agents to augment their ontologies with new knowledge, when agents have different domain models representing the same domain. Specifically, Bailin and Truszkowski's approach considers semantically equivalent representations, and Afsharchi et al. and Soh focus on the validation of the knowledge to be incorporated into the agent's ontology. These approaches augment an agent's ontology with one concept at a time, which increases the overhead cost of retrieving the information. In addition to this work, we presented an approach that reduces the cost associated with learning by augmenting a fragment into an ontology [11]. While the above discussed approaches allow agents to augment their ontologies, they do not prune them.

However, the Semantic Web community has produced methods that prune ontologies. An agent could apply the approaches of Eiter et al. [12] or Wang et al. [13] and [14] to prune its ontology, who provide algorithms to remove one concept from an ontology at a time. In contrast to the approach of [13], Eiter el al.'s approach requires axioms to first be translated from Description Logic (DL) syntax to rule representations, and translated back to DL syntax after the expansion of the rules and the removal of a concept has been performed. In contrast, Wang et al.'s approach can be applied to axioms without need of translation. [12]'s approach has restrictions which limit the use of this technique to OWL-Lite and subspecies of OWL-Lite. These approaches enable an agent to evaluate the knowledge and remove a single concept at a time.

In this work, we chose to use the technique presented by Wang et al. to remove concepts from our agent's ontologies because it can ensure the consistency of an ontology after the removal of a concept. However, while this work focuses on removing a single concept from the ontology, our approach focuses on selecting a set of concepts and remove them. Removing more than one concept at a time results in an overall smaller ontology and reduces the number of times that the forgetting approach needs to be used, resulting in an increase in performance.

3   RoboCup OWLRescue Framework

The RoboCup OWLRescue (RCOR) framework extends the RoboCup Rescue (RCR) platform, which models the effects of an earthquake on a virtual city's buildings, civilians, and roads [15]. In RCR, at the beginning of a simulation, buildings may have: collapsed, possibly with civilians buried inside; caused road blockages; and, ignited. There are three types of RCR agents with specific capabilities: ambulance teams recover buried civilians, and transfer them to refuges; fire teams extinguish fires, and police force teams clear blocked roads. The goal is to save the lives of as many civilians as possible, and to minimise the area of the city which is burnt. The performance of a team is evaluated using a formula which factors in the percentage of live civilians, the state of live civilians, and the average building damage. While this scenario provides a testbed for developing the co-ordination of agents, our extension aims to extend the variables associated with each target (civilians, buildings, and blockages) resulting in a set of possible actions an agent can take. Each action affects the outcome of the scenario.

Our RoboCup OWLRescue (RCOR) framework extends buildings to contain (possibly hazardous) chemicals, and extends civilians to have symptoms. The RCOR agents require different knowledge for each run because variables such as chemicals in buildings and civilians' symptoms are stochastic, and differ with each run. All agents have their own ontologies so that an agent can augment its ontology with information about its tasks from ontologies in the environment. These *environment ontologies* describe the available resources which can be used in the agents' decision making processes, and describe vehicles and their ability to deal with fires, building collapses and casualties. The RCOR agents access the environment ontologies by requesting information about concepts and receive fragments representing a desired concept. The agent can then augment its ontology with all the concepts or a selection of concepts depending on the agent's strategy. In order for an agent to retain its core knowledge, two ontologies are used, a Domain Ontology (DO) from which an agent cannot forget, and an Evolving Ontology (EO) which an agent learns in and forgets from. Both ontologies are used when deciding on the action to take. Each command centre is assigned a set of vehicles which it can allocate on a first-come, first-served basis to agents. Each agent is allocated a vehicle; if its vehicle does not have the necessary equipment for a task, it can then exchange it at a command centre.

The RCOR agents can learn about variables encountered while rescuing a target and alternative resources. For example, a police rescue agent can discover an appropriate construction vehicle which can remove a blockage from a road. It is beneficial for agents to augment their ontology so that they can successfully perform tasks that they could not complete before. In the RCR, a team of agents must complete a task within five seconds which represents one timestep in the simulation. Specifically, a timestep is the amount of time that each agent has to decide on its next action before the targets in the world are updated either with new targets or changes to existing targets. Thus an agent must spend its time efficiently performing actions. In order to do this, our agents maintain a relatively small ontology and send a minimal number of requests for information from the environment ontologies. The next section describes our forgetting approach.

4   The Forgetting Approach

When an ontology becomes too large to use given a specific timeframe, our approach: first evaluates the concepts in its ontology to select which concept to remove; second selects a fragment of the concept that is deemed to be the most irrelevant; and third removes the concept so that the ontology remains consistent.

In order to motivate forgetting concepts, we first consider costs associated with a large ontology. Using an ontology incurs costs with hosting, maintaining, and using it, and the larger the ontology, the greater the need for physical memory and time to access information. It is therefore beneficial to reduce the size of an ontology. We categorise three situations when forgetting concepts is beneficial:

1. **Performance:** If the performance of querying an ontology falls below required parameters, for example after new information has been learned, removing older less used information can result in performance gains.
2. **Specialisation:** In order to retain specialisation in an ontology, information that is unrelated to the domain can be removed. This can occur because the specialist domain of an ontology is predetermined, or because the specialist domain of an ontology changes over time.
3. **Relevance:** Concepts and relationships in an ontology can become outdated when superseded by information. Forgetting out of date concepts therefore keeps an ontology up to date. Depending on the scenario it might also be beneficial to utilise OWL's deprecation semantics to mark out of date concepts as obsolete.

In our RoboCup Rescue example, the agents decided to forget when they cannot complete their actions within a single timestep. In our scenario, we only consider removing concepts that have been learnt through participating in tasks because we do not want to change an agent's core knowledge. This is because fire brigade agents require a different core set of concepts than an ambulance team because of their specialisation, thus, we only remove concepts from an agent's EO. The following three sections describe how we enable an agent to automatically evaluate the concepts within its ontology, select a fragment to prune from the ontology, and remove the fragment. We use the following running example.

> A fire brigade is tasked with extinguishing a building. The chemicals in the building are particularly toxic and human exposure results in severe damage to airways. The fire brigade wants to be able to increase the amount of equipment it carries in its first aid kit so that it can treat affected civilians, and augments an ontology fragment with such equipment. During this augmentation the agent learns about intubation equipment, but is unable to use the equipment because it is not specialised to do so. Therefore, this knowledge is not used during the agent's lifetime. The agent's response time is waning and it decides to reduce the size of its ontology in order to reduce the cost of using it.

4.1   Evaluate Concepts

Once a task agent determines that it needs to contract its ontology, it decides which concepts it wants to remove. Our approach enables an agent to evaluate the concepts in its EO using two influential factors:

1. How recently and frequently the concept is used to answer queries: This approach aims to reduce the cost of acquiring regularly required concepts so we therefore adopt the Least Recently, Frequently Used value (LRFU) used in [17] and is used in caching scenarios to select concepts to remove from a query agent's ontology. Each time a concept is used the LRFU increases as does LRFUs of the concepts which are used to define it. The LRFU is normalised into a ranking so that it can be summed with the concept acquisition below.
2. The cost of the original acquisition of the concept: this cost is recorded in milliseconds and is recorded by our learning approach, in order to be used here. The acquisition value depends on the availability of the concept, and the network bandwidth available to transfer the fragment from another agent. The acquisition cost is normalised into a ranking, and is summed with the LRFU.

In order to indicate the usefulness of a concept, we sum these two factors to calculate a concept forgetting value (CFV) for each concept in an agent's EO (see Equation 1).

$$CFV = LRFU + AC \tag{1}$$

where CFV is the concept forgetting value of a concept in an agent's EO, the $LRFU$ is the LRFU value for a concept, and AC is the acquisition value of the concept. A low CFV weighting indicates that the concept has not been used recently, often, and was inexpensive time wise to acquire, and a high CFV weighting indicates that the concept is used recently, frequently and was expensive to acquire. A medium CFV weighting can indicate that LRFU is high and AC is low, or that AC is high and LRFU is low, and as such the likelihood of the concept being forgotten is lower than those with a low CFV weighting. In more detail, the LRFU is calculated for each concept in the agent's EO using Algorithm 1. This algorithm shows how an agent calculates the $LFRU$ value for each of its ontologies concepts, where $concept(EO)$ is a function that holds the set of concepts in an agent's EO, $T = \{< t_1, c_u1 >, \ldots, < t_n, c_un >\}$ is a set of tasks where each task is a tuple representing the task (t) and the set of concepts required to complete the task (c_u), all concepts in c_u are a subset of $concept(EO)$, and all concepts in the EO have a $LFRU$ weighting which is represented using a tuple $< c, LRFU >$. The LRFU weighting for each concept is calculated over time. After each time period each concept's LRFU is calculated, by increasing the value by 1 if it is used and decaying it exponentially when it is not, so that concepts not used recently have a lower value. In our RoboCup Rescue example, concepts' LRFU weightings are calculated each timestep and are represented by tasks in the Algorithm because an agent has to complete a task per timestep. Depending on the scenario, it may be appropriate to weight the AC or LRFU differently. For example if network bandwidth fluctuates, the acquisition cost is time-sensitive, and therefore it would be appropriate to weight it lower than LRFU. In our environment available bandwidth does not change, and therefore we do not apply weightings when calculating the CFV.

Algorithm 1. Algorithm calculating the LRFU for each concept in an agent's ontology

---

Ensure: $concepts(EO) \neq \emptyset$
Ensure: $T = \{< t_1, c_u 1 >, \ldots, < t_n, c_u n >\}$ /* T is the set of tasks, where tasks require a set of concepts to complete them. */
Ensure: $c_u \neq \emptyset$ /* the set of concepts used for current task */
Ensure: $c_u \subset concepts(EO)$
Ensure: $\forall c \in concepts(EO) = < c, lrfu >$
1. **for all** T **do**
2.   **for all** $c \in concepts(EO)$ **do**
3.   **if** $c \in c_u$ **then**
4. $c = < c, lrfu + 1 >$
5.   **else**
6. $c = < c, e^{-lrfu} >$
7.   **end if**
8.   **end for**
9. **end for**

Once a concept's LRFU factor has decayed so that the acquisition cost becomes more influential in the weighting, an agent can determine which concept from a set of concepts that have the same LRFU to forget. It is more likely that concepts will have different acquisition costs due to different agent's network location and bandwidth, than different a LRFU because concepts decay exponentially. Performance wise, it is better for the agent to forget concepts that are inexpensive to acquire because the cost of re-acquiring them is less, compared to concepts that are expensive to acquire. To summarise, the agent selects the concept with the lowest rating in its EO to remove. In our example (see Figure 1), the agent selects the concept labelled *endotrachealTubes*, which is a piece of intubation equipment, because it has the lowest weighting. In the next section, we describe how the agent removes a fragment representing the selected concept.

4.2   Select Concepts

Once the agent has selected a concept it desires to forget, it creates a fragment (made up of multiple concepts) representing that selected concept so that the agent can benefit performance wise from a smaller ontology. We also hypothesise that an agent can benefit from removing more than one concept at a time so that it can perform forgetting less often than forgetting methods that forget less concepts (as proposed by other state of the art approaches, see Section 2).

In order to select concepts to prune, the agent generates a fragment representing the selected concept and selects concepts with a similar CFV weighting to prune. The fragment is generated using the basic segmentation technique presented in [6], where the technique selects the target concept first, in our example (see Figure 1) the target concept is *endotrachealTubes*, and selects concepts by traversing the ontology's concept hierarchy upwards all the way to the root class. It then traverses downwards to the target's leaf classes. Additionally, any links across the hierarchy from any of the traversed classes are followed upwards but not downwards. Once there are no concepts to traverse, the traversed concepts form a fragment representing the target concept (described in more detail in [6]).

In order to detail how our agent selects the concepts to remove from the fragment, we formally introduce the components described above. Let: l be the capacity limit at which the agent is required to prune concepts from its EO; W be the set of weightings for the concepts contained in the EO, where $W = \{w : C \in concepts(EO) \land w = weight(c)\}$ and $c_1 \ldots c_n \in concepts(EO)$; f_{o_q,c_t} be the fragment representing the concept to be forgotten, where o_q is the query agent's ontology (where $o_q = DO \cup EO$) and c_t is the concept to be forgotten; $W_{f_{o_q,c_t}} = \{W : C \in f_{o_q,c_t} \land w = weight(c)\}$ be the set of concept weightings associated with the concepts contained in f_{o_q,c_t}, where $concepts(f_{o_q,c_t}) = \{c_1, \ldots c_n\}$. Using this formal notation we describe how we select the concepts to forget in Algorithm 2, which is run over all concepts in the EO.

Algorithm 2. Lowest Weighted Concept Selection Technique: This algorithm is used to select the concepts to be pruned from an agent's EO

Initialise: $c_t \leftarrow null$
Initialise: $conceptsToRemove \leftarrow \emptyset$
Initialise: $w_{c_t} \leftarrow +\infty$ /* the weight of the concept; calculated from the LRFU value and Acquisition Cost of c_t*/
1. {finds the concept with the lowest concept weighting}
2. **for** $\forall c \in concepts(EO)$ **do**
3.   **if** $w_c < w_{c_t}$ **then**
4. $c_t \leftarrow c$
5. $w_{c_t} \leftarrow w_c$
6.   **end if**
7. **end for**
8. $conceptsToRemove \leftarrow conceptsToRemove \cup \{c_t\}$
9. {finds all the concepts in the fragment with a similar concept weighting to c_t}
10. **for** $\forall c \in concepts(f_{o_q,c_t})$ **do**
11.   **if** $|w_c - w_{c_t}| \leq t$ **then**
12. $conceptsToRemove \leftarrow conceptsToRemove \cup \{c\}$
13.   **end if**
14. **end for**
15. **return** $conceptsToRemove$

In our example, Figure 1 shows the ontology fragment representing concept *endotrachealTubes* which has weight $w_{c_t} = 0.02$, thus our selection algorithm selects the concepts *endotrachealTubes*, *laryngoscopes*, *connellAnotomicMask*, and *intubationEquipment* to forget (these concepts have a shaded background). These concepts have not been used by the agent so they have a low CFV because the agent has not had specialist knowledge to use the equipment, because it specialises in extinguishing fires and only supports first response first aid and triage. Once our agents have selected the concepts they desire to remove they then remove these concepts from their ontology, this is described in the next section.

4.3   Remove Concepts

After the agent has selected the concepts that it desires to remove, the agent then prunes these from its ontology. In order to prune these concepts from an ontology we use the technique presented in [13] so that the ontology remains consistent. For example, we aim to remove concept B from $A \sqsubseteq B$, $B \sqsubseteq C$ which results in $A \sqsubseteq C$; and the removal of B from $A \sqsubseteq B$, $B \sqsubseteq C$ and $B \sqsubseteq D$ results in $A \sqsubseteq C$ and $A \sqsubseteq D$.

5   Evaluation

In order to evaluate our approach, *forget-fragment*, we compare the effectiveness of agents using different forgetting techniques (discussed below) to: rescue civilians; put out burning buildings; to evaluate the requirements of rescuing a target against its RoboCup score (see Section 2); and investigate the amount of time used to forget. Similar to the RCR Competition, our simulation allows a team of agents five seconds to complete an action in a timestep, and are given two thousand timesteps to save as many of the civilians and burning buildings as possible. The pseudo-code in Algorithm 3 provides the basic scenario of the agents in the RCOR.

In our experiments, we initialise a RCOR scenario where there are ten of each of the ambulance, fire brigade and police agents and they use the learning technique presented in [11] when they encounter unknown concepts or do not have the right equipment to rescue their target. This learning technique selects fragments from ontologies about a requested concept, and demonstrated the most efficient learning algorithm (in terms of resulting performance) compared to benchmark approaches. We compare our technique to the following approaches:

Forget Concept. This approach removes all concepts and relationships related to the selected concept, where eo^{-c_p} and c_p is the concept to be pruned from

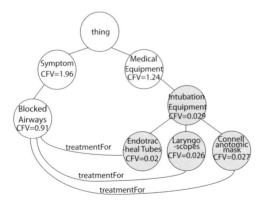

Fig. 1. Concepts selected to be forgotten, the curved lines represent relationships (domain and range restrictions) between concepts

Algorithm 3. Pseudo-code of the RoboCup simulator

Require: function: *contains(set, element)* returns true if *set* contains *element*
Require: *simulator* ← RoboCup rescue simulator
Require: *agent* ← RoboCup rescue agent
 1. *simulator.generateFires()*
 2. *simulator.generateBlockades()*
 3. *simulator.generateCivilians()*
 4. **for** *timestep* ∈ *timesteps* **do**
 5.   *target* = *getFirstTarget()*
 6.   *targetInfo* = *agent.getInformation(target)*
 7.   **if** ¬ *contains(agent.ontology, targetInfo)* **then**
 8.   *fragments* ← *requestFragements(targetInfo)*
 9.   *axioms* ← *selectAxioms(fragments)*
10.   *agent.ontology.learn(axioms)*
11.   **end if**
12.   **if** ¬ *contains(agent.vehicle, requiredEquipment)* **then**
13.   *travelToCentre()*
14.   *changeVehicle()*
15.   **end if**
16.   *rescue(target)*
17.   *simulator.update()*
18. **end for**

the ontology. This technique removes one concept at a time, and is comparable to the techniques presented by [12] and [13].

Forget Tree. This approach extends the above approach by selecting a subtree from the hierarchy of concepts in the agent's EO. The subtree is selected by comparing the weight used for each concept (see Algorithm 4), where $eo^{c_{tree}}$ and c_{tree} is a concept represented by the fragment (which is a subtree) being pruned from the ontology. Removing a connected subtree can result in removing a subtree, branch, or extraction of a subtree (shown in Figure 2).

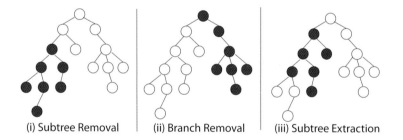

(i) Subtree Removal   (ii) Branch Removal   (iii) Subtree Extraction

Fig. 2. Subtree Removal, Branch Removal and Subtree Extraction, where the highlighted nodes are removed from the graph

Forget Redundant. This approach removes all concepts that are not used in future queries. We provide a list of the future queries to the agent at the start of the simulation, which have been recorded on a dummy run using the same random seed. This is the only agent that requires a complete list of future queries. This agent is not limited by a capacity.

Forget Nothing. This approach does not remove any concepts from the agent's ontology. Hence this agent is not limited by a capacity.

Algorithm 4. This algorithm is used to select the concepts which are connected by their concept weighting, to form a subtree, to be pruned from an agent's EO

Require: function: $getConceptWithMinimalWeight(set)$, returns tuple $< concept, weight >$ with the lowest weight in the set.
Require: $c_t \leftarrow null$, $w_{c_t} \leftarrow null$
Require: $conceptsToRemove \leftarrow \emptyset$
Require: $CH \leftarrow \{\emptyset\}$
1. $< c_t, w_{c_t} > =getConceptWithMinimalWeight(concepts(EO))$ $\{w_{c_t}$ is the lowest weight in EO $\}$
2. $\{$traverses the children of $c_t\}$
3. $CH \leftarrow children(c_t)$
4. **for** CH **do**
5.   **for** $ch \in CH$ **do**
6.   **if** $|w_c h - w_{c_t}| \leq t$ **then**
7. $conceptsToRemove \leftarrow conceptsToRemove \cup \{ch\}$
8. $CH \leftarrow CH \cup \{children(ch)\}$
9.   **end if**
10.   **end for**
11. **end for**
12. $P \leftarrow parents(c_t)$
13. **for** $\forall P$ **do**
14.   **for** $\forall p \in P$ **do**
15.   **if** $|w_p - w_{c_t}| \leq t$ **then**
16. $conceptsToRemove \leftarrow conceptsToRemove \cup \{p\}$
17. $P \leftarrow P \cup \{parents(p)\}$
18.   **end if**
19.   **end for**
20. **end for**
21. **return** $conceptsToRemove$

We put forward these four approaches: **forget-concept, forget-tree, forget-redundant, forget-nothing**, as benchmarks for the performance of our forgetting approach, **forget-fragment**. These agents adopt their behaviour defined by the sample package in RCR, which determines the agents' behaviour such as planning a path through the virtual city and which target to rescue first. We note this adopted behaviour is basic, whereby there are no algorithms used to co-ordinate agents' targets or to minimise path traversal. Our investigation consists of comparing how five different learning techniques perform given the same set of 200 scenarios, using the standard Kobe map provided with RCR. Each scenario is randomly generated by the RCR simulators. For our evaluation our framework includes the following environment ontologies:

1. **EAC Ontology:** This ontology describes the Emergency Action Code (EAC), which is a three character code displayed on all dangerous goods classed carriers. This ontology provides fire agents with information about the required equipment for attending burning targets, and is derived from the National Chemical Emergency Centre (NCEC) code list.

2. **HazChem Ontology:** This Hazardous Chemical (HazChem) ontology classifies chemicals using Hazardous Identification (ID) Numbers (HIN). Similar to the EAC Ontology, this ontology provides fire agents with information about the required equipment for attending burning targets, and is derived from the The National Institute for Occupational Safety and Health (NIOSH) HIN system.

3. **Chemical Ontology:** This ontology contains chemicals and their EAC and HIN classification. This ontology allows agents to use either standard provided by the EAC and HIN, and enables the agent to translate chemicals between both standards.

4. **Vehicle Ontology:** This ontology describes vehicles, their attributes, purpose, and manufacturer. In particular, this ontology provides information about the track type of a vehicle, and its capabilities, and is derived from vehicle categorisations from the Driver and Vehicle Licensing Agency (DVLA).

5. **HantsFireEngineFleet Ontology:** This ontology contains information about the fleet of fire engines in the county of Hampshire (UK). This information is derived from the Hampshire fire service website[1], which details vehicle types, their model, manufacturer, and registration numbers.

6. **Ambulance Ontology:** This ontology contains information about different types of ambulance, their attributes, and equipment and is derived from the standards of the Ontario Ministry of Health and Long-Term Care[2].

7. **ConstructionVehicles Ontology:** This ontology contains information about construction vehicles and their capacity, and is derived from information from the book "Fundamentals of Technical Rescue."[3]

8. **Triage Ontology:** This ontology describes the 5-Category Triage System and identifies symptoms for each category, and is derived from the Australian Ministry of Health guidelines[4].

9. **CSI Ontology:** This ontology contains information from the Chemical Sampling Information (CSI) of the US Department of Labor Occupational Safety and Health Administration. The CSI contains details about chemicals and their health effects on humans, and the organs affected.

10. **Treatment Ontology:** This ontology contains information about burns and broken bones, their symptoms, and their treatment. This information has been taken from the NHS website[5].

These ten ontologies have been chosen because they are representative of standard industry vocabularies for the domains of interest of RCR agents. This combination of ontologies covers the areas required by the RCOR extension and represent a realistic set of information that rescue agents would need to consult in real conditions. The number of concepts in each of the ontologies is given in Table 1. The next section presents the results and our analysis of our evaluation.

---

[1] Hampshire Fire and Rescue Service: http://www.hantsfire.gov.uk/theservice/sp-and-sr/fleetmanagement

[2] Ontario Ambulance Standards: http://www.health.gov.on.ca/english/providers/pub/ambul/equi pment/standard.pdf

[3] Fundamentals of Technical Rescue, International Association of Fire Chiefs: http://books.google.com/books?id=mLyYsT8YEWkC&pg=PT33

[4] Ministry of Health Triage Guidelines: http://www.moh.govt.nz/moh.nsf/indexmh/ed-about-triage

[5] NHS Health Information: http://www.nhs.uk/chq/pages/Category.aspx?CategoryID=72

Table 1. The number of concepts in each of the environment ontologies

Ontology	No. of Concepts
EAC Ontology	1906
Chemical Ontology	1800
HantsFireEngineFleet Ontology	745
ConstructionVehicles Ontology	114
CSI Ontology	3841
EAC Ontology	1906
Chemical Ontology	1800
HantsFireEngineFleet Ontology	745
ConstructionVehicles Ontology	114
CSI Ontology	3841

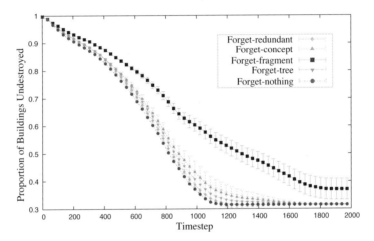

Fig. 3. Chart showing the percentage of buildings unburned for each forgetting approach

6   Results

We compare our results using standard measures of the RoboCup Rescue *scorevector* [18] scoring system: saved civilians, and buildings unburned. The agents using the *forget-fragment* approach outperform the other agents using benchmark approaches, by having the highest average number of civilians alive (99.3% more civilians saved compared to the next highest approach, *forget-tree*) and percentage of the city that is unburned at the end of the simulation (12% more than the next highest approach, *forget-tree*), see Table 2 and Figures 3 and 4.

The agents using our approach reduce the size of their ontologies by removing a fragment of a concept, instead of removing trees that contain a smaller number of concepts or removing a single concept. By removing a higher number of concepts from an ontology the agents forget less frequently, and ultimately spend less time deciding what to forget and forgetting than the other approaches, with the exception of the *forget-nothing* approach (*forget-tree* forgets 332% more times than *forget-fragment*, and 14% more concepts when it forgets), see Table 2 and

Table 2. Comparison of average results for each forgetting technique

Technique	Percentage of City Unburned	Percentage of Civilians in Refuge	Number of Times Forgot	Number of Concepts Forgotten per Forget
Forget-Concept	31.5	29.1	11.0	50.4
Forget-Tree	33.0	29.0	10.3	50.3
Forget-Redundant	31.5	19.8	50.3	215.1
Forget-Nothing	31.5	1.4	0.0	0.0
Forget-Fragment	37.1	58.0	3.1	44.1

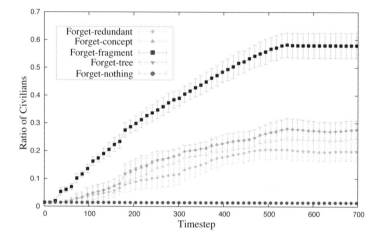

Fig. 4. Chart showing the percentage of civilians rescued for each forgetting approach, our results range from 0 - 700 timesteps; result after 700 timesteps are the same trend

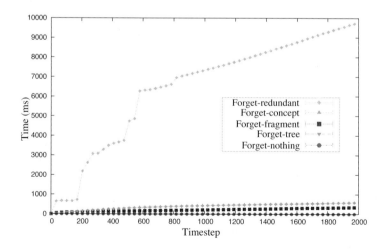

Fig. 5. Chart showing the time spent forgetting for each forgetting approach

Figure 5. We also note that despite our approach forgets fewer concepts than other approaches, it still outperforms them, because it spends less time forgetting, thus demonstrating the trade-off between spending time managing an ontology, and spending time querying a large ontology. Our approach helps agents save civilians at a faster rate than the other approaches, because the fire brigade agents put out more fires thus reducing the injuries to the civilians in the simulation. The *forget-tree* and *forget-concept* approaches' performances are similar; this is because they have a similar forgetting frequency, and thus spend a similar amount of time performing actions derived from their ontology. The *forget-nothing* approach is unable to submit any commands because it took too long to deliberate over its actions. Despite the *forget-redundant* approach having perfect foresight it spent too long forgetting because it forgot every unnecessary concept increasing the forgetting frequency compared to the *forget-fragment* approach.

7   Conclusions

In this paper we present a novel technique that can be used to remove a fragment from an ontology. Our technique is tested using an agent-based search and rescue domain, but is generalised and applicable to any scenario where an ontology evolves limitlessly over time. In order to support this contribution we have also developed a semantic extension to the RoboCup Rescue framework which enables its agents to evolve their ontologies, and a technique that rates all concepts in an ontology with a weighting. We have also implemented benchmark approaches which were used to compare the forgetting approaches. Our evaluation shows that our method saves 99.3% more civilians and 12.4% more city area, compared with the next best approach. For the future, we plan to investigate the benefits of using our technique in scenarios that require random and regularly used concepts. This investigation aims to explore the hypothesis that our approach will remove the concepts acquired from the random queries, while generally keeping the fragments required for the regularly repeated queries. We will also explore other motivations for an agent to forget concepts, for example agents that can predict future queries so that they can prioritise forgetting concepts which are unlikely to reoccur.

References

1. Berners-Lee, T., Hendler, J., Lassila, O.: The Semantic Web. Scientific American 284(5), 28–37 (2001)
2. Ashburner, M., Ball, C.A., Blake, J.A., Botstein, D., Butler, H., Cherry, J.M., Davis, A.P., Dolinski, K., Dwight, S.S., Eppig, J.T., et al.: Gene Ontology: tool for the unification of biology. Nature genetics 25(1), 25–29 (2000)
3. Sidhu, A.S., Dillon, T.S., Chang, E., Sidhu, B.S.: Protein ontology: vocabulary for protein data. IEEE Computer Society, Los Alamitos (2005)
4. Rogers, J., Rector, A.: The GALEN ontology. In: Medical Informatics Europe, pp. 174–178 (1996)

5. Stuckenschmidt, H., Klein, M.: Structure-based partitioning of large concept hierarchies. In: McIlraith, S.A., Plexousakis, D., van Harmelen, F. (eds.) ISWC 2004. LNCS, vol. 3298, pp. 289–303. Springer, Heidelberg (2004)
6. Seidenberg, J., Rector, A.: Web Ontology Segmentation: Analysis, Classification and Use. In: Proceedings of the 15th International Conference on WWW, NY, USA, pp. 13–22. ACM, New York (2006)
7. Bailin, S., Truszkowski, W.: Ontology Negotiation between Intelligent Information Agents. The Knowledge Engineering Review 17(01), 7–19 (2002)
8. Afsharchi, M., Far, B., Denzinger, J.: Ontology-guided Learning to Improve Communication between Groups of Agents. In: Proceedings of the 5th International Joint Conference on Autonomous Agents and Multiagent Systems, Hakodate, Japan, pp. 923–930 (2006)
9. Wiesman, F., Roos, N.: Domain Independent Learning of Ontology Mappings. In: Proceedings of the 3rd International Joint Conference on Autonomous Agents and Multiagent Systems, New York City, New York, USA, vol. 2, pp. 846–853 (2004)
10. Soh, L.: Multiagent, Distributed Ontology Learning. In: Working Notes of the 2nd AAMAS OAS Workshop, Bologna, Italy (2002)
11. Packer, H.S., Gibbins, N., Jennings, N.R.: Collaborative Learning of Ontology Fragments by Co-operating Agents. In: Web Intelligence-Intelligent Agent Technology International Conference, Toronto (2010)
12. Eiter, T., Ianni, G., Schindlauer, R., Tompits, H., Wang, K.: Forgetting in managing rules and ontologies. In: Proc. of the International Conference on Web Intelligence, pp. 411–419 (2006)
13. Wang, Z., Wang, K., Topor, R., Pan, J.: Forgetting Concepts in DL-Lite. In: Bechhofer, S., Hauswirth, M., Hoffmann, J., Koubarakis, M. (eds.) ESWC 2008. LNCS, vol. 5021, p. 245. Springer, Heidelberg (2008)
14. Wang, K., Wang, Z., Topor, R., Pan, J.Z., Antoniou, G.: Concept and Role Forgetting in ALC Ontologies. In: Bernstein, A., Karger, D.R., Heath, T., Feigenbaum, L., Maynard, D., Motta, E., Thirunarayan, K. (eds.) ISWC 2009. LNCS, vol. 5823, pp. 666–681. Springer, Heidelberg (2009)
15. Kitano, H., Tadokoro, S.: Robocup rescue: A grand challenge for multiagent and intelligent systems. AI Magazine 22, 1–39 (2001)
16. Wang, K., Sattar, A., Su, K.: A Theory of Forgetting in Logic Programming. In: Proc. of the National Conference on Artificial Intelligence, vol. 20(2), p. 682 (2005)
17. Lee, D., Choi, J., Choe, H., Noh, S.H., Min, S.L., Cho, Y.: Implementation and Performance Evaluation of the LRFU Replacement Policy, pp. 106–111 (1997)
18. Siddhartha, H., Sarika, R., Karlapalem, K.: Retrospective analysis of RoboCup rescue simulation agent teams. In: Proceedings of The 8th International Conference on Autonomous Agents and Multiagent Systems, vol. (2), pp. 1365–1366 (2009)

Linking and Building Ontologies of Linked Data

Rahul Parundekar, Craig A. Knoblock, and José Luis Ambite

University of Southern California,
Information Sciences Institute and Department of Computer Science
4676 Admiralty Way, Marina del Rey, CA 90292
{parundek,knoblock,ambite}@isi.edu

Abstract. The Web of Linked Data is characterized by linking structured data from different sources using equivalence statements, such as owl:sameAs, as well as other types of linked properties. The ontologies behind these sources, however, remain unlinked. This paper describes an extensional approach to generate alignments between these ontologies. Specifically our algorithm produces equivalence and subsumption relationships between classes from ontologies of different Linked Data sources by exploring the space of hypotheses supported by the existing equivalence statements. We are also able to generate a complementary hierarchy of derived classes within an existing ontology or generate new classes for a second source where the ontology is not as refined as the first. We demonstrate empirically our approach using Linked Data sources from the geospatial, genetics, and zoology domains. Our algorithm discovered about 800 equivalences and 29,000 subset relationships in the alignment of five source pairs from these domains. Thus, we are able to model one Linked Data source in terms of another by aligning their ontologies and understand the semantic relationships between the two sources.

1 Introduction

The last few years have witnessed a paradigm shift from publishing isolated data from various organizations and companies to publishing data that is *linked* to related data from other sources using the structured model of the Semantic Web. As the data being published on the Web of Linked Data grows, such data can be used to supplement one's own knowledge base. This provides significant benefits in various domains where it is used in the integration of data from different sources. A necessary step to publish data in the Web of Linked Data is to provide links from the instances of a source to other data 'out there' based on background knowledge (e.g. linking DBpedia to Wikipedia), common identifiers (e.g. ISBN numbers), or pattern matching (e.g. names, latitude, longitude and other information used to link Geonames to DBpedia). These links are often expressed by using *owl:sameAs* statements. Often, when such links between instances are asserted, the link between their corresponding concepts is not made. Such conceptual links would ideally help a consumer of the information (agent/human) to model data from other sources in terms of their own knowledge. This problem is widely known as ontology alignment [12], which is a form of schema alignment [16]. The advent of the Web of Linked Data warrants a renewed inspection of these methods.

P.F. Patel-Schneider et al. (Eds.): ISWC 2010, Part I, LNCS 6496, pp. 598–614, 2010.

Our approach provides alignments between classes from ontologies in the Web of Linked Data by examining their linked instances. We believe that providing ontology alignments between sources on the Web of Linked Data provides valuable knowledge in understanding and reusing such sources. Moreover, our approach can provide a more refined ontology for a source described with a simple ontology (like GEONAMES) by aligning it with a more elaborate ontology (like DBPEDIA). Alternatively, by aligning an ontology (like GEOSPECIES) with itself using the same approach, we are able to generate a hierarchy of derived classes, which provide class definitions complimentary to those already existing in the source.

The paper is organized as follows. First, we briefly provide background on Linked Data and describe the domains (geospatial, genetics and zoology) and data sources (LINKEDGEODATA, GEONAMES, DBPEDIA, GEOSPECIES, MGI, and GENEID) that we focus on in this paper. Second, we describe our approach to ontology alignment, which is based on defining *restriction classes* over the ontologies and comparing the extensions of these classes to determine the alignments. Third, we provide an empirical evaluation of the alignment algorithm on five pairs of sources: (LINKEDGEODATA-DBPEDIA, GEONAMES-DBPEDIA, GEOSPECIES-DBPEDIA, MGI-GENEID and GEOSPECIES-GEOSPECIES). Finally, we describe related and future work and discuss the contributions of this paper.

2   Linked Data Background and Sources

In this section, we provide a brief introduction to Linked Data and the three domains and six data sources that we consider.

The Linked Data movement, as proposed by Berners-Lee [6], aims to provide machine-readable connections between data in the Web. Bizer et al. [7] describe several approaches to publishing such Linked Data. Most of the Linked Data is generated automatically by converting existing structured data sources (typically relational databases) into RDF, using an ontology that closely matches the original data source. For example, GEONAMES gathers its data from over 40 different sources and it primarily exposes its data as a flat-file structure[1] that is described with a simple ontology [19]. Such an ontology might have been different if designed at the same time as the collection of the actual data. For example, all instances of GEONAMES have *geonames:Feature* as their only *rdf:type*, however, they could have been more effectively typed based on the *featureClass* and *featureCode* properties (cf. Section 3.1).

The links in the Web of Linked Data make the Semantic Web browsable and, moreover, increase the amount of knowledge by complementing data in a source with existing data from other sources. A popular way of linking data on the Web is the use of *owl:sameAs* links to represent *identity links* [14,8]. Instead of reusing existing URIs, new URIs are automatically generated while publishing linked data and an *owl:sameAs* link is used to state that two URI references refer to the same thing. Halpin et al. [14] distinguish four types of semantics for these links: (1) same thing as but different context, (2) same thing as but referentially opaque, (3) represents, and (4) very similar to.

---

[1] http://download.geonames.org/export/dump/readme.txt

For the purposes of this paper, we refrain from going into the specifics and use the term as asserting equivalence.

In this paper we consider six sources sources from three different domains (geospatial, zoology, and genetics), which are good representatives of the Web of Linked Data:

LINKEDGEODATA is a geospatial source with its data imported from the Open Street Map (OSM) [13] project containing about 2 billion triples. The data extracted from the OSM project was linked to DBPEDIA by expanding on the user created links on OSM to WIKIPEDIA using machine learning based on a heuristic on the combination of type information, spatial distance, and name similarity [4].

GEONAMES is a geographical database that contains over 8 million geographical names. The structure behind the data is the Geonames ontology [19], which closely resembles the flat-file structure. A typical individual in the database is an instance of type *Feature* and has a *Feature Class* (administrative divisions, populated places, etc.), a *Feature Code* (subcategories of *Feature Class*) along with latitude, longitude, etc. associated with it.

DBPEDIA is a source of structured information extracted from WIKIPEDIA containing about 1.5 million objects that are classified with a consistent ontology. Because of the vastness and diversity of the data in DBPEDIA, it presents itself as a hub for links in the Web of Linked Data from other sources [3]. We limit our approach to only the *rdf:type* assertions and info-box triples from DBPEDIA as they provide factual information. LINKEDGEODATA, GEONAMES are both linked to DBPEDIA using the *owl:sameAs* property asserting the equivalence of instances.

GEOSPECIES is an initiative intended to unify biological taxonomies and to overcome the problem of ambiguities in the classification of species.[2] GEOSPECIES is linked to DBPEDIA using the *skos:closeMatch* property.

Bio2RDF's MGI and GENEID. The Bio2RDF project aims at integrating mouse and human genomic knowledge by converting data from bioinformatics sources and publishing this information as Linked Data [5]. The Mouse Genome Informatics (MGI) database contains genetic, genomic, and biological data about mice and rats. This database also contains assertions to a gene in the National Center for Biotechnology Information - Entrez Gene database, which is identified with a unique GeneID.[3] The data from the MGI database and Entrez Gene was triplified and published by Bio2RDF on the Web of Linked Data[4], which we refer to as MGI and GENEID. We align these two sources using the *bio2RDF:xGeneID* link from MGI to GENEID.

3   Ontology Alignment Using Linked Data

An *Ontology Alignment* is "a set of correspondences between two or more ontologies," where a *correspondence* is "the relation holding, or supposed to hold according to a particular matching algorithm or individual, between entities of different ontologies" [12]. *Entities* here, can be classes, individuals, properties, or formulas.

---

[2] http://about.geospecies.org/
[3] http://www.ncbi.nlm.nih.gov/entrez/query/static/help/
    genehelp.html
[4] http://quebec.bio2rdf.org/download/data/

Our alignment algorithm uses data analysis and statistical techniques for matching the *classes* of two ontologies using what Euzenat et al. [12] classify as a *common extension comparison* approach for aligning the structure. This approach considers classes from different ontologies that have instances in common, and derives the alignment relationship between the classes based on the set containment relationships between the sets of instances belonging to each of the classes. Our approach is novel in two respects. First, we identify common instances by using the equivalence links in the Web of Linked Data. Specifically, we use the *owl:sameAs* property to link LINKEDGEODATA with DBPEDIA, and GEONAMES with DBPEDIA; the *skos:closeMatch* property to link GEOSPECIES with DBPEDIA,[5] and the *bio2rdf:xGeneID* property to link MGI with GENEID. Second, instead of limiting ourselves to the existing classes in an ontology, we overlay a richer space of class descriptions over the ontology and define alignments over these sets of new classes, as we describe next.

3.1 Restriction Classes

In the alignment process, instead of focusing only on classes defined by *rdf:type*, we also consider classes defined by conjunctions of property value restrictions (i.e, *hasValue* constraints in the Web Ontology Language), which we will call *restriction class*es in the rest of the paper. *Restriction classes* help us identify existing as well as derived set of classes in an ontology. A *restriction class* with only a single constraint on the *rdf:type* property gives us a class already present in the ontology, for example in LINKEDGEODATA the restriction (*rdf:type*=lgd:country) identifies the class Country. Using restrictions also helps us get a refined set of classes when the ontology of the source is rudimentary i.e. when there are little or no specializations of top level classes, as can be seen in the case of GEONAMES. In GEONAMES, the *rdf:type* for all instances is *Feature*. Thus, the ontology contains a single concept. Traditional alignments would then only be between the class *Feature* from GEONAMES and another class from DBPEDIA, for example *Place*. However, instances from GEONAMES have *featureCode* and *featureClass* properties. A restriction on the values of such properties gives us classes that we can effectively align with classes from DBPEDIA. For example, the *restriction class* defined by (*featureCode*=geonames:A.PCLI) (independent political entity) aligns with the class *Country* from DBPEDIA. Our algorithm defines restriction classes from the source ontologies and generates alignments between such restrictions classes using subset or equivalence relationships.

The space of *restriction classes* is simply the powerset of distinct property-value pairs occurring in the ontology. For example assume that the GEONAMES source had only three properties: *rdf:type, featureCode* and *featureClass*; and the instance *Saudi Arabia* had as corresponding values: geonames:Feature, geonames:A.PCLI, and geonames:A. Then this instance belongs to the *restriction class* defined by (*rdf:type*=geonames: Feature & *featureClass*=geonames:A). The other elements of the powerset also form such restriction classes as shown in Figure 1. It is evident that in order to consider all *restriction classes*, the algorithm would be exponential. We thus need some preprocessing that eliminates those properties that are not useful.

---

[5] Based on the 'Linked Open Data Cloud Connections' section in
   http://about.geospecies.org/

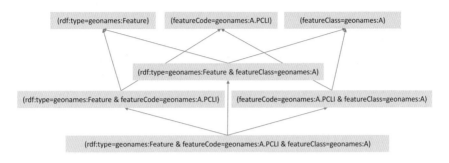

Fig. 1. Hierarchy showing how restriction classes are built

3.2   Pre-processing of the Data

Before we begin exploring alignments, we perform a simple pre-processing on the input sources in order to reduce the search space and optimize the representation. First,
for each pair of sources that we intend to align, we only consider instances that are
actually linked. For example, instances from DBPEDIA not relevant to alignments in the
geospatial domain (like People, Music Albums, etc.) are removed. This has the effect
of removing some properties from consideration. For example, when considering the
alignment of DBPEDIA to GEONAMES, the *dbpedia:releaseDate* property is eliminated
since the instances of type album are eliminated.

Second, in order to reduce the space of alignment hypotheses, we remove properties that cannot contribute to the alignment. Inverse functional properties resemble for-
eign keys in databases and identify an instance uniquely. Thus, if a *restriction class*
is constrained on the value of an inverse functional property, it would only have a
single element in it and not be useful. For example, consider the *wikipediaArticle*
property in GEONAMES, which links to versions of the same article in WIKIPEDIA in
different languages. The GEONAMES instance for the country Saudi Arabia[6] has links
to 237 different articles. Each of these, in turn, however could be used to identify
only *Saudi Arabia*. Similarly, in LINKEDGEODATA the 'georss:point' property from the
'http://www.georss.org/georss/' namespace contains a String representation of the latitude and longitude and would tend to be an inverse functional property. On the other
hand, the *addr:country* property in LINKEDGEODATA has a *range* of 2-letter country
codes that can be used to group instances into useful restriction classes.

Third, we transform the instance data of a source into a tabular form, which allows
us to load the data in a relational database and process it more efficiently. Specifically,
each instance is represented as a row in a table, each property occurring in the ontology
is a column, and the instance URI is the key. For example, the table for GEONAMES
contains 11 columns not including the identifier. We call this tuple representation of
an instance a *vector*. In cases of multivalued properties, the row is replicated in such a
way that each cell contains a single value but the number of rows equals the number
of multiple values. Each new row however, is still identified with the same URI, thus

---

[6] http://sws.geonames.org/102358/about.rdf

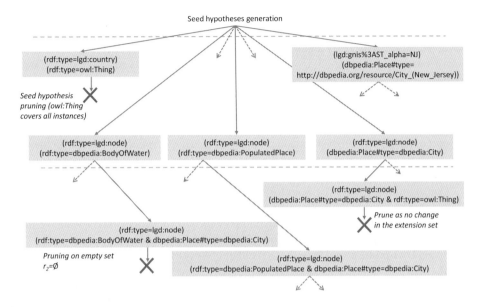

Fig. 2. Exploring and pruning the space of alignments

retaining the number of distinct individuals. In general, the total number of rows for each individual is the product of cardinalities of the value sets for each of its properties.

From these individual vectors, we then perform a join on the equivalence property (e.g. *owl:sameAs* property from LINKEDGEODATA to DBPEDIA) such that we get a combination of vectors from both ontologies. We call this concatenation of two vectors an *instance pair*.

3.3   Searching the Space of Ontology Alignments

An *alignment hypothesis* is a pair of restriction classes, one from each of the ontologies under consideration. The space of alignment hypotheses is combinatorial, thus our algorithm exploits the set containment property of the hypotheses in a top-down fashion along with several pruning features to manage the search space.

We describe the search algorithm that we use to build the alignments by example. Figure 2 shows a small subset of the search space, as explored by this algorithm while aligning LINKEDGEODATA with DBPEDIA. Each gray box represents a candidate hypothesis where the first line within it is the *restriction class* from the first source(O_1) and the second line is the *restriction class* from the second source (O_2). The levels in the exploration space, denoted by dashed horizontal lines, separate alignments where the one from a lower level contains a *restriction class* with one extra property-value constraint than its parent alignment (that is, it is a subclass by construction).

We first seed the space by computing all alignment hypotheses with a single property-value pair from each ontology, that is $[(p_i^1 = v_j^1)(p_k^2 = v_l^2)]$, as shown at the top of Figure 2. There are $O(n^2 m^2)$ seed hypotheses, where n is the larger of the number of properties in each source, and m is the maximum number of distinct values for any property. Then, we explore the hypotheses space by using a depth-first search. At each level we choose a property and add a property-value constraint to one of the *restriction classes* and thus explore all specializations. The *instance pairs* that support the new hypothesis are obtained by restricting the set of *instance pairs* of the current hypothesis with the additional constraint. In Figure 2, while adding a new constraint '*dbpedia:Place#type*=dbpedia:City' to the restriction (*rdf:type*=dbpedia:PopulatedPlace) while aligning it with (*rdf:type*=lgd:node), we take the intersection of the set of identifiers covered by [(*rdf:type*=dbpedia:PopulatedPlace) (*rdf:type*=lgd:node)] with the set of instances in DBPEDIA that have a value of 'dbpedia:City' for the property 'dbpedia:Place#type'.

Our algorithm prunes the search space in several ways. First, we prune those hypotheses with a number of supporting instance pairs less than a given threshold. For example, the hypothesis [(*rdf:type*=lgd:node) (*rdf:type*=dbpedia:BodyOfWater & *dbpedia:Place#type*=dbpedia:City)] is pruned since it has no support.

Second, we prune a seed hypothesis if either of its constituent *restriction classes* covers the entire set of instances from one of the sources, then the algorithm does not search children of this node, because the useful alignments will appear in another branch of the search space. For example, in the alignment between (*rdf:type*=lgd:country) from LINKEDGEODATA and (*rdf:type*=owl:Thing) from DBPEDIA in Figure 2, the *restriction class* (*rdf:type*=owl:Thing) covers all instances from DBPEDIA. The alignment of such a seed hypothesis will always be a subset relation. Moreover, each of its child hypotheses can also be explored through some other hypotheses that does not contain the non-specializing property-value constraint. For example, our algorithm will explore a branch with [(*rdf:type*=lgd:country) (*dbpedia:Place#type*=dbpedia:City)], where the restriction class from the second ontology actually specializes in the extensional sense (*rdf:type*=owl:Thing).

Third, if the algorithm constrains one of the restriction classes of an hypothesis, but the resulting set of *instance pairs* equals the set from the current hypothesis, then it means that adding the constraint did not really specialize the current hypothesis. Thus, the new hypothesis is not explored. Figure 2 shows such pruning when the constraint (*rdf:type*=owl:Thing) is added to the alignment [(*rdf:type*=lgd:node) (*dbpedia:Place#type* =dbpedia:City)].

Fourth, we prune hypotheses [$r_1 r_2$] where $r_1 \cap r_2 = r_1$ as shown in Figure 3(a). Imposing an additional restriction on r_1 to form r_1' would not provide any immediate specialization as the resultant hypothesis could be inferred from $r_1' \subset r_1$ and the current hypothesis. The same holds for the symmetrical case $r_1 \cap r_2 = r_2$.

Finally, to explore the space systematically the algorithm specializes the restriction classes in a lexicographic order. By doing this, we prune symmetric cases as shown by Figure 3(b). The effect of lexicographic selection of the property can also be seen in Figure 2 when the hypotheses [(*rdf:type*=lgd:node) (*rdf:type*=dbpedia:PopulatedPlace & *dbpedia:Place#type*=dbpedia:City)] is not explored through the hypothesis [(*rdf:type*= lgd:node) (*dbpedia:Place#type*=dbpedia:City)].[7]

---

[7] Note that the pruning from Figure 3(a) & (b) is not explicitly depicted in Figure 2.

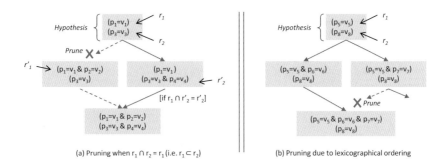

(a) Pruning when $r_1 \cap r_2 = r_1$ (i.e. $r_1 \subset r_2$)   (b) Pruning due to lexicographical ordering

Fig. 3. Pruning the hypotheses search space

3.4 Scoring Alignment Hypotheses

After building the hypotheses, we score each hypothesis to assign a degree of confidence for each alignment. Figure 4 illustrates the instance sets considered to score an alignment. For each hypothesis, we find the instances belonging to the *restriction class* r_1 from the first source and r_2 from the second source. We then compute the *image* of r_1 (denoted by $I(r_1)$), which is the set of instances from the second source that form *instance pairs* with instances in r_1, by following the *owl:sameAs* links. The dashed lines in the figure represent these *instance pairs*. All the pairs that match both restrictions r_1 and r_2 also support our hypothesis and thus are equivalent to the *instance pairs* corresponding to instances belonging to the intersection of the sets r_2 and $I(r_1)$. This set of *instance pairs* that support our hypothesis is depicted as the shaded region. We can now capture subset and equivalence relations between the *restriction classes* by set-containment relations from the figure. For example, if the set of *instance pairs* identified by r_2 are a subset of $I(r_1)$, then the set r_2 and the shaded region would be entirely contained in the $I(r_1)$.

We use two metrics P and R to quantify these set-containment relations. Figure 5 summarizes these metrics and also the different cases of intersection. In order to allow a certain margin of error induced by the dataset, we are lenient on the constraints and use the relaxed versions P' and R' as part of our scoring mechanism. For example, consider the alignment between the *restriction class* (*lgd:gnis%3AST_alpha*=NJ) from LINKED-GEODATA to the restriction (*dbpedia:Place#type*=http://dbpedia.org/resource/City_(New_Jersey)) from DBPEDIA shown in Figure 2. Based on the extension sets, our algorithm finds $|I(r_1)| = 39$, $|r_2| = 40$ and $|I(r_1) \cap r_2| = 39$. The value of R' therefore is 0.97 and that of P' is 1.0. Based on our margins, we hence assert the relation of the alignment as equivalent in an extensional sense.

3.5 Eliminating Implied Alignments

From the result set of alignments that pass our scoring thresholds, we need to only keep those that are not implied by other alignments. We hence perform a transitive reduction based on containment relationships to remove the implied alignments. Figure 6 explains

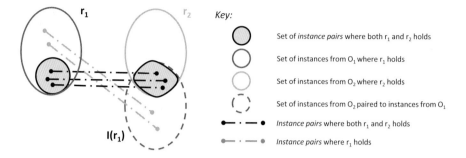

Fig. 4. Scoring of a Hypothesis

Set Representation	Relation	$P = \frac{|I(r_1) \cap r_2|}{|r_2|}$	$R = \frac{|I(r_1) \cap r_2|}{|r_1|}$	P'	R'
	Disjoint	= 0	= 0	≤ 0.01	≤ 0.01
	$r_1 \subset r_2$	< 1	= 1	> 0.01	≥ 0.90
	$r_2 \subset r_1$	= 1	< 1	≥ 0.90	> 0.01
	$r_1 = r_2$	= 1	= 1	≥ 0.90	≥ 0.90
	Not enough support	0 < P < 1	0 < R < 1	0.01 < P' < 0.90	0.01 < R' < 0.90

Fig. 5. Metrics

these reductions, where alignments between r_1 and r_2 and between r'_1 and r_2 are at different levels in the hierarchy such that r'_1 is a subclass of r_1 by construction (i.e., by conjoining with an additional property-value pair). Figure 6(a) through (i) depict the combinations of the equivalence and containment relations that might occur in the alignment result set. Solid arrows depict these containment relations. Arrows in both directions denote an equivalence of the two classes.

A typical example of the reduction is Figure 6(e) where the result set contains a relation such that $r_1 \subset r_2$ and $r'_1 \subset r_2$. Based on the implicit relation $r'_1 \subset r_1$, the relation $r'_1 \subset r_2$ can be eliminated (denoted with a cross). Thus, we only keep the relation $r_1 \subset r_2$ (denoted with a check). The relation $r_1 \subset r_2$ could alternatively be eliminated but instead we choose to keep the simplest alignment and hence remove $r'_1 \subset r_2$. Other such transitive relations and their reductions are depicted with a 'T' in box on the bottom-right corner.

Another case can be seen in Figure 6(d) where the subsumption relationships found in the alignment results can only hold if all the three classes r_1, r'_1 and r_2 are equivalent. These relations have a characteristic cycle of subsumption relationships. We hence need to correct our existing results by converting the subset relations into equivalences. This is depicted by an arrow with a dotted line in the figure. Other similar cases can be seen in Figure 6(a), (c) and (f) where the box on the bottom-right is has a 'C' (cycle).

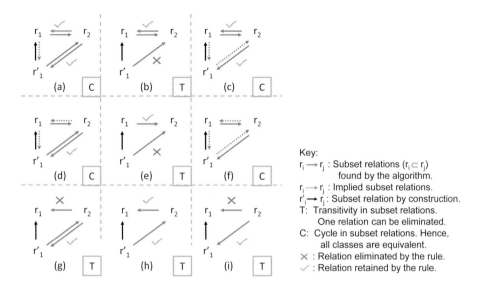

Fig. 6. Eliminating Implied Alignments

In such cases, we order the two equivalences such that the one with more support is said to be a 'better' match than the other (i.e. if $|I(r_1) \cap (r_2)| > |I(r'_1) \cap (r_2)|$, then $r_1 = r_2$ is a better match than $r'_1 = r_2$). The corrections in the result alignments based on transitive reductions may induce a cascading effect. Hence our algorithm applies the 'C' rules shown in Figure 6(a), (c), (d), (f) to identify equivalences until quiescence. Then it applies the 'T' rules to eliminate hypotheses that are not needed.

In sources like DBPEDIA an instance may be assigned multiple *rdf:type*s with values belonging to a single hierarchy of classes in the source ontology. This results in multiple alignments where relations were found to be implied based on the *rdf:type* hierarchy. Such alignments were also considered as candidates for cycle correction, equivalence ordering and elimination of implied subsumptions. We used the ontology files (RDF-S/OWL) provided by GEONAMES, LINKEDGEODATA, DBPEDIA AND GEOSPECIES as the source for the ontologies.

4   Empirical Evaluation

We evaluate our algorithm on the domain and sources described in Section 2. Table 1 shows the number of properties and instances in the original sources. For example, LINKEDGEODATA has 5,087 distinct properties and 11,236,351 instances.[8]

As described in Section 3.2, we consider only linked instances and remove properties that cannot generate useful restriction classes. This reduced dataset contains instances that reflect the practical usage of the equivalence links and properties relevant to the domain. In LINKEDGEODATA, most of the instances coming from OSM have a rudimentary type information (classified as 'lgd:node' or 'lgd:way') and are not linked to any

---

[8] Data and results available at:
http://www.isi.edu/integration/data/LinkedData

Table 1. Properties and instances in the original sources

Source	# properties	# instances
LinkedGeoData	5087	11236351
DBpedia	1043	1481003
Geonames	17	6903322
Geospecies	173	98330
MGI	24	153646
GeneID	32	4153014

instance from DBPEDIA. DBPEDIA similarly has instances not linked to LINKEDGEO-
DATA and they were removed as well.

Table 2 shows the results of pre-processing on the source pairs. The table lists the
number of properties and instances retained in either sources, the count of the number
of combinations of the *vectors* as a result of the join, and the count of the distinct
instance pairs as identified by the concatenation of their respective URIs. Our algorithm
processed this set of *instance pairs* for each source pair and generated alignments that
have a minimum support level of 10 *instance pairs*.

Table 2. Generation of instance pairs in pre-processing

Source 1	# properties after elimination	# instances after reduction	Source 2	# properties after elimination	# instances after reduction	# vector combin-ations	# distinct instance pairs
LinkedGeoData	63	23594	DBpedia	16	23632	329641	23632
Geonames	5	71114	DBpedia	26	71317	459716	71317
Geospecies	31	4997	Dbpedia	13	4982	289967	4998
MGI	7	31451	GeneID	4	47491	829454	47557
Geospecies	22	48231	Geospecies	22	48231	771690	48231

The alignment results after eliminating implied alignments, as described in
Section 3.5, are shown in Table 3. The table shows the two sources chosen for the
alignment and the count of the hypotheses classified as equivalent, $r_1 \subset r_2$ and $r_2 \subset r_1$
both before and after elimination.[9] Even though our algorithm provides for the correction and cascading of mislabeled equivalence relations, for all the source pairs that we
considered for alignment, such corrections did not arise. The number of equivalences
that our algorithm finds can be seen in Table 3 along with the count of equivalences that
were labeled as the best match in a hierarchy of equivalence relations. The procedure for
elimination of implied relations further prunes the results and helps the system focus on
the most interesting alignments. For example, in linking LINKEDGEODATA to DBPEDIA,
the 2528 ($r_1 \subset r_2$) relations were reduced to 1837 by removing implied subsumptions.
Similarly, in aligning GEOSPECIES with itself, we found 188 equivalence relations, 94
of which were unique due to the symmetrical nature of the hypotheses.

Since the subset and equivalence relationship our algorithm finds are based on extensional reasoning, they hold by definition. However, in the remainder of this section we

---

[9] The counts of any of the containment relations in the table do not include the logically implied
relations within the same source, that is, when r'_1 is a subset of r_1 by construction.

Table 3. Alignment results

Source 1 (O_1)	Source 2 (O_2)	$\#(r_1 = r_2)$ total	$\#(r_1 = r_2)$ best matches	$\#(r_1 \subset r_2)$ before	$\#(r_1 \subset r_2)$ after	$\#(r_2 \subset r_1)$ before	$\#(r_2 \subset r_1)$ after
LinkedGeoData	DBpedia	158	152	2528	1837	1804	1627
Geonames	DBpedia	31	19	809	400	1384	1247
Geospecies	DBpedia	509	420	9112	2294	6098	4455
MGI	GeneID	10	9	2031	1869	3594	2070
Geospecies	Geospecies	94	88	1550	1201	-	-

show some examples of the alignments discovered and discuss whether the extensional subset relationships correspond to the intuitive intensional interpretation. As we use an extensional approach as opposed to an intensional one, the results reflect the practical nature of the links between the datasets and the instances in these sources.

Table 4 provides an assessment of the experimental results by selecting some interesting alignment examples from the five source pairs. For each alignment, the table depicts the restrictions from the two sources, the values of the metrics used for hypotheses evaluation (P' and R'), the relation, and the support for that relation.

We refer to the row numbers from Table 4 as a shorthand for the alignments. For example alignment 1 refers to the alignment between the *restriction class* (*rdf:type*=lgd: node) from LINKEDGEODATA and the class (*rdf:type*=owl:Thing) from DBpedia classified as an equivalent relation. Alignments 1, 2, 3 and 5 are the simplest alignments found by our algorithm as they are constrained on values of only the *rdf:type* property. However, we are also able to generate alignments like 4, as shown in Figure 2. GEONAMES has a rudimentary ontology comprised of only a single *Feature* concept. Hence alignments between the restriction classes prove to be more useful. Alignments 6 and 7 suggest that such restrictions from GEONAMES are equivalent to existing concepts in DBPEDIA. Our algorithm is thus able to build a richer set of classes for GEONAMES. This ontology building can also be observed in GEOSPECIES in alignment 12. A more complicated and interesting set of relations is also found in alignments 8, 15, 17, 18, 20 and 22. For example, in alignment 8, pointing a web browser to 'http://sws.geonames.org/3174618/' confirms that for any instance in GEONAMES that has this URI as a parent feature, would also belong to the region of 'Lombardy' in DBPEDIA. In a similar way, 20 provides an alternate definition for a *restriction class* with another class in the same ontology and thus build complimentary descriptions to existing classes and thus reinforce it.

The alignments closely follow the ontological choices of the sources. For example, we could assume that alignment 11, mapping 'geonames:featureCode=T.MT' (Mountain) to 'rdf:type=dbpedia:Mountain', should be equivalent. Closer inspection of the GEONAMES dataset shows, however, that there are some places with *Feature Codes* like 'T.PK' (Peak), 'T.HLL' (Hill), etc. from GEONAMES whose corresponding instances in DBPEDIA are all typed 'dbpedia:Mountain'. This implies that the interpretation of the concept 'Mountain' is different in both the sources and only a subset relation holds. Alignments 16, 19 and 21 also express a similar nature of the classes. As our results follow the data in the sources, incompleteness in the data reflects closely on the alignments generated. Alignment 9 suggests *Schools* from GEONAMES is extensionally equivalent *Educational Institutions*. It should naturally follow that *Schools in the US* be a subset

610 R. Parundekar, C.A. Knoblock, and J.L. Ambite

Table 4. Example alignments from the LinkedGeoData-DBpedia, Geonames-DBpedia, Geospecies-DBpedia, MGI-GeneID & Geospecies-Geospecies datasets

#	LinkedGeoData restriction	DBpedia restriction	P'	R'	Relation	$|I(r_1) \cap r_2|$
1	rdf:type=lgd:node	rdf:type=owl:Thing	97.27	99.99	$r_1 = r_2$	22987
2	rdf:type=lgd:aerodrome	rdf:type=dbpedia:Airport	90.94	100	$r_1 = r_2$	251
3	rdf:type=lgd:island	rdf:type=dbpedia:Island	90.81	99.44	$r_1 = r_2$	178
4	lgd:gnis_%3AST_alpha=NJ	dbpedia:Place#type= http://dbpedia.org/resource/City_-(New_Jersey)	100	97.5	$r_1 = r_2$	39
5	rdf:type=lgd:village	rdf:type=dbpedia:PopulatedPlace	67.3	98.71	$r_1 \subset r_2$	14391

#	Geonames restriction	DBpedia restriction	P'	R'	Relation	$|I(r_1) \cap r_2|$
6	geonames:featureClass=geonames:P	rdf:type=dbpedia:PopulatedPlace	91.07	96.7	$r_1 = r_2$	54927
7	geonames:featureClass=geonames:H	rdf:type=dbpedia:BodyOfWater	98.49	91.88	$r_1 = r_2$	1959
8	geonames:parentFeature=http://sws.geonames.org/3174618/	dbpedia:City_region=http://dbpedia.org/resource/Lombardy	99.91	91.2	$r_1 = r_2$	1245
9	geonames:featureCode=geonames:S.SCH	rdf:type=dbpedia:EducationalInstitution	92.45	94.52	$r_1 = r_2$	380
10	geonames:featureCode=geonames:S.SCH & geonames:inCountry=geonames:US	rdf:type=dbpedia:EducationalInstitution	91.72	94.72	$r_1 = r_2$	377
11	geonames:featureCode=geonames:T.MT	rdf:type=dbpedia:Mountain	78.4	96.8	$r_1 \subset r_2$	1728

#	Geospecies restriction	DBpedia restriction	P'	R'	Relation	$|I(r_1) \cap r_2|$
12	geospecies:inKingdom=http://lod.geospecies.org/kingdoms/Aa	rdf:type=dbpedia:Animal	99.96	99.96	$r_1 = r_2$	3029
13	geospecies:hasOrderName=Lepidoptera	dbpedia:order=http://dbpedia.org/resource/Lepidoptera	100	99.42	$r_1 = r_2$	344
14	geospecies:hasOrderName=Lepidoptera	dbpedia:kingdom=http://dbpedia.org/resource/Animal & dbpedia:order=http://dbpedia.org/resource/Lepidoptera	100	97.68	$r_1 = r_2$	338
15	geospecies:hasGenusName=Falco	dbpedia:genus=http://dbpedia.org/resource/Falcon	100	90.9	$r_1 = r_2$	10
16	geospecies:hasOrderName=Primates	dbpedia:order=http://dbpedia.org/resource/Primates	100	40.22	$r_2 \subset r_1$	35

#	MGI restriction	GeneID restriction	P'	R'	Relation	$|I(r_1) \cap r_2|$
17	bio2rdf:subType=Pseudogene	bio2rdf:subType=pseudo	93.76	93.56	$r_1 = r_2$	5971
18	bio2rdf:subType=Pseudogene & mgi::genomeStart=17	geneid:chromosome=17 & bio2rdf:subType=pseudo	91.49	94.38	$r_1 \subset r_2$	269
19	bio2rdf:chromosomePosition=-1.00 & mgi::genomeStart=4	geneid:chromosome=4 & bio2rdf:subType=pseudo	97.07	14.79	$r_2 \subset r_1$	332

#	Geospecies restriction	Geospecies restriction	P'	R'	Relation	$|I(r_1) \cap r_2|$
20	geospecies:hasKingdomName=Animalia	geospecies:inKingdom=http://lod.geospecies.org/kingdoms/Aa	91.99	100	$r_1 = r_2$	563
21	geospecies:hasClassName=Insecta	geospecies:inClass= http://lod.geospecies.org/bioclasses/aQado	87.83	100	$r_1 \subset r_2$	195
22	geospecies:inFamily= http://lod.geospecies.org/families/amTJ9	geospecies:hasSubfamilyName=Sigmodontinae	100	37.03	$r_2 \subset r_1$	10

of *Educational Institutions*. However, as there are only 3 other *Schools* (outside the US), extensionally these classes are very close, as shown by alignment 10. This example illustrates that reasoning extensionally actually provides additional insight on the relationship between the sources. Alignments 13 and 14 show two equivalent alignments that have different support due to missing assertions in one of the ontologies (the property *dbpedia:kingdom* for all moths and butterflies).

Our approach makes an implicit 'closed-world' assumption in using the instances of a class to determine the relationships between the classes in different sources. We believe that this is an important feature of our approach in that it allows one to understand the relationships in the actual linked data and their corresponding ontologies. The alignments generated can be readily used for modeling and understanding the sources since we are modeling what the sources actually contain as opposed as to what an ontology disassociated from the data appears to contain based on the class name or description. Moreover, even if we delve into the open-world assumption of data, it would be very difficult to categorize the missing instances as either: (1) yet unexplored, (2) explored but purposefully classified as not belonging to the dataset, or (3) explored but not included in the dataset by mistake. Hence, our method provides a practical approach to understanding the relationships between sources.

In summary, our algorithm is able to find a significant number of interesting alignments, both equivalent and subset relationships, as well as build and refine the ontologies of real sources in the Web of Linked Data.

5   Related Work

There is a large body of literature on ontology matching [12]. Ontology matching has been performed based on *terminological* (e.g. linguistic and information retrieval techniques [11]), *structural* (e.g. graph matching [15]), and *semantic* (e.g. model-based) approaches or their combination. The FCA-merge algorithm [18] uses extensional techniques over common instances between two ontologies to generate a concept lattice in order to merge them and, thus, align them indirectly. This algorithm, however, relies on a domain expert (a user) to generate the merged ontology and is based on a single corpus of documents instead of two different sources, unlike our approach. A strong parallel to our work is found in Duckham et al. [10], which also uses an extensional approach for fusion and alignment of ontologies in the geospatial domain. The difference in our approach in comparison to their work (apart from the fact that it predates Linked Data) is that while their method fuses ontologies and aligns only existing classes, our approach is able to generate alignments between classes that are derived from the existing ontology by imposing restrictions on values of any or all of the properties not limited to the class *type*. The GLUE system [9] also uses an instance-based similarity approach to find alignments between two ontologies. It uses the labels of the classes that a concept belongs to along with the textual content of the attribute values of instances belonging to that concept to train a classifier and then uses it to classify instances of a concept from the other ontology as either belonging to the first concept or not. Similarly, it also tries to classify the concepts in the other direction. GLUE then hypothesizes alignments based on the probability distributions obtained from the classifications. Our approach, instead, relies on the links already present in the Web of Linked Data, which in some cases uses a much more sophisticated approach for finding instance equivalences.

Most of the work in information integration within the Web of Linked Data is in instance matching as explained in Bizer et al. [7]. Raimond et al. [17] use string and graph matching techniques to interlink artists, records, and tracks in two online music datasets (Jamendo and MusicBrainz) and also between personal music collections and the MusicBrainz dataset. Our approach solves a complimentary piece of the information integration problem on the Web of Linked Data by aligning ontologies of linked data sources. Schema matching in the Web of Linked Data has also been explored by Nikolov et al. [2], who use existing instance and schema-level evidence of Linked Data to augment instance mappings in those sources. First, instances from different sources are clustered together by performing a transitive closure on *owl:sameAs* links such that all instances in a cluster are equivalent. Class re-assignment is then performed by labeling each instance with all the other classes in the same cluster. Second, a similarity score is computed based on the size of the intersection sets and classes are labeled as equivalent. Finally, more equivalence links are generated based on the new class assignments. Our approach differs from this in the sense that, first, the class re-assignment step increases the coverage of a class. Such an assumption in aligning schemas would bias the extensional approach as it modifies the original extension of a class. Second, only existing classes are explored for similarity in that work and thus faces severe limitations with rudimentary ontologies like GEONAMES, where our approach performs well as it considers restriction classes.

6   Conclusion

The Web of Linked Data contains linked instances from multiple sources without the ontologies of the sources being themselves linked. It is useful to the consumers of the data to define the alignments between such ontologies. Our algorithm generates alignments, consisting of conjunctions of *restriction classes*, that define subsumption and equivalence relations between the ontologies. This paper focused on automatically finding alignments between the ontologies of geospatial, zoology and genetics data sources and building such ontologies using an extensional technique. However, the technique is general and can be applied to other Web of Linked Data data sources.

In our future work, we plan to improve the scalability of our approach, specifically, improve the performance of the algorithm that generates alignment hypotheses by using a more heuristic exploration of the space of alignments. The sizes of the sources in this paper were quite large (on the order of thousands of instances after preprocessing). Although we have fixed a minimum support size of ten instance pairs for a hypothesis, the effectiveness of the extensional approach needs to be verified when the sources are small (number of instances in the order of hundreds or less). We also plan to explore the integration of this work with our previous work on automatically building models of sources [1]. Linking the data from a newly discovered source with a known source already linked to an ontology will allow us to more accurately determine the classes of the discovered data. Finally, we plan to apply our alignment techniques across additional domains and to pursue in depth alignments in biomedical Linked Data.

Acknowledgements

This work was supported in part by the NIH through the following NCRR grant: the Biomedical Informatics Research Network (1 U24 RR025736-01), and in part by the Los Angeles Basin Clinical and Translational Science Institute (1 UL1 RR031986-01).

References

1. Ambite, J.L., Darbha, S., Goel, A., Knoblock, C.A., Lerman, K., Parundekar, R., Russ, T.: Automatically constructing semantic web services from online sources. In: Bernstein, A., Karger, D.R., Heath, T., Feigenbaum, L., Maynard, D., Motta, E., Thirunarayan, K. (eds.) ISWC 2009. LNCS, vol. 5823, pp. 17–32. Springer, Heidelberg (2009)
2. Andriy Nikolov, V.U., Motta, E.: Data Linking: Capturing and Utilising Implicit Schema Level Relations. In: International Workshop on Linked Data on the Web, Raleigh, North Carolina (2010)
3. Auer, S., Bizer, C., Kobilarov, G., Lehmann, J., Cyganiak, R., Ives, Z.: Dbpedia: A nucleus for a web of open data. In: Aberer, K., Choi, K.-S., Noy, N., Allemang, D., Lee, K.-I., Nixon, L.J.B., Golbeck, J., Mika, P., Maynard, D., Mizoguchi, R., Schreiber, G., Cudré-Mauroux, P. (eds.) ASWC 2007 and ISWC 2007. LNCS, vol. 4825, pp. 722–735. Springer, Heidelberg (2007)
4. Auer, S., Lehmann, J., Hellmann, S.: LinkedGeoData: Adding a Spatial Dimension to the Web of Data. In: Bernstein, A., Karger, D.R., Heath, T., Feigenbaum, L., Maynard, D., Motta, E., Thirunarayan, K. (eds.) ISWC 2009. LNCS, vol. 5823, pp. 731–746. Springer, Heidelberg (2009)
5. Belleau, F., Tourigny, N., Good, B., Morissette, J.: Bio2RDF: A Semantic Web atlas of post genomic knowledge about human and mouse, pp. 153–160. Springer, Heidelberg (2008)
6. Berners-Lee, T.: Design Issues: Linked Data (2009),
   http://www.w3.org/DesignIssues/LinkedData.html
7. Bizer, C., Cyganiak, R., Heath, T.: How to publish linked data on the web (2007),
   http://www4.wiwiss.fu-berlin.de/bizer/pub/LinkedDataTutorial/
8. Ding, L., Shinavier, J., Finin, T., McGuinness, D.L.: owl: sameAs and Linked Data: An Empirical Study. In: Second Web Science Conference, Raleigh, North Carolina (2010)
9. Doan, A., Madhavan, J., Domingos, P., Halevy, A.: Ontology matching: A machine learning approach. In: Handbook on Ontologies, pp. 385–516 (2004)
10. Duckham, M., Worboys, M.: An algebraic approach to automated geospatial information fusion. International Journal of Geographical Information Science 19(5), 537–557 (2005)
11. Euzenat, J.: An API for Ontology Alignment. In: McIlraith, S.A., Plexousakis, D., van Harmelen, F. (eds.) ISWC 2004. LNCS, vol. 3298, pp. 698–712. Springer, Heidelberg (2004)
12. Euzenat, J., Shvaiko, P.: Ontology matching. Springer, Heidelberg (2007)
13. Haklay, M.M., Weber, P.: OpenStreetMap: user-generated street maps
14. Halpin, H., Hayes, P.J.: When owl: sameAs isn't the same: An analysis of identity links on the semantic web. In: International Workshop on Linked Data on the Web, Raleigh, North Carolina (2010)
15. Melnik, S., Garcia-Molina, H., Rahm, E.: Similarity flooding: A versatile graph matching algorithm and its application to schema matching. In: International Conference on Data Engineering, San Jose, California, pp. 117–128 (2002)

16. Rahm, E., Bernstein, P.A.: A survey of approaches to automatic schema matching. VLDB Journal 10(4), 334–350 (2001)
17. Raimond, Y., Sutton, C., Sandler, M.: Automatic interlinking of music datasets on the semantic web. In: First Workshop on Linked Data on the Web, Beijing, China (2008)
18. Stumme, G., Maedche, A.: FCA-Merge: Bottom-up merging of ontologies. In: International Joint Conference on Artificial Intelligence, Seattle, Washington, pp. 225–234 (2001)
19. Vatant, B., Wick, M.: Geonames ontology, http://www.geonames.org/ontology/

A Feature and Information Theoretic Framework for Semantic Similarity and Relatedness

Giuseppe Pirró* and Jérôme Euzenat

INRIA Rhône-Alpes, Montbonnot, France
{Giuseppe.Pirro,Jerome.Euzenat}@inrialpes.fr

Abstract. Semantic similarity and relatedness measures between ontology concepts are useful in many research areas. While similarity only considers subsumption relations to assess how two objects are alike, relatedness takes into account a broader range of relations (e.g., part-of). In this paper, we present a framework, which maps the feature-based model of similarity into the information theoretic domain. A new way of computing IC values directly from an ontology structure is also introduced. This new model, called Extended Information Content (eIC) takes into account the whole set of semantic relations defined in an ontology. The proposed framework enables to rewrite existing similarity measures that can be augmented to compute semantic relatedness. Upon this framework, a new measure called FaITH (Feature and Information THeoretic) has been devised. Extensive experimental evaluations confirmed the suitability of the framework.

Keywords: Semantic Similarity, Feature Based Similarity, Ontologies.

1 Introduction

Semantic similarity and relatedness investigates how alike two or more objects are, and plays an important role in many contexts. Generally speaking, similarity allows to infer knowledge and categorize objects into kinds. This is important when either it is not possible to exactly state what properties are salient for an object, or when it is not easy to separate an object into distinct properties [5,26]. Semantic similarity has a long tradition in psychology and cognitive science where different models have been postulated. Among these, the *geometric* model enables to asses similarity between entities by considering them as points in a dimensionally organized metric space. The *feature-based* model, leverages features (i.e., characteristics) of the examined objects and assumes that similarity is a function of both common and distinctive features [24]. Recently, findings in information theory have been considered in computing similarity [19]. From a computer science perspective, similarity measures exploit some source of knowledge such as search engines [3] or ontologies such as WordNet [13]. More recently, similarity measures have been defined in Description Logics (DLs) [2,4]. In [4] several similarity measures are described, which take into account ontology

---

* This work was carried out during the tenure of an ERCIM "Alain Bensoussan" Fellowship Programme.

P.F. Patel-Schneider et al. (Eds.): ISWC 2010, Part I, LNCS 6496, pp. 615–630, 2010.

instances for assessing the similarity between two concepts. While similarity only considers subsumption relations to assess how two objects are alike, relatedness takes into account a broader range of relations (e.g., part-of). The work presented in this paper focuses on computing similarity and relatedness by exploiting the terminological definition of ontology concepts. We leave the investigation about how this method can be applied to DLs as a future work.

Computing semantic similarity between ontology concepts is an important issue since having many applications in different contexts including: Information Retrieval, to improve the performance of current search engines [8], ontology matching, to discover correspondences between entities belonging to different ontologies [16], semantic query routing, to choose among the set of possible peers only those relevant, bioinformatics to assess the similarity between proteins [25] just to cite a few. This paper presents a semantic similarity framework, which is based on two main pillars. One is the projection of the feature-based model of similarity into the information theoretical domain. The reason to combine these two models is twofold. On one hand, the feature-based model has a solid theoretical underpinning supported by several psychological studies [24] and is more flexible than other theoretical models (e.g., geometric). On the other hand, the information-theoretic formulation of similarity allows to compare concept features not by simply counting object properties but taking into account the informativeness of the concepts being compared. The second pillar, is a new way to obtain IC values called Extended Information content (*eIC*). *eIC* considers the whole set of semantic relations defined in an ontology and assigns a score of informativeness to each concept without referring to external corpora as usually done by traditional IC-based approaches where time expensive and corpus-dependent occurrence count has to be performed.

The generality of this framework enables to rewrite several existing similarity measures that can be augmented to compute semantic relatedness. This aspect has been investigated and resulted in an improvement of existing similarity measures as will be discussed in Section 4. Finally, a new measure called FaITH (Feature and Information Theoretic) has been designed, which is a versatile tool to compute both similarity and relatedness. Extensive experimental evaluation of similarity and relatedness show the suitability of the proposed framework and FaITH in particular.

The remainder of this paper is organized as follows. Section 2 provides some background and surveys on popular measures. Section 3 presents the new similarity framework and the logical path toward its definition; here the FaITH measure and the *eIC* are discussed. Section 4 presents an extensive evaluation campaign. Section 5 concludes the paper.

2   Definitions and Background

We consider an ontology O as a graph, where nodes represent concepts and edges represent relations between concepts. If we consider the hierarchical structure of the ontology, each concept can have a set of sub-concepts (its descendants) in the hierarchy. However, an ontology usually includes a broader set of semantic relation such as part-of. Figure 1 reports an excerpt of an ontology. Hereafter we will consider WordNet as reference ontology even if the same reasoning applies to any other ontology. In WordNet, the definition of a concept consists of its immediate superordinate(s) followed by a relative clause that describes how this concept differs from all others. For example *Fortified Wine* is distinguished from *Wine* because *"... alcohol (usually grape brandy)"* has been added just as the gloss accompanying its definition mentions.

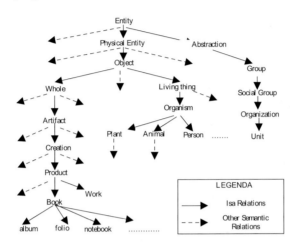

Fig. 1. An excerpt of ontology

An object *feature* (a concept in our case) can be seen as a property of the object. According to the definition above, concepts in the hierarchy inherit all the features of their superordinate even if they can have their own specific features. As an example, since *car* and *bicycle* both serve to transport people or objects, in other words they are both types of vehicles, they share all features pertaining to the concept *vehicle*. However, each concept has also its specific features as *steering wheel* for *car* and *pedal* for *bicycle*. Moreover, even if specialization relations constitute the majority in WordNet, there are other kinds of relations accompanying each definition that are useful to identify object features. For instance, *car* has a relation of type *part-of* with *engine* whereas *bicycle* has a *part-of* relation with *sprocket*. The use of immediate concept features can be seen as a special case of semantic neighbourhood with radius equals to 1.

Similarity or relatedness measures, by looking at the ontology structure or by exploiting some additional information, address the problem of assessing (typically in terms of a numerical score) how alike two concepts are. As an example of similarity and relatedness, *car* and *bicycle* are similar whereas *car* and *wheel* are related. The choice to focus either on similarity or relatedness depends on the particular application context, even though many approaches to compute relatedness are extensions of similarity measures [6,20]. The framework presented in this paper can be adopted to compute both similarity and relatedness.

2.1 State of the Art

Similarity measures can be divided into different and not necessarily disjoint categories. In this work we consider information-theoretic approaches, ontology-based approaches and hybrid approaches.

Information Theoretic Approaches. Information theoretic approaches employ the notion of Information Content (IC), which quantifies the informativeness of concepts. Early IC approaches [19,9,12] obtained IC values by associating probabilities to each concept in an ontology on the basis of its occurrences in large text corpora. In the specific case of hierarchical ontologies, these probabilities are cumulative as we travel up from specific concepts to more abstract ones. This means that every occurrence of a concept in a given corpus is also counted as an occurrence of each concept containing it. IC values are obtained by computing the negative likelihood of encountering a concept in a given corpus. Note that this method ensures that IC is monotonically decreasing as we move from the leaves of the taxonomy to its roots.

Resnik [19] was the first to leverage IC for the purpose of semantic similarity. The basic intuition behind the use of the negative likelihood is that the more probable a concept is of appearing in a corpus the less information it conveys, in other words, infrequent words are more informative than frequent ones. Once IC values are available for each concept in the considered ontology, semantic similarity can be calculated. Resnik's formula to compute similarity states that similarity depends on the amount of information two concepts c_1 and c_2 share, which is given by the Most Specific Common Abstraction ($msca(c_1, c_2)$), that is, the concept that subsumes the two concepts being compared.

Starting from Resnik's work, Jiang and Conrath [9] and Lin [12] proposed two measures, which calculate IC-values in the same manner as proposed by Resnik while correcting some problems with this similarity measure; if one were to calculate $sim_{res}(c_1, c_1)$ one would not obtain the maximal similarity value of 1, but instead the value given by IC(c_1). Besides, with Resnik's approach any two pairs of concepts having the same $msca$ have exactly the same semantic similarity; for instance, in the WordNet ontology, $sim_{res}(Horse, Plant)$ = $sim_{res}(Animal, Plant)$ because in each case the $msca(Horse, Plant)$ and $msca(Animal, Plant)$ is *Living Thing*. However, in this case the semantic leap is not the same.

The Lin measure considers the ratio between the amount of information needed to state the commonality between two concepts and the information needed to describe them as discussed in [12].

Ontology Based Approaches. As for ontology based approaches, the work by Rada et al. [18] is similar to the Resnik measure since it also computes the $msca(c_1, c_2)$, but instead of considering the IC as the value of similarity, it considers the number of links that were needed to attain the $msca(c_1, c_2)$. Obviously, the less the number of links separating the concepts the more similar they are. The work by Hirst et al., which actually measures relatedness, is similar to the previous one but it uses a wider set of relations coupled with rules restricting the way concepts are transversed [6]. Nonetheless, the intuition also in this case is that the number of links separating two concepts is inversely proportional to the degree of similarity.

Hybrid Approaches. Hybrid approaches usually combine multiple information sources. Li et al. [11] proposed to combine structural semantic information in a nonlinear model. The authors empirically defined a similarity measure that uses shortest path length, depth and local density in a taxonomy and combine them.

In [22] the *OSS* distance function, combining *a-priori* scores of concepts with distance, is proposed. *OSS* performs the following steps to assess similarity between two concepts c_1 and c_2: (i) computing the score of the concept c_2 from the concept c_1; (ii) computing how much score has been transferred between the concepts; (iii) transforming the transfer of score into a distance measure.

Our previous work [17], defined a similarity measure combining features and information content that adopts Tversky's contrast model. This measure treats similarity between identical concepts as a special case and can give as output negative values, which make difficult the interpretation of results. The differences with the present work are: i) this paper describes a general framework, which can be used to rewrite even existing similarity measures; ii) here a new similarity measure is proposed, which adopts a different representation of the feature-based model; iii) in this paper a new way to compute IC values is proposed, which enables to compute both semantic similarity and relatedness; iv) an extensive evaluation of relatedness is proposed for FaITH and several other measures.

2.2 Comparison among Measures

Each measure has its limitations. IC-based measures making use of corpora, though having a strong mathematical formalization, may sometimes fail to capture certain aspects of language. For instance, it is possible that corpus such as the British National Corpus, may not even mention certain words. Besides, values of IC are obtained through time intensive analysis of corpora and can heavily depend on the considered corpora (as discussed in Section 4). Ontology-based approaches require to work with consistent ontologies, that is, ontologies where distance between specific and more general concepts have the same interpretation. As an example it is obvious that the semantic leap between *Entity* and *Psychological Feature* is higher than that between *Canine* and *Dog* even if both couples are separated by one edge. Finally, hybrid approaches require the different information sources to be correctly "weighted". A common limitation of the considered approaches is that they can only compute either similarity or relatedness. The proposed framework, and in particular FaITH, are more flexible as the notion of *Extended Information Content* (*eIC*) can be exploited to compute both similarity and relatedness without depending on external corpora.

3   A Framework for Semantic Similarity and Relatedness

This section presents a new framework for computing semantic similarity and relatedness. After providing some preliminary definitions, the Tversky's formulation of similarity, which is based on a representation of concepts according to their features, is introduced. This will serve as a basis to motivate the present framework. In more detail, the proposed framework adopts a ratio-based formulation of the Tversky's model of similarity and projects it into the information-theoretic domain. Section 3.4 describes the *Extended Information Content (eIC)*, which can be used to compute relatedness between concepts. Note that the generality of this framework enables to rewrite several existing similarity measures, which can be augmented to compute relatedness.

3.1   Tversky's Feature-Based Model of Similarity

Amos Tversky, in his seminal work, proposed an alternative way to compute similarity by taking into account both common and distinguish "features "of the objects being compared. As an example of Tversky's formulation, *car* and *bicycle* both serve to transport people or objects (in other words they are both types of vehicles), then they share all features that pertain to the concept *vehicle*. However, each concept has also its specific features such as *steering wheel* for *car* or *pedal* for *bicycle*. Moreover, if we look beyond the hierarchical structure of their definitions we can find different kinds of relations with other concepts such as *engine part-of car* and *sprocket part-of bicycle*. The set of all relations can be exploited to further characterize concept features. Fig. 2 depicts an example of such reasoning. Early semantic similarity models, such as the geometric model,

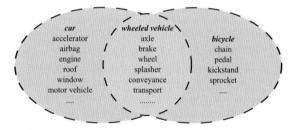

Fig. 2. An example of concept features

required to respect metric properties such as the triangle inequality or symmetry. Tversky's discussed several examples to support the idea that certain axioms, required by the geometric model, were not necessary in the process of similarity estimation. For instance, since Germany is judged to be more like Austria than Austria is to Germany [24] the symmetry property could not be respected in this case. According to the feature-based model, the similarity of a concept c_1 to a concept c_2 is a function of the features common to c_1 and c_2, those in c_1 but not in c_2 and those in c_2 but not in c_1. If we admit a function $\Psi(c)$ that yields the set of features relevant to c, Tversky's similarity model can be represented by the following equation, also known as *contrast model*:

$$sim_{tvr}(c_1, c_2) = \alpha F(\Psi(c_1) \cap \Psi(c_2)) - \beta F(\Psi(c_1) \setminus \Psi(c_2)) - \gamma F(\Psi(c_2) \setminus \Psi(c_1)). \quad (1)$$

where F is some function that reflects the salience of a set of features, and α, β and γ are parameters that provide for differences in focus on the different components. According to this model, features in common increase similarity whereas features that are unique to the two objects decrease similarity. However, note that the above formulation is not framed in information theoretic terms since it is based on sets of concept features.

3.2   A Ratio-Based Formulation of Tverky's Similarity Model

The difficulty with the contrast model described in equation (1) and discussed in our previous study [17] is that the more unique features a concept presents the lower the similarity. Moreover similarity values are not bounded between 0 and 1, which can make interpretation of results difficult. To overcome these issues, therefore, a *ratio model* is more appropriate since it is bounded between 0 and 1, irrespective of the size of the features being compared. Thus a more useful definition of feature-based similarity is:

$$sim_{tvr-ratio}(c_1, c_2) = \frac{F(\Psi(c_1) \cap \Psi(c_2))}{\beta F(\Psi(c_1) \setminus \Psi(c_2)) + \gamma F(\Psi(c_2) \setminus \Psi(c_1)) + F(\Psi(c_1) \cap \Psi(c_2))}. \tag{2}$$

Note that $\alpha = 1$ in the ratio model and then common features are maximally important in process of similarity estimation. At this point there are two main tasks we can perform:

1. Assess the degree to which concept c_1 and c_2 are similar to each other. In this case $\beta = \gamma$ since the similarity is not intended to be directional.
2. Assess the degree to which concept c_2 is similar to concept c_1. In this second task, similarity is directional and we are more interested in the features in c_1 than we are in the features unique to c_2. Here, β and γ do not need to be equal. This latter case is useful in many application contexts such as Information Retrieval (IR) or clustering where starting from a concept we are interested in finding what it is similar to.

Table 1 analyzes different scenarios obtained by manipulating the coefficients β and γ in equation (2). For the purpose of this paper, we consider $\beta = \gamma$ since we want to compute the similarity not directionally. Moreover, for the definition of the ratio based model described in equation (2) $\alpha = 1$, which maximizes the contribution of common features. We leave as future work the investigation of other values for these parameters in more targeted applications such as IR.

3.3   The FaITH Similarity Measure

This section describes the FaITH measure for semantic similarity and relatedness. The cornerstone of this measure is the $msca(c_1, c_2)$, which reflects the information shared by two concepts c_1 and c_2 in an ontology structure. In the information-theoretic domain, Resnik exploited the $msca(c_1, c_2)$ to assess the similarity between concepts. IC values are obtained by exploiting equation (3):

$$IC(c) = -log\ p(c). \tag{3}$$

Table 1. Possible scenarios obtained by manipulating equation (2)

Case	Coefficients	Description
Commonalities between c_1 and c_2	$\beta = \gamma = 0$	If there exists any commonality then $sim_{tvr'}(c_1, c_2) = 1$
Given c_1 assess to which degree c_2 is similar to it	$\beta = 1, \gamma = 0$	When the full set of features of c_1 are contained in c_2 then $sim_{tvr'}(c_1, c_2) = \frac{\alpha F(\Psi(c_1) \cap \Psi(c_2))}{\beta F(\Psi(c_1) \setminus \Psi(c_2)) + \alpha F(\Psi(c_1) \cap \Psi(c_2))}$
	$\beta = 0, \gamma = 1$	When the set of features of c_1 contains the features of c_2 then $sim_{tvr_r}(c_1, c_2) = \frac{\alpha F(\Psi(c_1) \cap \Psi(c_2))}{\gamma F(\Psi(c_2) \setminus \Psi(c_1)) + \alpha F(\Psi(c_1) \cap \Psi(c_2))}$
Given c_1 and c_2 assess to which degree they are similar to each other	$\beta = \gamma = 1$	Tversky's similarity is represented in terms of Tanimoto index.
	$\beta = \gamma = 0.5$	Tversky's similarity is represented in terms of Dice index.

where c is a concept and $p(c)$ is the probability of encountering c in a given corpus. Note that this method ensures that IC is monotonically decreasing as we move from the leaves of the taxonomy to its roots.

In Fig. 2, the $msca(car, bicycle)$ is *wheeled vehicle* and these two concepts share all the features belonging to their *msca*. In a feature-based formulation of similarity, the $msca(c_1, c_2)$ can be seen as the intersection of features from c_1 and c_2. Therefore, one can speculate that the function F, that reflects the saliency of features, can be substituted by the function IC in the information theoretic domain (this new IC is referred to as $IC_{features}$). Starting from this assumption, by looking at Fig. 2, it is immediate to infer that the set of features specific to car (resp. $bicycle$) is given by $IC_{features}(car) - IC_{features}(wheeled\_vehicle)$ (resp. $IC_{features}(bicycle) - IC_{features}(wheeled\_vehicle)$). These three analogies, generalized in Table 2, are the building blocks of the proposed framework.

Table 2. Mapping between feature-based and information theoretic similarity models

Description	Feature-based model	Information-theoretic model
Common features	$\Psi(c_1) \cap \Psi(c_2)$	$IC(msca(c_1, c_2))$
Features of c_1 alone	$\Psi(c_1) \setminus \Psi(c_2)$	$IC(c_1) - IC(msca(c_1, c_2))$
Features of c_2 alone	$\Psi(c_2) \setminus \Psi(c_1)$	$IC(c_2) - IC(msca(c_1, c_2))$

Moreover, as it will be discussed in Section 3.4, the way we compute the IC values for each concept (i.e., eIC) can take into account the different features of an object defined both in terms of the hierarchical structure and other kinds of semantic relations. By substituting the analogies from Table 2 in equation (2) the similarity measure called FaITH, reported in equation (4), is obtained.

$$sim_{FaITH}(c_1, c_2) = \frac{IC(msca(c_1, c_2))}{\beta(IC(c_1) - IC(msca(c_1, c_2))) + \gamma(IC(c_2) - IC(msca(c_1, c_2))) + IC(msca(c_1, c_2))}.$$

$$(4)$$

As we are concerned to compute how two concepts c_1 and c_2 are similar to each other we set the values of β and γ to 1 (see Table 1) thus obtaining:

$$sim_{FaITH}(c_1, c_2) = \frac{IC(msca(c_1, c_2))}{IC(c_1) + IC(c_2) - IC(msca(c_1, c_2))}. \tag{5}$$

Note that in the case of ontologies with multiple inheritance, the $msca(c_1, c_2)$ may be unique. In this case, FaITH considers the most informative $msca$ (i.e., the $msca$ with the highest information content).

3.4   Extended Information Content (eIC)

The proposed framework combines the feature and the information theoretic models of similarity. One of the main difficulty with this model is that IC values have to be derived by analyzing large corpora, which may not even contain certain specific words. In order to overcome this issue, the intrinsic IC formulation proposed in [23] is adopted. The *intrinsic* IC (iIC) for a concept c is defined as:

$$iIC(c) = 1 - \frac{log(sub(c) + 1)}{log(max_{con})}. \tag{6}$$

where the function sub returns the number of subconcepts of a given concept c. Note that concepts representing leaves in the taxonomy will have an IC of one, since they do not have hyponyms. The value of one states that a concept is maximally expressed and is not further differentiated. Moreover max_{con} is a constant that indicates the total number of concepts in the considered taxonomy.

However, since an ontology usually contains relations beyond inheritance also useful to assess to what extent two concepts are alike, the Extended Information Content (eIC) is introduced. eIC by investigating each kind of ontological relation between concepts provides a better indicator about the features of concepts and then can be used to compute relatedness. For instance, by only focusing on *isa* relations, in the example in Fig. 2 we would lose some important information (e.g., that *car* has *part-of engine* or that *bicycle* has as *part-of sprocket*) that can help to further characterize commonalities and differences between two concepts. For each concept, the coefficient EIC is defined as follows:

$$EIC(c) = \sum_{j=1}^{m} \frac{\sum_{k=1}^{n} iIC(c_k \in C_{R_j})}{|C_{R_j}|}. \tag{7}$$

This formula takes into account all the m kinds of relations that connect a given concept c with other concepts. Moreover, for all the concepts at the other end of a particular relation (i.e., each $c_k \in C_{R_j}$) the average iIC is computed. This enables to take into account the expressiveness of concepts to which a given concept is related in terms of their information content. The final value of *Extended Information Content* (eIC) is computed by weighting the contribution of the iIC and EIC coefficients thus leading to:

$$eIC(c) = \zeta iIC(c) + \eta EIC(c). \tag{8}$$

The two parameters ζ and η can be settled in order to give more or less emphasis to the hierarchical IC of the two concepts. At this point, we can rewrite equation (5) thus obtaining:

$$sim_{FaITH}(c_1, c_2) = \frac{eIC(msca(c_1, c_2))}{eIC(c_1) + eIC(c_2) - eIC(msca(c_1, c_2))}. \tag{9}$$

This similarity measure corrects some drawbacks of existing approaches. First, it exploits features of concepts, expressed in terms of IC, and not only their position in the ontology structure. Second, it corrects the problem with Resnik's measure, in fact, $sim_{FaITH}(c_1, c_1) = 1$. Finally, by taking into account relations beyond inheritance, FaITH allows to compute semantic relatedness.

4   Evaluation

This section discusses the evaluation of the FaITH similarity measure and its comparison w.r.t. the state of the art. In the first experiment we evaluated FaITH as a semantic similarity measure while in the second experiment we evaluated FaITH as a semantic relatedness measure as using the eIC formulation. Finally, in order to have an insight of how FaITH works with more domain-related ontologies, we performed an evaluation using couples of concepts taken from the MeSH biomedical ontology. In each experiment, we evaluated the performance of the different methods in two settings. The first one (denoted as $F + eIC$) by exploiting the proposed framework along with the eIC while the second one using the classical approach to compute IC without mapping features in the IC domain. In particular, the $SemCor(S)$ $Brown$ (B) and BNC (Bnc) text corpora, of increasing size, have been used to obtain IC values. For the Li measure we adopted the same optimal parameter values as indicated by authors in [11].

In order to have an idea of the improvement using the $F + eIC$ formulation we computed for each measure and corpus the loss (L) in performance, which represents how much the performance of a given measure decrease when using the classical IC formulation. Besides, for each evaluation, as statistical test of significance, we computed the $p\text{-}value$. The analyzed similarity measures have been implemented in the Java WordNet Similarity Library available upon request, along with the datasets, at `http://grid.deid.unical.it/similarity`.

4.1   Experiment 1: Evaluating FaITH on Similarity

In the first experiment, we evaluate the FaITH measure on a dataset collected by an online similarity experiment described in our previous work [17]. The dataset contains similarity judgments for 65 word pairs [21] (referred to as $S_{R\&G}$) which are commonly used, along with a subset of 28 word pairs [14] (referred to as $S_{M\&C}$), to measure accuracy of similarity measures. The word pairs in the dataset have been originally chosen to range from very similar (e.g., *car-automobile*) to *semantically unrelated* (e.g., *chord-smile*) as discussed in [21]. Figure 3 reports the ratings of similarity provided by both human participants and computational methods. Values of correlation (ρ) for the different measures are reported in Table 3. The

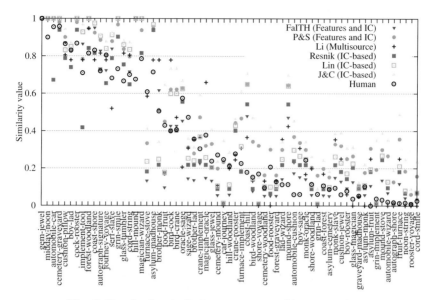

Fig. 3. Results for similarity measures and human ratings

first column indicates the correlation by using the IC formulation introduced in Section 3.4. Each other column considers the correlation according to one corpus and the loss as compared to the result in the first column. The second column, for instance, indicates that by using the *SemCor(S)* corpus, the correlation of the Resnik measure is 0.71, with a loss of 16.4 % .

Table 3. Correlation values with $F + eIC$ (ρ) and different corpora

	Correlation on $S_{M\&C}$				Correlation on $S_{R\&G}$			
	ρ	$\rho_S/\textbf{\textit{L}(\%)}$	$\rho_B/\textbf{\textit{L}(\%)}$	$\rho_{Bnc}/\textbf{\textit{L}(\%)}$	ρ	$\rho_S/\textbf{\textit{L}(\%)}$	$\rho_B/\textbf{\textit{L}(\%)}$	$\rho_{Bnc}/\textbf{\textit{L}(\%)}$
Lenght	0.61	0.61	0.61	0.61	0.58	0.58	0.58	0.58
Depth	0.84	0.84	0.84	0.84	0.80	0.80	0.80	0.80
Li	0.91	0.91	0.91	0.91	0.90	0.90	0.90	0.90
Resnik	0.85	0.71/16.4	0.73/14.5	0.75/11.8	0.87	0.83/5.1	0.84/4.1	0.85/2.4
Lin	0.87	0.69/**20.2**	0.74/**15.0**	0.75/**14.2**	0.89	0.75/**15.0**	0.79/**10.9**	0.80/**9.8**
J&C	0.88	0.72/17.8	0.80/9.3	0.81/8.1	0.87	0.82/6.3	0.83/5.4	0.84/4.0
P&S	0.91	0.86/4.7	0.87/4.2	0.89/2.3	0.90	0.87/3.7	**0.88/3.0**	0.89/1.8
FaITH	**0.92**	**0.87/5.5**	**0.88/4.6**	**0.90/2.4**	**0.91**	**0.88/3.3**	0.88/2.9	**0.90/1.0**

For the *Length* measure, lower values correspond to higher similarity values. For instance, the two word pairs (i.e., *gem-jewel* and *automobile-car*) have a length equal to zero since belonging to the same WordNet synset respectively and then are maximally similar according to the WordNet's design principle. On the other hand, examples of unrelated words are the couples *rooster-voyage* and *chord-smile* having a path length of 30. The *Depth* measure obtained a value of correlation of about 30% better than the *Path* measure. This measure assesses similarity by considering the depth of the $msca(c_1, c_2)$. Edge counting approaches reach the lowest correlation w.r.t. human ratings in both datasets. That is because these approaches work well only when the values computed have a "consistent interpretation", that is, when the length of the path (resp. depth of the $msca(c_1, c_2)$) between two general concepts and that between two specific ones express the same semantic leap, which is not the case of WordNet.

As for IC-based approaches, Resnik's measures obtained the lowest value of correlation. However, the usage of the $IC(msca(c_1, c_2))$ brings better results in terms of correlation as compared to path-based measures. The other two IC-based measures (i.e., Lin and J&C) obtained better results since considering the IC of the two concepts as well. As for hybrid approaches, the Li measure, which combines the depth of the $msca(c_1, c_2)$ and the length of the path between two concepts, obtained a higher value of correlation. However, note that this measure to correctly weights the contributions of the different information sources requires the tuning of two coefficients as described in [11]. The P&S measure, described in [17], obtained a remarkable value of correlation in both $S_{R\&G}$ and $S_{M\&C}$. However, the P&S formulation treats the computation of similarity between identical concepts as a special case as discussed in [17]. Moreover, in some cases, $sim_{P\&S(c_1, c_2)} < 0$, which makes the interpretation of results difficult.

Moreover, this measure is not as flexible as FaITH, which can be adopted to different contexts as discussed in Section 3.2. The FaITH measure obtained the best value of correlation in both $S_{R\&G}$ and $S_{M\&C}$. Note that in all cases the $F + eIC$ formulation brings better results. The loss L can reach the 20% and 15% with the Lin measure in $S_{R\&G}$ and $S_{M\&C}$ respectively. Moreover, using classical approaches the performance heavily depend on the adopted corpus even if it can be noted that larger corpora bring better results. The *p-values* in both evaluations are $p - value < 0.001$, which indicate that the results are significant. Finally, one note about the couple *car-journey*. The two words, even if generally related since a *car* can be the means to do a *journey*, are not *similar*. This is because similarity, which is a special case of relatedness, only considers the relations of hypernymy/hyponymy (i.e., isa). The FaITH measure assigned a similarity score of 0.007 to this couple while the J&C, Resnik, Lin and P&S assigned 0.346, 0.009, 0.013 and 0.233 respectively. In this case, the FaITH measure since giving the lowest value of similarity seems to better comply with the definition of similarity. In summary, our intuition to exploit a ratio-based representation of Tversky's similarity model and project it into the information theoretic domain is consistent.

4.2   Experiment 2: Evaluating FaITH on Relatedness

In this experiment, FaITH has been evaluated as a semantic relatedness measure by using the eIC formulation. For the evaluation, the WordSim353 dataset, which is a test collection for measuring word relatedness often used in the literature has been adopted. Further detail on the dataset are available in [1]. Even in this case, for each measure, the Pearson correlation coefficient w.r.t. human ratings of similarity has been computed. In this evaluation we compare FaITH with more relatedness measures. In particular, we also considered the Leacock & Chodorow (referred to as *Lch*) [10] and the Wu & Palmer (referred to as *Wup*) [27] measures. We also used a measure of relatedness between two words (referred to as *Ovp*), which assesses the overlap score between two concepts by augmenting glosses with glosses of related concepts [15]. The optimal values for the parameters ζ and η, experimental determined, are 0.4 and 0.6 respectively.

Table 4. Evaluation on relatedness

Measure	ρ	$\rho_S/L\%$	$\rho_B/L\%$	$\rho_{Bnc}/L\%$
Lch	0.36	0.36	0.36	0.36
Wup	0.32	0.32	0.32	0.32
Ovp	0.21	0.21	0.21	0.21
Resnik	0.40	0.36/**11.1**	0.36/**9.9**	0.38/5.4
Lin	0.404	0.37/7.9	0.378/6.4	0.38/5.7
J&C	0.40	0.38/4.0	0.38/2.8	0.39/1.8
P&S	0.41	0.38/5.4	0.38/5.1	0.39/4.7
FaITH	**0.43**	**0.40**/7.0	**0.40**/6.3	**0.40**/5.8

While similarity measures perform extremely well on small similarity datasets such as the M&C and R&G discussed in Section 4.1, their performance drastically decrease when applied to a larger dataset such as WordSim353. The values of correlation reported in Table 4 are related to the word pairs contained in WordNet. Note that for the *Lch*, *Wup* and *Ovp* measures the results are the same as they are not based on IC.

As can be observed, FaITH performs clearly better than the other measures, which substantiate our intuition of adopting the $F + eIC$ strategy. Besides, all the similarity measures perform worse when not using $F + eIC$. The loss (L) in performance is reported in Table 4. In particular, all the IC-based measures take advantage of this formulation, with the Resnik measure improving of about 11%. In the case of not adopting the $F + eIC$, correlation values heavily depend on the considered corpus. Overall, FaITH and the eIC formulation represent a promising technique to compute similarity and relatedness between words and help to augment and improve existing similarity measures.

4.3   Experiment 3: Evaluation on the MeSH Ontology

The MeSH Medical Subject Headings (MeSH) ontology is mainly a hierarchy of medical and biological terms. It consists of a controlled vocabulary and a *Tree*. The controlled vocabulary contains several different types of terms such as *Descriptors*, *Qualifiers*, *Publication Types*, *Geographics* and *Entry* terms. Entry terms are the synonyms or the related terms to descriptors. MeSH descriptors are organized in a tree, which defines the MeSH Concept Hierarchy. In the MeSH tree there are 15 categories each of which is further divided into subcategories. For each subcategory, its descriptors are arranged in a hierarchy from most general to most specific. This evaluation investigates how FaITH performs with domain

Table 5. Correlation with $F + iIC$

Measure	ρ
Resnik	0.72
Lin	0.71
J&C	0.71
Li	0.70
P&S	0.72
FaITH	**0.74**

Table 6. Evaluation on MeSH

Word 1	Word 2	Human	Resnik [19]	Lin [12]	J&C [9]	Li [11]	P&S [17]	FaITH
Antibiotics	Antibacterial Agents	0.93	1.00	1.00	1.00	0.99	1.00	1.00
Measles	Rubeola	0.91	0.92	1.01	1.00	0.99	1.03	1.00
Chicken Pox	Varicella	0.97	1.00	1.00	1.00	0.99	1.00	1.00
Down Syndrome	Trisomy 21	0.87	1.00	1.00	1.00	0.99	1.00	1.00
Seizures	Convulsions	0.84	0.88	1.04	0.90	0.81	1.10	0.99
Pain	Ache	0.87	0.86	1.00	1.00	0.99	1.00	0.95
Malnutrition	Nutritional Deficiency	0.87	0.62	1.00	1.00	0.98	1.00	0.87
Myocardial Ischemia	Myocardial Infarction	0.75	0.59	0.92	0.89	0.80	0.85	0.83
Hepatitis B	Hepatitis C	0.56	0.65	0.82	0.86	0.66	0.70	0.79
Pulmonary Valve Stenosis	Aortic Valve Stenosis	0.53	0.65	0.78	0.81	0.66	0.64	0.76
Psychology	Cognitive Science	0.59	0.68	0.77	0.81	0.80	0.63	0.75
Asthma	Pneumonia	0.37	0.51	0.79	0.87	0.52	0.66	0.75
Diabetic Nephropathy	Diabetes Mellitus	0.50	0.61	0.76	0.79	0.77	0.61	0.74
Hypothyroidism -	Hyperthyroidism	0.41	0.62	0.73	0.75	0.63	0.57	0.72
Sickle Cell Anemia	Iron Deficiency Anemia	0.44	0.60	0.72	0.79	0.36	0.56	0.71
Carcinoma	Neoplasm	0.75	0.25	0.68	0.85	0.45	0.46	0.65
Urinary Tract Infection	Pyelonephritis	0.65	0.47	0.58	0.67	0.42	0.42	0.60
Hyperlipidemia	Hyperkalemia	0.15	0.33	0.48	0.47	0.51	0.32	0.56
Lactose Intolerance	Irritable Bowel Syndrome	0.47	0.47	0.47	0.40	0.30	0.30	0.47
Adenovirus	Rotavirus	0.44	0.27	0.33	0.45	0.35	0.20	0.40
Vaccines	Immunity	0.59	0.00	0.00	0.52	0.00	0.00	0.34
Migraine	Headache	0.72	0.23	0.24	0.37	0.17	0.14	0.80
Bacterial Pneumonia	Malaria	0.15	0.00	0.00	0.20	0.13	0.00	0.22
AIDS	Congenital Heart Defects	0.06	0.00	0.00	0.27	0.10	0.00	0.18
Sarcoidosis	Tuberculosis	0.40	0.00	0.00	0.25	0.07	0.00	0.17
Anemia	Appendicitis	0.03	0.00	0.00	0.19	0.13	0.00	0.13
Meningitis	Tricuspid Atresia	0.03	0.00	0.00	0.19	0.13	0.00	0.13
Failure to Thrive	Malnutrition	0.62	0.00	0.00	0.18	0.13	0.00	0.12
Sinusitis	Mental Retardation	0.03	0.00	0.00	0.36	0.13	0.00	0.11
Hypertension	Kidney Failure	0.50	0.00	0.00	0.21	0.13	0.00	0.11
Breast Feeding	Lactation	0.84	0.00	0.00	0.04	0.08	0.00	0.03
Dementia	Atopic Dermatitis	0.06	0.00	0.00	0.16	0.10	0.00	0.00
Osteoporosis	Patent Ductus Arteriosus	0.15	0.00	0.00	0.03	0.10	0.00	0.00
Amino Acid Sequence -	AntiBacterial Agents	0.15	0.00	0.00	0.15	0.00	0.00	0.00
Otitis Media	Infantile Colic	0.15	0.00	0.00	0.07	0.08	0.00	0.00
Neonatal Jaundice	Sepsis	0.19	0.00	0.00	0.19	0.16	0.00	0.00

related ontologies. Similarly to the first evaluation, a dataset of human similarity judgments has been exploited (refer to [7] for further details). Results obtained by computational methods are compared with those provided by humans in Table 6 whereas, Table 5 reports values of correlations.

The P&S measure, which on WordNet similarity was the closest to FaITH, obtained even in this case a lower value of correlation. Note that the Li measure, which on WordNet obtained a remarkable value of correlation, obtained the lowest correlation on MeSH. We hypothesize that this can be due to two reasons. First, the Li measure depends on two parameters to correctly balance the contribution of the path between c_1 and c_2 to be compared and the depth of their $msca(c_1, c_2)$. Hence, it is possible that parameter values that achieved a good correlation in WordNet do not obtain the same (comparable) performance in MeSH. The second reason is related to the structure of the considered ontology. MeSH is a more domain-specific ontology than WordNet and therefore, in MeSH the combination of path and depth in a non linear function as suggested by the Li measure could not have the same consistent interpretation as in WordNet. The three information content measures obtained better correlation, with Resnik's measure showing a slightly higher level of correlation. This trend is in contrast with the results obtained by the same measure on WordNet where it obtained the lowest correlation both on the $M\&C$ and $R\&G$ datasets. This fact can be justified assuming the in MeSH the $msca(c_1, c_2)$ better expresses the amount of information shared by two terms. Finally, even on this dataset the FaITH measure obtained the highest correlation. In this case the value of correlation is lower than that obtained on WordNet. Results are significant due to the very low value of *p-value* (i.e., $p - value < 0.001$).

5   Concluding Remarks and Future Work

This paper described a new model of similarity combining features [24] and information-content [19]. In particular, by exploiting a ratio-based formulation of the feature model a family of similarity measures as reported in Table 2 has been defined. One of these measures, called FaITH, to quantify how two ontology concepts are similar to each other, has been presented. Another contribution of this paper is the definition of Extended Information Content (eIC) that enables to compute relatedness between concepts by taking into account relations beyond subsumption. The proposed framework enabled to rewrite existing IC-based measures with significant improvement in their performance.

There are at least two interesting strands for future research. One is how to extend the framework to Description Logics (DLs). The main aspect that should be addressed is how to express Extended Information Content values for concepts defined in DLs. Moreover, investigating how similarity depends on the expressiveness of the considered DL is another interesting concern.

The second aspect we want to address is how this strategy, and in particular FaITH, works in more targeted applications such as document clustering, information retrieval and query answering across ontologies.

References

1. Agirre, E., Alfonseca, E., Hall, K., Kravalova, J., Pasca, M., Soroa, A.: A Study on Similarity and Relatedness Using Distributional and WordNet-based Approaches. In: Proc. of NAACL-HLT (2009)
2. Borgida, A., Walsh, T., Hirsh, T.: Towards Measuring Similarity in Description Logics. In: Proc. of Description Logics (2005)
3. Danushka, B., Yutaka, M., Mitsuru, I.: Measuring Semantic Similarity Between Words using Web Search Engines. In: Proc. of WWW 2007, pp. 757–766 (2007)
4. D ' Amato, C.: Similarity-based Learning Methods for the Semantic Web. PhD Thesis, University of Bari (2007)
5. Son, J.Y., Goldstone, R.L.: The Transfer of Scientific Principles using Concrete and Idealized Simulation. The Journal of the Learning Sciences (14), 69–110 (2005)

6. Hirst, G., St-Onge, D.: Lexical Chains as Representations of Context for the Detection and Correction of Malapropisms. In: Fellbaum, C. (ed.) WordNet. An Electronic Lexical Database, ch. 13, pp. 305–332
7. Hliaoutakis, A.: Semantic Similarity Measures in MeSH Ontology and their Application to Information Retrieval on Medline, Technical report, Technical Univ. of Crete, Dept. of Electronic and Computer Engineering (2005)
8. Hliaoutakis, A., Varelas, G., Voutsakis, E., Petrakis, E.G.M., Milios, E.E.: Information Retrieval by Semantic Similarity. Int. J. SWIS 2(3), 55–73 (2006)
9. Jiang, J.J., Conrath, D.W.: Semantic Similarity based on Corpus Statistics and Lexical Taxonomy. In: Proc. of ROCLING X (1997)
10. Leacock, C., Chodorow, M.: Combining Local Context and WordNet Similarity for Word Sense Identification. In: Fellbaum, C. (ed.) WordNet. An Electronic Lexical Database, ch. 11, pp. 265–283
11. Li, Y., Bandar, A., McLean, D.: An Approach for Measuring Semantic Similarity between Words Using Multiple Information Sources. IEEE TKDE 15(4), 871–882
12. Lin, D.: An Information-theoretic Definition of Similarity. In: Proc. of Conf. on Machine Learning, pp. 296–304 (1998)
13. Miller, G.A.: WordNet an on-line Lexical Database. International Journal of Lexicography 3(4), 235–312 (1990)
14. Miller, G.A., Charles, W.G.: Contextual Correlates of Semantic Similarity. Language and Cognitive Processes (6), 1–28 (1991)
15. Banerjee, S., Pedersen, T.: Extended Gloss Overlaps as a Measure of Semantic Relatedness. In: Proc. of IJCAI, pp. 805–810 (2003)
16. Pirró, G., Ruffolo, M., Talia, D.: SECCO: On Building Semantic Links in Peer to Peer Networks. Journal on Data Semantics XII, 1–36 (2009)
17. Pirró, G.: A Semantic Similarity Metric Combining Features and Intrinsic Information Content. Data Knowl. Eng. 68(11), 1289–1308 (2009)
18. Rada, R., Mili, H., Bicknell, M., Blettner, E.: Development and Application of a measure on Semantic Nets. IEEE TSMC (19), 17–30 (1989)
19. Resnik, P.: Information Content to Evaluate Semantic Similarity in a Taxonomy. In: Proc. of IJCAI, pp. 448–453 (1995)
20. Rodriguez, M.A., Egenhofer, M.J.: Determining Semantic Similarity among Entity Classes from Different Ontologies. IEEE TKDE 15(2), 442–456 (2003)
21. Rubenstein, H., Goodenough, J.B.: Contextual Correlates of Synonymy. CACM 8(10), 627–633 (1965)
22. Schickel-Zuber, V., Faltings, B.: OSS: A Semantic Similarity Function based on Hierarchical Ontologies. In: IJCAI, pp. 551–556 (2007)
23. Seco, N., Veale, T., Hayes, J.: An Intrinsic Information Content measure for Semantic Similarity in WordNet. In: Proc. of ECAI 2004, pp. 1089–1090 (2004)
24. Tversky, A.: Features of Similarity. Psychological Review 84(2), 327–352 (1977)
25. Wang, J., Du, Z., Payattakool, R., Yu, P., Chen, C.: A New Method to Measure the Semantic Similarity of GO Terms. Bioinformatics 23(10), 1274–1281 (2007)
26. Watanable, S.: Knowing and Guessing: A Quantitative Study of Inference and Information. Wiley, Chichester (1969)
27. Wu, Z., Palmer, M.: Verb semantics and Lexical Selection. In: Proc. of FQAS ACL 1994, pp. 133–138 (1994)

Combining Approximation and Relaxation in Semantic Web Path Queries

Alexandra Poulovassilis and Peter T. Wood

London Knowledge Lab, Birkbeck, University of London, UK
{ap,ptw}@dcs.bbk.ac.uk

Abstract. We develop query relaxation techniques for regular path queries and combine them with query approximation in order to support flexible querying of RDF data when the user lacks knowledge of its full structure or where the structure is irregular. In such circumstances, it is helpful if the querying system can perform both approximate matching and relaxation of the user's query and can rank the answers according to how closely they match the original query. Our framework incorporates both standard notions of approximation based on edit distance and RDFS-based inference rules. The query language we adopt comprises conjunctions of regular path queries, thus including extensions proposed for SPARQL to allow for querying paths using regular expressions. We provide an incremental query evaluation algorithm which runs in polynomial time and returns answers to the user in ranked order.

1 Introduction

The volume of semistructured data available to users on the web continues to grow, increasingly in the form of RDF linked data. Given the complexity and heterogeneity of such data, users may not be aware of its full structure and need to be assisted by querying systems which do not require that users' queries necessarily match exactly the data structures being queried.

In this paper we consider general semistructured data modelled as a graph, with RDF linked data being a particular application of this model. We are interested in developing efficient algorithms which allow for both *approximate matching* and *relaxation* of users' queries on such data, with the answers to queries being returned to users in ranked order. We restrict our query language to that of *conjunctive regular path queries* [2]. A conjunctive regular path (CRP) query Q consisting of n conjuncts is of the form

$$(Z_1, \ldots, Z_m) \leftarrow (X_1, R_1, Y_1), \ldots, (X_n, R_n, Y_n)$$

where each X_i and Y_i, $1 \leq i \leq n$, is a variable or constant, each Z_i, $1 \leq i \leq m$, is a variable appearing in the body of Q, and each R_i, $1 \leq i \leq n$, is a regular expression over the alphabet from which edge labels in the graph are drawn.

The answer to a CRP query Q on a graph G, $Q(G)$, is defined as follows. For each conjunct (X_i, R_i, Y_i), $1 \leq i \leq n$, let r_i be a binary relation over the scheme

P.F. Patel-Schneider et al. (Eds.): ISWC 2010, Part I, LNCS 6496, pp. 631–646, 2010.

(X_i, Y_i). Let $t[X_i]$ and $t[Y_i]$ denote the first and second components, respectively, of any tuple $t \in r_i$. There is a tuple $t \in r_i$ if and only if there exists a path from node $t[X_i]$ to node $t[Y_i]$ in G such that $t[X_i] = X_i$ if X_i is a constant, $t[Y_i] = Y_i$ if Y_i is a constant, and the concatenation of the edge labels in the path satisfies the regular expression R_i. Then $Q(G) = \pi_{Z_1,...,Z_m}(r_1 \bowtie \cdots \bowtie r_n)$.

Using regular expressions to query data has been much studied, e.g. [2,17], as have approximate query matching techniques, e.g. [4,7,14,15,18]. In [13], we studied a combination of these and showed that approximate matching of CRP queries can be undertaken in polynomial time. The edit operations we allowed in approximate matching of queries were insertions, deletions, substitutions, transpositions and inversions of edge labels (corresponding to reverse traversal of edges) — each with an assumed edit cost of 1. Here, for simplicity of exposition, we exclude inversions and transpositions — we note though that the techniques we develop here extend straightforwardly to this more general case, and the query complexity results still hold.

Example 1. The L4All system allows users to create and maintain a chronological record of their learning, work and personal episodes — their "timelines" — with the aim of supporting lifelong learners in exploring learning opportunities and in planning and reflecting on their learning [3]. Figure 1 illustrates a fragment of data and metadata relating to a user's timeline (where sc denotes subclassOf). The episodes within a timeline have a start and an end date associated with them (for simplicity these are not shown). Episodes are ordered by their start date — as indicated by edges labelled next. There are several types of episode, e.g. University and Work. Associated with each type of episode are several properties — we show just two of these, qualif[ication] and job.

Suppose that Mary is studying for a BA in English and wishes to find out what possible future career choices there are for her. Timelines may have edges labelled prereq between episodes, indicating that the timeline's owner believes that undertaking an earlier episode was necessary in order for them to be able to proceed to or achieve a later episode. So Mary might pose this query, Q_1[1]:

```
(?E2,?P)<-(?E1,type,University),(?E1,qualif.type,EnglishStudies),
        (?E1,prereq+,?E2), (?E2,type,Work), (?E2,job.type,?P)
```

However, this will return no results relating to the timeline of Figure 1, even though it is evident that this contains information that would be relevant to Mary. This is because, in practice, users may or may not create prereq metadata relating to their timelines. If Mary chooses to allow replacement of the edge label prereq in her query by the label next, she can submit a variant of Q_1:

```
(?E2,?P)<-(?E1,type,University), (?E1,qualif.type,EnglishStudies),
        APPROX(?E1,prereq+,?E2),(?E2,type,Work),(?E2,job.type,?P)
```

The regular expression prereq+ can be approximated by the regular expression next.prereq* at edit distance 1 from prereq+. This allows the system to return

---
[1] In our assumed concrete syntax, variable names are preceded with '?'.

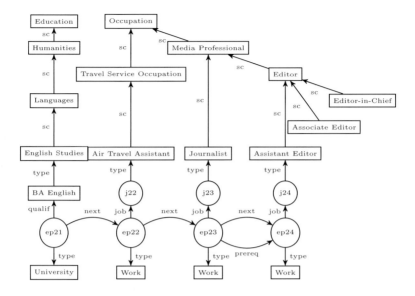

Fig. 1. A fragment of timeline data and metadata

the answer (ep22,AirTravelAssistant) at an edit distance 1 from Q_1. Mary may judge this not to be relevant and may seek further results, at a further level of approximation. The regular expression next.prereq* can be approximated by next.next.prereq*, now at edit distance 2 from Q_1, allowing the following answers (ep23,Journalist), (ep24,AssistantEditor) to be returned. Mary may judge both of these as being relevant, and she can then request the system to return the whole of this user's timeline for her to explore further.

Suppose now Mary knows she wants to become an Assistant Editor and would like to find out how she might achieve this, given that she's done an English degree. Mary might pose this query, Q_2:

```
(?E2,?P)<-(?E1,type,University), (?E1,qualif.type,EnglishStudies),
        APPROX (?E1,prereq+,?E2), (?E2,job.type,?P)
        APPROX (?E2,prereq+,?Goal), (?Goal,type,Work),
        (?Goal,job.type,AssistantEditor)
```

At distance 0 and 1 there are no results from the timeline of Figure 1. At distance 2, the answers (ep22,AirTravelAssistant), (ep23,Journalist) are returned, the second of which gives Mary potentially useful information.

Suppose Mary wants to know what other jobs, similar to an Assistant Editor, might be open to her. There are many categories of jobs classified under Media Professional but none of these will be matched by query Q_2 above. What she would like to pose instead (borrowing the 'RELAX' syntax of [12]) is query Q_3:

```
(?E2,?P)<-(?E1,type,University),(?E1,qualif.type,EnglishStudies),
        APPROX (?E1,prereq+,?E2),(?E2,job.type,?P)
        APPROX (?E2,prereq+,?Goal), (?Goal,type,Work),
```

```
RELAX (?Goal,job.type,AssistantEditor)
```

which would relax `Assistant Editor` to its parent concept `Editor`, matching jobs such as `Assistant Editor`, `Associate Editor`, `Editor-in-Chief` etc., as well as in parallel approximating the two instances of `prereq+`. Query results would be returned in increasing overall distance (relaxation and approximation) from the original query.

As a further extension, suppose another user, Joe, wants to know what jobs similar to being an Assistant Editor might be open to someone who has studied English or a similar subject at university. Subject disciplines are classified, e.g. `English Studies` under `Languages` which in turn is classified under `Humanities`. So Joe may pose query Q_4 which is identical to Q_3 above but with `RELAX` in front of (`?E1,qualif.type,EnglishStudies`).   □

In Section 2, we first consider computing approximate and relaxed answers for regular path queries consisting of a single conjunct. We show that in both cases answers can be computed in polynomial time in the size of the query and the input graph, and returned to the user in ranked order. Section 3 generalises to the case of multi-conjunct queries and shows that computation can still be achieved in polynomial time as long as the queries are acyclic and have a fixed number of head variables. In a multi-conjunct query, approximation and relaxation are combined by allowing each conjunct to be qualified by either an `APPROX` or a `RELAX` operator, as shown in the above example. Section 4 discusses related work. Section 5 presents our conclusions and future work.

2   Single-Conjunct Regular Path Queries

In this paper we consider a semistructured data model comprising a directed graph $G = (V, E)$ and an ontology $K = (V_K, E_K)$. V contains nodes representing entity instances or entity classes. E represents relationships between the members of V. Each node in V is labelled with a distinct constant. Each edge in E is labelled with a symbol drawn from a finite alphabet $\Sigma \cup \{\texttt{type}\}$. V_K contains nodes representing entity classes or properties. Each node in V_K is labelled with a distinct constant. We call a node in V_K representing an entity class a 'class node' and a node representing a property a 'property node'. So $V \cap V_K$ contains the set of class nodes of V. Each edge in E_K is labelled with a symbol drawn from $\{\texttt{sc}, \texttt{sp}, \texttt{dom}, \texttt{range}\}$. We assume that $\Sigma \cap \{\texttt{type}, \texttt{sc}, \texttt{sp}, \texttt{dom}, \texttt{range}\} = \emptyset$. We also assume that the set of labels of edges in E, except for the label `type`, is contained in the set of labels of property nodes in V_K. We observe that this general graph model encompasses RDF data, except that it does not allow for the representation of RDF's 'blank' nodes (but these are discouraged for linked data [10]). It also comprises a fragment of the RDFS vocabulary: `rdf:type`, `rdfs:subClassOf`, `rdfs:subPropertyOf`, `rdfs:domain`, `rdfs:range`, which we abbreviate by `type`, `sc`, `sp`, `dom`, `range`.

A *single-conjunct regular path query* Q over a graph G is of the form:

$$vars \leftarrow (X, R, Y) \tag{1}$$

where X and Y are constants or variables, R is a regular expression over $\Sigma \cup \{\texttt{type}\}$, and $vars$ is the subset of $\{X, Y\}$ that are variables.

A *regular expression* R over $\Sigma \cup \{\texttt{type}\}$ is defined as follows:

$$R := \epsilon \mid a \mid \texttt{type} \mid \_ \mid (R1 \cdot R2) \mid (R1|R2) \mid R^* \mid R^+$$

where ϵ is the empty string, a is any symbol in Σ, "$\_$" denotes the disjunction of all constants in $\Sigma \cup \{\texttt{type}\}$, and the operators have their usual meaning.

A *path* p in $G = (V, E)$ from $x \in V$ to $y \in V$ is a sequence of the form $(v_1, l_1, v_2, l_2, v_3, \ldots, v_n, l_n, v_{n+1})$, where $n \geq 0$, $v_1 = x$, $v_{n+1} = y$ and for each v_i, l_i, v_{i+1}, $v_i \xrightarrow{l_i} v_{i+1} \in E$. A path p *conforms* to a regular expression R if $l_1 \cdots l_n \in L(R)$, the language denoted by R.

Given a single-conjunct regular path query Q and graph G, let θ be a matching from variables and constants of Q to nodes of G, that maps each constant to itself. A tuple $\theta(vars)$ *satisfies* Q on G if there is a path from $\theta(X)$ to $\theta(Y)$ which conforms to R. The *answer* of Q on G is the set of tuples which satisfy Q on G. The answer can be found in polynomial time in the size of Q and G (from Lemma 1 in [17]).

Below we first briefly review approximate matching of single-conjunct regular path queries, from [13]. We then discuss relaxation of such queries based on information from the ontology K. Section 3 discusses combined approximation and relaxation for multi-conjunct queries.

2.1 Approximate Matching of Single-Conjunct Queries

The *edit distance* from a path p to a path p' is the minimum cost of any sequence of edit operations which transforms the sequence of edge labels of p to the sequence of edge labels of p' (note that edge labels are treated as atomic values and it is sequences of such labels that are transformed using edit operations). The edit operations that we consider here are insertions, deletions and substitutions of edge labels, each with an assumed edit cost of α, for some α.

The *edit distance* of a path p to a regular expression R is the minimum edit distance from p to any path that conforms to R. Given a matching θ from variables and constants of a query Q to nodes in a graph G, where constants must be matched to themselves, we say that the tuple $\theta(vars)$ has *edit distance* $edist(\theta, Q)$ to Q, and we define this to be the minimum edit distance to R of any path p from $\theta(X)$ to $\theta(Y)$ in G. Note that if p conforms to R, then $\theta(vars)$ has edit distance zero to Q.

The *approximate answer* of Q on G is a list of pairs $(\theta(vars), edist(\theta, Q))$, ranked in order of non-decreasing edit distance. The *approximate top-k answer* of Q on G comprises the first k tuples in the approximate answer of Q on G.

We now describe how the approximate answer can be computed in time polynomial in the size of R and G. The process is similar to that described in [13], but differs in a number of respects which are described below:

(i) We construct a *weighted* NFA M_R of size $O(R)$ to recognise $L(R)$, using Thompson's construction (which makes use of ϵ-transitions). M_R has set of states S, alphabet $\Sigma' = \Sigma \cup \{\text{type}\}$, transition relation δ, start state s_0, and final state s_f. Each transition is labelled with a label from Σ' and a weight, or cost, which is zero in M_R. If X (or, respectively, Y) in the query is a constant n, we annotate s_0 (s_f) with n; otherwise we annotate s_0 (s_f) with a wildcard symbol $*$ that matches any constant.

(ii) We now construct the *approximate automaton* A_R corresponding to M_R. A_R has the same set of states as M_R, with the following additional transitions:

- For each state $s \in S$ and label $a \in \Sigma$, there is a transition (s, a, α, s), where α is the cost of insertion.
- For each transition $(s, a, 0, t)$ in M_R where $a \in \Sigma$, there is a transition (s, ϵ, α, t), where α is the cost of deletion.
- For each transition $(s, a, 0, t)$ in M_R, where $a \in \Sigma$, and label $b \in \Sigma$ ($b \neq a$), there is a transition (s, b, α, t), where α is the cost of substitution.

Thus A_R has $O(|R| \cdot |\Sigma'|)$ transitions.

(iii) We form the weighted *product automaton*, H, of A_R with the graph $G = (V, E)$, viewing each node in V as both an initial and a final state. The states of H are of the form (s, n), $s \in S$ and $n \in V$.

(iv) To evaluate query Q, if X is a node v of G, we perform a shortest path traversal of H starting from the vertex (s_0, v). Whenever we reach a vertex (s_f, m) in H we output m, provided m matches the annotation on s_f. The distance of (v, m) to Q is given by the total cost of the shortest path from (s_0, v) to (s_f, m). If X is a variable, we perform such a traversal of H starting from vertex (s_0, v) for every node v of G.

This construction differs from that of [13] where the NFA for approximate matching of regular expression R was constructed using a number of copies of the NFA for recognising R, each corresponding to matching at a difference distance. Hence, in that NFA, distance was represented implicitly by the "copy number" of states, rather than explicitly using a weight as above. The use of annotations on states also does not appear in [13].

Proposition 1. *Let $G = (V, E)$ be a graph and Q be a single-conjunct query using regular expression R over alphabet Σ. The approximate answer of Q on G can be found in time $O(|R|^2 |V|(|\Sigma'||E| + |V| \log(|R||V|)))$.*

The proof follows by using Dijkstra's algorithm on the product automaton H, which can be shown to have $O(|R||V|)$ nodes and $O(|R||\Sigma'||E|)$ edges.

The above query evaluation can also be accomplished "on-demand" by incrementally constructing the edges of H as required, thus avoiding precomputation and materialisation of the entire graph H. This is performed by calling a function Succ with a node (s, n) of H. The function returns a set of transitions $\overset{a,d}{\to} (p, m)$, such that there is an edge in H from (s, n) to (p, m) with label a and cost d. We show Succ below, where the function $\text{nextStates}(A_R, s, a)$ returns the set of states in A_R that can be reached from state s on reading input a, along with the cost of reaching each. Note that we need either to remove ϵ-transitions from A_R (using a standard algorithm that potentially squares the

---

Procedure. Succ(s, n)

Input: state s of A_R and node n of G
Output: set of transitions which are successors of (s, n) in H
$W \leftarrow \emptyset$
for $(n, a, m) \in G$ *and* $(p, d) \in$ nextStates(A_R, s, a) **do**
$\quad \lfloor$ add $\xrightarrow{a,d} (p, m)$ to W
return W

---

size of A_R) or nextStates needs to repeatedly follow ϵ-transitions until it finds a non-ϵ-transition, while summing costs of transitions.

A set visited$_R$ is maintained, storing tuples of the form (v, n, s) representing the fact that node n of G was visited in state s having started the traversal from node v. Also maintained is a priority queue queue$_R$ containing quadruples of the form (v, n, s, d), ordered by increasing values of d, where d is the distance associated with visiting node n in state s having started from node v. We begin by enqueueing the initial quadruple $(v, v, s_0, 0)$, if X is some node v, or enqueueing a set of initial quadruples otherwise, one for each node v of G. We maintain a list answers$_R$ containing tuples of the form (v, n, d) where d is the smallest distance of this answer tuple to Q and ordered by non-decreasing value of d. This list is used to avoid returning again (v, n, d') for any $d' \geq d$.

We then call a procedure getNext to return the next query answer, in order of non-decreasing distance from Q. getNext repeatedly dequeues the first quadruple of queue$_R$, (v, n, s, d), adding (v, n, s) to visited$_R$, until queue$_R$ is empty. After dequeueing the quadruple (v, n, s, d), we enqueue $(v, m, s', d + d')$ for each transition $\xrightarrow{e, d'} (s', m)$ returned by Succ(s, n) such that $(v, m, s') \notin$ visited$_R$. If s is a final state, its annotation matches n, and the answer (v, n, d') has not been been generated before for some d', then the triple (v, n, d) is returned.

2.2 Ontology Relaxation of Single-Conjunct Regular Path Queries

In [12], we considered relaxation of conjunctive queries over RDF data, and the formalisation of relaxation using RDFS entailment with respect to an RDFS ontology K. We assumed that the predicates of triples in K are in the set {type, dom, range, sp, sc} and we adopted an operational semantics for the notion of RDFS *entailment*, denoted by $\models$ and characterised by the six rules shown in Fig. 2 (see [8,9] for details).

We assumed infinite sets I (IRIs) and L (RDF literals). The elements in $I \cup L$ are called RDF *terms*. A triple $(v_1, v_2, v_3) \in I \times I \times (I \cup L)$ is called an *RDF triple*. In such a triple, v_1 is called the *subject*, v_2 the *predicate* and v_3 the *object*. An *RDF graph* is a set of RDF triples.

For RDF graphs G_1 and G_2, we stated that $G_1 \models_{\text{rule}} G_2$ if G_2 can be derived from G_1 by iteratively applying the rules of Fig. 2. We used the notion of the *closure* of an RDF graph G [9], denoted cl(G), which is the closure of G under the rules. By a result from [9], RDFS entailment (for the fragment of RDFS we consider) can be characterized as follows: $G_1 \models_{\text{RDFS}} G_2$ if and only if $G_2 \subseteq$ cl(G_1).

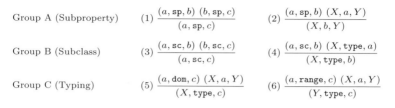

Fig. 2. RDFS Inference Rules

Given a set of variables V disjoint from the sets I and L, a *triple pattern* is a triple $(v_1, v_2, v_3) \in (I \cup V) \times (I \cup V) \times (I \cup V \cup L)$. A *graph pattern* P is a set of triple patterns. We denote the variables mentioned in P by $\texttt{var}(P)$.

A conjunctive query as considered in [12] is a rule whose body is a graph pattern. We investigated two broad classes of relaxations for such queries in that paper: *ontology relaxation* and *simple relaxation*. Ontology relaxation encompasses relaxations that are entailed using information from the ontology and are captured by the rules of Fig. 2; we note that when applying these rules to triple patterns, rather than (ground) triples, a, b and c must be instantiated to RDF terms, while X and Y can be instantiated to either RDF terms or variables. Simple relaxation consists of relaxations that can be entailed without an ontology, e.g. dropping triple patterns, replacing constants with variables, and breaking join dependencies.

In this paper, we extend the application of ontology relaxation from graph patterns to regular path queries, leaving consideration of simple relaxation to future work. Before proceeding further we introduce some assumptions and terminology.

We consider the cost of applying rule 2 or 4 to be β, and the cost of applying rule 5 or 6 to be γ. (Because queries and data graphs cannot contain sc and sp, rules 1 and 3 are inapplicable as far as relaxation is concerned.) We assume that the subgraphs of K induced by edges labelled sc and sp are acyclic; this ensures that the transitive reduction (see below) of each of these subgraphs is unique. We also assume that all the edges labelled with symbols from $\Sigma \cup \{\texttt{type}\}$ that are entailed by $G \cup K$ are included in G.

For each edge $(a, \texttt{type}, c)$ in G, we also add to G the "reverse" edge $(c, \texttt{type}^-, a)$. We do this because, while we do not consider reverse traversal of graph edges in general in this paper (leaving this as an area of further work), we do allow the reverse traversal of type edges, which we accommodate by generating reverse edges in G labelled $\texttt{type}^-$. We need these edges in order to accomodate Rule 6 of Fig. 2 without changing the position of the variable Y in the relaxed triple. This is because (as we will see below) in our context of relaxing regular path queries, the relaxed triples are generally part of a sequence of relaxed triples. Thus, we use the equivalent form of $(c, \texttt{type}^-, Y)$ for the relaxed triple inferred by Rule 6.

Finally, we assume that $K = \texttt{extRed}(K)$, where $\texttt{extRed}(K)$ is the *extended reduction* of K. Given ontology K, $\texttt{extRed}(K)$ can be computed as follows: (i) compute $\text{cl}(K)$; (ii) apply the rules of Fig. 3 in reverse until no longer applicable;

and (iii) apply rules 1 and 3 of Fig. 2 in reverse until no longer applicable. (Applying a rule in reverse means deleting the triple deduced by the rule.) Using this extended reduction allows us to perform what were termed *direct* relaxations in [12] which correspond to the "smallest' relaxation steps. This is necessary if we are to return query answers to users incrementally in order of increasing cost, which we discuss in more detail shortly.

$$(e1)\ \frac{(b, \mathrm{dom}, c)\ (a, \mathrm{sp}, b)}{(a, \mathrm{dom}, c)} \qquad (e2)\ \frac{(b, \mathrm{range}, c)\ (a, \mathrm{sp}, b)}{(a, \mathrm{range}, c)}$$

$$(e3)\ \frac{(a, \mathrm{dom}, b)\ (b, \mathrm{sc}, c)}{(a, \mathrm{dom}, c)} \qquad (e4)\ \frac{(a, \mathrm{range}, b)\ (b, \mathrm{sc}, c)}{(a, \mathrm{range}, c)}$$

Fig. 3. Additional rules used to compute the extended reduction of an RDFS ontology

Let t_1 and t_2 be triple patterns such that $t_1, t_2 \notin \mathrm{cl}(G \cup K)$, and $\mathrm{var}(t_2) = \mathrm{var}(t_1)$. We say that t_1 *relaxes to* t_2 (or t_2 is a *relaxation* of t_1), denoted $t_1 \leq t_2$[2], if $(\{t_1\} \cup G \cup K) \models_{\mathrm{rule}} t_2$. Let P_1 and P_2 be graph patterns such that for all $t_1 \in P_1$ and $t_2 \in P_2$, $t_1, t_2 \notin \mathrm{cl}(G \cup K)$ and $\mathrm{var}(P_2) = \mathrm{var}(P_1)$. We say that P_1 *relaxes to* P_2 (or P_2 is a *relaxation* of P_1), denoted $P_1 \leq P_2$, if for all $t_1 \in P_1$ there is a $t_2 \in P_2$ such that $t_1 \leq t_2$ and for all $t_2 \in P_2$ there is a $t_1 \in P_1$ such that $t_1 \leq t_2$. We note that the relaxation relation is reflexive and transitive.

Example 2. If we did not use the extended reduction of an ontology K, we could have the triples (a, dom, c), (a, dom, c') and (c, sc, c') in K. Given a conjunct (X, a, w), we could apply rule 5 in order to relax (X, a, w) to (X, type, c) with cost γ and to (X, type, c'), also with cost γ. However, the cost of relaxing (X, a, w) to (X, type, c') should really be $\gamma + \beta$, reflecting the cost of using rule 5 to relax (X, a, w) to (X, type, c) followed by the cost of using rule 4 to relax (X, type, c) to (X, type, c'). The extended reduction of K does not contain the triple (a, dom, c') because of applying rule e3 in reverse; hence, although the rules of Fig. 3 are not sound for RDFS entailment, using $\mathrm{extRed}(K)$ allows us finer control over computing the cost of various relaxations.   □

Given a query Q with a single conjunct (X, R, Y), let $q = l_1 l_2 \cdots l_n$ be a string in $L(R)$. We define a *triple form* of (Q, q) as a set of triple patterns

$$\{(X, l_1, W_1), (W_1, l_2, W_2), \ldots, (W_{n-1}, l_n, Y)\}$$

where $W_1, \ldots, W_{n-1}$ are variables not appearing in Q. Thus, a triple form of (Q, q) is a graph pattern which can be relaxed to another graph pattern.

Example 3. Let query Q contain the single conjunct $(X, R, 4)$, where X is a variable, 4 is a constant, and $R = (a \cdot b \cdot d)$. Assume that K contains the triples (d, sp, e), (e, dom, c) and (c, sc, c'). There is only a single $q \in L(R)$, namely $q = abd$. Consider the following triple form T of (Q, q)

$$\{(X, a, W_1), (W_1, b, W_2), (W_2, d, 4)\}$$

---

[2] For notational simplicity we assume that the parameters G and K are implicit.

and let P be the graph pattern

$$\{(X, a, W_1), (W_1, b, W_2), (W_2, \texttt{type}, c')\}$$

Then T relaxes to P since $(W_2, d, 4) \leq (W_2, \texttt{type}, c')$ by applying rules 2, 5 and 4. We also have that $(W_2, d, 4) \leq (W_2, e, 4)$ (by rule 2), $(W_2, e, 4) \leq (W_2, \texttt{type}, c)$ (by rule 5) and $(W_2, \texttt{type}, c) \leq (W_2, \texttt{type}, c')$ (by rule 4).

Note that, because of our requirement that variables be preserved when performing relaxation, rules 4, 5 and 6 can only be applied to the first or last triple pattern of a triple form of a string. So if, for example, $(b, \texttt{dom}, f) \in K$, the triple pattern (W_1, b, W_2) cannot be relaxed to $(W_1, \texttt{type}, f)$ by rule 5. □

We now define the *relaxed semantics* of such queries as follows. Let p be the path $(v_1, l_1, v_2, l_2, v_3, \ldots, v_n, l_n, v_{n+1})$, $n \geq 1$, in G. We define a *triple form* of p as a set of triple patterns

$$\{(v_1, l_1, W_1), (W_1, l_2, W_2), \ldots, (W_{n-1}, l_n, v_{n+1})\}$$

where $W_1, \ldots, W_{n-1}$ are variables. If p is of length zero, then p is of the form (v, ϵ, v) and the only triple form of p is also (v, ϵ, v).

Given a query Q of the form (1) and a graph G, let θ be a matching from variables and constants of Q to nodes of G such that θ maps each constant to itself. We denote $(\theta(X), R, \theta(Y))$ by $\theta(Q)$. Path p in G *r-conforms* to $\theta(Q)$ if there is a $q \in L(R)$, a triple form T_q of $(\theta(Q), q)$ and a triple form T_p of p such that $T_q \leq T_p$. A tuple $\theta(vars)$ *r-satisfies* Q on G if there is a path in G that r-conforms to $\theta(Q)$.

Note that a path in G can r-conform to a query on the basis of a triple pattern t relaxing to a triple pattern t' such that the constants in t and t' differ (due to applications of rules 5 and 6, provided Y is a constant). Hence relaxation of a conjunct induces a mapping on constants which may not be the identity.

We now consider the cost of applying relaxations in order to be able to return answers ordered by increasing cost. For this we need the notion of direct relaxation. In [12] we defined the *direct relaxation relation*, denoted by $\prec$, as the reflexive, transitive reduction of $\leq$. The *direct relaxations* of a triple pattern t (i.e., triple patterns t' such that $t \prec t'$) are the result of the smallest steps of relaxation. We write $t, o \vdash t'$ if t' can be derived from t and $o \in \text{cl}(G \cup K)$ by the application of a single rule from Fig. 2. We also write $t, o \vdash_i t'$ if rule i was the rule used in the derivation.

It is shown in [12] that a single application of each of the rules in Fig. 2 to a triple pattern t and a triple $o \in \texttt{extRed}(K)$ (where applicable) yields precisely the direct relaxations of t with respect to K. Given graph patterns P_1 and P_2, we say that P_1 directly relaxes to P_2, denoted $P_1 \prec P_2$, if $P_1 = \{t_1\} \cup P$ and $P_2 = \{t_2\} \cup P$, for some (possibly empty) graph pattern P, and $t_1 \prec t_2$; in other words, $t_1, o \vdash_i t_2$ for some triple $o \in \texttt{extRed}(K)$ and rule i. The *cost* of the direct relaxation is the cost of applying rule i. The cost of a sequence of direct relaxations is the sum of the costs of each relaxation in the sequence.

Given ontology $K = \texttt{extRed}(K)$, path p in G, matching θ, query Q as in (1), string $q \in L(R)$, triple form T_q for $(\theta(Q), q)$, triple form T_p for p such that

$T_q \leq T_p$ (so p r-conforms to $\theta(Q)$), the *relaxation distance* from p to $(\theta(Q), q)$ is the minimum cost of any sequence of direct relaxations which yields T_p from T_q. The cost of the empty sequence of direct relaxations (so that T_q is already a triple form of p) is zero. The *relaxation distance* from p to $\theta(Q)$ is the minimum relaxation distance from p to $(\theta(Q), q)$ for any string $q \in L(R)$.

Given graph G, query Q and matching θ, the *relaxation distance* of $\theta(Q)$, denoted $rdist(\theta, Q)$, is the minimum relaxation distance to $\theta(Q)$ from any path p that r-conforms to $\theta(Q)$. The *relaxed answer* of Q on G is a list of pairs $(\theta(vars), rdist(\theta, Q))$, where $\theta(vars)$ is an r-satisfying tuple, ranked in order of non-decreasing relaxation distance. The *relaxed top-k answer* of Q on G comprises the first k tuples in the relaxed answer of Q on G.

Example 4. Consider the conjunct $Q = $ (`?Goal,job.type,AssistantEditor`) from query Q_3 in Example 1. Suppose the graph G contains the triples (`ep24,job,j24`), (`j24,type,AssistantEditor`) shown in Fig. 1, and also the triples (`ep33,job,j33`), (`j33,type,AssociateEditor`) from another timeline. Path (`ep24,job,j24,type,AssistantEditor`) r-conforms to $\theta(Q)$ when θ(`?Goal`) = `ep24` with relaxation distance 0. Path (`ep33,job,j33,type,AssociateEditor`) r-conforms to $\theta(Q)$ when θ(`?Goal`) = `ep33` with relaxation distance β. So tuples (`ep24`) and (`ep33`) both r-satisfy Q on G.   □

2.3   Computing the Relaxed Answer

We now describe how the relaxed answer can be computed, starting from the weighted NFA M_R that recognises $L(R)$ which was described in Section 2.1.

In computing a relaxed answer, it is useful to be able to make (possibly partial) copies of states in an automaton. Given an automaton M with a set of states S and a state $s \in S$, a *clone* of s in M is a new state s' which is added to S such that s' is an initial or final state if s is, and s' has the same sets of incoming and outgoing transitions as s. An *incoming (outgoing) clone* of s is a new state s' such that s' is an initial or final state if s is, s' has the same set of incoming (outgoing) transitions as s, and has no outgoing (incoming) transitions.

Given a weighted automaton $M = (S, \Sigma', \delta, s_0, s_f)$ and ontology K such that $K = \texttt{extRed}(K)$, we construct as described below the *relaxed automaton* $M^K = (S', \Sigma', \tau, S_0, S_f)$ of M with respect to K. The set of states S' includes S as well as any new states defined below. S_0 and S_f are sets of initial and final states, respectively, with S_0 including s_0, S_f including s_f and both possibly including additional cloned states defined below. Each state in S_0 and S_f is annotated either with a constant or with the wildcard symbol $*$. The transition relation τ includes δ as well as any transitions added to τ by the process defined below. The process continues until no further changes to τ and S' occur.

- (rule 2) For each transition $(s, a, d, t) \in \tau$ and $(a, \texttt{sp}, b) \in K$, add the transition $(s, b, d + \beta, t)$ to τ.
- (rule 4 (i)) For each transition $(s, \texttt{type}, d, t) \in \tau$, $t \in S_f$ and $(c, \texttt{sc}, c') \in K$ such that t is annotated with c, (i) add an outgoing clone t' of t annotated with c' to S', and (ii) add the transition $(s, \texttt{type}, d + \beta, t')$ to τ.

- (rule 4 (ii)) For each transition $(s, \texttt{type}^-, d, t) \in \tau$, $s \in S_0$ and $(c, \texttt{sc}, c') \in K$ such that s is annotated with c, (i) add an incoming clone s' of s annotated with c' to S', and (ii) add the transition $(s', \texttt{type}^-, d + \beta, t)$ to τ.
- (rule 5) For each $(s, a, d, t) \in \tau$, $t \in S_f$ and $(a, \texttt{dom}, c) \in K$ such that t is annotated with a constant, (i) add an outgoing clone t' of t annotated with c to S', and (ii) add the transition $(s, \texttt{type}, d + \gamma, t')$ to τ.
- (rule 6) For each $(s, a, d, t) \in \tau$, $s \in S_0$ and $(a, \texttt{range}, c) \in K$ such that s is annotated with a constant, (i) add an incoming clone s' of s annotated with c to S', and (ii) add the transition $(s', \texttt{type}^-, d + \gamma, t)$ to τ.

Given a regular expression R and ontology $K = \texttt{extRed}(K)$, we denote by M_R^K the automaton obtained by first constructing the automaton M_R for R and then constructing the relaxed automaton of M_R with respect to K.

Example 5. Consider again conjunct $(X, R, 4)$, where $R = (a \cdot b \cdot d)$, and ontology $K = \{(d, \texttt{sp}, e), (e, \texttt{dom}, c), (c, \texttt{sc}, c')\}$ from Example 3. The relaxed automaton M_R^K initially comprises the states $\{s_0, s_1, s_2, s_f\}$ and the transitions labelled with cost zero between them, as shown in Fig. 4. Applying the transformation for rule 2 to the transition labelled $d, 0$ and the triple $(d, \texttt{sp}, e) \in K$, adds the transition labelled e, β from s_2 to s_f. Applying rule 5 to this transition and the triple $(e, \texttt{dom}, c) \in K$, adds the outgoing clone s'_f of s_f, annotated with c, as well as the transition labelled $\texttt{type}, \beta + \gamma$ from s_2 to s'_f. Applying rule 4(i) to this transition and the triple $(c, \texttt{sc}, c') \in K$, adds the outgoing clone s''_f of s'_f, annotated with c', as well as the transition labelled $\texttt{type}, 2\beta + \gamma$ from s_2 to s''_f.

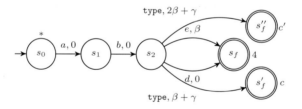

Fig. 4. Relaxed automaton M_R^K for conjunct $(X, (a \cdot b \cdot d), 4)$

Given a graph G, automaton M_R^K will match (i) paths labelled $a \cdot b \cdot d$ from any node to node 4 with distance 0, (ii) paths labelled $a \cdot b \cdot e$ from any node to node 4 with distance β, (iii) paths labelled $a \cdot b \cdot \texttt{type}$ from any node to node c with distance $\beta + \gamma$, and (iv) paths labelled $a \cdot b \cdot \texttt{type}$ from any node to node c' with distance $2\beta + \gamma$. □

Proposition 2. *Let Q be a query comprising a single conjunct (X, R, Y). Let $M_R^K = (S', \Sigma', \tau, S_0, S_f)$ be the relaxed automaton for regular expression R and ontology $K = \texttt{extRed}(K)$, where the ϵ-transitions have been removed from M_R^K. Let G be a graph and H be the product automaton of M_R^K and G. Let θ be a matching from Q to G such that $\theta(X) = v_0$ and $\theta(Y) = v_n$. (i) There is a path*

$r = (v_0, l_1, \ldots, l_n, v_n)$ in G that r-conforms to $\theta(Q)$ if and only if there is a path $p = ((s_0, v_0), (l_1, c_1), \ldots, (l_n, c_n), (s_n, v_n))$ in H, where $s_0 \in S_0$ and $s_n \in S_f$. (ii) Consider all paths of the form of p in (i). The relaxation distance from r to $(\theta(Q), q)$, where $q = l_1 \cdots l_n$, is given by the minimum value of $c_1 + \cdots + c_n$.

The proof of (i) follows from the fact that the rules used to add transitions to M_R^K correspond to direct relaxations applied to triples. The proof of (ii) follows from the definition of relaxation distance.

Proposition 3. *Given a query Q comprising a single conjunct (X, R, Y) and ontology $K = \texttt{extRed}(K)$, the relaxed automaton M_R^K has at most $O(|R||K|)$ states and $O(|R||K|^2)$ transitions.*

The proof follows from the fact that automaton M_R contains $O(|R|)$ states, for each of which we can potentially add $O(|K|)$ cloned states. Each of the rules adds no more than $O(|K|)$ transitions for each of the $O(|R||K|)$ states in M_R^K.

Proposition 4. *Let $G = (V, E)$ be a graph, Q be a single-conjunct query using regular expression R, and $K = \texttt{extRed}(K)$ be an ontology. The relaxed answer of Q on G can be found in time $O(|R|^2|K|^2|V|(|E| + |V|\log(|R||K||V|)))$.*

The proof follows from Propositions 2 and 3, along with using Dijkstra's algorithm on the product automaton H, which can be shown to have $O(|R||K||V|)$ nodes and $O(|R||K|^2|E|)$ edges.

In order to compute the relaxed answers incrementally, we can use the `getNext` function from Section 2.1 along with the same initialisation of program variables. The only difference is that the `Succ` function now uses the relaxed automaton M_R^K rather than approximate automaton A_R.

3   General Queries

Combining the possibility of approximating and relaxing query conjuncts, a general query Q is of the form

$$
\begin{aligned}
&(Z_1, \ldots, Z_m) \leftarrow (X_1, R_1, Y_1), \ldots, (X_j, R_j, Y_j), \\
&APPROX(X_{j+1}, R_{j+1}, Y_{j+1}), \ldots, APPROX(X_{j+k}, R_{j+k}, Y_{j+k}), \\
&RELAX(X_{j+k+1}, R_{j+k+1}, Y_{j+k+1}), \ldots, RELAX(X_{j+k+n}, R_{j+k+n}, Y_{j+k+n})
\end{aligned}
$$

where $j, k, n \geq 0$, the X_i and Y_i are constants or variables, the R_i are regular expressions, and each Z_i is one of $X_1, \ldots, X_{j+k+n}$ or $Y_1, \ldots, Y_{j+k+n}$. In the concrete syntax, conjuncts may be specified in any order.

Let θ be a matching from variables and constants of Q to nodes in graph G. The *distance* from θ to Q, $dist(\theta, Q)$, is defined as

$$
\begin{aligned}
&w_A(edist(\theta, (X_{j+1}, R_{j+1}, Y_{j+1})) + \cdots + edist(\theta, (X_{j+k}, R_{j+k}, Y_{j+k}))) + \\
&w_R(rdist(\theta, (X_{j+k+1}, R_{j+k+1}, Y_{j+k+1})) + \cdots + rdist(\theta, (X_{j+k+n}, R_{j+k+n}, Y_{j+k+n})))
\end{aligned}
$$

where the coefficients w_A and w_R are set according to the preferences of the user. For example, they can be set to the same value if the same "cost" is associated with query approximation and query relaxation, or to different relative values to penalise one or the other more. Let $\theta(Z_1, \ldots, Z_m) = (a_1, \ldots, a_m)$. We call θ a *minimum-distance* matching if for all matchings ϕ from Q to G such that $\phi(Z_1, \ldots, Z_m) = (a_1, \ldots, a_m)$, $dist(\theta, Q) \leq dist(\phi, Q)$.

The *answer* of Q on G is the list of pairs $(\theta(Z_1, \ldots, Z_m), dist(\theta, Q))$, for some minimum-distance matching θ, ranked in order of non-decreasing distance. The *top-k answer* of Q on G comprises the first k tuples in the answer of Q on G.

The query Q can be evaluated by joining the answers arising from the evaluation of each of its conjuncts. For each APPROXed or RELAXed conjunct we can use the techniques described in Sections 2.1 and 2.3, respectively, to incrementally compute a relation r_i with scheme (X_i, Y_i, ED, RD). If $i \leq j$, then $t[ED] = t[RD] = 0$. If $j < i \leq j + k$, then for any tuple $t \in r_i$, $t[RD] = 0$ and $t[ED]$ is the edit distance for that tuple. If $j + k < i \leq j + k + n$, then for any tuple $t \in r_i$, $t[ED] = 0$ and $t[RD]$ is the relaxation distance for that tuple.

To ensure polynomial-time evaluation, we require that the conjuncts of Q are *acyclic* [6]. Hence a query evaluation tree can be constructed for Q, consisting of nodes denoting join operators and nodes representing conjuncts of Q. Given that the answers for single conjuncts are ordered by non-decreasing distance, we can use a pipelined execution of any rank-join operator, such as the recent instance-optimal FRPA operator proposed in [5], to produce the answers to Q on graph G in order of non-decreasing distance.

Example 6. Consider query Q_4 from Example 1. Suppose graph G contains the triples shown in Fig. 1 and also the triples
(ep31,type,University),(ep31,qualif,BA History),
(ep32,type,Work),(ep32,job,j32),(j32,type,Writer)
(ep33,type,Work),(ep33,job,j33),(j33,type,AssociateEditor)
(BA History,type,History),(ep31,next,ep32),(ep32,next,ep33)
from another timeline. Suppose also that in the ontology, there are triples
(History,sc,Humanities) and (Writer,sc,MediaProfessional). We set the approximation cost $\alpha = 1$, the two relaxation costs $\beta = \gamma = 2$ and $w_A = w_R = 1$. Then, answers are produced for query Q_4 as shown in the table below:

?E1	?E1,RD	?E1,?E2,ED	?E2,?P	?E2,?Goal,ED	?Goal	?Goal,RD	?E2,P,D
ep21	*ep21,0*	ep23,ep24,0	ep22,AT	ep23,ep24,0	ep22	e24,0	*ep23,J,2*
ep31	**ep31,4**	ep21,ep22,1	ep23,J	ep21,ep22,1	ep23	**e33,2**	ep22,AT,6
		ep22,ep23,1	ep24,IE	ep22,ep23,1	ep24	e23,4	**ep32,W,8**
		ep31,ep32,1	**ep32,W**	ep31,ep32,1	ep32	e32,4	
		ep32,ep33,1	ep33,OE	**ep32,ep33,1**	**ep33**	e22,6	
		ep21,ep23,2		ep21,ep23,2			
		ep21,ep24,2		ep21,ep24,2			
		ep31,ep33,2		ep31,ep33,2			

The first seven columns refer to the answers produced for the individual conjuncts of Q_4. For brevity, we do not show the full four-attribute answer tuples, only the non-zero distances and the variable instantiations. We also abbreviate Air Travel Assistant by AT, Journalist by J, Writer by W, Assistant Editor by IE and Associate Editor by OE. The final column shows the overall query answers and distances. Tuples contributing to the first two answers are *italicised* and those contributing to the third answer are **bold**. □

4   Related Work

Various forms of query approximation and relaxation have been studied for a number of data models and query languages. For approximate querying, [14] considered querying semistructured data using flexible matchings which allow paths whose edge labels simply contain those appearing in the query to be matched. Such semantics can be captured by the edit operations of transposition and insertion. More generally, [7] used weighted regular transducers for performing transformations to regular path queries (but not CRP queries) to allow them to match semi-structured data approximately. The approximate queries of [18] are simply selections placed on attributes of form-based web data, where value constraints can be relaxed according to their perceived importance to the user.

In terms of query relaxation, work has been done on relaxing tree pattern queries for XML, recently in [16]. Relaxation of conjunctive queries on RDF is considered in [4,12]. Rewriting rules are used on query patterns in [4] to perform both query refinement by including user preferences as well as query relaxation. Building on the work of [12], [11] develops a similarity measure for relaxed queries in an attempt to improve the relevance of answers. Similarity-based querying was also the focus of iSPARQL [15], where resources (rather than paths connecting them) are compared using similarity measures. Flexible querying of RDF using SPARQL and preferences expressed as fuzzy sets is investigated in [1].

In contrast to all the above, our work combines within one framework both query approximation and query relaxation, and applies it to the more general query language of conjunctive regular path queries on graph-structured data.

5   Concluding Remarks

We have discussed query relaxation for conjunctive regular path queries, and have shown how this can be combined with query approximation in order to provide greater flexibility in the querying of complex, irregular seminstructured data sets. Using the techniques proposed here, users are able to specify approximations and relaxations to be applied to their original query, and the relative costs of these. Query results are returned incrementally, ranked in order of increasing 'distance' from the user's original query. We have presented polynomial-time algorithms for incrementally computing the top-k answers to such queries.

In practice, we expect that a visual query interface would be required, providing users with readily understandable options from which to select their query formulation, approximation and relaxation requirements, and set the relative cost associated with each operation they have selected. Our future work includes the design, prototyping and evaluation of such a query interface, or interfaces, and the empirical evaluation of our query processing algorithms, in domains such as querying of lifelong learners' metadata and heterogeneous medical data sets.

Another direction of ongoing research is to merge the APPROX and RELAX operations into one integrated 'FLEX' operation that applies concurrently both approximation and relaxation to a regular path query. For this, we are taking advantage of the common NFA-based approach that we have adopted.

References

1. Buche, P., Dibie-Barthélemy, J., Chebil, H.: Flexible SPARQL querying of web data tables driven by an ontology. In: Proc. FQAS, pp. 345–357 (2009)
2. Calvanese, D., Giacomo, G.D., Lenzerini, M., Vardi, M.Y.: Containment of conjunctive regular path queries with inverse. In: Proc. KR, pp. 176–185 (2000)
3. de Freitas, S., Harrison, I., Magoulas, G., Mee, A., Mohamad, F., Oliver, M., Papamarkos, G., Poulovassilis, A.: The development of a system for supporting the lifelong learner. British Journal of Educational Technology 37(6), 867–880 (2006)
4. Dolog, P., Stuckenschmidt, H., Wache, H., Diederich, J.: Relaxing RDF queries based on user and domain preferences. J. Intell. Inf. Syst. 33(3), 239–260 (2009)
5. Finger, J., Polyzotis, N.: Robust and efficient algorithms for rank join evaluation. In: Proc. ACM SIGMOD, pp. 415–428 (2009)
6. Gottlob, G., Leone, N., Scarcello, F.: The complexity of acyclic conjunctive queries. J. ACM 43(3), 431–498 (2001)
7. Grahne, G., Thomo, A.: Regular path queries under approximate semantics. Ann. Math. Artif. Intell. 46(1-2), 165–190 (2006)
8. Gutierrez, C., Hurtado, C., Mendelzon, A.O.: Foundations of semantic web databases. In: Proc. PODS, pp. 95–106 (2004)
9. Hayes, P. (ed.): RDF Semantics, W3C Recommendation, (February10, 2004)
10. Heath, T., Hausenblas, M., Bizer, C., Cyganiak, R.: How to publish linked data on the web (tutorial). In: Proc. ISWC (2008)
11. Huang, H., Liu, C., Zhou, X.: Computing relaxed answers on RDF databases. In: Bailey, J., Maier, D., Schewe, K.-D., Thalheim, B., Wang, X.S. (eds.) WISE 2008. LNCS, vol. 5175, pp. 163–175. Springer, Heidelberg (2008)
12. Hurtado, C.A., Poulovassilis, A., Wood, P.T.: Query relaxation in RDF. Journal on Data Semantics X, 31–61 (2008)
13. Hurtado, C.A., Poulovassilis, A., Wood, P.T.: Ranking approximate answers to semantic web queries. In: Aroyo, L., Traverso, P., Ciravegna, F., Cimiano, P., Heath, T., Hyvönen, E., Mizoguchi, R., Oren, E., Sabou, M., Simperl, E. (eds.) ESWC 2009. LNCS, vol. 5554, pp. 263–277. Springer, Heidelberg (2009)
14. Kanza, Y., Sagiv, Y.: Flexible queries over semistructured data. In: Proc. PODS, pp. 40–51 (2001)
15. Kiefer, C., Bernstein, A., Stocker, M.: The fundamentals of iSPARQL: A virtual triple approach for similarity-based semantic web tasks. In: Aberer, K., Choi, K.-S., Noy, N., Allemang, D., Lee, K.-I., Nixon, L.J.B., Golbeck, J., Mika, P., Maynard, D., Mizoguchi, R., Schreiber, G., Cudré-Mauroux, P. (eds.) ASWC 2007 and ISWC 2007. LNCS, vol. 4825, pp. 295–309. Springer, Heidelberg (2007)
16. Liu, C., Li, J., Yu, J.X., Zhou, R.: Adaptive relaxation for querying heterogeneous XML data sources. Information Systems 35(6), 688–707 (2010)
17. Mendelzon, A.O., Wood, P.T.: Finding regular simple paths in graph databases. SIAM J. Computing 24(6), 1235–1258 (1995)
18. Meng, X., Ma, Z.M., Yan, L.: Answering approximate queries over autonomous web databases. In: Proc. WWW, pp. 1021–1030 (2009)

EvoPat – Pattern-Based Evolution and Refactoring of RDF Knowledge Bases

Christoph Rieß, Norman Heino, Sebastian Tramp, and Sören Auer

AKSW, Institut für Informatik, Universität Leipzig, Pf 100920, 04009 Leipzig
{lastname}@informatik.uni-leipzig.de
http://aksw.org

Abstract. Facilitating the seamless evolution of RDF knowledge bases on the Semantic Web presents still a major challenge. In this work we devise *EvoPat* – a pattern-based approach for the evolution and refactoring of knowledge bases. The approach is based on the definition of *basic evolution patterns*, which are represented declaratively and can capture simple evolution and refactoring operations on both data and schema levels. For more advanced and domain-specific evolution and refactorings, several simple evolution patterns can be combined into a compound one. We performed a comprehensive survey of possible evolution patterns with a combinatorial analysis of all possible before/after combinations, resulting in an extensive catalog of usable evolution patterns. Our approach was implemented as an extension for the OntoWiki semantic collaboration platform and framework.

1 Introduction

The challenge of facilitating the smooth evolution of knowledge bases on the Semantic Web is still a major one. The importance of addressing this challenge is amplified by the shift towards employing agile knowledge engineering methodologies (such as Semantic Wikis), which particularly stress the evolutionary aspect of the knowledge engineering process.

The *EvoPat* approach is inspired by software refactoring. In software engineering, refactoring techniques are applied to improve software quality, to accommodate new requirements or to represent domain changes. The term refactoring refers to the process of making persistent and incremental changes to a system's internal structure without changing its observable behavior, yet improving the quality of its design and/or implementation [5]. Refactoring is based on two key concepts: *code smells* and *refactorings*. Code smells are an informal but still useful characterization of patterns of bad source code. Examples of code smells are "too long method" and "duplicate code". Refactorings are piecemeal transformations of source code which keep the semantics while removing (totally or partly) a code smell. For example, the "extract method" refactoring extracts a section of a "long method" into a new method and replaces it by a call to the new method, thus making the original method shorter (and clearer).

P.F. Patel-Schneider et al. (Eds.): ISWC 2010, Part I, LNCS 6496, pp. 647–662, 2010.

Compared to software source code refactoring, where refactorings have to be performed manually or with limited programmatic support, the situation in knowledge base evolution on the Semantic Web is slightly more advantageous. On the Semantic Web we have a unified data model, the RDF data model, which is the basis for both, data and ontologies. In this work we exploit the RDF data model by devising a pattern-based approach for the data evolution and ontology refactoring of RDF knowledge bases. The approach is based on the definition of *basic evolution patterns*, which are represented declaratively and can capture atomic evolution and refactoring operations on the data and schema levels. In essence, a basic evolution pattern consists of two main components: 1) a *SPARQL SELECT query template* for selecting objects, which will be changed and 2) a *SPARQL/Update query template*, which is executed for every returned result of the SELECT query. In order to accommodate more advanced and domain-specific data evolution and refactoring strategies, we define a compound evolution pattern as a linear combination of several simple ones.

To obtain a comprehensive catalog of evolution patterns, we performed a survey of possible evolution patterns with a combinatorial analysis of all possible before/after combinations. Starting with the basic constituents of a knowledge base (i. e. graphs, properties and classes), we consider all possible combinations of the elements potentially being affected by an evolution pattern and the prospective result after application of the evolution pattern. This analysis led to a comprehensive library of 24 basic and compound evolution patterns. The catalog is not meant to be exhaustive but covers the most common knowledge base evolution scenarios as confirmed by a series of interviews with domain experts and knowledge engineers. The EvoPat approach was implemented as an extension for the OntoWiki semantic collaboration platform and framework.

Compared to existing approaches for knowledge base evolution, our declarative, pattern-based approach has a number of advantages:

- EvoPat is a *unified method*, which works for both data evolution and ontology refactoring.
- The modularized, *declarative* definition of evolution patterns is relatively simple compared to an imperative description of evolution. It allows domain experts and knowledge engineers to amend the ontology structure and modify data with just a few clicks.
- Combined with our RDF representation of evolution patterns and their exposure on the Linked Data Web, EvoPat facilitates the development of an *evolution pattern ecosystem*, where patterns can be shared and reused on the Data Web.
- The declarative definition of bad smells and corresponding evolution patterns promotes the (semi-)automatic *improvement of information quality*.

This paper is structured as follows: We describe the evolution pattern concepts in Section 2 and survey possible evolution patterns in Section 3. We showcase our implementation in Section 4 while we present our work in the light of related approaches in Section 5 and conclude with an outlook on future work in Section 6.

2 Concepts

The EvoPat approach is based on the rationale of working as closely as possible with the RDF data model and the common ontology construction elements, i.e. classes, instances as well as datatype and object properties. With EvoPat we also aim at delegating bulk of the work during evolution processing to the underlying triple store. Hence, for the definition of evolution patterns we employ a combination of different SPARQL query templates. In order to ensure modularity and facilitate reusability of evolution patterns our definition of evolution patterns is twofold: *basic evolution patterns* accommodate atomic ontology evolution and data migration operations, while *compound evolution patterns* represent sequences of either basic or other compound evolution patterns in order to capture more complex and domain specific evolution scenarios. The application of a particular evolution pattern to a concrete knowledge base is performed with the help of the EvoPat *pattern execution algorithm.* In order to optimally assist a knowledge engineer we also define the concept of a *bad smell* in a knowledge base. We describe these individual EvoPat components in more detail in the remainder of this paper.

2.1 Evolution Pattern

Figure 1 describes the general composition of EvoPat evolution patterns. Bad smells (depicted in the lower left of Figure 1 have a number of basic or compound evolution patterns associated, which are triggered once a bad smell is traced. Basic and compound evolution patterns can be annotated with descriptive attributes, such as a label for the pattern, a textual description and other metadata such as the author of the pattern the creation date, revision etc.

Basic Evolution Pattern (BP). A basic evolution pattern consists of two main components: 1. a SPARQL SELECT query template for selecting objects, which will be changed and 2. a SPARQL/Update query template, which is executed for every returned result of the SELECT query. In addition, the placeholders contained in both query templates are typed in order to facilitate the classification and choreography of different evolution patterns. Please note, that in the following we will use the term variable for placeholders contained in SPARQL query templates. These should not be confused with variables contained in SPARQL graph patterns, which, however, do not play any particular role in this article. The following definition describes basic evolution patterns formally:

Definition 1 (Basic Evolution Pattern). *A basic evolution pattern is a tuple (V, S, U), where V is a set of typed variables, S is a SPARQL query template with placeholders for the variables from V, and U is a SPARQL/Update query template with placeholders referring to a result set which is generated by the SPARQL query template S.*

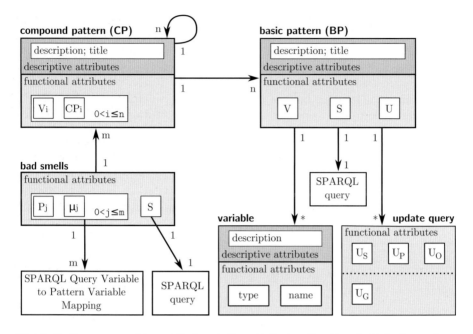

Fig. 1. Pattern composition with descriptive attributes, functional attributes and cardinality restrictions

```
1   V: dtProp type: PROPERTY
2      objProp type: PROPERTY
3      p type: TEMP
4      o type: TEMP
5   S: SELECT DISTINCT * WHERE {
6         %dtProp% %p% %o% .
7         FILTER (
8            !sameTerm(%p%, rdfs:range) &&
9            !sameTerm(%p%, rdf:type)
10        )
11     }
12  U: INSERT: %objProp% %p% %o% .
13     DELETE: %dtProp% %p% %o% .
```

Listing 1. Basic Evolution Pattern example: moving axioms from one property to another

Listing 1 shows a basic evolution pattern, which moves axioms from one property to another. Lines 1-4 define the typed variables used in the pattern. Lines 5-11 contain the SELECT query template, while lines 12-13 contain the SPAR-QL/Update query template to be executed for each result of the SELECT query.

Query preprocessor. In order to give a SPARQL query for previously unknown entities (since they are selected by the pattern SPARQL query), we introduce an extension to SPARQL that defines two additional types of variables and preprocessor functions:

- *Pattern variables* are enclosed in % characters and will be replaced with the corresponding entity. *Input variables* are defined by the user applying the pattern (e. g. on which entity the pattern is to operate.). *Temp variables* are variables to which query results from the pattern SPARQL query are bound. They can be used in the SPARQL/Update query of the same pattern to describe triple updates. In Listing 1, line 12 the variable %objProp% is used to bind the newly created object property.
- *Preprocessor functions* are a means of performing certain actions with the entities bound to a variable. If e. g. the user wants URIs of a certain format or change the datatype of a created literal value, those functions can be used. We provide a number of pre-defined functions for the most common use cases.

Compound Evolution Pattern (CP). Basic evolution patterns alone are not sufficient to cover arbitrary evolution scenarios. Especially on higher abstraction levels of represented domain knowledge, it is feasible to represent ontology changes on the same level of abstraction. To this end, we define compound evolution patterns, consisting of several evolution patterns that are subsequently applied to a knowledge base.

Definition 2 (Compound Evolution Pattern). *Let $0 < i \leq n$, P_i be (basic or compound) patterns and V_i the corresponding sets of unbound variables in P_i. A sequence $CP := (V_i, P_i)$ of patterns is called a* compound pattern *(CP).*

An example of a compound pattern for transforming a datatype property into an object property (including instance transformation) is given in listing 2. It consists of the following four basic sub patterns: moving property axioms, deleting datatype property, transforming instance data and creating object property.

```
1  // Sub pattern 1: (move axioms from dtProp to objProp)
2  V: dtProp type: PROPERTY
3     objProp type: PROPERTY
4     p type: TEMP
5     o type: TEMP
6  S: SELECT DISTINCT * WHERE {
7       %dtProp% %p% %o% .
8       FILTER (
9          !sameTerm(%p%,rdfs:range) &&
10         !sameTerm(%p%,rdf:type)
11      )
12  }
13 U: INSERT: %objProp% %p% %o% .
14    DELETE: %dtProp% %p% %o% .
```

```
15
16 // Sub pattern 2: (delete dtProp)
17 V: dtProp type: PROPERTY
18    p type: TEMP
19    o type: TEMP
20 S: SELECT DISTINCT * WHERE {
21      %dtProp% %p% %o% .
22    }
23 U: DELETE: %dtProp% %p% %o% .
24
25 // Sub pattern 3: (transform instance data)
26 V: dtProp type: PROPERTY
27    inst type: TEMP
28    o type: TEMP
29    objProp: PROPERTY
30 S: SELECT DISTINCT * WHERE {
31      %inst% %dtProp% %o% .
32    }
33 U: INSERT:
34    %inst% %objProp%getTempUri(getNamespace(%objProp%),%o%).
35    getTempUri(getNamespace(%objProp%),%o%) rdfs:label %o%.
36    DELETE: %inst% %dtProp% %o%
37
38 // Sub pattern 4: (create property)
39 V: objProp type: PROPERTY
40 S:
41 U: INSERT: %objProp% rdf:type owl:ObjectProperty .
```

Listing 2. Compound Evolution Pattern example: transforming a datatype into an object property while maintaining instance consistency

2.2 Evolution Pattern Processing

Algorithm 2.2 outlines the evolution pattern processing. The algorithm uses an evolution pattern P, a graph G and a set of variable bindings B as input. Depending on the type of pattern (basic or compound) the following steps are performed.

Basic pattern. If P is a basic pattern, the variables in the query are substituted with respect to their binding in B. Each of the update patterns contained in P is processed as follows:

1. If the update pattern sets an explicit graph, the active graph is set to that graph, else it is set to the default graph.
2. The variables in the update pattern are substituted according to B.
3. Changes are determined by executing the SPARQL query in P on G.
4. The changes are then applied to the active graph.

Compound pattern. Compound patterns are resolved to basic patterns. For each of the basic patterns the above steps are performed. The output of the algorithm is a set of changes on the respective graphs.

Algorithm 1. Pattern execution sequence

Require: Pattern P
Require: RDF graph G
Require: Variable bindings B
  if P is Basic Pattern **then**
    substitute variables in SPARQL Query according to B
    execute preprocessor functions in P
    $QR :=$ SPARQL query result of P on G
    for all update patterns of P as UP **do**
      if UP has graph **then**
        active graph $AG =$ graph of UP
      else
        active graph $AG =$ default graph G
      end if
      substitute variables in UP according to B
      generate changes CS of UP on AG with QR
      apply changes CS to AG
    end for
  else
    for all basic patterns in compound pattern P as SP **do** //maintain correct order
      execute Base Pattern SP //see above
    end for
  end if

2.3 Bad Smells

In order to assist knowledge engineers and domain experts as much as possible with the evolution of a knowledge base we also provide a formal definition for a bad smell in a certain knowledge base. In essence, a bad smell is represented via a SPARQL SELECT query, which detects a suspicious structure in a knowledge base. In most scenarios, there will be one (or multiple) evolution patterns addressing exactly the issue raised by a certain bad smell. Hence, we allow to assign one (or multiple) evolution patterns to the bad smell for resolving that issue. In order to further automatize the resolving of bad smells each evolution pattern can be assigned with a mapping from the bad smells result set to the variables used in the evolution patterns.

Definition 3 (Bad smell). *A bad smell is a tuple $(S, (P_i, \mu_i))$, where S is a SPARQL query and (P_i, μ_i) is a list of possible evolution patterns P_i for resolving the bad smell with an associated mapping μ_i, which maps results of S to the variables in P_i.*

```
1    SELECT ?s ?p ?o
2    WHERE {
3       ?s ?p ?o .
4       ?p a owl:DatatypeProperty .
5       ?p rdfs:range ?range .
6       FILTER (DATATYPE(?o) != ?range)
7    }
```

Listing 3. Bad smell example: selecting statements for which the datatype of the object doesn't match the `rdfs:range` of the property

An example of a bad smell is given in listing 3. It selects all statements whose object is a literal with a datatype that does not match the `rdfs:range` of the property of that statement. The result set from the bad smell query can be directly applied as input to a pattern that typecasts literal values to the correct datatype.

In certain cases a knowledge base evolution can be even performed completely automatically. This is the case if and only if both of the following conditions are met.

- The bad smell can only be resolved by exactly one evolution pattern and
- the mapping to the evolution pattern's variables is complete in the sense that all variables will be assigned values from the bad smell's query result set.

2.4 Serialization in RDF

To facilitate the exchange and reuse of previously defined evolution patterns we developed an RDF serialization, i.e. an RDF vocabulary for representing evolution patterns[1]. Together with an updated log publishing (such as e.g. proposed in [1]) on the Linked Data Web this facilitates the creation of an evolution ecosystem, where generic and domain specific evolution patterns are shared and reused and data cleansing and migration strategies can be also performed in network of linked knowledge bases.

3   Pattern Survey and Classification

In order to obtain a comprehensive catalog of evolution patterns we pursued a three-fold strategy: (1) we performed a comprehensive literature review, (2) we looked at all combinatorial combinations of before/after states and (3) we conducted a number of interviews with knowledge engineers and domain experts, which were involved in medium-scale knowledge base construction projects and retrospectively reviewed the evolution of these knowledge bases.

---

[1] The vocabulary for representing evolution patterns is available at:
http://ns.aksw.org/Evolution/

Table 1. Combinatorially possible before/after evolution states. C, P, G stand for class, property, graph respectively. The '+' indicates that multiple entities of the same type participate in the evolution pattern. Impossible combinations are blackened out.

	$\emptyset$	$C+$	$P+$	$G+$	PC	PG	CG
$\emptyset$	ok	ok	ok	ok			
$C+$	ok	ok	ok				ok
$P+$	ok	ok	ok		ok	ok	
$G+$	ok			ok			
PC			ok				
PG			ok				
CG		ok					

Literature review. Most work concerned with ontology evolution patterns identifies a number of useful patterns but gives only an informal description which cannot be used for implementing an evolution software system. In [10], evolution patterns that work on the ontology level are identified. A classification of evolution patterns in four levels of abstraction is presented in [7]. The levels identified by the authors helped us in our classification system. In the interviews we conducted, the need for representational changes was identified. Thus, we added another layer that deals with syntactic changes to resources (i. e. renaming a URI). The authors of [3] present a number of patterns with formally defined participants and execution steps. We extended the approach, providing a pattern behavior in the form of SPARQL/Update queries that can directly be built into Semantic Web applications.

Combinatorial analysis. In order to ensure, that we achieved a comprehensive coverage of all possible evaluation patterns we followed a combinatorial analysis. We considered all possible combinations of ontology construction elements (i. e. classes, properties and (sub-)graphs) which are potentially affected by the application of a basic evolution pattern and the possible combinations of remaining elements after the pattern has been applied. All possible combinations are displayed in Table 1. For each of the potentially possible combinations we performed an analysis whether evolution patterns actually exist in practice. The results of this analysis are also summarized in Table 2. Combinations where possible patterns can be represented as combinations of basic evolution patterns are marked with a white background. Those combinations were no basic evolution patterns exist are blackened out.

Interviews and retrospective coverage checks. In order to ground our findings from the literature review and combinatorial analysis, we had an in-depth look at several medium- to large-scale knowledge base construction projects. These included in particular the Vakantieland e-tourism knowledge base for the Netherlands [9], the Leipzig Professors Catalog [2] and the development of an ontology for the energy sector, which was performed by our industry partner Business

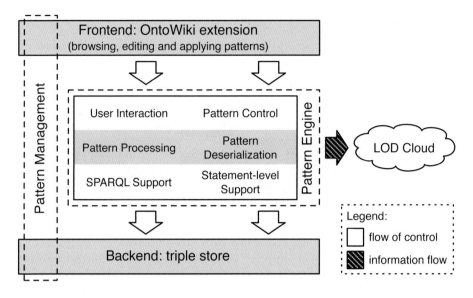

Fig. 2. System architecture with internal functional units and provided services. Patterns are exposed as Linked Data.

Intelligence GmbH. We also retrospectively reviewed the evolution of these knowledge bases and analyzed to what extend the previously defined evolution patterns would cover the found evolution steps.

4   Implementation

The EvoPat approach was implemented as an extension to OntoWiki – a tool for browsing and collaboratively editing RDF knowledge bases. It differs from other Semantic Wikis insofar as OntoWiki uses RDF as its natural data model instead of Wiki texts. Information in OntoWiki is always represented according to the RDF statement paradigm and can be browsed and edited by means of views. These views are generated automatically by employing ontology features such as class hierarchies or domain and range restrictions. OntoWiki adheres to the Wiki principles by striving to make the editing of information as simple as possible and by maintaining a comprehensive revision history. This history is also based on the RDF statement paradigm and allows to roll back prior change sets. OntoWiki has recently been extended to incorporate a number of Linked Data features, such as exposing all information stored in OntoWiki as Linked Data as well as retrieving background information from the Linked Data Web [6]. Apart from providing a comprehensive user interface, OntoWiki also contains a number of components for the rapid development of Semantic Web applications, such as the RDF API Erfurt[2], methods for authentication, access control, caching and various visualization components.

---
[2] http://aksw.org/Projects/Erfurt/

Table 2. Overview of valid evolution patterns on four levels of abstraction

Ontology level (OWL)		
Before	After	Description
$\emptyset$	$\emptyset$	Trivial empty pattern (no actions taken)
$\emptyset$	C+, P+ or G+	Creating class, property or graph
C+, P+ or G+	$\emptyset$	Deleting class, property or graph
C+	C+	Subclassing, union, merging, splitting classes
P+	P+	Property axioms (functional, symmetric, domain, range, etc.)
G+	G+	Graph merging and splitting, graph annotation
C+	P+	Remodeling from class membership to distinct property value
P+	C+	Remodeling from distinct property value to class membership
C+	CG	Class extraction from named graph
CG	C+	Merging classes into graph
P	PC	Converting datatype to object property
PC	P	Converting object to datatype property (incl. axioms)

Instance and data level (RDFS)		
Input	Output	Description
I^*	I	Instances merging
I^*, C^*	I^*	Instances reclassification
I^*, P, O	I^*	Adding data to instances
I^*, P, P^*	I^*	Generating data from existing instances data
I^*, L^*	I^*	Converting literal property values to resources
I^*, R^*	I^*	Converting resources to literal property values
$I^*(, P^*, O^*)$	I^*	Moving data (predicates and objects) from one instance to another

Entity level (RDF)		
Input	Function	Description
Literal, datatype	Setting datatype on literal	Datatype added, changed or removed
Literal, language	Setting language on literal	Literal language added, changed or removed
RegExp search/replace	regexp replace	Performs a regular expression search and replace on literal value

Syntactic/representational level (RDF/XML, N3, etc.)		
Input	Function	Description
URI, namespace	Set URI prefix	Changes prefixes for a resource
URI, local name	Set local name	Changes local name of a resource

The general architecture of the EvoPat extension is depicted in Figure 2. It consists of four distinct components. Core of the EvoPat implementation is the *pattern engine*, which in particular handles processing, storing, versioning and exposing evolution patterns as Linked Data on the data web. It interacts via SPARQL with a triple store representing the EvoPat *backend*. The EvoPat *frontend* facilitates the user friendly browsing/selection, configuration and application of evolution patterns. The *pattern management component* as a logical component spans several architectural layers. It implements the required APIs needed by the user interface and backend for managing patterns.

Different versions of ontologies resulting from applying evolution patterns can be managed through OntoWiki's versioning component. Similar to database transactions, the changes on the statement level that result from applying a certain evolution pattern can be grouped and versioned as a single change.

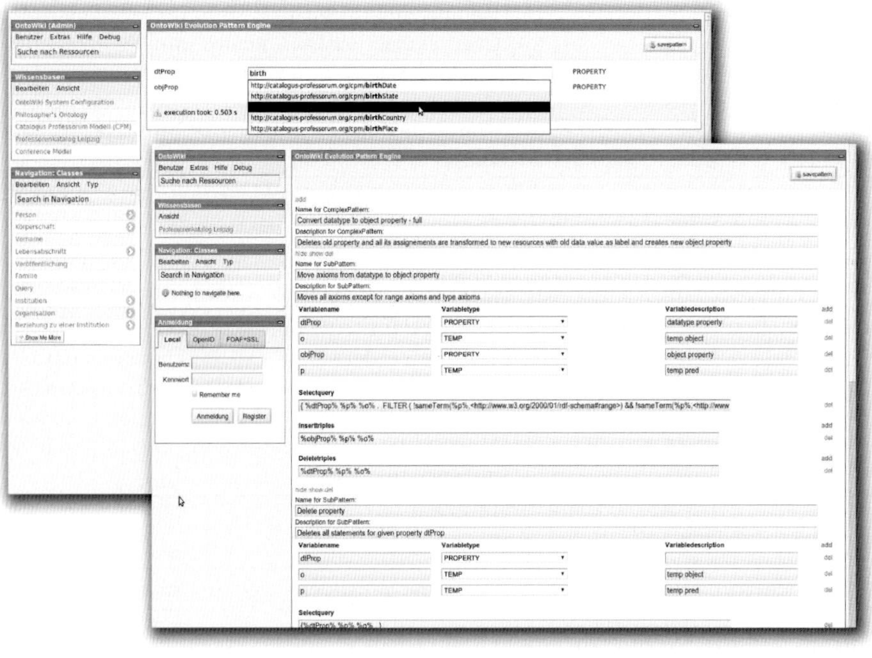

Fig. 3. EvoPat user interface showing pattern editor (right) and pattern execution view (left)

Figure 3 showcases the EvoPat user interface with the pattern editor and the pattern execution. The pattern editor allows to create basic and compound evolution patterns. A user friendly form is generated, where the descriptive attributes, the variables used in the pattern and the respective SPARQL SELECT and UPDATE queries can be filled in. For pattern execution (as shown in the upper left part of Figure 3), the EvoPat implementation generates a form based on the variables definition of the evolution pattern at hand. Employing the typing of the variable a type ahead search simplifies the selection of concrete values for the variables.

Scalability evaluation. One of the main goals of developing EvoPat was to push as much of the evolution pattern processing down to the triple store. In order to evaluate whether EvoPat lives up to this promise we evaluated the processing of selected evolution patterns with different knowledge base sizes. The results of the evaluation are summarized in Table 3. We used the Catalogus Professorum Lipiensis knowledge base and simply created three different versions of it in different sizes, by simply copying the data. The results of the performance evaluation show, that the evolution pattern processing grows linearly with the knowledge base size. As a consequence, EvoPat can be used with arbitrarily large knowledge bases, the performance of the evolution pattern processing primarily depends on the speed of the underlying triple store.

Table 3. Scalability evaluation with two compound patterns on Catalogus Professorum Lipsiensis. The benchmarks were performed in three different sizes of the original knowledge base: original size (150K triples), 3 × the size (450K triples), 5 × the size (750K triples). Figures are quoted for two patterns each KB size.

	pattern exec. [s]	affect. rsrc. [pcs]	throughput $[\frac{pcs}{s}]$
KB size: 1 × 150K triples			
Datatype to Object Property	8.593	1300	151.3
Class merging	5.949	1500	252.1
KB size: 3 × 150K triples			
Datatype to Object Property	24.813	3900	157.2
Class merging	17.753	4500	253.4
KB size: 5 × 150K triples			
Datatype to Object Property	39.822	6500	163.2
Class merging	30.603	7500	245.1

5   Related Work

Ontology evolution has constantly been under research during the past two decades. In recent years a ramp-up could be observed due to Semantic Web research activity, thus providing a more user-centric view on ontology evolution.

A comprehensive overview on the field of ontology change is given in [4]. The authors conduct an extensive literature review, extracting and defining common vocabulary as a base for discussion. They define ontology evolution as a "response to a change in the domain or conceptualization". The term ontology evolution, as used in this paper, covers what Flouris et al. refer to as ontology translation and by which they mean changes in the syntactical representation of the ontology (e. g. changing the URI of a resource).

To the best of our knowledge, there is no existing approach for formally specifying modular evolution patterns in a declarative manner. The most closely related approach in this regard is a categorization of pattern-based change operators in [7]. The paper defines four levels of abstraction of an ontology (element, element context, domain-specific and generic abstract level) to whose elements the said operators can be applied. Taking into account the Semantic Web infrastructure, our approach defines an additional level on the representation layer.

Stojanovic et al. in [12] define three requirements for ontology evolution: 1) ensuring consistency, 2) allowing the user supervision of evolution and 3) advice for continuous ontology refinement. In addition, the authors identify six phases of ontology evolution, namely 1) capturing, 2) representation, 3) semantics of change, 4) implementation, 5) propagation and 6) validation of changes. The KAON API[3], implementing the approach, also introduced by the authors. Furthermore, they identify the need for representing changes on different levels of granularity. To cope with different methods of applying changes to an ontology, they introduce basic evolution strategies, which define the steps of a complex evolution process. For a given change request there are usually more than on applicable strategy, resulting in different ontologies. Seen in a broader sense, these basic evolution strategies can be combined into so called advanced evolution strategies, of which they introduce four. Our compound patterns are similar in nature to Stojanovic's basic evolution strategies, but differ in the inclusion of explicit declarative semantics by means of SPARQL/Update queries.

An interesting approach to ontology evolution with particular respect to consistency management is given by Djedidi and Aufaure [3]. They propose a process model, an attached pattern and a versioning layer. If applying a change pattern results in a match to an inconsistency pattern, an alternative pattern is automatically applied by the proposed system. Furthermore, a quality assessment step is integrated into the process. The system can thus alleviate the need for user interaction by applying quality-improving patterns in an automated fashion.

Noy and Klein determine in [10] to what extent ontology evolution resembles schema evolution, which has been extensively researched in the database community. By arguing that different versions of an ontology have to be kept in parallel, they conclude that the traditional distinction between schema evolution and schema versioning is not applicable to ontology evolution and ontology versioning. Even though, EvoPat distinguishes between versioning and evolution, both subsystems are closely related and cannot be used exclusively. All evolutionary changes are automatically versioned and can be reverted at any time.

---

[3] http://kaon.semanticweb.org/developers

A declarative update language for RDF graphs, named RUL is defined in [8]. RUL is based on RQL and RVL and ensures consistency on the RDF and RDFS levels. It, therefore, contains *primitive, set-oriented* and *complex* updates as compositions of primitive or complex ones. Primitive RUL updates are similar in expressiveness to SPARQL 1.1 updates. Complex updates are expressed by means of fine-grained updates on class and property instance level. Our basic evolution patterns with variable placeholders are similar to the set-oriented RUL updates (i. e. repeating the same query for several bindings). Additionally, we, however, define a functional extension that allows for arbitrarily replacing entities in a preprocessor-like manner.

Finally, applying the software engineering concept of *code smell* [5] to ontologies has been inspired by the work of Rosenfeld et al. [11]. They use *bad smells* in a Semantic Wiki context for triggering refactoring operations.

6   Conclusion and Future Work

We introduced an approach to pattern-based evolution of RDF knowledge bases. By considering the complete stack of Semantic Web knowledge representation techniques including its syntactic infrastructure as opposed to just the ontology layer, our approach fulfills additional requirements identified for example in user interviews (cf. Section 3). We provide a concrete implementation that leverages the plug-in architecture of OntoWiki[4], our semantic collaboration platform and framework. Thus, our implementation can make use of existing functionality of the OntoWiki framework like versioning of RDF knowledge bases.

Currently, EvoPat only ensures consistency through the definition of consistency-preserving patterns by the knowledge engineer. User-defined patterns can, however, lead to inconsistent knowledge bases. An approach that ensures consistency by proposing only those patterns whose application will not result in an inconsistent ontology, would thus be desirable. A straightforward (but admittedly not very scalable) solution to this problem is to combined EvoPat with a reasoner and test the application of a pattern employing the reasoner before its actual application in order to ensure correctness.

As opposed to bad smells, which indicate modeling problems, a promising approach is also to share and reuse modeling best practices. A problem which has to be solved in this regard, is the formalization and elicitation of a user's modeling requirements. A related idea for future work is the consumption of Linked Data. Our current implementation publishes evolution patterns on the Data Web but makes no use of gathering further information about resources. Doing so, could deliver hints for the applicability of specific patterns.

In a number of application projects we learned, that a key factor for the success of a knowledge engineering project is the efficient co-design of knowledge-bases and knowledge-based applications. Through the declarative definition of evolution with EvoPat it becomes possible to (semi-)automatize this co-design, since a knowledge base refactoring can trigger code refactoring and vice versa.

---

[4] Online at `http://code.google.com/p/ontowiki/wiki/ExtensionCookbook`

References

1. Auer, S., Dietzold, S., Lehmann, J., Hellmann, S., Aumueller, D.: Triplify: light-weight linked data publication from relational databases. In: Quemada, J., León, G., Maarek, Y.S., Nejdl, W. (eds.) Proceedings of the 18th International Conference on World Wide Web, WWW 2009, Madrid, Spain, April 20-24, pp. 621–630. ACM, New York (2009)
2. Augustin, C., Kuchta, B., Morgenstern, U., Riechert, T.: Datenbank und web-site catalogus professorum lipsiensis. ein sozialstatistisches analyseinstrumentarium und seine repräsentation im netz. In: Schattkowsky, M., Metasch, F. (eds.) Biografische Lexika im Internet. Bausteine, vol. 14, pp. 167–184. TUDPress, Verlag der Wissenschaften GmbH, Dresden (2009)
3. Djedidi, R., Aufaure, M.-A.: ONTO-EVO^{A}L an Ontology Evolution Approach Guided by Pattern Modeling and Quality Evaluation. In: Link, S., Prade, H. (eds.) FoIKS 2010. LNCS, vol. 5956, pp. 286–305. Springer, Heidelberg (2010)
4. Flouris, G., Manakanatas, D., Kondylakis, H., Plexousakis, D., Antoniou, G.: Ontology change: classification and survey. Knowledge Eng. Review 23(2), 117–152 (2008)
5. Fowler, M.: Refactoring: Improving the Design of Existing Code. Addison-Wesley, Reading (1999)
6. Heino, N., Dietzold, S., Martin, M., Auer, S.: Developing Semantic Web Applications with the OntoWiki Framework. In: Networked Knowledge – Networked Media. Springer, Heidelberg (2009)
7. Javed, M., Abgaz, Y.M., Pahl, C.: A Pattern-Based Framework of Change Operators for Ontology Evolution. In: Meersman, R., Herrero, P., Dillon, T.S. (eds.) OTM 2009 Workshops. LNCS, vol. 5872, pp. 544–553. Springer, Heidelberg (2009)
8. Magiridou, M., Sahtouris, S., Christophides, V., Koubarakis, M.: RUL: A Declarative Update Language for RDF. In: Gil, Y., Motta, E., Benjamins, V.R., Musen, M.A. (eds.) ISWC 2005. LNCS, vol. 3729, pp. 506–521. Springer, Heidelberg (2005)
9. Martin, M.: Exploring the netherlands on a semantic path. In: Auer, S., Bizer, C., Müller, C., Zhdanova, A. (eds.) Proceedings of the 1st Conference on Social Semantic Web, Leipzig, Germany, GI-edn., LNI, vol. P-113, p. 179. Bonner Köllen Verlag (2007) ISSN 1617-5468
10. Noy, N.F., Klein, M.C.A.: Ontology Evolution: Not the Same as Schema Evolution. Knowl. Inf. Syst. 6(4), 428–440 (2004)
11. Rosenfeld, M., Fernández, A., Díaz, A.: Semantic Wiki Refactoring. A strategy to assist Semantic Wiki evolution. In: Proceedings of the Fifth Workshop on Semantic Wikis (SemWiki 2010), co-located with 7th European Semantic Web Conference, ESWC 2010 (2010)
12. Stojanovic, L., Maedche, A., Motik, B., Stojanovic, N.: User-Driven Ontology Evolution Management. In: Gómez-Pérez, A., Benjamins, V.R. (eds.) EKAW 2002. LNCS (LNAI), vol. 2473, p. 285. Springer, Heidelberg (2002)

How to *Reuse* a Faceted Classification and Put It on the *Semantic* Web

Bene Rodriguez-Castro, Hugh Glaser, and Leslie Carr

School of Electronics and Computer Science, University of Southampton,
Southampton SO17 1BJ, UK
{b.rodriguez,hg,lac}@ecs.soton.ac.uk
http://www.ecs.soton.ac.uk

Abstract. There are ontology domain concepts that can be represented according to multiple alternative classification criteria. Current ontology modeling guidelines do not explicitly consider this aspect in the representation of such concepts. To assist with this issue, we examined a domain-specific simplified model for facet analysis used in Library Science. This model produces a Faceted Classification Scheme (FCS) which accounts for the multiple alternative classification criteria of the domain concept under scrutiny. A comparative analysis between a FCS and the Normalisation Ontology Design Pattern (ODP) indicates the existence of key similarities between the elements in the generic structure of both knowledge representation models. As a result, a mapping is identified that allows to transform a FCS into an OWL DL ontology applying the Normalisation ODP. Our contribution is illustrated with an existing FCS example in the domain of "Dishwashing Detergent" that benefits from the outcome of this study.

Keywords: facet analysis, faceted classification, normalisation, ontology design pattern, ontology modeling.

1 Introduction

Ontologies remain as one of the key components needed for the realization of the Semantic Web vision. They bring with them a broad range of development activities that can be grouped into what it is referred to as Ontology Engineering. Ontology Engineering for the Semantic Web is a very active research area and has experienced remarkable advancements in recent years, although it is still relatively new compared to other engineering practices within Computer Science or other fields. A constant ongoing effort in Ontology Engineering deals with harnessing the field with sound development methodologies analogous to those successfully employed in Software Engineering for decades. One of the objectives of these methodologies is to address areas of the ontology development process vulnerable to ad-hoc practices that could potentially lead to unexpected or undesirable results in ontology artifacts.

This paper describes a specific, very recurrent modeling scenario in ontology development, subject to such vulnerability. The scenario consists of domain-specific

P.F. Patel-Schneider et al. (Eds.): ISWC 2010, Part I, LNCS 6496, pp. 663–678, 2010.

concepts that can be represented according to multiple alternative classification criteria. To the best of our knowledge, guidelines for the conceptualization and representation of domain-specific concepts prone to be described based on multiple (potentially alternative) classification criteria, has not been explicitly considered in the context of ontology modeling for the Semantic Web.

General examples of domain-specific concepts that exhibit the characteristics described abound, going from a "bibliographic reference", (which could be classified according to several criteria such as "subject", "author", "publication venue", etc.); to a "toy" (which could be classified based on "suitable age", "brand", "subject type", etc.). The list of examples can go on. We have seen in our own experience that lack of specific design guidelines leaves ample room for conceptual errors when trying to develop a simple domain-specific ontology model for such concepts. For example, common mistakes when trying to represent these concepts and their classification criteria are to use subsumption relations between classes when in fact a *part-of* relation would be in order, or to use subsumption to model relationships that are outside OWL DL expressivity altogether.

Other examples of domain-specific concepts that can fit into the modeling scenario described are particularly interesting because they are used in well-known ontology development literature using OWL. They include: "Wine" [1], "Person" (in the context of family history relations) [2], or "Pizza" [3]. However, in none of them, they refer explicitly to the various classification criteria of the domain concept that are considered implicitly, nor attempt to represent these criteria explicitly in the respective ontology models developed.

To assist with these issues, we aim to put forward an initial set of basic design guidelines to mitigate the opportunity for ad-hoc modeling decisions in the development of ontologies for the problem scenario described. To obtain the conceptual model of a domain-specific concept and its multiple classification criteria we examined a simplified model for facet analysis in the field of Library and Information Science [4]. The outcome of this facet analysis is a Faceted Classification Scheme (FCS) for the domain concept in question where in most cases a *facet* would correspond to a *classification criterion*. To obtain an ontology representation of the FCS, we examined the Normalisation Ontology Design Pattern (ODP) [5] [6] [7]. A comparative analysis between a FCS and the Normalisation ODP revealed the existence of key similarities between both knowledge representation paradigms. The similarities allowed us to identify a series of mappings to transform a FCS into an OWL ontology applying the Normalisation pattern. Moreover, the ontology model obtained through this process contains a valid OWL DL representation of the classification criteria involved in the characterization of the domain concept.

To illustrate our contribution, we used throughout the document an existing FCS example in the domain of "Dishwashing Detergent" [8]. In fact, there are aspects of the work presented in this paper that could be viewed as a follow-up to [8] in the context of the Semantic Web and we attempted to acknowledge that in our title.

There is an additional important use case worth highlighting for motivating the need of this work as well. That is the modeling of the concept "Fault" in the domain of resilient and dependable computer systems. The representation of "Fault" is part of an ontology featured in a web portal knowledge base (RKB-Explorer[1]) for the project ReSIST[2] (Resiliance for Survivability in Information Society Technologies) [9].

The rest of this paper is structure as follows: Section 2 describes the structure and elements of a generic FCS; Section 3 does likewise regarding the Normalisation ODP; Section 4 introduces the alignments identified between both knowledge representation paradigms to enable the transformation of a generic FCS into a normalised ontology; Section 5 provides a comparison to previous work closely related to our proposal; and finally, Section 6 concludes the paper with some final remarks.

2 Faceted Classification Scheme

This section remarks the main features of a FCS involved in the comparative analysis to the Normalisation ODP for a given domain of discourse, while a thorough overview of facet analysis and FCSs can be found in [4] [10]. The latter also explores how FCSs compare to other knowledge representation approaches in classification and provides an account of its strengths and limitations.

Denton [8](§ 0) characterized a FCS for a given domain as follows: "a set of mutually exclusive and jointly exhaustive categories, each made by isolating one perspective on the items (a facet), that combine to completely describe all the objects in question, and which users can use, by searching and browsing, to find what they need".

However, in order to develop a FCS it is required to go through the process of Facet Analysis. Vickery [8](§ 2.3) describes Facet Analysis as: "The essence of facet analysis is the sorting of terms in a given field of knowledge into homogeneous, mutually exclusive facets, each derived from the parent universe by a single characteristic of division".

The key to Facet Analysis and FCSs is the notion of *facet*. Spiteri [4] simplified existing principles used in established Universal FCSs in Library Science. A fundamental of such principles is introduced as follows: "The Principles of Homogeneity and Mutual Exclusivity state respectively that facets must be homogeneous and mutually exclusive, i.e., that the contents of any two facets cannot overlap, and that each facet must represent only one characteristic of division of the parent universe".

In this sense, each facet can be designed separately and it models the domain of discourse from a distinct aspect. Each facet consists of a terminology, a finite set of terms that exhaust the facet. This set of terms is also referred to as *foci*.

There are numerous types of FCSs that vary in complexity. For example, FCSs that include several subject fields containing multiple facets and subfacets

---

[1] http://www.rkbexplorer.com/
[2] http://www.resist-noe.org/

[11](§ 8, Fig. 1). However, the rest of this section characterizes the elements of a simple generic FCS that this paper will refer to hereafter.

2.1   Structure and Elements

Definition 1. *Elements of a simple generic Faceted Classification Scheme:*

- Target Domain Concept (TDC).
- Facets: Facet1, Facet2, ..., rest of facets.
- Terms or foci (organized by facets):
  - Facet1: F1Term1, F1Term2, ..., rest of terms in Facet1.
  - Facet2: F2Term1, F2Term2, ..., rest of terms in Facet2.
  - ... rest of terms by facet.
- Set of items (from the TDC) to classify: Item1, Item2, ..., rest of items.

The following notation is introduced to refer to the elements of a generic FCS in Def. 1:

- TDC denotes the domain or universe of discourse. The domain-specific concept targeted by the FCS.
- $Facet_i$ denotes one of the facets of the FCS.
- F_iTerm_j denotes one of the terms of $Facet_i$.
- $Item_x$ denotes one the items from the domain of discourse to be classified.

Example 1. The structure below recaps the final FCS developed for the "Dishwashing Detergent" domain example in [8](§ 2.4). The elements of the schema fit into the generic structure presented in Def. 1.

- The TDC element is populated with the domain "Dishwashing Detergent".
- $Facet_i$ elements are populated with the facets: "Agent", "Form", "Brand Name", "Scent", "Effect On Agent", and "Special Property".
- F_iTerm_j elements are populated with the terms or foci listed below (grouped by facet):
  - Agent: dishwasher, person.
  - Form: gel, gelpac, liquid, powder, tablet.
  - Brand Name: Cascade, [...], Palmolive, President's Choice, Sunlight.
  - etc.
- $Item_x$ elements are populated in this case with two example items to classify:
  - "President's Choice Antibacterial Hand Soap and Dishwashing Liquid".
  - "Palmolive Aroma Therapy, Lavender and Ylang Ylang".

3   Normalisation Ontology Design Pattern

This section highlights the main characteristics of the Normalisation ODP relevant to the comparative analysis to a FCS.

The Normalisation pattern is classified as a "Good Practice" ODP in the catalog of ODPs introduced in [6] [7] (available online[3]). It can be applied to any OWL DL ontology that consists of a polyhierarchy where some *semantic axes* can be pointed. Each of those axes will be a *module*. One of their most powerful features, is the ability of logical reasoners to link these independent ontology modules to allow them to be separately maintained, extended, and re-used.

The pattern also establishes a series of requirements that a normalised ontology should meet, some of which are summarized below:

- The essence for the normalisation proposal is that the primitive skeleton of the domain ontology should consist of disjoint homogeneous trees (also referred to as *modules*) [5].
- Each primitive class that is part of the primitive skeleton should only have a primitive parent, and primitive sibling classes should be disjoint, creating the *modules* [6](§ 4.3.2.1).
- This implies that for any two primitive concepts either one subsumes the other or they are disjoint. Assertion of multiple inheritance relations among primitive concepts are not allowed [5].
- Normalisation allows exactly one unlabelled flavour of *is-kind-of* link corresponding to the links declared in the primitive skeleton. All others are inferred by the reasoner [5].

3.1   Structure and Elements

There are several examples of the generic structure of the Normalisation ODP in the literature [6](§ 4.3.2.1), [7](§ 6.5.1, § A.13) and online[3]. Figure 1 presents the specific version of the generic structure that this paper will refer to hereafter, which preserves the required characteristics of the pattern. Every node of the owl:Thing tree in Fig. 1, denotes an owl:Class. The symbol "($\equiv$)" indicates that the corresponding node is a *defined* class. Otherwise, the node is a *primitive* class. Every node of the owl:topObjectProperty tree denotes an owl:ObjectProperty. Figure 2 depicts a further generalization of the structure in Fig. 1 and introduces the following notation:

- :TDC denotes a primitive class representing the domain concept being normalised.
- :$Module_i$ denotes a primitive class that represents one of the modules.
- :M_iClass_j denotes a primitive class that represents a subset of the module class :$Module_i$.

---

[3] `http://odps.sourceforge.net/` (§ Normalisation).

```
owl:Thing
   |-- :Module1
      |-- :M1Class1
      |-- :M1Class2
      |-- (... rest of subclasses of Module1)
   |-- Module2
      |-- :M2Class1
      |-- :M2Class2
      |-- (... rest of subclasses of Module2)
   |-- (... rest of modules and subclasses)
   |-- :TargetDomainConcept (or :TDC)
      |-- (≡) :M1Class1TDC
      |-- (≡) :M1Class2TDC
      |-- (≡) (... rest of defined classes based on Module1)
      |-- (≡) :M2Class1TDC
      |-- (≡) :M2Class2TDC
      |-- (≡) (... rest of defined classes based on Module2)
      |-- (≡) (... rest of defined classes based on subclasses of the rest of modules)
      |-- :SpecificTDC1
      |-- :SpecificTDC2
      |-- (... rest of specific items from the TDC to be represented and classified)

owl:topObjectProperty
   |-- :hasModule1
   |-- :hasModule2
   |-- (... rest of properties based on the rest of modules)
```

Fig. 1. Generic structure of the Normalisation ODP

- $:hasModule_i$ denotes an object property that links every module $:Module_i$ to the different subclasses of the target domain concept $:M_iClass_jTDC$ and $:SpecificTDC_x$.
- $:M_iClass_jTDC$ denotes a defined class that represents a subset of the target domain concept class $:TDC$. Every class $:M_iClass_jTDC$ is defined based on a *one-to-one* relationship to the single corresponding class $:M_iClass_j$ that it is derived from.
- $:SpecificTDC_x$ denotes a primitive class that represents a subset of the target domain concept class $:TDC$ and an entity from the domain to be classified. Every class $:SpecificTDC_x$ is described based on a *one-to-many* relationship to various classes $:M_iClass_j$ from potentially different modules. As a consequence of this one-to-many relationship, the classes $:SpecificTDC_x$ could introduce the polyhierarchy scenarios in the ontology model that the Normalisation ODP aims to manage.

3.2   Implementation

One of the main features of the Normalisation ODP is to enable a reasoner to mantain the subsumption relations between a class $:SpecificTDC_x$ and the various classes $:M_iClass_jTDC$ involved in its description. This feature is accomplished encoding the conditions of the subsumption relation as restrictions in the implementation of the classes $:M_iClass_jTDC$ and $:SpecificTDC_x$.

Definition 2. *The implementation of a generic defined class* $:M_iClass_jTDC$ *is given as follows:*

```
owl:Thing
    |-- :Module_i
       |-- :M_iClass_j
    |-- :TargetDomainConcept (or :TDC)
       |-- (≡) :M_iClass_jTDC
       |-- :SpecificTDC_x

owl:topObjectProperty
    |-- :hasModule_i
```

Fig. 2. Generic structure of the Normalisation ODP

```
:M_iClass_jTDC
    rdf:type owl:Class ;
    rdfs:subClassOf :TDC ;
    owl:equivalentClass [ rdf:type owl:Restriction ;
                          owl:onProperty :hasModule_i ;
                          owl:someValuesFrom :M_iClass_j ] .
```

This implementation indicates that:

- A $:M_iClass_jTDC$ class is equivalent to an anonymous class described by an existential property restriction.
- The restriction is on the object property $:hasModule_i$ associated to the module $:Module_i$ that subsumes the class $:M_iClass_j$.
- The filler of the restriction is the class $:M_iClass_j$ linked to the definition of $:M_iClass_jTDC$.

Definition 3. *The implementation of a generic class $:SpecificTDC_x$ is given as follows:*

```
:SpecificTDC_x
    rdf:type owl:Class ;
    rdfs:subClassOf :TDC ,
                    [ rdf:type owl:Restriction ;
                      owl:onProperty :hasModule_i ;
                      owl:someValuesFrom :M_iClass_j ] ,
                    [ ... rest of existential restrictions on :hasModule_i
                          for every class :M_iClass_j that participates
                          in the description of :SpecificTDC_x ] .
```

This representation indicates the following:

- A class $:SpecificTDC_x$ is subsumed by a variable number of anonymous classes. More specifically, one anonymous class for every class $:M_iClass_j$ of every module $:Module_i$ that is linked to the description of $:SpecificTDC_x$. Every anonymous class is represented by an existential property restriction such as:

- The restriction is on the object property $:hasModule_i$, associated to the module $:Module_i$ that subsumes the class $:M_iClass_j$.
- The filler of the restriction is the class $:M_iClass_j$, linked to the description of $:SpecificTDC_x$.

This implementation of the classes $:M_iClass_jTDC$ and $:SpecificTDC_x$ respectively, enable a reasoner to infer and maintain the subsumption relations between a given class $:SpecificTDC_x$ and the various classes $:M_iClass_jTDC$ that it is related to.

Specific examples of the Normalisation ODP in the literature [6]($\S$ 4.3.2.1), [7]($\S$ 6.5.1, $\S$ A.13) and online[3] demonstrate the features of the pattern in specific use case scenarios.

4   Alignment of a FCS to the Normalisation ODP

A comparative analysis between the main characteristics of a FCS and the Normalisation ODP presented in previous sections, indicates the existence of key similarities between the elements in the generic structures of both conceptual models.

One such key similarity lies in the notion of *facet* in FCSs and the notion of *module* (or *semantic axis*) in the Normalisation ODP. Both elements represent one perspective of the domain being modelled, a single characteristic of division, a single criterion of classification in their respective paradigm.

Another key similarity is linked to the requirement for facets in a FCS to be homogeneous and mutually exclusive and likewise the requirement of modules in the Normalisation ODP to be comprised of primitive classes arranged in a structure of disjoint homogeneous class trees.

These key similarities prompt us to identify a mapping between the elements of both conceptual models that allows to transform a FCS into a normalised ontology model. In this first approach, the mapping aims to keep the design choices of the resultant normalised ontology as simple and straight-forward as possible, without compromising any of the requirements and features of both FCSs and the normalisation mechanism. This approach might not be suitable for converting all possible schemas into a normalised ontology but it is an attempt to provide an initial set of basic design guidelines. These guidelines can be extended hereafter to support more complex cases of FCSs.

The main principle is to represent each *facet* as a independent *module* or *semantic axis*. Following this principle makes the application of the Normalisation ODP *almost* straight-forward. Moreover, the resultant ontology includes the representation of the multiple alternative classification criteria that were considered in the original FCS for the target domain concept.

Table 1 summarizes the alignment of the elements in the generic structure of both conceptual models. This alignment enables the conversion from a FCS to an OWL DL ontology by applying the Normalisation ODP.

- The first column (leftmost), contains the elements of a generic FCS as introduced in Sect. 2.1, Def. 1.

Table 1. Alignment of a Faceted Classification Scheme to the Normalisation ODP

Library Science	Ontology Modeling		
FCS	**Norm. ODP**	**FCS in Norm. ODP**	**OWL Implementation**
TDC		$:TDC$	owl:Class (primitive)
$Facet_i$	$:Module_i$	$:Facet_i$	owl:Class (primitive)
	$:hasModule_i$	$:hasFacet_i$	owl:ObjectProperty
F_iTerm_j	$:M_iClass_j$	$:F_iTerm_j$	owl:Class (primitive)
	$:M_iClass_jTDC$	$:F_iTerm_jTDC$	owl:Class (defined) ($\equiv$)
$Item_x$		$:SpecificTDC_x$	owl:Class (primitive)

- The second column contains the elements of the Normalisation ODP generic structure as introduced in Sect. 3.1, Fig. 2.
- The third column represents the selected OWL notation for the elements of a generic FCS in the context of the Normalisation ODP generic structure.
- The forth column (rightmost), indicates the OWL implementation chosen for every element. The selection complies with the requirements of the normalisation mechanism.

Based on the principle of representating each facet as a module, the underlying ideas behind the mappings in Table 1 can be outlined as follows:

- The target domain concept TDC represents the domain of discourse of both a FCS and the Normalisation ODP. The primitive class $:TDC$ fulfills that role in the normalised ontology.
- A facet $Facet_i$ from a generic FCS corresponds to a module $:Module_i$ in the Normalisation ODP, therefore it becomes a primitive class $:Facet_i$ in the normalised ontology model.
- A facet $Facet_i$ from a FCS also becomes an object property $:hasFacet_i$ in the normalised ontology, given that for every module $:Module_i$ in the Normalisation ODP, there is an object property $:hasModule_i$.
- From the relationship between facet and module, it follows that a facet term F_iTerm_j from a FCS maps to a module subclass $:M_iClass_j$ from the Normalisation ODP. Both elements represents the same notion in their respective conceptual models. A subvidision, a refinement of the facet or module that they complement respectively. Therefore, a facet term F_iTerm_j from a FCS becomes a primitive class $:F_iTerm_j$ in the normalised ontology.
- A facet term F_iTerm_j from a FCS also produces a defined class $:F_iTerm_jTDC$ in the normalised ontology, given that for every primitive class $:M_iClass_j$ in the Normalisation ODP, there is a corresponding defined class $:M_iClass_jTDC$.
- Every item $Item_x$ to be classified in the FCS aligns to a class $:Specific_xTDC$ that is automatically classified by a reasoner in the Normalization ODP. Therefore, every element $Item_x$ is represented as a primitive class $:SpecificTDC_x$ in the normalised ontology.

The rest of this section details the characteristics of the resultant normalised ontology model that is obtained by applying the Normalisation ODP to a generic

```
owl:Thing
   |-- :Facet_i
      |-- :F_iTerm_j
   |-- :TargetDomainConcept (or :TDC)
      |-- (≡) :F_iTerm_jTDC
      |-- :SpecificTDC_x

owl:topObjectProperty
   |-- :hasFacet_i
```

Fig. 3. Elements of a FCS placed into the Normalisation ODP generic structure

FCS. The application of the pattern is driven by the alignments summarized in Table 1. The process is illustrated using the example of the "Dishwashing Detergent" FCS presented in Sect. 2.1, Ex. 1.

4.1   Structure and Elements

Figure 3 depicts the placement of the elements of a generic FCS into the generic structure of the Normalisation ODP based on the structure of the pattern in Sect. 3.1, Fig. 2 and the corresponding mappings from Table 1.

Example 2. Now let us populate the generic ontology structure in Fig. 3 with the specific elements of the "Dishwashing Detergent" FCS example. Figure 4 presents the overall normalised ontology class diagram obtained.

It is important to note that the structure in Fig. 4 includes axioms to comply with the requirement already stated of the Normalization ODP. That is, the skeleton of primitive classes consists of disjoint homogeneous tress where each primitive class only has a primitive parent, and primitive sibling classes are disjoint, creating the modules. This normalization requirement complies as well with the FCS requirement of facets being homogeneous and mutually exclusive based on the alignments in Table 1.

4.2   Implementation

Defined Classes. The generic implementation of a defined class $:F_iTerm_jTDC$ in terms of FCS elements is straight-forward based on the definition of $:M_iClass_jTDC$ given in Sect. 3.2, Def. 2 and Table 1.

Example 3. Let us illustrate the implementation of a defined class in the "Dishwashing Detergent" FCS example. Consider the facet "Agent" which contains the terms "Person" and "Dishwasher". From Table 1, these FCS elements fit into the normalised ontology as follows:

- $:Facet_i$ is populated with :Agent.

```
owl:Thing
  |-- :Agent
     |-- :Person
     |-- :Dishwasher
  |-- :Form
     |-- :Gel
     |-- :Gelpac
     |-- (... rest of terms in the facet "Form")
  |-- :BrandName
     |-- :Cascade
     |-- :Electrasol
     |-- (... rest of terms in the facet "Brand Name")
  |-- :Scent
     |-- :GreenApple
     |-- :GreenTea
     |-- (... rest of terms in the facet "Scent")
  |-- :EffectOnAgent
     |-- :AromaTherapy
        |-- :Invigorating
        |-- :Relaxing
  |-- :SpecialProperty
     |-- :Antibacterial
  |-- :DishwashingDetergent (:TDC)
     |-- (≡) :ManualDishDetergent
     |-- (≡) :DishwasherDishDetergent
     |-- (≡) :GelDishDetergent
     |-- (≡) :GelpacDishDetergent
     |-- (≡) (... rest of subclasses for each term in the facet "Form")
     |-- (≡) :CascaseDishDetergent
     |-- (≡) :ElectrasolDishDetergent
     |-- (≡) (... rest of subclasses for each term in the facet "Brand Name")
     |-- (≡) :GreenAppleDishDetergent
     |-- (≡) :GreenTeaDishDetergent
     |-- (≡) (... rest of subclasses for each term in the facet "Scent")
     |-- (≡) :AromaTherapyDishDetergent
        |-- (≡) :InvigoratingDishDetergent
        |-- (≡) :RelaxingDishDetergent
     |-- (≡) :AntibacterialDishDetergent
     |-- :PresidentsPersonLiquidAntibacterial
     |-- :PalmoliveAromaTherapyLavenderYlangYlang
     |-- :SpecificDishDetergent3
     |-- (... rest of specific dish detergent classes :SpecificDishDetergent_x to classify)

owl:topObjectProperty
  |-- :hasAgent
  |-- :hasForm
  |-- :hasBrand
  |-- :hasScent
  |-- :hasEffectOnAgent
  |-- :hasSpecialProperty
```

Fig. 4. Normalised ontology structure of the "Dishwashing Detergent" FCS

- $:hasFacet_i$ is populated with :hasAgent.
- $:F_iTerm_j$ is populated with :Person and :Dishwasher respectively.
- $:F_iTerm_jTDC$ is populated with :ManualDishDetergent and :Dishwash-erDishDetergent respectively.

As an example, let us focus on the class :DishwasherDishDetergent. The implementation in the normalised ontology can be stated as follows:

```
:DishwasherDishDetergent
    rdf:type owl:Class ;
    rdfs:subClassOf :DishDetergent .
    owl:equivalentClass [ rdf:type owl:Restriction ;
                          owl:onProperty :hasAgent ;
                          owl:someValuesFrom :Dishwasher ] .
```

The implementation of the rest defined classes in the "Dishwashing Detergent" FCS shown in Fig. 4 follows the same rationale.

Classification Classes. The generic implementation of a class $:SpecificTDC_x$ in terms of FCS elements is straight-forward following the implementation of $:SpecificTDC_x$ given in Sect. 3.2, Def. 3 and Table 1.

Example 4. To illustrate the representation of a specific dishwashing detergent, let us reuse one of the classification examples presented in [8](§ 2.4). The item "President's Choice Antibacterial Hand Soap and Dishwashing Liquid" is classified in the cited reference, as follows: (Agent: person), (Form: liquid), (Brand Name: President's Choice), (Scent: none), (Effect on Agent: none) and (Special Property: antibacterial). From Table 1, the description of the example detergent reveals the following mappings:

- $:TDC$ is populated by :DishDetergent.
- $:SpecificTDC_x$ is populated by :PresidentsPersonLiquidAntibacterial.
- There are four existential restrictions. One per facet term involved in the description of the specific detergent at hand ("person", "liquid", "President's Choice", and "antibacterial"). Therefore, for each restriction:
  - $:hasFacet_i$ is populated with :hasAgent, :hasForm, :hasBrandName and :hasSpecialProperty respectively.
  - $:F_iTerm_j$ is populated with :Person, :Liquid, :PresidentsChoice and :Antibacterial respectively.

The implementation of this particular detergent in the normalised ontology can be stated as follows:

```
:PresidentsPersonLiquidAntibacterial
    rdf:type owl:Class ;
    rdfs:subClassOf :DishDetergent ,
                    [ rdf:type owl:Restriction ;
                      owl:onProperty :hasAgent ;
                      owl:someValuesFrom :Person ] ,
                    [ rdf:type owl:Restriction ;
                      owl:onProperty :hasForm ;
                      owl:someValuesFrom :Liquid ] ,
                    [ rdf:type owl:Restriction ;
                      owl:onProperty :hasBrandName ;
                      owl:someValuesFrom :PresidentsChoice ] ,
                    [ rdf:type owl:Restriction ;
                      owl:onProperty :hasSpecialProperty ;
                      owl:someValuesFrom :Antibacterial ] .
```

This description makes explicit the relationship between the specific detergent class and every term of every facet that participate in the facet classification of the item. Moreover, it enables a reasoner to infer that :PresidentsPersonLiquidAntibacterial is a subclass of the defined classes :ManualDishDetergent, :LiquidDishDetergent, :PresidentsChoiceDishDetergent and :AntibacterialDishDetergent.

A version of the complete normalised ontology model for the "Dishwashing Detergent" FCS example is available online[4] in RDF/XML format.

5   Relation to Other Methods

Previous work that defines mappings between different semantic models include [12]. The authors performs a rigorous and comprehensive comparative analysis between the primitive elements of three semantic models: the Semantic Web Ontology Language (OWL), the Relational Database Model (RDBM), and the Resource Space Model (RSM). Based on the identified mappings between every two models, a detailed set of criteria is provided to transform one of them to the other. The most relevant to us is the mapping between RSM and OWL because of its similarities with the conversion between a FCS and OWL that we propose here.

The RSM is defined as a semantic model for specifying, organizing and retrieving diverse multimedia resources by classifying their contents according to different partition methods and organizing them according to a multidimensional classification space. A FCS is also a multidimensional classification space and comparing the primitive elements of a FCS and a RSM the following mapping is instantly revealed:

- The domain or universe of discourse of the FCS (the target domain concept) corresponds to the overall resource space, the RS element in the RSM.
- A facet in the FCS corresponds to an axis X_i in the RSM.
- A facet term in the FCS corresponds to a coordinate C_i in the RSM.
- A facet is covered and exhausted by the set of terms associated to it in a FCS. The same principle holds in a RSM for an axis and the set of coordinates associated to it, $X_i = \langle C_{i1}, C_{i2}, ..., C_{in} \rangle$.
- An item to be classified by the FCS corresponds to a point p in the RSM.

These mappings show that a generic FCS can be converted into a RSM, which in turn can be converted into an OWL model using the RSM to OWL mappings in [12]. Now there are two possible paths to convert a FCS into an OWL model.

- Path 1: FCS to RSM via mappings above and RSM to OWL via mappings in [12]. Let us refer to this OWL model as O_1.
- Path 2: FCS to OWL via mappings presented in our paper using the Normalization ODP. Let us refer to this OWL model as O_2.

---

[4] http://purl.org/net/project/enakting/ontology/detergent_fcs_norm

There are important differences between the ontologies O_1 and O_2. An important difference is due to the RSM to OWL conversion in [12]. RSM describes mainly classification semantics and as the authors explain, this means that there is no semantic loss when converting from RSM to OWL but there might be semantic loss when transforming an OWL model that includes richer semantics into a RSM. This also means that, in terms of W3C standards, the expressivity level of the resultant OWL model O_1, will be within the RDF Schema or OWL Lite boundary.

On the other hand, the ontology O_2 is within OWL DL and presents richer OWL semantics than O_1, provided by the Normalization ODP. These additional OWL DL semantics in O_2 enable one of the main features of the normalization pattern such as the automatic classification and maintenance of complex subsumption relations by a reasoner. So while O_1 is a valid OWL description of the FCS that it is based on, O_2 using our proposed method provides additional semantics at the OWL DL level that support a richer description and additional features of the classification criteria considered in the initial FCS.

Additional research that made use of facet analysis in Library and Information Science to build computational ontologies includes [13]. Giunchiglia et al. introduces the concept of Faceted Lightweight Classification Ontology as "a lightweight (classification) ontology where each term and corresponding concept occurring in its node labels must correspond to a term and corresponding concept in the background knowledge, modeled as a faceted classification scheme".

Similarities to our approach include:

- The use of a FCS to model certain background knowledge and to derive and ontology based on it.
- Each concept in the ontology model obtained in our method also corresponds to a concept in the FCS.

There are important differences where our approach deviates from that in [13] probably due to the different type of problems that both are trying to address respectively. Giunchiglia et al. are trying to counteract the lack of interest and difficulties on the user side to build and reuse ontologies while our concern focuses on identifying explicit guidelines to represent the notion of multiple classification criteria in domain concepts. Additional differences include:

- The expressive level for the resultant ontology model in our method is OWL DL. In contrast, [13] focuses on lightweight classification ontologies which expressive level would loosely correspond to no more than RDF Schema in terms of W3C Standards. Key features provided by the Normalisation ODP found in our method, can not be implemented using solely RDF Schema semantics.
- The type of FCS used in [13] is based on a Universal Faceted Classification System. On the other hand, we have focused on simpler custom domain-specific FCSs to serve as an starting point for our initial proof of concept. This helped limiting the complexity of the classification criteria to consider and represent in the corresponding ontology.

6   Conclusions

This paper has presented an initial set of basic design guidelines to develop an ontology model within OWL DL that supports the representation of multiple alternative classification criteria of a specific domain concept.

A lack of explicit guidance in the ontology development literature on how to address this recurrent modeling scenario, leaves ample room for ad-hoc practices that can lead to unexpected or undesired results in ontology artifacts. In our attempt to mitigate this void, we examined a simplified procedure to develop a Faceted Classification Scheme (FCS) which contains the conceptualization of various classification criteria (facets) of a specific target domain concept. A series of mappings between the elements of a generic FCS and the Normalization Ontology Design Pattern (ODP) have been identified that allow us to convert a given FCS into an OWL DL ontology model following a consistent and systematic approach. The resultant ontology model includes the representation of the various classification criteria of the domain concept considered in the original FCS. An existing FCS example in the domain of "Dishwashing Detergent" is used to illustrate the main steps of our conversion procedure.

The guidelines presented in this first effort consider explicitly the conceptualization of existing classification criteria in the context of ontology modeling for the Semantic Web and provide a *partial* solution to the problem scenario described. They do not cover all existing types of generic structures of FCSs (which can be the aim of future work) and they do not eliminate all opportunities of potentially hazardous ad-hoc decisions in the development process. However, we believe the use of a consistent, systematic and fit-for-purpose approach allows to significantly reduced them.

Acknowledgments. This work was supported in part by the UK Engineering and Physical Sciences Research Council (EPSRC), in the context of the EnAKTing project under grant number EP/G008493/1. Additionally, many people have contributed directly and indirectly to this work and we thank them all. In particular, everyone that participated in the online discussion "The notion of a 'classification criterion' as a class"[5],[6] via the mailing lists: "ontolog-forum@ontolog.cim3.net" and "public-owl-dev@w3c.org"; and the British Chapter of the International Society for Knowledge Organization (ISKO).

References

1. Welty, C., McGuinness, D.L., Smith, M.K.: OWL web ontology language guide. W3C recommendation, W3C (February 2004),
   `http://www.w3.org/TR/2004/REC-owl-guide-20040210/`
2. Krötzsch, M., Patel-Schneider, P.F., Rudolph, S., Hitzler, P., Parsia, B.: OWL 2 web ontology language primer. Technical report, W3C (October 2009),
   `http://www.w3.org/TR/2009/REC-owl2-primer-20091027/`

---

[5] `http://lists.w3.org/Archives/Public/public-owl-dev/2010AprJun/0009.html`
[6] `http://ontolog.cim3.net/forum/ontolog-forum/2010-04/msg00051.html`

3. Horridge, M., Drummond, N., Jupp, S., Moulton, G., Stevens, R.: A practical guide to building owl ontologies using the protege-owl plugin and co-ode tools edition 1.2. Technical report, The University Of Manchester (March 2009)
4. Spiteri, L.: A simplified model for facet analysis: Ranganathan 101. Canadian Journal of Information and Library Science 23(1/2), 1–30 (1998)
5. Rector, A.L.: Modularisation of domain ontologies implemented in description logics and related formalisms including owl. In: Proceedings of the 2nd International Conference on Knowledge Capture, K-CAP 2003, pp. 121–128. ACM, New York (2003)
6. Egana-Aranguren, M.: Ontology Design Patterns for the Formalisation of Biological Ontologies. MPhil Dissertation, Bio-Health Informatics Group, School of Computer Science, University of Manchester (2005)
7. Egana-Aranguren, M.: Role and Application of Ontology Design Patterns in Bio-ontologies. PhD thesis, School of Computer Science, University of Manchester (2009)
8. Denton, W.: How to make a faceted classification and put it on the web (November 2003), http://www.miskatonic.org/library/facet-web-howto.html
9. Rodriguez-Castro, B., Glaser, H.: Whose "fault" is this? untangling domain concepts in ontology design patterns. In: Workshop on Knowledge Reuse and Reengineering over the Semantic Web in the 5th European Semantic Web Conference (June 2008)
10. Kwasnik, B.H.: The role of classification in knowledge representation and discovery. Library Trends 48(1) (1999)
11. Vickery, B.: Faceted classification for the web. Axiomathes 18(2), 145–160 (2008)
12. Zhuge, H., Xing, Y., Shi, P.: Resource space model, owl and database: Mapping and integration. ACM Trans. Internet Technol. 8(4), 1–31 (2008)
13. Giunchiglia, F., Dutta, B., Maltese, V.: Faceted lightweight ontologies. In: Borgida, A., Chaudhri, V.K., Giorgini, P., Yu, E.S.K. (eds.) Conceptual Modeling: Foundations and Applications. LNCS, vol. 5600, pp. 36–51. Springer, Heidelberg (2009)

OWL-POLAR: Semantic Policies for Agent Reasoning[*]

Murat Şensoy[1], Timothy J. Norman[1], Wamberto W. Vasconcelos[1], and Katia Sycara[1,2]

[1] Department of Computing Science, University of Aberdeen, AB24 3UE, Aberdeen, UK
{m.sensoy,t.j.norman,w.w.vasconcelos}@abdn.ac.uk
[2] Carnegie Mellon University, Robotics Institute, Pittsburgh, PA 15213, USA
katia@cs.cmu.edu

Abstract. Policies are declarations of constraints on the behaviour of components within distributed systems, and are often used to capture norms within agent-based systems. A few machine-processable representations for policies have been proposed, but they tend to be either limited in the types of policies that can be expressed or limited by the complexity of associated reasoning mechanisms. In this paper, we argue for a language that sufficiently expresses the types of policies essential in practical systems, and which enables both policy-governed decision-making and policy analysis within the bounds of decidability. We then propose an OWL-based representation of policies that meets these criteria using and a reasoning mechanism that uses a novel combination of ontology consistency checking and query answering. In this way, agent-based systems can be developed that operate flexibly and effectively in policy-constrained environments.

1 Introduction

In this paper, we present a novel and powerful OWL 2.0 [7] knowledge representation and reasoning mechanism for policies: OWL-POLAR (an acronym for OWL-based POlicy Language for Agent Reasoning). Policies (aka. norms) are system-level principles of ideal activity that are binding upon the components of that system. Depending on the nature of the system itself, policies may serve to control, regulate or simply guide the activities of components. In systems security, for instance, the aim is typically to control behaviour such that the system complies with the policies [18]. In real socio-technical systems, however, there are important limits to this and the aim is to develop effective sets of policies along with incentives to regulate behaviour [2]. In systems of autonomous agents, the term norm is most prevalent, but the concept and issues remain the same [4]; for example, norms are used to regulate the behaviour of agents representing disparate interests in electronic institutions [6]. The objective of this research is to capture the essential requirements of policy representation and reasoning. In meeting this objective three key requirements must be met:

---

[*] This research was sponsored by the U.S. Army Research Laboratory and the U.K. Ministry of Defence and was accomplished under Agreement Number W911NF-06-3-0001. The views and conclusions contained in this document are those of the author(s) and should not be interpreted as representing the official policies, either expressed or implied, of the U.S. Army Research Laboratory, the U.S. Government, the U.K. Ministry of Defence or the U.K. Government. The U.S. and U.K. Governments are authorized to reproduce and distribute reprints for Government purposes notwithstanding any copyright notation hereon.

P.F. Patel-Schneider et al. (Eds.): ISWC 2010, Part I, LNCS 6496, pp. 679–695, 2010.
© Springer-Verlag Berlin Heidelberg 2010

1. System/institutional policies must be machine understandable and underpinned by a clear interpretation.
2. The representation must be sufficiently expressive to capture the notion of a policy across domains.
3. Policies must be able to be effectively shared/interpreted at run-time.

The choice of OWL 2.0 as an underlying language addresses the first requirement, but in meeting the second two, we must clearly outline what is required of a policy language and what reasoning should be supported by it. The desiderata of a model of policies that motivates the language OWL-POLAR are as follows:

- **Representational adequacy.** Policies (or norms) must capture the distinction between activities that are required (obliged), restricted (prohibited) and, in some way, authorised but not necessarily expected (permitted) by some representational entity within the environment. It is essential to capture the authority from which the policy/norm comes, the subject (agent) to whom it applies, the object (activity) to which the policy/norm refers, and the circumstances within which it applies.
- **Supporting decisions.** Any reasoning mechanism that is driven/guided by policies must support both the determination of what policies/norms apply in a given situation, and what activities are warranted by the normative state of the agent if it were to comply with these policies.
- **Supporting analysis.** Any reasoning mechanism that is driven/guided by normative/policy constraints must support the assessment of policies in terms of: (i) whether a policy/norm is meaningful and (ii) whether norms conflict, and in what circumstances they do conflict.

With the introduction of data ranges within the OWL 2.0 specification [7], we believe that this desiderata of a model of policies can be met within the confines of OWL-DL. If this claim can be shown to be valid (as we aim to do within this paper), we believe that OWL-POLAR provides, for the first time, a sufficiently expressive policy language for which the key reasoning mechanisms required of such a language are decidable.

The paper is organised as follows: in Section 2 we formally specify the OWL-POLAR language within OWL-DL; in Section 3 we describe how a set of active policies may be computed, and how decisions about what activities are warranted by some set of policies may be made; then in Section 4 we present in detail the reasoning mechanisms that support the analysis of policies. OWL-POLAR is then compared to existing languages for policies in Section 5, and we present our conclusions in Section 6.

2 Semantic Representation of Policies

The proposed language for semantic representation of policies is based on OWL-DL [7]. An OWL-DL ontology $o = (TBox_o, ABox_o)$ consists of a set of axioms defining the classes and relations ($TBox_o$) as well as a set of assertional axioms about the individuals in the domain ($ABox_o$). Concept axioms have the form $C \sqsubseteq D$ where C and D are concept descriptions, and relation axioms are expressions of the form $R \sqsubseteq S$, where R and S are relation descriptions. The ABox contains concept assertions of the form $C(a)$ where C is a concept and a is an individual name, and relation assertions of the form $R(a, b)$, where R is a relation and a and b are individual names.

Conjunctive semantic formulas are used to express policies. A conjunctive semantic formula $F_v^o = \bigwedge_{i=0}^n \phi_i$ over an ontology o is a conjunction of atomic assertions ϕ_i, where $v = \langle ?x_0, \ldots, ?x_n \rangle$ represents a vector of variables used in these assertions. For the sake of convenience, we assume $\bigwedge_{i=0}^n \phi_i \equiv \{\phi_1, \ldots \phi_n\}$ in order to consider a conjunctive formula as a set of atomic assertions. Based on this, F_v^o can be considered as $T_v^o \cup R_v^o \cup C_v^o$, where T_v^o is a set of type assertions using the concepts from o, e.g., $\{student(?x_i), nurse(?x_j)\}$; R_v^o is set of of relation assertions using the relations from o, e.g., $\{marriedTo(?x_i, ?x_j)\}$; C_v^o is a set of constraint assertions on variables. Each constraint assertion is of the form $?x_i \triangleleft \beta$, where β is a constant and $\triangleleft$ is any of the symbols $\{>, <, =, \neq, \geq, \leq\}$. A constant is either a data literal (e.g, a numerical value) or an individual defined in o.

Variables are divided into two categories; data-type and object variables. A data-type variable refers to data values (e.g., integers) and can be used only once in R_v^o. On the other hand, an object variable refers to individuals (e.g., University_of_Aberdeen) and can be used freely many times in R_v^o. Equivalence and distinction between the values of object variables can be defined using OWL properties *sameAs* and *differentFrom* respectively, e.g., *owl:sameAs(?x,?y)*. In the rest of the paper, we use the symbols α, ρ, φ, and e as a short hand for semantic formulas.

Given an ontology o, a conditional policy is defined as $\alpha \longrightarrow N_{\chi:\rho}(a : \varphi)/e$, where

1. α, a conjunctive semantic formula, is the activation condition of the policy.
2. $N \in \{O, P, F\}$ indicates if the policy is an obligation, permission or prohibition.
3. χ is the policy addressee and ρ describes χ using only the *role* concepts from the ontology (e.g., $?x : student(?x) \wedge female(?x)$, where *student* and *female* are defined as sub-concepts of the *role* concept in the ontology). That is, ρ is of the form $\bigwedge_{i=0}^n r_i(\chi)$, where $r_i \sqsubseteq role$. Note that χ may directly refer to a specific individual (e.g., *John*) in the ontology or a variable.
4. $a : \varphi$ describes what is prohibited, permitted or obliged by the policy. Specifically, a is a variable referring to the action to be regulated by the policy and φ describes a as an action instance using the concepts and properties from the ontology (e.g., $?a : SendFileAction(?a) \wedge hasReceiver(?a, John) \wedge hasFile(?a, TechReport218.pdf)$, where *SendFileAction* is an *action* concept). Each action concept has only a number of functional relations (aka. functional properties) [7] and these relations are used while describing an instance of that action.
5. e defines the expiration condition.

Table 1 illustrates how a conditional policy can be represented using the proposed approach. The policy in the table states that a person is obliged to leave a location when there is a fire risk.

Table 1. A person has to leave a location when there is a fire risk

α	$Place(?b) \wedge hasFireRisk(?b, true) \wedge in(?x, ?b)$
N	O
$\chi : \rho$	$?x : Person(?x)$
$a : \varphi$	$?a : LeavingAction(?a) \wedge about(?a, ?b) \wedge hasActor(?a, ?x)$
e	$hasFireRisk(?b, false)$

Given a semantic representation for the state of the world, policies are used to reason about actions that are permitted, obliged or prohibited. Let Δ_o be a semantic representation for a state of the world based on an ontology o. Each state of the world is partially observable; hence Δ_o is a partial representation of the world. Δ_o itself is represented as an ontology composed of $(TBox_o, ABox_\Delta)$ where $ABox_\Delta$ is an extension of $ABox_o$.

3   Reasoning with Policies

When its activation conditions are satisfied, a conditional policy leads to an activated policy. Definition 1 summarizes how a conditional policy is activated using ontological reasoning over a state of the world. Here we use query answering to determine activated policies and reason about actions. The query answering mechanism we use in this work is DL-safe; i.e. variables are bound only to the named individuals, to guarantee decidability [8]. In this section, we address some of the key issues in supporting decisions governed by policies: activation and expiration, and reasoning about interactions between policies and actions.

Definition 1. Let Δ_o be a state of the world represented based on a domain ontology o. If there is a substitution σ such that $\Delta_o \vdash (\alpha \wedge \rho) \cdot \sigma$, but there is no substitution σ' such that $\Delta_o \vdash (e \cdot \sigma) \cdot \sigma'$, then the policy $(N_\chi (a : \varphi)) \cdot \sigma$ becomes active. This policy expires when there exists a substitution σ' such that $\Delta_o \vdash (e \cdot \sigma) \cdot \sigma'$.   ∎

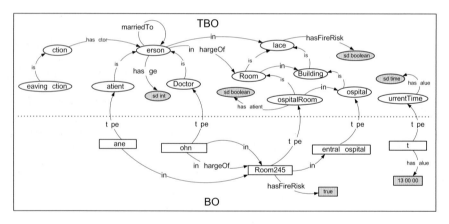

Fig. 1. A partial state of the world represented based on a domain ontology

3.1   Policy Activation

A policy is activated for a specific agent when the world state is such that the activation condition holds for that agent and the expiration condition does not hold, and expires when this latter condition holds. The above definition is rather standard [11], but we now describe how this is implemented efficiently through query answering. A conjunctive semantic formula can be trivially converted to a SPARQL query [15] and can be evaluated by OWL-DL reasoners with SPARQL-DL [17] support such as Pellet [17] to find a substitution for its variables satisfying a specific state of the world. Therefore, we can test $\Delta_o \vdash (\alpha \wedge \rho) \cdot \sigma$ by writing a query for $(\alpha \wedge \rho)$ and testing whether it is entailed by Δ_o or not. Consider the conditional policy in Table 1 and assume that we have the partially represented state of the world in Figure 1. We can write the semantic query in Figure 2 to find σ for the conditional policy. When we query the state of the world using SPARQL, each result in the result set provides a substitution σ; in our case, we have two σ values: {?x/John, ?b/Room245} and {?x/Jane, ?b/Room245}, representing that there is a fire risk in the room 245 of the Central Hospital and that John and Jane are in that room.

```
Query:

 q(?x, ?b):-
    Place(?b) ∧
    hasFireRisk(?b, true) ∧
    Person(?x) ∧
    in(?x,?b).
```

```
SPARQL SYNTAX:

 PREFIX example: <http://www.example.com/ns#>
 PREFIX rdf: <http://www.w3.org/...rdf-syntax-ns#>
 PREFIX xsd: <http://www.w3.org/2001/XMLSchema#>
 SELECT ?x ?b
 WHERE {
     ?b rdf:type example:Place.
     ?b example:hasFireRisk "true"^^xsd:boolean.
     ?x rdf:type example:Person.
     ?x example:in ?b.
 }
```

Fig. 2. Query for the activation of a policy

Now, using the computed σ values, we should try to find a σ' such that $\Delta_o \vdash (e \cdot \sigma) \cdot \sigma'$. In our case, for this purpose, we can use the semantic query "$q():-$ has-FireRisk(Room245, false)". When the SPARQL representation of this query is executed over the state of the world shown in Figure 1, it returns *false*; that is the RDF graph pattern represented by the query could not be found in the ontology. This means that the policy in Table 1 should be activated using the variable bindings in σ. The result is activations of $O_{John}(?a : LeavingAction(?a) \wedge about(?a, Room245))$ and $O_{Jane}(?a : LeavingAction(?a) \wedge about(?a, Room245))$. These policies mean that *John* and *Jane* are obliged to leave the *room 245*; the obligation expires when the fire risk is removed.

3.2 Reasoning about Actions

Let us assume that a specific action $a' : \varphi'$ will be performed by x, where a' is a URI referring to the action instance and φ' is a conjunctive semantic formula describing a' without using any variables. Let Δ_o be the current state of the world. We can test if the action a' is permitted, forbidden or prohibited in Δ_o. For this purpose, based on Δ_o, we create a "sandbox" (hypothetical) state of the world Δ'_o to make *what-if* reasoning [21], i.e., Δ'_o shows what happens if the action is performed. This is achieved by simply adding the described action instance to Δ_o, i.e., $\Delta'_o = \Delta_o \cup \varphi'$. For example, the state of the world in Figure 1 is extended using action instance *LeaveAct_1: LeavingAction(LeaveAct_1) $\wedge$ hasActor(LeaveAct_1,John) $\wedge$ about(LeaveAct_1,room245)*. The resulting state of the world is shown in Figure 3.

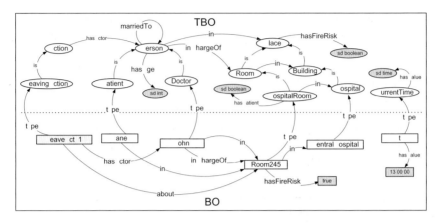

Fig. 3. The "sandbox" (hypothetical) state of the world

For each active policy $N_x(y : \varphi_y)$, we test the expiration conditions on Δ'_o as explained before. If the policy's expiration conditions are satisfied, we can conclude that the action $a' : \varphi'$ leads to the expiration of the policy. Otherwise, a semantic query Q of the form $q(\boldsymbol{v}_{\varphi_y}):- \varphi_y$ is created, where $\boldsymbol{v}_{\varphi_y}$ is the vector of variables in φ_y. Then, Δ'_o is queried with Q. Let the query return a result set rs; each result $r \in rs$ is a substitution such that $\Delta'_o \vdash \varphi_y \cdot r$. If $y \cdot r = a'$ for any such r, then a' is regulated by the policy. In this case, we can interpret the policy based on its modality as follows:

1. $N_x = O$: In this case, the policy represents an obligation; that is, x is obliged to perform a'. Performing a' will remove this obligation.
2. $N_x = P$: Performing a' is explicitly permitted.
3. $N_x = F$: Performing a' is prohibited.

After examining the active policies as described above, we can identify a number of possible normative positions with respect to the action instance a': (i) doing a' may be explicitly permitted if there is a policy permitting it; (ii) doing a' may be obligatory if there exists a policy obliging it; (iii) doing a' may be prohibited if there is a policy prohibiting it; and (iv) there may be a conflict in the normative position with respect to a' if it is either both prohibited and explicitly permitted, or both prohibited and obliged.

4   Reasoning about Policies

In this section, we demonstrate reasoning techniques to support the analysis of policies in terms of their meaningfulness (Section 4.2) and possibility of conflict (Section 4.3), and hence address our third desideratum. Prior to this, however, we propose methods for reasoning about semantic formulas to underpin our mechanisms for policy analysis.

4.1   Reasoning about Semantic Formulas

Here, we introduce methods for reasoning about semantic conjunctive formulas using query freezing and constraint transformation.

Conjunctive Queries. There is a relation between conjunctive formulas and conjunctive queries. A conjunctive semantic formula can trivially be converted into a conjunctive semantic query. For example, $A^o_{v_1}$ can be converted into the query $q_A():- A^o_{v_1}$. Therefore, we can use query reasoning techniques to reason about semantic formulas. For instance, in order to reason about the subsumption between semantic formulas, we can use query subsumption (containment).

In conjunctive query literature, in order to test whether q_A subsumes q_B, the standard technique of *query freezing* is used to reduce query containment problem to query answering in Description Logics [13,20]. For this purpose, we build a canonical ABox Φ_{q_B} from the query $q_B():- B^o_{v_2}$ in three steps. First, for each variable in v_2, we put a fresh individual into Φ_{q_B} using the type assertions about the variable. Note that this individual should not exist in o. Second, we add each individual appearing in q_B into Φ_{q_B}. This is done using the information about the individual from the $ABox_o$ (e.g., type assertions). Third, relationships between individuals and constants defined in q_B are inserted into Φ_{q_B}. As a result of this process, Φ_{q_B} contains a pattern that exists only in ontologies that satisfy q_B. We combine Φ_{q_B} and our $TBox_o$ to create a new canonical ontology, $o' = (TBox_o, \Phi_{q_B})$. Example 1 demonstrates a simple case. Based on [20,13], we conclude that $o \vdash q_B \sqsubseteq q_A$ if and only if o' entails q_A. In order to test whether o' entails q_A or not, we query o'. That is, o' entails q_A if there exists at least one match for q_A in o'. This can easily be achieved by converting q_A to SPARQL syntax and use Pellet's SPARQL-DL query engine to answer q_A on o' [17].

Example 1. Let query q_A be *q():- Person(?p) $\wedge$ marriedTo(?p,?x) $\wedge$ Patient(?x)* and query q_B be *q():- Doctor(?x) $\wedge$ marriedTo(?x,Jane) $\wedge$ hasChild(?x,?c)*. Then, Φ_{q_B} contains an individual x, which is created for the variable $?x$. The individual x is defined as of type *Doctor*. In Φ_{q_B}, we also have another individual *Jane*, which is defined in the original $ABox_o$ as an instance of the *Patient* class; we get all of its type assertions from the $ABox_o$. Then, we insert the object property *marriedTo* between the individuals x and *Jane*. Lastly, we create another individual c for the variable $?c$ in Φ_{q_B} and insert the *hasChild* object property between x and c. The resulting ontology is shown in Figure 4.

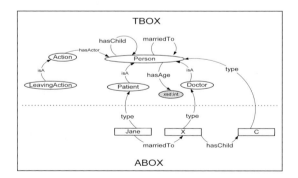

Fig. 4. The ontology created for q_B in Example 1

```
<owl:Class rdf:about="#AgeConst1">
 <rdfs:subClassOf>
    <owl:Restriction>
      <owl:onProperty rdf:resource="#hasAge"/>
        <owl:allValuesFrom>
         <rdfs:Datatype>
            <owl:onDataRange rdf:resource="&xsd;nonNegativeInteger"/>
            <xsd:minInclusive rdf:datatype="&xsd;int">10</xsd:minInclusive>
            <xsd:maxExclusive rdf:datatype="&xsd;int">20</xsd:maxExclusive>
         </rdfs:Datatype>
        </owl:allValuesFrom>
    </owl:Restriction>
 </rdfs:subClassOf>
</owl:Class>
```

Fig. 5. A concept named **AgeConst1** is created for *hasAge(?c, ?a)* $\wedge$ *?a $\geq$ 10* $\wedge$ *?a $\leq$ 20*

The query freezing method described above enables us to create a canonical ABox for a semantic conjunctive formula; this ABox represents a pattern which only exists in ontologies satisfying the semantic formula. On the other hand, this method assumes that variables in queries can be assigned fresh individuals in a canonical ABox. However, in OWL-DL, individuals can refer to objects, but not data values [19]. Therefore, the proposed query freezing method can be used to test for subsumption between q_A and q_B only if the variables in q_A and q_B refer to objects. A variable can refer to an object if it is used as the domain of an object or datatype property (e.g., *hasAge(?x,10)*) or if it is used as the range of an object property (e.g., *marriedTo(Jack,?x)*). Unfortunately, in many real-life settings, queries may have variables referring to data values with various constraints, which we refer to here as *datatype variables*. In these settings, the query freezing described above cannot be used to test subsumption. Example 2 illustrates a simple scenario.

Example 2. Let query q_A be *q():- Person(?p)* $\wedge$ *hasChild(?p,?c)* $\wedge$ *hasAge(?c,?y)* $\wedge$ *?y $\geq$ 12* $\wedge$ *?y $\leq$ 16* and query q_B be *q():- Doctor(?x)* $\wedge$ *marriedTo(?x,Jane)* $\wedge$ *hasChild(?x, ?c)* $\wedge$ *hasAge(?c,?a)* $\wedge$ *?a $\geq$ 10* $\wedge$ *?a $\leq$ 20*. In this example, the query freezing method cannot be used directly to test subsumption between q_A and q_B, because the variables *?y* and *?a* refer to data values, which cannot be represented by individuals in an OWL-DL ontology.

Constraint Transformation. Here, we propose *constraint transformation*. It is a pre-processing step which enables us to create a canonical ABox for semantic formulas with datatype variables. Note that a datatype variable is used in a semantic formula to constrain one datatype property, e.g., *?y* is used to constrain the *hasAge* datatype property in q_A of Example 2. Constraint transformation in contrast uses *data-ranges* introduced in OWL 2.0 [7] to transform each constrained datatype property to a named OWL class. As a result, datatype variables and related datatype properties and constraints are replaced with type assertions. This procedure is detailed in Algorithm 1.

The algorithm takes a conjunctive semantic formula F_v^o and the ontology o as inputs (line 1). F_v^o is of the form $T_v^o \cup R_v^o \cup C_v^o$, where T_v^o, R_v^o, and C_v^o are sets of type, relation and constraint assertions respectively. The outputs of the algorithm are the transformed

---

Algorithm 1. Constraint transformation

1: **Inputs:**   Formula $F_v^o \equiv T_v^o \cup R_v^o \cup C_v^o$,
                   Ontology $o \equiv (ABox_o, TBox_o)$
2: **Outputs:** Formula F_u^ϕ,
                   Ontology $\phi \equiv (ABox_o, TBox_\phi)$
3: **Initialization:** $F_u^\phi = T_v^o$, $TBox_\phi = TBox_o$
4: **for all** $(r(a,b) \in R_v^o)$ **do**
5:   **if** $(isDatatypeVariable(b))$ **then**
6:       $\gamma_d = getConstraints(b, C_v^o)$
7:       $c = createConcept(r, \gamma_d, TBox_\phi)$
8:       $\tau = createTypeAssertion(a,c)$
9:       $F_u^\phi = F_u^\phi \cup \tau$
10: **else**
11:      $F_u^\phi = F_u^\phi \cup r(a,b)$
12:      $\gamma_b = getConstraints(b, C_v^o)$
13:      **if** $(\gamma_b \neq \emptyset \; \& \; \neg(\gamma_b \subset F_u^\phi))$ **then**
14:          $F_u^\phi = F_u^\phi \cup \gamma_b$
15:      **end if**
16: **end if**
17: **end for**

---

semantic formula F_u^ϕ (containing no datatype variables) and the updated ontology ϕ (line 2). Initially, F_u^ϕ is set as equal to T_v^o and ϕ is the same as o (line 3). For each relation assertion $r(a,b)$ in R_v^o, we do the following (line 4). First, we check if b is a datatype variable (line 5). If so, this means that r is a datatype property with a variable in its range. In this case, we extract the set of constraints related to b from C_v^o, which is referred by γ_d (line 6). Based on r and γ_d, we create a concept c in $TBox_\phi$ using the *createConcept* function (line 7). This function works as follows:

1. If $\gamma_d \neq \emptyset$, then b implies some restrictions on the range of r. In this case, c should refer to objects that have the property r with the restrictions defined in γ_d on its range. While creating c in $TBox_\phi$, we use *data-ranges*[1] introduced in OWL 2.0 to restrict the range of r accordingly. For example, if $r(a,b)$ corresponds to $hasAge(?c, ?a)$ and $\gamma_d = \{?a \geq 10, ?a \leq 20\}$, then a concept named *AgeConst1* can be described as shown in Figure 5. For more sophisticated constraints, we create more complex class expressions using the OWL constructors *owl:unionOf*, *owl:intersectionOf*, and *owl:complementOf*.
2. If $\gamma_d = \emptyset$, then b has no constraints, which means that the data-range of b is equivalent to the range of its data-type (i.e., for *xsd:int*, the range is min inclusive -2147483648 and max inclusive 2147483647).

After creating the concept c in $TBox_\phi$, we create a type assertion τ to declare a as an instance of c (e.g., *AgeConst1(?c)*) (line 8). This type assertion is added to F_u^ϕ in order to substitute $r(a,b)$ and γ_d in F_v^o (line 9). On the other hand, if b is not a datatype variable (line 10), there are two possibilities: (1) r is a datatype property but b is not a variable, or (2) r is an object property. In both cases, we directly add $r(a,b)$ to F_u^ϕ (line 11). If b

---

[1] http://www.w3.org/TR/2008/WD-owl2-syntax-20081008/#Data_Ranges

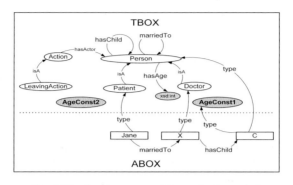

Fig. 6. The Ontology created for q_B in Example 2

has constraints defined in C_v^o, we extract these constraints and add them to F_u^ϕ if they are not already added (lines 12-15).

In order to test subsumption between q_A and q_B in Example 2, we should transform the bodies of these queries and update the ontology they are based on. For this purpose, we use constraint transformation twice. That is, we first update the ontology by adding the concept $AgeConst1$ to handle $hasAge(?c,?y) \land ?y \geq 10 \land ?y \leq 20$ and transform q_B to $q():- Doctor(?x) \land marriedTo(?x,Jane) \land hasChild(?x,?c) \land AgeConst1(?c)$. Then, we add concept $AgeConst2$ to the ontology to handle $hasAge(?c,?y) \land ?y \geq 12 \land ?y \leq 16$ and transform q_A to $q():- Person(?p) \land hasChild(?p,?c) \land AgeConst2(?c)$. After this preprocessing step, we use query freezing to test $q_B \sqsubseteq q_A$; the ontology with a canonical ABox created during query freezing is shown in Figure 6.

With these techniques in place, we are now in a position to address the issue of policy analysis supported by OWL-POLAR. It is descried in the following sections.

4.2   Idle Policies

A policy is *idle* if it is never activated or the policy's expiration condition is satisfied whenever the policy is activated. This condition is formally described in Definition 2. If a policy is idle, it cannot be used to regulate any action, because either it never activates or whenever it activates an obligation, permission, or prohibition about an action, the activated policy expires. While designing policies, we may take domain knowledge into account to avoid idle policies.

Definition 2. A policy $\alpha \longrightarrow N_{\chi:\rho}(a:\varphi)/e$ is an idle policy if it does not activate for any state of the world Δ_o or there is a substitution σ' such that $\Delta_o \vdash (e \cdot \sigma) \cdot \sigma'$, whenever there is a substitution σ such that $\Delta_o \vdash (\alpha \land \rho) \cdot \sigma$. ∎

Let us demonstrate idle policies with a simple example. Assume that object property *hasParent* is an inverse property of *hasChild*. Also, let us assume in the domain ontology, we have a SWRL rule such as $hasSponsor(?c, true) \leftarrow hasParent(?c,?p) \land hasAge(?c,?age) \land ?age < 18$, which means that children under 18 have a sponsor if they have a parent. Now, consider the policy in Table 2. This policy is activated when a person $?p$ has a child $?c$, which is a student under 18. The activated policy expires when

Table 2. A simple idle policy example

α	$hasChild(?p, ?c) \wedge Student(?c) \wedge hasAge(?c, ?age) \wedge ?age < 18$
N	O
$\chi : \rho$	$?p : Person(?p)$
$a : \varphi$	$?a : PayTuitionsOfStudent(?a) \wedge about(?a, ?c) \wedge hasActor(?a, ?p)$
e	$hasSponsor(?c, true)$

$?c$ has a sponsor. Interestingly, whenever the policy is activated, the domain knowledge implies that $?c$ has a sponsor. That is, whenever the policy is activated, it expires.

In order to detect idle policies, we reason about the activation and expiration conditions of policies. Specifically, a policy $\alpha \longrightarrow N_{\chi:\rho} (a : \varphi) / e$ is an idle policy if $(\alpha \wedge \rho)$ is unrealistic or implies e using the knowledge in the domain ontology. More formally, we can show that the policy is idle if we show $(\alpha \wedge \rho)$ never holds or $(\alpha \wedge \rho) \rightarrow e$. This can be achieved as follows. First, we freeze $(\alpha \wedge \rho)$ and create a canonical ontology o'. If the resulting o' is not a consistent ontology, then we can conclude that the policy is an idle policy, because $(\alpha \wedge \rho)$ never holds. Let o' be consistent and σ be a substitution denoting the mapping of variables in $(\alpha \wedge \rho)$ to the fresh individuals in o'. If there exists a substitution σ' such that $o' \vdash (e \cdot \sigma) \cdot \sigma'$, we conclude that $(\alpha \wedge \rho) \rightarrow e$. We can test $o' \vdash (e \cdot \sigma) \cdot \sigma'$ by querying o' with $q() : - (e \cdot \sigma)$.

4.3   Anticipating Conflicts between Policies

In many settings, policies may conflict. In the simplest case, one policy may prohibit an action while another requires it. There are, however, many less obvious interactions between policies that may lead to logical conflicts [9,16,12,5]. Further developing our earlier example, consider the policy presented in Table 3 that states that a doctor cannot leave a room with patients if he is in charge of the room. This policy conflicts with the policy in Table 1 under some specific conditions. For example, in the scenario described Figure 1, *room 245* of Central Hospital has a fire risk and *Dr. John* is in charge of the room, in which there are some patients. In this setting, the policy in Table 1 obligates Dr. John to leave the room while the policy in Table 3 prohibits this action until the room has no patient.

Table 3. A doctor cannot leave a room containing patients if he is in charge of the room

α	$Room(?r) \wedge hasPatient(?r, true) \wedge inChargeOf(?d, ?r)$
N	F
$\chi : \rho$	$?d : Doctor(?d)$
$a : \varphi$	$?x : LeavingAction(?x) \wedge about(?x, ?r) \wedge hasActor(?x, ?d)$
e	$hasPatient(?r, false)$

If we can determine possible logical conflicts while designing policies, we can create better policies that are less likely to raise conflicts at run time. Furthermore, we can use various conflict resolution strategies such as setting a priority ordering between the policies to solve conflicts [11,21,22], once we determine that two policies may conflict.

In this section, we propose techniques to anticipate possible conflicts between policies at design time. Suppose we have two non-idle policies $P_i = \alpha^i \longrightarrow A_{\chi^i:\rho^i} \left(a^i : \varphi^i \right) / e^i$ and $P_j = \alpha^j \longrightarrow B_{\chi^j:\rho^j} \left(a^j : \varphi^j \right) / e^j$. These policies are active for the same policy addressee in the same state of the world Δ if the following requirements are satisfied:

(i) $\Delta \vdash \left(\alpha^i \wedge \rho^i \right) \cdot \sigma_i$, but no σ_i' such that $\Delta \vdash \left(e^i \cdot \sigma_i \right) \cdot \sigma_i'$
(ii) $\Delta \vdash \left(\alpha^j \wedge \rho^j \right) \cdot \sigma_j$, but no σ_j' such that $\Delta \vdash \left(e^j \cdot \sigma_j \right) \cdot \sigma_j'$
(iii) $\chi^i \cdot \sigma_i = \chi^j \cdot \sigma_j$

The policies P_i and P_j conflict if the following requirements are also satisfied:

(iv) $(\varphi^i \cdot \sigma_i) \sqsubseteq (\varphi^j \cdot \sigma_j)$ or $(\varphi^j \cdot \sigma_j) \sqsubseteq (\varphi^i \cdot \sigma_i)$
(v) A conflicts with B. That is, $A \in \{P, O\}$ while $B \in \{F\}$ or vice versa.

We can use Algorithm 2 to test if it is possible to have such a state of the world where P_i conflicts with P_j. The first step of the algorithm is to test if A conflicts with B (line 2). If they are conflicting, we continue with testing the other requirements. We create a canonical state of the world Δ in which P_i is active by freezing $\left(\alpha^i \wedge \rho^i \right)$ with a substitution σ_i mapping the variables in $\left(\alpha^i \wedge \rho^i \right)$ to the fresh individuals in Δ. Given that $(\varphi^j \cdot \sigma) \sqsubseteq \varphi^j$ for any substitution σ mapping variables into individuals, the requirement (iv) implies that $(\varphi^i \cdot \sigma_i) \sqsubseteq \varphi^j$. We test this as follows. First, we create a canonical ontology o' by freezing $(\varphi^i \cdot \sigma_i)$ (line 4) and then query o' with φ^j (line 5). Each answer to this query defines a substitution σ_k mapping variables in φ^j into the terms in $(\varphi^i \cdot \sigma_i)$, so that $(\varphi^i \cdot \sigma_i) \sqsubseteq (\varphi^j \cdot \sigma_k)$. If φ^j does not have any variable but it repeats in o' as a pattern, the result set contains only one empty substitution. If the query fails, the result set is an empty set $(\emptyset)$, which means that it is not possible to have a σ_k such that $(\varphi^i \cdot \sigma_i) \sqsubseteq (\varphi^j \cdot \sigma_k)$. For each σ_k satisfying $(\varphi^i \cdot \sigma_i) \sqsubseteq (\varphi^j \cdot \sigma_k)$, we test

---

Algorithm 2. An algorithm to anticipate if P_i may conflict with P_j

1: **Inputs:**   Policy $P_i = \alpha^i \longrightarrow A_{\chi^i:\rho^i} \left(a^i : \varphi^i \right) / e^i$,
           Policy $P_j = \alpha^j \longrightarrow B_{\chi^j:\rho^j} \left(a^j : \varphi^j \right) / e^j$
2: **if** (($A \in \{O, P\}$ **and** $B \in \{F\}$) **or** ($A \in \{F\}$ **and** $B \in \{O, P\}$)) **then**
3:    $\langle \Delta, \sigma_i \rangle = freeze(\alpha^i \wedge \rho^i)$
4:    $\langle o', \_ \rangle \;\; = freeze(\varphi^i \cdot \sigma_i)$
5:    $rs \;\;\;\;\; = query(o', \varphi^j)$
6:    **for all** $(\sigma_k \in rs)$ **do**
7:       $\langle \Delta, \sigma_j \rangle = update(\Delta, \left(\alpha^j \wedge \rho^j \right) \cdot \sigma_k)$
8:       **if** $(isConsistent(\Delta))$ **then**
9:          **if** $(query(\Delta, e^i \cdot \sigma_i) = \emptyset$ **and** $query(\Delta, (e^j \cdot \sigma_k) \cdot \sigma_j) = \emptyset)$ **then**
10:             return **true**
11:          **end if**
12:       **end if**
13:    **end for**
14: **end if**
15: return **false**

the other requirements as follows. First, we update Δ by freezing $\left(\alpha^j \wedge \rho^j\right) \cdot \sigma_k$ without removing any individual from its existing $ABox$ (line 7). Note that as a result of this process, σ_j is the substitution mapping the variables in $\left(\alpha^j \wedge \rho^j\right) \cdot \sigma_k$ to the new fresh individuals in the updated Δ, so that $\chi^i \cdot \sigma_i = \left(\chi^j \cdot \sigma_k\right) \cdot \sigma_j$. We test the consistency of the resulting state of the world Δ (line 8). If this is not consistent, we can conclude that it is not possible to have a state of the world satisfying the requirements. If the resulting Δ is consistent, we check the expiration conditions of the policies. If both are active in the resulting state of the world (line 9), the algorithm returns *true* (line 10). If any of these requirements do not hold, the algorithm returns *false* (line 15).

As described above, the algorithm transforms the problem of anticipating conflict between two policies into an ontology consistency checking problem. To check the consistency of the constructed canonical state of the world Δ, we have used the Pellet [17] reasoner. This reasoner adopts the *open world assumption* and does not have Unique Name Assumption (UNA). Hence, it searches for a model[2] of Δ, also considering the possible overlapping between the individuals (i.e., individuals referring the same object). If there is no model of Δ, it is not possible to have a state of the world satisfying the requirements stated above. We should also note that, while anticipating the conflict, Algorithm 2 tests only the case $(\varphi^i \cdot \sigma_i) \sqsubseteq (\varphi^j \cdot \sigma_j)$. However, we also need to test $(\varphi^j \cdot \sigma_j) \sqsubseteq (\varphi^i \cdot \sigma_i)$ to capture the possibility of conflict. Therefore, if the algorithm returns *false*, we should swap the policies and run the algorithm again. If it returns *true* with the swapped policies, we can conclude that there is a state of the world where these policies may conflict.

To demonstrate the algorithm, let us use the policies presented in Tables 1 and 3 and refer to them as P_i and P_j respectively. In this example, P_j is a prohibition while P_i is an obligation, so the algorithm proceeds as follows (line 2). We create a canonical state of the world Δ by freezing $Person(?x) \wedge Place(?b) \wedge hasFireRisk(?b, true) \wedge in(?x, ?b)$ with a substitution $\sigma_i = \{?x/x, ?b/b\}$ (line 3). Now we create a canonical ontology a' by freezing $\varphi^i \cdot \sigma_i$ with substitution $\{?a/a\}$ (line 4). This ontology has the following $ABox$ assertions: $LeavingAction(a), about(a, b), hasActor(a, x)$. We query o' with $LeavingAction(?x) \wedge about(?x, ?r) \wedge hasActor(?x, ?d)$ (line 5). The result set is composed of only one substitution: $\sigma_k = \{?x/a, ?r/b, ?d/x\}$. The next step is to update Δ by freezing $Doctor(x) \wedge Room(b) \wedge hasPatient(b, true) \wedge inChargeOf(x, b)$ without removing the current $ABox$ of Δ (line 7). The resulting canonical state of the world is shown in Figure 7. Lastly, we check whether both policies remain in effect by checking their expiration conditions (line 9). In this example, we query Δ with $hasFireRisk(b, false)$ and $hasPatient(b, false)$. Both of these queries return $\emptyset$, hence we conclude that there is a state of the world where these policies conflict (line 10).

5   Related Work and Discussion

There have been several policy languages proposed that are built upon Semantic Web technologies. Rei [10] is a policy language based on OWL-Lite and Prolog. It allows

---

[2] A model of an ontology o is an interpretation of o satisfying all of its axioms [1].

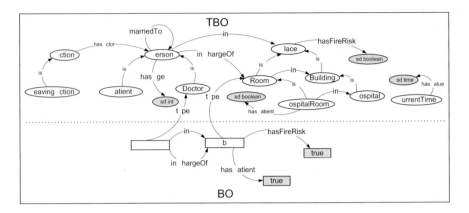

Fig. 7. The canonical state of the world where the policies of Table 1 and Table 3 conflict

logic-like variables to be used while describing policies. This gives it the flexibility to specify relations like *role value maps* that are not directly possible in OWL. The use of these variables, however, makes DL reasoning services (e.g., static conflict detection between policies) unavailable for Rei policies. KAoS [21] is, probably, the most developed language for describing policies that are built upon OWL. KAoS was originally designed to use OWL-DL to define actions and policies. This, however, restricts the expressive power to DL and prevents KAoS from defining policies in which one element of an action's context depends on the value of another part of the current context. For example, KAoS cannot be used to represent a policy like *two soldiers are allowed to communicate only if they are in the same team*. To handle such situations, KAoS has been enhanced with *role-value maps* using Stanford JTP, a general purpose theorem prover [21]. Unfortunately, subsumption reasoning is undecidable in the presence of arbitrary role-value-maps [1].

KAoS distinguishes between (positive and negative) obligation policies and (positive and negative) authorization policies. Authorization policies permit (positive) or forbid (negative) actions, whereas obligation policies require (positive) or do not require (negative) action. Thus the general types of policies that can be described are similar to those that we have discussed in this paper. Actions are also the object of a KAoS policy, and conditions on the application of policies can be described (context), although the subject (individual/role) of the policy is not explicit (it is, however, in Rei). In common with OWL-POLAR in its present form, KAoS does not capture the notion of the authority from which/whom a policy comes, but there is a notion of the priority of a policy which partially (although far from adequately) addresses this issue. Unlike OWL-POLAR, Rei and KAoS do not provide means to explicitly define expiration conditions of the policies.

Policy analysis within both KAoS and Rei is restricted to subsumption. A policy in KAoS is expressed as an OWL-DL class regulating an action, which is expressed as an OWL-DL class expression (e.g., using restrictions on properties such as *performedBy* and *hasDestination*). Two policies are regarded in conflict if their actions overlap (one subsumes another) while the modality of these policies conflict (e.g., negative vs. positive authorization). Similarly, if there exist two policies within Rei that overlap with respect to the agent and action concerned and they are obligued and prohibited, then a conflict is recognised. In such a situation, meta-policies are used to resolve the conflict. Policy conflicts can also be detected within the Ponder2 framework [18,23], where analysis is far more sophisticated than that developed for either KAoS or Rei, but analysis is restricted to design time. In general, different methods can be used to resolve conflicts between policies. This issue has been explored in detail elsewhere [11].

The expressiveness of OWL-POLAR is not restricted to DL. Using semantic conjunctive formulas, it allows variables to be used while defining policies. However, in semantic formulas, OWL-POLAR allows only object-type variables to be compared using *owl:sameAs* and *owl:differentFrom* properties. On the other hand, data-type variables can be used to define constraints on the datatype properties. In other words, semantic formulas are restricted to describe states of the world, each of which can be represented as an OWL-DL ontology. Therefore, when a semantic formula is frozen, the result is a canonical OWL-DL ontology. OWL-POLAR converts problems of reasoning with and about policies into query answering and ontology consistency checking problems. Then, it uses an off-the-shelf reasoner (Pellet) to solve these problems. It is known that consistency checking in OWL-DL is decidable [17], and query answering in OWL-DL has also been shown to be decidable under DL-safety restrictions [8].

Ontology languages like KAoS are built on OWL 1.0, which does not support dataranges. Therefore, while defining policies, they either do not allow complex constraints to be defined on datatype properties or use non-standard representations for these constraints, which prevents them from using the off-the-shelf reasoning technologies. The clear distinctions between OWL-POLAR and KAoS, however, are manifest in the fact that data ranges are exploited in OWL-POLAR to enable the expression of more complex constraints on policies, and the sophistication of the reasoning mechansims described in this paper.

To the best of our knowledge, OWL-POLAR is the first policy framework that formally defines and detects idle policies. Existing approaches like KAoS and Rei analyse policies only to detect some type of conflict, considering only subsumption between policies. On the other hand, OWL-POLAR provides advanced policy analysis support that is not limited to subsumption checking. Consider the following policies: (i) *Dogs are prohibited from entering to a restaurant*, and (ii) *A member of CSI team is permitted to enter a crime scene*. There is no subsumption relationship between these policies, and so KAoS and Rei could not detect a conflict. However, OWL-POLAR anticipates a conflict by composing a state of the world where these policies are in conflict, e.g., the crime scene is a restaurant and there is a dog in the CSI team.

Building upon this research, we plan to explore various extensions to OWL-POLAR. We will explore extending the representation of policies to include deadlines and penalties associated with their violation, along the lines of [3]. Another issue we would like to investigate concerns how policing mechanisms [14] could make use of our representation and associated mechanisms to foster welfare in societies of self-interested components/agents. We plan to enhance our representation so as to allow constraints over arbitrary terms (and not just $?x \lhd \beta$, β being a constant), possibly using constraint satisfaction mechanisms to deal with these. Two further extensions should address policies over many actions (as in, for instance, "ξ is obliged to perform φ_1 and φ_2") and

694   M. Şensoy et al.

disjunctions (as in, for instance, "ξ is obliged to perform φ_1 or φ_2"). Finally, we are exploring the use of OWL-POLAR in support of human decision-making, including joint planning activities in hybrid human-software agent teams.

6 Conclusions

Policies provide useful abstractions to constrain and control the behaviour of components in loosely coupled distributed systems. Policies, also called norms, help designers of large-scale, open, and heterogenous distributed systems (including multi-agent systems) to specify, in a concise fashion, acceptable (or policy-compliant) global and individual computational behaviours, thus providing guarantees for the system as a whole.

In this paper, we have presented a semantically-rich representation for policies as well as efficient mechanisms to reason with/about them. OWL-POLAR meets all the essential requirements of policies, as well as achieving an effective balance between expressiveness (realistic policies can be adequately represented) and computational complexity of associated reasoning for decision-making and analysis (reasoning with and about policies operate in feasible time).

References

1. Baader, F., Calvanese, D., McGuinness, D.L., Nardi, D., Patel-Schneider, P.F. (eds.): The Description Logic Handbook: Theory, Implementation and Applications. Cambridge University Press, Cambridge (2003)
2. Beautement, A., Pym, D.: Structured systems economics for security management. In: Proceedings of the Ninth Workshop on the Economics of Information Security, Harvard, USA (June 2010)
3. Boella, G., Broersen, J., Torre, L.: Reasoning about constitutive norms, counts-as conditionals, institutions, deadlines and violations. In: Bui, T.D., Ho, T.V., Ha, Q.T. (eds.) PRIMA 2008. LNCS (LNAI), vol. 5357, pp. 86–97. Springer, Heidelberg (2008)
4. Castelfranchi, C.: Modelling social action for AI agents. Artificial Intelligence 103, 157–182 (1998)
5. Elhag, A., Breuker, J., Brouwer, P.: On the formal analysis of normative conflicts. Information & Communications Technology Law 9(3), 207–217 (2000)
6. García-Camino, A., Rodríguez-Aguilar, J.A., Sierra, C., Vasconcelos, W.: Constraint rule-based programming of norms for electronic institutions. Autonomous Agents and Multi-Agent Systems 18(1), 186–217 (2009)
7. W. O. W. Group. OWL 2 web ontology language: Document overview (October 2009), http://www.w3.org/TR/owl2-overview
8. Haase, P., Motik, B.: A mapping system for the integration of owl-dl ontologies. In: Proceedings of the First International Workshop on Interoperability of Heterogeneous Information Systems, IHIS 2005, pp. 9–16. ACM, New York (2005)
9. Hill, H.: A functional taxonomy of normative conflict. Law and Philosophy 6(2), 227–247 (1987)
10. Kagal, L., Finin, T., Joshi, A.: A policy language for a pervasive computing environment. In: Proceedings of the 4th IEEE International Workshop on Policies for Distributed Systems and Networks, POLICY 2003, pp. 63–74 (2003)

11. Kollingbaum, M.J., Norman, T.J.: Norm adoption and consistency in the NoA agent architecture. In: Dastani, M.M., Dix, J., El Fallah-Seghrouchni, A. (eds.) PROMAS 2003. LNCS (LNAI), vol. 3067, pp. 169–186. Springer, Heidelberg (2004)

12. Lupu, E., Sloman, M.: Conflicts in policy-based distributed systems management. IEEE Transactions on Software Engineering 25(6), 852–869 (1999)

13. Motik, B.: Reasoning in Description Logics using Resolution and Deductive Databases. PhD thesis, Universitt Karlsruhe (TH), Karlsruhe, Germany (January 2006)

14. Patel, J., et al.: Agent-based virtual organisations for the grid. Int. Journal of Multi-Agent and Grid Systems 1(4), 237–249 (2005)

15. Prud'hommeaux, E., Seaborne, A.: SPARQL Query Language for RDF. Technical report, W3C (2006), http://www.w3.org/TR/rdf-sparql-query/

16. Sartor, G.: Normative conflicts in legal reasoning. Artificial Intelligence and Law 1(2), 209–235 (1992)

17. Sirin, E., Parsia, B., Grau, B.C., Kalyanpur, A., Katz, Y.: Pellet: A practical OWL-DL reasoner. Web Semant. 5(2), 51–53 (2007)

18. Sloman, M., Lupu, E.: Policy specification for programmable networks. In: Covaci, S. (ed.) IWAN 1999. LNCS, vol. 1653, pp. 73–84. Springer, Heidelberg (1999)

19. Smith, M.K., Welty, C., McGuinness, D.L.: OWL: Web ontology language guide (February 2004), http://www.w3.org/TR/owl-guide

20. Ullman, J.D.: Information integration using logical views. Theoretical Computer Science 239(2), 189–210 (2000)

21. Uszok, A., Bradshaw, J.M., Lott, J., Breedy, M., Bunch, L., Feltovich, P., Johnson, M., Jung, H.: New developments in ontology-based policy management: Increasing the practicality and comprehensiveness of KAoS. In: Proceedings of the 2008 IEEE Workshop on Policies for Distributed Systems and Networks, POLICY 2008, pp. 145–152 (2008)

22. Vasconcelos, W.W., Kollingbaum, M.J., Norman, T.J.: Normative conflict resolution in multiagent systems. Autonomous Agents and Multi-Agent Systems 19(2), 124–152 (2009)

23. Zhao, H., Lobo, J., Bellovin, S.M.: An algebra for integration and analysis of ponder2 policies. In: Proceedings of the 2008 IEEE Workshop on Policies for Distributed Systems and Networks, POLICY 2008, Washington, DC, USA, pp. 74–77. IEEE Computer Society, Los Alamitos (2008)

Query Strategy for Sequential Ontology Debugging

Kostyantyn Shchekotykhin and Gerhard Friedrich*

Universitaet Klagenfurt
Universitaetsstrasse 65-67
9020 Klagenfurt, Austria
{firstname.lastname}@ifit.uni-klu.ac.at

Abstract. Debugging is an important prerequisite for the wide-spread application of ontologies, especially in areas that rely upon everyday users to create and maintain knowledge bases, such as the Semantic Web. Most recent approaches use diagnosis methods to identify sources of inconsistency. However, in most debugging cases these methods return many alternative diagnoses, thus placing the burden of fault localization on the user. This paper demonstrates how the target diagnosis can be identified by performing a sequence of observations, that is, by querying an oracle about entailments of the target ontology. We exploit probabilities of typical user errors to formulate information theoretic concepts for query selection. Our evaluation showed that the suggested method reduces the number of required observations compared to myopic strategies.

1 Introduction

The application of semantic systems, including the Semantic Web technology, is largely based on the assumption that the development of ontologies can be accomplished efficiently even by every day users. However, studies in cognitive psychology, like [1], discovered that humans make systematic errors while formulating or interpreting logical descriptions. Results presented in [10,12] confirmed these observations regarding ontology development. Therefore it is essential to create methods that can identify and correct erroneous ontological definitions. Ontology debugging tools simplify the development of ontologies by localizing a set of axioms that should be modified in order to formulate the intended target ontology.

To debug an ontology a user must specify some requirements such as coherence and/or consistency. Additionally, one can provide test cases [3] which must be fulfilled by the target ontology $\mathcal{O}_t$. A number of ontology diagnosis methods have been developed [13,6,3] to pinpoint alternative sets of possibly faulty axioms (called a set of diagnoses). A user has to change at least all of the axioms of one diagnosis in order to satisfy all of the requirements and test cases.

However, the diagnosis methods can return many alternative diagnoses for a given set of test cases and requirements. A sample study of real-world inconsistent ontologies presented in Table 1 shows that even a small number of irreducible sets of axioms that are together inconsistent/incoherent (conflict sets) can be a source of a large number of diagnoses. For instance only 8 conflict sets in the Economy ontology resulted in

---

* The research project is funded by grants of the Austrian Science Fund (Project V-Know, contract 19996).

P.F. Patel-Schneider et al. (Eds.): ISWC 2010, Part I, LNCS 6496, pp. 696–712, 2010.

Table 1. Dianosis results for some real-world ontologies presented in [6]. #C/#P/#I are the numbers of concepts, properties, and individuals in an ontology. #CS/min/max are the number of conflict sets, their minimum and maximum cardinality. The same notation is used for diagnoses #D/min/max. These ontologies are available upon request.

	Ontology	Axioms	#C/#P/#I	#CS/min/max	#D/min/max	Domain
1.	Chemical	114	48/20/0	6/5/6	6/1/3	Chemical elements
2.	Sweet-JPL	2579	1537/121/50	8/1/13	13/8/8	Earthscience
3.	University	50	30/12/4	4/3/5	90/3/4	Training
4.	Tambis	596	395/100/0	7/3/9	147/3/7	Biological science
5.	Economy	1781	339/53/482	8/3/4	864/4/9	Mid-level
6.	Transport	1300	445/93/183	9/2/6	1782/6/9	Mid-level

864 diagnoses. In the case of Transportation ontology the diagnosis method was able to identify 1782 diagnoses. In such situations simple visualization of all alternative changes of the ontology is ineffective.

A possible solution would be to introduce an ordering using some preference criteria. For instance, Kalyanpur et al. [7] suggest measures to rank the axioms of a diagnosis depending on their structure, occurrence in test cases, etc. Only the top ranking diagnoses are then presented to the user. Of course this set of diagnoses will contain the target one only in the case when a faulty ontology, the given requirements and test cases, provide sufficient data to appropriate heuristics. However, in most debugging sessions a user has to provide additional information (e.g. in the form of tests) to identify the target diagnosis.

In this paper we present an approach to acquisition of additional information by generating a sequence of queries, which should be answered by some oracle such as a user, an information extraction system, etc. Our method uses each answer to a query to reduce the set of diagnoses until finally it identifies the target diagnosis. In order to construct queries we exploit the property that different diagnoses imply unequal sets of axioms. Consequently, we can differentiate between diagnoses by asking the oracle if the target ontology should imply an axiom or not. These axioms can be generated by classification and realization services provided in description logic reasoning systems [15,4]. In particular, the classification process computes a subsumption hierarchy (sometimes also called "inheritance hierarchy" of parents and children) for each concept name mentioned in a TBox. For each individual mentioned in an ABox, realization computes the atomic concepts (or concept names) of which the individual is an instance [15].

In order to generate the most informative query we exploit the fact that some diagnoses are more likely than others because of typical user errors. The probabilities of these errors can be used to estimate the change in entropy of the set of diagnoses if a particular query is answered. We select those queries which minimize the expected entropy, i.e. maximize the information gain. An oracle should answer these queries until a diagnosis is identified whose probability is significantly higher than those of all other diagnoses. This diagnosis is the most likely to be the target one.

We compare our entropy-based method with a greedy approach that selects those queries which try to cut the number of diagnoses in half as well as with a "random" strategy when the algorithm selects queries to be asked completely randomly. The evaluation was performed using the set of ontologies presented in Table 1 and generated examples. Its results show that on average the suggested entropy-based approach is at least 50% better than the greedy one.

The remainder of the paper is organized as follows: Section 2 presents two introductory examples as well as the basic concepts. The details of the entropy-based query selection method are given in Section 3. Section 4 describes the implementation of the approach and is followed by evaluation results in Section 5. The paper concludes with an overview of related work.

2   Motivating Examples and Basic Concepts

In order to explain the fundamentals of our approach let us introduce two examples.

Example 1. Consider a simple ontology $\mathcal{O}$ with the terminology $\mathcal{T}$:

$$ax_1 : A \sqsubseteq B \qquad ax_2 : B \sqsubseteq C \qquad ax_3 : C \sqsubseteq Q \qquad ax_4 : Q \sqsubseteq R$$

and the background theory $\mathcal{A} : \{A(w), \neg R(w)\}$. Let the user explicitly define that the two assertional axioms should be considered as correct.

The ontology $\mathcal{O}$ is inconsistent and the only irreducible set of axioms (minimal conflict set) that preserves the inconsistency is $CS : \{\langle ax_1, ax_2, ax_3, ax_4 \rangle\}$. That is one has to modify or remove the axioms of at least one diagnosis:

$$\mathcal{D}_1 : [ax_1] \quad \mathcal{D}_2 : [ax_2] \quad \mathcal{D}_3 : [ax_3] \quad \mathcal{D}_4 : [ax_4]$$

to restore the consistency of the ontology. However it is unclear, which diagnosis from the set $\mathbf{D} : \{\mathcal{D}_1 \ldots \mathcal{D}_4\}$ corresponds to the target one.

In order to focus on the essentials of our approach we employ the following simplified definition of diagnosis without limiting its generality. A more detailed version can be found in [3].

We allow the user to define a background theory (represented as a set of axioms) which is considered to be correct, a set of logical sentences which must be implied by the target ontology and a set of logical sentences which must *not* be implied by the target ontology. Following the standard definition of the diagnosis [11,8], we assume that each axiom $ax_j \in \mathcal{D}_i$ is faulty whereas each axiom $ax_k \notin \mathcal{D}_i$ is correct.

Definition 1. *Given a diagnosis problem* $\langle \mathcal{O}, B, T^{\models}, T^{\nvDash} \rangle$ *where $\mathcal{O}$ is an ontology, B a background theory, $T^{\models}$ a set of logical sentences which must be implied by the target ontology $\mathcal{O}_t$, and $T^{\nvDash}$ a set of logical sentences which must not be implied by the target ontology $\mathcal{O}_t$.*

A diagnosis is a set of axioms $\mathcal{D} \subseteq \mathcal{O}$ such that the set of axioms $\mathcal{O} \setminus \mathcal{D}$ can be extended by a logical description EX and $(\mathcal{O} \setminus \mathcal{D}) \cup B \cup EX \models t^{\models}$ for all $t^{\models} \in T^{\models}$ and $(\mathcal{O} \setminus \mathcal{D}) \cup B \cup EX \nvDash t^{\nvDash}$ for all $t^{\nvDash} \in T^{\nvDash}$.

A diagnosis $\mathcal{D}$ is minimal if there is no proper subset of the faulty axioms $\mathcal{D}' \subset \mathcal{D}$ such that $\mathcal{D}'$ is a diagnosis. The following proposition allows us to characterize diagnoses without the extension EX. The idea is to use the sentences which must be implied to approximate EX.

Corollary 1. *Given a diagnosis problem* $\langle \mathcal{O}, B, T^{\models}, T^{\nvDash} \rangle$, *a set of axioms $\mathcal{D} \subseteq \mathcal{O}$ is a diagnosis iff $(\mathcal{O} \setminus \mathcal{D}) \cup B \cup \{\bigwedge_{t^{\models} \in T^{\models}} t^{\models}\} \cup \neg t^{\nvDash}$ consistent for all $t^{\nvDash} \in T^{\nvDash}$.*

In the following we assume that a diagnosis always exists under the (reasonable) condition that the background theory together with the axioms in $T^\models$ and the negation of axioms in $T^\nvDash$ are mutually consistent. For the computation of diagnoses the set of conflicts is usually employed.

Definition 2. *Given a diagnosis problem* $\langle \mathcal{O}, B, T^\models, T^\nvDash \rangle$, *a conflict set* $CS \subseteq \mathcal{O}$ *is a set of axioms s.t. there is a* $t^\nvDash \in T^\nvDash$ *and* $CS \cup B \cup \{\bigwedge_{t^\models \in T^\models} t^\models\} \cup \neg t^\nvDash$ *is inconsistent.*

A conflict is the part of the ontology that preserves the inconsistency/incoherency. A minimal conflict CS has no proper subset which is a conflict. $\mathcal{D}$ is a (minimal) diagnosis iff $\mathcal{D}$ is a (minimal) hitting set of all (minimal) conflict sets [11].

In order to differentiate between the minimal diagnoses $\{\mathcal{D}_1 \dots \mathcal{D}_4\}$ an oracle can be queried for information about the entailments of the target ontology. For instance, in our example the ontologies $\mathcal{O}_i = \mathcal{O} \setminus \mathcal{D}_i$ have the following entailments $\mathcal{O}_1 : \emptyset$, $\mathcal{O}_2 : \{B(w)\}$, $\mathcal{O}_3 : \{B(w), C(w)\}$, and $\mathcal{O}_4 : \{B(w), C(w), Q(w)\}$ provided by the realization of the ontology. Based on these entailments we can ask the oracle whether the target ontology has to entail $Q(w)$ or not $(\mathcal{O}_t \nvDash Q(w))$. If the answer is *yes* (which we model with the boolean value 1), then $Q(w)$ is added to $T^\models$ and $\mathcal{D}_4$ is the target diagnosis. All other diagnoses are rejected because $(\mathcal{O} \setminus \mathcal{D}_i) \cup B \cup \{Q(w)\}$ for $i = 1, 2, 3$ is inconsistent. If the answer is *no* (which we model with the boolean value 0), then $Q(w)$ is added to $T^\nvDash$ and $\mathcal{D}_4$ is rejected as $(\mathcal{O} \setminus \mathcal{D}_4) \cup B \models Q(w)$ (rsp. $(\mathcal{O} \setminus \mathcal{D}_4) \cup B \cup \neg Q(w)$ is inconsistent) and we have to ask the oracle another question.

Property 1. Given a diagnosis problem $\langle \mathcal{O}, B, T^\models, T^\nvDash \rangle$, a set of diagnoses $\mathbf{D}$, and a set of logical sentences X representing the query $\mathcal{O}_t \models X$:

If the oracle gives the answer *1* then every diagnosis $\mathcal{D}_i \in \mathbf{D}$ is a diagnosis for $T^\models \cup X$ iff $(\mathcal{O} \setminus \mathcal{D}_i) \cup B \cup \{\bigwedge_{t^\models \in T^\models} t^\models\} \cup \{X\} \cup \neg t^\nvDash$ is consistent for all $t^\nvDash \in T^\nvDash$.

If the oracle gives the answer *0* then every diagnosis $\mathcal{D}_i \in \mathbf{D}$ is a diagnosis for $T^\nvDash \cup \{X\}$ iff $(\mathcal{O} \setminus \mathcal{D}_i) \cup B \cup \{\bigwedge_{t^\models \in T^\models} t^\models\} \cup \neg X$ is consistent.

Note, a set X corresponds to a logical sentence where all elements of X are connected by $\wedge$. This defines the semantic of $\neg X$.

As possible queries we consider sets of entailed concept definitions provided by a classification service and sets of individual assertions provided by realization. In fact, the intention of classification is that a model for a specific application domain can be verified by exploiting the subsumption hierarchy [2].

One can use different methods to select the best query in order to minimize the number of questions asked to the oracle. "Split-in-half" heuristic is one of such methods that prefers queries which remove half of the diagnoses from the set $\mathbf{D}$. To apply this heuristic it is essential to compute the set of diagnoses that can be rejected depending on the query outcome. For a query X the set of diagnoses $\mathbf{D}$ can be partitioned in sets of diagnoses $\mathbf{D}^X$, $\mathbf{D}^{\neg X}$ and $\mathbf{D}^\emptyset$ where

- for each $\mathcal{D}_i \in \mathbf{D}^X$ it holds that $(\mathcal{O} \setminus \mathcal{D}_i) \cup B \cup \{\bigwedge_{t^\models \in T^\models} t^\models\} \models X$
- for each $\mathcal{D}_i \in \mathbf{D}^{\neg X}$ it holds that $(\mathcal{O} \setminus \mathcal{D}_i) \cup B \cup \{\bigwedge_{t^\models \in T^\models} t^\models\} \models \neg X$
- $\mathbf{D}^\emptyset = \mathbf{D} \setminus (\mathbf{D}^X \cup \mathbf{D}^{\neg X})$

Given a diagnosis problem we say that the diagnoses in $\mathbf{D}^X$ predict 1 as a result of the query X, diagnoses in $\mathbf{D}^{\neg X}$ predict 0, and diagnoses in $\mathbf{D}^\emptyset$ do not make any predictions.

Property 2. Given a diagnosis problem $\langle \mathcal{O}, B, T^{\models}, T^{\not\models} \rangle$, a set of diagnoses **D**, and a query X:

If the oracle gives the answer *1* then the set of rejected diagnoses is $\mathbf{D}^{\neg X}$ and the set of remaining diagnoses is $\mathbf{D}^{X} \cup \mathbf{D}^{\emptyset}$.

If the oracle gives the answer *0* then the set of rejected diagnoses is $\mathbf{D}^{X}$ and the set of remaining diagnoses is $\mathbf{D}^{\neg X} \cup \mathbf{D}^{\emptyset}$.

For our first example let us consider three possible queries X_1, X_2 and X_3 (see Table 2). For each query we can partition a set of diagnoses **D** into three sets $\mathbf{D}^{X}$, $\mathbf{D}^{\neg X}$ and $\mathbf{D}^{\emptyset}$. Using this data and the heuristic given above we can determine that asking the oracle if $\mathcal{O}_t \models C(w)$ is the best query, as two diagnoses from the set **D** are removed regardless of the answer.

Let us assume that $\mathcal{D}_1$ is the target diagnosis, then an oracle will answer 0 to our question (i.e. $\mathcal{O}_t \not\models C(w)$). Given this feedback we can decide that $\mathcal{O}_t \models B(w)$ is the next best query, which is also answered with 0 by the oracle. Consequently, we identified that $\mathcal{D}_1$ is the only remaining minimal diagnosis. More generally, if n is the number of diagnoses and we can split the set of diagnoses in half by each query then the minimum number of queries is $log_2 n$. However, if the probabilities of diagnoses are known we can reduce this number of queries by using two effects: (1) We can exploit diagnoses probabilities to asses the probabilities of answers and the change in information content after an answer is given. (2) Even if there are multiple diagnoses in the set of remaining diagnoses we can stop further query generation if one diagnosis is highly probable and all other remaining diagnoses are highly improbable.

Table 2. Possible queries in Example 1

Query	$\mathbf{D}^{X}$	$\mathbf{D}^{\neg X}$	$\mathbf{D}^{\emptyset}$
$X_1 : \{B(w)\}$	$\{\mathcal{D}_2, \mathcal{D}_3, \mathcal{D}_4\}$	$\{\mathcal{D}_1\}$	$\emptyset$
$X_2 : \{C(w)\}$	$\{\mathcal{D}_3, \mathcal{D}_4\}$	$\{\mathcal{D}_1, \mathcal{D}_2\}$	$\emptyset$
$X_3 : \{Q(w)\}$	$\{\mathcal{D}_4\}$	$\{\mathcal{D}_1, \mathcal{D}_2, \mathcal{D}_3\}$	$\emptyset$

Example 2. Consider an ontology $\mathcal{O}$ with the terminology $\mathcal{T}$:

$$ax_1 : A_1 \sqsubseteq A_2 \sqcap M_1 \sqcap M_2 \qquad ax_4 : M_2 \sqsubseteq \forall s.A \sqcap C$$
$$ax_2 : A_2 \sqsubseteq \neg\exists s.M_3 \sqcap \exists s.M_2 \qquad ax_5 : M_3 \equiv B \sqcup C$$
$$ax_3 : M_1 \sqsubseteq \neg A \sqcap B$$

and the background theory $\mathcal{A} : \{A_1(w), A_1(u), s(u,w)\}$. The ontology is inconsistent and includes two minimal conflict sets: $\{\langle ax_1, ax_3, ax_4 \rangle, \langle ax_1, ax_2, ax_3, ax_5 \rangle\}$. To restore consistency, the user should modify all axioms of at least one minimal diagnosis:

$$\mathcal{D}_1 : [ax_1] \quad \mathcal{D}_2 : [ax_3] \quad \mathcal{D}_3 : [ax_4, ax_5] \quad \mathcal{D}_4 : [ax_4, ax_2]$$

Following the same approach as in the first example, we compute entailments for each ontology $\mathcal{O}_i = \mathcal{O} \backslash \mathcal{D}_i$ for all minimal diagnoses $\mathcal{D}_i \in \mathbf{D}$. To construct a query we select a $\mathbf{D}^{X} \subset \mathbf{D}$ and determine the common set X of concept instantiations and concept subsumption axioms, which are entailed by each $\mathcal{O}_i = \mathcal{O} \backslash \mathcal{D}_i$, where $\mathcal{D}_i \in \mathbf{D}^{X}$. If the set X is empty, the query is rejected. For each accepted query the remaining diagnoses $\mathcal{D}_j \in \mathbf{D} \backslash \mathbf{D}^{X}$ are partitioned into three sets $\mathbf{D}^{X}$, $\mathbf{D}^{\neg X}$, and $\mathbf{D}^{\emptyset}$ as defined above. If

Table 3. Possible queries in Example 2

Query	$\mathbf{D^X}$	$\mathbf{D^{\neg X}}$	$\mathbf{D^\emptyset}$
$X_1 : \{B \sqsubseteq M_3\}$	$\{\mathcal{D}_1, \mathcal{D}_2, \mathcal{D}_4\}$	$\{\mathcal{D}_3\}$	$\emptyset$
$X_2 : \{B(w)\}$	$\{\mathcal{D}_3, \mathcal{D}_4\}$	$\{\mathcal{D}_2\}$	$\{\mathcal{D}_1\}$
$X_3 : \{M_1 \sqsubseteq B\}$	$\{\mathcal{D}_1, \mathcal{D}_3, \mathcal{D}_4\}$	$\{\mathcal{D}_2\}$	$\emptyset$
$X_4 : \{M_1(w), M_2(u)\}$	$\{\mathcal{D}_2, \mathcal{D}_3, \mathcal{D}_4\}$	$\{\mathcal{D}_1\}$	$\emptyset$
$X_5 : \{A(w)\}$	$\{\mathcal{D}_2\}$	$\{\mathcal{D}_3, \mathcal{D}_4\}$	$\{\mathcal{D}_1\}$
$X_6 : \{M_2 \sqsubseteq D\}$	$\{\mathcal{D}_1, \mathcal{D}_2\}$	$\emptyset$	$\{\mathcal{D}_3, \mathcal{D}_4\}$
$X_7 : \{M_3(u)\}$	$\{\mathcal{D}_4\}$	$\emptyset$	$\{\mathcal{D}_1, \mathcal{D}_2, \mathcal{D}_3\}$

the the ontology $\mathcal{O}_j = \mathcal{O} \setminus \mathcal{D}_j$ is inconsistent with X then we add $\mathcal{D}_j$ to the set $\mathbf{D^{\neg X}}$. In the case when $\mathcal{O}_j \cup \{\neg X\}$ is inconsistent $\mathcal{D}_j$ is added to $\mathbf{D^X}$. Otherwise we add $\mathcal{D}_j$ to the set $\mathbf{D^\emptyset}$.

For instance, ontologies $\mathcal{O}_i = \mathcal{O} \setminus \mathcal{D}_i$ obtained for diagnoses $\mathcal{D}_2$, $\mathcal{D}_3$ and $\mathcal{D}_4$ have the following set of common entailments:

$$X_4' : \{A_1 \sqsubseteq A_2, A_1 \sqsubseteq M_1, A_1 \sqsubseteq M_2, A_2(u), M_1(u), M_2(u), A_2(w), M_1(w)\} \quad (1)$$

Since the set X_4' is not empty it is considered as the query and the set $\mathbf{D^X}$ includes three elements $\{\mathcal{D}_2, \mathcal{D}_3, \mathcal{D}_4\}$. The ontology $\mathcal{O} \setminus \mathcal{D}_1 \cup \{X_4'\}$ is inconsistent therefore the set $\mathbf{D^{\neg X}} = \{\mathcal{D}_1\}$ and the set $\mathbf{D^\emptyset} = \emptyset$. However, a query need not include all of these axioms. If a query X' partitions the set of diagnoses into $\mathbf{D^X}$, $\mathbf{D^{\neg X}}$ and $\mathbf{D^\emptyset}$ and there exists an irreducible set $X \subset X'$ which preserves the partition then it is sufficient to query X. In our example, the set X_4' can be reduced to its subset X_4 : $\{M_1(w), M_2(u)\}$. If there are multiple subsets that preserve the partition we select one with minimal cardinality. For query generation we investigate all possible subsets of $\mathbf{D}$. This is feasible since we consider only the n most probable minimal diagnoses (e.g. $n = 12$) during query generation and selection.

The possible queries presented in Table 3 partition the set of diagnoses $\mathbf{D}$ in a way that makes the application of myopic strategies, such as split-in-half, inefficient. A greedy algorithm based on such a heuristic would select the first query X_1 as the next query, since there is no query that cuts the set of diagnoses in half. If $\mathcal{D}_4$ is the target diagnosis then X_1 will be positively evaluated by an oracle (see Fig. 1). On the next iteration the algorithm would also choose a suboptimal query since there is no partition that divides the diagnoses $\mathcal{D}_1$, $\mathcal{D}_2$, and $\mathcal{D}_4$ into two equal groups. Consequently,

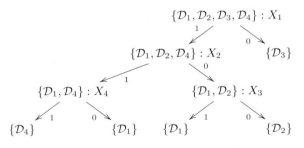

Fig. 1. Greedy algorithm

it selects the first untried query X_2. The oracle answers positively, and the algorithm identifies query X_4 to differentiate between $\mathcal{D}_1$ and $\mathcal{D}_4$.

However, in real-world settings the assumption that all axioms fail with the same probability is rarely the case. For example, Roussey et al. [12] present a list of "anti-patterns". Each anti-pattern is a set of axioms, like $\{C1 \sqsubseteq \forall R.C2, C1 \sqsubseteq \forall R.C3, C2 \equiv \neg C3\}$, that correspond to a minimal conflict set. The study performed by the authors shows that such conflict sets occur often in practice and therefore can be used to compute probabilities of diagnoses.

The approach that we follow in this paper was suggested by Rector at al. [10] and considers the syntax of the description logics, such as quantifiers, conjunction, negation, etc., rather than axioms to describe a failure pattern. For instance, if a user modifies a quantifier of one of the roles to restore coherency, then we can assume that axioms including universal quantifier are more probable to fail than the other ones. In [10] the authors report that in most cases inconsistent ontologies were created because users (a) mix up $\forall r.S$ and $\exists r.S$, (b) mix up $\neg \exists r.S$ and $\exists r.\neg S$, (c) mix up $\sqcup$ and $\sqcap$, (d) wrongly assume that classes are disjoint by default or overuse disjointness, (e) wrongly apply negation. Observing that misuses of quantifiers are more likely than other failure patterns one might find that the axioms ax_2 and ax_4 are more likely to be faulty than ax_3 (because of the use of quantifiers), whereas ax_3 is more likely to be faulty than ax_5 and ax_1 (because of the use of negation). Therefore, diagnosis $\mathcal{D}_2$ is the most probable one, followed closely by $\mathcal{D}_4$ although it is a double fault diagnosis. $\mathcal{D}_1$ and $\mathcal{D}_3$ are significantly less probable because ax_1 and ax_5 have a significantly lower fault probability than ax_3. A detailed justification based on probability is given in the next section.

Taking into account the information about user faults provided in [10], it is almost useless to ask query X_1 because it is highly probable that the target diagnosis is either $\mathcal{D}_2$ or $\mathcal{D}_4$ and therefore it is highly probable that the oracle will respond with 1. Instead, asking X_3 is more informative because given any possible answer we can exclude one of the highly probable diagnoses, i.e. either $\mathcal{D}_2$ or $\mathcal{D}_4$. If the oracle responds to X_3 with 0 then $\mathcal{D}_2$ is the only remaining diagnosis. However, if the oracle responds with 1, diagnoses $\mathcal{D}_4$, $\mathcal{D}_3$, and $\mathcal{D}_1$ remain, where $\mathcal{D}_4$ is significantly more probable compared to diagnoses $\mathcal{D}_3$ and $\mathcal{D}_1$. We can stop, since the difference between the probabilities of the diagnoses is high enough such that $\mathcal{D}_1$ can be accepted as the target diagnosis. In other situations additional questions may be required. This strategy can lead to a substantial reduction in the number of queries compared to myopic approaches as we will show in our evaluation.

Note that in real-world application scenarios failure patterns and their probabilities can be discovered by analyzing actions of a user in an ontology editor, like Protégé, while debugging an ontology or just repairing an inconsistency/incoherency. In this case it is possible to "personalize" the debugging algorithm such that it will prefer user-specific faults.

3 Entropy-Based Query Selection

To select the best query we make the assumption that knowledge is available about the a-priori failure probabilities in specifying axioms. Such probabilities can be estimated either by studies such as [10,12] or can be personalized by observing the typical failures of specific users working with an ontology development tool. In the last case an ontology editor should just save logs of debugging sessions, as well as user actions taken to restore the consistency/coherency of an ontology. Such observations can be then used to identify typical failures of a particular user. Using observations about failure patterns, for instance obtained from an ontology editor as described above, we can calculate the initial probability of each axiom $p(ax_i)$ containing a failure. If no information about failures is available then the debugger can initialize all probabilities $p(ax_i)$ with some small number.

Given the failure probabilities $p(ax_i)$ of axioms, the diagnosis algorithm first calculates the a-priori probability $p(\mathcal{D}_j)$ that $\mathcal{D}_j$ is the target diagnosis. Since all axioms fail independently, this probability can be computed as [8]:

$$p(\mathcal{D}_j) = \prod_{ax_n \in \mathcal{D}_j} p(ax_n) \prod_{ax_m \notin \mathcal{D}_j} 1 - p(ax_m) \tag{2}$$

The prior probabilities for diagnoses are then used to initialize an iterative algorithm that includes two main steps: (a) selection of the best query and (b) update of the diagnoses probabilities given the query feedback.

According to information theory the best query is the one that, given the answer of an oracle, minimizes the expected entropy of a the set of diagnoses [8]. Let $p(X_i = v_{ik})$ where $v_{i0} = 0$ and $v_{i1} = 1$ be the probability that query X_i is answered with either 0 or 1. Let $p(\mathcal{D}_j|X_i = v_{ik})$ be the probability of diagnosis $\mathcal{D}_j$ after the oracle answers $X_i = v_{ik}$. The expected entropy after querying X_i is:

$$H_e(X_i) = \sum_{k=0}^{1} p(X_i = v_{ik}) \times - \sum_{\mathcal{D}_j \in \mathbf{D}} p(\mathcal{D}_j|X_i = v_{ik}) \log_2 p(\mathcal{D}_j|X_i = v_{ik})$$

The query which minimizes the expected entropy is the best one based on a one-step-look-ahead information theoretic measure. This formula can be simplified to the following score function [8] which we use to evaluate all available queries and select the one with the minimum score to maximize information gain:

$$sc(X_i) = \sum_{k=0}^{1} p(X_i = v_{ik}) \log_2 p(X_i = v_{ik}) + p(\mathbf{D_i^\emptyset}) + 1 \tag{3}$$

where $\mathbf{D_i^\emptyset}$ is the set of diagnoses which do not make any predictions for the query X_i. $p(\mathbf{D_i^\emptyset})$ is the total probability of the diagnoses that predict no value for the query X_i. Since, for a query X_i the set of diagnoses $\mathbf{D}$ can be partitioned into the sets $\mathbf{D^{X_i}}$, $\mathbf{D^{\neg X_i}}$ and $\mathbf{D_i^\emptyset}$, the probability that an oracle will answer a query X_i with either 1 or 0 can be computed as:

$$p(X_i = v_{ik}) = p(\mathbf{S_{ik}}) + p(\mathbf{D_i^\emptyset})/2 \tag{4}$$

where $\mathbf{S_{ik}}$ corresponds to the set of diagnoses that predicts the outcome of a query, e.g. $\mathbf{S_{i0}} = \mathbf{D^{\neg X_i}}$ for $X_i = 0$ and $\mathbf{S_{i1}} = \mathbf{D^{X_i}}$ in the other case. Under the assumption that *both outcomes are equally likely* the probability that a set of diagnoses $\mathbf{D_i^\emptyset}$ predicts $X_i = v_{ik}$ is $p(\mathbf{D_i^\emptyset})/2$.

Since by Definition 1 each diagnosis is a unique partition of all axioms in an ontology $\mathcal{O}$ into correct and faulty, we consider all diagnoses as mutually exclusive events.

Therefore the probabilities of their sets can be calculated as:

$$p(\mathbf{D}_i^{\emptyset}) = \sum_{\mathcal{D}_j \in \mathbf{D}_i^{\emptyset}} p(\mathcal{D}_j) \qquad p(\mathbf{S}_{ik}) = \sum_{\mathcal{D}_j \in \mathbf{S}_{ik}} p(\mathcal{D}_j)$$

Given the feedback v of an oracle to the selected query X_s, i.e. $X_s = v$ we have to update the probabilities of the diagnoses to take the new information into account. The update is made using Bayes' rule for each $\mathcal{D}_j \in \mathbf{D}$:

$$p(\mathcal{D}_j|X_s = v) = \frac{p(X_s = v|\mathcal{D}_j)p(\mathcal{D}_j)}{p(X_s = v)} \qquad (5)$$

where the denominator $p(X_s = v)$ is known from the query selection step (Equation 4) and $p(\mathcal{D}_j)$ is either a prior probability (Equation 2) or is a probability calculated using Equation 5 during the previous iteration of the debugging algorithm. We assign $p(X_s = v|\mathcal{D}_j)$ as follows:

$$p(X_s = v|\mathcal{D}_j) = \begin{cases} 1, & \text{if } \mathcal{D}_j \text{ predicted } X_s = v; \\ 0, & \text{if } \mathcal{D}_j \text{ is rejected by } X_s = v; \\ \frac{1}{2}, & \text{if } \mathcal{D}_j \in \mathbf{D}_s^{\emptyset} \end{cases}$$

Example 1 (continued). Suppose that the debugger is not provided with any information about possible failures and therefore it assumes that all axioms fail with the same probability $p(ax_i) = 0.01$. Using Equation 2 we can calculate probabilities for each diagnosis. For instance, $\mathcal{D}_1$ suggests that only one axiom ax_1 should be modified by the user. Hence, we can calculate the probability of diagnosis D_1 as follows $p(\mathcal{D}_1) = p(ax_1)(1 - p(ax_2))(1 - p(ax_3))(1 - p(ax_4)) = 0.0097$. All other minimal diagnoses have the same probability, since every other minimal diagnosis suggests the modification of one axiom. To simplify the discussion we only consider minimal diagnoses for the query selection. Therefore, the prior probabilities of the diagnoses can be normalized to $p(\mathcal{D}_j) = p(\mathcal{D}_j)/\sum_{\mathcal{D}_j \in \mathbf{D}} p(\mathcal{D}_j)$ and are equal to 0.25.

Given the prior probabilities of the diagnoses and a set of queries (see Table 2) we evaluate the score function (Equation 3) for each query. E.g. for the first query $X_1 : \{B(w)\}$ the probability $p(\mathbf{D}^{\emptyset}) = 0$ and the probabilities of both the positive and negative outcomes are: $p(X_1 = 1) = p(\mathcal{D}_2) + p(\mathcal{D}_3) + p(\mathcal{D}_4) = 0.75$ and $p(X_1 = 0) = p(\mathcal{D}_1) = 0.25$. Therefore the query score is $sc(X_1) = 0.1887$.

The scores computed during the initial stage (see Table 4) suggest that X_2 is the best query. Taking into account that $\mathcal{D}_1$ is the target diagnosis the oracle answers 0 to the

Table 4. Expected scores for queries ($p(ax_i) = 0.01$)

Query	Initial score	$X_2 = 1$
$X_1 : \{B(w)\}$	0.1887	**0**
$X_2 : \{C(w)\}$	**0**	1
$X_3 : \{Q(w)\}$	0.1887	1

Table 5. Expected scores for queries ($p(ax_1) = 0.025$, $p(ax_2) = p(ax_3) = p(ax_4) = 0.01$)

Query	Initial score
$X_1 : \{B(w)\}$	**0.250**
$X_2 : \{C(w)\}$	0.408
$X_3 : \{Q(w)\}$	0.629

Table 6. Probabilities of diagnoses after answers

Answers	$\mathcal{D}_1$	$\mathcal{D}_2$	$\mathcal{D}_3$	$\mathcal{D}_4$
Prior	0.0970	0.5874	0.0026	0.3130
$X_3 = 1$	0.2352	0	0.0063	0.7585
$X_3 = 1, X_4 = 1$	0	0	0.0082	0.9918
$X_3 = 1, X_4 = 1, X_1 = 1$	0	0	0	1

Table 7. Expected scores for queries

Queries	Initial	$X_3 = 1$	$X_3 = 1, X_4 = 1$
$X_1 : \{B \sqsubseteq M_3\}$	0.974	0.945	**0.931**
$X_2 : \{B(w)\}$	0.151	0.713	1
$X_3 : \{M_1 \sqsubseteq B\}$	**0.022**	1	1
$X_4 : \{M_1(w), M_2(u)\}$	0.540	**0.213**	1
$X_5 : \{A(w)\}$	0.151	0.713	1
$X_6 : \{M_2 \sqsubseteq D\}$	0.686	0.805	1
$X_7 : \{M_3(u)\}$	0.759	0.710	0.970

query. The additional information obtained from the answer is then used to update the probabilities of diagnoses using the Equation 5. Since $\mathcal{D}_1$ and $\mathcal{D}_2$ predicted this answer, their probabilities are updated, $p(\mathcal{D}_1) = p(\mathcal{D}_2) = 1/p(X_2 = 1) = 0.5$. The probabilities of diagnoses $\mathcal{D}_3$ and $\mathcal{D}_4$ which are rejected by the outcome are also updated, $p(\mathcal{D}_3) = p(\mathcal{D}_4) = 0$.

On the next iteration the algorithm recomputes the scores using the updated probabilities. The results show that X_1 is the best query. The other two queries X_2 and X_3 are irrelevant since no information will be gained if they are performed. Given the negative feedback of an oracle to X_1, we update the probabilities $p(\mathcal{D}_1) = 1$ and $p(\mathcal{D}_2) = 0$. In this case the target diagnosis $\mathcal{D}_1$ was identified using the same number of steps as the split-in-half heuristic.

However, if the first axiom is more likely to fail, e.g. $p(ax_1) = 0.025$, then the first query will be $X_1 : \{B(w)\}$ (see Table 5). The recalculation of the probabilities given the negative outcome $X_1 = 0$ sets $p(\mathcal{D}_1) = 1$ and $p(\mathcal{D}_2) = p(\mathcal{D}_3) = p(\mathcal{D}_4) = 0$. Therefore the debugger identifies the target diagnosis only in one step.

Example 2 (continued). Suppose that in ax_4 the user specified $\forall s.A$ instead of $\exists s.A$ and $\neg \exists s.M_3$ instead of $\exists s.\neg M_3$ in ax_2. Therefore $\mathcal{D}_4$ is the target diagnosis. Moreover, the debugger is provided with observations of three types of failures: (1) conjunction/disjunction occurs with probability $p_1 = 0.001$, (2) negation $p_2 = 0.01$, and (3) restrictions $p_3 = 0.05$. Using the probability addition rule for non-mutually exclusive events we can calculate the probability of the axioms containing an error: $p(ax_1) = 0.0019$, $p(ax_2) = 0.1074$, $p(ax_3) = 0.012$, $p(ax_4) = 0.051$, and $p(ax_5) = 0.001$. These probabilities are exploited to calculate the prior probabilities of the diagnoses (see Table 6) and to initialize the query selection process.

On the first iteration the algorithm determines that X_3 is the best query and asks an oracle whether $\mathcal{O}_t \models M_1 \sqsubseteq B$ is true or not (see Table 7). The obtained information is then used to recalculate the probabilities of the diagnoses and to compute the next best

---

Algorithm 1. Ontology debugging algorithm

Input: ontology $\mathcal{O}$, set of background axioms B, set of fault probabilities for axioms FP,
maximum number of most probable minimal diagnoses n, acceptance threshold σ

Output: a diagnosis $\mathcal{D}$

1 $DP \leftarrow \emptyset; DS \leftarrow \emptyset; T^{\models} \leftarrow \emptyset; T^{\not\models} \leftarrow \emptyset; \mathbf{D} \leftarrow \emptyset; s \leftarrow 0;$

2 **while** belowThreshold$(DP, \sigma) \wedge s \neq 1$ **do**

3      $\mathbf{D} \leftarrow$ getDiagnoses(HS-Tree$(\mathcal{O}, B \cup T^{\models}, T^{\not\models}, n));$

4      $DS \leftarrow$ computeDataSet$(DS, \mathbf{D});$

5      $DP \leftarrow$ computePriors$(\mathbf{D}, FP);$

6      $DP \leftarrow$ uptateProbablities$(DP, DS, T^{\models}, T^{\not\models});$

7      $s \leftarrow$ getMinimalScore$(DS, DP);$

8      $\langle X, \mathbf{D^X}, \mathbf{D^{\neg X}} \rangle \leftarrow$ selectQuery$(DS, s);$

9   **if** getAnswer$(\mathcal{O}_t \models X)$ **then** $T^{\models} \leftarrow T^{\models} \cup X;$

10   **else** $T^{\not\models} \leftarrow T^{\not\models} \cup \neg X;$

11 **return** mostProbableDiagnosis$(\mathbf{D}, DP);$

---

query X_4, and so on. The query process stops after the third query, since $\mathcal{D}_4$ is the only diagnosis that has the probability $p(\mathcal{D}_4) > 0$.

Given the feedback of the oracle $X_4 = 1$ for the second query, the updated probabilities of the diagnoses show that the target diagnosis has a probability of $p(\mathcal{D}_4) = 0.9918$ whereas $p(\mathcal{D}_3)$ is only 0.0082. In order to reduce the number of queries a user can specify a threshold, e.g. $\sigma = 0.95$. If the probability of some diagnosis is greater than this threshold, the query process stops and returns the most probable diagnosis. Note, that even after the first answer $X_3 = 1$ the most probable diagnosis $\mathcal{D}_3$ is three times more likely than the second most probable diagnosis $\mathcal{D}_1$. Given such a great difference we could suggest to stop the query process after the first answer. Thus, in this example the debugger requires less queries than the split-in-half heuristic.

4   Implementation Details

The ontology debugger (Algorithm 1) takes an ontology $\mathcal{O}$ as input. Optionally, a user can provide a set of axioms B that are known to be correct, a set FP of fault probabilities for axioms $ax_i \in \mathcal{O}$, a maximum number n of most probable minimal diagnoses that should be considered by the algorithm, and a diagnosis acceptance threshold σ. The fault probabilities of axioms are computed as described by exploiting knowledge about typical user errors. Parameters n and σ are used to speed up the computations. In Algorithm 1 we approximate the set of the n most probable diagnoses with the set of the n most probable *minimal* diagnoses, i.e. we neglect non-minimal diagnoses which are more probable than some minimal ones. This approximation is correct, under a reasonable assumption that probability of each axiom $p(ax_i) < 0.5$. In this case for every non-minimal diagnosis ND, a minimal diagnosis $\mathcal{D} \subset ND$ exists which from Equation 2 is more probable than ND. Consequently the query selection algorithm operates on the set of minimal diagnoses instead of all diagnoses (including non-minimal ones). However, the algorithm can be adapted with moderate effort to also consider non-minimal diagnoses.

We implemented the computation of diagnoses following the approach proposed by Friedrich et al. [3]. The authors employ the combination of two algorithms, QUICKX-PLAIN [5] and HS-TREE [11]. The latter is a search algorithm that takes an ontology $\mathcal{O}$, a set of correct axioms, a set of axioms $T^{\not\models}$ which must not be implied by the target ontology, and the maximal number of most probable minimal diagnoses n as an input. HS-TREE implements a breadth-first search strategy to compute a set of minimal hitting sets from the set of all minimal conflicts in $\mathcal{O}$. As suggested in [3] it ignores all branches of the search tree that correspond to hitting sets inconsistent with at least one element of $T^{\not\models}$. HS-TREE terminates if either it identifies the n most probable minimal diagnoses or there are no further diagnoses which are more probable than the already computed ones. Note, HS-TREE often calculates only a small number of minimal conflict sets in order to generate the n most probable minimal hitting sets (i.e. minimal diagnoses), since only a subset of all minimal diagnoses is required.

The search algorithm computes minimal conflicts using QUICKXPLAIN. This algorithm, given a set of axioms AX and a set of correct axioms B returns a minimal conflict set $CS \subseteq AX$, or $\emptyset$ if axioms $AX \cup B$ are consistent. Minimal conflicts are computed on-demand by HS-TREE while exploring the search space. The set of minimal hitting sets returned by HS-TREE is used by GETDIAGNOSES to create a set $\mathbf{D}$ with at most n minimal diagnoses.

At the beginning of the main loop the algorithm calls COMPUTEDATASET function to generate a set of ontologies $\mathbf{O} : \{\mathcal{O}_i\}$ for each diagnosis $\mathcal{D}_i \in \mathbf{D}$ by removing all elements of a diagnosis from $\mathcal{O}$. The algorithm uses this set to generate data sets like the ones presented in Tables 2 and 3. For each ontology $\mathcal{O}_i \in \mathbf{O}$ the algorithm gets a set of entailments from the reasoner and associates them with the corresponding diagnosis $\mathcal{D}_i$. The algorithm uses the set of diagnoses/entailments pairs to compute the set of queries. For each query X_i it partitions the set $\mathbf{D}$ into $\mathbf{D}^{X_i}$, $\mathbf{D}^{\neg X_i}$ and $\mathbf{D}_i^{\emptyset}$, as defined in Section 2. Then X_i is iteratively reduced by applying QUICKXPLAIN such that sets $\mathbf{D}^{X_i}$ and $\mathbf{D}^{\neg X_i}$ are preserved.

In the next step COMPUTEPRIORS computes prior probabilities for a set of diagnoses given the fault probabilities of the axioms contained in FP. To take past answers into account the algorithm updates the prior probabilities of the diagnoses by evaluating Equation 5 for each diagnosis in $\mathbf{D}$ (UPDATEPROBABILITIES). All data required for the update is stored in sets DS, $T^{\models}$, and $T^{\not\models}$.

The function GETMINIMALSCORE evaluates the scoring function (Equation 3) for each element of DS and returns the minimal score.

In the query-selection phase the algorithm selects a set of axioms that should be evaluated by an oracle. SELECTQUERY retrieves a triple $\langle X, \mathbf{D}^X, \mathbf{D}^{\neg X} \rangle \in DS$ that corresponds to the best (minimal) score s. The set of axioms X is then presented to the oracle. If there are multiple queries with a minimal score SELECTQUERY returns the triple where X has the smallest cardinality in order to reduce the answering effort.

Depending on the answer of the oracle, the algorithm extends either set $T^{\models}$ or $T^{\not\models}$. This is done to exclude corresponding diagnoses from the results of HS-TREE in further iterations. Note, the algorithm can be easily extended to allow the oracle to reject a query if the answer is unknown. In this case the algorithm proceeds with the next best query until no further queries are available.

The algorithm stops if there is a diagnosis probability above the acceptance threshold σ or if no query can be used to differentiate between the remaining diagnoses (i.e. all scores are 1). The most probable diagnosis is then returned to the user. If it is impossible to differentiate between a number of highly probable minimal diagnoses, the algorithm returns a set that includes all of them.

5   Evaluation

The evaluation of our approach was performed using generated examples and real-world ontologies presented in Table 1. We employed generated examples to perform controlled experiments where the number of minimal diagnoses and their cardinality could be varied to make the identification of the target diagnosis more difficult. The main goal of the experiment using ontologies is to demonstrate applicability of our approach in the real-world settings.

For the first test we created a generator which takes a consistent and coherent ontology, a set of fault patterns together with their probabilities, the minimum number of minimal diagnoses m, and the required minimum cardinality of these minimal diagnoses $|\mathcal{D}_t|$ as inputs. The output was an alteration of the input ontology for which at least the given number of minimal diagnoses with the required cardinality exist. In order to introduce inconsistencies and incoherences, the generator applied fault patterns randomly to the input ontology depending on their probabilities.

In this experiment we took five fault patterns from a case study reported by Rector at al. [10] and assigned fault probabilities according to their observations of typical user errors. Thus we assumed that in cases (a) and (b) (see Section 2, when an axiom includes some roles (i.e. property assertions), axiom descriptions are faulty with a probability of 0.025, in cases (c) and (d) 0.01 and in case (e) 0.001. In each iteration the generator randomly selected an axiom to be altered and applied a fault pattern to this axiom. Next it selected another axiom using the concept taxonomy and altered it correspondingly to introduce an incoherency/inconsistency. The fault patterns were randomly selected in each step using the probabilities given above.

For instance, given the description of a randomly selected concept A and the fault pattern "misuse of negation", we added the construct $\sqcap \neg X$ to the description of A, where X is a new concept name. Next, we randomly selected concepts B and S such that $S \sqsubseteq A$ and $S \sqsubseteq B$ and added $\sqcap X$ to the description of B. During the generation process, we applied the HS-TREE algorithm after each introduction of a in-coherency/inconsistency to control two parameters: the minimum number of minimal diagnoses in the ontology and their minimum cardinality. The generator continued to introduce incoherences/inconsistencies until the specified parameter values were reached. For instance, if the minimum number of minimal diagnoses equals to $m = 6$ and their cardinality to $|\mathcal{D}_t| = 4$, then the generated ontology will include at least 6 diagnoses of cardinality 4 and some additional number of diagnoses of higher cardinalities.

The resulting faulty ontology as well as the fault patterns and their probabilities were inputs for the ontology debugger. The acceptance threshold σ was set to 0.95 and the number of most probable minimal diagnoses n was set to 12. One of the minimal diagnoses with the required cardinality was randomly selected as the target diagnosis.

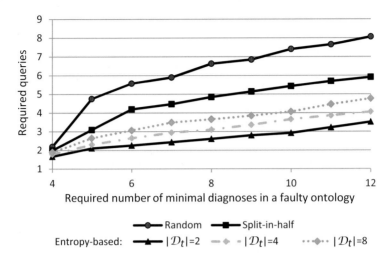

Fig. 2. Number of queries required to select the target diagnosis $\mathcal{D}_t$ with threshold $\sigma = 0.95$. Random and "split-in-half" are shown for the cardinality of minimal diagnoses $|\mathcal{D}_t| = 2$.

Note, the target ontology is not equal to the original ontology, but rather is a corrected version of the altered one, in which the faulty axioms were repaired by replacing them with their original (correct) versions according to the target diagnosis. The tests were done on ontologies bike2 to bike9, bcs3, galen and galen2 from Racer's benchmark suite[1].

The average results of the evaluation performed on each test suite (depicted in Fig. 2) show that the entropy-based approach outperforms the split-in-half method described in Section 2 as well as random query selection by more than 50% for the $|\mathcal{D}_t| = 2$ case due to its ability to estimate the probabilities of diagnoses. On average the algorithm required 8 seconds to generate a query. Figure 2 also shows that the cardinality of the target diagnosis increases as the number of required queries increases. This holds for the random and split-in-half methods (not depicted) as well. However, the entropy-based approach is still better than the split-in-half method even for diagnoses with increasing cardinality. The approach required more queries to discriminate between high cardinality diagnoses because the prior probabilities of these diagnoses tend to converge.

In the tests performed on the real-world ontologies we initialized the input parameters n and σ of Algorithm 1 with the same values as in the test with generated examples. Also we used the same five fault patterns together with their probabilities as given above. Before the experiment each ontology was analyzed by the HS-TREE algorithm and all minimal diagnoses of these ontologies were identified. In each test for a given ontology we selected randomly one of its minimal diagnoses as the target one and applied our approach using both split-in-half and entropy-based strategies. The evaluation of queries was done automatically by verifying if a query is also entailed by the target ontology obtained by removing all axioms of the target diagnosis from the input ontology. For

---

[1] http://www.racer-systems.com/products/download/benchmark.phtml

Table 8. Number of queries required to identify a target diagnosis

		Split-in-half			Entropy-based		
	Ontology	min	max	avg	min	max	avg
1.	Chemical	3	4	3	1	3	2
2.	Sweet-JPL	4	5	4	1	4	2
3.	University	7	9	8	2	7	4
4.	Tambis	8	10	8	2	7	5
5.	Economy	10	12	11	3	10	6
6.	Transport	11	14	12	4	11	7

Table 9. Time in seconds required to calculate 12 first and all minimal diagnoses as well as an average time used to generate a query

	Ontology	Diagnoses		Query
		12	all	avg
1.	Chemical	0,97	1,39	1,50
2.	Sweet-JPL	31,97	36,47	5,48
3.	University	0,27	0,61	1,12
4.	Tambis	80,29	286,11	3,91
5.	Economy	8,33	55,70	1,87
6.	Transport	6,70	99,02	2,39

each ontology we performed 20 tests and on each iteration the target diagnosis was randomly reselected.

The results of this experiment are presented in Tables 8 and 9 and show that in terms of queries, the entropy-based approach outperformed split-in-half. As the number of diagnoses grew we observed that the difference between the two strategies increased. In the best case for the entropy-based strategy, when the target diagnoses were assigned a high a-priori fault probability, the number of queries was usually twice as low as required by the split-in-half strategy. Also in the worst case, when the target diagnoses were assigned a low a-priori fault probability, the entropy-based strategy performed better than split-in-half, because it was able to adapt the a-posteriori fault probabilities using Bayes rule and the oracle's feedback to queries. In this case the entropy-based strategy corresponds to active learning [14] applied to learn fault probabilities which is not exploited in the split-in-half strategy. The more queries are asked, the better the entropy-based method can predict the target diagnosis.

6   Related Work

To the best of our knowledge no sequential ontology debugging methods (neither employing split-in-half nor entropy-based methods) have been proposed to debug faulty ontologies so far. Diagnosis methods for ontologies are introduced in [13,6,3]. Ranking of diagnoses and proposing a target diagnosis is presented in [7]. This method uses a number of measures such as: (a) the frequency with which an axiom appears in conflict sets, (b) impact on an ontology in terms of its "lost" entailments when some axiom is modified or removed, (c) ranking of test cases, (d) provenance information about the axiom, and (e) syntactic relevance. All these measures are evaluated for each axiom in a conflict set. The scores are then combined in a rank value which is associated with the corresponding axiom. These ranks are then used by a modified HS-TREE algorithm that identifies diagnoses with a minimal rank. In this work no query generation and selection strategy is proposed if the target diagnosis cannot be determined reliably with the given a-priori knowledge. In our work additional information is acquired until the target diagnosis can be identified with confidence. In general, the work of [7] can be combined with the one presented in this paper as axiom ranks can be taken into account together with other observations while calculating the prior probabilities of the diagnoses.

The idea of selecting the next best query based on the expected entropy was exploited in the generation of decisions trees [9] and further refined for selecting measurements in the model-based diagnosis of circuits [8]. We extended these methods to query selection in the domain of ontology debugging.

7   Conclusions

In this paper we presented an approach to the sequential diagnosis of ontologies. We showed that the axioms generated by classification and realization can be used to build queries which differentiate between diagnoses. To rank the utility of these queries we employ knowledge about typical user errors in ontology axioms. Based on the likelihood of an ontology axi om containing an error we predict the information gain produced by a query result, enabling us to select the next best query according to a one-step-lookahead entropy-based scoring function. We outlined the implementation of a sequential debugging algorithm and compared our proposed method with a split-in-half strategy. Our experiments showed a significant reduction in the number of queries required to identify the target diagnosis.

References

1. Ceraso, J., Provitera, A.: Sources of error in syllogistic reasoning. Cognitive Psychology 2(4), 400–410 (1971)
2. Baader, F., Calvanese, D., McGuinness, D.L., Nardi, D., Patel-Schneider, P.F. (eds.): The Description Logic Handbook, 2nd edn. Cambridge University Press, New York (2007)
3. Friedrich, G., Shchekotykhin, K.: A General Diagnosis Method for Ontologies. In: Gil, Y., Motta, E., Benjamins, V.R., Musen, M.A. (eds.) ISWC 2005. LNCS, vol. 3729, pp. 232–246. Springer, Heidelberg (2005)
4. Haarslev, V., Müller, R.: RACER System Description. In: Goré, R.P., Leitsch, A., Nipkow, T. (eds.) IJCAR 2001. LNCS (LNAI), vol. 2083, pp. 701–705. Springer, Heidelberg (2001)
5. Junker, U.: QUICKXPLAIN: Preferred Explanations and Relaxations for Over-Constrained Problems. In: Association for the Advancement of Artificial Intelligence (AAAI 2004), pp. 167–172. AAAI, Menlo Park (2004)
6. Kalyanpur, A., Parsia, B., Horridge, M., Sirin, E.: Finding all Justifications of OWL DL Entailments. In: Aberer, K., Choi, K.-S., Noy, N., Allemang, D., Lee, K.-I., Nixon, L.J.B., Golbeck, J., Mika, P., Maynard, D., Mizoguchi, R., Schreiber, G., Cudré-Mauroux, P. (eds.) ASWC 2007 and ISWC 2007. LNCS, vol. 4825, pp. 267–280. Springer, Heidelberg (2007)

7. Kalyanpur, A., Parsia, B., Sirin, E., Cuenca-Grau, B.: Repairing Unsatisfiable Concepts in OWL Ontologies. In: Sure, Y., Domingue, J. (eds.) ESWC 2006. LNCS, vol. 4011, pp. 170–184. Springer, Heidelberg (2006)
8. de Kleer, J., Williams, B.C.: Diagnosing multiple faults. Artificial Intelligence 32(1), 97–130 (1987)
9. Quinlan, J.R.: Induction of Decision Trees. Machine Learning 1(1), 81–106 (1986)
10. Rector, A., Drummond, N., Horridge, M., Rogers, J., Knublauch, H., Stevens, R., Wang, H., Wroe, C.: OWL Pizzas: Practical Experience of Teaching OWL-DL: Common Errors & Common Patterns. In: Motta, E., Shadbolt, N.R., Stutt, A., Gibbins, N. (eds.) EKAW 2004. LNCS (LNAI), vol. 3257, pp. 63–81. Springer, Heidelberg (2004)
11. Reiter, R.: A Theory of Diagnosis from First Principles. Artificial Intelligence 23, 57–95 (1987)
12. Roussey, C., Corcho, O., Vilches-Blázquez, L.M.: A catalogue of OWL ontology antipatterns. In: 5th International Conference On Knowledge Capture (K-CAP-2009), pp. 205–206. ACM, New York (2009)
13. Schlobach, S., Huang, Z., Cornet, R., Harmelen, F.: Debugging Incoherent Terminologies. Journal of Automated Reasoning 39(3), 317–349 (2007)
14. Settles, B.: Active Learning Literature Survey. Computer sciences technical report 1648, University of Wisconsin-Madison (2009)
15. Sirin, E., Parsia, B., Grau, B.C., Kalyanpur, A., Katz, Y.: Pellet: A practical OWL-DL reasoner. Journal of Web Semantics: Science, Services and Agents on the World Wide Web 5(2), 51–53 (2007)

Preference-Based Web Service Composition: A Middle Ground between Execution and Search

Shirin Sohrabi and Sheila A. McIlraith

Department of Computer Science, University of Toronto, Toronto, Canada
{shirin,sheila}@cs.toronto.edu

Abstract. Much of the research on automated Web Service Composition (WSC) relates it to an AI planning task, where the composition is primarily done offline prior to execution. Recent research on WSC has argued convincingly for the importance of optimizing quality of service, trust, and user preferences. While some of this optimization can be done offline, many interesting and useful optimizations are data-dependent, and must be done following execution of at least some information-gathering services. In this paper, we examine this class of WSC problems, attempting to balance the trade-off between offline composition and online information gathering with a view to producing high-quality compositions efficiently and without excessive data gathering. Our investigation is performed in the context of the semantic web employing an existing preference-based Hierarchical Task Network WSC system. Our experiments illustrate the potential improvement in both the quality and speed of composition generation afforded by our approach.

1 Introduction

Web Service Composition (WSC) requires a computer program to automatically select, integrate, and invoke multiple web services in order to achieve a user-defined objective. It is an example of the more general task of composing business processes or component software. Automated WSC is motivated by the need to improve the efficiency of composing and integrating services. A number of Business Process Management (BPM) systems exist to help organizations optimize business performance by discovering, managing, composing, and integrating business processes, including SAP's NetWeaver, and IBM's WebSphere and BPM Suite. With the advent of cloud computing, an increasing number of small- and medium-sized businesses are attempting to blend cloud services from multiple providers. Performing such integration and interoperation manually is costly and time consuming. Automated WSC and semantic integration address this emerging challenge [14]. For the purposes of this paper, we illustrate concepts in terms of the intuitive but over-used travel domain, however compelling examples exist in sectors such as Banking and Finance, Government, Healthcare and Life Sciences, Insurance, Retail, and Supply Chain Management. Many of these applications exploit extensive internet- or intranet-accessible data and will directly benefit from the work described here.

P.F. Patel-Schneider et al. (Eds.): ISWC 2010, Part I, LNCS 6496, pp. 713–729, 2010.

A popular approach to WSC is to characterize it as an Artificial Intelligence (AI) planning task and to solve it as such (e.g., [13,15,4]). In previous work (e.g., [15,20,19]) we have argued that for a number of WSC problems it is desirable to specify a **flexible workflow**, generic procedure, or composition template that specifies the basic steps of the composition at an abstract level, but has sufficient flexibility to support their customization for different stakeholders, scenarios, and applications. To this end, we have specified flexible workflows using Golog (e.g., [15,20]), or alternatively Hierarchical Task Networks (HTNs) [19], and developed associated machinery for WSC. We are not alone is proposing such a vision. Others have similarly used HTNs (e.g., [17]) and finite state automata (e.g., [5]) to specify composition objectives with varying flexibility.

While customization of flexible workflows can take the form of hard constraints imposed by the specific application scenario and its stakeholders, in cases where such customizing constraints are conflicting, some form of prioritization is required. Similarly, in cases where customizations are desirable but not mandatory, customizations can be specified as preferences. This observation has led us to characterize the WSC task as a preference-based planning (PBP) task where actions (services, service parameters, and/or data) are selected not only to achieve the composition objective but to produce compositions that are of high quality with respect to quality of service, trust, or other composition-, service-, or data-oriented user preferences (e.g., [20,11,19,1,10,21]).

Previous work on preference-based WSC (and indeed much of the work on WSC without preferences) has assumed that all the information required to generate the composition is on hand at the outset, and as such, composition is done offline followed by subsequent execution of the composition, perhaps in association with execution monitoring. However, this is not realistic in many settings. Consider the task of travel planning or any other multi-step purchasing process on the Web. A good part of the composition for these domains involves data gathering, followed by generation of an optimized composition with respect to that data and other criteria. Indeed many of the choice points relating to the composition require data acquired at execution time.

To address this, most current WSC systems will acquire all the information required for the composition prior to initiating composition generation. This can result in a lot of unnecessary data access. Further, it results in an enormous search space for a planner. Most state-of-the-art planners require actions to be grounded. However, unlike typical planning applications, many WSC applications are data-intensive, which results in an enormous number of ground actions and a huge search space. While this space may still be manageable for computing *a* composition, to compute *an optimal* composition, and to guarantee optimality, the entire search space must be searched, at least implicitly. This has the effect that most data-intensive WSC tasks that involve optimization of data (like picking preferred flights) will not scale using conventional PBP techniques.

Consider a flexible workflow that describes the travel domain in terms of the tasks of booking transportation and booking accommodations, with varying options for their realization. We add to this the following preferences: *If destination is more than 500 km away, book a flight, otherwise I prefer to rent a car; I prefer to fly with a Star Alliance carrier; I prefer to book cars with Avis, and if not Budget; I prefer to book a Hilton hotel, and if not a Sheraton.* A naive PBP would access all the flight, car, hotel, etc. information prior to composition and create grounded actions (e.g., *book-car(Avis,Pria,Daily,$39,...)*) for each data instance, resulting in a huge set of actions. In order to guarantee optimality of a composition, one needs to guarantee that all compositions were considered, which would (naively) involve considering all combinations of flight-hotel and/or car-hotel. However, there is clearly a smarter way to do this. In particular, either flight information or car rental information (but not both) need to be considered, depending on the distance to destination. Further, the choice of airline is independent of the choice of hotel, so optimality can be guaranteed by optimizing these choices independently. These simple, intuitive observations provide motivation for the work presented here.

In this paper, we investigate the class of WSC problems that endeavour to generate high-quality compositions through optimization of service and data selection. We attempt to balance the trade-off between offline composition and online information gathering with a view to producing high-quality compositions. Our objective is to minimize data access and to make optimization as efficient as possible by exploiting the independence of ground actions within the search space. Finally we wish to ensure that our techniques will maintain the guarantees a more naive approach would afford, including guarantees regarding the soundness of our compositions and their optimality.

Our investigation is performed in the context of our existing preference-based HTN WSC system, HTNWSC-P [19]. We propose a means of analyzing a WSC problem in order to identify places where optimization can be localized while preserving global optimality. Further, building on previous work that addresses the problem of information gathering (e.g., [15,9]), we propose a middle-ground execution engine that executes information-gathering services, as needed, while only *simulating* the execution of world-altering services. In doing so, the HTN WSC engine is able to benefit from the further knowledge afforded by information-gathering while still supporting backtrack search, by not actually or not necessarily executing world-altering services. We illustrate the effectiveness of our approach through experimentation.

2 Background and Preliminaries

The setting for this work is the semantic web. We assume that both the Web services and our composition template are described in OWL-S, an ontology for describing Web services [12]. We use an OWL-S to HTN translator to translate the OWL-S process descriptions and composition template to an HTN domain description and initial task network, respectively. Customization of the composition template is specified in PDDL3, the Planning Domain Definition Language, which provides a means of specifying preferences for planning domains [6]. Web service compositions now take the form of plans, and optimized compositions take the form of optimized PBPs. In order to compute such PBPs, we exploit our previous work [18], which uses state-of-the-art heuristic search techniques to generate optimized PBPs from HTN specifications. Note throughout this paper we distinguish between information-gathering actions – actions that collect data, and world-altering actions – actions that effect change in the world.

HTN Planning: Hierarchical Task Network (HTN) planning [7] is a popular and widely used planning paradigm that has been employed for WSC (e.g., [17,11]). Given an initial state, an initial task network (the objective to be achieved), and a domain description comprising a set of operators and *methods* – a description of how tasks can be decomposed, an HTN planner constructs a plan by repeatedly decomposing tasks into smaller and smaller subtasks until a primitive decomposition of the initial task network is found. In the travel domain, the initial task network is the single task *arrange-travel*. This task can be decomposed into arranging transportation, accommodations, local transportation, activities, tours, and entertainment. Basic definitions are taken from [7].

Definition 1 (HTN Planning Problem). *An HTN planning problem is a 3-tuple* $\mathcal{P} = (s_0, w_0, D)$ *where* s_0 *is the initial state,* w_0 *is the initial task network, and* D *is the HTN planning domain which consists of a set of operators and methods.*

An operator is a primitive action, described by its name, preconditions and effects. In the travel domain, ignoring the parameters, operators might include: *book-hotel* and *book-flight*. A *task* consists of a task symbol and a list of arguments. A task is primitive if its task symbol is an operator name and its parameters match, otherwise it is *nonprimitive*. *arrange-transportation* and *arrange-activity* are nonprimitive tasks, while *book-tour* and *book-car* are primitive.

A method, m, is a 4-tuple $(name(m),\ task(m), subtasks(m),\ constr(m))$ corresponding to the method's name, a nonprimitive task and the method's task network, comprising subtasks and constraints. Method m is relevant for a task t if there is a substitution σ such that $\sigma(t) = task(m)$. Several methods can be relevant to a particular nonprimitive task t, leading to different decompositions of t. In our example, the method with *name by-air-trans* can be used to decompose the *task arrange-trans* into the *subtasks* of booking a flight and paying, with the constraint (*constr*) that the booking precede payment.

Definition 2 (Task Network). *A task network is a pair* $w=(U,\ C)$ *where* U *is a set of task nodes and* C *is a set of constraints. The constraints normally considered are of type precedence constraint, before-constraint, after-constraint or between-constraint.*

Definition 3 (Plan). $\pi = o_1 o_2 \ldots o_k$ *is a plan for HTN planning program* $\mathcal{P} = (s_0, w_0, D)$ *if there is a primitive decomposition,* w, *of* w_0 *of which* π *is an instance.*

Specifying User Preferences and Constraints: Customizing preferences and constraints are specified in a version of PDDL3 that we have augmented to express preferences over how HTN tasks are parameterized and decomposed as well as preferences over service (i.e., task) properties [19,18]. This allows us to combine optimization of service selection (such as quality of service) with optimization of the composition. This augmented version of PDDL3 supports specification of temporally extended preferences via a subset of Linear Temporal Logic (LTL). *always, sometime, sometime-before* are among the supported constructs. **occ**(a) refers to the occurrence of a primitive task, while **initiate**(x) and **terminate**(x) refer to the initiation and termination of a nonprimitive task or method. To specify preferences over non-functional properties of services such as trust, reliability, and reputation, we associate a unique id with each task via the predicate *isAssociatedWith* and augment the domain with additional predicates for these properties The constructs described above are used to describe desirable properties of plans. These properties (called preferences) are then aggregated together into an objective function. Some simplified examples follow.

```
(preference p1   (sometime (initiate (book-flight AirCanada Eco Direct))))
(preference p2   (always (not (occ (pay MasterCard)))))
(preference p3   (imply (hasBookedCar ?Z) (sometime (occ (pay ?Z AE)))))
```

p1 states that at some point the user books a direct economy flight with Air Canada, **p2** states that the user never pays by Mastercard, and **p3** states that if a car is booked, at some point the user pays with their American Express (AE).

The quality of a plan is measured by the value of a PDDL3 **metric function** – an objective function over preferences that can either be maximized or minimized. The PDDL3 function `is-violated` takes as input a preference name and returns the *number of times* the corresponding preference is violated. The example metric function below stipulates that it is to be minimized. As such, the lower its value, the higher the quality of the plan. The violation of individual preferences can be weighted to reflect their relative importance. E.g.,

```
(:metric minimize (+ (* 2 (is-violated p1))   (* 1 (is-violated p2))))
```

specifies that it is twice as important to satisfy **p1** as to satisfy preference **p2**. Note that since the metric function is a weighted sum of individual preference formulae, by trying to minimize its value, it automatically deals with inconsistent preferences. Hence, an appropriate trade-off between inconsistent preferences is made so that the metric function can be optimized.

Definition 4 (Preference-based HTN Planning). *An HTN planning problem with user preferences is described as a 4-tuple $\mathcal{P} = (s_0, w_0, D, \preceq)$ where $\preceq$ is a preorder between plans. A plan π is a solution to $\mathcal{P}$ if and only if: π is a plan for $\mathcal{P}' = (s_0, w_0, D)$ and there does not exist a plan π' for $\mathcal{P}'$ such that π' is more preferred than π.*

The $\preceq$ relation can be defined in many ways (e.g., $\preceq$ can be quantitatively defined using a metric function). Note, from now on we will refer to the metric function as M, and use $M(N)$ to denote the value of the metric in a search node N (a search node contains the current state, task network, and partial plan).

3   Decoupling Data Optimization from Search

Given the HTN domain description of a WSC problem, the initial task network, and the customizing constraints and preferences, we are interested in generating a high-quality (ideally optimal) composition. Unfortunately, unlike the task of generating a composition, its optimization requires considering all alternative compositions, at least implicitly. And even in the case where the composition can be decomposed into independent subproblems, customizing preferences and constraints over the composition can introduce new inter-dependencies.

In previous work [20,19] we proposed an algorithm based on planning with heuristic search that employs a best-first, forward search strategy capable of computing an optimal composition. We elaborate on the algorithm in Section 5. Here we consider how to exploit this algorithm in data-intensive settings where the search space can be prohibitively large.

As noted earlier, data acquired via information gathering is typically encoded as parameters of the actions that act on that data. E.g., the *book-flight* action would be parameterized by the data associated with a flight, such as airline, origin, destination, fare class, etc. State-of-the-art planning algorithms require actions/operators to be grounded. As such, in data-intensive settings, there can be an enormous number of ground actions and as a consequence an enormous search space to explore. Consider a simplified version of the task of booking a flight, a hotel, a car, and booking a tour for a vacation. Assume that these four tasks can be performed in any order and are completely independent of each other. Given 20 possible flights, 10 hotels, 10 types of car, 5 tours of the city, and 4! ways in which the booking of these items can be performed, there are 20*10*10*5*4! different compositions that need to be explored (at least implicitly) to determine the optimal composition. Using the algorithm proposed in our previous work, some of these combinations will be eliminated by our exploitation of state-of-the-art heuristic search and sound pruning – a means of pruning partial plans that have no prospect of producing a plan that is superior to the current best plan. Nevertheless, the algorithm is still doing a lot of unnecessary search.

From our experience with WSC applications that involve preferences, we observe that most of the search time is spent on resolving the optimization that relates to the data that we have collected. We henceforth refer to this type of optimization as **data optimization**. We observe that just as the subtasks afford a degree of independence in many WSC scenarios, so too do the different data choices, and that this independence allows us to perform some optimization *locally*, external to the composition process, or even arbitrarily (if they don't matter) while still guaranteeing that the choice does not eliminate the globally optimal solution. For example, in our simplified scenario we can select the best car, best flight, best hotel, and best tour independently of each other. And in doing so, we can reduce the search space to (20+10+10+5)*4!. More generally, if we are able to identify that subset of the data that is relevant to the optimization of the composition and attempt to localize its optimization then we can significantly streamline our search.

In what follows, we elaborate on the exploitation of three scenarios: (1) a data choice must be done in concert with the composition but choosing the optimal data can be localized; (2) a data choice can be optimized in isolation of the composition generation process; and (3) a data choice is irrelevant to the optimization of the composition and can be made arbitrarily. We begin by defining the notion of localized data optimization and identify conditions under which it retains the possibility of finding the optimal solution.

Definition 5 (Localized Data Optimization with respect to an Operator). *Let $\mathcal{P}'$ be an information-gathering HTN planning problem with preferences, following Definitions 4 and 8. Let N be a search node that represents a partial plan, and let O be the world-altering operator that is to be applied next in our search – the operator that extends the partial plan currently under consideration. Let $N_1...N_k$ be different nodes that result from different possible groundings of O from node N. Localized data optimization for O selects node N_i, $1 \leq i \leq k$ if $M(N_i) \leq M(N_j)$, $\forall 1 \leq j \leq k$, where $M(N)$ is the metric value of search node N.*

According to the above definition, the node with the least metric value is selected when localized data optimization for an operator is performed. The question is when is such a strategy sound, i.e., when can we do such a local selection without eliminating the overall best solution? For example, assume a best flight among all available flights is selected, but the selected flight arrives at night preventing the planner from booking an activity for that day. In such situations, even though the selected flight is the best flight choice among all available flights in isolation (or locally), because of the interactions among operators within and between tasks, this choice is not the best choice for the composition.

Definition 6 (Sound Localized Data Optimization with respect to an Operator). *Let $\mathcal{P}'$, N, O, $N_1...N_k$ be as in Definition 5. Localized data optimization with respect to O is said to be sound if there does not exist a plan extending any node N_j, $1 \leq j \leq k$ that would result in a better metric value than any plan extending the node N_i that is selected via localized data optimization. Hence, if there exists an optimal plan π from extending the partial plan in node N, π is not achievable from extending any of the nodes N_j and is only achievable from N_i.*

This definition has important implications. If localized data optimization is sound, then all nodes N_j can be pruned from the search space because we know the optimal plan cannot be reached by extending any of these nodes. Now that we know the condition under which localized data optimization is sound, we need to discuss how such an operator can be identified. Doing so involves analyzing the structure of the planning problem to identify operators that are completely independent and have no interactions with the rest of the planning problem including (1) the operators and methods in the domain, (2) the user preferences, and (3) the hard constraints, assuming for simplicity that there are no indirect effects that we have to worry about. The following is a syntactic criterion that can be used to identify operators whose grounding choices will have no impact on the rest of the decisions made during the generation of a composition.

Definition 7 (Non-interacting Operator with respect to the Domain). *An operator O is said to be non-interacting with respect to the domain if (1) no predicate in the precondition of O or in the condition of the conditional effect statement of O appears in the effect of any other operator in the domain, and (2) there is no predicate in the effect of O that appears in the precondition (or in the condition of the conditional effect statement) of any other operators or methods[1] of the planning problem.*

Intuitively this definition says that nothing affects the execution or outcome of this operator. Returning to our example, if the flight booking operator changes anything that is a precondition of another operator, then the flight booking operator interacts with that operator. E.g., if the flight booking operator has the effect of depleting available monetary funds, precluding the booking of a particular hotel, or if it results in arrival at a time that impacts the booking of a tour, then it is considered to interact with other aspects of the problem.

The above condition can be easily checked as a preprocessing step by analyzing the domain definition. However, syntactically identifying how preferences play a role in data interactions is more difficult, particularly when trajectory preferences – preferences expressed in a subset of LTL – are involved. One way to identify interacting operators with respect to the preferences, is to determine whether the operator's add effects – the positive effects of an operator – appear in any preference formulae. More specifically to enforce non-interaction, we need to ensure that the add effects of the operator never appear in the "*b* part" of preference formulae, where the "*b* part" is as follows: (*sometime-after* b a) (*always* (*imply* b a)) or (*sometime* (*imply* b a)). This is because the "*b* part" is the condition that if true requires the preference formula to be true, and in particular necessitates the "*a* part" holding. Thus, if the "*b* part" refers to an add effect of a world-altering operator for which localized data optimization is performed, and the "*a* part" is hard or impossible to achieve then the choice made in the data optimization interacts with a choice that has to be made later.

Theorem 1 (Criterion for Sound Localized Data Optimization). *If an operator O is non-interacting with respect to the planning domain, user preferences, and hard constraints then performing localized data optimization on this operator is sound.*

To this point we have defined the notion of localized data optimization and identified some syntactic criteria that will ensure its soundness. Before concluding, we informally discuss two further cases. We observe that in some instances the optimization of data can be completely separated or decoupled from the dynamics of the composition problem and the optimal data choice can be determined as a separate process. For example, if a user's sole preference is to book the

---

[1] Precondition for a method can be specified as a before constraint.

cheapest car, then the identification of what car to book can be performed in isolation of the generation of the composition altogether. Further, some data choices have no effect at all on the quality of the composition and as such can be made arbitrarily. For example, if the user does not care what car they rent, then the choice of rental car can be made arbitrarily. In both of these cases, the search space can exclude consideration of the different data values by insertion of a single placeholder value. Execution of the information-gathering service can be delayed until after composition, and the placeholder resolved at that time.

4 Middle-Ground Execution

For many WSC problems it is impractical, and often impossible to reduce the WSC problem to a planning problem with complete initial state – i.e., for which all the information necessary to generate a composition (and in our case to optimize it) is known prior to commencement of the search for a composition. In the travel domain this would necessitate collecting data relating to all the different modes of transportation, means of accommodation, etc. The space of ground actions would be enormous and the planning and optimization task unsolvable. However, one can instead imagine gathering information as it becomes necessary to choice points in the generation and optimization of the composition, and using this to inform the search for different compositions. In this section, we investigate how to perform information gathering in this manner.

The problem of gathering information during composition has been examined in several research papers (e.g., [15,17,8]). McIlraith and Son in [15] describe a middle-ground interpreter that collects relevant information, but only simulates the effects of world-altering actions. Their interpreter works under the **Invocation and Reasonable Persistence (IRP) Assumption** that (1) assumes all information gathering actions can be executed by the middle-ground interpreter and (2) assumes that the gathered information persists for a reasonable period of time, and none of the actions in the composition cause this assumption to be violated. Kuter et al. in [8] take a similar approach but their work focuses on dealing with services that do not return a result (if any) immediately. They provide a Query Manager that allows the planner to continue search without waiting for all of the information-gathering services to return data. They also assume that the information-gathering services are executable (similar to condition 1 of IRP) but they allow the planner itself to change the gathered information during planning (a variant of condition 2 of IRP). More recently, Au et al [2] proposed an approach to relaxing the IRP assumption, however their approach does not seem amenable to generating optimized compositions.

Our translation builds on the work by Sirin et al. [17]. We encode each OWL-S atomic process as an HTN operator and each OWL-S composite process as an HTN method. Similarly, we assume that all atomic processes are either information gathering or world altering and distinguish our set of planning operators accordingly. The fidelity of our translation relies on the IRP assumption, i.e., none of the actions in the HTN or any exogenous action can violate the assumption. To improve the efficiency of the system by avoiding multiple calls to the same source with the same parameters, we implement a caching system similar to [17]. However instead of using a monitoring system we modify the translation of information-gathering atomic processes into HTN operators (this operator has preconditions that externally call information-gathering sources and add the return response) to explicitly encode the caching for the gathered information, and to reflect the different courses of action that must be followed. We consider the following 3 cases in our translation:

1. cannot delay the call and are calling the information source for the first time, so call the information source and cache the gathered information.
2. cannot delay the call and have already called the information source once, so use the cached information.
3. can delay the call to the information source, so use a placeholder data value.

Our translation relies on the use of a SHOP2-based HTN planner; it exploits SHOP2's features to perform runtime binding of variables and to make external procedure calls to invoke services. The full translation is excluded for space.

In Section 3, we discussed circumstances where data optimization can be performed in isolation of the generation of the composition. This can occur when the data is irrelevant to the optimization of the composition. i.e., it is not mentioned in any preferences, or when the data choice does not does not interact with the dynamics of the composition. For example, consider the book-hotel service and the information-gathering service that gathers information regarding available hotels. If the user has no preference regarding the choice of hotel, then it is efficient to delay the execution of this information-gathering service and the arbitrary selection of a hotel until after the composition is generated. To implement this, we identify these data and associated services a prior and modify the translation to remove the execution of the information-gathering service and to replace occurrences of the data with placeholders. The information-gathering service is then executed following composition generation and the placeholder replaced with an appropriate choice.

Similar to [8], let X be a set of information-gathering services available during planning. Then we represent the body of information that can be obtained from services in X as $\delta(X)$. More specifically, $\delta(X)$ represents all possible bindings of the predicates that appear in the `output` or the `postcondition` of the OWL-S descriptions of the services in X. Note that we operate under the IRP assumption, and more specifically, we assume that the results returned from these sources will not change during the planning step.

Definition 8 (Information-Gathering HTN Planning Problem). *An information-gathering HTN planning problem $\mathcal{P}'$ is a 3-tuple (s'_0, w'_0, D') where s'_0 is what is known of the initial state, and w'_0 and D' are generated following our modified OWL-S to HTN translator, described above. Assuming the IRP assumption holds for our planning problem, we define a corresponding HTN planning problem $\mathcal{P} = (s_0, w_0, D)$ where s_0 is a consistent complete initial state such that $s'_0 \cup \delta(X) \subseteq s_0$, w_0 is the initial task network, D is an HTN planning domain, and where w_0 and D are generated using the original OWL-S to HTN translator described in [17].*

From Definitions 3 and 8, a plan for the information-gathering HTN planning problem is a primitive decomposition of the task network w_0'. To find such a decomposition, some information-gathering operators, as dictated by the methods and operators of the domain, have to be applied to collect the relevant information needed to successfully decompose w_0'. These operators interact with the information sources and add new information to the state of the planning problem. The following theorem establishes soundness of our approach.

Theorem 2. *Let $\mathcal{P}$ and $\mathcal{P}'$ be corresponding planning problems as defined above. π is a plan for $\mathcal{P}'$ if and only if π is a plan for $\mathcal{P}$.*

The above theorem states that if a plan can be found in the information-gathering problem $\mathcal{P}'$ the same plan can be found from the corresponding complete problem $\mathcal{P}$, and vice versa. This holds by looking at the relevant search space. The following corollary immediately follows. Recall π is an optimal plan for $\mathcal{P}$ if there is no other plan of superior quality.

Corollary 1. *Let $\mathcal{P}$ and $\mathcal{P}'$ be corresponding planning problems as described in Theorem 1. π is an optimal plan for $\mathcal{P}'$ if and only if π is an optimal plan for $\mathcal{P}$.*

5   Computing a Preferred Composition

In this section, we address the problem of computing a most preferred composition by using AI planning techniques to help guide the construction of the composition. Our algorithm performs best-first, incremental search and uses state-of-the-art heuristics developed in [18]. The search is performed in a series of *episodes*, each of which returns a plan with better quality than the previous plan. The search in each episode performs branch-and-bound pruning, that is we prune nodes from the search space if provably there does not exist a plan extending this node with a better metric value than the one found in the previous episode. In addition, we perform sound localized data optimization on some already identified non-interacting operators. The two important heuristics we use are the Optimistic Metric Function (OM) and the Lookahead Metric Function (LA). The OM function estimates optimistically the metric value resulting from the current node. LA function estimate the metric of the *best successor* to the current node. In short, it first solves the current node up to a certain depth, and then it computes a single decomposition for each of the resulting nodes and returns the best metric value among all the fully decomposed nodes.

Our algorithm is outlined in Figure 1. The algorithm takes as input an information-gathering HTN planning problem (s_0', w_0', D'), a metric function METRICFN, and a heuristic function HEURISTICFN. The nodes are of the form $\langle s, w, partialP \rangle$, where s is a plan state, w is a task network, and $partialP$ is a partial plan. This means w remains to be decomposed in state s and state s is reached from s_0' by performing the sequence of actions $partialP$. The algorithm keeps the elements of *frontier* sorted according to the function HEURISTICFN.

function HTNWSC(s_0', w_0', D', METRICFN,HEURISTICFN)
  frontier ← $\langle s_0', w_0', \emptyset \rangle$, *bestMetric* ← worst case upper bound   ▷ initialization
  while *frontier* is not empty **do**
    current ← *frontier*'s first element   ▷ best element since *frontier* is always sorted
    $\langle s, w, partialP \rangle$ ← *current*   ▷ establish the current values for s, w, and *partialP*
    lbound ← METRICBOUNDFN(s)   ▷ estimating the lower bound for s
    if *lbound* < *bestMetric* **then**   ▷ pruning suboptimal partial plans
      if $w = \emptyset$ and *current*'s metric < *bestMetric* **then**
        Output plan *partialP*, *bestMetric* ← METRICFN(s)
      succ ← successors of *current*
      if possible to perform sound localized data optimization **then**
        succ ← the best node among successors of *current*   ▷ pruning other nodes
      frontier ← merge *succ* into *frontier*

Fig. 1. A sketch of our HTN WSC algorithm

The HEURISTICFN function we use is a *prioritized sequence* of our heuristics (i.e., when comparing two nodes we look at the value of their heuristics in sequence to break ties when needed). We use a variable *bestMetric* that stores the metric value of the best plan found so far. This variable is initialized to a worst case upper bound. In each iteration of the while loop, the algorithm extracts the first element from the *frontier* and initializes the *current*. Then, it estimates a lowerbound, *lbound*, using the function METRICBOUNDFN and prunes nodes with a *lbound* greater than or equal to *bestMetric*. If *current* corresponds to a plan (i.e., w is empty), *bestMetric* is updated, and the plan is returned.

All successors to *current* are computed using the Partial-order Forward Decomposition procedure (PFD) [7]. If computing a successor to *current* implies picking a primitive task to decompose next and it is possible to perform sound localized data optimization for the operator that accomplishes this task, then data optimization on this node will select the best successor according to METRICFN and replace *succ* with the selected node[2]. The resulting *succ* is then merged into the *frontier*. Note that *succ* will have only one element if the algorithm chose to perform localized data optimization, that is all other nodes will get pruned from the search space. The search terminates when *frontier* is empty.

Optimality and Pruning. The search space for computing the preferred composition is significantly reduced by the flexible workflow captured in the structure of the HTN, by pruning performed from incremental search, and by the localized data optimization. So, under sound pruning we can guarantee that by exhausting the search space, an optimal plan can be found. We use the OM function to estimate the lower bound. Baier et al. [3] show that the OM function provides sound pruning under certain conditions.

---

[2] There are some subtleties, not discussed here, that ensure all appropriate grounding choices are considered and evaluated.

% of Identified Non-Inteferences	Case 1 Time(sec)	Case 2 Time(sec)	Case 3 Time(sec)	Case 4 Time(sec)	Average STI	Average PMI
0%	128	131	136	277	1.00	50.89%
20%	80	80	88	221	1.51	50.89%
40%	41	39	50	178	2.69	50.89%
60%	29	29	40	119	3.66	50.89%
80%	23	23	33	89	4.62	50.89%
100%	17	18	30	30	7.14	44.91%

Fig. 2. Time comparison between the four cases that found the optimal plan even without localized data optimization. STI is the search time improvement between each case and the no data optimization case (i.e., 0% case). PMI is the percent metric improvement. i.e., the percent difference between the metric of the first and the last plan returned relative to the first plan.

Proposition 1. *The OM function provides sound pruning if the metric function is non-decreasing in the number of satisfied preferences, non-decreasing in plan length, and independent of other state properties. A metric is non-decreasing in plan length if one cannot make a plan better by increasing its length only.*

Theorem 3. *If the OM function used to calculate the lower bound provides sound pruning, and any localized data optimization performed is sound, then the last plan returned, if any from the algorithm, is optimal.*

The proof follows from the proof of optimality for the **HPlan-P** planner [3] using Definition 7 and Theorem 1.

6   Implementation and Evaluation

We implemented our proof-of-concept WSC engine with two modules: a preprocessor and a preference-based HTN planner. The preprocessor reads PDDL3 problems and generates an HTN planning problem. Additionally, it finds noninteracting operators, making it possible to perform sound localized data optimization on this selection. Our implementation builds on **HTNPlan-P** [18], itself a modification of the LISP version of **SHOP2** [16], that implements the algorithm and heuristics described above. We have three main objectives in our experimental evaluation: (1) to measure the search time gain as well as the quality improvement by performing localized data optimization, (2) to see if performing localized data optimization helps in finding the optimal plan, (3) to investigate if the improvement (both time and quality) depends on other dimensions of search such as the heuristics used or the difficulty of the domain.

We use the travel domain described in this paper as our benchmark. We created 8 problem sets each with 6 different instances (we have 48 instances in total). In half of the problem sets we allowed interleaving of tasks and in the other half we did not. An example of interleaving is one that allows booking an accommodation when a transportation is booked, but not paid for (i.e., the transportation task is not done yet). Furthermore, the problem sets within the allowed (or not allowed) interleaving group differ in the difficulty of their top-level task. In the easiest case, the order of the execution of all tasks in *arrange-travel* (e.g., *arrange-trans*, *arrange-acc*, and *arrange-activity*) was known, and in the hardest case, these tasks could be carried out in any order. As explained earlier, if there are n tasks and they can be carried out in any order, then in the worst case there are $n!$ different combinations to evaluate in order to find the optimal composition. Finally in each problem set we know the number of non-interacting operators, but intentionally select the percentage of the one identified from this range [0, 20, 40, 60, 80, 100]. So in the 0% case none of the non-interacting operators are identified, hence, no localized data optimization can be performed, on the other hand in the 100% case all of the non-interacting operators are known, and localized data optimization is performed whenever possible. We used a 60 minute time out and a limit of 1 GB per process.

We ran all of the instances in two modes, one that makes use of the LA heuristic and one that does not. To compare the relative performance between the two modes, we averaged the percent metric difference of the final plan (relative to the worst plan) for all our 48 instances. This difference is 43% indicating that not surprisingly, using the LA heuristic greatly improves the quality of search. In particular, without the use of the LA heuristic, an optimal plan was not found in any of the instances. However, when the LA heuristic was used, many instances found an optimal plan. In particular, in four of the problem sets (we named them cases 1-4 in Figure 2), an optimal plan was found even without any localized data optimization The result shows (see Figure 2) that as the percentage of identified non-interacting operators increases (i.e., more localized data optimization is done), the time it took to find the optimal plan decreases. We averaged this improvement and show it in the STI column (search time improvement with respect to the 0% case). This column shows that optimal plans are found for example, 2.69 times faster than the 0% case in the 40% case, and 7.14 times faster in the 100% case. Also recall that our algorithm is incremental, performing search in a series, each one returning a better-quality plan than the last. To see how effective this approach is, we calculated the *percent metric improvement* (PMI), i.e., the percent difference between the metric of the first and the last plan returned relative to the first plan. The result shows that the incremental approach improves the quality of the plan almost by 50%.

Finally, we looked at the other four cases where without localized data optimization an optimal plan was not found. Out of these, in two, an optimal plan was found in the 100% case and this was found 3.5 times faster than the time it took to find a non-optimal plan in the 0% case. This suggests that doing localized data optimization for these harder problem sets is helpful. In the remaining two cases, an optimal plan was not found even with optimization. This is not surprising, since the search space in these sets is very large, and pruning even though helpful, is not able to exhaust the search space; in these cases interleaving was allowed and the top level tasks were unordered. However, we observed that with optimization, the quality of the final plan was improved by 10%, and the time spend on finding this better quality plan was 5 times faster.

7   Summary and Related Work

A significant number of WSC problems involve both optimization of the composition and the collection of information. Work on preference-based WSC has begun to address this problem but much of the work has ignored the critical information-gathering component, assuming that all information is given a priori. In this paper, we are motivated by the observation that even though some classes of WSC problems can be addressed without the need for any execution during the composition phase, without explicit consideration of the data, and without consideration of preferences that distinguish high-quality solutions, many interesting and useful compositions must be done hand in hand with the data collection and optimization. Specifically this is done following execution of some information-gathering services. The main contributions of this paper include: identification of a way to exploit structure in the preference specification and domain in order to generate compositions more efficiently by performing what we call *localized data optimization*, identification of a condition where performing localized data optimization is sound, development of an execution engine for preference-based WSC that interleaves online information gathering with offline search as deemed necessary, and identification of a case where we could prove the optimality of resulting compositions. To assess the effectiveness of our approach to WSC, we performed experiments to evaluate the performance of our system. We showed that our approach to data optimization has the potential to greatly improve the quality of compositions and the speed with which they are generated. While the focus of this paper was reasonably narrow, the problem it presents and the advances it makes are important first steps in addressing a broad and important problem.

While no other WSC planners can perform true preference-based planning, **SHOP2** [16] and **enquirer** [8] handle some simple user constraints. The **scup** prototype PBP planner in [11] is related but there are several differences to our work. In particular, their preferences are pre-processed into task networks and conflicting user preferences are detected and removed prior to invocation of their planner. Further, they do not consider handling regulations and are not able to specify preferences over the quality of services.

Another body of related work is the research on quality-driven WSC (e.g., [10,21,1]). This research addresses the problem of run-time service selection based on the functional (e.g., input and output matching) and non-functional (e.g.., reliability, availability, and reputation) properties of a service. This is addressed by encoding the problem as an optimization problem that can be solved using for example: Integer Programming (e.g., [21]), Mixed Integer Programming (e.g., [1]) or Genetic Algorithms (e.g., [10]). Our work differs in many ways. In particular, in our framework we are able to find a composition that is optimal with respect to the user's preferences some of which are over the entire composition, and we can do so while interleaving execution and search. Further, we are concerned with optimizing the selection of data within the services in addition to the selection of services themselves based on their quality.

Acknowledgements. We gratefully acknowledge funding from the Natural Sciences and Engineering Research Council of Canada (NSERC) and the Ontario Ministry of Innovations Early Researcher Award (ERA).

References

1. Alrifai, M., Risse, T.: Combining global optimization with local selection for efficient QoS-aware service composition. In: Proc. of the 18th Int'l World Wide Web Conference (WWW 2009), pp. 881–890 (2009)
2. Au, T.C., Nau, D.S.: Reactive query policies: A formalism for planning with volatile external information. In: Proc. of the IEEE Symposium on Computational Intelligence and Data Mining (CIDM), pp. 243–250 (2007)
3. Baier, J.A., Bacchus, F., McIlraith, S.A.: A heuristic search approach to planning with temporally extended preferences. Artificial Intelligence 173(5-6), 593–618 (2009)
4. Bertoli, P., Kazhamiakin, R., Paolucci, M., Pistore, M., Raik, H., Wagner, M.: Continuous orchestration of Web services via planning. In: Proc. of the 19th Int'l Conference on Automated Planning and Scheduling (ICAPS), pp. 18–25 (2009)
5. Calvanese, D., Giacomo, G.D., Lenzerini, M., Mecella, M., Patrizi, F.: Automatic service composition and synthesis: the Roman Model. IEEE Data Eng. Bull. 31(3), 18–22 (2008)
6. Gerevini, A., Haslum, P., Long, D., Saetti, A., Dimopoulos, Y.: Deterministic planning in the fifth international planning competition: PDDL3 and experimental evaluation of the planners. Artificial Intelligence 173(5-6), 619–668 (2009)
7. Ghallab, M., Nau, D., Traverso, P.: Hierarchical Task Network Planning. In: Automated Planning: Theory and Practice. Morgan Kaufmann, San Francisco (2004)
8. Kuter, U., Sirin, E., Nau, D.S., Parsia, B., Hendler, J.A.: Information gathering during planning for Web service composition. In: McIlraith, S.A., Plexousakis, D., van Harmelen, F. (eds.) ISWC 2004. LNCS, vol. 3298, pp. 335–349. Springer, Heidelberg (2004)
9. Kuter, U., Sirin, E., Parsia, B., Nau, D.S., Hendler, J.A.: Information gathering during planning for Web service composition. J. Web Sem. 3(2-3), 183–205 (2005)
10. Lécué, F.: Optimizing QoS-aware semantic Web service composition. In: Bernstein, A., Karger, D.R., Heath, T., Feigenbaum, L., Maynard, D., Motta, E., Thirunarayan, K. (eds.) ISWC 2009. LNCS, vol. 5823, pp. 375–391. Springer, Heidelberg (2009)
11. Lin, N., Kuter, U., Sirin, E.: Web service composition with user preferences. In: Bechhofer, S., Hauswirth, M., Hoffmann, J., Koubarakis, M. (eds.) ESWC 2008. LNCS, vol. 5021, pp. 629–643. Springer, Heidelberg (2008)
12. Martin, D., Burstein, M., McDermott, D., McIlraith, S., Paolucci, M., Sycara, K., McGuinness, D., Sirin, E., Srinivasan, N.: Bringing semantics to Web services with OWL-S. World Wide Web Journal 10(3), 243–277 (2007)
13. McDermott, D.V.: Estimated-regression planning for interactions with Web services. In: Proc. of the 6th Int'l Conference on Artificial Intelligence Planning and Scheduling (AIPS), pp. 204–211 (2002)
14. McDougall, P.: IBM eyes plug-and-play cloud framework, informationWeek (July 8, 2010)
15. McIlraith, S., Son, T.: Adapting Golog for composition of semantic Web services. In: Proc. of the 8th Int'l Conference on Knowledge Representation and Reasoning (KR), pp. 482–493 (2002)

16. Nau, D.S., Au, T.C., Ilghami, O., Kuter, U., Murdock, J.W., Wu, D., Yaman, F.:
    SHOP2: An HTN planning system. Journal of Artificial Intelligence Research 20,
    379–404 (2003)
17. Sirin, E., Parsia, B., Wu, D., Hendler, J., Nau, D.: HTN planning for Web service
    composition using SHOP2. J. Web Sem. 1(4), 377–396 (2005)
18. Sohrabi, S., Baier, J.A., McIlraith, S.A.: HTN planning with preferences. In: Proc.
    of the 21st Int'l Joint Conference on Artificial Intelligence (IJCAI), pp. 1790–1797
    (2009)
19. Sohrabi, S., McIlraith, S.A.: Optimizing Web service composition while enforcing
    regulations. In: Bernstein, A., Karger, D.R., Heath, T., Feigenbaum, L., Maynard,
    D., Motta, E., Thirunarayan, K. (eds.) ISWC 2009. LNCS, vol. 5823, pp. 601–617.
    Springer, Heidelberg (2009)
20. Sohrabi, S., Prokoshyna, N., McIlraith, S.A.: Web service composition via generic
    procedures and customizing user preferences. In: Cruz, I., Decker, S., Allemang,
    D., Preist, C., Schwabe, D., Mika, P., Uschold, M., Aroyo, L.M. (eds.) ISWC 2006.
    LNCS, vol. 4273, pp. 597–611. Springer, Heidelberg (2006)
21. Zeng, L., Benatallah, B., Ngu, A.H.H., Dumas, M., Kalagnanam, J., Chang,
    H.: QoS-aware middleware for web services composition. IEEE Trans. Software
    Eng. 30(5), 311–327 (2004)

A Self-Policing Policy Language

Sebastian Speiser and Rudi Studer

Karlsruhe Institute of Technology (KIT), Karlsruhe, Germany
Institute of Applied Informatics and Formal Description Methods (AIFB)
`firstname.lastname@kit.edu`

Abstract. Formal policies allow the non-ambiguous definition of situations in which usage of certain entities are allowed, and enable the automatic evaluation whether a situation is compliant. This is useful for example in applications using data provided via standardized interfaces. The low technical barriers of integrating such data sources is in contrast to the manual evaluation of natural language policies as they currently exist. Usage situations can themselves be regulated by policies, which can be restricted by the policy of a used entity. Consider for example the Google Maps API, which requires that applications using the API must be available without a fee, i.e. the application's policy must not require a payment. In this paper we present a policy language that can express such constraints on other policies, i.e. a self-policing policy language. We validate our approach by realizing a use case scenario, using a policy engine developed for our language.

1 Introduction

Policies are declarative descriptions of constraints and conditions that apply to some entity (the policy subject). Formal languages allow non-ambiguous policies, that can be automatically evaluated by computers. Many existing policy languages represent essentially an implicit access control matrix [1]. While this is sufficient for applications such as rights management for local file systems, there are entities that still impose constraints on their use after initial access was granted. This often applies to data representing factual information or creative works. Examples include images that require attribution of their creator, or real-time stock quotes that can only be published for a fee. Generally such policies classify usage situations into compliant or non-compliant. Conditions, required to be fulfilled by compliant situations, may restrict the policy of the situation. Consider for example the Google Maps API, which requires that applications using the API must be made available to the public without a fee. This is basically a constraint in the API's policy, which restricts the application's policy to not grant exclusive access to paying users. There exist approaches to usage restrictions, but our work is to the best of our knowledge the first self-policing policy language, in the sense that it can express restrictions on other policies.

Today, vast amounts of data are published on the Internet with standardized interfaces, e.g. as Web services or as Linked Data[1]. This imposes only low technical

---

[1] `http://linkeddata.org`

P.F. Patel-Schneider et al. (Eds.): ISWC 2010, Part I, LNCS 6496, pp. 730–746, 2010.

barriers to the use and reuse of data in new ways and their composition into new applications or data sources. In contrast the policies regulating their allowed uses are either not made explicit at all [2], or published in natural language, in form of terms and conditions. The former case makes it impossible, the latter case a manual and very tedious task to evaluate if a given usage situation is compliant or not. This may lead to frequent violations of usage restrictions, not because of ill will, but convenience. Evidence for this assumption is delivered e.g. by Seneviratne et al. who discovered that around 70%-90% of the reuses of Flickr images with a Creative Commons attribution license actually violate the license [3]. Formal policies are required to build tools that help users to check compliance of their data usages with the same ease as just using the data.

Restrictions on other policies include testing if one policy is contained in another. The resulting query containment problem is undecidable for many policy languages (e.g. in the presence of general negations and disjunctions). This means that these languages cannot simply be extended with self-policing conditions. Another difficulty is that simple query containment may not work, as restrictions have to apply to policies with subjects that are unknown at specification time. Therefore a policy structure is required that separates identifying applicable policy subjects and required compliance conditions. Other restrictions include checking if a partial situation description is sufficient for fulfilling a policy, possibly under further restrictions on aspects not specified in the partial description. Such restrictions need novel algorithms. Another requirement for the policy language is usability for the policy specifiers. Two enabling properties for usability are an intuitive policy structure and the reuse of policy conditions.

The rest of the paper is structured as follows. After introducing a use case in Section 2 for further motivation and evaluation of the approach, and presenting preliminaries in Section 3, the following contributions are presented:

- A policy model with formal semantics based on unions of conjunctive queries and RDFS (see Section 4.1).
- A model for structuring policies to improve usability and enable the reuse of policy parts (comparable to the Creative Commons building blocks, such as (non-)commercial use). The structure is based on RDF and RIF and is provided with rules that map it to our policy model (see Sections 4.2 and 4.3).
- Formal definitions of useful types of policy restrictions and their integration into policy conditions (see Section 5).

We evaluate the approach by implementing a policy engine and applying it to policies realizing the use cases. This is described in Section 6 together with some performance experiments. In Section 7 the policy language is compared to existing work. In Section 8 we conclude and give an outlook to future work.

2   Use Case and Requirements

The policy language presented in this paper is thought to be applicable to different application scenarios. However for further motivating the features of the language and validating how they fulfill concrete requirements, we describe a specific application and a concrete use case in this section. The application area we deal with is the use of services and data in dynamic and composed documents. Another thinkable application would be expressing right restrictions of music pieces that also affect the right restrictions of a musical work that samples the original piece.

Dynamic and composed documents are an approach for integrating data and functionalities that are provided over standardized interfaces, e.g. as Web services or as Linked Data. Dynamic document compositions specify links to resources and how the obtained data is combined to form a final document. An example for such a composition is a dynamic PHP page, that reads stock quotes from a Web service and displays them in a human-friendly way. Both the Web service and the PHP page can be equipped with a policy restricting who can access them. The Web service could also have a clause that requires that Web pages displaying its result, have to have the same access restrictions as the service.

Fig. 1. Use case scenario

For realizing the policies we use an abstract model of service and data usages that is the base for policy conditions. The policy-aware composition tool, as visualized in Figure 1, mediates between the concrete document composition (e.g. the PHP page) and its abstract description in terms of the usage model. The policy engine classifies the composition according to the policies of the used services and returns the result to the composition tool. In future work the

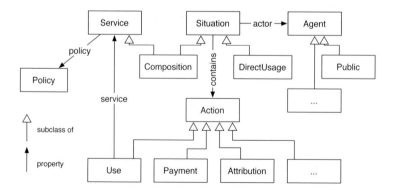

Fig. 2. Conceptual Model of Use Case for Policy Conditions

classification will be accompanied with a justification that helps to fix problems, if a situation is non-compliant.

The abstract conceptual model for compositions and service usages is visualized in Figure 2. Situations contain actions, that can be for example uses of services, payments or attributions. A situation can either be a direct usage, meaning that the actions are executed and the result directly used, or a composition, meaning that the result of the situation is again provided as a service. Situations are conducted by an agent, which can be optionally classified in subclasses. Services (including compositions) have a policy regulating their allowed uses.

In Section 6 we will show how our policy language can be used to model the following representative examples:

- The terms and conditions of the Google Maps API[2], which require (besides other clauses) that "Your Maps API Implementation must be generally accessible to users without charge."
- A service in a company internal scenario delivers confidential information, thus it can only be accessed by managers; the same must hold for compositions using the service.
- A service provider offers two stock quote services: one with real-time quotes that requires a payment, and one with delayed quotes that only requires an attribution. A service user is searching for stock quote services that can be used without payments.

3   Preliminaries

We choose RDF Schema (RDFS [4]) as data model for situation descriptions, as it provides desirable modeling features, but still has decidable algorithms for conjunctive query answering and containment. Modeling features of RDFS that are useful for describing usage situations include: (i) the use of URIs for individuals and classes, allowing heterogeneous actors and extensibility of situation models, (ii) class memberships and subclasses, e.g. an action belonging to a credit card payment class, fulfills the requirement of a general payment action, and (iii) subproperties, e.g. two actions in an application that always occur together (subproperty) are also related by a property describing actions that can possibly occur together (superproperty).

Let $I, B, L,$ and V be disjoint infinite sets of IRIs, blank nodes, literals and variables. In the following $P(S)$ denotes the powerset of S.

Definition 1. *An RDF graph is a finite set of triples, defined as $r \in P((I \cup B) \times I \times (I \cup B \cup L))$.*

In Section 4, we introduce our policy model, which is based on conjunctive queries (CQs), as defined in the following.

---

[2] http://code.google.com/apis/maps/terms.html

Definition 2. *A conjunctive query $cq = (x, t)$ is a pair of head variables $x \subset V$ and a finite set of triple patterns $t \in P((I \cup V) \times I \times (I \cup V \cup L))$. We denote as $V_t = \{v \in V \mid \exists p, o\ (v, p, o) \in t \lor \exists s, p\ (s, p, v) \in t\}$ the set of all variables in a set of triple patterns t.*

Let M be the set of all function $\mu : I \cup L \cup V \to I \cup L$, s.t. $\forall a : (a \in I \cup L \to \mu(a) = a)$. As an abbreviation we also apply a function $\mu \in M$ to a set S ($\mu(S) = \{\mu(s) \mid s \in S\}$), to a triple or triple pattern $t = (s, p, o)$ ($\mu(t) = (\mu(s), \mu(p), \mu(o))$) or to sets of triples or triple patterns.

Definition 3. *The result set for a conjunctive query $cq = (x, t)$ applied to a RDF graph r is defined as $Q_{cq}(r) = \{x' \in (I \cup L)^{|x|} \mid \exists \mu \in M\ \mu(x) = x' \land \mu(t) \subseteq r\}$.*

Definition 4. *A union of conjunctive queries (UCQ) is a set CQ of conjunctive queries with the same head predicate. We define $Q_{CQ}(r) = \bigcup_{cq \in CQ} Q_{cq}(r)$.*

We assume that the we can evaluate queries on a RDF graph that is the fixpoint according to RDFS semantics for the properties and classes used in the queries, i.e. all implicit properties and class memberships are materialized.

In Section 5, we discuss restrictions on policies, which are partially defined using query containment. Query containment of a query CQ_1 in a query CQ_2, denoted as $CQ_1 \sqsubseteq CQ_2$, means that for every possible RDF graph r, every result of CQ_1 is also a result of CQ_2, i.e. $CQ_1(r) \subseteq CQ_2(r)$.

Definition 5. *A function $h : (I \cup L \cup B \cup V) \to (I \cup L \cup B \cup V)$ is a containment mapping from $cq_2 = (x_2, t_2)$ to $cq_1 = (x_1, t_1)$, if the following conditions hold:*
- *$\forall x \in (I \cup L) : h(x) = x$*
- *$\forall x \in x_2 : h(x) \in x_1$*
- *$\forall (s, p, o) \in t_2 : (p = $ `rdf:type` $\to$*
  $\exists (s', p', o') \in t_1 : h(s) = s' \land p' = $ `rdf:type` $\land o'$ `rdfs:subClassOf` $o)$
- *$\forall (s, p, o) \in t_2 : (p \neq $ `rdf:type` $\to$*
  $\exists (s', p', o') \in t_1 : h(s) = s' \land h(o) = o' \land p'$ `rdf:subPropertyOf` $p)$

Note that `rdfs:subClassOf` *and* `rdfs:subPropertyOf` *are both reflexive.*

Definition 6. *A CQ cq_1 is contained in a CQ cq_2, if and only if there exists a containment mapping h from cq_2 to cq_1 (see[5, p. 882]).*

For showing query containment of a UCQ CQ_1 in another UCQ CQ_2, it is sufficient to show containment on the component CQs, i.e. $CQ_1 \sqsubseteq CQ_2 \leftarrow \forall cq_1 \in CQ_1 \exists cq_2 \in CQ_2 : cq_1 \sqsubseteq cq_2$ (see [5, p. 904]).

4   Policy Model

As mentioned in the introduction, we want a policy to describe the circumstances in which it is allowed to use the entity that is the subject of the policy. We distinguish between policy applicability and compliance. Applicability describes the situations, which are regulated by a policy, i.e. considered a use of the policy subject. If a situation is not applicable it is trivially compliant, otherwise only if the situation fulfills the corresponding conditions.

This corresponds to a goal-based policy as defined by Kephart and Walsh in [6], as only the desired states are specified. Such policies are on a higher conceptual level than action-based policies, which specify for every situation what has to be done next. The notions are based on the classification of agents according to Russel and Norvig [7]. To arrive at a compliant state based on a goal policy, algorithms are needed that help to determine the needed actions, respectively situation modifications. In Section 4.2 we further elaborate on this aspect, after we describe in Section 4.1 the used formalisms for modeling descriptions and policy conditions.

4.1   Formal Policy Model

The sets of situations that are applicable, respectively compliant for a given policy, are described by conjunctive queries. CQs allow the declarative specification of properties that a situation must fulfill, using predicates (i.e. RDF properties and classes) on variables and constants which are connected by conjunctions.

Consider for example a policy that requires either a payment by credit card or if the usage is for scientific purposes, then an attribution of the service provider is sufficient. In order to avoid having two different policies, we define policy compliance conditions to be UCQs. Formally we define: a policy $P = (id, cq_a, CQ_c)$, where $id \in I$ is the IRI representing the policy entity, cq_a is a CQ defining the applicable policy subjects, and CQ_c is a UCQ defining the compliant policy subjects.

We define the two properties `applicable` and `compliant` with domain of policy subjects and range of policies. The extensions of these properties are defined in the following way for all policies $P = (id, cq_a, CQ_c)$ and all potential policy subjects s in an RDF graph r:

$$s \ \texttt{applicable} \ id \leftrightarrow (s) \in Q_{cq_a}(r), \text{ and}$$
$$s \ \texttt{compliant} \ id \leftrightarrow (s) \in Q_{cq_a}(r) \land \exists cq \in CQ_c : (s) \in Q_{cq}(r).$$

For the representation of such policies we employ the RIF-Core Dialect [8], to define a policy as a group of conjunctive rules, using RIF's annotation to link it to the policy entity. The RIF documents specify in [9] how RIF frame formulas of the form `s[p->o]` correspond to RDF triple (patterns) of the form `s' p' o'`.

Note that the policies do not support negation. This means that for example it is not possible to check that there is no activity with commercial purpose, instead such an absence has to be stated and required explicitly. Approaches like scoped negation (cf. [10]) make it possible to combine negation as failure with RDF's open world assumption. However, negation together with hierarchical predicates as introduced in Section 4.2 generally leads to undecidability of query containment.

4.2   Policy Structure

Unions of conjunctive queries (UCQs) provide a nice formal model of policies that is suitable for evaluation. However specifying them can introduce redundancy in the likely case that several alternatives of a union share common conditions. Furthermore UCQs lack an hierarchical structure which eases the specification and maintainability of policies. Therefore we allow not only the use of frame formulas in conditions that can be directly mapped to triple patterns but also the use of predicates with arbitrary arity that are themselves again defined as UCQs. This essentially means that policies can be specified as non-recursive datalog programs, which can always be expanded to UCQs using only base predicates (i.e. RDF class memberships and properties).

Note that such predicates can also be defined externally, which enables reuse of conditions across policies from different specifiers. This is comparable to the Creative Commons approach, where certain standard terms are defined that can be used to define custom policies (cf. [11]). As rules are identified by IRIs, they can be described not only by their formal definition as RIF documents but also by a legal or layman description, if the IRI is resolved by a Web browser (recognized by the `Accept` header of the HTTP request).

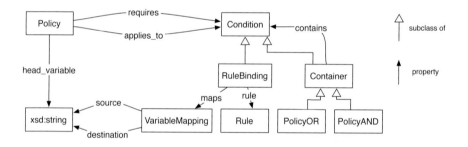

Fig. 3. Visualization of Policy Structures

In the following we define a conceptual model of combining policies from predicates defined by UCQs. It is based on an RDF model that refers by IRIs to rules defined in RIF. The model is visualized in Figure 3. A `Policy` `applies_to` subjects that are answers to the applicability query, which is defined by a `Condition`, which is either a `RuleBinding`, a conjunction of other `Conditions` (i.e. a `PolicyAND` container), or a disjunction of `Conditions` (i.e. a `PolicyOR` container). Furthermore a policy `requires` a condition, which represents the validity test, and has a `head_variable` which defines the policy subject in the conditions. Both `PolicyAND` and `PolicyOR` `contain` a number of conditions. `RuleBindings` refer by the `rule` property to an IRI which is the id of a group in a RIF document that defines the corresponding predicate. Note that by resolving the IRI we expect a RIF representation containing this group. The metadata of the group specifies via `defines_predicate` the head predicate of the rules. Furthermore a `RuleBinding` maps a number of `VariableMapping`s each with a `source` variable name of the defined predicate that is mapped to the `destination`, which is either a variable in the policy condition or a string representation of an IRI.

Such an hierarchical policy definition with simple boolean operators to combine basic conditions is a more user friendly way to specify policies, which is already familiar from filter creation in many email programs. Furthermore the structuring allows users to group conditions in sensible blocks, which can be exploited for giving justifications of (mainly negative) policy decisions. Due to the use of IRIs and metadata, the rules and policy parts can be annotated with further useful and human-readable information. See for example the work by Kagal et al. [12] for a policy engine that exploits policy structures for human-friendly justifications.

4.3   Mapping the Policy Structure to the Formal Model

The mapping from the proposed structural model to a policy's normal form (i.e. its UCQ as defined in Section 4) is defined in a bottom-up way. The most basic part is a rule defining a predicate based only on RDF properties. Using RIF presentation syntax (cf. [13]) it is expressed in the following way (`p:` is used in the following for the namespace of the policy vocabulary):

```
(* "RULEID"^^rif:iri
    "RULEID"^^rif:iri[p:defines_predicate->"PREDICATE"^^rif:iri] *)
Group (
    Forall ?h1 ... ?hn (
        "PREDICATE"^^rif:iri(?h1 ... ?hn) :-
            Exists ?e1 ... ?em (
            And( s1[p1->o1]

                  ...
                  sk[pk->ok])))
    Forall ?h1 ... ?hn (
        "PREDICATE"^^rif:iri(?h1 ... ?hn) :-
            Exists ?e1 ... ?em (
            And( s'1[p'1->o'1]

                  ...
                  s'l[p'l->o'l])))))
```

This maps to a union of conjunctions of the following form:

$$CQ_{RULEID} = \Big\{ \big((h_1,\ldots,h_n), \{(s_1,p_1,o_1),\ldots,(s_k,p_k,o_k)\}\big),\ldots,$$
$$\big((h_1,\ldots,h_n), \{(s'_1,p'_1,o'_1),\ldots,(s'_l,p'_l,o'_l)\}\big)\Big\},$$

where $(s_1,p_1,o_1),\ldots,(s'_l,p'_l,o'_l) \in I\cup V \times I \times I\cup V\cup L$. Note that it is also possible to use other (non-recursive) RIF predicates instead of only RDF properties. In this case, we assume that the IRI of the used predicate resolves to a RIF document that defines the corresponding UCQ. In this way a rule definition can always be expanded to a union of conjunctions in terms of simple RDF properties.

738   S. Speiser and R. Studer

The rules are used in our policy model by `RuleBindings`, which has the general form:

```
RB a p:RuleBinding;
    p:rule RULEID;
    p:maps MAP1;  p:maps ...;  p:maps MAPN.
```

We define a function $f_{\text{MAP}} : V \cup I \cup L \to V \cup I \cup L$ for each variable mapping `MAP = (source, destination)` in the following way:

$$f_{\text{MAP}}(x) = \begin{cases} \text{destination}, & \text{if } x = \text{source} \\ x, & \text{otherwise.} \end{cases}$$

We also use these functions when applied to UCQs with the meaning that it is applied to all variables, IRIs and literals in the UCQ. Thus, we can define the UCQ of the rule binding `RB` in the following way:

$$CQ_{\text{RB}} = f_{\text{MAP1}}(f_{...}(f_{\text{MAPN}}(CQ_{\text{RULEID}}))).$$

The mapping for both AND and OR containers are defined by treating them as binary operators. Due to the associativity of these operators, the mapping naturally applies also to containers with more components.

Conditions (e.g. rule bindings) are used in `PolicyAND` containers of the following form:

```
AND a p:PolicyAND;
    p:contains C1;
    p:contains C2.
```

The corresponding UCQ is obtained by creating the union of the conjunctions for each pair of alternatives of the two components. More formally:

$$CQ_{\text{AND}} = \bigcup_{(x_1,t_1) \in CQ_{\text{C1}}} \bigcup_{(x_2,t_2) \in CQ_{\text{C2}}} \{(x_1 \cup x_2, t_1 \cup t_2)\}.$$

For a `PolicyOR` container of the following form

```
OR a p:PolicyOR;
    p:contains C1;
    p:contains C2.
```

we define the UCQ as the union of the two components: $CQ_{\text{OR}} = CQ_{\text{C1}} \cup CQ_{\text{C2}}$. Finally we define the mapping for a `Policy` object to the formal model. Given the following representation

```
POL a p:Policy;
    p:head_variable HV;
    p:applies_to CA;
    p:requires CR.
```
, we define a policy $P_{\text{POL}} = (\text{POL}, CQ_{\text{CA}}, CQ_{\text{CR}})$.

5   Restrictions on Policies

Policies classify policy subjects into compliant, non-compliant, and inapplicable categories. If we want to ensure that certain kinds of policy subjects are always, respectively never, compliant with a policy, we have to restrict the policy with regard to a specification of the policy subject. Specifying restrictions on policies is useful for several tasks, as outlined in the following:

- Searching for an entity with a policy that allows certain situations, e.g. searching for a service that can be used without a payment.
- Validation of a policy, i.e. ensuring that it fulfills test restrictions.
- Comparison to other policies, e.g. to a previous version, in order to see, if the policy is stricter or more lax.
- Policing other policies, if the compliance of a policy subject depends on restrictions of the subject's policy.

Independent of their application, we found the three types of restrictions particularly useful, which are listed in the following, including examples of their use:

1. **Required for compliance:** is it necessary that a policy subject fulfills certain conditions in order to be compliant. If we specify the required conditions themselves as a policy, this restriction is equivalent to asking if all compliant subjects of the restricted policy are also compliant with the restricting policy. This can be solved by checking query containment of the policies.

   Examples for such restrictions are: (i) a policy must always require a payment, or (ii) a policy must restrict data access to a certain class of users.

2. **Not required for compliance:** is it possible that a policy subject is compliant without necessarily fulfilling certain conditions. This restriction is basically just the negation of the previous one, and thus can also be checked by query containment.

   Examples are: (i) can a subject be compliant without having a payment, (ii) can a service be used without being a registered user?

3. **Sufficient for compliance:** can a partially described subject be compliant by adding only further restrictions that do not affect the given description?

   An example for this restriction is: can a situation, where data is provided to the general public, be compliant? This is true if the policy does not further restrict the data recipient, but it may for example require a payment.

The presented restrictions rely on query containment, i.e. comparison of policies. As we want to compare policies that generally can apply to different subjects (i.e. the subjects of the restricted and the restricting policy), we define the comparisons in terms of the compliance conditions of policies, and dismiss the applicability conditions. This is one of the reasons for separating applicability and compliance, besides avoiding redundancy as applicability is part of every policy alternative (i.e. conjunction in the policy's UCQ).

In order to use the above defined restrictions in policy conditions, we introduce three RDF properties and formally define their extensions:

- `req_for_comp` (see 1. "required for compliance"),
- `not_req_for_comp` (see 2. "not required for compliance", needed as we do not support negation), and
- `sufficient_for_comp` (see 3., "sufficient for compliance").

As discussed above we can reduce the first two restrictions to query containment in the following way:

P2 `req_for_comp` P1 $\leftrightarrow P_{P1} = (P1, cq_a^1, CQ_c^1) \wedge P_{P2} = (P2, cq_a^2, CQ_c^2) \wedge CQ_c^1 \sqsubseteq CQ_c^2$

P2 `not_req_for_comp` P1 $\leftrightarrow \neg$(P2 `req_for_comp` P1)

The `sufficient_for_comp` property is defined between policies specifying the sufficient condition and a target policy. The sufficient condition policy $P_s = (id^s, cq_a^s, CQ_c^s)$ should only consist of a single acyclic conjunctive query ([14], also called tree queries [15]) $CQ_c^s = \{cq_s(x_s, t_s)\}$, whereas the target policy $P_t = (id^t, cq_a^t, CQ_c^t)$ can be a union of CQs. P_s is sufficient for P_t if there exists one policy alternative $cq_t \in CQ_c^t$ for which it is sufficient. Finding out, if $cq_s = (x_s = \{hv_s\}, t_s)$ is sufficient for $cq_t = (x_t = \{hv_t\}, t_t)$ can be done by doing a tree traversal of cq_s according to the following recursive condition:
$suff(cq_s, cq_t) \leftrightarrow is\_suff(hv_s, cq_s, cq_t, \{(hv_s, hv_t)\})$, where:

$$is\_suff(n, cq_s, cq_t, \mu)$$
$$= \Big(\forall c \in \{c \mid (n, \texttt{rdf:type}, c) \in t_s\} :$$
$$\forall c' \in \{c' \mid (\mu(n), \texttt{rdf:type}, c') \in t_t\} : c \ \texttt{rdfs:subClassOf} \ c'\Big) \wedge$$

$$\Big(\forall p \in \{p \mid \exists o : (n, p, o) \in t_s\} :$$
$$\forall p' \in \{p' \mid \exists o' : (\mu(n), p', o') \in t_t\} :$$
$$((p' \ \texttt{rdfs:subPropertyOf} \ p) \rightarrow (p \ \texttt{rdfs:subPropertyOf} \ p'))\Big) \wedge$$
$$\Big(\forall (p, o) \in \{(p, o) \mid (n, p, o) \in t_s\} :$$
$$\forall (p', o') \in \{(p', o') \mid (\mu(n), p', o') \in t_t\} :$$
$$((p \ \texttt{rdfs:subPropertyOf} \ p') \wedge \{(x, x') \in \mu \mid x = o\} = \emptyset \rightarrow$$
$$is\_suff(o, cq_s, cq_t, \mu \cup \{(o, o')\})\Big)\Big) \wedge$$
$$\Big(\forall (s, p) \in \{(s, p) \mid (s, p, n) \in t_s\} :$$
$$\forall (s', p') \in \{(s', p') \mid (s', p', \mu(n)) \in t_t\} :$$
$$((p \ \texttt{rdfs:subPropertyOf} \ p') \wedge \{(x, x') \in \mu \mid x = s\} = \emptyset \rightarrow$$
$$is\_suff(s, cq_s, cq_t, \mu \cup \{(s, s')\})\Big)\Big).$$

The definition of is_suff is divided into four conditions. The first condition checks for a node mapping, that there are no stricter class requirements in the target policy than in the sufficiency condition. The second condition checks that no stricter property requirements occur (i.e. every mapping to a subproperty must be an equivalent property). The third and fourth conditions follow the patterns connected to a node (depending on its position as a subject or object) and recursively apply is_suff to the newly mapped variables. As we required the sufficiency condition to be an acyclic conjunctive query and only patterns are followed that map previously unmapped variables, the recursion will always come to an end.

The proposed policy restriction properties are defined to have special interpretations, as defined in this section. As the definitions for query containment and sufficiency rely on normal RDFS interpretations of properties, the restriction properties cannot be freely used in policies occuring in restriction conditions. Specifically the current definitions do not support restriction properties in sufficiency conditions (i.e. S in S `sufficient_for_comp` P) and containing policies (i.e. P2 in P2 `(not_)req_for_comp` P1). Note that the properties can occur in contained policies, as they just reduce the set of compliant subjects and thus can be ignored.

6   Evaluation

In Section 2 we presented a use case and three concrete examples. We modeled the use case using our policy language and tested it with a prototypical implementation of a policy engine, that we developed. In the following we present and discuss interesting aspects of the example policies. The full examples in RDF and RIF, as well as the policy engine and its source code are available online[3]. At the end of the section, we elaborate on the performance of the policy engine.

Policy for Google Maps API. In subsequent descriptions we use N3-syntax for RDF and the abstract syntax for RIF. The URI prefix `p:` stands for the policy vocabulary, and `m:` points to the conceptual model for compositions and service usages. In the following we show the description of the maps policy, the `policy:` prefix points to a RIF file containing maps policy rules, and `generalrules:` refers to a RIF file describing general rules that can be reused by different policy specifiers.

```
@prefix gm: <http://example.org/googlemapsapi#> .
@prefix policy: <http://example.org/gmpolicy#> .

gm:policy a p:Policy;
  p:head_variable "situation";
  p:applies_to [a p:RuleBinding;
                p:rule policy:apprule];
  p:requires [a p:PolicyAND;
    p:contains :RegisteredUser;
    p:contains [a p:PolicyOR;
      p:contains [a p:PolicyAND;
        p:contains <http://example.org/nopaymentreq#NoPaymentReqCondition>;
        p:contains :AvailForPublic];
      p:contains [a p:RuleBinding; p:rule generalrules:DirectUse]]].

:RegisteredUser a p:RuleBinding;
  p:rule policy:GMRegisteredUser.
:AvailForPublic a p:RuleBinding;
  p:rule policy:AvailForPublicRule.
```

---

[3] http://code.google.com/p/seppl/

The policy defines that applicability is determined by the `apprule` and requires for compliance that (i) the actor of an applicable situation is a registered user (rule `GMRegisteredUser`), and (ii) that the situation is either a direct use (rule `generalrules:DirectUse`), or a composition that has a policy which makes it available to the public (rule `AvailForPublicRule`) and does not require a payment (link to external rule binding, reusing this common condition).

The `apprule` specifies that situations are applicable to this policy, if it uses the Google Maps API (defined as `gm:service a m:Service`):

```
(* policy:apprule
   policy:apprule[p:defines_predicate -> policy:apprulepred *)
Group (
  Forall ?situation (
    policy:apprulepred(?situation)  :- Exists ?usage (
           And ( ?situation[rdf:type -> m:Situation]
                 ?situation[m:contains -> ?usage]
                 ?usage[rdf:type -> m:Usage]
                 ?usage[m:service -> gm:service] ) ) ) )
```

The `AvailForPublicRule` has the following rule body:

```
And ( ?situation[rdf:type -> m:Composition]
      ?situation[m:policy -> ?policy]
      gm:AvailableForPublicPolicy[p:sufficient_for_comp -> ?policy] )
```

This means that another policy is described (`gm:AvailableForPublicPolicy`) that defines a partial situation description which must be sufficient for fulfilling the policy of a composition which is using the API. The partial situation is defined by a binding of a rule with the following body:

```
And ( ?situation[m:actor -> ?actor]
      ?actor[rdf:type -> m:Public] )
```

The partial situation is thus only sufficient if the composition's policy allows access by actors without requiring them to belong to any other class than `m:Public`.

Confidential company internal service. The policy of the confidential service requires two rules: (i) one checking if the actor of the using situation is a manager, and (ii) one that checks if the policy of a using composition is contained in a policy restricting access to managers. The second rule ensures that if the service is used in a composition, then the composition inherits the access restrictions. For the realization of this rule, the `p:req_for_comp` property was used.

Stock quotes service. The policies of the stock quote services are rather straightforward, one checking for a payment and the other one for an attribution. The search process is realized in the following way: (i) the user creates a policy that requires a situation that contains a payment, (ii) he asks the policy engine to check for both stock quote services if their policy is not contained in his policy. The engine answers the request by using the `p:not_req_for_comp` property, which only holds for the delayed stock quote service.

Performance. Compliance checking using our policy language corresponds to answering unions of conjunctive queries. Conjunctive query answering is known to be NP-complete [16] for relational databases. This result can be transferred to RDFS knowledge bases with a materialized fixpoint, where the properties can be treated as relations. However, in our approach special properties exist that check restrictions on policies. The evaluation, if two instances are related by such a property involves checking query containment, which for positive conjunctive queries is equivalent to query answering and thus also NP-complete. Thus in combination this means a complexity of up to $\Sigma_2 P$ (i.e. NP with an NP oracle) for policy evaluation. The theoretical complexity relates to the size of the queries defining the policies.

For testing what the theoretical complexity means for practical purposes we conducted some performance measurements using our (non-optimized) policy engine. We created for both the maps API policy and the confidential service policy each three situation descriptions: one that is compliant, one that is non-compliant and one that is not applicable. We measured the classification time on a laptop with an Intel Core2Duo 2.4GHz processor and 4 GB of main memory. Furthermore we measured the search time for determining for the real-time and the delayed stock quote services if they do not require payments. The results are shown in Table 1.

Table 1. Results of the Performance Experiments

Task/Policy	time non-compliant	time compliant	time not-applicable
Maps API	0.69 s	0.68 s	0.60 s
Confidential Service	0.53 s	0.54 s	0.38 s
Policy search	0.35 s	0.34 s	n/a

Even with our prototypical policy engine the time required for performing policy checks are all well below 1 second. With further optimizations (e.g. caching formal representations of policies instead of parsing them again for every policy action), it seems feasible to integrate real-time compliance checking in a policy-aware composition tool.

7   Related Work

XACML is a widely-used industry standard for policies [17], but lacks a formal, declaratively defined semantics for its very extensive condition model, which includes XPath queries, string and date comparisons, arithmetic functions, logical negation and regular expressions besides others. Especially negation in combination with arbitrary XPath queries leads to undecidability of query containment. Another difference to our work is that XACML focuses specifically on access control policies, whereas our proposed policy language is suitable for usage control, which does not only check if initial access to data or services is allowed, but also restricts the ongoing usage afterwards.

WS-Policy provides a standard that can be used to specify policies that express requirements and capabilities in systems based on Web services [18]. The policy language itself is not especially targeted at Web services and can be extended by custom policy assertions, which are basic conditions that can be combined to form policies (similar to using our containers). The standard is based on XML and syntactic matching and is thus, in contrast to our approach, not suitable for heterogeneous environments where different vocabularies are mixed. There exist however several extensions to WS-Policy, which link assertions to OWL concepts (e.g. [19,20]). Such policies are thus based on description logics and therefore restricted to conditions with tree structures. The same restriction applies to KAoS, an early semantic policy framework [21]. Our approach uses conjunctive queries and thus can express non-tree conditions.

Accountability in RDF (AIR) is a policy language that comes with an engine that supports RDFS models, and an extensive justification framework [12]. It is based on N3 syntax, and supports quantified variables, as well as if-then-else statements. The "else" path is followed if the condition does not hold, which means that the language supports negation on non-atomic conditions. Therefore query containment on AIR policies is not decidable and thus the policy restrictions presented in this paper cannot be easily integrated into AIR. For future work it is certainly interesting to see which features of AIR and our policy language can be fruitfully combined. Especially interesting is an adaption of AIR's justification framework, for which we already laid out the foundation by associating policy rules with RIF metadata.

Another semantic policy language is Protune [22]. It is based on logic programming rules, including negation. Its main focus is not on the classification of situations, but on trust negotiation, which includes the execution of actions. It includes the explanation facility ProtuneX [23], which supports decision justifications and different kind of policy queries, such as how-to queries that tell a user what is needed to fulfill a policy. None of these queries can however be integrated into the conditions of other policies, which is a key feature of our policy language. Bonatti and Mogavero present a restricted version of Protune (e.g. no negation) for which they show decidability of policy comparison, i.e. query containment [24]. Their work does not support integration of the comparisons into policy conditions, and does not treat the "sufficient for compliance" restriction introduced in this paper.

8   Conclusions and Future Work

We presented a policy language with the novel capability to express restrictions on other policies given in the same language. The policy language has formal semantics defined in terms of conjunctive queries over RDFS data. Furthermore we described a concrete representation format being based on the W3C standards RDF and RIF. We motivated the need for our self-policing policy language by a use case about composed documents, including a real-world service, namely the Google Maps API.

We implemented a policy engine for our language and used it to model the use case. We conducted first performance measurements. The results show that the language and engine can effectively represent the required policies. As next steps we plan to develop a justification framework for the language and based thereon build a policy-aware composition tool.

Furthermore we plan to extend the expressivity of the policy language in one of the following possible directions:

- a more expressive data model, i.e. using one of the OWL 2 profiles, instead of RDFS,
- allow some limited negation (e.g. only on basic patterns),
- allow viral policies, in the meaning that restricting policies can also include conditions using the special policy restriction properties.

We currently evaluate, which of these extensions are most desirable in terms of required expressivity and preservation of decidability.

Acknowledgments. The authors wish to thank Markus Krötzsch and Andreas Harth for the useful discussions. This work was supported by the European project SOA4All.

References

1. Lampson, B.W.: Protection. In: Proc. Fifth Princeton Symposium on Information Sciences and Systems, pp. 437–443. Princeton University, Princeton (March 1971); reprinted in Operating Systems Review 8 (1), 18 – 24 (January 1974)
2. Dodds, L.: Rights Statements on the Web of Data. Nodalities Magazine (9) (2010), http://www.talis.com/nodalities/pdf/nodalities_issue9.pdf
3. Seneviratne, O., Kagal, L., Berners-Lee, T.: Policy Aware Content Reuse on the Web. In: Bernstein, A., Karger, D.R., Heath, T., Feigenbaum, L., Maynard, D., Motta, E., Thirunarayan, K. (eds.) ISWC 2009. LNCS, vol. 5823, pp. 553–568. Springer, Heidelberg (2009)
4. W3C: RDF Vocabulary Description Language 1.0: RDF Schema. W3C Recommendation (2004), http://www.w3.org/TR/rdf-schema/
5. Ullman, J.D.: Principles of Database and Knowledge-Base Systems, vol. II. Computer Science Press, Rockville (1989)
6. Kephart, J.O., Walsh, W.E.: An Artificial Intelligence Perspective on Autonomic Computing Policies. In: IEEE Workshop on Policies for Distributed Systems and Networks, POLICY (2004)
7. Russel, S., Norvig, P.: Artificial Intelligence: A Modern Approach, 2nd edn. Prentice Hall, Englewood Cliffs (2003)
8. W3C: RIF Core Dialect. W3C Recommendation (2010), http://www.w3.org/TR/rif-core/
9. W3C: RIF RDF and OWL Compatibility. W3C Recommendation (2010), http://www.w3.org/TR/rif-rdf-owl/
10. Polleres, A., Feier, C., Harth, A.: Rules with Contextually Scoped Negation. In: Sure, Y., Domingue, J. (eds.) ESWC 2006. LNCS, vol. 4011, pp. 332–347. Springer, Heidelberg (2006)

11. Abelson, H., Adida, B., Linksvayer, M., Yergler, N.: ccREL: The Creative Commons Rights Expression Language. W3C Submission (2008)
12. Kagal, L., Hanson, C., Weitzner, D.: Using Dependency Tracking to Provide Explanations for Policy Management. In: IEEE Workshop on Policies for Distributed Systems and Networks, POLICY (2008)
13. W3C: RIF Basic Logic Dialect. W3C Recommendation (2010), http://www.w3.org/TR/rif-bld/
14. Gottlob, G., Leone, N., Scarcello, F.: The complexity of acyclic conjunctive queries. Journal of the ACM 48(3), 431–498 (2001)
15. Goodman, N., Shmueli, O.: Tree queries: a simple class of relational queries. ACM Trans. Database Syst. 7(4), 653–677 (1982)
16. Chandra, A.K., Merlin, P.M.: Optimal implementation of conjunctive queries in relational data bases. In: Annual ACM Symposium on Theory of Computing (1977)
17. OASIS: eXtensible Access Control Markup Language (XACML) Version 2.0. OASIS Standard (2005), http://docs.oasis-open.org/xacml/2.0/
18. W3C: Web Services Policy 1.5 - Framework. W3C Recommendation (2007), http://www.w3.org/TR/ws-policy/
19. Verma, K., Akkiraju, R., Goodwin, R.: Semantic matching of web service policies. In: Semantic and Dynamic Web Processes (SDWP) In Conjunction with the Third International Conference on Web Services, ICWS 2005 (2005)
20. Kolovski, V., Parsia, B.: WS-Policy and Beyond: Application of OWL Defaults to Web Service Policies. In: Semantic Web Policy Workshop (SWPW) at 5th International Semantic Web Conference, ISWC (2006)
21. Uszok, A., Bradshaw, J., Jeffers, R., Suri, N., Hayes, P., Breedy, M., Bunch, L., Johnson, M., Kulkarni, S., Lott, J.: KAoS Policy and Domain Services: Toward a Description-Logic Approach to Policy Representation, Deconfliction, and Enforcement. In: IEEE Workshop on Policies for Distributed Systems and Networks, POLICY (2003)
22. Bonatti, P.A., De Coi, J.L., Olmedilla, D., Sauro, L.: A Rule-based Trust Negotiation System. IEEE Transactions on Knowledge and Data Engineering (2010)
23. Bonatti, P.A., Olmedilla, D., Peer, J.: Advanced Policy Explanations on the Web. In: European Conference on Artificial Intelligence, ECAI (2006)
24. Bonatti, P.A., Mogavero, F.: Comparing Rule-Based Policies. In: IEEE Workshop on Policies for Distributed Systems and Networks, POLICY (2008)

Completeness Guarantees for Incomplete Reasoners

Giorgos Stoilos, Bernardo Cuenca Grau, and Ian Horrocks

Oxford University Computing Laboratory
Wolfson Building, Parks Road, Oxford, UK

Abstract. We extend our recent work on evaluating incomplete reasoners by introducing *strict testing bases*. We show how they can be used in practice to identify ontologies and queries where applications can exploit highly scalable incomplete query answering systems while enjoying completeness guarantees normally available only when using computationally intensive reasoning systems.

1 Introduction

A key application of OWL ontologies is ontology-based data access [12,9,3,2,7,11], where an ontology is used to support query answering against distributed and/or heterogeneous data sources. The ontology provides the vocabulary used to formulate queries, and a conceptual model (or schema) that is used in computing query answers. In a Semantic Web setting, a typical scenario would involve the use of an OWL ontology to answer SPARQL queries over RDF datasets.

Unfortunately, when using an expressive ontology language such as OWL, computing query answers can be very costly, and in a (Semantic) Web setting, datasets may be extremely large. There has therefore been a growing interest in the development of query answering systems that are highly scalable in practice, but that are not guaranteed to be *complete* in all cases; i.e., for some combinations of query, ontology and dataset, they will not compute all query answers. Most such systems (e.g., Oracle's Semantic Data Store, Sesame, Jena, HAWK, OWLim, Minerva, and Virtuoso) are based on database or RDF triple store technologies; others are based on approximate reasoning techniques [10,5].

Although the scalability of such systems is attractive, application developers face two main difficulties when using them. Firstly, incomplete query answers may not be acceptable in a given application; and secondly, even if some incompleteness is acceptable, it may be important to know just how incomplete answers are likely to be, and to compare the scalability-completeness trade-off offered by different systems. One way to address these issues is via empirical testing, e.g., checking the answers given by query answering systems w.r.t. a particular ontology, dataset and query, and although primarily intended for performance testing, benchmark suites such as LUBM [4] have sometimes been used for this purpose. However, this kind of testing has serious limitations: results are specific to a given query, ontology and dataset, and may tell us nothing about

P.F. Patel-Schneider et al. (Eds.): ISWC 2010, Part I, LNCS 6496, pp. 747–763, 2010.

the behaviour of the system more generally; and, in order to determine the system's degree of completeness, we need to already know, or be able to compute, exact answers to the given queries.

In our recent work [13], we addressed these issues by introducing the notion of a *Testing Base* (TB). For a given ontology and query, a TB is a set of datasets such that, for any "well behaved" query answering system, if the system is complete for each dataset in the TB, then it will be complete for *any* dataset. As well as providing a quantitative measure of completeness, which we call the *completeness degree*, TBs thus allow us to identify circumstances under which a completeness guarantee can be provided even when the system being used is incomplete in general. This is very useful in practice given that in many applications the ontology and (kinds of) query are fixed at design time, or change relatively infrequently, whereas the data is typically unknown and/or frequently changing. Unfortunately, we were unable to devise a practical algorithm for computing TBs; instead, we devised an algorithm that efficiently computes an approximation of a TB. This algorithm can be used to approximate the completeness degree, but it cannot be used to provide completeness guarantees.

In this paper we extend our previous work in several directions. Most importantly, we define the notion of a *Strict Testing Base*; we show that strict TBs are typically much smaller than TBs, prove that they can be used to provide the same completeness guarantee as TBs, and present an efficient algorithm for computing them. This algorithm can thus be used to identify ontologies and queries where applications can exploit highly scalable incomplete systems while enjoying completeness guarantees normally available only when using computationally intensive reasoning systems—i.e., they can have *the best of both worlds*.

Additionally, we propose four properties that any "reasonable" measure of completeness should ideally enjoy, and we show that while completeness degree w.r.t. a TB satisfies all of these properties, completeness degree w.r.t. a strict TB satisfies only two of them, albeit the most important two.

Finally, our preliminary evaluation, which includes the LUBM ontology and queries as well as (a version of) the Galen ontology of clinical terms, suggests not only that strict TBs are easy to compute in practice, but also that completeness guarantees can often be provided for realistic ontologies and queries.

2   Preliminaries

Description Logics. We assume that the reader is familiar with the basics of DL syntax, semantics and standard reasoning problems [1], and we use standard notions of a TBox $\mathcal{T}$ (the terminology, or conceptual schema) and an ABox $\mathcal{A}$ (the assertions, or data). In the context of ontology-based data access, the ontology may be thought of as consisting only of a TBox, with the data being stored in the sources. From an OWL point of view, however, we can treat the contents of the sources as ABox assertions, and the ontology as being the union of the TBox and ABox, i.e., $\mathcal{O} = \mathcal{T} \cup \mathcal{A}$. To avoid conflating the schema (TBox) and the data (ABox), we consider only fragments of the DLs underpinning OWL

DL and OWL 2 that do not provide for *nominals* (i.e., that do not allow ABox individuals to be used to define TBox concepts); we also assume (without loss of generality) that ABoxes contain only *atomic assertions*—that is, each assertion of the form $C(a)$ or $R(a, b)$ in $\mathcal{A}$ must be such that C and R are atomic.

Queries. We use the standard notions of term, (function-free) atom and variable. A datalog clause is an expression $H \leftarrow B_1 \wedge \ldots \wedge B_n$ where H (the *head*) is a (possibly empty) atom, $B_1 \wedge \ldots \wedge B_n$ (the *body*) is a conjunction of atoms, and each variable in the head also occurs in the body. A union of conjunctive queries (UCQ) is a tuple $u = \langle Q_P, P \rangle$ with Q_P a query predicate and P a finite set of datalog clauses such that Q_P is the only predicate occurring in head position in P and the body of each clause in P does not contain Q_P. We denote with $\mathsf{var}(q)$ the set of variables in q and say that a variable is distinguished if it appears in the head. Finally, q is a conjunctive query (CQ) if it is a UCQ and P has one clause. If $q = \langle Q_P, P \rangle$ is a CQ, we often abuse notation and write $q = P$; if u is a UCQ with $P = \{P_1, \ldots, P_n\}$, we write $u = \{q_1, \ldots, q_n\}$ with $q_i = P_i$ a CQ.

A tuple of constants $\vec{a}$ is a certain answer of a UCQ $q = \langle Q_P, P \rangle$ with respect to $\mathcal{O} = \mathcal{T} \cup \mathcal{A}$ iff $\mathcal{O} \cup P \models Q_P(\vec{a})$, where P is seen as a set of universally quantified implications with first-order semantics. The set of certain answers of q w.r.t. $\mathcal{O} = \mathcal{T} \cup \mathcal{A}$ is (equivalently) denoted as either $\mathsf{cert}(q, \mathcal{T}, \mathcal{A})$ or $\mathsf{cert}(q, \mathcal{O})$. Clearly, the set of certain answers satisfies the following useful properties:

1. *Monotonicity*: $\mathsf{cert}(q, \mathcal{O}) \subseteq \mathsf{cert}(q, \mathcal{O}')$ for each $\mathcal{O}$, $\mathcal{O}'$ and q with $\mathcal{O} \subseteq \mathcal{O}'$.
2. *Invariance under isomorphisms*: For each pair of *isomorphic ABoxes* $\mathcal{A}$ and $\mathcal{A}'$ (i.e., identical modulo renaming of individuals), $\mathsf{cert}(q, \mathcal{T}, \mathcal{A})$ and $\mathsf{cert}(q, \mathcal{T}, A')$ are also identical modulo the same renaming.

UCQ Rewritings. Intuitively, a *UCQ rewriting* for a TBox $\mathcal{T}$ and a CQ q is a UCQ that extends q with the information from $\mathcal{T}$ that is relevant to answering the query. Formally, a *UCQ rewriting* for $\mathcal{T}$ and q is a UCQ u such that, for each ABox $\mathcal{A}$ where $\mathcal{T} \cup \mathcal{A}$ is consistent, the following properties hold:

1. (Soundness:) For each $q' \in u$, we have that $\mathsf{cert}(q', \emptyset, \mathcal{A}) \subseteq \mathsf{cert}(q, \mathcal{T}, \mathcal{A})$.
2. (Completeness:) $\mathsf{cert}(q, \mathcal{T}, \mathcal{A}) \subseteq \bigcup_{q' \in u} \mathsf{cert}(q', \emptyset, \mathcal{A})$.

Several well-known techniques can be used to reduce the size of (U)CQs. A CQ q is reduceable if it contains distinct body atoms that are unifiable. A reduction q' of q is obtained by applying the most general unifier θ to the body of q. A *condensation reduction* $\mathsf{cond}(u)$ is a UCQ obtained from u by ensuring that no two queries q, q' exist such that q' subsumes q and q' is a reduction of q. Finally, a *subsumption reduction* $\mathsf{sub}(u)$ is a UCQ obtained from u by ensuring that no two queries q, q' in the reduction are such that q' subsumes q.

Justifications. Finally, our framework in [13] relies on the well-established notion of a *justification* for an entailment (see e.g., [6]). In the case of CQ answering, a justification for a CQ q and a tuple $\vec{a} \in \mathsf{cert}(q, \mathcal{O})$ in a consistent ontology $\mathcal{O}$ is an ontology $J \subseteq \mathcal{O}$ such that $\vec{a} \in \mathsf{cert}(q, J)$ and $\vec{a} \notin \mathsf{cert}(q, J')$ for each $J' \subset J$.

3   A Framework for Evaluating Completeness

In this section we present our revised and extended framework for evaluating the completeness of Semantic Web CQ answering systems.

3.1   CQ Answering Algorithms

Our framework adopts a rather general notion of a CQ answering algorithm. This allows us to abstract from the specifics of implemented systems and establish general results that hold for any system satisfying certain basic properties.

Definition 1. *A CQ answering algorithm* ans *for a DL $\mathcal{L}$ is a procedure that, for each $\mathcal{L}$-ontology $\mathcal{O} = \mathcal{T} \cup \mathcal{A}$ and CQ $q = \langle Q_P, P \rangle$ computes in a finite number of steps a set* ans$(q, \mathcal{O})$ *of tuples of constants of the same arity as Q_P.*

- *It is* sound *if* ans$(q, \mathcal{O}) \subseteq$ cert$(q, \mathcal{O})$ *for each $\mathcal{O}$ and q.*
- *It is* complete *if* cert$(q, \mathcal{O}) \subseteq$ ans$(q, \mathcal{O})$ *for each $\mathcal{O}$ and q.*
- *It is* faithful *if it satisfies the same monotonicity and invariance under isomorphisms properties as* cert.
- *It is* compact *if for each consistent $\mathcal{O}$, each q, and each $\vec{a} \in$ cert$(q, \mathcal{O}) \cap$ ans$(q, \mathcal{O})$, there exists a justification J for $q, \vec{a}$ in $\mathcal{O}$ such that $\vec{a} \in$ ans(q, J).*

Intuitively, ans is faithful if it implements the semantics of CQ answering in a "reasonable" way; in particular, the set of computed query answers for a fixed query can only grow if new axioms are added to the ontology (*monotonicity*) and the algorithm should be robust under trivial isomorphic renamings of individuals in the ABox (*invariance under isomorphisms*). Most of the results in our framework require ans to be at least *sound* and *faithful*, which we believe to be reasonable requirements that are satisfied by most if not all existing incomplete reasoners. For some of our results, however, *compactness* is also an issue. Intuitively, ans is compact if, whenever it correctly computes a certain answer $\vec{a}$ for some query q and ontology $\mathcal{O}$, then it will also compute $\vec{a}$ for q and some minimal subset of $\mathcal{O}$ that is sufficient to derive $\vec{a}$.

Consider an (incomplete) algorithm ans that, given $\mathcal{O} = \mathcal{T} \cup \mathcal{A}$ and q, ignores $\mathcal{T}$ and answers q only w.r.t. $\mathcal{A}$. Clearly, ans is sound and faithful. Furthermore, it is compact since, for each consistent $\mathcal{O} = \mathcal{T} \cup \mathcal{A}$ and certain answer $\vec{a} \in$ ans$(q, \mathcal{O})$, there is a minimal subset $\mathcal{A}'$ of $\mathcal{A}$ (a justification) that is sufficient to derive $\vec{a}$.[1] Suppose, however, that in order to handle atomic implications of the form $A \sqsubseteq B$, ans is extended as follows: it selects from $\mathcal{T}$ the set $\mathcal{T}'$ of atomic implications, extends $\mathcal{A}$ to $\mathcal{A}'$ by adding assertions implied by $\mathcal{T}'$ (e.g., adding $B(a)$ if $A(a) \in \mathcal{A}'$ and $A \sqsubseteq B \in \mathcal{T}'$), and uses $\mathcal{A}'$ to answer queries as before. Assume, however, that ans contains a bug, and only adds $B(a)$ if both $A(a)$ and $C(a)$ occur in $\mathcal{A}'$, for $C \neq A$ a (fixed) atomic concept. Despite the bug, the algorithm is still sound and faithful, but it is not compact. To see this, consider $\mathcal{T} = \{A \sqsubseteq B\}$ and q asking for the instances of B. For $\mathcal{A} = \{A(a), C(a)\}$ we have

---

[1] Recall that we are assuming that TBoxes do not contain nominals.

that $\mathsf{cert}(q, \mathcal{T}, \mathcal{A}) = \mathsf{ans}(q, \mathcal{T}, \mathcal{A}) = \{a\}$. However, the only relevant justification is $J = \mathcal{T} \cup \mathcal{A}_J$ for $\mathcal{A}_J = \{A(a)\}$; but $a \notin \mathsf{ans}(q, J)$, and thus ans is not compact.

We believe that compactness is also a reasonable property to expect from a CQ answering algorithm, and that non-compactness is likely to be indicative of some "oddity" in the algorithm, as in the above example.

3.2   Testing Bases

Next, we briefly recapitulate from [13] the central notion of a *testing base*: a collection of minimal ABoxes (called *testing units*) which can produce an answer to q w.r.t. some minimal subset of $\mathcal{T}$. To check completeness, a testing base must include all "relevant" testing units.

Definition 2. *An ABox $\mathcal{A}$ is a* testing unit *for a CQ q and TBox $\mathcal{T}$ if $\mathcal{T} \cup \mathcal{A}$ is consistent and there exists a tuple $\vec{a} \in \mathsf{cert}(q, \mathcal{T}, \mathcal{A})$ such that $\mathcal{A}$ is the ABox part of some justification for $q, \vec{a}$ in $\mathcal{T} \cup \mathcal{A}$. A* testing base *(TB) for $q, \mathcal{T}$ is a finite set $\mathbf{B}$ of testing units for $q, \mathcal{T}$ such that for each testing unit $\mathcal{A}$ for q and $\mathcal{T}$, there is some $\mathcal{A}' \in \mathbf{B}$ such that $\mathcal{A}'$ is isomorphic to $\mathcal{A}$. A testing base is* minimal *if no two ABoxes in it are isomorphic.*

Consider, as a running example, the following TBox $\mathcal{T}$ stating that everyone taking a maths course is a student and every instance of the relation "takes calculus course" is also an instance of "takes maths course"; consider also the following query q asking for the set of students taking a maths course.

$$\mathcal{T} = \{\exists \mathsf{takesMathCo}.\top \sqsubseteq \mathsf{St}, \mathsf{takesCalcCo} \sqsubseteq \mathsf{takesMathCo}\}$$
$$q = Q_P(x) \leftarrow \mathsf{St}(x) \wedge \mathsf{takesMathCo}(x, y)$$

By Definition 2, the following ABoxes are testing units for $q, \mathcal{T}$, and the set $\mathbf{B} = \{\mathcal{A}_1, \ldots, \mathcal{A}_8\}$ is a minimal TB for $q, \mathcal{T}$:

$$\mathcal{A}_1 = \{\mathsf{takesMathCo}(a, b)\} \quad \mathcal{A}_2 = \{\mathsf{St}(a), \mathsf{takesMathCo}(a, b)\}$$
$$\mathcal{A}_3 = \{\mathsf{takesMathCo}(a, a)\} \quad \mathcal{A}_4 = \{\mathsf{St}(a), \mathsf{takesMathCo}(a, a)\}$$
$$\mathcal{A}_5 = \{\mathsf{takesCalcCo}(a, b)\} \quad \mathcal{A}_6 = \{\mathsf{St}(a), \mathsf{takesCalcCo}(a, b)\}$$
$$\mathcal{A}_7 = \{\mathsf{takesCalcCo}(a, a)\} \quad \mathcal{A}_8 = \{\mathsf{St}(a), \mathsf{takesCalcCo}(a, a)\}$$

As shown in [13], TBs provide the following completeness guarantee for any CQ answering algorithm ans that is sound and faithful: if ans correctly computes the set of certain answers for each ABox in a TB, then it will also compute the set of certain answers for *any* ABox that is consistent with the TBox. In our example, this means that we only need to check whether $\mathsf{ans}(q, \mathcal{T}, \mathcal{A}_i) = \{a\}$ for each $\mathcal{A}_i \in \{\mathcal{A}_1, \ldots, \mathcal{A}_8\}$ in order to determine if ans will compute the set of certain answers of q w.r.t. $\mathcal{T}$ and any ABox that is consistent with $\mathcal{T}$.

3.3   Strict Testing Bases

Intuitively, to check whether each $\mathcal{A}_i$ in our running example is a testing unit, one would need to compute all justifications for q and each certain answer a in

$\mathcal{T} \cup \mathcal{A}_i$, and then check whether $\mathcal{A}_i$ is the ABox part of one of them. This may be infeasible in practice, as we may need to consider all possible subsets of $\mathcal{T}$.

In this paper, we address this issue by investigating the notion of a *strict testing unit*—a minimal ABox that can produce an answer to q w.r.t. $\mathcal{T}$.

Definition 3. *An ABox $\mathcal{A}$ is a* strict testing unit *for a CQ q and TBox $\mathcal{T}$ if $\mathcal{T} \cup \mathcal{A}$ is consistent and there exists a tuple $\vec{a} \in \mathsf{cert}(q, \mathcal{T}, \mathcal{A})$ such that $\vec{a} \notin \mathsf{cert}(q, \mathcal{T}, \mathcal{A}')$ for each $\mathcal{A}' \subset \mathcal{A}$.*

To check whether $\mathcal{A}$ is a strict testing unit, we only need to find a certain answer that is lost when removing any assertion from $\mathcal{A}$. Furthermore, it can be easily shown that each strict testing unit for q and $\mathcal{T}$ is also a testing unit for q and $\mathcal{T}$, and in our running example only the testing units $\mathcal{A}_1, \mathcal{A}_3, \mathcal{A}_5$, and $\mathcal{A}_7$ are strict. The notion of a strict testing unit leads to that of a *strict testing base*.

Definition 4. *A* strict testing base $\mathbf{B}_s$ *for $q, \mathcal{T}$ is a finite set of strict testing units for $q, \mathcal{T}$ such that, for each strict testing unit $\mathcal{A}$ for $q, \mathcal{T}$, there is some $\mathcal{A}' \in \mathbf{B}_s$ such that $\mathcal{A}'$ is isomorphic to $\mathcal{A}$. Finally, a strict testing base is* minimal *if no two ABoxes in it are isomorphic.*

Given any TB $\mathbf{B}$, we can always construct a strict one $\mathbf{B}_s$ by removing from $\mathbf{B}$ the testing units that are not strict, and hence $\mathbf{B}_s$ is likely to be smaller than $\mathbf{B}$ (in our example, $\mathbf{B}_s = \{\mathcal{A}_1, \mathcal{A}_3, \mathcal{A}_5, \mathcal{A}_7\}$ is a strict and minimal TB).

We next present our main result in this section: although strict TBs are smaller than TBs, they provide *exactly the same* completeness guarantees.

Theorem 1. *Let ans be a sound and faithful CQ answering algorithm for $\mathcal{L}$. Let q be a CQ, $\mathcal{T}$ an $\mathcal{L}$-TBox and $\mathbf{B}_s$ a strict TB for q, $\mathcal{T}$. The following property ($\Diamond$) holds for any ABox $\mathcal{A}'$ s.t. $\mathcal{T} \cup \mathcal{A}'$ is consistent: If $\mathsf{ans}(q, \mathcal{T}, \mathcal{A}) = \mathsf{cert}(q, \mathcal{T}, \mathcal{A})$ for each $\mathcal{A} \in \mathbf{B}_s$, then $\mathsf{ans}(q, \mathcal{T}, \mathcal{A}') = \mathsf{cert}(q, \mathcal{T}, \mathcal{A}')$.*

Proof. By contradiction, let $\mathsf{ans}(q, \mathcal{T}, \mathcal{A}) = \mathsf{cert}(q, \mathcal{T}, \mathcal{A})$ for each $\mathcal{A} \in \mathbf{B}_s$ and assume there exists $\mathcal{A}'$ s.t. $\mathcal{T} \cup \mathcal{A}'$ is consistent but $\mathsf{ans}(q, \mathcal{T}, \mathcal{A}') \neq \mathsf{cert}(q, \mathcal{T}, \mathcal{A}')$. Since ans is sound, $\mathsf{ans}(q, \mathcal{T}, \mathcal{A}') \neq \mathsf{cert}(q, \mathcal{T}, \mathcal{A}')$ iff $\mathsf{cert}(q, \mathcal{T}, \mathcal{A}') \nsubseteq \mathsf{ans}(q, \mathcal{T}, \mathcal{A}')$. Hence, let $\vec{a} \in \mathsf{cert}(q, \mathcal{T}, \mathcal{A}')$ be s.t. $\vec{a} \notin \mathsf{ans}(q, \mathcal{T}, \mathcal{A}')$. Since $\vec{a} \in \mathsf{cert}(q, \mathcal{T}, \mathcal{A}')$ and $\mathcal{L}$ does not provide for nominals, there is a minimal (w.r.t. set inclusion), non-empty $\mathcal{A}_{min} \subseteq \mathcal{A}'$ s.t. $\vec{a} \in \mathsf{cert}(q, \mathcal{T}, \mathcal{A}_{min})$. But then, $\mathcal{A}_{min}$ is a strict testing unit by Definition 3. Since $\mathbf{B}_s$ is a strict TB, there exists $\mathcal{A}'_{min} \in \mathbf{B}_s$ isomorphic to $\mathcal{A}_{min}$. Finally, since $\mathcal{A}_{min} \subseteq \mathcal{A}'$ and ans is monotonic and invariant under isomorphisms, we have that $\vec{a} \in \mathsf{ans}(q, \mathcal{T}, \mathcal{A}')$, which is a contradiction. $\square$

Thus, given our example $\mathcal{T}$ and q, to check whether a sound and faithful reasoner correctly computes $\mathsf{cert}(q, \mathcal{T}, \mathcal{A})$ for any ABox $\mathcal{A}$, we only need to check whether it returns all the certain answers w.r.t. $\mathcal{A}_1, \mathcal{A}_3, \mathcal{A}_5$ and $\mathcal{A}_7$.

3.4   Existence and Size of Strict Testing Bases

In [13] we showed that, unfortunately, there exist CQs and ontologies written in rather simple ontology languages for which a TB does not exist, because infinitely many testing units would be needed. As already discussed, a strict TB exists whenever a TB does. The converse, however, may not hold, and hence our non-existence results from [13] do not transfer directly to strict TBs. The following example shows a TBox and a CQ for which there is a strict TB containing just one ABox with a single assertion, but for which no TB exists.

Example 1. Consider the following TBox and query:

$$\mathcal{T} = \{A \sqsubseteq \exists R.B, \exists R.B \sqsubseteq B\}; \quad q = Q_P(x) \leftarrow A(x) \wedge B(x)$$

The set $\mathbf{B}_s = \{\{A(a)\}\}$ is a strict TB. However, for any value of n, the ABox $\mathcal{A}_n = \{A(a), R(a, b_1), \ldots, R(b_{n-1}, b_n), B(b_n)\}$ is a testing unit (it is the ABox part of a justification J for the certain answer a in $\mathcal{T} \cup \mathcal{A}_n$, whose TBox part is $\mathcal{T}_J = \{\exists R.B \sqsubseteq B\}$), and $\mathcal{A}_i$ and $\mathcal{A}_j$ are non-isomorphic for any $i \neq j$. Thus no TB exists, because from Definition 2 a TB must be a finite set of testing units.

Although a strict TB may exist even if no TB does, it may not be possible in general to guarantee the existence of one. For instance, if we modify $\mathcal{T}$ from Example 1 to be $\mathcal{T} = \{\exists R.B \sqsubseteq B\}$, no strict TB exists for the same reason that no TB does. The proof of Theorem 2 is identical to the one in [13] for TBs.

Theorem 2. *Let $\mathcal{L}$ be $\mathcal{EL}$, or $\mathcal{FL}_0$, or a DL allowing for transitivity axioms. There is a CQ q and a $\mathcal{L}$-TBox $\mathcal{T}$ for which no strict testing base exists.*

In cases when a TB $\mathbf{B}$ does exist (see Section 4), the corresponding strict TB $\mathbf{B}_s$ is likely to be much smaller. A natural question is how small $\mathbf{B}_s$ can be in comparison to $\mathbf{B}$. We next provide an example of an *exponential reduction in size*.

Example 2. Consider the following TBox and query:

$$\mathcal{T} = \{B \sqsubseteq A_i \mid 1 \leq i \leq n\} \cup \{A_1 \sqcap \ldots \sqcap A_n \sqsubseteq C\}; \quad q = Q_P(x) \leftarrow C(x)$$

Let $\mathbf{B}_s$ and $\mathbf{B}$ be as follows, where $\mathcal{A} = \{A_1(a), \ldots, A_n(a)\}$, $\mathcal{B} = \{B(a)\}$, and $\wp(\mathcal{A})$ is the power set of $\mathcal{A}$:

$$\mathbf{B}_s = \{\mathcal{B}, \mathcal{A}, \{C(a)\}\}; \quad \mathbf{B} = \mathbf{B}_s \cup \bigcup_{\mathcal{A}' \in \wp(\mathcal{A}) \backslash \mathcal{A}} \{\mathcal{B} \cup \mathcal{A}'\}$$

The set $\mathbf{B}_s$ with three testing units is a strict and minimal TB for q, $\mathcal{T}$. Also, given any $\mathcal{A}' \subset \mathcal{A}$, we have that $\mathcal{B} \cup \mathcal{A}'$ is a testing unit since it is the ABox of a justification with $\mathcal{T}' = \{B \sqsubseteq A_j \mid 1 \leq j \leq n, A_j(a) \notin \mathcal{A}'\} \cup \{A_1 \sqcap \ldots \sqcap A_n \sqsubseteq C\}$. Therefore, $\mathbf{B}$ is a minimal TB containing $2^n + 1$ testing units.

Although strict TBs can be exponentially smaller than TBs, this is not always the case. The following example shows that an exponential blowup w.r.t. the size of the TBox may not be avoidable when computing strict and minimal TBs.

Example 3. For $n \geq 1$, consider the TBox $\mathcal{T}_n$ consisting of the following axioms for each $0 \leq j < i \leq n$:[2]

---

[2] A similar TBox was used in [8] for a different purpose.

$$\overline{X}_0 \sqcap \ldots \sqcap \overline{X}_n \sqsubseteq \bot; \qquad\qquad X_0 \sqcap \ldots \sqcap X_n \sqsubseteq B;$$
$$\exists R.\overline{X}_0 \sqsubseteq X_0; \qquad\qquad \exists R.X_0 \sqsubseteq \overline{X}_0;$$
$$\exists R.(\overline{X}_i \sqcap X_0 \sqcap \ldots \sqcap X_{i-1}) \sqsubseteq X_i; \qquad \exists R.(X_i \sqcap X_0 \sqcap \ldots \sqcap X_{i-1}) \sqsubseteq \overline{X}_i;$$
$$\exists R.(\overline{X}_i \sqcap \overline{X}_j) \sqsubseteq \overline{X}_i; \qquad\qquad \exists R.(X_i \sqcap \overline{X}_j) \sqsubseteq X_i.$$

and the query $q = Q_P(x) \leftarrow B(x)$. Intuitively, $\mathcal{T}_n$ implements the incrementation of an n-bit counter along an R-chain. For each $1 \le k < 2^{n+1}$, let Z_k be of the form $Z_k = \sqcap_{0 \le i \le n} Y_i$ with $Y_i \in \{X_i, \overline{X}_i\}$ s.t. the binary number obtained by replacing each Y_i in the chain $Y_0 \ldots Y_n$ with 1 if $Y_i = X_i$ and 0 otherwise is precisely the binary encoding of k. Then, for each $2 \le j < 2^{n+1}$, the following ABox $\mathcal{A}_j$ is a strict testing unit (and is not isomorphic to any $\mathcal{A}_{j'}$ with $j' \ne j$):

$$\mathcal{A}_j = \{R(a_0, a_1), \ldots, R(a_{j-1}, a_j), Z_j(a_j)\}$$

Existence of a strict TB is ensured by the axiom $\overline{X}_0 \sqcap \ldots \sqcap \overline{X}_n \sqsubseteq \bot$, which precludes the computation of an infinite number of (non-isomorphic) strict testing units by "appending" relevant R-chains an arbitrary number of times (recall that a strict testing unit must be consistent with $\mathcal{T}_n$). It can easily be verified that a strict and minimal TB must contain exponentially many testing units w.r.t. n.

3.5 Measuring the Degree of Completeness

In this section, we turn our attention to measuring quantitatively "how complete" a sound and faithful reasoner is for a fixed query q and TBox $\mathcal{T}$, when completeness guarantees are not provided. To this end, we next introduce the notion of *completeness degree*, which in its most general form can be defined as follows.[3]

Definition 5. *Let* ans *be a sound and faithful CQ answering algorithm for $\mathcal{L}$ a DL, q a CQ, $\mathcal{T}$ an $\mathcal{L}$-TBox and $\mathbf{A}$ a non-empty set of ABoxes such that, for each $\mathcal{A} \in \mathbf{A}$, $\mathcal{T} \cup \mathcal{A}$ is consistent and $\mathrm{cert}(q, \mathcal{T}, \mathcal{A}) \ne \emptyset$. The completeness degree δ of* ans *for q, $\mathcal{T}$ and $\mathbf{A}$ is defined as follows (where $\sharp \mathbf{S}$ denotes the number of elements in a set $\mathbf{S}$):*

$$\delta_{\mathbf{A}}(\mathsf{ans}, q, \mathcal{T}) = \frac{1}{\sharp \mathbf{A}} \times \sum_{\mathcal{A} \in \mathbf{A}} \frac{\sharp \mathsf{ans}(q, \mathcal{T}, \mathcal{A})}{\sharp \mathrm{cert}(q, \mathcal{T}, \mathcal{A})}$$

Therefore, $\delta_{\mathbf{A}}$ represents the proportion of certain answers w.r.t. ABoxes in $\mathbf{A}$ that ans is able to compute correctly. The specific properties of $\delta_{\mathbf{A}}$, however, will obviously depend on the particular set of ABoxes under consideration. Intuitively, in order to obtain a reasonable measure of completeness for $\mathcal{T}$ and q, the set $\mathbf{A}$ should be chosen such that the following basic properties are satisfied:

1. If ans misses a certain answer for some (arbitrary) ABox consistent with the TBox, then $\delta_{\mathbf{A}}(\mathsf{ans}, q, \mathcal{T})$ should be smaller than one.

---

[3] The notion given here slightly differs from the one in our previous work.

2. If ans correctly computes some certain answer for some (arbitrary) ABox consistent with the TBox, then $\delta_{\mathbf{A}}(\mathsf{ans}, q, \mathcal{T})$ should be larger than zero.
3. If each certain answer computed by ans is also computed by ans′, then $\delta_{\mathbf{A}}(\mathsf{ans}', q, \mathcal{T})$ should be at least as large as $\delta_{\mathbf{A}}(\mathsf{ans}, q, \mathcal{T})$.
4. If Property 3 holds and, in addition, there is an (arbitrary) ABox $\mathcal{A}$ consistent with $\mathcal{T}$ for which ans′ computes a certain answer that ans fails to compute, then $\delta_{\mathbf{A}}(\mathsf{ans}', q, \mathcal{T})$ should be strictly larger than $\delta_{\mathbf{A}}(\mathsf{ans}, q, \mathcal{T})$.

Consider our running example CQ q, TBox $\mathcal{T}$, TB $\mathbf{B} = \{\mathcal{A}_1, \ldots, \mathcal{A}_8\}$ and strict TB $\mathbf{B}_s = \{\mathcal{A}_1, \mathcal{A}_3, \mathcal{A}_5, \mathcal{A}_7\}$. An algorithm a_1 that ignores $\mathcal{T}$ and simply answers q w.r.t. the data would only compute the correct answers for $\mathcal{A}_2$ and $\mathcal{A}_4$; hence, $\delta_{\mathbf{B}}(\mathsf{a}_1, q, \mathcal{T}) = 0.25$, whereas $\delta_{\mathbf{B}_s}(\mathsf{a}_1, q, \mathcal{T}) = 0$. An algorithm a_2 that handles role inclusions but not existential quantification would only compute the correct answers for $\mathcal{A}_2$, $\mathcal{A}_4$, $\mathcal{A}_6$ and $\mathcal{A}_8$; thus, $\delta_{\mathbf{B}}(\mathsf{a}_2, q, \mathcal{T}) = 0.5$, but we again have $\delta_{\mathbf{B}_s}(\mathsf{a}_2, q, \mathcal{T}) = 0$. Finally, a complete algorithm would compute the correct answers for all ABoxes; hence, $\delta_{\mathbf{B}}(\mathsf{a}_3, q, \mathcal{T}) = \delta_{\mathbf{B}_s}(\mathsf{a}_3, q, \mathcal{T}) = 1$, as desired.

Our example suggests that by choosing a (possibly strict) TB, we can guarantee Properties 1 and 3. Indeed, this can be shown in general as a direct consequence of Theorem 1.

Proposition 1. *The following properties hold for* ans *and* ans′ *sound and faithful, q a CQ, $\mathcal{T}$ a TBox and $\mathbf{B}_s$ a strict TB for $q, \mathcal{T}$:*

1. *If $\mathsf{cert}(q, \mathcal{T}, \mathcal{A}) \neq \mathsf{ans}(q, \mathcal{T}, \mathcal{A})$ for some $\mathcal{A}$ s.t. $\mathcal{T} \cup \mathcal{A}$ is consistent, then $\delta_{\mathbf{B}_s}(\mathsf{ans}, q, \mathcal{T}) < 1$.*
2. *If $\mathsf{ans}(q, \mathcal{T}, \mathcal{A}) \subseteq \mathsf{ans}'(q, \mathcal{T}, \mathcal{A})$ for each $\mathcal{A}$, $\delta_{\mathbf{B}_s}(\mathsf{ans}, q, \mathcal{T}) \leq \delta_{\mathbf{B}_s}(\mathsf{ans}', q, \mathcal{T})$.*

Our running example also illustrates an important advantage of using TBs over strict TBs for measuring completeness degrees, namely that Properties 2 and 4 fail if δ is measured in terms of $\mathbf{B}_s$, but hold if δ is measured w.r.t. $\mathbf{B}$. We finally show that Properties 2 and 4 always hold for TBs provided that the relevant CQ answering algorithm is also compact.

Proposition 2. *The following properties hold for* ans *and* ans′ *sound, faithful and compact, q a CQ, $\mathcal{T}$ a TBox and $\mathbf{B}$ a TB for $q, \mathcal{T}$.*

1. *If $\vec{a} \in \mathsf{ans}(q, \mathcal{T}, \mathcal{A})$ for some tuple $\vec{a}$ and some ABox $\mathcal{A}$ s.t. $\mathcal{T} \cup \mathcal{A}$ is consistent, then $\delta_{\mathbf{B}}(\mathsf{ans}, q, \mathcal{T}) > 0$.*
2. *If $\mathsf{ans}(q, \mathcal{T}, \mathcal{A}) \subseteq \mathsf{ans}'(q, \mathcal{T}, \mathcal{A})$ for each ABox $\mathcal{A}$, and there exists $\mathcal{A}'$ and $\vec{a} \in \mathsf{ans}'(q, \mathcal{T}, \mathcal{A}')$ s.t. $\vec{a} \notin \mathsf{ans}(q, \mathcal{T}, \mathcal{A}')$, then $\delta_{\mathbf{B}}(\mathsf{ans}, q, \mathcal{T}) < \delta_{\mathbf{B}}(\mathsf{ans}', q, \mathcal{T})$.*

Proof. 1. Suppose that such tuple $\vec{a}$ and ABox $\mathcal{A}$ exist. Since ans is sound, $\vec{a} \in \mathsf{cert}(q, \mathcal{T}, \mathcal{A})$. Since ans is also compact, there is a justification $J = \mathcal{T}_J \cup \mathcal{A}_J$ for q, $\vec{a}$ in $\mathcal{T} \cup \mathcal{A}$ such that $\vec{a} \in \mathsf{ans}(q, \mathcal{T}_J, \mathcal{A}_J)$. But then, $\mathcal{A}_J$ is a testing unit by Definition 2. Since $\mathbf{B}$ is a testing base, there exists an ABox $\mathcal{A}' \in \mathbf{B}$ that is isomorphic to $\mathcal{A}_J$ with $\vec{a}'$ the tuple obtained after the corresponding renaming of $\vec{a}$. Finally, since ans is invariant under isomorphisms and monotonic, we clearly have $\vec{a}' \in \mathsf{ans}(q, \mathcal{T}, A')$ and hence $\delta_{\mathbf{B}}(\mathsf{ans}, q, \mathcal{T}) > 0$.

2. By Proposition 1, $\delta_{\mathbf{B}}(\mathsf{ans}, q, \mathcal{T}) \leq \delta_{\mathbf{B}}(\mathsf{ans}', q, \mathcal{T})$. The property then follows from the following statement, which we show next: there exist $\mathcal{A}_{min} \in \mathbf{B}$ and $\vec{b} \in \mathsf{ans}'(q, \mathcal{T}, \mathcal{A}_{min})$ such that $\vec{b} \notin \mathsf{ans}(q, \mathcal{T}, \mathcal{A}_{min})$. Since ans' is compact, there is a justification $J = \mathcal{T}_J \cup \mathcal{A}_J$ for q, $\vec{a}$ in $\mathcal{T} \cup \mathcal{A}'$ s.t. $\vec{a} \in \mathsf{ans}'(q, \mathcal{T}_J, \mathcal{A}_J)$. By definition of a TB, there exists $\mathcal{A}_{min} \in \mathbf{B}$ isomorphic to $\mathcal{A}_J$ with $\vec{b}$ the result of renaming $\vec{a}$ accordingly. By monotonicity and invariance under isomorphisms of ans', $\vec{b} \in \mathsf{ans}'(q, \mathcal{T}, \mathcal{A}_{min})$. But then, $\vec{b} \notin \mathsf{ans}(q, \mathcal{T}, \mathcal{A}_{min})$ since otherwise by monotonicity and invariance under isomorphisms of ans we have $\vec{a} \in \mathsf{ans}(q, \mathcal{T}, \mathcal{A}')$, which is a contradiction.   □

4   Computing Strict Testing Bases

In our previous work [13], we identified sufficient conditions for a TB to exist. We showed that it is always possible to construct a TB for $\mathcal{T}, q$ whenever there exists a UCQ rewriting for q and each subset $\mathcal{T}'$ of $\mathcal{T}$. The connection between the existence of UCQ rewritings and of TBs is relevant for practice: on the one hand, UCQ rewritings are guaranteed to exist if $\mathcal{T}$ is expressed in the DLs underpinning the QL profile of OWL 2, and they may also exist even if $\mathcal{T}$ is in other fragments of OWL 2 (such as the $\mathcal{EL}$ profile); on the other hand, there are currently a number of implemented algorithms for computing UCQ rewritings (e.g., those implemented in the systems QuOnto and REQUIEM).

Roughly speaking, the algorithm for computing a TB for $\mathcal{T}$ and q proceeds as follows: first, for each subset $\mathcal{T}'$ of $\mathcal{T}$ it computes a UCQ rewriting $u_{\mathcal{T}'}$; second, for each such $u_{\mathcal{T}'}$ it constructs a fixed set of individuals whose cardinality is bounded by $\mathsf{var}(u_{\mathcal{T}'})$; finally, it computes the required testing units by instantiating each $u_{\mathcal{T}'}$ with a *valid instantiation*—a (maximal) subset of the mappings from the variables of each CQ in $u_{\mathcal{T}'}$ to individuals satisfying certain properties.

This naive algorithm is not practical since it may need to examine an exponential number of subsets of $\mathcal{T}$. This is required to ensure, on the one hand, that a TB exists and, on the other hand, that all relevant testing units are computed via a valid instantiation. For example, the CQ $Q_P(x) \leftarrow A(x)$ is a UCQ rewriting for the query q and TBox $\mathcal{T}$ from Example 1, but no TB exists for $q, \mathcal{T}$. The algorithm from [13] rejects the input $q, \mathcal{T}$ because it additionally considers the subset $\mathcal{T}' = \{\exists R.B \sqsubseteq B\}$ of $\mathcal{T}$ and finds that no UCQ rewriting exists for $q, \mathcal{T}'$.

We next show that a *strict* TB can be computed solely from a UCQ rewriting u for $\mathcal{T}$ and q, and hence computing a UCQ rewriting for each of the (exponentially many) subsets of $\mathcal{T}$ is no longer required. Indeed, we can compute the strict TB $\mathbf{B}_s = \{A(a)\}$ for q and $\mathcal{T}$ from Example 1 by just computing and instantiating the UCQ rewriting $Q_P(x) \leftarrow A(x)$.

We start by recapitulating the notions from [13] of an *instantiation* of a CQ and a *valid instantiation* for a UCQ.

Definition 6. *Let q be a CQ, B_q the body atoms in q, and π a mapping from all variables of q to individuals. The following ABox is an* instantiation *of q:*

$$\mathcal{A}_{\pi}^q := \{A(\pi(x)) \mid A(x) \in B_q\} \cup \{R(\pi(x), \pi(y)) \mid R(x, y) \in B_q\}$$

Let $u = \{q_1, \ldots, q_n\}$ be a UCQ and assume w.l.o.g. that $\mathsf{var}(q_i) \cap \mathsf{var}(q_j) = \emptyset$ for $i \neq j$. Let $\mathsf{ind} = \{a_1, \ldots, a_m\}$ be a set of individuals s.t. $m = \sharp\mathsf{var}(u)$ and let Π_{q_i} be the set of all mappings from $\mathsf{var}(q_i)$ to ind. A set $\Pi_u = \Pi_{q_1}^u \uplus \ldots \uplus \Pi_{q_n}^u$ with $\Pi_{q_i}^u \subseteq \Pi_{q_i}$ is a valid instantiation of u if it is a maximal subset of $\Pi_{q_1} \uplus \ldots \uplus \Pi_{q_n}$ with the following property:

> (*): for each $q_i, q_j \in u$ and $\pi \in \Pi_{q_i}^u$, there is no $\pi' \in \Pi_{q_j}$ s.t. π and π' map the distinguished variables in q_i and q_j identically and $\mathcal{A}_{\pi'}^{q_j} \subset \mathcal{A}_{\pi}^{q_i}$.

Intuitively, when instantiating a CQ q in a rewriting u using a valid instantiation, Property (*) from Definition 6 ensures that there is no "smaller" instantiation of a (possibly different) CQ q' from u. In our running example about students and math courses we have that $u = \{q, q_1, q_2, q_3\}$, with q_1, q_2 and q_3 given as follows, is a UCQ rewriting of q and $\mathcal{T}$:

$$q_1 = Q_P(x_1) \leftarrow \mathsf{St}(x_1) \wedge \mathsf{takesCalcCo}(x_1, y_1)$$
$$q_2 = Q_P(x_2) \leftarrow \mathsf{takesMathCo}(x_2, y_2)$$
$$q_3 = Q_P(x_3) \leftarrow \mathsf{takesCalcCo}(x_3, y_3)$$

For $\mathsf{ind} = \{a_i, b_i \mid 0 \leq i \leq 3\}$, we have that $\Pi = \{x_i \mapsto a_i, y_i \mapsto b_i \mid 2 \leq i \leq 3\}$ is a valid instantiation; in contrast, $\Pi' = \{x_1 \mapsto a_1, y_1 \mapsto b_1\}$ is not valid since the mapping $\pi = \{x_3 \mapsto a_1, y_3 \mapsto b_1\}$ leads to a smaller ABox.

Our previous example clearly shows that non-valid instantiations can lead to ABoxes that are not strict testing units. Furthermore, each ABox obtained by instantiating a CQ in a UCQ rewriting using a valid instantiation is indeed a strict testing unit, as shown by the following lemma.

Lemma 1. *Let u be a UCQ rewriting for $\mathcal{T}$ and q, and let Π_u be a valid instantiation. Then, for each $q_i \in u$ and each $\pi \in \Pi_{q_i}^u$ such that $\mathcal{T} \cup \mathcal{A}_{\pi}^{q_i}$ is consistent, we have that $\mathcal{A}_{\pi}^{q_i}$ is a strict testing unit for $\mathcal{T}, q$.*

Proof. Let π map the distinguished variables of q_i to $\vec{a}$. Then, $\vec{a} \in \mathsf{cert}(q_i, \emptyset, \mathcal{A}_{\pi}^{q_i})$. Since $q_i \in u$, soundness of UCQ rewritings implies $\vec{a} \in \mathsf{cert}(q, \mathcal{T}, \mathcal{A}_{\pi}^{q_i})$. To show that $\mathcal{A}_{\pi}^{q_i}$ is a strict testing unit, it suffices to show that $\vec{a} \notin \mathsf{cert}(q, \mathcal{T}, \mathcal{A}_{\pi}^{q_i} \setminus \alpha)$ for each $\alpha \in \mathcal{A}_{\pi}^{q_i}$. By the contrapositive of the completeness property of UCQ rewritings it suffices to show that $\forall q_j \in u$, $\vec{a} \notin \mathsf{cert}(q_j, \emptyset, \mathcal{A}_{\pi}^{q_i} \setminus \alpha)$. But, this is ensured by Property (*) of Definition 6, as we show next.

By contradiction. For some $q_j \in u$ assume that $\vec{a} \in \mathsf{cert}(q_j, \emptyset, \mathcal{A}_{\pi}^{q_i} \setminus \alpha)$.[4] Then, by the semantics of CQ answering there is a mapping σ from the variables in q_j to the individuals in $\mathcal{A}_{\pi}^{q_i} \setminus \alpha$ that maps the distinguished variables of q_j to $\vec{a}$ and such that $\mathcal{A}_{\sigma}^{q_j} \subseteq \mathcal{A}_{\pi}^{q_i} \setminus \alpha$. Hence, $\mathcal{A}_{\sigma}^{q_j} \subset \mathcal{A}_{\pi}^{q_i}$; however, this contradicts the assumption that $\pi \in \Pi_{q_i}^u$, since Property (*) in Definition 6 would fail. $\square$

To compute strict TBs, we need to consider in the worst case all the possible ABoxes that can be obtained by instantiating a given UCQ rewriting using a given valid instantiation, as shown by the following theorem.

---

[4] Note that q_j could be q_i.

Algorithm 1. Compute a strict testing base

Algorithm: tb(u)
Input: a UCQ rewriting u for $\mathcal{T}$, q
1 Compute $u' := \mathsf{sub}(\mathsf{cond}(u))$
2 Construct $\mathsf{ind} := \{a_1, \ldots, a_n\}$ for $n = \sharp\mathsf{var}(u')$
3 Initialize $\mathsf{Out} := \emptyset$
4 **For each** $q_i \in u'$
    For each $\pi : \mathsf{var}(q_i) \mapsto \mathsf{ind}$
        If $\mathcal{T} \cup \mathcal{A}^{q_i}_{\pi}$ is consistent **then**
            $\mathsf{Out} := \mathsf{Out} \cup \{\mathcal{A}^{q_i}_{\pi}\}$
            For each $q_j \in u'$
                If $\mathsf{Sig}(q_j) \subseteq \mathsf{Sig}(q_i)$ **and** there exists $\pi' : \mathsf{var}(q_j) \mapsto \mathsf{ind}$ s.t. π, π' map
                distinguished vars. in q_i and q_j identically and $\mathcal{A}^{q_j}_{\pi'} \subset \mathcal{A}^{q_i}_{\pi}$ **then**
                $\mathsf{Out} := \mathsf{Out} \setminus \{\mathcal{A}^{q_i}_{\pi}\}$
3 **Return** Out

Theorem 3. *Let u be a UCQ rewriting for $\mathcal{T}$, q and let Π_u be a valid instantiation. The following set is a strict testig base for $\mathcal{T}$ and q.*

$$\mathbf{B}_s = \{\mathcal{A}^{q_j}_{\pi} \mid q_j \in u, \pi \in \Pi^u_{q_j}, \mathcal{T} \cup \mathcal{A}^{q_j}_{\pi} \ consistent\}$$

Proof. By Lemma 1, $\mathbf{B}_s$ only contains strict testing units. We show that for each strict testing unit $\mathcal{A}$, there exists $q_j \in u$ and a mapping $\pi \in \Pi^u_{q_j}$ s.t. $\mathcal{A}^{q_j}_{\pi}$ is isomorphic to $\mathcal{A}$ and thus $\mathbf{B}_s$ is a strict TB. $\mathcal{A}$ being a strict testing unit implies that $\mathcal{T} \cup \mathcal{A}$ is consistent and there exists $\vec{a} \in \mathsf{cert}(q, \mathcal{T}, \mathcal{A})$ s.t. $\vec{a} \notin \mathsf{cert}(q, \mathcal{T}, \mathcal{A}')$ for each $\mathcal{A}' \subset \mathcal{A}$. Since $\vec{a} \in \mathsf{cert}(q, \mathcal{T}, \mathcal{A})$ and $\mathcal{T} \cup \mathcal{A}$ is consistent, the completeness property of rewritings implies that $q_j \in u$ exists s.t. $\vec{a} \in \mathsf{cert}(q_j, \emptyset, \mathcal{A})$. Hence, there exists π from $\mathsf{var}(q_j)$ to individuals in $\mathcal{A}$ that maps the distinguished variables in q_j to $\vec{a}$. Since $\mathcal{A}$ is minimal, there is no other π' that maps q_j or any other $q_i \in u$ to a strict subset of $\mathcal{A}$ and s.t. it maps their distinguished variables to $\vec{a}$. Thus, by maximality, there exists $\pi \in \Pi^u_{q_j}$ s.t. $\mathcal{A}^{q_j}_{\pi}$ is isomorphic to $\mathcal{A}$. □

According to Theorem 3, an algorithm for computing a strict TB for $\mathcal{T}$, q must use a valid instantiation to compute all testing units. A naive implementation would check Property ($*$) from Definition 6 by performing a number of ABox containment tests that is exponential in the number of query variables.

We next present a practical algorithm for computing a strict TB. Algorithm 1 takes a UCQ rewriting u for $\mathcal{T}$ and q (computed using any state of the art rewriting algorithm), and implements several optimisations aimed at reducing the number of ABox containment tests needed to check Property ($*$).

First, as in our practical algorithm from [13], Algorithm 1 uses condensation and subsumption to reduce the size of the input rewriting. This can avoid (possibly exponentially) many tests when checking Property ($*$). For instance, subsumption would eliminate the queries q and q_1 from the rewriting for our running example, thus discarding each of their instantiations.

Finally, Algorithm 1 only checks the containment of an instantiation of the form $\mathcal{A}_{\pi'}^{q_j}$ in an instantiation of the form $\mathcal{A}_{\pi}^{q_i}$ whenever all the body predicates in q_j occur also in q_i. For instance, consider the following TBox and CQ:

$$\mathcal{T} = \{C \sqsubseteq \exists R.\top\} \quad q = Q_P(x) \leftarrow R(x,y) \wedge R(y,z)$$

Given $u = \{q, q_1\}$ with $q_1 = Q_P(x) \leftarrow R(x,y) \wedge C(y)$, neither cond nor sub removes any query. Algorithm 1, however, will not perform any test of the form $\mathcal{A}_{\pi'}^{q_1} \subset \mathcal{A}_{\pi}^{q}$ since q_1 mentions the predicate C, which is not mentioned in q.

As shown by the following theorem, none of these optimisations results in a loss of relevant strict testing units, and the output of Algorithm 1 is a strict TB.

Theorem 4. *Algorithm 1 computes a strict TB for q and $\mathcal{T}$.*

Proof. The application of cond and sub on a UCQ rewriting preseves the soundness and completeness properties, and $u' = \mathsf{sub}(\mathsf{cond}(u))$ is also a UCQ rewriting. We show that each $\mathcal{A}_{\pi}^{q_i} \in \mathsf{Out}$ is a strict testing unit. To this end, we show that π belongs to $\Pi_{q_i}^{u'}$ with $\Pi_{u'}$ a valid instantiation of u' w.r.t. ind. This is so unless the following condition holds: there exists $q_j \in u'$ with $\mathsf{Sig}(q_j) \nsubseteq \mathsf{Sig}(q_i)$ and $\pi' : \mathsf{var}(q_j) \mapsto \mathsf{ind}$ s.t. π and π' map the distinguished variables in q_i and q_j identically and $\mathcal{A}_{\pi'}^{q_j} \subset \mathcal{A}_{\pi}^{q_i}$. This condition, however, cannot hold: q_j has an atom X not occurring in q_i and hence for each π' the ABox $\mathcal{A}_{\pi'}^{q_j}$ has an assertion involving X which cannot occur in $\mathcal{A}_{\pi}^{q_i}$. To show that Out is a strict TB, let $\Pi_{u'}$ be a valid instantiation of u' w.r.t. ind. By Theorem 3 we show that for arbitrary $q_i \in u'$ and $\pi \in \Pi_{q_i}^{u'}$, we have $\mathcal{A}_{\pi}^{q_i} \in \mathsf{Out}$. This is the case because Algorithm 1 considers all possible queries q_i, q_j and all possible mappings from variables in those queries to individuals in ind, and it only excludes from Out ABoxes corresponding to instantiations violating Property $(*)$ from Definition 6. □

We conclude by briefly comparing Algorithm 1 with the TB approximation algorithm from our previous work. In [13], we showed that the approximation algorithm produces only testing units, but not necessarily all those needed to obtain a TB. In fact, the approximation algorithm produces only strict testing units, but not necessarily all those needed to obtain a strict TB, and hence it cannot be used to provide completeness guarantees for sound and faithful CQ answering algorithms. Algorithm 1, in contrast, produces strict TBs, and so can be used to provide such guarantees. Furthermore, as we show in the next section, the computation of strict TBs is computationally feasible in practice.

5   Implementation and Evaluation

We have implemented Algorithm 1 in our prototype tool SyGENiA,[5] which uses REQUIEM[6] for the computation of the UCQ rewritings, and used it to evaluate

---

[5] http://code.google.com/p/sygenia/
[6] http://www.comlab.ox.ac.uk/projects/requiem/home.html

Table 1. Generation times of strict TBs for each LUBM query (in sec.)

Q1	Q2	Q3	Q4	Q5	Q6	Q7	Q8	Q9	Q10	Q11	Q12	Q13	Q14
1.3	1.1	0.09	6.3	0.25	2.1	0.6	7	2.7	6	0.06	0.1	0.1	0.05

Table 2. LUBM queries for which completeness can be guaranteed

System	Completeness Guarantee	Completeness w.r.t. LUBM dataset
Jena Max	Q1-Q14	Q1-Q14
OWLim	Q1-Q5, Q7, Q9, Q11-Q14	Q1-Q14
Minerva	Q1-Q4, Q9, Q11, Q14	Q1-Q14
Jena Mini/Micro	Q1-Q3, Q5, Q11, Q13-Q14	Q1-Q5, Q11, Q13-Q14
Sesame	Q1, Q3, Q11, Q14	Q1-Q5, Q11, Q14

four systems: Sesame 2.3-prl,[7] OWLim 2.9.1,[8] Minerva v1.5,[9] and Jena v2.6.3[10] in each of its three variants (Micro, Mini and Max).

We first ran SyGENiA over the LUBM TBox and queries and computed a strict TB for each query—each LUBM query leads to a UCQ rewriting w.r.t. the TBox, and hence a strict TB is guaranteed to exist.[11] Table 1 presents the generation time for each of these strict TBs. We then used these strict TBs to compute the corresponding completeness degrees for each evaluated system.

In contrast to our previous work, Algorithm 1 ensures that each computed collection of datasets is a strict TB, and hence we can provide completeness guarantees in practice. This is illustrated in Table 2 where, for each system, we list the queries for which testing using strict TBs shows that it is complete for any dataset, and the queries for which it is complete w.r.t. the LUBM dataset.

Our results show that Jena Max is the only system that is guaranteed to be complete for all 14 LUBM queries regardless of the dataset—that is, it behaves exactly like a complete OWL reasoner w.r.t. the LUBM queries and TBox. Furthermore, as already noted in our previous work, completeness w.r.t. the LUBM benchmark is no guarantee of completeness in general; for example, OWLim and Minerva are both complete w.r.t. LUBM (and even w.r.t. to the more expressive UOBM benchmark), but for some queries they were found to be incomplete w.r.t. to our datasets. OWLim is, however, guaranteed to be complete for all LUBM queries that do not involve reasoning with existential quantifiers—a feature not supported by the system. Minerva, which uses a DL reasoner to classify the ontology and explicate subsumption between atomic concepts, is still guaranteed to be complete for only 8 queries; this is because our datasets reveal missing answers that depend on subsumptions between *complex concepts* that are not

---

[7] http://www.openrdf.org/

[8] http://www.ontotext.com/owlim/

[9] http://www.alphaworks.ibm.com/tech/semanticstk

[10] http://jena.sourceforge.net/

[11] Since REQUIEM does not currently support individuals in the queries or transitivity in the TBox, we have replaced the individuals in queries by distinguished variables and dispensed with the only transitivity axiom in the LUBM TBox.

Table 3. Completeness degrees for Jena Mini/Micro

Datasets	Q4	Q6	Q7	Q8	Q9	Q10	Q12
LUBM	1	.83	.87	.83	.64	.83	0
SyGENiA	.68	.003	.04	.058	0	.001	.25

Table 4. Completeness analysis on Galen

	Sesame	OWLim	Jena Mini	Minerva
Q1	∼0	.84	∼0	.97
Q2	.07	.83	.07	.96
Q3	.01	.84	∼0	.96
Q4	.01	.77	.01	1

pre-computed by the system. Jena Mini and Micro are guaranteed to be complete for 7 queries. Surprisingly, Jena Mini behaved exactly like Jena Micro, despite the fact that, in theory, Jena Mini can handle a larger fragment of OWL; these differences are, however, not revealed by the structure of the LUBM TBox and queries. Finally, Sesame is only guaranteed to be complete for 4 of the queries.

Concerning completeness degrees, the values we have obtained using strict TBs are in line with those from our previous work. However, as discussed in Section 3.5, a completeness degree value smaller than one should be interpreted with caution when using strict TBs, especially in the case of very small values. For instance, consider the values obtained for Jena Mini/Micro given in Table 3. When using strict TBs, the completeness degree for query $Q9$ is 0%, but the system is clearly able to correctly compute certain answers for some ABoxes (e.g., the LUBM dataset). As already discussed, this is because completeness degree measures based on strict TBs fail to satisfy properties 2 and 4 of Proposition 2.

Finally, we have considerd a small version of Galen (an expressive ontology with complex structure used in medical applications) and four queries asking respectively for the instances of the concepts HaemoglobinConcentrationProcedure, PlateletCountProcedure, LymphocyteCountProcedure, and HollowStructure. Each of these queries has a UCQ rewriting that can be computed using REQUIEM. Thus, a strict TB exists for each of them and can be computed in times ranging from 2 seconds to 1 minute. Our results are summarised in Table 4.

We could not run Jena Max since Galen makes heavy use of existential restrictions, which (according to the Jena documentation) might cause problems. Among the other systems, Minerva exhibited the best behavior: it was the only one for which completeness could be guaranteed for at least one query, and it exhibited a high completeness degree for the remaining three queries; this is because Minerva pre-computes many subsumption relationships between atomic concepts that depend on existential restrictions, which most other systems do not handle. Jena Mini and Sesame were surprisingly incomplete, although as already discussed, values close to zero should be interpreted with caution.

G. Stoilos, B. Cuenca Grau, and I. Horrocks

6 Conclusion and Future Work

In this paper we have extended in several important ways our prior work on completeness evaluation of Semantic Web reasoners. Most importantly, we have introduced the notion of a strict testing base, studied its formal properties, and shown that it can be used to identify circumstances in which completeness guarantees can be provided for reasoners that are incomplete in general. Finally, we have proposed a practical algorithm for the generation of strict testing bases, implemented it in the SyGENiA tool, and used SyGENiA to evaluate several incomplete reasoners, using both the LUBM benchmark, and the Galen ontology.

Our results suggest not only that strict testing bases are relatively easy to compute in practice, but also that completeness guarantees can often be provided for realistic ontologies and queries. The main limitation of strict testing bases is that the associated completeness degree fails to satisfy certain desirable properties. An interesting problem for future work is to try to design a practical algorithm that can be used to provide an accurate measure of completeness degree.

Acknowledgments. Supported by the EU project SEALS (FP7-ICT-238975). B. Cuenca Grau is supported by a Royal Society University Research Fellowship.

References

1. Baader, F., McGuinness, D., Nardi, D., Patel-Schneider, P.: The Description Logic Handbook: Theory, implementation and applications. Cambridge Uni. Press, New York (2002)
2. Calvanese, D., De Giacomo, G., Lembo, D., Lenzerini, M., Rosati, R.: Tractable reasoning and efficient query answering in description logics: The DL-Lite family. J. of Automated Reasoning 39(3), 385–429 (2007)
3. Glimm, B., Horrocks, I., Lutz, C., Sattler, U.: Conjunctive query answering for the description logic $\mathcal{SHIQ}$. In: Proc. of IJCAI 2007 (2007)
4. Guo, Y., Pan, Z., Heflin, J.: LUBM: A Benchmark for OWL Knowledge Base Systems. Journal of Web Semantics 3(2), 158–182 (2005)
5. Hitzler, P., Vrandecic, D.: Resolution-based approximate reasoning for OWL DL. In: Gil, Y., Motta, E., Benjamins, V.R., Musen, M.A. (eds.) ISWC 2005. LNCS, vol. 3729, pp. 383–397. Springer, Heidelberg (2005)
6. Kalyanpur, A., Parsia, B., Horridge, M., Sirin, E.: Finding all justifications of OWL DL entailments. In: Aberer, K., Choi, K.-S., Noy, N., Allemang, D., Lee, K.-I., Nixon, L.J.B., Golbeck, J., Mika, P., Maynard, D., Mizoguchi, R., Schreiber, G., Cudré-Mauroux, P. (eds.) ASWC 2007 and ISWC 2007. LNCS, vol. 4825, pp. 267–280. Springer, Heidelberg (2007)
7. Lutz, C., Toman, D., Wolter, F.: Conjunctive query answering in the description logic $\mathcal{EL}$ using a relational database system. In: Proc. of IJCAI 2009 (2009)
8. Lutz, C., Wolter, F.: Conservative extensions in the lightweight description logic $\mathcal{EL}$. In: Pfenning, F. (ed.) CADE 2007. LNCS (LNAI), vol. 4603, pp. 84–99. Springer, Heidelberg (2007)
9. Ortiz, M., Calvanese, D., Eiter, T.: Characterizing data complexity for conjunctive query answering in expressive description logics. In: Proc. of AAAI 2006 (2006)

10. Pan, J.Z., Thomas, E.: Approximating OWL-DL Ontologies. In: Proc. of AAAI 2007, pp. 1434–1439 (2007)
11. Pérez-Urbina, H., Horrocks, I., Motik, B.: Efficient query answering for OWL 2. In: Bernstein, A., Karger, D.R., Heath, T., Feigenbaum, L., Maynard, D., Motta, E., Thirunarayan, K. (eds.) ISWC 2009. LNCS, vol. 5823, pp. 489–504. Springer, Heidelberg (2009)
12. Poggi, A., Lembo, D., Calvanese, D., Giacomo, G.D., Lenzerini, M., Rosati, R.: Linking data to ontologies. J. Data Semantics 10, 133–173 (2008)
13. Stoilos, G., Cuenca Grau, B., Horrocks, I.: How incomplete is your semantic web reasoner? In: Proc. of AAAI 2010 (2010)

Signal/Collect: Graph Algorithms for the (Semantic) Web

Philip Stutz[1], Abraham Bernstein[1], and William Cohen[2]

[1] DDIS, Department of Informatics, University of Zurich, Zurich, Switzerland
[2] Machine Learning Department, Carnegie Mellon University, Pittsburgh, PA
{stutz,bernstein}@ifi.uzh.ch,
wcohen@cs.cmu.edu

Abstract. The Semantic Web graph is growing at an incredible pace, enabling opportunities to discover new knowledge by interlinking and analyzing previously unconnected data sets. This confronts researchers with a conundrum: Whilst the data is available the programming models that facilitate scalability and the infrastructure to run various algorithms on the graph are missing.

Some use MapReduce – a good solution for many problems. However, even some simple iterative graph algorithms do not map nicely to that programming model requiring programmers to shoehorn their problem to the MapReduce model.

This paper presents the Signal/Collect programming model for synchronous and asynchronous graph algorithms. We demonstrate that this abstraction can capture the essence of many algorithms on graphs in a concise and elegant way by giving Signal/Collect adaptations of various relevant algorithms. Furthermore, we built and evaluated a prototype Signal/Collect framework that executes algorithms in our programming model. We empirically show that this prototype transparently scales and that guiding computations by scoring as well as asynchronicity can greatly improve the convergence of some example algorithms. We released the framework under the Apache License 2.0 (at http://www.ifi.uzh.ch/ddis/research/sc).

1 Introduction

The Semantic Web confronts researchers and practitioners with increasing data set sizes. One approach to deal with this problem is to hope for the computational capabilities of computers to grow faster than the datasets relying on Moore's Law [1]. This approach is somewhat impractical as it makes current work rather tedious and relies on the hope that Moore's Law will be sustainable and will outpace the growth of data—both of which are unsure prospects. As a consequence, many researchers have tried to use parallelism to improve the performance of Semantic Web computational tasks. Hereby, they used two avenues of investigation. On the one hand, they have tried to use distributed computing programming models such as MapReduce [2] to achieve their goals [3,4]. This,

P.F. Patel-Schneider et al. (Eds.): ISWC 2010, Part I, LNCS 6496, pp. 764–780, 2010.

usually, requires to cumbersomely shoehorn their computation to the programming model. In the case of MapReduce the typed graphs of the Semantic Web need to be inconveniently mapped to the key/value-pair programming model. On the other hand, many have used low-level distributed computing primitives such as message passing interfaces [5], clusters [6], or distributed hash trees [7], which requires the building of the whole infrastructure for Semantic Web graph processing from scratch—a tedious task.

This paper proposes a *novel and scalable programming model for typed graphs*. The core idea lies in the realization, that most computations on Semantic Web data involve the passage of (1) some kind of information between the resources (or vertices) along the properties (or edges) of the RDF graph and (2) some computation at the vertices of the RDF graph. Specifically, we propose a programming model where vertices send *signals* along the property-defined edges of a compute graph and then a *collect* function gathers the incoming signals at the vertices to perform some computation. Given the two core elements we call our model SIGNAL/COLLECT.

This programming model allows an elegant and concise definition of programming tasks on typed graphs that a suitable execution framework can process transparently in a distributed fashion. In some cases the framework can exploit asynchronous execution to further speed up the accomplishment of the task. As such SIGNAL/COLLECT provides a natural programming model for the Semantic Web, which can serve as an alternative to paradigms such as MapReduce.

Given the above, the contributions of this paper are the following: First, we introduce an *elegant and concise programming model for Semantic Web computing tasks*. We show the elegance and conciseness by presenting the model and providing a number of typical algorithm examples. Second, we empirically show that a simple *execution framework is able to transparently parallelize* SIGNAL/COLLECT *computations* and can be *simply initialized with SPARQL queries*. Third, we empirically show that the exploitation of *asynchronous execution can further increase the performance and convergence* of an already parallel algorithm.

The remainder of the paper is organized as follows: Section 2 formally introduces the programming model and its extensions. Section 3 describes a number of increasingly complex graph algorithms to illustrate the elegance and conciseness of the programming model. We then introduce an actual implementation and evaluate the scalability, the impact of guiding computations by scoring and asynchronous computations in Sections 4 and 5. We close with a discussion of the related and future work.

2   The Signal/Collect Programming Model

The general intuition behind our SIGNAL/COLLECT programming model is that all computations are executed on a compute graph, where the vertices are the computational units that interact by the means of signals that flow along the edges. All computations in the vertices are accomplished by collecting the incoming signals and performing some computation on them employing, possibly, some vertex-state, and then signaling their neighbors in the compute graph.

To give a more concrete example: imagine a graph with RDFS classes as vertices and edges from superclasses to subclasses (i.e., `rdfs:subClassOf` triples). Every vertex has a set of superclasses as state, which initially only contains itself. Now all the superclasses send their own states as signals to their subclasses, which collect those signals by setting their own new state to the union of the old state and all signals received. It is easy to imagine how these steps, when repeatedly executed, iteratively compute the transitive closure of the `rdfs:subClassOf` relationship in the vertex states.

SIGNAL/COLLECT provides an elegant and concise abstraction for describing such graph-based algorithms. So far, however, we glossed over a number of important details which we elaborate in this section. We will introduce a formal definition of the basic structures of the SIGNAL/COLLECT programming model and continue by specifying the synchronous/asynchronous execution modes for computations as well as extending the basic model to support them.

2.1   A Formal Definition of the Signal/Collect Structures

The basis for any SIGNAL/COLLECT computation is the *compute graph*

$$G = (V, E),$$

where V is the set of vertices and E the set of edges in G. Every *vertex* $\mathbf{v} \in V$ has the following attributes:

v.id a unique id.

v.state the current vertex state which represents computational intermediate results.

v.outgoingEdges a list of all edges $\mathbf{e} \in E$ with e.source $= \mathbf{v}$.

v.signalMap a map with the ids of vertices as keys and signals as values. Every key represents the id of a neighboring vertex and its value represents the most recently received signal from that neighbor. We will use the alias v.signals to refer to the list of values in v.signalMap.

v.uncollectedSignals a list of signals that arrived since the collect operation was last executed on this vertex.

Every *edge* $\mathbf{e} \in E$ has the following attributes:

e.source the source vertex

e.sourceId id of the source vertex

e.targetId id of the target vertex

The default vertex type also defines an abstract **collect** function and the default edge type defines an abstract **signal** function. To specify an algorithm in the SIGNAL/COLLECT programming model the default types have to be extended with implementations of those functions. The **collect** function calculates a new vertex state, while the **signal** function calculates the signal that will be sent along an edge.

We have now defined the basic structures of the programming model. In order to completely define a SIGNAL/COLLECT computation we still need to describe how to execute computations on them.

2.2   The Computation Model and Extensions

In this section we will specify how both synchronous and asynchronous computations are executed in the SIGNAL/COLLECT programming model. Also we will provide extensions to the core model.

We will use the attribute `target` on edges to denote the target vertex, but this is only to specify the behavior without having to describe how signals are relayed.

We first define an additional attribute `lastSignalState` and two additional functions on all vertices $v \in V$, which will enable us to describe computations in SIGNAL/COLLECT:

v.executeSignalOperation

> `lastSignalState := state`
> **for all** (e ∈ outgoingEdges) **do**
> > `e.target.uncollectedSignals.append(e.signal)`
> > `e.target.signalMap.put(e.sourceId, e.signal)`
> **end for**

v.executeCollectOperation

> `state := collect`
> `uncollectedSignals := ` *Nil*

With these functions we are now able to describe a synchronous SIGNAL/COLLECT execution.

Synchronous Execution. A synchronous computation is specified in Algorithm 1. Its parameter num_iterations defines the number of iterations (computation steps the algorithm is going to perform. Everything inside the inner loops can be executed in parallel, with a global synchronization between the signaling and collecting phases, which is similar to computations in Pregel [8]. This parallel programming model is more generally referred to as Bulk Synchronous Parallel (BSP).

Algorithm 1. Synchronous execution of SIGNAL/COLLECT

> **for** i ← 1..num_iterations **do**
> > **for all** v ∈ V **parallel do**
> > > `v.executeSignalOperation`
> > **end for**
> > **for all** v ∈ V **parallel do**
> > > `v.executeCollectOperation`
> > **end for**
> **end for**

This specification allows the efficient execution of algorithms, where every vertex is equally involved in all steps of the computation. However, in many algorithms only a subset of the vertices is involved in each part of the computation. In order to be able to define a computational model that enables us to guide the computation and give priority to more "important" operations, we will introduce scoring.

Extension 1: Score-Guided Execution. In order to enable the scoring (or prioritizing) of signal/collect operations, we need to extend the core structures of the SIGNAL/COLLECT programming model. This is why we define two additional functions on all vertices $v \in V$:

v.scoreSignal : Double
   is a function that calculates a number that reflects how important it is for this vertex to signal. The result of this function is only allowed to change when the v.state changes. Its default implementation returns 0 if state = lastSignalState and 1 otherwise. This captures the intuition that it is desirable to inform the neighbors iff the state has changed since they were informed last.

v.scoreCollect : Double
   is a function that calculates a number that reflects how important it is for this vertex to collect. The result of this function is only allowed to change when uncollectedSignals changes. Its default implementation returns uncollectedSignals.size. This captures the intuition that the more new information is available, the more important it is to update the state.

The default implementations can be overridden with functions that capture the algorithm-specific notion of "importance" more accurately.

Now that we have extended the basic model with scoring we specify a score-guided synchronous execution of a SIGNAL/COLLECT computation in Algorithm 2. There are three parameters that influence when the algorithm stops:

Algorithm 2. Score-guided synchronous execution of SIGNAL/COLLECT

```
done := false
while iterations < num_iterations and !done do
    done := true
    iterations := iterations +1
    for all v ∈ V parallel do
        if (v.signalScore > signal_threshold) then
            done := false
            v.executeSignalOperation
        end if
    end for
    for all v ∈ V parallel do
        if (v.collectScore > collect_threshold) then
            done := false
            v.executeCollectOperation
        end if
    end for
end while
```

---

signal_threshold and collect_threshold which set a minimum level of "importance" for operations that get executed and num_iterations, which limits the number of computation steps. This means that the algorithm is guaranteed to stop, either because the maximum number of iterations is reached or because there are no operations anymore that score higher than the threshold. If the second condition is fulfilled, we say that the algorithm has converged.

Asynchronous Execution. We can now also define an asynchronous execution which gives no guarantees about the order of execution or the ratio of signal/-collect operations. We referred to the first two execution modes as synchronous because they guarantee that all vertices are in the same "loop" at the same time. In a synchronous execution it can never happen that one vertex executes a signal operation while another vertex is executing a collect operation, because the switch from one phase to the other is globally synchronized. In an asynchronous computation, in contrast, no such guarantees exist.

Algorithm 3. Asynchronous execution of SIGNAL/COLLECT

---

ops := 0
while [ops < num_ops] ∧ [∃v ∈ V((v.signalScore > signal_threshold) ∨ (v.collectScore > collect_threshold))] **do**
    S := radomly choose subset of V
    for all v ∈S **parallel do**
        Randomly call either
            v.executeSignalOperation or v.executeCollectOperation
            assuming respective threshold is reached
        ops := ops + 1
    end for
end while

As shown in Algorithm 3 there are again three parameters that influence when the asynchronous algorithm stops: signal_threshold and collect_threshold, which have the same function as in the synchronous case and num_ops which instead of the number of iterations limits the number of operations executed. Again this guarantees that an asynchronous execution terminates, either because the maximum number of operations is exceeded or because it converged. The purpose of Algorithm 3 is not to be executed directly, but to specify what kind of restrictions are guaranteed (by an execution environment) during asynchronous execution. This freedom is useful, because if an algorithm no longer has to maintain the execution order of operations then one is able use different scheduling strategies for those operations.

Extension 2: Scheduled Asynchronous Operations. As an extension to the asynchronous execution we can define operation schedulers that optimize certain measures. For example we can define an eager scheduler (see Algorithm 4 in Section 4) that will execute the signal operation of a vertex immediately after the collect operation of that same vertex. This allows other vertices to receive those signals sooner. Another example is a scheduler that gives priority to signals that have high scores by only executing signals with at least an average score. Depending on the algorithm this can result in fewer operations being executed, which, depending on the operation costs, impact of scheduling on convergence and the cost of scheduling itself, can pay off.

Extension 3: Multiple Vertex/Edge Types. Some algorithms, for example operating on OWL ontologies in RDF or bipartite factor graphs, require several kinds of vertices and edges with different associated functions for signaling, collecting, etc. This is why a compute graph can contain vertices and edges that have different types.

Extension 4: Result Processing. Results are processed by a `result Processing` function defined on the default vertex type. The default implementation does nothing and is meant to be overridden. This function gets executed on all vertices once the computation has ended.

Extension 5: Weights. The model supports weights on edges and the vertices keep track of the sum of weights of outgoing edges. It is also possible to extend the default edge/vertex type with labels or whatever additional attributes or functions should be required.

Extension 6: Conditional Edges & Computation Phases. Edges can be extended to only send a signal when certain conditions have been met. A possible condition is that a source vertex has received a convergence signal from an aggregation vertex, which can be used to trigger a next computation phase in the target vertices. Another use for this feature is to avoid sending the same signal repeatedly.

Feature: Aggregation. Sometimes it is desirable to aggregate over the state of multiple vertices to, for example, obtain a global convergence criterium. This can be easily achieved by introducing an aggregation vertex that receives a signal from all the vertices it needs to aggregate.

We have now specified the structures and execution model of the SIGNAL/COLLECT programming model. In the next section we show the usefulness of this programming model by giving implementations for various algorithms.

3   Algorithms in Signal/Collect

We argue that the SIGNAL/COLLECT programming model is a useful abstraction. There are many important algorithms that can be expressed in a concise and elegant way, which proves that this abstraction captures the essence of many computations on graphs indeed.

We demonstrate these characteristics of the SIGNAL/COLLECT programming model by giving examples (written in Scala-like[1] pseudocode, where the initialization of class variables is passed in parentheses) of several interesting algorithms expressed in SIGNAL/COLLECT. Note that not all algorithms work with all execution modes. Vertex coloring for example does not converge without score-guidance.

Single-source shortest path. Here the vertex states represent the shortest currently known path from the path-source, edge weights are used to represent distance. The signals represent the total path length of the shortest currently known path from the path-source to e.target that passes through e. In the Semantic Web context this algorithm can be used to compute the Rada shortest path distance along subclass vertices, which is sometimes used to denote the similarity between two classes.

```
class Location(id: Any, initalState: Int) extends Vertex {
  def collect: Integer = min(state, min(uncollectedSignals))
}
class Path(sourceId: Any, targetId: Any) extends Edge {
  def signal: Integer = source.state+weight
}
```

RDFS subclass inference. A vertex represents an RDFS class. The vertex state represents the set of currently known superclasses of the given vertex. As the edges just signal the set of currently known superclasses of the class represented by the source vertex, one can simply use the predefined StateForwarder edge, which has the signal function: def signal = source.state. The compute graph can be built with edges from vertices representing super-classes to vertices representing sub-classes, which can easily be done with a SPARQL query (see full PageRank example code in Figure 1, Section 4).

```
class RdfsClass(id: String, initialState=Set(iri)) extends Vertex {
  def collect: Set[String] = state ∪ ⋃          s
                                    s∈uncollectedSignals
}
```

Vertex coloring. The following algorithm solves the vertex coloring problem by assigning to each vertex a random color. The vertices keep switching to different random colors wherever conflicts with neighbors remain. The predefined StateForwarder edges are used again to signal the color of a vertex to its neighbors.

```
class Colored(id: Any, initialState=randomColor) extends Vertex {
  def collect: Int = {
```

[1] http://www.scala-lang.org/

```
      if (signals.contains(state)) randomColorExcept(state) else state
  }
}
```

PageRank [9]. The vertex state represents the current pagerank of a vertex.
The signals represent the rank transferred from e.source to e.target.

```
class Document(id: Any, initialState=0.15) extends Vertex {
  def collect: Double = 0.15+0.85*    ∑         s
                                  s∈v.signalMap.values
}
class Citation(sourceId: Any, targetId: Any) extends Edge {
  def signal: Double = weight*source.state
                       ──────────────────────
                       source.sumOfOutWeights
}
```

Loopy Belief Propagation [10]. Loopy belief propagation subsumes inferencing on Relational Probabilistic Models. These can be used to combine logical
and probabilistic inference on Semantic Web data—a highly desirable goal.

Because of space constraints we just convey the intuition: A factor graph can
be defined in SIGNAL/COLLECT with two vertex types Factor and Variable and
edge types FactorToVariable and VariableToFactor. Loopy belief propagation
is a message passing algorithm on a factor graph where messages are passed
back and forth between factor and variable vertices. Those messages in turn
are calculated from the messages received by the respective factor/variable. In
the simplest adaptation we put the code that computes those messages into the
signal functions of the edges. These functions can directly calculate the new
outgoing signals from the received signals in the signalMap of the source vertex.

All current evidence also indicates (but we have not tried it yet) that SIG-
NAL/COLLECT can straightforwardly implement the general sum-product (GSP)
algorithm [11]. According to Kschischang et al. [11] a wide variety of algorithms,
such as the forward/backward algorithm, the Viterbi algorithm, the iterative
turbo decoding algorithm, Pearls belief propagation algorithm for Bayesian networks, the Kalman filter, and certain fast Fourier transform (FFT) algorithms
can be derived as specific instances of the GSP algorithm.

In this subsection we demonstrated that the SIGNAL/COLLECT programming
model is a useful abstraction by giving examples of several interesting algorithms,
which we were able to express in a concise and elegant way. Note that whilst not
all of them are initially recognizable as typical Semantic Web approaches they
all provide important functionality.

In the next section we are going to evaluate the properties of a prototype of
the SIGNAL/COLLECT framework which can execute algorithms such as the ones
we just described.

4   The Signal/Collect Framework — An Implementation

The SIGNAL/COLLECT framework provides an execution platform for algorithms specified according to the SIGNAL/COLLECT programming model. This is analogous to the Hadoop MapReduce framework which executes algorithms expressed in the MapReduce programming model. The framework has been implemented in Scala—a fusion of the object-oriented and functional programming paradigms running on the Java Virtual Machine. We released the framework under the Apache License 2.0 (http://www.ifi.uzh.ch/ddis/research/sc).

Parallel Computations: The current implementation of the framework can parallelize computations exploiting multiple processor cores of one computer and shared memory for efficient signal passing. To that end we assign the vertices to worker threads that are each responsible for a part of the graph. We use a hash function on the vertex ids for the mapping of vertices to workers, edges are always assigned to the same worker as their source vertex.

We implemented the synchronous computation similar to [8] with a master that orchestrates the synchronized execution of signal/collect steps for all worker threads.

Asynchronous Scheduling: In an asynchronous computation every worker decides on its own which operations to execute. For this purpose every worker has a scheduler that determines the order in which the signal/collect operations get executed. We experimented with different schedulers for the signal/collect operations in the asynchronous case. Every computation in asynchronous mode starts with one synchronous score-guided signal/collect step, as this improved performance for the algorithms we analyzed. After that a scheduler takes over. The "eager" scheduler (Algorithm 4) for example tries to have a vertex signal as soon as possible after collecting.

Algorithm 4. "Eager" scheduler: tries to signal right after collection

---

    for all v $\in V$ **do**
        if (v.collectScore > collect_threshold) **then**
            v.executeCollectOperation
            if (v.signalScore > signal_threshold) **then**
                v.executeSignalOperation
            end if
        end if
    end for

We also experimented with other schedulers that, for example, only execute signal operations that score above or equal to the average signal score ("above average" scheduler).

Specifying Graphs: The PageRank example. In order to specify compute graphs one needs to define the necessary elements of the SIGNAL/COLLECT programming model. Consider the SIGNAL/COLLECT implementation of the

PageRank algorithm optimized with residual scoring on SwetoDblp[2] citations shown in Figure 1. First, the Figure specifies the PageRank algorithm by defining both the `collect` and `signal` functions. Second, the `Algorithm` object initializes a score-guided and synchronous compute-graph by iterating over the answers of a SPARQL query and then executes the algorithm with a signal threshold of 0 using `computeGraph.execute(.)`. Note that this is the executable code and not simplified pseudocode.

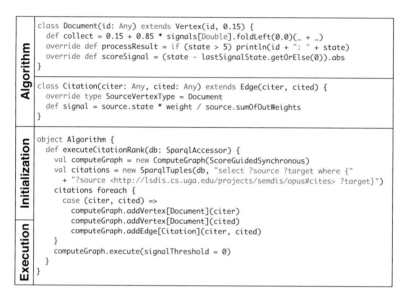

Fig. 1. Complete implementation of the PageRank algorithm on citations, including residual scoring and result processing. Written in Scala and executable as-is on the framework.

5   Evaluation

Having established the elegance and conciseness of the SIGNAL/COLLECT programming model by example in Sections 3/4 we can now turn to validate the second and third of our claims: the scalability/transparency of our parallelization framework and the ability of our programming model to exploit score-guidance and asynchronous execution to further improve performance.

5.1   Scalability

To establish the ability of the SIGNAL/COLLECT framework to transparently scale we evaluated its performance when running the single-source shortest path algorithm ("above average" score-guided asynchronously) with a varying number of worker threads on a computer with two quad-core Intel® Xeon® X5570

---

[2] http://knoesis.wright.edu/library/ontologies/swetodblp

processors (turbo boost & hyper-threading disabled to reduce confounding effects) and 72 GB RAM.

Figure 2(a) shows the average, fastest and slowest running-times over 10 executions, while Figure 2(b) shows that the performance scaled almost linearly with the number of worker threads used. Bearing the limitations discussed in Section 5.4 below we have, hence, established that *given the right algorithm and graph our programming model and framework can provide excellent scalability.*

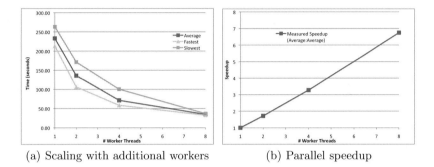

(a) Scaling with additional workers   (b) Parallel speedup

Fig. 2. Scalability of Signal/Collect: Single-source shortest path on a randomly generated graph with a log-normal distribution of out-degrees. The graph had 1 million vertices, 94 million edges and a longest path of 5. Results of 10 executions for each number of worker threads.

5.2   Score-Guided Computations

In order to evaluate the impact of guiding computations by scoring and, hence, establish its usefulness we ran PageRank (as shown in Figure 1) with and without score-guiding (residual scoring, signal_threshold=0.001) on two different graphs. The hardware we used for all further evaluations was a MacBook Pro i7 2.66 GHz (2 cores, hyper-threading enabled) with 8 GB RAM running four worker threads. Also we used a newer version of the framework than in the previous experiment. Figures 3 and 4 show the averages over 10 executions for each algorithm and the error bars indicate min/max values.

These results show that score-guided execution in general performed very well on the less densely connected citation graph, where some parts of the graph probably converged faster than others. On the densely connected generated graph the synchronous version performed comparably to the score-guided algorithms. We can conclude that given a suitable combination of algorithm and graph, *score-guidance can improve convergence significantly by focusing the computation only on the parts of the graph that still require it.*

5.3   Asynchronous vs. Synchronous

The results in Figure 4 suggest that the performance of the asynchronous version is highly dependent on the scheduling. For this combination of algorithm and

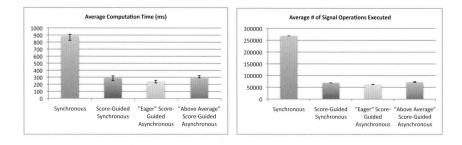

Fig. 3. PageRank on SwetoDblp citations, 22 387 vertices (publications) connected by 112 303 edges (citations)

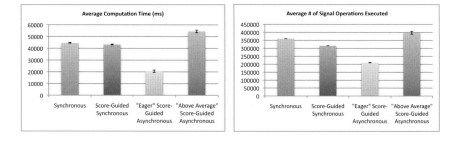

Fig. 4. PageRank on a generated graph with log-normal distributed out-degrees (drawn from $e^{\mu+\sigma N}$, where $\mu = 4$, $\sigma = 1.3$ and N is drawn from a standard normal distribution). The generated graph has 100 000 vertices connected by 1 284 495 edges.

graph the "eager" asynchronous version performed well and outperformed the synchronous approach, while the "above average" asynchronous scheduler performed poorly. Hence, more evaluations are required to determine which combinations of scheduling algorithms and graph algorithms/structures work well.

To establish that for some algorithms the asynchronous version outperforms the synchronous we ran the vertex coloring algorithm introduced in Section 3 on a generated graph. The results in Table 1 show that the asynchronous version converges quickly for some problems, where the synchronous version fails to converge (within a reasonable amount of time). Other algorithms share this property: Koller and Friedman note that some asynchronous loopy belief propagation computations converge where the synchronous computations keep oscillating. They summarize [12, p. 408]: *"In practice an asynchronous message passing scheduling works significantly better than the synchronous approach. Moreover, even greater improvements can be obtained by scheduling messages in a guided way."*

5.4 Limitations—Threats to Validity

The main limitation of the evaluations above is that our current SIGNAL/COLLECT framework only runs on a single machine using shared memory for signaling. It is not entirely clear how the overhead of signaling across the network with

Table 1. Vertex coloring on a generated graph with log-normal distributed out-degrees (drawn from $e^{\mu + \sigma N}$, where $\mu = 1$, $\sigma = 0.2$ and N is drawn from a standard normal distribution). The generated graph has 100 000 vertices connected by 554 118 edges. The table shows the average time (in milliseconds) over 10 executions it took to find a vertex coloring with the given number of colors. When the algorithm failed to converge in less than a minute (on average) the time was noted as "did not converge" (d.n.c).

Number of colors	5	6	7	8	9	10	11
"Eager" Score-Guided Asynchronous	d.n.c	12870	1690	1392	1243	1218	1046
Score-Guided Synchronous	d.n.c	d.n.c	d.n.c	d.n.c	d.n.c	2856	1876

the involved bandwidth and latency implications would impact the scalability of the prototype system. We do not expect this limitation to have an impact on the evaluations of score-guided and asynchronous execution.

In terms of scalability, we only ran our experiment on one large graph with an algorithm that has a very simple interaction pattern and many more vertices than worker threads and many edges per vertex. For a refined evaluation we need to investigate the impact of different graph structures and interaction scenarios.

Note also, that we only ran our second experiment on one algorithm. Before analyzing the impact of all important factors (algorithm, graph, scoring functions/thresholds, number of worker threads, asynchronous operation scheduling, etc.) it is difficult to make a general statement about the trade-offs involved with regard to guided vs. unguided and synchronous vs. asynchronous computations.

6   Related Work

Many general programming models for distributed computing have been presented. Most notable is the MapReduce [2] programming model, which is based on parallel operations on sets of key-value pairs. The Hadoop MapReduce framework[3] has been used by [3,4] for scalable RDFS/OWL reasoning. The big disadvantage of the MapReduce model is that it is based on key-value pairs requiring a translation of Semantic Web tasks to this abstraction. Also, the programming model was not designed with iterated executions in mind and if it is used iteratively, the model is limited to synchronous execution.

Most closely related to our programming model is Pregel [8]: a system developed by Google for large-scale graph processing. It has been shown to scale to graphs with billions of vertices/edges via distribution to thousands of commodity PCs. Its limitations are that it only handles synchronous computations, can only support graphs with one kind of vertex sharing a single "compute" function, and edges are not first class citizens. As we have seen in our evaluation, score-guided asynchronous computations are essential for some graph computations. Pregel's limitation to one vertex type makes the implementation of algorithms employing multiple kinds of vertices rather tedious.

---

[3] http://hadoop.apache.org/mapreduce

The concurrency model of the asynchronous SIGNAL/COLLECT computation was inspired by the actor formalism [13], in which many processor objects take part in a computation and can only influence each other via messaging. This bears a lot of similarity to vertices in SIGNAL/COLLECT, which do local computations and can only influence each other via signaling.

We did not find any other programming models specialized for parallel iterated computation on typed graphs (such as the Semantic Web). However, in addition to the use of generic distributed computing frameworks many have implemented their own distributed systems for Semantic Web tasks [6,7]. Weaver and Hendler [5], e.g., present an RDFS closure using MPI—a low-level message passing interface. Oren et al. [14] have implemented a distributed reasoner using their own low-level primitives. We believe that most of these solutions could profit from our generic framework.

7   Limitations, Future Work and Conclusions

In order to master the onslaught of data the Semantic Web is in dire need of distributed computation paradigms. Current paradigms either have the problem that their programming model does not lend itself naturally to the typed graph based Semantic Web computation tasks or provide only low-level functionality requiring the tedious implementation of the whole functionality for every algorithm. This paper presented a novel, distributed, and scalable computing model for typed graphs called SIGNAL/COLLECT. We showed a framework that can be used to elegantly and concisely specify and execute a number of computations that are typical for the Semantic Web fully incorporating Semantic Web techniques such as SPARQL (to initialize the graph). We also showed that the programming model allows for scalable implementations given suitable algorithms and graphs. Lastly, we showed that the support for asynchronous execution of graph algorithms enables the convergence for some algorithms that will not converge in the synchronous case.

Whilst these results are remarkable SIGNAL/COLLECT is still at its beginning. First, we need to find the limitations of the programming model. Although it is suitable for computations on graphs it is, obviously, not quite as suitable for computations on lists. Second, we need to extend the framework for distribution and explore heuristics for the distribution of vertices in the compute graph to compute nodes. This is a non-trivial problem as signals transmitted across the network will incur significant latencies compared to signals transmitted in a shared memory setting. Consequently, the algorithms need to be robust against signal latency variance. Third, for the use outside research we need to build a framework that provides typical middle-ware services (such as distributed filesystem access). We plan to investigate each of these areas in the future.

The Semantic Web is growing and so are the needs for processing its RDF-based data. Many have approached the call for processing these large-sized RDF graph data sets. Researchers have developed stores (or data bases) that scale to disk, have explored various means for computing the logical closure, and built large-scale systems. In order for large-scale processing of these data to go main-stream we need elegant programming models that allow for the concise formulation of a large amount of Semantic Web tasks. SIGNAL/COLLECT is such a programming model that, we believe, can serve the function as a general purpose Semantic Web infrastructure. As such, it has the potential to bring distributed computing transparently to the Semantic Web and become a major building block for future Semantic Web applications.

Acknowledgemement. We would like to thank Stefan Schurgast for using early prototypes of the framework and providing valuable feedback on its usage.

References

1. Moore, G.E.: Cramming more components onto integrated circuits. Electronics 38(8) (1965)
2. Dean, J., Ghemawat, S.: Mapreduce: simplified data processing on large clusters. In: Proceedings of the 6th Conference on Symposium on Opearting Systems Design & Implementation, OSDI 2004, Berkeley, CA, USA, USENIX Association, p. 10 (2004)
3. Urbani, J., Kotoulas, S., Oren, E., van Harmelen, F.: Scalable distributed reasoning using mapreduce. In: Bernstein, A., Karger, D.R., Heath, T., Feigenbaum, L., Maynard, D., Motta, E., Thirunarayan, K. (eds.) ISWC 2009. LNCS, vol. 5823, pp. 634–649. Springer, Heidelberg (2009)
4. Urbani, J., Kotoulas, S., Maassen, J., van Harmelen, F., Bal, H.E.: Owl reasoning with webpie: Calculating the closure of 100 billion triples. In: Aroyo, L., Antoniou, G., Hyvönen, E., ten Teije, A., Stuckenschmidt, H., Cabral, L., Tudorache, T. (eds.) ESWC 2010. LNCS, vol. 6088, pp. 213–227. Springer, Heidelberg (2010)
5. Weaver, J., Hendler, J.: Parallel materialization of the finite rdfs closure for hundreds of millions of triples. In: Bernstein, A., Karger, D.R., Heath, T., Feigenbaum, L., Maynard, D., Motta, E., Thirunarayan, K. (eds.) ISWC 2009. LNCS, vol. 5823, pp. 682–697. Springer, Heidelberg (2009)
6. Harth, A., Umbrich, J., Hogan, A., Decker, S.: Yars2: A federated repository for querying graph structured data from the web. In: Aberer, K., Choi, K.-S., Noy, N., Allemang, D., Lee, K.-I., Nixon, L.J.B., Golbeck, J., Mika, P., Maynard, D., Mizoguchi, R., Schreiber, G., Cudré-Mauroux, P. (eds.) ASWC 2007 and ISWC 2007. LNCS, vol. 4825, pp. 211–224. Springer, Heidelberg (2007)
7. Aberer, K., Cudré-Mauroux, P., Hauswirth, M., Pelt, T.V.: Gridvine: Building internet-scale semantic overlay networks. In: McIlraith, S.A., Plexousakis, D., van Harmelen, F. (eds.) ISWC 2004. LNCS, vol. 3298, pp. 107–121. Springer, Heidelberg (2004)
8. Malewicz, G., Austern, M.H., Bik, A.J.C., Dehnert, J.C., Horn, I., Leiser, N., Czajkowski, G.: Pregel: a system for large-scale graph processing. In: Elmagarmid, A.K., Agrawal, D. (eds.) SIGMOD Conference, pp. 135–146. ACM, New York (2010)
9. Page, L., Brin, S., Motwani, R., Winograd, T.: The PageRank citation ranking: Bringing order to the Web. Technical report, Stanford Digital Library Technologies Project (1998)
10. Bishop, C.M.: Pattern Recognition and Machine Learning (Information Science and Statistics), 1st edn. Springer, Heidelberg (October 2007)

11. Kschischang, F., Frey, B., Loeliger, H.: Factor graphs and the sum-product algorithm. IEEE Transactions on Information Theory 47(2), 498–519 (2001)
12. Koller, D., Friedman, N.: Probabilistic Graphical Models: Principles and Techniques. MIT Press, Cambridge (January 2009)
13. Hewitt, C., Bishop, P., Steiger, R.: A universal modular actor formalism for artificial intelligence. In: Proceedings of the 3rd International Joint Conference on Artificial intelligence, IJCAI 1973, pp. 235–245. Morgan Kaufmann Publishers Inc., San Francisco (1973)
14. Oren, E., Kotoulas, S., Anadiotis, G., Siebes, R., ten Teije, A., van Harmelen, F.: Marvin: Distributed reasoning over large-scale semantic web data. Web Semantics: Science, Services and Agents on the World Wide Web 7(4), 305–316 (2009); Semantic Web challenge 2008

Summary Models for Routing Keywords to Linked Data Sources

Thanh Tran, Lei Zhang, and Rudi Studer

Institute AIFB, Karlsruhe Institute of Technology, Germany
{dtr,lzh,studer}@kit.edu

Abstract. The proliferation of linked data on the Web paves the way to a new generation of applications that exploit heterogeneous data from different sources. However, because this Web of data is large and continuously evolving, it is non-trivial to identify the relevant link data sources and to express some given information needs as structured queries against these sources. In this work, we allow users to express needs in terms of simple keywords. Given the keywords, we define the problem of finding the relevant sources as the one of keyword query routing. As a solution, we present a family of summary models, which compactly represents the Web of linked data and allows to quickly find relevant sources. The proposed models capture information at different levels, representing summaries of varying granularity. They represent different trade-offs between effectiveness and efficiency. We provide a theoretical analysis of these trade-offs and also, verify them in experiments carried out in a real-world setting using more than 150 publicly available datasets.

1 Introduction

The Web is no longer only a collection of textual documents but also a *Web of linked data*. One prominent project which largely contributes to this development is the Linking Open Data project. Collectively, linked data comprises hundreds of sources containing over 13.1 billions RDF triples, which are connected by 142 millions links (November 2009, http://linkeddata.org/).

This development offers new opportunities for addressing complex information needs. Instead of documents, complex results ranging over different sources of linked data can be returned to Web users. To exploit this, users can specify complex queries using structured query languages such as SPARQL[1]. While such a query language is powerful, it requires users to know not only the query syntax and semantics but also the schema as well as the underlying data.

Problem. So far, these requirements have proven to be a large burden. Given the amount of linked data is large and continuously evolving, it is inherently difficult to know what is in there (i.e., the data and the schema) and to formulate the corresponding structured queries for addressing some given information needs. Hence, it is desirable to have a mechanism, which allows users to express

---

[1] http://www.w3.org/TR/rdf-sparql-query/

P.F. Patel-Schneider et al. (Eds.): ISWC 2010, Part I, LNCS 6496, pp. 781–797, 2010.

information needs in their own words. Another aspect of dealing with the large Web of linked data is scalability. Processing the needs against the entire Web might be too time consuming and not needed, especially when users are interested in and want to choose some particular sources of information. Processing against a relevant subset of linked data identified by the user is more scalable and possibly the only practical solution for the large Web of linked data. Concerning these problems, the question we deal with is *given the needs expressed by users as sets of keywords, are there corresponding answers in linked data and what combination of data sources shall be used to produce them?*.

Existing Work. In the Semantic Web community, there exists a large body of work on processing queries against RDF and linked data. Given structured queries, RDF stores such as RDF-3X [5] and YAR2 [3] can compute structured results and in the context of linked data query processing [2], can also identify relevant sources. They however do not apply when the information need is provided as keywords. While keyword search is supported by some Semantic Web search engines such as SWSE [1] and Sig.ma [9], they are limited to processing simple list of keywords that refer to entities. This work deals with *complex information needs*, which may involve complex results providing information about sets of entities and relations between them, i.e., result tuples that may form *graphs*. Further, the aim is not to directly compute results but to quickly identify and let users and system focus on the combination of sources that produce non-empty results.

To this end, work in information retrieval (IR) and database research dealing with keyword search constitutes the starting point. Keyword search has become the most widely used IR paradigm on the Web, enabling lay users without knowledge of the schema and data to search for a priori unknown documents. This kind of schema-agnostic search is not limited to textual data but can also be used for querying structured data. In database research, solutions have been proposed that allow for the retrieval of the most relevant, possibly graph-structured results [4,6,7]. Unlike IR approaches, which consider only keywords for finding matching documents (we called *keyword coverage*), database approaches also use structure information. Possible join sequences in the data are explored to ensure that matching result tuples not only "contain" the keywords but also, represent meaningful connections between these keywords (called *structure coverage*). Given the data graph in Fig. 1a and the query "Stanford, John, Award" for instance, an IR-style approach might return none (AND-semantics) or all the entities $uni1, per2, ...$ in the graph because they all partially match the keyword query (OR-semantics) whereas the DB approach would return the subgraph that connects $uni1$ with $per2, per1, per3$ and $prize1$. However, computing complex results in this way is expensive, especially in a multi-source setting like the linked data Web. Authors of state-of-the-art work explicitly considered only the setting where "number of databases that can be dealt with is up to the tens" [6].

Thus, database researchers started to look at a problem we consider most related, namely the of finding the single most relevant databases [11,10]. They recognized the fact that the computational complexity resulting from a large-scale setting can be partially addressed when allowing users to choose and retrieve answers from only some particular databases. Given a set of keywords, the goal is to find and rank the single most relevant databases that contain the answers. Following this line, we propose specific solutions for the linked data context. The differences to this work called *database selection* will be discussed in detail throughout the paper.

Contributions. While existing approaches select single databases, we deal with the Web of linked data where results are not bounded by a single source but may encompass several linked data sources. Instead of computing the most relevant single sources, we extend the work in [11,10] to compute the most relevant combinations of sources. The goal is to produce *keyword routing plans* which capture combinations of sources that contain non-empty results. This novel *keyword query routing* problem raises additional challenges. Most notably, query keywords may be covered by several linked sources, resulting in a large search space. The size of this search space grow exponentially with the number of sources and their associated links. Targeting this problem of *scale*, we report the following contributions in this paper:

- We propose *solutions for keyword query routing* which enable the exploitation of linked data. Without putting any burden on the users, this kind of approaches help to find relevant sources containing complex answers to ad-hoc information needs in the large and evolving Web of linked data.
- We propose a *multi-level relationship graph* to capture the search space of the keyword query routing problem. Based on this, we elaborate on *a family of summary models*, which compactly represent the Web of linked data. These models capture information at different levels, representing summaries of different granularities. In a theoretical analysis, we prove that finer grained models can improve the result quality. This however, comes at the expense of higher complexity. Thus, the models represent different trade-offs between effectiveness and efficiency.
- In the experiments, we investigate these trade-offs by analyzing the precision and the processing time needed using different models. The experiments were carried out in a real-world setting using more than 150 publicly available datasets, and an open-source implementation we made available at http://code.google.com/p/rdfstores/. Results of using summaries are promising. While the "best" one shall be determined w.r.t a concrete application, there is one model that seems to represent the most practical trade-off: the D-KERG model, which summarizes elements according to sources, produces results in less than 10ms, out of which every second is a valid one.

Outline. Section 2 introduces the readers to the concepts of linked data and keyword query routing. The search space and its summary models are presented in Section 3. Strategies for computing routing plans using these models and ramifications for result quality and performance are discussed in Section 4. Evaluation results are provided in Section 5 before we conclude in Section 6.

2   Preliminary

In this section, we discuss the underlying data and problem.

2.1   Web of Linked Data

Linked data can be conceived as a set of data graphs, each represents a particular source. As a working definition, we present a simple graph-based model of linked data called the *Web graph*. In that model, we distinguish between the *Web data graph* representing relationships between individual data elements, the *Web schema graph*, which captures information about group of elements, and the *Web source graph* that contains information at the level of data sources.

Definition 1 (Web Graph). *The Web of linked data is modeled as a* Web Graph $\mathcal{W}^*(\mathcal{G}^*, \mathcal{M}^*, \mathcal{N}^*, \mathcal{E}^*)$ *where* $\mathcal{G}^*$ *denotes the set of* data graphs, $\mathcal{M}^*$ *is the set of edges also called mappings or* links, *which establish connections between elements of two different graphs,* $\mathcal{N}^*$ *is the set of all nodes and* $\mathcal{E}^*$ *is the set of all edges, i.e.,* $\mathcal{G}^* = \{g_1(\mathcal{N}^*{}_1, \mathcal{E}^*{}_1), g_2(\mathcal{N}^*{}_2, \mathcal{E}^*{}_2), \ldots, g_n(\mathcal{N}^*{}_n, \mathcal{E}^*{}_n)\}$, $\mathcal{N}^* = \bigcup_{i=1}^{n} \mathcal{N}^*{}_i$, $\mathcal{M}^* = \{m(n_i, n_j) | n_i \in \mathcal{N}^*{}_i, n_j \in \mathcal{N}^*{}_j, \mathcal{N}^*{}_i, \mathcal{N}^*{}_j \subseteq \mathcal{N}^*, i \neq j\}$ *and* $\mathcal{E}^* = \bigcup_{i=1}^{n} \mathcal{E}^*{}_i \cup \mathcal{M}^*$. *We use* $\mathcal{W}(\mathcal{G}, \mathcal{M}, \mathcal{N}, \mathcal{E})$ *to distinguish the* Web data graph *from the* Web schema graph $\mathcal{W}'(\mathcal{G}', \mathcal{M}', \mathcal{N}', \mathcal{E}')$ *and the* Web source graph $\mathcal{W}''(\mathcal{N}'', \mathcal{E}'')$. *We have* $n \in \mathcal{N}$ *representing a data element,* $n' \in \mathcal{N}'$ *stands for a group of elements, and* $n'' \in \mathcal{N}''$ *denotes a data source. For simplicity, we use* $n \in n'$ *to denote that an element* n *belongs to the group* n' *and* $n, n' \in n''$ *to assert the element* n *and the group* n' *belongs to the source* n''. *Elements in* $\mathcal{N}$ *and* $\mathcal{N}'$ *are labeled, i.e., there is a function* $label : \mathcal{N} \cup \mathcal{N}' \mapsto 2^{\mathcal{V}}$ *that associates an element with a set of labels drawn from* $\mathcal{V}$, *the vocabulary of words. We have* $m(n'_i, n'_j) \in \mathcal{M}'$ *iff there is* $m(n_i, n_j) \in \mathcal{M}$ *where* $n_i \in n'_i$ *and* $n_j \in n'_j$. *Analogously,* $e(n''_i, n''_j) \in \mathcal{E}''$ *iff there is* $m(n'_i, n'_j) \in \mathcal{M}'$ *where* $n'_i \in n''_i$ *and* $n'_j \in n''_j$. *We use the Web graph* $\mathcal{W}^*(\mathcal{G}^*, \mathcal{M}^*, \mathcal{N}^*, \mathcal{E}^*)$ *to refer to the union set of elements of the Web data graph, the Web schema graph and the Web source graph.*

This is a simple model of linked data that omits details not necessary for this work. In particular, data elements may correspond to RDF resources, blank nodes or literals. Schema elements might stand for classes or data types. For keyword query routing, these distinctions are not relevant but the fact that the elements can be recognized via their labels. While different kinds of links can be established, the ones frequently found are *sameAs* links, which denote that two RDF resources or two classes are the same. There is also no need to distinguish the types of links. Only the fact that sources can be reached via some kinds of link $m \in \mathcal{M}^*$ matters. An example of this model is illustrated in Figs. 1.

2.2   Keyword Query Routing

Given the need expressed as keywords, we aim to identify sources containing results. A DB-style result to a keyword query is typically a Steiner graph, which in the linked data scenario, may combine data from several sources:

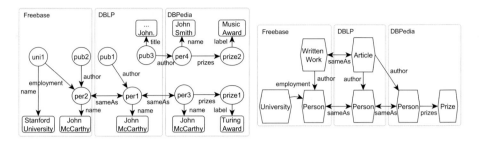

Fig. 1. (a) A Web data graph (left) and (b) its Web schema graph (right)

Definition 2 (Keyword Query Result). *A Web data graph $W(\mathcal{G}, \mathcal{M}, \mathcal{N}, \mathcal{E})$ contains a result for a query $\mathcal{K} = \{k_1, k_2, \ldots, k_{|\mathcal{K}|}\}$ if there is subgraph also called* Steiner graph $W_{\mathcal{K}}(\mathcal{G}_{\mathcal{K}}, \mathcal{M}_{\mathcal{K}}, \mathcal{N}_{\mathcal{K}}, \mathcal{E}_{\mathcal{K}})$*, where for all $k_i \in \mathcal{K}$, there is an $n_{\mathcal{K}}^{\mathcal{M}} \in \mathcal{N}_{\mathcal{K}}^{\mathcal{M}} \subseteq \mathcal{N}_{\mathcal{K}} \subseteq \mathcal{N}$ with a label that matches k_i ($\mathcal{N}_{\mathcal{K}}^{\mathcal{M}}$ is called the set of* keyword elements*), and there is path $n_i \rightsquigarrow n_j$ for all $n_i, n_j \in \mathcal{N}_{\mathcal{K}}^{\mathcal{M}}$. In a d-max* Steiner graph, *the length of the paths $n_i \rightsquigarrow n_j$ is d-max or less.*

Typical for keyword search is the pragmatic assumption that users are only interested in compact results such that a threshold d_{max} can be used to constrain the connections to be considered. Thus, instead of general Steiner graphs, keyword search solutions proposed so far and the work presented here consider *d-max Steiner graphs* as results. For our example query "Stanford, John, Award", we have $\mathcal{N}_{\mathcal{K}}^{\mathcal{M}} = \{uni1, per2, per1, per3, prize1\}$; the subgraph that connects these keyword elements is a 1-Steiner graph because the maximum distance between keyword elements is 1; and since there are no other elements between keyword elements, $\mathcal{N}_{\mathcal{K}}^{\mathcal{M}} = \mathcal{N}_{\mathcal{K}}$.

Definition 3 (Keyword Routing Plan). *Given the Web data graph $W = (\mathcal{G}, \mathcal{M}, \mathcal{N}, \mathcal{E})$ and a set of keyword queries $\mathcal{SK}$, the mapping $\mu : \mathcal{SK} \mapsto 2^{\mathcal{G}}$ that associates a query with a set of data graphs is called a* keyword routing plan $\mathcal{RP}$. *A plan $\mathcal{RP} = \{g_1, \ldots, g_{|\mathcal{K}|}\}$ for a query $\mathcal{K} \in \mathcal{SK}$ is considered* valid *when there is a combination of data graphs $g_i \in \mathcal{RP}$ that produces non-empty results for $\mathcal{K}$.*

A valid plan in our example is $\mathcal{RP} = \{Freebase, DBLP, DBPedia\}$. Note that validity does not imply relevance. That is, a valid plan ensures that results can be produced, but for the users, these results may differ in relevance. A proper account of relevance and the ranking of routing plans based on the relevance of their results go beyond the scope of this paper, which is focused on *efficiency* aspects of computing valid plans. We assume a fixed ranking function, which equally applies to all summaries discussed in this paper. We refer the interested readers to our report [8], which discusses relevance and the ranking function.

3   Summary Models for Keyword Query Routing

We now discuss the most related work in detail and introduce the models we use for keyword query routing.

3.1   Keyword Query Routing Search Space

For database selection, the search space is composed of a set of databases. The idea behind previous work [11,10] is to model every database using a keyword relationship model. A keyword relationship $\langle k_i, k_j \rangle$ is a pair of keywords, which can be connected via a sequence of join operations, i.e., there exists two data elements $n_i \rightsquigarrow n_j$ that contain k_i and k_j. For instance, $\langle Stanford, Award \rangle$ is a keyword relationship because there is a path between FB:$uni1$ and DBP:$prize1$ in Fig. 1a. The state-of-the-art [10] employs a keyword relationship graph (KRG), with keywords being nodes and keyword relationships being edges. A database is relevant when all pairs of query keywords match some edges of the KRG.

In our example, we have the keyword pairs $(Stanford, John)$, $(Stanford, Award)$ and $(John, Award)$. It is clear that when using keyword relationships in every source to form separate KRGs, none of them matches all the 3 keyword pairs. To match the pair $(Stanford, Award)$, relationships across sources from $Freebase$ to $DBPedia$ have to be incorporated into in the model. In keyword query routing, the search space does not comprise single databases but constitutes one integrated Web data graph. Instead of computing a set of summary models, this problem requires the construction of one *integrated summary model*. It shall allow for answers capturing relationships across sources. Thus, not only single sources but also combinations of sources might be relevant. Another aspect not addressed by current work is efficiency. Instead of capturing all possible relationships, we aim to use a more *compact* representations of the search space.

We conceive the search space as a *multi-level inter-relationship graph* (MIRG), as illustrated in Fig. 2. For clarity, this figure does not show the labels and also, omits some data and schema elements of our running example. At the lowest level, it models relationships between keywords. In the upper-levels, there is the Web data graph $\mathcal{W}$ followed by $\mathcal{W}'$ and $\mathcal{W}''$. Elements and relationships at the upper level represent sets of elements and sets of relationships at the lower level: a node at the source level represents a set of schema elements; every schema node represents a set of data elements; and every data element n is composed of a set of keywords K. We say $k \in K$ is mentioned in n, denoted $mentionedIn(k, n)$.

3.2   Summary Models

Thus, MIRG provides different perspectives on the search space and different views on the data. The lower levels capture more fine-grained views of the data. In order to extend the KRG [10] to deal with keyword query routing, the keyword level and keyword relationships at this level that also capture links between sources have to be taken into account. We will now discuss such an extension of

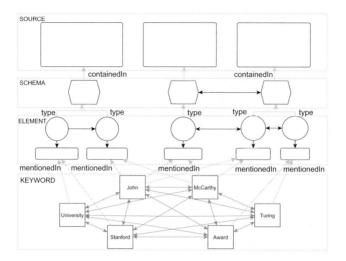

Fig. 2. Multi-level inter-relationship graph

the KRG, and introduce further summary models that capture relationships at different levels of granularity. Examples of the models are shown in Fig. 3.

Definition 4 (Keyword Sets). *The* keyword sets *(KS) of a Web graph* $\mathcal{W}^*(\mathcal{G}^*, \mathcal{M}^*, \mathcal{N}^*, \mathcal{E}^*)$ *is* $\mathcal{W}_{\mathcal{K}}^{KS} = \mathcal{N}_{\mathcal{K}}^{KS}$, *where* $\mathcal{N}_{\mathcal{K}}^{KS}$ *stands for all the keywords that are mentioned in elements of the graphs* $\mathcal{G}^*$. *Every* $n_k^{KS} \in \mathcal{N}_{\mathcal{K}}^{KS}$ *is in fact a tuple* $(k, \mathcal{G}_k)$ *that represents a keyword* k *and the graphs* $\mathcal{G}_k \subset \mathcal{G}^*$ *mentioning* k.

This is a simple model that contains only keywords but no relationships between them. It captures all nodes at the keyword level of MIRG.

Definition 5 (Element-level KERG). *An* element-level keyword-element relationship graph *(E-KERG) of a Web graph* $\mathcal{W}^*(\mathcal{G}^*, \mathcal{M}^*, \mathcal{N}^*, \mathcal{E}^*)$ *is a tuple* $\mathcal{W}_{\mathcal{K}} = (\mathcal{N}_{\mathcal{K}}, \mathcal{E}_{\mathcal{K}})$. *Every keyword-element* $n_{\mathcal{K}} \in \mathcal{N}_{\mathcal{K}}$ *is a tuple* $(n, g, \mathcal{K})$ *where* $n \in \mathcal{N}$ *is the corresponding element node it represents,* $g \in \mathcal{G} \subset \mathcal{G}^*$ *is the data graph containing* n, *and* $\mathcal{K}$ *is the set of all keywords that are mentioned in* n, *i.e.,* $K = \{k | mentionedIn(k, n)\}$. *There is a relationship* $e_{\mathcal{K}} = (\langle k_i, n_{\mathcal{K}_i}(n_i, g_i, K_i) \rangle, \langle k_j, n_{\mathcal{K}_j}(n_j, g_j, K_j) \rangle) \in \mathcal{E}_{\mathcal{K}}$, *iff* $mentionedIn(k_i, n_i)$, $mentionedIn(k_j, n_j)$, *and* $n_i \leftrightsquigarrow n_j$.

This can be seen as an extension of the KRG because it captures all keywords and relationships. As shown in Fig. 3a, it also represents the data elements in which the keywords are mentioned. Hence, we use "keyword-element" to make clear that a node captures both the data element and its keywords. This model captures elements at the keyword and element level of the MIRG.

Definition 6 (Schema-level KERG). *A* schema-level keyword-element relationship graph *(S-KERG) is a tuple* $\mathcal{W}'_{\mathcal{K}} = (\mathcal{N}'_{\mathcal{K}}, \mathcal{E}'_{\mathcal{K}})$. *It captures elements*

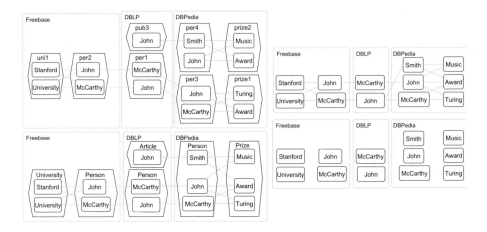

Fig. 3. (a)The 1-E-KERG (top left), (b) the 1-S-KERG (bottom left), (c) the 1-D-KERG (top right) and the (d) KS for our running example.

at the keyword and schema level of the MIRG. For a keyword-element node $n'_\mathcal{K}(n', g, \mathcal{K}) \in \mathcal{N}'_\mathcal{K}$, we have $n' \in \mathcal{N}'$ being a schema-level node, $g \in \mathcal{G}' \subset \mathcal{G}^*$ is the schema graph containing n', and $\mathcal{K}$ comprises keywords that are mentioned in the elements $n \in n'$, i.e., $\mathcal{K} = \{k|n \in n', mentionedIn(k, n)\}$. There is a relationship $e'_\mathcal{K} = (\langle k_i, n'_{\mathcal{K}_i}(n'_i, g_i, K_i)\rangle, \langle k_j, n'_{\mathcal{K}_j}(n'_j, g_j, K_j)\rangle) \in \mathcal{E}'_\mathcal{K}$, iff $mentionedIn(k_i, n_i)$, $mentionedIn(k_j, n_j)$, $n_i \in n'_i$, $n_j \in n'_j$, and $n_i \rightsquigarrow n_j$.

As opposed to E-KERG, this one is indeed a summary model because it clusters two element-level relationships $(\langle k_i, n_{\mathcal{K}_i}(n_i, g_i, K_i)\rangle, \langle k_j, n_{\mathcal{K}_j}(n_j, g_j, K_j)\rangle)$ and $(\langle k_v, n_{\mathcal{K}_v}(n_v, g_v, K_v)\rangle, \langle k_w, n_{\mathcal{K}_w}(n_w, g_w, K_w)\rangle)$ to one schema-level relationship when they capture the same keyword relationships (i.e., $k_i = k_v$ and $k_j = k_w$) between the same classes (i.e, $n'_i = n'_v$ and $n'_j = n'_w$). For instance, $(\langle John, (Person, DBPedia, \{Smith, John, McCarthy\})\rangle, \langle Award, (Prize, DBPedia, \{Music, Award, Turing\})\rangle)$ in Fig. 3b is an aggregation of the relationships $(\langle John, (per4, DBPedia, \{Smith, John\})\rangle, \langle Award, (prize2, DBPedia, \{Music, Award\})\rangle)$ and $(\langle John, (per3, DBPedia, \{McCarthy, John\})\rangle, \langle Award, (prize1, DBPedia, \{Turing, Award\})\rangle)$ in Fig. 3a. These E-KERG relationships are aggregated because they represent the same relationships $(John, Award)$ between the classes $(Person, Prize)$.

Definition 7 (Source-level KERG). *A source-level keyword-element relationship graph (D-KERG) is a tuple $\mathcal{W}''_\mathcal{K} = (\mathcal{N}''_\mathcal{K}, \mathcal{E}''_\mathcal{K})$. For a keyword-element node $n''_\mathcal{K}(n'', \mathcal{K}) \in \mathcal{N}''_\mathcal{K}$, we have $n'' \in \mathcal{N}''$ being a source-level node, i.e., a graph, and $\mathcal{K}$ is the set of all keywords that are mentioned in elements of the graph n''. There is a relationship $e''_\mathcal{K} = (\langle k_i, n''_{\mathcal{K}_i}(n''_i, K_i)\rangle, k_j, n''_{\mathcal{K}_j}(n''_j, K_j)\rangle) \in \mathcal{E}''_\mathcal{K}$, iff k_i is mentioned in some elements n_i of the graph n''_i, k_j mentioned in some elements n_j of the graph n''_j, and $n_i \rightsquigarrow n_j$.*

Thus, this model is conceptually similar to S-KERG but aggregate elements at the level of sources. It combines schema-level relationships when they capture the same keyword relationships between the same sources. As shown in Fig. 3b, there are only distinct keyword relationships in S-KERG. Thus, no further aggregation is needed in this case.

As keyword search results, we consider d-max Steiner graphs where paths between keyword elements are of length d_{max} or less. Accordingly, we actually employ a d_{max}-$KERG$ versions where the maximum distance to be considered between n_i and n_j is d_{max} ($n_i \leftrightsquigarrow^{d_{max}} n_j$). Note that the summaries illustrated in Figs. 3 resemble the structure of the underlying data and schema graphs because relationships in the summaries in fact correspond to graph edges, i.e., only paths with length 1 are considered such that $d_{max} = 1$ (1-KERG models). Clearly, a higher value for d_{max} would result in a blowup of paths. In particular, the E-KERG model would contain much more relationships than there are edges in the data graph. Hence, summarizing relationships is essential for efficient keyword query routing.

3.3   Computing Summary Models

The computation of d_{max}-KERG models is performed in three steps. Firstly, the relationships between entities are computed for various distances within a threshold d_{max}. Then, connected term pairs are extracted based on the computed relationships. They are used for computing E-KERG. For computing S-KERG and D-KERG, term pairs are further grouped according to schema and source-level elements, respectively.

All information are finally stored in an specialized index that enables the lookup of keyword-element relationships, given a pair of keywords. In particular, for (k_i, k_j), we have (1) I_{E-KERG} returning the relationships $e_\mathcal{K} = (\langle k_i, n_{\mathcal{K}_i} \rangle, \langle k_j, n_{\mathcal{K}_j} \rangle)$, (2) I_{S-KERG} returning $e'_\mathcal{K} = (\langle k_i, n'_{\mathcal{K}_i} \rangle, \langle k_j, n'_{\mathcal{K}_j} \rangle)$ and (3) I_{D-KERG} returning $e''_\mathcal{K} = (\langle k_i, n''_{\mathcal{K}_i} \rangle, \langle k_j, n''_{\mathcal{K}_j} \rangle)$. Also, we construct a KS model based on keywords extracted from the data graphs and build the index (4) I_{KS}, which returns the elements n_k^{KS}, given the keyword k.

4   Computing Keyword Routing Plans

For computing valid query routing plans, the idea behind existing work on keyword search [4,6,7] and database selection [11,10] applies: we search for Steiner graphs to discover sources that produce answers. Recall that a Steiner graph is basically a graph that connects keyword elements. The existence of such a graph indicates that there are answers to the keyword query. In our approach, the search is not performed directly on the Web data graph but on the summary models. Specifically, we search for Steiner graphs in either (1) $\mathcal{W}_\mathcal{K}^{ks}$, (2) $\mathcal{W}_\mathcal{K}$, (3) $\mathcal{W}'_\mathcal{K}$ or (4) $\mathcal{W}''_\mathcal{K}$. Since KS (1) do not capture relationships, the results that can be derived from it do not completely adhere to the notion of Steiner graph. Also for the KERG models (2-4), Steiner graphs that can be computed are different in granularities. We will now elaborate on strategies for searching Steiner graphs using different summaries.

4.1   Routing Plan Computation Using KS

Using the KS model and its index, routing plans can be computed as follows:

- Given the keyword query $\mathcal{K} = \{k_1, k_2, \ldots, k_i\}$, retrieve the elements $n_{k_i}^{KS}(k_i, \mathcal{G}_{k_i})$ for every $k_i \in \mathcal{K}$ using the I_{KS} index.
- For every $n_{k_i}^{KS}$ retrieved before, put the sources $\mathcal{G}_{k_i}$ that is associated with $n_{k_i}^{KS}$ into the set of relevant sources $\mathcal{G}_{\mathcal{K}}$.
- Compute all $|\mathcal{K}|$-combinations for the set $\mathcal{G}_{\mathcal{K}}$.
- Output these combinations as the set of routing plans $\mathcal{SRP}$.

Intuitively speaking, this procedure simply retrieves sources that cover the keywords and in order to cover all $|\mathcal{K}|$ query keywords, it uses $|\mathcal{K}|$-combinations of these sources as routing plans.

4.2   Routing Plan Computation Using KERGs

Since KERG models capture relationships, we retrieve data for pair of keywords (k_i, k_j), instead of single keywords. Retrieved data are joined to compute Steiner graphs. It is necessary to ensure that all keyword elements in a Steiner graph are pairwise connected through a path of length d_{max} or less. Thus, it is necessary to join all possible keyword pairs. Given a query with three keywords k_1, k_2, k_3 for instance, we need to retrieve keyword elements and perform the joins $(k_1, k_2) \bowtie_{k_2} (k_2, k_3) \bowtie_{k_3, k_1} (k_1, k_3)$ to verify that (1) the elements n_2 matching k_2 are connected with both n_1 that match k_1 and n_3 that match k_3 over a distance of d_{max} or less (by means of the first join $((k_1, k_2) \bowtie_{k_2} (k_2, k_3)$ on k_2), (2) the elements n_3 just found to be connected with n_2, are also connected with n_1 (by means of the second join on k_3), and (3) the n_1 found to be connected with n_2, is also connected with n_3 (by means of the third join on k_1). The complete procedure can be summarized as follows:

- Given the keyword query $\mathcal{K} = \{k_1, k_2, \ldots, k_i\}$, compute all 2-combinations of $\mathcal{K}$ to get all possible keyword pairs, resulting in a total of $N = |K|(|K| - 1)/2$ different pairs. Subsequently, retrieve relationships for these pairs and perform joins according to a random[2] or alternatively, optimized order.
- In particular, inputs for every keyword pair are obtained using the underlying index. Given (k_1, k_2) and I_{E-KERG} for instance, relationships of the form $e_\mathcal{K} = (\langle k_1, n_{\mathcal{K}_1}(n_1, g_1, K_1)\rangle, \langle k_2, n_{\mathcal{K}_2}(n_2, g_2, K_2)\rangle)$ are retrieved. Joining this with the next inputs retrieved for (k_2, k_3) for instance, ensures that n_2 is connected with both n_1 and n_3.
- Processing the entire join sequence of keyword pairs yields a set of graphs. Depending on the underlying summary model, these graphs capture Steiner graphs at the source, schema or element level.

---

[2] For this work, we omit the aspect of join order optimization and simply generate a random order for joining keyword pairs.

– Some resulting graphs might be indistinguishable in terms of the sources and connections between sources they represent. Keep only one of those because the other does not contain additional information.
– Extract sources associated with elements of the graphs to obtain combination of sources, i.e., the routing plans $\mathcal{RP}$.

This procedure is the same for all KERGs. Given that the underlying data contain results, we provide proofs in the report [8] to show that applying this procedure on the S-KERG summary will yield routing plans, i.e., when Steiner graphs can be found for $\mathcal{K}$ in the data, then there will be corresponding graphs that can be found in the summary. Thus, given $\mathcal{K}$, the procedure will output a non-empty set of $\mathcal{RP}$ if $\mathcal{W}$ contains a result for $\mathcal{K}$. In the same manner, it is straightforward to show that E-KERG and D-KERG can provide this guarantee. However, we show formally in [8] that the other way around is not true, i.e., the graphs derived from the summary are not necessarily valid such that there might be no corresponding Steiner graph in the data. Thus, the fact that a routing plan can be derived from the summaries does not guarantee there exists a result for $\mathcal{K}$. This formal result is interesting because it makes clear that while the use of summaries might be required to obtain the desired performance, it has consequences on the result validity. In particular, it implies that the more compact the summary, the more likely that plans computed from it are not valid. We will now discuss the intuition behind this formal result.

4.3 Result Validity

A graph derived from a summary does not always have a corresponding Steiner graph in the data, unless we use E-KERG. This model makes a difference because it in fact captures all nodes and paths in the Web data graph. In particular, when a d_{max}^{sum}-E-KERG is used, d_{max}^{data}-Steiner graphs are keyword query answers, and $d_{max}^{sum} = d_{max}^{data}$, then E-KERG captures all the paths in the data that are relevant for Steiner graph computation, i.e., all paths up to length d_{max}^{sum}. For every path $n_i \leadsto^{d_{max}} n_j$ in the data, there is a one-to-one corresponding relationship $e_{\mathcal{K}} = (\langle k_i, n_{\mathcal{K}_i}(n_i, g_i, K_i)\rangle, \langle k_j, n_{\mathcal{K}_j}(n_j, g_j, K_j)\rangle)$ in E-KERG. This one-to-one correspondence of paths constitutes the base argument, which can be extended inductively to show that there is also a one-to-one correspondence of graphs such that a graph derived from E-KERG always has a corresponding Steiner graph in the data, and vice versa.

A S-KERG however, combines two edges $e(n_{i_1}, n_{j_1})$ and $e(n_{i_2}, n_{j_2})$ (paths) in the data graph to one single relationship $e'_{\mathcal{K}} = (\langle k_i, n'_{\mathcal{K}_i}(n'_i, g_i, K_i)\rangle, \langle k_j, n'_{\mathcal{K}_j}(n'_j, g_j, K_j)\rangle)$ iff $n_{i_1}, n_{i_2} \in n'_i$, $n_{j_1}, n_{j_2} \in n'_j$, n_{i_1}, n_{i_2} mention k_i, and n_{j_1}, n_{j_2} mention k_j. Thus, for an element in the data, there is always a counterpart in S-KERG, which is however a grouping of elements, constituting a one-to-many correspondence. Through this grouping, we loose detailed information about elements in the group. That is, for the pair of keyword (k_i, k_j), S-KERG captures the corresponding connection from the element group n'_i to n'_j but can no longer tell for instance, whether this represents a connection between n_{i_1} to n_{j_1} or n_{i_1} to n_{j_2}. In other words, it can be inferred from S-KERG that n_{i_1} is connected with n_{j_2} even though such a connection does not exist in the data. With respect to our example, a graph can be derived from S-KERG that covers the keywords *Stanford*, *John* and *Music*. It is clear from Fig. 3a that there is no Steiner graph corresponding to this. The problem here is that S-KERG does not distinguish the John McCarthy connected with "Stanford" from the John Smith connected with "Music". Thus, it incorrectly infers the connection "Stanford" and "Music".

The same arguments can be applied to D-KERG. The difference is that the grouping in D-KERG is even more coarse-grained. Two edges $e(n_{i_1}, n_{j_1})$ $e(n_{i_2}, n_{j_2})$ are aggregated to one single relationship in D-KERG, when n_{i_1} and n_{i_2} mention the same keyword k_i, n_{j_1} and n_{j_2} mention k_j, and they belong to the same data source, i.e., $n_{i_1}, n_{i_2} \in n_i''$ and $n_{j_1}, n_{j_2} \in n_j''$. Note that with S-KERG, the incorrect inference mentioned before would not occur when John Smith is in a different class than John McCarthy. They would not have been aggregated to one single node in S-KERG. This however happens with D-KERG. No matter the classes they belong to, these elements would be aggregated to one single node when they are in the same data source. With respect to our example, D-KERG makes one additional false inference: it does not distinguish the person "John" connected with "Stanford" from yet another "John" connected with "Music", which is an article (see that John as an article and John as a person in Fig. 3b is aggregated to one element in Fig. 3c).

Compared to the KERG models, KS does not capture relationships between keywords at all. Given two keywords k_i, k_j, the sources which cover these keywords can be derived from KS, e.g. the graphs n_i'', n_j''. However, this does not imply there exist two elements $n_i \in n_i''$ and $n_j \in n_j''$, and $n_i \leftrightsquigarrow n_j$. More generally, a combination of sources derived from KS covers all keywords but does not ensure that elements matching these keywords are connected, and thus, does not necessarily correspond to a Steiner graph.

In summary, the percentage of valid plans for D-KERG is less or equal that for S-KERG, which in turn is less or equal that for E-KERG. When d_{max}^{sum} value of E-KERG is sufficiently large to cover all paths relevant for Steiner graph computation, i.e., $d_{max}^{sum} = d_{max}^{data}$, this percentage is 100 for E-KERG. By chance, the percentage of valid plans for KS might be higher than that for the summary models but in general, is expected to be less (because relationships between elements are not considered).

4.4 Complexity

Using KS, complexity is $O(input_{max}^{|K|-1})$, where $input_{max}$ denotes the largest number of elements that can be obtained for a keyword k_i. This is because for computing the combination of sources for a 2-keyword query $K = \{k_i, k_j\}$, we have to union every element retrieved for k_i with every other retrieved for k_j (Cartesian product), thus requiring $|input_i| \times |input_j|$ time and space. For queries with $|K|$ keywords, we have to combine elements retrieved for one keyword k_i

with elements retrieved for every other keyword $k_j \in K, k_j \neq k_i$. Thus, $|K| - 1$ combinations of input sets of maximum size $input_{max}$ have to be performed.

With KERG models, retrieved elements have to be joined. While in practice, this operation can be performed more efficiently using special indexes and join implementation, this operation in worst case, also requires $|input_i| \times |input_j|$ time and space. Inputs are retrieved not for every k_i but for all possible pair of keywords. This results in complexity $O(input_{max}^{C(\mathcal{K},2)-1})$, where $input_{max}$ here refers to the largest number of relationships that can be obtained for a keyword pair, and $C(\mathcal{K}, 2)$ is the number of 2-combinations of the set $\mathcal{K}$, denoting the number of joins that have to be processed.

While the number of operations are same for all KERG models, the size of $input_{max}$ varies. Clearly, the more coarse-grained the grouping, i.e., the higher the number of elements aggregated to one group at the summary level, the smaller will be $input_{max}$. In particular, we have $input_{max}(D\text{-}KERG) \leq input_{max}(S\text{-}KERG) \leq input_{max}(E\text{-}KERG)$. How much smaller a KERG summary is compared to one other depends on the data. In the extreme case where every data element mentions only distinct terms, i.e., does not share terms with one other, all KERG models are actually equal in size.

While KERG models require joins on input sets to be performed $C(\mathcal{K}, 2) - 1$ times, KS only needs $|K| - 1$ combinations of input sets. However, the advantage of using KERG is that the size of the input sets that have to be processed is expected to be smaller. This is not only due to the effect of summarization. For all KERG models, inputs are retrieved for keyword pair while for KS, inputs are retrieved using single keywords. Two keywords are more selective than one keyword, thus more likely result in smaller input.

5   Evaluation

We implemented our approach for keyword query routing in Java using JDK 1.6 on top of MySQL 5.1. The experiments were conducted on a commodity PC with 2.5GHz Intel Core, 4GB of RAM and 500GB HDD SATA II 7200rpm, running on Windows 7. As discussed, while KRG [10] is limited to the problem of database selection, E-KERG can be seen as an extension that captures the ideas behind KRG. KS represents a naive baseline. The goal was to assess the performance of routing plan computation and the validity of results that can be achieved with S-KERG and D-KERG, compared to the baselines E-KERG and KS.

5.1   Data Preprocessing

We employed a chunk of RDF data part of the Billion Triple Challenge dataset[3]. It contains about 10M RDF triples that are from 154 different data sources, linked via 500K mappings.

---

[3] http://vmlion25.deri.ie/index.html

In total, the number of distinct terms extracted from all sources was 121,434. We measured the number of elements in KS and the number of KERG relationships. This was done for different settings of d_{max} to investigate the changes in the number of relationships as longer distances are considered. KS contains 804,528 elements. For $d_{max} = 0, 1, 2, 3, 4$, E-KERG contains 2.4M, 7.7M, 364M, 616M and 889M relationships, S-KERG contains 1.8M, 5.1M, 144M, 215M and 312M relationships and D-KERG contains 1.7M, 4.7M, 141M, 203M and 279M relationships. Clearly, there were more relationships in KERG models than elements in KS. The number of relationships increases with d_{max}. The increase was particularly sharp (one order of magnitude) when changing d_{max} from 1 to 2.

Similar results were obtained for index size. The E-KERG index was the largest. As an average over different settings for d_{max}, S-KERG was about 36%, D-KERG was about 32% and KS was less than 1% the size of E-KERG. For $d_{max} = 2$ for instance, the sizes for E-KERG, S-KERG, D-KERG and KS were 8694MB, 3438MB, 3279MB and 22 MB, respectively.

Larger indexes required more building times. The times for building the S-KERG indexes for $d_{max} = 4, 3, 2, 1$ for instance, were 846 Min, 583 Min, 339 Min and 27 Min, respectively.

Our report [8] provides a breakdown of the results into 6 categories of datasets that vary in size. According to these results, both index size and building time increased with the size of the dataset. However, there is no strict correlation because there are cases where relatively small datasets resulted in large indexes. Rather, structural density was the dominant factor. Large index and high building costs were obtained for datasets which exhibit large number of links to other datasets, and contain nodes with large in- and outdegree.

5.2   Query Processing

For the experiment, we used a set of 30 keyword queries. All queries are valid, i.e., they produce non-empty keyword answers (4-Steiner graphs to be precise). For each query, at least two data sources contribute to the answers. One example submitted by participants is "Rudi AIFB ISWC2008". The sources containing partial answers to this are *uni-karlsruhe.de* and *semanticweb.org*. Other examples are "Town River America", "Markus Denny Semantic Wikis" and "Beijing Conference Database 2007". All queries can be found in our report [8].

Validity of Routing Plans. To investigate the validity, we use precision at k ($P@k$) to measure the percentage of plans that are valid out of the top-k plans returned by the system. For instance, $P@10$ is 1 when every plan in the top-10 list returned by the system, produces at least one keyword query result.

Fig. 4a shows $P@5$ for the settings $d_{max} = 0, 1, 2, 3, 4$. These values represent the average computed for all 30 queries. Using E-KERG, precision was up to 100 percent, i.e., for $d_{max}^{sum} = d_{max}^{data} = 4$. With $P@5$ being always above 0.6 when $d_{max} > 1$, S-KERG and D-KERG also achieved relatively good results. $P@5$ for KS was only 6%. Clearly, d_{max} had a positive effect. More valid plans were

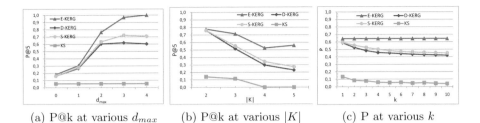

(a) P@k at various d_{max} (b) P@k at various $|K|$ (c) P at various k

Fig. 4. Validity of the plans measured using $P@k$

computed when a higher value was used for d_{max}. However, using $d_{max} = 4$ instead of 3 did not yield clear improvement.

Fig. 4b shows the effect of query length $|K|$. Quite clear, queries with larger number of keywords resulted in lower precision. It dropped as low as 0.23 when using D-KERG for queries with 5 keywords.

Fig. 4c shows that as more results from the system were taken into account (larger k), precision decreased. The decrease is small for k values larger than 3.

Experimental results thus correspond to the analysis we presented before: KS is the model that produces only very few valid plans. This result was improved by one order of magnitude when relationships between keywords were used. The more fine-grained a model captures the relationships, the larger was the percentage of valid plans. Even a summary at the level of sources produced reasonably high quality results, i.e., every second plan was a valid one.

Performance. Performance is measured as the average response time for computing routing plans. Fig. 5a shows the performance for queries at various settings using different values for d_{max}. This parameter had no effect on the KS's results but clearly influenced the performance achieved with KERG summaries. Times increased with higher values for d_{max}. While this increase was sharp for E-KERG and S-KERG, time performance of D-KERG was relatively stable. In particular, time required by D-KERG was no more than 10ms on average.

Expectedly, more time was needed when the number of query keywords increases, as illustrated in Fig. 5b. It seems that all the other models had poor performance w.r.t complex queries but D-KERG. In particular, E-KERG is no longer affordable for queries with more than 2 keywords because it needed more than 100s to produce results. While the times shown are the actual times obtained for the other models, only the lower bound was shown for E-KERG. This is because we applied a timeout of 6min. Fig. 5c shows the exact times obtained for E-KERG and the queries that had to be aborted due to timeout. For $d_{max} = 4$ for instance, 1 out of every three queries was aborted.

Less expected, Fig. 5a+ 5b show that KS did not achieve good performance. It needed more than 30s on average, up to 100s for queries with 5 keywords.

This can be explained using the theoretical result achieved in the previous section. Namely, the poor performance of KS indicates that the number of elements (see $input_{max}$ in Section 4.4) retrieved for single keywords must have been much

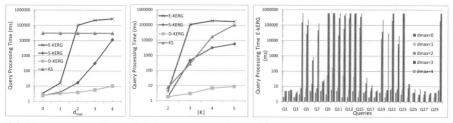

(a) Times at var. d_{max} (b) Times at var. $|K|$ (c) Times for E-KERG at var. d_{max}

Fig. 5. Processing times

larger than for two keywords. In other words, keyword pairs proved to be the much more selective queries. Considering relationships between keywords thus did not only improve result validity but also performance.

6   Conclusion

We presented a solution to the novel problem of keyword query routing. It helps users without knowledge of the evolving linked data and schema to find combination of sources that contain answers corresponding to their needs. This solution also partially addresses the aspect of efficiency as queries can be then evaluated against the relevant sources identified by the user, instead of using the entire Web of linked data.

We have proposed a family of summary models. Through theoretical and experimental analysis, we showed that it is important to capture keyword relationships. Compared to the KS model representing the naive baseline that stores only single keywords, the KERG models relying on relationships could produce a much larger number of valid results, i.e., improved precision by more than one order of magnitude when compared to the naive baseline represented by KS. Further, finding out which *relationships* are covered as opposed to single keywords resulted in less intermediate results to be processed. Thus, using relationships also has a positive effect on performance.

We could also show that *summarizing relationships* is essential for dealing with the large-scale linked data Web. Using a fine-grained E-KERG model representing an extension of work in database selection that captures all relationships in the data, precision was up to 100%, but response time was too high. While specific requirements shall determine what is the "best" model, it seems that D-KERG which summarizes at the level of sources represents the most practical trade-off. It produced results in less than 10ms out of which every second one was valid.

As future work, we will combine the proposed work on query routing with query processing to obtain a scalable procedure for computing relevant sources as well as retrieving the final answers from them.

Acknowledgements. Research reported in this paper was supported by the German Federal Ministry of Education and Research (BMBF) under the iGreen (grant 01A08005) and CollabCloud project (grant 01IS0937A-E).

References

1. Harth, A., Hogan, A., Delbru, R., Umbrich, J., O'Riain, S., Decker, S.: Swse: Answers before links! In: Semantic Web Challenge (2007)
2. Harth, A., Hose, K., Karnstedt, M., Polleres, A., Sattler, K.-U., Umbrich, J.: Data summaries for on-demand queries over linked data. In: WWW, pp. 411–420 (2010)
3. Harth, A., Umbrich, J., Hogan, A., Decker, S.: Yars2: A federated repository for querying graph structured data from the web. In: Aberer, K., Choi, K.-S., Noy, N., Allemang, D., Lee, K.-I., Nixon, L.J.B., Golbeck, J., Mika, P., Maynard, D., Mizoguchi, R., Schreiber, G., Cudré-Mauroux, P. (eds.) ASWC 2007 and ISWC 2007. LNCS, vol. 4825, pp. 211–224. Springer, Heidelberg (2007)
4. Liu, F., Yu, C.T., Meng, W., Chowdhury, A.: Effective keyword search in relational databases. In: SIGMOD Conference, pp. 563–574 (2006)
5. Neumann, T., Weikum, G.: The rdf-3x engine for scalable management of rdf data. VLDB J. 19(1), 91–113 (2010)
6. Sayyadian, M., LeKhac, H., Doan, A., Gravano, L.: Efficient keyword search across heterogeneous relational databases. In: ICDE, pp. 346–355 (2007)
7. Tran, T., Wang, H., Rudolph, S., Cimiano, P.: Top-k exploration of query candidates for efficient keyword search on graph-shaped (rdf) data. In: ICDE, pp. 405–416 (2009)
8. Tran, T., Zhang, L.: Keyword query routing. Technical report, Karlsruhe Institute of Technology (2010), http://www.aifb.uni-karlsruhe.de/WBS/dtr/papers/kqueryrouting.pdf
9. Tummarello, G., Cyganiak, R., Catasta, M., Danielczyk, S., Delbru, R., Decker, S.: Sig.ma: live views on the web of data. In: WWW, pp. 1301–1304 (2010)
10. Vu, Q.H., Ooi, B.C., Papadias, D., Tung, A.K.H.: A graph method for keyword-based selection of the top-k databases. In: SIGMOD Conference, pp. 915–926 (2008)
11. Yu, B., Li, G., Sollins, K.R., Tung, A.K.H.: Effective keyword-based selection of relational databases. In: SIGMOD Conference, pp. 139–150 (2007)

Declarative Semantics for the Rule Interchange Format Production Rule Dialect

Carlos Viegas Damásio, José Júlio Alferes, and João Leite

CENTRIA, Dep. Informática, FCT/Universidade Nova de Lisboa, Portugal
{cd,jja,jleite}@di.fct.unl.pt

Abstract. The Rule Interchange Format Production Rule Dialect (RIF-PRD) is a W3C Recommendation to define production rules for the Semantic Web, whose semantics is defined operationally via labeled terminal transition systems.

In this paper, we introduce a declarative logical characterization of the full default semantics of RIF-PRD based on Answer Set Programming (ASP), including matching, conflict resolution and acting.

Our proposal to the semantics of RIF-PRD enjoys several features. Being based on ASP, it enables a straightforward integration with Logic Programming rule based technology, namely for reasoning and acting with ontologies. Then, its full declarative logical character facilitates the investigation of formal properties of RIF-PRD itself. Furthermore, it turns out that our characterization based on ASP is flexible enough so that new conflict resolution semantics for RIF-PRD can easily be defined and encoded. Finally, it immediately serves as the declarative specification of an implementation, whose prototype we developed.

1 Introduction

In this paper we present a sound and complete declarative semantical characterization of the Production Rule Dialect of the Rule Interchange Format (RIF-PRD) [6] – including matching, conflict resolution and acting – based on Answer-Set Programming [11], accompanied by a prototypical implementation. While contributing to a better understanding of RIF-PRD, our proposal brings greater flexibility to RIF-PRD as it facilitates integration with other rule based technologies and is easily extensible e.g. with other conflict resolution strategies.

The W3C Rule Interchange Format (RIF) exists to enable interoperability among rule languages in general, allowing rules written for one application to be published, shared, and re-used in other applications and other rule engines. Whereas the core dialect of RIF [3] is designed to support the interchange of definite Horn rules without function symbols ("Datalog"), the Production Rule Dialect of RIF (RIF-PRD) [6] extends it to deal with production rules, and is currently a W3C Recommendation. Production rules can be seen as condition-action rules, and are particularly useful to specify behaviors and support the separation of business logic from business objects. According to RIF-PRD, the condition part of production rules is like the condition part of logic rules (as

P.F. Patel-Schneider et al. (Eds.): ISWC 2010, Part I, LNCS 6496, pp. 798–813, 2010.

covered by RIF-Core and its basic logic dialect extension, RIF-BLD [4]). Actions can assert facts, modify facts, retract facts, and have other side-effects, unlike conclusions of logic rules, which contain only a logical statement.

The following are examples of production rules taken from [6], about the status of customers, and corresponding discounts at checkout:

−**Gold rule:** *"Silver"* customers with shopping carts worth at least $2,000 are awarded the *"Gold"* status.

−**Discount rule:** *"Silver"* and *"Gold"* customers are awarded a 5% discount on the total worth of their shopping cart.

−**New customer and widget rule:** *"New"* customers who buy a widget are awarded a 10% discount on the total worth of their shopping carts, but loose any voucher they may have been awarded.

−**Unknown status rule:** a message must be printed, identifying any customer whose status is unknown (that is, neither *"New"*, *"Bronze"*, *"Silver"* nor *"Gold"*), and the customer must be assigned the status *"New"*.

RIF-PRD specifies an abstract syntax and associates the abstract constructs with normative semantics and a normative XML concrete syntax. It also specifies a presentation syntax that provides a more succinct representation of production rules. For example, the third rule above can be represented as follows [6]:

```
Forall ?cust such that (And( ?cust # ex1:Customer
                                    ?cust[status->"New"] ) )
    (If Exists ?cart ?item ( And ( ?customer[shoppingCart->?cart]
                                    ?cart[containsItem->?item]
                                    ?item # ex1:Widget ) ) )
    Then Do( (?s ?cust[shoppingCart->?s])
             (?val ?s[value->?val])
             (?voucher ?cust[voucher->?voucher])
             Retract( ?cust[voucher->?voucher] )    Retract( ?voucher )
             Modify( ?s[value->func:numeric-multiply(?val 0.90)] ) ) ) )
```

The RIF-PRD operational semantics for production rules and rule sets is based on labeled terminal transition systems [14] where state transitions result from executing the action part of instantiated rules, according to the loop: **(Match)**: the rules are instantiated based on the definition of the rule conditions and the current state of the data source; **(Conflict resolution)**: a decision algorithm, often called the conflict resolution strategy, is applied to select which rule instance will be executed; **(Act)**: the state of the data source is changed, by executing the selected rule instance's actions. If a terminal state has not been reached, the control loops back to the first step (Match).

An important part of the control loop that governs the semantics concerns the conflict resolution strategy used to select one of the several available rules for execution. Strategies are denoted by keywords (of type `rif:IRI`), that are attached to rule sets permitting that production rule producers and consumers agree on a different semantics. RIF-PRD also prescribes a normative strategy, *forward chaining* denoted by `rif:forwardChaining`, which eliminates rules from a conflict set (a set of applicable rules) based on the following ordered criteria:

1.Refraction: eliminate rules that were already applied and whose conditions for application haven't changed since;

2.Priority: eliminate rules with lower priority;

3.Recency: eliminate rules that have been applicable for longer.

At the end of the application of these criteria, RIF-PRD prescribes that one of the remaining rules be chosen "in some way" (e.g. randomly).

The RIF-PRD W3C Recommendation is a crucial and significant step in standardizing the syntax and semantics of production rules, enabling their interoperability among rule languages in general, and not limited to the Web. However, there are some issues that require further attention, and some steps that need to be taken, in order to provide a better understanding and greater flexibility of RIF-PRD. One important component missing in [6] is a purely logical declarative semantics for RIF-PRD, which would serve as a counterpart to the operational semantics provided. Such a semantics would provide a better understanding and further insights into RIF-PRD, while facilitating the integration of production rules with declarative rules and Logic Programming rule based technology in general, useful e.g. for reasoning and acting with ontologies.

Another issue that needs further attention is that of providing alternatives to the default conflict resolution strategy. Though RIF-PRD foresees the specification of different conflict resolution strategies, there is no indication in [6] as to how such alternative strategies could be specified in a way that facilitates their shared understanding by document producers and consumers. We believe that any such strategy, including the one normatively specified by RIF-PRD, should be defined by a set of rules which precisely defines its meaning. In this case, the keyword for the strategy could be a URI for the set of rules which precisely defines the strategy.

In this paper, we present a sound and complete declarative semantical characterization of RIF-PRD – including matching, conflict resolution and acting – based on Answer-Set Programming (ASP) [11], that addresses these outstanding issues. As suggested by RIF-PRD designers, we assume RIF-Core strong safeness [3] in order to guarantee finite grounding in forward chaining mode.

ASP is a form of declarative programming, similar in syntax to traditional logic programming and close in semantics to non-monotonic logic, that is now widely recognized as a valuable tool for knowledge representation and reasoning. On the one hand, ASP is fully declarative in the sense that the program specifications resemble the problem specifications, the semantics is very intuitive, and there is extensive theoretical work that facilitates proving several properties of answer-set programs. On the other hand, ASP is very expressive, allowing for compact representations of all NP and coNP problems, or even more complex ones if disjunctive programs are used [7]. Other important characteristics of ASP include the use of default negation to allow for reasoning with assumptions and incomplete knowledge, as well as the existence of a number of well studied extensions such as preferences, revision, abduction, etc. More relevant for this work, are the recent results on $MKNF^+$ hybrid knowledge bases where a faithful, tight and flexible integration of description logics and rules has been achieved [13]. The integration of rules with ontologies is also possible with dl-programs [8]. Finally, there are very efficient ASP solvers available (e.g. Clingo, DLV, Smodels, etc.).

Our proposal enjoys the following features that address the mentioned issues:

- Being based on ASP, it paves the way to a direct integration with Logic Programming based technology, viz. for reasoning and acting with ontologies;
- Being fully declarative, it facilitates the investigation of further formal properties of RIF-PRD, e.g. using the approach followed in [5];
- Enjoying the expressivity of ASP, it is flexible enough so that conflict resolution strategies for RIF-PRD are easily defined and encodable;
- Benefiting from the existence of efficient ASP solvers, it can be directly and efficiently implemented – which we have done using iClingo [9], and is, to the best of our knowledge, the first implementation of RIF-PRD.

The remainder of this paper is structured as follows: in Sect. 2 we review ASP; in Sect. 3 we present a sound and complete translation of RIF-PRD rule sets into ASP; in Sect. 4 we address the specification of conflict resolution strategies in ASP, illustrating with a sound and complete encoding of *forward chaining*, the RIF-PRD normative strategy; we conclude in Sect. 5.

2   Answer Set Programming

In this Section we start by describing the syntax and semantics of Answer-set Programming, before we introduce *iClingo*[9], an incremental answer-set system. We follow the presentation in [9], with some modifications.

The language is built from a set $\mathcal{F}$ of constants and function symbols (including the natural numbers and usual arithmetic operators), a set $\mathcal{V}$ of variable symbols, and a set $\mathcal{P}$ of predicate symbols (including the binary equality and inequality predicates, and ordinary arithmetic comparison operators). We assume that $\mathcal{V}$ contains a distinguished parameter symbol κ (varying over natural numbers). The set $\mathcal{T}$ of terms is the smallest set containing $\mathcal{V}$ and all expressions of the form $f(t_1, ..., t_n)$, where $f \in \mathcal{F}$ and $t_i \in \mathcal{T}$ for $0 \leq i \leq n$. The set $\mathcal{A}$ of atoms contains all expressions of the form $p(t_1, ..., t_n)$, where $p \in \mathcal{P}$ and $t_i \in \mathcal{T}$ for $1 \leq i \leq n$. A literal is an atom a or its (default) negation **not** a. Given a set L of literals, let $L^+ = \{a \in \mathcal{A} \mid a \in L\}$ and $L^- = \{a \in \mathcal{A} \mid \textbf{not}\ a \in L\}$. A logic program over $\mathcal{A}$ is a set of rules of the form $a_0 \leftarrow a_1, ..., a_m, \textbf{not}\ a_{m+1}, ..., \textbf{not}\ a_n$, where $a_i \in \mathcal{A}$ for $0 \leq i \leq n$. For a rule r of the form above, let $head(r) = a$ be the head of r, $body(r) = \{a_1, ..., a_m, \textbf{not}\ a_{m+1}, ..., \textbf{not}\ a_n\}$ be the body of r, and $atom(r) = \{head(r)\} \cup body(r)^+ \cup body(r)^-$. For a program P, let $head(P) = \{head(r) \mid r \in P\}$ and $atom(P) = \bigcup_{r \in P} atom(r)$. Given an expression $e \in \mathcal{T} \cup \mathcal{A}$, let $var(e)$ denote the set of all variables occurring in e, and given a rule r, let $var(r)$ denote the set of all variables occurring in r. Expression $e \in \mathcal{T} \cup \mathcal{A}$ is ground if $var(e) = \emptyset$. The ground instantiation of a program P is defined as $grd(P) = \{r\theta \mid r \in P, \theta : var(r) \rightarrow \mathcal{U}\}$ where $\mathcal{U} = \{t \in \mathcal{T} \mid var(t) = \emptyset\}$. Similarly, $grd(\mathcal{A}) = \{a \in \mathcal{A} \mid var(a) = \emptyset\}$.

A set $M \subseteq grd(\mathcal{A})$ is an answer set [11,1] of a program P over A if M is the $\subseteq$-smallest model of $\{head(r) \leftarrow body(r)^+ \mid r \in grd(P), body(r)^- \cap M = \emptyset\}$. The set of answer-sets of P is denoted by $AS(P)$. The semantics of integrity constraints is given through a program transformation where an integrity constraint of the form $\leftarrow a_1, ..., a_m, \mathbf{not}\ a_{m+1}, ..., \mathbf{not}\ a_n$ is a shorthand for the rule $a' \leftarrow a_1, ..., a_m, \mathbf{not}\ a_{m+1}, ..., \mathbf{not}\ a_n, \mathbf{not}\ a'$ where a' is a new atom.

2.1   *iClingo*

Real-world applications such as planning or model checking include a parameter encoding the size of a solution. In Answer Set Programming (ASP), essentially a propositional formalism, this is dealt with by considering one problem instance after another by gradually increasing the bound on the solution size. In most cases, Answer-Set Programming systems simply produce a ground set of rules for each problem instance, incurring in a high efficiency cost.

iClingo[1] [9] is an incremental ASP (iASP) system where both the grounder as well as the solver are implemented in a stateful way, interleaving grounding and solving within incremental computations. Both the grounder and the solver maintain their previous states while increasing an incremental parameter. At each incremental step, the grounder just produces ground rules generated from the current program slice, i.e. generated by instantiating the incremental parameter with the current value. Such ground program slices are gradually passed to the solver that accumulates ground rules and computes answer sets for them.

In the context of *iClingo*, the concept of a *(parametrized) domain description* is introduced, as being a triple $\langle B, S[\kappa], Q[\kappa] \rangle$ of logic programs where $S[\kappa]$ and $Q[\kappa]$ contain a (single) parameter κ ranging over the natural numbers. The base program B describes static knowledge, independent of parameter κ. Program $S[\kappa]$ contains knowledge that accumulates with increasing values of κ. Program $Q[\kappa]$ contains knowledge that is specific for each value of κ. Given a *domain description* $\Pi = \langle B, S[\kappa], Q[\kappa] \rangle$ and an integer $i \geq 1$, let $P[i] = B \cup \left(\bigcup_{1 \leq j \leq i} S[j] \right) \cup Q[i]$, and $AS(\Pi_i)$ denote $AS(P[i])$, $min(\Pi)$ denote the minimum integer such that $AS(\Pi_i) \neq \emptyset$, and $AS(\Pi)$ denote $AS\left(\Pi_{min(\Pi)}\right)$. The goal is then to determine $AS(\Pi)$. *iClingo* accepts *domain descriptions* Π[2] and computes $AS(\Pi)$ by incrementally constructing and solving for $P[i]$. Detailed information regarding the implementation of *iClingo* can be found in [9].

---

[1] *iClingo* is part of *Potassco*, a set of tools for Answer Set Programming developed at the University of Potsdam, and available at http://potassco.sourceforge.net

[2] Function symbols with non-zero arity may lead to logic programs over an infinite Herbrand base. To maintain decidability at each iteration, it is important to restrict the language to fragments for which finite equivalent ground programs are guaranteed to exist. Level-restricted (or λ-restricted) logic programs [10] constitute such a fragment, where finiteness is guaranteed by the requirement that any variable in a rule be bound to a finite set of ground terms via a predicate not subject to positive recursion through that rule.

3   Fact Bases, States, Conditions and Rules

In this Section we synthetically overview some of the main concepts of the Production Rule dialect of RIF [6] and provide a mapping of RIF-PRD initial states (*fact base*) and rule sets into iASP which is sound and complete wrt. the possible traces of execution of the rules on the initial state. For now, we do not consider the inclusion of a conflict resolution strategy – it will be dealt with in Sect. 4.

RIF-PRD defines rules with action heads for performing changes over a set of facts (i.e. an extensional logic database) dependent on logical conditions over a logical state derived from this set of facts. The underlying logical language is constructed from a first-order alphabet.

3.1   Atomic Formulas and Conditions

RIF-PRD defines the notion of term as in ASP, except for the introduction of the special list term which, for all purposes in the rest of this paper, can be seen as an ordinary complex term. Terms are used to construct atomic formulas.

Definition 1 (RIF-PRD term and atomic formulas). *A term is either an arbitrary constant* c, *an arbitrary variable* ?V, *a lists of ground terms* List(g_1 ...g_n), *or a (complex) positional terms* f(t_1 ...t_n) *formed from a constant* f *and a sequence of arbitrary term arguments* t_1 ...t_n *with* $n \geq 1$.

Given arbitrary terms t, s, *and* p_i, t_i *where* $1 \leq i \leq n$, *atomic formulas are ordinary atoms (i.e. positional terms), equality of terms (*t=s*), membership of object* t *in class* s *(*t#s*), subclass relation (*t##s*), frames (*t[p_1->t_1 ... p_n->t_n]*), or externally defined terms (* External(t)*).*

In RIF-PRD, there is no syntactical distinction between positional terms and ordinary atoms. Equality is used to check if two terms are identical, while membership atomic formulas t#s are used to represent that the object denoted by term t belongs to the class denoted by s. A subclass atomic formula t##s expresses that t is a subclass of s. A frame term t[p_1->t_1 ... p_n->t_n] roughly states that the object denoted by term t has for each property p_i the value t_i. Externally defined terms are used for representing built-in functions, e.g. to perform numerical operations. Condition formulas are to be used in the antecedents of production rules to define conditions for their applicability, corresponding syntactically to a fragment of first-order logic without universal quantifiers.

Definition 2 (RIF-PRD condition formulas). *Condition formulas are inductively defined from atomic formulas, conjunction* And(ϕ_1 ...ϕ_n) *and disjunction* Or(ϕ_1 ...ϕ_n) *of conditional formulas, negation* Not(ϕ) *or existential quantification* Exists ?v_1 ...?v_m (ϕ), *where* ϕ, ϕ_1 ...ϕ_n *are condition formulas and* ?v_1 ...?v_m *are variables.*

3.2   Fact Bases and States

The knowledge dynamics is captured by a set of ground atomic formulas – the *fact base* – which changes through the addition and removal of atomic formulas.

The execution of a RIF-PRD production rule system starts with an initial fact base, and proceeds by updating it step by step. At a given step of the execution κ a fact base will be encoded in iASP by a set of facts of the form $\mathtt{fact}(\varphi',\ \kappa)$ where φ' is the translation of the RIF-PRD ground atomic formula φ.

Definition 3 (Translation of atomic formulas). *An atomic RIF-PRD formula φ is translated into the iASP term φ' as follows:*
- *A positional atom, an equality or an externally defined term φ is mapped into itself;*
- *A membership atomic formula* $\mathtt{t\#s}$ *is mapped into term* $isa(t, s)$;
- *A subclass atomic formula* $\mathtt{t\#\#s}$ *is mapped into term* $sub(t, s)$;
- *A frame atomic formula* $\mathtt{s[p\text{->}o]}$ *is mapped into term* $frame(s, p, o)$.

This representation assumes that a ground frame $\mathtt{t[p_1\text{->}t_1\ \ldots\ p_n\text{->}t_n]}$ is represented by the set of facts $frame(t, p_1, t_1), \ldots, frame(t, p_n, t_n)$. For simplicity of presentation, externally defined formulas are mapped into themselves. However, a concrete implementation should implement these resorting to their own built-ins; this is ignored in the translation.

Definition 4 (Fact bases translation). *Consider an initial fact base Φ.*
- *Program $\pi_{\mathtt{INIT}}(\Phi)$ is formed by $fact(\varphi, 0)$, for each $\varphi \in \Phi$.*
- *Program $\pi_{\mathtt{FLUENT}}(\Phi)$ is formed by $fluent(\varphi)$, for each formula φ that may occur in a fact base.*
- *Program $\pi_{\mathtt{CHANGE}}[\kappa]$ is formed by the rules:*
$$fact(F, \kappa) \leftarrow fluent(F), fact(F, \kappa - 1), \mathbf{not}\ retract(F, \kappa - 1).$$
$$fact(F, \kappa) \leftarrow fluent(F), assert(F, \kappa - 1).$$

$\pi_{\mathtt{INIT}}$ collects the initial fact base which will be updated using the rules in $\pi_{\mathtt{CHANGE}}[\kappa]$. The first rule states that fluents which are not retracted in the previous step remain in the fact base (inertia), while the second states that fluents asserted in the previous step will be added. Notice that the things which can be added or deleted are collected in program $\pi_{\mathtt{FLUENT}}$. For simplicity, the definition of predicate $fluent/1$ is extensional but could also be defined intensionally by rules. Also note that by RIF-Core strong safeness at each step there may exist only a finite number of alternatives which can be dealt with in practice. Another essential use of predicate $fluent/1$ is to ground variables in the final iASP domain description.

Definition 5 (States translation). *Program $\pi_{\mathtt{STATES}}[\kappa]$ is formed by the rules:*

$$state(F, \kappa) \leftarrow fact(F, \kappa).$$
$$state(F, \kappa) \leftarrow fact(F, 0), \mathbf{not}\ fluent(F).$$

$$state(isa(O1, C2), \kappa) \leftarrow fluent(isa(O1, C1)), fluent(sub(C1, C2)),$$
$$state(isa(O1, C1), \kappa), state(sub(C1, C2), \kappa).$$
$$state(sub(C1, C3), \kappa) \leftarrow fluent(sub(C1, C2)), fluent(sub(C2, C3)),$$
$$state(sub(C1, C2), \kappa), state(sub(C2, C3), \kappa).$$

The first rule includes in the state of step κ the fact base of κ. The second states that any non-fluent (static) fact holding at the initial fact base also holds at step κ. According to RIF-PRD semantics the set of initial facts can be arbitrarily ground atomic formula but actions are syntactically limited to specific types of formula (e.g. it is impossible to change subclass atomic formulas). The third rule captures class inheritance while the last one expresses transitivity of the subclass relationship, imposed to any state by the semantics of RIF-PRD.

Conditions are matched to a given state. However, the case of non-atomic formulas introduces extra complexity:

Definition 6 (Conditions translation). *Let Φ be an arbitrary condition formula and κ an execution step. Define condition iASP formula Φ' and program $\pi_{\text{COND}}^{\Phi}[\kappa]$ inductively as follows:*

- *If Φ is an atomic formula φ then $\Phi'[\kappa] = state(\varphi', \kappa)$ and $\pi_{\text{COND}}^{\Phi}[\kappa] = \{\}$;*
- *If $\Phi = \text{And}(\phi_1 \ldots \phi_n)$ then $\Phi'[\kappa] = (\phi'_1, \ldots, \phi'_n)$ and $\pi_{\text{COND}}^{\Phi}[\kappa] = \bigcup_{1 \leq i \leq n} \pi_{\text{COND}}^{\phi_i}[\kappa]$;*
- *If $\Phi = \text{Or}(\phi_1 \ldots \phi_n)$ then $\Phi'[\kappa] = or_{\Phi}(X_1, \ldots, X_m, \kappa)$ where $?X_1, \ldots ?X_m$, are the free variables of Φ and or_{Φ} is a new predicate symbol, and $\pi_{\text{COND}}^{\Phi}[\kappa] = \bigcup_{1 \leq i \leq n} \left(\pi_{\text{COND}}^{\phi_i}[\kappa] \cup \{or_{\Phi}(X_1, \ldots, X_m, \kappa) \leftarrow \phi'_i[\kappa]\} \right)$;*
- *If $\Phi = \text{Exists } ?V_1 \ldots ?V_n \ (\phi)$ then $\Phi'[\kappa] = exists_{\Phi}(X_1, \ldots, X_m, \kappa)$ where $?X_1, \ldots ?X_m$, are the free variables of Φ and $exists_{\Phi}$ is a new predicate symbol, and $\pi_{\text{COND}}^{\phi}[\kappa] = \pi_{\text{COND}}^{\phi}[\kappa] \cup \{exists_{\Phi}(X_1, \ldots, X_m, \kappa) \leftarrow \phi'[\kappa]\}$;*
- *If $\Phi = \text{Not}(\phi)$ then $\Phi'[\kappa] = \textbf{not } arg_{\Phi}(X_1, \ldots, X_m, \kappa)$ where $?X_1, \ldots ?X_m$, are the free variables of Φ and arg_{Φ} is a new predicate symbol, and $\pi_{\text{COND}}^{\Phi}[\kappa] = \pi_{\text{COND}}^{\phi}[\kappa] \cup \{arg_{\Phi}(X_1, \ldots, X_m, \kappa) \leftarrow \phi'[\kappa]\}$;*

Basically, this transformation applies Lloyd-Topor's transformation [12] to obtain the corresponding normal rules capturing the conditional formula, taking into account what is true in the current step. Mark that both a (conjunctive) goal $\Phi'[\kappa]$ and a program $\pi_{\text{COND}}^{\Phi}[\kappa]$ is returned for each condition formula Φ. Additional details and justification of this process can be found in [1].

3.3   Actions and Rules

The RIF-PRD language defines several atomic actions for updating the fact base, and these will be used to define the effects of RIF-PRD production rules.

Definition 7 (RIF-PRD atomic actions). *An atomic action is a simple construct that represents an atomic transaction.*

1. *Assert fact: If Φ is a positional atom, a frame or a membership atomic formula in the RIF-PRD condition language, then $\text{Assert}(\Phi)$ is an atomic action.*
2. *Retract fact: If Φ is a positional atom or a frame in the RIF-PRD condition language, then $\text{Retract}(\Phi)$ is an atomic action.*
3. *Retract all slot values: If o and s are terms in the RIF-PRD condition language, then $\text{Retract}(o\ s)$ is an atomic action.*
4. *Retract object: If t is a term in the RIF-PRD condition language, then $\text{Retract}(t)$ is an atomic action.*
5. *Execute: if Φ is a positional atom in the RIF-PRD condition language, then $Execute(\Phi)$ is an atomic action.*

The arguments of the action are dubbed the target of the action.

The effects of RIF-PRD atomic actions are captured by our translation using the following iASP rules.

Definition 8 (Effects of actions). *Program* $\pi_{\text{ACTIONS}}[\kappa]$ *is:*

$assert(F, \kappa) \leftarrow action(assert(F), \kappa).$
$retract(F, \kappa) \leftarrow action(retract(F), \kappa).$

$retract(isa(O, C), \kappa) \leftarrow action(retract\_object(O), \kappa), fact(isa(O, C), \kappa).$
$retract(frame(O, S, V), \kappa) \leftarrow action(retract\_object(O), \kappa), fact(frame(O, S, V), \kappa).$

$retract(frame(O, S, V), \kappa) \leftarrow action(retract\_slots(O, S), \kappa), fact(frame(O, S, V), \kappa).$

Note that the *execute* actions do not have an effect in the fact base and should be interpreted externally. The first two rules of program $\pi_{\text{ACTIONS}}[\kappa]$ apply when an assert (resp. retract) action occurs at step κ, whose effects in the fact base have been defined previously in program π_{CHANGE}. The next two rules translate a retract object action into a set of simultaneous retracts, while the last one takes care of the retract all slots action. The interaction of rules with the fact base is performed via the $action/2$ predicate to be defined subsequently.

Actions are combined sequentially into action blocks, allowing binding patterns for binding variables occurring in the actions. Additionally, RIF-PRD defines a compound **Modify** frame action which can be substituted by a sequence of a retract all slot values followed by an assert; it is assumed that such a replacement has been performed.

Definition 9 (Action variable declaration and action blocks). *An action variable declaration is a pair* (?V b) *where* ?V *is a variable and* b *is binding having one of the forms:* New() *for generating a new identifier, or a frame* o[s->?V] *where* o *and* s *are ground terms. If* (?V$_1$ b$_1$), ..., (?V$_n$ b$_n$), $n \geq 0$, *are action variable declarations, and if* a$_1$, ..., a$_m$, $m \geq 1$, *are simple actions, then* Do((?V$_1$ b$_1$) ...(?V$_n$ b$_n$) a$_1$ ...a$_m$) *denotes an action block.*

Finally, the RIF Production Rules are captured by the following definition. Mark that well-formedness conditions are imposed to rules and conditions, which we are ignoring in this summary presentation.

Definition 10 (RIF production rule). *A rule can be one of:*
- *An (unconditional) action block* Do((?V$_1$ b$_1$) ...(?V$_n$ b$_n$) a$_1$...a$_m$).
- *A conditional action block* If Φ Then Do((?V$_1$ b$_1$) ...(?V$_n$ b$_n$) a$_1$...a$_m$), *where Φ is a condition formula and the conclusion is an action block.*
- *A quantified rule* Forall ?V$_1$...?V$_n$ such that (p$_1$...p$_m$) (r), *where each* p$_i$ *is a conditional formula (a pattern) and r is a RIF Production rule.*

Without loss of generality we assume that quantified rules have only one level of universal quantification, i.e. the rule r is limited to be a conditional action block since it is always possible to write quantified rules in this way, by variable renaming and appending patterns.

Definition 11 (Translation of a RIF production rule). *Let ri be a RIF production rule and let id be a unique identifier assigned to that rule (i.e. its "name"). Program $\pi_{\text{RULE}}^{ri}[\kappa]$ is constructed as follows:*

- If ri is $\text{Do}((?V_1\ b_1)\ldots(?V_n\ b_n)\ a_1\ldots a_m\ )$ *then include in* $\pi_{\text{RULE}}^{ri}[\kappa]$ *the fact* $fireable(rule(id, subs), \kappa)$.
- If ri is $\text{If}\ \Phi\ \text{Then}\ \text{Do}(\ (?V_1\ b_1)\ \ldots(?V_n\ b_n)\ a_1\ \ldots a_m\ )$ *then include* $\pi_{\text{COND}}^{\Phi}[\kappa]$ *in* $\pi_{\text{RULE}}^{ri}[\kappa]$, *and the following rule where* $?X_1, \ldots, ?X_l$ *are the free variables of* ri: $fireable(rule(id, subs(X_1, \ldots, X_l), \kappa)) \leftarrow \Phi'$.
- If ri is $\text{Forall}\ ?V_1\ldots?V_n\ \text{such that}\ (p_1\ldots p_m)\ (\text{If}\ \Phi\ \text{Then}\ \text{Do}(B))$ *then treat this as the conditional action block* $\text{If}\ \text{And}(p_1\ldots p_m\ \Phi)\ \text{Then}\ \text{Do}(B)$.

Additionally, from the action block $\text{Do}((?V_1\ b_1)\ldots(?V_n\ b_n)\ a_1\ldots a_m\ )$ *in the conclusion of* ri *add to program* $\pi_{\text{RULE}}^{ri}[\kappa]$, *for each* $1 \leq j \leq m$, *the rule:*

$$action(a_j', \kappa + j) \leftarrow instance(id, subs(V_1, \ldots, V_n, X_1, \ldots, X_l), \kappa).$$

Finally, include in $\pi_{\text{RULE}}^{ri}[\kappa]$ *the rule below, where* $bind_{v_i}$ *is* $state(frame(o, s, V_i), \kappa)$ *if* $b_i = \text{o}[\text{s->}?V_i]$. *Otherwise* $b_i = \text{New}()$, *and let* $bind_{v_i}$ *be* $V_i = obj(id, i, \kappa)$ *with obj an arbitrary but fixed constant symbol.*

$$instance(id, subs(V_1, \ldots, V_n, X_1, \ldots, X_l), \kappa) \leftarrow picked(rule(id, subs(X_1, \ldots, X_l)), \kappa),$$
$$bind_{V_1}, \ldots, bind_{V_n}.$$

Predicate $fireable(rule(id, subs(\ldots)), \kappa)$ holds in step κ whenever the rule identified by id has a condition true, and thus may be applied. The complex term $sub(\ldots)$ keeps the substitution of variables for which the condition matches state κ, and is also used to distinguish between different matching instances of the same rule. If the rule is picked for execution then $picked(rule(id, subs(\ldots)), \kappa)$ will hold and consequently action a_j will be executed in step $k + j$ with the action instance (i.e. substitution of variables) collected in auxiliary predicate $instance/3$.

Example 1. Consider the rule presented in the introduction of this paper. Its encoding into iASP as constructed by π_{RULE} transformation is shown below, following the usual answer-set convention of variables beginning with upper-case and, to simplify the presentation, the constants belonging to namespace $ex1$ are represented using CURIE notation:

$$fireable(rule(widget, subs(Cust)), \kappa) \leftarrow state(isa(Cust, ex1:Customer), \kappa),$$
$$state(frame(Cust, status, ``New"), \kappa), exists_1(Cust, \kappa).$$
$$exists_1(Cust, \kappa) \leftarrow state(frame(Cust, shoppingCart, Cart), \kappa),$$
$$state(frame(Cart, containsItem, Item), \kappa), state(isa(Item, ex1:Widget), \kappa).$$

$$action(retract(frame(Cust, voucher, Voucher)), \kappa + 1) \leftarrow$$
$$instance(widget, subs(Cust, S, Val, Voucher), \kappa).$$
$$action(retract\_object(Voucher)), \kappa + 2) \leftarrow$$
$$instance(widget, subs(Cust, S, Val, Voucher), \kappa).$$
$$action(retract\_slots(S, value), \kappa + 3) \leftarrow$$
$$instance(widget, subs(Cust, S, Val, Voucher), \kappa).$$
$$action(assert(frame(S, value, Val * 90/100)), \kappa + 4) \leftarrow$$
$$instance(widget, subs(Cust, S, Val, Voucher), \kappa).$$

$$instance(widget, subs(Cust, S, Val, Voucher), \kappa) \leftarrow$$
$$picked(rule(widget, subs(Cust), \kappa), state(frame(Cust, shoppingCart, S), \kappa),$$
$$state(frame(S, value, Val), \kappa), state(frame(Cust, voucher, Voucher), \kappa).$$

It is clear from the example that the fireable conditions are not yet connected to the rules performing the actions, which will be tackled next. First, it is necessary to pick one rule for execution from the pickable ones (i.e. the ones which fire and can be executed). This is straightforward to encode:

Definition 12 (Pick rule). *Program* $\pi_{\texttt{PICK}}[\kappa]$ *is formed by:*

$picked(Rule, \kappa) \leftarrow pickable(Rule, \kappa), \textbf{not } picked\_other(Rule, \kappa), \textbf{not } transitional(k).$
$picked\_other(Rule, \kappa) \leftarrow pickable(Other, \kappa), pickable(Rule, \kappa), Rule! = Other,$
$\qquad\qquad\qquad\qquad\qquad\qquad\qquad\qquad\qquad\qquad\qquad\qquad picked(Other, \kappa).$
$picked(\kappa) \leftarrow picked(Rule, \kappa). \qquad\qquad transitional(\kappa) \leftarrow action(A, \kappa).$

The execution of RIF-PRD proceeds by first picking one rule, then performing its actions sequentially, then picking another rule, performing its actions, etc. . . . The steps in which the fact base is being updated are dubbed "transitional" in the RIF-PRD recommendation. The first two rules in $\pi_{\texttt{PICK}}[\kappa]$ choose exactly one alternative (i.e. a rule) from the pickable rules, when κ is not a transitional step. If no strategy is defined, the general operational semantics prescribes that all fireable rules are pickable, which can be captured by the program $\pi_{\texttt{ONE}}[\kappa]$ with the single rule $pickable(Rule, \kappa) \leftarrow fireable(Rule, \kappa)$. Computation terminates in a non-transitional step where no rule is picked. This is captured by $\pi_{\texttt{HALT}}[\kappa]$, which ends our translation of a RIF-PRD rule set, summarized in Def. 14.

Definition 13 (Termination). *Program* $\pi_{\texttt{HALT}}[\kappa]$ *is defined by:*
$\qquad\qquad \leftarrow \textbf{not } final(\kappa).$
$\qquad\qquad final(\kappa) \leftarrow \textbf{not } transitional(\kappa), \textbf{not } picked(\kappa).$

Definition 14 (Rule set translation). *The translation of a RIF-PRD rule set RS with initial fact base w and set of fluents F is the iASP domain specification* $\Pi_{\texttt{RULESET}}(RS, w) = \langle B_{\texttt{RS}}(w), S_{\texttt{RS}}(RS)[\kappa], Q_{\texttt{RS}}[\kappa] \rangle$ *where:*

$$B_{\texttt{RS}}(w) \quad = \pi_{\texttt{INIT}}(w) \cup \pi_{\texttt{FLUENT}}(F)$$
$$S_{\texttt{RS}}(RS)[\kappa] = \pi_{\texttt{CHANGE}}[\kappa] \cup \pi_{\texttt{STATES}}[\kappa] \cup \pi_{\texttt{ACTION}}[\kappa] \cup \pi_{\texttt{PICK}}[\kappa] \cup \pi_{\texttt{ONE}}[\kappa] \cup \bigcup_{ri \in RS} \pi_{\texttt{RULE}}^{ri}[\kappa]$$
$$Q_{\texttt{RS}}[\kappa] \quad = \pi_{\texttt{HALT}}[\kappa]$$

An advantage of this encoding is that all possible "traces" of execution can be generated by the iASP system, where each different trace corresponds to an answer set. Formally[3]:

Theorem 1 (Correctness of translation). *Let RS be a rule set and w an initial fact base. Then*[4]*:*
Soundness: *If* $M \in AS(\Pi_{\texttt{RULESET}}(RS, w)_n)$ *and* $(c_1, \ldots, c_m)$ *is the increasing sequence of integers such that* $transitional(c_j) \notin M, 1 \leq j \leq m$, *then, for every* $i : 1 \leq i \leq m - 1$ $(State^i(M), Picked^i(M), State^{i+1}(M)) \in \rightarrow_{PRD}$,

---

[3] Lack of space prevents us from presenting the proofs of theorems.

[4] $\rightarrow_{PRD}$ stands for the transition system which serves as the basis for defining the semantics of RIF-PRD, $ConflictSet(RS, s_i)$ the set of all applicable rules in state s_i. Lack of space prevents us from presenting the semantics of RIF-PRD, which is available in [6].

where $State^i(M)$ denotes the set of formulae Φ such that $state(\Phi', c_i) \in M$ and $Picked^i(M)$ the name of the (only) rule R such that $picked(R, c_i) \in M$.

Completeness: If $(s_1, \ldots, s_m)$ is a sequence of non-transitional states such that $w = s_1$, and for each pair (s_i, s_{i+1}) there exists a rule $r \in ConflictSet(RS, s_i)$ such that $(s_i, r, s_{i+1}) \in \rightarrow_{PRD}$, then, there exists $M \in AS\left(\Pi_{\text{RULESET}}(RS, w)_n\right)$ for some $n \geq m$ such that the sequence of integers $(c_1, \ldots, c_m)$, constructed from M as above, is such that $State^i(M) = s_i$, for all $1 \leq i \leq m$.

4   Conflict Resolution Strategies

For selecting (ideally one) among these possible executions (or traces), as mentioned in the Introduction RIF-PRD foresees the existence of *conflict resolution strategies*. Each of the strategies is denoted by a keyword (of type `rif:IRI`), that is attached to the rule set. The current version of RIF-PRD prescribes a normative strategy, *forward chaining*, denoted by `rif:forwardChaining`, and anticipates the specification of additional keywords, each corresponding to an additional strategy for selecting rules in conflict. Furthermore, it also allows for the inclusion of other keywords, not specified in the RIF-PRD specification, in which case it is the responsibility of the producers and consumers of those documents to agree on the strategy denoted by the keywords.

Our stance is that any conflict resolution strategy should be defined by a set of rules, including those normatively specified by RIF-PRD, which precisely defines its meaning. In this case, the keyword for the strategy could be a URI for the set of rules which precisely defines the strategy. In this section we show that iASP, along with the translation defined in the previous section, is expressive enough to specify conflict resolution strategies. In particular, we show how to specify conflict resolution strategies, and illustrate by precisely characterizing the `rif:forwardChaining` strategy.

4.1   General Definition of Strategies

A conflict resolution strategy is defined in [6] by an algorithm that, in a series of steps, selects from the set of all fireable rules in some state, a subset of (pickable) rules from which one is finally picked for execution. For example, the `rif:forwardChaining` strategy can be summarized as the following algorithm:

Definition 15 (Forward chaining algorithm). *Given a conflict set (i.e. a set of fireable rules):*

1. *Remove all rules which where previously applied and, since their last application, the conditions that made them applicable haven't changed – refraction.*
2. *The remaining rules are ordered by decreasing priority, and only the rule instances with the highest priority are kept. Recall that in RIF-PRD every rule is assigned a priority which is a natural number.*
3. *The remaining rules are ordered by decreasing recency, and only the most recent rule instances are kept. Here, a rule is more recent than another if it is (consecutively) applicable for less prior states than the other.*

Each of these steps applies one strategy element (refraction, priority and recency). In [6], a fourth (tie-break) element is considered, to be applied after these 3, stating that one of the remaining rules should be picked in some "implementation specific way" [6]. Here we do not need to consider this last step. On the one hand, the translation is such that each answer set is guaranteed to reflect the application of a single rule at each state. On the other hand, the existence of more than one answer set reflects the fact that there may be more than one pickable rule at some state after the application of these 3 strategy elements. As a result of the translation, each answer set encodes one possible sequence of application of rules, and one can either consider all resulting answer-sets, or arbitrarily pick one of them.

For encoding such a strategy in a set of iASP rules, to be added to the domain description obtained from the translation of the previous section, we first need to replace the rule of $\pi_{\text{ONE}}[\kappa]$ which specified that all fireable rules are pickable, by a set of general rules allowing for restrictions on pickable rules. Accordingly, a rule is pickable if it is fireable and it is not rejected by one of the strategy elements:

Definition 16 (Strategy). *Program* $\pi_{\text{STRATEGY}}[\kappa]$ *is formed by the rules*

$$pickable(Rule, \kappa) \leftarrow fireable(Rule, \kappa), \textbf{not } rejected(Rule, \kappa).$$
$$rejected(Rule, \kappa) \leftarrow rejected(Rule, \kappa, S), st\_element(S).$$

Note that, without any defined strategy, $\pi_{\text{STRATEGY}}[\kappa]$ has exactly the same effect as $\pi_{\text{ONE}}[\kappa]$. In fact, if there are no rules for neither $rejected/3$ nor $st\_element/1$, $rejected(Rule, \kappa)$ is false in all answer-sets for all rules and κ, and so pickable is true for all fireable rules, as is the case in $\pi_{\text{ONE}}[\kappa]$.

Strategy elements are identified by a name. Then, for each strategy, facts to specify the order of application of the elements must be added. For example, for `rif:forwardChaining` the specification of the order of elements is as follows:

$$st\_element(refraction, 1). \quad st\_element(priority, 2). \quad st\_element(recency, 3).$$

For referring to the element without its order of application, the following rule is also needed $st\_element(S) \leftarrow st\_element(S, \_)$.

In general, for the definition of conflict resolution strategies, a predicate is needed to indicate whether a rule is active when a given strategy element is being applied. For example, in `rif:forwardChaining`, if a rule is removed by refraction, then that rule should no longer be available for consideration (i.e. active) when considering the priority-element. The specification of this predicate is quite straightforward: a rule is inactive if there is a strategy element prior in the application order which rejected it, and active otherwise.

Definition 17 (Active Rules). *Program* $\pi_{\text{ACTIVE}}[\kappa]$ *is defined by*

$$inactive(Rule, \kappa, N) \leftarrow st\_element(\_, N), st\_element(S, N1), N1 < N,$$
$$rejected(Rule, \kappa, S).$$
$$active(Rule, \kappa, N) \quad \leftarrow \textbf{not } inactive(Rule, \kappa, N), st\_element(\_, N).$$

The iASP domain description associated with a RIF-PRD rule set becomes:

Definition 18 (RIF-PRD domain description). *The RIF-PRD iASP domain description of a rule set RS with initial fact base w and fluents F is $\Pi_{\mathrm{RS}}(RS, w) = \langle B_{\mathrm{RS}}(w), S_{\mathrm{RS}}(RS)[\kappa], Q_{\mathrm{RS}}[\kappa] \rangle$ with $B_{\mathrm{RS}}(w)$ and Q_{RS} as in Def. 14, and*

$$S_{\mathrm{RS}}(RS)[\kappa] = \pi_{\mathrm{CHANGE}}[\kappa] \cup \pi_{\mathrm{STATES}}[\kappa] \cup \pi_{\mathrm{ACTION}}[\kappa] \cup \pi_{\mathrm{PICK}}[\kappa] \cup$$
$$\cup \bigcup\nolimits_{ri \in RS} \pi_{\mathrm{RULE}}^{ri}[\kappa] \cup \pi_{\mathrm{STRATEGY}}[\kappa] \cup \pi_{\mathrm{ACTIVE}}[\kappa]$$

Theorem 2. *Theorem 1 holds if we replace $\Pi_{\mathrm{RULESET}}(RS, w)$ with $\Pi_{\mathrm{RS}}(RS, w)$.*

4.2 Defining One Specific Strategy

To completely specify one conflict resolution strategy, we add facts defining the strategy elements and their application order (as above for `rif:forwardChaining`) and define, for each element, which rules are rejected. Below we show how this can be done for each of the elements in the `rif:forwardChaining` algorithm.

Refraction. Once a rule is picked at some state, then it is rejected by refraction from that state onwards, for as long as the rule remains fireable. The test for the rule being fireable is only done in states when the system is not being updated.

$rejected(Rule, \kappa, refraction) \leftarrow fireable(Rule, \kappa), picked(Rule, \kappa - 1).$
$rejected(Rule, \kappa, refraction) \leftarrow rejected(Rule, \kappa - 1, refraction), transitional(\kappa).$
$rejected(Rule, \kappa, refraction) \leftarrow fireable(Rule, \kappa), rejected(Rule, \kappa - 1, refraction)$
$\qquad\qquad\qquad\qquad\qquad \mathbf{not}\ transitional(\kappa).$

Priority. All rules for which there is another (different) active fireable rule with a strictly higher priority should be rejected. We do not need to test that rejected rules are active (i.e. not rejected by a previous strategy element), since according to $\pi_{\mathrm{STRATEGY}}[\kappa]$ a rejected rule is never pickable.

$rejected(rule(Id, Var), \kappa, priority) \leftarrow fireable(rule(Id, Var), \kappa),$
$\qquad fireable(rule(Id2, Var2), \kappa), Id! = Id2, priority(Id, P), priority(Id2, P2),$
$\qquad P < P2, active(rule(Id2, Var2), \kappa, N), strategy(priority, N).$

Recency. A rule is rejected if there is a more recent one also active and fireable. We use an auxiliary predicate (*recency/3*) that, for each rule instance and state κ, determines the number of consecutive states before κ that the instance has been fireable. Then, a rule is rejected if there is another one which is more recent. Predicate $state(K)$ is just used for grounding, and is true for any state K.

$rejected(rule(Id, Var), \kappa, recency) \leftarrow fireable(Rule, \kappa),\ fireable(Other, \kappa),$
$\qquad Rule! = Other, recency(Rule, TR, \kappa), recency(Other, TO, \kappa), TO < TR,$
$\qquad state(TR), state(TO), active(Other, \kappa, N),\ st\_element(recency, N).$

$recency(Rule, \kappa, \kappa) \leftarrow fireable(Rule, \kappa), not\ fireable(Rule, \kappa - 1).$
$recency(Rule, K, \kappa) \leftarrow recency(Rule, K, \kappa - 1), transitional(\kappa), state(K).$
$recency(Rule, K, \kappa) \leftarrow fireable(Rule, \kappa), recency(Rule, K, \kappa - 1),$
$\qquad\qquad\qquad\qquad \mathbf{not}\ transitional(\kappa), state(K).$

The set composed by all rules described in this subsection is meant to encode the `rif:forwardChaining`, and we denote it by $\pi_{\mathrm{rif:fC}}[\kappa]$.

The next theorem shows in which terms the encoding is correct with respect to the RIF-PRD `rif:forwardChaining` as described in [6]:

Theorem 3 (Correctness for `rif:forwardChaining`). *Let RS be a rule set, w an initial fact base, and $\langle B_{\mathsf{RS}}(w), S_{\mathsf{RS}}(RS)\,[\kappa], Q_{\mathsf{RS}}\,[\kappa]\rangle$ the corresponding iASP domain description as in Def. 18. Let LS be the* `rif:forwardChaining` *strategy of definition 15, and H the halting test that halts whenever no rule is picked. Let $\Pi_{\mathtt{rif:fC}}(RS, w) = \langle B_{\mathsf{RS}}(w), S_{\mathsf{RS}}(RS)\,[\kappa] \cup \pi_{\mathtt{rif:fC}}[\kappa], Q_{\mathsf{RS}}\,[\kappa]\rangle$. Then*[5]:

Soundness: *if $M \in AS\left(\Pi_{\mathtt{rif:fC}}(RS, w)\right)$, then there exists a state s_f such that $Eval(RS, LS, H, w) \rightarrow^*_{PRD} s_f$ and where s_f is the set of all formulae Φ such that $state(\Phi', min\left(\Pi_{\mathtt{rif:fC}}(RS, w)\right)) \in M$.*

Completeness: *if $Eval(RS, LS, H, w) \rightarrow^*_{PRD} s_f$, then there exists an M such that $M \in AS\left(\Pi_{\mathtt{rif:fC}}(RS, w)\right)$ and $\forall \Phi \in s_f, state(\Phi', min\left(\Pi_{\mathtt{rif:fC}}(RS, w)\right)) \in M$.*

One can impose other conflict resolution strategies, by specifying different rejection rules. For example, `rif:forwardChaining` behaves in a depth-first manner, in that it always selects the rule that has been more recently applied. Imposing a breadth-first strategy can be accomplished by simply changing "$TO < TR$" into "$TO > TR$" in the rule defining the rejection by recency, thus obtaining $\pi_{\mathtt{st:breadth}}[\kappa]$. Also note that `rif:forwardChaining` does not behave in a purely depth-first manner since it only applies recency after removing rules with less priority. For a strategy where a depth-first behavior is more important than complying with the declared priority of rules, one can simply change the facts that impose the order in the application of strategy elements, e.g. by including the facts $st\_element(priority, 3)$ and $st\_element(recency, 2)$ instead.

5   Conclusions

In this paper, we presented a declarative logical characterization of RIF-PRD through a sound and complete transformation into ASP, which can be seen as an equivalent alternative to the transitional semantics proposed in [6], giving further insights into RIF-PRD and providing for an immediate implementation using *iASP*, which we have developed using iClingo[9]. This transformation considers not only the RIF-PRD rule sets and their transitions, but also the conflict resolution strategies which are essential to select among applicable rules. We have illustrated how the default normative strategy – *forward chaining* – is encodable in ASP, and have shown that ASP provides an appropriate language in which to precisely define alternative non-standard conflict resolution strategies, which are also foreseen in [6], facilitating their development and unambiguous sharing, due to the simple, expressive and well known semantics of ASP. The work in [2] uses the Situation Calculus, although without handling the idiosyncrasies of RIF-PRD. A Situation Calculus based approach like the one in [2]

---

[5] $Eval(RS, LS, H, w)$ is the input function of the RIF-PRD production rule system that is responsible for choosing one among the rules in the conflict set and for the halting conditions. $\rightarrow^*_{PRD}$ is the transitive closure of $\rightarrow_{PRD}$.

could have been followed, although with extra complexity introduced by the situation terms which would not be easily handled by answer set solvers. A critique of the Situation Calculus is made in [5], where it is shown how to capture the semantics of rule production systems in μ-calculus and FPL. This work captures a result equivalent to our Theorem 1, thus not handling other conflict resolution strategies. We expect to use the work of [5] to study the formal properties of our translation. An implementation using an external DL reasoner is underway to assess the practicality of our approach, namely by comparing with more traditional approaches like CLIPS or JESS.

References

1. Baral, C.: Knowledge Representation, Reasoning and Declarative Problem Solving. Cambridge University Press, Cambridge (2003)
2. Baral, C., Lobo, J.: Characterizing production systems using logic programming and situation calculus,
   http://www.public.asu.edu/?cbaral/papers/char-prod-systems.ps
3. Boley, H., Hallmark, G., Kifer, M., Paschke, A., Polleres, A., Reynolds, D. (eds.): RIF Core Dialect. W3C Recommendation (June 22, 2010),
   http://www.w3.org/TR/2010/REC-rif-core-20100622/
4. Boley, H., Kifer, M. (eds.): RIF Basic Logic Dialect. W3C Recommendation (June 22, 2010), http://www.w3.org/TR/2010/REC-rif-bld-20100622/
5. de Bruijn, J., Rezk, M.: A logic based approach to the static analysis of production systems. In: Polleres, A. (ed.) RR 2009. LNCS, vol. 5837, pp. 254–268. Springer, Heidelberg (2009)
6. de Sainte Marie, C., Hallmark, G., Paschke, A. (eds.): RIF Production Rule Dialect. W3C Recommendation (June 22, 2010),
   http://www.w3.org/TR/2010/REC-rif-prd-20100622/
7. Eiter, T., Gottlob, G.: Expressiveness of stable model semantics for disjunctive logic programs with functions. Journal of Logic Programming 33(2), 167–178 (1997)
8. Eiter, T., Ianni, G., Lukasiewicz, T., Schindlauer, R., Tompits, H.: Combining answer set programming with description logics for the semantic web. Artificial Intelligence 172(12-13), 1495–1539 (2008)
9. Gebser, M., Kaminski, R., Kaufmann, B., Ostrowski, M., Schaub, T., Thiele, S.: Engineering an incremental asp solver. In: Garcia de la Banda, M., Pontelli, E. (eds.) ICLP 2008. LNCS, vol. 5366, pp. 190–205. Springer, Heidelberg (2008)
10. Gebser, M., Schaub, T., Thiele, S.: Gringo: A new grounder for answer set programming. In: Baral, C., Brewka, G., Schlipf, J. (eds.) LPNMR 2007. LNCS (LNAI), vol. 4483, pp. 266–271. Springer, Heidelberg (2007)
11. Gelfond, M., Lifschitz, V.: Logic programs with classical negation. In: Procs. of ICLP 1990, pp. 579–597. MIT Press, Cambridge (1990)
12. Lloyd, J.W., Topor, R.W.: Making prolog more expressive. Journal of Logic Programming 1(3), 225–240 (1984)
13. Motik, B., Rosati, R.: Reconciling description logics and rules. J. ACM 57(5) (2010)
14. Plotkin, G.D.: A structural approach to operational semantics. Journal of Logic and ALgebraic Programming 60-61, 17–139 (2004)

Measuring the Dynamic Bi-directional Influence between Content and Social Networks

Shenghui Wang and Paul Groth

VU University Amsterdam
De Boelelaan 1081a, 1081 HV, Amsterdam, The Netherlands
{swang,pgroth}@few.vu.nl

Abstract. The Social Semantic Web has begun to provide connections between users within social networks and the content they produce across the whole of the Social Web. Thus, the Social Semantic Web provides a basis to analyze both the communication behavior of users together with the content of their communication. However, there is little research combining the tools to study communication behaviour and communication content, namely, social network analysis and content analysis. Furthermore, there is even less work addressing the longitudinal characteristics of such a combination. This paper presents a general framework for measuring the dynamic bi-directional influence between communication content and social networks. We apply this framework in two use-cases: online forum discussions and conference publications. The results provide a new perspective over the dynamics involving both social networks and communication content.

1 Introduction

Does an informative post on a microblogging service lead to a user gaining followers? If a user is popular in a social network, will their new status updates be widely quoted? If a researcher identifies a new topic one year, does that result in the research having more coauthors the next? As an increasing amount of content is mediated through social networks, these types of questions are of great interest, in particular, to developers, social scientists, and business that aim to understand the link between content generation and social connection. A key aspect to answering these questions is to understand how the relationships between users influence the content of their communication and vice versa.

In this paper, we extend our work in [26] by proposing a general framework for measuring such influence over time. In our approach, we translate both user relationships and content into two corresponding networks: a social network and a content networks. The networks are then characterized using common network properties such as (in-/out-)degree and betweenness centrality. The influence is then measured using a set of multilevel time-series regression models producing what we term an influence network showing how these variables impact each other in time. Additionally, our Influence Framework can integrate other network properties tailored to a given problem domain.

P.F. Patel-Schneider et al. (Eds.): ISWC 2010, Part I, LNCS 6496, pp. 814–829, 2010.

The use of the Influence Framework is facilitated by the emergence of Semantic Web technologies not only to represent relationships between users on the Social Web but also to link to the content those users exchange. For example, the Semantically Interlinked Online Communities (SIOC) ontology is for the representation of the content of discussions but is explicitly intertwined with the Friend of a Friend (FOAF) ontology that is used to represent personal relationships. Because the Semantic Web provides these explicit links, it is easier to obtain the input data sets required by our Influence Framework. Thus, as more Social Web content is made available using Semantic Web standards, the Framework can be used to investigate a wider variety of content and social networks. Later, we show how the Influence Framework can be applied to networks obtained by querying the Semantic Web Dog Food dataset [24] as well as networks extracted from a Dutch political forum. The ability to study the connection between people through their objects was posited as a key benefit to the Social Semantic Web [5]. This work is an example of where these benefits are coming to fruition.

In summary, the contributions of this paper are as follows:

- A general framework for measuring the bi-directional influence between networks of people and the content associated with those people.
- A multilevel time-series regression model for measuring the longitudinal influences between the network properties of content and social networks.
- The generation of influence networks for both Dutch political forums and the World Wide Web conference series, which provide new material for social scientists to investigate these domains.

The rest of this paper is organized as follows. We begin by presenting the Influence Framework and its constituent parts. This is followed by a discussion of the application of the Framework to two use cases: one studying a conference series and the other studying data from a Dutch political forum. Related work is then discussed followed by a conclusion.

2   Influence Framework

The Influence Framework is a three stage framework for measuring the influence between (and within) user relationships and the content they communicate. While such measures of influence are clearly possible to perform on a case-by-case basis, a key realization in this work is that by representing content and user relationships as networks, standard network properties can provide a good initial insight into influence in different domains. We note that influence is a time-dependent notion and thus our framework requires time series data.

The three stages of the framework are:

1. Network Generation
2. Measuring Network Properties
3. Time Series Analysis

We now discuss each of these stages.

2.1 Network Generation

The first stage of the framework is to generate a series of both content and social networks as well as bindings between those networks. The starting point is information about a set of actors who interact over time, *e.g.* , participants in online discussions, scientists who co-author, *etc.* . From these data sets, *a series of social networks* representing the interaction of these actors over time can be produced. Then, a corpus of content related to each actor produced over time is needed *e.g.* , the textual content of online discussions a participant posted, the abstract a scientist wrote, the movies a star acted in, *etc.* . This content corpus should also have the property that pieces of content are somehow similar across a group of actors. Based on some similarity measure between content at each time step, a *series of content networks* can be generated. A key artifact for the framework is documentation of the relationship between actors and the content they produce at each time step. We term these *bindings*.

The network generation stage is perhaps the most domain specific part of the framework as a decision must be made about which content and which sort of user relationship should be represented in the network. Furthermore, many domains have different data formats requiring specialized programs to generate the needed networks. This is where Social Semantic Web technologies are particularly important. By providing common query interfaces and data representations, the extraction of these networks is significantly easier as demonstrated in Section 3.1.

2.2 Measuring Network Properties

Once the content networks and social networks have been produced, the properties of those networks that are of interest need to be defined (as variables) and then measured. The necessary requirement of these properties is that they vary over time. Because the content and social relationships are defined as networks, common network properties can be measured first. For a graph $G = (V, E)$ with a set of vertices $V = \{v_1, \ldots, v_n\}$ and a set of edges $E = \{e_{ij} \mid 1 \leqslant i, j \leqslant n\}$, the common network properties suggested are:

Degree centrality. For a given vertex v_i, its degree centrality is equal to the degree of v_i divided by the maximum possible degree. That is, the degree centrality $C_D(v_i)$ for vertex v_i is:

$$C_D(v_i) = \frac{deg(v_i)}{n-1}$$

In a directed network, two separate measures of degree centrality, namely **in-degree** and **out-degree**, should be measured instead.

Betweenness centrality. The betweenness centrality of a vertex is defined as the fraction of all shortest paths that pass through it over all shortest paths in the network. That is,

$$C_B(v_i) = \sum_{\substack{v_s \neq v_i \neq v_t \in V \\ v_s \neq v_t}} \frac{\sigma_{st}(v_i)}{\sigma_{st}}$$

where σ_{st} is the number of shortest paths from v_s to v_t (v_s, $v_t \in V$) and $\sigma_{st}(v_i)$ is the number of shortest paths from v_s to v_t that pass through v_i.

Clustering coefficients. Our analysis is at the vertex level, therefore, we measure the local clustering coefficient of a vertex which quantifies how close its neighbors are to being a clique (complete graph). It is measured as the proportion of links between the vertices within its neighbourhood divided by the number of links that could possibly exist between them. Let N_i be the neighbourhood of vertex v_i, *i.e.*, its immediately connected neighbours. For directed graphs, the local clustering coefficient of vertex v_i is given as

$$CC(v_i) = \frac{|\{e_{jk}\}|}{k_i(k_i - 1)} : v_j, v_k \in N_i, e_{jk} \in E.$$

While for undirected graphs, it is defined as

$$CC(v_i) = \frac{2|\{e_{jk}\}|}{k_i(k_i - 1)} : v_j, v_k \in N_i, e_{jk} \in E.$$

A higher $CC(v_i)$ means the neighbours of v_i are more densely connected.

It is important to note that while these network properties can be measured for every graph, their underlying meaning with respect to the social reality needs to be defined on a per domain basis.

While these measures are a useful start, any network property that varies over time is allowable within the Influence Framework. Later, we show how other more domain specific network properties can be used to gain additional insight into the influence between content and social networks.

The output of this stage is a table mapping each actor to values for each property at each time step.

2.3   Multilevel Time-Series Regression Models

Our Framework aims to model the longitudinal influences between network properties derived from both social and content networks. The output of Stage 2 provides data at successive time steps spaced at uniform time intervals, which form a *time series*. Thus, we need to apply *time series analysis* to extract meaningful statistics of the data in order to better understand the underlying forces and structures that produced the observed data. By fitting to a time series model, we can proceed to forecasting and predicting the forthcoming data [30]. When modeling variations in the level of a process, one of the typical methods is to use the *autoregressive* (AR) models.

Let $\mathbf{X}$ be a time series: $\mathbf{X} = \{\mathbf{x}^{(1)}, \mathbf{x}^{(2)}, \dots\}$, where $\mathbf{x}^{(t)}$ is the data observation at time t. Here, $\mathbf{x}^{(t)}$ is a vector, *i.e.* $\mathbf{x}^{(t)} = (x_1^{(t)}, x_2^{(t)}, \dots, x_m^{(t)})^T$, where m is the total number of variables we are modelling and each $x_i^{(t)}$, $i = 1, \dots, m$, is a variable we are interested in, such as the betweenness and degree centrality of a node in the social network or the centrality values of certain political or scientific topics. The $AR(p)$ model is defined as

$$x^{(t)} = a + \sum_{j=1}^{p} b_j \, x^{(t-j)} + \varepsilon^{(t)}, \tag{1}$$

where $b_1, \ldots, b_p$ are the parameters of the model, a is a constant and $\varepsilon^{(t)}$ is the noise with Gaussian distribution. In this paper, we opt for a simple model for each variable x_i independently, which only includes the values from the last time-point as independent variables, *i.e.* , an AR(1)-process:

$$x_i^{(t)} = a_i + b_{1i} \, x_1^{(t-1)} + \cdots + b_{mi} \, x_m^{(t-1)} + \varepsilon_i^{(t)}, \tag{2}$$

where $\varepsilon_i^{(t)}$ is Gaussian noise with zero mean and variance σ_ε^2.

In these models, each variable $x_i^{(t)}$ at time t is modelled as a linear combination of the predictor variables at time $t - 1$, each weighted by a coefficient that quantifies how variation in the predictor variable at time $t - 1$ is related to the variation of the predicted variable at time t. Such coefficients or *effects* can tell us the influence among different variables over time.

Generally, the above mentioned variables are referred to in statistics as *units of analysis*. In social reality, these variables are often from different levels, which are frequently hierarchically nested. For example, when studying the research achievements, attributes of individual researchers, research groups, faculties and the universities as a whole can all be important units of measures. This stage applies the above introduced regressive model to study the influence between variables, and the resulting coefficients are also called *fixed* effects. However, there exist variations among different actors, *i.e.* , *random* effects (actor-level errors). Therefore, such single-level statistical methods are no longer appropriate to study these so-called *complex data sets* [31]. We thus need to apply *multilevel analysis* to examine both fixed and random effects of variables measured at different levels [13,31].

Formally, we define $\mathbf{x}_p^{(t)} = (x_{1,p}^{(t)}, \ldots, x_{m,p}^{(t)})^T$, a vector containing the variables for actor p at time t. We can then rewrite equation (2) as

$$x_{i,p}^{(t)} = a_i + \mathbf{b}_i^T \, \mathbf{x}_p^{(t-1)} + \varepsilon_i^{(t)} + \mathbf{c}_{i,p}^T \, \mathbf{x}_p^{(t-1)} + \varepsilon_{i,p}^{(t)}, \tag{3}$$

where $\mathbf{b}_i = (b_{i1}, \ldots, b_{im})^T$ and $\mathbf{c}_i = (c_{i1}, \ldots, c_{im})^T$ are the fixed-effect coefficients and random-effects coefficients respectively.

In order to compare the resulting fixed effects to each other, all variables in the random effects regression equations need to be linearly transformed into standardised values, *i.e.* , subtraction of their mean, division by their standard deviation. In this way, the fixed effects can be interpreted as the effect of one standard deviation of change in the independent variable on the number of standard deviations change in the dependent variable.

The output of this stage is the set of statistics generated in fitting the regression models as well as a diagram, called an *influence network*, that shows the statistically significant effects between variables.

3   Use Cases

We now present two use cases applying the Influence Framework. First, a simple use case based on existing Semantic Web data is discussed. It analyses the influence between co-authorship and the topics addressed at a conference. The second use case looks at the influence of social status of forum participants and their focus on particular political parties. This use case is then extended to consider newly defined variables to answer specific questions of the domain.

3.1   Influence between Co-authors of Academic Papers and the Topics They Address

Data Collection. The World Wide Web Conference is the preeminent conference on Web Technologies covering both advances in academia and industry. We obtained a corpus of metadata about this conference from the Semantic Web Dogfood repository [24]. The metadata covers the conference program including paper metadata (*e.g.*, authors, paper titles, keywords, *etc.*) and organization metadata (*e.g.*, program committee members, collocated workshops, *etc.*). Importantly for use with the Influence Framework, the metadata spans four years of the conference from 2007 to 2010 using generally the same schema. The data was downloaded in bulk and loaded into separate RDF stores for each year.

Generating Social Networks. We chose the co-author network as the social network of interest. For every year, we retrieved the co-author pairs for each article using the SPARQL query shown in Figure 1. From these results, we built a weighted undirected graph for each year where nodes are authors, edges are shared authorship of an article and the weights on edges are the number of co-authorships between the two linked authors. For wider coverage, we did not distinguish between paper types that is a workshop, main track, or poster paper are all considered equal for the purposes of co-authorship.

```
PREFIX rdf: <http://www.w3.org/1999/02/22-rdf-syntax-ns#>
PREFIX rdfs: <http://www.w3.org/2000/01/rdf-schema#>
PREFIX foaf: <http://xmlns.com/foaf/0.1/>
PREFIX swrc: <http://swrc.ontoware.org/ontology#>

SELECT ?author ?coauthor ?article WHERE {
        ?article swrc:author ?author.
        ?article swrc:author ?coauthor
}
```

Fig. 1. Query to extract co-author pairs

For each year, we measured the degree and betweenness centrality of each of the authors. The degree centrality represents how activity the author is in coauthoring with others. Clustering coefficient provides a measure for how closely knit a group is. In this case, it provides a measure of whether authors write with the same set of other authors. For example, one can imagine that the authors from the same department may form a cluster within the co-author network.

Generating Content Networks. Here, we are interested in the topics under discussion at the conference in each year. To obtain those topics, we use author assigned keywords as proxies for those topics. This is common practice within the bibliometrics community [3]. Similar to the co-author network, we retrieved the keywords for each article in the conference via a SPARQL query. To improve overlap between keywords assigned by different authors, keywords containing more than one word were split into separate words and then stemmed. Stemming allows keywords such as ontologies and ontology to be treated the same. Based on the stemmed keywords, a weighted undirected graph is built, where a node is a keyword and an edge is the co-occurrence between two keywords in the set of keywords for an article. Edges are weighted by the number of co-occurrences. A graph is produced for each year.

As prescribed by the Influence Framework, we then compute several common network metrics. Again, the degree provides information about the popularity of a given topic. The betweenness centrality provides information about whether a keyword is a bridge between two other keywords (*i.e.* topics).

Binding Social Content Networks. We bind the two networks together via the papers within the conference. Thus, we know which author discusses a topic and what topics are associated with particular authors via their connection to papers.

Influence Network. For this use case, we use five network measures.

- Three social network properties: degree centrality, betweenness centrality, and clustering coefficient.
- Two content-wise properties: degree centrality, betweenness centrality.

The units of analysis are all year × participant combinations. The multilevel time-series regression models are then constructed to to study the influence network between topics of a conference and the co-authorship of papers.

Figure 2 shows the resulting influence network. This network only shows effects which are statistically significant. Note, when reading such an influence network, the edges are directional in time. For example, in Figure 2, the edge between degree in the content network and clustering coefficient in the social network, should be read as the degree at some time t has large negative effect on the clustering coefficient in time $t + 1$.

The network suggests a number of avenues for investigation. First, there is strong negative effect between the degree centrality of a topic (*i.e.* , keyword) on itself, which suggests that a popular topic one year is likely to be less popular the next. Degree centrality of a topic also has strong negative effects on the degree centrality and clustering coefficient for an author. One interpretation of this

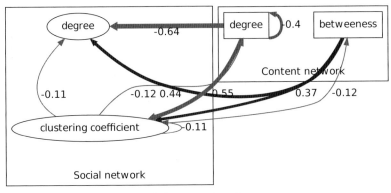

Influence network

Fig. 2. Influence network for WWW conference

result is that after a burst of collaboration on a hot topic, the topic becomes less exciting and the collaboration between authors around it dies down. There are strong positive effects of the betweenness centrality of a topic and the subsequent degree centrality and clustering coefficient of an author. A possible explanation for these effects is that if a topic bridges the gap between other topics in one conference year, it is likely to become the focus for new collaborations between authors concentrating on these normally separate topics. Such new collaborations would then come to the foreground in the next conference year.

3.2   Influence between Social Status of Online Forum Participants and Their Political Attention

Data Collection. Our data is collected from the biggest and one of the oldest Dutch forums, NL.politiek, which is entirely devoted to politics. This forum has more than 40,000 participants. Our dataset contains all the postings from October 2003 to December 2008, in total more than 1.1 million postings.

Generating Social Networks. All postings were divided into weekly subsets. In each subset, all postings were grouped by their threads and ranked based on their time stamps. Each thread corresponded to a mini discussion network, where the participants reacted to others by replying to their postings. Formally, a mini discussion network (*i.e.* , a thread) is a graph $G = (V, E)$, where V is a set of participants in this thread, and E the weighted and directed connections between the participants. There is a directed link (v_i, v_j) if participant v_i replied at least once to one of the postings of participant v_j. The frequency of the occurrence of such replying action was considered as the weight of the link, $w(v_i, v_j)$. Note, online participants often post more than once in the same thread, replying to previous postings which may include their own. Therefore, such networks can be reflexive.

We then aggregated all the mini discussion networks within one week into a bigger network, producing a series of 259 weekly social networks where 21,127 participants are involved. We note that the extraction of these networks would have been greatly simplified if they had been represented using SIOC, for example.

For each week, we measured the in-/out-degree and betweenness centrality of all participants. In this setting, the in-degree centrality of a participant indicates the degree of *popularity* he has in the online community. The out-degree centrality indicates how active one participant is. The betweenness centrality is an indicator of the *mediating/brokerage* role of a participant. A high betweenness centrality suggests that the participant connects separate communities. The brokerage role of the persons with a higher betweenness centrality is the key to understand the structural hole theory of organisational communication [9].

Generating Content Networks. In this use case, we are interested in the attention to the political parties that online participants have when they discuss in the forum. We thus extract the co-occurrence of parties as the content network. Since co-occurrence is symmetric, the content networks are therefore undirected. In the content network, the vertices are 19 Dutch parties, *i.e.*, $V = \{p_1, \ldots, p_{19}\}$. At the weekly basis, for each party p_i, we gathered a set of postings where the party was mentioned,[1] noted as S_{p_i}, $i = 1, \ldots, 19$. The weight of the edge (p_i, p_j) is calculated as the Jaccard similarity coefficient between two sets S_{p_i} and S_{p_j}, that is,

$$w(p_i, p_j) = \frac{S_{p_i} \cap S_{p_j}}{S_{p_i} \cup S_{p_j}}$$

In this way, we also extracted 259 weekly content networks. We then measured the betweenness and degree centrality of each party in each week. These centrality can tell us how one party's popularity and breakage role evolves over time. When a party has a higher degree centrality, then this party is more often mentioned while other parties are being discussed, *i.e.*, this party is more relevant or important. A party with a higher betweenness centrality is more often mentioned as a reference while more than two parties are mentioned.

Binding Social and Content Networks. We bind two networks based on *who talked when, about what*. For each participant, we counted how many times he talked about one or more of the 19 Dutch parties in a particular week, noted as $\{O_{p_1}, \ldots, O_{p_{19}}\}$. Then the degree centrality of this participant in terms of his discussion content is calculated as

$$cdc = \sum_{1}^{19} O_{p_1} \times dc(p_i)$$

---

[1] This is done through the AmCAT tool (http://content-analysis.org/) which uses a dictionary of keywords to signify an occurrence of a party when one of its keywords is used in the posting.

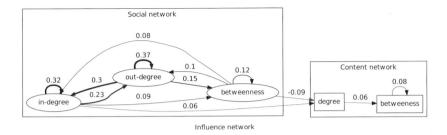

Fig. 3. Standard influence network

where $dc(p_i)$ is the degree centrality of Party p_i in the extracted content network of this week. The betweenness centrality in terms of the content, cbc, is calculated similarly.

Influence Network. Similar to the conference case, we have five standard network variables to model:

- Three social network properties: sbc (betweenness centrality), $sidc$ (in-degree centrality) and $sodc$ (out-degree centrality)
- Two content-wise properties: cbc (betweenness centrality) and cdc (degree centrality)

The units of analysis are all the week × participant combinations. We built the multilevel time-series regression models as introduced in Section 2.3 to study the influence network among political attention and social status in the online community.

There are 1762 participants have posted more than 10 postings during the whole period of time. Therefore, the Figure 3 is based on 433,453 observations from these 1762 participants. The value on the links are the fixed effects, with the critical value $p < 0.05$.

Not surprisingly, the in and out degree centrality have positive effect upon each other and to themselves. When a participant is more active, they are also likely to be more popular and more active in the social network, and vice versa. Also, the two degree centralities and the betweenness centrality have positive effects on each other with the similar strength. Once a participant gains a relatively strong brokerage role, they are more likely to maintain this role, by continuing to react to others, which consequently causes more people to reply to them. Looking at the effects between social network and the content network, the in-degree centrality (*i.e.*, the popularity of a participant) has a positive effect on the degree centrality of the content. This suggests that when a popular participant talks about certain parties, these parties are likely to become popular in the next week. When a participant becomes a broker, they tend to communicate with different opinion-holders, therefore they discuss more parties instead of only popular ones. This might be the reason for the negative effect from the social betweenness centrality to the content degree centrality. However, this may also be because of the correlations between these fixed effects, which needs to be further investigated.

3.3   Influence between User-Defined Content Variables with Social Network Properties

Content networks can be extracted in a manner that is more suitable to specific problems within a domain. Communication scientists are interested in not only the attention that the online forum participants pay to the political parties, but also the degree to which they follow the agenda of the mass media. Online discussions are expected to be more emotional and more aggressive (negativity, hatred, disgust, in short *flaming*) as compared to the news from the mass media [25]. It is natural for the communication scientists to ask to which degree emotions and aggression are expressions of autonomous or even anarchistic of online participants, to which degree they are caused by the news content in the mass media, as the classic theory of agenda setting would suggest, and to which degree they reflect depersonalised, scale-free properties of the social network of online participants that can be predicted from the previous state of their social networks.

Data Collection. We further collected newspaper articles from five biggest Dutch national newspapers. The selected national newspapers (Telegraaf, NRC Handelsblad, Algemeen Dagblad, de Volkskrant and Trouw) represent mainstream politics in the Netherlands. These newspapers reach one third of the Dutch population (official figures in 2008, http://www.cebuco.nl). The newspaper articles were retrieved from the LexisNexis archive,[2] each of which mentioned at least one political actor (*e.g.* , Dutch politicians or parties). We took a random subset of newspaper articles published between 2006 and 2008. Therefore, we also take 157 weekly social networks in these three years into the analysis.

Extracting Content Variables. In this paper, we focus on two aspects related to the forum content. The first aspect is related to the agenda setting [29,33]. The agenda setting hypothesis maintains that the participants in the online environment will take over the issue agenda from the mass media in a top-down fashion. An alternative hypothesis is that the mass media nowadays take over the topics raised in online discussion forums, in order to express and disseminate the opinions of their audience to decision-makers in business and politics, or in order to keep their audience in competitive media markets. Here, we are interested in whether the social status of the participants is influenced by the extent to which they follow mass media. Therefore, we use a list of political issues and measure weekly the attention to these issues (the frequencies of occurrence of these issues) in the newspaper articles and online discussions, respectively. Then a correlation is calculated between these two lists of the attention, which gives the first content

---

[2] http://www.lexisnexis.com/

variable *NewspaperContagion*. A higher *NewspaperContagion* indicates that the participant more strongly follows the agenda of the newspapers.

Another interesting aspect is the above-mentioned emotion expressed in the forum discussions. We would like to check whether the amount of emotion expressed in the online discussion influences the social status of the participants and his willingness to following the mass media. Starting from Brouwers thesaurus for Dutch [7], a list of keywords was developed for each emotion. Similar to measuring the attention to political parties, the frequencies of occurrence of these keywords were also measured. We separated the emotion of disgust and hate as a separate variable as they are the major emotions the communication scientists are studying [25]. Therefore, we have two other content variables: *DisgustHate* and *OtherEmotions*.

Influence Network. The five variables we investigate are

- Two network properties: *IPopularity* (=indegree centrality) and *CBetweenness* (betweenness centrality)
- Three communication contents: *DisgustHate*, *OtherEmotions* and *NewspaperContagion*.

Similar multilevel time-series regression models were built to study the influence between these variables. The resulting influence network is shown in Figure 4, based on 171,756 observations from 1101 participants.[3]

Similar to Figure 3, the betweenness centrality and popularity (in-degree centrality) have strong positive effects on themselves and each other. As we can see, a popular member or a brokerage member has a strong tendency to express emotions in their postings, and such emotional expressions also increase their social status. Especially the social popularity and the usage of the language of disgust and hate have impressively strong effects on each other. It may be the case that online participants who feel that they are in the centre of the debate, as measured by a high popularity and betweenness centrality, feel unhindered or even obliged to use rather crude words to maintain their position.

In our dataset, there seems no significant effect from the degree one follows the mass media to the aptitude for flaming and blaming, which is suggested on the basis of the classic agenda setting theory [21].

The decision of following newspaper agenda is influenced by the previous popularity in the community and also the expression of disgust and hate. This finding corresponds with earlier findings that especially citizens who are preoccupied with negativity will like the current type of news, especially men like negative news [12]. A new finding is that a high popularity in an online discussion forum also contributes to taking over agenda cues from the mass media. Apparently popular participants feel inclined to follow the news and to take over the news agenda. This corresponds with the old idea that opinion leaders in a group tend to follow the mass media closely.

---

[3] Again, only these 1101 participants have more than 9 postings within these three years.

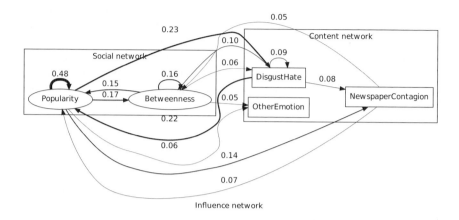

Fig. 4. Regression model of user-defined variables

4   Related Work

Social network analysis (SNA) has recently become a popular topic of study in organisation studies, communication studies, information science, *etc.* It views social relationships in terms of network theory consisting of nodes and ties. Using graph algorithms, SNA characterises the structure of social networks, strategic positions in these networks, specific sub-networks and decompositions of people and activities [28]. SNA has been applied not only to Web 2.0 platforms such as Facebook [1] and wikis [32], but also directly to the whole Web, the blogsphere, ontologies and the Semantic Web [16,17,15]. Recently, Semantic Web techniques have been adopted to facilitate standard SNA procedures [23,20,10].

On the other hand, content analysis is a research tool which has been used since the mid-1950's to determine the presence of certain words or concepts within texts [4,18]. By quantifying and analysing the presence, meanings and relations of such words and concepts, social scientists can make inferences about the content of the texts. As it is applicable to any piece of writing or recorded communication, it has been widely used in many fields, such as media studies, literature, sociology and political science [14,8,34]. Recently, many efforts have been focused on automated content analysis, such as [2], which to a large degree improves the access to large corpora.

These two classes of analysis have been investigated and applied in a rather parallel style. Only until recently, social scientists started to combine social network analysis and content analysis, such as the *discourse network analysis* in [19], and the work in [27]. This paper is the first to combine these two kinds of analysis in the Semantic Web context.

Another focus of our paper is on the longitudinal analysis over content and social networks. Recognised as a *Holy Grail* for network researchers, there has been a large degree of focus on the analysis of social networks over time [22]. However, there has not been much work with respect to the longitudinal analysis on the combination of social and content networks. The closest work is that of Gloor et al, who use network analysis over social networks and corresponding content to identify trends , however, they concentrate on a time dependent betweenness measure and do not provide a general framework for a variety of network properties [11]. Our previous work in [26] is extended in this paper by providing a general framework which is suitable for the analysis of the longitudinal influence between social networks and communication content in the Semantic Web context.

5   Conclusion

In this paper, we presented a general framework for analyzing the dynamic bi-directional influence between social relationships and the content produced with respect to those relationships. The Influence Framework leverages a key insight that by representing both social relationships and content as networks, common network properties can be used to bootstrap the analysis of influence. Based on these properties, the framework applies a time-series regression model to generate influence network diagrams representing the statistically significant effects of these properties. We applied our framework to two domains, dutch politics and a conference series, resulting in interesting conclusions about the influence of media on political forum participants and the impact of topics on academic collaboration. The data was acquired from both a web crawl and a Semantic Web source, we note that the acquisition of networks was easier using the Semantic Web data source. *To the best of our knowledge, this is the first work that combines longitudinal social network analysis and content analysis in the context of the Semantic Web.* In future work, we aim to expand the integration with Semantic Web data sources by providing reusable modules for widely used ontologies such as SIOC. Additionally, we aim to provide a service allowing others to more easily apply this framework to their own data sources.

By linking across both content and social networks, the Social Semantic Web is providing a new data source for understanding the relationship between users and the content that they produce [6]. The framework described in this paper provides a new tool for analyzing these relationships from a longitudinal perspective.

References

1. Ackland, R.: Social network services as data sources and platforms for e-researching social networks. Social Science Computer Review 27, 481–492 (2009)
2. van Atteveldt, W., Kleinnijenhuis, J., Ruigrok, N.: Parsing, semantic networks, and political authority using syntactic analysis to extract semantic relations from dutch newspaper articles. Political Analysis 16(4), 428–446 (2008)
3. Becker, H.A., Sanders, K.: Innovations in meta-analysis and social impact analysis relevant for tech mining. Technological Forecasting and Social Change 73(8), 966–980 (2006); tech Mining: Exploiting Science and Technology Information Resources
4. Berelson, B.: Content Analysis in Communication Research. Free Press, New York (1952)

5. Bojars, U., Breslin, J.G., Peristeras, V., Tummarello, G., Decker, S.: Interlinking the social web with semantics. IEEE Intelligent Systems 23, 29–40 (2008)
6. Bojrs, U., Breslin, J.G., Finn, A., Decker, S.: Using the semantic web for linking and reusing data across web 2.0 communities. Web Semant. 6(1), 21–28 (2008)
7. Brouwers, L.: Het juiste woord. Standaard betekeniswoordenboek der Nederlandse taal, 7de druk, bewerkt door F. Claes. Antwerpen: Standaard Uitgeverij (1989)
8. Budge, I., Klingemann, H.D., Volkens, A., Bara, J., Tanenbaum, E.: Mapping Policy Preferences. In: Estimates for Parties, Electors and Governments 1945-1998. Oxford University Press, Oxford (2001)
9. Burt, R.S.: Structural Holes: The Social Structure of Competition. Harvard University Press, Cambridge (1992)
10. Ereteo, G., Buffa, M., Gandon, F., Corby, O.: Analysis of a real online social network using semantic web frameworks. In: Bernstein, A., Karger, D.R., Heath, T., Feigenbaum, L., Maynard, D., Motta, E., Thirunarayan, K. (eds.) ISWC 2009. LNCS, vol. 5823, pp. 180–195. Springer, Heidelberg (2009)
11. Gloor, P.A., Krauss, J., Nann, S., Fischbach, K., Schoder, D.: Web science 2.0: Identifying trends through semantic social network analysis. In: CSE (4), pp. 215–222. IEEE Computer Society, Los Alamitos (2009)
12. Grabe, M., Kamhawi, R.: Hard wired for negative news? gender differences in processing broadcast news. Communication Research 33(5), 346–369 (2006)
13. Hayes, A.F.: A Primer on Multilevel Modeling. Human Communication Research 4, 385–410 (2006)
14. Holsti, O.R.: Content Analysis for the Social Sciences and Humanities. Addison-Wesley, Reading (1969)
15. Hoser, B., Hotho, A., Jäschke, R., Schmitz, C., Stumme, G.: Semantic network analysis of ontologies. In: Sure, Y., Domingue, J. (eds.) ESWC 2006. LNCS, vol. 4011, pp. 514–529. Springer, Heidelberg (2006)
16. Jamali, M., Abolhassani, H.: Different aspects of social network analysis. In: IEEE/WIC/ACM International Conference on Web Intelligence, Hong Kong, pp. 66–72 (2006)
17. Kim, H.M., Biehl, M., Buzacott, J.A.: M-ci2: Modelling cyber interdependencies between critical infrastructures. In: Proceedings of 3rd IEEE International Conference on Industrial Informatics, pp. 644–648 (2005)
18. Krippendorff, D.K.H.: Content Analysis: An Introduction to Its Methodology. Sage Publications, Inc., Thousand Oaks (2003)
19. Leifeld, P., Haunss, S.: A comparison between political claims analysis and discourse network analysis: The case of software patents in the european union. In: MPI Collective Goods Preprint. 2010/21 (May 2010)
20. Martin, M.S., Gutierrez, C.: Representing, querying and transforming social networks with rdf/sparql. In: Aroyo, L., Traverso, P., Ciravegna, F. (eds.) Semantic Web: Research and Applications, pp. 293–307 (2009)
21. McCombs, M., Shaw, D.: The agenda-setting function of mass media. Public Opinion Quarterly (1972)
22. McCulloh, I., Carley, K.: Longitudianl dynamic network analysis, using the over time viewer feature in ora. Tech. rep., Institute for Software Research, School of Computer Science, Carnegie Mellon University (2009)
23. Mika, P.: Flink: Semantic web technology for the extraction and analysis of social networks. Journal of Web Semantics 3, 211–223 (2005)

24. Möller, K., Heath, T., Handschuh, S., Domingue, J.: Recipes for semantic web dog food: the ESWC and ISWC metadata projects. In: Aberer, K., Choi, K.-S., Noy, N., Allemang, D., Lee, K.-I., Nixon, L.J.B., Golbeck, J., Mika, P., Maynard, D., Mizoguchi, R., Schreiber, G., Cudré-Mauroux, P. (eds.) ASWC 2007 and ISWC 2007. LNCS, vol. 4825, pp. 802–815. Springer, Heidelberg (2007)

25. Oegema, D., Kleinnijenhuis, J., Anderson, K., Van Hoof, A.: Flaming and blaming: The influence of mass media content on interactions in on-line discussions. In: Konijn, E., Tanis, M., Utz, S. (eds.) Mediated Interpersonal Communication. Erlbaum, Mahwah (2008)

26. Oegema, D., Wang, S., Kleinnijenhuis, J.: Dynamics of online discussions about politics: a function of structural network properties, mass media attention or emotional utterances? In: Proceedings of the WebSci10: Extending the Frontiers of Society On-Line. US, Raleigh (April 2010)

27. Oliver, A.L., Montgomery, K.: Using field-configuring events for sense-making: A cognitive network approach. Journal of Management Stuidies 45, 1147–1167 (2008)

28. Scott, J.: Social Network Analysis: A Handbook, 2nd edn. Sage, Newberry Park (2000)

29. Severin, W., Tankard, J.: Communication theories. Pearson, New York (2010)

30. Shumway, R.H.: Applied Statistical Time Series Analysis. Prentice Hall Series in Statistics. Prentice Hall, Englewood Cliffs (1988)

31. Snijders, T.A.B., Bosker, R.J.: Multilevel analysis: an introduction to basic and advanced multilevel modeling. Sage, Thousand Oaks (1999)

32. Tomasev, N., Mladenic, D.: Semantic web wiki: Social network analysis of page editing. In: LuzarStiffler, V., Jarec, I., Bekic, Z. (eds.) Proceedings of the ITI 2009 31st International Conference on Information Technology Interfaces, pp. 505–510 (2009)

33. Walgrave, S., Van Aelst, P.: The contingency of the mass medias political agenda setting power: Toward a preliminary theory. Journal of Communication 56, 88–190 (2006)

34. Wimmer, R.D., Dominick, J.R.: Mass Media Research: An Introduction, 8th edn. Wadsworth, Belmont (2005)

Author Index

Printing: Mercedes-Druck, Berlin
Binding: Stein + Lehmann, Berlin